EIGHTH EDITION

Environmental Science

A GLOBAL CONCERN

EIGHTH EDITION

Environmental Science

A GLOBAL CONCERN

William P. Cunningham
University of Minnesota

MaryAnn Cunningham
Vassar College

Barbara Woodworth Saigo
St. Cloud State University
Saiwood Biology Resources
Saiwood Publications

 **Higher Education**

Boston   Burr Ridge, IL   Dubuque, IA   Madison, WI   New York   San Francisco   St. Louis
Bangkok   Bogotá   Caracas   Kuala Lumpur   Lisbon   London   Madrid   Mexico City
Milan   Montreal   New Delhi   Santiago   Seoul   Singapore   Sydney   Taipei   Toronto

Higher Education

ENVIRONMENTAL SCIENCE: A GLOBAL CONCERN
EIGHTH EDITION

Published by McGraw-Hill, an business unit of The McGraw-Hill Companies, Inc., 1221 Avenue of the Americas, New York, NY, 10020. Copyright © 2005, 2003, 2001, 1999, 1997 by The McGraw-Hill Companies, Inc. All rights reserved. No part of this publication may be reproduced or distributed in any form or by any means, or stored in a database or retrieval system, without the prior written consent of The McGraw-Hill Companies, Inc., including, but not limited to, in any network or other electronic storage or transmission, or broadcast for distance learning.

Some ancillaries, including electronic and print components, may not be available to customers outside the United States.

 This book is printed on recycled, acid-free paper containing 10% postconsumer waste.

1 2 3 4 5 6 7 8 9 0 / 0 9 8 7 6 5 4 3

ISBN 0-07-243956-4

Publisher: *Margaret J. Kemp*
Senior developmental editor: *Kathleen R. Loewenberg*
Executive marketing manager: *Lisa L. Gottschalk*
Lead project manager: *Joyce M. Berendes*
Lead production supervisor: *Sandy Ludovissy*
Lead media project manager: *Judi David*
Senior media technology producer: *Jeffry Schmitt*
Senior coordinator of freelance design: *Michelle D. Whitaker*
Cover/interior designer: *Jamie E. O'Neal*
Cover image: *"Tiny Figure" by Dettifoss/Bryan & Cherry Alexander Photography*
Senior photo research coordinator: *Lori Hancock*
Photo research: *Connie Mueller*
Supplement producer: *Brenda A. Ernzen*
Compositor: *Precision Graphics*
Typeface: *10/12 Times Roman*
Printer: *Quebecor World Versailles Inc.*

Interior design image credits:
Title Page: *Digital Vision, PhotoDisc;* Preface Header: *PhotoDisc, Digital Vision, Artville;* Brief Contents Header: *Corbis, Artville;* Contents Header: *Corbis, Digital Vision;* Glossary/Credits/Index Headers: *Corbis, Digital Vision;* What Do You Think? Icon: *Getty Images;* Exploring Science Icon: *Getty Images, Corbis;* Case Study Icon: *Artville;* What Can You Do? Icon: *Getty Images;* Tables: *Corbis;* Key Concepts: *Corbis*

Library of Congress Cataloging-in-Publication Data
Cunningham, William P.
    Environmental science : a global concern. — 8th ed. / William P. Cunningham, Mary Ann Cunningham, Barbara Woodworth Saigo.
    p.   cm.
Includes index.
ISBN 0-07-243956-4
1. Environmental sciences. I. Cunningham, Mary Ann. II. Saigo, Barbara Woodworth. III. Title.

GE105.C86   2005
363.7—dc22   2003025301
                                                    CIP

www.mhhe.com

CONTENTS IN BRIEF

CONTENTS

PART ONE

PRINCIPLES FOR UNDERSTANDING
OUR ENVIRONMENT

PART TWO

PEOPLE IN THE ENVIRONMENT

PART FOUR

PHYSICAL RESOURCES
AND ENVIRONMENTAL SYSTEMS

PART FIVE

ISSUES AND POLICY

PREFACE

We face a rising epidemic of global environmental problems: global warming, diminishing biodiversity, growing shortages in freshwater supplies, long range transport of air pollutants and accumulation of persistent organic compounds in food webs, to mention just a few. To combat these problems and to find ways to prevent others from occurring, we need an environmentally-informed citizenry. The purpose of this book is to provide an interesting, accessible introduction to environmental science for students from a variety of backgrounds. Combining a broad, inter-disciplinary approach that includes both natural sciences and human dimensions of environmental issues, this book integrates information from many different areas in a way that is accessible and useful to students from any field of study.

AUDIENCE

This book is intended for use in a one- or two-semester course in environmental science, human ecology, or environmental studies at the college or advanced placement high school level. Because most students who will use this book are freshmen or sophomore non-science majors, we have tried to make the text readable and accessible without technical jargon or a presumption of prior science background. At the same time, enough data and depth are presented to make this book suitable for many upper-division classes and a valuable resource for students who will keep it in their personal libraries after their formal studies are completed.

SUSTAINABILITY

An overarching theme in this book is sustainability: can we find ways to meet our present needs without compromising the ability of future generations to meet their own needs. Can we live on renewable energy sources and the surplus produced by biogeo-chemical cycles without damaging the productive capacity of our environment? The concepts of inherent values, ethical rights, stewardship, and equity between generations and between people living under different conditions now all play important roles in our consideration of how natural resources should be managed. Consequently, ethics, philosophy and environmental worldviews are among the first topics we discuss in this book.

"This text is excellent as it provides a balanced view of renewable energy sources, taking into account both the advantages and disadvantages of the available technologies."

Lawrence Roberge
Goodwin College

CRITICAL THINKING

Critical thinking is another central theme in this book. Environmental science is a complex field, one in which a large number of special interests, contradictory data, and conflicting interpretations battle for our attention. How can we decide what to believe when apparently equally eminent experts hold diametrically opposed opinions on controversial topics? Perhaps the most valuable skill any student can gain from the study of environmental science is the ability to think purposively, analytically, and clearly about evidence. To understand the complexity and conflicting interpretations of environmental problems, students need a number of skills. They need to be able to identify and evaluate biases, recognize and assess assumptions, and understand conceptual frameworks. They must also learn to acknowledge and clarify uncertainties, equivocations, and contradictions in arguments. Reaching satisfactory conclusions about environmental dilemmas isn't just a matter of logic and rationality; we also need open-mindedness, skepticism, independence, and an ability to empathize with others. We discuss these skills in the introductory chapter of this book and then model their application in boxed readings, case studies, and questions at the ends of each chapter.

"Objectivity, readability, and visual presentations all combine to make this text stand out from all of the others out there. The authors' thoroughness and objective treatment of the topics are genuine strengths of this textbook."

Ned Knight
Linfield College

BALANCED VIEW

In every edition of this textbook, we have tried to pull together and summarize the most important current environmental information, and to explain the context and significance of scientific evidence. There's a temptation, in discussing environmental conditions to focus on extremes. While acknowledging problems, we also are careful to describe good news, progress towards sustainability, and the many ways individuals can make positive contributions toward environmental protection. Because science is always conditional, and there can be many ways to interpret data, we also present a balanced view that recognizes uncertainties and conflicting interpretations. At the same time, we stress that scientific consensus does emerge on major issues. We feel it is essential that students understand the need for differing interpretations of evidence and also recognize the value of general agreement among scholars.

"The voice of the Cunningham text is more optimistic than the book we are currently using."

Susan Brydon Golz
Rockland Community College

We hope you will find this book a valuable source of information about our global environment, as well as an inspiration for solutions to the dilemmas we face. Everyone has a role to play in this endeavor. Whether as students, educators, researchers, activists, or consumers, each of us can find ways to contribute in solving our common problems.

GLOBAL CONCERN

We live in an increasingly interconnected world. An awareness of international events, population trends, health conditions, and environmental quality are essential for educated citizens. The coal burned in China, the nuclear waste dumped in the ocean by Russia, or the pesticides used on farms in Central America affects all of us. This text has set the standard in the market for incorporating a worldview of environmental issues into each chapter with discussions in the text, photos, examples used, boxed readings, and data.

"Seldom have I seen such a good, succinct explanation of historical trends in world temperature means and why Milankovitch cycles occur."

David A. Francko
Miami University (Ohio)

UNIQUE "HOW TO STUDY" CHAPTER

Our first chapter provides information that most students need but that is rarely discussed in introductory texts: how to study, how to prepare for tests, critical thinking, concept maps, and why environmental science is exciting and important. These topics are presented in the beginning of the book so students can begin to use them immediately. This is the kind of information that most of us cover in the first lecture of a class. No other textbook goes into the depth on the fundamentals of critical thinking theory and application found here.

"What a novel idea! Many of our students come into the course with a circumscribed background in science, and this section answers many questions that are foremost in their minds. I believe that this chapter does a wonderful job of opening the idea of active self-learners to them and importantly describes the techniques needed to make this transition."

Glenn Wehner
Truman State University

NEW TO THIS EDITION

The eighth edition has undergone a major revision and reorganization reflecting both the wealth of new information available and valuable suggestions for improvement by a large number of reviewers who have been kind enough to read the text carefully and give us their detailed comments. Among these changes are:

- Updated art program with 129 new photographs and over 100 new or revised pieces of line art, including 50 new, realistic, 3D drawings.

"The photographs are good and generally have short and to-the-point captions. There are a number of very good illustrations of which I have not seen this type of before in any other text."

Patricia Smith
Valencia Community College

- New Key Concepts boxes to help students keep track of major points.
- New Exploring Science boxes to emphasize important scientific questions and help students understand how science works.
- New large fold-out piece featuring full-color physiographic and political world maps.
- New bulleted list format for Chapter Summaries so that students will recognize major issues more clearly.
- Updated graphs and tables with new data or better presentations.
- Revised Chapter Objectives and Questions for Review and Discussion to reflect new and revised material. To help students study effectively, all these elements follow the chapter organization more closely than before.
- New brief list of Selected Readings in each chapter to suggest some especially valuable sources for further study. We also have a much more extensive reading list on the Online Learning Center with roughly 100 citations from recent literature per chapter.

- Moved chapters 8 (Ecological Economics) and 10 (Environmental Policy and Law) at the suggestion of several reviewers, from the middle to the end of the book. These chapters can now serve as a capstone for previous discussions.

Visit www.mhhe.com/environmentalscience and click on this text's title to access a detailed list of changes for each chapter.

LEARNING AIDS

This text is designed to be useful as a self-education tool for students. To facilitate studying and encourage higher-level thinking, each chapter begins with a set of **Objectives** based on major concepts that students should master. The **Learning Online** section lists important chapter topics for which there are hyperlinks available on the accompanying website.

New **"Exploring Science"** boxes focus on the science behind the story. **Case Studies, "What Do You Think?"** essays, many with **"Ethical Considerations"** attached, also give students real-life examples to evaluate. All of these boxed readings are carefully planned to build upon chapter content and encourage students to practice critical thinking skills and formulate reasoned opinions.

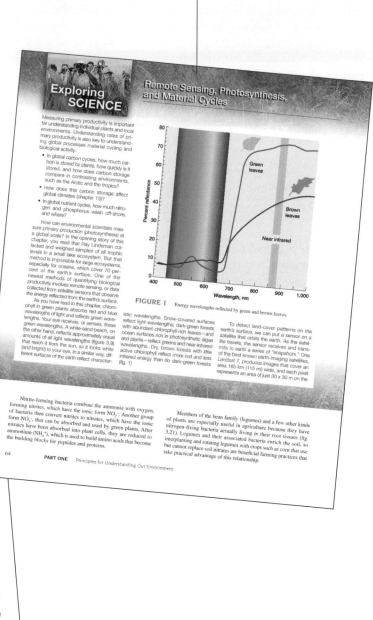

A short **Opening Story,** taken from recent news events, sets the subject in context and illuminates the importance of the material to be discussed. **Key Terms,** indicated by boldface type, are defined in the context where they are first used, and are also defined in the **Glossary** for quick reference.

The **"What Can You Do?"** listings help students to learn that small, individual steps can make a real difference in affecting our environment.

What can you do?

Lowering Our Forest Impacts

Americans throw away 30 million trees' worth of newspaper every year. Your habits and purchases affect the health of world forests. Here are some ways you can make a difference.

- Reuse and recycle paper. Make double-sided copies. Save office paper and use the back for scratch paper. Buy recycled paper.
- Use e-mail. Store information in digital form, and only print messages you really need to keep.
- If you build, conserve wood. Use wafer board, particle board, laminated beams or other composites rather than plywood and timbers made from old-growth trees.
- Buy products made from "good wood" or other certified sustainably-harvested wood.
- Don't patronize fast-food restaurants that purchase beef from cattle grazing on deforested rainforest land. Don't buy coffee, bananas, pineapples or other cash crops if their production contributes to forest destruction.
- Do buy Brazil nuts, cashews, mushrooms, rattan furniture, and other non-timber forest products harvested sustainably by local people from intact forests. Remember that tropical rainforest is not the only biome under attack. Contact the Taiga Rescue Network (www.sll.fi/TRN/Taiga News) for information about boreal forests.
- Stay informed about resource and land-use policies, and let your elected representatives know what you think.

Opponents of forest thinning also worry that it is another disguise for below-cost timber sales. A recent Forest Service study found that the cost of thinning the 1.6 million acres of forest in the Klamath Mountains of southeastern Oregon would be $2.7 billion, more than $1,685 per acre, and more than the entire fire-fighting budget for 2004.

Sustainable Forestry and Non-Timber Forest Products

Creative solutions to forest management problems are available. In both temperate and tropical regions, scores of certification programs are being developed to identify sustainably produced wood products. One organization that is currently active in 40 countries is the Forest Stewardship Council (FSC). The FSC works to set standards for certification. SmartWood, a program of the Rainforest Alliance, is the most extensive certification program. This organization works with both tropical and temperate forest products companies. One of the promising movements in North American forestry is the development of cooperatives and networks among private landowners. In the United States alone, there are more than 9 million owners of small (less than 100 acres) forest lands. Groups such as the Community Forestry Resource Center are sharing information and resources to assist in sustainable management of small working forests like these.

Consumer preferences play a role in forest protection (see this chapter's opening story). In 2003, Home Depot adopted a policy of buying wood products only from suppliers committed to environmentally friendly logging and lumber practices. The retailer sells about $5 billion of wood products each year. The number of vendors providing Home Depot with products certified by the FSC grew from 5 in 1999 to 40 in 2000. In addition, Home Depot says that nearly all of the cedar it now buys comes from second- or third-generation forests, rather than old-growth. It also has cut purchases of Indonesian luan wood by 70 percent, because much of it is illegally logged. Staples, an office supply retailer with more than 1,000 stores, announced that it would increase the average amount of recycled content in its paper products from less than 10 percent to more than 30 percent.

Logging is not the only way to make a living in a forest. Increasingly, non-timber forest products are seen as an alternative to timber production. In the United States alone, a $3 billion natural plants industry depends on healthy forests. Non-timber forest products have been around for centuries: latex (rubber), chicle (gum), nuts, and many other products have long been gathered sustainably from tropical forests (fig. 12.20). Medicinal plants, fruits,

FIGURE 12.20 Non-timber forest products, such as natural rubber, can provide an income without destroying forests.

New **Three-Dimensional Art** has transformed this eighth edition and raised it to a new standard, providing students with images that are more realistic and identifiable. For example, life-like images of wolves, hares, Inuit people, and other organisms involved in the artic food web allow the students to more accurately visualize the connections between these various components.

"These are great illustrations, much improved over the common diagrammatic-flow representations used in most texts."

David I. Johnson
Michigan State University

At the end of each chapter, a bulleted **Summary** and a set of **Questions for Review** provide an opportunity for students to test their understanding of the material just covered, while **Questions for Critical Thinking** are designed to stimulate creative, analytical thinking and to serve as a springboard for class discussions. **Web Exercises** make use of current data on the Internet and ask students to perform activities such as graphing data, comparing maps, and using live GIS sources to learn about environmental issues and information sources.

USEFUL SUPPLEMENTS

- **Digital Content Manager (DCM) CD-ROM.** This multimedia collection of visual resources allows instructors to utilize artwork from the text in multiple formats to create customized classroom presentations, visually based tests and/or quizzes, dynamic course website content, or attractive printed support materials (see fold-out piece for more information).

- **Instructor's Testing and Resource CD-ROM.** This cross-platform CD-ROM provides a computerized test bank utilizing Brownstone Diploma@ testing software to quickly create customized exams. The user-friendly program allows instructors to search for questions by topic, format, or difficulty level; edit existing questions or add new ones; and scramble questions and answer keys for multiple versions of the same test.

- **Transparencies.** A set of 100 transparencies is available to users of the text. These acetates include key figures from the text, including new art from this edition.

- **Online Learning Center** (www.mhhe.com/ environmentalscience/). This comprehensive website offers numerous resources for both students and instructors.

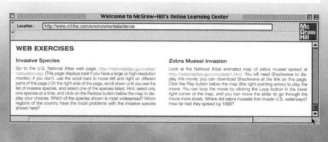

Student Resources—Everything you need in one place:

—Practice quizzing

—How to study tips

—Hyperlinks on chapter topics

—Web exercises

—Guide to electronic research

—Regional Perspectives (case studies)

—Environmental issues world map

—Key-term flashcards

—How to Contact Your Elected Officials

—Further readings

—Metric equivalents and conversion tables

—Career information

—PowerWeb's hundreds of current articles and daily news items have been integrated into each chapter on the OLC

—Access Science offers the advantage of an online, interactive encyclopedia

Instructor Resources—In addition to <u>all of the above</u>, you'll receive:

—Supplements resource chart for each chapter

—Questions for eInstruction

—Answers to web exercises

—Additional case studies

—Answers to critical thinking questions

—PageOut (create your own course website)

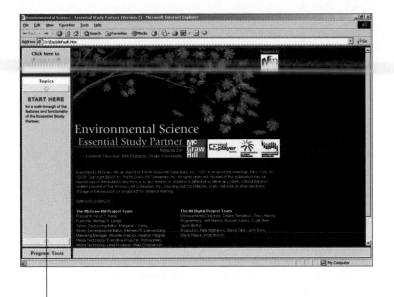

- **Environmental Science Essential Study Partner CD-ROM.**
A complete, interactive student study tool, this CD features
animations, videos, and learning activities. From quizzes to
interactive diagrams, you'll find that there has never been a
better study partner to ensure the mastery of core concepts.
Best of all, it's available FREE with a new textbook purchase
in an optional package.

PACKAGING OPPORTUNITIES AND RELATED TITLES

McGraw-Hill offers many different packaging opportunities that
not only provide your students with valuable environmental-related
material, but also a substantial cost savings. Ask your McGraw-Hill
sales representative for information on discounts and special ISBNs
for ordering a package that contains one or more of the following:

Annual Edition: Environment 04/05

This 23rd edition is a compilation of current articles from the best
of the public press. The selections explore the global environment,
the world's population, energy, the biosphere, natural resources,
and pollution.

Interactive World Issues CD-ROM

This CD explores environmental issues that affect various geographic regions. For example, you'll visit Oregon and investigate
water rights of the Columbia River. Listen to Native Americans
whose living depends on salmon fishing and then to the farmers
who need water to irrigate their crops. Additional case studies discuss migration in Mexico, apartheid in South Africa, population
issues in China, and farming in urban Chicago.

New!! Exploring Environmental Science with GIS

This short book provides exercises for students and instructors who
are new to GIS, but are familiar with the Windows operating system. The exercises focus on improving analytical skills, understanding spatial relationships, and understanding the nature and
structure of environmental data. Because the software used is distributed free of charge, this text is appropriate for courses and
schools that are not yet ready to commit to the expense and time
involved in acquiring other GIS packages.

Student Interactive CD-ROM

This CD is packaged complimentary with every new copy of Cunningham et al: Environmental Science, 8th edition. The CD-ROM
features chapter-based quizzes, chapter-based text web exercises,
student tutorial, animations and PowerPoints of all the images found
in the textbook.

Taking Sides: Clashing Views on Controversial Environmental Issues, Revised 10th Edition

This represents the arguments of leading environmentalists, scientists, and policymakers. The issues reflect a variety of viewpoints
and are staged as "pro" and "con" debates. Issues are organized
around four core areas: general philosophical and political issues,
the environment and technology, disposing of wastes, and the environment and the future.

Field and Laboratory Activities for Environmental Science, 7th Edition by Enger and Smith

The major objectives of this manual are to provide students with
hands on experiences that are relevant, easy-to-understand and
applicable to the student's life, presented in an interesting, informative format. Ranging from field and lab experiments to con-
ducting social and personal assessments of the environmental
impact of human activities, the manual presents something for
everyone, regardless of the budget or facilities of each class. These
labs are grouped by categories that can be used in conjunction with
any introductory environmental textbook.

Sources: Notable Selections in Environmental Studies, 2nd Edition

This volume brings together primary source selections of enduring
intellectual value—classic articles, book excerpts, and research
studies—that have shaped environmental studies and our contemporary understanding of it. The book includes carefully edited
selections for the works of the most distinguished environmental
observers, past and present. Selections are organized topically

around the following major areas of study: energy, environmental degradation, population issues and the environment, human health and the environment, and environment and society.

Student Atlas of Environmental Issues by Allen

The Student Atlas of Environmental Issues is an invaluable pedagogical tool for exploring the human impact on the air, waters, biosphere, and land in every major world region. This informative resource provides a unique combination of maps and data helping students understand the dimensions of the world's environmental problems and the geographical basis of these problems.

You Can Make a Difference: Be Environmentally Responsible, 2nd Edition by Getis

This book is organized around the three parts of the biosphere: land, water, and air. Each section contains descriptions of the environmental problems associated with that part of the biosphere. Immediately following each problem or "challenge" are suggested ways that individuals can help solve or alleviate them. This book has been written to provide the reader with some easy and practical ways to protect the Earth and to help understand why the task is so important.

ACKNOWLEDGMENTS

We're indebted to all the students and teachers who have sent helpful suggestions, corrections, and recommendations for improving this book. Unfortunately, space doesn't permit inclusion of all the excellent ideas that were provided. All have been saved, however, and will be helpful in future editions. We hope that those who read this edition will offer their advice and insights as well. Little of the vast range of material in this book represents our own personal research. All of us owe a great debt to the many scholars whose work forms the basis of our understanding of environmental science. We stand on the shoulders of giants. If errors persist in spite of our best efforts to root them out, we accept responsibility and ask for your indulgence.

We want to express our appreciation to the entire McGraw-Hill book team for their wonderful work in putting together this edition. Kathy Loewenberg oversaw the developmental stages and has made many creative contributions to the book. Joyce Berendes, as production project manager, kept everything running smoothly and has been extremely tolerant and accommodating even when some of us have missed deadlines. Cathy Conroy did an excellent job of copyediting and spotting errors/inconsistencies. Connie Mueller and Lori Hancock found superb photographs. The folks at Precision Graphics did an excellent job of composition and page layout. Marge Kemp has continued to support this project over the years with enthusiasm and creative ideas.

We especially want to thank our distinguished panel of advisors who helped select this edition's cover, and more importantly, guided the amazing new art program through development. We're very grateful for their thoughtful and timely comments on such critical illustrations.

Board of Advisors
Dawn Ford, *University of Tennessee*
Dan Gleason, *Georgia Southern University*
Peggy Green, *Broward Community College*
David Johnson, *Michigan State University (East Lansing)*
Lissa Leege, *Georgia Southern University*
Stacy Smith, *Lexington Community College*
Ed Standora, *Buffalo State College*

We also gratefully acknowledge the constructive criticism of the many colleagues who provided reviews of this, and the previous, edition of the book. They include:

Reviewers for the Eighth Edition
C. Marjorie Aelion
    University of South Carolina
James R. Albanese
    State University of New York at Oneonta
M. Lizabeth Allyn
    Penn State York
Robb A. Bajema
    Aquinas College
Christine Baumann Feurt
    University of New England
Jerry Beilby
    Northwestern College
Brian L. Bingham
    Western Washington University
Richard J. Bryant
    Southwestern Oklahoma State University
Susan Brydon Golz
    Rockland Community College
Dan Buresh
    Sitting Bull College
Ray D. Burkett
    Southwest Tennessee Community College
Lawrence D. Cahoon
    University of North Carolina at Wilmington
William A. Calder
    University of Arizona
Winifred Caponigri
    Holy Cross College
Raymond A. Catalano
    California University of Pennsylvania

Amy B. Chan Hilton
Florida A & M University

Mingteh Chang
Stephen F. Austin State University

David T. Corey
Midlands Technical College

Anne M. Cummings
Pikes Peak Community College

Randi Darling
Westfield State College

James N. DeVries
Lancaster Bible College

Ronald E. D'Orazio
Ellsworth Community College

Leslie E. Dorworth
Illinois–Indiana Sea Grant College

L. Donald Duke
University of California - Los Angeles

David S. Duncan
University of South Florida

David J. Eisenhour
Morehead State University

David L. Evans
Pennsylvania State University

Edwin M. Everham III
Florida Gulf Coast University

Anne M. Falke
Worcester State College

Qinguo Fan
University of Massachusetts Dartmouth

David G. Fisher
Maharishi University of Management

Malcolm Fitz Patrick
Worcester Polytechnic Institute

Catherine Folio
Brookdale Community College

Carl F. Friese
University of Dayton

Allan A. Gahr
Gordon College

Lesley Garner
University of West Alabama

Rodney G. Handy
Western Kentucky University

Gregory S. Holden
Colorado School of Mines

Robert E. Holtz
Concordia University

Frank Huang
New Mexico Tech

John C. Inman
Presbyterian College

Dan F. Ippolito
Anderson University

Marlo G. Johansen
Gavilan College

Kristen A. Keteles
University of Central Arkansas

Vishnu R. Khade
Eastern Connecticut State University

Carol A. Kimmons
University of Tennessee at Chattanooga

Eric C. Kindahl
Hood College

Mark E. Knauss
Shorter College

Ned J. Knight
Linfield College

Mark Kozubowski
Bethany College

Steve LaDochy
California State University-Los Angeles

Robert W. Ling Jr.
Kankakee Community College

Chris Lobban
University of Guam

Peter Lortz
North Seattle Community College

Judy Ann Lowman
Chaffey College

Dorothy May
Park University

Emmanuel K. Mbobi
Kent State University Stark Campus

Kathy McCann Evans
Reading Area Community College

Alan McInTosh
University of Vermont

Michael J. McLeod
Belmont Abbey College

Sheila G. Miracle
Southeast Community College

Dusan Miskovic
Northwood University

Thomas E. Murray
Elizabethtown College

James L. Nation
University of Florida

Victor I. Okereke
State University of New York – Morrisville

Carl S. Oplinger
Muhlenberg College

Mark A. Ouimette
Hardin-Simmons University

Jon K. Piper
Bethel College

Richard Puetz
Illinois Valley Community College

Kathleen L. Purvis
The Claremont Colleges

Jodie Ramsay
Northern State University

Lakshmi N. Reddi
Kansas State University

Samuel K. Riffell
Michigan State University–East Lansing

Lawrence F. Roberge
Lesley University

Lynette Rushton
South Puget Sound Community College

May Linda Samuel
Benedict College

Bradley A. Sarchet
Colby-Sawyer College

Rick Schmude
College of Lake County

R.P. Sinha
Elizabeth City State University

Jerry M. Skinner
Keystone College

Edwin J. Skoch
John Carroll University

William A. Smith
Charleston Southern University

Patricia L. Smith
Valencia Community College–Orlando

Ravi Srinivas
University of St. Thomas–Houston

Richard T. Stevens
Cleveland State Community College

Thomas M. Tharp
Purdue University

Teresa A. Thomas
Southwestern College at Chula Vista

Anne Todd Bockarie
Philadelphia University

John C. Tucker
University of Tennessee at Chattanooga

Robert F. Volp
Murray State University

Carl Waltz
Gwynedd Mercy College

Thomas Waring
Montana Tech

Phillip L. Watson
Ferris State University

Harold J. Webster
Penn State DuBois

Glenn R. Wehner
Truman State University

Richard Wilke
University of Wisconsin–Stevens Point

Danielle M. Wirth
Des Moines Area Community College

David R. Yesner
University of Alaska–Anchorage

Reviewers for the Seventh Edition:

Ghulam Sediq Aasef
Kaskaskia College

David Arieti
Waubonsee College

Lisa R. Arnold
South Georgia College

David Bass
University of Central Oklahoma

Sharmistha Basu-Dutt
State University of West Georgia

R. P. Benard
American International College

Bruce Bennett
Community College of Rhode Island

David Bixler
Chaffey College

Del Blackburn
Clark College

Dorothy F. Boorse
Gordon College

Fred J. Brenner
Grove City College

Joel G. Burken
University of Missouri-Rolla

William A. Calder
University of Arizona

Catherine W. Carter
Georgia Perimeter College

Richard Clements
Chattanooga State Technical Community College

Terence H. Cooper
University of Minnesota

William C. Culver
St. Petersburg Junior College

Roy G. Darville
East Texas Baptist University

Linda M. Desmarteau
University of Washington, Tacoma

Jean W. Dupon
Menlo College

David A. Easterla
Northwest Missouri State University

Kathy McCann Evans
Reading Area Community College

David G. Fisher
Maharishi University of Management

Chris Fox
Community College of Baltimore County-Catonsville

Heather Gallacher
Cleveland State University

Jianbang Gan
Tuskegee University

Sandi Gardner
Triton College

J. Phil Gibson
Agnes Scott College

Joseph W. Goy
Harding University

W. David Hacker
New Mexico Highlands University

Gregory J. Haenel
Elon College

Mark F. Hammer
Wayne State College

Stephen Herr
Oral Roberts University

Graham C. Hickman
Texas A&M University-Corpus Christi

Robert E. Hoitz
Concordia University

Jean R. Hushagen
Bismarck State College

Gina Johnston
California State University, Chico

J. Timothy Kimmel
Barton County Community College

John C. Kinworthy
Concordia University

Peter Kish
Southwestern Oklahoma State University

Penelope M. Koines
University of Maryland, College Park

John Koscelny
Coffeyville Community College

Bennett D. Kottler
Southern Connecticut State University

Thomas A. Kreiling
William Rainey Harper College

John F. Logue
University of South Carolina Sumter

David A. Lovejoy
Westfield State College

Paul E. Lutz
Lenoir-Rhyne College

Michael J. Manetas
Humboldt State University

Heidi Marcum
Baylor University

Bernard A. Marcus
Genesee Community College

Dorothy G. May
Park University

Dave Mense
Pitt Community College

Gary L. Miller
University of North Carolina-Asheville

David M. Myton
Lake Superior State University

Muthena Naseri
Moorpark College

Melvin L. Northup
Grand Valley State University

Chuks A. Ogbonnaya
Mountain Empire College

Joyce H. Ownbey
Sacramento City College

Jeff Port
Ottawa University

Brian C. Reeder
Morehead State University

Learning to Learn

What kind of world do you want to live in? Demand that your teachers teach you what you need to know to build it.

Peter Kropotkin

OBJECTIVES

After studying this introduction, you should be able to:

*   form a plan to organize your efforts and become a more effective and efficient student.
*   make an honest assessment of the strengths and weaknesses of your current study skills.
*   assess what you need to do to get the grade you want in this class.
*   set goals, schedule your time, and evaluate your study space.
*   use this textbook effectively, practice active reading, and prepare for exams.
*   be prepared to apply critical and reflective thinking in environmental science.
*   understand the advantages of concept mapping and use it in your studying.

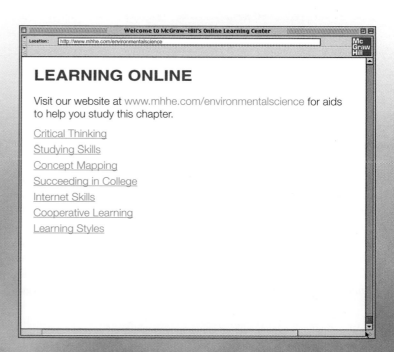

LEARNING ONLINE

Visit our website at www.mhhe.com/environmentalscience for aids to help you study this chapter.

Critical Thinking
Studying Skills
Concept Mapping
Succeeding in College
Internet Skills
Cooperative Learning
Learning Styles

Photo: Burmese schoolchildren begin a life-long process of learning. © Corbis/Volume 198.

Why Study Environmental Science?

Welcome to environmental science. We hope you'll enjoy learning about the material presented in this book, and that you'll find it both engaging and useful. There should be something here for just about everyone, whether your interests are in basic ecology, natural resources, or the broader human condition. You'll see, as you go through the book, that it covers a wide range of topics. It defines our environment, not only the natural world, but also the built world of technology, cities, and machines, as well as human social or cultural institutions. All of these interrelated aspects of our life affect us, and, in turn, are affected by what we do.

You'll find that many issues discussed here are part of current news stories on television or in newspapers. Becoming an educated environmental citizen will give you a toolkit of skills and attitudes that will help you understand current events and be a more interesting person. Because this book contains information from so many different disciplines, you will find connections here with many of your other classes. Seeing material in an environmental context may assist you in mastering subject matter in many courses, as well as in life after you leave school.

One of the most useful skills you can learn in any of your classes is critical thinking—a principal topic of this chapter. Much of the most important information in environmental science is highly contested. Facts vary depending on when and by whom they were gathered. For every opinion there is an equal and opposite opinion. How can you make sense out of this welter of ever-changing information? The answer is that you need to develop a capacity to think independently, systematically, and skillfully to form your own opinions (fig. I.1). These qualities and abilities can help you in many aspects of life. Throughout this book you will find "What Do You Think?" boxes that invite you to practice your critical and reflective thinking skills.

There is much to be worried about in our global environment. Evidence is growing relentlessly that we are degrading our environment and consuming resources at unsustainable rates. Biodiversity is disappearing at a pace unequaled since the end of the age of dinosaurs 65 million years ago. Irreplaceable topsoil erodes from farm fields, threatening global food supplies. Ancient forests are being destroyed to make newsprint and toilet paper. Rivers and lakes are polluted with untreated sewage, while soot and smoke obscure our skies. Even our global climate seems to be changing to a new regime that could have catastrophic consequences.

At the same time, we have better tools and knowledge than any previous generation to do something about these crises. Worldwide public awareness of—and support for—environmental protection is

FIGURE I.1 What does it all mean? Studying environmental science gives you an opportunity to develop creative, reflective, and critical thinking skills. © Corbis/Volume 198.

at an all-time high. Over the past 50 years, human ingenuity and enterprise have brought about a breathtaking pace of technological innovations and scientific breakthroughs. We have learned to produce more goods and services with less material. The breathtaking spread of communication technology makes it possible to share information worldwide nearly instantaneously. Since World War II, the average real income in developing countries has doubled; malnutrition has declined by almost one-third; child death rates have been halved; average life expectancy has increased by 30 percent; and the percentage of rural families with access to safe drinking water has risen from less than 10 percent to almost 75 percent.

The world's gross domestic product has increased more than tenfold over the past five decades, but the gap between the rich and poor has grown ever wider. More than a billion people now live in abject poverty without access to adequate food, shelter, medical care, education, and other resources required for a healthy, secure life. The challenge for us is to spread the benefits of our technological and economic progress more equably and to find ways to live sustainably over the long run without diminishing the natural resources and vital ecological services on which all life depends. We've tried to strike a balance in this book between enough doom and gloom to give you a realistic view of our problems, and enough positive examples to give hope that we can discover workable solutions.

What would it mean to become a responsible environmental citizen? What rights and privileges do you enjoy as a member of the global community? What duties and responsibilities go with that citizenship? In many chapters of this book you will find practical advice on things you can do to conserve resources and decrease adverse environmental impacts. Ethical perspectives are an important part of our relationship to the environment and the other people with whom we share it. The discussion of ethical principles and worldviews in chapter 2 is a key section of this book. We hope you'll think about the ethics of how we treat our common environment.

Clearly, to become responsible and productive environmental citizens each of us needs a basis in scientific principles, as well as some insights into the social, political, and economic systems that impact our global environment. We hope this book and the class you're taking will give you the information you need to reach those goals. As the noted Senegalese conservationist and educator, Baba Dioum, once said, "in the end, we will conserve only what we love, we will love only what we understand, and we will understand only what we are taught."

HOW CAN I GET AN A IN THIS CLASS?

"What have I gotten myself into?" you are probably wondering as you began to read this book. "Will environmental science be worth my while? Do I have a chance to get a good grade?" The answers to these questions depend, to a large extent, on you and how you decide to apply yourself. Expecting to be interested and to do either well or poorly in your classes often turns out to be a self-fulfilling prophecy. As Henry Ford once said, "If you think you can do a thing, or think you can't do a thing, you're right." Cultivating good study skills can help you to reach your goals and make your experience in environmental science a satisfying and rewarding one. The purpose of this introduction is to give you some tips to help you get off to a good start in studying. You'll find that many of these techniques are also useful in other courses and after you graduate, as well.

Environmental science, as you can see by skimming through the table of contents of this book, is a complex, transdisciplinary field that draws from many academic specialties. It is loaded with facts, ideas, theories, and confusing data. It is also a dynamic, highly contested subject. Topics such as environmental contributions to cancer rates, potential dangers of pesticides, or when and how much global warming may be caused by human activities are widely disputed. Often you will find distinguished and persuasive experts who take completely opposite positions on any particular question. It will take an active, organized approach on your part to make sense of the vast amount of information you'll encounter here. And it will take critical, thoughtful reasoning to formulate your own position on the many controversial theories and ideas in environmental science. Learning to learn will help you keep up-to-date on important issues after you leave this course. Becoming educated voters and consumers is essential for a sustainable future.

Develop Good Study Habits

Many students find themselves unprepared for studying in college. In a survey released in 2003 by the Higher Education Research Institute, more than two-thirds of high school seniors nationwide reported studying outside of class less than one hour per day. Nevertheless, because of grade inflation, nearly half those students claim to have an A average. It comes as a rude shock to many to discover that the study habits they developed in high school won't allow them to do as well—or perhaps even to pass their classes—in college. Many will have to triple or even quadruple their study time. In addition, they need urgently to learn to study more efficiently and effectively.

What are your current study skills and habits? Making a frank and honest assessment of your strengths and weaknesses will help

TABLE I.1	Assess Your Study Skills

Rate yourself on each of the following study skills and habits on a scale of 1 (excellent) to 5 (needs improvement). If you rate yourself below 3 on any item, think about an action plan to improve that competence or behavior.

_____ How strong is your commitment to be successful in this class?

_____ How well do you manage your time (e.g., do you always run late or do you complete assignments on time)?

_____ Do you have a regular study environment that reduces distraction and encourages concentration?

_____ How effective are you at reading and note-taking (e.g., do you remember what you've read; can you decipher your notes after you've made them)?

_____ Do you attend class regularly and listen for instructions and important ideas? Do you participate actively in class discussions and ask meaningful questions?

_____ Do you generally read assigned chapters in the textbook before attending class or do you wait until the night before the exam?

_____ Are you usually prepared before class with questions about material that needs clarification or that expresses your own interest in the subject matter?

_____ How do you handle test anxiety (e.g., do you usually feel prepared for exams and quizzes or are you terrified of them? Do you have techniques to reduce anxiety or turn it into positive energy)?

_____ Do you actively evaluate how you are doing in a course based on feedback from your instructor and then make corrections to improve your effectiveness?

_____ Do you seek out advice and assistance outside of class from your instructors or teaching assistants?

you set goals and make plans for achieving them during this class. Answer the questions in table I.1 as a way of assessing where you are as you begin to study environmental science and where you need to work to improve your study habits.

One of the first requirements for success is to set clear, honest, attainable goals for yourself. Are you willing to commit the time and effort necessary to do well in this class? Make goals for yourself in terms that you can measure and in time frames within which you can see progress and adjust your approach if it isn't taking you where you want to go. Be positive but realistic. It's more effective to try to accomplish a positive action than to avoid a negative one. When you set your goals, use proactive language that states what you want rather than negative language about what you're trying to avoid. It's good to be optimistic, but setting impossibly high standards will only lead to disappointment. Be objective about the obstacles you face and be willing to modify your goals if necessary. As you gain more experience and information, you may need to adjust your expectations either up or down. Take stock from time to time to see whether you are on track to accomplish what you expect from your studies. In environmental planning, this is called adaptive management.

One of the most common mistakes many of us make is to procrastinate and waste time. Be honest, are you habitually late for meetings or in getting assignments done? Do you routinely leave your studying until the last minute and then frantically cram the night before your exams? If so, you need to organize your schedule so that you can get your work done and still have a life. Make a study schedule for yourself and stick to it. Allow enough time for sleep, regular meals, exercise, and recreation so that you will be rested, healthy, and efficient when you do study. Schedule regular study times between your classes and work. Plan some study times during the day when you are fresh; don't leave all your work until late night hours when you don't get much done. Divide your work into reasonable sized segments that you can accomplish on a daily basis. Plan to have all your reading and assignments completed several days before your exams so you will have adequate time to review and process information. Carry a calendar so you will remember appointments and assignments.

Establish a regular study space in which you can be effective and productive. It might be a desk in your room, a carrel in the library, or some other quiet, private environment. Find a place that works for you and be disciplined about sticking to what you need to do. If you get in the habit of studying in a particular place and time, you will find it easier to get started and to stick to your tasks. Many students make the mistake of thinking that they can study while talking to their friends or watching TV. They may put in many hours but not really accomplish much. On the other hand, some people think most clearly in the anonymity of a crowd. The famous philosopher, Immanuel Kant, found that he could think best while wandering through the noisy, crowded streets of Königsberg, his home town.

How you behave in class and interact with your instructor can have a big impact on how much you learn and what grade you get. Make an effort to get to know your instructor. She or he is probably not nearly as formidable as you might think. Sit near the front of the room where you can see and be seen. Pay attention and ask questions that show your interest in the subject matter. Practice the skills of good note-taking (table I.2). Attend every class and arrive on time. Don't fold up your papers and prepare to leave until after the class period is over. Arriving late and leaving early says to your instructor that you don't care much about either the class or your grade. If you think of yourself as a good student and act like one, you may well get the benefit of the doubt when your grade is assigned.

Practice active, purposeful learning. It isn't enough to passively absorb knowledge provided by your instructor and this textbook. You need to actively engage the material in order to really understand it. The more you invest yourself in the material, the easier it will be to comprehend and remember. It is very helpful to have a study buddy with whom you can compare notes and try out ideas (fig. I.2). You will get a valuable perspective on whether you're getting the main points and understanding an adequate amount by comparing. It's an old adage that the best way to learn something is to teach it to someone else. Take turns with your study buddy explaining the material you're studying. You may think you've mastered a topic by quickly skimming the text but you're

TABLE I.2	Learning Skills—Taking Notes

1. Identify the important points in a lecture and organize your notes in an outline form to show main topics and secondary or supporting points. This will help you follow the sense of the lecture.
2. Write down all you can. If you miss something, having part of the notes will help your instructor identify what you've missed.
3. Leave a wide margin in your notes in which you can generate questions to which your notes are the answers. If you can't write a question about the material, you probably don't understand it.
4. Study for your test under test conditions by answering your own questions without looking at your notes. Cover your notes with a sheet of paper on which you write your answers, then slide it to the side to check your accuracy.
5. Go all the way through your notes once in this test mode, then go back to review those questions you missed.
6. Compare your notes and the questions you generated with those of a study buddy. Did you get the same main points from the lecture? Can you answer the questions someone else has written?
7. Review your notes again just before test time, paying special attention to major topics and questions you missed during study time.

Source: Dr. Melvin Northrup, Grand Valley State University.

likely to find that you have to struggle to give a clear description in your own words. Anticipating possible exam questions and taking turns quizzing each other can be a very good way to prepare for tests.

Recognize and Hone Your Learning Styles

Each of us has ways that we learn most effectively. Discovering techniques that work for you and fit the material you need to learn is an important step in reaching your goals. Do any of the following fit your preferred ways of learning?

- **Visual Learner:** understands and remembers best by reading, looking at photographs, figures, and diagrams. Good with maps and picture puzzles. Visualizes image or spatial location for recall. Uses flash cards for memorization.
- **Verbal Learner:** understands and remembers best by listening to lectures, reading out loud, and talking things through with a study partner. May like poetry and word games. Memorizes by repeating item verbally.
- **Logical Learner:** understands and remembers best by thinking through a subject and finding reasons that make sense. Good at logical puzzles and mysteries. May prefer to find patterns and logical connections between items rather than memorize.
- **Active Learner:** understands and remembers best those ideas and skills linked to physical activity. Takes notes, makes lists,

FIGURE I.2 Cooperative learning, in which you take turns explaining ideas and approaches with a friend, can be one of the best ways to comprehend material. © Corbis/Volume 198.

uses cognitive maps. Good at working with hands and learning by doing. Remembers best by writing, drawing, or physically manipulating items.

The list above represents only a few of the learning styles identified by educational psychologists. How can you determine which approaches are right for you? Think about the one thing in life that you most enjoy and in which you have the greatest skills. What hobbies or special interests do you have? How do you learn new material in that area? Do you read about a procedure in a book and then do it, or do you throw away the manual and use trial and error to figure out how things work? Do you need to see a diagram or a picture before things make sense, or are spoken directions most memorable and meaningful for you? Some people like to learn by themselves in a quiet place where there are no distractions, while others need to discuss ideas with another person to feel really comfortable about what they're learning.

Sometimes you have to adjust your preferred learning style to the specific material you're trying to master. You may be primarily an oral learner, but if what you need to remember for a particular subject is spatial or structural, you may need to try some visual learning techniques. Memorizing vocabulary items might be best accomplished by oral repetition, while developing your ability to work quantitative problems should be approached by practicing analytical or logical skills.

Use This Textbook Effectively

An important part of productive learning is to read assigned material in a purposeful, deliberate manner. Ask yourself questions as you read. What is the main point being made here? Does the evidence presented adequately support the assertions being made? What personal experience have you had or what prior knowledge can you bring to bear on this question? Can you suggest alternative explanations for the phenomena being discussed? What addi-

tional information would you need in order to make an informed judgment about this subject and how might you go about obtaining that information or making that judgment?

A study technique developed by Frances Robinson and called the SQ3R method (table I.3) can be a valuable aid in improving your reading comprehension. Start your study session with a *survey* of the entire chapter or section you are about to read so you'll have an idea of how the whole thing fits together. What are the major headings and subdivisions? Notice that there is usually a hierarchical organization that gives you clues about the relationship between the various parts. This survey will help you plan your strategy for approaching the material. Next, *question* what the main points are likely to be in each of the sections. Which parts look most important or interesting? Ask yourself where you should invest the most time and effort. Is one section or topic likely to be more relevant to your particular class? Has your instructor emphasized any of the topics you see? Being alert for important material can help you plan the most efficient way to study.

After developing a general plan, begin *active reading* of the text. Read in small segments and stop frequently for reflection and to make notes. Don't fall into a trance in which the words swim by without leaving any impression. Highlight or underline the main points but be careful that you don't just paint the whole page yellow. If you highlight too much, nothing will stand out. Try to distinguish what is truly central to the argument being presented. Make brief notes in the margins that identify main points. This can be very helpful in finding important sections or ideas when you are reviewing. Check your comprehension at the end of each major section. Ask yourself: Did I understand what I just read? What are the main points being made here? Does this relate to my own personal experiences or previous knowledge? Are there details or ideas that need clarification or elaboration?

As you read, stop periodically to *recite* the information you've just acquired. Summarize the information in your own words to be

TABLE I.3 The SQ3R Method for Studying Texts

SURVEY
Preview the information to be studied before reading.

QUESTION
Ask yourself critical questions about the content of what you are reading.

READ
Conduct the actual reading in small segments.

RECITE
Stop periodically to recite to yourself what you have just read.

REVIEW
Once you have completed the section, review the main points to make sure you remember them clearly.

FIGURE I.3 Cooperative learning is an important part of mastering environmental science. You never grasp material as clearly as when you explain it to someone else. © Photodisc.

sure that you really understand and are not just depending on rote memory. This is a good time to have a study group (fig. I.3). Taking turns to summarize and explain material really helps you internalize it. If you don't have a study group and you feel awkward talking to yourself, you can try writing your summary. Finally, *review* the section. Did you miss any important points? Do you understand things differently the second time through? This is a chance to think critically about the material. Do you agree with the conclusions suggested by the authors? Can you think of alternative explanations for the same evidence? As you review each section, think about how this may be covered on the test. Put yourself in the position of the instructor. What would be some good questions based on this material? Don't try to memorize everything but try to anticipate what might be the most important points.

After class, compare your lecture notes with your study notes. Do they agree? If not, where are the discrepancies? Is it possible that you misunderstood what was said in class, or does your instructor differ with what's printed in the textbook? Are there things that your instructor emphasized in lecture that you missed in your pre-class reading? This is a good time to go back over the readings to reinforce your understanding and memory of the material.

Will This Be on the Test?

Students often complain that test results don't adequately reflect what they know and how much they've learned in studying. It may well be that test questions won't cover what you think is important

or use a style that appeals to you, but you'll probably be more successful if you adapt yourself to the realities of your instructor's test methods rather than trying to force your instructor to accommodate to your preferences. One of your first priorities in studying, therefore, should be to learn your instructor's test style. Are you likely to have short-answer objective questions (multiple choice, true or false, fill in the blank) or does your instructor prefer essay questions? If you have an essay test, will the questions be broad and general or more analytical? You should develop a very different study strategy depending on whether you are expected to remember and choose between a multitude of facts and details, or whether you will be asked to write a paragraph summarizing some broad topic.

Organize the ideas you're reading and hearing in lecture. This course will probably include a great deal of information. Unless you have a photographic memory, you won't be able to remember every detail. What's most important? What's the big picture? If you see how pieces of the course fit together, it will all make more sense and be easier to remember. As you read and review, ask yourself what might be some possible test questions in each section. If you're likely to have factual questions, what are the most significant facts in the material you've read? Memorize some benchmark figures. Just a few will help a lot. Pay special attention to tables, graphs, and diagrams. They were chosen because they illustrate important points.

You probably won't be expected to remember all the specific numbers in this book but you probably should know orders of magnitude. The world population is about six *billion* people, not thousands, millions, or trillions. Highlight facts and figures in your lecture notes about which your instructor seemed especially interested. There is a good chance you'll see those topics again on a test. It often helps to remember facts and figures if you can relate them to some other familiar example. The United States, for instance, has about 275 million residents. The European Union is slightly larger, India is almost four times and China is nearly five times as large. Be sure you're familiar with the bold-face key terms in the textbook. Vocabulary terms make good objective questions. The questions for review at the end of each chapter generally cover objective material that makes good short-answer questions.

A number of strategies can help you be successful in test-taking. Look over the whole test at the beginning and answer the questions you know well first, then tackle the harder ones. On multiple choice tests, find out whether there is a penalty for guessing. Use the process of elimination to narrow down the possible choices and improve the odds for guessing. Often you can get hints from the context of the question or from other similar questions. Notice that the longest or most specific answer often is right while those that are vague or general are more likely wrong. Be alert for absolutes (such as always, never, all) which could indicate wrong choices. Qualifiers (such as sometimes, may, or could) on the other hand, often point to correct answers. Exactly opposite answers may indicate that one of them is correct.

If you anticipate essay questions, practice writing one- or two-paragraph summaries of major points in each chapter. Develop your ability to generalize and to make connections between important facts and ideas. Notice that questions for review at the end of each chapter are open-ended topics that can work well either for discussion groups or as questions for an essay test. You'll have a big advantage on a test if you have some carefully thought out arguments for and against the major ideas presented in each chapter. If you don't have any idea what a particular essay question means, you often can make a transition to something you do understand. Look for a handle that links the question to a topic you are prepared to answer. Even if you have no idea what the question means, make an educated guess. You might get some credit. Anything is better than a zero. Sometimes if you explain your answer, you'll get at least some points. "If the question means such and such, then the answer would be _____" may get you partial credit.

Does your instructor like thought questions? Does she/he expect you to be able to interpret graphs or to draw inferences from a data table? Might you be asked to read a paragraph and describe what it means or relate it to other cases you've covered in the class? If so, you should practice these skills. Making up and sharing these types of questions with your study group can greatly increase your understanding of the material as well as improve your performance on exams. Notice that the end of every chapter in this book includes a list of questions for critical and reflective thinking that makes excellent essay questions. Writing a paragraph answer for each of these questions could be a very good way to study for an essay test.

Concentrate on positive attitudes and building confidence before your tests. If you have fears and test anxiety, practice relaxation techniques and visualize success. Be sure you are rested and well prepared. You certainly won't do well if you're sleep-deprived and a bundle of nerves. Often the worst thing you can do is to stay up all night to cram your brain with a jumble of data. Being able to think clearly and express yourself well may count much more than knowing a pile of unrelated facts. Review your test when it is returned to learn what you did well and where you need to improve. Ask your instructor for pointers on how you might have answered the questions better. Carefully add your score to be sure you got all the points you deserve. Sometimes graders make simple mathematical errors in adding up points.

THINKING ABOUT THINKING

Perhaps the most valuable skill you can learn in any of your classes is the ability to think clearly, creatively, and purposefully. In a rapidly moving field such as environmental science, facts and explanations change constantly. It's often said that in six years approximately half the information you learn from this class will be obsolete. During your lifetime you will probably change careers four to six times. Unfortunately, we don't know which of the ideas we now hold will be outdated or what qualifications you will need for those future jobs. Developing the ability to learn new skills, examine new facts, evaluate new theories, and formulate your own interpretations is essential to keep up in a changing world. In other words, you need to learn how to learn on your own.

Even in our everyday lives most of us are inundated by a flood of information and misinformation. Competing claims and contradictory ideas battle for our attention. The rapidly growing complexity of our world and our lives intensifies the difficulties in knowing what to believe or how to act. Consider how the communication revolution has brought us computers, e-mail, cell phones, mobile faxes, pagers, the World Wide Web, hundreds of channels of satellite TV, and direct mail or electronic marketing that overwhelm us with conflicting information. We have more choices than we can possibly manage, and know more about the world around us than ever before but, perhaps, understand less. How can we deal with the barrage of often contradictory news and advice that inundates us?

To complicate our difficulty in knowing what to believe, distinguished authorities disagree vehemently about many important topics. A law of environmental science might be that for any expert there is always an equal and opposite expert. How can you decide what is true and meaningful in such a welter of confusing information? Is it simply a matter of what feels good at the moment or supports our preconceived notions? Or are there ways to use logical, orderly, creative thinking procedures to reach decisions?

By now, most of us know not to believe everything we read or hear (fig. I.4). "Tastes great . . . Low, low sale price . . . Vote for me . . . Lose 30 pounds in 3 weeks . . . You may already be a winner . . . Causes no environmental harm . . . I'll never lie to you . . . Two out of three doctors recommend . . ." More and more of the information we use to buy, elect, advise, judge, or heal has

FIGURE I.4 "There is absolutely no cause for alarm at the nuclear plant!" © Tribune Media Services, Inc. All Rights Reserved. Reprinted with permission.

been created not to expand our knowledge but to sell a product or advance a cause. It would be unfortunate if we become cynical and apathetic due to information overload. It does make a difference what we think and how we act.

Approaches to Truth and Knowledge

A number of skills, attitudes, and approaches can help us evaluate information and make decisions. **Analytical thinking** asks, "How can I break this problem down into its constituent parts?" **Creative thinking** asks, "How might I approach this problem in new and inventive ways?" **Logical thinking** asks, "How can orderly, deductive reasoning help me think clearly?" **Critical thinking** asks, "What am I trying to accomplish here and how will I know when I've succeeded?" **Reflective thinking** asks, "What does it all mean?" In this section, we'll look more closely at critical and reflective thinking as a foundation for your study of environmental science. We hope you will apply these ideas consistently as you read this book.

As figure I.5 suggests, critical thinking is central in the constellation of thinking skills. It challenges us to examine theories, facts, and options in a systematic, purposeful, and responsible manner. It shares many methods and approaches with other methods of reasoning but adds some important contextual skills, attitudes, and dispositions. Furthermore, it challenges us to plan methodically and to assess the process of thinking as well as the implications of our decisions. Thinking critically can help us discover hidden ideas and means, develop strategies for evaluating reasons and conclusions in arguments, recognize the differences between facts and values, and avoid jumping to conclusions. Professor Karen J. Warren of Macalester College identifies ten steps in critical thinking (table I.4).

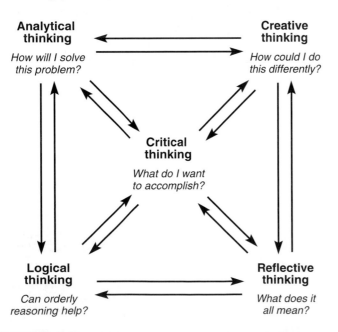

FIGURE I.5 Different approaches to thinking are used to solve different kinds of problems or to study alternate aspects of a single issue.

TABLE I.4	Steps in Critical Thinking

1. What is the *purpose* of my thinking?
2. What precise *question* am I trying to answer?
3. Within what *point of view* am I thinking?
4. What *information* am I using?
5. How am I *interpreting* that information?
6. What *concepts* or ideas are central to my thinking?
7. What *conclusions* am I aiming toward?
8. What am I taking for granted; what *assumptions* am I making?
9. If I accept the conclusions, what are the *implications*?
10. What would the *consequences* be, if I put my thoughts into action?

Source: Courtesy of Karen Warren, Philosophy Department, Macalester College, St. Paul, MN.

Notice that many critical thinking processes are self-reflective and self-correcting. This form of thinking is sometimes called "thinking about thinking." It is an attempt to plan rationally how to analyze a problem, to monitor your progress while you are doing it, and to evaluate how your strategy worked and what you have learned when you are finished. It is not critical in the sense of finding fault, but it makes a conscious, active, disciplined effort to be aware of hidden motives and assumptions, to uncover bias, and to recognize the reliability or unreliability of sources.

What Do I Need to Think Critically?

Certain attitudes, tendencies, and dispositions are essential for well-reasoned analysis. Professor Karen Warren suggests the following list:

- *Skepticism and independence.* Question authority. Don't believe everything you hear or read—including this book; even experts sometimes are wrong.

- *Open-mindedness and flexibility.* Be willing to consider differing points of view and to entertain alternative explanations. Try arguing from a viewpoint different from your own. It will help you identify weaknesses and limitations in your own position.

- *Accuracy and orderliness.* Strive for as much precision as the subject permits or warrants. Deal systematically with parts of a complex whole. Be disciplined in the standards you apply.

- *Persistence and relevance.* Stick to the main point. Don't allow diversions or personal biases to lead you astray. Information may be interesting or even true, but is it relevant?

- *Contextual sensitivity and empathy.* Consider the total situation, relevant context, feelings, level of knowledge, and sophistication of others as you evaluate information. Imagine being in someone else's place to try to understand how they feel.

- *Decisiveness and courage.* Draw conclusions and take a stand when the evidence warrants doing so. Although we often wish for more definitive information, sometimes a well-reasoned but conditional position has to be the basis for action.

- *Humility.* Realize that you may be wrong and that reconsideration may be called for in the future. Be careful about making absolute declarations; you may need to change your mind someday.

While critical thinking shares many of the orderly, systematic approaches of formal logic, it also invokes traits like empathy, sensitivity, courage, and humility. Formulating intelligent opinions about some of the complex issues you'll encounter in environmental science requires more than simple logic. Developing these attitudes and skills is not easy or simple. It takes practice. You have to develop your mental faculties just as you need to train for a sport. Traits such as intellectual integrity, modesty, fairness, compassion, and fortitude are not things you can use only occasionally. They must be cultivated until they become your normal way of thinking.

Applying Critical Thinking

We all use critical or reflective thinking at times. Suppose a television commercial tells you that a new breakfast cereal is tasty and good for you. You may be suspicious and ask yourself a few questions. What do they mean by good? Good for whom or what? Does "tasty" simply mean more sugar and salt? Might the sources of this information have other motives in mind besides your health and happiness? Although you may not have been aware of it, you already have been using some of the techniques of critical analysis. Working to expand these skills helps you recognize the ways information and analysis can be distorted, misleading, prejudiced, superficial, unfair, or otherwise defective.

Here are some steps in critical thinking:

1. *Identify and evaluate premises and conclusions in an argument.* What is the basis for the claims made here? What evidence is presented to support these claims and what conclusions are drawn from this evidence? If the premises and evidence are correct, does it follow that the conclusions are necessarily true?

2. *Acknowledge and clarify uncertainties, vagueness, equivocation, and contradictions.* Do the terms used have more than one meaning? If so, are all participants in the argument using the same meanings? Are ambiguity or equivocation deliberate? Can all the claims be true simultaneously?

3. *Distinguish between facts and values.* Are claims made that can be tested? (If so, these are statements of fact and should be able to be verified by gathering evidence.) Are claims made about the worth or lack of worth of something? (If so, these are value statements or opinions and probably cannot be verified objectively.) For example, claims of what we *ought* to do to be moral or righteous or to respect nature are generally value statements.

4. *Recognize and assess assumptions.* Given the backgrounds and views of the protagonists in this argument, what underlying reasons might there be for the premises, evidence, or conclusions presented? Does anyone have an "axe to grind" or a personal agenda in this issue? What do they think you know, need, want, or believe? Is there a subtext based on race, gender, ethnicity, economics, or some belief system that distorts this discussion?

5. *Distinguish the reliability or unreliability of a source.* What makes the experts qualified in this issue? What special knowledge or information do they have? What evidence do they present? How can we determine whether the information offered is accurate, true, or even plausible?

6. *Recognize and understand conceptual frameworks.* What are the basic beliefs, attitudes, and values that this person, group, or society holds? What dominating philosophy or ethics control their outlook and actions? How do these beliefs and values affect the way people view themselves and the world around them? If there are conflicting or contradictory beliefs and values, how can these differences be resolved?

Some Clues for Unpacking an Argument

In logic, an argument is made up of one or more introductory statements (called **premises**), and a **conclusion** that supposedly follows logically from the premises. Often in ordinary conversation, different kinds of statements are mixed together, so it is difficult to distinguish between them or to decipher hidden or implied meanings. Social theorists call the process of separating and analyzing textual components *unpacking*. Applying this type of analysis to an argument can be useful.

An argument's premises are usually claimed to be based on facts; conclusions are usually opinions and values drawn from, or used to interpret, those facts. Words that often introduce a premise include: *as, because, assume that, given that, since, whereas,* and *we all know that . . .* Words that generally indicate a conclusion or statement of opinion or values include: *and, so, thus, therefore, it follows that, consequently, the evidence shows,* and *we can conclude that.*

For instance, in the example we used earlier, the television ad might have said: "*Since* we all need vitamins, and *since* this cereal contains vitamins, *consequently* the cereal must be good for you." Which are the premises and which is the conclusion? Does one necessarily follow from the other? Remember that even if the facts in a premise are correct, the conclusions drawn from them may not be. Information may be withheld from the argument such as the fact that the cereal is also loaded with unhealthy amounts of sugar.

Avoiding Logical Errors and Fallacies

Formal logic catalogs a large number of fallacies and errors that invalidate arguments. Although we don't have room here to include all of these fallacies and errors, it may be helpful to review a few of the more common ones.

- *Red herring:* Introducing extraneous information to divert attention from the important point.

- *Ad hominem attacks:* Criticizing the opponent rather than the logic of the argument.
- *Hasty generalization:* Drawing conclusions about all members of a group based on evidence that pertains only to a selected sample.
- *False cause:* Drawing a link between premises and conclusions that depends on some imagined causal connection that does not, in fact, exist.
- *Appeal to ignorance:* Because some facts are in doubt, therefore a conclusion is impossible.
- *Appeal to authority:* It's true because _____ says so.
- *Begging the question:* Using some trick to make a premise seem true when it is not.
- *Equivocation:* Using words with double meanings to mislead the listener.
- *Slippery slope:* A claim that some event or action will cause some subsequent action.
- *False dichotomy:* Giving either/or alternatives as if they are the only choices.

Avoiding these fallacies yourself or being aware of them in another's argument can help you be more logical and have more logical and reasonable discussions.

Using Critical Thinking in Environmental Science

As you go through this book, you will have many opportunities to practice critical thinking skills. Every chapter includes many facts, figures, opinions, and theories. Are all of them true? No, probably not. They were the best information available when this text was written, but much in environmental science is in a state of flux. Data change constantly as does our interpretation of them. Do the ideas presented here give a complete picture of the state of our environment? Unfortunately, they probably don't. No matter how comprehensive our discussion is of this complex, diverse subject, it can never capture everything worth knowing, nor can it reveal all possible points of view.

When reading this text, try to distinguish between statements of fact and opinion. Ask yourself if the premises support the conclusions drawn from them. Although we have tried to be fair and even-handed in presenting controversies, we, like everyone, have biases and values—some of which we may not even recognize—that affect how we see issues and present arguments. Watch for cases in which you need to think for yourself and utilize your critical and reflective thinking skills to find the truth.

CONCEPT MAPS

Concept mapping is a learning strategy that many students find useful in understanding complex ideas and clarifying ambiguous relationships. Creating a graphic representation of a topic often can help you visualize key concepts and organize your knowledge more clearly than will other methods of study. You may find that the physical process of drawing a map of a topic engages a different part of your brain than does ordinary reading or taking notes. Taking time to think carefully about what is most important about a particular topic will help you remember it better at a later date. It can also point out weaknesses in your understanding as well as areas in which you need more study. Practicing this technique as you think about environmental science can help you comprehend difficult material and prepare for exams.

From ancient times, people have made maps to help them understand and remember important aspects of the world around them. Maps integrate and summarize knowledge. They suggest linkages that we may not have seen before, and they suggest routes for further exploration. But maps can never record every possible detail of the world or the ideas they represent. Only the most useful and meaningful information is put onto the map so that important points can easily be seen and remembered. No map is ever complete and finished. As we learn more, we revise our maps, correcting errors, adding new information, removing unnecessary features, and refining the presentation. The act of drawing a map exposes doubtful knowledge and shows us where we need more data and a clearer understanding.

If you ask different people to construct a map of the same area of a city, you would probably get very different results. A young child, for example, might draw in only the locations of her school, home, and playground. A commuter from the suburbs, on the other hand, might show the locations of the major office buildings, filling stations, freeway ramps, and the cheapest parking lots in that same city. Neither of these representations is wrong, they just emphasize the aspects of greatest importance to each person.

When we think about maps, we usually visualize physical features such as mountains, lakes, roads, buildings, and so forth. But we can also create maps of biodiversity, magnetic fields, economic flows, cultures, language families, or anything else that can be presented in graphical form. A **concept map** is a two-dimensional representation of the relationship between key ideas. It shows us how we think and suggests affinities and associations that might not otherwise be obvious. At first glance, a concept map looks like a flow chart in which key terms are placed in boxes connected by directional arrows. Based on educational psychology theories of how we organize information, concept maps are hierarchical, with broader, more general items at the top and more specific topics arranged in a cascade below them. They are metacognitive tools that empower the learner to take charge of his/her own learning in a highly organized and meaningful manner.

How Do I Create a Concept Map?

To create a concept map, start with what you already know. Build from what's familiar. What are the key components or ideas in the topic you're trying to understand? Place each concept in its own individual circle, box, or other geometrical shape. You might want to use different shapes to indicate relative levels or types of ideas.

What do you think?

Don't Believe Everything You See on the Internet

The Internet (the World Wide Web, e-mail, newsgroups, etc.) provides a wonderful resource for students of environmental science. You can find a vast amount of information to update or supplement topics in this textbook. An Internet search that takes just seconds often will provide you with hundreds, even thousands, of sources for almost any subject you can imagine. The chances are that you've already gotten material for a term paper or class project using the Internet. But you have to be very careful about evaluating the reliability of information that you obtain from these sources. Skepticism and critical thinking skills should be part of every Internet session.

One of the best things about the Internet is also its greatest problem: Anyone with access to a server can put up anything they want. There is no editorial control. On the positive side, alternative groups that ordinarily wouldn't have access to mainstream media can have a voice. People with unusual interests can find each other to share ideas and information. It also means, however, that anyone can post rumors, myths, speculation, or outright lies anonymously and inexpensively. You can find conspiracy theories, vicious racist diatribes, recipes for building bombs, groups convinced they've seen Elvis at the local mall, or people who've been abducted by little green men in space ships. Just about any crazy idea that you can imagine is intermixed with real news and valuable information. How can you know what, or whom, to believe in an information free-for-all?

One of the first questions you should ask is what information you have about the source of a particular story. Is it a person or group whose history and interests you know anything about? What motivations or interest would they be likely to have? Think about why this person or group might be taking this particular stand. What special interest might they be promoting? A good way to evaluate the reliability of a source on the Internet is to look at the kind of language they use. Is it moderate, balanced reporting, or is it a shrill, strident, call-to-action? Does it use broad generalizations and stock char-

"On the Internet, nobody knows you're a dog." Drawing by P. Steiner; © 1993, The New Yorker Magazine, Inc.

acterizations to denounce its opponents, or is it a well-reasoned analysis? Is the purpose of the piece to inform or outrage you? Sometimes outrage is appropriate, but be aware if your emotions are being manipulated.

What sources are quoted or what links to other pages are referenced in the information you have found? Be aware that someone may claim to have obtained data from a well-known source but it may not be true. Check the original, claimed source to see if they have the same story. Knowing the affiliations of a group can help you assess what their agenda and funding sources may be. An important check on any information you find on the Internet is to ask whether it is corroborated by other independent sources. It may well be that the mainstream media have chosen to ignore or suppress a particular story. The Internet may be the only way to find out what's happening. But be wary of claims that aren't supported by any other evidence.

Beware of deceptive names that portray groups as something they aren't. Anti-

environmental groups often hide behind innocuous-sounding names. Take the Information Council for the Environment, for example. Run by a public relations firm, it represents coal companies and mining firms trying to convince the government and the public that global warming is a myth. Similarly, Citizens for the Environment claims to be a grassroots citizen group but has no citizen members. Instead, it represents transnational corporations in supporting unfettered free enterprise. And the Environmental Conservation Organization, a "wise use" group organized by real estate developers, primarily opposes wetland regulations and land use planning. You can find more about these and many other similar organizations at the Clearinghouse on Environmental Advocacy and Research.

Finally, don't take any single source or study as the last word on a subject. Think of it as one rung on a ladder leading to the truth. Continuing to search for an ever closer approximation to the truth takes effort, but it's important.

Connect concept boxes with directional arrows to show relationships. Label each arrow with descriptive terms so that your diagram can be read as a statement or proposition by following interconnections from the top down. In figure I.6, for example, you can read the proposition that "concept maps are used to develop learning strategies that lead to knowledge acquisition that contributes to class performance and determines your grade" as one set of associations. Following another pathway, you can see that "values and beliefs affect learning strategies which help discover key concepts that clarify concept maps."

As you can see in this example, branches to one side or the other of the key concepts show related ideas. Where appropriate, cross links or bridges can connect branches of your map. Linking arrows can be bidirectional to indicate mutual interactions, but be careful not to make everything connected to everything else. Focus on the most important concepts and the most significant relationships. View this as an exercise in discrimination. Don't try to make your map perfect. Sometimes working briskly will help you cut away the superfluous while fostering creativity and synthesis. The point is not to create a work of art but to organize, discover, and understand central meanings. The map helps you learn how to learn.

A concept map can show just a small part or a subset of a broader field of knowledge. The top or key concept in one map may be subsumed into a lower position in a map with a different focus. A small branch in a general map could be expanded into a much more specific map of its own.

Remember that concept maps are works in progress. There are no right or wrong maps. Each one represents one possible way of understanding a particular set of ideas by an individual at the time the map was made. Expect to do several iterations, right from the outset. There are probably as many ways of representing a collection of concepts as there are concepts in the collection, but some—typically those discovered by a process of trial and error—are more elegant and easier to understand than others.

The benefits of mapping are mainly to the individual making the map. The process of simplifying concepts and arranging them on a page forces you to think about what's most important. It helps you clarify your thoughts and understandings and makes learning more meaningful. A concept map can be a heuristic device: i.e., a process in which you make discoveries and uncover meanings

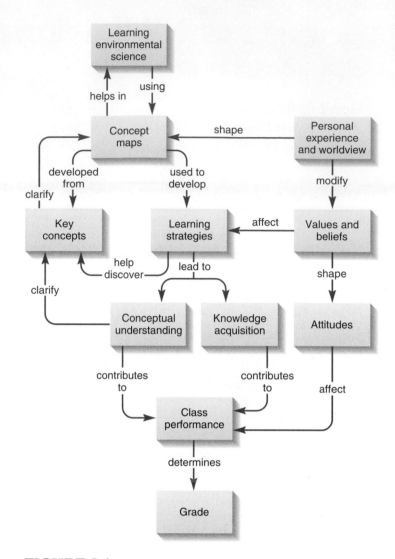

FIGURE I.6 How concept map use can affect your grade.

through trial and error. Although you benefit most directly from making your own map, it can be instructive to compare your map to that of a fellow student to see how her or his take on a topic compares to yours.

Summary

- Whether you find environmental science interesting and useful depends largely on your own attitudes and efforts. Developing good study habits, setting realistic goals for yourself, taking the initiative to look for interesting topics, finding an appropriate study space, and working with a study partner can both make your study time more efficient and improve your final grade.

- Each of us has his or her own learning style. You may understand and remember things best if you see them in writing, hear them spoken by someone else, reason them out for yourself, or learn by doing. By determining your preferred style, you can study in the way that is most comfortable and effective for you.

- Many students find the SQ3R method a helpful study technique. The acronym stands for study, question, read, recite, and review.

- Knowing what kinds of test questions you will be asked can aid you in knowing how to study. Some simple analytic keys can help you choose the best answer in objective tests.

- Writing short essays, such as those that answer the critical and reflective thinking questions at the end of every chapter in this book, can be a great help if you are likely to have this kind of test.

- Practicing drawing the concept maps described at the end of this chapter may help you analyze and organize complex topics.

- Critical thinking is a conscious effort to "think about thinking." It challenges you to think about what you are trying to accomplish in a systematic, purposeful, and responsible manner. It also asks how you will know when you have reached an adequate conclusion.

- While critical thinking shares many of the skills and techniques of formal logic (as well as drawing on analytical, creative, and reflective thinking), it also introduces attitudes such as open-mindedness, flexibility, skepticism, independence, persistence, relevance, contextual sensitivity, empathy, decisiveness, courage, and humility that will help you reach understanding in the complex and uncertain issues you will encounter in this course.

- One of the best ways to practice critical thinking skills is by evaluating the reliability of information you see on the Internet. Although the World Wide Web is a wonderful source of information, everything you see there should be approached with a sense of skepticism and caution.

Questions for Review

1. Which study skills in table I.1 do you need to improve?

2. Describe some ways you can avoid procrastination and keep on schedule.

3. List four learning styles. Which fits you best?

4. Describe the SQ3R study techniques.

5. What are ten steps in critical thinking?

6. Name (and describe) seven attributes or dispositions essential for critical thinking.

7. List some adverbs or adverbial phrases that introduce premises and conclusions.

8. Distinguish between an *ad hominem* attack and an appeal to ignorance.

9. Describe three questions you'd ask to evaluate the reliability of Internet information.

Questions for Critical Thinking

1. What is critical thinking? Why is it important?

2. Suppose your preferred learning style doesn't match the teaching style of your instructor. How might you find a common ground?

3. You may find that you fit two or more of the learning profiles described in this chapter. How would you decide which is most effective or appropriate for you?

4. Why is critical thinking important in environmental science?

5. What is the difference between critical and reflective thinking?

6. Suppose you see a claim on television or the Internet, how would you evaluate its reliability?

7. Why are empathy and contextual sensitivity important in critical thinking?

8. If some facts in this book are vague or doubtful, why mention them at all? How would you decide which "facts" to use and which to ignore?

9. Why are concept maps usually hierarchical and linked by active verbs?

Key Terms

active learner 4
analytical thinking 8
concept map 10
conclusion 9
creative thinking 8
critical thinking 8

logical learner 4
logical thinking 8
premises 9
reflective thinking 8
verbal learner 4
visual learner 4

Further Readings

Clabaugh, Gary K., and Edward G. Rozycki. 1997. *Analyzing Controversy: An Introductory Guide.* McGraw-Hill/Dushkin.

Giere, Ronald N. 1997. *Understanding Scientific Reasoning,* 4th ed. Holt, Rinehart, and Winston.

Green, Gordon W. 1993. *Getting Straight A's.* Lyle Stuart Publisher.

Pauk, Walter. 2000. *How to Study in College,* 7th ed. Houghton Mifflin Co.

Paul, Kevin. 2002. *Study Smarter, Not Harder (Self-Counsel Reference Series).* Self Counsel Press.

Paul, Richard, and Linda Elder. 2001. *Critical Thinking: Tools for Taking Charge of Your Learning and Your Life.* Prentice Hall.

Paul, Richard, and Linda Elder. 2003. *The Miniature Guide to Scientific Thinking.* Foundation for Critical Thinking.

Robinson, Adam. 1993. *What Smart Students Know: Maximum Grades, Optimum Learning, Minimum Time.* Crown Publisher.

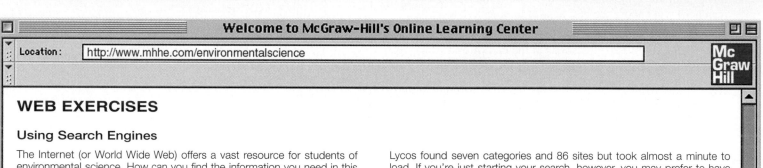

Welcome to McGraw-Hill's Online Learning Center

Location: http://www.mhhe.com/environmentalscience

WEB EXERCISES

Using Search Engines

The Internet (or World Wide Web) offers a vast resource for students of environmental science. How can you find the information you need in this maze of information? Search engines offer an excellent way to find and filter the myriad sources on the Web. Probably you have already used at least one of these sources, but are you aware how many of these valuable tools are now available and how different they are?

Go to www.surffast.com/ to find an extensive list of search engines. Test the same term (perhaps global warming or biodiversity losses, for example) on a half dozen of them and compare how fast they respond, how many hits they find and what the differences are between the information they offer. Try some hierarchical search engines (those that organize information in categories) such as Yahoo and Lycos. They often have annoying ads that slow your search, but they allow you to browse related themes and show you similar words or phrases that could be related to your topic.

Also try some meta-search engines such as Dogpile.com, Google.com, or Profusion.com. These meta-engines examine other search engines to give you a much broader database. They also tend to find more resources. In a test to see what it would find on "global warming," Google searched 13 billion web pages and found 311,000 hits in 0.05 second. By contrast,

Lycos found seven categories and 86 sites but took almost a minute to load. If you're just starting your search, however, you may prefer to have fewer sites and more selectivity in finding useful ones. The Lycos Environment News, for instance, lets you browse current news stories and may reveal topics that you didn't even know existed.

Many search engines allow you to type in a question in ordinary English: "Where can I find an article on global warming?" This may take you to places that you'd rather not go, however. The more specific and limited your query, the more likely you'll be to get useful results. Some search engines allow you to use Boolean terms such as AND, OR, BUT NOT. If you enclose a phrase in quotation marks it will direct the search to that exact phrase including all the words in it. AND (in caps) will limit the search to both words (global AND warming), while OR (global OR warming) will find sources that have either word. The phrase, BUT NOT, excludes terms that you know you don't want.

You'll probably find that searching for the same words in different systems gives you very different results. When you're doing research, it pays to use more than one search engine to make sure that you aren't missing important perspectives and information.

Understanding Our Environment

1

*We travel together, passengers on a little space ship, dependent upon its vulnerable reserves of air and soil,
all committed for our safety to its security and peace; preserved from annihilation only by the care,
the work, and I will say, the love that we give to our fragile craft.*

Adlai Stevenson

OBJECTIVES

After studying this chapter, you should be able to:

- define the term *environment* and identify some important
  environmental concerns that we face today.
- discuss the history of conservation and the different attitudes toward
  nature revealed by utilitarian conservation and biocentric
  preservation.
- briefly describe some major environmental dilemmas and issues that
  shape our current environmental agenda.
- understand the connection between poverty and environmental
  degradation, as well as the division between the wealthy,
  industrialized countries and the poorer, developing countries of the
  world.
- define the term *sustainable development* and describe some of its
  requirements.

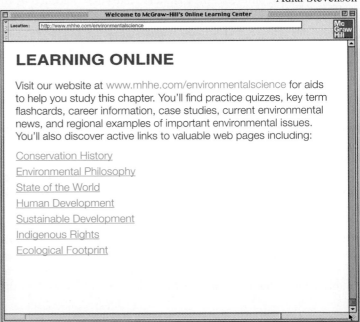

Welcome to McGraw-Hill's Online Learning Center

Location: http://www.mhhe.com/environmentalscience

LEARNING ONLINE

Visit our website at www.mhhe.com/environmentalscience for aids
to help you study this chapter. You'll find practice quizzes, key term
flashcards, career information, case studies, current environmental
news, and regional examples of important environmental issues.
You'll also discover active links to valuable web pages including:

Conservation History
Environmental Philosophy
State of the World
Human Development
Sustainable Development
Indigenous Rights
Ecological Footprint

Photo: Mount Kilimanjaro rises above the tropical savanna of Kenya and Tanzania. Glaciologists warn
that by 2015, all permanent ice on the mountaintop will probably be melted. © Corbis/Volume 35.

Measuring Sustainability and Ecological Footprints

Can the earth sustain our current lifestyles, and will there be adequate natural resources for future generations? These questions are among the most important in environmental science today. We depend on nature for food, water, energy, oxygen, waste disposal and other life-support services. Sustainability implies that we cannot turn our resources into waste faster than nature can recycle and replenish the supplies on which we depend. It also recognizes that degrading ecological systems ultimately threatens everyone's well-being. Although we may be able to overspend nature's budget temporarily, future generations will have to pay the debts we leave them. Living sustainably means meeting our own vital needs without compromising the ability of future generations to meet their own needs.

How can we evaluate our ecological impacts? Redefining Progress, a nongovernmental environmental organization, has developed a measure called the "ecological footprint" to compute the demands placed on nature by individuals and nations. A simple questionnaire of 16 items gives a rough estimate of our personal footprint. A more complex assessment of 60 categories involved in primary commodities (such as milk, wood, or metal ores), as well as the manufactured products derived from them, gives a measure of national consumption patterns.

Ecological footprint assessments are based on estimates of the biologically productive area (representing a supply of ecological services) needed for all the different human demands on nature. This isn't an easy calculation to make. There are vast differences between both the productive and absorptive capacity of diverse biomes as well as the varied impacts of human activities. Still, general estimates of the area required to provide essential services using current technology can give us a useful picture of how sustainably we're living.

According to Redefining Progress, the average world citizen has an ecological footprint of 2.3 global hectares (5.6 acres), while the biologically productive space available is only 1.9 global hectares (ha) per person. The unbalance is far more pronounced in some of the richer countries. The average resident of the United States, for example, lives at a consumption level that requires 9.6 ha of bioproductive land. If everyone in the world were to adopt this same U.S. lifestyle, we'd need about four more planets to support us all. Leaving room for the other species with which we share resources further reduces our available space. According to the World Commission on Environment and Development, at least 12 percent of all ecosystem types should be preserved for biodiversity protection. You can check your own ecological footprint by going to www.redefiningprogress.org/.

Of course, not everyone lives at a U.S. level of consumption. Residents of Germany and Japan, for example, use about half as much resources as the average American, while people in Bangladesh use 16 times less. If all of us lived in the style of Bangladeshis, the earth could support four times as many people as it does now. On the other hand, we don't have to be deprived and miserable to live sustainably. Changing to renewable, nonpolluting energy sources such as solar or wind, reusing or recycling our wastes, and adopting conservation designs in housing, transportation, food production, and other areas could dramatically cut our ecological impacts without detracting from the quality of our lives. How we might make this transition is a challenge we hope you'll keep in mind as you read the rest of this book.

WHAT IS ENVIRONMENTAL SCIENCE?

Humans have always inhabited two worlds. One is the natural world of plants, animals, soils, air, and water that preceded us by billions of years and of which we are a part. The other is the world of social institutions and artifacts that we create for ourselves using science, technology, and political organization. Both worlds are essential to our lives, but integrating them successfully causes enduring tensions.

Where earlier people had limited ability to alter their surroundings, we now have power to extract and consume resources, produce wastes, and modify our world in ways that threaten both our continued existence and that of many organisms with which we share the planet. To ensure a sustainable future for ourselves and future generations, we need to understand something about how our world works, what we are doing to it, and what we can do to protect and improve it.

Environment (from the French *environner:* to encircle or surround) can be defined as (1) the circumstances or conditions that surround an organism or group of organisms, or (2) the complex of social or cultural conditions that affect an individual or community. Since humans inhabit the natural world as well as the "built" or technological, social, and cultural world, all constitute important parts of our environment (fig. 1.1).

Environmental science, then, is the systematic study of our environment and our proper place in it. A relatively new field, environmental science is highly interdisciplinary, integrating natural sciences, social sciences, and humanities in a broad, holistic study of the world around us. In contrast to more theoretical disciplines, environmental science is mission-oriented. That is, it seeks new, valid, contextual knowledge about the natural world and our impacts on it, but obtaining this information creates a responsibility to get involved in trying to do something about the problems we have created.

As distinguished economist Barbara Ward pointed out, for an increasing number of environmental issues, the difficulty is not to identify remedies. Remedies are now well understood. The problem is to make them socially, economically, and politically acceptable. Foresters know how to plant trees, but not how to establish conditions under which villagers in developing countries can manage plantations for themselves. Engineers know how to control pollution, but not how to persuade factories to install the necessary equipment. City planners know how to build housing and design safe drinking water systems, but not how to make them affordable

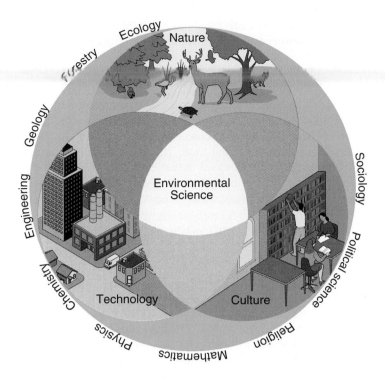

FIGURE 1.1 The intersections of the natural, cultural, and technological worlds outline the province of environmental science. Many disciplines contribute to our understanding and management of our environment.

for the poorest members of society. The solutions to these problems increasingly involve human social systems as well as natural science.

Criteria for environmental literacy suggested by the National Environmental Education Advancement Project in Wisconsin include: awareness and appreciation of the natural and built environment; knowledge of natural systems and ecological concepts; understanding of current environmental issues; and the ability to use critical-thinking and problem-solving skills on environmental issues. These are good overall goals to keep in mind as you study this book.

Chapter 2 looks more closely at science as a way of knowing, environmental ethics, and other tools that help us analyze and understand the world around us. For the remainder of this chapter, we'll complete our overview with a short history of environmental thought and a survey of some important current issues that face us.

A BRIEF HISTORY OF CONSERVATION AND ENVIRONMENTALISM

Although many early societies had negative impacts on their surroundings, others lived in relative harmony with nature. In modern times, however, growing human populations and the power of our technology have heightened our concern about what we are doing to our environment. We can divide conservation history and environmental activism into at least four distinct stages: (1) pragmatic

resource conservation, (2) moral and aesthetic nature preservation, (3) a growing concern about health and ecological damage caused by pollution, and (4) global environmental citizenship. Each era focused on different problems and each suggested a distinctive set of solutions. These stages are not necessarily mutually exclusive, however; parts of each persist today in the environmental movement and one person may embrace them all simultaneously.

Historic Roots of Nature Protection

Recognizing human misuse of nature is not unique to modern times. Plato complained in the fourth century B.C. that Greece once was blessed with fertile soil and clothed with abundant forests of fine trees. After the trees were cut to build houses and ships, however, heavy rains washed the soil into the sea, leaving only a rocky "skeleton of a body wasted by disease." Springs and rivers dried up while farming became all but impossible. Many classical authors regarded Earth as a living being, vulnerable to aging, illness, and even mortality. Periodic threats about the impending death of nature as a result of human misuse have persisted into our own time. Many of these dire warnings have proven to be premature or greatly exaggerated, but others remain relevant to our own times. As Mostafa K. Tolba, former Executive Director of the United Nations Environment Programme has said, "The problems that overwhelm us today are precisely those we failed to solve decades ago."

Some of the earliest scientific studies of environmental damage were carried out in the eighteenth century by French and British colonial administrators who often were trained scientists and who considered responsible environmental stewardship as an aesthetic and moral priority, as well as an economic necessity. These early conservationists observed and understood the connection between deforestation, soil erosion, and local climate change. The pioneering British plant physiologist, Stephen Hales, for instance, suggested that conserving green plants preserved rainfall. His ideas were put into practice in 1764 on the Caribbean island of Tobago, where about 20 percent of the land was marked as "reserved in wood for rains."

Pierre Poivre, an early French governor of Mauritius, an island in the Indian Ocean, was appalled at the environmental and social devastation caused by destruction of wildlife (such as the flightless dodo) and the felling of ebony forests on the island by early European settlers. In 1769, Poivre ordered that one-quarter of the island was to be preserved in forests, particularly on steep mountain slopes and along waterways. Mauritius remains a model for balancing nature and human needs. Its forest reserves shelter a larger percentage of its original flora and fauna than most other human-occupied islands.

Pragmatic Resource Conservation

Many historians consider the publication of *Man and Nature* in 1864 by geographer George Perkins Marsh as the wellspring of environmental protection in North America. Marsh, who also was a lawyer, politician, and diplomat, traveled widely around the

Mediterranean as part of his diplomatic duties in Turkey and Italy. He read widely in the classics (including Plato) and personally observed the damage caused by the excessive grazing by goats and sheep and by the deforesting of steep hillsides. Alarmed by the wanton destruction and profligate waste of resources still occurring on the American frontier in his lifetime, he warned of its ecological consequences. Largely as a result of his book, national forest reserves were established in the United States in 1873 to protect dwindling timber supplies and endangered watersheds.

Among those influenced by Marsh's warnings were President Theodore Roosevelt and his chief conservation advisor, Gifford Pinchot (fig. 1.2). In 1905, Roosevelt, who was the leader of the populist, progressive movement, moved the Forest Service out of the corruption-filled Interior Department into the Department of Agriculture. Pinchot, who was the first native-born professional forester in North America, became the founding head of this new agency. He put resource management on an honest, rational, and scientific basis for the first time in our history. Together with naturalists and activists such as John Muir, William Brewster, and George Bird Grinnell, Roosevelt and Pinchot established the framework of our national forest, park, and wildlife refuge systems, passed game protection laws, and tried to stop some of the most flagrant abuses of the public domain. In 1908, Pinchot organized and chaired the White House Conference on Natural Resources, perhaps the most prestigious and influential environmental meeting ever held in the United States.

The basis of Roosevelt's and Pinchot's policies was pragmatic **utilitarian conservation.** They argued that the forests should be saved "not because they are beautiful or because they shelter wild creatures of the wilderness, but only to provide homes and jobs for people." Resources should be used "for the greatest good, for the greatest number for the longest time." "There has been a fundamental misconception," Pinchot said, "that conservation means nothing but husbanding of resources for future generations. Nothing could be further from the truth. The first principle of conservation is development and use of the natural resources now existing on this continent for the benefit of the people who live here now. There may be just as much waste in neglecting the development and use of certain natural resources as there is in their destruction." This pragmatic approach still can be seen in the multiple use policies of the Forest Service.

Moral and Aesthetic Nature Preservation

John Muir (fig. 1.3), geologist, author, and first president of the Sierra Club, strenuously opposed Pinchot's influence and policies. Muir argued that nature deserves to exist for its own sake, regardless of its usefulness to us. Aesthetic and spiritual values formed the core of his philosophy of nature protection. This outlook has been called **biocentric preservation** because it emphasizes the fundamental right of other organisms to exist and to pursue their own interests. Muir wrote: "The world, we are told, was made for

FIGURE 1.2 Gifford Pinchot, first chief of the U.S. Forest Service and founder of the utilitarian conservation movement.
Courtesy Grey Towers National Historic Landmark.

FIGURE 1.3 President Teddy Roosevelt (*left*) and naturalist John Muir pose at Glacier Point in Yosemite National Park during a historical meeting in 1903.
Courtesy of the Bancroft Library University of California, Berkeley.

FIGURE 1.4 The National Park Service has generally attempted to preserve pristine wilderness—such as this area in Redwood National Park—along the altruistic preservation principles proposed by John Muir.
Courtesy David Swanlund, Save-The-Redwoods League.

FIGURE 1.5 Rachel Carson's book *Silent Spring* was a landmark in modern environmental history. She alerted readers to the dangers of indiscriminate pesticide use.
© AP/Wide World Photos.

man. A presumption that is totally unsupported by the facts . . . Nature's object in making animals and plants might possibly be first of all the happiness of each one of them. . . . Why ought man to value himself as more than an infinitely small unit of the one great unit of creation?"

Muir, who was an early explorer and interpreter of the Sierra Nevada Mountains in California, fought long and hard for establishment of Yosemite and Kings Canyon National Parks. The National Park Service, established in 1916, was first headed by Muir's disciple, Stephen Mather, and has always been oriented toward preservation of nature in its purest state (fig. 1.4). It has often been at odds with Pinchot's utilitarian Forest Service. Environmental ethics is discussed further in chapter 2.

Modern Environmentalism

The undesirable effects of pollution probably have been recognized at least as long as those of forest destruction. In 1273, King Edward I of England threatened to hang anyone burning coal in London because of the acrid smoke it produced. In 1661, the En-

glish diarist John Evelyn complained about the noxious air pollution caused by coal fires and factories and suggested that sweet-smelling trees be planted to purify city air. Increasingly dangerous smog attacks in Britain led, in 1880, to formation of a national Fog and Smoke Committee to combat this problem.

The tremendous industrial expansion during and after the Second World War added a new set of concerns to the environmental agenda. *Silent Spring,* written by Rachel Carson (fig. 1.5) and published in 1962, awakened the public to the threats of pollution and toxic chemicals to humans as well as other species. The movement she engendered might be called **environmentalism** because its concerns are extended to include both environmental resources and pollution. Among the pioneers of this movement were activist David Brower and scientist Barry Commoner. Brower, while executive director of the Sierra Club, Friends of the Earth, and, more recently, Earth Island Institute, introduced many of the techniques of modern environmentalism, including litigation, intervention in regulatory hearings, book and calendar publishing, and using mass media for publicity campaigns. Commoner, who was trained as a molecular biologist, has been a leader in analyzing the links between science, technology, and society. Both activism and research remain hallmarks of the modern environmental movement.

Under the leadership of a number of other brilliant and dedicated activists and scientists, the environmental agenda was expanded in the 1960s and 1970s to include issues such as human population growth, atomic weapons testing and atomic power, fossil fuel extraction and use, recycling, air and water pollution, wilderness protection, and a host of other pressing problems that are addressed in this textbook. Environmentalism has become well established on the public agenda since the first national Earth Day in 1970. A majority of Americans now consider themselves environmentalists, although there is considerable variation in what that term means.

Global Concerns

Increased opportunities to travel, as well as greatly expanded international communications, now enable us to know about daily events in places unknown to our parents or grandparents. We have become, as Marshal McLuhan announced in the 1960s, a global village. As in a village, we are all interconnected in various ways. Events that occur on the other side of the globe have profound and immediate effects on our lives.

Photographs of the earth from space (fig. 1.6) provide a powerful icon for the fourth wave of ecological concern that might be called **global environmentalism.** These photos remind us how small, fragile, beautiful, and rare our home planet is. We all share a common environment at this global scale. As our attention shifts from questions of preserving particular landscapes or preventing pollution of a specific watershed or airshed, we begin to worry about the life-support systems of the whole planet.

Minnesota geologist Roger Hooke estimates that current human earth-moving activities now rival those of natural geological forces. In addition, we are changing planetary weather systems and atmospheric chemistry, reducing the natural variety of organisms, and degrading ecosystems in ways that could have devastating effects, both on humans and on all other life-forms. Protecting our environment has become an international cause and it will take international cooperation to bring about many necessary changes.

Among the leaders of this worldwide environmental movement have been British economist Barbara Ward, French/American scientist René Dubos, Norwegian Prime Minister Gro Harlem Brundtland, and Canadian diplomat Maurice Strong. All have been central in major international environmental conventions, such as the 1972 UN Conference on the Human Environment in Stockholm or the 1992 UN "Earth Summit" on Environment and Development in Rio de Janeiro. Once again, new issues have become part of the agenda as our field of vision widens. We have begun to appreciate the links between poverty, injustice, oppression, and exploitation of humans and our environment. We will discuss human development in more detail later in this chapter.

CURRENT CONDITIONS

As you probably already know, many environmental problems now face us. Before surveying them in the following section, we should pause for a moment to consider the extraordinary natural world that we inherited and that we hope to pass on to future generations in as good—perhaps even better—a condition than when we arrived.

A Marvelous Planet

Imagine that you are an astronaut returning to Earth after a long trip to the moon or Mars. What a relief it would be to come back to this beautiful, bountiful planet (fig. 1.7) after experiencing the hostile, desolate environment of outer space. Although there are dangers and difficulties here, we live in a remarkably prolific and hospitable world that is, as far as we know, unique in the universe. Compared to the conditions on other planets in our solar system, temperatures on the earth are mild and relatively constant. Plentiful supplies of clean air, fresh water, and fertile soil are regenerated endlessly and spontaneously by geological and biological cycles (discussed in chapters 3 and 4).

Perhaps the most amazing feature of our planet is the rich diversity of life that exists here. Millions of beautiful and intriguing species populate the earth and help sustain a habitable environment. This vast multitude of life creates complex, interrelated communities where towering trees and huge animals live together with, and depend upon, tiny life-forms such as viruses, bacteria, and fungi. Together all these organisms make up delightfully diverse, self-sustaining communities, including dense, moist forests, vast sunny savannas, and richly colorful coral reefs. From time to time,

FIGURE 1.6 We live in a bountiful and beautiful world. Ours is a unique and irreplaceable planet on whose life-sustaining systems we are totally dependent. © Corbis/Volume 262.

about 85 million more to the world every year. While demographers report a transition to slower growth rates in most countries, present trends project a population between 8 and 10 billion by 2050. As the opening story of this chapter illustrates, questions remain about whether resources and ecological services can support that many humans.

Water may well be the most critical resource in the twenty-first century. Already at least 1.1 billion people lack access to safe drinking water, and twice that many don't have modern sanitation. Polluted water and lack of sanitation are estimated to contribute to the ill health of more than 1.2 billion people annually, including the death of 15 million children per year. About 40 percent of the world population lives in countries where water demands now exceed supplies, and by 2025 the UN projects that as many as three-fourths of us could live under similar conditions. Water wars may well become the major source of international conflict in coming decades.

Over the past century, global food production has more than kept pace with human population growth, but there are worries about whether we will be able to maintain this pace. Soil scientists report that about two-thirds of all agricultural lands show signs of degradation. Biotechnology and intensive farming techniques responsible for much of our recent production gains often are too expensive for poor farmers. Can we find ways to produce the food we need without further environmental degradation? And will that food be distributed equitably? In a world of food surpluses, more than 800 million people were chronically undernourished in 2002, and at least 60 million faced acute food shortages due to bad weather or politics.

How we obtain and use energy is likely to play a crucial role in our environmental future. Fossil fuels (oil, coal, and natural gas) presently provide around 80 percent of the energy used in industrialized countries (fig. 1.8). Supplies of these fuels are diminishing, however, and problems associated with their acquisition and use— air and water pollution, mining damage, shipping accidents, and

FIGURE 1.7   We are fortunate to live in a beautiful, bountiful world. It will take wisdom, care, and hard work to keep it this way. © Corbis Royalty Free Website.

we should pause to remember that, in spite of the challenges and complications of life on earth, we are incredibly lucky to be here. We should ask ourselves: what is our proper place in nature? What *ought* we do and what *can* we do to protect the irreplaceable habitat that produced and supports us? These are some of the central questions of environmental science.

Environmental Dilemmas

In preparation for the 2002 World Summit on Sustainable Development in Johannesburg, South Africa, the United Nations released a sobering assessment of the state of our global environment. With a current population of more than 6 billion humans, we're adding

FIGURE 1.8   Fossil fuels supply about 85 percent of world commercial energy. They also produce a large percentage of all air pollutants and greenhouse gases, and contribute to economic and political instability. © Corbis Royalty Free Website.

Exploring SCIENCE

What's Happening to Frogs?

Around the world, scientists are finding increasing evidence of striking deformities in frogs, toads, salamanders, and their amphibian kin. How can we evaluate the causes of a widespread environmental problem such as this?

The first report of frog abnormalities to attract widespread attention came from Minnesota schoolchildren on a summer fieldtrip to a marsh in 1995. Of the 22 frogs they caught that day, more than half had either too few or too many legs. The students' questions about these deformities led to discovery that similar problems also occurred in places throughout North America. Investigators report that as much as 80 percent of frog or salamander species in some local populations have abnormal or missing limbs, digits, or eyes. When dissected, many of these animals are found to have internal problems as well, including defective digestive systems and abnormal reproductive organs. Altogether more than 60 species in 46 states have been found to have some or all of these anomalies.

Several hypotheses have been proposed to explain amphibian malformations. Chemical contamination, ultraviolet (UV) solar radiation, and parasite infection are among the leading candidates. To test the proposal that pesticides, industrial chemicals, or other environmental contaminants might be responsible, frogs were grown in the laboratory in the presence of a variety of toxic substances. Many different developmental abnormalities occurred. Field studies, however, have failed to show a clear correlation between any of these contaminants and frog abnormalities in wild populations. It's possible that high toxin levels at a crucial developmental stage—just as frog eggs are hatching, for example—might cause permanent physical deformities even though pollutants may be degraded or diluted to lower levels by the time researchers collect water samples.

Similarly, lab studies demonstrate that high levels of UV exposure can result in many of the same deformities observed in nature. We know that UV radiation has increased because of stratospheric ozone depletion, but it's difficult to determine how much UV exposure wild frogs are getting. Both tadpoles and adults take shelter under debris or aquatic vegetation. In exposed areas, they may be more active at night than during the day. In addition, UV is absorbed by organic material, so that the exposure to animals in the water may be much less than that at the surface.

Currently, the leading candidate for amphibian malformalities seems to be parasite infections. Deformed animals often have cysts of a invasive trematode fluke named *Ribeiroia ondatrae*. The fluke burrows into tadpoles and can cause formation of either extra, absent, or deformed limbs. Infecting tadpoles with living trematodes resulted in many of the same defects found in nature. Although first observed in California, *Ribeiroia* has also been found in Wisconsin, Illinois, Pennsylvania, New York, and Minnesota, and appears to be spreading rapidly.

Fertilizer runoff and other water pollutants may play a role in *Ribeiroia* infections. Extra nutrients stimulate algal blooms and aquatic vegetation growth that provide food for snails, the intermediate host for the par-

An alarming number of deformed frogs have been found in recent years. What causes these abnormalities is not yet known. Are they a warning of pollution or other serious environmental problems? Courtesy of the Minnesota Pollution Control Agency.

asite. Endocrine disrupters can weaken immune systems and make amphibians more susceptible to infection. Still, like other suspected causes of amphibian problems, *Ribeiroia* field studies have produced mixed results. Parasites aren't always found in sites with frog abnormalities, and they haven't been shown to produce all the types of defects found in nature.

It seems likely that different combinations of chemical, biological, and physical factors are probably responsible for causing the deformities observed in wild amphibian populations. Malformations undoubtedly play a role in the widespread decline of many species of frogs, toads, salamanders, and newts from wetlands around the world. It's important for us to continue this research to understand what's happening to species that serve as indicators of more widespread environmental problems.

---

geopolitics—may limit what we do with remaining reserves. Cleaner renewable energy resources—solar power, wind, geothermal, and biomass—together with conservation, could give us cleaner, less destructive options if we invest in appropriate technology.

Burning fossil fuels, making cement, cultivating rice paddies, clearing forests, and other human activities release carbon dioxide and other so-called "greenhouse gases" that trap heat in the atmosphere. Over the past 200 years, atmospheric CO_2 concentrations have increased about 30 percent. By 2100, if current trends continue, climatologists warn that mean global temperatures will probably warm 1.5° to 6°C (2.7°–11°F). Global climate change already is affecting a wide variety of biological species (fig. 1.9). Further warming is likely to cause increasingly severe weather events including droughts in some areas and floods in others. Melting alpine glaciers and snowfields could threaten water supplies on which millions of people depend. Rising sea levels already are flooding low-lying islands and coastal regions, while habitat losses and climatic changes are affecting many biological species.

In 1997, 160 nations meeting in Kyoto, Japan, agreed to roll back greenhouse gas emissions, but the United States, which produces about 25 percent of all greenhouse gases, has refused to ratify this treaty. Criticizing the U.S. stand on climate change, British Prime Minister Tony Blair said environmental degradation in general and climate change in particular are "just as devastating in their potential impact" as weapons of mass destruction and terrorism. "There will be no genuine security," he warned, "if the planet is ravaged by climate change."

Air quality has worsened dramatically in many areas. Over southern Asia, for example, satellite images recently revealed a 3-km (2-mile)-thick toxic haze of ash, acids, aerosols, dust, and photochemical products regularly covers the entire Indian subcontinent for much of the year. Nobel laureate Paul Crutzen estimates that at least 3 million people die each year from diseases triggered by air pollution. Worldwide, the United Nations estimates that more than 2 billion metric tons of air pollutants (not including carbon dioxide or wind-blown soil) are emitted each year. Air pollution no longer is merely a local problem. Mercury, polychlorinated biphenyls (PCB), DDT and other long-lasting pollutants accumulate in arctic ecosystems and native people after being transported by air currents from industrial regions thousands of kilometers to the south. And during certain days, as much as 75 percent of the smog and particulate pollution recorded on the west coast of North America can be traced to Asia.

Biologists report that habitat destruction, overexploitation, pollution, and introduction of exotic organisms are eliminating species at a rate comparable to the great extinction that marked the end of the age of dinosaurs. The UN Environment Programme reports that over the past century, more than 800 species have disappeared and at least 10,000 species are now considered threatened. This includes about half of all primates and freshwater fish together with around 10 percent of all plant species (fig. 1.10). More than three-quarters of all global fisheries are overfished or harvested at their biological limit. At least half of the forests existing before the introduction of agriculture have been cleared, and much of the diverse "old growth" on which many species depend for habitat, is rapidly being cut and replaced by secondary growth or monoculture.

Finding solutions to these problems requires good science as well as individual and collective actions. Becoming educated about our global environment is the first step in understanding how to control our impacts on it. We hope this book will help you in that quest.

Signs of Hope

The dismal litany of problems facing us seems overwhelming, doesn't it? Is there hope that we can find solutions to these dilemmas? We think so. As you will see in subsequent chapters in this book, progress has been made in many areas by reducing pollution and curbing wasteful resource use. Many cities in Europe and North America, for example, are cleaner and much more livable now than they were a century ago. Population has stabilized in most industrialized countries and even in some very poor countries

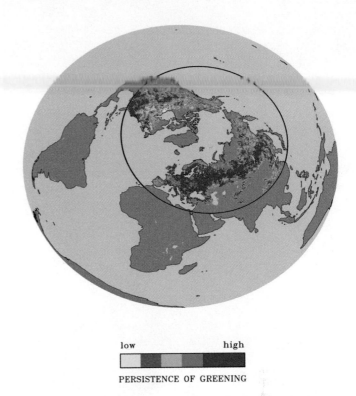

low   high

PERSISTENCE OF GREENING

FIGURE 1.9   Satellite images and surface temperature data show that polar regions, especially in Eurasia, are becoming green earlier and staying green longer than ever in recorded history. This appears to be evidence of a changing global climate. *Source:* NASA, 2002.

where social security and democracy have been established. Over the last 20 years, the average number of children born per woman worldwide has decreased from 6.1 to 3.4. By 2050, the UN Population Division predicts that all developed countries and 75 percent of the developing world will experience a below-replacement fertility rate of 2.1 children per woman. This prediction suggests that the world population will stabilize at about 8.9 billion rather than 9.3 billion, as previously estimated.

The incidence of life-threatening infectious diseases has been reduced sharply in most countries during the past century, while life expectancies have nearly doubled on average. In 2003, the entire European region was declared free of polio, and volunteers vaccinated all of India's 165 million children under the age of five against this crippling disease in just six days. Since 1990, more than 800 million people have gained access to improved water supplies and modern sanitation. In spite of population growth that added nearly a billion people to the world during the 1990s, the number facing food insecurity and chronic hunger during this period actually declined by about 40 million.

Deforestation has slowed in Asia, from more than 8 percent during the 1980s to less than 1 percent in the 1990s. Nature preserves and protected areas have increased nearly fivefold over the past 20 years, from about 2.6 million km^2 to about 12.2 million km^2. This represents only 8.2 percent of all land area—less than the 12 percent thought necessary to protect a viable sample of the world's biodiversity—but is a dramatic expansion nonetheless.

FIGURE 1.10 At least half of all primates are considered threatened or endangered. Hunting and habitat destruction are the biggest problems. © Corbis/Volume 244.

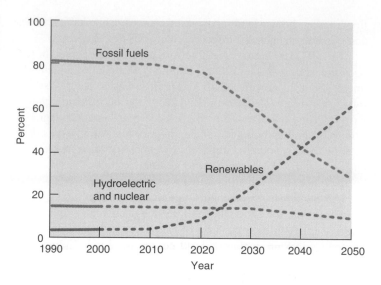

FIGURE 1.11 A possible energy future. Global warming and other environmental problems may require that we switch from our current dependence on fossil fuels to renewable sources such as wind and solar energy. *Source:* World Bank, 2000.

Dramatic progress is being made in a transition to renewable energy sources. The European Union has announced a goal of obtaining 22 percent of its electricity and 12 percent of all energy from renewable sources by 2010. In 2003, Prime Minister Tony Blair laid out ambitious plans to fight global warming by cutting carbon dioxide emissions in Britain by 60 percent through energy conservation and a switch to renewables. If nonpolluting, sustainable energy technology is made available to the world's poorer countries, it may be possible to promote human development while simultaneously reducing environmental damage (fig 1.11).

Democracy and the environment were a major focus of the 2002 World Summit on Sustainable Development. Over the past two decades, the world has made dramatic progress in opening up political systems and expanding political freedoms. During this time, some 81 countries took significant steps toward democracy. Currently, nearly three-quarters of the world's 200 countries now hold multiparty elections. At least 60 developing countries claim to be transferring decision-making authority to local units of government. Of course, decentralization doesn't always guarantee better environmental stewardship, but it puts people with direct knowledge of local conditions in a position of power rather than distant elites or bureaucrats.

Of the 500 or so international environmental agreements now in effect, 150 are global treaties, while others have a more limited set of parties. Some, such as the Montreal Protocol on Protecting Stratospheric Ozone, or the Convention on International Trade in Endangered Species, have had notable success. Among the most important challenges in planning and policy identified by the United Nations Environment Program are (1) building capacity for public participation, (2) recognizing the standing of all stakeholders in environmental decisions, (3) insisting on sustainability in all economic sectors, and (4) measuring progress in governance as a key environmental indicator.

RICH/POOR: A DIVIDED WORLD

We live in a world of haves and have-nots: a few of us live in increasing luxury while many others lack the basic necessities for a decent, healthy, productive life. The World Bank estimates that more than 2.8 billion people (nearly half the world population) live on less than $2 (U.S.) per day. At least 1.4 billion of the poorest of the poor live in **extreme poverty,** scraping by on less than $1 (U.S.) per day. Most in this category lack access to an adequate diet, decent housing, basic sanitation, clean water, education, medical care, and other essentials for a humane existence. The proportion living under these bitter conditions has declined steadily since 1990, but because of population growth, the total number living in extreme poverty (excluding China) has actually risen. The greatest progress in human development has been in East Asia, where nearly 200 million fewer people live in poverty now than in 1990.

The plight of the poor is not just a humanitarian concern. Policymakers are becoming aware that eliminating poverty and

FIGURE 1.12 While many of us live in luxury, more than a billion people lack access to food, housing, clean water, sanitation, education, medical care, and other essentials for a healthy, productive life. Often the poorest people are both the victims and agents of environmental degradation as they struggle to survive. Helping them meet their needs is not only humane, it is essential to protect our mutual environment. © The McGraw-Hill Companies, Inc./Barry W. Barker, photographer.

protecting our common environment are inextricably interlinked (fig. 1.12). In opening the World Summit on Sustainable Development in 2002, President Thabo Mbeki of South Africa said, "A global human society based on poverty for many and prosperity for a few, characterized by islands of wealth, surrounded by a sea of poverty is unsustainable."

The poor often are both the victims and agents of environmental degradation, forced to meet short-term survival needs at the cost of long-term sustainability. The bushmeat trade in Central Africa, for example, is estimated to consume more than one million metric tons of forest animals, including elephants, gorillas, chimpanzees, monkeys and other threatened or endangered species. Desperate for croplands to feed themselves and their families, many poor farmers move into virgin forests or cultivate steep, erosion-prone hillsides where soil nutrients are exhausted after only a few years. Others migrate to the grimy, crowded slums and ramshackle shantytowns that now surround most major cities in the developing world. With few options in energy sources, technology, or waste disposal, the residents often foul the air they breathe and the water on which they depend for washing and drinking.

Human Development

Where do the rich and poor live? About one-fifth of the world's population lives in the 20 wealthiest countries, where the average per capita income is $25,958 (U.S.) per year. Most of these countries are in North America or Western Europe, but Japan, Singapore, Australia, New Zealand, the United Arab Emirates, and Israel also fall into this group. By contrast, the 20 poorest countries, where the average per capita income is only $735 (U.S.) per year, are all in sub-Saharan Africa (fig. 1.13).

These national statistics, however, hide the full extent of the inequality. The 200 richest people in the world have a combined wealth of $1 trillion (U.S.). This is more than the total assets of the poorest half—some 3 billion people—of the world's population. Even some of the most impoverished countries have a wealthy elite—usually the ruling class—who enjoy a life of opulent luxury. And almost every country, even the richest, such as the United States and Canada, has poor people. No doubt, everyone reading this book knows about homeless people or others who lack resources for a safe, productive life. The U.S. Census Bureau reported in 2002 that 32.9 million Americans were living in poverty, defined as an income of less than $18,104 (U.S.) per year for a family of four. By that definition, however, nearly 90 percent of the world lives in poverty.

Of course, cash flow isn't the only way to measure quality of life. A subsistence farmer in a remote area might have little contact with the monetary economy, but may have a healthy diet, a beautiful environment, and a rewarding cultural, intellectual, and emotional life. The United Nations Development Program has devised a measure called the **human development index (HDI)** to evaluate real quality of life. Based on factors such as life expectancy, child survival, adult literacy, childhood education, gender equity, access to clean water and sanitation, and level of income, this index gives an indication of both the real progress in quality of life as well as suggestions for improvement (table 1.1).

The HDI gives some interesting insights into quality of life. While the lowest HDI is found in the poorest countries of sub-Saharan Africa, and the top rankings are found in Northern Europe

Key Concepts

- Environmental science is interdisciplinary, integrating natural and social sciences with humanities to study both the natural world and the social, cultural, and technological world that humans create.

- We face many serious environmental problems, but progress is being made in many areas toward finding sustainable solutions to these dilemmas.

- The human development index evaluates quality of life based on social and environmental factors as well as income.

- Sustainable development means meeting the needs of the present without compromising the ability of future generations to meet their own needs. It implies progress in real human well-being that can last for many generations.

FIGURE 1.13 Three-quarters of the world's poorest nations are in Africa. Millions of people lack adequate food, housing, medical care, clean water, and safety. The human suffering engendered by this poverty is tragic. © Norbert Schiller/The Image Works.

TABLE 1.1	Average Indicators of Quality of Life for the Twenty Richest and Poorest Countries	
INDICATOR	POOR COUNTRIES	RICH COUNTRIES
GNP/capita	$736	$25,958
Life expectancy	45 years	74 years
Infant mortality*	100	5
Safe drinking water	44%	99%
Adult literacy	38%	99%
Annual population growth rate	2.6%	0.3%

*per 1,000 live births
Source: UNDP Human Development Indicators, 2002.

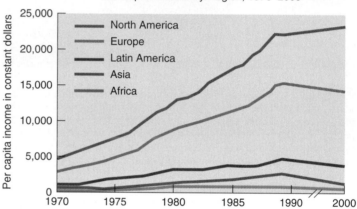

Growing Disparities in Incomes among Regions
Per Capita Income by Region, 1970–2000

- North America
- Europe
- Latin America
- Asia
- Africa

FIGURE 1.14 Although the percentage of the world's population living in poverty has decreased slightly over the past 30 years, the relative gap between rich and poor nations has increased sharply.
Source: United Nations (UN), *Critical Trends: Global Change and Sustainable Development* (UN, New York, 2001), p. 58.

and North America, some very rich countries have relatively low HDI while others with a moderate economy, but a more equitable distribution of wealth and power, have a much higher rank. Egalitarian Sweden, for instance, where the average income is $22,636 (U.S.) per year ranks fourth in the world, while the United States, with $31,872 per person, ranks sixth. Luxembourg, with an income nearly twice that of Sweden, ranks only twelfth in HDI. Similarly, Costa Rica, with an average per capita income of only $8,860 (U.S.) per year, but a high level of democracy and stability, is in the high human development category. Saudi Arabia, on the other hand, where the average income is $10,815 (U.S.) but there is little political freedom, ranks only in the middle of the medium development group.

The disparities between rich and poor increased dramatically over the past three decades (fig. 1.14). This gap isn't just academic.

Saudi Arabia, where ordinary citizens often feel oppressed and alienated, was home to 15 of the 19 terrorists who carried out the September 11, 2001 attacks on the United States. An open letter from 100 Nobel Prize winners warned that the most profound danger to world peace in the coming years will stem not from the irrational acts of states or individuals but from the legitimate anger of the world's dispossessed and marginalized people.

A Fair Share of Resources?

The affluent lifestyle that many of us in the richer countries enjoy consumes an inordinate share of the world's natural resources and produces a shockingly high proportion of pollutants and wastes. The United States, for instance, with less than 5 percent of the total population, consumes about one-quarter of most commercially

TABLE 1.2	The United States, with 4.5 Percent of the World's Population	
CONSUMES	**PRODUCES**	
26% of all oil	50% of all toxic wastes	
24% of aluminum	26% of nitrogen oxides	
20% of copper	25% of sulfur oxides	
19% of nickel	22% of chlorofluorocarbons	
13% of steel	26% of carbon dioxide	

Source: World Resources Institute 1998–99.

FIGURE 1.15   "And may we continue to be worthy of consuming a disproportionate share of this planet's resources." © The New Yorker Collection 1992 Lee Lorenz from cartoonbank.com. All Rights Reserved.

traded commodities and produces a quarter to half of most industrial wastes (table 1.2).

To get an average American through the day takes about 450 kg (nearly 1,000 lbs) of raw materials, including 18 kg of fossil fuels, 13 kg of other minerals, 12 kg of farm products, 10 kg of wood and paper, and 450 liters (119 gal) of water. Every year we throw away some 160 million tons of garbage, including 50 million tons of paper, 67 billion cans and bottles, 25 billion styrofoam cups, 18 billion disposable diapers, and 2 billion disposable razors (fig. 1.15).

This profligate resource consumption and waste disposal strains the life-support systems of the earth on which we all depend. If everyone in the world tried to live at consumption levels approaching those of the United States, the results would be disastrous. Unless we find ways to curb our desires and produce the things we truly need in less destructive ways, the sustainability of human life on our planet is questionable.

Economic Progress

Over the past 50 years, human ingenuity and enterprise have brought about a breathtaking pace of technological innovations and scientific breakthroughs. The world's gross domestic product increased more than tenfold during that period, from $2 trillion to $22 trillion per year. While not all that increased wealth was applied to human development, there has been significant progress in increasing general standard of living nearly everywhere. In 1960, for instance, nearly three-quarters of the world's population lived in abysmal conditions (HDI below 0.5). Now, less than one-third are still at this low level of development.

Since World War II, average real income in developing countries has doubled; malnutrition declined by almost one-third; child death rates have been reduced by two-thirds; average life expectancy increased by 30 percent. Overall, poverty rates have decreased more in the last 50 years than in the previous 500. Nonetheless, while general welfare has increased, so has the gap between rich and poor worldwide. In 1960, the income ratio between the richest 20 percent of the world and the poorest 20 percent was 30 to 1. In 2000, this ratio was 100 to 1. Because

perceptions of poverty are relative, people may feel worse off compared to their rich neighbors than development indices suggest they are.

SUSTAINABLE DEVELOPMENT

By now, it is clear that security and living standards for the world's poorest people are inextricably linked to environmental protection. One of the most important questions in environmental science is how we can continue improvements in human welfare within the limits of the earth's natural resources. A possible solution to this dilemma is **sustainable development,** a term popularized by *Our Common Future,* the 1987 report of the World Commission on Environment and Development, chaired by Norwegian Prime Minister Gro Harlem Brundtland (and consequently called the Brundtland Commission). In the words of this report, sustainable development means "meeting the needs of the present without compromising the ability of future generations to meet their own needs."

Another way of saying this is that we are dependent on nature for food, water, energy, fiber, waste disposal, and other life-support services. We can't deplete resources or create wastes faster than nature can recycle them if we hope to be here for the long term. Development means improving people's lives. Sustainable development, then, means progress in human well-being that can be extended or prolonged over many generations rather than just a few years. To be truly enduring, the benefits of sustainable development must be available to all humans rather than to just the members of a privileged group.

To many economists, it seems obvious that economic growth is the only way to bring about a long-range transformation to more advanced and productive societies and to provide resources to improve the lot of all people. As former President John F. Kennedy said, "A rising tide lifts all boats." But economic growth is not sufficient in itself to meet all essential needs. As the Brundtland Commission pointed out, political stability, democracy, and equitable economic distribution are needed to ensure that the poor will get a fair share of the benefits of greater wealth in a society.

Can Development Be Truly Sustainable?

Many ecologists regard "sustainable" growth of any sort as impossible in the long run because of the limits imposed by nonrenewable resources and the capacity of the biosphere to absorb our wastes. Using ever-increasing amounts of goods and services to make human life more comfortable, pleasant, or agreeable must inevitably interfere with the survival of other species and, eventually, of humans themselves in a world of fixed resources. But, supporters of sustainable development assure us, both technology and social organization can be managed in ways that meet essential needs and provide long-term—but not infinite—growth within natural limits, if we use ecological knowledge in our planning.

While economic growth makes possible a more comfortable lifestyle, it doesn't automatically result in a cleaner environment. As figure 1.16 shows, people will purchase clean water and sanitation if they can afford to do so. For low-income people, however, more money tends to result in higher air pollution because they can afford to burn more fuel for transportation and heating. Given enough money, people will be able to afford both convenience *and* clean air. Some environmental problems, such as waste generation and carbon dioxide emissions, continue to rise sharply with increasing wealth because their effects are diffuse and delayed. If we are able to sustain economic growth, we will need to develop personal restraint or social institutions to deal with these problems.

Some projects intended to foster development have been environmental, economic, and social disasters. Large-scale hydropower projects, like that in the James Bay region of Quebec or the Brazilian Amazon that were intended to generate valuable electrical power, also displaced indigenous people, destroyed wildlife, and poisoned local ecosystems with acids from decaying vegetation and heavy metals leached out of flooded soils. Similarly, introduction of "miracle" crop varieties in Asia and huge grazing projects in Africa financed by international lending agencies crowded out wildlife, diminished the diversity of traditional crops, and destroyed markets for small-scale farmers.

Other development projects, however, work more closely with both nature and local social systems. Socially conscious businesses and environmental, nongovernmental organizations sponsor ventures that allow people in developing countries to grow or make high-value products—often using traditional techniques and designs—that can be sold on world markets for good prices (fig. 1.17). Pueblo to People, for example, is a nonprofit organization that buys textiles and crafts directly from producers in Latin America. It sells goods in America, with the profits going to community development projects in Guatemala, El Salvador, and Peru. It also informs customers in wealthy countries about the conditions in the developing world.

As the economist John Stuart Mill wrote in 1857, "It is scarcely necessary to remark that a stationary condition of capital and pop-

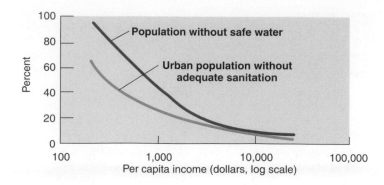

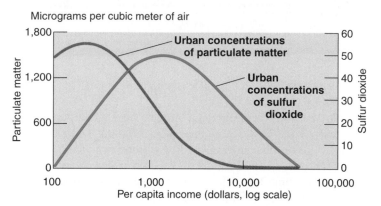

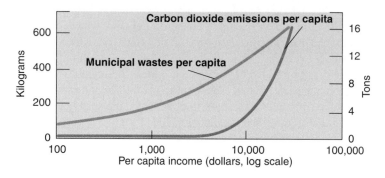

FIGURE 1.16 Environmental indicators show different patterns as average national income increases. When people have more money they invariably will purchase clean water and better sanitation. Rising income may temporarily produce increased urban air pollution (for example, particulates and sulfur dioxide) as people burn more fuel; eventually, however, people can afford both clean air and the benefits of technology. Some environmental problems such as waste generation and carbon dioxide emissions rise sharply with increasing wealth because of increased demands for goods and services without recognition of adverse environmental consequences. *Source:* World Bank, *World Development Report,* 1996.

ulation implies no stationary state of human improvement. There would be just as much scope as ever for all kinds of mental culture and moral and social progress; as much room for improving the art of living and much more likelihood of its being improved when minds cease to be engrossed by the art of getting on." Somehow, in our rush to exploit nature and consume resources, we have forgotten this sage advice.

FIGURE 1.17 A Mayan woman from Guatemala weaves on a back-strap loom. A member of a women's weaving cooperative, she sells her work to nonprofit organizations in the United States at much higher prices than she would get at the local market. © The McGraw-Hill Companies, Inc./Barry W. Barker, photographer.

The 20:20 Compact for Human Development

At the 1995 United Nations Summit for Social Development in Copenhagen, a world social charter was passed that calls on all nations to ensure basic human needs for everyone. Among the goals in this world action plan to vanquish poverty and injustice are:

- universal primary education—for girls as well as for boys;

- adult illiteracy rates to be halved—with the female rate to be no higher than the male one;

- elimination of severe malnutrition;

- family planning services for all who wish them;

- safe drinking water and sanitation for all;

- credit for all—to ensure self-employment opportunities.

How much will this cost? A rough estimate provided by the United Nations Development Agency is that an additional $30 to $40 billion (U.S.) per year is needed to meet these targets. This is a large amount but not an impossible one. Developing countries

FIGURE 1.18 Every year, military spending equals the total income of half the world's people. The cost of a single large aircraft carrier equals ten years of human development aid given by all the world's industrialized countries. © Stone/Getty Images.

now devote, on average, only $57 billion per year (13 percent of their national budgets) to basic human needs. Military spending by these countries, in contrast, averages $125 billion per year. If weapons purchases were cut in half, human development could be doubled.

An innovative suggestion called the 20:20 Compact was offered by the World Social Summit. Donor countries now allocate, on average, only 7 percent of their aid to humanitarian concerns. If they were to shift 20 percent of aid to social development, it would provide an additional $12 billion per year. Similarly, if developing countries would earmark 20 percent of their budgets to human priority concerns, $88 billion per year would be available. The $100 billion raised by this compact is dwarfed by the one-trillion-dollar annual total world military spending, but it would be three times the minimum needed for the human development agenda (fig. 1.18).

Indigenous People

Often at the absolute bottom of the social strata, whether in rich or poor countries, are the indigenous or native peoples who are generally the least powerful, most neglected groups in the world. Typically descendants of the original inhabitants of an area taken over by more powerful outsiders, they are distinct from their country's dominant language, culture, religion, and racial communities. Of the world's nearly 6,000 recognized cultures, 5,000 are indigenous ones that account for only about 10 percent of the total world population. In many countries, these indigenous people are repressed by traditional caste systems, discriminatory laws, economics, or prejudice. Unique cultures are disappearing along with biological diversity as natural habitats are destroyed to satisfy industrialized world appetites for resources. Traditional ways of life are disrupted further by dominant Western culture sweeping around the globe.

At least half of the world's 6,000 distinct languages are dying because they are no longer taught to children. When the last few elders who still speak the language die, so will the culture that was

FIGURE 1.19 Do indigenous people have unique knowledge about nature and inalienable rights to traditional territories?
© William P. Cunningham.

Highest cultural diversity		Highest biological diversity
Nigeria	Indonesia	Madagascar
Cameroon	New Guinea	South Africa
Australia	Mexico	Malaysia
Congo	China	Cuba
Sudan	Brazil	Peru
Chad	United States	Ecuador
Nepal	Philippines	New Zealand

FIGURE 1.20 Cultural diversity and biodiversity often go hand in hand. Seven of the countries with the highest cultural diversity in the world are also on the list of "megadiversity" countries with the highest number of unique biological organisms. (Listed in decreasing order of importance.) *Source:* Norman Myers, Conservation International and Cultural Survival Inc., 2002.

its origin. Lost with those cultures will be a rich repertoire of knowledge about nature and a keen understanding about a particular environment and a way of life (fig. 1.19).

Nonetheless, in many places, the 500 million indigenous people who remain in traditional homelands still possess valuable ecological wisdom and remain the guardians of little-disturbed habitats that are the refuge for rare and endangered species and undamaged ecosystems. Author Alan Durning estimates that indigenous homelands harbor more biodiversity than all the world's nature reserves and that greater understanding of nature is encoded in the languages, customs, and practices of native people than is stored in all the libraries of modern science. Interestingly, just 12 countries account for 60 percent of all human languages (fig. 1.20). Seven of those are also among the "megadiversity" countries that contain more than half of all unique plant and animal species. Conditions that support evolution of many unique species seem to favor development of equally diverse human cultures as well.

Recognizing native land rights and promoting political pluralism is often one of the best ways to safeguard ecological processes and endangered species. As the Kuna Indians of Panama say, "Where there are forests, there are native people, and where there are native people, there are forests." A few countries, such as Papua New Guinea, Fiji, Ecuador, Canada, and Australia acknowledge indigenous title to extensive land areas.

In other countries, unfortunately, the rights of native people are ignored. Indonesia, for instance, claims ownership of nearly three-quarters of its forest lands and all waters and offshore fishing rights, ignoring the interests of indigenous people who have lived in these areas for millennia. Similarly, the Philippine government claims possession of all uncultivated land in its territory, while Cameroon and Tanzania recognize no rights at all for forest-dwelling pygmies who represent one of the world's oldest cultures.

Summary

- Your "ecological footprint" estimates your impact on the earth and the amount of biologically productive land needed to support your current lifestyle. If everyone in the world lived at the same standard as those in the United States, we'd need four more planets to sustain us all.

- Environmental science is the systematic study of both the natural world and the social, cultural, and technological worlds created by humans.

- Nature protection has deep roots reaching back into ancient history. We can divide recent conservation history into at least four distinct stages: pragmatic (or utilitarian) resource conservation, moral or aesthetic nature preservation, modern environmentalism with its concern for chemical pollution, and global environmentalism.

- We live on a marvelous, bountiful planet that, as far as we know, is unique in the universe. We should ask ourselves: what ought we do or what can we do to protect the habitat that produced and supports us?

- Many environmental dilemmas now face us. Water shortages, food production, energy supplies, the effects of burning fossil fuels, global climate change, and biodiversity losses all are serious concerns.

- There are reasons, however, for hope. Population growth is slowing, many terrible diseases have been conquered, progress is being made in a transition to renewable energy, and some countries are reducing their greenhouse gas emissions.

- At least 1.4 billion people live in extreme poverty on less than $1 (U.S.) per day. All too often the poor are both the victims and agents of environmental degradation, forced to meet short-term survival at the cost of long-term sustainability.

- The affluent lifestyle of the richest people in the world consumes an inordinate share of the world's natural resources and produces a high proportion of pollution and wastes.

- Sustainable development means meeting the needs of the present without compromising the ability of future generations to meet their own needs. We must live within the means of nature if we hope to survive for the long term.

- Indigenous or native peoples are generally among the poorest and most oppressed of any group. Nevertheless, they often possess valuable ecological knowledge and remain the guardians of nature when allowed to do so. Recognizing their rights can be a good way to protect natural resources and environmental quality.

Questions for Review

1. Define *environment* and *environmental science*.
2. Describe four stages in conservation history and identify leaders associated with each stage.
3. List six environmental dilemmas that we now face and describe how each concerns us.
4. Define the *human development index* and compare some indicators of quality of life between the richest and poorest nations.
5. Why should we be concerned about the plight of the poor? How do they affect us?
6. What is sustainable development?
7. Identify six goals for human development.
8. What benefit to us would there be in protecting the rights of indigenous people?

Questions for Critical Thinking

1. What resource uses are most strongly represented in the ecological footprint? What are the best and worst features of this assessment?
2. What are the fundamental differences between utilitarian conservation and altruistic preservation? Which do you favor? Why?
3. Do the issues discussed in this chapter as global environmentalism belong in an environmental *science* text? Why would anyone ask this question?
4. Some people argue that we can't afford to be generous, tolerant, fair, or patient. There isn't enough to go around as it is, they say. What questions would you ask such a person?
5. Some people claim that we live in a world of bounty. They believe there would be plenty for all if we just shared equitably. What questions would you ask such a person?
6. Around 200 million children are forced into dangerous, degrading labor each year. Is it our business what goes on in other countries?

7. What would it take for human development to be truly sustainable? What does sustainable mean to you?

8. Are there enough resources in the world for 8 or 10 billion people to live decent, secure, happy, fulfilling lives? What do those terms mean to you? Try to imagine what they mean to others in our global village.

9. What responsibilities do we have to future generations? What have they done for us? Why not use whatever resources we want right now?

10. Do you believe that indigenous people have special knowledge about nature or unique rights to their native land? Why or why not?

Key Terms

biocentric preservation 18
environment 16
environmentalism 19
environmental science 16
extreme poverty 24

global environmentalism 20
human development index (HDI) 25
sustainable development 27
utilitarian conservation 18

Further Readings

Blaustein, A. R., and P. T. T. Johnson. 2003. Explaining frog deformities. *Scientific American* 288(2):60–65.

Brown, L., et al. 2003. *State of the World 2003.* Washington, D.C.: Worldwatch Institute.

Folke, Carl, et al. 2002. *Resilience and Sustainable Development: Building Adaptive Capacity in a World of Transformations.* International Council for Science. (www.icgu.org/library/WSS-REP/vol3.pdf)

Gewin, V. 2002. Ecosystem health: The state of the planet. *Nature* 417:112–13.

International Council for Science, 2002. *Report of the Scientific and Technological Community to the World Summit on Sustainable Development.* (www.sustainabilityscience.org/keydocs/)

Sachs, Jeffery, Andrew D. Mellinger, and John L. Gallup. 2001. The geography of poverty and wealth. *Scientific American* 284(3):70–73.

Sanderson E. W., et al. 2002. The human footprint and the last of the wild. *Bioscience* 52(10):891–904.

Wackernagel, Mathis, et al. 2002. Tracking the ecological overshoot of the human economy. *Proc. Natl. Acad. Sci. USA* 99(14):9266–71.

World Resources Institute. 2003. *World Resources 2003–2004: A Guide to World Resources 2002–2004: Decisions for the Earth: Balance, Voice, and Power.* New York: Oxford University Press.

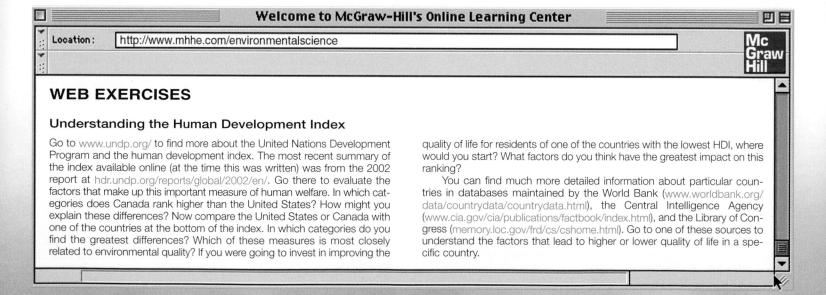

Welcome to McGraw-Hill's Online Learning Center

Location: http://www.mhhe.com/environmentalscience

WEB EXERCISES

Understanding the Human Development Index

Go to www.undp.org/ to find more about the United Nations Development Program and the human development index. The most recent summary of the index available online (at the time this was written) was from the 2002 report at hdr.undp.org/reports/global/2002/en/. Go there to evaluate the factors that make up this important measure of human welfare. In which categories does Canada rank higher than the United States? How might you explain these differences? Now compare the United States or Canada with one of the countries at the bottom of the index. In which categories do you find the greatest differences? Which of these measures is most closely related to environmental quality? If you were going to invest in improving the

quality of life for residents of one of the countries with the lowest HDI, where would you start? What factors do you think have the greatest impact on this ranking?

You can find much more detailed information about particular countries in databases maintained by the World Bank (www.worldbank.org/data/countrydata/countrydata.html), the Central Intelligence Agency (www.cia.gov/cia/publications/factbook/index.html), and the Library of Congress (memory.loc.gov/frd/cs/cshome.html). Go to one of these sources to understand the factors that lead to higher or lower quality of life in a specific country.

Environmental Philosophy, Ethics, and Science

A thing is right when it tends to preserve the integrity, stability, and beauty of the biotic community.
It is wrong when it tends otherwise.

Aldo Leopold

OBJECTIVES

After studying this chapter, you should be able to:

- understand some principles of environmental philosophy, ethics, and science.
- compare and contrast how different ethical perspectives shape our view of nature and our role in it.
- describe how religious and cultural traditions influence our perceptions of nature.
- explain anthropocentrism, biocentrism, utilitarianism, and ecofeminism, and what each says about relationships between humans and nature.
- summarize the methods, applications, and limitations of the scientific method.

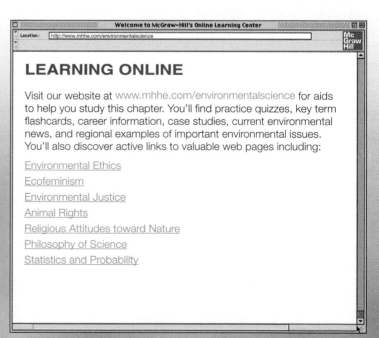

Welcome to McGraw-Hill's Online Learning Center

Location: http://www.mhhe.com/environmentalscience

LEARNING ONLINE

Visit our website at www.mhhe.com/environmentalscience for aids to help you study this chapter. You'll find practice quizzes, key term flashcards, career information, case studies, current environmental news, and regional examples of important environmental issues. You'll also discover active links to valuable web pages including:

Environmental Ethics

Ecofeminism

Environmental Justice

Animal Rights

Religious Attitudes toward Nature

Philosophy of Science

Statistics and Probability

Photo: Hammerhead sharks and other marine top predators are declining rapidly. © Dennis Scott/Corbis.

Sharkless Seas?

Every year, humans kill about 100 million sharks, skates, and rays, about half of them caught as unwanted "by-catch" while fishing for other species like swordfish and tuna. Sharks are particularly sensitive to overfishing because they grow slowly, mature late, and have few young in each generation. As top predators, sharks play a crucial role in the health of the ocean ecosystems. Researchers warn that shark extermination could lead to major ecological changes in the ocean.

Shark populations are thought to be plummeting around the world. In the North Atlantic, for example, Canadian researchers estimate that all coastal shark species plunged an average of 61 percent between 1986 and 2000. Some of the largest species had the biggest declines. Hammerhead sharks (see p. 33) dropped by 89 percent, while great white and thresher sharks fell 80 percent. In 2003, the UN Convention on International Trade in Endangered Species, or CITES, restricted commercial trade in whale and basking sharks, the largest fish in the world, because of worries that the species may be threatened with extinction.

Key factors in the recent increase in shark deaths are the rapid emergence of a middle class in China and the traditional Asian taste for shark fin soup. This expensive delicacy, which can sell for as much as $100 per bowl in Hong Kong restaurants, is a conspicuous way of showing increased affluence. Rising demand means that a single large fin from a whale shark can now sell for more than $10,000 in Asian markets. In the 1990s, rapidly increasing prices boosted the shark fin traffic in Hawaii more than 20-fold: from an estimated 200,000 pounds to 4.5 million pounds (41,000 kg to 2 million kg) per year.

Because the prices for shark meat are low, many fishing crews cut off the fins of any shark they catch—often while the animal is still alive—and simply throw the body overboard (fig. 2.1). That finless shark is eaten by another, bleeds to death, or drowns because it can no longer swim. A fishing trip is much more profitable if you fill up the hold with valuable fins rather than whole fish. Many conservationists consider this practice cruel, wasteful, and highly unethical.

In response to a public outcry, the United States and a few other nations have banned shark finning within their territorial waters. It is also illegal to bring fins into American ports without the corresponding carcasses. Certain trade groups and some other nations criticize these rules arguing that ethical considerations shouldn't apply to sharks, which are "merely fish." They consider such restrictions an imperialistic attempt to force American ideals on the rest of the world.

FIGURE 2.1   Cutting fins off living sharks is regarded by some people as both unethical and wasteful. © Jeffrey L. Rotman/Corbis.

And they worry that if these rules prevail, other regulations on the way they catch and kill other species might soon follow.

What do you think? Do sharks and other fish have intrinsic rights? Do we have an obligation to treat them humanely, or can they be harvested, like potatoes, without any thought to whether they suffer? As far as we know, sharks aren't as intelligent or self-conscious as humans. Is it enough that they are alive and can experience pain to grant them moral status? Every year sharks attack people without any apparent ethical qualms. Does this give us the right to attack them in return? Even if we consider sharks dangerous, the numbers seem out of balance. In 2002, only 76 unprovoked shark attacks occurred worldwide, resulting in five human deaths. These are far fewer human deaths than from dog bites every year, and very different from the 100 million or so sharks and their kin that we kill annually.

Sharks are superbly adapted to life in the ocean. For some people, killing such a wonderfully evolved animal is akin to destroying a magnificent work of art. Do beauty or excellence in design give some species special ethical value? These questions illustrate some of the moral dilemmas involved in our interactions with nature. Do other species deserve moral consideration? If so, which ones, and how much? How do western notions of morality and ethics compare to those of other cultures? These are some of the topics we'll examine in this chapter.

ENVIRONMENTAL ETHICS AND PHILOSOPHY

Ethics is a branch of philosophy concerned with **morals** (the distinction between right and wrong) and **values** (the ultimate worth of actions or things). Ethics evaluates the relationships, rules, principles, or codes that require or forbid certain conduct. Most Western ethicists consider the roots of their field to be the famous questions posed by Socrates and the Greek philosophers 2,500 years ago: "What is the good life? How ought we, as moral beings, behave?"

Environmental ethics asks about the moral relationships between humans and the world around us. Do we have special duties, obligations, or responsibilities to other species or to nature in general? Are there ethical principles that constrain how we use resources or modify our environment? If so, what are the foundations of those constraints and how do they differ from principles governing our relations to other humans? How are our obligations and responsibilities to nature weighed against human values and interests? Do some interests or values supercede others?

Are There Universal Ethical Principles?

The first question considered in ethics is whether *any* moral laws are objectively valid and independent of cultural context, history, or situation. How the question is answered depends largely on the philosophical disposition of the respondent. **Universalists,** such as Plato (427–347 B.C.E.) and Kant (1724–1804), assert that the fundamental principles of ethics are universal, unchanging, and eternal (fig. 2.2). These rules of right and wrong are valid regardless of our interests, attitudes, desires, or preferences. Some believe these rules are revealed by God, while others maintain they can be discovered through reason and knowledge.

Relativists, such as Plato's opponents, the Sophists, claim that moral principles always are relative to a particular person, society, or situation. Although there may be right and wrong—or at least better and worse—things to do, relativists assert that no transcendent, absolute principles apply regardless of circumstances. In this view, ethical values always are contextual. Friedrich Nietzsche's famous aphorism, "there are no facts, only interpretations" has been the banner for recent generations of relativists.

Nihilists, such as Schopenhauer (1788–1860), claim that the world makes no sense at all. Everything is completely arbitrary, and there is no meaning or purpose in life other than the dark, instinctive, unceasing struggle for existence. According to this view, there is no reason to behave morally. Only power, strength, and sheer survival matter. "Might is right; eat or be eaten." There is no such thing as a "good" life: we live in a world of uncertainty, pain, and despair. Nevertheless, Schopenhauer enjoyed living in a comfortable, civilized society where rules of normative behavior and good conduct prevailed.

Utilitarians hold that an action is right that produces the greatest good for the greatest number of people. This philosophy is usually associated with the English philosopher Jeremy Bentham (1748–1832); but something very similar was suggested by Plato, Socrates, Aristotle, and others. Bentham was an eccentric genius and a hedonist who equated goodness with happiness, and happiness with pleasure. He regarded pleasure as the only thing worth having in its own right. Thus, the good life is one of maximum pleasure. Insofar as people are moral animals, in his view, we should act to produce the greatest pleasure for the greatest number. To do so is good; not to do so is wrong.

Utilitarianism was modified and made less hedonistic by Bentham's brilliant protégé, John Stuart Mill (1806–1873). Mill believed that pleasures of the intellect are superior to pleasures of the body. He held that the greatest pleasure is to be educated and to act according to enlightened, humanitarian principles. This enlightened form of utilitarianism inspired Gifford Pinchot and the early conservationists (see chapter 1), who argued that the purpose of conservation is to protect resources for the "greatest good for the greatest number *for the longest time.*"

Although utilitarianism remains widely popular today, it has drawbacks. It can, for instance, be used to justify reprehensible acts. If ten thousand Romans greatly enjoyed watching a few Christians being eaten by lions, did that make it the right thing to do? Does the pleasure of the tormentors outweigh the suffering of victims? Most of us would conclude that it does not. Justice, freedom, morals, and loyalty take precedence over pleasure, or even happiness, although it could be argued that furthering moral ends and right action ought to bring the greatest happiness in the long run. A utilitarian calculation of costs and benefits of environmental projects remains the ultimate justification for much government policy (see chapter 23).

Values, Rights, and Obligations

For many philosophers, only humans are **moral agents,** beings capable of acting morally or immorally and who can—and should—accept responsibility for their acts. Capacities that enable humans to form moral judgments include moral deliberation, the resolve to carry out decisions, and the responsibility to hold oneself answerable for failing to do what is right.

Of course, not all humans have all these capacities all the time. Children and those who are mentally retarded, mentally ill, or for

FIGURE 2.2   Plato and Aristotle debate moral philosophy in a painting by Raphael. Plato (*left*) motions upward, indicating a transcendent, universal moral truth, while Aristotle (*right*) motions downward to suggest grounded, situational ethics. © Courtesy Monumenti Musei E Gallerie Pontificie.

some other reason, lacking a full use of reason, are not regarded as moral agents. If a child murders someone, we don't hold the child responsible. Nonetheless, she/he still has rights. Children are considered **moral subjects,** beings who are not moral agents themselves but who have moral interests of their own and can be treated rightly or wrongly by others.

Historically, the idea that weaker members of society deserve equal treatment with those who are stronger was not a universally held opinion. In many societies, women, children, outsiders, serfs, and others were treated as property by those who were more powerful. The Greek philosophers, for example, accepted slavery as natural and necessary. Gradually, we have come to believe that all humans have certain inalienable rights: life, liberty, and the pursuit of happiness, for example. No one can ethically treat another human as a mere object for their own pleasure, gratification, or profit. This gradually widening definition of whom we consider ethically significant is called **moral extensionism.**

Do Other Animals Have Rights?

Perhaps the most important question in environmental ethics is whether moral extensionism encompasses nonhumans. Do other species have rights as well? Are they moral agents or at least moral subjects? For many philosophers, the answer is no. Reason and consciousness—or at least a potential thereof—are essential for moral considerability. René Descartes (1596–1650), for instance, claimed that animals are mere automata (machines) and can neither reason nor feel pain (fig. 2.3). In this view, sharks and other creatures are simply objects that we can treat in any way we choose.

FIGURE 2.3 Do other organisms have inherent values and rights? These chickens are treated as if they are merely egg-laying machines. Many people argue that we should treat them more humanely.
© Grant Heilman.

Animal rights proponents, on the other hand, disagree, claiming that while animals may not have the same self-consciousness as humans, they are intelligent and clearly have feelings. As sentient (perceptive) beings, they deserve ethical treatment. But what about nonsentient beings? Should we extend moral consideration to bugs, plants, rocks, even landscapes? Some people think so. Let's see why.

Intrinsic and Instrumental Values

Rather than couch ethics strictly in terms of rights, some philosophers prefer to consider values. Value is a measure of the worth of something. But value can be either inherent or conferred. All humans, we believe, have **inherent value**—an intrinsic or innate worth—simply because they are human. They deserve moral consideration no matter who they are or what they do. Tools, on the other hand, have conferred, or **instrumental value.** They are worth something only because they are valued by someone who matters. If I hurt you without good reason, I owe you an apology. If I borrow your car and smash it into a tree, however, I don't owe the car an apology. I owe *you* the apology—or reimbursement—for ruining your car. The car is valuable only because you want to use it. It doesn't have inherent values or rights of itself.

Do Nonsentient Things Have Inherent Value?

How does this apply to nonhumans? Domestic animals clearly have an instrumental value because they are useful to their owners. But some philosophers would say they also have inherent values and interests. By living, breathing, struggling to stay alive, the animal carries on its own life independent of its usefulness to someone else. Some would draw a line of moral considerability at the limit of sentience (the ability to perceive sensations such as pain). Others argue that it is in the best interest of bugs, worms, and plants to be treated well, even if they aren't aware of it. In this perspective, just being alive gives things inherent value.

Some people believe that even nonliving things also have inherent worth. Rocks, rivers, mountains, landscapes, and certainly the earth itself, have value. These things were in existence before we came along and we couldn't re-create them if they are altered or destroyed. In a landmark 1969 court case, the Sierra Club sued the Disney Corporation on behalf of the trees, rocks, and wildlife of Mineral King Valley in the Sierra Nevada Mountains (fig. 2.4) where Disney wanted to build a ski resort. The Sierra Club argued that it represented the interests of beings that could not speak for themselves in court.

A legal brief entitled *Should Trees Have Standing?,* written for this case by Christopher D. Stone, proposed that organisms as well as ecological systems and processes should have standing (or rights) in court. After all, corporations—such as Disney—are treated as persons and given legal rights even though they are really only figments of our imagination. Why shouldn't nature have similar standing? The case went all the way to the Supreme Court but was overturned on a technicality. In the meantime, Disney lost interest in the project and the ski resort was never built. What do you think? Where would you draw the line of what deserves moral considerability? Are there ethical limits on what we can do to nature?

FIGURE 2.4 Mineral King Valley at the southern border of Sequoia National Park was the focus of an important environmental law case in 1969. The Disney Corporation wanted to build a ski resort here, but the Sierra Club sued to protect the valley on behalf of the trees, rocks, and native wildlife. © Carr Clifton.

RELIGIOUS AND CULTURAL PERSPECTIVES

Historian Lynn White, Jr., said, "What people do about their ecology depends on what they think about themselves in relation to the things around them." Often our ideas about our proper role in the world are based less on reason and logic than on culture, intuition, and religion. In this section, we'll look at some major religious or moral philosophies and what they tell us about our proper relationship with nature.

Buddhism, Shamanism and Nature-Based Religions

In many cultures, traditional religions hold that all creatures are imbued with souls. Even rocks, rivers, mountains, and phenomena such as storms are manifestations of sacred spirits that deserve reverence and respect. Before one cuts a tree, dams a brook, or kills

an animal, the particular spirit or deity in charge of that situation must be honored and placated. Shamans also communicate with these spirits to foretell the future, cure diseases, and control the forces of nature. Taboos that limit when, how, or where people could travel, live, or harvest natural resources have played important roles in preserving biodiversity and ecological integrity in many places. Native Americans in the Pacific Northwest, for example, were forbidden to fish on certain days during the salmon run, thus ensuring that enough fish would make it to their spawning grounds to produce future generations.

Buddhism, Shintoism, and Taoism share much of this reverence for nature. All living things are held to be worthy of respect, a philosophy known as **biocentrism.** Many Buddhists, for example, refrain from killing spiders, flies, rodents, and other creatures that many people might regard as pests (fig. 2.5). For some people, this reverence for nature is focused on a particular ancient tree, uniquely shaped rock, or sacred place. It doesn't necessarily protect all trees, rocks, or wildlife. Japan, for example, where Buddhism and Shintoism are widespread, has been very careful to protect its own forests, but has been among the biggest markets for tropical lumber from other countries. In China, as the opening story of this

FIGURE 2.5 Avalokiteshvara, the manifestation of the compassionate Buddha, seeks harmony between all beings. © William P. Cunningham.

chapter shows, a taste for rare wildlife is threatening the survival of some species. And some predominantly Buddhist countries such as Cambodia, Thailand, and Burma have destroyed large forest areas and decimated wildlife populations. Perhaps the most successful application of Buddhist principles in nature protection is the tiny mountain state of Bhutan, where more than 90 percent of the original forest cover and biodiversity are reported to be preserved.

Christianity, Judaism, and Islam

Drawing on common historic origins, various branches of Christianity, Judaism, and Islam share many beliefs about nature and our role in it. Often this role has been thought to be one of mastery, control, and special privilege in which we are called to exploit nature in whatever manner we wish. In an influential 1967 paper entitled *The Historic Roots of Our Ecological Crisis,* Lynn White, Jr., traced the view that justifies human domination of nature to the biblical injunction to "Be fruitful, and multiply, and replenish the earth, and subdue it: and have dominion over the fish of the sea, and over the fowl of the air, and over every living thing that moveth upon the earth" (Genesis 1:28). This view of creation, White claimed, destroyed pagan animism and made it possible to exploit nature with indifference to feelings of other species (fig. 2.6). A belief in humans as masters of the world with a unique set of rights and values is termed **anthropocentrism.**

Although many people agree with White's analysis, others argue that this passage is translated and interpreted inaccurately. Genesis is really intended, they claim, to teach us to love and nurture creation rather than dominate and exploit it. They point out that Judeo-Christian and Islamic traditions also have a long history of practicing **stewardship,** or responsibility to care for or manage a particular place. Humility and reverence are essential in this worldview, where we are seen as partners in nature rather than masters of it. As caretakers of resources, stewards see themselves working together with the rest of nature to sustain life and

make the world a better place. According to the Islamic scholar, Dr. Abubakr Ahmed Bagader, everything in the universe has value and purpose. "God's wisdom grants human beings stewardship (khilafah) on the earth. . . As such man is only a manager of the earth and not a proprietor; a beneficiary and not a disposer or ordainer. . . He is thus entrusted with its maintenance and care, and must use it as a trustee."

In recent years, a number of Christian, Jewish, and Islamic groups have played important roles in nature protection. The National Religious Partnership for the Environment, for instance, is made up of representatives of most major denominations in North America. "Although we may disagree on how the earth's creatures were created," they say, "we really need to work together to save what's left." Acting on this belief, a coalition of evangelical Christians has been instrumental in opposing Congressional attacks on the U.S. Endangered Species Act. Similarly, the Central Conference of American Rabbis declared that desecration of the headwaters Redwood Forest in California breaks our covenant with the Creator.

Religious concern extends beyond our treatment of plants and animals. Pope John Paul II and Orthodox Patriarch Bartholomew called on countries bordering the Black Sea to stop pollution, saying "to commit a crime against nature is a sin." And in 2002, the Evangelical Environmental Network ran ads asking, "What Would Jesus Drive?" The ads explained, "Of all the choices we make as consumers, the cars we drive have the single biggest impact on all of God's creation. Car pollution causes illness and death, and most afflicts the elderly, poor, sick and young. It also contributes to global warming, putting millions at risk from drought, flood, hunger and homelessness. Transportation is now a moral choice and an issue for Christian reflection. . . We call upon America's automobile industry to manufacture more fuel-efficient vehicles. And we call on Christians to drive them."

Ecofeminism

Many feminists believe that neither Eastern nor Western religious traditions are sufficient to solve environmental problems or to tell us how we ought to behave as moral agents. They argue that all these philosophies come out of patriarchal systems based on domination and duality that assigns prestige and importance to some things but not others. In a patriarchal worldview, men are superior to women, minds are better than bodies, and culture is higher than nature. Feminists contend that domination, exploitation, and mistreatment of women, children, minorities, and nature are intimately connected and mutually reinforcing. They reject all "isms" of domination: sexism, racism, classism, heterosexism, and speciesism.

Ecofeminism, a pluralistic, nonhierarchical, relationship-oriented philosophy that suggests how humans could reconceive themselves and their relationships to nature in nondominating ways, is proposed as an alternative to patriarchal systems of domination. It is concerned not so much with rights, obligations, ownership, and responsibilities as with care, appropriate reciprocity, and kinship. This worldview promotes a richly textured understanding or sense of what human life is and how this understanding can shape people's encounters with the natural world (fig. 2.7).

FIGURE 2.6 Do humans have a right to use or destroy other species or natural objects in any way we choose? Do we have duties, obligations, or responsibilities toward nature? © Inga Spence/Tom Stack & Associates.

FIGURE 2.7 Ecofeminism reexamines our relations with other humans and with nature. It calls for a new ethic of care, reciprocity, and kinship. © The McGraw-Hill Companies, Inc./Barry W. Barker, photographer.

Among ecofeminist leaders are Karen Warren, Vandana Shiva, Carolyn Merchant, Rosemary Ruether, and Ynestra King.

According to ecofeminist philosophy, when people see themselves as related to others and to nature, they will see life as bounty rather than scarcity, as cooperation rather than competition, and as a network of personal relationships rather than isolated egos. Like the postmodernists, ecofeminists reject the view of a single, ahistoric, context-free, neutral observation stance. Instead, they favor multiple understandings, complex relationships, and "embodied objectivity."

In *Healing the Wounds* Ynestra King wrote, "We can use [ecofeminism] as a vantage point for creating a different kind of culture and politics that would integrate intuitive, spiritual, and rational forms of knowledge, embracing both science and magic insofar as they enable us to transform the nature-culture distinction and to envision and create a free, ecological society."

ENVIRONMENTAL JUSTICE

People of color in the United States and around the world are subjected to a disproportionately high level of environmental health risks in their neighborhoods and on their jobs. Minorities, who tend to be poorer and more disadvantaged than other residents, work in the dirtiest jobs where they are exposed to toxic chemicals and other hazards. More often than not they also live in urban ghettos, barrios, reservations, and rural poverty pockets that have shockingly high pollution levels and are increasingly the site of unpopular industrial facilities, such as toxic waste dumps, landfills, smelters, refineries, and incinerators. **Environmental justice** combines civil rights with environmental protection to demand a safe, healthy, life-giving environment for everyone.

Among the evidence of environmental injustice is the fact that three out of five African-Americans and Hispanics, and nearly half of all Native Americans, Asians, and Pacific Islanders live in communities with one or more uncontrolled toxic waste sites, incinerators, or major landfills. A recent Greenpeace study found that minorities make up twice as large a population share in communities with these locally unwanted land uses (**LULUs**) as in communities without them. And the inequities are growing. Whereas in 1980 the average minority population near a landfill or hazardous waste facility was about 22 percent, in 1994 it was 36 percent.

Although it is difficult to distinguish between race, class, historical patterns of ethnic groups, economic disparities, and other social factors in these disputes, racial origins often seem to play a role in exposure to environmental hazards. Simple correlation doesn't prove causation, still, while poor people in general are more likely to live in polluted neighborhoods than rich people, the discrepancy between the pollution exposure of middle class blacks and middle class whites is even greater than the difference between poorer whites and blacks. Where upper class whites can "vote with their feet" and move out of polluted and dangerous neighborhoods, blacks and other minorities are restricted by color barriers and prejudice (overt or covert) to the less desirable locations (fig. 2.8).

FIGURE 2.8

Poor people and people of color often live in the most dangerous and least desirable places. Here children play next to a chemical refinery in Texas City, Texas.

Environmental Racism

Racial prejudice is a belief that someone is inferior merely because of their race. Racism is prejudice with power. **Environmental racism** is inequitable distribution of environmental hazards based on race. Evidence of environmental racism can be seen in lead poisoning in children. The Federal Agency for Toxic Substances and Disease Registry considers lead poisoning to be the number one environmental health problem for children in the United States. Some 4 million children—many of whom are African American, Latino, Native American, or Asian, and most of whom live in inner-city areas—have dangerously high lead levels in their bodies. This lead is absorbed from old lead-based house paint, contaminated drinking water from lead pipes or lead solder, and soil polluted by industrial effluents and automobile exhaust. The evidence of racism is that at every income level, whether rich or poor, black children are two to three times more likely than whites to suffer from lead poisoning.

Dumping Across Borders

Because of their quasi-independent status, most Native-American reservations are considered sovereign nations that are not covered by state environmental regulations. Court decisions holding that reservations are specifically exempt from hazardous waste storage and disposal regulations have resulted in a land rush of seductive offers from waste disposal companies to Native-American reservations for onsite waste dumps, incinerators, and landfills. The short-term economic incentives can be overwhelming for communities in which adult unemployment runs between 60 and 80 percent. Uneducated, powerless people often can be tricked or intimidated into signing environmentally and socially disastrous contracts. Nearly every tribe in America has been approached with proposals for some dangerous industry or waste facility.

The practice of targeting poor communities of color in the Third or Fourth World for waste disposal and/or experimentation with risky technologies has been described as **toxic colonialism.** Internationally, the trade in toxic waste has mushroomed in recent years as wealthy countries have become aware of the risks of industrial refuse. Poor, minority communities at home and abroad are being increasingly targeted as places to dump unwanted wastes. Although a treaty regulating international shipping of toxics was signed by 105 nations in 1989, millions of tons of toxic and hazardous materials continue to move—legally or illegally—from the richer countries to the poorer ones every year. This issue is discussed further in chapter 23.

Another form of toxic colonialism is the flight of polluting industries from developed nations and states where control requirements are stringent to less developed areas where regulations are lax and local politicians are easily co-opted. For example, more than 2,000 *maquiladoras,* or assembly plants operated by American, Japanese, or other foreign companies, are now located along the United States/Mexico border to take advantage of favorable import quotas, low wages, and weak pollution-control laws.

FIGURE 2.9   Living conditions in the *colonias* or unplanned settlements along the U.S./Mexican border often are substandard. Is this evidence of racism or simply a result of poverty and poor planning?
© Sam Kittner.

Mexican laborers work under appalling conditions in some of these factories, assembling imported components into consumer goods to be exported to the United States. Although the jobs are demeaning, dangerous, and pay far less than the minimum U.S. wage, large populations have been attracted to squalid shantytowns along the border where industrial effluents poison the air and water (fig. 2.9).

The U.S. Environmental Justice Act was established in 1992 to identify areas threatened by the highest levels of toxic chemicals, assess health effects caused by emissions of those chemicals, and ensure that groups or individuals residing within those areas have opportunities and resources to participate in public discussions concerning siting and cleanup of industrial facilities. Perhaps we need something similar worldwide.

SCIENCE AS A WAY OF KNOWING

Science, derived from "knowing" in Latin, is a process for producing knowledge. It depends on making precise observations of natural phenomena and on formulating rational theories to make sense out of those observations. Science rests on the assumptions that the world is knowable and that we can learn about how things work through careful empirical study and logical analysis. Moreover, we presume science can help us find practical solutions for many problems because it provides information about both materials and mechanisms in the world around us (table 2.1).

An important value of scientific thinking is that it reduces our tendency to rely on emotional reactions and unexamined assumptions. In the Middle Ages, the ultimate sources of knowledge about

TABLE 2.1	Some Basic Scientific Assumptions

1. The world is knowable. With careful, impartial observation and logical analysis, we can make sense of the fundamental processes and laws that shape our environment.

2. Basic patterns that describe events in the natural world are uniform throughout time and space. The forces at work now are the same as those that shaped the world in the past and will continue to do so in the future.

3. Where two equally plausible explanations for a phenomenon are possible, we should choose the simpler one. (Also known as the law of **parsimony,** or Ockham's razor, after the English philosopher who first proposed this rule.)

4. Change in knowledge is inevitable because new evidence may challenge prevailing theories. No matter how well one theory explains a set of observations, it is possible that another theory may fit just as well or better, or may fit a still wider range of data.

5. Although new facts can disprove existing theories, science can never provide absolute proof that a theory is correct. Every theory should be considered only conditionally or provisionally correct until contrary evidence is found.

6. Even if there is no way to secure complete and absolute truth, increasingly accurate approximations can be made to account for the world and how it works.

7. Science can help find practical solutions for many problems because it provides information about both mechanisms and processes in the world around us.

Key Concepts

- Environmental ethics asks about moral relationships between humans and the world around us.

- Intrinsic or inherent values and rights apply to things or persons simply because of what or who they are. Instrumental values apply to things that have worth to a moral agent (who has rights).

- Science can prove a hypothesis wrong, but can rarely prove something unquestionably true.

- Deductive reasoning starts with a general principle and predicts specific results. Inductive reasoning starts with observations and suggests principles to explain them.

matters such as how crops grow or how diseases spread were religious authorities or cultural traditions. While these sources may have provided useful insights in many cases, there was no way to test their explanations independently and objectively. They were right because custom, politics, or theology said so. The benefit of scientific thinking is that it searches for testable evidence: if you suspect that a disease spreads through contaminated water, you can close off access to that water source and see if the disease stops spreading.

Ideally, scientists are skeptical. They don't accept proposed explanations until there is substantial evidence to support them. Even then, every explanation is considered only provisionally true because there is always a possibility that some additional evidence may appear to disprove it. Scientists also try to be methodical, rigorous, and unbiased. Because bias and methodical errors are hard to avoid, scientific tests are subject to review by informed peers, who can help evaluate results and conclusions.

Modern science has its roots in antiquity. Greek philosophers such as Pythagoras (about 520 B.C.E.), Socrates (464–399 B.C.E.), and Aristotle (380–320 B.C.E.) laid the foundations for inquiry and logic. Arabic mathematicians and astronomers added to this store of knowledge and reintroduced it to the Western world at the end of the Middle Ages. Much of what we now know as the scientific worldview and scientific method, however, originated in seventeenth-

and eighteenth-century Europe. Among the scholars who pioneered empirical, experimental approaches to science were Galileo Galilei (1564–1642), a pioneer in physics and astronomy; philosophers such as René Descartes (1596–1650) and Francis Bacon (1561–1626), who emphasized the use of objective observation and inductive reasoning; and Isaac Newton (1642–1727), who invented calculus and studied optics and gravitation. The insights and methods introduced by these scientific pioneers laid the foundation for much of the material progress that we now enjoy.

Cooperation and Insight in Science

Good science rarely is carried out by a single individual working in isolation. Instead, a community of scientists collaborates in a cumulative, self-correcting process. Ideas and information are freely exchanged, debated, and tested to find the most accurate ones. Results are compared and interpretations are challenged in a democratic, impartial way. You often hear about big breakthroughs and dramatic discoveries that change our understanding overnight, but these are rare in ordinary science, Instead, many people work on different aspects of a common problem, each adding small insights into how a system works (fig. 2.10).

Confusing or erroneous results often misdirect researchers temporarily until mistakes are recognized and the search for truth gets back on the right track. One of the main benefits of doing research work is the realization you get of how tedious and contradictory science often is. If you're working at the cutting edge of either your theories or your instruments, you may well get wrong answers as often as you get right ones. Making sense of a baffling mass of information is one of the greatest challenges that scientists face.

Scientific Design

Scientists demand lots of evidence before they are willing to accept the accuracy of any data or the usefulness of a particular interpretation. One of the most important tests of any data is its

FIGURE 2.10 Careful data recording is important in science.
© William P. Cunningham.

reproducibility. Making an observation or obtaining a particular result just once doesn't count for much. You have to produce the same result consistently to be sure that your first outcome wasn't a fluke. Even more importantly, you ought to be able to describe the conditions under which you made the observation or obtained the result in sufficient detail so that someone else can reproduce your findings.

It's often difficult to study natural systems because so many important factors vary simultaneously. To avoid this problem, scientists design **controlled studies** in which comparisons are made between experimental and control populations that are identical (as far as possible) in every factor except the one variable being studied. Studies must be designed very carefully so that experimenters don't inadvertently treat the experimental and control groups differently or that data coders don't unconsciously let bias creep into their scoring. To avoid experimenter effects, **blind experiments** are designed in which those carrying out the experiment don't know until after data have been gathered and analyzed. In medical research involving human subjects, you have to be careful that expectations of the subjects and the experimenters don't bias results. In a **double-blind design,** neither the subject (participant) nor those administering treatment and analyzing results know who is receiving the experimental or the control treatments until after the experiment is completed.

Researchers try very hard to be accurate, but it's important to avoid unnecessary or meaningless precision. Suppose you've had a fresh snowfall in your area, and you want to know its depth. You have a ruler marked in millimeters. The first measurement you make is 24 mm, but the surface of the snow isn't even. By moving your ruler slightly, you can get measurements of 21 mm and 25 mm. What's the true depth of the snowfall? If you average these three measurements, you get 23.3333 mm, but this gives a false notion of precision. The last three digits aren't **significant numbers.** Your ruler isn't capable of measuring tenths of a millimeter, let alone thousandths or ten-thousandths. Nor is the surface of the snow even enough to make such a precise measurement. The most accurate answer is an approximation of about 23 mm.

Deductive and Inductive Reasoning

In 1919, Sir Arthur Eddington led an expedition to Principe Island off the coast of West Africa to photograph stars during a solar eclipse. These photographs were a landmark in modern science because they were the first successful empirical test of Einstein's theory of general relativity. Einstein's calculations predicted that photons of light should be affected by gravity. Thus, light coming from distant stars should be deflected if it passes close to our sun's surface. These stars should appear shifted slightly in position during the day, when the sun is present, compared to the night, when it is not. Normally, you can't make such an observation because the stars close to the sun are invisible in the daytime. During an eclipse, however, it is possible to photograph stars and compare their apparent day and night locations. This is what Eddington did, and his corroboration of Einstein's prediction was a triumphant moment in modern physics, one that helped Einstein's theory of relativity become widely accepted.

This series of logical steps is **deductive reasoning.** Starting with a general principle (the theory of general relativity), a testable prediction is derived about a specific case (that certain stars should appear closer together in the daytime than at night). You might think of this as "top-down" reasoning—proceeding from the general to the specific. Although deduction is often regarded as a scientific ideal, in many cases we don't have a general theory to explain certain phenomena in nature. Instead, we depend on "bottom-up" **inductive reasoning** in which we study specific examples and try to discover patterns and derive general explanations from collected observations.

An example of inductive reasoning is seen in the brilliant work of Austrian zoologist Karl Von Frisch, who discovered how honeybees communicate. Although people had noticed for many years that bees seem to have an uncanny ability to find their way to nectar-bearing flowers, no theoretical explanation gave any insight into this phenomenon. Through meticulous, methodical observations, Von Frisch discovered and described the dances by which foraging honeybees (*Apis mellifera*) indicate the direction and distance of food sources to their hive mates. Thus, he created the theory based on his observations. In 1973, Von Frisch shared the Nobel Prize in medicine with Konrad Lorenz and Nikolaas Tinbergen who made similar discoveries in animal behavior through equally detailed and painstaking observations. While we still don't have any widely accepted laws of animal behavior or communication, these studies have given us fascinating insights into the lives of other organisms.

Hypotheses and Theories

You may already be using scientific technique without being aware of it. Suppose your flashlight doesn't work. The problem could be in the batteries, the bulb, or the switch—or all of them could be faulty. How can you distinguish between these possibilities? If you change all the components at once, you may have a working flashlight, but you won't know which was the faulty part. A series of methodical steps to test each component can be helpful.

Starting with the observation that your flashlight doesn't work, you focus on the component you think most likely to be defective. Perhaps that might be the batteries. You propose a **hypothesis,** or a conditional explanation that can be verified or falsified. In this case, the hypothesis might be, "The batteries are dead." From this tentative assumption, you make a prediction: "If I replace the dead batteries with fresh ones, there should be light." If your hypothesis was correct, when you perform the experiment (replacing the batteries), you will see the predicted result. If not, you reject that hypothesis and form a new one: perhaps the bulb is burned out. By formulating testable hypotheses and evaluating each component in turn, you should finally be able to isolate the problem and find a way to solve it.

What you have just employed in this example is the scientific method of observations, systematic testing, and interpretation of results shown in figure 2.11. The results of one experiment may give us information that leads to further hypotheses and additional experiments. In each case, prior knowledge and experience help us design experiments and interpret results. Eventually, with evidence from a group of related investigations, we may be able to formulate a theory to explain a set of general principles.

Logically, you can show a hypothesis to be wrong, but you can almost never show it to be unquestionably true. The philosopher Ludwig Wittgenstein gave the following example to illustrate this principle. Suppose all the swans you have ever seen are white. You might assume that *all* swans are white. This hypothesis could be tested by examining a large number of swans. If you never find a black one, you might tentatively conclude that your hypothesis is right. But even if you look at a million white swans, there could still be a black one somewhere (actually, there are black swans in Australia) that would refute your conjecture. Thus, you could show your hypothesis is wrong, but you could never be absolutely sure that it is correct, no matter how much evidence you collect.

As a result of this uncertainty, scientific evidence should always be regarded as provisional. Additional information might come along to undermine the observations we have made thus far. When an explanation has been supported, however, by a large number of tests, and a majority of experts in a given field have reached a general consensus that it is the best description or explanation available, we call it a **scientific theory.** Note that the use of this term by scientists is very different from the way it is understood by the general public. To many people, a theory is speculative and unsupported by facts. To a scientist, it means just the opposite;

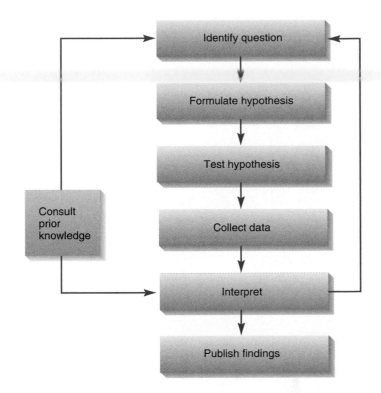

FIGURE 2.11   Ideally, scientific investigation follows a series of logical, orderly steps to formulate and test hypotheses.

while all explanations are tentative and open to revision and correction, one that counts as a scientific theory is supported by an overwhelming body of data and experience, and is generally accepted in the scientific community, at least for the present.

Modeling and Natural Experiments

Not every area that we might want to study scientifically is open to simple, direct experiments like those described above (fig. 2.12). A geologist, for instance, might want to study mountain building, or an ecologist might like to know how species coevolve, but neither scientist would be able to re-create the millions of years and complex environmental conditions required for these processes to occur. Similarly, a toxicologist might want to know about the effects of toxins but couldn't deliberately poison people, no matter how useful that information might be. One way to gather information in fields such as this is to look for historic evidence or what might be called "natural experiments" for support or contradiction of their ideas. Can you find examples that would support or refute a given hypothesis from what has already happened? The geologist, for example, might propose the hypothesis that mountain building always involves folding and uplifting of surface strata. If she can find counterexamples, this hypothesis could be discarded.

Another way to gather information about environmental systems is to use models. A model could be a substitute organism or physical mock-up that simulates a real system, or it could be a set of mathematical equations, often calculated in a computer program, that represents the phenomenon you want to study. For instance, the

FIGURE 2.12 Researchers use a directional antenna to track radio-collared wildlife. Experiments are difficult under these circumstances.
© Courtesy Dr. David Mech.

toxicologist might feed the toxin under question to rats, and then attempt to extrapolate or infer how their reaction is related to humans. The geologist might build a scale model of a mountain range to see what forces affect it. Mathematical models are often especially useful where there are simultaneous variables. Input conditions can be manipulated (initial population size, population growth rate, and so on) to test their effect on resulting conditions. Models, however, represent researchers' assumptions about how a system works, so models run by different people can produce contradictory results. Global warming models, for example, disagree on the amount of climate change expected, and the rate of change. However, models do provide heuristic information—suggestions of how things might be—and they allow us to test our understanding of the relationships in a system. Even though climate models disagree on the degree of change, nearly all agree that temperatures will increase. Models have become an important scien-

tific tool for exploring complex systems. In chapter 6 we will look at a few simple population models, and how these models help us understand population dynamics.

Statistics and Probability

You will encounter many numbers in this book. It is important to think carefully about what they mean and what they don't mean. Quantitative data can be precise, easily compared, and provide good benchmarks to measure change. But you have to look carefully at what numbers mean, not just take them at face value. Scientists use statistical methods to evaluate the reliability and significance of numerical data. An example with which you're already familiar is to compute the average (or mean) of a group of similar measurements. Standard deviation is a calculation that tells you how tightly the numbers in a data set are clustered around the mean. It allows you to evaluate how much variance there is within the set. We can also use statistical analysis to calculate the margin of error or confidence level that a particular data set is, or is not, different from mere random chance. A standard accepted in most scientific studies is that data should have at least a 95 percent confidence interval. This means that the margin of error is no more than 5 percent, or that if you repeat the measurement 20 times, at least 19 of those measurements would be the same.

Probability is another important concept in science. Probability, risk, and chance are all ideas that try to measure how likely something is to occur. They never tell what *will* happen, only what *might* happen. If you have a 20 percent chance of catching a cold this month, that means that about 20 out of every 100 people will probably get a cold. You are relatively unlikely to get one, but you might. We rarely can be 100 percent sure of anything, but scientists consider a 95 percent probability—or confidence level—that an event is not a random chance and may occur with a high degree of certainty.

A related question concerns appropriate sample size. How many individuals would you need to examine to have a reliable representation of a population? Is it better to sample small areas in many different places or one larger area? How many times do you need to replicate your investigation to be sure that you have an accurate result (fig. 2.13)? All of these questions depend on the variables that you intend to study as well as the accuracy level you hope to attain. Statistics is an important tool in both planning and evaluating scientific studies.

Intuition and Inspiration

Many people—even scientists—fail to recognize the role that human factors such as creativity, insight, aesthetics, and luck play in research. Some of our most important discoveries were made not because of superior scientific method and objective detachment, but because the investigators were passionately interested in their topic and pursued hunches and intuitions that appeared unreasonable to fellow scientists. A good example is Barbara McClintock, the geneticist who discovered that genes in corn can move and recombine spontaneously. Her years of experience in corn breeding and an uncanny ability to recognize patterns in what

FIGURE 2.13 How many measurements are needed for acceptable accuracy and significance? © Ron Nichols/URCS/USDA.

seemed to be random variation, gave her an intuitive understanding that it took other investigators years to accept.

Paradigms and Scientific Consensus

The notion of paradigm shifts has been an important one since it was introduced in the 1960s by the science historian Thomas Kuhn. **Paradigms,** according to Kuhn, are overarching models of the world that guide our interpretation of events. Two centuries ago, Noah's Flood was considered by many people to be one of the principal events that shaped our world. Geological evidence was interpreted in terms of the flood and its impacts. Now geologists interpret the world's history in terms of tectonic plates that have rearranged themselves repeatedly over billions of years. Tectonic theory also gives biogeographers and ecologists entirely different explanations for evidence such as distribution of fossils on different continents. Paradigm shifts like these even guide the sorts of questions we ask. Today few people would search for evidence that the Flood redistributed fish species, but they might look for evidence of how mountain building separated species into different watersheds.

The shift from one guiding paradigm to another occurs when a majority of scientists accept that the old explanation no longer explains new observations very well. This shift can be contentious and political, because whole careers, based on one sort of research and explanation, can be undermined by a new model. Sometimes this revolution is accomplished rather quickly. Quantum mechanics and Einstein's theory of relativity, for example, overturned classical physics in only about 30 years. In other cases, it may take centuries for new ideas to supplant old ones. Often a whole generation of scholars has to die before new paradigms can be accepted.

One of the things you should do as you study this book and think about environmental science is to try to identify some of the paradigms that guide our investigations and our explanations today. This is one of the skills involved in critical thinking, discussed in the introductory chapter of this book.

Pseudoscience and Baloney Detection

In every controversial topic, a few contrarian scientists always champion a view diametrically opposed to the majority position of the scientific community. There are several reasons to adopt a contradictory view. One might be sincerely convinced, for instance, that the majority opinion is wrong. Often the most brilliant and original thinkers are regarded as hopelessly off track by conventional scientists. Others may be motivated simply by a desire to consider all possibilities in case unexpected evidence emerges.

Some people also may be tempted to take an opposing viewpoint because they know it will generate publicity and even, perhaps, generous payments from those who benefit from countertestimony. In many court cases, for example, opposing sides are able to produce paid "experts" who are willing to present a partisan agenda rather than objective science. A common tactic is to use scientific uncertainty as an excuse to postpone or reverse an action or policy that a vast majority of the scientific community believes to be prudent. Opponents will claim that the evidence doesn't constitute conclusive proof that the action needs to be taken, or that the policy is based only on a scientific "theory" and therefore has no validity. As we've discussed earlier, theory describes an entirely different level of certainty in science than it does in common usage. In addition, while scientists usually avoid claims that a theory is unquestionably true, an overwhelming body of evidence is usually required before a consensus is reached on the provisional validity of a particular position.

Harvard's Edward O. Wilson writes, "We will always have contrarians whose sallies are characterized by willful ignorance, selective quotations, disregard for communications with genuine experts, and destructive campaigns to attract the attention of the media rather than scientists. They are the parasite load on scholars who earn success through the slow process of peer review and approval." How can we identify "junk science" that presents bogus analysis dressed up in quasi-scientific jargon but that really represents a perversion of objective inquiry? The astronomer Carl Sagan proposed a "Baloney Detection Kit" containing the questions in table 2.2.

TABLE 2.2 Questions for Baloney Detection
1. How reliable are the sources of this claim? Is there reason to believe that they might have an agenda to pursue in this case?
2. Have the claims been verified by other sources? What data are presented in support of this opinion?
3. What position does the majority of the scientific community hold in this issue?
4. How does this claim fit with what we know about how the world works? Is this a reasonable assertion or does it contradict established theories?
5. Are the arguments balanced and logical? Have proponents of a particular position considered alternate points of view or only selected supportive evidence for their particular beliefs?
6. What do you know about the sources of funding for a particular position? Are they financed by groups with partisan goals?
7. Where was evidence for competing theories published? Has it undergone impartial peer review or it is only in proprietary publication?

Summary

- Sharks are disappearing from many places in the ocean because of overharvesting. Many people regard shark finning as wasteful, cruel, and unethical.

- Environmental ethics asks about moral relationships between humans and the world around us. How ought we, as moral beings, behave?

- Utilitarians believe in actions that bring the greatest good to the greatest number. Biocentrism teaches us that all living things deserve respect and have inherent value and rights.

- Anthropocentrism claims that humans have unique, intrinsic rights and values that entitle us to use (or abuse) the world and its resources however we choose, while stewardship sees our proper role as caretakers of our environment.

- Ecofeminism sees a link between violence against oppressed people and environmental abuse. It advocates pluralistic, non-hierarchical relationships between people and their environment.

- Environmental racism results in inequitable distribution of access to resources and exposure to hazards based on race.

- Science depends on making accurate observations and then formulating rational theories and tests to understand natural phenomena. Explanations in science are generally considered to be only provisionally true because there is always a possibility that some additional evidence may be found that disproves what we now believe to be correct. Science can prove a hypothesis wrong, but can rarely prove something unquestionably true.

- Science assumes that the world is knowable, that the forces at work now are the same as those in the past, and that where equally plausible explanations exist, we should choose the simpler one.

- One of the most important tests of scientific data is reproducibility. If you and others can't repeat an observation, it doesn't count for much.

- Deductive reasoning starts with a general principle and predicts specific results. Inductive reasoning starts with observations and suggests principles to explain them.

- A hypothesis is a testable, provisional explanation. In science, a theory is a general principle supported by an overwhelming body of data widely accepted by the scientific community.

- A paradigm is an all-encompassing model that explains how the world works. It determines not only how we understand phenomena, but how we think about them, what language we use, and what is legitimate to ask about them. From time to time, our paradigms shift as new evidence becomes available, but it's usually very difficult to overturn entrenched worldviews.

Questions for Review

1. Why are sharks disappearing in many places?
2. Define universalist, nihilist, relativist, utilitarian, and ecofeminist ethics.
3. Describe the differences between moral agents, moral subjects, and those who fail to qualify for moral consideration.
4. What are inherent and instrumental values? Who has them? Why?
5. Give two examples in which religious groups are working to protect environmental resources.
6. Compare and contrast stewardship, anthropocentric, biocentric, and environmental justice worldviews. Which is closest to your own views?
7. What are environmental racism and toxic colonialism?
8. Draw a diagram showing the scientific method and describe, in your own words, what it means.
9. Explain the idea of a *paradigm* and why it is important.
10. Not every science is experimental. Explain how geologists or evolutionary biologists might test their hypotheses.

Questions for Critical Thinking

1. Review a topic in this chapter—ecofeminism or animal rights for example—and arrange the arguments for or against this idea as a series of short statements. Determine whether each is a statement of fact or opinion.
2. How would you verify or disprove fact statements given in question 1? Do the opinions or conclusions reached in this argument *necessarily* follow from the facts given? What alternatives could be proposed?
3. Reflect on the paradigms, values, beliefs, contextual perspective that you bring to the discussion above. Does it coincide with those in the textbook? What different interpretation would your perspective impose on the argument?
4. Try to put yourself in the place of a person from a minority community, an underdeveloped nation, or a developing country in discussing questions of environmental justice and environmental quality. What preconceptions, values, beliefs, and contextual perspective would you bring to the issue? What would you ask for from the majority society?
5. What is your environmental ethic? What experiences, cultural background, education, religious beliefs help shape your worldview?

6. Try some role-playing with a classmate, friend, or family member. Take the position of a universalist, biocentrist, utilitarian, anthropocentrist, or ecofeminist and debate the merits of cutting trees in the National Forest. On what points would you agree and where would you disagree in this issue?

7. This chapter addresses many social issues and theories. Are these areas appropriate for an environmental science textbook? Why or why not? Try answering this question from the perspective of Gifford Pinchot, John Muir, the ecofeminist, or the person of color whom you identified in earlier questions. Would they agree?

Key Terms

anthropocentrism 38	moral extensionism 36
biocentrism 37	moral subjects 36
blind experiments 42	morals 34
controlled studies 42	nihilists 35
deductive reasoning 42	paradigms 45
double-blind design 42	parsimony 41
ecofeminism 38	relativists 35
environmental ethics 34	reproducibility 42
environmental justice 39	science 40
environmental racism 40	scientific theory 43
hypothesis 43	significant numbers 42
inductive reasoning 42	stewardship 38
inherent value 36	toxic colonialism 40
instrumental value 36	universalists 35
LULUs 39	utilitarians 35
moral agents 35	values 34

Further Readings

Barrett, C. B., and R. Grizzle. 1999. A holistic approach to sustainability based on pluralism stewardship. *Environmental Ethics* 21(1):23–42.

Callicott, J. Baird. 1999. *Beyond the Land Ethic: More Essays in Environmental Philosophy.* State Univ. of New York Press.

Des Jardins, Joseph R. 1999. *Environmental Ethics: Concepts, Policy, and Theory.* McGraw-Hill Co.

Jamieson, Dale. 2003. *A Companion to Environmental Philosophy.* Blackwell Publishers.

Kellert, Stephen R., and Timothy J. Farnham. 2002. *The Good in Nature and Humanity: Connecting Science, Religion and Spirituality with the Natural World.* Island Press.

Kuhn, Thomas S. 1996. *The Structure of Scientific Revolutions,* 3rd ed. Univ. of Chicago Press.

Light, Andrew, and Holmes Ralston III. 2002. *Environmental Ethics: An Anthology.* Blackwell Publishing.

Popper, Karl R. 1992. *Conjectures and Refutations: The Growth of Scientific Knowledge,* 5th ed. Routledge Press.

Zimmerman, Michael E., J. Baird Callicott, John Clark, George Sessions, and Karen Warren, eds. 2000. *Environmental Philosophy: From Animal Rights to Radical Ecology,* 3rd ed. Prentice Hall.

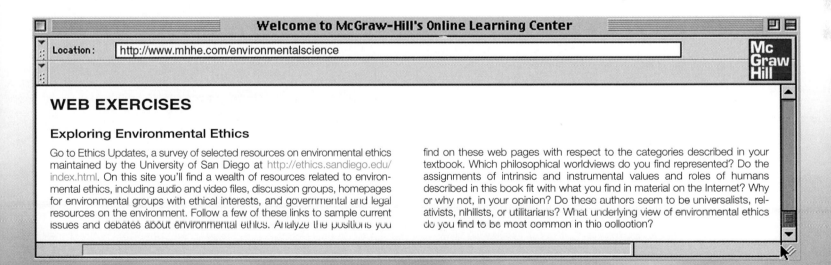

Welcome to McGraw-Hill's Online Learning Center

Location: http://www.mhhe.com/environmentalscience

WEB EXERCISES

Exploring Environmental Ethics

Go to Ethics Updates, a survey of selected resources on environmental ethics maintained by the University of San Diego at http://ethics.sandiego.edu/index.html. On this site you'll find a wealth of resources related to environmental ethics, including audio and video files, discussion groups, homepages for environmental groups with ethical interests, and governmental and legal resources on the environment. Follow a few of these links to sample current issues and debates about environmental ethics. Analyze the positions you find on these web pages with respect to the categories described in your textbook. Which philosophical worldviews do you find represented? Do the assignments of intrinsic and instrumental values and roles of humans described in this book fit with what you find in material on the Internet? Why or why not, in your opinion? Do these authors seem to be universalists, relativists, nihilists, or utilitarians? What underlying view of environmental ethics do you find to be most common in this collection?

3

Matter, Energy, and Life

To the best of our knowledge, our Sun is the only star proven to grow vegetables.

Philip Scherrer, Stanford University

OBJECTIVES

After studying this chapter, you should be able to:

- describe matter, atoms, and molecules and give simple examples of the role of four major kinds of organic compounds in living cells.
- define *energy* and explain the difference between kinetic and potential energy.
- understand the principles of conservation of matter and energy and appreciate how the laws of thermodynamics affect living systems.
- know how photosynthesis captures energy for life and how cellular respiration releases that energy to do useful work.
- define *species, populations, communities,* and *ecosystems* and understand the ecological significance of these levels of organization.
- discuss food chains, food webs, and trophic levels in biological communities and explain why there are pyramids of energy, biomass, and numbers of individuals in the trophic levels of an ecosystem.
- recognize the unique properties of water and explain why the hydrologic cycle is important to us.
- compare the ways that water, carbon, nitrogen, sulfur, and phosphorus cycle within ecosystems.

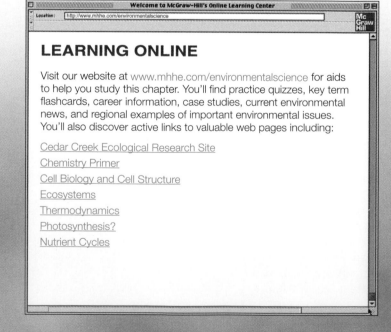

LEARNING ONLINE

Visit our website at www.mhhe.com/environmentalscience for aids to help you study this chapter. You'll find practice quizzes, key term flashcards, career information, case studies, current environmental news, and regional examples of important environmental issues. You'll also discover active links to valuable web pages including:

Cedar Creek Ecological Research Site

Chemistry Primer

Cell Biology and Cell Structure

Ecosystems

Thermodynamics

Photosynthesis?

Nutrient Cycles

Photo: Cedar Bog Lake, the site of the first detailed, quantitative study of matter and energy cycling in an ecosystem. © William P. Cunningham.

Measuring Energy Flows in Cedar Bog Lake

In 1938, a young graduate student named Ray Lindeman began his Ph.D. research on a small, marshy pond in Minnesota called Cedar Bog Lake. His pioneering work helped reshape the way ecologists think about the sytems they study. At the time, most ecologists were concerned primarily with descriptive histories and classifications of biological communities. A typical lake study might classify the taxonomy and life histories of resident species and describe the lake's stage in development from open water to marsh and then to forest. Blind in one eye, Lindeman couldn't do the microscopy necessary to identify the many species of algae, protozoans, and other aquatic organisms in the lake. Instead, following the ideas of two contemporary English ecologists, Charles Elton and A. G. Tansley, he concentrated on biological communities as systems and looked at broad categories of feeding relationships, for which he coined the term *trophic levels* (from the Greek word for eating) (fig. 3.1). Aided by his wife, Eleanor, Lindeman spent many hours collecting samples of aquatic plants and algae, grazing and predatory zooplankton and fish, and the benthic (bottom-dwelling) worms, insect larvae, crustaceans, and sediment. Back in the laboratory, he measured the plants' photosynthetic rates, the animals' respiration rates, and the total energy content of organic compounds in each of the different trophic levels.

Describing the system in terms of energy flows was a radical departure from ecological methods at the time. Lindeman made a careful balance sheet of the total energy content in the biomass at each trophic level, the energy used in respiration, and the energy content of organic matter deposited in the sediment. To his surprise, he found that each successive feeding level contained only about 10 percent of the energy captured by the level below it. The remainder is lost as heat or deposited in sediments, he argued, because of the work that organisms perform and the inefficiency of biological energy transformations. In his dissertation, Lindeman showed that energy represents a common denominator that allows us to sum up all the processes of production and consumption by the myriad organisms in a biological community.

Lindeman also broke from standard procedure by representing the relationships in his study lake as a mathematical model: He used a series of equations to describe thermodynamic relationships and the efficiency of energy capture and transfer. Ironically, Lindeman's most important paper was rejected by the journal *Ecology* as being too theoretical and too quantitative. It was only after the intercession

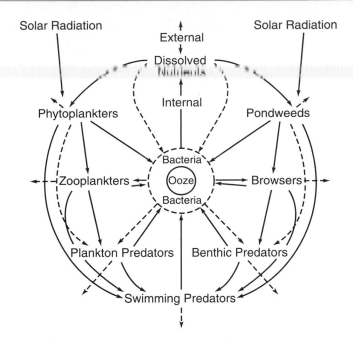

FIGURE 3.1   Feeding relationships in Cedar Bog Lake.
After R. Lindeman, 1942. The trophic-dynamic aspect of ecology. *Ecology* 23(4):399–417.

of G. Evelyn Hutchinson from Yale, with whom Lindeman had a post-doctoral fellowship after finishing his studies at Minnesota, that his mathematical model and energy analysis of Cedar Bog Lake was finally published. Unfortunately, Ray Lindeman died of liver failure before his article appeared. It has since become a landmark in ecological history.

In the years since Lindeman's work, the idea of taking a systemic view of a biological community together with its physical and inorganic environment has become standard in ecology. Energy flows and nutrient cycles are central to the way we understand the workings of ecological systems. Constructing quantitative models to describe, explain, and explore ecological processes has become routine. In this chapter, we will investigate the ways living things use energy and matter and the ways these flows create relationships in ecosystems. (See Lindeman, R. L. 1942. The trophic-dynamic aspect of ecology. *Ecology* 23:399–418.)

ELEMENTS OF LIFE

What substances are we made of, and how do we get the energy we need to function? These questions are at the core of **ecology,** the scientific study of relationships between organisms and their environment. In this chapter we'll introduce a number of concepts that are essential to understanding how living things function in their environment.

As an introduction to principles of ecology, this chapter first reviews the nature of matter and energy, then explores the ways organisms acquire and use energy and chemical elements. Then we'll investigate feeding relationships among organisms—the ways

that energy and nutrients are passed from one living thing to another—forming the basis of ecosystems. Finally, we'll review some of the key substances that cycle through organisms, ecosytems, and our environment.

In chapters 4 and 5 we'll continue investigating concepts of ecology: general organization of ecosystems by environmental types and landscapes, and principles of population growth.

In a sense, every organism is a chemical factory that captures matter and energy from its environment and transforms them into structures and processes that make life possible. To understand how these processes work, we will begin with some of the fundamental properties of matter and energy.

Matter, Atoms, Molecules, and Compounds

Everything that takes up space and has mass is **matter.** Matter exists in three distinct states—solid, liquid, and gas—due to differences in the arrangement of its constitutive particles. Water, for example, can exist as ice (solid), as liquid water, or as water vapor (gas).

Under ordinary circumstances, matter is neither created nor destroyed but rather is recycled over and over again. Some of the molecules that make up your body probably contain atoms that once made up the body of a dinosaur and most certainly were part of many smaller prehistoric organisms, as chemical elements are used and reused by living organisms. Matter is transformed and combined in different ways, but it doesn't disappear; everything goes somewhere. These statements paraphrase the physical principle of **conservation of matter.**

How does this principle apply to human relationships with the biosphere? Particularly in affluent societies, we use natural resources to produce an incredible amount of "disposable" consumer goods. If everything goes somewhere, where do the things we dispose of go after the garbage truck leaves? As the sheer amount of "disposed-of stuff" increases, we are having greater problems finding places to put it. Ultimately, there is no "away" where we can throw things we don't want any more.

Matter consists of **elements,** which are substances that cannot be broken down into simpler forms by ordinary chemical reactions. Each of the 115 known elements (92 natural, plus 23 created under special conditions) has distinct chemical characteristics. Just four elements—oxygen, carbon, hydrogen, and nitrogen—are responsible for more than 96 percent of the mass of most living organisms.

All elements are composed of discrete units called **atoms,** which are the smallest particles that exhibit the characteristics of the element. Atoms are composed of positively charged protons, negatively charged electrons, and electrically neutral neutrons. Protons and neutrons, which have approximately the same mass, are clustered in the nucleus in the center of the atom (fig. 3.2). Electrons, which are tiny in comparison to the other particles, rotate around the nucleus at the speed of light.

Each element has a characteristic number of protons per atom, called its **atomic number.** The number of neutrons in different atoms of the same element can vary slightly. Thus, the atomic mass, which is the sum of the protons and neutrons in each nucleus, also can vary. We call forms of a single element that differ in atomic mass **isotopes.** For example, hydrogen, the lightest element, normally has only one proton (and no neutrons) in its nucleus. A small percentage of hydrogen atoms have one proton and one neutron. We call this isotope deuterium (^{2}H). An even smaller percentage of natural hydrogen called tritium (^{3}H) has one proton plus two neutrons. Isotopes are of interest in environmental science because they are often unstable—that is, they may spontaneously emit particles. Radioactive waste, and nuclear energy, result from unstable isotopes of elements such as uranium and plutonium.

Chemical Bonds

Atoms often join to form **compounds,** or substances composed of different kinds of atoms (fig. 3.3). A pair or group of atoms that can exist as a single unit is known as a **molecule.** Some elements commonly occur as molecules, such as molecular oxygen (O_2) or molecular nitrogen (N_2), and some compounds can exist as molecules, such as glucose ($C_6H_{12}O_6$). In contrast to these molecules, sodium chloride (NaCl, table salt) is a compound that cannot exist as a single pair of atoms. Instead it occurs in a large mass of Na and Cl atoms or as two ions, Na^+ and Cl^-, in solution. Most molecules consist of only a few atoms. Others, such as deoxyribonucleic acid (DNA), which contains the genetic information in your cells, can include billions of atoms and be 2 meters long.

When ions with opposite charges form a compound, the electrical attraction holding them together is an *ionic bond.* Sometimes

FIGURE 3.2 As difficult as it may be to imagine when you look at a solid object, all matter is composed of tiny, moving particles, separated by space and held together by energy. It is hard to capture these dynamic relationships in a drawing. This model represents carbon-12, with a nucleus of six protons and six neutrons; the six electrons are represented as a fuzzy cloud of potential locations rather than as individual particles.

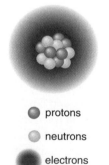

- ● protons
- ● neutrons
- ● electrons

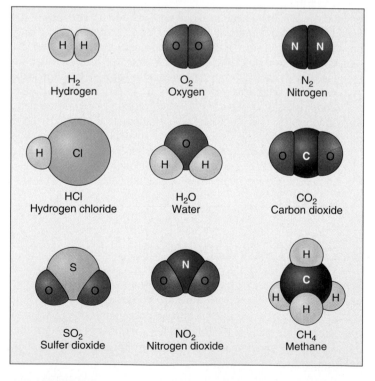

FIGURE 3.3 These common molecules, with atoms held together by covalent bonds, are important components of the atmosphere or important pollutants.

atoms form bonds by *sharing* electrons. For example, two hydrogen atoms can bond by sharing a single electron—it orbits the two hydrogen nuclei equally and holds the atoms together. Such electron-sharing bonds are known as *covalent bonds*. Carbon (C) can form covalent bonds simultaneously with four other atoms, so carbon can create complex structures such as sugars and proteins. Atoms in covalent bonds do not always share electrons evenly. An important example in environmental science is the covalent bonds in water (H_2O). The oxygen atom attracts the shared electrons more strongly than do the two hydrogen atoms. Consequently, the hydrogen portion of the molecule has a slight positive charge, while the oxygen has a slight negative charge. These charges create a mild attraction between water molecules, so that water tends to be somewhat cohesive. This fact helps explain some of the remarkable properties of water (Case Study, p. 52).

When an atom gives up one or more electrons, we say it is *oxidized* (because it is very often oxygen that takes the electron, as bonds are formed with this very common and highly reactive element). When an atom gains electrons, we say it is *reduced.* Chemical reactions necessary for life involve oxidation and reduction: oxidation of sugar and starch molecules, for example, is an important part of how you gain energy from food.

Forming bonds requires energy. Breaking bonds generally releases energy. For example, burning wood in your fireplace breaks the bonds of molecules in the wood, releasing heat. Some energy is used in forming new molecules, such as carbon dioxide, but the net result is a release of energy (heat). Generally, some energy input (activation energy) is needed to initiate these reactions. In your fireplace, a match might provide the needed activation energy. In your car, a spark from the battery provides activation energy to initiate the oxidation (burning) of gasoline.

Ions, Acids, and Bases

Atoms frequently gain or lose electrons, acquiring a negative or positive electrical charge. Charged atoms (or combinations of atoms) are called **ions.** Negatively charged ions (with one or more extra electrons) are *anions.* Positively charged ions are *cations.* A hydrogen (H) atom, for example, can give up its sole electron to become a hydrogen ion (H^+). Chlorine (Cl) readily gains electrons, forming chlorine ions (Cl^-).

Substances that readily give up hydrogen ions in water are known as **acids.** Hydrochloric acid, for example, dissociates in water to form H^+ and Cl^- ions. In later chapters, you may read about acid rain (which has an abundance of H^+ ions), acid mine drainage, and many other environmental problems involving acids. In general, acids cause environmental damage because the H^+ ions react readily with living tissues (such as your skin or tissues of fish larvae) and with nonliving substances (such as the limestone on buildings, which erodes under acid rain).

Substances that readily bond with H^+ ions are called **bases** or alkaline substances. Sodium hydroxide (NaOH), for example, releases hydroxide ions (OH^-) that bond with H^+ ions in water. Bases can be highly reactive, so they also cause significant environmental problems. Acids and bases can also be essential to living things: the

acids in your stomach dissolve food, for example, and acids in soil help make nutrients available to growing plants.

We describe the strength of an acid and base by its **pH,** the negative logarithm of its concentration of H^+ ions (fig. 3.4). Acids have a pH below 7; bases have a pH greater than 7. A solution of exactly pH 7 is "neutral." Because the pH scale is logarithmic, pH 6 represents *ten times* more hydrogen ions in solution than pH 7.

A solution's pH can be neutralized by adding buffers, or substances that accept or release hydrogen ions. In the environment, for example, alkaline rock can buffer acidic precipitation, decreasing its acidity. Lakes with acidic bedrock, such as granite, are especially vulnerable to acid rain because they have little buffering capacity.

Organic Compounds

Organisms use some elements in abundance, others in trace amounts, and others not at all. Certain vital substances are concentrated within cells, while others are actively excluded. Carbon is a particularly important element because chains and rings of carbon atoms form the skeletons of **organic compounds,** the material of which biomolecules, and therefore living organisms, are made.

The four major categories of organic compounds in living things ("bio-organic compounds") are lipids, carbohydrates, proteins, and nucleic acids. Lipids (including fats and oils) store

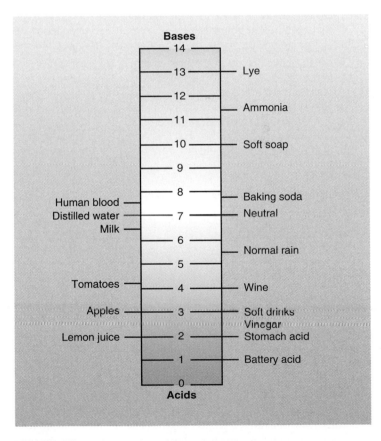

FIGURE 3.4   The pH scale. The numbers represent the negative logarithm of the hydrogen ion concentration in water.

Case Study

A "Water Planet"

If travelers from another solar system were to visit our lovely, cool, blue planet, they might call it Aqua rather than Terra because of its outstanding feature: the abundance of streams, rivers, lakes, and oceans of liquid water. Our planet is the only place we know where water exists as a liquid in any appreciable quantity. Water covers nearly three-fourths of the earth's surface and moves around constantly via the hydrologic cycle (discussed in chapter 15) that distributes nutrients, replenishes freshwater supplies, and shapes the land. Water makes up 60 to 70 percent of the weight of most living organisms. It fills cells, giving form and support to tissues. Among water's unique, almost magical qualities, are the following:

1. Water molecules are polar, that is, they have a slight positive charge on one side and a slight negative charge on the other side. Therefore, water readily dissolves polar or ionic substances, including sugars and nutrients, and carries materials to and from cells.

2. Water is the only inorganic liquid that exists under normal conditions at temperatures suitable for life. Most substances exist as either a solid or a gas, with only a very narrow liquid temperature range. Organisms synthesize organic compounds such as oils and alcohols that remain liquid at ambient temperatures and are therefore extremely valuable to life, but the original and predominant liquid in nature is water.

3. Water molecules are cohesive, tending to stick together tenaciously. You have experienced this property if you have ever done a belly flop off a diving board. Water has the highest surface tension of any common, natural liquid. Water also adheres to surfaces. As a result, water is subject to *capillary action:* it can be drawn into small channels. Without capillary action, movement of water and nutrients into groundwater reservoirs and through living organisms might not be possible.

4. Water is unique in that it expands when it crystallizes. Most substances shrink as they change from liquid to solid. Ice floats because it is less dense than liquid water. When temperatures fall below freezing, the surface layers of lakes, rivers, and oceans cool faster and freeze before deeper water. Floating ice then insulates underlying layers, keeping most water bodies liquid (and aquatic organisms alive) throughout the winter in most places. Without this feature, many aquatic systems would freeze solid in winter.

5. Water has a high heat of vaporization, using a great deal of heat to convert from liquid to vapor. Consequently, evaporating water is an effective way for organisms to shed excess heat. Many animals pant or sweat to moisten evaporative cooling surfaces. Why do you feel less comfortable on a hot, humid day than on a hot, dry day? Because the

Surface tension is demonstrated by the resistance of a water surface to penetration, as when it is walked upon by a water strider. © G. I. Bernard/ Animals Animals/Earth Scenes.

water vapor–laden air inhibits the rate of evaporation from your skin, thereby impairing your ability to shed heat.

6. Water also has a high specific heat; that is, a great deal of heat is absorbed before it changes temperature. The slow response of water to temperature change helps moderate global temperatures, keeping the environment warm in winter and cool in summer. This effect is especially noticeable near the ocean, but it is important globally.

All these properties make water a unique and vitally important component of the ecological cycles that move materials and energy and make life on Earth possible.

---

energy for cells, and they provide the core of cell membranes and other structures. Many hormones are also lipids. Lipids do not readily dissolve in water, and their basic structure is a chain of carbon atoms with attached hydrogen atoms. This structure makes them part of the family of hydrocarbons (fig. 3.5a). Carbohydrates (including sugars, starches, and cellulose) also store energy and provide structure to cells. Like lipids, carbohydrates have a basic structure of carbon atoms, but hydroxide (OH) groups replace half the hydrogen atoms in their basic structure, and they usually consist of long chains of ring-shaped structures. Glucose (fig. 3.5b) is an example of a very simple sugar.

Proteins are composed of chains of subunits called amino acids (fig. 3.5c). Folded into complex three-dimensional shapes, proteins provide structure to cells and are used for countless cell functions. Enzymes, such as those that release energy from lipids and carbohydrates, are proteins. Proteins also help identify disease-causing microbes, make muscles move, transport oxygen to cells, and regulate cell activity. Nucleic acids, including DNA and RNA, store the genetic information that direct the processes of life. These are extremely complex molecules, with a backbone of sugars and phosphate groups. Many subunits, known as nucleotides (fig. 3.5d) may be attached to these phosphate groups.

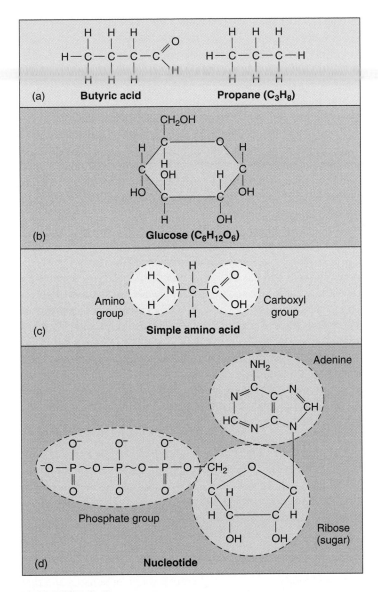

FIGURE 3.5 The four major groups of organic molecules are based on repeating subunits of these carbon-based structures. Basic structures are shown for (*a*) butyric acid (a building block of lipids) and a hydrocarbon, (*b*) a simple carbohydrate, (*c*) a protein, and (*d*) a nucleic acid.

Cells: The Fundamental Units of Life

All living organisms are composed of **cells,** minute compartments within which the processes of life are carried out (fig. 3.6). Microscopic organisms such as bacteria, some algae, and protozoa are composed of single cells. Most higher organisms are multicellular, usually with many different cell varieties. Your body, for instance, is composed of several trillion cells of about two hundred distinct types. Every cell is surrounded by a thin but dynamic membrane of lipid and protein that receives information about the exterior world and regulates the flow of materials between the cell and its environment. Inside, cells are subdivided into tiny organelles and subcellular particles that provide the machinery for life. Some of these organelles store and release energy. Others

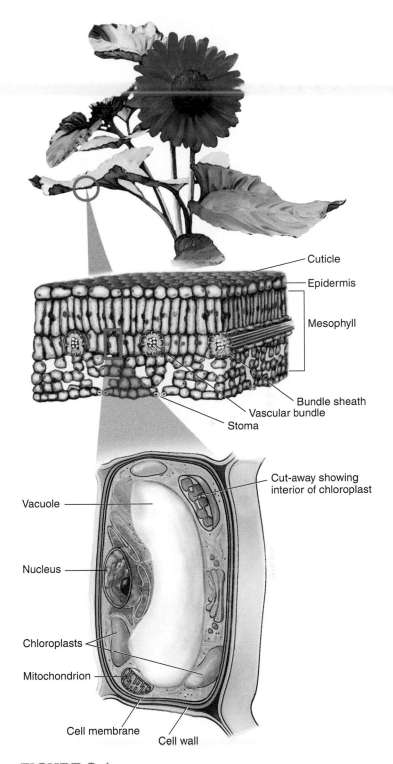

FIGURE 3.6 Plant tissues and a single cell's interior. Cell components include a cellulose cell wall, a nucleus, a large, empty vacuole, and several chloroplasts, which carry out photosynthesis. © E. H. Newcomb.

manage and distribute information. Still others create the internal structure that gives the cell its shape and allows it to fulfill its role.

All of the chemical reactions required to create these various structures, provide them with energy and materials to carry out their functions, dispose of wastes, and perform other functions of life at the cellular level are carried out by a special class of proteins called **enzymes.** Enzymes are molecular catalysts, that is, they regulate chemical reactions without being used up or inactivated in the process. Think of them as tools: like hammers or wrenches, they do their jobs without being consumed or damaged as they work. There are generally thousands of different kinds of enzymes in every cell, all necessary to carry out the many processes on which life depends. Altogether, the multitude of enzymatic reactions performed by an organism is called its **metabolism.**

ENERGY

If matter is the material of which things are made, energy provides the force to hold structures together, tear them apart, and move them from one place to another. In this section we will look at some fundamental characteristics of these components of our world.

Energy Types and Qualities

Energy is the ability to do work such as moving matter over a distance or causing a heat transfer between two objects at different temperatures. Energy can take many different forms. Heat, light, electricity, and chemical energy are examples that we all experience. The energy contained in moving objects is called **kinetic energy.** A rock rolling down a hill, the wind blowing through the trees, water flowing over a dam (fig. 3.7), or electrons speeding around the nucleus of an atom are all examples of kinetic energy. **Potential energy** is stored energy that is latent but available for use. A rock poised at the top of a hill and water stored behind a dam are examples of potential energy. **Chemical energy** stored in the

FIGURE 3.7 Water stored behind this dam represents potential energy. Water flowing over the dam has kinetic energy, some of which is converted to heat. © Carl Purcell/Photo Researchers, Inc.

food that you eat and the gasoline that you put into your car are also examples of potential energy that can be released to do useful work. Energy is often measured in units of heat (calories) or work (joules). One joule (J) is the work done when one kilogram is accelerated at one meter per second per second. One calorie is the amount of energy needed to heat one gram of pure water one degree Celsius. A calorie can also be measured as 4.184 J.

Heat describes the energy that can be transferred between objects of different temperature. When a substance absorbs heat, the kinetic energy of its molecules increases, or it may change state: a solid may become a liquid, or a liquid may become a gas. We sense change in heat content as change in temperature (unless the substance changes state).

An object can have a high heat content but a low temperature, such as a lake that freezes slowly in the fall. Other objects, like a burning match, have a high temperature but little heat content. Heat storage in lakes and oceans is essential to moderating climates and maintaining biological communities. Heat absorbed in changing states is also critical. As you will read in chapter 15, evaporation and condensation of water in the atmosphere helps distribute heat around the globe.

Energy that is diffused, dispersed, and low in temperature is considered low-quality energy because it is difficult to gather and use for productive purposes. The heat stored in the oceans, for instance, is immense but hard to capture and use, so it is low quality. Conversely, energy that is intense, concentrated, and high in temperature is high-quality energy because of its usefulness in carrying out work. The intense flames of a very hot fire or high-voltage electrical energy are examples of high-quality forms that are valuable to humans. Many of our alternative energy sources (such as wind) are diffuse compared to the higher-quality, more concentrated chemical energy in oil, coal, or gas.

Thermodynamics and Energy Transfers

Atoms and molecules cycle endlessly through organisms and their environment, but energy flows in a one-way path. A constant supply of energy—nearly all of it from the sun—is needed to keep biological processes running. Energy can be used repeatedly as it flows through the system, and it can be stored temporarily in the chemical bonds of organic molecules, but eventually it is released and dissipated.

The study of thermodynamics deals with how energy is transferred in natural processes. More specifically, it deals with the rates of flow and the transformation of energy from one form or quality to another. Thermodynamics is a complex, quantitative discipline, but you don't need a great deal of math to understand some of the broad principles that shape our world and our lives.

The **first law of thermodynamics** states that energy is *conserved;* that is, it is neither created nor destroyed under normal conditions. Energy may be transformed, for example, from the energy in a chemical bond to heat energy, but the total amount does not change.

The **second law of thermodynamics** states that, with each successive energy transfer or transformation in a system, less

Key Concepts

- The law of conservation of matter states that matter is always conserved, never destroyed, under normal conditions.

- The first law of thermodynamics states that energy is neither created nor destroyed as it moves through natural systems.

- The second law of thermodynamics states that energy is constantly degraded to lower forms (entropy increases) as it is used.

- Photosynthesis captures about 1 percent of the solar energy that reaches plants to create energy-rich molecules from sunlight, carbon dioxide, and water. Cellular respiration decomposes energy-rich molecules to provide energy to cells.

- Material cycles describe the movement of materials and elements, such as water, nitrogen, carbon, and phosphorus, through living organisms and their environment.

energy is available to do work. That is, energy is degraded to lower-quality forms, or it dissipates and is lost, as it is used. When you drive a car, for example, the chemical energy of the gas is degraded to kinetic energy and heat, which dissipates, eventually, to space. The second law recognizes that disorder, or **entropy,** tends to increase in all natural systems. Consequently, there is always less *useful* energy available when you finish a process than there was before you started. Because of this loss, everything in the universe tends to fall apart, slow down, and get more disorganized.

How does the second law of thermodynamics apply to organisms and biological systems? Organisms are highly organized, both structurally and metabolically. Constant care and maintenance is required to keep up this organization, and a constant supply of energy is required to maintain these processes. Every time some energy is used by a cell to do work, some of that energy is dissipated or lost as heat. If cellular energy supplies are interrupted or depleted, the result—sooner or later—is death.

ENERGY FOR LIFE

Ultimately, nearly all life on Earth depends on the sun for energy to create cells and carry out life processes. There are a few exceptions: some bacteria can capture energy by oxidizing inorganic compounds such as hydrogen sulfide (chemosynthesis), and this process can support entire living systems deep in the ocean, beyond the reach of sunlight (fig. 3.8). Most living things, however, depend on the sun for energy. In this section we will look at how green plants capture solar energy and use it to create organic molecules that are essential for life.

Solar Energy: Warmth and Light

Our sun is a star, a fiery ball of exploding hydrogen gas. Its thermonuclear reactions emit powerful forms of radiation, including potentially deadly ultraviolet and nuclear radiation (fig. 3.9), yet life here is nurtured by, and dependent upon, this searing, energetic source. Solar energy is essential to life for two main reasons.

First, the sun provides warmth. Most organisms survive within a relatively narrow temperature range. In fact, each species has its own range of temperatures within which it can function normally. At high temperatures (above 40°C), biomolecules begin to break down or become distorted and nonfunctional. At low temperatures (near 0°C), some chemical reactions of metabolism occur too slowly to enable organisms to grow and reproduce. Other planets in our solar system are either too hot or too cold to support life as we know it. The earth's water and atmosphere help to moderate, maintain, and distribute the sun's heat.

Second, organisms depend on solar radiation for life-sustaining energy, which is captured by green plants, algae, and some bacteria in a process called **photosynthesis.** Photosynthesis converts radiant energy into useful, high-quality chemical energy in the bonds that hold together organic molecules.

How much of the available solar energy is actually used by organisms? The amount of incoming solar radiation is enormous, about 1,372 watts/m^2 at the top of the atmosphere (1 watt = 1 J per second). However, more than half of the incoming sunlight may be reflected or absorbed by atmospheric clouds, dust, and gases. In particular, harmful, short wavelengths are filtered out by gases (such as ozone) in the upper atmosphere; thus, the atmosphere is a valuable shield, protecting life-forms from harmful doses of ultraviolet and other forms of radiation. Even with these energy reductions, however, the sun provides much more energy than biological systems can harness, and more than enough for all our energy needs if technology could enable us to tap it efficiently.

Of the solar radiation that does reach the earth's surface, about 10 percent is ultraviolet, 45 percent is visible, and 45 percent is infrared. Most of that energy is absorbed by land or water or is

FIGURE 3.8 Tube worms in a deep-ocean thermal vent ecosystem, which gathers energy from sulfur-based chemical reactions (chemosynthesis). Aside from these communities, all life on Earth depends on photosynthesis for energy. NOAA (http://oceanexplorer.noaa.gov).

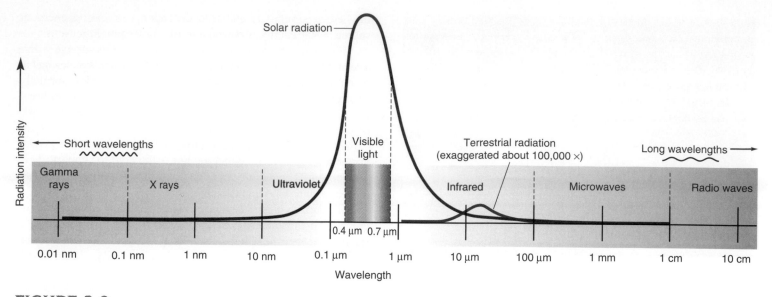

FIGURE 3.9   The electromagnetic spectrum. Our eyes are sensitive to light wavelengths, which make up nearly half the energy that reaches the earth's surface (represented by the area under the curve). Photosynthesizing plants also use the most abundant solar wavelengths. The earth reemits lower-energy, longer wavelengths, mainly the infrared part of the spectrum.

reflected into space by water, snow, and land surfaces. (Seen from outer space, Earth shines about as brightly as Venus.)

Of the energy that reaches the earth's surface, photosynthesis uses only certain wavelengths, mainly red and blue light. (Most plants reflect green wavelengths, so that is the color they appear to us.) Half of the energy plants absorb is used in evaporating water. In the end, only 1 to 2 percent of the sunlight falling on plants is available for photosynthesis. This small percentage represents the energy base for virtually all life in the biosphere!

How Does Photosynthesis Capture Energy?

Photosynthesis occurs in tiny membranous organelles called chloroplasts that reside within plant cells (see fig. 3.6). The most important key to this process is chlorophyll, a unique green molecule that can absorb light energy and use it to create high-energy chemical bonds in compounds that serve as the fuel for all subsequent cellular metabolism. Chlorophyll doesn't do this important job all alone, however. It is assisted by a large group of other lipid, sugar, protein, and nucleotide molecules. Together these components carry out two interconnected cyclic sets of reactions (fig. 3.10).

Photosynthesis begins with a series of steps called light-dependent reactions: these occur only while the chloroplast is receiving light. Enzymes split water molecules and release molecular oxygen (O_2). This is the source of all the oxygen in the atmosphere on which all animals, including you, depend for life. The light-dependent reactions also create mobile, high-energy molecules (adenosine triphosphate, or ATP, and nicotinamide adenine dinucleotide phosphate, or NADPH), which provide energy for the next set of processes, the light-independent reactions. As

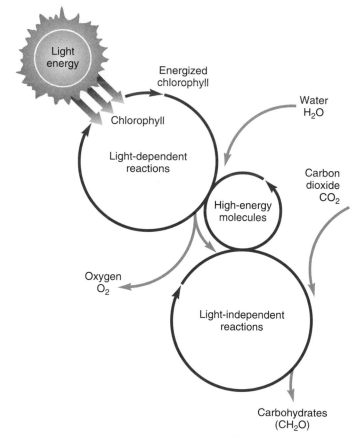

FIGURE 3.10   Photosynthesis involves a series of reactions in which chlorophyll captures light energy and forms high-energy molecules, ATP and NADPH. Light-independent reactions then use energy from ATP and NADPH to fix carbon (from air) in organic molecules.

their name implies, these reactions do not use light directly. Here, enzymes extract energy from ATP and NADPH to add carbon atoms (from carbon dioxide) to simple sugar molecules, such as glucose. These molecules provide the building blocks for larger, more complex organic molecules.

In most temperate-zone plants, photosynthesis can be summarized in the following equation:

$$6H_2O + 6CO_2 + \text{solar energy} \xrightarrow[\text{chlorophyll}]{} C_6H_{12}O_6 \text{ (sugar)} + 6O_2$$

We read this equation as "water plus carbon dioxide plus energy produces sugar plus oxygen." The reason the equation uses six water and six carbon dioxide molecules is that it takes six carbon atoms to make the sugar product. If you look closely, you will see that all the atoms in the reactants balance with those in the products. This is an example of conservation of matter.

You might wonder how making a simple sugar benefits the plant. The answer is that glucose is an energy-rich compound that serves as the central, primary fuel for all metabolic processes of cells. The energy in its chemical bonds—the ones created by photosynthesis—can be released by other enzymes and used to make other molecules (lipids, proteins, nucleic acids, or other carbohydrates), or it can drive kinetic processes such as movement of ions across membranes, transmission of messages, changes in cellular shape or structure, or movement of the cell itself in some cases. This process of releasing chemical energy, called **cellular respiration,** involves splitting carbon and hydrogen atoms from the sugar molecule and recombining them with oxygen to re-create carbon dioxide and water. The net chemical reaction, then, is the reverse of photosynthesis:

$$C_6H_{12}O_6 + 6O_2 \longrightarrow 6H_2O + 6CO_2 + \text{released energy}$$

Note that in photosynthesis, energy is *captured,* while in respiration, energy is *released.* Similarly, photosynthesis *consumes* water and carbon dioxide to *produce* sugar and oxygen, while respiration does just the opposite. In both sets of reactions, energy is stored temporarily in chemical bonds, which constitute a kind of energy currency for the cell.

We animals don't have chlorophyll and can't carry out photosynthetic food production. We do have the components for cellular respiration, however. In fact, this is how we get all our energy for life. We eat plants—or other animals that have eaten plants—and break down the organic molecules in our food through cellular respiration to obtain energy (fig. 3.11). In the process, we also consume oxygen and release carbon dioxide, thus completing the cycle of photosynthesis and respiration. Later in this chapter we will see how these feeding relationships work.

FROM SPECIES TO ECOSYSTEMS

While cellular and molecular biologists study life processes at the microscopic level, ecologists study interactions at the species, population, biotic community, or ecosystem level. In Latin, *species* lit-

erally means *kind.* In biology, **species** refers to all organisms of the same kind that are genetically similar enough to breed in nature and produce live, fertile offspring. There are several qualifications and some important exceptions to this definition of species (especially among bacteria and plants), but for our purposes this is a useful working definition.

Populations, Communities, and Ecosystems

A **population** consists of all the members of a species living in a given area at the same time. Chapter 6 deals further with population growth and dynamics. All of the populations of organisms living and interacting in a particular area make up a **biological community.** What populations make up the biological community of which you are a part? The population sign marking your city limits announces only the number of humans who live there, disregarding the other populations of animals, plants, fungi, and microorganisms that are part of the biological community within the city's boundaries. Characteristics of biological communities are discussed in more detail in chapter 4.

An ecological system, or **ecosystem,** is composed of a biological community and its physical environment. The environment includes abiotic factors (nonliving components), such as climate,

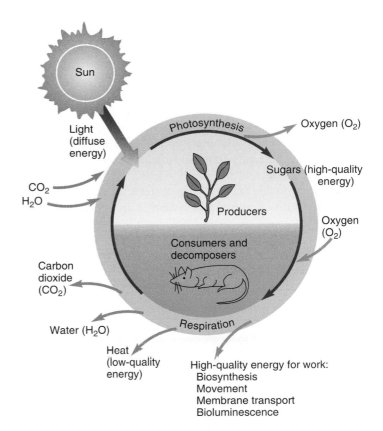

FIGURE 3.11 Energy exchange in ecosystems. Plants use sunlight, water, and carbon dioxide to produce sugars and other organic molecules. Consumers use oxygen and break down sugars during cellular respiration. Note that the laws of thermodynamics and conservation of matter can be observed in these processes.

water, minerals, and sunlight, as well as biotic factors, such as organisms, their products (secretions, wastes, and remains), and effects in a given area. It is useful to think about the biological community and its environment together, because energy and matter flow through both. Understanding how those flows work is a major theme in ecology.

For simplicity's sake, we think of ecosystems as fixed ecological units with distinct boundaries. If you look at a patch of woods surrounded by farm fields, for instance, a relatively sharp line separates the two areas, and conditions such as light levels, wind, moisture, and shelter are quite different in the woods than in the fields around them. Because of these variations, distinct populations of plants and animals live in each place. By studying each of these areas, we can make important and interesting discoveries about who lives where and why and about how conditions are established and maintained there.

The division between the fields and woods is not always clear, however. Air, of course, moves freely from one to another, and the runoff after a rainfall may carry soil, leaf litter, and even live organisms between the areas. Birds may feed in the field during the day but roost in the woods at night, giving them roles in both places. Are they members of the woodland community or the field community? Is the edge of the woodland ecosystem where the last tree grows, or does it extend to every place that has an influence on the woods?

As you can see, it may be difficult to draw clear boundaries around communities and ecosystems. To some extent we define these units by what we want to study and how much information we can handle. Thus, an ecosystem might be as large as a whole watershed or as small as a pond or even the surface of your skin. Even though our choices may be somewhat arbitrary, we still can make useful discoveries about how organisms interact with each other and with their environment within these units. The woods are, after all, significantly different from the fields around them.

Like the woodland we just considered, most ecosystems are open, in the sense that they exchange materials and organisms with other ecosystems. A stream ecosystem is an extreme example. Water, nutrients, and organisms enter from upstream and are lost downstream. The species and numbers of organisms present may be relatively constant, but they are made up of continually changing individuals. Other ecosystems are relatively closed, in the sense that very little enters or leaves them. A balanced aquarium is a good example of a closed ecosystem. Aquatic plants, animals, and decomposers can balance material cycles in the aquarium if care is taken to balance their populations. Because of the second law of thermodynamics, however, every ecosystem must have a constant inflow of energy and a way to dispose of heat. Thus, at least with regard to energy flow, every ecosystem is open.

Many ecosystems have mechanisms that maintain composition and functions within certain limits. A forest tends to remain a forest, for the most part, and to have forestlike conditions if it isn't disturbed by outside forces. Some ecologists suggest that ecosystems—or perhaps all life on the earth—may function as superorganisms because they maintain stable conditions and can be resilient to change.

Food Chains, Food Webs, and Trophic Levels

Photosynthesis (and rarely chemosynthesis) is the base of all ecosystems. Organisms that photosynthesize, mainly green plants and algae, are therefore known as **producers.** One of the major properties of an ecosystem is its **productivity,** the amount of **biomass** (biological material) produced in a given area during a given period of time. Photosynthesis is described as *primary productivity* because it is the basis for almost all other growth in an ecosystem. Manufacture of biomass by organisms that eat plants is termed *secondary productivity.* A given ecosystem may have very high total productivity, but if decomposers decompose organic material as rapidly as it is formed, the *net primary productivity* will be low.

Think about what you have eaten today and trace it back to its photosynthetic source. If you have eaten an egg, you can trace it back to a chicken, which ate corn. This is an example of a **food chain,** a linked feeding series. Now think about a more complex food chain involving you, a chicken, a corn plant, and a grasshopper. The chicken could eat grasshoppers that had eaten leaves of the corn plant. You also could eat the grasshopper directly—some humans do. Or you could eat corn yourself, making the shortest possible food chain. Humans have several options of where we fit into food chains.

In ecosystems, some consumers feed on a single species, but most consumers have multiple food sources. Similarly, some species are prey to a single kind of predator, but many species in an ecosystem are beset by several types of predators and parasites. In this way, individual food chains become interconnected to form a **food web.** Figure 3.12 shows feeding relationships among some of the larger organisms in a woodland and lake community. If we were to add all the insects, worms, and microscopic organisms that belong in this picture, however, we would have overwhelming complexity. Perhaps you can imagine the challenge ecologists face in trying to quantify and interpret the precise matter and energy transfers that occur in a natural ecosystem!

An organism's feeding status in an ecosystem can be expressed as its **trophic level** (from the Greek *trophe,* food). In our first example, the corn plant is at the producer level; it transforms solar energy into chemical energy, producing food molecules. Other organisms in the ecosystem are **consumers** of the chemical energy harnessed by the producers. An organism that eats producers is a primary consumer. An organism that eats primary consumers is a secondary consumer, which may, in turn, be eaten by a tertiary consumer, and so on. Most terrestrial food chains are relatively short (seeds → mouse → owl), but aquatic food chains may be quite long (microscopic algae → copepod → minnow → crayfish → bass → osprey). The length of a food chain also may reflect the physical characteristics of a particular ecosystem. A harsh arctic landscape has a much shorter food chain than a temperate or tropical one (fig. 3.13).

Organisms can be identified both by the trophic level at which they feed and by the *kinds* of food they eat (fig. 3.14). **Herbivores** are plant eaters, **carnivores** are flesh eaters, and **omnivores** eat both plant and animal matter. What are humans? We are natural

FIGURE 3.12 Each time an organism feeds, it becomes a link in a food chain. In an ecosystem, food chains become interconnected when predators feed on more than one kind of prey, thus forming a food web. The arrows in this diagram and in figure 3.13 indicate the direction in which matter and energy are transferred through feeding relationships. Only a few representative relationships are shown here. What others might you add?

omnivores, by history and by habit. Tooth structure is an important clue to understanding animal food preferences, and humans are no exception. Our teeth are suited for an omnivorous diet, with a combination of cutting and crushing surfaces that are not highly adapted for one specific kind of food, as are the teeth of a wolf (carnivore) or a horse (herbivore).

One of the most important trophic levels is occupied by the many kinds of organisms that remove and recycle the dead bodies and waste products of others. **Scavengers** such as crows, jackals, and vultures clean up dead carcasses of larger animals. **Detritivores** such as ants and beetles consume litter, debris, and dung, while **decomposer** organisms such as fungi and bacteria complete the final breakdown and recycling of organic materials. It could be argued that these microorganisms are second in importance only to producers, because without their activity nutrients would remain locked up in the organic compounds of dead organisms and discarded body wastes, rather than being made available to successive generations of organisms.

Ecological Pyramids

If we arrange the organisms in a food chain according to trophic levels, they often form a pyramid with a broad base representing

primary producers and only a few individuals in the highest trophic levels. This pyramid arrangement is especially true if we look at the energy content of an ecosystem (fig. 3.15). True to the second principle of thermodynamics, less food energy is available to the top trophic level than is available to preceding levels. For example, it takes a huge number of plants to support a modest colony of grazers such as prairie dogs. Several colonies of prairie dogs, in turn, might be required to feed a single coyote. And a very large top carnivore like a tiger may need a home range of hundreds of square kilometers to survive.

Why is there so much less energy in each successive level in figure 3.15? In the first place, some of the food that organisms eat is undigested and doesn't provide usable energy. Much of the energy that is absorbed is used in the daily processes of living or lost as heat when it is transformed from one form to another and thus isn't stored as biomass that can be eaten.

Furthermore, predators don't operate at 100 percent efficiency. If there were enough foxes to catch all the rabbits available in the summer when the supply is abundant, there would be too many foxes in the middle of the winter when rabbits are scarce. A general rule of thumb is that only about 10 percent of the energy in one consumer level is represented in the next higher level (fig. 3.16). The amount of energy available is often expressed in biomass. For

FIGURE 3.13　Harsh environments tend to have shorter food chains than environments with more favorable physical conditions. Compare the arctic food chains depicted here with the longer food chains in the food web in figure 3.12.

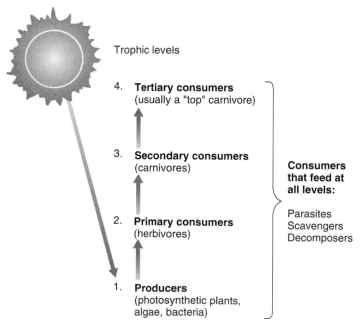

Trophic levels

4. **Tertiary consumers**
   (usually a "top" carnivore)

3. **Secondary consumers**
   (carnivores)

2. **Primary consumers**
   (herbivores)

1. **Producers**
   (photosynthetic plants,
   algae, bacteria)

**Consumers
that feed at
all levels:**

Parasites
Scavengers
Decomposers

FIGURE 3.14　Organisms in an ecosystem may be identified by how they obtain food for their life processes (producer, herbivore, carnivore, omnivore, scavenger, decomposer, reducer) or by consumer level (producer; primary, secondary, or tertiary consumer) or by trophic level (1st, 2nd, 3rd, 4th).

example, it generally takes about 100 kg of clover to make 10 kg of rabbit and 10 kg of rabbit to make 1 kg of fox.

The total number of organisms and the total amount of biomass in each successive trophic level of an ecosystem also may form pyramids (fig. 3.17) similar to those describing energy content. The relationship between biomass and numbers is not as dependable as energy, however. The biomass pyramid, for instance, can be inverted by periodic fluctuations in producer populations (for example, low plant and algal biomass present during winter in temperate aquatic ecosystems). The numbers pyramid also can be inverted. One coyote can support numerous tapeworms, for example. Numbers inversion also occurs at the lower trophic levels (for example, one large tree can support thousands of caterpillars).

MATERIAL CYCLES AND LIFE PROCESSES

To our knowledge, the earth is the only planet in our solar system that provides a suitable environment for life as we know it. Even our nearest planetary neighbors, Mars and Venus, do not meet these requirements. Maintenance of these conditions requires a constant recycling of materials between the biotic (living) and abiotic (nonliving) components of ecosystems.

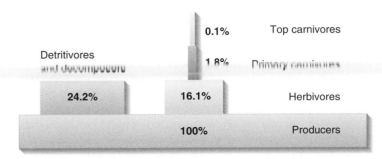

FIGURE 3.15 A classic example of an energy pyramid from Silver Springs, Florida. The numbers in each bar show the percentage of the energy captured in the primary producer level that is incorporated into the biomass of each succeeding level. Detritivores and decomposers feed at every level but are shown attached to the producer bar because this level provides most of their energy.

Source: Howard T. Odum, "Trophic Structure and Productivity of Silver Springs, Florida," in *Ecological Monographs*, 27:55–112, 1957, Ecological Society of America.

The Hydrologic Cycle

The path of water through our environment, known as the **hydrologic cycle**, is perhaps the most familiar material cycle, and it is discussed in greater detail in chapter 17 (fig. 3.18). Most of the earth's water is stored in the oceans, but solar energy continually evaporates this water, and winds distribute water vapor around the globe. Water that condenses over land surfaces, in the form of rain, snow, or fog, supports all terrestrial (land-based) ecosystems. Living organisms emit the moisture they have consumed through respiration and perspiration. Eventually this moisture reenters the atmosphere or enters lakes and streams, from which it ultimately returns to the ocean again.

As it moves through living things and through the atmosphere, water is responsible for metabolic processes within cells, for maintaining the flows of key nutrients through ecosystems, and for global-scale distribution of heat and energy (chapter 15). Water performs countless services because of its unusual properties. Water is so important that when astronomers look for signs of life on distant planets, traces of water are the key evidence they seek.

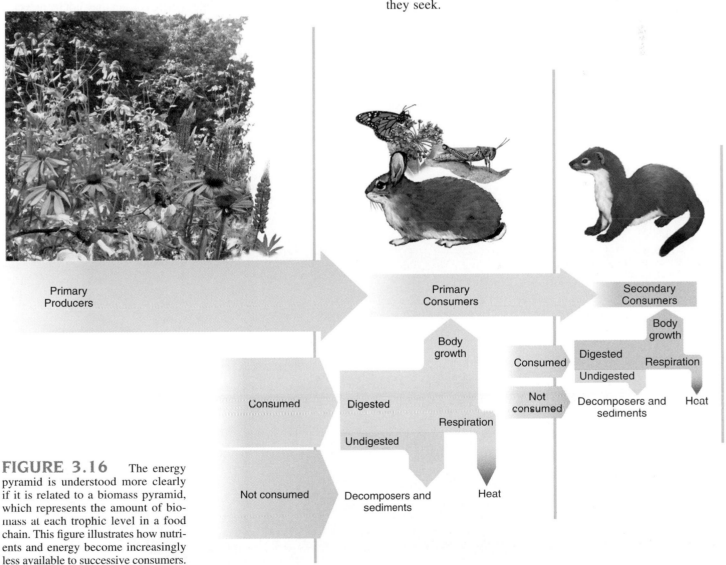

FIGURE 3.16 The energy pyramid is understood more clearly if it is related to a biomass pyramid, which represents the amount of biomass at each trophic level in a food chain. This figure illustrates how nutrients and energy become increasingly less available to successive consumers.

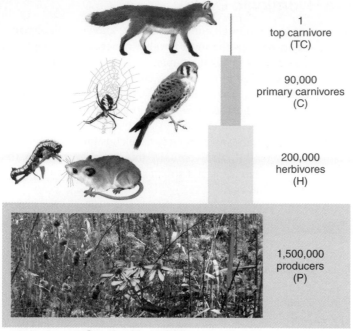

1
top carnivore
(TC)

90,000
primary carnivores
(C)

200,000
herbivores
(H)

1,500,000
producers
(P)

Grassland in summer

FIGURE 3.17 Usually, smaller organisms are eaten by larger organisms and it takes numerous small organisms to feed one large organism. The classic study represented in this pyramid shows numbers of individuals at each trophic level per 1,000 m² of grassland, and reads like this: to support one individual at the top carnivore level, there were 90,000 primary carnivores feeding upon 200,000 herbivores that in turn fed upon 1,500,000 producers.

The Carbon Cycle

Carbon serves a dual purpose for organisms: (1) it is a structural component of organic molecules, and (2) the energy-holding chemical bonds it forms represent energy "storage." The **carbon cycle** begins with the intake of carbon dioxide (CO_2) by photosynthetic organisms (fig. 3.19). Carbon (and hydrogen and oxygen) atoms are incorporated into sugar molecules during photosynthesis. Carbon dioxide is eventually released during respiration, closing the cycle. The carbon cycle is of special interest because biological accumulation and release of carbon is a major factor in climate regulation (see Exploring Science, p. 64).

The path followed by an individual carbon atom in this cycle may be quite direct and rapid, depending on how it is used in an organism's body. Imagine for a moment what happens to a simple sugar molecule you swallow in a glass of fruit juice. The sugar molecule is absorbed into your bloodstream where it is made available to your cells for cellular respiration or for making more complex biomolecules. If it is used in respiration, you may exhale the same carbon atom as CO_2 the same day.

Can you think of examples where carbon may not be recycled for even longer periods of time, if ever? Coal and oil are the compressed, chemically altered remains of plants or microorganisms that lived millions of years ago. Their carbon atoms (and hydrogen, oxygen, nitrogen, sulfur, etc.) are not released until the coal and oil are burned. Enormous amounts of carbon also are locked up as calcium carbonate ($CaCO_3$), used to build shells and skeletons of marine organisms from tiny protozoans to corals. Most of these

Movement of moist air
from ocean to land
40,000 km³

Precipitation
over land
111,000 km³

Transpiration
from vegetation
41,000 km³

Precipitation
over ocean
385,000 km³

Evaporation from soil,
streams, rivers and lakes
30,000 km³

Evaporation
from ocean
425,000 km³

Percolation
through
porous rock
and soil to
groundwater

FIGURE 3.18 The hydrologic cycle. Most exchange occurs with evaporation from oceans and precipitation back to oceans. About one-tenth of water evaporated from oceans falls over land, is recycled through terrestrial systems, and eventually drains back to oceans in rivers.

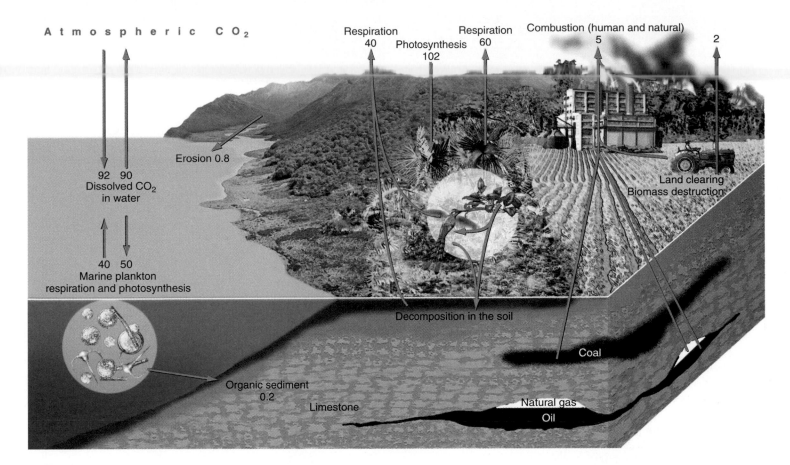

FIGURE 3.19   The carbon cycle. Numbers indicate approximate exchange of carbon in gigatons (Gt) per year. Natural exchanges are balanced, but human sources produce a net increase of CO_2 in the atmosphere. *Source:* Intergovernmental Panel for Climate Change (IPCC), 2002.

deposits are at the bottom of the oceans. The world's extensive surface limestone deposits are biologically formed calcium carbonate from ancient oceans, exposed by geological events. The carbon in limestone has been locked away for millennia, which is probably the fate of carbon currently being deposited in ocean sediments. Eventually, even the deep ocean deposits are recycled as they are drawn into deep molten layers and released via volcanic activity. Geologists estimate that every carbon atom on the earth has made about thirty such round trips over the last 4 billion years.

How does tying up so much carbon in the bodies and by-products of organisms affect the biosphere? Favorably. It helps balance CO_2 generation and utilization. Carbon dioxide is one of the so-called greenhouse gases because it absorbs heat radiated from the earth's surface, retaining it instead in the atmosphere. This phenomenon is discussed in more detail in chapter 15. Photosynthesis and deposition of $CaCO_3$ remove atmospheric carbon dioxide; therefore, vegetation (especially large forested areas such as the boreal forests) and the oceans are very important **carbon sinks** (storage deposits). Cellular respiration and combustion both release CO_2, so they are referred to as carbon sources of the cycle.

Presently, natural fires and human-created combustion of organic fuels (mainly wood, coal, and petroleum products) release huge quantities of CO_2 at rates that seem to be surpassing the pace of CO_2 removal. Scientific concerns over the linked problems of increased atmospheric CO_2 concentrations, massive deforestation, and reduced productivity of the oceans due to pollution are discussed in chapters 15 and 16.

The Nitrogen Cycle

Organisms cannot exist without amino acids, peptides, nucleic acids, and proteins, all of which are organic molecules containing nitrogen. The nitrogen atoms that form these important molecules are provided by producer organisms. Plants assimilate (take up) inorganic nitrogen from the environment and use it to build their own protein molecules, which are eaten by consumer organisms, digested, and used to build their bodies. Even though nitrogen is the most abundant gas (about 78 percent of the atmosphere), however, plants cannot use N_2, the stable diatomic (2-atom) molecule in the air.

Where and how, then, do green plants get *their* nitrogen? The answer lies in the most complex of the gaseous cycles, the **nitrogen cycle.** Figure 3.20 summarizes the nitrogen cycle. The key natural processes that make nitrogen available are carried out by nitrogen-fixing bacteria (including some blue-green algae or cyanobacteria). These organisms have a highly specialized ability to "fix" nitrogen, meaning they change it to less mobile, more useful forms by combining it with hydrogen to make ammonia (NH_3).

Measuring primary productivity is important for understanding individual plants and local environments. Understanding rates of primary productivity is also key to understanding global processes material cycling and biological activity:

- In global carbon cycles, how much carbon is stored by plants, how quickly is it stored, and how does carbon storage compare in contrasting environments, such as the Arctic and the tropics?

- How does this carbon storage affect global climates (chapter 15)?

- In global nutrient cycles, how much nitrogen and phosphorus wash off-shore, and where?

How can environmental scientists measure primary production (photosynthesis) at a global scale? In the opening story of this chapter, you read that Ray Lindeman collected and weighed samples of all trophic levels in a small lake ecosystem. But that method is impossible for large ecosystems, especially for oceans, which cover 70 percent of the earth's surface. One of the newest methods of quantifying biological productivity involves remote sensing, or data collected from satellite sensors that observe the energy reflected from the earth's surface.

As you have read in this chapter, chlorophyll in green plants *absorbs* red and blue wavelengths of light and *reflects* green wavelengths. Your eye receives, or senses, these green wavelengths. A white-sand beach, on the other hand, reflects approximately equal amounts of all light wavelengths (figure 3.9) that reach it from the sun, so it looks white (and bright!) to your eye. In a similar way, different surfaces of the earth reflect character-

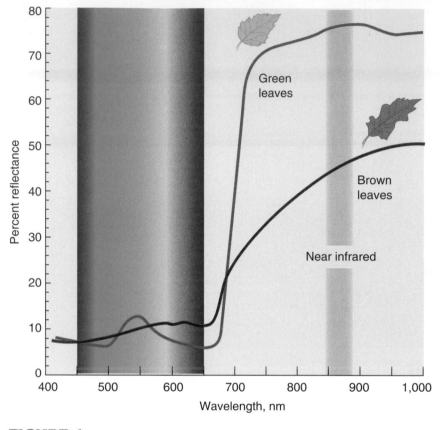

FIGURE 1   Energy wavelengths reflected by green and brown leaves.

istic wavelengths. Snow-covered surfaces reflect light wavelengths; dark-green forests with abundant chlorophyll-rich leaves—and ocean surfaces rich in photosynthetic algae and plants—reflect greens and near-infrared wavelengths. Dry, brown forests with little active chlorophyll reflect more red and less infrared energy than do dark-green forests (fig. 1)

To detect land-cover patterns on the earth's surface, we can put a sensor on a satellite that orbits the earth. As the satellite travels, the sensor receives and transmits to earth a series of "snapshots." One of the best known earth-imaging satellites, *Landsat 7,* produces images that cover an area 185 km (115 mi) wide, and each pixel represents an area of just 30 x 30 m on the

Nitrite-forming bacteria combine the ammonia with oxygen, forming nitrites, which have the ionic form NO_2^-. Another group of bacteria then convert nitrites to nitrates, which have the ionic form NO_3^-, that can be absorbed and used by green plants. After nitrates have been absorbed into plant cells, they are reduced to ammonium (NH_4^+), which is used to build amino acids that become the building blocks for peptides and proteins.

Members of the bean family (legumes) and a few other kinds of plants are especially useful in agriculture because they have nitrogen-fixing bacteria actually living *in* their root tissues (fig. 3.21). Legumes and their associated bacteria enrich the soil, so interplanting and rotating legumes with crops such as corn that use but cannot replace soil nitrates are beneficial farming practices that take practical advantage of this relationship.

ground. *Landsat* orbits approximately from pole to pole, so as the earth spins below the satellite, it captures images of the entire surface every 16 days. Another satellite, *SeaWIFS,* was designed mainly for monitoring biological activity in oceans (fig. 2). *SeaWIFS* follows a path similar to *Landsat*'s but it revisits each point on the earth every day and produces images with a pixel resolution of just over 1 km.

Since satellites detect a much greater range of wavelengths than our eyes can, they are able to monitor and map chlorophyll abundance. In oceans, this is a useful measure of ecosystem health, as well as carbon dioxide uptake. By quantifying and mapping primary production in oceans, climatologists are working to estimate the role of ocean ecosystems in moderating climate change: for example, they can estimate the extent of biomass production in the cold, oxygen-rich waters of the North Atlantic (fig. 2). Oceanographers can also detect near-shore areas where nutrients washing off the land surface fertilize marine ecosystems and stimulate high productivity, such as near the mouth of the Amazon or Mississippi Rivers. Monitoring and mapping these patterns helps us estimate human impacts on nutrient flows (figs. 3.20, 3.22) from land to sea.

>01 .02 .03 .05 .1 .2 .3 .5 1 2 3 5 10 15 20 30 50
Ocean: Chlorophyll *a* Concentration (mg/m³)

Maximum   Minimum
Land: Normalized Difference Land Vegetation Index

FIGURE 2 *SeaWIFS* image showing chlorophyll abundance in oceans and plant growth on land (normalized difference vegetation index). The *SeaWIFS* Project, NASA/Goddard Space Flight Center and ORBIMAGE.

Nitrogen reenters the environment in several ways. The most obvious path is through the death of organisms. Their bodies are decomposed by fungi and bacteria, releasing ammonia and ammonium ions, which then are available for nitrate formation. Organisms don't have to die to donate proteins to the environment, however. Plants shed their leaves, needles, flowers, fruits, and cones; animals shed hair, feathers, skin, exoskeletons, pupal cases, and silk. Animals also produce excrement and urinary wastes that contain nitrogenous compounds. Urinary wastes are especially high in nitrogen because they contain the detoxified wastes of protein metabolism. All of these by-products of living organisms decompose, replenishing soil fertility.

How does nitrogen reenter the atmosphere, completing the cycle? Denitrifying bacteria break down nitrates into N_2 and

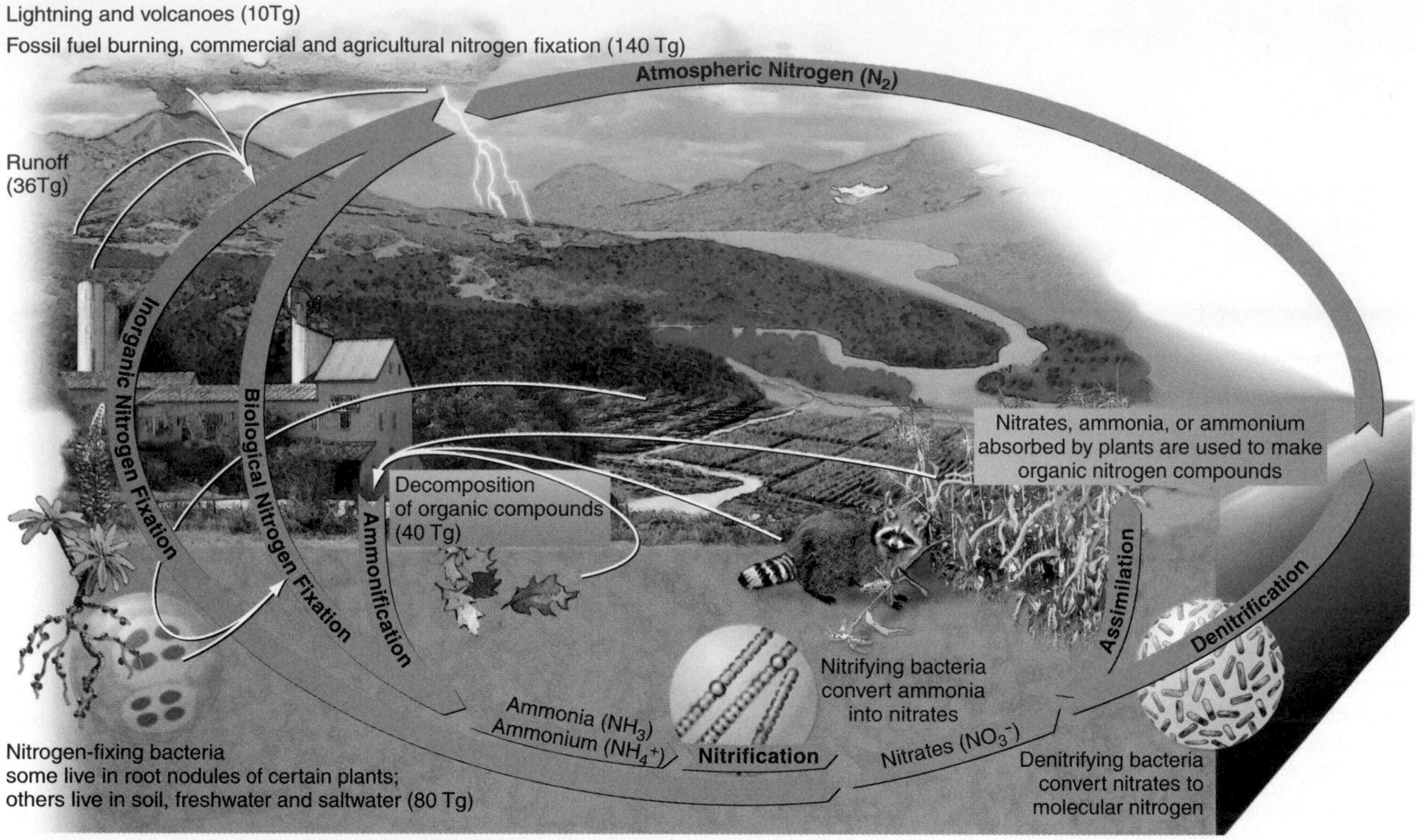

Lightning and volcanoes (10Tg)

Fossil fuel burning, commercial and agricultural nitrogen fixation (140 Tg)

Atmospheric Nitrogen (N₂)

Runoff (36Tg)

Inorganic Nitrogen Fixation

Biological Nitrogen Fixation

Ammonification

Decomposition of organic compounds (40 Tg)

Nitrates, ammonia, or ammonium absorbed by plants are used to make organic nitrogen compounds

Assimilation

Denitrification

Nitrifying bacteria convert ammonia into nitrates

Ammonia (NH₃)
Ammonium (NH₄⁺)

Nitrification

Nitrates (NO₃⁻)

Denitrifying bacteria convert nitrates to molecular nitrogen

Nitrogen-fixing bacteria some live in root nodules of certain plants; others live in soil, freshwater and saltwater (80 Tg)

FIGURE 3.20   The nitrogen cycle. Human sources of nitrogen fixation (conversion of molecular nitrogen to ammonia or ammonium) are now about 50 percent greater than natural sources. Bacteria convert ammonia to nitrates, which plants use to create organic nitrogen. Eventually, nitrogen is stored in sediments or converted back to molecular nitrogen (1 Tg = 10^{12} g). Galloway and Cowling, 2002.

FIGURE 3.21   The roots of this adzuki bean plant are covered with bumps called nodules. Each nodule is a mass of root tissue containing many bacteria that help to convert nitrogen in the soil to a form the bean plants can assimilate and use to manufacture amino acids.
© David Dennis/Tom Stack & Associates.

nitrous oxide (N_2O), gases that return to the atmosphere; thus, denitrifying bacteria compete with plant roots for available nitrates. However, denitrification occurs mainly in waterlogged soils that have low oxygen availability and a high amount of decomposable organic matter. These are suitable growing conditions for many wild plant species in swamps and marshes, but not for most cultivated crop species, except for rice, a domesticated wetlands grass.

In recent years, humans have profoundly altered the nitrogen cycle. By using synthetic fertilizers, cultivating nitrogen-fixing crops, and burning fossil fuels, we have more than doubled the amount of nitrogen cycled through our environment. This excess nitrogen input is causing serious loss of soil nutrients such as calcium and potassium, acidification of rivers and lakes, and rising atmospheric concentrations of the greenhouse gas, nitrous oxide. It also encourages the spread of weeds into areas such as prairies occupied by native plants adapted to nitrogen-poor environments. In coastal areas, blooms of toxic algae and dinoflagellates result from excess nitrogen carried by rivers from farmlands and cities.

The Phosphorus Cycle

Minerals become available to organisms after they are released from rocks. Two mineral cycles of particular significance to organisms are phosphorus and sulfur. Why do you suppose phosphorus is a primary ingredient in fertilizers? At the cellular level, energy-rich, phosphorus-containing compounds are primary participants in energy transfer reactions, as we have discussed. The amount of available phosphorus in an environment can, therefore, have a dramatic effect on productivity. Abundant phosphorus stimulates lush plant and algal growth, making it a major contributor to water pollution.

The **phosphorus cycle** (fig. 3.22) begins when phosphorus compounds are leached from rocks and minerals over long periods of time. Because phosphorus has no atmospheric form, it is usually transported in aqueous form. Inorganic phosphorus is taken in by producer organisms, incorporated into organic molecules, and then passed on to consumers. It is returned to the environment by decomposition. An important aspect of the phosphorus cycle is the very long time it takes for phosphorus atoms to pass through it. Deep sediments of the oceans are significant phosphorus sinks of extreme longevity. Phosphate ores that now are mined to make detergents and inorganic fertilizers represent exposed ocean sediments that are millennia old. You could think of our present use of phosphates, which are washed out into the river systems and eventually the oceans, as an accelerated mobilization of phosphorus from source to sink. Aquatic ecosystems often are

dramatically affected in the process because excess phosphates can stimulate explosive growth of algae and photosynthetic bacteria populations, upsetting ecosystem stability. Notice also that in this cycle, as in the others, the role of organisms is only one part of a larger picture.

The Sulfur Cycle

Sulfur plays a vital role in organisms, especially as a minor but essential component of proteins. Sulfur compounds are important determinants of the acidity of rainfall, surface water, and soil. In addition, sulfur in particles and tiny air-borne droplets may act as critical regulators of global climate. Most of the earth's sulfur is tied up underground in rocks and minerals such as iron disulfide (pyrite) or calcium sulfate (gypsum). This inorganic sulfur is released into air and water by weathering, emissions from deep seafloor vents, and by volcanic eruptions (fig. 3.23).

The **sulfur cycle** is complicated by the large number of oxidation states the element can assume, including hydrogen sulfide (H_2S), sulfur dioxide (SO_2), sulfate ion (SO_4^{-2}), and sulfur, among others. Inorganic processes are responsible for many of these transformations, but living organisms, especially bacteria, also sequester sulfur in biogenic deposits or release it into the

FIGURE 3.22   The phosphorus cycle. Natural movement of phosphorus is slight, involving recycling within ecosytems and some erosion and sedimentation of phosphorus-bearing rock. Use of phosphate (PO_4^{-3}) fertilizers and cleaning agents increases phosphorus in aquatic systems, causing eutrophication. Units are teragrams (Tg) phosphorus per year.

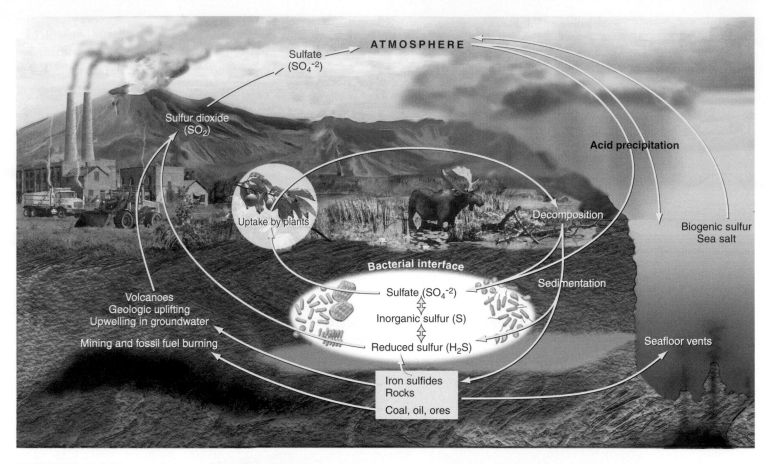

FIGURE 3.23 The sulfur cycle. Sulfur is present mainly in rocks, soil, and water. It cycles through ecosystems when it is taken in by organisms. Combustion of fossil fuels causes increased levels of atmospheric sulfur compounds, which create problems related to acid precipitation.

environment. Which of the several kinds of sulfur bacteria prevail in any given situation depends on oxygen concentrations, pH, and light levels.

Human activities also release large quantities of sulfur, primarily through burning fossil fuels. Total yearly anthropogenic sulfur emissions rival those of natural processes, and acid rain caused by sulfuric acid produced as a result of fossil fuel use is a serious problem in many areas (see chapter 16). Sulfur dioxide and sulfate aerosols cause human health problems, damage buildings and vegetation, and reduce visibility. They also absorb UV radiation and create cloud cover that cools cities and may be offsetting greenhouse effects of rising CO_2 concentrations.

Summary

- Ecology, the study of relationships between organisms and their environment, involves investigating how organisms acquire and use energy, nutrients, and water from their environment. Required conditions for life include chemical elements, availability of liquid water, and moderate temperatures.

- Matter—the substance of which things are made—is not created or destroyed under normal circumstances. It is recycled endlessly through organisms and their environment. This principle is known as the *conservation of matter*.

- Elements are identified by their atomic number, the number of protons in their nuclei. Atoms can have variable numbers of neutrons (isotopes) or electrons (ions). Four elements, oxygen, carbon, hydrogen, and nitrogen, make up more than 96 percent of most living organisms.

- Acids are substances that readily release hydrogen ions in solution; bases are substances that readily bond with hydrogen atoms. Acids and bases play important roles in life processes, and they are common sources of environmental problems, such as acid mine drainage. Buffers, substances that absorb or release hydrogen ions, neutralize acids and bases.

- Ionic or covalent bonds join atoms to produce compounds or molecules. Breaking bonds releases energy, including that used by living cells.

- Water's remarkable properties make it both the medium for life processes and a key factor in our global climate.

- Four major classes of organic compounds, hydrocarbons, carbohydrates, proteins, and nucleic acids, provide structure, function, and energy for cells.

- Energy cannot be created or destroyed (*the first law of thermodynamics*), but it is constantly degraded to less useful forms—entropy increases—as it is used (*the second law of thermodynamics*).

- Photosynthesis is the conversion of solar energy to chemical bonds in organic molecules, using carbon dioxide (from air) and water. Chlorophyll, the green pigment in plants, provides the energy for nearly all living things on Earth. Photosynthesis involves a complex series of reactions, some light-dependent and others not dependent on light. Cellular respiration is the process of breaking down organic molecules to release energy.

- *Species* are often (but not always) defined as all organisms that can breed and produce fertile offspring in nature. A population is all the members of a species that live in an area, and a biological community comprises all the populations in an area. An ecosystem is a biological community and its physical environment. Ecosystems are generally open systems: they exchange materials and energy with their surroundings.

- Energy and nutrients flow through food webs or trophic levels in an ecosystem. Productivity is a measure of how rapidly an ecosystem accumulates biomass, or biological material. Organisms in a food web can be identified as primary producers or consumers; they can also be identified as herbivores, carnivores, omnivores, scavengers, detritivores, or decomposers. The concept of ecological pyramids is that there are many more individuals, and more biomass, at lower trophic levels that at higher trophic levels.

- Materials, such as water, carbon, nitrogen, sulfur, and phosphorus, also cycle through organisms and their environment. Substances are stored in rocks, air, water, and living things for varying lengths of time. One of the major impacts of humans on our environment has been to alter the rates of material cycling.

Questions for Review

1. Define *atom* and *element*. Are these terms interchangeable?

2. Your body contains vast numbers of carbon atoms. How is it possible that some of these carbon atoms may have been part of the body of a prehistoric creature?

3. What are six characteristics of water that make it so valuable for living organisms and their environment?

4. In the biosphere, matter follows a circular pathway while energy follows a linear pathway. Explain.

5. The oceans store a vast amount of heat, but (except for climate moderation) this huge reservoir of energy is of little use to humans. Explain the difference between high-quality and low-quality energy.

6. Ecosystems require energy to function. Where does this energy come from? Where does it go? How does the flow of energy conform to the laws of thermodynamics?

7. Heat is released during metabolism. How is this heat useful to a cell and to a multicellular organism? How might it be detrimental, especially in a large, complex organism?

8. Photosynthesis and cellular respiration are complementary processes. Explain how they exemplify the laws of conservation of matter and thermodynamics.

9. What do we mean by carbon-fixation or nitrogen-fixation? Why is it important to humans that carbon and nitrogen be "fixed"?

10. The population density of large carnivores is always very small compared to the population density of herbivores occupying the same ecosystem. Explain this in relation to the concept of an ecological pyramid.

11. A species is a specific kind of organism. What general characteristics do individuals of a particular species share? Why is it important for ecologists to differentiate among the various species in a biological community?

Questions for Critical Thinking

1. When we say that there is no "away" where we can throw things we don't want anymore, are we stating a premise or a conclusion? If you believe this is a premise, supply the appropriate conclusion. If you believe it is a conclusion, supply the appropriate premises. Does the argument change if this statement is a premise or a conclusion?

2. Suppose one of your classmates disagrees with the statement above, saying, "Of course there is an 'away.' It's anywhere out of *my* ecosystem." How would you answer?

3. A few years ago, laundry detergent makers were forced to reduce or eliminate phosphorus. Other cleaning agents (such as dishwasher detergents) still contain substantial amounts of phosphorus. What information would make you change your use of nitrogen, phosphorus, or other useful pollutants?

4. The first law of thermodynamics is sometimes summarized as "you can't get something for nothing." The second law is summarized as "you can't even break even." Explain what these phrases mean. Is it dangerous to oversimplify these important concepts?

5. The ecosystem concept revolutionized ecology by introducing holistic systems thinking as opposed to individualistic life history studies. Why was this a conceptual breakthrough?

6. Defining a *species* has become increasingly tricky for ecologists. What are some problems with the simple definition given in the summary? How might you modify and improve that definition?

7. If ecosystems are so difficult to delimit, why is this such a persistent concept? Can you imagine any other ways to define or delimit environmental investigation?

8. The properties of water are so unique and so essential for life as we know it that some people believe it proves that our planet was intentionally designed for our existence. What would an environmental scientist say about this belief?

9. Choose one of the material cycles (carbon, nitrogen, phosphorus, or sulfur) and identify the components of the cycle in which you participate. For which of these components would it be easiest to reduce your impacts?

Key Terms

acid 51
atomic number 50
atoms 50
base 51
biological community 57

biomass 58
carbon cycle 62
carbon sink 63
carnivores 58
cells 53

cellular respiration 57
chemical energy 54
compound 50
conservation of matter 50
consumers 58
decomposers 59
detritivores 59
ecology 49
ecosystem 57
elements 50
energy 54
entropy 55
enzymes 54
first law of thermodynamics 54
food chain 58
food web 58
heat 54
herbivores 58
hydrologic cycle 61
ions 51

isotopes 50
kinetic energy 54
matter 50
metabolism 54
molecule 50
nitrogen cycle 63
omnivores 58
organic compounds 51
pH 51
phosphorus cycle 67
photosynthesis 55
population 57
potential energy 54
producer 58
productivity 58
scavengers 59
second law of thermodynamics 54
species 57
sulfur cycle 67
trophic level 58

Further Readings

Beardsley, Tim. 1997. When nutrients turn noxious: A little nitrogen is nice, but too much is toxic. *Scientific American* 276(6): 24–25.

Daily, Gretchen, ed. 2002. *The New Economy of Nature.* Island Press.

Emsley, John. 2001. *Nature's Building Blocks: A–Z Guide to the Elements.* Oxford University Press.

Galloway, James N., and Ellis B. Cowling. 2002. Reactive nitrogen and the world: 200 years of change. *AMBIO* 31(7):501.

Lindeman, Ray L. 1942. The trophic-dynamic aspect of ecology. *Ecology* 23:399–418.

Smith, Stephen V., et al. 2003. Humans, hydrology, and the distribution of inorganic nutrient loading to the ocean. *BioScience* 53(3):235–45.

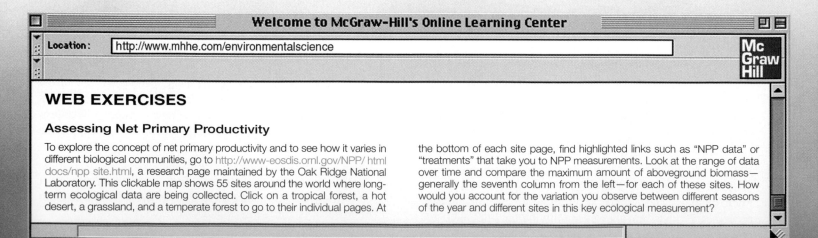

Welcome to McGraw-Hill's Online Learning Center

Location: http://www.mhhe.com/environmentalscience

WEB EXERCISES

Assessing Net Primary Productivity

To explore the concept of net primary productivity and to see how it varies in different biological communities, go to http://www-eosdis.ornl.gov/NPP/ html docs/npp site.html, a research page maintained by the Oak Ridge National Laboratory. This clickable map shows 55 sites around the world where long-term ecological data are being collected. Click on a tropical forest, a hot desert, a grassland, and a temperate forest to go to their individual pages. At the bottom of each site page, find highlighted links such as "NPP data" or "treatments" that take you to NPP measurements. Look at the range of data over time and compare the maximum amount of aboveground biomass—generally the seventh column from the left—for each of these sites. How would you account for the variation you observe between different seasons of the year and different sites in this key ecological measurement?

Biological Communities and Species Interactions

4

Any species of bug is an irreplaceable marvel, equal to the works of art which we religiously preserve in our museums.

Claude Levi-Strauss

OBJECTIVES

After studying this chapter, you should be able to:

- describe how environmental factors determine which species live in a given ecosystem and where or how they live.
- understand how random genetic variation and natural selection lead to evolution, adaptation, niche specialization, and partitioning of resources in biological communities.
- compare and contrast interspecific predation, competition, symbiosis, commensalism, mutualism, and coevolution.
- discuss productivity, diversity, complexity, and structure of biological communities and how these characteristics might be connected to resilience and stability.
- explain how ecological succession results in ecosystem development and allows one species to replace another.
- give some examples of exotic species introduced into biological communities and describe the effects such introductions can have on indigenous species.

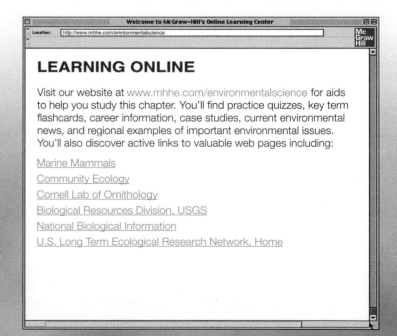

Welcome to McGraw~Hill's Online Learning Center

Location: http://www.mhhe.com/environmentalscience

LEARNING ONLINE

Visit our website at www.mhhe.com/environmentalscience for aids to help you study this chapter. You'll find practice quizzes, key term flashcards, career information, case studies, current environmental news, and regional examples of important environmental issues. You'll also discover active links to valuable web pages including:

Marine Mammals
Community Ecology
Cornell Lab of Ornithology
Biological Resources Division, USGS
National Biological Information
U.S. Long Term Ecological Research Network, Home

Photo: The bears, birds, and fish of the McNeil River, Alaska, form an interconnected biological community together with terrestrial and aquatic plants and invertebrates. © John Warden/Stone/Getty Images.

Why Trees Need Salmon

Ecologists have long known that salmon in the Pacific Northwest need clear streams to breed, and that clear streams need healthy forests. Surprising new evidence now indicates that forests themselves need salmon to remain healthy.

Pacific salmon (*Onchorhyncus spp.*) are anadromous: they hatch in freshwater lakes and streams, spend much of their lives at sea, then return to the stream where they were born to breed and die. To reproduce successfully, these fish require clear, cold, shaded streams and a clean gravel riverbed. When forests are cut, sediment washes down hillsides and into streams, clogging gravel streambeds and suffocating eggs. Sediment also absorbs sunlight, warming water and reducing oxygen saturation in the water. Lower oxygen level reduces survival rates of eggs and young fish.

Every year, as millions of fish return to spawn and die, they provide a banquet for bears, eagles, and other species that gorge themselves on the fat-rich fish. Ecologist Thomas Reimchen has found that bears fishing in British Columbia's rivers can catch 500 fish in a six-week salmon migration season (about 12 fish per bear per day). He also estimates that a bear gets 70 percent of its annual protein intake from fish. But bears also drag tons of fish up on shore and leave half-eaten carcasses strewn about the forest floor. Riemchen calculates that these scattered fish fertilize the forest at a rate of about 120 kg of nitrogen per acre. British Columbia's rainforests, with at least 30,000 fishing bears, may receive 60 million kg of salmon each year. Nitrogen is often a limiting nutrient for rainforest vegetation. Between one-quarter and one-half of the nitrogen in a towering Sitka Spruce or Douglas Fir may derive from salmon carcasses.

In addition to fertilizing trees, salmon carcasses provide food for insects and other scavengers. Birds and other predators consume these insects, and nutrients from salmon thus work through the entire forest ecosystem.

In a separate study, ecologists Robert Naiman and James Helfield found that trees along salmon-rich rivers can grow up to three times as fast as trees along streams without salmon. This is important, they point out, because salmon stocks are dwindling throughout the Pacific Northwest. In Washington, Oregon, and California, salmon populations have fallen by 90 percent from their historic numbers. Because of this close relationship, they argue, forest management and fish management need to be integrated. Each population—rainforest trees and ocean-going fish—affects the stability of the other.

Apparently, salmon need healthy forests, and forests need healthy salmon populations. Stream ecosystems need standing trees to retain soil and provide shade. So healthy streams depend on fish, just as the fish depend on the streams. As this case shows, links among organisms in an ecosystem are intricate, often subtle, and essential for ecological stability. Relationships between apparently separate environments, such as rivers and forests, can be equally important. In this chapter we'll explore some of these relationships among organisms and between organisms and their environment.

To read more:

Reimchen, T. E., D. Mathewson, M. D. Hocking, J. Moran, and D. Harris. 2003. Isotopic evidence for enrichment of salmon-derived nutrients in vegetation, soil and insects in riparian zones in coastal British Columbia. *American Fisheries Society Symposium* 34:59–69.

Helfield J. M., and R. J. Naiman. 2001. Effects of salmon-derived nitrogen on riparian forest growth and implications for stream productivity. *Ecology* 82(9):2403–09.

WHO LIVES WHERE, AND WHY?

"Why" questions often are the stimulus for scientific research, but the research itself centers on "how" questions. Why, we wonder, does a particular species live where it does? More to the point, how is it *able* to live there? How does it deal with the physical resources of its environment and are some of its techniques unique? How does it interact with the other species present? And what gives one species an edge over another species in a particular habitat?

In this section we will examine some specific ways organisms are limited by the physical aspects of their environment. We then will discuss how members of a biological community interact, pointing out a few of the difficulties ecologists encounter when they attempt to discern patterns and make generalizations about community interactions and organization.

Critical Factors and Tolerance Limits

Every living organism has limits to the environmental conditions it can endure. Temperatures, moisture levels, nutrient supply, soil and water chemistry, living space, and other environmental factors must be within appropriate levels for life to persist. In 1840, Justus von Liebig proposed that the single factor in shortest supply relative to demand is the critical determinant in the distribution of that species. Ecologist Victor Shelford later expanded this principle of limiting factors by stating that each environmental factor has both minimum and maximum levels, called **tolerance limits,** beyond which a particular species cannot survive (fig. 4.1) or is unable to reproduce. The single factor closest to these survival limits, he postulated, is the critical limiting factor that determines where a particular organism can live.

At one time, ecologists accepted this concept so completely that they called it Liebig's or Shelford's law and tried to identify unique factors limiting the growth of every population of plants and animals. For many species, however, we find that the interaction of several factors working together, rather than a single limiting factor, determines biogeographical distribution. If you have ever explored the rocky coasts of New England or the Pacific Northwest, for instance, you probably have noticed that mussels and barnacles endure extremely harsh conditions but generally are sharply limited to an intertidal zone where they grow so thickly that they often completely cover the substrate. No single factor determines this distribution. Instead, a combination of temperature extremes, drying time between tides, salt concentrations, competitors, and food availability limits the number and location of these animals.

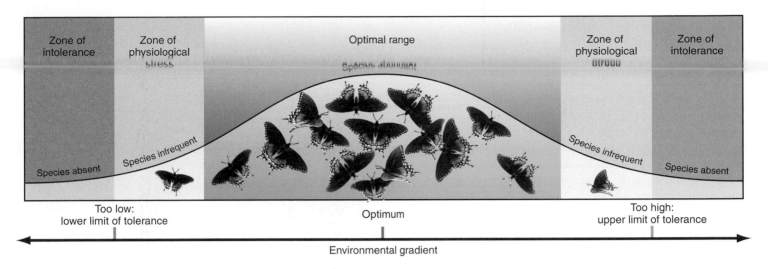

Zone of intolerance — Zone of physiological stress — Optimal range — Zone of physiological stress — Zone of intolerance

Species abundant

Species absent — Species infrequent — Species infrequent — Species absent

Too low: lower limit of tolerance — Optimum — Too high: upper limit of tolerance

Environmental gradient

FIGURE 4.1 The principle of tolerance limits states that for every environmental factor, an organism has both maximum and minimum levels beyond which it cannot survive. The greatest abundance of any species along an environmental gradient is around the optimum level of the critical factor most important for that species. Near the tolerance limits, abundance decreases because fewer individuals are able to survive the stresses imposed by limiting factors.

For other organisms, there may be a specific *critical factor* that, more than any other, determines the abundance and distribution of that species in a given area. A striking example of cold intolerance as a critical factor is found in the giant saguaro cactus (*Carnegiea gigantea*), which grows in the dry, hot Sonoran desert of southern Arizona and northern Mexico (fig. 4.2). Saguaros are extremely sensitive to low temperatures. A single exceptionally cold winter night with temperatures below freezing for 12 hours or more will kill growing tips on the branches. Young saguaros are more susceptible to frost damage than adults, but seedlings typically become established under the canopy of small desert trees such as mesquite that shield the young cacti from the cold night sky. Unfortunately, the popularity of grilling with mesquite wood has caused extensive harvesting of the nurse trees that once sheltered small saguaros, adversely affecting reproduction of this charismatic species.

Animal species, too, exhibit tolerance limits that often are more critical for the young than for the adults. The desert pupfish (*Cyprinodon*), for instance, occurs in small isolated populations in warm springs in the northern Sonoran desert. Adult pupfish can survive temperatures between 0°C and 42°C (a remarkably high temperature for a fish) and are tolerant to an equally wide range of salt concentrations. Eggs and juvenile fish, however, can only live between 20°C and 36°C and are killed by high salt levels. Reproduction, therefore, is limited to a small part of the range of adult fish, which is often restricted anyway by the size of the small springs and desert seeps in which the species lives.

Sometimes the requirements and tolerances of species are useful indicators of specific environmental characteristics. The presence or absence of such species can tell us something about the community and the ecosystem as a whole. Locoweeds, for example, are small legumes that grow where soil concentrations of selenium are high. Because selenium is often found with uranium deposits, locoweeds have an applied economic value as **environmental indicators.** Such indicator species also may demonstrate the effects of human activities. Lichens and eastern white pine are less restricted in habitat than locoweeds, but are indicators of air pollution because they are extremely sensitive to sulfur dioxide and acid precipitation. Bull thistle is a weed that grows on disturbed soil but is not eaten by cattle; therefore, an abundant population of bull thistle in a pasture is a good indicator of overgrazing. Similarly, anglers know that trout species require clean, well-oxygenated water, so the presence or absence of trout can be an indicator of water quality.

FIGURE 4.2 Saguaro cacti, symbolic of the Sonoran desert, are an excellent example of distribution controlled by a critical environmental factor. Extremely sensitive to low temperatures, saguaros are found only where minimum temperatures never dip below freezing for more than a few hours at a time. © William P. Cunningham.

Natural Selection, Adaptation, and Evolution

How is it that mussels have developed the ability to endure pounding waves, daily exposure to drying sun and wind, and seasonal threats of extreme cold or hot temperatures? What enables desert pupfish to tolerate hot, mineral-laden springs? How does the saguaro survive in the harsh temperatures and extreme dryness of the desert? We commonly say that each of these species is "adapted" to its special set of conditions, but what does that mean? In this section, we will examine one of the most important concepts in biology: how species acquire traits that allow them to live in unique ways in particular environments.

In common use, to *adapt* means to modify slightly, usually temporarily. We use the term *adapt* in two ways. One is a limited range of *physiological modifications* (called acclimation) available to individual organisms. If you keep house plants inside all winter, for example, and then put them out in full sunlight in the spring, they get sunburned. If the damage isn't too severe, your plants will probably grow new leaves with a thicker cuticle and denser pigments that protect them from the sun. But this change isn't permanent. Another winter inside will make them just as sensitive to the sun as before. Furthermore, the changes they acquire are not passed on to their offspring.

In biological terms, adaptation refers specifically to inherited traits that gradually change a *population* or a species, not an individual. These inherited traits allow a species to live in a particular environment. This process is explained by the theory of **evolution,** developed by Charles Darwin and Alfred Wallace. According to this theory, species change gradually through competition for scarce resources and **natural selection,** a process in which those members of a population that are best suited for a particular set of environmental conditions will survive and produce offspring more successfully than their ill-suited competitors.

Natural selection acts on preexisting genetic diversity created by a series of small, random mutations (changes in genetic material) that occur spontaneously in every population. These mutations produce a variety of traits, some of which are more advantageous than others in a given situation. Where resources are limited or environmental conditions place some selective pressure on a population, individuals with those advantageous traits become more abundant in the population, and the species gradually evolves or becomes better suited to that particular environment. Although each change may be very slight, many mutations over a very long time have produced the incredible variety of different life-forms that we observe in nature (fig. 4.3).

The variety of finches observed by Charles Darwin on the Galápagos Islands is a classic example of speciation driven by availability of different environmental opportunities (fig. 4.4). Originally derived from a single seed-eating species that somehow crossed the thousands of kilometers from the mainland, the finches have evolved into a dozen or more distinct species that differ markedly in appearance, food preferences, and habitats they occupy. Fruit eaters have thick parrot-like bills; seed eaters have heavy, crushing bills; insect eaters have thin probing beaks to catch

FIGURE 4.3   Giraffes don't have long necks because they stretch to reach tree-top leaves, but those giraffes that happened to have longer necks got more food and had more offspring, so the trait became fixed in the population. © Corbis/Volume 6.

their prey. One of the most unusual species is the woodpecker finch, which pecks at tree bark for hidden insects. Lacking the woodpecker's long tongue, however, the finch uses a cactus spine as a tool to extract bugs.

The amazing variety of colors, shapes, and sizes of dogs, cats, rabbits, fish, flowers, vegetables, and other domestic species is evidence of deliberate selective breeding. The various characteristics of these organisms arose through mutations. We simply kept the ones we liked. Note that sexual reproduction helps to redistribute genetic material in new and novel combinations that greatly increase the variation and diversity we see in both wild and domestic species. Organisms that reproduce asexually can evolve, but often do so very slowly.

What environmental factors cause selective pressure and influence fertility or survivorship in nature? They include (1) physiological stress due to inappropriate levels of some critical environmental factor, such as moisture, light, temperature, pH, or specific nutrients; (2) predation, including parasitism and disease; (3) competition; and (4) chance. In some cases the organisms that survive environmental catastrophes or find their way to a new habitat where they start a new population may simply be lucky rather than more fit or better suited to subsequent environmental conditions than their less fortunate contemporaries.

Be sure you understand that while selection affects individuals, evolution and adaptation work at the population level. Individuals don't evolve; species do. Each individual is locked in by genetics to a particular way of life. Most plants, animals, or microbes have relatively limited ability to modify their physical makeup or behavior to better suit a particular environment. Over time, however, random genetic changes and natural selection can change an entire population.

Given enough geographical isolation or selective pressure, the members of a population become so different from their ancestors that they may be considered an entirely new species that has

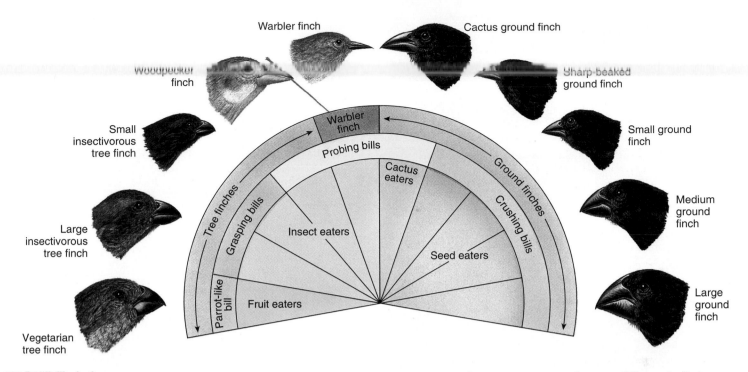

FIGURE 4.4 Some species of Galápagos Island finches. Although all are descendents of a common ancestor, they now differ markedly in appearance, habitat, and feeding behavior. Ground finches (*upper right*) eat cactus leaves; warbler finches (*upper left*) eat insects; others eat seeds or have mixed diets. The woodpecker finch (*upper left*) pecks tree bark as do woodpeckers, but lacks a long tongue. Instead, it uses cactus spines as tools to extract insects.
Source: Peter H. Raven and George B. Johnson, *Biology,* 4th edition. Copyright © 1996 McGraw Hill Company, Inc. All rights reserved. Reprinted by permission.

replaced the original one. Alternatively, isolation of population subsets by geographical or behavioral factors that prevent exchange of genetic material can result in branching off of new species that coexist with their parental line. Suppose that two populations of the same species become separated by a body of water, a desert, or a mountain range that they cannot cross. Over a very long time—often millions of years—random mutations and different environmental pressures may cause the populations to evolve along such dissimilar paths that they can no longer interbreed successfully even if the opportunity to do so arises. They have now become separate species as in the case of the Galápagos finches. The barriers that divide subpopulations are not always physical. In some cases, behaviors such as when and where members of a population feed, sleep, or mate—or how they communicate—may separate them sufficiently for divergent evolution and speciation to occur even though they occupy the same territory.

Natural selection and adaptation can cause organisms with a similar origin to become very different in appearance and develop different habits over time, but they can also result in unrelated organisms coming to look and act very much alike. We call this latter process *convergent* evolution. The cactus-eating Galápagos finches (fig. 4.4), for example, look and act very much like parrots even though they are genetically very dissimilar. The features that enable parrots to eat fruit successfully work well for these finches also.

A common mistake is to believe that organisms develop certain characteristics because they want or need them. This is incorrect. A duck doesn't have webbed feet because it wants to swim or needs to swim in order to eat; it has webbed feet because some ancestor happened to have a gene for webbed feet that gave it some advantage over other ducks in its particular pond and because those genes were passed on successfully to its offspring. A variety of different genetic types are always present in any population, and natural selection simply favors those best suited for particular conditions. Whether there is a purpose or direction to this process is a theological question rather than a scientific one and is beyond the scope of this book.

The Ecological Niche

Habitat describes the place or set of environmental conditions in which a particular organism lives. A more functional term, the **ecological niche,** is a description of either the role played by a species in a biological community or the total set of environmental factors that determine species distribution. Niches as community roles—describing how a species obtains food, what relationships it has with other species, and the services it provides its community, for example—were first described by the British ecologist, Charles Elton in 1927. Thirty years later, the American limnologist G. E. Hutchinson proposed a more biophysical definition of this concept. Every species, he pointed out, has a range of physical and chemical conditions (temperature, light levels, acidity, humidity, salinity, etc.) as well as biological interactions (predators and prey present, defenses, nutritional resources available, etc.) within which it can exist. Figure 4.1, for example, shows the abundance of a hypothetical species along a single factor gradient. If it were possible to

FIGURE 4.5 The giant panda feeds exclusively on bamboo. Although its teeth and digestive system are those of a carnivore, it is not a good hunter, and has adapted to a vegetarian diet. In the 1970s, huge acreages of bamboo flowered and died, and many pandas starved.
© William P. Cunningham.

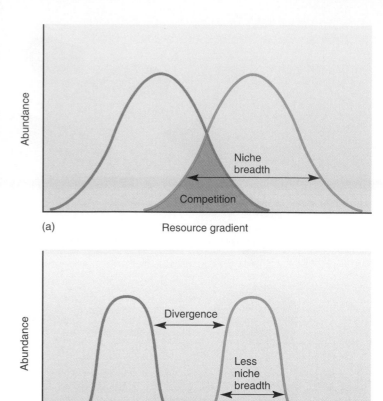

FIGURE 4.6 Resource partitioning and niche specialization caused by competition. Where niches of two species overlap along a resource gradient, competition occurs (shaded area in (*a*)). Individuals occupying this part of the niche are less successful in reproduction so that characteristics of the population diverge to produce more specialization, narrower niche breadth, and less competition between species (*b*).

graph simultaneously all of the factors that affect a particular species, a multidimensional space would result that describes the ecological niche available to that species.

The idea of niches can be further defined in terms of *fundamental niche* and *realized niche*. A species' fundamental niche is the full range of resources or habitat it *could* exploit if there were no competition with other species. A species' realized niche, the resources or habitat it actually uses, may be much less than its fundamental niche.

Some species, like raccoons or coyotes, are generalists that eat a wide variety of food and live in a broad range of habitats (including urban areas). Others, such as the panda (fig. 4.5), are specialists that occupy a very narrow niche. Specialists often tend to be rarer than generalists and less resilient to disturbance or change.

A few species such as elephants, chimpanzees, and baboons learn how to behave from their social group and can invent new ways of doing things when presented with new opportunities or challenges. Most organisms, however, are limited by genetically determined physical structure and instinctive behavior to established niches.

Over time, though, niches can evolve, just as physical characteristics do. The law of competitive exclusion states that no two species will occupy the same niche and compete for exactly the same resources in the same habitat for very long. Eventually, one group will gain a larger share of resources while the other will

either migrate to a new area, become extinct, or change its behavior or physiology in ways that minimize competition. We call this latter process of niche evolution **resource partitioning** (fig. 4.6). It can produce high levels of specialization that allow several species to utilize different parts of the same resource and coexist within a single habitat (fig. 4.7).

Niche specialization also can create behavioral separation that allows subpopulations of a single species to diverge into separate species. Why doesn't this process continue until there is an infinite number of species? The answer is that a given resource can be partitioned only so far. Populations must be maintained at a minimum size to avoid genetic problems and to survive bad times. This puts an upper limit on the number of different niches—and therefore the number of species—that a given community can support.

Perhaps you haven't thought of time as an ecological factor, but niche specialization in a community is a 24-hour phenomenon. Swallows and insectivorous bats both catch insects, but some insect species are active during the day and others at night, providing noncompetitive feeding opportunities for day-active swallows and night-active bats.

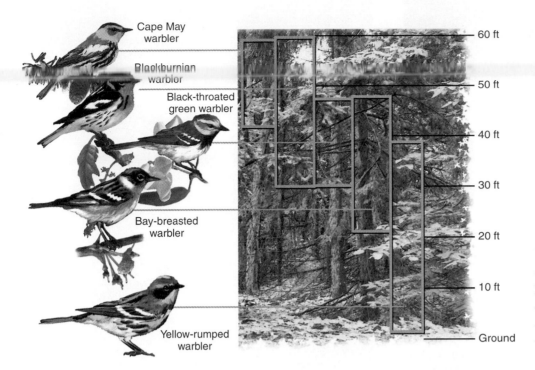

FIGURE 4.7 Resource partitioning and the concept of the ecological niche are demonstrated by several species of wood warblers that use different strata of the same forest. This is a classic example of the principle of competitive exclusion.
Source: Original observation by R. H. MacArthur.

Labels on figure: Cape May warbler, Blackburnian warbler, Black-throated green warbler, Bay-breasted warbler, Yellow-rumped warbler; 60 ft, 50 ft, 40 ft, 30 ft, 20 ft, 10 ft, Ground

SPECIES INTERACTIONS

Predation and competition for scarce resources are major factors in evolution and adaptation. Not all biological interactions are competitive, however. Organisms also cooperate with, or at least tolerate, members of their own species as well as individuals of other species in order to survive and reproduce. In this section, we will look more closely at the different interactions within and between species that shape biological communities.

Exploitation: Predation and Parasitism

All organisms need food to live. Producers make their own food, and consumers eat organic matter created by other organisms. In most communities, as we saw in chapter 3, photosynthetic organisms are the producers. Consumers include herbivores, carnivores, omnivores, scavengers, and decomposers. With which of these categories do you associate the term *predator?* Ecologically, the term has a much broader meaning than you might expect. A **predator** in an ecological sense, is an organism that feeds directly upon another living organism, whether or not it kills the prey to do so (fig. 4.8). By this definition herbivores, carnivores, and omnivores that feed on live prey are predators, but scavengers, detritivores, and decomposers that feed on dead things are not.

Predatory relationships can be complex, as in the case of marine shellfish. Many crustaceans, mollusks, and worms release eggs directly into the water, and the eggs and free-living larval and juvenile stages are part of the floating community, or **plankton** (fig. 4.9). Planktonic animals feed upon each other and are food for successively larger carnivores, including small fish. As prey species mature, their predators change. Barnacle larvae are planktonic and are eaten by fish. Adult barnacles, on the other hand,

FIGURE 4.8 Insect herbivores are predators as much as are lions and tigers. In fact, insects consume the vast majority of biomass in the world. Complex patterns of predation and defense have often evolved between insect predators and their plant prey. © Ray Coleman/Photo Researchers, Inc.

build hard shells that protect them from fish but can be crushed by limpets and other mollusks. Predators also may change their feeding targets. Adult frogs, for instance, are carnivores, but the tadpoles of most species are grazing herbivores. Sorting out the trophic levels in these communities can be very difficult.

Predation is an important factor in evolution. Predators prey most successfully on the slowest, weakest, least fit members of their target population, thus reducing competition, preventing excess population growth, allowing successful traits to become dominant in the prey population, and making the prey population stronger and healthier. As the poet Robinson Jeffers said, "What but the wolf's tooth whittled so fine/The fleet limbs of the antelope?"

FIGURE 4.9 Microscopic plants and animals form the basic levels of many aquatic food chains and account for a large percentage of total world biomass. Many oceanic plankton are larval forms that have habitats and feeding relationships very different from their adult forms. © D. P. Wilson/Photo Researchers, Inc.

Prey species have evolved many protective or defensive adaptations to avoid predation. In plants, for instance, this often takes the form of thick bark, spines, thorns, or chemical defenses. Animal prey may become very adept at hiding, fleeing, or fighting back against predators. Predators, in turn, evolve mechanisms to overcome the defenses of their prey. This process in which species exert selective pressure on each other is called **coevolution.**

Parasites are organisms that feed on a host, or take resources from it, without killing the host. Some parasites do little damage: a mosquito takes blood but usually causes little damage. Others cause significant harm and may eventually kill a host. **Pathogens** (disease-causing organisms) are often considered parasites. Your immune system is an evolved defense against pathogens in our environment.

Keystone Species

A **keystone species** is a species or group of species whose feeding activity has an inordinate influence on the structure of its community. Originally, keystone species were thought to be top predators, such as wolves, whose presence limits the abundance of herbivores and thereby reduces their grazing or browsing on plants. Recently, it has been recognized that less conspicuous species also play essential community roles. Certain tropical figs, for example, bear during seasons when no other fruit is available for frugivores (fruit-eating animals). If these figs were removed, many animals would starve to death during periods of fruit scarcity. With those animals gone, many other plant species that depend on them at other times of the year for pollination and seed-dispersal would disappear as well.

Even microorganisms can play vital roles. In some forest ecosystems, mycorrhizae (fungi associated with tree roots) are essential for mineral mobilization and absorption. If the fungi die, so do the trees and many other species that depend on a healthy forest community. Rather than being a single species, mycorrhizae are actually a group of species that together fulfill a keystone function.

Often a number of species are intricately interconnected in biological communities so that it is difficult to tell which is the

FIGURE 4.10 Giant kelp is a massive alga that forms dense "forests" off the Pacific coast of California. It is a keystone species in that it provides food, shelter, and structure essential for a whole community. Removal of sea otters allows sea urchin populations to explode. When the urchins destroy the kelp, many other species suffer as well. © Randy Morse/Tom Stack & Associates.

essential key. A classic example is in the Pacific kelp forests, where towering columns of kelp (algae) shelter myriad fish, shellfish, and mammals (fig. 4.10). The sheltering kelp could be regarded as the key to community structure. Sea urchins, however, feed on the kelp and determine their number and distribution while sea otters regulate urchins and kelp provides a resting place for dozing otters. Which of these species is the most important? Each depends on and affects the others. Perhaps we should think in terms of a "keystone set" of organisms in some ecosystems. (See "Oreas, Otters, Urchins, and Kelp: Disrupting a Marine Food Web," on the Online Learning Center in the chapter 4 Case Studies.)

Competition

Competition is another kind of antagonistic relationship within a community. For what do organisms compete? To answer this question, think again about what all organisms need to survive: energy and matter in usable forms, space, and specific sites for life activities (What Do You Think? p. 80). Plants compete for growing space for root and shoot systems so they can absorb and process sunlight, water, and nutrients. Animals compete for living, nesting, and feeding sites, as well as for food, water, and mates. Competition among members of the same species is called **intraspecific competition,** whereas competition between members of different species is called **interspecific competition.**

You can observe interspecific competition if you look closely at a patch of weeds growing on good soil early in the summer. First of all, many weedy species attempt to crowd out their rivals by producing prodigious numbers of seeds. After the seeds germinate, the plants race to grow the tallest, cover the most ground, and get the most sun. You may observe several strategies to do this. For example, vines don't build heavy stems of their own; they simply climb up over their neighbors to get to the light.

Species also race to new territory. Plants with highly mobile seeds can reach and colonize open ground ahead of other species (fig. 4.11). Some plants secrete substances that inhibit the growth of seedlings near them, including their own and those of other species. This strategy is particularly significant in deserts where water is a limiting factor.

We often think of competition among animals as a bloody battle for resources. A famous Victorian description of the struggle for survival was "nature red in tooth and claw." In fact, a better metaphor is a race. Have you ever noticed that birds always eat fruits and berries just before they are ripe enough for us to pick? Having a tolerance for bitter, unripe fruit gives them an advantage in the race for these food resources. Many animals tend to avoid fighting if possible. It's not worth getting injured. Most confrontations are more noise and show than actual fighting.

FIGURE 4.11 Dandelions and other opportunistic species generally produce many highly mobile offspring. © William P. Cunningham.

Intraspecific competition can be especially intense because members of the same species have the same space and nutritional requirements; therefore, they compete directly for these environmental resources. How do plants cope with intraspecific competition? The inability of seedlings to germinate in the shady conditions created by parent plants acts to limit intraspecific competition by favoring the mature, reproductive plants.

Symbiotic relationships often enhance the survival of one or both partners. Symbiotic relationships often entail some degree of coadaptation or coevolution of the partners, shaping—at least in part—their structural and behavioral characteristics. An interesting case of mutualistic coadaptation is seen in Central and South American swollen thorn acacias and their symbiotic ants. Acacia ant colonies live within the swollen thorns on the acacia tree branches and feed on two kinds of food provided by the trees: nectar produced in glands at the leaf bases and special protein-rich structures produced on leaflet tips. The acacias thus provide shelter and food for the ants. Although they spend energy to provide these services, the trees are not physically harmed by ant feeding.

Animals also have developed adaptive responses to intraspecific competition. Two major examples are varied life cycles and territoriality. The life cycles of many invertebrate species have juvenile stages that are very different from the adults in habitat and feeding. Compare a leaf-munching caterpillar to a nectar-sipping adult butterfly or a planktonic crab larva to its bottom-crawling adult form. In these examples, the adults and juveniles of each species do not compete because they occupy different ecological niches.

You may have observed robins chasing other robins during the mating and nesting season. Robins and many other vertebrate species demonstrate **territoriality,** an intense form of intraspecific competition in which organisms define an area surrounding their home site or nesting site and defend it, primarily against other members of their own species. Territoriality helps to allocate the resources of an area by spacing out the members of a population. It also promotes dispersal into adjacent areas by pushing grown offspring outward from the parental territory.

Territory size depends on the size of the species and the resources available. A pair of robins might make do with a suburban yard, but a large carnivore like a tiger may need thousands of square kilometers.

Symbiosis

In contrast to predation and competition, symbiotic interactions between organisms can be nonantagonistic. **Symbiosis** is the intimate living together of members of two or more species. **Commensalism** is a type of symbiosis in which one member clearly benefits and the other apparently is neither benefited nor harmed. Cattle often are accompanied by cattle egrets, small white shore birds who catch insects kicked up as the cattle graze through a field. The birds benefit while the cattle seem indifferent. Many of the mosses, bromeliads, and other plants growing on trees in the moist tropics are also considered to be commensals (fig. 4.12). These epiphytes get water from rain and nutrients from leaf litter and dust

What do you think?

Understanding Competition

Ecology is a relatively young science. Consequently, many ecological processes are incompletely understood. How a community comes to have its particular organization is one area of uncertainty. Some ecologists feel that physical factors are the most important determinants in community organization, while others feel that interspecific competition is most important.

How can we find out which view is correct? Ecologists employ the scientific method, as described in chapter 2, to better understand community dynamics. This process is mostly refined common sense and its basic elements can be useful in everyday life.

Once ecologists have decided on the concept to be investigated, they look for a specific situation that can either be observed or manipulated to provide relevant information. For example, ecologist Richard Karban was interested in how competition affected a community. He learned that larvae of two insect species, the meadow spittlebug and the calendula plume moth, both feed and develop on the seaside daisy, a common beach plant on the American west coast. The specific question to be investigated was: Does competition affect these two insect species, therefore impacting community organization?

Competition might reduce survival rate, larval growth, or both. Karban's procedure involved setting up four groups of plants at Bodega Bay, CA: one got both spittlebugs and moths, another got only spittlebugs, another only moths, and a fourth had neither. He compared survival rates of spittlebugs and moths when competitors were present and absent.

There are three important general considerations in designing scientific investigations:

1. Things need to be organized in such a way that the outcome can clearly be linked to a particular cause. In other words, differences in insect survival rates need to be clearly attributable to competition and not to other factors. Karban accomplished this by making his plant/insect groups as uniform as possible, except for the presence or absence of competitors. He eliminated genetic differences between plants by using plants from the same clone. He was careful to put the same numbers of insects on each plant to eliminate animal density as a factor, and so on.

2. The data collected must be a reliable representation of the larger situation and not simply the result of chance. This is usually accomplished by replicating the procedure many times. Instead of setting up just a few plants with one or both insects present, Karban set up 30 plants with each treatment. The procedure was repeated a second year. This gave him a cumulative total of 60 plants that had just spittlebugs, 60 plants having just moths, and 60 plants each having both or neither spittlebugs and moths. With such a large number of replications it was highly likely that differences in survival rates were, in fact, the result of competition and not simply chance occurrences.

3. Finally, conclusions must be justified by the data. Karban's statistical analysis revealed that spittlebug persistence was nearly 40 percent higher when the plume moths were absent. Plume moth persistence was not significantly affected by spittlebug presence, however.

His overall conclusion was:

Evidence from this and other studies supports the contention that interspecific com-

Spittlebugs produce mounds of foam under which they hide from predators while feeding on host plants. © Milton Tierney/Visuals Unlimited.

petition can play an important role in influencing densities of plant-feeding insects.

Notice the caution expressed in these words. He did not claim to have proven anything. Instead, his study "*supports* the contention." Second, he states competition "*can* play an important role," instead of using stronger language. And finally, he restricts these conclusions to plant-feeding insects. Karban carefully avoids drawing conclusions beyond the realm supported by his data.

Based on a healthy skepticism, clarity of language, critical evaluation of relationships and information, and caution in coming to judgment, critical thinking in science has been a very successful tool in enhancing understanding.

---

fall, and often neither help nor hurt the trees on which they grow. In a way, the robins and sparrows that inhabit suburban yards are commensals with humans.

Lichens are a combination of a fungus and a photosynthetic partner, either an alga or a cyanobacterium. Their association is a type of symbiosis called **mutualism,** in which both members of the partnership benefit (fig. 4.13). Some ecologists believe that cooperative,

mutualistic relationships may be more important in evolution than we have commonly thought. Aggressive interactions often are dangerous and destructive, while cooperation and compromise may have advantages that we tend to overlook. Survival of the fittest often may mean survival of those organisms that can live best with one another.

What do the acacias get in return and how does the relationship relate to community dynamics? Ants tend to be aggressive defenders of their home areas, and acacia ants are no exception. They drive off herbivorous insects that attempt to feed on their home acacia, thus reducing predation. They also trim away vegetation that grows around their home tree, thereby reducing competition. This is a fascinating example of how a symbiotic relationship fits into community interactions. It is also an example of coevolution based on mutualism rather than competition or predation.

Defensive Mechanisms

Many species of plants and animals have toxic chemicals, body armor, and other ingenious defensive adaptations to protect themselves from competitors or predators. Arthropods, amphibians, snakes, and some mammals, for instance, produce noxious odors or poisonous secretions to induce other species to leave them alone. Plants also produce a variety of chemical compounds that make them unpalatable or dangerous to disturb. Perhaps you have brushed up against poison ivy or stinging nettles in the woods or you have encountered venomous insects or snakes and appreciate the wisdom of leaving them alone. Often, species possessing these chemical defenses will evolve distinctive colors or patterns to warn potential enemies (fig. 4.14).

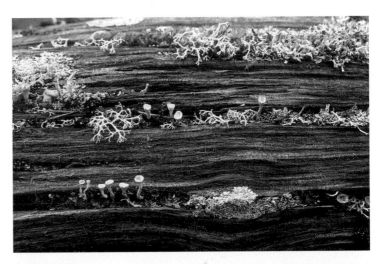

FIGURE 4.13 Lichens, such as the various species growing on this log, are a combination of algae and fungi in a classic example of mutualistic symbiosis. © William P. Cunningham.

FIGURE 4.12 Plants compete for light and growing space in this Indonesian rainforest. Epiphytes, such as the ferns and bromeliads shown here, find a place to grow in the forest canopy by perching on the limbs of large trees. This may be a commensal relationship if the epiphytes don't hurt their hosts. Sometimes, however, the weight of epiphytes breaks off branches and even topples whole trees. © William P. Cunningham.

FIGURE 4.14 Poison arrow frogs of the family Dendrobatidae use brilliant colors to warn potential predators of the extremely toxic secretions from their skin. Native people in Latin America use the toxin on blowgun darts. © Michael Fogden/Animals Animals/Earth Scenes.

FIGURE 4.15   An example of Batesian mimicry. The dangerous wasp (*left*) has bold yellow and black bands to warn predators away. The much rarer longhorn beetle (*right*) has no poisonous stinger, but looks and acts like a wasp and thus avoids predators as well. © Edward S. Ross.

FIGURE 4.16   This highly camouflaged scorpion fish lies in wait for its unsuspecting prey. Natural selection and evolution have created the elaborate disguise seen here. © Brian Parker/Tom Stack & Associates.

Sometimes species that actually are harmless will evolve colors, patterns, or body shapes that mimic species that are unpalatable or poisonous. This is called **Batesian mimicry** after the English naturalist H. W. Bates, who described it in 1857. Wasps, for example, often have bold patterns of black and yellow stripes to warn off potential predators. The rarer longhorn beetle (fig. 4.15), although it has no stinger, looks and acts much like wasps and thus avoids predators as well. Another form of mimicry, called **Müllerian mimicry,** named for the German biologist Fritz Müller, who described it in 1878, involves two species, both of which are unpalatable or dangerous and have evolved to look alike. When predators learn to avoid either species, both benefit.

Species also evolve amazing abilities to avoid being discovered. You very likely have seen examples of insects that look exactly like

dead leaves or twigs to hide from predators. Predators also use camouflage to hide as they lie in wait for their prey. The scorpion fish (fig. 4.16) blends in remarkably well with its surroundings as it waits for smaller fish to come within striking distance. Not all cases of mimicry are to avoid or carry out predation, however. Some tropical orchids have evolved flower structures that look exactly like female flies. Males attempting to mate unwittingly carry away pollen.

COMMUNITY PROPERTIES

The processes and principles that we have studied thus far in this chapter—tolerance limits, species interactions, resource partitioning, evolution, and adaptation—play important roles in determining the characteristics of populations and species. In this section we will look at some fundamental properties of biological communities and ecosystems—productivity, diversity, complexity, resilience, stability, and structure—to learn how they are affected by these factors.

Productivity

A community's **primary productivity** is the rate of biomass production, an indication of the rate of solar energy conversion to chemical energy. The energy left after respiration is net primary production. Photosynthetic rates are regulated by light levels, temperature, moisture, and nutrient availability. Figure 4.17 shows approximate productivity levels for some major ecosystems. As you can see, tropical forests, coral reefs, and estuaries (bays or inundated river valleys where rivers meet the ocean) have high levels of productivity because they have abundant supplies of all these resources. In deserts, lack of water limits photosynthesis. On the arctic tundra or in high mountains, low temperatures inhibit plant growth. In the open ocean, a lack of nutrients reduces the ability of algae to make use of plentiful sunshine and water.

Some agricultural crops such as corn (maize) and sugar cane grown under ideal conditions in the tropics approach the productivity levels of tropical forests. Because shallow water ecosystems such as coral reefs, salt marshes, tidal mud flats, and other highly productive aquatic communities are relatively rare compared to the vast extent of open oceans—which are effectively biological deserts—marine ecosystems are much less productive on average than terrestrial ecosystems.

Even in the most photosynthetically active ecosystems, only a small percentage of the available sunlight is captured and used to make energy-rich compounds. Between one-quarter and three-quarters of the light reaching plants is reflected by leaf surfaces. Most of the light absorbed by leaves is converted to heat that is either radiated away or dissipated by evaporation of water. Only 0.1 to 0.2 percent of the absorbed energy is used by chloroplasts to synthesize carbohydrates.

In a temperate-climate oak forest, only about half the incident light available on a midsummer day is absorbed by the leaves. Ninety-nine percent of this energy is used to evaporate water. A large oak tree can transpire (evaporate) several thousand liters of water on a warm, dry, sunny day while it makes only a few kilograms of sugars and other energy-rich organic compounds.

Abundance and Diversity

Abundance is an expression of the total number of organisms in a biological community, while **diversity** is a measure of the number of different species, ecological niches, or genetic variation present. The abundance of a particular species often is inversely related to the total diversity of the community. That is, communities with a very large number of species often have only a few members of any given species in a particular area. As a general rule, diversity decreases but abundance within species increases as we go from the equator toward the poles. The arctic has vast numbers of insects such as mosquitoes, for example, but only a few species. The tropics, on the other hand, have vast numbers of species—some of which have incredibly bizarre forms and habits—but often only a few individuals of any particular species in a given area.

Consider bird populations. Greenland is home to 56 species of breeding birds, while Colombia, which is only one-fifth the size of Greenland, has 1,395. Why are there so many species in Colombia and so few in Greenland?

Climate and history are important factors. Greenland has such a harsh climate that the need to survive through the winter or escape to milder climates becomes the single most important critical factor that overwhelms all other considerations and severely limits the ability of species to specialize or differentiate into new forms. Furthermore, because Greenland was covered by glaciers until about 10,000 years ago, there has been little time for new species to develop.

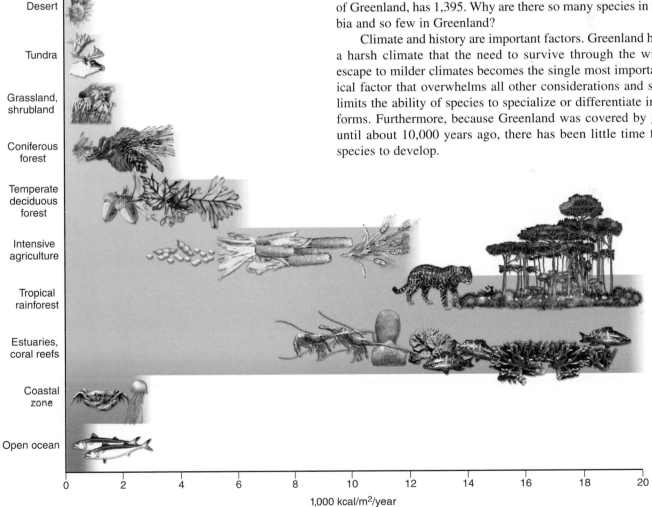

FIGURE 4.17 Relative biomass accumulation of major world ecosystems. Only plants and some bacteria capture solar energy. Animals consume biomass to build their own bodies.

Many areas in the tropics, by contrast, have relatively abundant rainfall and warm temperatures year-round so that ecosystems there are highly productive. The year-round dependability of food, moisture, and warmth supports a great exuberance of life and allows a high degree of specialization in physical shape and behavior. Coral reefs are similarly stable, productive, and conducive to proliferation of diverse and amazing life-forms. The enormous abundance of brightly colored and fantastically shaped fish, corals, sponges, and arthropods in the reef community is one of the best examples we have of community diversity.

Productivity is related to abundance and diversity, both of which are dependent on the total resource availability in an ecosystem as well as the reliability of resources, the adaptations of the member species, and the interactions between species. You shouldn't assume that all communities are perfectly adapted to their environment. A relatively new community that hasn't had time for niche specialization, or a disturbed one where roles such as top predators are missing, may not achieve maximum efficiency of resource use or reach its maximum level of either abundance or diversity.

Complexity and Connectedness

Community complexity and connectedness generally are related to diversity and are important because they help us visualize and understand community functions. **Complexity** in ecological terms refers to the number of species at each trophic level and the number of trophic levels in a community. A diverse community may not be very complex if all its species are clustered in only a few trophic levels and form a relatively simple food chain.

By contrast, a complex, highly interconnected community (fig. 4.18) might have many trophic levels, some of which can be compartmentalized into subdivisions. In tropical rainforests, for instance, the herbivores can be grouped into "guilds" based on the specialized ways they feed on plants. There may be fruit eaters, leaf nibblers, root borers, seed gnawers, and sap suckers, each composed of species of very different size, shape, and even biological kingdom, but that feed in related ways. A highly interconnected community such as this can form a very elaborate food web.

Resilience and Stability

Many biological communities tend to remain relatively stable and constant over time. An oak forest tends to remain an oak forest, for example, because the species that make it up have self-perpetuating mechanisms. We can identify three kinds of stability or resiliency in ecosystems: *constancy* (lack of fluctuations in composition or functions), *inertia* (resistance to perturbations), and *renewal* (ability to repair damage after disturbance).

In 1955, Robert MacArthur, who was then a graduate student at Yale, proposed that the more complex and interconnected a community is, the more stable and resilient it will be in the face of disturbance. If many different species occupy each trophic level, some can fill in if others are stressed or eliminated by external forces, making the whole community resistant to perturbations and able to recover relatively easily from disruptions. This theory has been controversial, however. Some studies support it, while others do not. For example, Minnesota ecologist David Tilman, in studies of native prairie and recovering farm fields, found that plots with high diversity were better able to withstand and recover from drought than those with only a few species.

On the other hand, in a diverse and highly specialized ecosystem, removal of a few keystone members can eliminate many other associated species. Eliminating a major tree species from a tropical forest, for example, may destroy pollinators and fruit distributors as well. We might replant the trees, but could we replace the whole web of relationships on which they depend? In this case, diversity has made the forest less resilient rather than more.

Diversity is widely considered important and has received a great deal of attention. In particular, human impacts on diversity are a primary concern of many ecologists (Case Study, p. 87).

Edges and Boundaries

An important aspect of community structure is the boundary between one habitat and its neighbors. We call these relationships **edge effects.** Sometimes, the edge of a patch of habitat is relatively sharp and distinct. In moving from a woodland patch into a grassland or cultivated field, you sense a dramatic change from the cool, dark, quiet forest interior to the windy, sunny, warmer, open space of the field or pasture (fig. 4.19). In other cases, one habitat type intergrades very gradually into another, so there is no distinct border.

Ecologists call the boundaries between adjacent communities **ecotones.** A community that is sharply divided from its neighbors is called a closed community. In contrast, communities with gradual or indistinct boundaries over which many species cross are called open communities. Often this distinction is a matter of degree or perception. As we saw earlier in this chapter, birds might feed in fields or grasslands but nest in the forest. As they fly back and forth, the birds interconnect the ecosystems by moving energy and material from one to the other, making both systems relatively open. Furthermore, the forest edge, while clearly different from the open field, may be sunnier and warmer than the forest interior, and may have a different

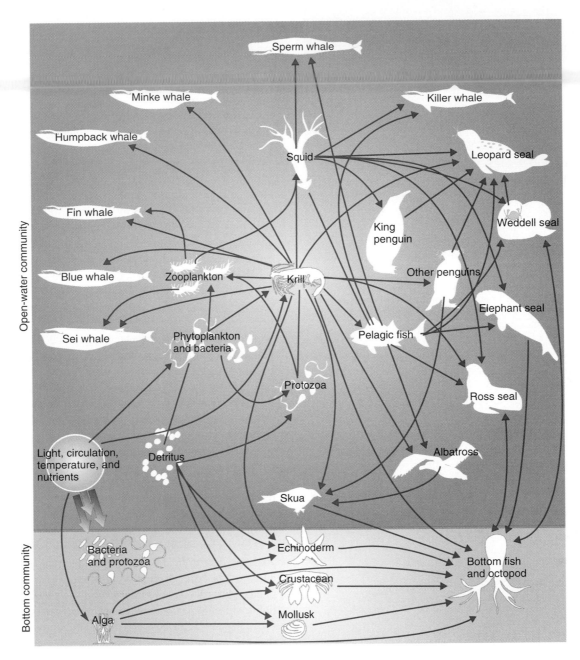

FIGURE 4.18 A complex and highly interconnected community can have many species at each trophic level and many relationships, as illustrated by this Antarctic marine food web.

combination of plant and animal species than either field or forest "core."

Depending on how far edge effects extend from the boundary, differently shaped habitat patches may have very dissimilar amounts of interior area (fig. 4.20). In Douglas fir forests of the Pacific Northwest, for example, increased rates of blowdown, decreased humidity, absence of shade-requiring ground cover, and other edge effects can extend as much as 200 m into a forest. A 40-acre block (about 400 meters square) surrounded by clear-cut would have essentially no true core habitat at all.

Many popular game animals, such as white-tailed deer and pheasants that are adapted to human disturbance, often are most plentiful in boundary zones between different types of habitat. Game managers once were urged to develop as much edge as possible to promote large game populations. Today, however, most wildlife conservationists recognize that the edge effects associated with habitat fragmentation are generally detrimental to biodiversity. Preserving large habitat blocks and linking smaller blocks with migration corridors may be the best ways to protect rare and endangered species (see chapter 13).

FIGURE 4.19 Ecological edges are known as ecotones. Temperature, wind, and humidity differ at the edges in a landscape. Edge conditions do extend into patches of habitat. Small or linear fragments may be mostly edge. © Corbis/Volume 262.

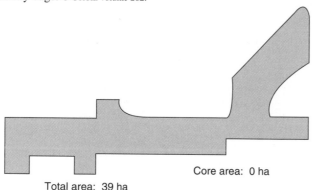

Total area: 39 ha   Core area: 0 ha

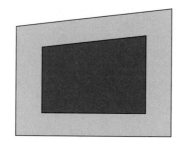

Total area: 47 ha   Core area: 20 ha

FIGURE 4.20 Shape can be as important as size in small preserves. While these areas are close to the same size, no place in the top figure is far enough from the edge to have characteristics of core habitat, while the bottom patch has a significant core.

COMMUNITIES IN TRANSITION

So far our view of communities has focused on the day-to-day interactions of organisms with their environments, set in a context of survival and selection. In this section, we'll step back to look at some transitional aspects of communities, including where communities meet and how communities change over time.

Ecological Succession

Biological communities have a history in a given landscape. The process by which organisms occupy a site and gradually change environmental conditions by creating soil, shade, shelter, or increasing humidity is called ecological succession or development. **Primary succession** occurs when a community begins to develop on a site previously unoccupied by living organisms, such as an island, a sand or silt bed, a body of water, or a new volcanic flow (fig. 4.21). **Secondary succession** occurs when an existing community is disrupted and a new one subsequently develops at the site. The disruption may be caused by some natural catastrophe, such as fire or flooding, or by a human activity, such as deforestation, plowing, or mining. Both forms of succession usually follow an orderly sequence of stages as organisms modify the environment in ways that allow one species to replace another.

In primary succession on a terrestrial site, the new site first is colonized by a few hardy **pioneer species,** often microbes, mosses, and lichens that can withstand harsh conditions and lack of resources. Their bodies create patches of organic matter in which protists and small animals can live (fig. 4.22). Organic debris accumulates in pockets and crevices, providing soil in which seeds can become lodged and grow. We call this process of environmental modification by organisms **ecological development** or facilitation. The community of organisms often becomes more diverse and increasingly competitive as development continues and new niche opportunities appear. The pioneer species gradually disappear as the environment changes and new species combinations replace the preceding community. In a global sense, the gradual changes brought about by living organisms have created many of the conditions that make life on earth possible. You could consider evolution to be a very slow, planetwide successional and developmental process.

Examples of secondary succession are easy to find. Observe an abandoned farm field or clear-cut forest (fig. 4.23) in a temperate climate. The bare soil first is colonized by rapidly growing annual plants (those that grow, flower, and die the same year) that have light, wind-blown seeds and can tolerate full sunlight and exposed soil. They are followed and replaced by perennial plants (those that live for several to many years), including grasses, various nonwoody flowering plants, shrubs, and trees. As in primary succession, plant species progressively change the environmental conditions. Biomass accumulates and the site becomes richer, better able to capture and store moisture, more sheltered from wind and climate change, and biologically more complex. Species that cannot survive in a bare, dry, sunny, open area find shelter and food as the field turns to prairie or forest.

Eventually, in either primary or secondary succession, a community often develops that resists further change. Ecologists call this a **climax community** because it appears to be the culmination of the successional process. An analogy is often made between community succession and organism maturation. Beginning with a primitive or juvenile state and going through a complex developmental process, each progresses until a complex, stable, and mature form is reached. It's dangerous to carry this analogy too far, however, because no mechanism is known to regulate communities in

Where Have All the Songbirds Gone?

Every June, some 2,200 amateur ornithologists and bird watchers across the United States and Canada join in an annual bird count called the Breeding Bird Survey. Organized in 1966 by the U.S. Fish and Wildlife Service to follow bird population changes, this survey has discovered some shocking trends. While birds such as robins, starlings, and blackbirds that prosper around humans have increased their number and distribution over the past 30 years, many of our most colorful forest birds have declined severely. The greatest decreases have been among the true songbirds such as thrushes, orioles, tanagers, catbirds, vireos, buntings, and warblers. These long-distance migrants nest in northern forests but spend the winters in South or Central America or in the Caribbean Islands. Scientists call them neotropical migrants.

In many areas of the eastern United States and Canada, three-quarters or more of the neotropical migrants have declined significantly since the survey was started. Some that once were common have become locally extinct. Rock Creek Park in Washington, D.C., for instance, lost 75 percent of its songbird population and 90 percent of its long-distance migrant species in just 20 years. Nationwide, cerulean warblers, American redstarts, and ovenbirds declined about 50 percent in the single decade of the 1970s. Studies of radar images from National Weather Service stations in Texas and Louisiana suggest that only about half as many birds fly across the Gulf of Mexico each spring now compared to the 1960s. This could mean a loss of about half a billion birds in total.

What causes these devastating losses? Destruction of critical winter habitat is clearly a major issue. Birds often are much more densely crowded in the limited areas available to them during the winter than they are on their summer range. Unfortunately, forests throughout Latin America are being felled at an appalling rate. Central America, for instance, is losing about 1.4 million hectares (2 percent of its forests or an area about the size of Yellowstone National Park) each year. If this trend continues, there will be essentially no intact forest left in much of the region in 50 years.

But loss of tropical forests is not the only threat. Recent studies show that fragmentation of breeding habitat and nesting fail-

ures in the United States and Canada may be just as big a problem for woodland songbirds. Many of the most threatened species are adapted to deep woods and need an area of 10 hectares (25 acres) or more per pair to breed and raise their young. As our woodlands are broken up by roads, housing developments, and shopping centers, it becomes more and more difficult for these highly specialized birds to find enough contiguous woods to nest successfully.

Predation and nest parasitism also present a growing threat to many bird species. In human-dominated landscapes, raccoons, opossums, crows, bluejays, squirrels, and house cats thrive. They are protected from larger predators like wolves or owls and find abundant supplies of food and places to hide. Cats are a particular problem. By some estimates, there are 100 million feral cats in the United States, and 73 million pet cats. A comparison of predation rates in the Great Smoky Mountain National Park and in small rural and suburban woodlands shows how devastating predators can be. In a 1,000-hectare study area of mature, unbroken forest in the national park, only one songbird nest in fifty was raided by predators. By contrast, in plots of 10 hectares or less near cities, up to 90 percent of the nests were raided.

Nest parasitism by brown-headed cowbirds is one of the worst threats for woodland songbirds. Rather than raise their young themselves, cowbirds lay their eggs in the nests of other species. The larger and more aggressive cowbird young either kick their foster siblings out of the nest, or claim so much food that the others starve. Well adapted to live around humans, there are now about 150 million cowbirds in the United States.

A study in southern Wisconsin found that 80 percent of the nests of woodland species were raided by predators and that three-quarters of those that survived were

This thrush has been equipped with a lightweight radio transmitter and antenna so that its movements can be followed by researchers. Courtesy Dr. David Mech.

invaded by cowbirds. Another study in the Shawnee National Forest in southern Illinois found that 80 percent of the scarlet tanager nests contained cowbird eggs and that 90 percent of the wood thrush nests were taken over by these parasites. The sobering conclusion of this latter study is that there probably is no longer any place in Illinois where scarlet tanagers and wood thrushes can breed successfully.

What can we do about this situation? Elsewhere in this book, we discuss sustainable forestry and economic development projects that could preserve forests at home and abroad. Preserving corridors that tie together important areas also will help. In areas where people already live, clustering of houses protects remaining woods. Discouraging the clearing of underbrush and trees from yards and parks leaves shelter for the birds.

Could we reduce the number of predators or limit their access to critical breeding areas? Would you accept fencing or trapping of small predators in wildlife preserves? How would you feel about a campaign to keep house cats inside during the breeding season?

Ethical Considerations

Some wildlife managers are already trapping cowbirds. The Kirtland's warbler is one of the rarest songbirds in the United States. It nests only in young, fire-maintained jackpine forests in Michigan. Controlled burning to maintain habitat for this endangered species was started in the 1960s, but the population continued to decline. Studies showed that 90 percent of the nests were being parasitized by cowbirds. Since 1972, refuge managers have trapped and killed some 7,000 cowbirds each year to protect the warblers. In the past two decades, the number of breeding pairs of warblers has risen from about 150 to nearly 400. Would it be possible to do something similar on a nationwide scale? Could we trap and kill 150 million cowbirds? How much should we reduce one species to save another? What do you think?

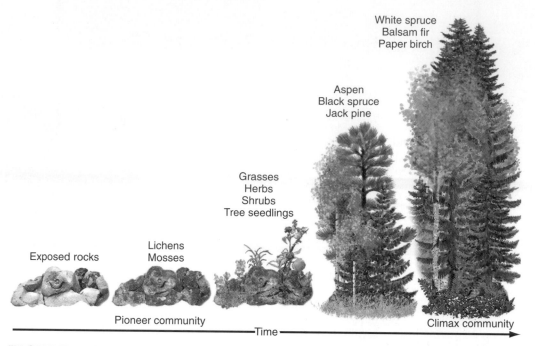

White spruce
Balsam fir
Paper birch

Aspen
Black spruce
Jack pine

Grasses
Herbs
Shrubs
Tree seedlings

Lichens
Mosses

Exposed rocks

Pioneer community ────Time────► Climax community

FIGURE 4.21 One example of primary succession, shown in five stages (*left to right*). Here, bare rocks are colonized by lichens and mosses, which trap moisture and build soil for grasses, shrubs, and eventually trees.

FIGURE 4.22 Primary succession occurs where there had been no living things, as on this lava in Hawaii. Fungi, algae, and bacteria grew here first, providing rooting material for these ferns. © William P. Cunningham.

the same way that genetics and physiology regulate development of the body.

The concept of succession to a climax community was first championed by the pioneer biogeographer F. E. Clements. He viewed this process as being like a parade or relay, in which species replace each other in predictable groups and in a fixed, regular order, and as being driven almost entirely by climate. This community-unit theory was opposed by Clements's contemporary,

H. A. Gleason, who saw community history as a much more individualistic and random process driven by many environmental factors. He argued that temporary associations are formed according to the conditions prevailing at a particular time and the species available to colonize a given area. You might think of the Gleasonian model as a time-lapse movie of a busy railroad station. Passengers come and go; groups form and then dissipate. Patterns and assemblages that seem significant to us may not mean much in the long run.

The process of succession may not be as deterministic as we once thought, yet mature or highly developed ecological communities may tend to be resilient and stable over long periods of time because they can resist or recover from external disturbances. Many are characterized by high species diversity, narrow niche specialization, well-organized community structure, good nutrient conservation and recycling, and a large amount of total organic matter. Community functions, such as productivity and nutrient cycling, tend to be self-stabilizing or self-perpetuating. What once were regarded as "final" climax communities, however, may still be changing. It's probably more accurate to say that the rate of succession is so slow in a climax community that, from the perspective of a single human lifetime, it appears to be stable.

Some landscapes never reach a stable climax in the traditional sense because they are characterized by, and adapted to, periodic disruption. They are called **equilibrium communities** or **disclimax communities.** Grasslands, the chaparral shrubland of California, and some kinds of coniferous forests, for instance, are shaped and maintained by periodic fires that have long been a part of their history. They are, therefore, often referred to as **fire-climax communities** (fig. 4.24). Plants in these communities are adapted to resist fires,

FIGURE 4.23   This area was once a cool, shady black spruce stand. The forest floor was covered by a deep, moist layer of sphagnum moss. Clear-cutting and burning have turned it into a dry, sunny, barren ground on which few of the former residents can survive. Secondary succession will probably restore previous conditions if the climate doesn't change and further disturbance is prevented. © William P. Cunningham.

Working Locally for Ecological Diversity

You might think that diversity and complexity of ecological systems are too large or too abstract for you to have any influence. But you can contribute to a complex, resilient, and interesting ecosystem, whether you live in the inner city, a suburb, or a rural area.

- Keep your cat indoors. As discussed in the Case Study (p. 87), our lovable domestic cats are also very successful predators. Migratory birds, especially those nesting on the ground, have not evolved defenses against these predators.

- Plant a butterfly garden. Use native plants that support a diverse insect population. Native trees with berries or fruit also support birds. (Be sure to avoid non-native invasive species: see chapter 11.) Allow structural diversity (open areas, shrubs, and trees) to support a range of species.

- Join a local environmental organization. Often, the best way to be effective is to concentrate your efforts close to home. City parks and neighborhoods support ecological communities, as do farming and rural areas. Join an organization working to maintain ecosystem health: start by looking for environmental clubs at your school, park organizations, a local Audubon chapter, or a local Nature Conservancy branch.

- Take walks. The best way to learn about ecological systems in your area is to take walks and practice observing your environment. Go with friends and try to identify some of the species and trophic relationships in your area.

- Live in town. Suburban sprawl consumes wildlife habitat and reduces ecosystem complexity by removing many specialized plants and animals. Replacing forests and grasslands with lawns and streets is the surest way to simplify, or eliminate, ecosystems.

FIGURE 4.24   This lodgepole pine forest in Yellowstone National Park was once thought to be a climax forest, but we now know that this forest must be constantly renewed by periodic fire. It is an example of an equilibrium, or disclimax, community. © William P. Cunningham.

reseed quickly after fires, or both. In fact, many of the plant species we recognize as dominants in these communities require fire to eliminate competition, to prepare seedbeds for germination of seedlings, or to open cones or thick seed coats. Without fire, community structure may be quite different.

Introduced Species and Community Change

Succession requires the continual introduction of new community members and the disappearance of previously existing species. New species move in as conditions become suitable; others die or move out as the community changes. New species also can be introduced after a stable community already has become established. Some cannot compete with existing species and fail to become established. Others are able to fit into and become part of the community, defining new ecological niches. If, however, an introduced species preys upon or competes more successfully with one or more populations that are native to the community, the entire nature of the community can be altered.

Human introductions of Eurasian plants and animals to non-Eurasian communities often have been disastrous to native species because of competition or overpredation. Oceanic islands offer classic examples of devastation caused by rats, goats, cats, and pigs liberated from sailing ships. All these animals are prolific, quickly developing large populations. Goats are efficient, non-specific herbivores; they eat nearly everything vegetational, from grasses and herbs to seedlings and shrubs. In addition, their sharp hooves are hard on plants rooted in thin island soils. Rats and pigs are opportunistic omnivores, eating the eggs and nestlings of seabirds that tend to nest in large, densely packed colonies, and digging up sea turtle eggs. Cats prey upon nestlings of both ground- and tree-nesting birds. Native island species are particularly vulnerable because they have not evolved under circumstances that required them to have defensive adaptations to these predators (What Can You Do? p. 89).

Sometimes we introduce new species in an attempt to solve problems created by previous introductions but end up making the situation worse. In Hawaii and on several Caribbean Islands, for instance, mongooses were imported to help control rats that had escaped from ships and were destroying indigenous birds and devastating plantations (fig. 4.25). Since the mongooses were diurnal (active in the day), however, and rats are nocturnal, they tended to ignore each other. Instead, the mongooses also killed native birds and further threatened endangered species. Our lessons from this and similar introductions have a new technological twist. Some of the ethical questions currently surrounding the release of geneti-

FIGURE 4.25 Mongooses were released in Hawaii in an effort to control rats. The mongooses are active during the day, however, while the rats are night creatures, so they ignored each other. Instead, the mongooses attacked defenseless native birds and became as great a problem as the rats. © Gerard Lacz/Peter Arnold, Inc.

cally engineered organisms are based on concerns that they are novel organisms, and we might not be able to predict how they will interact with other species in natural ecosystems—let alone how they might respond to natural selective forces. It is argued that we can't predict either their behavior or their evolution.

Summary

- Organisms are adapted to live within certain ranges of environmental conditions. Tolerance limits are the maximum or minimum conditions, such as temperature or moisture, that an organism can survive. Since many environmental factors affect survival, it is useful to consider critical factors that limit a species' growth or expansion.

- Evolution is gradual change of organisms by *natural selection*. Natural selection refers to a higher rate of survival and reproduction among individuals that happen to have advantageous traits. Environmental conditions can exert *selective pressure* by making some traits more advantageous than others.

- An ecological niche is usually described as its ecological role in a community; a niche can also be the place or set of environmental conditions in which an organism lives. Generalist species can occupy a range of habitats and ecological roles or environmental conditions. Highly specialized species occupy narrower niches.

- Resource partitioning occurs when species adapt to use a single resource differently.

- Species interact in many ways. Some general classes of interaction include predation, parasitism, symbiosis, and competition. All of these interactions can exert selective pressure, as organisms develop defenses against predators or parasites, as they develop traits that improve competitiveness, or as they develop mutually beneficial interactions. Both interspecific (between species) and intraspecific (within a species) competition can lead to changes in traits or behavior.

- Defensive mechanisms can include Batesian mimicry, in which a harmless species looks like a dangerous one, and Müllerian mimicry, in which two dangerous species look like each other, and thus both discourage predation.

- Primary productivity, or the rate of biomass accumulation, is a basic characteristic of communities. Abundance and species diversity are also important characteristics.

- Complexity refers to the number of species at each trophic level and the number of trophic levels in a community. Many ecologists believe that complexity contributes to stability in an ecosystem, or resilience to abrupt change such as fire, flood, or drought. Others believe that complex communities can be less resilient than simple ones.

- Edges, where contrasting conditions meet, are important features in biological communities. Ecotones, or zones of transition, have great diversity. Edges also reduce habitat quality for interior species.

- Primary succession occurs when pioneer species occupy areas previously lacking living things. Secondary succession occurs when an existing community is disrupted and a new, different community develops.

- The idea of a climax community is a stable community that appears to be the culmination of successional processes. A contrasting idea is that species occur individualistically, each according to its ability to colonize an area.

- Introduced species are one of the greatest modern threats to biological diversity and ecosystem complexity. When introduced species are free of predators, they can become abundant and cause significant damage to ecosystems.

Questions for Review

1. Explain how tolerance limits (fig. 4.2) to environmental factors determine distribution of a highly specialized species such as the saguaro cactus. Compare this to the distribution of a generalist species such as cowbirds or starlings. What would the curve in fig. 4.1 look like for one of these species?

2. Productivity, diversity, complexity, resilience, and structure are exhibited to some extent by all communities and ecosystems. Describe how these characteristics apply to the ecosystem in which you live.

3. Resource partitioning (figs. 4.6, 4.7) is an important adaptive strategy. Explain resource partitioning, and think of an example in your local area.

4. Define keystone species and explain their importance in community structure and function.

5. All organisms within a biological community interact with each other. The most intense interactions often occur between individuals of the same species. What concept discussed in this chapter can be used to explain this phenomenon?

6. Relationships between predators and prey play an important role in the energy transfers that occur in ecosystems. They also influence the process of natural selection. Explain how predators affect the adaptations of their prey. This relationship also works in reverse. How do prey species affect the adaptations of their predators?

7. Competition for a limited quantity of resources occurs in all ecosystems. This competition can be interspecific or intraspecific. Explain some of the ways an organism might deal with these different types of competition.

8. Each year fires burn large tracts of forestland. Describe the process of succession that occurs after a forest fire destroys an existing biological community. Is the composition of the final successional community likely to be the same as that which existed before the fire? What factors might alter the final outcome of the successional process? Why may periodic fire be beneficial to a community?

9. Which world ecosystems are most productive in terms of biomass (fig. 4.17)? Which are least productive? What units are used in this figure to quantify biomass accumulation?

10. Discuss the dangers posed to existing community members when new species are introduced into ecosystems. What type of organism would be most likely to survive and cause problems in a new habitat?

Questions for Critical Thinking

1. Ecologists debate whether biological communities have self-sustaining, self-regulating characteristics or are highly variable, accidental assemblages of individually acting species. What outlook or worldview might lead scientists to favor one or the other of these theories?

2. The concepts of natural selection and evolution are central to how most biologists understand and interpret the world, and yet the theory of evolution is contrary to the beliefs of many religious groups. Why do you think this theory is so important to science and so strongly opposed by others? What evidence would be required to convince opponents of evolution?

3. What is the difference between saying that a duck has webbed feet because it needs them to swim and saying that a duck is able to swim because it has webbed feet?

4. The concept of keystone species is controversial among ecologists because most organisms are highly interdependent. If each of the trophic levels is dependent on all the others, how can we say one is most important? Choose an ecosystem with which you are familiar and decide whether it has a keystone species or keystone set.

5. Some scientists look at the boundary between two biological communities and see a sharp dividing line. Others looking at the same boundary see a gradual transition with much intermixing of species and many interactions between communities. Why are there such different interpretations of the same landscape?

6. The absence of certain lichens is used as an indicator of air pollution in remote areas such as national parks. How can we be sure that air pollution is really responsible? What evidence would be convincing?

7. We tend to regard generalists or "weedy" species as less interesting and less valuable than rare and highly specialized endemic species. What values or assumptions underlie this attitude?

8. What part of this chapter do you think is most likely to be challenged or modified in the future by new evidence or new interpretations?

Key Terms

abundance 83
Batesian mimicry 82
climax community 86
coevolution 78
commensalism 79
complexity 84
disclimax communities 88
diversity 83
ecological development 86
ecological niche 75
ecotones 84
edge effects 84

environmental indicators 73
equilibrium communities 88
evolution 74
fire-climax communities 88
habitat 75
interspecific competition 79
intraspecific competition 79
keystone species 78
Müllerian mimicry 82
mutualism 80
natural selection 74
parasites 78

pathogens 78
pioneer species 86
plankton 77
predator 77
primary productivity 82
primary succession 86

resource partitioning 76
secondary succession 86
symbiosis 79
territoriality 79
tolerance limits 72

Further Readings

Botkin, Daniel B. 1989. *Discordant Harmonies: A New Ecology for the Twenty-First Century*. Oxford University Press.

Ehrlich, Paul R., and Peter H. Raven. 1967. Butterflies and plants: A study in coevolution. *Evolution* 18:586–608.

Gleason, Henry A. 1926. The individualistic concept of the plant association. *Bulletin of the Torrey Botanical Club* 53:7–26.

MacArthur, R. H. 1958. Population ecology of some warblers of Northeastern coniferous forests. *Ecology* 39:599–619.

Paracer, S., and V. Ahmadjian. 2000. *Symbiosis: An Introduction to Biological Associations*. Oxford University Press.

Simberloff, D. 1997. Flagships, umbrellas, and keystones: Is single-species management passé in the landscape era? *Biological Conservation* 83:247–57.

Tilman, David, et al. 1997. The influence of functional diversity and composition on ecosystem processes. *Science* 277(5330): 1300–02.

Welcome to McGraw-Hill's Online Learning Center

Location: http://www.mhhe.com/environmentalscience

McGraw Hill

WEB EXERCISES

Project FeederWatch

The FeederWatch Program coordinated by the Cornell Laboratory of Ornithology is an excellent example of citizen science. Thousands of volunteers collect data on bird frequency and distribution from backyard feeders throughout winter months. The data are displayed on innovative animated maps that allow you to view dynamic information about a given species in a particular region or state over time. Go to: http://birds. cornell.edu/PFWMaproom/pfwmaproom.html to find a species and location that interests you; then consider the following questions:

1. Does it surprise you that this species does or doesn't occur in your area?

2. How would you account for the patterns you see on the map? Is it possible that the results show a bias in data collection rather than a real variation in distribution of the species?

3. Some species display seasonal movements. Can you detect a pattern in changing distribution of the species you've chosen during the time shown? How would you account for the pattern (or the lack of a pattern) you observe?

Trophic Cascades in Aquatic Food Webs

Ecological relationships can affect physical qualities in our environment. To understand how this occurs, go to http://www.mhhe.com/environmental science. Click on the title of your textbook to take you to the Online Learning Center, and then click on the student edition. Click on "Regional Case Studies" on the left-hand navigational menu. Scroll down to the North region to find a case study titled "Food Web Control of Primary Production in Lakes." Read the text and study the graphics to answer the following questions.

1. Explain the three graphs. Why does an increase in game fish (piscivores) cause a decrease in phytoplankton (algae) in a lake?

2. If you were designing a test of this hypothesis, how would you regulate piscivore biomass experimentally?

3. What would you use as a control in your study?

4. What do the authors mean by top down and bottom up controls?

5. Why do they call this a trophic cascade?

Alien Invaders: When Weeds Do Good and Bad Things

On the same regional perspectives page, look at the first case study in the Southwest Region. You can also find an interesting international case study about water hyacinth on the USGS Eros site at http://edcsnw3.cr.usgs.gov/ip/hyacinth/hyacinth.html. Look at the Winam Gulf study for some impressive images of how this plant can clog lakes and waterways.

1. When and why was water hyacinth introduced into the United States?

2. Where did it come from?

3. How fast does it spread?

4. Why is it a problem?

5. What possible benefits does it convey?

6. How is it controlled?

7. Drawing on what you've learned about community interactions in this chapter, why is this plant so aggressive and so successful in its new home?

Biomes: Global Patterns of Life

What is the use of a house if you haven't got a tolerable planet to put it on?

Henry David Thoreau

OBJECTIVES

After studying this chapter, you should be able to:

- recognize the characteristics of major aquatic and terrestrial biomes and understand the most important factors that determine the distribution of each type.
- describe ways in which humans disrupt or damage each of these ecosystem types.
- understand how and why marine environments vary with depth and distance from shore.
- identify and describe a variety of nearshore environments and wetland environments.
- summarize the overall patterns of human disturbance of world biomes as well as some specific, important examples of losses obscured by broad aggregate categories.

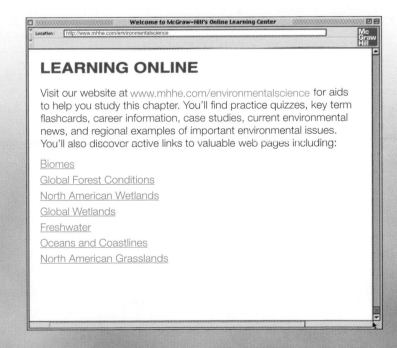

LEARNING ONLINE

Visit our website at www.mhhe.com/environmentalscience for aids to help you study this chapter. You'll find practice quizzes, key term flashcards, career information, case studies, current environmental news, and regional examples of important environmental issues. You'll also discover active links to valuable web pages including:

Biomes
Global Forest Conditions
North American Wetlands
Global Wetlands
Freshwater
Oceans and Coastlines
North American Grasslands

Photo: A taiga biome transitions to alpine tundra. What environmental conditions control the distribution of these communities? © Corbis/Volume 156.

Integrity, Stability, and Beauty of the Land

In 1935, pioneering wildlife ecologist Aldo Leopold bought 32 ha (80 acres) of worn out, sandy farmland on the banks of the Wisconsin River not far from his home in Madison. Originally intended to be merely a hunting camp, the farm quickly became a year-round retreat from the city, as well as a laboratory in which Leopold could test his theories about conservation, environmental ethics, and ecologically based land management. A dilapidated chicken shack, the only remaining building from the original farm, was remodeled into a rustic cabin (fig. 5.1). The whole Leopold family participated in tree planting, bird watching, gardening, and exploring nature.

The old farm was not pristine wilderness nor were the Leopolds merely spectators. They regarded themselves as participating citizens of the land community, seeking to restore it to ecological health and beauty. Planting as many as 6,000 trees and bushes each spring, they practiced "wild husbandry," using axes and shovels to reverse the abuses of previous owners and to revitalize the land through active management, care, and understanding. "Conservation," Leopold wrote, "is the positive exercise of skill and insight, not merely a negative exercise of abstinence or caution."

While building this relationship with the land—by which he meant all the plants and animals as well as the nonliving components of the landscape—Leopold mused on the ethics and meaning of conservation and the proper role of humans in nature. The first part of his much-beloved *Sand County Almanac* is a collection of essays about experiments and experiences at the farm. All of us, he claimed, should choose a piece of land on which we can practice stewardship and develop a sense of place. It doesn't have to be a beautiful place. In fact, it might be best to adopt a weedy, unwanted patch that needs our love and care. Both we and the land benefit from such connectedness, he maintained.

Leopold's essay on "The Land Ethic" is a cornerstone of the conservation movement and one of the most eloquent statements of environmental philosophy in American nature writing. In it, Leopold wrote,

FIGURE 5.1 Aldo Leopold's Sand County farm in central Wisconsin served as a refuge from the city and as a laboratory to test theories about land conservation, environmental ethics, and ecologically based land management. © William P. Cunningham.

"We abuse the land because we regard it as a commodity belonging to us. When we see land as a community to which we belong, we may begin to use it with love and respect. . . . A land ethic, then, reflects the existence of an ecological conscience, and this in turn reflects a conviction of individual responsibility for the health of the land. Health is the capacity of the land for self-renewal. Conservation is our effort to understand and preserve this capacity. . . . A thing is right when it preserves the integrity, stability, and beauty of the biotic community. It is wrong when it does otherwise."

Leopold's Sand County farm gave him an intimate understanding of the natural community in which he lived. This farm also shows how resilient natural systems can be with thoughtful management. In this chapter, we'll explore the major biological communities and the ways humans have altered them.

TERRESTRIAL BIOMES

Although all local environments are unique, it is helpful to understand them in terms of a few general groups with similar climate conditions, growth patterns, and vegetation types. We call these broad types of biological communities **biomes.** Understanding the global distribution of biomes, and knowing the differences in what grows where and why, is essential to the study of global environmental science. Biological productivity—and ecosystem resilience—varies greatly from one biome to another. Human use of biomes depends largely on those levels of productivity. Our ability to restore ecosystems, a topic of chapter 13, and nature's ability to restore itself, depend largely on biome conditions. Clear-cut forests regrow relatively quickly in New England, but very slowly in Siberia, where current logging is expanding. Some grasslands rejuvenate quickly after grazing, and some are slower to recover. Why these differences? The sections that follow seek to answer this question.

Temperature and precipitation are among the most important determinants in biome distribution on land (fig. 5.2). If we know the general temperature range and precipitation level, we can predict what kind of biological community is likely to occur there, in the absence of human disturbance. Landforms, especially mountains, and prevailing winds also exert important influences on biological communities.

It is helpful to understand biomes in terms of their global distribution (fig. 5.3). For example, a band of boreal (northern) forests crosses Canada and Siberia, tropical forests occur near the equator, and expansive grasslands lie near—or just beyond—the tropics. As you look at this map, which biomes do you think are most heavily populated by humans? Why?

In this chapter, we'll examine the major terrestrial biomes, then we'll investigate ocean and freshwater communities and environments. Ocean environments are important because they cover two-thirds of the earth's surface, provide food for much of human-

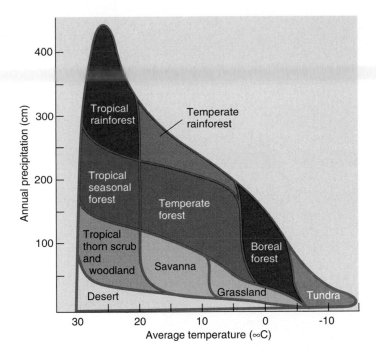

FIGURE 5.2 Biomes most likely to occur in the absence of human disturbance or other disruptions, according to average annual temperature and precipitation. *Note:* this diagram does not consider soil type, topography, wind speed, or other important environmental factors. Still, it is a useful general guideline for biome location.

Source: Communities and Ecosystems, 2/e by R. H. Whitaker, © 1975. Reprinted by permission of Prentice Hall, Upper Saddle River, New Jersey.

ity, and help regulate our climate through photosynthesis. Wetlands are often small, but they have great influence on environmental health, biodiversity, and water quality. In chapter 12, we'll look at how we use these communities; and in chapter 13, we'll see how we preserve, manage, and restore them.

Deserts

You may think of deserts as barren and biologically impoverished. Their vegetation is sparse, but it can be surprisingly diverse, and most desert plants and animals are highly adapted to survive long droughts, extreme heat, and often extreme cold. **Deserts** occur where precipitation is rare and unpredictable, usually with less than 30 cm of rain per year. Adaptations to these conditions include water-storing leaves and stems, thick epidermal layers to reduce water loss, and salt tolerance. Many desert plants are drought-deciduous, that is, they lose their leaves during the dry season. Most desert plants also bloom and set seed quickly when a spring rain does fall.

Warm, dry, high-pressure climate conditions (see chapter 15) create desert regions at about 30° north and south. Extensive deserts occur in continental interiors (far from oceans, which evaporate the moisture for most precipitation) of North America, Central Asia, Africa, and Australia (fig. 5.4). The rain shadow of the Andes produces the world's driest desert in coastal Chile. Deserts can also be cold. Antarctica is a desert. Some inland valleys apparently get almost no precipitation at all.

Like plants, animals in deserts are specially adapted. Many are nocturnal, spending their days in burrows to avoid the sun's heat and desiccation. Pocket mice, kangaroo rats, and gerbils can get most of their moisture from seeds and plants. Desert rodents also have highly concentrated urine and nearly dry feces that allow them to eliminate body waste without losing precious moisture.

Deserts are more vulnerable than you might imagine. Sparse, slow-growing vegetation is quickly damaged by off-road vehicles. Desert soils recover slowly. Tracks left by army tanks practicing in California deserts during World War II can still be seen today.

Deserts are also vulnerable to overgrazing. In Africa's vast Sahel (the southern edge of the Sahara Desert), livestock are destroying much of the plant cover. Bare, dry soil becomes drifting sand, and restabilization is extremely difficult. Without plant roots and organic matter, the soil loses its ability to retain what rain does fall, and the land becomes progressively drier and more bare. Similar depletion of dryland vegetation is happening in many desert areas, including Central Asia, India, and the American Southwest and Plains states.

Grasslands: Prairies and Savannas

Grasslands occur where there is enough rain to support abundant grass but not enough for forests (fig. 5.5). Usually grasslands are a complex, diverse mix of grasses and flowering herbaceous plants, generally known as forbs. Black-eyed susan and purple coneflower are forbs that may be familiar to you. In drier grasslands, grasses and forbs may be less than half a meter tall. In wetter areas, grasses can exceed 2 m. Where scattered trees occur in a grassland, we call it a savanna.

Like desert plants, grassland and savanna vegetation is adapted to survive drought and extreme heat or cold. Most have deep, long-lived roots that seek groundwater and that persist when leaves and stems above ground die back. These deep roots, and annual accumulation of dead leaves on the surface, produce thick, organic-rich soils in many grasslands. Most grasslands are also adapted to survive fire. Fresh green shoots appear quickly after a fire, and migratory grazers, such as American bison, African wildebeest, and Central Asian horses, thrive on this new growth.

Historically, the greatest threat to grasslands was conversion of the rich soils to farmland. The tallgrass prairies of the central Unites States and Canada are almost completely converted to corn, soy, and other crops. Remaining grasslands are mostly too dry for good farmland, so the greatest risk today is from overgrazing. As in a desert, excessive grazing eventually kills even deep-rooted plants. As groundcover diappears, soil erosion results, and unpalatable weeds, such as cheatgrass or leafy spurge, spread. An additional threat to remaining grasslands is fire suppression, which allows trees to encroach on former grasslands.

Tundra

Where temperatures are below freezing most of the year, only small, hardy vegetation can survive. **Tundra,** a treeless landscape that occurs at high latitudes or on mountaintops, has a growing season of only two to three months, and it may have frost any month

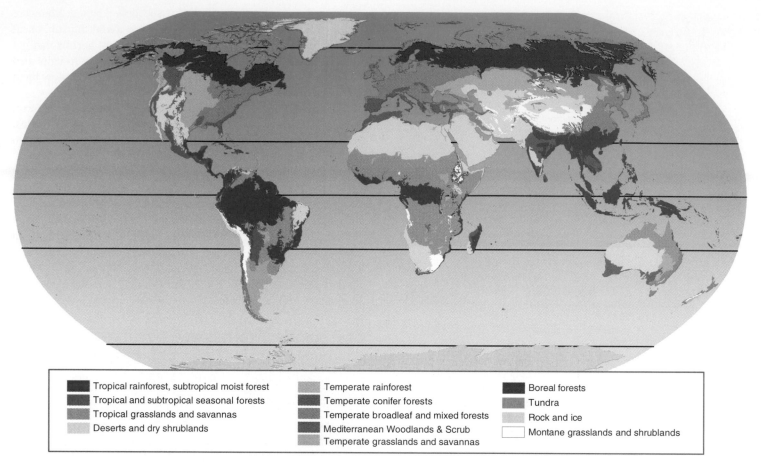

Tropical rainforest, subtropical moist forest	Temperate rainforest	Boreal forests
Tropical and subtropical seasonal forests	Temperate conifer forests	Tundra
Tropical grasslands and savannas	Temperate broadleaf and mixed forests	Rock and ice
Deserts and dry shrublands	Mediterranean Woodlands & Scrub	Montane grasslands and shrublands
	Temperate grasslands and savannas	

FIGURE 5.3   Major world biomes. Compare this map to figure 5.2 for generalized temperature and moisture conditions that control biome distribution. Also compare it to the satellite image of biological productivity (fig. 5.12). *Source:* WWF Ecoregions.

FIGURE 5.4   World deserts (*bottom*) and a desert landscape in Joshua Tree National Park. Like most desert plants, these Joshua trees (really members of the lily family) are adapted to conserve water and repel enemies. © William P. Cunningham.

of the year. Some people consider tundra a variant of grasslands because it has no trees; others consider it a very cold desert because water is unavailable (frozen) most of the year.

Arctic tundra is an expansive biome that has low productivity because it receives little light and has a short growing season. During midsummer, however, 24-hour sunshine supports a burst of plant growth and an explosion of insect life. Tens of millions of waterfowl, shorebirds, terns, and songbirds migrate to the arctic every year to feast on the abundant invertebrate and plant life and to raise their young on the brief bounty. These birds then migrate to wintering grounds, where they may be eaten by local predators—effectively they carry energy and protein from high latitudes to low latitudes. Arctic tundra is essential for global biodiversity, especially for birds.

Alpine tundra, occurring on or near mountaintops, has environmental conditions and vegetation similar to arctic tundra (fig. 5.6). These areas have a short, intense growing season. Often one sees a splendid profusion of flowers in alpine tundra: this is because everything must flower at once in order to produce seeds in a few weeks before the snow comes again. Many alpine tundra plants also have deep pigmentation and leathery leaves to protect against the strong ultraviolet light in the thin mountain atmosphere.

FIGURE 5.5 Grasslands occur at midlatitudes on all continents. Kept open by extreme temperatures, dry conditions, and periodic fires, grasslands can have surprisingly high plant and animal diversity. © Mary Ann Cunningham.

FIGURE 5.6 Tundra, with its short, slow-growing plant life, occurs at high altitudes (alpine tundra) or high latitudes (arctic tundra), where temperatures are cool and frost is possible even in summer. © Corbis/Volume 156.

Compared to other biomes, tundra has relatively low diversity. Dwarf shrubs, such as willows, sedges, grasses, mosses, and lichens tend to dominate the vegetation. Migratory musk-ox, caribou, or alpine mountain sheep and mountain goats can live on the vegetation because they move frequently to new pastures.

Because these environments are too cold for most human activities, they are not as badly threatened as other biomes. There are important problems, however. Global climate change may be altering the balance of some tundra ecosystems, and air pollution

from distant cities tends to accumulate at high latitudes (see chapter 15). In eastern Canada, coastal tundra is being badly depleted by overabundant populations of snow geese, whose numbers have exploded due to winter grazing on the rice fields of Arkansas and Louisiana. Oil and gas drilling—and associated truck traffic—threatens tundra in Alaska and Siberia. Clearly, this remote biome is not independent of human activities at lower latitudes.

Conifer Forests

Conifer (cone-bearing) forests occur in a wide range of temperate, or midlatitude, regions. Many grow where moisture is limited by sandy soil, and their thin, waxy leaves (needles) help them reduce moisture loss. In the United States, **southern pine forests** are one of the most important forest resources. These forests grow quickly, and they tolerate weathered, nutrient-poor soils. Bird and mammal diversity in these forests can be extremely high. Because these forests grow quickly in the warm, moist southern climate, though, many have been converted to plantations, with little plant or animal diversity.

Conifer needles can also survive the harsh winter of northern forests and mountains, so conifers tend to dominate high-latitude and high-altitude forests. The **boreal forest,** or northern conifer forest, stretches in a broad band around the world between about 45° and 60° north (fig. 5.7). Dominant trees are pines, hemlocks, spruce, cedar, and fir. Some deciduous trees are also mixed in, such as maples, birch, aspen, and alder. In Siberia, Canada, and the western United States, conifer forests are also a key resource, on which large, regional economies depend.

The extreme, ragged edge of the boreal forest, where forest gradually gives way to tundra, is known by its Russian name, **taiga.**

FIGURE 5.7 Boreal forest. At the northern limit of the boreal forest in Alaska, we find small, widely spaced black spruce intermixed with willows and heather on a wet peatland. © William P. Cunningham.

The extreme cold and short summer limits the growth rate of trees here. A 10 cm diameter tree may be a century or two old in the far north.

The coniferous forests of the Pacific coast represent yet another special set of environmental circumstances. Mild year-round temperatures and abundant rainfall, up to 250 cm (100 in.) per year, result in luxuriant plant growth and giant trees such as the California redwoods, the largest trees in the world and the largest organism of any kind known to have ever existed. Redwoods once grew along the Pacific coast from California to Oregon, but logging has reduced them to a few small fragments.

The wettest coastal forests are known as the **temperate rainforest,** a cool, rainy forest often enshrouded in fog (fig. 5.8). Condensation in the canopy (leaf drip) is a major form of precipitation in the understory.

Because conifer forests are widespread and the trees are often large, they have been one of our most important natural resources. Remaining fragments of ancient forests are important areas of biodiversity, especially in North America and Europe. Recent battles over old-growth conservation (chapter 12) are focused mainly on these forests. Forest management policies are one of the more important, and perennial, political issues in the United States and Canada, and these issues are emerging in Russia.

FIGURE 5.8 Temperate rainforests, with towering conifers and a wet understory, are a very rainy variant of temperate coniferous biomes.
© Corbis/Volume 90.

Broad-Leaved Deciduous Forests

Broad-leaf forests occur throughout the world where rainfall is plentiful. In midlatitudes, these forests are **deciduous,** that is, they lose their leaves in winter (fig. 5.9). At warmer latitudes, broad-leaved trees may lose their leaves in a dry season, or they may retain most leaves most of the year. Southern live oaks and cypresses, for example, are broad-leaved evergreen trees.

Although these forests have a dense canopy in summer, they have a diverse understory that blooms in spring, before the trees leaf out. Spring ephemeral (short-lived) plants produce lovely flowers, and vernal pools support amphibians and insects. The middle layers and understory of these forests also harbor a great diversity of North American songbirds.

In North America, deciduous forests once covered most of what is now the eastern half of the United States and southern Canada. Most of western Europe was once deciduous forest. Most of this forest was cleared a thousand years ago. When European settlers first came to North America, they quickly settled and cut most of these forests for timber, firewood, industrial uses, and to make farmland.

Deciduous forests can regrow readily because they occupy a moist, moderate climate. But most of these forests have been occupied so long that human impacts are extensive, and most native species are at least somewhat threatened. Currently, the greatest threat to broad-leaved deciduous forests is in eastern Siberia, where deforestation is proceeding rapidly. Siberia may have the highest deforestation rate in the world. As forests disappear, so do Siberian tigers, bears, cranes, and a host of other endangered species.

FIGURE 5.9 Temperate deciduous forests lose their leaves, and often change to lovely colors, as freezing weather approaches.
© William P. Cunningham.

Mediterranean/Chaparral/Thorn Scrub

Often, dry environments support drought-adapted shrubs and trees, as well as grass. These mixed environments can be highly variable. They can also be very rich biologically. Such conditions are often described as Mediterranean (with hot, dry summers and cool, moist winters). Evergreen shrubs with small, leathery, sclerophyllous (hard, waxy) leaves form dense thickets. Scrub oaks, drought-resistant pines, or other small trees often cluster in sheltered valleys. Periodic fires burn fiercely in this fuel-rich plant assemblage and are a major factor in plant succession. Annual spring flowers often bloom profusely, especially after fires. In California, this landscape is called **chaparral,** Spanish for thicket. Some typical animals include jackrabbits, kangaroo rats, mule deer, chipmunks, lizards, and many bird species. Very similar landscapes are found along the Mediterranean coast as well as southwestern Australia, central Chile, and South Africa. Although this biome doesn't cover a very large total area, it contains a high number of unique species and is often considered a "hot-spot" for biodiversity. It also is highly desired for human habitation, often leading to conflicts with rare and endangered plant and animal species.

Areas that are drier year-round, such as the African Sahel (edge of the Sahara Desert), northern Mexico, or the American Intermountain West (or Great Basin) tend to have a more sparse, open shrubland, characterized by sagebrush (*Artemisia* sp.), chamiso (*Adenostoma* sp.), or saltbush (*Atriplex* sp.). In Africa, acacias and other spiny plants dominate this landscape, giving it the name **thorn scrub.** Some typical animals of this biome in America are a wide variety of snakes and lizards, rodents, birds, antelope, and mountain sheep. In Africa, this landscape is home to gazelle, rhinos, and giraffes, and many other species (fig. 5.10).

Tropical Moist Forests

The humid tropical regions of South and Central America, Africa, Southeast Asia, and some of the Pacific Islands support one of the most complex and biologically rich biome types in the world (fig. 5.11). Although there are several kinds of moist tropical forests, they share common attributes of ample rainfall and uniform temperatures. Cool **cloud forests** are found high in the mountains where fog and mist keep vegetation wet all the time. **Tropical rainforests** occur where rainfall is abundant—more than 200 cm (80 in.) per year—and temperatures are warm to hot year-round.

The soil of both these tropical moist forest types tends to be old, thin, acidic, and nutrient-poor, yet the number of species present can be mind-boggling. For example, the number of insect species in the canopy of tropical rainforests has been estimated to be in the millions! It is estimated that one-half to two-thirds of all species of terrestrial plants and insects live in tropical forests.

The nutrient cycles of these forests also are distinctive. Almost all (90 percent) of the nutrients in the system are contained in the bodies of the living organisms. This is a striking contrast to temperate forests, where nutrients are held within the soil and made available for new plant growth. The luxuriant growth in tropical rainforests depends on rapid decomposition and recycling of dead

FIGURE 5.10   The thorny acacias in this mixed grassland–thorn scrub environment support abundant migratory grazers. © Corbis/Volume 6.

FIGURE 5.11   Tropical rainforests, like this one in Costa Rica, support a luxuriant profusion of life-forms. The canopies of tall trees harbor epiphytes and vines. Little light reaches the forest floor. © William P. Cunningham.

organic material. Leaves and branches that fall to the forest floor decay and are incorporated almost immediately back into living biomass.

When the forest is removed for logging, agriculture, and mineral extraction, the thin soil cannot support continued cropping and cannot resist erosion from the abundant rains. And if the cleared area is too extensive, it cannot be repopulated by the rainforest community. Rapid deforestation is occurring in many tropical areas as people move into the forests to establish farms and ranches, but the land soon loses its fertility.

Tropical Seasonal Forests

Many areas in India, Southeast Asia, Australia, West Africa, the West Indies, and South America have tropical regions characterized by distinct wet and dry seasons instead of uniform heavy rainfall throughout the year, although temperatures are hot year-round. These areas have produced communities of **tropical seasonal forests:** semievergreen or partly deciduous forests tending toward open woodlands and grassy savannas dotted with scattered, drought-resistant tree species.

Tropical dry forests have typically been more attractive than wet forests for human habitation and have suffered greater degradation. Clearing a dry forest with fire is relatively easy during the dry season. Soils of dry forests often have higher nutrient levels and are more agriculturally productive than those of a rainforest. Finally, having fewer insects, parasites, and fungal diseases than a wet forest makes a dry or seasonal forest a healthier place for humans to live. Consequently, these forests are highly endangered in many places. Less than 1 percent of the dry tropical forests of the Pacific coast of Central America or the Atlantic coast of South America, for instance, remain in an undisturbed state.

MARINE ECOSYSTEMS

The biological communities in oceans and seas are poorly understood, but they are probably as diverse and as complex as terrestrial biomes. In this section, we will explore a few facets of these fascinating environments. Oceans cover nearly three-fourths of the earth's surface, and they contribute in important, although often unrecognized, ways to terrestrial ecosystems (see chapter 4, opening essay). Like land-based systems, most marine communities depend on photosynthetic organisms. Often it is algae, coral, or tiny, free-floating photosynthetic plants (**phytoplankton**) that support a marine food web, rather than the trees and grasses we see on land. In oceans, photosynthetic activity tends to be greatest near coastlines, where nitrogen, phosphorus, and other nutrients wash off-shore and fertilize primary producers. Ocean currents also contribute to the distribution of biological productivity, as they transport nutrients and phytoplankton far from shore (fig. 5.12).

As plankton, algae, fish, and other organisms die, they sink toward the ocean floor. Deep-ocean ecosystems, consisting of crabs, filter-feeding organisms, strange phosphorescent fish, and many other life-forms, often rely on this "marine snow" as a primary nutrient source. Surface communities also depend on this material. Upwelling currents circulate nutrients from the ocean floor back to the surface. Along the coasts of South America, Africa, and Europe, these currents support rich fisheries.

Vertical stratification is a key feature of aquatic ecosystems. Light decreases rapidly with depth, and communities below the photic zone (light zone, often reaching about 20 m deep) must rely on energy sources other than photosynthesis to persist. Temperature also decreases with depth. Deep-ocean species often grow slowly in part because metabolism is reduced in cold conditions. In contrast, warm, bright, near-surface communities such as coral reefs and estuaries are among the world's most biologically productive environments. Temperature also affects the amount of oxygen and other elements that can be absorbed in water. Cold water holds abundant oxygen, so productivity is often high in cold oceans, as in the North Atlantic, North Pacific, and Antarctic.

Ocean systems can be described by depth and proximity to shore (fig. 5.13). In general, **benthic** communities occur on the bottom, and **pelagic** (from "sea" in Greek) zones are the water column. The epipelagic zone (*epi* = on top) has photosynthetic organisms. Below this are the mesopelagic (*meso* = medium), and bathypelagic (*bathos* = deep) zones. The deepest layers are the abyssal zone (to 4,000 m) and hadal zone (deeper than 6,000 m). Shorelines are known as littoral zones, and the area exposed by low tides is known as the intertidal zone. Often there is a broad, relatively shallow region along a continent's coast, which may reach a few kilometers or hundreds of kilometers from shore. This undersea area is the continental shelf.

FIGURE 5.12 Satellite measurements of chlorophyll levels in the oceans and on land. Dark green to blue land areas have high biological productivity. Dark blue oceans have little chlorophyll and are biologically impoverished. Light green to yellow ocean zones are biologically rich. Courtesy SeaWifs/NASA.

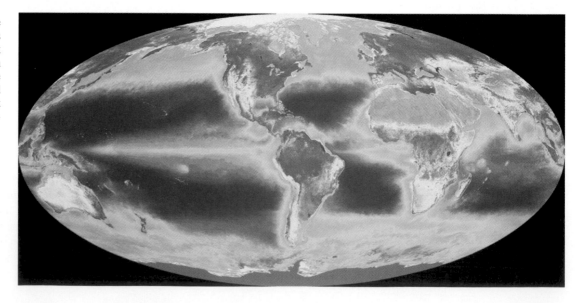

The Open Ocean

The open ocean is often referred to as a biological desert because it has relatively low productivity. Fish and plankton (small, floating organisms) abound in regions such as the equatorial Pacific and Antarctic oceans, where currents carry nutrients far from shore (fig. 5.12). Another notable exception, the Sargasso Sea in the western Atlantic, is known for its free-floating mats of brown algae. These algae mats support a phenomenal diversity of animals, including sea turtles, fish, and even eels that hatch amid the algae, then eventually migrate up rivers along the Atlantic coasts of North America and Europe.

Deep-sea thermal vent communities are another remarkable type of marine system (see fig. 3.8) that was completely unknown until 1977 explorations with the deep-sea submarine *Alvin*. These communities are based on microbes that capture chemical energy, mainly from sulfur compounds released from thermal vents—jets of hot water and minerals on the ocean floor. Magma below the ocean crust heats these vents. Tube worms, mussels, and microbes on these vents are adapted to survive both extreme temperatures, often above 350°C (700°F), and the intense water pressure at depths of 7,000 m (20,000 f) or more. Oceanographers have discovered thousands of different types of organisms, most of them microscopic, in these communities.

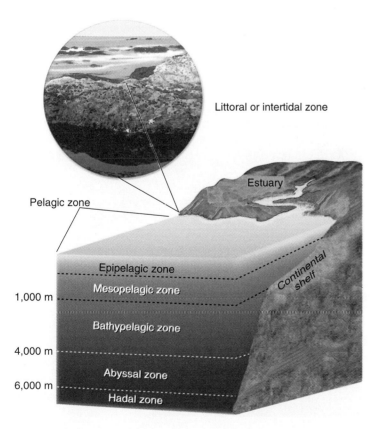

Littoral or intertidal zone

Pelagic zone

Estuary

Epipelagic zone

Mesopelagic zone

Continental shelf

1,000 m

Bathypelagic zone

4,000 m

Abyssal zone

6,000 m

Hadal zone

FIGURE 5.13 Light penetrates only the top 10–20 m of the ocean. Below this level, temperatures drop and pressure increases. Nearshore environments include the intertidal zone and estuaries.

Shallow Coasts: Coral Reefs and Mangroves

Coral reefs are among the best-known marine ecosystems. **Reefs** consist of colonies of minute, colonial animals that live symbiotically with photosynthetic algae. The calcium-rich skeletons of the coral provide structure to the reef, and the algae provide energy and nutrients. Fish, worms, crustaceans, and myriad other life-forms subsist on, and hide in, the coral. The coral structure also provides shelter and hiding places for small fish and fish larvae, so coral reefs provide a nursery for countless species of fish on which humans rely for food. In addition to being highly productive, coral reefs are among the most diverse and colorful ecosystems in the world (fig. 5.14).

Reefs are also among the most endangered biological communities. Because coral reefs rely on photosynthesis, they only occur in shallow, clear water, where sunlight is abundant. Often they occur along the edges of submerged banks or shelves, where upwelling currents, which carry nutrients, and sunlight are available. Because they are usually near shore, they are easily damaged when onshore activities, such as road construction, agriculture, or urban development, produce sediment that clouds water and clogs reef systems. Destructive fishing practices are widely used to catch colorful fish, including dynamite and cyanide (chapter 13), and these have destroyed large portions of reefs in Southeast Asia.

Sea-grass beds, or eel-grass beds, often occupy shallow, warm, sandy areas near coral reefs. Like reefs, these communities support rich communities of grazers, from snails and worms to turtles and manatees. Also like reefs, these environments are easily smothered by sediment originating from onshore agriculture and development.

Mangroves are trees that grow in salt water. They occur along calm, shallow, tropical coastlines. Mangrove forests or swamps help stabilize shorelines, and they are also critical nurseries for fish, shrimp, and other commercial species. Like coral reefs, mangroves

FIGURE 5.14 Coral reefs, with their complex structure and high rate of biological productivity, harbor some of the most diverse communities in the world. © Corbis/Royalty Free Website.

line tropical and subtropical coastlines, where they are vulnerable to development, sedimentation, and overuse. Unlike reefs, mangroves provide commercial timber, and they can be clear-cut to make room for aquaculture (fish farming) and other activities. Ironically, mangroves provide the protected spawning beds for most of the fish and shrimp farmed in these ponds. As mangroves become increasingly threatened in tropical countries, villages relying on fishing for income and sustenance are seeing reduced catches and falling income.

Tidal Environments and Barrier Islands

Among the many fascinating coastal communities are those occupying tidal zones. The rocky intertidal communities of the Pacific Northwest are excellent for "tide-pooling," or exploring the strange and beautiful animals and algae in the small pools left by a receding tide (fig. 5.15). These environments are distinctive largely because species are adapted to survive both the hammering of waves at high tide and the heat and desiccation of sunshine when the tide is out.

Estuaries are bays where rivers empty into the sea, mixing fresh water with salt water. Usually calm, warm, and nutrient-rich, estuaries are biologically diverse and productive (fig. 5.16). Rivers provide nutrients and sediments, and a muddy bottom supports emergent plants (whose leaves emerge from the water surface) as well as the young forms of crustaceans, such as crabs and shrimp, and molluscs, such as clams and oysters. Nearly two-thirds of all marine fish and shellfish rely on estuaries and saline wetlands for spawning and juvenile development. Estuaries and coastal wetlands also help stabilize shorelines and reduce storm damage inland.

Barrier islands are low, narrow, sandy islands that form parallel to a coastline (fig. 5.17). They occur where the continental shelf is shallow and rivers or coastal currents provide a steady source of sediments. They protect brackish (moderately salty), inshore lagoons and salt marshes from storms, waves, and tides. One of the world's most extensive sets of barrier islands lines the Atlantic coast from New England to Florida, as well as along the Gulf coast of Texas. Composed of sand that is constantly reshaped by wind and waves, these islands can be formed or removed by a single violent storm. Because they are mostly beach, barrier islands are also popular places for real estate development. About 20 percent of the barrier island surface in the United States has been developed. Barrier islands are also critical to preserving coastal shorelines, settlements, estuaries, and wetlands.

Unfortunately, human occupation often destroys the value that attracts us there in the first place. Barrier islands and beaches are dynamic environments, and sand is hard to keep in place. Wind and wave erosion is a constant threat to beach developments. Walking or driving vehicles over dune grass destroys the stabilizing vegetative cover and accelerates, or triggers, erosion. Cutting roads through the dunes further destabilizes these islands, making them increasingly vulnerable to storm damage. A single storm in 1962 caused $300 million in property damage along the East Coast and left hundreds of beach homes tottering into the sea (fig. 5.18). As barrier islands recede, inland waters and shorelines become increasingly exposed to ocean waves and wind.

FIGURE 5.15 A fascinating variety of life-forms has adapted to life in the rocky intertidal zone. © William P. Cunningham.

FIGURE 5.16 A complex of small channels forms in this calm, protected salt marsh, which is also an estuary—the tidal area where a river enters the ocean. © Andrew Martinez/Photo Researchers, Inc.

FIGURE 5.17   A barrier island, Assateague, along the Maryland–Virginia coast. Grasses cover and protect dunes, which keep ocean waves from disturbing the bay, salt marshes, and coast at right. Roads cut through the dunes expose them to erosion. *Source:* USGS.

FIGURE 5.18   Winter storms have eroded the beach and undermined the foundations of homes on this barrier island. Breaking through protective dunes to build such houses damages sensitive plant communities and exposes the whole island to storm sand erosion. Coastal zone management attempts to limit development on fragile sites.
© Stephen Rose/Gamma Liason International.

Because of these problems, we spend billions of dollars each year building protective walls and barriers, pumping sand onto beaches from off-shore, and moving sand from one beach area to another. Insurance for beach structures is expensive, but many people can afford to pay high premiums for a coastal view and access to the water.

FRESHWATER ECOSYSTEMS

Freshwater environments are far less extensive than marine environments, but they are centers of biodiversity. Most terrestrial communities rely, to some extent, on freshwater environments. In deserts, isolated pools, streams, and even underground water systems, support astonishing biodiversity as well as provide water to land animals. In Arizona, for example, most birds gather in trees and bushes surrounding the few available rivers and streams.

Lakes

Freshwater lakes, like marine environments, have distinct vertical zones (fig. 5.19). Near the surface a subcommunity of plankton, mainly microscopic plants, animals, and protists (single-celled organisms such as amoebae), float freely in the water column. Insects such as water striders and mosquitoes also live at the air-water interface. Fish move through the water column, sometimes near the surface, and sometimes at depth.

Finally, the bottom, or *benthos,* is occupied by a variety of snails, burrowing worms, fish, and other organisms. These make up the benthic community. Oxygen levels are lowest in the benthic environment, mainly because there is little mixing to introduce oxygen to this zone. Anaerobic bacteria (not using oxygen) may live in low-oxygen sediments. In the littoral zone, emergent plants such as cattails and rushes grow in the bottom sediment. These plants create important functional links between layers of an

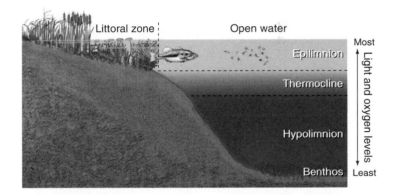

FIGURE 5.19   The layers of a deep lake are determined mainly by gradients of light, oxygen, and temperature. The epilimnion is affected by surface mixing from wind and thermal convections, while mixing between the hypolimnion and epilimnion is inhibited by a sharp temperature and density difference at the thermocline.

aquatic ecosystem, and they may provide the greatest primary productivity to the system.

Lakes, unless they are shallow, have a warmer upper layer that is mixed by wind and warmed by the sun. This layer is the *epilimnion.* Below the epilimnion is the hypolimnion (*hypo* = below), a colder, deeper layer that is not mixed. If you have gone swimming in a moderately deep lake, you may have discovered the sharp temperature boundary, known as the **thermocline,** between these layers. Below this boundary, the water is much colder. This boundary is also called the mesolimnion.

Local conditions that affect the characteristics of an aquatic community include (1) nutrient availability (or excess) such as nitrates and phosphates; (2) suspended matter, such as silt, that affects light penetration; (3) depth; (4) temperature; (5) currents; (6) bottom characteristics, such as muddy, sandy, or rocky floor; (7) internal currents; and (8) connections to, or isolation from, other aquatic and terrestrial systems (fig. 5.20).

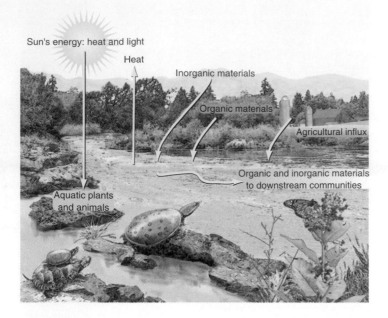

FIGURE 5.20   The character of freshwater ecosystems is greatly influenced by the immediately surrounding terrestrial ecosystems, and even by ecosystems far upstream or far uphill from a particular site.

Wetlands

Wetlands are shallow ecosystems in which the land surface is saturated or submerged at least part of the year. Wetlands have vegetation that is adapted to grow under saturated conditions. These legal definitions are important because although wetlands make up only a small part of most countries, they are disproportionately important in conservation debates and are the focus of continual legal disputes around the world and in North America. Beyond these basic descriptions, defining wetlands is a matter of hot debate. How often must a wetland be saturated, and for how long? How large must it be to deserve legal protection? Answers can vary, depending on political, as well as ecological, concerns.

These relatively small systems support rich biodiversity, and they are essential for both breeding and migrating birds (fig. 5.21). Although wetlands occupy less than 5 percent of the land in the United States, the Fish and Wildlife Service estimates that one-third of all endangered species spend at least part of their lives in wetlands. Wetlands retain storm water and reduce flooding by slowing the rate at which rainfall reaches river systems. Flood-water storage is worth $3 billion to $4 billion per year in the United States. As water stands in wetlands, it also seeps into the ground, replenishing groundwater supplies. Wetlands filter, and even purify, urban and farm runoff, as bacteria and plants take up nutrients and contaminants in water. They are also in great demand for filling and development. They are often near cities or farms, where land is valuable, and once drained, wetlands are easily converted to more lucrative uses.

Wetlands are described by their vegetation. **Swamps** are wetlands with trees (fig. 5.22). **Marshes** are wetlands without trees

FIGURE 5.21   Small wetlands are essential sources of biodiversity. Here a Western grebe sits on its floating nest. Courtesy of USDA Natural Resources Conservation Service.

(fig. 5.23). **Bogs** are areas of saturated ground, and usually the ground is composed of deep layers of accumulated, undecayed vegetation known as peat. **Fens** are similar to bogs except that they are mainly fed by groundwater, so that they have mineral-rich water and specially adapted plant species. Bogs are fed mainly by precipitation. Swamps and marshes have high biological productivity. Bogs and fens, which are often nutrient-poor, have low biological productivity. They may have unusual and interesting species, though, such as sundews and pitcher plants, which are adapted to capture nutrients from insects rather than from soil.

The water in marshes and swamps usually is shallow enough to allow full penetration of sunlight and seasonal warming. These mild conditions favor great photosynthetic activity, resulting in high productivity at all trophic levels. In short, life is abundant and varied. Wetlands are major breeding, nesting, and migration staging areas for waterfowl and shorebirds.

Wetlands may gradually convert to terrestrial communities as they fill with sediment, and as vegetation gradually fills in toward

FIGURE 5.22 Forested wetlands, such as the Okefenokee Swamp in Georgia, can be amazingly diverse, complex, and productive. Myriad life-forms coexist in exuberant abundance, interlinked by manifold relationships. Who lives where, and why, are central ecological questions. © John D. Cunningham/Visuals Unlimited.

FIGURE 5.23 Wetlands provide irreplaceable ecological services. They are among the most diverse and productive of all ecosystems. Many species spend at least part of their life cycles in freshwater or saltwater wetlands. © Fred Whitehead/Animals Animals/Earth Scenes.

Key Concepts

- Biomes are generalized communities controlled, at a regional scale, by precipitation and temperature.

- Plants and animals have characteristic adaptations suited to biome conditions.

- Marine and freshwater environments differ according to depth, which controls light, temperature, oxygen level, and pressure, as well as distance from shore.

- Wetlands are primary centers of terrestrial biodiversity, and they perform essential ecosystem services.

the center. Often this process is accelerated by increased sediment loads from urban development, farms, and roads. Wetland losses are one of the areas of greatest concern among biologists.

HUMAN DISTURBANCE

Humans have become dominant organisms over most of the earth, damaging or disturbing more than half of the world's terrestrial ecosystems to some extent. By some estimates, humans preempt about 40 percent of the net terrestrial primary productivity of the biosphere either by consuming it directly, by interfering with its production or use, or by altering the species composition or physical processes of human-dominated ecosystems. Conversion of natural habitat to human uses is the largest single cause of biodiversity losses.

Researchers from the environmental group Conservation International have attempted to map the extent of human disturbance of the natural world (fig. 5.24). The greatest impacts have been in Europe, parts of Asia, North and Central America, and islands such as Madagascar, New Zealand, Java, Sumatra, and those in the Caribbean. Data from this study are shown in table 5.1.

Temperate broad-leaved forests are the most completely human-dominated of any major biome. The climate and soils that support such forests are especially congenial for human occupation. In eastern North America or most of Europe, for example, only remnants of the original forest still persist. Regions with a Mediterranean climate generally are highly desired for human habitation. Because these landscapes also have high levels of biodiversity, conflicts between human preferences and biological values frequently occur (fig. 5.25).

Temperate grasslands, temperate rainforests, tropical dry forests, and many islands also have been highly disturbed by human activities. If you have traveled through the American corn-belt states such as Iowa or Illinois, you have seen how thoroughly former prairies have been converted to farmlands. Intensive cultivation of this land exposes the soil to erosion and fertility losses (see chapter 9). Islands, because of their isolation, often have high numbers of endemic species. Many islands, such as Madagascar, Haiti, and Java have lost more than 99 percent of their original land cover.

Tundra and Arctic deserts are the least disturbed biomes in the world. Harsh climates and unproductive soils make these areas unattractive places to live for most people. Temperate conifer forests also generally are lightly populated and large areas remain in a relatively natural state. However, recent expansion of forest harvesting in Canada and Siberia may threaten the integrity of this

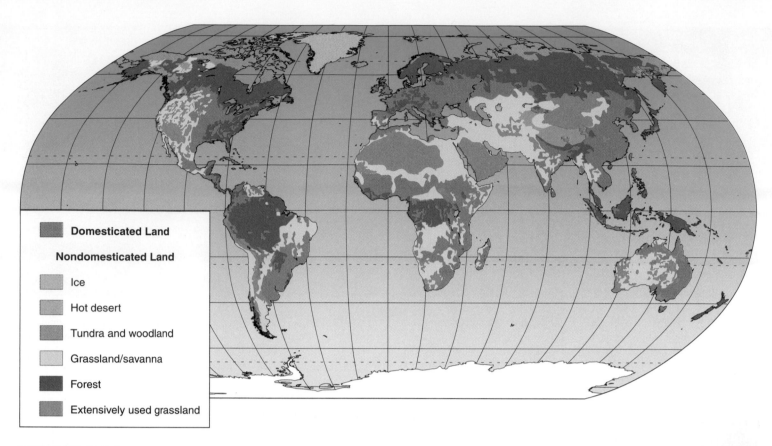

FIGURE 5.24 Domesticated land has replaced much of the earth's original land cover. *Source:* United Nations Environment Programme, *Global Environment Outlook, 1997.*

TABLE 5.1 Human Disturbance

BIOME	TOTAL AREA (10⁶ KM²)	% UNDISTURBED HABITAT	% HUMAN DOMINATED
Temperate broad-leaved forests	9.5	6.1	81.9
Chaparral and thorn scrub	6.6	6.4	67.8
Temperate grasslands	12.1	27.6	40.4
Temperate rainforests	4.2	33.0	46.1
Tropical dry forests	19.5	30.5	45.9
Mixed mountain systems	12.1	29.3	25.6
Mixed island systems	3.2	46.6	41.8
Cold deserts/semideserts	10.9	45.4	8.5
Warm deserts/semideserts	29.2	55.8	12.2
Moist tropical forests	11.8	63.2	24.9
Tropical grasslands	4.8	74.0	4.7
Temperate coniferous forests	18.8	81.7	11.8
Tundra and arctic desert	20.6	99.3	0.3

Note: *Where undisturbed and human-dominated areas do not add up to 100 percent, the difference represents partially disturbed lands.*

Source: *Hannah, Lee, et al., "Human Disturbance and Natural Habitat: A Biome Level Analysis of a Global Data Set," in* Biodiversity and Conservation, *1995, Vol. 4:128–155.*

biome. Large expanses of tropical moist forests still remain in the Amazon and Congo basins but in other areas of the tropics such as West Africa, Madagascar, Southeast Asia, and the Indo-Malaysian peninsula and archipelago, these forests are disappearing at a rapid rate (see chapter 12).

As mentioned earlier, wetlands have suffered severe losses in many parts of the world. About half of all original wetlands in the United States have been drained, filled, polluted, or otherwise degraded over the past 250 years. In the prairie states, small potholes and seasonally flooded marshes have been drained and converted to croplands on a wide scale. Iowa, for example, is estimated to have lost 99 percent of its presettlement wetlands (fig. 5.26). Similarly, California has lost 90 percent of the extensive marshes and deltas that once stretched across its central valley. Wooded swamps and floodplain forests in the southern United States have been widely disrupted by logging and conversion to farmland.

Similar wetland disturbances have occurred in other countries as well. In New Zealand, over 90 percent of natural wetlands have been destroyed since European settlement. In Portugal, some 70 percent of freshwater wetlands and 60 percent of estuarine habitats have been converted to agriculture and industrial areas. In Indonesia, almost all the mangrove swamps that once lined the coasts of Java have been destroyed, while in the Philippines and Thailand, more than two-thirds of coastal mangroves have been cut down for firewood or conversion to shrimp and fish ponds.

Slowing this destruction, or even reversing it, is a challenge that we will discuss in chapter 13.

FIGURE 5.25   "There was an environment before human beings and there will be an environment after human beings. So what's the big deal? © Peter Kohlsaat/Modern Times Syndicate, 1990.

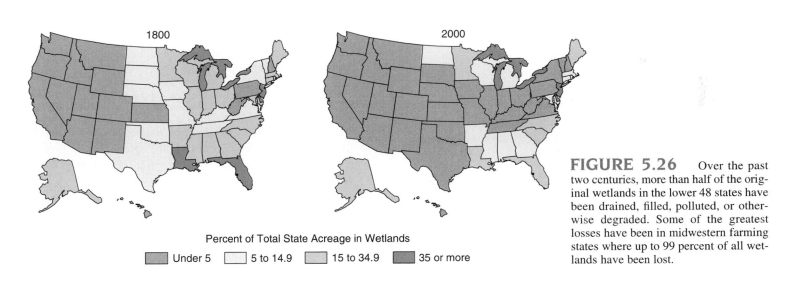

Percent of Total State Acreage in Wetlands

Under 5   5 to 14.9   15 to 34.9   35 or more

FIGURE 5.26   Over the past two centuries, more than half of the original wetlands in the lower 48 states have been drained, filled, polluted, or otherwise degraded. Some of the greatest losses have been in midwestern farming states where up to 99 percent of all wetlands have been lost.

Summary

- Biomes are major ecosystem types characterized by similar climates, soil conditions, and biological communities. Temperature and precipitation largely determine what kind of biome would occur in a place, in the absence of human disturbance.

- Understanding biome distribution is important in environmental science because resource availability and biological productivity differ greatly among biomes. Consequently, humans use biomes differently, and ecological resilience differs among biomes.

- Among the major terrestrial (land-based) biomes are deserts, tundra, grasslands, temperate deciduous forest, temperate coniferous forest, tropical moist forest, and tropical seasonal forest.

- Grasslands, shrublands, savannas, and thorn scrub are among the biomes occurring where temperatures are extreme and precipitation is too slight to support forests. Frequent fires and drought tend to keep trees from growing.

- Forests, consisting of trees and the complex, layered communities on the ground and in the soil, occur where there is enough moisture to support trees and tree seedlings. Tropical (low-latitude) rainforests and temperate (midlatitude) rainforests are extremely tall because of abundant rainfall. Tropical forests have relatively poor soils, so cutting them removes most organic matter from the forest.

- Grasslands have been lost through conversion to farmlands, overgrazing, and fire suppression. Forests have been lost through logging and conversion to agriculture.

- Marine environments can be described by depth and by proximity to shore. Shallow, light-rich environments are usually the most biologically productive because they include many photosynthetic organisms. Near-shore and onshore environments include intertidal zones, estuaries, coral reefs, and mangrove swamps.

- Marine environments play key roles in global systems because of photosynthetic activity and biomass production.

- Freshwater aquatic systems, including lakes, rivers, and wetlands, are major centers of biodiversity in terrestrial systems.

- Wetlands provide essential ecosystem services, including floodwater storage and water purification. Even so, wetlands continue to be converted in North America and worldwide.

- Human alteration of ecosystems has been most extensive in forests, grasslands, and wetlands. Many islands have lost nearly all their original habitat. Arctic environments are the least directly altered by human actions, although indirect effects are extensive.

Questions for Review

1. Who was Aldo Leopold and why is he considered important in the history of American conservation?

2. Throughout the central portion of North America is a large biome once dominated by grasses. Describe how physical conditions and other factors control this biome.

3. What is taiga and where is it found? Why might logging in taiga be more disruptive than in southern coniferous forests?

4. Why are tropical moist forests often less suited for agriculture and human occupation than tropical deciduous forests?

5. Find out the annual temperature and precipitation conditions where you live (fig. 5.2). Which biome type do you occupy?

6. Describe four different kinds of wetlands and explain why they are important sites of biodiversity and biological productivity.

7. Forests differ according to both temperature and precipitation. Name and describe a biome that occurs in (a) hot, (b) cold, (c) wet, and (d) dry climates (one biome for each climate).

8. How do physical conditions change with depth in marine environments?

9. Describe four different coastal ecosystems.

10. Define a landscape. Describe the major ecological and cultural features of the landscape in which you live.

Questions for Critical Thinking

1. What physical and biological factors are most important in shaping your biological community? How do the present characteristics of your area differ from those 100 or 1,000 years ago?

2. Forest biomes frequently undergo disturbances such as fire or flooding. As more of us build homes in these areas, what factors should we consider in deciding how to protect people from natural disturbances?

3. Often humans work to preserve biomes that are visually attractive. What biomes might be lost this way? Is this a problem?

4. What do you think Aldo Leopold meant by integrity, stability, and beauty of the land?

5. Disney World in Florida wants to expand onto a wetland. It has offered to buy and preserve a large nature preserve in a different area to make up for the wetland it is destroying. Is that reasonable? What conditions would make it reasonable or unreasonable?

6. Suppose further that the wetland being destroyed in question 5 and its replacement area both contain several endangered species (but different ones). How would you compare different species against each other? How many plant or insect species would one animal species be worth?

7. Historically, barrier islands have been hard to protect because links between them and inshore ecosystems are poorly recognized. What kinds of information would help a community distant from the coast commit to preserving a barrier island?

8. Real estate development is a main reason for both wetland losses and barrier island destruction. Whose responsibility is it to curtail these losses? Why?

Key Terms

barrier islands 102
benthic 100
biomes 94
bogs 104
boreal forest 97
chaparral 99
cloud forest 99
conifer 97
coral reefs 101
deciduous 98
deserts 95
estuaries 102
fens 104
grasslands 95
mangroves 101

marshes 104
pelagic 100
phytoplankton 100
reefs 101
southern pine forests 97
swamps 104
taiga 97
temperate rainforest 98
thermocline 103
thorn scrub 99
tropical rainforest 99
tropical seasonal forests 100
tundra 95
wetlands 104

Further Readings

Brown, James H., and Mark V. Lomolino. 1998, *Biogeography*. Sinauer Associates.

Cunningham, S., and J. Read. 2003. Comparison of temperate and tropical rainforest tree species: Growth responses to temperature. *Journal of Biogeography* 30(1):143–54.

MacDonald, Glenn. 2002. *Biogeography: Introduction to Space, Time, and Life*. Wiley.

Marcus C. Öhman, Mats Björk, Elisabeth Kessler, and Julius Francis. 2002. Bridging science and management in the Western Indian Ocean. *AMBIO: A Journal of the Human Environment* 31(7):501.

Schneider, Stephen H., and Terry Root, eds. 2001. *Wildlife Responses to Climate Change*. Island Press.

Soule, M. E., and G. H. Orians. 2001. *Conservation Biology: Research Priorities for the Next Decade*. Island Press.

Stephen V. Smith, et al. 2003. Humans, hydrology, and the distribution of inorganic nutrient loading to the ocean. *Bioscience* 53(5):235–46.

Stickney, R. R., ed. 2002. *Responsible Marine Aquaculture*. Oxford Univ. Press.

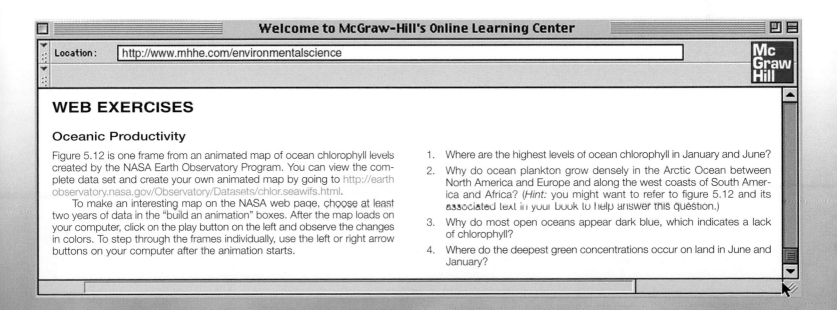

Welcome to McGraw-Hill's Online Learning Center

Location: http://www.mhhe.com/environmentalscience

WEB EXERCISES

Oceanic Productivity

Figure 5.12 is one frame from an animated map of ocean chlorophyll levels created by the NASA Earth Observatory Program. You can view the complete data set and create your own animated map by going to http://earthobservatory.nasa.gov/Observatory/Datasets/chlor.seawifs.html.

To make an interesting map on the NASA web page, choose at least two years of data in the "build an animation" boxes. After the map loads on your computer, click on the play button on the left and observe the changes in colors. To step through the frames individually, use the left or right arrow buttons on your computer after the animation starts.

1. Where are the highest levels of ocean chlorophyll in January and June?
2. Why do ocean plankton grow densely in the Arctic Ocean between North America and Europe and along the west coasts of South America and Africa? (*Hint*: you might want to refer to figure 5.12 and its associated text in your book to help answer this question.)
3. Why do most open oceans appear dark blue, which indicates a lack of chlorophyll?
4. Where do the deepest green concentrations occur on land in June and January?

6

Population Biology

Nature teaches more than she preaches.

John Burroughs

OBJECTIVES

After studying this chapter, you should be able to:

- appreciate the potential of exponential growth.
- draw a diagram of J and S curves and explain what they mean.
- describe environmental resistance and discuss how it can lead to logistic or stable growth.
- define *fecundity, fertility, birth rates, life expectancy, death rates,* and *survivorship.*
- compare and contrast density-dependent and density-independent population processes.
- explain the theory of island biogeography and what it says about populations of rare and endangered species in fragmented habitats.
- apply the concepts of genetic drift, founder effects, and demographic bottlenecks to minimum viable population sizes.
- illustrate how metapopulation analysis is important in conservation biology.

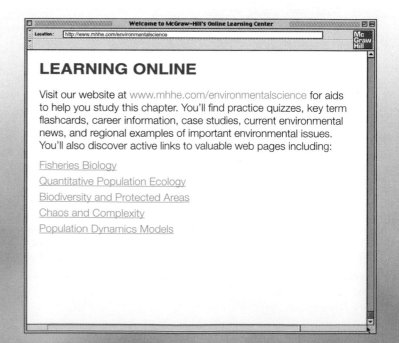

LEARNING ONLINE

Visit our website at www.mhhe.com/environmentalscience for aids to help you study this chapter. You'll find practice quizzes, key term flashcards, career information, case studies, current environmental news, and regional examples of important environmental issues. You'll also discover active links to valuable web pages including:

Fisheries Biology

Quantitative Population Ecology

Biodiversity and Protected Areas

Chaos and Complexity

Population Dynamics Models

Photo: Climate change threatens the survival of Emperor penguins in Antarctica. © Corbis/Royalty Free Website.

How Many Fish in the Sea?

When John Cabot discovered Newfoundland in 1497, cod (*Gadus morhua*) were so abundant that sailors claimed they could walk on the backs of the teeming fish. Growing up to 2 meters (6 feet) long, weighing as much as 100 kg (220 lbs) and living up to 25 years, cod have been a major food resource for Europeans for more than 500 years. Because the firm, white flesh of the cod has little fat, it can be salted and dried to produce a long-lasting food that can be stored or shipped to distant markets.

No one knows how many cod there may have once been in the ocean. Coastal people recognized centuries ago that huge schools would gather to spawn on shoals and rocky reefs from Massachusetts around the North Atlantic to the British Isles. In 1990, Canadian researchers watched on sonar as a school estimated to contain several hundred million fish spawned on the Gorges Bank off Newfoundland. Because a single mature female cod can lay up to 10 million eggs in a spawning, a school like this—only one of many in the ocean—might have produced a quadrillion eggs.

It seems that such an abundant and fecund an animal could never be threatened by humans. In 1883, Thomas Huxley, the eminent biologist and friend of Charles Darwin, said, "I believe that the cod fishery . . . and probably all the great sea fisheries are inexhaustible . . . Nothing we do seriously affects the number of fish." But in Huxley's time, most cod were caught on handlines by fishermen in small wooden dories. He couldn't have imagined the size and efficiency of modern fishing fleets. Following World War II, fishing boats grew larger, more powerful, and more numerous, while their fish-finding and harvesting technology grew tremendously more effective.

Modern trawlers now pull nets with mouths large enough to engulf a dozen jumbo jets at a time. Heavy metal doors, connected by a thick metal chain, hold the net down on the ocean floor, where it crushes bottom-dwelling organisms and reduces habitat to rubble. One oceanographer compared the process to harvesting forest mushrooms with a bulldozer. A single pass of the trawler not only can scoop up millions of fish, it leaves a devastated community that may take decades to repair.

It's difficult to know how many fish are in the ocean. We can't see them easily, and often we don't even know where they are. Our estimates of population size often are based on the harvest brought in by fishing boats. Biologists warn that many marine species are overfished and in danger of catastrophic population crashes. Fishermen claim that scientific models don't represent reality.

In 1991, the Canadian government predicted that Newfoundland fishermen would catch 350,000 metric tons of cod. The actual harvest that year was only 22,000 tons even though more boats were trying harder than ever before to catch fish. Other ground (bottom-dwelling) fish showed similar declines (fig. 6.1). Something was seriously amiss.

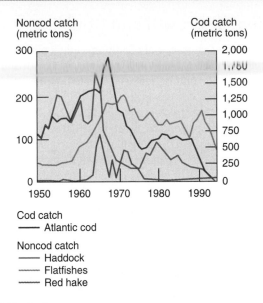

FIGURE 6.1 Commercial harvests in the Northwest Atlantic of some important ground (bottom) fish, 1950–1995. *Source:* World Resources Institute, 2000.

In an effort to save the few remaining fish, Canada banned most cod fishing within 320 km (200 mi) of its coastline. More than 40,000 people lost their jobs, and many fishing towns were decimated. Despite these limits, however, Canadian cod stocks have not recovered. In 2003, the Canadian government banned all cod fishing in the Gulf of St. Lawrence and in the Atlantic Ocean northeast of Newfoundland and Labrador. And after decades of debate about how many fish exist and how many can be safely caught, the European Union agreed on a similar fishing ban throughout the North Atlantic.

Whether cod populations will ever recover remains to be seen. It appears that overharvesting may have irreversibly disrupted marine ecosystems and food webs. On the Gorges Bank off the coast of Newfoundland, trillions of tiny tentacled organisms called hydroids now prey on both the organisms that once fed young cod as well as the juvenile cod themselves. Although hydroids have probably always been present, they once were held in check by adult fish. Now not enough fish survive to regulate hydroid populations. Is this a shift to a permanent new state, or just a temporary situation?

This case study illustrates some of the complexities and importance of population biology. How can we predict the impacts of human actions and environmental change on different kinds organisms? What are acceptable harvest limits and minimum viable population sizes? In this chapter, we'll look at some of the factors that effect population dynamics of biological organisms.

DYNAMICS OF POPULATION GROWTH

Why do some populations grow to enormous size while others do not? What causes abrupt population increases or decreases? How many individuals does it take to make a viable population? What roles do competition, genetic diversity, and ecological adaptations play in maintaining or reducing wild species? These are some of the questions asked by population biologists. In this chapter, we will survey the factors that lead to population growth and decline. We also will consider the mechanisms that regulate total population size, distribution, and age structure. One of the most important concepts in population biology is that both density-dependent and density-independent factors play roles in determining how big a particular population in a given habitat can be.

Exponential Growth and Doubling Times

Given optimum environmental conditions, many biological species have the capacity for **exponential growth,** or growth at a constant *rate* of increase per unit of time. It is called exponential because the rate of increase can be expressed as a constant *fraction,* or exponent, by which the existing population is *multiplied.* This pattern also is called **geometric growth** because the sequence of growth follows a geometric pattern of increase, such as 2, 4, 8, 16, and so on. By contrast, a pattern of growth that increases at a constant *amount* per unit of time is called **arithmetic growth.** The sequence in this case might be 1, 2, 3, 4 or 1, 3, 5, 7. Notice in these examples that a constant amount is *added* to the population (fig. 6.2).

Organisms are able to grow exponentially, given an appropriate environment, because of the amplifying power of biological reproduction. Most parents have the ability to produce multiple offspring over their lifetime. If even a few of those offspring survive to reproductive age, the population will grow.

As you can see in figure 6.2, the number of individuals added to a population at the beginning of a geometric growth curve is rather small. But within a very short time, the numbers begin to increase quickly because a given percentage becomes a much larger amount as the population gets bigger. The growth curve produced by this constant rate of unfettered growth is called a **J curve** because of its shape. At 1 percent per year, a population or a bank

TABLE 6.1	Doubling Times at Various Compound Interest Rates

ANNUAL % INCREASE	DOUBLING TIME (YEARS)
0.1	700
0.5	140
1.0	70
2.0	35
5.0	14
7.5	9
10.0	7
100.0	0.7

account doubles in roughly 70 years (table 6.1). A useful rule of thumb is that if you divide 70 by the annual percentage growth, you will get the approximate doubling time in years. As a result, a population growing at 35 percent doubles every two years. You can also apply this rule to calculate doubling time of human populations. Countries growing at 4 percent per year will *double* their populations in 17.5 years. A country growing at a rate of 0.1 percent annually will double its population in 700 years. We will use this rule to discuss human population growth in chapter 7.

Biotic Potential and Carry Capacity

Many species have amazingly high reproductive rates that give them the potential to produce enormous populations very quickly, given unlimited resources and freedom from limiting factors (fig. 6.3). We call the maximum reproductive rate of an organism its **biotic potential.** Table 6.2 shows the potential number of offspring that a single female housefly (*Musca domestica*) and her offspring could produce in a year. The result is astounding. Each female fly lays an average of 120 eggs in each generation. The eggs hatch and mature into sexually active adults and lay their own eggs in 56 days. In one year (seven generations), if all its offspring survived long enough to reproduce, a single female could be the ancestor of 5.6 *trillion* flies. If this rate of reproduction continued for ten years, the whole earth would be covered several meters deep with houseflies! Fortunately, this has not happened because of factors that limit the reproductive success of houseflies. This example, however, illustrates the potential for biological populations to increase rapidly.

Carrying capacity is defined as the maximum number of individuals of any species that can be supported by a particular ecosystem on a long-term basis. As environmental conditions in the ecosystem change, so, too, does the carrying capacity.

Population Oscillations and Irruptive Growth

In the real world there are limits to growth. When a population exceeds the carrying capacity of its environment or some other limiting factor comes into effect, death rates begin to surpass birth

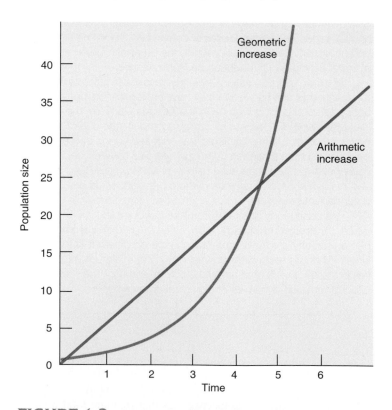

FIGURE 6.2 Arithmetic and geometric growth curves. Note that the geometric or exponential curve grows more slowly at first, but then accelerates past the arithmetic curve, which grows at a steady incremental pace throughout.

What do you think?

Too Many Deer?

A century ago, few Americans had ever seen a wild deer. Uncontrolled hunting and habitat destruction had reduced the deer population to about 500,000 animals nationwide. Some states had no deer at all. To protect the remaining deer, laws were passed in the 1920s and 1930s to restrict hunting, and the main deer predators—wolves and mountain lions—were exterminated throughout most of their former range.

As Americans have moved from rural areas to urban centers, forests have regrown, and deer populations have undergone explosive growth. Maturing at age two, a female deer can give birth to twin fauns every year for a decade or more. Increasing more than 20 percent annually, a deer population can double in just three years, an excellent example of irruptive, exponential growth.

Wildlife biologists estimate that the contiguous 48 states now have a population of more than 30 million white-tailed deer (*Odocoileus virginianus*), probably triple the number present in pre-Columbian times. Some areas have as many as 200 deer per square mile (77/km²). At this density, woodland plant diversity is generally reduced to a few species that deer won't eat. Most deer, in such conditions, suffer from malnourishment, and many die every year of disease and starvation. Other species are diminished as well. Many small mammals and ground-dwelling birds begin to disappear when deer populations reach 25 animals per square mile. At 50 deer per square mile, most ecosystems are seriously impoverished.

The social costs of large deer populations are high. In Pennsylvania alone, where deer numbers are now about 500 times greater than a century ago, deer destroy about $70 million worth of crops and $75 million worth of trees annually. Every year some 40,000 collisions with motor vehicles cause $80 million in property damage. Deer help spread Lyme disease, and, in many states, chronic wasting disease is found in wild deer herds. Some of the most heated criticisms of current deer management policies are in the suburbs. Deer love to browse on the flowers, young trees, and ornamental bushes in suburban yards. Heated disputes often arise between those who love to watch deer and their neighbors who want to exterminate them all.

In remote forest areas, many states have extended hunting seasons, increased the bag limit to four or more animals, and encouraged hunters to shoot does (females) as well as bucks (males). Some hunters criticize these changes because they believe that fewer deer will make it harder to hunt successfully and less likely that they'll find a trophy buck. Others, however, argue that a healthier herd and a more diverse ecosystem is better for all concerned.

In urban areas, increased sport hunting usually isn't acceptable. Wildlife biologists argue that the only practical way to reduce deer herds is culling by professional sharpshooters. Animal rights activists protest lethal control methods as cruel and inhumane. They call instead for fertility controls or trap and transfer programs. Birth control works in captive populations but is expensive and impractical with wild animals. Trapping, also, is expensive, and there's rarely anyplace willing to take surplus animals.

This case study shows that carrying capacity can be more complex than the maximum number of organisms an ecosystem can support. While it may be possible for 200 deer to survive in a square mile, there's an ecological carrying capacity lower

A white-tailed deer (Odocoileus virginianus).
© Corbis/Volume 72.

than that if we consider the other species dependent on that same habitat. There's also an ethical carrying capacity if we don't want to see animals suffer from malnutrition and starve to death every winter. And there's a cultural carrying capacity if we consider the tolerable rate of depredation on crops and lawns or an acceptable number of motor vehicle collisions.

If you were a wildlife biologist charged with managing the deer herd in your state, how would you reconcile the different interests in this issue? How would you define the optimum deer population, and what methods would you suggest to reach this level? What social or ecological indicators would you look for to gauge whether deer populations are excessive or have reached an appropriate level?

---

rates. The growth curve becomes negative rather than positive, and the population decreases as fast as, or faster than, it grew. We call this the population crash or **dieback.**

The extent to which a population exceeds the carrying capacity of its environment is called **overshoot,** and the severity of the dieback is generally related to the extent of the overshoot. This pattern of **population explosion** followed by a **population crash** is called **irruptive** or **Malthusian growth.** It is named after the eighteenth-century economist Thomas Malthus, who concluded that human populations tend to grow until they exhaust their resources and become subject to famine, disease, or war. Malthus and his theories are discussed further in chapter 7.

Populations may go through repeated oscillating cycles of exponential growth and catastrophic crashes, as shown in figure 6.4.

FIGURE 6.3 Reproduction gives many organisms the potential to expand populations explosively. The cockroaches in this kitchen could have been produced in only a few generations. A single female cockroach can produce up to 80 eggs every six months. This exhibit is in the Smithsonian Institute's National Museum of Natural History.
Courtesy National Museum of Natural History, Smithsonian Institution.

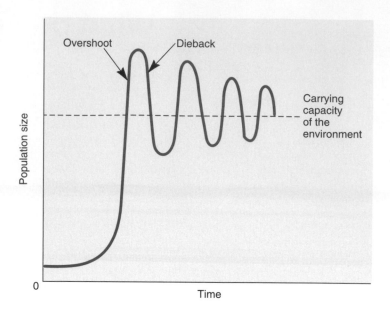

FIGURE 6.4 Population oscillations. Some species demonstrate a pattern of cyclic overshoot and dieback.

TABLE 6.2	Biotic Potential of Houseflies (*Musca domestica*) in One Year

Assume that a female lays 120 eggs per generation, half of the offspring are females, and all offspring live long enough to reproduce.

DAYS	TOTAL POPULATION
56	120
112	7,200
168	432,000
224	25,920,000
280	1,555,200,000
336	93,312,000,000
392	5,598,720,000,000

Source: E. J. Kormondy, Concepts of Ecology, 3d ed., © 1985 Harper & Row Publishers, Inc.

These cycles may be very regular if they depend on a few simple factors, such as the seasonal light- and temperature-dependent bloom of algae in a lake. They also may be very irregular if they depend on complex environmental and biotic relationships that control cycles, such as the population explosions of migratory locusts in the desert or tent caterpillars and spruce budworms in northern forests.

Growth to a Stable Population

Not all biological populations go through these cycles of irruptive population growth and catastrophic decline. The growth rates of many species are regulated by both internal and external factors so that they come into equilibrium with their environmental resources. These species may grow exponentially when resources are unlimited, but their growth slows as they approach the carrying capacity of the environment. This pattern is called **logistic growth,** a mathematical description of its constantly changing rate.

Together, factors that tend to reduce population growth rates are called **environmental resistance.** In later sections of this chapter, we will look in more detail at these factors and how they limit growth and regulate population size. First, we will see how logistic growth compares to the Malthusian growth and what kinds of organisms we are talking about in these different patterns.

How does the growth curve of a stable population differ from the J curve of an exploding population? Figure 6.5 shows an idealized comparison between biotic potential and sustainable growth. The J curve on the left in this figure represents the growth without restraint that we just discussed. It rises rapidly toward the maximum biotic potential of the species. The curve to the right represents logistic growth. We call this pattern an **S curve** because of its shape. It is also called a sigmoidal curve (for the Greek letter sigma). The area between these curves is the cumulative effect of environmental resistance. Note that the resistance becomes larger and the rate of logistic growth becomes smaller as the population approaches the carrying capacity of the environment.

Strategies of Population Growth

There appear to be evolutionary advantages, as well as disadvantages, in both Malthusian and logistic growth patterns. Although we should avoid implying intention in natural systems where the controlling forces may be entirely mechanistic, it sometimes helps us see these advantages in terms of "strategies" of adaptation and "logic" in different modes of reproduction.

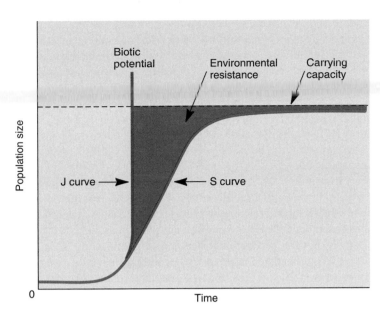

FIGURE 6.5 Idealized J and S population curves. The J curve represents theoretical unlimited growth. The S curve represents population growth and stabilization in response to environmental resistance.

TABLE 6.3 Characteristics of Contrasting Reproductive Strategies

EXTERNALLY CONTROLLED GROWTH	INTRINSICALLY CONTROLLED GROWTH
1. Short life	1. Long life
2. Rapid growth	2. Slower growth
3. Early maturity	3. Late maturity
4. Many small offspring	4. Fewer large offspring
5. Little parental care or protection	5. High parental care and protection
6. Little investment in individual offspring	6. High investment in individual offspring
7. Adapted to unstable environment	7. Adapted to stable environment
8. Pioneers, colonizers	8. Later stages of succession
9. Niche generalists	9. Niche specialists
10. Prey	10. Predators
11. Regulated mainly by extrinsic factors	11. Regulated mainly by intrinsic factors
12. Low trophic level	12. High trophic level

Malthusian "Strategies"

Organisms with Malthusian growth patterns often tend to occupy low trophic levels in their ecosystems (chapter 3), or to be pioneers in succession (chapter 4). As generalists or opportunists, they move quickly into disturbed environments, grow rapidly, mature early, and produce many offspring. They usually do little to care for their offspring or protect them from predation. They depend on sheer numbers and effective dispersal mechanisms to ensure that some offspring survive to adulthood (table 6.3). They have little investment in individual offspring, using their energy to produce vast numbers instead.

Many insects, rodents, marine invertebrates, parasites, and annual plants (especially the ones we consider weeds) follow this reproductive strategy. Their numbers generally are limited by predators or other controlling factors in the environment. They reproduce at the maximum rate possible to offset these losses. If the external factors that normally control their populations are inoperative, they tend to rapidly overshoot the carrying capacity of the environment and then die back catastrophically, as we have just seen. Among this group are the weeds, pests, or other species we consider nuisances that reproduce profusely, adapt quickly to environmental change, and survive under a broad range of conditions.

Logistic "Strategies"

While environmental resistance is a factor in controlling population growth in all species, those exhibiting logistic growth tend to grow more slowly and are more likely to be regulated by intrinsic characteristics than those with Malthusian patterns. These organisms are usually larger, live longer, mature more slowly, produce fewer offspring in each generation, and have fewer natural predators than

the species below them in the ecological hierarchy. Some typical examples of this strategy are wolves, elephants, whales, and primates. Each of these species provides more care and protection for its offspring than do "lower" organisms.

Elephants, for instance, are not reproductively mature until they are 18 to 20 years old. During youth and adolescence, a young elephant is part of a complex extended family that cares for it, protects it, and teaches it how to behave. A female elephant normally conceives only once every four or five years after she matures. The gestation period is about 18 months; thus, an elephant herd doesn't produce many babies in a given year. Since they have few enemies (except humans) and live a long life (often 60 or 70 years), however, this low reproductive rate produces enough elephants to keep the population stable, given appropriate environmental conditions.

An important underlying question to much of the discussion in this book is which of these strategies humans follow. Do we more closely resemble wolves and elephants in our population growth, or does our population growth pattern more closely resemble that of moose and rabbits? Will we overshoot the carrying capacity of our environment (or are we already doing so), or will our population growth come into balance with our resources?

FACTORS THAT INCREASE OR DECREASE POPULATIONS

Now that you have seen population dynamics in action, let's focus on what happens *within* populations, which are, after all, made up of individuals. In this section, we will discuss how new members are added to and old members removed from populations. We also will examine the composition of populations in terms of age classes and introduce terminology that will apply in subsequent chapters.

Natality, Fecundity, and Fertility

Natality is the production of new individuals by birth, hatching, germination, or cloning, and is the main source of addition to most biological populations. Natality is usually sensitive to environmental conditions so that successful reproduction is tied strongly to nutritional levels, climate, soil or water conditions, and—in some species—social interactions between members of the species. The maximum rate of reproduction under ideal conditions varies widely among organisms and is a species-specific characteristic. We already have mentioned, for instance, the differences in natality between several different species.

Fecundity is the physical ability to reproduce, while **fertility** is a measure of the actual number of offspring produced. Because of lack of opportunity to mate and successfully produce offspring, many fecund individuals may not contribute to population growth. Human fertility often is determined by personal choice of fecund individuals.

Immigration

Organisms are introduced into new ecosystems by a variety of methods. Seeds, spores, and small animals may be floated on winds or water currents over long distances. This is a major route of colonization for islands, mountain lakes, and other remote locations. Sometimes organisms are carried as hitchhikers in the fur, feathers, or intestines of animals traveling from one place to another. They also may ride on a raft of drifting vegetation. Some animals travel as adults—flying, swimming, or walking. In some ecosystems, a population is maintained only by a constant influx of immigrants. Salmon, for instance, must be important predators in some parts of the ocean, but their numbers are maintained only by recruitment from mountain streams thousands of kilometers away.

Mortality and Survivorship

An organism is born and eventually it dies; it is mortal. **Mortality,** or death rate, is determined by dividing the number of organisms that die in a certain time period by the number alive at the beginning of the period.

Since the number of survivors is more important to a population than is the number that died, mortality is often better expressed in terms of **survivorship** (the percentage of a cohort that survives to a certain age) or **life expectancy** (the probable number of years of survival for an individual of a given age). Table 6.4 shows U.S. life expectancies at different ages. Notice that for each year you survive, your life expectancy increases. If you were one year old in 2000, for example, you could expect to live, on average, 76.7 years more. But if you had already reached age 75 that year, rather than having only 1.7 more years to live, you could expect to survive another 11.2 years, on average. How could this be? At age 1,

many in your cohort were likely to die early for one reason or another. By age 75, most of those individuals are already dead, giving the rest of you a longer average life probability. Even at age 100, long past your starting life expectancy, you would still have 2.7 more years to live, on average.

Life expectancies in the United States rose dramatically during the twentieth century. In 1900, the average life expectancy was only 47.3 years. By 2003, this number had risen to 77.4 years. There are striking differences, however, between sexes, races, and economic classes (fig. 6.6). In 2000, white males and black females had essentially the same life expectancies (74.7 years). White females, by contrast, could expect to live, on average, to 79.9, while black males had a life expectancy of only 67.8 years. A number of factors underlie these differences including genetics, access to medical care, occupations, diet, and behavior.

Life span is the longest period of life reached by a given type of organism. The process of living entails wear and tear that eventually overwhelm every organism, but maximum age is dictated primarily by physiological aspects of the organism itself. There is an enormous difference in life span between different species. Some microorganisms live their whole life cycles in a matter of hours or minutes. Bristlecone pine trees in the mountains of California, on the other hand, have life spans up to 4,600 years. The maximum life span for humans appears to be about 120 years.

TABLE 6.4 Remaining Years of Life in the U.S. in 2000 (all races)

AGE	YEARS REMAINING
1	76.7
5	72.4
10	67.4
15	62.5
20	57.7
25	53.0
30	48.2
35	43.5
40	38.8
45	34.3
50	29.8
55	25.5
60	21.5
65	17.7
70	14.3
75	11.2
80	9.0
85	6.7
90	4.9
95	3.6
100	2.7

Source: *National Vital Statistics Report 50(3), March 2002.*

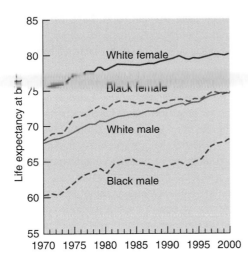

FIGURE 6.6 Life expectancy at birth by race and sex, 1970–2000.
Source: National Vital Statistics Reports 51(3):2002.

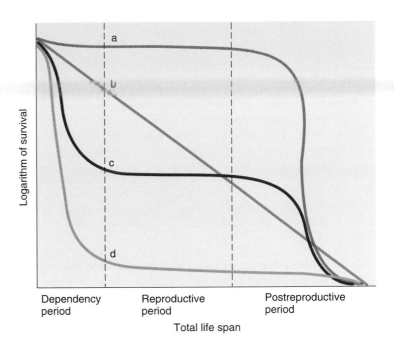

FIGURE 6.7 Four basic types of survivorship curves for organisms with different life histories. Curve (*a*) represents organisms such as humans or whales, which tend to live out the full physiological life span if they survive early growth. Curve (*b*) represents organisms such as sea gulls with a fairly constant mortality at all age levels. Curve (*c*) represents such organisms as white-tailed deer that have high mortality rates in early and late life. Curve (*d*) represents such organisms as clams and redwood trees with a high mortality rate early in life but live a full life if they reach adulthood.

Most organisms do not live anywhere near the maximum life span for their species. The major factors in early mortality are predation, parasitism, disease, accidents, fighting, and environmental influences, such as climate and nutrition. Important differences in relative longevity among different species are reflected in the survivorship curves shown in figure 6.7.

Four general patterns of survivorship can be seen in this idealized figure. Curve (*a*) is the pattern of organisms that tend to live their full physiological life span if they reach maturity and then have a high mortality rate when they reach old age. This pattern is typical of many large mammals, such as whales, bears, and elephants (when not hunted by humans), as well as humans in developed countries. Interestingly, some very small organisms, including predatory protozoa and rotifers (small, multicellular, freshwater animals), have similar survivorship curves even though their maximum life spans may be hundreds or thousands of times shorter than those of large mammals. In general, curve (*a*) is the pattern for top consumers in an ecosystem, although many annual plants have a similar survivorship pattern.

Curve (*b*) represents the survivorship pattern for organisms for which the probability of death is generally unrelated to age. Sea gulls, for instance, die from accidents, poisoning, and other factors that act more or less randomly. Their mortality rate is generally constant with age, and their survivorship curve is a straight line.

Curve (*c*) is characteristic of many songbirds, rabbits, members of the deer family, and humans in less-developed countries (see chapter 7). They have a high mortality early in life when they are more susceptible to external factors, such as predation, disease, starvation, or accidents. Adults in the reproductive phase have a high level of survival. Once past reproductive age, they become susceptible again to external factors and the number of survivors falls quite rapidly.

Curve (*d*) is typical of organisms at the base of a food chain or those especially susceptible to mortality early in life. Many tree species, fish, clams, crabs, and other invertebrate species produce

a very large number of highly vulnerable offspring, few of which survive to maturity. Those individuals that do survive to adulthood, however, have a very high chance of living most of the maximum life span for the species.

Emigration

Emigration, the movement of members out of a population, is the second major factor that reduces population size. The dispersal factors that allow organisms to migrate into new areas are important in removing surplus members from the source population. Emigration can even help protect a species. For instance, if the original population is destroyed by some catastrophic change in their environment, their genes still are carried by descendants in other places. Many organisms have very specific mechanisms to facilitate migration of one or more of each generation of offspring.

FACTORS THAT REGULATE POPULATION GROWTH

So far, we have seen that differing patterns of natality, mortality, life span, and longevity can produce quite different rates of population growth. The patterns of survivorship and age structure created by these interacting factors not only show us how a population is growing but also can indicate what general role that species plays in its ecosystem. They also reveal a good deal about how that species is likely to respond to disasters or resource bonanzas in its environment. But what factors *regulate* natality, mortality, and the other components of population growth? In this section, we will look at some of the mechanisms that determine how a population grows.

Various factors regulate population growth, primarily by affecting natality or mortality, and can be classified in different ways. They can be *intrinsic* (operating within individual organisms or between organisms in the same species) or *extrinsic* (imposed from outside the population). Factors can also be either **biotic** (caused by living organisms) or **abiotic** (caused by nonliving components of the environment). Finally, the regulatory factors can act in a *density-dependent* manner (effects are stronger or a higher proportion of the population is affected as population density increases) or *density-independent* manner (the effect is the same or a constant proportion of the population is affected regardless of population density).

In general, biotic regulatory factors tend to be density-dependent, while abiotic factors tend to be density-independent. There has been much discussion about which of these factors is most important in regulating population dynamics. In fact, it probably depends on the particular species involved, its tolerance levels, the stage of growth and development of the organisms involved, the specific ecosystem in which they live, and the way combinations of factors interact. In most cases, density-dependent and density-independent factors probably exert simultaneous influences. Depending on whether regulatory factors are regular and predictable or irregular and unpredictable, species will develop different strategies for coping with them.

Density-Independent Factors

In general, the factors that affect natality or mortality independently of population density tend to be abiotic components of the ecosystem. Often weather (conditions at a particular time) or climate (average weather conditions over a longer period) are among the most important of these factors. Extreme cold or even moderate cold at the wrong time of year, high heat, drought, excess rain, severe storms, and geologic hazards—such as volcanic eruptions, landslides, and floods—can have devastating impacts on particular populations.

Abiotic factors can have beneficial effects as well, as anyone who has seen the desert bloom after a rainfall can attest. Fire is a powerful shaper of many biomes. Grasslands, savannas, and some montane and boreal forests often are dominated—even created—by periodic fires. Some species, such as jack pine and Kirtland's warblers, are so adapted to periodic disturbances in the environment that they cannot survive without them.

In a sense, these density-independent factors don't really regulate population *per se,* since regulation implies a homeostatic feedback that increases or decreases as density fluctuates. By definition, these factors operate without regard to the number of organisms involved. They may have such a strong impact on a population, however, that they completely overwhelm the influence of any other factor and determine how many individuals make up a particular population at any given time.

Density-Dependent Factors

Density-dependent mechanisms tend to reduce population size by decreasing natality or increasing mortality as the population size increases. Most of them are the results of interactions *between* populations of a community (especially predation), but some of them are based on interactions *within* a population.

Interspecific Interactions

As we discussed in chapter 4, a predator feeds on—and usually kills—its prey species. While the relationship is one-sided with respect to a particular pair of organisms, the prey species as a whole may benefit from the predation. For instance, the moose that gets eaten by wolves doesn't benefit individually, but the moose *population* is strengthened because the wolves tend to kill old or sick members of the herd. Their predation helps prevent population overshoot, so the remaining moose are stronger and healthier.

Sometimes predator and prey populations oscillate in a sort of synchrony with each other as is shown in figure 6.8, which shows the number of furs brought into Hudson Bay Company trading posts in Canada between 1840 and 1930. As you can see, the numbers of Canada lynx fluctuate on about a ten-year cycle that is similar to, but slightly out of phase with, the population peaks of snowshoe hares. Although there are some doubts now about how and where these data were collected, this remains a classic example of population dynamics. When prey populations (hares) are abundant, predators (lynx) reproduce more successfully and their population grows. When hare populations crash, so do the lynx. This predator-prey oscillation is known as the Lotka-Volterra model after the scientists who first described it mathematically.

Not all interspecific interactions are harmful to one of the species involved. Mutualism and commensalism, for instance, are interspecific interactions that are beneficial or neutral in terms of population growth (chapter 4).

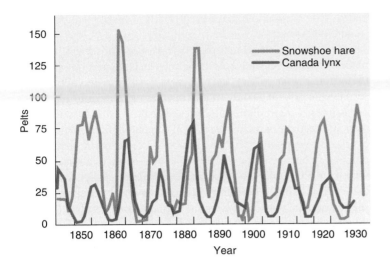

FIGURE 6.8 Oscillations in the populations of snowshoe hare and lynx in Canada suggest the close interdependency of this prey-predator relationship. These data are based on the number of pelts received by the Hudson Bay Company. Both predator and prey show about a ten-year cycle in population growth and decline.

Source: D. A. MacLulich, *Fluctuations in the Numbers of the Varying Hare (Lepus americus).* Toronto: University of Toronto Press, 1937, reprinted 1974.

FIGURE 6.9 Animals often battle over resources. This conflict can induce stress and affect reproductive success. © Art Wolfe/Stone/Getty Images.

Intraspecific Interactions

Individuals within a population also compete for resources. When population density is low, resources are likely to be plentiful and the population growth rate will approach the maximum possible for the species, assuming that individuals are not so dispersed that they cannot find mates. As population density approaches the carrying capacity of the environment, however, one or more of the vital resources becomes limiting. The stronger, quicker, more aggressive, more clever, or luckier members get a larger share, while others get less and then are unable to reproduce successfully or survive.

Territoriality is one principal way many animal species control access to environmental resources. The individual, pair, or group that holds the territory will drive off rivals if possible, either by threats, displays of superior features (colors, size, dancing ability), or fighting equipment (teeth, claws, horns, antlers). Members of the opposite sex are attracted to individuals that are able to seize and defend the largest share of the resources. From a selective point of view, these successful individuals presumably represent superior members of the population and the ones best able to produce offspring that will survive.

Stress and Crowding

Stress and crowding also are density-dependent population control factors. When population densities get very high, organisms often exhibit symptoms of what is called stress shock or **stress-related diseases.** These terms describe a loose set of physical, psychological, and/or behavioral changes that are thought to result from the stress of too much competition and too close proximity to other members of the same species. There is a considerable con-

TABLE 6.5	The Influence of Density on Fecundity in the House Mouse (*Mus musculus*)			
	SPARSE	MEDIUM	DENSE	VERY DENSE
Average number/m³	34	118	350	1,600
Average percentage pregnant	58.3	49.4	51.0	43.4
Average number per litter	6.2	5.7	5.6	5.1

Source: C. Southwick, "Population Characteristics of House Mice Living in English Corn Ricks: Density Relationships," Proc. Zool. Soc. London, 131:163–175, 1958, The Zoological Society of London.

troversy about what causes such changes and how important they are in regulating natural populations. The strange behavior and high mortality of arctic lemmings or hares during periods of high population density may be a manifestation of stress shock (fig. 6.9). On the other hand, they could simply be the result of malnutrition, infectious disease, or some other more mundane mechanism at work.

Some of the best evidence for the existence of stress-related disease comes from experiments in which laboratory animals, usually rats or mice, are grown in very high densities with plenty of food and water but very little living space (table 6.5). A variety of symptoms are reported, including reduced fertility, low resistance to infectious diseases, and pathological behavior. Dominant animals seem to be affected least by crowding, while subordinate animals—the ones presumably subjected to the most stress in intraspecific interactions—seem to be the most severely affected.

CONSERVATION BIOLOGY

Small, isolated populations can undergo catastrophic declines due to environmental change, genetic problems, or stochastic (random or unpredictable) events. A critical question in conservation biology is the minimum population size of a rare and endangered species required for long-term viability. In this section, we'll look at some factors that limit species and genetic diversity. We'll also consider the interaction of collections of subpopulations of species in fragmented habitats.

Island Biogeography

In a classic 1967 study, R. H. MacArthur and E. O. Wilson asked why it is that small islands far from the mainland generally have far fewer species than larger or nearer islands. Their theory of **island biogeography** explains that diversity in isolated habitats is a balance between colonization and extinction rates. An island far from a population source has a lower colonization rate for terrestrial species because it is harder for organisms to reach (fig. 6.10). At the same time, the limited habitat on a small island can support fewer individuals of any given species. This creates a greater probability that a species could go extinct due to natural disasters, diseases, or demographic factors such as imbalance between sexes in a particular generation. Larger islands, or those closer to the mainland, on the other hand, are more likely to be colonized or to retain those species already successfully established. Thus they tend to have greater diversity than smaller, more remote places.

Island biogeographical effects have been observed in many places. Cuba, for instance, is 100 times as large and has about 10 times as many amphibian species as its Caribbean neighbor, Monserrat. Similarly, in a study of bird species on the California Channel Islands, Jared Diamond observed that on islands with fewer than 10 breeding pairs, 39 percent of the populations went extinct over an 80-year period. In contrast, only 10 percent of populations numbering between 10 and 100 pairs went extinct, and no species with over 1,000 pairs disappeared over this time (fig. 6.11). This theory of a balance between colonization and extinction is now seen to explain species dynamics in many small, isolated habitat fragments whether on islands or not.

Conservation Genetics

Genetics plays an important role in the survival or extinction of small, isolated populations. In large populations, genetic variation tends to persist in what is called a Hardy-Weinberg equilibrium, named after the scientists who first described why this occurs. If mating is random, no mutations (changes in genetic material) occur, and there is no gene in-flow or selective pressure for or against particular traits, random distribution of gene types (alleles) will occur during gamete formation and sexual reproduction. That is, different alleles will be distributed in the offspring in the same ratio they occur in the parents, and genetic diversity is preserved.

In a large population, these conditions for maintaining genetic equilibrium are generally operative. The addition or loss of a few individuals or appearance of new genotypes makes little difference in the total gene pool, and genetic diversity is relatively constant. In small, isolated populations, however, immigration, mortality, mutations, or chance mating events involving only a few individuals can greatly alter the genetic makeup of the whole pop-

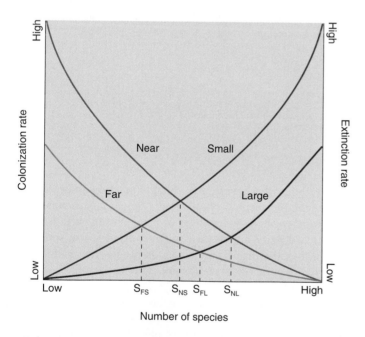

FIGURE 6.10   Predicted species richness on an island resulting from a balance between colonization (immigration) and extinction by natural causes. This island biogeography theory of MacArthur and Wilson (1967) is used to explain why large islands near a mainland (S_{NL}) tend to have more species than small, far islands (S_{FS}). *Source:* Based on MacArthur and Wilson, *The Theory of Island Biogeography,* 1967, Princeton University Press.

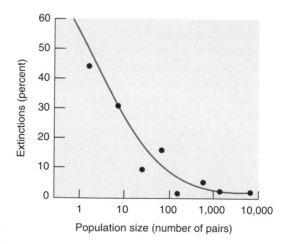

FIGURE 6.11   Extinction rates of bird species on the California Channel Islands as a function of population size over 80 years. *Source:* H. L. Jones and J. Diamond, "Short-term-base Studies of Turnover in Breeding Bird Populations on the California Coast Island," in *Condor,* vol. 78: 526–549, 1976.

ulation. We call the gradual changes in gene frequencies due to random events **genetic drift.**

For many species, loss of genetic diversity causes a number of harmful effects that limit adaptability, reproduction, and species survival. A **founder effect** or **demographic bottleneck** occurs when just a few members of a species survive a catastrophic event or colonize new habitat geographically isolated from other members of the same species. Any deleterious genes present in the founders will be overrepresented in subsequent generations (fig. 6.12). Inbreeding, mating of closely related individuals, also makes expression of rare or recessive genes more likely.

Some species seem not to be harmed by inbreeding or lack of genetic diversity. The northern elephant seal, for example, was reduced by overharvesting a century ago, to fewer than 100 individuals. Today there are more than 150,000 of these enormous animals along the Pacific coast of Mexico and California. No marine mammal is known to have come closer to extinction and then made such a remarkable recovery. All northern elephant seals today appear to be essentially genetically identical and yet they seem to have no apparent problems. Although interpretations of their situation are controversial, in highly selected populations, where only the most fit individuals reproduce, or in which there are few deleterious genes, inbreeding and a high degree of genetic identity may not be such a negative factor.

Cheetahs, also appear to have undergone a demographic bottleneck sometime in the not-too-distant past. All the male cheetahs throughout the world appear to be nearly genetically identical, suggesting that they all share a single male ancestor (fig. 6.13). This lack of diversity is thought to be responsible for an extremely low fertility rate, a high abundance of abnormal sperm, and low survival rate for offspring, all of which threatens the survival of the species.

Population Viability Analysis

Conservation biologists use the concepts of island biogeography, genetic drift, and founder effects to determine **minimum viable population size,** or number of individuals needed for long-term survival of rare and endangered species. A classic example is that of the grizzly bear (*Ursus arctos horribilis*) in North America. Before European settlement, grizzlies roamed from the Great Plains west to California and north to Alaska. Hunting and habitat destruction reduced the number of grizzlies from an estimated 100,000 in 1800 to less than 1,000 animals in six separate subpopulations that now occupy less than 1 percent of the historic range. Recovery target sizes—based on estimated environmental carrying capacities—call for fewer than 100 bears for some subpopulations. Conservation genetics predicts that a completely isolated population of 100 bears cannot be maintained for more than a few generations. Even the 200 bears now in Yellowstone National Park will be susceptible to genetic problems if completely isolated. Interestingly, computer models suggest that translocating only two unrelated bears into small populations every generation (about ten years) could greatly increase population viability.

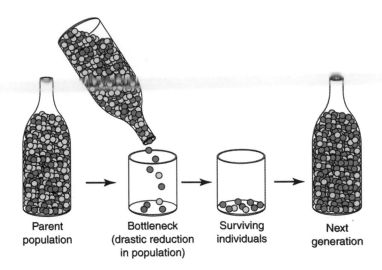

FIGURE 6.12   Genetic drift: the bottleneck effect. The parent population contains roughly equal numbers of blue and yellow individuals. By chance, the few remaining individuals that comprise the next generation are mostly blue. The bottleneck occurs because so few individuals form the next generation, as might happen after an epidemic or catastrophic storm.

FIGURE 6.13   Sometime in the past, cheetahs underwent a severe population crash. Now all male cheetahs in the world are nearly genetically identical, and deformed sperm, low fertility levels, and low infant survival are common in the species. © The McGraw-Hill Companies, Inc./ Barry Barker, photographer.

Metapopulations

For mobile organisms, separated populations can have gene exchange if suitable corridors or migration routes exist. A **metapopulation** is a collection of populations that have regular or intermittent gene flow between geographically separate units (fig. 6.14). For example, the Bay checkerspot butterfly (*Euphydrays editha bayensis*) in California exists in several distinct habitat patches. Individuals occasionally move among these patches, mating with existing animals or recolonizing empty habitats. Thus, the apparently separate groups form a functional metapopulation.

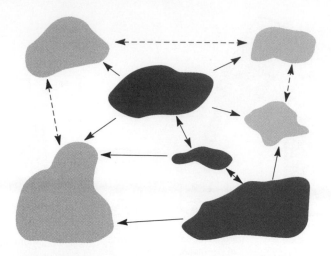

FIGURE 6.14 A metapopulation is composed of several local populations linked by regular (*solid arrows*) or occasional (*dashed lines*) gene flows. Source populations (*dark*) provide excess individuals, which emigrate to and colonize sink habitats (*light*).

A "source" habitat, where birth rates are higher than death rates, produces surplus individuals that can migrate to new locations within a metapopulation. "Sink" habitats, on the other hand, are places where mortality exceeds birth rates. Sinks may be spatially larger than sources but because of unfavorable conditions, the species would disappear in the sink habitat if it were not periodically replenished from a source population. In general, the larger a reserve is, the better it is for endangered species. Sometimes, however, adding to a reserve can be negative if the extra area is largely sink habitat. Individuals dispersing within the reserve may settle in unproductive areas if better habitat is hard to find. Recent studies using a metapopulation model for spotted owls predict just such a problem for this species in the Pacific Northwest.

Some conservation biologists argue that we ought to try to save every geographically distinct population or "evolutionarily significant unit (ESU)" possible in order to preserve maximum genetic diversity. Paul Ehrlich and Gretchen Daily estimate there may be an average of 220 ESU for every one of the 10 to 50 billion species in the world. Saving all of them would be a gargantuan task.

Summary

- Population biology plays an important role in determining how ecosystems work and how we can sustainably harvest wild species like cod or deer.

- Given optimum conditions, populations of many organisms can grow exponentially; that is, they can expand at a constant *rate* per unit of time.

- We describe the rapidly rising curve of exponential growth as a J curve.

- Without some external limits, exponential growth will cause a population to overshoot the carrying capacity of its environment and then crash catastrophically. "Weedy" organisms, or those at the lower trophic levels of a food web often exhibit extrinsically regulated growth.

- Top predators, such as tigers or lions, generally have intrinsic factors that allow them to grow to a stable population size at or near the carrying capacity of their ecosystem.

- The most important components of population dynamics are natality, fertility, fecundity, life span, longevity, survivorship, immigration, and emigration.

- The most important biological factors in population dynamics generally are competition, predation, and disease, all of which are usually density dependent. The most important abiotic factors in population dynamics generally are weather and climate, both of which are usually density independent.

- Island biogeography proposes that small islands or habitat fragments that are far from a colonization source will have little biodiversity, while larger islands or habitats closer to a colonization source will generally have more biodiversity.

- Small, isolated populations often are subject to genetic drift, that is, a gradual change in gene frequencies and composition.

- A population arising from only a few founder individuals can undergo a demographic bottleneck in which genetic diversity is reduced and reproductive or developmental problems can arise.

- A metapopulation is a network of populations having regular or intermittent gene flow between geographically separate units. Source habitats provide surplus individuals to replenish populations in sink habitats where mortality exceeds natality.

Questions for Review

1. What caused the sudden collapse of Atlantic cod populations?

2. Explain exponential and arithmetic growth.

3. Given a growth rate of 3 percent per year, how long will it take for a population of 100,000 individuals to double? How long will it take to double when the population reaches 10 million?

4. What is environmental resistance? How does it affect populations?

5. What is the difference between fertility and fecundity?

6. Describe the four major types of survivorship patterns and explain what they show about the role of the species in an ecosystem.

7. What are the main interspecific population regulatory interactions? How do they work?

8. What is island biogeography and why is it important in conservation biology?

9. Why does genetic diversity tend to persist in large populations, but gradually drift or shift in small populations?

10. Draw a diagram showing gene flow between source and sink habitat in a metapopulation. Explain.

Questions for Critical Thinking

1. Compare the advantages and disadvantages to a species that result from exponential or logistic growth. Why do you think hares have evolved to reproduce as rapidly as possible, while lynx appear to have intrinsic or social growth limits?

2. Are humans subject to environmental resistance in the same sense that other organisms are? How would you decide whether a particular factor that limits human population growth is ecological or social?

3. Species differ greatly in birth and death rates, survivorship, and life spans. There must be advantages and disadvantages in living longer or reproducing more quickly. Why hasn't evolution selected for the most advantageous combination of characteristics so that all organisms would be more or less alike?

4. Abiotic factors that influence population growth tend to be density-independent, while biotic factors that regulate population growth tend to be density-dependent. Explain.

5. Some people consider stress and crowding studies of laboratory animals highly applicable in understanding human behavior. Other people question the cross-species transferability of these results. What considerations would be important in interpreting these experiments?

6. What implications (if any) for human population control might we draw from our knowledge of basic biological population dynamics?

7. If you were a wildlife manager charged with regulating an urban deer herd, how would you bring hunters, animal rights activists, home owners, and deer feeders together to reach a management strategy?

Key Terms

abiotic 118
arithmetic growth 112
biotic 118
biotic potential 112
carrying capacity 112
demographic bottleneck 121
dieback 113
emigration 117
environmental resistance 114
exponential growth 112
fecundity 116
fertility 116
founder effect 121
genetic drift 121
geometric growth 112
irruptive or Malthusian growth 113

island biogeography 120
J curve 112
life expectancy 116
life span 116
logistic growth 114
metapopulation 121
minimum viable population size 121
mortality 116
natality 116
overshoot 113
population crash 113
population explosion 113
S curve 114
stress-related diseases 119
survivorship 116

Further Readings

Case, Ted J., Martin L. Cody, and Exequiel Ezcurra, eds. 2002. *Island Biogeography in the Sea of Cortez II.* Oxford Univ. Press.

Dobbs, David. 2000. *The Great Gulf: Fishermen, Scientists, and the Struggle to Revive the World's Greatest Fishery.* Island Press.

Hanski, Iikka. 1999. *Metapopulation Ecology.* Oxford Univ. Press.

Kurlansky, Mark. 1997. *Cod: A Biography of the Fish That Changed the World.* Walker & Co.

MacArthur, Robert H., and Edward O. Wilson. 2001. *The Theory of Island Biogeography.* Princeton Univ. Press.

Meffe, G. K., and C. R. Carroll. 1997. *Principles of Conservation Biology,* 2nd ed. Macmillan.

Safina, Carl. 1995. The world's imperiled fish. *Scientific American* 273(5):46–53.

Williams, Ted. 2003. Wanted: More hunters. *Audubon* 104(2):43–51.

Welcome to McGraw-Hill's Online Learning Center

Location: http://www.mhhe.com/environmentalscience

Mc Graw Hill

WEB EXERCISES

Exponential and Logistic Growth

Find the Excel file Population_Growth.xls in the OLC, chapter 6, under Web Exercises. This file models and graphs growth since 1950, based on actual data, exponential growth, and logistic growth. Review these growth patterns in your text, then experiment with growth rates as follows:

1. Raise the growth rate (r), then lower it, and watch how the curve changes. At 4 percent (comparable to the fastest growing countries), how many years does it take for the 1950 population to double? How many years at 3.5 percent? 0.5 percent? What rate brings the population closest to the projected curve for 2050?

2. Click on the Census Data tab (bottom of sheet). What is the range of actual rates? How have they changed over time?

3. Now look at the logistic curve. The key difference here is that growth rate decreases as the population approaches K. Set r to 1.4. Set K to 100. Do the logistic and exponential curves differ much? Reduce K to 80, then to 60, 40, and 20. When does the logistic curve fall below the projected curve? As implemented here, is the carrying capacity an abrupt population ceiling, or is it a density-dependent growth control?

4. Enter what you think would be a reasonable carrying capacity. What number have you entered? What population does this produce in 2050? How much does your restricted population vary from the U.S. Census' projection? What does K mean for humans, as compared to other populations?

7

Human Populations

Every child comes with the message that God is not yet tired of the man.

—Rabindranath Tagore—

OBJECTIVES

After studying this chapter, you should be able to:

- trace the history of human population growth.
- summarize Malthusian and Marxian theories of limits to growth as well as why technological optimists and supporters of social justice oppose these theories.
- explain the process of demographic transition and why it produces a temporary population surge.
- understand how changes in life expectancy, infant mortality, women's literacy, standards of living, and democracy affect population changes.
- evaluate pressures for and against family planning in traditional and modern societies.
- compare modern birth control methods and prepare a personal family planning agenda.

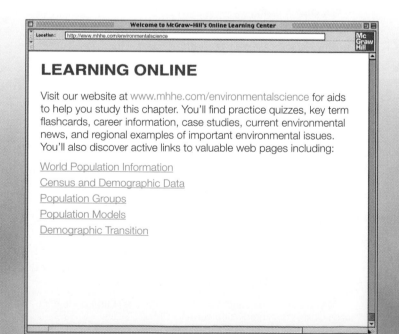

LEARNING ONLINE

Visit our website at www.mhhe.com/environmentalscience for aids to help you study this chapter. You'll find practice quizzes, key term flashcards, career information, case studies, current environmental news, and regional examples of important environmental issues. You'll also discover active links to valuable web pages including:

World Population Information
Census and Demographic Data
Population Groups
Population Models
Demographic Transition

Photo: More than a billion people now live in India. © The McGraw-Hill Companies, Inc./Barry Barker, photographer.

A Billion People and Growing

In 1999, having added more than 180 million people in just a decade, India reached a population of 1 billion humans. If current growth rates persist, India will have 1.63 billion residents in 2050, and will surpass China as the world's most populous country. How will the country, which already has more than a quarter of its population living in abject poverty, feed, house, educate, and employ all those being added each year? And what's the best way to slow this rapid growth? The fierce debate now taking place about how to control India's population has ramifications for the rest of the world as well.

On one side of this issue are those who believe that the best way to reduce the number of children born is poverty eradication and progress for women. Drawing on social justice principles established at the 1994 UN Conference on Population and Development in Cairo, some argue that responsible economic development, a broad-based social welfare system, education and empowerment of women, and high-quality health care—including family planning services—are essential components of population control. Without progress in these areas, they believe, efforts to provide contraceptives or encourage sterilization are futile.

On the other side of this debate are those who contend that, while social progress is an admirable goal, India doesn't have the time or resources to wait for an indirect approach to population control. The government must push aggressively, they argue, to reduce births now or the population will be so huge and its use of resources so great that only mass starvation, class war, crime, and disease will be able to bring it down to a manageable size.

Unable to reach a consensus on population policy, the Indian government decided in 2000 to let each state approach the problem in its own way. Some states have chosen to focus on social justice, while others have adopted more direct, interventionist policies.

The model for the social justice approach is the southern state of Kerala, which achieved population stabilization in the mid-1980s, the first Indian state to do so. Although still one of the poorest places in the world, economically, Kerala's fertility rate is comparable to that of many industrialized nations, including the United States. Both women and men have a nearly 100 percent literacy rate and share affordable and accessible health care, family planning, and educa-tional opportunities; therefore, women have only the number of children they want, usually two. The Kerala experience suggests that increased wealth isn't a prerequisite for zero population growth.

Taking a far different path to birth reduction is the nearby state of Andra Pradesh, which reached a stable growth rate in 2001. Boasting the most dramatic fertility decline of any large Indian state, Andra Pradesh has focused on targeted, strongly enforced sterilization programs. The poor are encouraged—some would say, compelled—to be sterilized after having only one or two children. The incentives include cash payments. You might receive 500 rupees—equivalent to $11 (U.S.) or a month's wages for an illiterate farm worker—if you agree to have "the operation." In addition, participants are eligible for better housing, land, wells, and subsidized loans.

The pressure to be sterilized is overwhelmingly directed at women, for whom the procedure is major abdominal surgery. Sterilizations often are done by animal husbandry staff and carried out in government sterilization camps. This practice raises troubling memories of the 1970s for many people, when then-Prime Minister Indira Gandhi suspended democracy and instituted a program of forced sterilization of poor people. There were reports at the time of people being rounded up like livestock and castrated or neutered against their will.

While many feminists and academics regard Andra Pradesh's policies as appallingly intrusive and coercive to women and the poor, the state has successfully reduced population growth. By contrast, the hugely populous northern states of Uttar Pradesh and Bihar have seen slightly increased growth rates over the past two decades to a current rate above 2.5 percent per year. How will they slow this exponential growth, and what might be the social and environmental costs of not doing so?

India's population problems introduce several of the important themes of this chapter. What are the trends in human populations around the world, what do those trends mean for our resources and environment, and what is the best way to approach population planning? Keep in mind, as you read this chapter, that resource limits aren't simply a matter of total number of people, they also depend on consumption levels and the types of technology used to produce the things we use and consume.

POPULATION GROWTH

Every second, on average, four or five children are born somewhere on the earth. In that same second, two other people die. This difference between births and deaths means a net gain of roughly 2.5 more humans per second in the world population. This means we are adding 9,000 per hour, 217,000 per day, or about 79 million more people per year. By 2003, the world population stood at about 6.3 billion, making us the most numerous vertebrate species on the planet. For the families to whom these children are born, this may well be a joyous and long-awaited event (fig. 7.1). But is a continuing increase in humans good for the planet in the long run?

Many people worry that overpopulation will cause—or perhaps already is causing—resource depletion and environmental degradation that threaten the ecological life-support systems on which we all depend. These fears often lead to demands for imme-diate, worldwide birth control programs to reduce fertility rates and to eventually stabilize or even shrink the total number of humans.

Others believe that human ingenuity, technology, and enterprise can extend the world carrying capacity and allow us to overcome any problems we encounter. From this perspective, more people may be beneficial rather than disastrous. A larger population means a larger workforce, more geniuses, more ideas about what to do. Along with every new mouth comes a pair of hands. Proponents of this worldview—many of whom happen to be economists—argue that continued economic and technological growth can both feed the world's billions and enrich everyone enough to end the population explosion voluntarily. Not so, counter many ecologists. Growth is the problem; we must stop both population and economic growth.

Yet another perspective on this subject derives from social justice concerns. In this worldview, there are sufficient resources for

FIGURE 7.1 A Mayan family in Guatemala with four of their six living children. Decisions on how many children to have are influenced by many factors, including culture, religion, need for old age security for parents, immediate family finances, household help, child survival rates, and power relationships within the family. Having many children may not be in the best interest of society at large, but may be the only rational choice for individual families. Courtesy John W. Cunningham.

TABLE 7.1	World Population Growth and Doubling Times	
DATE	POPULATION	DOUBLING TIME
5000 B.C.	50 million	?
800 B.C.	100 million	4,200 years
200 B.C.	200 million	600 years
A.D. 1200	400 million	1,400 years
A.D. 1700	800 million	500 years
A.D. 1900	1,600 million	200 years
A.D. 1965	3,200 million	65 years
A.D. 2000	6,100 million	51 years
A.D. 2050 (estimate)	9,300 million	140 years

Source: *Population Reference Bureau and United Nations Population Division.*

everyone. Current shortages are only signs of greed, waste, and oppression. The root cause of environmental degradation, in this view, is inequitable distribution of wealth and power rather than population size. Fostering democracy, empowering women and minorities, and improving the standard of living of the world's poorest people are what are really needed. A narrow focus on population growth only fosters racism and an attitude that blames the poor for their problems while ignoring the deeper social and economic forces at work.

Whether human populations will continue to grow at present rates and what that growth would imply for environmental quality and human life are among the most central and pressing questions in environmental science. In this chapter, we will look at some causes of population growth as well as how populations are measured and described. Family planning and birth control are essential for stabilizing populations. The number of children a couple decides to have and the methods they use to regulate fertility, however, are strongly influenced by culture, religion, politics, and economics, as well as basic biological and medical considerations. We will examine how some of these factors influence human demographics.

Human Population History

For most of our history, humans have not been very numerous compared to other species. Studies of hunting and gathering societies suggest that the total world population was probably only a few million people before the invention of agriculture and the domestication of animals around 10,000 years ago. The larger and more secure food supply made available by the agricultural revolution allowed the human population to grow, reaching perhaps 50 million people by 5000 B.C. For thousands of years, the number of humans increased very slowly. Archaeological evidence and historical descriptions suggest that only about 300 million people were living at the time of Christ (table 7.1).

Until the Middle Ages, human populations were held in check by diseases, famines, and wars that made life short and uncertain for most people (fig. 7.2). Furthermore, there is evidence that many early societies regulated their population size through cultural taboos and practices such as infanticide. Among the most destructive of natural population controls were bubonic plagues that periodically swept across Europe between 1348 and 1650. During the worst plague years (between 1348 and 1350), it is estimated that at least one-third of the European population perished. Notice, however, that this did not retard population growth for very long. In 1650, at the end of the last great plague, there were about 600 million people in the world.

As you can see in figure 7.2, human populations began to increase rapidly after A.D. 1600. Many factors contributed to this rapid growth. Increased sailing and navigating skills stimulated commerce and communication between nations. Agricultural developments, better sources of power, and better health care and hygiene also played a role. We are now in an exponential or J curve pattern of growth (What Do You Think? p. 128).

It took all of human history to reach 1 billion people in 1804, but little more than 150 years to reach 3 billion in 1960. To go from 5 to 6 billion took only 12 years. Another way to look at population growth is that the number of humans tripled during the twentieth century. Will it do so again in the twenty-first century? If it does, will we overshoot the carrying capacity of our environment and experience a catastrophic dieback similar to those described in chapter 6? As you will see later in this chapter, there is evidence that population growth already is slowing, but whether we will reach equilibrium soon enough and at a size that can be sustained over the long term remains a difficult but important question.

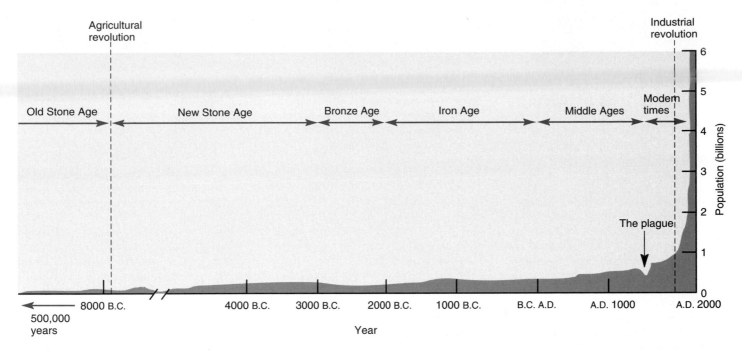

FIGURE 7.2   Human population levels through history. Since about A.D. 1000, our population curve has assumed a J shape. Are we on the upward slope of a population overshoot? Will we be able to adjust our population growth to an S curve? Or can we just continue the present trend indefinitely?

LIMITS TO GROWTH: SOME OPPOSING VIEWS

As with many topics in environmental science, people have widely differing opinions about population and resources. Some believe that population growth is the ultimate cause of poverty and environmental degradation. Others argue that poverty, environmental degradation, and overpopulation are all merely symptoms of deeper social and political factors. In this section, we will examine some opposing worldviews and their implications.

Malthusian Checks on Population

In 1798, the Rev. Thomas Malthus wrote *An Essay on the Principle of Population* to refute the views of progressives and optimists—including his father—who were inspired by the egalitarian principles of the French Revolution to predict a coming utopia. The younger Malthus argued that human populations tend to increase at an exponential or compound rate while food production either remains stable or increases only slowly. The result, he predicted, is that human populations inevitably outstrip their food supply and eventually collapse into starvation, crime, and misery. Malthus's theory might be summarized by the graphic in figure 7.3a.

According to Malthus, the only ways to stabilize human populations are "positive checks," such as disease or famines that kill people, or "preventative checks," including all the factors that prevent human birth. Among the preventative checks he advocated were "moral restraint," including late marriage and celibacy until a couple can afford to support children. Many social scientists and

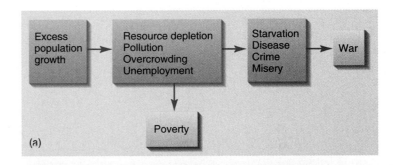

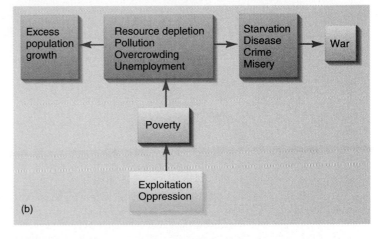

FIGURE 7.3   (a) Thomas Malthus argued that excess population growth is the ultimate cause of many other social and environmental problems. (b) Karl Marx argued that oppression and exploitation are the real causes of poverty and environmental degradation. Population growth in this view is a symptom or result of other problems, not the source.

What do you think?

Looking for Bias in Graphs

Graphs, pictorial representations of information, can help others understand what we have to say. They are particularly helpful in conveying patterns of relationship. Why are graphs used so widely and why do they make such powerful impressions on us?

Not only are images expressive, but graphs come equipped with substantiating data. It is one thing to read that "the human population is rising explosively," but it is considerably more compelling to see the J curve pictured together with concrete numbers as evidence. The combination of picture and numbers often conveys powerful impressions beyond the numbers themselves.

An important question addressed in this chapter is: How serious is the worldwide population problem? Does it demand action? The following graphs present information relevant to human population studies, yet the impressions left with the reader vary considerably, depending on the type of data used and the design of the graph.

Reexamine figure 7.2, which presents an historical perspective on the growth in human numbers. What conclusions do you draw from it? Does it suggest we are expe-riencing an explosive rise in human numbers or that the current increase is unprecedented in our species history? Does it also imply, perhaps more indirectly, that the earth's population problem is serious, that there is an urgency to respond? In light of concepts of environmental resistance, carrying capacity, and overshoot and dieback, many people would conclude that the current growth, highlighted by the graph, is unsustainable.

Now examine figure 1.

It plots the same variables as figure 7.2, but suggests that population is rising at a modest rate. It does not produce the sense of explosiveness and urgency as figure 7.2. How can the impressions be so different?

Notice that the graph in figure 1 covers a much shorter time period. This greatly changes the time interval lengths on the horizontal scale. In figure 7.2, one millimeter represents about 50 years, but less than one year in figure 1. This changes the line's slope from nearly vertical in figure 7.2 to a modest incline in figure 1. Slope impacts the visual impression created by a graph, and therefore on its interpretation. How could you change the time axis in figure 1 to flatten the slope even more?

Next examine figure 2. This graph plots the stabilization ratio rather than population size over time. This graph reveals that the rate of growth for most regions except Africa has been in decline since about 1970. By itself, does the graph's downward-trending line suggest there really isn't much of a population problem, or at least not much reason for concern?

All three graphs present information relevant to a discussion of worldwide population growth, yet each gives a different impression of the seriousness of the problem. So, how can we analyze graphs and their messages in a thoughtful way? There is no single formula applicable to all situations, certainly, but a few questions are worth keeping in mind.

Is the time frame of reference appropriate or is it too restricted to allow a valid, comprehensive assessment of the issue? Are the unit intervals on the graph appropriately sized? Is the impression created by the graph real or simply an artifact of the graph's format? Do the data shown in a graph present only a partial, perhaps misleading, view of the whole?

Because graphs can create powerful impressions, they need to be interpreted with care.

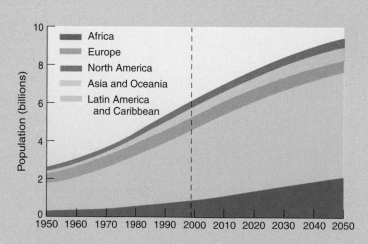

FIGURE 1   The UN Population Division projects continued population growth, although at a gradually slowing rate over the next 50 years. *Source:* World Resources Institute, *World Resources 1998–99.*

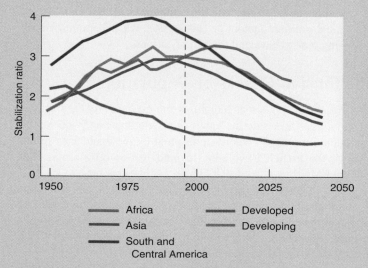

FIGURE 2   The stabilization ratio is measured by dividing crude birth rate by crude death rate. A ratio of 1 indicates zero population growth.
Source: United Nations (UN) Population Division, World Population Prospects, 1950–2050 (The 1996 Revision), on diskette (UN, New York, 1996).

biologists have been influenced by Malthus. Charles Darwin, for instance, derived his theories about the struggle for scarce resources and survival of the fittest after reading Malthus's essay.

If Malthus's views of the consequences of exponential population growth were dismal, the corollary he drew was even more bleak. He believed that most people are too lazy and immoral to regulate birth rates voluntarily. Consequently, he opposed efforts to feed and assist the poor in England because he feared that more food would simply increase their fertility and thereby perpetuate the problems of starvation and misery.

Not surprisingly, Malthus's ideas provoked a great social and economic debate. Karl Marx was one of his most vehement critics, claiming that Malthus was a "shameless sycophant of the ruling classes." According to Marx, population growth is a symptom rather than a root cause of poverty, resource depletion, pollution, and other social ills. The real causes of these problems, he believed, are exploitation and oppression (fig. 7.3b). Marx argued that workers always provide for their own sustenance given access to means of production and a fair share of the fruits of their labor. According to Marxians, the way to slow population growth *and* to alleviate crime, disease, starvation, misery, and environmental degradation is through social justice.

Malthus and Marx Today

Both Marx and Malthus developed their theories about human population growth in the nineteenth century when understanding of the world, technology, and society were much different than they are now. Still, the questions they raised are relevant today. While the evils of racism, classism, and other forms of exploitation that Marx denounced still beset us, it is also true that at some point available resources must limit the numbers of humans that the earth can sustain.

Those who agree with Malthus, that we are approaching—or may already have surpassed—the carrying capacity of the earth are called **neo-Malthusians.** In their view, we should address the issue of surplus population directly by making birth control our highest priority. An extreme version of this worldview is expressed by Cornell University entomologist David Pimentel, who claims that the "optimum human population" would be about 2 billion, or about the number living in 1950. He believes this would allow everyone to enjoy a standard of living equal to the average European today.

Neo-Marxians, on the other hand, believe that only eliminating oppression and poverty through technological development and social justice will solve population problems. Claims of resource scarcity, they argue, are only an excuse for inequity and exclusion. If distribution of wealth and access to resources were more fair, they believe, there would be plenty for everyone.

Perhaps a compromise position between these opposing viewpoints is that population growth, poverty, and environmental degradation all are interrelated. No factor exclusively causes any other, but each influences and, in turn, is influenced by the others.

Can Technology Make the World More Habitable?

Technological optimists argue that Malthus was wrong in his predictions of famine and disaster 200 years ago because he failed to account for scientific progress. In fact, food supplies have increased faster than population growth since Malthus's time. There have been terrible famines in the past two centuries, but they were caused more by politics and economics than lack of resources or sheer population size. Whether this progress will continue remains to be seen, but technological advances have increased human carrying capacity more than once in our history.

The burst of growth of which we are a part, was stimulated by the scientific and industrial revolutions. Progress in agricultural productivity, engineering, information technology, commerce, medicine, sanitation, and other achievements of modern life have made it possible to support approximately 1,000 times as many people per unit area as was possible 10,000 years ago.

Much of our growth in the past 300 years has been based on availability of easily acquired natural resources, especially cheap, abundant fossil fuels. Whether we can develop alternative, renewable energy sources in time to avert disaster when current fossil fuels run out is a matter of great concern (chapter 19).

Can More People Be Beneficial?

There can be benefits as well as disadvantages in larger populations. More people mean larger markets, more workers, and efficiencies of scale in mass production of goods. Often a larger population produces more ideas and faster innovation, and sometimes these innovations actually increase or extend resources, or reduce our environmental impacts. Recent innovations in green design, for example, allow us to produce less waste and consume fewer materials, if we choose to. New homes, with insulation, efficient lighting, and tight windows, can be far more efficient than traditional homes. Many new cars use only a fraction of the oil and gas of those a generation ago; new farming techniques save soil and water while producing more food.

Economist Julian Simon was one of the most outspoken champions of this rosy view of human history. People, he argued, are the "ultimate resource" and there is no evidence that pollution, crime, unemployment, crowding, the loss of species, or any other resource limitations will worsen with population growth. This outlook is shared by leaders of many developing countries who insist that instead of being obsessed with population growth, we should focus on the inordinate consumption of the world's resources by people in richer countries. What constitutes a resource and which resources might limit further human population growth are questions we will return to in subsequent chapters in this book. For now, we will move on to discuss how people are counted.

HUMAN DEMOGRAPHY

Demography is derived from the Greek words *demos* (people) and *graphos* (to write or to measure). It encompasses vital statistics about people, such as births, deaths, and where they live, as well as total population size. In this section, we will survey ways human populations are measured and described, and discuss demographic factors that contribute to population growth.

How Many of Us Are There?

On October 12, 1999, the United Nations officially declared that the human population had reached 6 billion. The date is only an estimate, however. Even in this age of information technology and communication, counting the number of people in the world is like shooting at a moving target. Some countries have never even taken a census, and those that have been done may not be accurate. Governments may overstate or understate their populations to make their countries appear larger and more important or smaller and more stable than they really are. Individuals, especially if they are homeless, refugees, or illegal aliens, may not want to be counted or identified.

We really live in two very different demographic worlds. One of these worlds is poor, young, and growing rapidly. It is occupied by the vast majority of people who live in the less-developed countries of Africa, Asia, and Latin America. These countries represent 80 percent of the world population but more than 90 percent of all projected growth (fig. 7.4). In countries such as Yemen and Niger, the average age is less than 15, the current doubling time is only 20 years, and the average person lives on less than $1 (U.S.) per day.

Some countries in the developing world have experienced amazing growth rates and are expected to reach extraordinary population sizes by the middle of the twenty-first century. Table 7.2 shows the ten largest countries in the world, arranged by their projected size in 2050. Note that, while China reached 1.28 billion people in 2002, India also passed 1 billion and is expected to have

TABLE 7.2	The World's Largest Countries		
IN 2002		**IN 2050**	
COUNTRY	POPULATION (IN MILLIONS)	COUNTRY	POPULATION (IN MILLIONS)
China	1,281	India	1,628
India	1,050	China	1,394
United States	287	United States	413
Indonesia	217	Pakistan	332
Brazil	174	Indonesia	316
Russia	144	Nigeria	304
Pakistan	144	Brazil	247
Bangladesh	134	Bangladesh	205
Nigeria	130	Congo, Dem. Rep. of	182
Japan	127	Ethiopia	173
Mexico	102	Mexico	151
Germany	82	Philippines	146
Philippines	80	Vietnam	117
Vietnam	80	Egypt	115
Egypt	71	Russia	102

Source: *Population Reference Bureau 2003.*

the largest population in a few decades because its population control programs have been less successful than China's. Nigeria, which had only 33 million residents in 1950, is forecast to have more than 300 million in 2050. Ethiopia, with about 18 million people 50 years ago, is likely to grow at least tenfold over a century. In many of these countries, rapid population growth is a serious problem. Overall, the population of less-developed countries is projected to rise from 5 billion in 2001 to 8.2 billion in 2050. Just six countries (India, China, Pakistan, Nigeria, Bangladesh, and Indonesia) account for almost half this growth.

The other demographic world is made up of the richer countries of North America, Western Europe, Japan, Australia, and New Zealand. This world is wealthy, old, and mostly shrinking. Italy, Germany, Hungary, and Japan, for example, all have negative growth rates. The average age in these countries is now 40, and life expectancy of their residents is expected to exceed 90 by 2050. With many couples choosing to have either one or no children, the populations of these countries are expected to decline significantly over the next century. Japan, which has 126 million residents now, is expected to shrink to about 100 million by 2050. Europe, which now makes up about 12 percent of the world population, will constitute less than 7 percent in 50 years, if current trends continue. Even the United States and Canada would have nearly stable populations if immigration were stopped.

It isn't only wealthy countries that have declining populations. Russia, for instance, is now declining by nearly 1 million people per year as death rates have soared and birth rates have plummeted. A collapsing economy, hyperinflation, crime, corruption, and despair have demoralized the population. Horrific

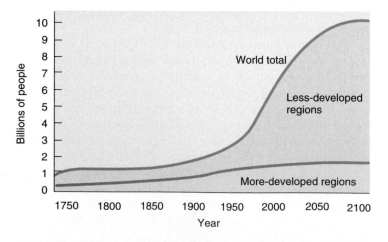

FIGURE 7.4 Estimated human population growth, 1750–2100, in less-developed and more-developed regions. More than 90 percent of all growth in this century and projected for the next is in the less-developed countries. *Source:* World Resources Institute, 2000.

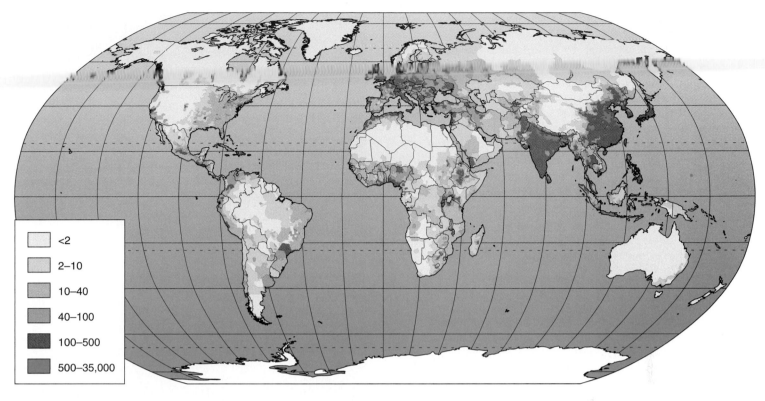

FIGURE 7.5 Population density in persons per square kilometer. *Source:* World Bank, 2000.

Legend:
- <2
- 2–10
- 10–40
- 40–100
- 100–500
- 500–35,000

pollution levels left from the Soviet era, coupled with poor nutrition and health care, have resulted in high levels of genetic abnormalities, infertility, and infant mortality. Abortions are twice as common as live births, and the average number of children per woman is now 1.3, one of the lowest in the world. Death rates, especially among adult men, have risen dramatically. Male life expectancy dropped from 68 years in 1990 to 58 years in 2000. After having been the fourth largest country in the world in 1950, Russia is expected to have a smaller population than Vietnam by 2050. Other former Soviet states are experiencing similar declines. Estonia, Bulgaria, Georgia, and Ukraine, for example, now have negative growth rates and are expected to lose about 40 percent of their population in the next 50 years.

The situation is even worse in many African countries, where AIDS and other communicable diseases are killing people at a terrible rate. In Zimbabwe, Botswana, Zambia, and Namibia, for example, up to 36 percent of the adult population have AIDS or are HIV positive. Health officials predict that more than two-thirds of the 15-year-olds now living in Botswana will die of AIDS before age 50. Without AIDS, the average life expectancy is estimated to be 69.7 years. Now, with AIDS, Botswana's life expectancy has dropped to only 36.1 years. Altogether, Africa's population is expected to be nearly 200 million lower in 2050 than it would have been without AIDS. Overall, however, because of high fertility rates, Africa is expected to grow by at least 1 billion people over the next century.

Figure 7.5 shows human population distribution around the world. Notice the high densities supported by fertile river valleys of the Nile, Ganges, Yellow, Yangtze, and Rhine Rivers and the well-watered coastal plains of India, China, and Europe. Historic factors, such as technology diffusion and geopolitical power, also play a role in geographic distribution.

Fertility and Birth Rates

As we pointed out in chapter 6, fecundity is the physical ability to reproduce, while fertility describes the actual production of offspring. Those without children may be fecund but not fertile. The most accessible demographic statistic of fertility is usually the **crude birth rate,** the number of births in a year per thousand persons. It is statistically "crude" in the sense that it is not adjusted for population characteristics such as the number of women in reproductive age.

The **total fertility rate** is the number of children born to an average woman in a population during her entire reproductive life. Upper-class women in seventeenth- and eighteenth-century England, whose babies were given to wet nurses immediately after birth and who were expected to produce as many children as possible, often had 25 or 30 pregnancies. The highest recorded total fertility rates for working-class people is among some Anabaptist agricultural groups in North America who have averaged up to 12 children per woman. In most tribal or traditional societies, food shortages, health problems, and cultural practices limit total fertility to about six or seven children per woman even without modern methods of birth control.

Fertility is usually calculated as births per woman because, in many cases, it is difficult to establish paternity. Some demographers argue, nevertheless, that we should pay more attention to birth rates per male, because in some cultures men have far more children, on average, than do women. In Cameroon, for instance, due to multiple marriages, extramarital affairs, and a high rate of female mortality, men are estimated to have 8.1 children in their lifetime, while women average only 4.8.

Zero population growth (ZPG) occurs when births plus immigration in a population just equal deaths plus emigration. It takes several generations of replacement level fertility (where people just replace themselves) to reach ZPG. Where infant mortality rates are high, the replacement level may be five or more children per couple. In the more highly developed countries, however, this rate is usually about 2.1 children per couple because some people are infertile, have children who do not survive, or choose not to have children.

Fertility rates have declined dramatically in every region of the world except Africa over the past 50 years (fig. 7.6). Only a few decades ago, total fertility rates above 6 were common in many countries. The average family in Mexico in 1975, for instance, had 7 children. By 2000, however, the average Mexican woman had only 2.5 children. According to the World Health Organization, 61 out of the world's 190 countries are now at or below a replacement rate of 2.1 children per couple. The greatest fertility reduction has been in Southeast Asia, where rates have fallen by more than half. Most of this decrease has occurred in just the past few decades and, contrary to what many demographers expected, some of the poorest countries in the world have been remarkably successful in lowering growth rates. Bangladesh, for instance, reduced its fertility rate from 6.9 in 1980, to 3.3 children per woman in 2002.

China's one-child-per-family policy decreased the fertility rate from 6 in 1970 to 1.8 in 1990. This policy, however, has sometimes resulted in abortions, forced sterilizations, and even infanticide. Another adverse result is that the only children (especially boys) allowed to families may grow up to be spoiled "little emperors" who have an inflated impression of their own importance (fig.

FIGURE 7.7 China's one-child-per-family policy has been remarkably successful in reducing birth rates. It may, however, have created a generation of "little emperors," since parents and grandparents focus all their attention on an only child. © William P. Cunningham.

7.7). Furthermore, there may not be enough workers to maintain the army, sustain the economy, or support retirees when their parents reach old age.

Although the world as a whole still has an average fertility rate of 2.8, growth rates are now lower than at any time since World War II. If fertility declines like those in Bangladesh and China were to occur everywhere in the world, our total population could begin to decline by the end of the twenty-first century. Interestingly, Spain and Italy, although predominately Roman Catholic, have fertility rates (1.2 children per woman) far below replacement. And Iran, which has an Islamic government, has seen birth rates fall by more than half in less than 20 years (Case Study, p. 133).

Mortality and Death Rates

A traveler to a foreign country once asked a local resident, "What's the death rate around here?" "Oh, the same as anywhere," was the reply, "about one per person." In demographics, however, **crude death rates** (or crude mortality rates) are expressed in terms of the number of deaths per thousand persons in any given year. Countries in Africa where health care and sanitation are limited may have mortality rates of 20 or more per 1,000 people. Wealthier countries generally have mortality rates around 10 per 1,000. The number of deaths in a population is sensitive to the age structure

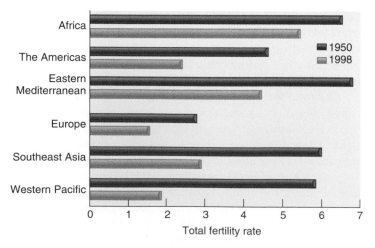

FIGURE 7.6 Declines in fertility rates by region, 1950 and 1998. *Source:* Data from World Health Organization, *World Health Report 1999.*

Case Study

Family Planning in Iran

After the Islamic Revolution in 1979, Iran had one of the world's highest population growth rates. In spite of civil war, large-scale emigration, and economic austerity, the country surged from 34 million to 63 million in just 20 years. A crude birth rate of 43.4/1,000 people and a total fertility rate of 5.1 per woman during this time resulted in annual population growth of 3.9 percent and a doubling time of less than 18 years. Religious authorities exhorted couples to have as many children as Allah would give them. Any mention of birth control or family planning (other than to have as many children as possible) was forbidden, and the marriage age for girls was dropped to 9 years old. When a devastating war with Iraq in the 1980s killed at least 1 million young soldiers, producing more children to rebuild the army became a civic as well as religious duty.

In the late 1990s, however, the Iranian government became aware of the costs of such rapid population growth. With religious moderates gaining greater political power, public policy changed abruptly. Now, the Iranian government is spending millions of dollars to lower birth rates. Couples must pass a national family planning course before they are allowed to marry. While it took a few years to convince people that this change will be long-lasting, most Iranian citizens are now eager for access to birth-control information. Family planning classes are sought out both by engaged couples and those already married. A wide range of birth control methods are available. Implantable or injectable slow-release hormones, condoms, intrauterine devices (IUDs), pills, and male or female sterilization are free to all. Billboards, newspapers, television, and even water towers advertise this national program. Religious leaders have issued a *fatwah,* or command, that all faithful Muslims participate in family planning.

As a consequence, Iran has been remarkably successful in stemming its population growth. Between 1986 and 1996, the fertility rates for urban residents dropped almost by half, to less than three children per woman, and the crude birth rate dropped from 43 to 18 per 1,000 people. By 2000, the average annual growth rate had fallen to 1.4 percent. While the population is still increasing, another decade of such progress would bring the country to a stable or even declining rate of growth.

Several societal changes have contributed to this rapid birth reduction. While the minimum marriage age has been returned to 15, couples are encouraged to wait until at least age 20 to begin their families. The educational benefits of concentrating the family resources on just one or two children are being promoted. Although women's roles are still highly restricted in the Islamic Republic, greater gender equity has given women more control over their reproductive lives. Access to modern, information-age jobs gives people an incentive to seek out education both for themselves and their children.

The demographic transition hasn't spread to all levels of Iranian society. Rural families, ethnic minorities, and some urban poor still tend to have many children. Still, this example of how quickly both ideals of the perfect family size and information about modern birth control can spread through a society—even a highly religious, fundamentalist one—is encouraging for what might be accomplished worldwide in a surprisingly short time.

of the population. Rapidly growing, developing countries such as Libya or Costa Rica have lower crude death rates (4 per 1,000) than do the more-developed, slowly growing countries, such as Denmark (12 per 1,000). This is because there are proportionately more youths and fewer elderly people in a rapidly growing country than in a more slowly growing one.

Population Growth Rates

Crude death rate subtracted from crude birth rate gives the **natural increase** of a population. We distinguish natural increase from the **total growth rate,** which includes immigration and emigration, as well as births and deaths. Both of these growth rates are usually expressed as a percent (number per hundred people) rather than per thousand. A useful rule of thumb is that if you divide 70 by the annual percentage growth, you will get the approximate doubling time in years. Palestine, for example, which is growing 3.5 percent per year, is doubling its population every 20 years. The United States, which has a natural increase rate of 0.6 percent per year, is doubling in 116.7 years. Actually, because of immigration, U.S. total growth is considerably faster than natural increase. Spain and

the United Kingdom, with natural increase rates of 0.1 percent, are doubling in about 700 years. Russia, on the other hand, with a growth rate of –0.8 percent, will lose about 30 percent of its population in the next 50 years. The world growth rate is now 1.3 percent, which means that the population will double in about 54 years if this rate persists.

Life Span and Life Expectancy

Life span is the oldest age to which a species is known to survive. Although there are many claims in ancient literature of kings living a millennium or more, the oldest age that can be certified by written records was that of Jeanne Louise Calment of Arles, France, who was 122 years old at her death in 1997. The aging process is still a medical mystery, but it appears that cells in our bodies have a limited ability to repair damage and produce new components. At some point they simply wear out, and we fall victim to disease, degeneration, accidents, or senility.

Life expectancy is the average age that a newborn infant can expect to attain in any given society. It is another way of expressing the average age at death. For most of human history, we believe that life expectancy in most societies has been between 35 and 40 years. This doesn't mean that no one lived past age 40, but rather that so many deaths at earlier ages (mostly early childhood) balanced out those who managed to live longer.

Declining mortality, not rising fertility, is the primary cause of most population growth in the past 300 years. Crude death rates began falling in western Europe during the late 1700s. Most of this advance in survivorship came long before the advent of modern medicine and is due primarily to better food and better sanitation.

The twentieth century has seen a global transformation in human health unmatched in history. This revolution can be seen in the dramatic increases in life expectancy in most places. Worldwide, the average life expectancy has risen from about 40 to 67 years over the past century. Table 7.3 shows gains in some selected countries. Globally, the number of people over 60 years old is expected to triple, increasing from 600 million today to nearly 2 billion in 2050. The oldest old (over 80 years) is projected to grow five-fold to about 400 million in that same period.

The greatest progress in life expectancy has been in developing countries. Take the case of Nicaragua, for example. In 1900, the average Nicaraguan man could expect to live only 29 years, while the average woman would reach just 33 years. By 2002, although Nicaragua had an annual per capita income less than $2,100 (U.S.), the average life expectancy for both men and women had more than doubled and was close to that of countries with ten times its income level. Longer lives were due primarily to better nutrition, improved sanitation, clean water, and education rather than miracle drugs or high-tech medicine. While the gains were not as great for the already industrialized countries, residents of the United States, Italy, and Japan, for example, now live about half-again as long as they did at the beginning of the century.

As figure 7.8 shows, there is a good correlation between annual income and life expectancy up to about $4,000 (U.S.) per person. Beyond that level—which is generally enough for adequate food, shelter, and sanitation for most people—life expectancies level out at about 75 years for men and 85 for women.

Large discrepancies in how benefits of modernization and social investment are distributed within countries are revealed in

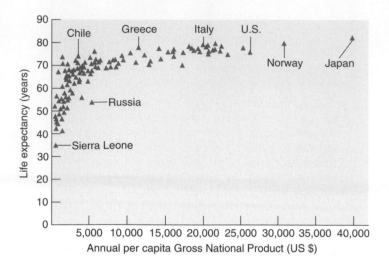

FIGURE 7.8 As incomes rise, so does life expectancy up to about $4,000 (U.S.). Russia is an exception, with a life expectancy nearly 20 years less than that of Chile, even though their GNP is about the same. *Source:* The World Bank, *World Development Indicators 1997,* the World Bank, Washington, D.C., 1997.

differential longevity of various groups. The greatest life expectancy reported anywhere in the United States is for women in Stearns County, Minnesota, who live to an average age of 84. By contrast, Native American men on Pine Ridge Indian Reservation in neighboring South Dakota, live, on average, only to age 45. Only a few countries in Africa have a lower life expectancy. Stearns County is populated mainly by prosperous German-Catholic farmers. The Pine Ridge Reservation is the poorest area in America with an unemployment rate near 75 percent and high rates of poverty, alcoholism, drug use, and cultural alienation. Similarly, African-American men in Washington, D.C., live, on average, only 57.9 years, or less than men in Lesotho or Swaziland.

Some demographers believe that life expectancy is approaching a plateau, while others predict that advances in biology and medicine might make it possible to live 150 years or more. If our average age at death approaches 100 years, as some expect, society will be profoundly affected. In 1970 the median age in the United States was 30. By 2100 the median age could be over 60. If workers continue to retire at 65, half of the population could be unemployed, and retirees might be facing 35 or 40 years of retirement. We may need to find new ways to structure and finance our lives.

Living Longer: Demographic Implications

A population that is growing rapidly by natural increase has more young people than does a stationary population. One way to show these differences is to graph age classes in a histogram as shown in figure 7.9. In Niger, which is growing at a rate of 3.5 percent per year, 47.8 percent of the population is in the prereproductive category (below age 15). Even if total fertility rates were to fall abruptly, the total number of births, and population size, would continue to grow for some years as these young people enter reproductive age. This phenomenon is called population momentum.

TABLE 7.3	Life Expectancy at Birth for Selected Countries in 1900 and 2000			
	1900		2000	
COUNTRY	MALES	FEMALES	MALES	FEMALES
India	22.6	23.3	60.3	60.5
Japan	42.4	43.7	77.4	84.2
Russia	30.9	33.0	61.7	73.6
Sweden	56.6	59.5	77.0	82.1
United States	45.6	48.3	74.7	79.3

Source: *Population Reference Bureau, 2002.*

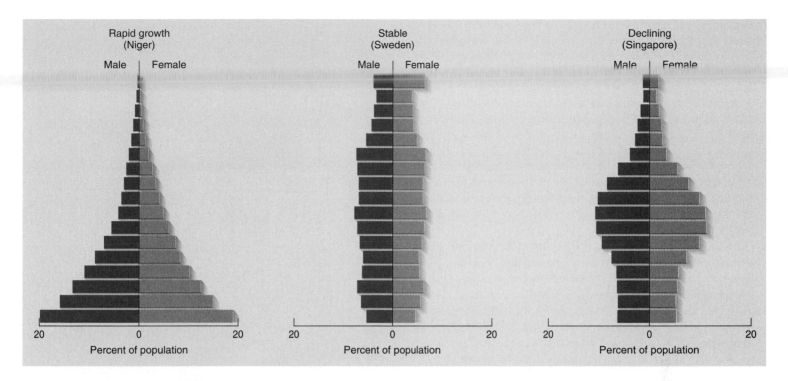

FIGURE 7.9   Age structure graphs for rapidly growing, stable, and declining populations. *Source:* U.S. Census Bureau, 2003.

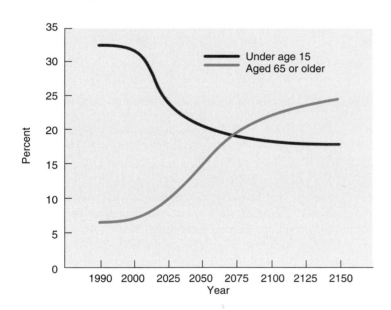

FIGURE 7.10   Changing age structure of world population. In the twenty-first century, children under 15 years of age will make up a smaller percentage of world population, while people over 65 years old will make up a rapidly rising share of the population.

By contrast, a country with a stable population, like Sweden, has nearly the same number in each age cohort. A population that has recently entered a lower growth rate pattern, such as Singapore, has a bulge in the age classes for the last high-birth-rate generation. Notice that there are more females than males in the older age group in Singapore because of differences in longevity between the sexes. The United States has a high percentage of retired people because of long life expectancy.

Both rapidly growing countries and slowly growing countries can have a problem with their **dependency ratio,** or the number of nonworking compared to working individuals in a population. In Mexico, for example, each working person supports a high number of children. In the United States, by contrast, a declining working population is now supporting an ever larger number of retired persons and there are dire predictions that the social security system will soon be bankrupt. This changing age structure and shifting dependency ratio are occurring worldwide (fig. 7.10). By 2050 the UN predicts there will be two older persons for every child in the world.

Emigration and Immigration

Humans are highly mobile, so emigration and immigration play a larger role in human population dynamics than they do in those of many species. Currently, about 800,000 people immigrate legally to the United States each year, but many more enter illegally. Western Europe receives about 1 million applications each year for asylum from economic chaos and wars in former socialist states and the Middle East. The United Nations High Commission on Refugees reported that in January 2002 there were 19.8 million refugees who had left their countries for political or economic reasons, while about 20 million more were displaced persons in their own countries.

The more-developed regions are expected to gain about 2 million immigrants per year for the next 50 years. Without migration, the population of the wealthiest countries would already be declining and would be more than 126 million less than the current 1.2 billion by 2050. The 2000 census showed that 35 million U.S. residents (12.5 percent of the total population) classify themselves as Hispanic or Latino. They now constitute the largest U.S. minority.

Immigration is a controversial issue in many countries. "Guest workers" often perform heavy, dangerous, or disagreeable work that citizens are unwilling to do. Many migrants and alien workers are of a different racial or ethnic background than the majority in their new home. They generally are paid low wages and given substandard housing, poor working conditions, and few rights. Local residents often complain that immigrants take away jobs, overload social services, and ignore established rules of behavior or social values.

Some countries encourage, or even force, internal mass migrations as part of a geopolitical demographic policy. In the 1970s, Indonesia embarked on an ambitious "transmigration" plan to move 65 million people from the overcrowded islands of Java and Bali to relatively unpopulated regions of Sumatra, Borneo, and New Guinea. Attempts to turn rainforest into farmland had disastrous environmental and social effects, however, and this plan was greatly scaled back. China has announced a plan to move up to 100 million people to a sparsely populated region along the Amur River in Heilongjang.

POPULATION GROWTH: OPPOSING FACTORS

A number of social and economic pressures affect decisions about family size, which in turn affects the population at large. In this section we will examine both positive and negative pressures on reproduction.

Pronatalist Pressures

Factors that increase people's desires to have babies are called **pronatalist pressures.** Raising a family may be the most enjoyable and rewarding part of many people's lives. Children can be a source of pleasure, pride, and comfort. They may be the only source of support for elderly parents in countries without a social security system. Where infant mortality rates are high, couples may need to have many children to be sure that at least a few will survive to take care of them when they are old. Where there is little opportunity for upward mobility, children give status in society, express parental creativity, and provide a sense of continuity and accomplishment otherwise missing from life. Often children are valuable to the family not only for future income, but even more as a source of current income and help with household chores. In much of the developing world, children as young as 6 years old tend domestic animals and younger siblings, fetch water, gather firewood, and help grow crops or sell things in the marketplace (fig. 7.11). Parental desire for children rather than an unmet need for contraceptives may be the most important factor in population growth in many cases.

FIGURE 7.11 Children in rural areas can help with many household chores, such as tending livestock or caring for younger children. © William P. Cunningham.

Society also has a need to replace members who die or become incapacitated. This need often is codified in cultural or religious values that encourage bearing and raising children. In some societies, families with few or no children are looked upon with pity or contempt. The idea of deliberately controlling fertility may be shocking, even taboo. Women who are pregnant or have small children are given special status and protection. Boys frequently are more valued than girls because they carry on the family name and are expected to support their parents in old age. Couples may have more children than they really want in an attempt to produce a son.

Male pride often is linked to having as many children as possible. In Niger and Cameroon, for example, men, on average, want 12.6 and 11.2 children, respectively. Women in these countries consider the ideal family size to be only about one-half that desired by their husbands. Even though a woman might desire fewer children, however, she may have few choices and little control over her own fertility. In many societies, a woman has no status outside of her role as wife and mother. Without children, she has no source of support.

Birth Reduction Pressures

In more highly developed countries, many pressures tend to reduce fertility. Higher education and personal freedom for women often result in decisions to limit childbearing. The desire to have children is offset by a desire for other goods and activities that compete with childbearing and childrearing for time and money. When women have opportunities to earn a salary, they are less likely to stay home and have many children. Not only are the challenge and variety of a career attractive to many women, but the money that they can earn outside the home becomes an important part of the family budget. Thus, education and socioeconomic status are usually inversely related to fertility in richer countries. In developing countries, however, fertility is likely to increase as educational levels and socioeconomic status rise. With higher income, families are better able to afford the children they want; more money means that women are likely to be healthier, and therefore better able to conceive and carry a child to term.

In less-developed countries where feeding and clothing children can be a minimal expense, adding one more child to a family usually doesn't cost much. By contrast, raising a child in the United States can cost hundreds of thousands of dollars by the time the child is through school and is independent. Under these circumstances, parents are more likely to choose to have one or two children on whom they can concentrate their time, energy, and financial resources.

Figure 7.12 shows U.S. birth rates between 1910 and 2000. As you can see, birth rates have fallen and risen in a complex pattern. The period between 1910 and 1930 was a time of industrialization and urbanization. Women were getting more education than ever before and entering the workforce. The Great Depression in the 1930s made it economically difficult for families to have children, and birth rates were low. The birth rate increased at the beginning of World War II (as it often does in wartime). For reasons that are unclear, a higher percentage of boys are usually born during war years.

At the end of the war, there was a "baby boom" as couples were reunited and new families started. This high birth rate per-sisted through the times of prosperity and optimism of the 1950s, but began to fall in the 1960s. Part of this decline was caused by the small number of babies born in the 1930s. This meant fewer young adults to give birth in the 1960s. Part was due to changed perceptions of the ideal family size. Whereas in the 1950s women typically wanted four children or more, in the 1970s the norm dropped to one or two (or no) children. A small "echo boom" occurred in the 1980s as people born in the 1960s began to have babies, but changing economics and attitudes seem to have permanently altered our view of ideal family size in the United States.

Birth Dearth?

Most European countries now have birth rates below replacement rates, and Italy, Russia, Austria, Germany, Greece, and Spain are experiencing negative rates of natural population increase. Asia, Japan, Singapore, and Taiwan are also facing a "child shock" as fertility rates have fallen well below the replacement level of 2.1 children per couple. There are concerns in all these countries about falling military strength (lack of soldiers), economic power (lack of workers), and declining social systems (not enough workers and taxpayers) if low birth rates persist or are not balanced by immigration.

Economist Ben Wattenberg warns that this "birth dearth" might seriously erode the powers of Western democracies in world affairs. He points out that Europe and North America accounted for 22 percent of the world's population in 1950. By the 1980s, this number had fallen to 15 percent, and by the year 2030, Europe and North America probably will make up only 9 percent of the world's population. Germany, Hungary, Denmark, and Russia now offer incentives to encourage women to bear children. Japan offers financial support to new parents, and Singapore provides a dating service to encourage marriages among the upper classes as a way of increasing population.

On the other hand, since Europeans and North Americans consume so many more resources per capita than most other people in the world, a reduction in the population of these countries will do more to spare the environment than would a reduction in population almost anywhere else.

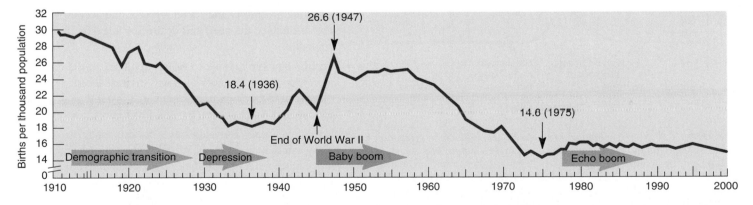

FIGURE 7.12 Birth rates in the United States, 1910–2000. The falling birth rate from 1910 to 1929 represents a demographic transition from an agricultural to an industrial society. Note that this decline occurred before the start of the Great Depression. The baby boom following World War II lasted from 1945 to 1957. A much smaller "echo boom" occurred around 1980 when the baby boomers started to reproduce, but it produced far fewer births than anticipated. *Sources:* Population Reference Bureau, Inc., and U.S. Bureau of the Census.

DEMOGRAPHIC TRANSITION

In 1945, demographer Frank Notestein pointed out that a typical pattern of falling death rates and birth rates due to improved living conditions usually accompanies economic development. He called this pattern the **demographic transition** from high birth and death rates to lower birth and death rates. Figure 7.13 shows an idealized model of a demographic transition. This model is often used to explain connections between population growth and economic development.

Development and Population

The left side of figure 7.13 represents the conditions in a premodern society. Food shortages, malnutrition, lack of sanitation and medicine, accidents, and other hazards generally keep death rates in such a society around 35 per 1,000 people. Birth rates are correspondingly high to keep population densities relatively constant. As economic development brings better jobs, medical care, sanitation, and a generally improved standard of living, death rates often fall very rapidly. Birth rates may actually rise at first as more money and better nutrition allow people to have the children they always wanted. Eventually, however, birth rates fall as people see that all their children are more likely to survive and that the whole family benefits from concentrating more resources on fewer children. Note that populations grow rapidly during the time that death rates have already fallen but birth rates remain high. Depending on how long it takes to complete the transition, the population may go through one or more rounds of doubling before coming into balance again.

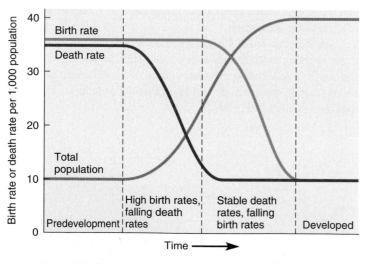

FIGURE 7.13 Theoretical birth, death, and population growth rates in a demographic transition accompanying economic and social development. In a predevelopment society, birth and death rates are both high, and total population remains relatively stable. During development, death rates tend to fall first, followed in a generation or two by falling birth rates. Total population grows rapidly until both birth and death rates stabilize in a fully developed society.

The right-hand side of each curve in figure 7.13 represents conditions in developed countries, where the transition is complete and both birth rates and death rates are low, often, a third or less than those in the predevelopment era. The population comes into a new equilibrium in this phase, but at a much larger size than before. Most of the countries of northern and western Europe went through a demographic transition in the nineteenth or early twentieth century similar to the curves shown in this figure.

Many of the most rapidly growing countries in the world, such as Kenya, Yemen, Libya, and Jordan, now are in the middle phase of this demographic transition. Their death rates have fallen close to the rates of the fully developed countries, but birth rates have not fallen correspondingly. In fact, both their birth rates and total population are higher than those in most European countries when industrialization began 300 years ago. The large disparity between birth and death rates means that many developing countries now are growing at 3 to 4 percent per year. Such high growth rates in the Third World could boost total world population to 9 billion or more before the end of the twenty-first century. This raises what may be the two most important questions in this entire chapter: Why are birth rates not yet falling in these countries, and what can be done about it?

An Optimistic View

Some demographers claim that a demographic transition already is in progress in most developing nations. Problems in taking censuses and a normal lag between falling death and birth rates may hide this for a time, but the world population should stabilize sometime in the next century. Some evidence supports this view. As we mentioned earlier in this chapter, fertility rates have fallen dramatically nearly everywhere in the world over the past half century.

Some countries have had remarkable success in population control. In Thailand, Indonesia, Colombia, and Iran, for instance, total fertility dropped by more than half in 20 years (Case Study, p. 133). Morocco, Dominican Republic, Jamaica, Peru, and Mexico all have seen fertility rates fall between 30 percent and 40 percent in a single generation. The following factors contribute to stabilizing populations:

- Growing prosperity and social reforms that accompany development reduce the need and desire for large families in most countries.

- Technology is available to bring advances to the developing world much more rapidly than was the case a century ago, and the rate of technology transfer is much faster than it was when Europe and North America were developing.

- Less-developed countries have historic patterns to follow. They can benefit from our mistakes and chart a course to stability more quickly than they might otherwise do.

- Modern communications (especially television) have caused a revolution of rising expectations that act as a stimulus to spur change and development.

A Pessimistic View

Economist Lester Brown of the Worldwatch Institute takes a more pessimistic view. He warns that many of the poorer countries of the world appear to be caught in a "demographic trap" that prevents them from escaping from the middle phase of the demographic transition. Their populations are now growing so rapidly that human demands exceed the sustainable yield of local forests, grasslands, croplands, or water resources. The resulting resource shortages, environmental deterioration, economic decline, and political instability may prevent these countries from ever completing modernization. Their populations may continue to grow until catastrophe intervenes.

Many people argue that the only way to break out of the demographic trap is to immediately and drastically reduce population growth by whatever means are necessary. They argue strongly for birth control education and bold national policies to encourage lower birth rates. Some agree with Malthus that helping the poor will simply increase their reproductive success and further threaten the resources on which we all depend. Author Garret Hardin described this view as lifeboat ethics. "Each rich nation," he said, "amounts to a lifeboat full of comparatively rich people. The poor of the world are in other much more crowded lifeboats. Continu-ously, so to speak, the poor fall out of their lifeboats and swim for a while, hoping to be admitted to a rich lifeboat, or in some other way to benefit from the goodies on board. . . . We cannot risk the safety of all the passengers by helping others in need. What happens if you share space in a lifeboat? The boat is swamped and everyone drowns. Complete justice, complete catastrophe."

A Social Justice View

A third view is that **social justice** (a fair share of social benefits for everyone) is the real key to successful demographic transitions. The world has enough resources for everyone, but inequitable social and economic systems cause maldistributions of those resources. Hunger, poverty, violence, environmental degradation, and overpopulation are symptoms of a lack of social justice rather than a lack of resources. Although overpopulation exacerbates other problems, a narrow focus on this factor alone encourages racism and hatred of the poor. A solution for all these problems is to establish fair systems, not to blame the victims. Small nations and minorities often regard calls for population control as a form of genocide. Figure 7.14 expresses the opinion of many people in less-developed countries about the relationship between resources and population.

FIGURE 7.14 Controlling our population and resources—there may be more than one side to the issue.
Used with permission of the Asian Cultural Forum on Development.

An important part of this view is that many of the rich countries are, or were, colonial powers, while the poor, rapidly growing countries were colonies. The wealth that paid for progress and security for developed countries was often extracted from colonies, which now suffer from exhausted resources, exploding populations, and chaotic political systems. Some of the world's poorest countries such as India, Ethiopia, Mozambique, and Haiti had rich resources and adequate food supplies before they were impoverished by colonialism. Those of us who now enjoy abundance may need to help the poorer countries not only as a matter of justice but because we all share the same environment.

An Ecojustice View

In addition to considering the rights of fellow humans, we should also consider those of other species. Rather than ask what is the maximum number of humans that the world can possibly support, perhaps we should think about the needs of other creatures. As we convert natural landscapes into agricultural or industrial areas, species are crowded out that may have just as much right to exist as we do. Perhaps we should seek the optimum number of people at which we can provide a fair and decent life for all humans while causing the minimum impact on nonhuman neighbors.

Women's Rights and Fertility

Opportunities for education and paying jobs are critical factors in fertility rates (fig. 7.15). Child survival also is crucial in stabilizing population. When infant and child mortality rates are high, as they are in much of the developing world, parents tend to have high numbers of children to ensure that some will survive to adulthood. There has never been a sustained drop in birth rates that was not first preceded by a sustained drop in infant and child mortality. One of the most important distinctions in our demographically divided world is the high infant mortality rates in the less-developed countries. Better nutrition, improved health care, simple oral rehydration therapy, and immunization against infectious diseases (chapter 8) have brought about dramatic reductions in child mortality rates, which have been accompanied in most regions by falling birth rates. It has

Key Concepts

- The total fertility rate is the number of children an average woman will have in a lifetime.

- The replacement fertility rate (or zero population growth rate) is approximately 2.1 children per woman.

- A demographic transition usually is defined as a change from high birth and death rates to lower birth and death rates. It usually (but not always) accompanies urbanization, industrialization, and higher standards of living.

- A social justice view holds that gender equity including education and economic security for women, access to modern contraception and family planning services, maternal and infant health care and nutrition, protection from violence, and reduction in sexually transmitted diseases all are essential for stabilizing human populations and achieving sustainable development.

been estimated that saving 5 million children each year from easily preventable communicable diseases would avoid 20 or 30 million extra births.

Increasing family income does not always translate into better welfare for children since men in many cultures control most financial assets. Often the best way to improve child survival is to ensure the rights of mothers. Land reform, political rights, opportunities to earn an independent income, and improved health status of women often are better indicators of total fertility and family welfare than rising GNP.

FAMILY PLANNING AND FERTILITY CONTROL

Family planning allows couples to determine the number and spacing of their children. It doesn't necessarily mean fewer children—people may use family planning to have the maximum number of children possible—but it does imply that the parents will control

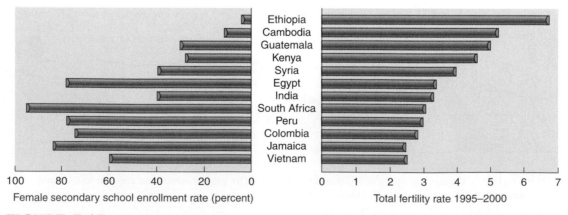

FIGURE 7.15 Total fertility declines as women's education increases. *Source:* Worldwatch Institute 2003.

their reproductive lives and make rational, conscious decisions about how many children they will have and when those children will be born, rather than leaving it to chance. As the desire for smaller families becomes more common, birth control becomes an essential part of family planning in most cases. In this context, **birth control** usually means any method used to reduce births, including celibacy, delayed marriage, contraception, methods that prevent implantation of embryos, and induced abortions.

Traditional Fertility Control

Evidence suggests that people in every culture and every historic period have used a variety of techniques to control population size. Studies of hunting and gathering people, such as the !Kung or San of the Kalahari Desert in southwest Africa, indicate that our early ancestors had stable population densities, not because they killed each other or starved to death regularly, but because they controlled fertility.

For instance, San women breast-feed children for three or four years. When calories are limited, lactation depletes body fat stores and suppresses ovulation. Coupled with taboos against intercourse while breast-feeding, this is an effective way of spacing children. Other ancient techniques to control population size include celibacy, folk medicines, abortion, and infanticide. We may find some or all of these techniques unpleasant or morally unacceptable, but we shouldn't assume that other people are too ignorant or too primitive to make decisions about fertility.

Current Birth Control Methods

Modern medicine gives us many more options for controlling fertility than were available to our ancestors (fig. 7.16). Some of these techniques are safer, easier, or more pleasant to use. The major categories of birth control techniques include (1) avoidance of sex during fertile periods [celibacy; using changes in body temperature or cervical mucus color and viscosity to judge when ovulation will occur]; (2) mechanical barriers that prevent contact between sperm and egg [condoms, spermicides, diaphragm, cervical cap, and vaginal sponge]; (3) surgical methods that prevent release of sperm or egg [sterilization: tubal ligation or use of the Filshie clip in females, vasectomy in males]; (4) chemicals that prevent maturation or release of sperm or eggs or implantation of the embryo in the uterus [the pill: estrogen and progesterone, progesterone alone for females; gossypol for males]; (5) physical barriers to implantation [IUD]; and (6) abortion.

Norplant, the trade name for flexible, matchstick-sized, silicon-rubber implants containing a slow-release analog of progesterone, was approved for use in the United States. The implants are inserted under the skin where they will release hormones for up to five years. Depo-Provera, an injectable progesterone analog, is given by injections four times a year. Both implants and injections have very low failure rates (0.3 percent compared to about 1 percent for oral contraceptives) and eliminate the need to keep track of and take daily pills. They also can be used without knowledge of one's partner, who may oppose birth control. Both injections

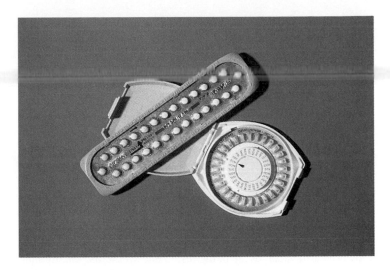

FIGURE 7.16   Modern birth control methods can be highly effective when used properly. © The McGraw-Hill Companies, Inc./Bob Coyle, photographer.

and implants cause problems for some women who experience increased vaginal bleeding or absence of menstrual periods. Norplant may be linked to ovarian cysts and the inserts have been difficult to remove.

Condom use has more than doubled over the past 20 years, from about 3.5 million users in 1980 to 8 million in 1999. While condoms have about a 10 percent failure rate when used alone, their effectiveness can be increased by combining them with spermicidal creams. Condoms are important protection against sexually transmitted diseases.

None of these methods is perfect and none suits every contraceptive need. Many require careful, conscientious use. Which choice is best for you depends on your life situation and your plans for the future.

For nearly 20 years, the French drug, RU486 (mifepristone or mifegyne), has been used as a "morning after" pill in Europe. RU486 blocks the effects of progesterone in maintaining the lining of the uterine wall. It is usually administered together with misoprostol (a prostaglandin analog), which causes uterine contraction and expulsion of the fetus. Interestingly, RU486 also appears to have promise in treating breast cancer, brain cancer, diabetes, and hypertension. RU486 has been approved for use in the United States, but restrictions on its use may make it unavailable for many women.

Some drugs already on the market for other uses have been shown to be as safe and effective as RU486. Methotrexate and misoprostol are administered a week apart to induce abortion. "Emergency contraception," taken within 72 hours of unprotected sex, uses oral hormones similar to standard birth control pills to prevent implantation of an egg.

New Developments in Birth Control

More than 100 new contraceptive methods are now being studied, and some appear to have great promise. In the past two years, the

U.S. Food and Drug Administration (FDA) approved five new birth control products:

- Ensure, a springlike device that blocks the fallopian tubes in a noninvasive alternative to tubal ligation.
- Mirena, a hormone-releasing intrauterine device that can stay in place for five years.
- Lunelle, a hormone shot administered monthly.
- NuvaRing, a hormonal vaginal ring changed every three weeks.
- Ortho Evra, a hormone patch changed weekly.

A group of drugs known as gonadotropin releasing-hormone agonists show promise in suppressing egg and sperm development. Some antipregnancy vaccines (immunization against chorionic gonadotropin—a hormone required to maintain the uterine lining) and antisperm vaccines are being tested that would use the immune system to prevent fertilization or embryonic implantation, but they are years away from the market. Clinical trials of hormone injections (progestin and testosterone) and calcium channel blockers (also used to treat high blood pressure) have shown some promise in suppressing sperm production but it may be years before they are ready to market or before men will be willing to take them. A drug known as N-butyldeoxynorjirimycin or NB-DJN prevents sperm production in mice with no observed side effects. Whether it will prove effective in humans remains to be seen.

THE FUTURE OF HUMAN POPULATIONS

How many people will be in the world a century from now? Most demographers believe that world population will stabilize sometime during the next century. The total number of humans, when we reach that equilibrium, is likely to be somewhere around 8 to 10 billion people, depending on the success of family planning programs and the multitude of other factors affecting human populations. Figure 7.17 shows three scenarios projected by the UN Population Division in its 2003 revision. The optimistic (low) projection shows that world population might reach about 8 billion in 2050, and then fall back to about where it is today. The medium projection suggests that growth might continue to around 9.3 billion in 2050, and then stabilize. The most pessimistic projection assumes a constant rate of growth (no change from present) to 25 billion people by 2150.

Which of these scenarios will we follow? As you have seen in this chapter, population growth is a complex subject. To accomplish a stabilization or reduction of human populations will require substantial changes from business as usual.

An encouraging sign is that worldwide contraceptive use has increased sharply in recent years. About half of the world's married couples used some family planning techniques in 2000, compared to only 10 percent 30 years earlier, but another 100 million couples say they want, but do not have access to, family planning. Contraceptive use varies widely by region, with high levels in Latin America and East Asia but relatively low use in much of Africa.

Figure 7.18 shows the unmet need for family planning among married women in some representative countries. When people in developing countries are asked what they want most, men say they want better jobs, but the first choice for a vast majority of women is family planning assistance. In general, a 15 percent increase in contraceptive use equates to about one fewer birth per woman per lifetime. In Chad, for example, where only 4 percent of all women use contraceptives, the average fertility is 6.6 children per woman. In Columbia, by contrast, where 77 percent of the women who would prefer not to be pregnant use contraceptives, the average fertility is 2.6.

In July 2002, U.S. President George Bush announced that he would not release the $34 million dollars appropriated by Congress

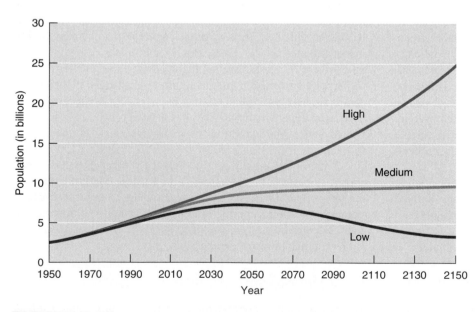

FIGURE 7.17   Estimated and projected world population, 1950 to 2150, with different fertility levels. *Source:* United Nations Population Division, 2003.

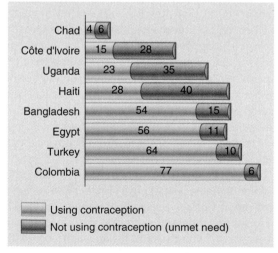

FIGURE 7.18   Unmet need for family planning in selected countries. Globally, more than 100 million women in developing countries would prefer to avoid pregnancy but do not have access to family planning.
Source: Population Reference Bureau, 2003.

for the United Nations Population Fund (UNFPA). He claimed that the fund, by working in China, tacitly supports the forced abortions reported to be part of that country's one-child policy. Ms. Thoraya Obaid, executive director of UNFPA, replied that "UNFPA has not, does not, and will not ever condone or support coercive activities of any kind, anywhere." She also said, "The denial of these funds will, unfortunately, significantly affect millions of women and children worldwide for whom the life-saving services provided by the UNFPA will have to be discontinued." She estimated that the $34 million withheld by the United States could have prevented 2 million unwanted pregnancies, 800,000 abortions, 4,700 maternal deaths, 60,000 cases of serious maternal illness, and more than 77,000 infant and child deaths. The United States is the only country ever to deny funding to UNFPA for nonbudgetary reasons.

The UNFPA is the world's largest international source of funding for population and reproductive health programs. Since it began operations in 1969, the Fund has provided nearly $6 billion in assistance to developing countries. UNFPA works with governments and nongovernmental organizations in over 140 countries to help women, men, and young people plan their families and avoid accidental pregnancies, undergo pregnancy and childbirth safely, avoid sexually transmitted diseases including HIV/AIDS, and combat violence against women. Access to contraception and education, together with maternal and child health care, are mainly responsible for plummeting birth rates that have slowed global population growth. Since the UNFPA began its work, access to voluntary family planning programs in developing countries has increased and fertility has fallen by half, from six children per woman to three. Nearly 60 percent of married women in developing countries are now able to practice modern contraception methods, compared with 10–15 percent when the organization was founded.

The UNFPA takes the position that reproductive health is a fundamental human right. The association's work is guided by the goals adopted by 179 governments at the International Conference on Population and Development in Cairo in 1994, which said that

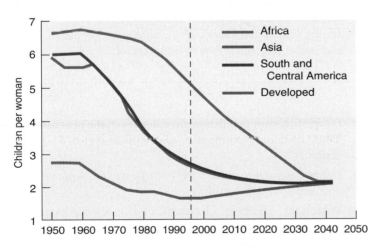

FIGURE 7.19   Fertility declines, real and projected, 1950–2050.
Source: U.N. Population Division, *World Population Prospects, 1996.*

The "34 Million Friends" of the UNFPA

When Jane Roberts of Redlands, California, sat down at her computer to express her outrage at the U.S. decision to withdraw $34 million from UNFPA, she had no idea that a campaign had been born. She was also unaware that Lois Abraham of Taos, New Mexico, was doing the same thing.

Both women sent emails to friends, urging them to donate one dollar or more to UNFPA to help bridge the 12.5 percent funding gap caused by the U.S. withdrawal of funds in July 2002. They reasoned that if 34 million people would give $1 each, the U.S. contribution could be restored. So they started a grass-roots movement, called the "34 Million Friends" campaign.

Through the power of electronic networking, Jane and Lois reached out to individuals and organizations across the country, including Planned Parenthood, Sierra Club, National Women's Federation, and Rotary. The campaign has been a resounding success. In its first year, the friends raised more than $1 million for the UNFPA. Donations are from men and women, within and outside the United States, who firmly believe that all people should have access to family planning and reproductive health services.

If you'd like to learn more about this organization, and how you can support their mission, visit their website: http://www.unfpa.org/support/friends/34million.htm.

meeting people's needs for education and health, including reproductive health, is a prerequisite of sustainable development.

In a related incident, the United States attempted to block a plan of action proposed at the UN-sponsored Asian and Pacific Population Conference in Bangkok, Thailand, in December 2002. The plan, which was based on the principles of the 1994 Cairo conference, was intended to reduce poverty in the region by concentrating on 12 areas including family planning, gender equality, and combating HIV/AIDS. The United States objected to the use of the terms "reproductive health services" and "reproductive rights," asserting they could be read as advocating abortion and condom use by adolescents. Other nations, however, voted 32 to 1 to accept the plan without change. Ms. Thoraya Obaid of the UNFPA welcomed passage of the plan, and said the United States would have an opportunity to express its concerns in a separate document.

If you agree with the goals and policies of the UNFPA, you can support its work with an individual donation (What Can You Do? p. 143).

The World Health Organization estimates that nearly 1 million conceptions occur daily around the world as a result of some 100 million sex acts. At least half of those conceptions are unplanned or unwanted. Still, birth rates already have begun to fall in East Asia and Latin America (fig. 7.19). Similar progress is expected in South Asia in a few years. Only Africa will probably continue to grow in the twenty-first century.

Deep societal changes are often required to make family planning programs successful. Among the most important of these are

(1) improved social, educational, and economic status for women (birth control and women's rights are often interdependent); (2) improved status for children (fewer children are born as parents come to regard them as valued individuals rather than possessions); (3) acceptance of calculated choice as a valid element in life in general and in fertility in particular (belief that we have no control over our lives discourages a sense of responsibility); (4) social security and political stability that give people the means and the confidence to plan for the future; (5) knowledge, availability, and use of effective and acceptable means of birth control.

Concerted efforts to bring about these changes can be effective. Twenty years of economic development and work by voluntary family planning groups in Zimbabwe, for example, have lowered total fertility rates from 8.0 to 5.5 children per woman on average. Surveys showed that desired family sizes have fallen nearly by half (9.0 to 4.6) and that nearly all women and 80 percent of men in Zimbabwe use contraceptives. If similar progress can be sustained elsewhere, human populations may be restrained after all. The choices all of us make—both in our private lives and in public policies—will determine our future course.

Summary

- Human populations have grown at an unprecedented rate over the past three centuries. By 2003, the world population stood at 6.3 billion people.

- If the current growth rate of 1.3 percent per year persists, the population will double in 54 years. Most of that growth will occur in the less-developed countries of Asia, Africa, and Latin America. There is a serious concern that the number of humans in the world and our impact on the environment will overload the life-support systems of the earth.

- The crude birth rate is the number of births in a year divided by the average population. A more accurate measure of growth is the general fertility rate, which takes into account the age structure and fecundity of the population.

- The crude birth rate minus the crude death rate gives the rate of natural increase. When this rate reaches a level at which people are just replacing themselves, zero population growth is achieved.

- The change from high birth and death rates that accompanies industrialization is called a demographic transition. Many developing countries have already begun this transition. Death rates have fallen, but birth rates remain high.

- Some demographers believe that as infant mortality drops and economic development progresses so that people in these countries can be sure of a secure future, they will complete the transition to a stable population. Others fear that excessive population growth and limited resources will catch many of the poorer countries in a demographic trap that could prevent them from ever achieving a stable population or a high standard of living.

- While larger populations bring many problems, they also may be a valuable resource of energy, intelligence, and enterprise that will make it possible to overcome resource limitation problems. A social justice view argues that a more equitable distribution of wealth might reduce both excess population growth and environmental degradation.

- We have many more options now for controlling fertility than were available to our ancestors.

- Sometimes successful family planning requires deep cultural changes such as improved social, educational, and economic status for women; higher values on individual children; accepting responsibility for our own lives; social security and political stability that give people the means and confidence to plan for the future; and knowledge, availability, and use of effective and acceptable means of birth control.

Questions for Review

1. At what point in history did the world population pass its *first* billion? What factors restricted population before that time, and what factors contributed to growth after that point?

2. How might growing populations be beneficial in solving development problems?

3. Why do some economists consider human resources more important than natural resources in determining the future of a country?

4. Where will most population growth occur in the next century? What conditions contribute to rapid population growth in some countries?

5. Define *crude birth rate, total fertility rate, crude death rate,* and *zero population growth.*

6. What is the difference between life expectancy and longevity?

7. What is dependency ratio, and how might it affect the United States in the future?

8. What pressures or interests make people want or not want to have babies?

9. Describe the conditions that lead to a demographic transition.

10. Describe the major choices in modern birth control.

Questions for Critical Thinking

1. What do you think is the optimum human population? The maximum human population? Are the numbers different? If so, why?

2. Some people argue that technology can provide solutions for environmental problems; others believe that a "technological fix" will make our problems worse. What personal experiences or worldviews do you think might underlie these positions?

3. Karl Marx called Thomas Malthus a "shameless sycophant of the ruling classes." Why would the landed gentry of the eighteenth century be concerned about population growth of the lower classes? Are there comparable class struggles today?

4. Try to imagine yourself in the position of a person your age in a developing world country. What family planning choices and pressures would you face? How would you choose among your options?

5. Some demographers claim that population growth has already begun to slow; others dispute this claim. How would you evaluate the competing claims of these two camps? Is this an issue of uncertain facts or differing beliefs? What sources of evidence would you accept as valid?

6. What role do race, ethnicity, and culture play in our immigration and population policies? How can we distinguish between prejudice and selfishness on one hand and valid concerns about limits to growth on the other?

7. President Bush says the UNFPA condones abortion by working in countries where abortion is allowed. Do you agree? Does giving family planning advice constitute support for abortion? Why or why not?

Key Terms

birth control 141
crude birth rate 131
crude death rates 132
demographic transition 138
demography 130
dependency ratio 135
family planning 140
life expectancy 133

natural increase 133
neo Malthusian 129
pronatalist pressures 136
social justice 139
total fertility rate 131
total growth rate 133
zero population growth (ZPG) 132

Further Readings

Ashford, Lori. 2003. *Unmet Need for Family Planning: Recent Trends and Their Implications for Programs.* Population Reference Bureau. Published online at http://www.prb.org.

Dasgupta, Partha S. 1995. Population, poverty and the local environment. *Scientific American* (February 1995), pp. 40–45.

Ehrlich, Paul R. 2000. *Human Nature: Genes, Cultures, and the Human Prospect.* Island Press.

Hartmann, Betsy. 1999. Population, environment, and security: A new trinity. In *Dangerous Intersections: Feminist Perspectives on Population, Environment, and Development.* Jael Silliman and Ynestra King, eds. South End Press, pp. 1–23.

Pimentel, David, et al. 1999. Will limits of the earth's resources control human numbers? *Environment, Development and Sustainability.* 1:19–39.

Sen, Amaryta. 1994. Population and reasoned agency: Food, fertility, and economic development. In *Population, Economic Development, and the Environment.* Kerstin Lindahl-Kiessling and Hans Landberg, eds. Oxford University Press, pp. 51–78.

Welcome to McGraw-Hill's Online Learning Center

Location: http://www.mhhe.com/environmentalscience

McGraw Hill

WEB EXERCISES

Exploring Growth Factors in Population Data

How and why populations grow are key questions in environmental science. In this exercise, you examine and graph current world population data to explore which factors most strongly correlate with birth rates. Go to this book's online learning center (www.mhhe.com/environmentalscience) and click on Chapter 7/Web Exercises.

There you will find an Excel data file named popdata.xls. Double-click on the file name to copy it to your hard disk. If you have Excel on your computer, you should be able to open the data file by double-clicking on it. (Other spreadsheet programs can also read this file, but you must open it from within your program, not by double-clicking.)

1. This file contains population data for the countries of the world, sorted by the United Nations Human Development Index (HDI) rank. First look at the top 20 countries. Where are they? What is the range of income levels (in GNP per capita) of the top 20? What is the range of income for the bottom 20 countries?

2. Now make an X,Y scatter graph of Adult Literacy and Birth Rate. (Detailed instructions for making graphs in Excel are included at the far right side of the spreadsheet page [column N].) How would you describe the relationship between these variables? How would you explain this relationship? Keep this graph in your spreadsheet while you make three more scatter graphs: (1) GNP per Capita and Birth Rate, (2) Life Expectancy and Birth Rate, and (3) Infant Mortality and Birth Rate. Describe the trends you observe, and explain what they mean.

3. How would you compare the relative amount of scatter in each of your graphs? Why do some curves slope from right to left, while others slope in the opposite direction? If you draw a line through the middle of the dot cluster, some curve smoothly, while others seem to have a break or inflection point. How would you interpret these patterns?

4. Try changing the shape of your graphs. (See instructions on the right side of the spreadsheet to do this.) Would manipulating the graph shape change how people interpret it? Is this ethical?

5. Now make a dot graph of GNP Per Capita and Adult Literacy. Is there a linear relationship between the two variables? Why or why not?

8

Environmental Health and Toxicology

To wish to become well is a part of becoming well.

Seneca

OBJECTIVES

After studying this chapter, you should be able to:

* define *health* and *disease* and describe how global disease burden is now changing.
* identify some major infectious organisms and hazardous agents that cause environmental diseases.
* identify examples of emergent human and ecological diseases.
* distinguish between toxic and hazardous chemicals and between chronic and acute exposures and responses.
* compare factors that affect toxin movement and persistence in the environment as well as routes of entry and effects in our bodies.
* evaluate the major environmental risks we face and how risk assessment and risk acceptability are determined.

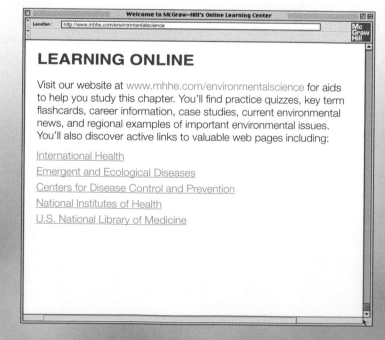

Photo: Hong Kong residents protect themselves from Severe Acute Respiratory Syndrome (SARS), a highly infectious, pneumonia-like emergent disease. *Source:* AP Photo/Anat Givon.

The Cough Heard Round the World

Early in 2003, news began to trickle out of China that a very infectious "atypical pneumonia" was spreading rapidly in hospitals around Guangzhou (formerly known as Canton) in Guangdong Province. Symptoms included fever, chills, headaches, muscle pains, and a dry cough, but in many patients, especially the elderly, the disease would quickly turn into a deadly pneumonia. In February of 2003, a doctor from Guangzhou, who had contracted this disease from his patients, traveled to Hong Kong. There he passed the infection to other travelers, who carried what is now know as Severe Acute Respiratory Syndrome (SARS) to Beijing, Canada, Taiwan, Singapore, and Vietnam.

Within six months, SARS spread to 31 countries around the world, where more than 8,500 probable cases and 812 deaths were reported. Fear of SARS traveled even faster and farther than the disease itself. Rumors multiplied across the Internet as conferences and sporting events were canceled, factories closed, and tourism to China fell by as much as 85 percent after the epidemic was revealed. The economic impact of SARS is estimated to be at least $30 billion (U.S.) in 2003 alone. By July 2003, the World Health Organization (WHO) declared the outbreak contained, but suggested that the world remain vigilant for further infections.

The rapid transmission of this disease shows how interconnected we all are. A virus can travel in just a few days anywhere a plane flies. One flight attendant is thought to have been the source of infection for 160 people in seven countries. SARS also points to the need for better communication and identification of new diseases. Because the Chinese government hid the extent and severity of the disease for several months, patients with this highly infectious disease mingled with the general hospital population, spreading the illness. Medical professionals, not knowing how serious the infection was, treated patients without wearing protective clothing. Major hospitals in Beijing, Taipei, Hong Kong, and Singapore were closed because so many staff members were sick, making treatment of the contagion even more difficult.

While globalization helped spread SARS, it also helped in rapid recognition and treatment of the disease. It took centuries to discover the cause of cholera. Identifying the virus that causes AIDS took two years. But within weeks after the WHO issued its first warning about SARS, an electron microscopist at the Centers for Disease Control in Atlanta found a new coronavirus in cell cultures infected with tissue from SARS patients. Less than a month later, labs from Vancouver to Singapore were sequencing its RNA. In May 2003, Scientists at Hong Kong University announced they had found coronavirus nearly identical to those from SARS patients in civets, badgers, and raccoon dogs being sold in Guangdong meat markets.

Wild species had been suspected as a source of the SARS virus since some of the first infections occurred among chefs and animal merchants. Exotic animals are regarded as delicacies in southern China, where they are featured at banquets and dinners at expensive restaurants. In April of 2003, Chinese police raided animal markets and hotels and restaurants, seizing 838,500 animals including many rare and endangered species. The emergence of SARS may reduce animal smuggling, but the existence of a reservoir of this virus in the wild may mean that it will be impossible to completely eradicate the disease.

SARS deaths, so far, are relatively insignificant compared to the 3 million people who die from AIDS or the 1 million who die from malaria every year, but we tend to fear new risks with unknown causes, while ignoring more routine but perhaps more dangerous risks that we believe we can control. Still, the emergence of SARS reminds us how susceptible we remain to infectious diseases in an interconnected world. Dr. Jong-wook Lee, the newly elected head of the WHO said, "SARS is the first new disease of the twenty-first century, but it will not be the last." The U.S. Institute of Medicine warns that gaps in our defenses against biological assault from both terrorists and natural sources makes us vulnerable to other deadly epidemics. In this chapter, we'll look at some principles of environmental health to help you understand some of the risks we face and what we might do about them.

ENVIRONMENTAL HEALTH

What is health? The WHO defines **health** as a state of complete physical, mental, and social well-being, not merely the absence of disease or infirmity. By that definition, we all are ill to some extent. Likewise, we all can improve our health to live happier, longer, more productive, and more satisfying lives if we think about what we do.

What is disease? A **disease** is an abnormal change in the body's condition that impairs important physical or psychological functions. Diet and nutrition, infectious agents, toxic substances, genetics, trauma, and stress all play roles in **morbidity** (illness) and **mortality** (death). **Environmental health** focuses on external factors that cause disease, including elements of the natural, social, cultural, and technological worlds in which we live. Figure 8.1 shows some major environmental disease agents as well as the media through which we encounter them. Ever since the publication of Rachel Carson's *Silent Spring* in 1962, the discharge, movement, fate, and effects of synthetic chemical toxins have been a special focus of environmental health. Later in this chapter, we'll study these topics in detail. First, however, let's look at some of the major causes of illness worldwide.

Global Disease Burden

In the past, health organizations have focused on the leading causes of death as the best summary of world health. Mortality data, however, fail to capture the impacts of nonfatal outcomes of disease and injury, such as dementia or blindness, on human well-being. When people are ill, work isn't done, crops aren't planted or harvested, meals aren't cooked, and children can't study and learn. Health agencies now calculate **disability-adjusted life years (DALYs)** as a measure of disease burden. DALYs combine premature deaths and loss of a healthy life resulting from illness or disability. This is an attempt to evaluate the total cost of disease, not simply how many people die. Clearly, many more years of expected life are lost when a child dies of neonatal tetanus than when an 80-year-old dies

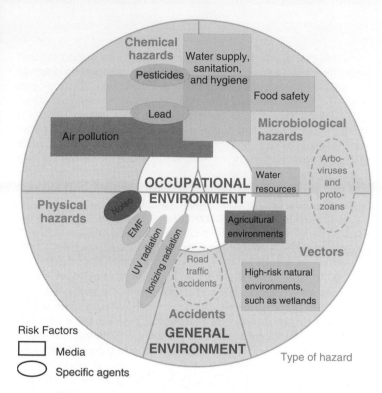

Risk Factors

▢ Media

⬭ Specific agents

FIGURE 8.1 Major sources of environmental health risks.
Source: WHO, 2002.

of pneumonia. Similarly, a teenager permanently paralyzed by a traffic accident will have many more years of suffering and lost potential than will a senior citizen who has a stroke. According to the WHO, chronic diseases now account for nearly 60 percent of the 56.5 million total deaths worldwide each year and about half of the global disease burden.

The world is now undergoing a dramatic epidemiological transition. Chronic conditions, such as cardiovascular disease and cancer, no longer afflict only wealthy people. Marvelous progress in eliminating communicable diseases, such as smallpox, polio, and malaria, is allowing people nearly everywhere to live longer. As chapter 7 points out, over the past century the average life expectancy worldwide has risen by about two-thirds. In some poorer countries, like India, life expectancies nearly tripled in the twentieth century. Although the traditional killers in developing countries—infections, maternal and perinatal (birth) complications, and nutritional deficiencies—still take a terrible toll, diseases such as depression and heart attacks that once were thought to occur only in rich countries are rapidly becoming the leading causes of disability and premature death everywhere.

In 2020, the WHO predicts heart disease, which was fifth in the list of causes of global disease burden a decade ago, will be the leading source of disability and deaths worldwide (table 8.1). Most of that increase will be in the poorer parts of the world where people are rapidly adopting the lifestyles and diet of the richer countries. Similarly, global cancer rates will increase by 50 percent. By 2020, it's expected that 15 million people will have cancer and 9 million will die from it.

Ask American women what disease they're most afraid of and a majority will probably answer breast cancer. What many don't realize is that cardiovascular disease is the leading cause of death among U.S. women. While focusing on the 40,000 women who die each year from breast cancer we aren't aware of the 500,000 who die in that same time from heart attacks and strokes. For reasons we don't fully understand, women are far more likely than men to be disabled or die as a result of heart attack or strokes. Forty-six percent of women are seriously disabled by their first heart attack, for instance, compared to only 22 percent of men. Smoking, diabetes, high blood pressure, high cholesterol, excess weight, and lack of physical activity increase cardiovascular disease risks in both women and men, but women are less likely than men to be aware of the importance of these health factors.

Taking disability as well as death into account in our assessment of disease burden reveals the increasing role of mental health as a worldwide problem. WHO projections suggest that psychiatric and neurological conditions could increase their share of the global burden from 10 percent currently to 15 percent of the total load by 2020. Again, this isn't just a problem of the developed world. Depression is expected to be the second largest cause of all years lived with disability worldwide, as well as the cause of 1.4 percent of all deaths. For women in both developing and developed regions, depression is the leading cause of disease burden, while suicide, which often is the result of untreated depression, is the fourth largest cause of female deaths.

Notice in table 8.1 that diarrhea, which was the second leading cause of disease burden in 1990, is expected to be ninth on the list in 2020, while measles and malaria are expected to drop out of the top 15 causes of disability. Tuberculosis, which is becoming

TABLE 8.1 Leading Causes of Global Disease Burden

RANK	1990	RANK	2020
1	Pneumonia	1	Heart disease
2	Diarrhea	2	Depression
3	Perinatal conditions	3	Traffic accidents
4	Depression	4	Stroke
5	Heart disease	5	Chronic lung disease
6	Stroke	6	Pneumonia
7	Tuberculosis	7	Tuberculosis
8	Measles	8	War
9	Traffic accidents	9	Diarrhea
10	Birth defects	10	HIV/AIDS
11	Chronic lung disease	11	Perinatal conditions
12	Malaria	12	Violence
13	Falls	13	Birth defects
14	Iron anemia	14	Self-inflicted injuries
15	Malnutrition	15	Respiratory cancer

Source: World Health Organization, 2002.

resistant to antibiotics and is spreading rapidly in many areas (especially in the former Soviet Union), is the only infectious disease whose ranking is not expected to change over the next 20 years. Traffic accidents are now soaring as more people drive. War, violence, and self-inflicted injuries similarly are becoming much more important health risks than ever before.

Chronic obstructive lung diseases (e.g., emphysema, asthma, and lung cancer) are expected to increase from eleventh to fifth in disease burden by 2020. A large part of the increase is due to rising use of tobacco in developing countries, sometime called "the tobacco epidemic." Every day about 100,000 young people—most of them in poorer countries—become addicted to tobacco. At least 1.1 billion people now smoke, and this number is expected to increase at least 50 percent by 2020. If current patterns persist, about 500 million people alive today will eventually be killed by tobacco. This is expected to be the biggest single cause of death worldwide (because illnesses such as heart attack and depression are triggered by multiple factors). In 2003, the World Health Assembly adopted a historic tobacco-control convention that requires countries to impose restrictions on tobacco advertising, establish clean indoor air controls, and clamp down on tobacco smuggling. Dr. Gro Harlem Brundtland, former director-general of the WHO, predicted that the convention, if passed, could save billions of lives.

Emergent and Infectious Diseases

Although the ills of modern life have become the leading killers almost everywhere in the world, communicable diseases still are responsible for about one-third of all disease-related mortality. Diarrhea, acute respiratory illnesses, malaria, measles, tetanus, and a few other infectious diseases kill about 11 million children under age five every year in the developing world. Better nutrition, clean water, improved sanitation, and inexpensive inoculations could eliminate most of those deaths (fig. 8.2).

A wide variety of **pathogens** (disease-causing organisms) afflict humans including viruses, bacteria, protozoans (single-celled animals), parasitic worms and flukes (fig. 8.3). Probably the greatest loss of life from an individual disease in a single year was the great influenza epidemic of 1918. Nearly every year a new version of the flu virus appears, but most are merely unpleasant rather than fatal for healthy adults. In 1918, however, a particularly virulent flu spread rapidly across the globe, killing somewhere around 30 million people in less than 12 months. This was more than the total number killed in all the battles of World War I, which was raging at the time.

Giardia, a parasitic intestinal protozoan, is thought to be the largest single cause of diarrhea in the United States. It is spread from human feces through food and water. You may think of this as a wilderness disease, but researchers report that the best place to find the pathogens is day care centers and nursery schools. At any given time, somewhere around 2 billion people—one third of the world population—suffer from worms, flukes, and other internal parasites. Guinea worms (*Dracunculus medinesis*) are round worms transmitted via contaminated water. After a year of migrat-

FIGURE 8.2   At least 3 million children die every year from easily preventable diseases. This billboard in Guatemala encourages parents to have their children vaccinated against polio, diphtheria, TB, tetanus, pertussis (whooping cough), and scarlet fever. © William P. Cunningham.

ing through the body, the adult, which can be 1 meter long, emerges to lay eggs (fig. 8.3c). A campaign to eradicate this scourge has eliminated it from most countries.

Malaria is one of the most prevalent remaining infectious diseases. Every year, about 300 million new cases of this disease occur, and about 1 million die from it. The territory infected by this disease is expanding as global climate change allows mosquito vectors to move into new territory. Simply providing insecticide-treated bed nets and a few dollars worth of chloroquine pills could prevent tens of millions of cases of this debilitating disease every year. Tragically, some of the countries where malaria is most widespread tax both bed nets and medicine as luxuries, placing them out of reach for ordinary people.

Emergent diseases are those not previously known or that have been absent for at least 20 years. The story of SARS that introduced this chapter is a good example of an emergent disease. Although coronaviruses have long been known to cause a variety of diseases—some lethal—in animals, and two members of this virus family cause about 30 percent of all human colds, the particularly virulent form that appears to have jumped from wild animals to humans in southern China had been previously unknown to science. Figure 8.4 shows some recent outbreaks of emergent diseases around the world. Ebola, which can kill 90 percent of those infected, is thought to have spread to humans in central Africa when they ate infected monkeys or chimpanzees.

Some health experts warn that West Nile virus may be more deadly for the United States than SARS. West Nile belongs to a family of mosquito-transmitted viruses that cause encephalitis (brain

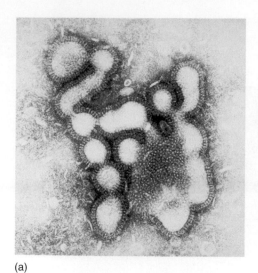

(a)

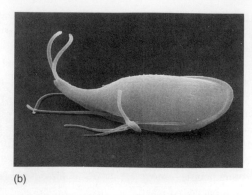

(b)

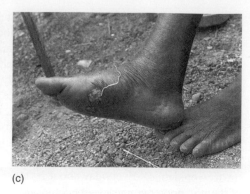

(c)

FIGURE 8.3 (*a*) A group of influenza viruses magnified about 300,000 times. (*b*) *Giardia,* a parasitic intestinal protozoan, magnified about 10,000 times. (*c*) A guinea worm emerges from the foot of an infected person. (*a*) © Corbis/Volume40. (*b*) Courtesy of Stanley Erlandsen, University of Minnesota. (*c*) Courtesy Donald R. Hopkins.

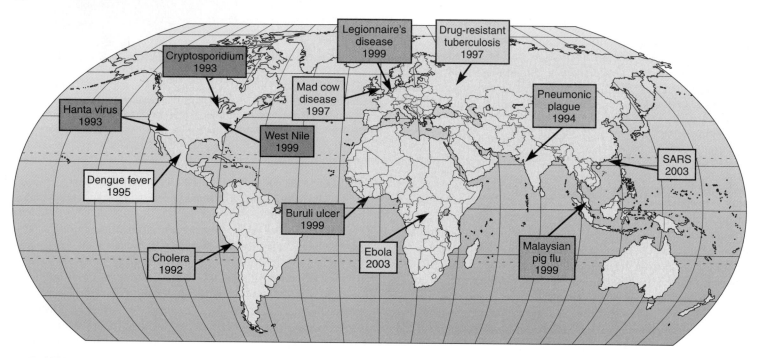

FIGURE 8.4 Some recent outbreaks of highly lethal, infectious diseases. Why are supercontagious organisms emerging in so many different places? *Source:* U.S. Centers for Disease Control and Prevention.

inflammation). Although recognized in Africa in 1937, the West Nile virus was absent from North America until 1999, when it apparently was introduced by a bird or mosquito from the Middle East. The disease spread rapidly from New York, where it was first reported, throughout the eastern United States in only two years (fig. 8.5). The virus infects 230 species of animals, including 130 bird species. In 2002, more than 4,000 people contracted the disease and 284 died in the United States. There's a worry that if the disease reaches Hawaii, it could completely eradicate rare species such as the Hawaiian crow and several types of honeycreeper.

The largest recent death toll from an emergent disease is HIV/AIDS. Although virtually unknown 15 years ago, acquired immune deficiency syndrome has now become the fifth greatest cause of contagious deaths. The WHO estimates that 60 million people are now infected with the human immune-deficiency virus, and that 3 million die every year from AIDS complications. Although two-thirds of all current HIV infections are now in sub-Saharan Africa, the disease is spreading rapidly in South and East Asia. Over the next 20 years, there could be an additional 65 million AIDS deaths. In Botswana, health officials estimate about 40 percent of all adults are HIV positive and that two-thirds of all current 15-year-olds will die of AIDS before age 50. As chapter 7 points out, without AIDS, the life expectancy in Botswana would be expected to be 69.7 years. With AIDS, Botswana's average life expectancy is now 36.1 years. Worldwide, more than 14 million children—the equivalent of every child under age five in America—

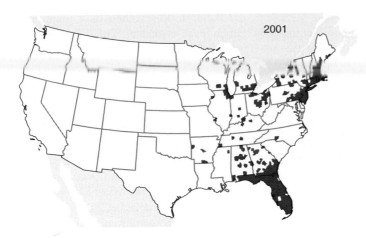

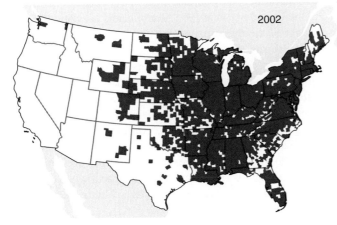

FIGURE 8.5 The spread of West Nile virus in birds, 2001–2002. *Source:* CDC and USGS.

have lost one or both parents to AIDS. The economic costs of treating patients and lost productivity from premature deaths resulting from this disease are estimated to be at least $35 billion (U.S.) per year or about one-tenth of the total GDP of sub-Saharan Africa.

Go to the Online Learning Center at www.mhhe.com/environmentalscience for related essays about Ebola fever, eradicating Guinea worms, and biological terrorism.

Funding Health Care

The heaviest burden of illness is borne by the poorest people who can afford neither a healthy environment nor adequate health care. Women in sub-Saharan Africa, for example, suffer six times the disease burden per 1,000 population as do women in most European countries. The WHO estimates that 90 percent of all disease burden occurs in developing countries where less than one-tenth of all health care dollars is spent. The group Medecins Sans Frontieres (MSF, or Doctors Without Borders) calls this the 10/90 gap. While wealthy nations pursue drugs to treat baldness and obesity, depression in dogs, and erectile dysfunction, billions of people are sick or dying from treatable infections and parasitic diseases to which little attention is paid. Worldwide, only 2 percent of the

people with AIDS have access to modern medicines. Every year, some 600,000 infants acquire HIV—almost all of them through mother-to-child transmission during birth or breast-feeding. Antiretroviral therapy costing only a few dollars can prevent most of this transmission. The Bill and Melinda Gates Foundation has pledged $200 million for medical aid to developing countries to help fight AIDS, TB, and malaria.

Dr. Jeffrey Sachs of the Columbia University Earth Institute says that disease is as much a cause as a consequence of poverty and political unrest, yet the world's richest countries now spend just $1 per person per year on global health. He predicts that raising our commitment to about $25 billion annually (about 0.1 percent of the annual GDP of the 20 richest countries) would not only save about 8 million lives each year, but would boost the world economy by billions of dollars. There also would be huge social benefits for the rich countries in not living in a world endangered by mass social instability, the spread of pathogens across borders, and the spread of other ills such as terrorism and drug trafficking caused by social problems. Sachs also argues that reducing disease burden would help reduce population growth. When parents believe their offspring will survive, they have fewer children and invest more in food, health, and education for smaller families.

The United States is among the least generous of the world's rich countries, donating only about 12 cents per $100 of GDP on international development aid. Could this country do better? During this time of fear of terrorism and rising anti-American feelings around the globe, it's difficult to interest legislators in international aid, and yet, helping to reduce disease might win the United States more friends and make the nation safer than buying more bombs and bullets. Improved health care in poorer countries may also help prevent the spread of emergent diseases like SARS in a globally interconnected world.

At the 2003 meeting of the Global Health Council, epidemiologists noted that almost all of the 2.2 billion people expected to be added to the world population in the next 30 years will live in megacities of the developing world. The economic and environmental conditions in those cities will have a profound impact on global disease burden. Madagascar President Marc Ravalomanana urged conference attendees—and the world—to address the "lethal disease of poverty." More discussion of urban areas and their problems is presented in chapter 22.

Ecological Diseases

Humans aren't the only ones to suffer from new and devastating diseases. Domestic animals and wildlife also experience sudden and widespread epidemics. In 1998, for example, a distemper virus killed half the seals in western Europe. It's thought that toxic pollutants and hormone-disrupting environmental chemicals might have made seals and other marine mammals susceptible to infections. In 2002, more dead seals were found in Denmark, raising fears that distemper might be reappearing.

Chronic wasting disease (CWD) is spreading through deer and elk populations in North America (fig. 8.6). Caused by a strange protein called a prion, CWD is one of a family of irreversible,

FIGURE 8.6 Wild elk and deer are widely affected by chronic wasting disease, which originated with domestic herds. © Corbis/Volume 72.

degenerative neurological diseases known as transmissible spongiform encephalopathies (TSE) that include mad cow disease in cattle, scrapie in sheep, and Creutzfelt-Jacob disease in humans. CWD probably started when elk ranchers fed contaminated animal by-products to their herds. Infected animals were sold to other ranches, and now the disease has spread to wild populations. First recognized in 1967 in Saskatchewan, CWD has been identified in wild deer populations and ranch operations in at least eight American states.

The Canadian government is estimated to have spent $65 million in an attempt to stop the spread of CWD. In 2002, Wisconsin encouraged hunters to kill some 20,000 deer in an area near Madison in an effort to contain the disease. No humans are known to have contracted TSE from deer or elk, but there is a concern that we might see something like the mad cow disaster that inflicted Europe in the 1990s. At least 100 people died, and nearly 5 million European cattle and sheep were slaughtered in an effort to contain that disease.

In 1995, California residents noticed that the oak trees were dying from a fast-spreading new disease known as sudden oak death syndrome (SODS), The disease is caused by funguslike organism known as *Phytophthora ramorum,* part of a group known as water molds. Possibly imported with Asian rhododendrons shipped from European plant nurseries, this organism is a close relative of the pathogen that caused Ireland's potato blight and another species that has wiped out vast swaths of forest in western Australia. Recently, foresters reported that Douglas fir, one of the nation's most economically important timber species, and California coast redwood also are infected with SODS. In addition to being beautiful trees and dominant members of forest communities, each of these species provides commercial products worth more than a billion dollars per year in the United States (fig. 8.7).

Starting in the early 1970s, an illness called black-band disease has been attacking corals throughout the Caribbean. A cyanophyte alga (*Phormidium corallyticum*) actually kills the coral polyps. As the black ring of dead polyps spreads through the colony, it leaves behind a bleached coral skeleton. Researchers have found pathogenic bacteria from human feces associated with dying corals and think that they may play a role in triggering the algal attack. Coliform bacteria are present on reefs far from any human occupation. These pathogens may be carried by dust storms from as far away as Africa. Coral subjected to environmental stressors such as nutrient imbalance or elevated seawater temperatures may be especially susceptible to infection.

In 2003, a naturally occurring but deadly toxin produced by sea algae killed record numbers of dolphins and sea lions along sections of California's southern coast. More than 1,000 marine mammals were found stranded or dead on state beaches. Hundreds of seabirds, including endangered brown pelicans, grebes, and loons, also were affected. The animals are being poisoned by domoic acid, a nerve toxin produced by certain algae. First detected on the West Coast of the United States in 1991, domoic acid is a naturally occurring product of the diatom species *Pseudo-nitzschia.* Marine animals and seabirds are poisoned by eating small fish that have consumed diatoms. Filter-feeding animals such as mussels also feed on the toxin-laced algae. Humans who eat toxin-contaminated fish or shellfish contract an illness called amnesic shellfish poisoning (ASP). Symptoms include vomiting, nausea, diarrhea, and abdominal cramps. In more severe cases, confusion, seizures, cardiac arrhythmia, and coma can occur. People poisoned with very high doses of the toxin can die. The exact cause of the toxin increase is a mystery, but scientists speculate that the algae may be thriving on nutrients from agricultural runoff or sewage. Weather patterns could also play a role.

One thing all of these diseases have in common is human-made environmental changes that stress biological communities and upset normal ecological relationships. How many other ways might we be altering the world around us, and what might the consequences be both for ourselves and other species?

FIGURE 8.7 The pathogen that causes sudden oak death also infects redwoods, Douglas fir, and dozens of other species. © Corbis/Volume 63.

Antibiotic and Pesticide Resistance

Malaria, the most deadly of all insect-borne illnesses, is an example of the return of a disease that once was thought to be nearly vanquished. Malaria now claims about a million lives every year—90 percent are in Africa, and most of them children. With the advent of modern medicines and pesticides, malaria had nearly been wiped out in many places but recently has come roaring back. The protozoan parasite that causes the disease is now resistant to most antibiotics, while the mosquitoes that transmit it have developed resistance to many insecticides. Spraying of DDT in India and Sri Lanka, for instance, reduced malaria from millions of infections per year to only a few thousand in the 1950s and 1960s. Now South Asia is back to its pre-DDT level of about half a million new cases of malaria every year. Other places that never had cases of malaria now have them as a result of climate change and habitat alteration.

Why have vectors such as mosquitoes, and pathogens such as the malaria parasite become resistant to pesticides and antibiotics? Part of the answer is natural selection and the ability of many organisms to evolve rapidly. Another factor is the human tendency to use control measures carelessly. When we discovered that DDT and other insecticides could control mosquito populations, we spread them everywhere. This not only harmed wildlife and beneficial insects, but it created perfect conditions for natural selection. Many pests and pathogens were exposed only minimally to control measures, allowing those with natural resistance to survive and spread their genes through the population (fig. 8.8). After repeated cycles of exposure and selection, many microorganisms and their vectors are insensitive to almost all our weapons against them.

As chapter 9 discusses, raising huge numbers of cattle, hogs, and poultry in densely packed barns and feedlots is another reason for widespread antibiotic resistance in pathogens. Confined animals are dosed constantly with antibiotics and steroid hormones to keep them disease-free and to make them gain weight faster. More than half of all antibiotics used in the United States each year is fed to livestock. A significant amount of these antibiotics and hormones are excreted in urine and feces, which are spread, untreated, on the land or discharged into surface water where they contribute further to the evolution of supervirulent pathogens.

At least half of the 100 million antibiotic doses prescribed for humans every year in the United States are unnecessary or are the wrong ones. Furthermore, many people who start a course of antibiotic treatment fail to carry it out for the time prescribed. For your own health and that of the people around you, if you are taking an antibiotic, follow your doctor's orders and don't stop taking the medicine as soon as you start feeling better.

TOXICOLOGY

Toxicology is the study of **toxins** (poisons) and their effects, particularly on living systems. Because many substances are known to be poisonous to life (whether plant, animal, or microbial), toxicology is a broad field, drawing from biochemistry, histology, pharmacology, pathology, and many other disciplines. Toxins damage or kill living organisms because they react with cellular components to disrupt metabolic functions. Because of this reactivity, toxins often are harmful even in extremely dilute concentrations. In some cases billionths, or even trillionths of a gram can cause irreversible damage.

All toxins are hazardous, but not all hazardous materials are toxic. Some substances, for example, are dangerous because they're flammable, explosive, acidic, caustic, irritants, or sensitizers. Many of these materials must be handled carefully in large doses or high concentrations, but can be rendered relatively innocuous by dilution, neutralization, or other physical treatment. They don't react with cellular components in ways that make them poisonous at low concentrations.

(a) Mutation and selection create drug-resistant strains

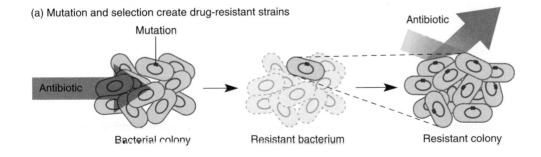

Antibiotic

Mutation

Antibiotic

Bacterial colony   Resistant bacterium   Resistant colony

FIGURE 8.8 How microbes acquire antibiotic resistance. Random mutations make a few cells resistant. When challenged by antibiotics, only those cells survive to give rise to a resistant colony. Sexual reproduction (conjugation) or plasmid transfer moves genes from one strain or species to another.

(b) Conjugation transfers drug resistance from one strain to another

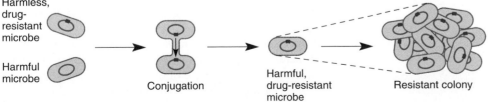

Harmless, drug-resistant microbe

Harmful microbe

Conjugation

Harmful, drug-resistant microbe

Resistant colony

Environmental toxicology or ecotoxicology specifically deals with the interactions, transformation, fate, and effects of natural and synthetic chemicals in the biosphere including individual organisms, populations, and whole ecosystems. In aquatic systems the fate of the pollutants is primarily studied in relation to mechanisms and processes at interfaces of the ecosystem components. Special attention is devoted to the sediment/water, water/organisms, and water/air interfaces. In terrestrial environments, the emphasis tends to be on effects of metals on soil fauna community and population characteristics.

Table 8.2 lists the top 20 toxins compiled by the U.S. Environmental Protection Agency from the 275 substances regulated by the Comprehensive Environmental Response, Compensation, and Liability Act (CERCLA), commonly known as the Superfund Act. These materials are listed in order of assessed importance in terms of human and environmental health.

Allergens are substances that activate the immune system. Some allergens act directly as **antigens;** that is, they are recognized as foreign by white blood cells and stimulate the production of specific antibodies (proteins that recognize and bind to foreign cells or chemicals). Other allergens act indirectly by binding to and changing the chemistry of foreign materials so they become antigenic and cause an immune response.

Formaldehyde is a good example of a widely used chemical that is a powerful sensitizer of the immune system. It is directly allergenic and can also trigger reactions to other substances. Widely

TABLE 8.2	Top 20 Toxic and Hazardous Substances

1. Arsenic
2. Lead
3. Mercury
4. Vinyl Chloride
5. Polychlorinated Biphenyls (PCBs)
6. Benzene
7. Cadmium
8. Benzo(a)pyrene
9. Polycyclic aromatic hydrocarbons
10. Benzo(b)fluoranthene
11. Chloroform
12. DDT
13. Aroclor 1254
14. Aroclor 1260
15. Trichloroethylene
16. Dibenz(a,h)anthracene
17. Dieldrin
18. Chromium, Hexavalent
19. Chlordane
20. Hexachlorobutadiene

Source: *U.S. EPA, 2003.*

What can you do?

Tips for Staying Healthy

- Eat a balanced diet with plenty of fresh fruits, vegetables, legumes, and whole grains. Wash fruits and vegetables carefully, they may well have come from a country where pesticide and sanitation laws are lax.

- Use unsaturated oils such as olive or canola rather than hydrogenated or semisolid fats such as margarine.

- Cook meats and other foods at temperatures high enough to kill pathogens; clean utensils and cutting surfaces; store food properly.

- Wash your hands frequently. You transfer more germs from hand to mouth than any other means of transmission.

- When you have a cold or flu, don't demand antibiotics from your doctor—they aren't effective against viruses.

- If you're taking antibiotics, continue for the entire time prescribed—quitting as soon as you feel well is an ideal way to select for antibiotic-resistant germs.

- Practice safe sex.

- Don't smoke and avoid smoky places.

- If you drink, do so in moderation. Never drive when your reflexes or judgment are impaired.

- Exercise regularly: walk, swim, jog, dance, garden. Do something you enjoy that burns calories and maintains flexibility.

- Get enough sleep. Practice meditation, prayer, or some other form of stress reduction.

- Make a list of friends and family who make you feel more alive and happy. Spend time with one of them at least once a week.

used in plastics, wood products, insulation, glue, and fabrics, formaldehyde concentrations in indoor air can be thousands of times higher than in normal outdoor air. Some people suffer from what is called **sick building syndrome:** headaches, allergies, chronic fatigue, and other symptoms caused by poorly vented indoor air contaminated by mold spores, carbon monoxide, nitrogen oxides, formaldehyde, and other toxins released from carpets, insulation, plastics, building materials, and other sources (fig. 8.9). The Environmental Protection Agency estimates that poor indoor air quality may cost the nation $60 billion a year in absenteeism and reduced productivity.

Immune system depressants are pollutants that suppress the immune system rather than activate it. Little is known about how this occurs or which chemicals are responsible. Immune system failure is thought to have played a role, however, in widespread deaths of seals in the North Atlantic and of dolphins in the Mediterranean. These dead animals generally contain high levels of pesticide residues, polychlorinated biphenyls (PCBs), and other contaminants that are suspected of disrupting the immune system and making it susceptible to a variety of opportunistic infections.

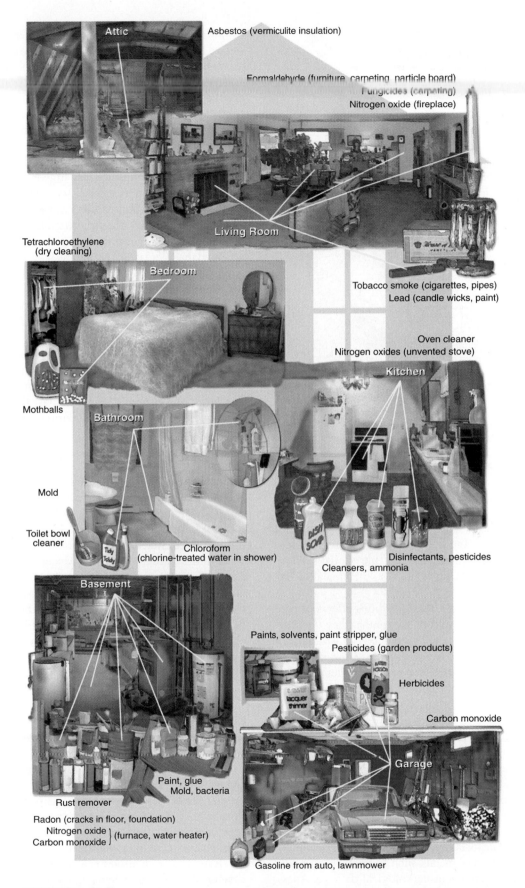

Asbestos (vermiculite insulation)

Formaldehyde (furniture, carpeting, particle board)
Fungicides (carpeting)
Nitrogen oxide (fireplace)

Attic

Living Room

Tetrachloroethylene
(dry cleaning)

Bedroom

Tobacco smoke (cigarettes, pipes)
Lead (candle wicks, paint)

Oven cleaner
Nitrogen oxides (unvented stove)

Mothballs

Kitchen

Bathroom

Mold

Toilet bowl
cleaner

Chloroform
(chlorine-treated water in shower)

Disinfectants, pesticides
Cleansers, ammonia

Basement

Paints, solvents, paint stripper, glue
Pesticides (garden products)

Herbicides

Carbon monoxide

Paint, glue
Mold, bacteria

Garage

Rust remover

Radon (cracks in floor, foundation)
Nitrogen oxide ⎫
Carbon monoxide ⎬ (furnace, water heater)

Gasoline from auto, lawnmower

FIGURE 8.9 Some sources of toxic and hazardous substances in a typical home.

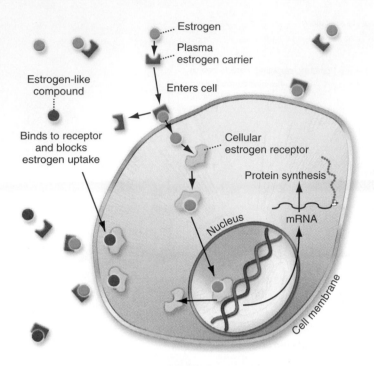

FIGURE 8.10 Steroid hormone action. Plasma hormone carriers deliver regulatory molecules to the cell surface, where they cross the cell membrane. Intracellular carriers deliver hormones to the nucleus, where they bind to and regulate expression of DNA.

Endocrine disrupters are chemicals that disrupt normal hormone functions. Hormones are chemicals released into the blood stream by cells in one part of the body to regulate development and function of tissues and organs elsewhere in the body (fig. 8.10). You undoubtedly have heard about sex hormones and their powerful effects on how we look and behave, but these are only one example of the many regulatory hormones that rule our lives. Some other powerful hormones include thyroxin, insulin, adrenalin, and endorphins.

We now know that some of the most insidious effects of persistent chemicals such as dioxins and PCBs are that they interfere with normal growth, development, and physiology of a variety of animals—including humans—at very low doses. In some cases, picogram concentrations (trillionths of a gram per liter) may be enough to cause developmental abnormalities in sensitive organisms. Because these chemicals often cause sexual dysfunction (reproductive health problems in females or feminization of males, for example), these chemicals are sometimes called environmental estrogens or androgens. They are just as likely, however, to disrupt functions of other important regulatory molecules as they are to obstruct sex hormones.

Neurotoxins are a special class of metabolic poisons that specifically attack nerve cells (neurons). The nervous system is so important in regulating body activities that disruption of its activities is especially fast-acting and devastating. Different types of neurotoxins act in different ways. Heavy metals such as lead and mercury kill nerve cells and cause permanent neurological damage. Anesthetics (ether, chloroform, halothane, etc.) and chlorinated hydrocarbons (DDT, Dieldrin, Aldrin) disrupt nerve cell membranes necessary for nerve action. Organophosphates (Malathion, Parathion) and carbamates (carbaryl, zeneb, maneb) inhibit acetylcholinesterase, an enzyme that regulates signal transmission between nerve cells and the tissues or organs they innervate (for example, muscle). Most neurotoxins are both extremely toxic and fast-acting.

Mutagens are agents, such as chemicals and radiation, that damage or alter genetic material (DNA) in cells. This can lead to birth defects if the damage occurs during embryonic or fetal growth. Later in life, genetic damage may trigger neoplastic (tumor) growth. When damage occurs in reproductive cells, the results can be passed on to future generations. Cells have repair mechanisms to detect and restore damaged genetic material, but some changes may be hidden, and the repair process itself can be flawed. It is generally accepted that there is no "safe" threshold for exposure to mutagens. Any exposure has some possibility of causing damage.

Teratogens are chemicals or other factors that specifically cause abnormalities during embryonic growth and development. Some compounds that are not otherwise harmful can cause tragic problems in these sensitive stages of life. Perhaps the most prevalent teratogen in the world is alcohol. Drinking during pregnancy can lead to **fetal alcohol syndrome**—a cluster of symptoms including craniofacial abnormalities, developmental delays, behavioral problems, and mental defects that last throughout a child's life. Even one alcoholic drink a day during pregnancy has been associated with decreased birth weight.

Carcinogens are substances that cause **cancer,** invasive, out-of-control cell growth that results in malignant tumors. Cancer rates rose in most industrialized countries during the twentieth century, and cancer is now the second leading cause of death in the United States, killing more than half a million people in 2002. According to the American Cancer Society, 1 in 2 males and 1 in 3 females in the United States will have some form of cancer in their lifetime. Some authors blame this cancer increase on toxic synthetic chemicals in our environment and diet. Others argue that it is attributable mainly to lifestyle (smoking, sunbathing, alcohol) or simply living longer. The U.S. EPA estimates that 200 million U.S. residents live in areas where the combined lifetime cancer risk from environmental carcinogens exceeds 1 in 100,000, or ten times the risk normally considered acceptable.

Diet

Diet also has an important effect on health. For instance, there is a strong correlation between cardiovascular disease and the amount of salt and animal fat in one's diet.

Fruits, vegetables, whole grains, complex carbohydrates, and dietary fiber (plant cell walls) often have beneficial health effects. Certain dietary components, such as pectins; vitamins A, C, and E; substances produced in cruciferous vegetables (cabbage, broccoli, cauliflower, brussels sprouts); and selenium, which we get from plants, seem to have anticancer effects.

Eating too much food is a significant dietary health factor in developed countries and among the well-to-do everywhere. Sixty percent of all U.S. adults are now considered overweight, and the

 TABLE 8.4	National Health Recommendations and Diet Goals

1. Balance the food you eat with physical activity to maintain or improve your weight.
2. Choose a diet with plenty of grain products, vegetables, and fruits.
3. Choose a diet low in fat, saturated fat, and cholesterol.
4. Eat a variety of foods.
5. Choose a diet moderate in salt and sodium.
6. Choose a diet moderate in sugars.
7. If you drink alcoholic beverages, do so in moderation.

Source: *U.S. Department of Health and Human Services, 1995.*

 TABLE 8.5	Factors in Environmental Toxicity

FACTORS RELATED TO THE TOXIC AGENT
1. Chemical composition and reactivity
2. Physical characteristics (such as solubility, state)
3. Presence of impurities or contaminants
4. Stability and storage characteristics of toxic agent
5. Availability of vehicle (such as solvent) to carry agent
6. Movement of agent through environment and into cells

FACTORS RELATED TO EXPOSURE
1. Dose (concentration and volume of exposure)
2. Route, rate, and site of exposure
3. Duration and frequency of exposure
4. Time of exposure (time of day, season, year)

FACTORS RELATED TO ORGANISM
1. Resistance to uptake, storage, or cell permeability of agent
2. Ability to metabolize, inactivate, sequester, or eliminate agent
3. Tendency to activate or alter nontoxic substances so they become toxic
4. Concurrent infections or physical or chemical stress
5. Species and genetic characteristics of organism
6. Nutritional status of subject
7. Age, sex, body weight, immunological status, and maturity

worldwide total of obese or overweight people is estimated to be over 1 billion. Every year in the United States, 300,000 deaths are linked to obesity.

The U.S. Centers for Disease Control in Atlanta warn that one in three U.S. children will become diabetic unless many more people start eating less and exercising more. The odds are worse for Black and Hispanic children: nearly half of them are likely to develop the disease. And among the Pima tribe of Arizona, nearly 80 percent of all adults are diabetic. Some goals for reducing obesity and other diet-related problems are shown in table 8.4. More information about food and its health effects is available in chapter 9.

MOVEMENT, DISTRIBUTION, AND FATE OF TOXINS

There are many sources of toxic and hazardous chemicals in the environment and many factors related to each chemical itself, its route or method of exposure, and its persistence in the environment, as well as characteristics of the target organism (table 8.5), that determine the danger of the chemical. We can think of both individuals and an ecosystem as sets of interacting compartments between which chemicals move, based on molecular size, solubility, stability, and reactivity (fig. 8.11). The dose (amount), route of entry, timing of exposure, and sensitivity of the organism all play important roles in determining toxicity. In this section, we will consider some of these characteristics and how they affect environmental health.

Solubility and Mobility

Solubility is one of the most important characteristics in determining how, where, and when a toxic material will move through the environment or through the body to its site of action. Chemicals can be divided into two major groups: those that dissolve more readily in water and those that dissolve more readily in oil. Water-soluble compounds move rapidly and widely through the environ-

ment because water is ubiquitous. They also tend to have ready access to most cells in the body because aqueous solutions bathe all our cells. Molecules that are oil- or fat-soluble (usually organic molecules) generally need a carrier to move through the environment, into, and within, the body. Once inside the body, however, oil-soluble toxins penetrate readily into tissues and cells because the membranes that enclose cells are themselves made of similar oil-soluble chemicals. Once they get inside cells, oil-soluble materials are likely to be accumulated and stored in lipid deposits where they may be protected from metabolic breakdown and persist for many years.

Exposure and Susceptibility

Just as there are many sources of toxins in our environment, there are many routes for entry of dangerous substances into our bodies (fig. 8.12). Airborne toxins generally cause more ill health than any other exposure source. We breathe far more air every day than the volume of food we eat or water we drink. Furthermore, the lining of our lungs, which is designed to exchange gases very efficiently, also absorbs toxins very well. Still, food, water, and skin contact also can expose us to a wide variety of toxins. The largest exposures for many toxins are found in industrial settings, where workers may encounter doses thousands of times higher than would be found anywhere else. The European Agency for Safety and Health at Work warns that 32 million people (20 percent of all employees)

Air
Photolysis
Oxidation
Precipitation

Source
Industry
Agriculture
Domestic, etc.

Biota
Metabolism
Storage
Excretion

Soil and sediment
Photolysis and metabolism
Evaporation

Water
Hydrolysis
Oxidation
Microbial degradation
Evaporation
Sedimentation

FIGURE 8.11 Movement and fate of chemicals in the environment. Processes that modify, remove, or sequester compounds are shown in parentheses. Toxins also move directly from a source to soil and sediment.

Bioaccumulation and Biomagnification

Cells have mechanisms for **bioaccumulation,** the selective absorption and storage of a great variety of molecules. This allows them to accumulate nutrients and essential minerals, but at the same time, they also may absorb and store harmful substances through these same mechanisms. Toxins that are rather dilute in the environment can reach dangerous levels inside cells and tissues through this process of bioaccumulation.

The effects of toxins also are magnified in the environment through food webs. **Biomagnification** occurs when the toxic burden of a large number of organisms at a lower trophic level is accumulated and concentrated by a predator in a higher trophic level. Phytoplankton and bacteria in aquatic ecosystems, for instance, take up heavy metals or toxic organic molecules from water or sediments (fig. 8.13). Their predators—zooplankton and small fish—collect and retain the toxins from many prey organisms, building up higher concentrations of toxins. The top carnivores in the food chain—game fish, fish-eating birds, and humans—can accumulate such high toxin levels that they suffer adverse health effects (chapter 18). One of the first known examples of bioaccumulation and biomagnification was DDT, which accumulated through food chains so that by the 1960s it was shown to be interfering with reproduction of peregrine falcons, brown pelicans, and other predatory birds at the top of their food chains (chapter 10).

Persistence

Some chemical compounds are very unstable and degrade rapidly under most environmental conditions so that their concentrations decline quickly after release. Most modern herbicides and pesticides, for instance, quickly lose their toxicity. Other substances are more persistent and last for years or even centuries in the

in the European Union are exposed to unacceptable levels of carcinogens and other toxins in their workplace.

Condition of the organism and timing of exposure also have strong influences on toxicity. Healthy adults, for example, may be relatively insensitive to doses that would be very dangerous for young children or for someone already weakened by disease (What Do You Think? p. 160). Similarly, exposure to a toxin may be very dangerous at certain stages of developmental or metabolic cycles, but may be innocuous at other times. A single dose of the notorious teratogen thalidomide, for example, taken in the third week of pregnancy (a time when many women aren't aware they're pregnant) can cause severe abnormalities in fetal limb development. A complication in measuring toxicity is that great differences in sensitivity exist between species. Thalidomide was tested on a number of laboratory animals without any deleterious effects. Unfortunately, however, it is a powerful teratogen in humans.

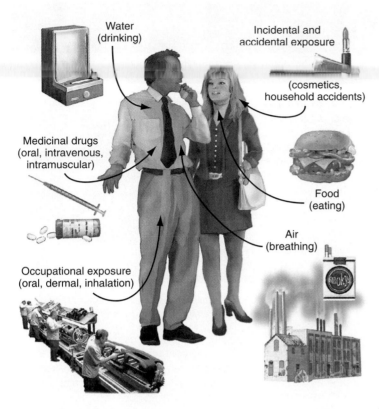

FIGURE 8.12 Routes of exposure to toxic and hazardous environmental factors.

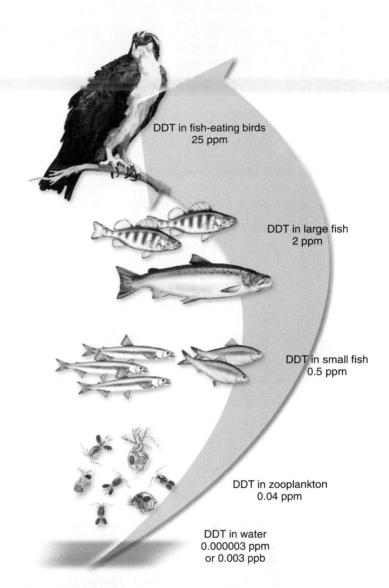

DDT in fish-eating birds
25 ppm

DDT in large fish
2 ppm

DDT in small fish
0.5 ppm

DDT in zooplankton
0.04 ppm

DDT in water
0.000003 ppm
or 0.003 ppb

FIGURE 8.13 Bioaccumulation and biomagnification. Organisms lower on the food chain take up and store toxins from the environment. They are eaten by larger predators, who are eaten, in turn, by even larger predators. The highest members of the food chain can accumulate very high levels of the toxin.

environment. PVC plastics, chlorinated hydrocarbon pesticides, and asbestos are valuable because they are resistant to degradation, among other properties. This stability, however, also causes problems because these materials persist in the environment and have unexpected effects far from the sites of their original use.

Some **persistent organic pollutants (POPs)** have become extremely widespread, being found now from the tropics to the arctic. They often accumulate in food webs and reach toxic concentrations in long-living top predators such as humans, sharks, raptors, swordfish, and bears. Some of greatest current concerns are:

- Polybrominated diphenyl ethers (PBDE). Widely used as flame-retardants in textiles, foam in upholstery, and plastic in appliances and computers, this compound was first reported accumulating in women's breast milk in Sweden in the 1990s. It was subsequently found in humans and other species everywhere from Canada to Israel. Nearly 150 million metric tons (330 million lbs) of PBDEs are used every year worldwide. The toxicity and environmental persistence of PBDE is much like that of PCBs, to which it is closely related chemically. The dust at ground zero in New York City after September 11 was heavily laden with PBDE. The European Union has already banned this compound.

- Perfluorooctane sulfonate (PFOS) and perfluorooctanoic acid (PFOA, also known as C8) are members of a chemical family used to make nonstick, waterproof, and stain-resistant products such as Teflon, Gortex, Scotchguard, and Stainmaster.

Industry makes use of their slippery, heat-stable properties to manufacture everything from airplanes and computers to cosmetics and household cleaners. Now these chemicals—which are reported to be infinitely persistent in the environment—are found throughout the world, even the most remote and seemingly pristine sites. Almost all Americans have one or more perfluorinated compounds in their blood. Heating some nonstick cooking pans above 500°F (260°C) can release enough PFOA to kill pet birds. This chemical family has been shown to cause liver damage as well as various cancers and reproductive and developmental problems in rats. Exposure may be especially dangerous to women and girls, who may be 100 times more sensitive than men to these chemicals.

What do you think?

Children's Health

Increasing evidence shows that children are much more vulnerable than adults to environmental toxins. Pound for pound, children drink more water, eat more food, and breathe more air than do adults. Constantly putting fingers, toys, and other objects into their mouths increases children's exposure to toxins in dust or soil. Compared to adults, children generally have less-developed protective mechanisms such as immune systems or metabolic processes to degrade or excrete toxins. Rapid growth and development in childhood makes their metabolism and maturation highly susceptible to disruption. And because children have more future years of life remaining than do most adults, they have more time to develop chronic diseases, like cancer, that may require decades to be expressed.

Childhood cancer incidence has risen 1 percent per year since the early 1970s. Asthma also is up sharply. Some 6.3 million children now suffer from this disease, a four-fold increase over the past 30 years. Asthmatic attacks, which cause blockage of the airways and result in coughing, wheezing, and difficulty breathing, are now the fourth-ranking cause of childhood hospitalization. Inner-city children seem to be particularly at risk for this disease. One study found that 25 percent of all children in New York City's central Harlem suffer from asthma. Scientists believe genetic predisposition plays an important role in this disease, but that environmental exposure to factors such as dust, cockroaches and mites, cigarette smoke, mold and mildew, and diesel engine fumes also act as triggers.

Currently, about 430,000 American children between the ages of one and five have elevated blood lead levels that can cause irreversible disabilities such as lower IQ and neurological damage. Autism and attention-deficit hyperactivity disorder (ADHD) are reported to have increased sharply in the past 20 years, although increasing awareness and diagnosis of these conditions may be partly responsible for this increase. A growing percentage of boys with reproductive tract defects also have been reported.

A recently recognized environmental hazard for children is chromated copper arsenate (CCA)-treated lumber often used to build rot-resistant decks, sidewalks, and playground equipment. All three metals are toxic. Arsenic, the worst of the three, can leach out of weathered wood and has been shown to cause muscle damage as well as lung, bladder, and skin cancer. CCA-treated wood is being phased out in the United States, but not fast enough according to some groups.

In 2003, the U.S. EPA issued its first guidelines to define the greater risks that children face from environmental toxins. According to these proposed rules, regulators should assume that children have ten times the exposure risk of adults to cancer-causing chemicals. Some health groups criticized these guidelines as not being strong enough, pointing out that certain chemicals such as vinyl chloride (used in PVC plastics), diethylnitrosamine (from tobacco smoke), and polychlorinated biphenyls (PCBs, in electrical transformers) can be up to 65 times more potent in infants and toddlers than in adults.

Should all arsenic-treated lumber be removed from places where children play? © William P. Cunningham.

To study connections between environmental exposures and potential health effects issues, the U.S. government has started a landmark National Children's Study, which will follow 100,000 children from birth to age 18. In establishing this study, former EPA Administrator Christie Whitman said, "Children represent 25 percent of our population, but 100 percent of the future."

What do you think? What would be the best way to safeguard children from environmental hazards? Would you ban CCA-treated lumber completely, or just remove it from playgrounds and other places in which children might be exposed? How would you distinguish between the effects of poverty and environmental hazards for inner-city children? Might it be more effective to focus on better housing, nutrition, and education than protecting them from environmental chemicals? What research would you request if you were an environmental regulator?

- Phthalates (pronounced *thalates*) are found in cosmetics, deodorants, and many plastics (such as soft polyvinyl chloride or PVC) used for food packaging, children's toys, and medical devices. Some members of this chemical family are known to be toxic to laboratory animals, causing kidney and liver damage and possibly some cancers. In addition, many phthalates act as endocrine hormone disrupters, and have been linked to reproductive abnormalities and decreased fertility. A correlation has been found between phthalate levels in urine and low sperm numbers and decreased sperm motility in men. Nearly everyone in the United States has phthalates in their body at levels reported to cause these problems. While not yet conclusive, these results could help explain a 50-year decline in semen quality in most industrialized countries.

- Bisphenol A (BPA), a prime ingredient in polycarbonate plastic (commonly used for products ranging from water bottles to tooth-protecting sealants), has been widely found in humans, even those who have no known exposure to these chemicals. So far, there is little direct evidence linking BPA exposure to human health risks, but studies in animals have found that the chemical can cause abnormal chromosome numbers, a condition called aneuploidy, which is the leading cause of miscarriages and several forms of mental retardation. It also is an environmental estrogen and may alter sexual development in both males and females.

- Atrazine, one of the most widely used herbicides in the Midwestern corn belt, was shown to cause abnormal development and sexual dysfunction in frogs. Farmers who use Atrazine heavily have high rates of certain lymphomas, but these farmers are generally exposed to other toxic compounds as well. A study of farm families in northwestern Minnesota found considerably higher rates of birth defects than in urban families. "The data is associative," said Dr. Vince Gary, who directed the research. No definitive cause and effect can be shown with any particular pesticide, but the sensitivity of other species raises concerns.

Further discussion of POPs can be found in chapter 10.

Chemical Interactions

Some materials produce *antagonistic* reactions. That is, they interfere with the effects or stimulate the breakdown of other chemicals. For instance, vitamins E and A can reduce the response to some carcinogens. Other materials are *additive* when they occur together in exposures. Rats exposed to both lead and arsenic show twice the toxicity of only one of these elements. Perhaps the greatest concern is synergistic effects. **Synergism** is an interaction in which one substance exacerbates the effects of another. For example, occupational asbestos exposure increases lung cancer rates 20-fold. Smoking increases lung cancer rates by the same amount. Asbestos workers who also smoke, however, have a 400-fold increase in cancer rates. How many other toxic chemicals are we exposed to that are below threshold limits individually but combine to give toxic results?

MECHANISMS FOR MINIMIZING TOXIC EFFECTS

A fundamental concept in toxicology is that every material can be poisonous under some conditions, but most chemicals have some safe level or threshold below which their effects are undetectable or insignificant. Each of us consumes lethal doses of many chemicals over the course of a lifetime. One hundred cups of strong coffee, for instance, contain a lethal dose of caffeine. Similarly, one hundred aspirin tablets, or 10 kilograms (22 lbs) of spinach or rhubarb, or a liter of alcohol would be deadly if consumed all at once. Taken in small doses, however, most toxins can be broken down or excreted before they do much harm. Furthermore, damage they cause can be repaired. Sometimes, however, mechanisms that protect us from one type of toxin or at one stage in the life cycle become deleterious with another substance or in another stage of development. Let's look at how these processes help protect us from harmful substances as well as how they can go awry.

Metabolic Degradation and Excretion

Most organisms have enzymes that process waste products and environmental poisons to reduce their toxicity. In mammals, most of these enzymes are located in the liver, the primary site of detoxification of both natural wastes and introduced poisons. Sometimes, however, these reactions work to our disadvantage. Compounds, such as benzopyrene, for example, that are not toxic in their original form are processed by these same enzymes into cancer-causing carcinogens. Why would we have a system that makes a chemical more dangerous? Evolution and natural selection are expressed through reproductive success or failure. Defense mechanisms that protect us from toxins and hazards early in life are "selected for" by evolution. Factors or conditions that affect postreproductive ages (like cancer or premature senility) usually don't affect reproductive success or exert "selective pressure."

We also reduce the effects of waste products and environmental toxins by eliminating them from our body through excretion. Volatile molecules, such as carbon dioxide, hydrogen cyanide, and ketones are excreted via breathing. Some excess salts and other substances are excreted in sweat. Primarily, however, excretion is a function of the kidneys, which can eliminate significant amounts of soluble materials through urine formation. Accumulation of toxins in the urine can damage this vital system, however, and the kidneys and bladder often are subjected to harmful levels of toxic compounds. In the same way, the stomach, intestine, and colon often suffer damage from materials concentrated in the digestive system and may be afflicted by diseases and tumors.

Repair Mechanisms

In the same way that individual cells have enzymes to repair damage to DNA and protein at the molecular level, tissues and organs that are exposed regularly to physical wear-and-tear or to toxic or hazardous materials often have mechanisms for damage repair. Our skin and the epithelial linings of the gastrointestinal tract, blood vessels, lungs, and urogenital system have high cellular reproduction rates to replace injured cells. With each reproduction cycle, however, there is a chance that some cells will lose normal growth controls and run amok, creating a tumor. Thus any agent, such as smoking or drinking, that irritates tissues is likely to be carcinogenic. And tissues with high cell-replacement rates are among the most likely to develop cancers.

MEASURING TOXICITY

Almost 500 years ago, the Swiss scientist Paracelsus said "the dose makes the poison," by which he meant that almost everything is toxic at some level. This remains the most basic principle of toxicology. Sodium chloride (table salt), for instance, is essential for human life in small doses. If you were forced to eat a kilogram of salt all at once, however, it would make you very sick. A similar amount injected into your bloodstream would be lethal. How a material is delivered—at what rate, through which route of entry, and in what medium—plays a vitally important role in determining toxicity.

This does not mean that all toxins are identical, however. Some are so poisonous that a single drop on your skin can kill you. Others require massive amounts injected directly into the blood to be lethal. Measuring and comparing the toxicity of various materials is difficult because not only do species differ in sensitivity, but individuals within a species respond differently to a given exposure. In this section, we will look at methods of toxicity testing and at how results are analyzed and reported.

Animal Testing

The most commonly used and widely accepted toxicity test is to expose a population of laboratory animals to measured doses of a specific substance under controlled conditions. This procedure is expensive, time-consuming, and often painful and debilitating to the animals being tested. It commonly takes hundreds—or even thousands—of animals, several years of hard work, and hundreds of thousands of dollars to thoroughly test the effects of a toxin at very low doses. More humane toxicity tests using computer simulation of model reactions, cell cultures, and other substitutes for whole living animals are being developed. However, conventional large-scale animal testing is the method in which we have the most confidence and on which most public policies about pollution and environmental or occupational health hazards are based.

In addition to humanitarian concerns, there are several problems in laboratory animal testing that trouble both toxicologists and policymakers. One problem is differences in sensitivity to a toxin of the members of a specific population. Figure 8.14 shows a typical dose/response curve for exposure to a hypothetical toxin. Some individuals are very sensitive to the toxin, while others are insensitive. Most, however, fall in a middle category forming a bell-shaped curve. The question for regulators and politicians is whether we should set pollution levels that will protect everyone, including the most sensitive people, or only aim to protect the average person. It might cost billions of extra dollars to protect a very small number of individuals at the extreme end of the curve. Is that a good use of resources?

Dose/response curves are not always symmetrical, making it difficult to compare toxicity of unlike chemicals or different species of organisms. A convenient way to describe toxicity of a chemical is to determine the dose to which 50 percent of the test population is sensitive. In the case of a lethal dose (LD), this is called the **LD50** (fig. 8.15).

Unrelated species can react very differently to the same toxin, not only because body sizes vary but also because of differences in physiology and metabolism. Even closely related species can have very dissimilar reactions to a particular toxin. Hamsters, for instance, are nearly 5,000 times less sensitive to some dioxins than are guinea pigs. Of 226 chemicals found to be carcinogenic in either rats or mice, 95 caused cancer in one species but not the other. These variations make it difficult to estimate the risks for humans since we don't consider it ethical to perform controlled experiments in which we deliberately expose people to toxins.

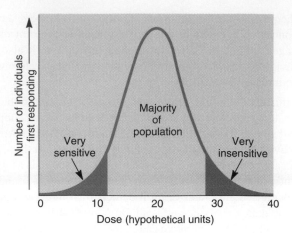

FIGURE 8.14 Probable variations in sensitivity to a toxin within a population. Some members of a population may be very sensitive to a given toxin, while others are much less sensitive. The majority of the population falls somewhere between the two extremes.

Toxicity Ratings

It is useful to group materials according to their relative toxicity. A moderate toxin takes about one gram per kilogram of body weight (about two ounces for an average human) to make a lethal dose. Very toxic materials take about one-tenth that amount, while extremely toxic substances take one-hundredth as much (only a few drops) to kill most people. Supertoxic chemicals are extremely potent; for some, a few micrograms (millionths of a gram—an amount invisible to the naked eye) make a lethal dose. These materials are not all synthetic. One of the most toxic chemicals known, for instance, is ricin, a protein found in castor bean seeds. It is so toxic that 0.3 billionths of a gram given intravenously will generally kill a mouse. If aspirin were this toxic, a single tablet, divided evenly, could kill 1 million people.

Many carcinogens, mutagens, and teratogens are dangerous at levels far below their direct toxic effect because abnormal cell growth exerts a kind of biological amplification. A single cell, perhaps altered by a single molecular event, can multiply into millions of tumor cells or an entire organism. Just as there are different levels of direct toxicity, however, there are different degrees of carcinogenicity, mutagenicity, and teratogenicity. Methanesulfonic acid, for instance, is highly carcinogenic, while the sweetener saccharin is a suspected carcinogen whose effects may be vanishingly small.

Acute versus Chronic Doses and Effects

Most of the toxic effects that we have discussed so far have been **acute effects.** That is, they are caused by a single exposure to the toxin and result in an immediate health crisis of some sort. Often, if the individual experiencing an acute reaction survives this immediate crisis, the effects are reversible. **Chronic effects,** on the other hand, are long lasting, perhaps even permanent. A chronic effect can result from a single dose of a very toxic substance, or it can be the result of a continuous or repeated sublethal exposure.

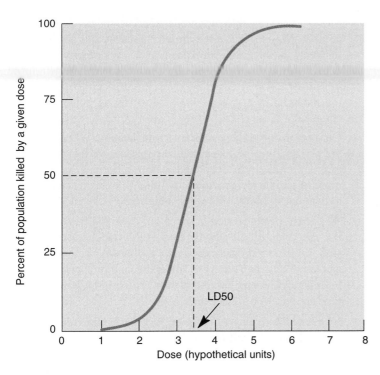

FIGURE 8.15 Cumulative population response to increasing doses of a toxin. The LD50 is the dose that is lethal to half the population.

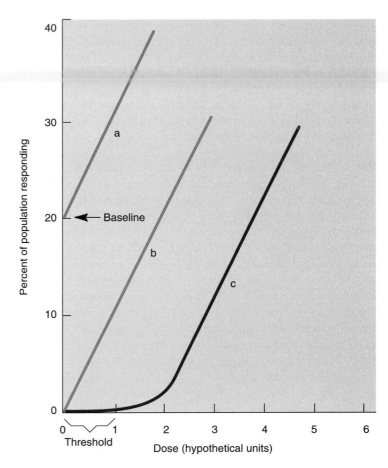

FIGURE 8.16 Three possible dose-response curves at low doses. (*a*) Some individuals respond, even at zero dose, indicating that some other factor must be involved. (*b*) Response is linear down to the lowest possible dose. (*c*) Threshold must be passed before any response is seen.

We also describe long-lasting *exposures* as chronic, although their effects may or may not persist after the toxin is removed. It usually is difficult to assess the specific health risks of chronic exposures because other factors, such as aging or normal diseases, act simultaneously with the factor under study. It often requires very large populations of experimental animals to obtain statistically significant results for low-level chronic exposures. Toxicologists talk about "megarat" experiments in which it might take a million rats to determine the health risks of some supertoxic chemicals at very low doses. Such an experiment would be terribly expensive for even a single chemical, let alone for the thousands of chemicals and factors suspected of being dangerous.

An alternative to enormous studies involving millions of animals is to give massive amounts—usually the maximum tolerable dose—of a toxin being studied to a smaller number of individuals and then to extrapolate what the effects of lower doses might have been. This is a controversial approach because it is not clear that responses to toxins are linear or uniform across a wide range of doses.

Figure 8.16 shows three possible results from low doses of a toxin. Curve (*a*) shows a baseline level of response in the population, even at zero dose of the toxin. This suggests that some other factor in the environment also causes this response. Curve (*b*) shows a straight-line relationship from the highest doses to zero exposure. Many carcinogens and mutagens show this kind of response. Any exposure to such agents, no matter how small, carries some risks. Curve (*c*) shows a threshold for the response where some minimal dose is necessary before any effect can be observed. This generally suggests the presence of some defense mechanism

that prevents the toxin from reaching its target in an active form or repairs the damage that it causes. Low levels of exposure to the toxin in question may have no deleterious effects, and it might not be necessary to try to keep exposures to zero.

Which, if any, environmental health hazards have thresholds is an important but difficult question. The 1958 Delaney Clause to the U.S. Food and Drug Act forbids the addition of *any* amount of known carcinogens to food and drugs, based on the assumption that any exposure to these substances represents unacceptable risks. This standard was replaced in 1996 by a *no reasonable harm* requirement, defined as less than one cancer for every million people exposed over a lifetime. This change was supported by a report from the National Academy of Sciences concluding that synthetic chemicals in our diet are unlikely to represent an appreciable cancer risk. We will discuss risk analysis in the next section.

Detection Limits

You may have seen or heard dire warnings about toxic materials detected in samples of air, water, or food. A typical headline announced recently that 23 pesticides were found in 16 food samples. What does that mean? The implication seems to be that any amount of dangerous materials is unacceptable and that counting the numbers of compounds detected is a reliable way to establish danger. We have seen, however, that the dose makes the poison. It matters not only what is there, but how much, where it is located, how accessible it is, and who is exposed. At some level, the mere presence of a substance is insignificant.

Toxins and pollutants may seem to be more widespread now than in the past, and this is surely a valid perception for many substances. The daily reports we hear of new materials found in new places, however, are also due, in part, to our more sensitive measuring techniques. Twenty years ago, parts per million were generally the limits of detection for most chemicals. Anything below that amount was often reported as zero or absent rather than more accurately as undetected. A decade ago, new machines and techniques were developed to measure parts per billion. Suddenly, chemicals were found where none had been suspected. Now we can detect parts per trillion or even parts per quadrillion in some cases. Increasingly sophisticated measuring capabilities may lead us to believe that toxic materials have become more prevalent. In fact, our environment may be no more dangerous; we are just better at finding trace amounts.

RISK ASSESSMENT AND ACCEPTANCE

Risk is the possibility of suffering harm or loss. **Risk assessment** is the scientific process of estimating the threat that particular hazards pose to human health. This process includes risk identification, dose response assessment, exposure appraisal, and risk characterization. In hazard identification, scientists evaluate all available information about the effects of a toxin to estimate the likelihood that a chemical will cause a certain effect in humans. The best evidence comes from human studies, such as physician case reports. Animal studies are also used to assess health risks. Risk assessment for identified toxicity hazards (for example, lead) includes collection and analysis of site data, development of exposure and risk calculations, and preparation of human health and ecological impact reports. Exposure assessment is the estimation or determination of the magnitude, frequency, duration, and route of exposure to a possible toxin. Toxicity assessment weighs all available evidence and estimates the potential for adverse health effects to occur.

Understanding Risks

A number of factors influence how we perceive relative risks associated with different situations.

- People with social, political, or economic interests—including environmentalists—tend to downplay certain risks and emphasize others that suit their own agendas. We also tend to tolerate risks that we choose—such as driving, smoking, or overeating—while objecting to risks we cannot control—such as potential exposure to slight amounts of toxic substances.

- Most people have difficulty understanding and believing probabilities. We feel that there must be patterns and connections in events, even though statistical theory says otherwise. If the coin turned up heads last time, we feel certain that it will turn up tails next time. In the same way, it is difficult to understand the meaning of a 1-in-10,000 risk of being poisoned by a chemical.

- Our personal experiences often are misleading. When we have not personally experienced a bad outcome, we feel it is more rare and unlikely to occur than it actually may be. Furthermore, the anxieties generated by life's gambles make us want to deny uncertainty and to misjudge many risks.

- We have an exaggerated view of our own abilities to control our fate. We generally consider ourselves above-average drivers, safer than most when using appliances or power tools, and less likely than others to suffer medical problems, such as heart attacks. People often feel they can avoid hazards because they are wiser or luckier than others.

- News media give us a biased perspective on the frequency of certain kinds of health hazards, overreporting some accidents or diseases while downplaying or underreporting others. Sensational, gory, or especially frightful causes of death like murders, plane crashes, fires, or terrible accidents occupy a disproportionate amount of attention in the public media. Heart diseases, cancer, and stroke kill nearly 15 times as many people in the United States as do accidents and 75 times as many people as do homicides, but the emphasis placed by the media on accidents and homicides is nearly inversely proportional to their relative frequency compared to either cardiovascular disease or cancer. This gives us an inaccurate picture of the real risks to which we are exposed.

- We tend to have an irrational fear or distrust of certain technologies or activities that leads us to overestimate their dangers. Nuclear power, for instance, is viewed as very risky, while coal-burning power plants seem to be familiar and relatively benign; in fact, coal mining, shipping, and combustion cause an estimated 10,000 deaths each year in the United States. An old, familiar technology seems safer and more acceptable than does a new, unknown one.

Accepting Risks

How much risk is acceptable? How much is it worth to minimize and avoid exposure to certain risks? Most people will tolerate a higher probability of occurrence of an event if the harm caused by that event is low. Conversely, harm of greater severity is acceptable only at low levels of frequency. A 1-in-10,000 chance of being killed might be of more concern to you than a 1-in-100 chance of being injured. For most people, a 1-in-100,000 chance of dying from some event or some factor is a threshold for changing what we do. That is, if the chance of death is less than 1 in 100,000, we are not likely to be worried enough to change our ways. If the risk is greater, we will probably do something about it. The Environmental Protection Agency generally assumes that a risk of 1 in 1 million is acceptable for most environmental hazards. Critics of this policy ask, acceptable to whom?

For activities that we enjoy or find profitable, we are often willing to accept far greater risks than this general threshold. Conversely, for risks that benefit someone else we demand far higher protection. For instance, your chance of dying in a motor vehicle accident in any given year is about 1 in 5,000, but that doesn't deter many people from riding in automobiles. Your lifetime chance of dying from lung cancer if you smoke one pack of cigarettes per day is about 1 in 4. By comparison, the risk from drinking water with the EPA limit of trichloroethylene is about 1 in 10 million. Strangely, many people demand water with zero levels of trichloroethylene, while continuing to smoke cigarettes.

Table 8.6 lists lifetime odds of dying from a few leading diseases and accidents. These are statistical averages, of course, and there clearly are differences in where one lives or how one behaves that affect the danger level of these activities. Although the average lifetime chance of dying in an automobile accident is 1 in 100, there clearly are things you can do—like wearing a seatbelt, following safety rules, and avoiding risky situations—that improve your odds. Still, it is interesting how we readily accept some risks while shunning others.

Our perception of relative risks is strongly affected by whether risks are known or unknown, whether we feel in control of the out-

come, and how dreadful the results are. Risks that are unknown or unpredictable and results that are particularly gruesome or disgusting seem far worse than those that are familiar and socially acceptable.

Studies of public risk perception show that most people react more to emotion than statistics. We go to great lengths to avoid some dangers while gladly accepting others. Factors that are involuntary, unfamiliar, undetectable to those exposed, catastrophic, or that have delayed effects or are a threat to future generations are especially feared while those that are voluntary, familiar, detectable, or immediate cause less anxiety. Even though the actual number of deaths from automobile accidents, smoking, or alcohol, for instance, are thousands of times greater than those from pesticides, nuclear energy, or genetic engineering, the latter preoccupy us far more than the former.

ESTABLISHING PUBLIC POLICY

Risk management combines principles of environmental health and toxicology together with regulatory decisions based on socioeconomic, technical, and political considerations (fig. 8.17). The biggest problem in making regulatory decisions is that we are usually dealing with many sources of harm to which we are exposed, often without being aware of them. It is difficult to separate the effects of all these different hazards and to evaluate their risks accurately, especially when the exposures are near the threshold of measurement and response. In spite of often vague and contradictory data, public policymakers must make decisions.

The case of the sweetener saccharin is a good example of the complexities and uncertainties of risk assessment in public health. Studies in the 1970s at the University of Wisconsin and the Canadian Health Protection Branch suggested a link between saccharin

TABLE 8.6	Lifetime Chances of Dying in the U.S.	
SOURCE	**ODDS (1 IN x)**
Heart disease	2
Cancer	3
Smoking	4
Lung disease	15
Pneumonia	30
Automobile accident	100
Suicide	100
Falls	200
Firearms	200
Fires	1,000
Airplane accident	5,000
Jumping from high places	6,000
Drowning	10,000
Lightning	56,000
Hornets, wasps, bees	76,000
Dog bite	230,000
Poisonous snakes, spiders	700,000
Botulism	1 million
Falling space debris	5 million
Drinking water with EPA limit of trichloroethylene	10 million

Source: U.S. National Safety Council, 2003.

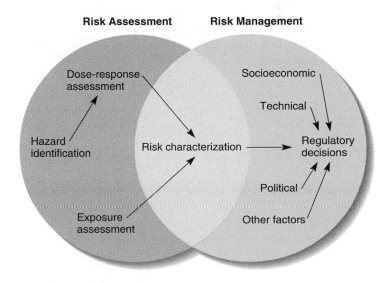

FIGURE 8.17 Risk assessment organizes and analyzes data to determine relative risk. Risk management sets priorities and evaluates relevant factors to make regulatory decisions.

Source: D. E. Patton, "USEPA's Framework for Ecological Risk Assessment" in *Human Ecological Risk Assessment*, Vol. 1 No. 4.

and bladder cancer in male rats. Critics of these studies pointed out that humans would have to drink 800 cans of diet soda *per day* to get a saccharin dose equivalent to that given to the rats. Furthermore, they argued this response may be unique to male rats. In 2000, the U.S. Department of Health concluded a study that found no association between saccharin and cancer in humans. Congress ordered that all warnings be removed from saccharin-containing products. Still, some groups like the Center for Science in the Public Interest consider this sweetener dangerous and urge us to avoid it if possible.

In setting standards for environmental toxins, we need to consider (1) combined effects of exposure to many different sources of damage, (2) different sensitivities of members of the population, and (3) effects of chronic as well as acute exposures. Some people argue that pollution levels should be set at the highest amount that does *not* cause measurable effects. Others demand that pollution be reduced to zero if possible, or as low as is technologically feasible. It may not be reasonable to demand that we be protected from every potentially harmful contaminant in our environment, no matter how small the risk. As we have seen, our bodies have mechanisms that enable us to avoid or repair many kinds of damage so that most of us can withstand some minimal level of exposure without harm (fig. 8.18).

On the other hand, each challenge to our cells by toxic substances represents stress on our bodies. Although each individual stress may not be life-threatening, the cumulative effects of all the environmental stresses, both natural and human-caused, to which we are exposed may seriously shorten or restrict our lives. Furthermore, some individuals in any population are more susceptible to those stresses than others. Should we set pollution standards so that no one is adversely affected, even the most sensitive individuals, or should the acceptable level of risk be based on the average member of the population?

Finally, policy decisions about hazardous and toxic materials also need to be based on information about how such materials affect the plants, animals, and other organisms that define and maintain our environment. In some cases, pollution can harm or destroy whole ecosystems with devastating effects on the life-supporting cycles on which we depend. In other cases, only the most sensitive species are threatened. Table 8.7 shows the Environmental Protection Agency's assessment of relative risks to human welfare. This ranking reflects a concern that our exclusive focus on reducing pollution to protect human health has neglected risks to natural ecological systems. While there have been many benefits from a case-by-case approach in which we evaluate the health risks of individual chemicals, we have often missed broader ecological problems that may be of greater ultimate importance.

FIGURE 8.18   "Do you want to stop reading those ingredients while we're trying to eat?"
Reprinted with permission of the *Star-Tribune,* Minneapolis-St. Paul.

TABLE 8.7   **Relative Risks to Human Welfare**

RELATIVELY HIGH-RISK PROBLEMS
Habitat alteration and destruction
Species extinction and loss of biological diversity
Stratospheric ozone depletion
Global climate change

RELATIVELY MEDIUM-RISK PROBLEMS
Herbicides/pesticides
Toxics and pollutants in surface waters
Acid deposition
Airborne toxics

RELATIVELY LOW-RISK PROBLEMS
Oil spills
Groundwater pollution
Radionuclides
Thermal pollution

Source: *Environmental Protection Agency.*

Summary

- Health is a state of physical, mental, and social well-being, not merely the absence of disease or infirmity. Environmental health focuses on health risks in the natural, social, cultural, and technological worlds in which we live.

- The world is now undergoing a dramatic epidemiological transition. The traditional killers—infectious diseases, maternal and perinatal complications, and nutritional deficiencies—still take a terrible toll in the developing world, but health problems such as heart attack, depression, and traffic accidents, once thought to occur only in rich countries, are now becoming the leading killers everywhere as people live longer and adopt Western lifestyles and diets.

- Disability-adjusted life years (DALYs) measure the disease burden arising both from premature deaths and the loss of healthy life resulting from illness and disability.

- Emergent diseases are those not previously known or that have been absent for at least 20 years. Some examples are SARS (severe acute respiratory syndrome), West Nile virus, and AIDS (acquired immune deficiency syndrome). AIDS is currently the most deadly of these diseases, infecting about 65 million people and killing some 3 million every year.

- Only 10 percent of all medical expenditures go to the major diseases that affect 90 percent of the world population. Currently, the United States donates only 0.01 percent of its GDP for health care in the developing world. Investing more would not only save millions of lives, but could boost the world economy significantly.

- CWD (chronic wasting disease), a rapidly spreading brain disease in deer, is an emergent ecological disease. Some other examples are sudden oak death in California, black band coral disease in the Caribbean, and domoic acid poisoning, which is killing marine mammals in the Pacific.

- Toxins are poisons. They can be very specific because they interact with and disrupt the metabolic machinery that keeps cells alive. Some materials are so supertoxic that a single molecule can cause cell death.

- Allergens, carcinogens, mutagens, neurotoxins, teratogens, and endocrine disrupters all are examples of toxins that cause specific health problems. Diet, also, is an important health factor.

- Solubility, persistence, chemical reactivity, and bioaccumulation all are important factors in toxicity as are timing, dose, and route of exposure. Characteristics and condition of the target organism also are very influential in determining effects of toxins.

- Children are much more sensitive to most toxins than are adults. Pound for pound, children drink more water, eat more food, and breathe more air than adults. How best to protect children from environmental hazards is a difficult, but important, question.

- Some persistent organic pollutants (POPs) such as PBDE, PFOS, PFOA, and BPA are now found throughout the world. Every human has them in his or her blood. The health effects of these compounds are not yet known, but there are concerns about chronic exposures.

- We depend on animal testing for much of our assessment of toxins, but great differences in sensitivity between species makes risk evaluation difficult. Just because something can be detected, doesn't mean that it's present in a dangerous form or concentration.

- Our perception of risk is strongly influenced by emotion and factors such as whether the hazard is voluntary, familiar, has a lag before its effects are known, and whether those at risk benefit from the source of exposure.

- Health experts generally regard a 1 in 1 million risk of death to be acceptable, but some people ask, "Acceptable to whom?" We often disregard familiar but serious risks, while demanding protection from other, highly improbable risks.

Questions for Review

1. What is SARS? How is it thought to have started?

2. Define the terms *disease* and *health*.

3. What were some of the most serious diseases in the world in 1990? How is this list expected to change in the next 20 years?

4. What are emergent diseases? Give a few examples, and describe their cause and effects.

5. How do bacteria acquire antibiotic resistance? What might we do about this?

6. What is the difference between toxic and hazardous? Give some examples of materials in each category.

7. How do the physical and chemical characteristics of materials affect their movement, persistence, distribution, and fate in the environment?

8. What is the difference between acute and chronic toxicity?

9. Define *carcinogenic, mutagenic, teratogenic,* and *neurotoxic.*

10. What are the relative risks of smoking, driving a car, and drinking water with the maximum permissible levels of trichloroethylene? Are these relatively equal risks?

Questions for Critical Thinking

1. What consequences (positive or negative) do you think might result from defining health as a state of complete physical, mental, and social well-being? Who might favor or oppose such a definition?

2. Do rich countries bear any responsibilities if the developing world adopts unhealthy lifestyles or diets? What could (or should) we do about it?

3. Why do we spend more money on heart or cancer research than childhood illnesses?

4. How much do you think the United States should donate for health care in developing countries? What evidence would you offer to support your position?

5. Some people seem to have a poison paranoia about synthetic chemicals. Why do we tend to assume that natural chemicals are benign while industrial chemicals are evil?

6. Analyze the claim that we are exposed to thousands of times more natural carcinogens in our diet than industrial ones. Is this a good reason to ignore pollution?

7. Are good health and a clean environment a basic human right or merely something for which we should strive?

8. What are the premises in the discussion of assessing risk? Could conflicting conclusions be drawn from the facts presented in this section? What is your perception of risk from your environment?

9. Should pollution levels be set to protect the average person in the population or the most sensitive? Why not have zero exposure to all hazards?

10. What level of risk is acceptable to you? Are there some things for which you would accept more risk than others?

Key Terms

acute effects 162
allergens 154
antigens 154
bioaccumulation 158

biomagnification 158
cancer 156
carcinogens 156
chronic effects 162

disability-adjusted life years (DALY) 147
disease 147
emergent disease 149
endocrine disrupters 156
environmental health 147
fetal alcohol syndrome 156
health 147
LD50 162
morbidity 147
mortality 147

mutagens 156
neurotoxins 156
pathogens 149
persistent organic pollutants (POPs) 159
risk 164
risk assessment 164
sick building syndrome 154
synergism 161
teratogens 156
toxins 153

Further Readings

Calabrese, E. J., and L. A. Baldwin. 2003. Toxicology rethinks its central belief. *Nature* 421:691–92.

Daszak, P., et al. 2000. Emerging infectious diseases of wildlife— Threats to biodiversity and human health. *Science* 287:443–49.

Donnelly, C. A., et al. 2003. Epidemiological determinants of spread of causal agent of severe acute respiratory syndrome in Hong Kong. *Lancet* 361:111–34.

Hughes, J. M. 2001 Emerging infectious diseases: A CDC perspective. *Emerging Infectious Diseases* 7(3) Supplement:494–96.

Klaassen, Curtis D., ed. 2001. *Casarett & Doull's Toxicology: The Basic Science of Poisons,* 6th ed. McGraw-Hill Co.

Koop, C. Everett, et al. 2002. *Critical Issues in Global Health.* Jossey-Bass.

Landrigan, Phillip, et al. 2002. *Raising Healthy Children in a Toxic World: 101 Smart Solutions for Every Family.* Rodale Press.

Miller, Judith, Stephen Engelberg, and William J. Broad. 2001. *Germs: Biological Weapons and America's Secret War.* Simon & Schuster.

Slovic, Paul. 2000. *The Perception of Risk.* Earthscan Publications.

Woloshin, S., et al. 2002. Risk charts: Putting cancer in context. *J. of the National Cancer Institute* 94(11):799–84.

World Health Organization. 2003. *World Health Report.* Oxford University Press.

Yam, Philip. 2003. Shoot this deer. *Scientific American* 288(6): 38–43.

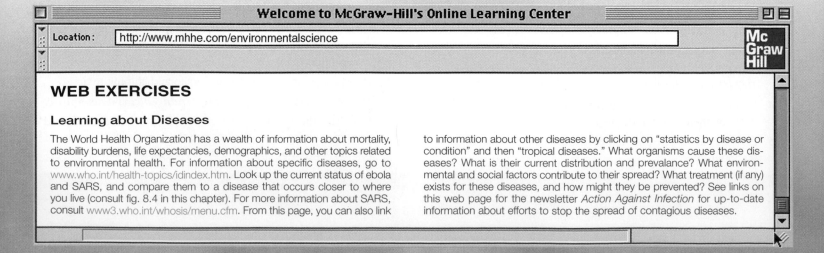

Welcome to McGraw-Hill's Online Learning Center

Location: http://www.mhhe.com/environmentalscience

WEB EXERCISES

Learning about Diseases

The World Health Organization has a wealth of information about mortality, disability burdens, life expectancies, demographics, and other topics related to environmental health. For information about specific diseases, go to www.who.int/health-topics/idindex.htm. Look up the current status of ebola and SARS, and compare them to a disease that occurs closer to where you live (consult fig. 8.4 in this chapter). For more information about SARS, consult www3.who.int/whosis/menu.cfm. From this page, you can also link to information about other diseases by clicking on "statistics by disease or condition" and then "tropical diseases." What organisms cause these diseases? What is their current distribution and prevalance? What environmental and social factors contribute to their spread? What treatment (if any) exists for these diseases, and how might they be prevented? See links on this web page for the newsletter *Action Against Infection* for up-to-date information about efforts to stop the spread of contagious diseases.

Food and Agriculture

We abuse the land because we regard it as a commodity belonging to us.
When we see land as a community to which we belong, we may begin to use it with love and respect.

Aldo Leopold

OBJECTIVES

After studying this chapter, you should be able to:

- describe world food supplies and some causes of chronic hunger and acute food shortages in the midst of growing food surpluses.
- differentiate between famine and chronic undernutrition and understand the relation between natural disasters and social or economic forces in triggering food shortages.
- explain some major human nutritional requirements as well as the consequences of deficiencies in those nutrients.
- sketch the role of living organisms, physical forces, and other factors in creating and maintaining fertile soil.
- differentiate between the sources and effects of land degradation including erosion, nutrient depletion, waterlogging, and salinization.
- explain the need for water, energy, and nutrients for sustained crop production as well as some limits on our use of these resources.
- analyze some of the promises and perils of genetic engineering.
- recognize the potential for low-input, sustainable, regenerative agriculture.

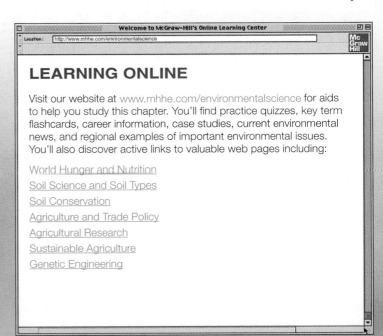

LEARNING ONLINE

Visit our website at www.mhhe.com/environmentalscience for aids to help you study this chapter. You'll find practice quizzes, key term flashcards, career information, case studies, current environmental news, and regional examples of important environmental issues. You'll also discover active links to valuable web pages including:

World Hunger and Nutrition
Soil Science and Soil Types
Soil Conservation
Agriculture and Trade Policy
Agricultural Research
Sustainable Agriculture
Genetic Engineering

Blue Revolution

Seafood, including fin fish, shellfish, and crustaceans, is prized by many people, and it provides the main animal protein source for nearly 1 billion residents of developing countries. Populations of wild fish are declining, however, in many parts of the world. The UN Food and Agriculture Organization (FAO) estimates that three-quarters of all marine stocks are being exploited at or beyond sustainable yields. Furthermore, indiscriminate and destructive harvest techniques are destroying habitat and disrupting marine communities in ways that may make it difficult to rebuild populations. Canadian fisheries experts report that only 10 percent of large predators such as sharks, marlin, tuna, and swordfish remain in the ocean. In the North Atlantic, once immense groundfish populations (cod, halibut, flounder, and skate) have been exploited to near extinction.

How can we replace this valuable protein source? Aquaculture, or fish farming, which now produces about one-third of the seafood eaten by humans every year, is the fastest growing sector of our food producing system. Economist Lester Brown predicts that fish farms may be the feedlots of the future. Beef production, he reports, which now provides about 55 million metric tons per year, has grown by about half over the past 30 years. Aquaculture, in that same time, has grown nearly tenfold, from about 3 million metric tons per year to more than 30 million metric tons. If these trends continue, fish farming may soon overtake cattle ranching as the largest source of animal protein in the human diet. Some marine biologists expect this "blue revolution" in aquaculture to be as important as the "green revolution" in terrestrial crops that greatly reduced chronic hunger and malnutrition worldwide over the past 30 years.

Many seafood species can be raised in captivity more efficiently than can birds or mammals. It takes about 7 kg of feed, for example, to produce 1 kg of beef. By contrast, only 1.6 kg of feed will yield 1 kg gain in catfish. Although at least 220 species of aquatic organisms are raised for food, a dozen or so dominate world output. China, where fish farming began more than 3,000 years ago, is by far the world leader in this field, producing 71 percent of the total volume and 50 percent of the total value of all aquaculture every year. About 6.5 million ha of Chinese ponds, lakes, reservoirs, and rice paddies are used for seafood cultivation. Most of China's aquaculture is integrated with agriculture, enabling farmers to use agricultural wastes, such as rice straw and pig manure, to fertilize ponds.

Not all fish farming, however, is environmentally or socially benign. Much of the production, especially that destined for markets in richer countries, is in high-value carnivorous marine species such as salmon, cod, and shrimp. Huge volumes of wild-caught fish such as anchovies are processed into fish meal and fish oil to feed these carnivores, which may take more protein to produce a kilogram of edible product than do cattle, hogs, or chickens. Fish meal can be contaminated

FIGURE 9.1 Pens for fish-rearing in Thailand.
Source: Food and Agriculture Organization (FAO).

with PCBs and other dioxin-like compounds from the fish oil it contains. According to the Environmental Working Group, farmed salmon are likely the most PCB-contaminated protein source in the U.S. food supply. In samples from U.S. grocery stores, farmed salmon were found to have 16 times as much PCBs as wild fish and 4 times as much as beef. Fish meal production also depletes food for wild marine species such as cod, seals, and seabirds. Furthermore, many fish farms depend on wild-caught rather than hatchery-reared young to stock captive-rearing operations. This practice can severely deplete natural communities.

Fish feces and uneaten food falling from net pens can seriously pollute near-shore waters (fig. 9.1). Dense populations of captive species in ponds or pens often require use of antibiotics and disinfectants, which also can contaminate surrounding ecosystems if they are discharged into surface waters. Non-native species and genetically modified organisms can escape from captivity to hybridize with or outcompete native wild fish. And destruction of hundreds of thousands of hectares of mangrove forests and coastal wetlands—especially for shrimp ponds—threatens nurseries important for many marine species. Fish ponds created on former agricultural land can have high levels of pesticides, PCBs, and other toxic contaminants.

Still, there is great promise for ecologically, socially, and economically sustainable fish farming to produce valuable, high-quality protein to feed growing human populations. In this chapter, we'll look further at food supplies, soil science, farming practices, and public policies designed to insure adequate nutrition and food security for everyone.

FOOD AND NUTRITION

Despite dire predictions that runaway population growth would soon lead to terrible famines (see chapter 7), world food supplies have more than kept up with increasing human numbers over the past two centuries. The past 40 years have seen especially encouraging strides in reducing world hunger. While population growth averaged 1.7 percent per year during that time, world food production increased an average of 2.2 percent. Increased use of irrigation, improved crop varieties, more readily available fertilizers, and distribution systems to transport food from regions with surpluses to those in need have brought improved nutrition to billions of people. In this section, we'll look at the causes and effects of remaining chronic and acute food shortages, as well as recommendations for a balanced, healthful diet.

Chronic Hunger and Food Security

In 1960, nearly 60 percent of the residents of developing countries were considered **chronically undernourished,** meaning their diet didn't provide the 2,200 kcal per day, on average, considered necessary for a healthy productive life. Today, despite the fact that their population has doubled over the past 40 years, the proportion in these countries suffering from chronic caloric deficiency has fallen to less than 15 percent.

The UN Food and Agriculture Organization (FAO) expects agricultural production to continue to grow over the next few decades. Where the current world food supply would be sufficient, if equitably distributed, to provide an average of 2,800 kcal per person per day, the FAO predicts that by 2030, there will be enough food available to supply 3,050 kcal per day to everyone, or about 30 percent more than most of us need. In countries such as the United States, the problem already has become what to do with surplus food. Farmers in these countries are paid billions of dollars per year not to grow crops.

Still, in a world of surplus food, some 815 million people don't have enough to eat. Figure 9.2 shows countries with the highest risk of food shortages. As you can see, most of sub-Saharan Africa, South and Southeast Asia, and parts of Latin America fall in this category. Ninety-five percent of the chronically undernourished are in developing countries, but an increasing number are in transition countries (primarily states of the former Soviet Union and its allies undergoing a change from socialism to market economies) where bad weather, poor management, and social crises have resulted in sharply falling agricultural production (fig. 9.3). Even in the richest countries, where excess calories are the greatest problem for the majority, some 11 million people don't have enough to eat.

Poverty is the greatest threat to **food security,** or the ability to obtain sufficient food on a day-to-day basis. The 1.4 billion people in the world who live on less than $1 per day all too often can't buy the food they need and don't have access to resources to grow it for themselves. Food security occurs at multiple scales. In the poorest countries, hunger may affect nearly everyone. In other countries, although the average food availability may be satisfactory, some individual communities or families may not have enough to eat. And within families, males often get both the largest share as well as the most nutritious food, while women and children—who need food most—all too often get the poorest diet. At least 6 million children under 5 years old die every year of diseases exacerbated by hunger and malnutrition. Providing a healthy diet might eliminate as much as 60 percent of all premature deaths worldwide.

Hungry people can't work their way out of poverty, the Nobel Prize-winning economist Robert Fogel points out. He estimates that in 1790, about 20 percent of the population of England and France was effectively excluded from the labor force because they were too weak and hungry to work. Improved nutrition, he calculates, accounted for about half of all European economic growth during the nineteenth century. Since many developing countries are as poor now (in relative terms) as Britain and France were in 1790, his analysis suggests that reducing hunger could yield more than $120 billion (U.S.) in economic growth produced by longer, healthier, more productive lives for several hundred million people.

The 2003 UN World Food Summit reaffirmed the goal set by previous conventions of reducing the number of chronically undernourished people from 800 million to 400 million by 2015. We aren't on track to meet that goal, but some countries have made impressive progress. China, alone, has reduced its number of undernourished people by 74 million over the past decade. Indonesia,

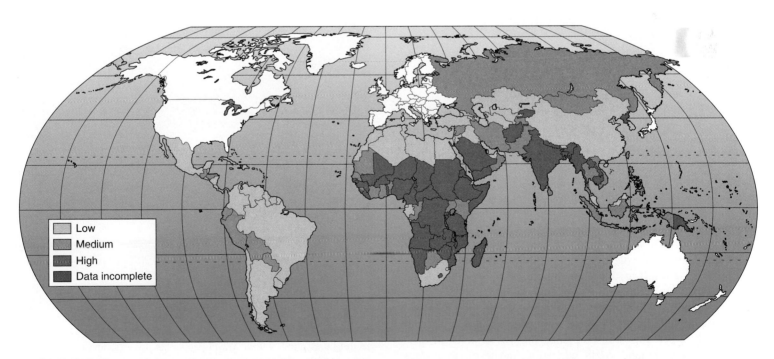

FIGURE 9.2   Countries with populations at risk of inadequate nutrition. The United States, Canada, Europe, Japan, and Australia have little risk.
Source: Food and Agriculture Organization (FAO) 2002.

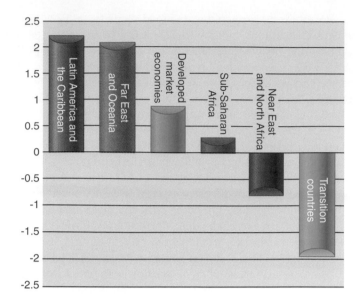

FIGURE 9.3 Changes in agricultural production in 2000. Transition nations are states from the former Soviet Union undergoing a change from socialism to capitalism. *Source:* Food and Agriculture Organization (FAO) 2002.

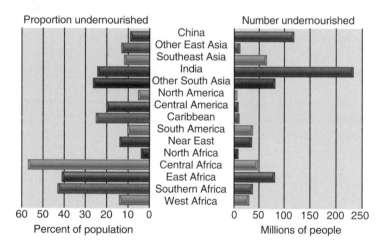

FIGURE 9.4 Number and percent of population chronically undernourished by region in the developing world. *Source:* Food and Agriculture Organization (FAO) 2002.

Vietnam, Thailand, Nigeria, Ghana, and Peru each reduced chronic hunger by about 3 million people. In 47 other countries, however, the numbers of chronically underfed people have increased. India now has by far the largest number of persistently hungry people in the world, while central Africa has the highest percentage (fig. 9.4).

Recognizing the role of women in food production is an important step toward food security for all. Throughout the developing world, women do 50 to 70 percent of all farm work but control only a tiny fraction of the land and rarely have access to capital or developmental aid. In Nigeria, for example, home gardens occupy only 2 percent of all cropland, but provide half the food families eat. Making land, credit, education, and access to markets available to women could contribute greatly to family nutrition.

Acute Food Shortages

The chronic hunger and malnutrition described in the previous section is silent and generally invisible, affecting individuals, families, and communities, often in the midst of general plenty, but **famines** are characterized by large-scale food shortages, massive starvation, social disruption, and economic chaos. Starving people eat their seed grain and slaughter their breeding stock in a desperate attempt to keep themselves and their families alive. Even if better conditions return, they have sacrificed their productive capacity and will take a long time to recover. Famines are characterized by mass migrations as starving people travel to refugee camps in search of food and medical care (fig. 9.5). Many die on the way or fall prey to robbers.

In 2002, the FAO reported that 67 million people in 32 countries needed emergency food aid. What causes these emergencies? Environmental conditions are usually the immediate trigger, but politics and economics are often equally important. Bad weather, insect outbreaks, and other natural disasters cause crop failures and create food shortages. But the Nobel Prize-winning work of Harvard economist Amartya K. Sen shows that these factors have often been around for a long time, and local people usually have adaptations to get through hard times if they aren't thwarted by inept or corrupt governments and greedy elites. National politics, however, together with commodity hoarding, price gouging, poverty, wars, landlessness, and other external factors often conspire so that poor people can neither grow their own food nor find jobs to earn money to buy the food they need. Professor Sen points out that armed conflict and political oppression almost always are at the root of famine. No democratic country with a relatively free press, he says, has ever had a major famine.

FIGURE 9.5 Children wait for their daily ration of porridge at a feeding station in Somalia. When people are driven from their homes by hunger or war, social systems collapse, diseases spread rapidly, and the situation quickly becomes desperate. © Norbert Schiller/The Image Works.

The aid policies of rich countries often serve more to get rid of surplus commodities and make us feel good about our generosity than to get at the root causes of starvation. But herding people into feeding camps often is the worst thing to do for them. The stress of getting there kills many of them, and the crowding and lack of sanitation in the camps exposes them to epidemic diseases. There are no jobs in the refugee camps, so people can't support themselves if they try. Social chaos and family breakdown expose those who are weakest to robbery and violence. Having left their land and tools behind, people can't replant crops when the weather returns to normal.

Malnutrition and Obesity

In addition to energy (calories), we also need specific nutrients in our diet, such as proteins, vitamins, and certain trace minerals. You might have more than enough calories and still suffer from **malnourishment,** a nutritional imbalance caused by a lack of specific dietary components or an inability to absorb or utilize essential nutrients. Those of us in richer countries often eat too much meat, salt, and saturated fat, and too little fiber, vitamins, trace minerals, and other components lost from highly processed foods. On average, we consume about one-third more calories than we need and get too little exercise. Food provides solace in stressful lives, and we're constantly bombarded with advertising tempting us to consume more. According to the U.S. Surgeon General, 62 percent of all adult Americans are overweight, up from 40 percent only a decade ago, and about one-third of us are seriously overweight or **obese**—generally considered to be a body mass greater than 30 kg/m², or roughly 30 pounds above normal for an average person.

Being overweight substantially raises your risk of hypertension, diabetes, heart attacks, stroke, gallbladder disease, osteoarthritis, respiratory problems, and certain cancers. Every year, about 300,000 in the United States die from illnesses related to obesity. This number is close to the 400,000 annual deaths from diseases related to smoking. Paradoxically, food insecurity and poverty can contribute to obesity. In one study, more than half of the women who reported not having enough to eat were overweight compared to one-third of food-secure women. Lack of time for cooking and access to health food choices along with ready availability of fast-food snacks and calorie-laden drinks lead to dangerous dietary imbalances for many people.

This trend isn't limited to richer countries. Obesity is spreading around the world (fig. 9.6). For the first time in history, there are probably more overweight people (more than 1 billion) than underweight. As chapter 8 points out, western diets and lifestyles are being adopted by many in the developing world, and diseases such as heart attack, stroke, diabetes, and depression that once were thought to inflict only wealthy nations are now becoming the most prevalent causes of death and disability everywhere.

Many poor people can't afford meat, fruits, and vegetables that would provide a balanced diet. The FAO estimates that nearly 3 billion people (half the world population) suffer from vitamin, mineral, or protein deficiencies. This results in devastating illnesses and deaths as well as reduced mental capacity, developmental abnormalities, and stunted growth. Altogether, these problems bring an incalculable loss of human potential and social capital.

Anemia (low hemoglobin levels in the blood, usually caused by dietary iron deficiency) is the most common nutritional problem in the world. According to the FAO, more than 2 billion people (52 percent are pregnant women and 39 percent are children under age 5) suffer from iron deficiencies. The problem is most severe in India, where it's estimated that 80 percent of all pregnant women are anemic. Anemia increases the risk of maternal deaths from hemorrhage in childbirth and affects childhood development. Red meat, eggs, legumes, and green vegetables all are good sources of dietary iron.

Iodine is essential for synthesis of thyroxin, an endocrine hormone that regulates metabolism and brain development, among other things. Chronic iodine deficiency causes goiter (a swollen thyroid gland), stunted growth, and reduced mental ability. The FAO estimates that 740 million people—mainly in South and Southeast Asia—suffer from iodine deficiency and that 177 million children have stunted growth and development. Adding a few pennies worth of iodine to our salt has largely eliminated this problem in developed countries.

Starchy foods like maize (corn), polished rice, and manioc (tapioca), which form the bulk of the diet for many poor people, tend to be low in several essential vitamins as well as minerals. According to the FAO, vitamin A deficiencies affect 100–140 million children at any given time. At least 350,000 go blind every year from the effects of this vitamin shortage. Folic acid (found in dark green, leafy vegetables) is essential for early fetal development. Folic acid deficiencies have been linked to neurological problems in babies including microencephaly (an abnormally small head) or even anencephaly (lacking a brain).

Protein also is essential for normal growth and development. The two most widespread human protein deficiency diseases are kwashiorkor and marasmus. **Kwashiorkor** is a West African word meaning "displaced child." (A young child is displaced—and deprived of nutritious breast milk—when a new baby is born.) This

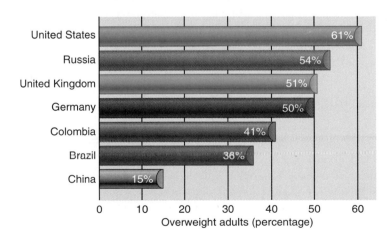

FIGURE 9.6 While nearly a billion people are chronically undernourished, people in wealthier countries are at risk from eating too much. *Source:* Worldwatch Institute 2001.

FIGURE 9.7 Marasmus is caused by combined energy (calorie) and protein deficiencies. Children with marasmus have the wizened look and dry, flaky skin of an old person. © Scott Daniel Peterson/Gamma Liaison International.

condition most often occurs in young children who eat mainly cheap starchy food and don't get enough good-quality protein. Children with kwashiorkor often have reddish-orange hair, puffy, discolored skin, and a bloated belly. **Marasmus** (from the Greek "to waste away") is caused by a diet low in both calories and protein. A child suffering from severe marasmus is generally thin and shriveled, like a tiny, very old starving person (fig. 9.7). Children with both these deficiencies have low resistance to infections and are likely to suffer from stunted growth, mental retardation, and other developmental problems. Altogether, the FAO estimates that the annual losses due to deaths and diseases caused by calorie and nutrient deficiencies are equivalent to 46 million years of productive life (see more in chapter 8 on disability-adjusted life years).

Eating a Balanced Diet

What's the best way to be sure you're getting a healthy diet? Generally, it isn't necessary to take synthetic dietary supplements. Eating a good variety of foods should give you all the nutrients you need. For years, Americans were advised to eat daily servings of four major food groups: meat, dairy products, grains, and fruits and vegetables. These recommendations were revised in 1992 to emphasize only sparing servings of meat, dairy, fats, and sweets.

Some nutritionists believe that the 1992 recommendations for 6 to 11 servings a day of bread, cereal, rice, and pasta still provides too many simple sugars (or starches that are quickly converted to sugar). Based on observations of health effects of Mediterranean

diets as well as a long-term study of 140,000 U.S. health professionals, Drs. Walter Willett and Meir Stampfer of Harvard University have recommended a new dietary pyramid (fig. 9.8). Both red meat and starchy food such as white rice, white bread, potatoes, and pasta should be eaten sparingly. Nuts, legumes (beans, peas, and lentils), fruit, vegetables, and whole grain foods form the basis of this diet. Unsaturated plant oils should make up 30 to 40 percent of dietary calories, according to this view. Trans fat (the kind found in hydrogenated margarine), on the other hand, is not recommended at all. Combined with regular, moderate exercise, this food selection should provide most people with all the nutrition they need.

KEY FOOD SOURCES

Of the thousands of edible plants and animals in the world, only about a dozen types of seeds and grains, three root crops, twenty or so common fruits and vegetables, six mammals, two domestic fowl, and a few fish and other forms of marine life make up almost all of the food humans eat. Table 9.1 shows annual production of some important foods in human diets. In this section, we will highlight the characteristics of some important food sources.

Major Crops

The three crops on which humanity depends for the majority of its nutrients and calories are wheat, rice, and maize (called corn in the United States). Together, nearly 2,100 million metric tons of these three grains are grown each year. Wheat and rice are especially important since they are the staple foods for most of the 5 billion people in the developing countries of the world. These two grass species supply around 60 percent of the calories consumed directly by humans.

Potatoes, barley, oats, and rye are staples in mountainous regions and high latitudes (northern Europe, north Asia) because they grow well in cool, moist climates. Cassava, sweet potatoes, and other roots and tubers grow well in warm, wet areas and are staples in Amazonia, Africa, Melanesia, and the South Pacific. Sorghum and millet are drought resistant and are staples in the dry regions of Africa.

Fruits, vegetables, and vegetable oils make a surprisingly large contribution to human diets. They are especially welcome because they typically contain high levels of vitamins, minerals, dietary fiber, and complex carbohydrates.

Meat and Dairy

Protein-rich foods are prized by people nearly everywhere. In the past, the rich countries consumed a vast majority of the meat, dairy, and high-quality seafood traded internationally. The four-fifths of the world's people in less-developed countries generally raised 60 percent of the 3 billion domestic ruminants and 6 billion poultry in the world but consumed only one-fifth of all commercial animal products. The FAO reports, however, that as incomes rise in developing countries, food choices throughout the world are shifting

FIGURE 9.8 A new food pyramid proposed by Drs. Walter Willett and Meir Stampfer deemphasizes consumption of red meat, butter, white rice, potatoes, and pasta, and suggests a diet rich in whole grains, unsaturated plant oils, fruits, and vegetables.

Source: Willett and Stampfer, 2002.

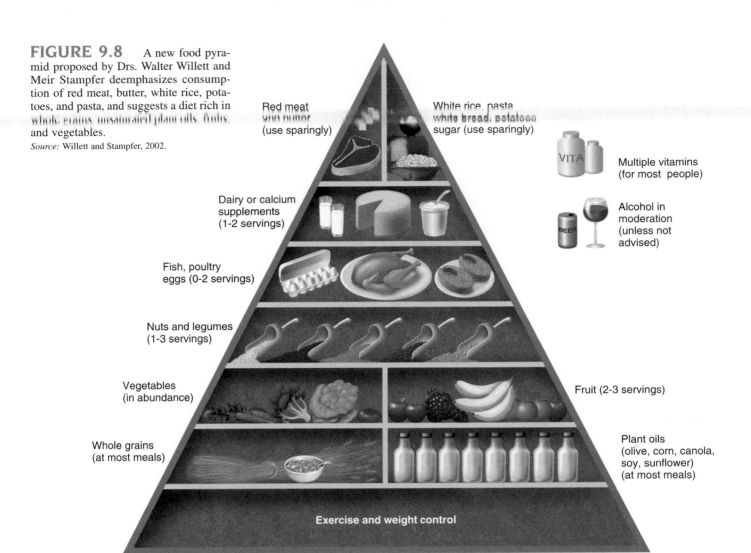

Red meat and butter (use sparingly)

White rice, pasta, white bread, potatoes, sugar (use sparingly)

Multiple vitamins (for most people)

Dairy or calcium supplements (1-2 servings)

Alcohol in moderation (unless not advised)

Fish, poultry eggs (0-2 servings)

Nuts and legumes (1-3 servings)

Vegetables (in abundance)

Fruit (2-3 servings)

Whole grains (at most meals)

Plant oils (olive, corn, canola, soy, sunflower) (at most meals)

Exercise and weight control

TABLE 9.1 **Some Important Food Sources**

CROP	2002 YIELD (MILLION METRIC TONS)
Wheat	603
Rice (paddy)	593
Maize (corn)	933
Potatoes	308
Barley and oats	168
Soybeans	177
Cassava and sweet potato	015
Sugar (cane and beet)	141
Pulses (beans, peas)	54
Oil seeds	324
Vegetables and fruits	905
Meat and milk	735
Fish and seafood	140

Source: Food and Agriculture Organization (FAO), 2003.

---

toward higher-quality and more expensive foods. Meat consumption in developing countries, for example, has risen from only 10 kg per person annually in the 1960s to 26 kg currently (fig. 9.9). By 2030, average annual meat consumption in those countries is expected to be nearly 40 kg per person. Far more cereal grain will be needed to raise livestock if this trend continues.

Most of the livestock grown in North America are confined in concentrated animal feeding operations for part or all of their lifetime. A diet rich in grain, oil, and protein fattens animals quickly and produces meat preferred by many consumers. Globally, some 660 million metric tons of cereals are used as livestock feed each year, representing just over a third of the world total cereal use. It's often pointed out that we could feed about ten times as many people if we ate grain directly rather than feeding it to livestock. Surprisingly, the FAO claims that using cereals as animal feed does not contribute to hunger and undernutrition. Given current agricultural economics, if these cereals were not used as animal feed, they would probably not be produced at all, and thus would not be available as food. According to this view, if everyone became a vegetarian, the lack of demand for cereals for livestock production would simply lead to lower crop production, not more food for the hungry.

FIGURE 9.9 Meat and dairy consumption have quadrupled in the past 40 years, and China represents about 40 percent of that increased demand. © William P. Cunningham.

The rapid proliferation of large-scale animal confinement operations raises a number of social and environmental questions. Up to 10,000 hogs enclosed in a giant barn, or 100,000 cattle in a single feedlot complex can cause serious local air and water pollution (fig. 9.10). Animal waste often is stored in enormous open lagoons, which can leak or rupture, contaminating both surface waters and groundwater supplies. When Hurricane Floyd dumped torrential rains on North Carolina in 1999, an estimated 10 million m³ (2.5 billion gal) of hog and poultry waste overflowed from storage lagoons into local rivers. The high density of animals in these facilities and the rich diet they're fed to speed weight gain requires a constant use of antibiotics and growth hormones. More than half of all antibiotics used in the United States are administered to livestock, mostly in confinement facilities. A big part of the rapid rise in antibiotic resistant pathogens is due to this massive and constant dosing with drugs.

Seafood

As the opening story of this chapter illustrates, the 140 million metric tons of seafood eaten every year is an important part of our diet. Seafood provides about 15 percent of all animal protein eaten by humans, and is the main animal protein source for about 1 billion people in developing countries. Unfortunately, overharvesting and habitat destruction threaten most of the world's wild fisheries. Annual catches of ocean fish rose by about 4 percent annually between 1950 and 1988. Since 1989, however, 13 of 17 major marine fisheries have declined dramatically or become commercially unsustainable. According to the United Nations, three-quarters of the world's edible ocean fish, crustaceans, and mollusks are declining and in urgent need of managed conservation.

The problem is too many boats using efficient but destructive technology to exploit a dwindling resource base. Boats as big as ocean liners travel thousands of kilometers and drag nets large enough to scoop up a dozen jumbo jets, sweeping a large patch of ocean clean of fish in a few hours. Long-line fishing boats set cables up to 10 km long with hooks every 2 meters that catch birds, turtles, and other unwanted "by-catch" along with targeted species. Trawlers drag heavy nets across the bottom, scooping up everything indiscriminately and reducing broad swaths of habitat to rubble. One marine biologist compared the technique to harvesting forest mushrooms with a bulldozer. In some operations, up to 15 kg of dead and dying by-catch is dumped back into the ocean for every kilogram of marketable food. The FAO estimates that operating costs for the 4 million boats now harvesting wild fish exceed sales by $50 billion (U.S.) per year. Countries subsidize fishing fleets to preserve jobs and to ensure access to this valuable resource.

Aquaculture is providing an increasing share of the world's seafood. Fish can be grown in farm ponds that take relatively little space but are highly productive. As mentioned earlier in this chapter, cultivation of high-value carnivorous species such as salmon, sea bass, and tuna threaten wild stocks exploited to stock captive operations or to provide fish food. Building coastal fish-rearing ponds causes destruction of hundreds of thousands of hectares of mangrove forests and wetlands that serve as irreplaceable nurseries for marine species. Net pens anchored in nearshore areas allow spread of diseases, escape of exotic species, and release of feces, uneaten food, antibiotics, and other pollutants into surrounding ecosystems.

Polyculture systems of mixed species of herbivores or filter feeders can alleviate many aquaculture problems. Raising species in enclosed, land-based ponds or warehouses can eliminate the pollution problems associated with net pens in lakes or oceans. In China, for example, most fish are raised in ponds or rice paddies (fig. 9.11). One ecologically balanced system uses four carp species that feed at different levels of the food chain. The grass carp, as its name implies, feeds largely on vegetation, while the common carp is a bottom feeder, living on detritus that settles to the bottom. Silver

FIGURE 9.10 Up to 100,000 cattle may be confined in a single feedlot operation. Note the animals in the background standing on a manure mound. © Corbis/Volume 102.

FIGURE 9.11 Fish ponds are interspersed with mulberry plantations in China. Dead tree leaves fertilize the ponds and silkworms provide fish food, while enriched pond water irrigates the trees. Food and Agriculture Organization photo/H. Zhang.

carp and bighead carp are filter feeders that feed on phytoplankton and zooplankton, respectively. Agricultural wastes such as manure, dead silkworms, and rice straw fertilize ponds and encourage phytoplankton growth. All these carp species are considered dangerous invasive species in North America. Still, these integrated polyculture systems typically boost fish yields per hectare by 50 percent or more compared with monoculture farming. Of the top ten seafood species recommended by the Monterey Aquarium as ecologically acceptable, six are farmed.

FARM POLICY

Much of the increase in food production over the past 50 years has been fueled by public support for agricultural education, research, and development projects such as irrigation systems, transportation networks, and crop insurance. While helping local farmers, agricultural subsidies also can distort markets and cripple production in developing countries. The World Bank estimates that rich countries pay their own farmers $350 billion per year, or nearly seven times as much as all developmental aid to poor countries. A typical cow in Europe enjoys annual subsidies three times the average yearly income for African farmers. In the past, European countries have had the highest aid per farm, but recent policy reforms have promised to decouple farm support from crop production.

Farm subsidies in many countries are protected by powerful political and economical interests. The most recent U.S. Farm Bill, for example, promised $180 billion in payments over the next decade for American farmers. Most of this aid goes to heavily supported crops such as corn, wheat, cotton, rice, and soybeans along with certain specially protected commodities such as milk, sugar, and peanuts. For some crops, this bill represented an 80 percent increase in support. Legislators claimed that subsidies are intended to preserve "family farms," but analysis of previous payments shows that 10 percent of all farms received 70 percent of all support. One giant farm in Arkansas, for example, received $38 million in payments between 1996 and 2001.

Agricultural subsidies encourage surpluses and allow American farmers to sell their products overseas at as much as 20 percent below the actual cost of production. This makes imported food cheaper than locally grown crops in many developing countries, and destroys the livelihoods of millions of indigenous farmers. The FAO argues that ending distorting financial support in the richer countries would have a far more positive impact on food supplies and livelihoods in the developing world than any aid program. What do you think? Does agriculture deserve special protections that we don't extend to other sectors of our economy? How might we protect our own farms without threatening those of our global neighbors?

SOIL: A RENEWABLE RESOURCE

Growing the food and fiber needed to support human life is a complex enterprise that requires knowledge from many different fields and cooperation from many different groups of people. In this section, we'll survey some of the principles of soil science and look at some of the inputs necessary for continued agricultural production.

Of all the earth's crustal resources, the one we take most for granted is soil. We are terrestrial animals and depend on soil for life, yet most of us think of it only in negative terms. English is unique in using "soil" as an interchangeable word for earth and excrement. "Dirty" has a moral connotation of corruption and impurity. Perhaps these uses of the word enhance our tendency to abuse soil without scruples; after all, it's only dirt.

The truth is that **soil** is a marvelous substance, a living resource of astonishing beauty, complexity, and frailty. It is a complex mixture of weathered mineral materials from rocks, partially decomposed organic molecules, and a host of living organisms. It can be considered an ecosystem by itself. Soil is an essential component of the biosphere, and it can be used sustainably, or even enhanced, under careful management.

Key Concepts

- Chronic undernourishment comes from a diet providing less than the 2,200 kcal per day, on average, considered necessary by the FAO for a healthy productive life.

- Food security is the ability to obtain sufficient food on a day-to-day basis. Even in communities or families where, on average, there would be enough for everyone, those with lowest status (women and children) may not get the food they need for a healthy life.

- Being overweight substantially raises your risk of hypertension, diabetes, heart attacks, strokes, and other diseases. Currently, more than 60 percent of Americans are considered overweight and this problem is spreading around the world.

- Many current farming practices degrade soil and deplete resources, but soil conservation, agroecology, regenerative farming, and sustainable agriculture all offer ways to grow the food we need on a sustainable basis.

There are at least 15,000 different soil series or types in the United States and many thousands more worldwide. They vary because of the influences of parent material, time, topography, climate, and organisms on soil formation. There are young soils that, because they have not weathered much, are rich in soluble nutrients. There are old soils, like the red soils of the tropics, from which rainwater has washed away most of the soluble minerals and organic matter, leaving behind clay and rust-colored oxides.

To understand the potential for feeding the world on a sustainable basis we need to know how soil is formed, how it is being lost, and what can be done to protect and rebuild good agricultural soil. With careful husbandry, soil can be replenished and renewed indefinitely. Many farming techniques deplete soil nutrients, however, and expose the soil to the erosive forces of wind and moving water. As a result, in many places we are essentially mining this resource and using it much faster than it is being replaced.

Building good soil is a slow process. Under the best circumstances, good topsoil accumulates at a rate of about 10 tons per hectare (2.5 acres) per year—enough soil to make a layer about 1 mm deep when spread over a hectare. Under poor conditions, it can take thousands of years to build that much soil. Perhaps one-third to one-half of the world's current croplands are losing topsoil faster than it is being replaced. In some of the worst spots, erosion carries away about 2.5 cm (1 in.) of topsoil per year. With losses like that, agricultural production has already begun to fall in many areas.

Soil Composition

Most soil is about half mineral. The rest is plant and animal residue, air, water, and living organisms. The mineral particles are derived either from the underlying bedrock or from materials transported and deposited by glaciers, rivers, ocean currents, windstorms, or landslides. The weathering processes that break rocks down into soil particles are described in chapter 14.

Particle sizes affect the characteristics of the soil (fig. 9.12). The spaces between sand particles give sandy soil good drainage and usually allow it to be well aerated, but also cause it to dry out quickly when rains are infrequent. Tight packing of small particles in silty or clay soils makes them less permeable to air and water than sandy soils. Tiny capillary spaces between the particles, on the other hand, store water and mineral ions better than more porous soils. Because clay particles have a proportionately large surface area and a high ionic charge, they stick together tenaciously, giving clay its slippery plasticity, cohesiveness, and impermeability. Soils with a high clay content are called "heavy soils," in contrast to easily worked "light soils" that are composed mostly of sand or silt. Varying proportions of these mineral particles occur in each soil type (fig. 9.13). Farmers usually consider sandy loam the best soil type for cultivating crops.

The organic content of soil can range from nearly zero for pure sand, silt, or clay, to nearly 100 percent for peat or muck, which is composed mainly of partly decomposed plant material. Much of the organic material in soil is **humus,** a sticky, brown, insoluble residue from the partially decomposed bodies of dead plants and animals. Humus is the most significant factor in the development

Sand (0.05 - 2mm)

Silt (0.02 - 0.05mm)

Clay (less than 0.02 mm)

FIGURE 9.12 Relative sizes of soil particles magnified about 100-fold.

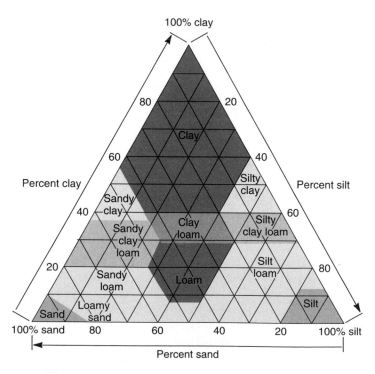

FIGURE 9.13 Soil texture is determined by the percentages of clay, silt, and sand particles in the soil. Soils with the best texture for most crops are loams, which have enough larger particles (sand) to be loose, yet enough smaller particles (silt and clay) to retain water and dissolved mineral nutrients. *Source:* Soil Conservation Service.

of "structure," a description of how the soil particles clump together. Humus coats mineral particles and holds them together in loose crumbs, giving the soil a spongy texture that holds water and nutrients needed by plant roots, and maintains the spaces through which delicate root hairs grow.

Soil Organisms

Without soil organisms, the earth would be covered with sterile mineral particles far different from the rich, living soil ecosystems on which we depend for most of our food. The activity of the myriad organisms living in the soil help create structure, fertility, and tilth (condition suitable for tilling or cultivation) (fig. 9.14).

Soil organisms usually stay close to the surface, but that thin living layer can contain thousands of species and billions of individual organisms per hectare. Algae live on the surface, while bacteria and fungi flourish in the top few centimeters of soil. A single gram of soil (about one-half teaspoon) can contain hundreds of millions of these microscopic cells. Algae and blue-green bacteria capture sunlight and make new organic compounds. Bacteria and fungi decompose organic detritus and recycle nutrients that plants can use for additional growth. The sweet aroma of freshly turned soil is caused by actinomycetes, bacteria that grow in funguslike strands and give us the antibiotics streptomycin and tetracyclines.

Mycorrhizal symbiosis is an association between the roots of most plant species and certain fungi. The plant provides organic compounds to the fungus, while water and inorganic nutrients are absorbed from the soil by the fungus and transferred to the plant. As a result, micorrhizal plants often grow better—especially in infertile soil—than those lacking fungal partners. There can be 20 m of fungal strands in a gram of soil.

Roundworms, segmented worms, mites, and tiny insects swarm by the thousands in that same gram of soil from the surface. Some of them are herbivorous, but many of them prey upon one another. Soil roundworms (nematodes) attack plant rootlets and can cause serious crop damage. A carnivorous fungus snares nematodes with tiny loops of living cells that constrict like a noose when a worm blunders into it. Burrowing animals, such as gophers, moles, insect larvae, and worms, tunnel deeper in the soil, mixing and aerating it. Plant roots also penetrate lower soil levels, drawing up soluble minerals and secreting acids that decompose mineral particles. Fallen plant litter adds new organic material to the soil, returning nutrients to be recycled.

FIGURE 9.14 Soil ecosystems include numerous consumer organisms, as depicted here: (1) snail, (2) termite, (3) nematodes and nematode-killing constricting fungus, (4) earthworm, (5) wood roach, (6) centipede, (7) carabid (ground) beetle, (8) slug, (9) soil fungus, (10) wireworm (click beetle larva), (11) soil protozoan, (12) sow bug, (13) ant, (14) mite, (15) springtail, (16) pseudoscorpion, and (17) cicada nymph.

Soil Profiles

Most soils are stratified into horizontal layers called **soil horizons** that reveal much about the history and usefulness of the soil. The thickness, color, texture, and composition of each horizon are used to classify the soil. Together these horizons make up a **soil profile** (fig. 9.15).

The soil surface is often covered with a layer of leaf litter, crop residues, or other fresh or partially decomposed organic material. This organic layer is know as the O horizon. Below this layer is the A horizon or **topsoil,** composed of mineral particles mixed with organic material. The A horizon can range from several meters thick under virgin prairie to almost nothing in dry deserts. The O and A horizons contain most of the living organisms and organic material in the soil, and it is in these layers that most plants spread their roots to absorb water and nutrients. An eluviated (leached) E horizon often lies below the A horizon. This layer is depleted of clays and soluble nutrients, which are removed by rainwater seeping down through the soil. These clays and nutrients generally accumulate in the B horizon, or **subsoil.** The B horizon may have a dense or clayey texture because of the accumulated clays. In desert regions, a dense, impermeable "hardpan" layer of accumulated minerals or salts may develop on the B horizon. This hardpan blocks plant root growth and prevents water from draining properly.

Beneath the subsoil is the parent material, or C horizon, made of weathered rock fragments with very little organic material. Weathering in the C horizon, largely accomplished by rain water mixed with organic compounds from plants, produces new soil particles and allows downward expansion of the soil profile. Parent material can be sand, solid rock, or other material. About 70 percent of the parent material in the United States was transported to its present site by glaciers, wind, or water and is not directly related to the underlying bedrock.

Soil Types

Soils are classified according to their structure and composition. In the United States, soils are classified into 12 *soil orders.* The richest farming soils, with thick, organic-rich A horizons are *mollisols* (formed under grasslands) and *alfisols* (formed under deciduous forests). Both of these develop where rainfall and temperatures are moderate. *Spodosols* also develop in temperate climates, but usually under pine forests, where acidic needle litter causes a characteristic whitish, ashy-looking E horizon. In hot and rainy environments, *oxisols* and *ultisols* develop. These soils are severely depleted of nutrients and tend to be reddish because iron-rich minerals make up much of the remaining soil material. *Aridosols* form in arid environments, have little organic material, and often contain accumulated salt or hardpan layers. Some soil types are defined mainly by parent material: *andisols* develop from volcanic material, and *vertisols* develop from clay-rich material such as lake beds or shale bedrock. *Histosols* are formed of waterlogged, incompletely decayed plant material. Other soils are largely defined by degree of development; *entisols* and *inceptisols* have little or no horizon development. This may be because these soils are on recently exposed parent material (such as glacial debris) or because they are in an environment where soil development is extremely slow because of slight rainfall or organic activity. A relatively new soil order is *gelisols,* soils in permafrost areas. These soils occur only at high latitudes or high altitudes.

WAYS WE USE AND ABUSE SOIL

Only about 11 percent of the earth's land area (14.66 million km^2 out of a total of 132.4 million km^2) is currently in agricultural production. Perhaps four times as much land could potentially be converted to cropland, but much of this land serves as a refuge for cultural or biological diversity or suffers from constraints, such as steep slopes, shallow soils, poor drainage, tillage problems, low nutrient levels, metal toxicity, or excess soluble salts or acidity, that limit the types of crops that can be grown there (fig. 9.16).

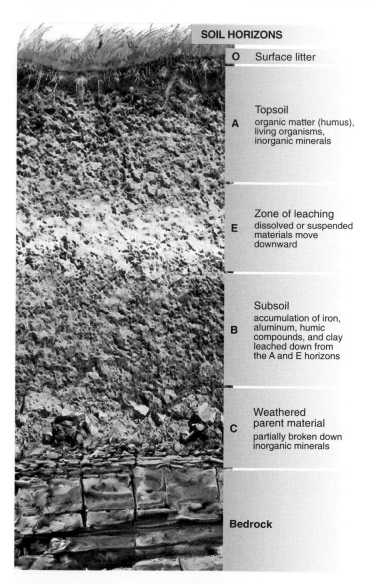

SOIL HORIZONS

O   Surface litter

A   Topsoil
organic matter (humus), living organisms, inorganic minerals

E   Zone of leaching
dissolved or suspended materials move downward

B   Subsoil
accumulation of iron, aluminum, humic compounds, and clay leached down from the A and E horizons

C   Weathered parent material
partially broken down inorganic minerals

Bedrock

FIGURE 9.15   Soil profile showing possible soil horizons. The actual number, composition, and thickness of these layers varies in different soil types.

FIGURE 9.16 In many areas, soil or climate constraints limit agricultural production. These hungry goats in Sudan feed on a solitary Acacia shrub. *Source:* Food and Agriculture Organization (FAO) photo/R. Faidutti.

Land Resources

Arable land is unevenly distributed around the world. In parts of Canada, the United States, and Europe, temperate climates, abundant water, and high soil fertility produce high crop yields that contribute to high standards of living. Other parts of the world, although rich in land area, lack suitable soil, topography, water, or climate to sustain our levels of productivity.

Worldwide, the cropland available for agriculture is shrinking. In 1970, there was a global average of 0.38 ha per person. By 2002, this had declined to 0.21 ha per person. The FAO forecasts that by 2030 the land supply will shrink to 0.16 ha per person. In Asia, cropland will be even more scarce—0.09 ha (0.22 acre) per person—30 years from now. If you live on a typical quarter-acre suburban lot, look at your yard and imagine feeding yourself for a year on what you could produce there.

In the developed countries, 95 percent of recent agricultural growth in the twentieth century has come from improved crop varieties or increased fertilization, irrigation, and pesticide use, rather than from bringing new land into production. In fact, less land is being cultivated now than 100 years ago in North America, or 600 years ago in Europe. As more effective use of labor, fertilizer, and water and improved seed varieties have increased in the more developed countries, productivity per unit of land has increased, and marginal land has been retired, mostly to forests and grazing lands. In many developing countries, land continues to be cheaper than other resources, and new land is still being brought under cultivation, mostly at the expense of forests and grazing lands. Still, at least two-thirds of recent production gains have come from new crop varieties and more intense cropping rather than expansion into new lands.

The largest increases in cropland over the last 30 years occurred in South America and Oceania where forests and grazing lands are rapidly being converted to farms. Many developing countries are reaching the limit of lands that can be exploited for agriculture without unacceptable social and environmental costs, but others still have considerable potential for opening new agricultural lands. East Asia, for instance, already uses about three-quarters of its potentially arable land. Most of its remaining land has severe restrictions for agricultural use. Further increases in crop production will probably have to come from higher yields per hectare. Latin America, by contrast, uses only about one-fifth of its potential land, and Africa uses only about one-fourth of the land that theoretically could grow crops. However, there would be serious ecological trade-offs in putting much of this land into agricultural production.

While land surveys tell us that much more land in the world *could* be cultivated, not all of that land necessarily *should* be farmed. Much of it is more valuable in its natural state. The soils over much of tropical Asia, Africa, and South America are old, weathered, and generally infertile. Most of the nutrients are in the standing plants, not in the soil. In many cases, clearing land for agriculture in the tropics has resulted in tragic losses of biodiversity and the valuable ecological services that it provides. Ultimately, much of this land is turned into useless scrub or semidesert.

On the other hand, there are large areas of rich, subtropical grassland and forest that are well watered, have good soil, and could become productive farmland without unduly reducing the world's biological diversity. Argentina, for instance, has pampas grasslands about twice the size of Texas that closely resemble the American Midwest a century ago in climate and potential for agricultural growth. Some of this land could probably be farmed with relatively little ecological damage if it were done carefully.

Land Degradation

Agriculture both causes and suffers from environmental degradation. The International Soil Reference and Information Centre in the Netherlands estimates that every year 3 million ha (7.4 million acres) of cropland are ruined by erosion, 4 million ha are turned into deserts, and 8 million ha are converted to nonagricultural uses such as homes, highways, shopping centers, factories, reservoirs, etc. Over the past 50 years, some 1.9 *billion* ha of agricultural land (an area greater than that now in production) have been degraded to some extent. About 300 million ha of this land are strongly degraded (deep gullies, severe nutrient depletion, crops grow poorly, restoration is difficult and expensive), while 910 million ha—about the size of China—are moderately degraded. Nearly 9 million ha of former croplands are so degraded that they no longer support any crop growth at all. The causes of this extreme degradation vary: In Ethiopia it is water erosion, in Somalia it is wind, and in Uzbekistan salt and toxic chemicals are responsible.

Definitions of degradation are based on both biological productivity and our expectations about what the land should be like. Often this is a subjective judgment and it is difficult to distinguish between human-caused deterioration and natural processes like drought. We generally consider the land degraded when the soil is impoverished or eroded, water runs off or is contaminated more than is normal, vegetation is diminished, biomass production is decreased, or wildlife diversity diminishes. On farmlands this results in lower crop yields. On ranchlands it means fewer livestock can be supported per unit area. On nature reserves it means fewer species.

The amount and degree of land degradation varies by region and country. About 20 percent of land in Africa and Asia is degraded, but most is in either the light or moderate category. In Central America and Mexico, by contrast, 25 percent of all vegetated land suffers moderate to extreme degradation. Figure 9.17 shows some areas of greatest concern for soil degradation, as well as the prime mechanisms for this problem. Water and wind erosion provide the motive force for the vast majority of all soil degradation, worldwide. Chemical deterioration includes nutrient depletion, salinization (salt accumulation), acidification, and pollution. Physical deterioration includes compaction by heavy machinery or trampling by cattle, waterlogging—water accumulation—from excess irrigation and poor drainage, and laterization—solidification of iron and aluminum-rich tropical soil when exposed to sun and rain.

Erosion: The Nature of the Problem

Erosion is an important natural process, resulting in the redistribution of the products of geologic weathering, and is part of both soil formation and soil loss. The world's landscapes have been sculpted by erosion. When the results are spectacular enough, we enshrine them in national parks as we did with the Grand Canyon. Where erosion has worn down mountains and spread soil over the plains, or deposited rich alluvial silt in river bottoms, we gladly farm it. Erosion is a disaster only when it occurs in the wrong place at the wrong time.

In some places, erosion occurs so rapidly that anyone can see it happen. Deep gullies are created where water scours away the soil, leaving fenceposts and trees sitting on tall pedestals as the land erodes away around them. In most places, however, erosion is more subtle. It is a creeping disaster that occurs in small increments. A thin layer of topsoil is washed off fields year after year until eventually nothing is left but poor-quality subsoil that requires more and more fertilizer and water to produce any crop at all.

The net effect, worldwide, of this general, widespread topsoil erosion is a reduction in crop production equivalent to removing about 1 percent of world cropland each year. Many farmers are able to compensate for this loss by applying more fertilizer and by bringing new land into cultivation. Continuation of current erosion rates, however, could reduce agricultural production by 25 percent in Central America and Africa and 20 percent in South America by the year 2020. The total annual soil loss from croplands is thought to be 25 billion metric tons. About twice that much soil is lost from rangelands, forests, and urban construction sites each year.

In addition to reduced land fertility, this erosion results in sediment-loading of rivers and lakes, siltation of reservoirs, smothering of wetlands and coral reefs, and clogging of water intakes and waterpower turbines.

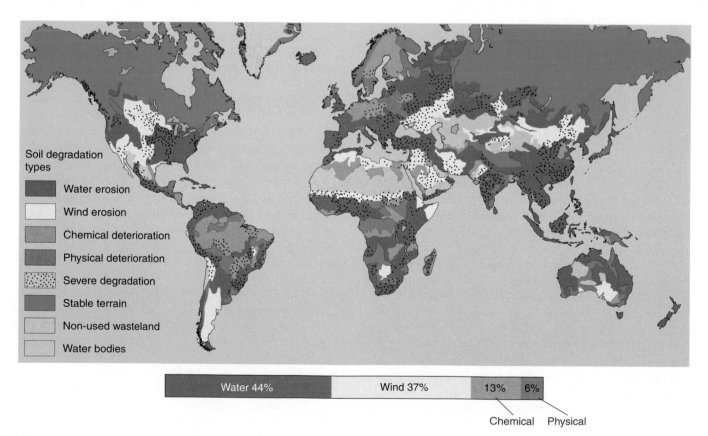

FIGURE 9.17 Causes and location of global soil degradation. *Source:* Food and Agriculture Organization (FAO), 2003.

Mechanisms of Erosion

Wind and water are the main agents that move soil around. A thin layer taken off the land surface is called **sheet erosion**. When little rivulets of running water gather together and cut small channels in the soil, the process is called **rill erosion** (fig. 9.18). When rills enlarge to form bigger channels or ravines that are too large to be removed by normal tillage operations, we call the process **gully erosion** (fig. 9.19). Streambank erosion refers to the washing away of soil from the banks of established streams, creeks, or rivers, often as a result of removing trees and brush along streambanks and by cattle damage to the banks.

Most soil loss on agricultural land is sheet or rill erosion. Large amounts of soil can be transported a little bit at a time without being very noticeable. A farm field can lose 20 metric tons of soil per hectare during winter and spring runoff in rills so small that they are erased by the first spring cultivation. That represents a loss of only a few millimeters of soil over the whole surface of the field, hardly apparent to any but the most discerning eye. But it doesn't take much mathematical skill to see that if you lose soil twice as fast as it is being replaced, eventually it will run out.

Wind can equal or exceed water in erosive force, especially in a dry climate and on relatively flat land. When plant cover and surface litter are removed from the land by agriculture or grazing, wind lifts loose soil particles and sweeps them away. Windborne dust is sometimes transported from one continent to another. Scientists in Hawaii can tell when spring plowing begins in China because dust from Chinese farmland is carried by winds all the way across the Pacific Ocean. Similarly, summer dust storms in the Sahara Desert of North Africa carry about 1 billion tons of soil in massive airborne dust plumes over the Atlantic and Mediterranean every year. This dust creates a hazy atmosphere over islands in the Caribbean Sea, 5,000 km (3,000 mi) away and has measurable regional climatic effects. Similarly, it has been estimated that winds blowing over the Mississippi River basin have 1,000 times the soil-carrying capacity of the river itself.

Some of the highest erosion rates in the world occur in the United States and Canada. The U.S. Department of Agriculture reports that 69 million hectares (170 million acres) of U.S. farmland and range are eroding at rates that reduce long-term productivity. Five tons per acre (11 metric tons per hectare) is generally considered the maximum tolerable rate of soil loss because that is generally the highest rate at which soil forms under optimum conditions. Some farms lose soil at twice that rate or more.

Intensive farming practices are largely responsible for this situation. Row crops, such as corn and soybeans, leave soil exposed for much of the growing season (fig. 9.20). Deep plowing and heavy herbicide applications create weed-free fields that look neat but are subject to erosion. Because big machines cannot easily follow contours, they often go straight up and down the hills, creating ready-made gullies for water to follow. Farmers sometimes plow through grass-lined watercourses and have pulled out windbreaks and fencerows to accommodate the large machines and to get every last square meter into production. Consequently, wind and water carry away the topsoil.

Pressed by economic conditions, many farmers have abandoned traditional crop rotation patterns and the custom of resting land as pasture or fallow every few years. Continuous monoculture cropping can increase soil loss tenfold over other farming patterns. A soil study in Iowa showed that a three-year rotation of corn, wheat, and clover lost an average of only 6 metric tons per hectare. By comparison, continuous wheat production on the same land caused nearly four times as much erosion, and continuous corn cropping resulted in seven times as much soil loss as the rotation with wheat and clover.

Erosion Hotspots

Data on soil condition and soil erosion often is incomplete, but it is evident that many places have problems as severe as, or perhaps worse than ours. China, for example, has a large area of loess (windblown silt) deposits on the North China Plain that once was

FIGURE 9.18   Runoff from a rainstorm erodes soil from an Iowa cornfield. Photo by Lynn Betts, courtesy of USDA Natural Resources Conservation Service.

FIGURE 9.19   Uncontrolled erosion has cut a deep gully through this Kansas pasture. Photo by Jeff Vanuga, courtesy of USDA Natural Resources Conservation Service.

FIGURE 9.20 Annual row crops leave soil bare and exposed to erosion for most of the year, especially when fields are plowed immediately after harvest, as this one always is. © William P. Cunningham.

covered by forest and grassland. The forests were cut down and the grasslands were converted to cropland. This plateau is now scarred by gullies 30 to 40 m deep, and the soil loss is thought to be at least 480 metric tons per hectare per year. This would be equivalent to 3 cm of topsoil per year.

One way to estimate soil loss is to measure the sediment load carried by rivers draining an area. The highest concentration of sediment in any river is in the Huang (Yellow) River that originates in the loess plateau of China. Although its drainage basin is only one-fifth as big as that of the Mississippi River, the Huang carries more than four times as much soil each year. This suggests that the average soil loss *per hectare* in China may be 20 times that in the United States.

The Mississippi River carries enough fertilizer every year to create a "dead zone" in the Gulf of Mexico that can be as large as 57,000 km². Algal growth stimulated by high nitrogen in runoff from farms and cities depletes oxygen within this zone to levels that are lethal for most marine life. A task force recommended a 20 to 30 percent decrease in nitrogen loading to reduce the size and effects of this zone. Similar hypoxic zones occur near mouths of many other rivers that drain agricultural areas.

OTHER AGRICULTURAL RESOURCES

Soil is only part of the agricultural resource picture. Agriculture is also dependent upon water, nutrients, favorable climates to grow crops, productive crop varieties, and upon the mechanical energy to tend and harvest them.

Water

All plants need water to grow. Agriculture accounts for the largest single share of global water use. About two-thirds of all fresh water withdrawn from rivers, lakes, and groundwater supplies is used for irrigation (chapter 17). Although estimates vary widely (as do definitions of irrigated land), about 15 percent of all cropland, worldwide, is irrigated.

Some countries are water rich and can readily afford to irrigate farmland, while other countries are water poor and must use water very carefully. The efficiency of irrigation water use is rather low in most countries. High evaporative and seepage losses from unlined and uncovered canals often mean that in some places, up to 80 percent of water withdrawn for irrigation never reaches its intended destination (chapter 17). Farmers often tend to over-irrigate because water prices are relatively low and because they lack the technology to meter water and distribute just the amount needed. In the United States and Canada, many farmers are adopting water-saving technologies such as drip irrigation or downward-facing sprinklers (fig. 9.21).

Excessive use not only wastes water; it often results in **waterlogging.** Waterlogged soil is saturated with water, and plant roots die from lack of oxygen. **Salinization,** in which mineral salts accumulate in the soil, occurs particularly when soils in dry climates are irrigated with saline water. As the water evaporates, it leaves behind a salty crust on the soil surface that is lethal to most plants. Flushing with excess water can wash away this salt accumulation but the result is even more saline water for downstream users.

Worldwide, irrigation problems are a major source of land degradation and crop losses. The Worldwatch Institute reports that 60 million ha (150 million acres) of cropland have been damaged by salinization and waterlogging. Water conservation techniques can greatly reduce problems arising from excess water use. Conservation also makes more water available for other uses or for expanded crop production where water is in short supply (chapter 17).

FIGURE 9.21 Downward-facing sprinklers on this center-pivot irrigation system deliver water more efficiently than upward-facing sprinklers. © Corbis/Volume 13.

Fertilizer

In addition to water, sunshine, and carbon dioxide, plants need small amounts of inorganic nutrients for growth. The major elements required by most plants are nitrogen, potassium, phosphorus, calcium, magnesium, and sulfur. Calcium and magnesium often are limited in areas of high rainfall and must be supplied in the form of lime. Lack of nitrogen, potassium, and phosphorus even more often limits plant growth. Adding these elements in fertilizer usually stimulates growth and greatly increases crop yields. A good deal of the doubling in worldwide crop production since 1950 has come from increased inorganic fertilizer use. In 1950, the average amount of fertilizer used was 20 kg per hectare. In 1990, this had increased to an average of 91 kg per hectare worldwide.

Farmers may overfertilize because they are unaware of the specific nutrient content of their soils or the needs of their crops. While European farmers use more than twice as much fertilizer per hectare as do North American farmers, their yields are not proportionally higher. Phosphates and nitrates from farm fields and cattle feedlots are a major cause of aquatic ecosystem pollution. Nitrate levels in groundwater have risen to dangerous levels in many areas where intensive farming is practiced. Young children are especially sensitive to the presence of nitrates. Using nitrate-contaminated water to mix infant formula can be fatal for newborns.

What are some alternative ways to fertilize crops? Manure and green manure (crops grown specifically to add nutrients to the soil) are important natural sources of soil nutrients. Nitrogen-fixing bacteria living symbiotically in root nodules of legumes are valuable for making nitrogen available as a plant nutrient (chapter 3). Interplanting or rotating beans or some other leguminous crop with such crops as corn and wheat are traditional ways of increasing nitrogen availability.

There is considerable potential for increasing world food supply by increasing fertilizer use in low-production countries if ways can be found to apply fertilizer more effectively and reduce pollution. Africa, for instance, uses an average of only 19 kg of fertilizer per hectare, or about one-fourth of the world average. It has been estimated that the developing world could at least triple its crop production by raising fertilizer use to the world average.

Energy

Farming as it is generally practiced in the industrialized countries is highly energy-intensive. Fossil fuels supply almost all of this energy. Between 1920 and 1980, direct energy use on farms rose as gasoline and diesel fuels were consumed by increasing mechanization of agriculture. An even greater increase in indirect energy use, in the form of synthetic fertilizers, pesticides, and other agricultural chemicals, also occurred, especially after World War II.

After crops leave the farm, additional energy is used in food processing, distribution, storage, and cooking. It has been estimated that the average food item in the American diet travels 2,000 km between the farm that grew it and the person who consumes it. The energy required for this complex processing and distribution system may be five times as much as is used directly in farming. Alto-

FIGURE 9.22   Oil from fuel crops like these sunflowers can be burned directly in diesel engines. © Corbis/Volume 90.

gether the food system in the United States consumes about 16 percent of the total energy we use. Most of our foods require more energy to produce, process, and get to market than they yield when we eat them.

Farmers could assist in moving to a renewable energy future by growing energy crops that can be converted into biofuels. Many Midwestern states have encouraged construction of corn- or soy-based ethanol factories in recent years. Mixing ethanol with gasoline helps reduce air pollution, and having a new market helps struggling farmers, but these programs aren't truly renewable if it takes more petroleum energy to grow and process the crop than is available in the final product. Plant oils from sunflowers, soybeans, corn, and other oil seed crops can be burned directly in most diesel engines, and may be closer than ethanol to a net energy gain (fig 9.22). Fast-growing trees, kenaf, switch grass, cattails, and other energy crops may be grown on marginal land and burned in power plants. Many Great Plains farmers are finding that their most lucrative crop may be the wind (chapter 20).

NEW CROPS AND GENETIC ENGINEERING

Although at least 3,000 species of plants have been used for food at one time or another, most of the world's food now comes from only 16 species. Many new or unconventional varieties might be valuable human food supplies, however, especially in areas where conventional crops are limited by climate, soil, pests, or other problems. The FAO predicts that 70 percent of future production growth will come from higher yields and new crop varieties rather than expansion of arable lands. Among the plants now being investigated as potential additions to our crop roster is the winged bean (fig. 9.23), a perennial plant that grows well in hot climates where other legumes will not grow. It is totally edible (pods, mature seeds, shoots, flowers, leaves, and tuberous roots), resistant to diseases,

FIGURE 9.23 Winged beans bear fruit year-round in tropical climates and are resistant to many diseases that prohibit growing other bean species. Whole pods can be eaten when they are green, or dried beans can be stored for later use. It is a good protein source in a vegetarian diet. © William P. Cunningham.

and enriches the soil. Another promising crop is tricale, a hybrid between wheat (*Triticum*) and rye (*Secale*) that grows in light, sandy, infertile soil. It is drought resistant, has nutritious seeds, and is being tested for salt tolerance for growth in saline soils or irrigation with seawater. Some traditional crop varieties grown by Native Americans, such as tepary beans, amaranth, and Sonoran panicgrass are being collected by seed conservator Gary Nabhan both as a form of cultural revival for native people and as a possible food crop for harsh environments.

Green Revolution

So far, the major improvements in farm production have come from technological advances and modification of a few well-known species. Yield increases often have been spectacular. A century ago, when all maize (corn) in the United States was open-pollinated, average yields were about 25 bushels per acre. In 1999, average yields from hybrid maize were around 130 bushels per acre, and under optimum conditions, 250 bushels per acre are possible. Most of this gain was accomplished by conventional plant breeding: geneticists laboriously hand-pollinating plants, moving selected genes from one variety to another.

Starting about 50 years ago, agricultural research stations began to breed tropical wheat and rice varieties that would provide food for growing populations in developing countries. The first of the "miracle" varieties was a dwarf, high-yielding wheat devel-

oped by Norman Borlaug (who received a Nobel Peace Prize for his work) at a research center in Mexico (fig 9.24). At about the same time, the International Rice Institute in the Philippines developed dwarf rice strains with three or four times the production of varieties in use at the time. The dramatic increases obtained as these new varieties spread around the world has been called the **green revolution.** It is one of the main reasons that world food supplies have more than kept pace with the growing human population over the past few decades.

Most green revolution breeds really are "high responders," meaning that they yield more than other varieties if given optimum levels of fertilizer, water, and protection from pests and diseases (fig. 9.25). Under suboptimum conditions, on the other hand, high responders may not produce as well as traditional varieties. Poor farmers who can't afford the expensive seed, fertilizer, and water

FIGURE 9.24 Semi-dwarf wheat (*right*), bred by Norman Borlaug, has shorter, stiffer stems and is less likely to lodge (fall over) when wet than its conventional cousin (*left*). This "miracle" wheat responds better to water and fertilizer, and has played a vital role in feeding a growing human population. © William P. Cunningham.

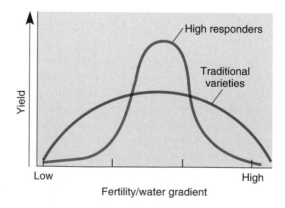

FIGURE 9.25 Green revolution miracle crops are really high responders, meaning that they have excellent yields under optimum conditions. For poor farmers who can't afford the fertilizer and water needed by high responders, traditional varieties may produce better yields.

required to become part of this movement, usually are left out of the green revolution. In fact, they may be driven out of farming altogether as rising land values and falling commodity prices squeeze them from both sides.

Genetic Engineering

Genetic engineering involves removing genetic material from one organism and splicing it into the chromosomes of another (fig. 9.26). This new technology has the potential to greatly increase both the quantity and quality of our food supply. It is now possible to build entirely new genes, and even new organisms, taking a bit of DNA from here, a bit from there, even synthesizing artificial DNA sequences to create desired characteristics in **genetically modified organisms (GMOs).**

Proponents predict dramatic benefits from genetic engineering. Research is now underway to improve yields and create crops that resist drought, frost, or diseases. Other strains are being developed to tolerate salty, waterlogged, or low-nutrient soils, allowing degraded or marginal farmland to become productive. All of these could be important for reducing hunger in developing countries. Plants that produce their own pesticides might reduce the need for toxic chemicals; while engineering for improved protein or vitamin content could make our food more nutritious. Attempts to remove specific toxins or allergens from crops also could make our food safer. Crops such as bananas and potatoes have been altered to contain oral vaccines that can be grown in developing countries where refrigeration and sterile needles are unavailable. Plants have been engineered to make industrial oils and plastics. Animals, too, are being genetically modified to grow faster, gain weight on less food, and produce pharmaceuticals such as insulin in their milk. It may soon be possible to create animals with human cell-recognition factors that could serve as organ donors.

Opponents worry that moving genes willy-nilly could create a host of problems, some of which we can't even imagine. GMOs themselves might escape and become pests or they might interbreed with wild relatives. In either case, we may create superweeds or reduce native biodiversity. Constant presence of pesticides in plants could accelerate pesticide resistance in insects or leave toxic residues in soil or our food. Genes for toxicity or allergies could be transferred along with desirable genes, or novel toxins could be created as genes are mixed together. This technology may be available only to the richest countries or the wealthiest corporations, making family farms uncompetitive and driving developing countries even further into poverty.

Both the number of GMO crops and the acreage devoted to growing them is increasing rapidly. Between 1987 and 2002, more than 9,000 field tests for some 750 different crop varieties were carried out in the United States (fig. 9.27). Hawaii is the most popular site for GMO testing, with about 1,500 field releases over the past 14 years. Illinois, Iowa, California, and Puerto Rico each have had more than 1,000 tests during that time. Worldwide, 53 million ha (131 million acres) were planted with GMO crops in 2002. The United States accounted for 68 percent of that acreage, followed by Argentina with 23 percent. Canada, Australia, Mexico, China, and South Africa together make up 9 percent of all transgenic cropland.

Some transgenic crops have reached mainstream status. Currently about 82 percent of all soybeans, 71 percent of the cotton, and one-quarter of all maize (corn) grown in the United States are GMOs. You've already probably eaten some genetically modified food. Estimates are that at least 60 percent of all processed food in America contains GMO ingredients. Since the United States, Argentina, and Brazil account for 90 percent of international trade in maize and soybeans (in which GMOs often are mixed with other grains), a large fraction of the world most likely has been exposed to some GMO products.

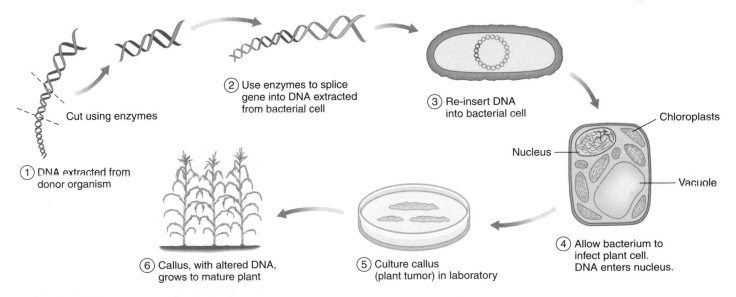

FIGURE 9.26 One method of gene transfer, using an infectious, tumor-forming bacterium such as agrobacterium. Genes with desired characteristics are cut out of donor DNA and spliced into bacterial DNA using special enzymes. The bacteria then infect plant cells and carry altered DNA into cells' nuclei. The cells multiply, forming a tumor, or callus, which can grow into a mature plant.

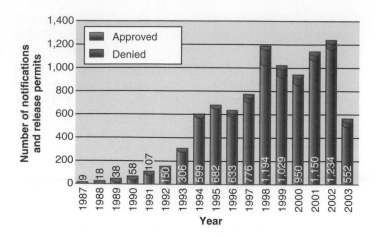

FIGURE 9.27 Transgenic crop field releases in the United States 1987–2003 (2003 data only through June).

Source: Information Systems for Biotechnology, Virginia Tech University.

Pest Resistance and Weed Control

Biotechnologists recently have created plants with genes for endogenous insecticides. *Bacillus thuringiensis* (Bt), a bacterium, makes toxins lethal to Lepidoptera (butterfly family) and Coleoptera (beetle family). The genes for some of these toxins have been transferred into crops such as maize (to protect against European cut worms), potatoes (to fight potato beetles), and cotton (for protection against boll weevils). This allows farmers to reduce insecticide spraying. Arizona cotton farmers, for example, report reducing their use of chemical insecticides by 75 percent. Small farms in India report an 80 percent yield increase with Bt cotton compared to neighboring plots growing conventional cotton.

Entomologists worry that Bt plants churn out toxin throughout the growing season, regardless of the level of infestation, creating perfect conditions for selection of Bt resistance in pests. The effectiveness of this natural pesticide—one of the few available to organic growers—is likely to be destroyed within a few years. One solution is to plant at least a part of every field in non-Bt crops that will act as a refuge for nonresistant pests. The hope is that interbreeding between these "wild-type" bugs and those exposed to Bt will dilute out recessive resistance genes. Deliberately harboring pests and letting them munch freely on crops is something that many farmers find hard to do. In addition, devoting a significant part of their land to nonproductive crops lowers the total yield and counteracts the profitability of engineered seed.

There also is a concern about the effects on nontarget species. In laboratory tests, about half of a group of monarch butterfly caterpillars died after being fed on plants dusted with pollen from Bt corn. Under field conditions, however, it was difficult to show any harm to butterflies. Critics still worry, however, that there may be unexpected problems associated with this new technology.

The other major group of transgenic crops are engineered to tolerate high doses of herbicides. Currently these crops occupy 20 million hectares worldwide, or about three-quarters of all genetically engineered acreage. The two main products in this category are Monsanto's "Roundup Ready" crops—so-called because they can withstand treatment with Monsanto's best-selling herbicide, Roundup (glyphosate)—and AgrEvo's "Liberty Link" crops, which resist that company's Liberty (glufosinate) herbicide. Because crops with these genes can grow in spite of high herbicide doses, farmers can spray fields heavily to exterminate weeds. This practice allows for conservation tillage and leaving more crop residue on fields to protect topsoil from erosion, which are both good ideas, but it also means using much more herbicide in higher doses than might otherwise be done.

Is Genetic Engineering Safe?

Ever since scientists discovered how to move genes from one organism to another, critics have worried about irresponsible use of this technology or unforeseen consequences arising from novel combinations of genetic material. Environmental and consumer groups have campaigned against transgenic organisms, labeling them "Frankenfoods." Industry groups, on the other hand, describe their critics as Neo-Luddites, who blindly oppose any new technology. In 2002, while millions of its people faced famine, Zambia's government refused to accept thousands of tons of genetically modified maize from the United States, claiming that it might be unsafe for human consumption. Most European nations have bans on genetically engineered crops. In 2003, the United States filed a suit at the World Trade Organization claiming that European policies constitute an unwarranted restraint on trade. How can we decide what to believe in this welter of claims and counterclaims?

At the base of some people's unease with genetic modification is a feeling that it simply isn't right to mess with nature. It doesn't seem proper to recombine genes any way we want. Doing so raises specters of gruesome monsters stitched together from spare parts, like Mary Shelley's Frankenstein. Is this merely a fear of science, or is it a valid ethical issue?

The U.S. Food and Drug Administration declined to require labeling of foods containing GMOs, saying that these new varieties are "substantially equivalent" to related varieties bred via traditional practices. After all, proponents say, we have been moving genes around for centuries through plant and animal breeding. All domesticated organisms should be classified as genetically modified, some people argue. Biotechnology may be a more precise way of creating novel organisms than normal breeding procedures. We're moving only a few genes—or even part of one gene—at a time with biotechnology, rather than hundreds of unknown genes through classical techniques.

What if GMOs escape and interbreed with native species? In a ten-year study of genetically modified crops, a group from Imperial College, London, concluded that GMOs tested so far do not survive well in the wild, and are no more likely to invade other habitats than their unmodified counterparts. Other scientists counter that some GMO crops already have been shown to spread their genes to nearby fields. Normal rapeseed (canola oil) varieties in Canada were found to contain genes from genetically modified varieties in nearby fields. It isn't clear, however, if the genes involved will have any adverse effect or whether they will persist

Shade-Grown Coffee and Cocoa

Has it ever occurred to you that your purchases of coffee and chocolate may be contributing to the protection or destruction of tropical forests? Both coffee and cocoa are examples of food products grown exclusively in developing countries but consumed almost entirely in the wealthy nations (vanilla and bananas are some other examples). Coffee grows in cool, mountain areas of the tropics, while cocoa is native to the warm, moist lowlands. Both are small trees of the forest understory, adapted to low light levels.

Until a few decades ago, most of the world's coffee and cocoa were grown under a canopy of large forest trees. Recently, however, new varieties of both crops have been developed that can be grown in full sun. Yields for sun-grown crops are higher because more coffee or cocoa trees can be crowded into these fields, and they get more solar energy than in a shaded plantation.

There are costs, however, in this new technology. Sun-grown trees die earlier from the stress and diseases common in these fields. Furthermore, ornithologists have found that the number of bird species can be cut in half in full-sun plantations, and the number of individual birds may be reduced by 90 percent. Shade-grown coffee and cocoa generally require fewer pesticides (or sometimes none) because the birds and insects residing in the forest canopy eat many of the pests. Shade-grown plantations also need less chemical ferilizer because many of the plants in these complex forests add nutrients to the soil. In addition, shade-grown crops rarely need to be irrigated because heavy leaf fall protects the soil, while forest cover reduces evaporation.

Currently, about 40 percent of the world's coffee and cocoa plantations have been converted to full-sun varieties and another 25 percent are in process. Traditional techniques for coffee and cocoa production are worth preserving. Thirteen of the world's 25 biodiversity hot spots occur in coffee or cocoa regions. If all the 20 million ha (49 million acres) of coffee and cocoa plantations in these areas are converted to monocultures, an incalculable number of species will be lost.

The Brazilian state of Bahia is a good example of both the ecological importance of these crops and how they might help preserve forest species. At one time, Brazil produced much of the world's cocoa, but in the early 1900s, the crop was introduced into West Africa. Now Côte d'Ivoire alone grows more than 40 percent of the world total, and the value of Brazil's harvest has dropped by 90 percent. Côte d'Ivoire is aided in this competition by a labor system that reportedly includes widespread child slavery. Even adult workers in Côte d'Ivoire get only about $165 per year (if they get paid at all) compared to a minimum wage of $850 per year in Brazil. As African cocoa production ratchets up, Brazilian landowners are converting their plantations to pastures or other crops.

The area of Bahia where cocoa was once king is part of Brazil's Atlantic forest, one of the most threatened forest biomes in the world. Only 8 percent of this forest remains undisturbed. Although cocoa plantations don't represent the full diversity of intact forests, they protect a surprisingly large sample of what once was there. And shade-grown cocoa can provide an economic rationale for preserving that biodiver-

Cocoa pods grow directly on the trunk and large branches of cocoa trees. © William P. Cunningham.

sity. Brazilian cocoa will probably never compete with that from other areas for lowest cost. There is room in the market, however, for specialty products. If consumers were willing to pay a small premium for organic, fair-trade, shade-grown chocolate and coffee it might provide the incentive needed to preserve biodiversity. Wouldn't you like to know that your chocolate or coffee wasn't grown with child slavery, and is helping protect plants and animal species that might otherwise go extinct?

---

for many generations. In 2001, researchers reported finding traces of genetically modified corn in wild relatives in Mexico, which supposedly has banned planting of GMOs, especially in regions where species were thought to have originated. This report was later withdrawn due to criticism of the techniques used to detect genetic markers, but it gave an interesting insight into the politics and sociology of science.

The first genetically modified animal designed to be eaten by humans is an Atlantic salmon containing extra growth hormone genes from an oceanic pout. The greatest worry from this fish is not that it will introduce extra hormones into our diet—that's already being done by chickens and beef that get extra growth hormone via injections or their diet—but rather the ecological effects if the fish escape from captivity. The transgenic fish grow seven times faster and are more attractive to the opposite sex than a normal salmon. If they escape from captivity, they may outcompete already endangered wild relatives for food, mates, and habitat. Fish farmers say they will grow only sterile females and will keep them in secure net pens. Opponents point out that salmon frequently escape from aquaculture operations, and that if just a few fertile transgenic fish break out it could be catastrophic for wild stocks.

Many people argue that we should take a better-safe-than-sorry "precautionary approach" that errs—if at all—on the side of safety. In 2002, the 5,000-member Society of Toxicology issued a position paper concluding that, "Based on available tests, there's no reason to suspect that transgenic plants differ in any substantive way from traditional varieties." On the other hand, a panel of biologists and agricultural scientists convened by the U.S. National Academy of Sciences urged the government to more carefully, and more publically, review the potential environmental impacts of genetically modified plants and animals before approving them for commercial use.

Finally, there are social and economic implications of GMOs. Will they help feed the world or will they lead to a greater consolidation of corporate power and economic disparity? Might higher yields and fewer losses to pests and diseases allow poor farmers in developing countries to stop using marginal land and avoid cutting down forests to create farmland? Is this simply a technological fix or could it help promote agricultural sustainability? Critics suggest that there are simpler and cheaper ways other than high-tech crop varieties to provide vitamin A to children in developing countries or to increase the income of poor, rural families. Adding a cow or a fishpond or training people in water harvesting or regenerative farming techniques (as we'll discuss in the next section) may have a longer-lasting impact than selling them expensive new seeds.

On the other hand, if we hope to reduce malnutrition and feed 8 billion people in 50 years, maybe we need all the tools we can get. Where do you stand in this debate? What additional information would you need to reach a reasoned judgment about the risks and benefits of this new technology?

SUSTAINABLE AGRICULTURE

How, then, shall we feed the world? Can we make agriculture compatible with sustainable ecological and social systems? Having discussed some of the problems that beset modern agriculture, we will now consider some suggested ways to overcome problems and make farming and food production just and lasting enterprises. This goal is usually termed **sustainable agriculture,** regenerative farming, or agroecology, all of which aim to produce food and fiber on a sustainable basis and repair the damage caused by destructive practices. Some alternative methods are developed through scientific research; others are discovered in traditional cultures and practices nearly forgotten in our mechanization and industrialization of agriculture.

Soil Conservation

With careful husbandry, soil is a renewable resource that can be replenished and renewed indefinitely. Since agriculture is the area in which soil is most essential and also most often lost through erosion, agriculture offers the greatest potential for soil conservation and rebuilding. Some rice paddies in Southeast Asia, for instance, have been farmed continuously for a thousand years without any apparent loss of fertility. The rice-growing cultures that depend on these fields have developed management practices that return organic material to the paddy and carefully nurture the soil's ability to sustain life.

While American agriculture hasn't reached that level of sustainability, there is evidence that soil conservation programs are having a positive effect. In one Wisconsin study, erosion rates in one small watershed were 90 percent less in 1975–1993 than they were in the 1930s. Among the most important elements in soil conservation are land management, ground cover, climate, soil type, and tillage system.

Managing Topography

Water runs downhill. The faster it runs, the more soil it carries off the fields. Comparisons of erosion rates in Africa have shown that a 5 percent slope in a plowed field has three times the water runoff volume and eight times the soil erosion rate of a comparable field with a 1 percent slope. Water runoff can be reduced by leaving grass strips in waterways and by **contour plowing,** that is, plowing across the hill rather than up and down. Contour plowing is often combined with **strip farming,** the planting of different kinds of crops in alternating strips along the land contours (fig. 9.28). When one crop is harvested, the other is still present to protect the soil and keep water from running straight downhill. The ridges created by cultivation make little dams that trap water and allow it to seep into the soil rather than running off. In areas where rainfall is very heavy, intersecting, or "tied," ridges are often useful. This method involves a series of ridges running at right angles to each other, so that water runoff is blocked in all directions and is encouraged to soak into the soil.

Terracing involves shaping the land to create level shelves of earth to hold water and soil. The edges of the terrace are planted with soil-anchoring plant species. This is an expensive procedure, requiring either much hand labor or expensive machinery, but makes it possible to farm very steep hillsides. The rice terraces in China and the Philippines have transformed whole mountain ranges into constructed landscapes (fig. 9.29). They are considered one of the wonders of the world.

Planting **perennial species** (plants that grow for more than two years) is the only suitable use for some lands and some soil types. Establishing forest, grassland, or crops such as tea, coffee, or other crops that do not have to be cultivated every year may be necessary to protect certain unstable soils on sloping sites or watercourses (low areas where water runs off after a rain).

Providing Ground Cover

Annual row crops such as corn or beans generally cause the highest erosion rates because they leave soil bare for much of the year (table 9.2). Often, the easiest way to provide cover that protects soil from erosion is to leave crop residues on the land after harvest. They not only cover the surface to break the erosive effects of wind and water, but they also reduce evaporation and soil temperature in hot climates and protect ground organisms that help aerate and rebuild soil. In some experiments, 1 ton of crop residue per acre (0.4 ha) increased water infiltration 99 percent, reduced runoff 99 percent, and reduced erosion 98 percent. Leaving crop residues on the field also can increase disease and pest problems, however, and may require increased use of pesticides and herbicides.

FIGURE 9.28 Countour plowing and strip cropping help prevent soil erosion on hilly terrain as well as creating a beautiful landscape. Photo by Lynn Betts, courtesy of USDA Natural Resources Conservation Service.

FIGURE 9.29 Rice terraces on Java, Indonesia. Some rice paddies have been cultivated for hundreds or even thousands of years without any apparent loss of productivity. © The McGraw Hill Companies, Inc./Barry Barker, photographer.

Where crop residues are not adequate to protect the soil or are inappropriate for subsequent crops or farming methods, such **cover crops** as rye, alfalfa, or clover can be planted immediately after harvest to hold and protect the soil. These cover crops can be plowed under at planting time to provide green manure. Another method is to flatten cover crops with a roller and drill seeds through the residue to provide a continuous protective cover during early stages of crop growth.

TABLE 9.2	Soil Cover and Soil Erosion	
CROPPING SYSTEM	AVERAGE ANNUAL SOIL LOSS (TONS/HECTARE)	PERCENT RAINFALL RUNOFF
Bare soil (no crop)	41.0	30
Continuous corn	19.7	29
Continuous wheat	10.1	23
Rotation: corn, wheat, clover	2.7	14
Continuous bluegrass	0.3	12

Source: *Based on 14 years' data from Missouri Experiment Station, Columbia, MO.*

In some cases, interplanting of two different crops in the same field not only protects the soil but also is more efficient use of the land, providing double harvests. Native Americans and pioneer farmers, for instance, planted beans or pumpkins between the corn rows. The beans provided nitrogen needed by the corn, pumpkins crowded out weeds, and both crops provided foods that nutritionally balance corn. Traditional swidden (slash-and-burn) cultivators in Africa and South America often plant as many as 20 different crops together in small plots. The crops mature at different times so that there is always something to eat, and the soil is never exposed to erosion for very long.

Mulch is a general term for a protective ground cover that can include manure, wood chips, straw, seaweed, leaves, and other natural products. For some high-value crops, such as tomatoes, pineapples, and cucumbers, it is cost-effective to cover the ground with heavy paper or plastic sheets to protect the soil, save water, and prevent weed growth. Israel uses millions of square meters of plastic mulch to grow crops in the Negev Desert.

Reduced Tillage

Farmers have traditionally used a moldboard plow to till the soil, digging a deep trench and turning the topsoil upside down. In the 1800s, it was shown that tilling a field fully—until it was "clean"—increased crop production. It helped control weeds and pests, reducing competition; it brought fresh nutrients to the surface, providing a good seedbed; and it improved surface drainage and aerated the soil. This is still true for many crops and many soil types, but it is not always the best way to grow crops. We are finding that less plowing and cultivation often makes for better water management, preserves soil, saves energy, and increases crop yields.

There are several major **reduced tillage systems.** *Minimum till* involves reducing the number of times a farmer disturbs the soil by plowing, cultivating, etc. This often involves a disc or chisel plow rather than a traditional moldboard plow. A chisel plow is a curved chisel-like blade that doesn't turn the soil over but creates ridges on which seeds can be planted. It leaves up to 75 percent of plant debris on the surface between the rows, preventing erosion (fig. 9.30). *Conserv-till* farming uses a coulter, a sharp disc like a pizza cutter, which slices through the soil, opening up a furrow or slot just wide enough to insert seeds. This disturbs the soil very little

FIGURE 9.30 In ridge tilling, a chisel plow is used to create ridges on which crops are planted and shallow troughs filled with crop residue. Less energy is used in plowing and cultivation, weeds are suppressed, and moisture is retained by the ground cover, or crop residue, left on the field. Courtesy of Dave Hanson, College of Agriculture, University of Minnesota.

and leaves almost all plant debris on the surface. *No-till* planting is accomplished by drilling seeds into the ground directly through mulch and ground cover. This allows a cover crop to be interseeded with a subsequent crop.

Farmers who use these conservation tillage techniques often must depend on pesticides (insecticides, fungicides, and herbicides) to control insects and weeds. Increased use of toxic agricultural chemicals is a matter of great concern. Massive use of pesticides is not, however, a necessary corollary of soil conservation. It is possible to combat pests and diseases with integrated pest management that combines crop rotation, trap crops, natural repellents, and biological controls (chapter 10).

Low-Input Sustainable Agriculture

In contrast to the trend toward industrialization and dependence on chemical fertilizers, pesticides, antibiotics, and artificial growth factors common in conventional agriculture, some farmers are going back to a more natural, agroecological farming style. Finding that they can't—or don't want to—compete with factory farms, these folks are making money and staying in farming by returning to small-scale, low-input agriculture. The Minar family, for instance, operate a highly successful 150-cow dairy operation on 97 ha (240 acres) near New Prague, Minnesota. No synthetic chemicals are used on their farm. Cows are rotated every day between 45 pastures or paddocks to reduce erosion and maintain healthy grass. Even in the winter, livestock remain outdoors to avoid the spread of diseases common in confinement (fig. 9.31). Antibiotics are used only to fight diseases. Milk and meat from this operation are marketed through co-ops and a community-supported agriculture (CSA) program. Sand Creek, which flows across the Minar land, has been shown to be cleaner when leaving the farm than when it enters.

Similarly, the Franzen family, who raise livestock on their organic farm near Alta Vista, Iowa, allow their pigs to roam in lush pastures, where they can supplement their diet of corn and soybeans with grasses and legumes. Housing for these happy hogs is in spacious, open-ended hoop structures. As fresh layers of straw are added to the bedding, layers of manure beneath are composted, breaking down into odorless organic fertilizer.

Low-input farms such as these typically don't turn out the quantity of meat or milk that their intensive agriculture neighbors do, but their production costs are lower, and they get higher prices for their crops, so that the all-important net gain is often higher. The Franzens, for example, calculate that they pay 30 percent less for animal feed, 70 percent less for veterinary bills, and half as much for buildings and equipment as their neighboring confinement operations. And on the Minar's farm, erosion after an especially heavy rain was measured to be 400 times lower than a conventional farm nearby.

Preserving small-scale, family farms also helps preserve rural culture. As Marty Strange of the Center for Rural Affairs in Nebraska asks, "Which is better for the enrollment in rural schools, the membership of rural churches, and the fellowship of rural communities—two farms milking 1,000 cows each or twenty farms milking 100 cows each?" Family farms help keep rural towns alive by purchasing machinery at the local implement dealer, gasoline at the neighborhood filling station, and groceries at the mom-and-pop grocery store.

One way to support local agriculture is to shop at a farmer's market. The produce is fresh, and profits go directly to the person who grows the crop. A local food co-op or owner-operated grocery store also is likely to buy from local farmers and to feature pesticide-free foods. Many co-ops and buyers' associations sign contracts directly with producers to grow the types of food they want to eat. This benefits both parties. Producers are guaranteed a local market for organic food or specialty items. Consumers can be assured of quality and can even be involved in production.

Many large supermarkets also now carry organic produce and other foods. If yours doesn't, why not ask the manager to look into it? Organic foods are becoming increasingly accepted—and profitable—as their benefits become more widely understood.

FIGURE 9.31 On the Minar family's 230-acre dairy farm near New Prague, MN, cows and calves spend the winter outdoors in the snow, bedding down on hay. Dave Minar is part of a growing counterculture that is seeking to keep farmers on the land and bring prosperity to rural areas. © Tom Sweeny/Minneapolis Star Tribune.

Agroecology, or sustainable farming using ecological knowledge, also can be applied effectively in poorer countries. Brian Haliweil, of the Worldwatch Institute, reports that farmers in the Guatemalan Highlands raised their crop yields tenfold without the use of chemical fertilizers or pesticides. Encouraged by World Neighbors, a nongovernmental development organization, to find innovative, local solutions to their problems, the farmers came up with low-cost techniques such as planting cover crops and grass strips to reduce erosion, or rotating corn with beans, peas, and other legumes to add nitrogen to the soil. Corn harvests increased from 0.4 tons to 4.5 tons per hectare, yields comparable to many richer countries.

One of the largest experiments in low-input farming is currently taking place in Cuba, where loss of aid from the former Soviet Union coupled with a trade embargo imposed by the United States have forced farmers to turn to organic, nonmechanized agriculture (chapter 10).

Summary

- Fish farming offers a means of replacing diminishing catches of wild seafood. There are, however, environmental and social concerns about the effects of some types of aquaculture. Nevertheless, if done right, fish farming can provide valuable, high-quality protein to the human diet.

- Forty years ago, 60 percent of the developing world was considered undernourished, meaning their diet didn't provide the average 2,200 kcal per day considered necessary for a healthy productive life. Today, despite the fact that world population has doubled since 1960, less than 15 percent of the population suffers from chronic caloric deficiency. However, that means that 815 million people don't have enough to eat.

- Poverty is the greatest threat to food security, or the ability to obtain sufficient food on a day-to-day basis. The 1.4 billion people in the world who live on less than $1 per day all too often can't buy the food they need and don't have access to resources to grow it for themselves. Even in communities or families where, on average, there would be enough for everyone, those with lowest status (women and children) may not get the food they need for a healthy life.

- An epidemic of obesity is spreading around the world as more of us eat too much meat, salt, and saturated fat and get too little exercise. Being overweight raises the risk of hypertension, diabetes, heart attacks, stroke, and many other diseases, which are becoming the leading causes of death and disability everywhere.

- A few crop species provide almost all the food humans eat. Wheat, rice, and maize supply the majority of the nutrients and calories for the vast majority of the world. Meat, dairy, and seafood consumption are rising rapidly as more people can afford these foods. Growing animals in dense concentrations in confinement operations can cause serious environmental and social problems.

- Soil is a marvelous, complex substance. Thousands of specific soil types exist in the world, having arisen from different parent material under diverse ecological conditions. Some are fertile, tillable, and wonderfully suited for agriculture. Others may need a great deal of husbandry to become useful.

- Large areas of the world suffer soil degradation caused by erosion, nutrient depletion, salinization, waterlogging, or other symptoms of abuse. It's estimated that 25 billion metric tons of soil are lost from cropland every year due to wind and water erosion. This erosion causes pollution and siltation of rivers, lakes, reservoirs, wetlands, and coastal ocean areas.

- The FAO predicts that most future production growth will come from higher yields and new crop varieties rather than expansion of arable lands. Genetic engineering involves removing genetic material from one organism and splicing it into the chromosomes of another. This new technology has the potential to greatly increase both the quantity and quality of our food supply, but there are worries about both the ecological safety and possible health effects of genetic modification of organisms.

- Sustainable agriculture, agroecology, or regenerative farming all aim to produce food and fiber on a sustainable basis, and to repair damage caused by destructive practices. Soil conservation provides techniques to preserve, protect, and rebuild this precious resource. Small-scale, low-input farming offers an alternative to industrial, chemically intensive agriculture. Even in very poor countries, using ecological knowledge and local initiatives can increase yields and improve profits.

Questions for Review

1. How many people in the world are chronically undernourished and how many die each year from starvation and nutritionally related diseases?

2. What is the relation between poverty and food security?

3. Define malnutrition and obesity. How many Americans are considered overweight?

4. How have recommendations for a balanced diet changed over the years?

5. What do we mean by the green and blue revolutions?

6. Why are wild fish populations endangered? How might aquaculture contribute to our food supply?

7. What is the composition of soil? What is humus? Why are soil organisms so important?

8. What are four kinds of erosion? Why is erosion a problem?

9. What is genetic engineering, or biotechnology, and how might they help or hurt agriculture and the environment?

10. What is sustainable agriculture?

Questions for Critical Thinking

1. Why might commercial fishing companies argue that the oceans have plenty of fish? If you were a marine resource manager, what information would you need to set sustainable harvest limits?

2. Suppose that a seafood company wants to start a fish farming operation in a lake near your home. What regulations or safeguards would you want to see imposed on its operation? How would you weigh the possible costs and benefits of this operation?

3. Debate the claim that famines are caused more by human actions (or inactions) than by environmental forces. What is the critical element or evidence in this debate?

4. Should farmers be forced to use ecologically sound techniques that serve farmers' best interests in the long run, regardless of short-term consequences? How could we mitigate hardships brought about by such policies?

5. In a crisis, when small farms are in danger of being lost, should preserving the soil be the farmer's first priority?

6. Should we encourage (and subsidize) the family farm? What are the advantages and disadvantages (economic and ecological) of the small farm and the corporate farm?

7. Should we try to increase food production on existing farmland, or should we sacrifice other lands to increase farming areas?

8. Some rice paddies in Southeast Asia have been cultivated continuously for a thousand years or more without losing fertility. Could we, and should we, adapt these techniques to our own country?

9. Do you think that agribusinesses should be allowed to insert lethal "terminator" genes in their crop varieties? Why or why not?

Key Terms

anemia 173
chronically undernourished 171
contour plowing 190
cover crops 191
famines 172
food security 171
genetically modified organisms (GMOs) 187
genetic engineering 187
green revolution 186
gully erosion 183
humus 178
kwashiorkor 173
malnourishment 173
marasmus 174
micorrhizal symbiosis 179

mulch 191
obese 173
perennial species 190
reduced tillage systems 191
rill erosion 183
salinization 184
sheet erosion 183
soil 177
soil horizons 180
soil profile 180
strip farming 190
subsoil 180
sustainable agriculture 190
terracing 190
topsoil 180
waterlogging 184

Further Readings

Bailey, Britt, and Marc Lappé. 2002. *Engineering the Farm: The Social and Ethical Aspects of Agricultural Biotechnology.* Island Press.

Brown, Kathryn. 2001. Seeds of concern. *Scientific American* 284(4):50–57.

Ellis, Richard. 2003. *The Empty Ocean.* Island Press.

Food and Agriculture Organization (UN). 2002. *The State of Food Insecurity in the World.* Published by the Food and Agriculture Organization of the United Nations.

Halweil, Brian. 2002. Farming in the public interest. *State of the World 2002.* W.W. Norton & Co. for the Worldwatch Institute.

Helvarg, David. 2003. The last fish. *Earth Island Journal* 18(1): 26–30.

Jackson, Dana L., and Laura L. Jackson. 2002. *Farm as Natural Habitat: Reconnecting Food Systems with Ecosystems.* Island Press.

Tillman, David, et al. 2002. Agricultural sustainability and intensive production practices. *Nature* 418:671–77.

Willett, Walter C., and Meir J. Stampfer. 2003. Rebuilding the food pyramid. *Scientific American* 288(1):64–71.

Welcome to McGraw-Hill's Online Learning Center

Location: http://www.mhhe.com/environmentalscience

Mc Graw Hill

WEB EXERCISES

Looking at Soils

Go the U.S. Department of Agriculture, Natural Resources Conservation Service Soils page at http://soils.usda.gov/. Click on "Soil Education" to read about soil types and soil classification. Then go to the Photo Gallery page to find the "State Soil" for your area. (If you live outside the United States, look at a place familiar to you.) Is that soil widespread or localized? Do you suppose climate conditions or geology help explain the formation and distribution of that soil type? What color is it? Why? Can you distinguish the different horizons? What are the distinguishing characteristics of the soil, according to the descriptive text? What influence do you think the presence of this soil type may have had on agricultural and economic development in your area? Under Dominant Soil Orders, on the Photo Gallery page, you can see soil maps. How widespread is the order of your state soil?

Pest Control

It ain't the things we know that cause all the trouble; it's the things we think we know that ain't so.

Will Rogers

OBJECTIVES

After studying this chapter, you should be able to:

- define the major types of pesticides and describe the pests they are meant to control.
- outline the history of pest control, including the changes in pesticides that have occurred in the last half of the twentieth century.
- appreciate the benefits of pest control.
- relate some of the problems of pesticide use.
- explain some alternative methods of pest control.
- discuss pesticide regulation and the special concerns about the effects of pesticides on children.

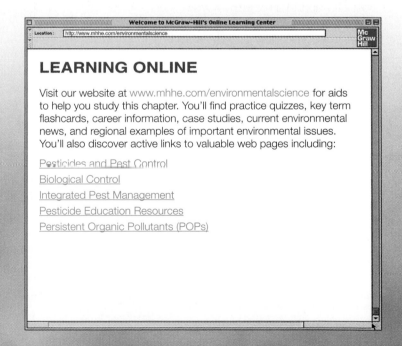

LEARNING ONLINE

Visit our website at www.mhhe.com/environmentalscience for aids to help you study this chapter. You'll find practice quizzes, key term flashcards, career information, case studies, current environmental news, and regional examples of important environmental issues. You'll also discover active links to valuable web pages including:

Pesticides and Pest Control
Biological Control
Integrated Pest Management
Pesticide Education Resources
Persistent Organic Pollutants (POPs)

Photo: A helicopter sprays pesticides on a farm field. © Inga Spence/Tom Stack & Associates.

The Promise and Perils of DDT

In 1939, Paul Müller of Geigy Pharmaceuticals in Switzerland discovered the amazing insecticidal properties of dichloro-diphenyl-trichloroethane (DDT). Inexpensive, stable, easily applied, and highly effective, this compound seemed ideal for crop protection and disease prevention. It is remarkably lethal to a wide variety of insects but relatively nontoxic to mammals. Where other control processes act slowly, and must be started before a crop is planted, DDT can save a harvest even when pests already are well established. A single application can produce 90 percent pest mortality.

Mass production of DDT began during World War II, when the Allied Armies used it to protect both troops and civilian populations against insect-borne diseases. Manufacture of this broad-spectrum insecticide soared from a few kilograms to thousands of metric tons in less than a decade. It was sprayed on crops and houses, dusted on people and livestock, and used to combat insects nearly everywhere (fig. 10.1). It was spectacularly effective in controlling diseases. Venezuela, for instance, had 8 million cases of malaria in 1945, but only 800 in 1955. Similarly, Taiwan went from 1 million malaria cases in 1945 to only nine a decade later. In 1948, Müller was awarded the Nobel Prize in medicine and physiology for discovery of DDT's beneficial properties.

By the 1960s, however, evidence began to accumulate that indiscriminate use of DDT and other persistent industrial toxins were having unexpected wildlife effects. Peregrine falcons, bald eagles, osprey, brown pelicans, shrikes, and several other predatory bird species were disappearing from former territories in eastern North America. Studies revealed that eggs laid by these birds had thin, fragile shells that broke before hatching. DDT and its degradation product, DDE, had concentrated millions of times through food webs until reaching toxic concentrations in top trophic levels such as these birds (see fig. 8.13).

In 1962, biologist Rachel Carson published *Silent Spring,* warning that pollution and persistent industrial toxins such as DDT posed a threat to wildlife, and perhaps to humans. Industry vilified Carson, calling her a "simplistic nature worshiper intent on subverting the continuing progress of science central to the development of the nation." In 1968, however, in a landmark court case, DDT was banned in Wisconsin. Within a few years, nearly all developed countries had banned most DDT applications. Many tropical countries, however, continued to permit its use for agriculture and disease control into the 1980s. They argued that without DDT and other similar chlorinated hydrocarbons, they would lose massive amounts of food and experience devastating epidemics.

Banning of DDT in the United States has led to substantial recovery of many predatory bird species. By the mid-1970s, peregrine fal-

FIGURE 10.1 Before we realized the toxicity of DDT, it was sprayed freely on people to control insects as shown here at Jones Beach, New York, in 1948. © Bettmann/Corbis.

cons had completely disappeared from the United States east of the Mississippi River. Captive breeding and successful reintroduction programs have rebuilt populations so there are now between 2,000 and 5,000 breeding falcon pairs in North America. In 1994, peregrines were removed from the endangered list and bald eagles were downgraded from endangered to threatened.

In spite of having been restricted from most uses for two decades or more, DDT and DDE (its major breakdown product) continue to be present in soils, food, humans, and wildlife throughout the world—even in places like the Arctic far from where it was ever used. We are now learning that these compounds—along with many other pesticides and synthetic industrial chemicals—can disrupt endocrine hormone functions and interfere with developmental processes at concentrations far below those causing immediate toxicity.

We see, in the example of DDT, some of the dilemmas in modern pest control. Chemical pesticides offer quick, effective, and relatively inexpensive relief from many pests that threaten our health and food supply, but they also can have unexpected effects, often far from their site or time of initial application. In this chapter, we'll study the major types of pests and pesticides together with some of the benefits and problems involved in our battle against pests.

PESTS AND PESTICIDES

A pest is something or someone that annoys us, detracts from some resource that we value, or interferes with a pursuit that we enjoy. In this chapter, we will concentrate on **biological pests,** organisms that reduce the availability, quality, or value of resources useful to humans. What's annoying or undesirable depends, of course, on your perspective. The mosquitoes that swarm in clouds over a marsh in the summer may be irritating to us, but they are an essential food source for birds and bats that feed on them. You may regard dandelions in your yard as tenacious and obnoxious weeds, but in some countries dandelions are cultivated as beautiful flowers and as a food source. Of the millions of species of organisms only about 100 plants, animals, fungi, and microbes cause 90 percent of all crop damage worldwide.

Insects tend to be the most frequent pests, in part, because they make up at least three-quarters of all species on the earth. Most pest organisms tend to be generalists, the opportunistic species that reproduce rapidly, migrate quickly into disturbed areas, and that are pioneers in ecological succession. They compete aggressively against more specialized endemic species and can often take over a biotic community, especially where humans have disrupted natural conditions and created an opening into which they can slip. Most Americans are familiar with dandelions, ragweed, English sparrows, starlings, European pigeons, and other "weedy" species that survive well in urban habitats. Chapter 11 describes how exotic aliens brought in by humans are crowding out native species in many places.

A **pesticide** is a chemical that kills pests. We generally think of toxic substances in this category, but chemicals that drive away pests or prevent their development are sometimes included as well. Pest control can also include activities such as killing pests by burning crop residues or draining wetlands to eliminate breeding sites. A broad-spectrum pesticide that kills a wide range of living organisms is called a **biocide** (fig. 10.2). Fumigants, such as ethylene dibromide or dibromochloropropane, used to protect stored grain or sterilize soil fall into this category. Generally, we prefer narrower spectrum agents that attack a specific type of pest: **herbicides** kill plants; **insecticides** kill insects; **fungicides** kill fungi; acaricides kill mites, ticks, and spiders; nematicides kill nematodes (microscopic roundworms); rodenticides kill rodents; and avicides kill birds. Pesticides can also be defined by their method of dispersal (fumigation, for example) or their mode of action, such as an ovicide, which kills the eggs of pests. In a sense, the antibiotics used in medicine to fight infections are pesticides as well.

Early Pest Controls

Using chemicals to control pests may well have been among our earliest forms of technology. People in every culture have known that salt, smoke, and insect-repelling plants can keep away bothersome organisms and preserve food. The Sumerians controlled insects and mites with sulfur 5,000 years ago. Chinese texts 2,500 years old describe mercury and arsenic compounds used to control body lice and other pests. Greeks and Romans used oil sprays, ash and sulfur ointments, lime, and other natural materials to protect themselves, their livestock, and their crops from a variety of pests.

In addition to these metals and inorganic chemicals, people have used organic compounds, biological controls, and cultural practices for a long time. Alcohol from fermentation and acids in pickling solutions prevent growth of organisms that would otherwise ruin food. Spices were valued both for their flavors and because they deterred spoilage and pest infestations. Romans burned fields and rotated crops to reduce crop diseases. They also employed cover crops to reduce weeds. The Chinese developed plant-derived insecticides and introduced predatory ants in orchards to control caterpillars 1,200 years ago. Many farmers still use ducks and geese to catch insects and control weeds (fig. 10.3).

Current Pesticide Use

According to the EPA, total pesticide use in the United States amounts to about 5.3 billion pounds (2.4 million metric tons) per year. Roughly half of that amount is chlorine and hypochlorites

FIGURE 10.2 Synthetic chemicals can eliminate pests quickly and efficiently, but what are the long-term costs to us and to our environment? © William P. Cunningham.

FIGURE 10.3 Geese make good biological control agents. They eat weeds, grass, and insects but leave many crops alone. Their droppings enrich the soil, their down can be used to make garments and pillows, and a goose dinner makes a welcome protein source for many people. © Steve Wilson/Entheos.

(bleach) used for water purification (fig. 10.4). Eliminating pathogens from drinking water prevents a huge number of infections and deaths, but there's concern that using so much chlorine and hypochlorite to do so may be creating other chronic health risks. The next largest category is conventional pesticides: primarily insecticides, herbicides, and fungicides.

Specialty biocides, such as preservatives used in adhesives and sealants, paints and coatings, leather, petroleum products, and plastics as well as recreational and industrial water treatment amount to some 300 million pounds per year, although they represent only

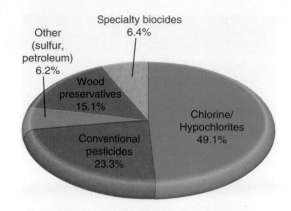

FIGURE 10.4 Of the 5.3 billion pounds of pesticides used in the United States each year, nearly half is chlorine/hypochlorite disinfectant. Specialty biocides include other antiseptics and sanitizers. *Source:* U.S. EPA 2000.

about 6 percent of total pesticide use. The "other" category in figure 10.4 includes sulfur, oil, and chemicals used for insect repellents (such as DEET) and moth control. Wood preservatives represent just 15 percent of total pesticide consumption, but can be especially dangerous to our health because they tend to be both highly toxic and very long lasting. We'll discuss some of the concerns about these chemicals later in this chapter.

Roughly 80 percent of all conventional pesticides applied in the United States are used in agriculture or food storage and shipping. According to CropLife America, some 90 million ha of crops in the United States—including 96 percent of all corn (maize) and one-third of all soybeans—are treated with herbicides every year. In addition, 25 million ha of agricultural fields and 7 million ha of parks, lawns, golf courses, and other lands are treated with insecticides and fungicides. By some accounts, cotton has the highest rate of insecticide application of any crop, while golf courses often have higher rates of application of all conventional pesticides per unit area than any farm fields.

Homes and gardens account for only about 8 percent of total annual pesticide use in the United States, but three-quarters of all American homes use some type of pesticide, amounting to 20 million applications per year. Often people use much larger quantities of chemicals in their homes, yards, or gardens than farmers would to eradicate the same species in their fields. Storage and accessibility of toxins in homes also can be a problem. As we'll discuss later in this chapter, children's exposure to toxins in their home may be of greater concern than pesticide residues in food.

Information on global pesticide use is spotty and uncertain. The U.S. EPA estimates that world usage of conventional pesticides alone amounts to some 5.7 billion pounds (2.6 million metric tons) per year. In addition, millions of metric tons of so-called "inert" ingredients are added to pesticide formulations as carriers, stabilizers, emulsifiers, etc. Some of these materials are dangerous in their own right, and should be regulated more carefully than they often are. The 1.24 billion pounds of conventional pesticides used annually in the United States accounts for about 20 percent of total world consumption (fig. 10.5). Good data for water purification compounds, wood preservatives, and specialty biocides are lacking for many countries.

Pesticide Types

One way to classify pesticides is by their chemical structure. This is useful because environmental properties—such as stability, solubility, and mobility—and toxicological characteristics of members of a particular chemical group are often similar.

Inorganic pesticides include compounds of arsenic, sulfur, copper, lead, and mercury. These broad-spectrum poisons are generally highly toxic and essentially indestructible, remaining in the environment forever. Seeds are sometimes coated with a mercury or arsenic powder to deter insects and rodents during storage or after planting. Handling such seeds with bare hands can be very dangerous for farmers or gardeners. They are generally neurotoxins and even a single dose can cause permanent damage.

Natural organic pesticides, or "botanicals," generally are extracted from plants. Some important examples are nicotine and nicotinoid alkaloids from tobacco; rotenone from the roots of derris and cubé plants; pyrethrum, a complex of chemicals extracted from the daisylike *Chrysanthemum cinerariaefolium* (fig. 10.6); and turpentine, phenols, and other aromatic oils from conifers. All are toxic to insects, but nicotine is also toxic to a broad spectrum of organisms including humans. Rotenone is commonly used to kill fish. Turpentine, phenols, and other natural hydrocarbons are effective pesticides, but synthetic forms such as pentachlorophenol are more stable and more toxic than natural forms. They penetrate surfaces well and are used to prevent wood decay.

Fumigants are generally small molecules such as carbon tetrachloride, carbon disulfide, ethylene dichloride, ethylene dibromide, methylene bromide, and dibromochloropropane that gasify easily and penetrate rapidly into a variety of materials. They are used to sterilize soil and prevent decay or rodent and insect infestation of stored grain. Because these compounds are extremely dangerous for workers who apply them, many have been restricted or banned altogether.

Chlorinated hydrocarbons, or organochlorines such as DDT, chlordane, aldrin, dieldrin, toxaphene, paradichlorobenzene (mothballs), and lindane, are synthetic organic insecticides that inhibit

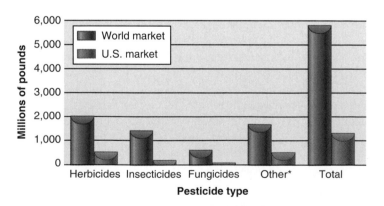

FIGURE 10.5 World and U.S. "conventional" pesticide use in 1999 in millions of pounds of active ingredients. U.S. consumption of these chemicals is about 20 percent of world total.
Source: U.S. EPA 2002.

FIGURE 10.6 Chrysanthemum flowers are a source of pyrethrum, a natural insecticide. © William P. Cunningham.

Carbamates, or urethanes, such as carbaryl (Sevin), aldicarb (Temik), aminocarb (Zineb), carbofuran (Baygon), and Mirex share many organophosphate properties, including mode of action, toxicity, and lack of environmental persistence and low bioaccumulation. Carbamates generally are extremely toxic to bees and must be used carefully to prevent damage to these beneficial organisms.

Halogenated pyrroles are a new class of compounds based on a naturally occurring microbial toxin. The first of these compounds to be released commercially is chlorfenapyr, which is marketed as Pirate. It was hailed as a replacement for organophosphates like malthion, but has been shown to reduce egg laying and egg survival in ducks at very low concentrations. Opponents argue that it may be the new DDT and call for its withdrawal.

Microbial agents and **biological controls** are living organisms or toxins derived from them used in place of pesticides. Bacteria such as *Bacillus thuringiensis* or *Bacillus popilliae* kill caterpillars or beetles by producing a toxin that ruptures the digestive tract lining when eaten. Parasitic wasps such as the tiny *Trichogramma* genus attack moth caterpillars and eggs, while lacewings and ladybugs control aphids. Viral diseases also have been used against specific pests.

nerve membrane ion transport and block nerve signal transmission. They are fast acting and highly toxic in sensitive organisms. Toxaphene, for instance, kills goldfish at 5 parts per billion (5 µg/liter), one of the highest toxicities for any compound in any organism. Chlorinated hydrocarbons may persist in soil for decades, become concentrated through food chains, and are stored in fatty tissues of a variety of organisms. The chloriphenoxy herbicides 2,4 D and 2,4,5 T have hormone-like growth-regulating properties and are selective for broad-leaved flowering plants. Because these compounds are so toxic and so long-lasting, many chlorinated hydrocarbons have been banned for most uses over much of the world.

Organophosphates, such as parathion, malathion, dichlorvos, chlorpyrifos, dimethyldichlorovinylphosphate (DDVP), and tetraethylpyrophosphate (TEPP), are an outgrowth of nerve gas research during World War II. They generally inhibit cholinesterase, an enzyme essential for removing excess neurotransmitter from synapses in the peripheral nervous system. Many organophosphates are extremely toxic to mammals, birds, and fish (generally 10 to 100 times more poisonous than most chlorinated hydrocarbons). A single drop of TEPP on your skin, for example, can be lethal. Because they are quickly degraded, they are much less persistent in the environment than organochlorines, generally lasting only a few hours or a few days. These compounds are very dangerous for workers who often are sent into fields too soon after they have been sprayed (fig. 10.7).

Glyphosate (N-phosphonomethylglycine) is a special organophosphate that does not react with the central nervous system as most others in this group do. Instead, it is a broad spectrum, non-selective herbicide used to kill weeds. Monsanto markets glyphosate as Roundup, and has genetically engineered soy beans, corn, wheat, and other crops to be Roundup resistant. This allows no-till agriculture, but encourages use of the herbicide.

PESTICIDE BENEFITS

Like all organisms, humans compete with other species for food and shelter and struggle to protect ourselves from diseases and predators. Pesticides are important weapons in this fight for survival.

FIGURE 10.7 The United Farm Workers of America claims that 300,000 farmworkers in the United States suffer from pesticide-related illnesses each year. Worldwide, the WHO estimates that 25 million people suffer from pesticide poisoning and 20,000 die each year from improper use or storage of pesticides. © Corbis/Volume 14.

Disease Control

Insects and ticks serve as vectors in the transmission of a number of disease-causing pathogens and parasites. Consider malaria, for example. About 300 million people suffer from this disease at any given time, and about 1 million die each year from *Plasmodium* protozoa spread to humans by *Anopheles* mosquitoes (fig. 10.8). It is estimated that insecticidal mosquito control has prevented at least 50 million deaths from malaria over the past 50 years. Sri Lanka is a classic example of pesticide benefits. In the early 1950s, more than 2 million cases of malaria were reported in Sri Lanka each year. After DDT spraying began in 1954, new malaria cases almost completely disappeared. When DDT spraying was discontinued in 1964, however, malaria reappeared almost immediately.

Within three years the annual incidence was more than 1 million cases per year. The Sri Lankan government resumed DDT spraying in 1968 and continues limited use of this insecticide despite environmental concerns. They concluded that the reductions in medical expenses, social disruption, pain and suffering, and lost work resulting from malaria outweighed the direct costs of insecticide spraying by a thousand to one.

Some other diseases spread by biting insects include yellow fever and related viral diseases such as encephalitis and West Nile, also carried by mosquitoes, and trypanosomiasis or sleeping sickness caused by protozoa transmitted by the tsetse fly. Onchocerciasis (river blindness) and filariasis (one form of which is elephantiasis, fig. 10.9) are caused by tiny worms spread by biting flies that afflict hundreds of millions of people in tropical countries. All of these terrible diseases can be reduced by judicious use of pesticides. If you were faced with a choice of going blind before age 30 because of masses of worms accumulating in your eyeballs or a small chance of cancer due to pesticide exposures if you live to age 50 or 60, which would you choose?

FIGURE 10.8   Malaria, spread by the *Anopheles* mosquito, is one of the largest causes of human disease and premature death in the world. By controlling mosquitoes, pesticides save millions of lives per year in tropical countries. © Stephen Dalton/Photo Researchers, Inc.

FIGURE 10.9   Elephantiasis is caused by parasitic worms (filaria) that block lymph vessels and cause fluid accumulation in various parts of the body. © SPL/Photo Researchers, Inc.

Crop Protection

Although reliable data on crop losses are difficult to obtain, it is thought that plant diseases, insect and bird predation, and competition by weeds reduce crop yields worldwide by at least one-third. Postharvest losses to rodents, insects, and fungi may be as high as another 20 to 30 percent. Without modern chemical pesticides, these losses might be much higher. Although we said in chapter 9 that there is more than enough food in the world to adequately feed everyone now living—if food were equitably distributed—this most certainly would not be true, given current intensive farming practices, if modern chemical pesticides were unavailable.

A commonly quoted estimate is that farmers save $3 to $5 for every $1 spent on pesticides. This means lower costs and generally more reliable quality for consumers. In some cases, insects and fungal diseases cause only small losses in terms of the total crop quantity, but the cosmetic damage they cause greatly reduces the economic value of crops. For example, although codling moth larvae consume very little of the apples they infest, the brown trails they leave as they crawl through the fruit and the possibility of biting into a living worm often make an unsprayed crop unsalable. Which is the more worrisome risk for you—consuming a bug or consuming pesticides?

CropLife America calculates that prohibiting pesticides in the United States would result in a $21 billion per year loss in food and fiber production. Without herbicides, they conclude, crop yields would be reduced up to 67 percent, soil erosion would increase by at least 1 billion metric tons per year (because of more cultivation to remove weeds), and it would take up to 70 million additional farm laborers to remove weeds by hand. This assumes, however, that we would continue growing the same single-species crops year after year without crop rotation or other sustainable agriculture practices.

PESTICIDE PROBLEMS

While synthetic chemical pesticides have brought us valuable economic and social benefits, they also cause a number of serious problems. In this section we will examine some of the worst of those problems.

Effects on Nontarget Species

It is estimated that up to 90 percent of the pesticides we use never reach their intended targets (fig. 10.10). Many beneficial organisms are poisoned unintentionally as a result. For instance, about 20 percent of all honeybee colonies in the United States are destroyed each year and another 15 percent are damaged by pesticide spray drift or residues on the flowers they visit. Direct losses to beekeepers amount to several million dollars per year. Losses to crops the bees would have pollinated may be ten times higher.

In some cases, the effects of poisoning nontarget species are immediate and unmistakable. In one episode in 1972, a single application of the insecticide Azodrin to combat potato aphids on a farm in Dade County, Florida, killed 10,000 migrating robins in three days. Similarly, a 1991 derailment of a Southern Pacific tanker car on a tricky canyon bridge just north of Dunsmuir, California, dumped 75,000 liters (20,000 gallons) of highly toxic metam sodium herbicide into the Sacramento River. The entire river ecosystem—including aquatic plants, insects, amphibians, and at least 100,000 trout—was completely wiped out for 45 kilometers (27 miles) downstream.

In other cases, the effects are more difficult to pin down, although not less serious. Insecticide spraying in Canadian forests has been linked to dramatic declines (up to 77 percent) in Atlantic salmon. The chemical suspected in this case is 4-nonylphenol, a powerful endocrine-hormone disrupter (Exploring Science, p. 202).

Pesticide Resistance and Pest Resurgence

Pesticides almost never kill 100 percent of a target species even under the most ideal conditions. As we discussed in chapter 4, every population contains some diversity in tolerance to adverse environmental factors. The most resistant members of a population survive pesticide treatment and produce more offspring like themselves with genes that enable them to withstand further chemical treatment (fig. 10.11). Because most pests propagate rapidly and produce many offspring, the population quickly rebuilds with pesticide-resistant individuals. We call this phenomenon **pest resurgence** or rebound. The Worldwatch Institute reports that at least 1,000 insect pest species and another 550 or so weeds and plant pathogens worldwide have developed chemical resistance (fig. 10.12). Of the 25 most serious insect pests in California, three-quarters are now resistant to one or more insecticides. This resistance means that it takes constantly increasing doses to get the same effect or that farmers who are caught on a **pesticide treadmill** must constantly try newer and more toxic chemicals in an attempt to stay ahead of the pests. Cornell University entomologist David Pimentel reports that a larger percentage of crops are lost now to insects, diseases, and weeds than in 1944

FIGURE 10.10 This machine sprays insecticide on orchard trees—and everything nearby. Up to 90 percent of pesticides applied in this fashion never reach target organisms. © Joe Munroe/Photo Researchers, Inc.

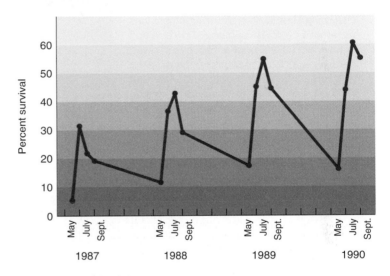

FIGURE 10.11 Survival of tobacco budworms to a constant dose of pyrethroid at different times of the year. Highest survival percentages are in July of each year, but maximum survival grows as worms become resistant. *Source:* S. Micinski, et al., *Resistant Pest Management Newsletter,* Vol. 3, No. 2, July 1991.

despite the use of millions of tons more pesticides. Some people worry that we are losing the battle to control pests.

One of the most ominous developments in this race to find effective pesticides is that we increasingly find pests that are resistant to chemicals to which they have never been exposed. Apparently, genes for pesticide resistance are being transferred from one species to another by means of vectors such as viruses and plasmids (naked pieces of virus-like DNA). Often a whole cluster of genes jump between species so that multiple chemical-tolerance is inherited before a particular pest is ever exposed to any of the chemicals. Widespread use of crops genetically engineered to produce pesticides endogenously (chapter 9) is likely to cause even more pesticide resistance.

What might alligators in Florida, seals in the North Sea, salmon in the Great Lakes, and you have in common? All are at the top of their respective food chains and all appear to be accumulating threatening levels of toxic environmental chemicals in their body tissues. One of the most serious effects of those chemicals is that many disrupt endocrine hormones that regulate many important bodily functions, including reproductive and immune systems. Evidence for this seems quite convincing in some wildlife populations, but whether it also is true for humans is one of the most contentious and important questions in environmental toxicology today.

One of the first examples of hormone-disrupting chemicals in the environment was a dramatic decline in alligators a decade ago in Florida's Lake Apopka. Surveys showed that 90 percent of the alligator eggs laid each year were infertile and that of the few that hatched, only about half survived more than two weeks. Male hatchlings had shrunken penises and unusually low levels of the male hormone testosterone. Female alligators, meanwhile, had highly elevated estrogen levels and abnormal ovaries. The explanation seems to be that a DDT spill in the lake in the 1980s, along with pesticide-laden runoff from adjacent farm fields, has led to high levels of DDE (a persistent breakdown product of DDT) in the reptiles' tissues and eggs. Because of a similarity in chemical structure, DDE appears to interfere with the action of androgens and estrogens, the normal sex hormones (see fig. 8.10).

Researchers have begun to suspect that mysterious outbreaks of health and reproductive problems and immune system failures in other wildlife populations may have similar origins. A USGS survey of 139 U.S. rivers, for instance, found that 40 percent had hormonally active industrial chemicals. In many rivers, fish that appeared superficially to be females were actually genetically males. In some rivers and lakes, all male fish are feminized. Recent evidence suggests that atrazine converts male frogs into hermaphrodites and makes both sexes more susceptible to parasites and limb deformities.

Humans may be affected as well. DDE and other compounds have been correlated with elevated cases of estrogen-sensitive breast and vaginal cancers in women, with developmental disorders in children, and

A shocking decrease in fertility and hatchling survival of alligators in Lake Apopka has been linked to pesticide pollution. Are humans affected by similar toxins? © Howard Suzuki, Aquatic Life Sculptures/ Discover Magazine.

possibly with falling sperm counts and low fertility in men.

Good evidence exists from controlled laboratory experiments that rats and mice exposed *in utero* or through mother's milk to very low levels of estrogen-like compounds develop physical, reproductive, and behavioral problems. Among some pesticides shown to have hormone-disrupting effects are the herbicides alachlor and atrazine, the fungicides mancozeb and benomyl, and the insecticides carbaryl, clicofol, endosulfan, methomyl, methoxychlor, parathion, linuron, chlorfenapyr, and sythetic pyrethroids.

We know that some of these chemicals act as synthetic hormones, others are antagonists that block normal hormone function. Furthermore, there can be striking synergy between some compounds. When endosulfan and DDT or chlordane are applied together, for example, the combination is 1,600 times more estrogenic than either chemical alone.

The question is whether specific chemical exposures are linked to particular human health problems. Many of these compounds are hundreds or thousands of times less active than normal hormones, leading skeptics to doubt that they have any noticeable effects except in people exposed to extremely high levels. Furthermore, we may have protective mechanisms that are lacking in highly

inbred laboratory rodents, and we can eat a highly varied diet that includes protective factors as well as toxins.

A wide variety of industrial products—including phthalates in plastics and cosmetics, polychlorinated biphenyls (PCBs), dioxins, chlorinated solvents, and metals such as lead, mercury, arsenic, and cadmium—also can disrupt hormone actions. Every human in the world now carries dozens of these chemicals in his or her body from before birth to death. Determining which of these toxins—or which combination—might be responsible for a particular disease or developmental abnormality may never be possible.

The bottom line is that we don't know (and we may never know for sure) whether falling sperm counts, increasing cancers, birth defects, immune diseases, and behavioral disorders in humans are caused by hormonally active, bioaccumulative, environmental chemicals. In 1996, the EPA ordered pesticide manufacturers to begin testing for disrupting effects. Given the continuing uncertainty about the dangers we face, what more do you think we should do? Is this threat serious enough to warrant drastic steps to reduce our risk? If you were head of the Environmental Protection Agency or the Food and Drug Administration, how much certainty would you demand before acting to protect our environment and ourselves from this ominous threat?

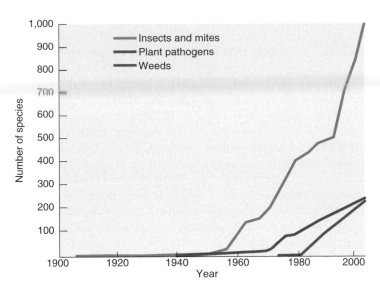

FIGURE 10.12 Many pests have developed resistance to pesticides. Because insecticides were the first class of pesticide to be used widely, selection pressures led insects to show resistance early. More recently, plant pathogens and weeds are also becoming insensitive to pesticides. *Source:* Worldwatch Institute, 2003.

In a way, every pesticide has a very limited useful life span before target species become resistant to it or the pesticide builds up intolerable environmental concentrations. We may have made a big mistake in broadcasting pesticides recklessly and extravagantly. DDT, for instance, is such a helpful insecticide that we should have used it sparingly and carefully, so that it would still be effective against the worst insect pests. It has been spread so widely that 50 of the 60 malaria-carrying mosquitoes are now resistant to it and the environmental side effects outweigh its benefits. You can think of the useful life of DDT as a nonrenewable resource that we squandered.

Creation of New Pests

Often the worst side effect of broadcast spraying a pesticide is that we kill beneficial predators that previously kept a number of pests under control. Many of the agricultural pests that we want to eliminate, for instance, are herbivores such as aphids, grasshoppers, or moth larvae that eat crop plants. Under natural conditions, their populations are kept under control by predators such as wasps, lady bugs, and praying mantises. As we discussed in chapter 3, there are generally far fewer predators in a food web than the species they prey upon. When we use a broad-spectrum pesticide, higher trophic levels are more likely to be knocked out than lower ones. This means that species that previously were insignificant can be released from natural controls and suddenly become major pests.

Consider the case of the Canete Valley in Peru. Before DDT was introduced in 1949, cotton yields were about 500 kg per hectare. By 1952 yields had risen to nearly 750 kg per hectare, but DDT-resistant boll weevils also had appeared (fig. 10.13). Toxaphene replaced DDT, but within two years it also became ineffective against boll weevils, which rebounded to higher levels than ever.

Even worse, *Heliothis* worms—which had not previously been a problem—began increasing rapidly. The wasps that earlier had kept both organisms in check were poisoned by the increasing pesticide doses. By 1955, cotton yields were down to 330 kg per hectare, one third less than before any pesticides were used.

Similarly, in California in the 1990s, cotton yields plunged by 20 percent despite a doubling of the number of insecticide applications and a 43 percent increase in pesticide costs. With the advent of chemical pest controls, farmers tend to abandon mixed crops, rotation regimes, and other traditional methods of management. This creates a greater potential for pest damage and pest resistance and thus makes it economical to use even higher levels of pesticide treatment. The application of large amounts of pesticides over wide expanses of land, together with changes in the ways the land is managed, have undermined the biological and ecological forces and interactions that previously governed population dynamics among species. A lack of understanding of the ways pesticides adversely affect beneficial organisms is one reason pest managers stick with chemical-based systems despite declining pesticide efficiency.

Persistence and Mobility in the Environment

The qualities that make DDT and other chlorinated hydrocarbons so effective—stability, high solubility, and high toxicity—also make them environmental nightmares. Because they persist for years, even decades in some cases, and move freely through air, water, and soil, they often show up far from the point of original application. Some of these compounds have been discovered far from any possible source and long after they most likely were used. Because they have an affinity for fat, many chlorinated hydrocarbons are bio-concentrated and stored in the bodies of predators—such as porpoises, whales, polar bears, trout, eagles, ospreys, and

FIGURE 10.13 The cotton boll weevil caused terrible devastation in southern U.S. cotton fields in the 1930s and 1940s. DDT and other insecticides controlled this pest briefly, but the beetle quickly became resistant. © Norm Thomas/Photo Researchers, Inc.

humans—that feed at the top of food webs. In a study of human pesticide uptake and storage, Canadian researchers found the levels of chlorinated hydrocarbons in the breast milk of Inuit mothers living in remote arctic villages were five times that of women from Canada's industrial region some 2,500 km (1,600 mi) to the south. Inuit people have the highest levels of these persistent pollutants of any human population except those contaminated by industrial accidents.

These compounds accumulate in polar regions by what has been called the "grasshopper effect," in which they evaporate from water and soil in warm areas and then condense and precipitate in colder regions. In a series of long-distance hops, they eventually collect in polar regions, where they accumulate in top predators. Polar bears, for instance, have been shown to have concentrations of certain chlorinated compounds 3 billion times greater than the seawater around them. In Canada's St. Lawrence estuary, beluga (white whales), which suffer from a wide range of infectious diseases and tumors thought to be related to environmental toxins, have such high levels of chlorinated hydrocarbons that their carcasses must be treated as toxic waste.

Although DDT hasn't been used legally in the United States or Canada for more than 20 years, in 1999 researchers reported discovery of p,p′-DDE, a DDT breakdown by-product, in the amniotic fluid of 30 percent of a sample of pregnant women in Los Angeles, California. A growing baby is surrounded by amniotic fluid during its nine months in the mother's womb. Retrospective studies have shown that mothers with high DDT or DDE levels in their blood are more likely to have premature or low-birthweight babies, and that these children have lower fertility rates as adults than those not exposed to pesticides in the womb.

Atrazine and alochlor are the most widely used herbicides in North America. More than 96 percent of the maize (corn) and soybean crops in the United States are treated with these compounds. The 31 million kg (70 million lbs) of atrazine used every year in the United States makes it the biggest commercial pesticide. Surveys of wells have found that up to 30 percent of all community wells and as much as 60 percent of all private wells in the mid-

western U.S. corn belt are contaminated with atrazine and alochlor. In a related study, researchers measured levels of these two compounds in rain falling over much of Europe that would be illegal if it were supplied as drinking water.

Because these **persistent organic pollutants (POPs)** are so long-lasting and so dangerous, a meeting of 127 countries agreed in 2001 to a global ban on the worst of them including aldrin, clordane, dieldrin, DDT, endrin, hexachlorobenzene, neptachlor, mirex, toxaphene, polychlorinated biphenyls (PCBs), dioxins, and furans. Most of this "dirty dozen" had been banned or severely restricted in developed countries for years. However, their production continued. Between 1994 and 1996, U.S. ports shipped more than 100,000 tons of POPs each year. Most of this was sent to developing countries where regulations were lax. Ironically, many of these pesticides returned to the United States on bananas and other imported crops. According to the 2001 POPs treaty, eight of the dirty dozen were banned immediately; PCBs, dioxins, and furans will be phased out; and use of DDT, still allowed for limited uses such as controlling malaria, must be publically registered in order to permit monitoring. The POPs treaty has been hailed as a triumph for environmental health and international cooperation. Unfortunately, other compounds—perhaps just as toxic—have been introduced to replace POPs.

Human Health Problems

Pesticide effects on human health can be divided into two categories: (1) acute effects, including poisoning and illnesses caused by relatively high doses and accidental exposures, and (2) chronic effects suspected to include cancer, birth defects, immunological problems, endometriosis, neurological problems, Parkinson's disease, and other chronic degenerative diseases.

The World Health Organization (WHO) estimates that 25 million people suffer pesticide poisoning and at least 20,000 die each year (fig. 10.14). At least two-thirds of this illness and death results from occupational exposures in developing countries where people use pesticides without proper warnings or protective clothing. A tragic example of occupational pesticide exposure is found among workers in the Latin American flower industry. Fueled by the year-round demand in North America for fresh vegetables, fruits, and flowers, a booming export trade has developed in countries such as Guatemala, Colombia, Chile, and Ecuador. To meet demands in North American markets for perfect flowers, table grapes and other produce, growers use high levels of pesticides, often spraying daily with fungicides, insecticides, nematicides, and herbicides. Working in warm, poorly ventilated greenhouses with little protective clothing, the workers—70 to 80 percent of whom are women—find it hard to avoid pesticide contact. Nearly two-thirds of nearly 9,000 workers surveyed in Colombia experienced blurred vision, nausea, headaches, conjunctivitis, rashes, and asthma. Although harder to document, they also reported serious chronic effects such as stillbirths, miscarriages, and neurological problems.

Long-term health effects caused by chronic pesticide exposures are difficult to document conclusively. As we discussed in chapter 8, isolating a specific pesticide-related disease from the myriad of

FIGURE 10.14 Handling pesticides requires protective clothing and an effective respirator. Pesticide applicators in tropical countries, however, often can't afford these safeguards or can't bear to wear them because of the heat. Photo by Tim McCabe, courtesy of USDA Natural Resources Conservation Service.

FIGURE 10.15 Representative examples of drawings of people by 4- and 5-year-old Yaqui children relatively unexposed to pesticides (foothills) and those heavily exposed (valley).

Source: From E. A. Guillette, et al., "An Anthropological Approach to the Evaluation of Preschool Children Exposed to Pesticides in Mexico" in *Environmental Health Perspective,* 106(6):347–353, 1998. U.S. Dept. of Health and Human Services.

other risks we encounter is complex. For example, numerous studies have shown that farmers who use pesticides have elevated rates of prostate cancer, non-Hodgkin's lymphoma, Parkinson's, and several other chronic diseases, but distinguishing between pesticide effects, smoking, exposure to gasoline and solvents, and other health risks is complex. Similarly, children of farmers who apply herbicides and fungicides have higher rates of birth defects, but it's difficult to know which of the mixture of chemicals used by these farmers may be responsible.

In an ongoing study of prenatal exposure to PCBs and other persistent contaminants, researchers from Wayne State University in Detroit, Michigan, have documented significant learning and attention problems in children whose mothers ate Lake Michigan fish regularly in the years prior to pregnancy. At age 11, the most highly exposed children had difficulty paying attention, suffered from poorer short- and long-term memory, were twice as likely to be below average in reading comprehension, and were three times as likely to have low IQ scores as their age-matched peers.

Researchers in a similar study in Mexico showed striking differences in behavior and motor skills in children exposed to pesticides compared to peers with minimal pesticide risk. Two groups of Yaqui children from northwestern Mexico were tested in this study. The children were similar in every respect except for their pesticide exposure. Children who lived in the foothills come from ranches where pesticides are used sparingly, if at all. Children from the Yaqui valley, live in a farming area where pesticide use has been heavy for the past 50 years. Samples of mother's milk and umbilical cord blood taken from valley families contained high amounts of several persistent pesticides including aldrin, endrin, dieldrin, heptachlor, and DDE. In tests designed to measure growth and development, the valley children fell far behind their foothill-dwelling peers (fig. 10.15). Farm children also exhibited diminished memory, decreased physical stamina, higher irritability, a greater tendency to fight, and poorer eye-hand coordination.

Farm families aren't the only ones to suffer from pesticide exposure. The 63 million pounds of pesticides used annually in American homes and gardens pose a threat to families, especially those with young children. In California, for example, children from homes fumigated by professional bug exterminators have been found to be three times as likely to have acute lymphocytic leukemia as those from homes where no pesticides were used (fig. 10.16). One recent study concluded that more than 1 million children under age five in the United States may exceed the safe daily dose of organophosphate insecticides. Children are much more susceptible to toxic chemicals than adults because—pound for pound—they eat more food, drink more water, and breathe more air. Children spend more time playing on carpets, grass, or in dirt where pesticides may accumulate. They put their fingers in their mouths more often, and children's rapid growth and development make them vulnerable to toxic effects (see What Do You Think? p. 160). Many teachers believe that the sudden rise in autism and hyperactivity/attention deficit disorder in America may be linked to environmental toxins.

All of us now have traces of persistent pollutants in our bodies. In its "Body Burden Study," the Environmental Working Group found detectable levels of 91 different industrial chemicals per person on average (161 different chemicals in all) in a sample

FIGURE 10.16 Our approach to nature is to beat it into submission, but at what cost? Ziggy © ZIGGY & FRIENDS, INC. Reprinted with permission of UNIVERSAL PRESS SYNDICATE. All rights reserved.

of people none of whom had any known occupational exposure. Of these chemicals, 76 are suspected to be cancer-causing, 94 are known to be neurotoxins in laboratory animals, and 79 are suspected of causing birth defects (the total adds up to more than 161 because some have multiple effects). Of course, some materials can be detected at levels far below those known to have toxic effects, but the presence of so many different compounds in ordinary people suggests the ubiquity of these chemicals in our environment. Still, there are steps you can take to reduce your exposure to pesticides (What Can You Do? p. 206).

ALTERNATIVES TO CURRENT PESTICIDE USES

Can we avoid using toxic pesticides or lessen their environmental and human-health impacts? In many cases, improved management programs can cut pesticide use between 50 and 90 percent without reducing crop production or creating new diseases. Some of these techniques are relatively simple and save money while maintaining disease control and yielding crops with just as high quality and quantity as we get with current methods. In this section, we will examine behavioral changes, biological controls, and integrated pest management systems that could substitute for current pest-control methods.

Behavioral Changes

Crop rotation (growing a different crop in a field each year in a two- to six-year cycle) keeps pest populations from building up. For instance, a soybean/corn/hay rotation is effective and eco-

What can you do?

Controlling Pests

Based on the principles of Integrated Pest Management, the U.S. EPA has released a helpful and informative guide to pest control. Among their recommendations:

1. *Identify the pest problem.* What kinds of pests and how many do you have? Many free resources are available from your library or County Cooperative Extension Service to help you understand what you face and how best to deal with it.

2. *Decide how much pest control is necessary.* Does your lawn really need to be totally weed free? Could you tolerate some blemished fruits and vegetables, or could you replace some of the plants you now grow with ones less sensitive to pests?

3. *Eliminate pest sources.* Remove food, water, and habitat that encourage pest growth from your house or yard. Block off hiding places or routes of entry into your home. Don't plant the same crops in the same spot year after year.

4. *Develop a healthy, weed-resistant yard.* Make sure your soil has the right pH balance, key nutrients, and good texture. Grow a type of grass or cover crop that does well in your climate. Set realistic weed and pest controls.

5. *Use biological controls.* Beneficial predators such as purple martins and bats eat insects; lady beetles (ladybugs) and their larvae eat aphids, mealybugs, whiteflies, and mites. Other beneficial predators include spiders, centipedes, dragonflies, wasps, and ants. Planting a border of marigolds may keep insects or herbivores away from your crops or flowers.

6. *Use simple manual methods.* Cultivate your garden and hand-pick weeds and pests from your garden. Use a flyswatter. Set traps to control rats, mice, and some insects. Mulch to reduce weed growth.

7. *Use chemical pesticides carefully.* If you decide that the best solution is chemical, choose the right pesticide product, read safety warnings and handling instructions, buy the amount you need, store the product safely, and dispose of excess properly.

8. *Evaluate the results.* Compare pretreatment and posttreatment conditions. Is there clear evidence of pest reduction? Are the benefits of short-term chemical pesticide control worth the risks? Would other treatments, including prevention of pest buildup work as well?

Source: Citizen's Guide to Pest Control and Pesticide Safety: EPA 730-K-95-001.

nomical against white-fringed weevils. Mechanical cultivation can substitute for herbicides but increases erosion. Flooding fields before planting or burning crop residues and replanting with a cover crop can suppress both weeds and insect pests. Habitat diversification, such as restoring windbreaks, hedgerows, and ground cover on watercourses, not only prevents soil erosion but also provides perch areas and nesting space for birds and other predators that eat insect pests. Growing crops in areas where pests are absent makes good sense. Adjusting planting times can avoid pest outbreaks, while switching from huge monoculture fields to mixed polyculture (many crops grown together) makes it more difficult for pests to find the crops they like. Tillage at the right time can greatly reduce pest populations. For instance, spring or fall plowing can help control overwintering corn earworms.

Biological Controls

Biological controls such as predators (wasps, ladybugs, praying mantises; fig. 10.17) or pathogens (viruses, bacteria, fungi) can control many pests more cheaply and safely than broad-spectrum, synthetic chemicals. *Bacillus thuringiensis* or Bt, for example, is a naturally occurring bacterium that kills the larvae of lepidopteran (butterfly and moth) species but is generally harmless to mammals. A number of important insect pests such as tomato hornworm, corn rootworm, cabbage loopers, and others can be controlled by spraying bacteria on crops. Larger species are effective as well. Ducks, chickens, and geese, among other species, are used to rid fields of both insect pests and weeds. These biological organisms are self-reproducing and often have wide prey tolerance. A few mantises or ladybugs released in your garden in the spring will keep producing offspring and protect your fruits and vegetables against a multitude of pests for the whole growing season.

Herbivorous insects also have been used to control weeds. For example, the prickly pear cactus was introduced to Australia about 150 years ago as an ornamental plant. This hardy cactus escaped from gardens and found an ideal home in the dry soils of the outback. It quickly established huge, dense stands that dominated 25 million ha (more than 60 million acres) of grazing land. A natural predator from South America, the cactoblastis moth, was introduced into Australia in 1935 to combat the prickly pear. Within a few years, cactoblastis larvae had eaten so much prickly pear that the cactus has become rare and is no longer economically significant.

Some plants make natural pesticides and insect repellents. The neem tree (*Azadirachta indica*) is native to India, but is now grown in many tropical countries (fig. 10.18). The leaves, bark, roots and flowers all contain compounds that repel insects and can be used to combat a number of crop pests and diseases.

Genetics and bioengineering can help in our war against pests. Traditional farmers have long known to save seeds of disease-resistant crop plants or to breed livestock that tolerate pests well. Modern science can speed up this process through selection regimes or by using biotechnology to transfer genes between closely related or even totally unrelated species. (Often we don't know, however what unintended consequences, such as super weeds or new pests, might result from these activities.) Insect pest reproduction has sometimes been reduced by releasing sterile males. Screwworms, for example, are the flesh-eating larvae of flies that lay their eggs in scratches or skin wounds of livestock. They were a terrible problem for ranchers in Texas and Florida in the 1950s, but release of massive numbers of radiation-sterilized males disrupted reproduction and eliminated this pest in Florida. In Texas, where flies continue to cross the border from Mexico, control has been more difficult, but continual vigilance keeps the problem manageable.

Other promising approaches are to use hormones that upset development or sex attractants to bait traps containing toxic pesticides. Many municipalities control mosquitoes with these techniques rather than aerial spraying of insecticides because of worries about effects on human health. Briquettes saturated with insect juvenile hormone are scattered in wetlands where mosquitoes

FIGURE 10.17 The praying mantis looks ferocious and is an effective predator against garden pests, but it is harmless to humans. They can even make interesting and useful pets. © William P. Cunningham.

FIGURE 10.18 A Nigerian woman examines a neem tree, the leaves, seeds, and bark of which provide a natural insecticide. Food and Agriculture Organization (FAO) photo/W. Ciesla.

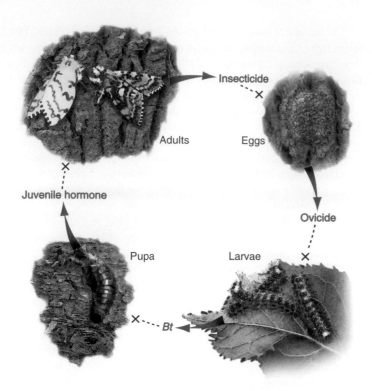

FIGURE 10.19 Different strategies can be used to control pests at various stages of their life cycles. *Bacillus thuringiensis* (Bt) kills caterpillars when they eat leaves with these bacteria on the surface. Releasing juvenile hormone in the environment prevents maturation of pupae. Predators attack at all stages.

FIGURE 10.20 This machine, nicknamed the "salad vac," vacuums bugs off crops as an alternative to treating them with toxic chemicals. © W. C. Wood/Tanimura & Antle, Inc.

breed. The presence of even minute amounts of this hormone prevent larvae from ever turning into biting adults (fig. 10.19). Unfortunately, many beneficial insects as well as noxious ones are affected by the hormone, and birds or amphibians that eat insects may be adversely affected when their food supply is reduced. Some communities that formerly controlled mosquitoes have abandoned these programs, believing that having naturally healthy wetlands is worth getting a few bites in the summer.

Integrated Pest Management

Integrated pest management (IPM) is a flexible, ecologically based pest-control strategy that uses a combination of techniques applied at specific times, aimed at specific crops and pests. It often uses mechanical cultivation and techniques such as vacuuming bugs off crops as an alternative to chemical application (fig. 10.20). IPM doesn't give up chemical pest controls entirely but rather tries to use the minimum amount necessary only as a last resort and avoids broad-spectrum, ecologically disruptive products. IPM relies on preventive practices that encourage growth and diversity of beneficial organisms and enhance plant defenses and vigor. Careful, scientific monitoring of pest populations to determine **economic thresholds,** the point at which potential economic damage justifies pest control expenditures, and the precise time, type, and method of pesticide application is critical in IPM.

Trap crops, small areas planted a week or two earlier than the main crop, are also useful. This plot matures before the rest of the field and attracts pests away from other plants. The trap crop then is sprayed heavily with enough pesticides so that no pests are likely to escape. The trap crop is destroyed so that workers will not be exposed to the pesticide and consumers will not be at risk. The rest of the field should be mostly free of both pests and pesticides.

IPM programs are already in use all over the United States on a variety of crops. Massachusetts apple growers who use IPM have cut pesticide use by 43 percent in the past ten years while maintaining per-acre yields of marketable fruit equal to that of farmers who use conventional techniques. Some of the most dramatic IPM success stories come from the developing world. Cuba, for example (Case Study, p. 210), has turned almost entirely to organic farming. In Brazil, pesticide use on soybeans has been reduced up to 90 percent with IPM. In Costa Rica, use of IPM on banana plantations has eliminated pesticides altogether in one region. In Africa, mealybugs were destroying up to 60 percent of the cassava crop (the staple food for 200 million people) before IPM was introduced in 1982. A tiny wasp that destroys mealybug eggs was discovered and now controls this pest in over 65 million ha (160 million acres) in 13 countries.

A successful IPM program that could serve as a model for other countries is found in Indonesia, where brown planthoppers developed resistance to virtually every insecticide and threatened the country's hard-won self-sufficiency in rice. In 1986, President Suharto banned 56 of 57 pesticides previously used in Indonesia and declared a crash program to educate farmers about IPM and the dangers of pesticide use. Researchers found that farmers were spraying their fields habitually—sometimes up to three times a week—regardless of whether fields were infested. By allowing natural predators to combat pests and spraying only when

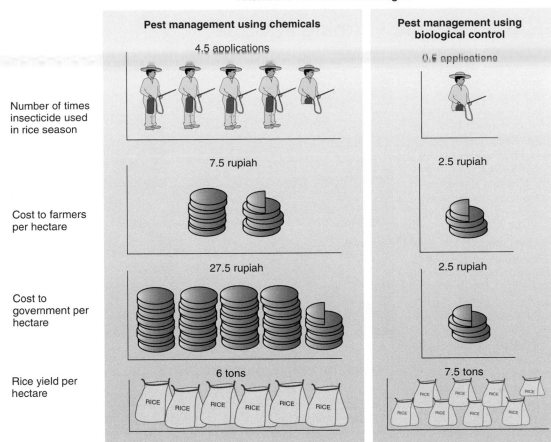

Alternative Pest Control Strategies

Pest management using chemicals	Pest management using biological control

Number of times insecticide used in rice season

4.5 applications / 0.5 applications

Cost to farmers per hectare

7.5 rupiah / 2.5 rupiah

Cost to government per hectare

27.5 rupiah / 2.5 rupiah

Rice yield per hectare

6 tons / 7.5 tons

FIGURE 10.21   Indonesia has one of the world's most successful integrated pest management (IPM) programs. Switching from toxic chemicals to natural pest predators has saved money while also increasing rice production. *Source:* Tolba, et al., *World Environment, 1972–1992*, p. 307, Chapman & Hall, 1992 United Nations Environment Programme.

absolutely necessary with chemicals specific for planthoppers, Indonesian farmers using IPM have had higher yields than their neighbors using normal practices and they cut pesticide costs by 75 percent. In 1988, only two years after its initiation, the program was declared a success. It is being extended throughout the whole country. Since nearly half the people in the world depend on rice as their staple crop, this experiment could have important implications elsewhere (fig. 10.21).

While IPM can be a good alternative to chemical pesticides, it also presents environmental risks in the form of exotic organisms. Wildlife biologist George Boettner of the University of Massachusetts reported in 2000 that biological controls of gypsy moths, which attack fruit trees and ornamental plants, have also decimated populations of native North American moths. *Compsilura* flies, introduced in 1905 to control the gypsy moths, have a voracious appetite for other moth caterpillars as well. One of the largest North American moths, the Cecropia moth (*Hyalophora cecropia*), with a 15 cm wingspan, was once ubiquitous in the eastern United States, but it is now rare in regions where *Compsilura* flies were released.

REDUCING PESTICIDE EXPOSURE

The total numbers and amounts of different chemicals to which we are potentially exposed is overwhelming. The 2.4 million metric tons of pesticides used in the United States every year contain about 600 active ingredients combined with more than 1,200 presumably inactive carriers, solvents, preservatives, and other ingredients in about 25,000 commercial products. Less than 10 percent of the active pesticide ingredients have been subjected to a full battery of ten chronic health-effect tests, and studies of the inactive ingredients have started only recently (fig. 10.22). Of the 321 pesticides screened so far, the EPA reports that 146 are probable human carcinogens. Since 1972, only 40 pesticides have been banned. Critics charge that this slow pace puts all of us at risk and that assessment and regulation must proceed more quickly.

Regulating Pesticides

Three federal agencies share responsibility for regulating pesticides used in food production in the United States: the Environmental

Case Study

Organic Farming in Cuba

The biggest experiment in low-input, sustainable agriculture in world history is occurring now in Cuba. The sudden collapse of the socialist bloc, upon which Cuba had been highly dependent for trade and aid, has forced an abrupt and difficult conversion from conventional agriculture to organic farming on a nationwide scale. Methods developed in Cuba could help other countries find ways to break their dependence on synthetic pesticides and fossil fuels.

Between the Cuban revolution in 1959 and the breakdown of trading relations with the Soviet Union in 1989, Cuba experienced rapid modernization, a high degree of social equity and welfare, and a strong dependence on external aid. Cuba's economy was supported during this period by the most modern agricultural system in Latin America. Farming techniques, levels of mechanization, and output often rivaled those in the United States. The main crop was sugarcane, almost all of which was grown on huge state farms and sold to the former Soviet Union at premium prices. More than half of all food eaten by Cubans came from abroad, as did most fertilizers, pesticides, fuel, and other farm inputs on which agricultural production depended.

Under the theory of comparative advantage, it seemed reasonable for Cuba to rely on international trade. With the collapse of the socialist bloc, however, Cuba's economy also fell apart. In 1990, wheat and grain imports decreased by half and other foodstuffs declined even more. At the same time, fertilizer, pesticide, and petroleum imports were down 60 to 80 percent. Farmers faced a dual challenge: how to produce twice as much food using half the normal inputs.

The crisis prompted a sudden turn to a new model of agriculture. Cuba was forced to adopt sustainable, organic farming practices based on indigenous, renewable resources. Typically, it takes three to five years for a farmer in the United States to make the change from conventional to organic farming profitable. Cuba, however, didn't have that long; it needed food immediately.

This organic, communal garden in Havana, Cuba provides most of the fresh produce for the housing complex surrounding it. Courtesy Bill Wilcke.

Cuba's agricultural system is based on a combination of old and new ideas. Broad community participation and use of local knowledge is essential. Scientific, adaptive management is another key. Diverse crops suitable to local microclimates, soil types, and human nutritional needs have been adopted. Natural, renewable energy sources such as wind, solar, and biomass fuels are being substituted for fossil fuels. Oxen and mules have replaced some 500,000 tractors idled by lack of fuel.

Soil management is vital for sustainable agriculture. Organic fertilizers substitute for synthetic chemicals. Livestock manure, green manure crops, composted municipal garbage, and industrial-scale cultivation of high-quality humus in earthworm farms all replenish soil fertility. In 1995 more than 100,000 metric tons of worm compost were produced and spread on fields.

Pests are suppressed by crop rotation and biological controls rather than chemical pesticides. For example, the parasitic fly (*Lixophaga diatraeae*) controls sugarcane borers; wasps in the genus *Trichogramma* feed on the eggs of grain weevils; while the predatory ant (*Pheidole megacephala*) attacks sweet potato weevils. Pest control also involves innovative use of biopesticides, such as *Bacillis thuringiensis*, that are poisonous or repellent to crop pests. Finally, integrated pest management includes careful monitoring of crops and measures to build populations of native beneficial organisms and to enhance the vigor and defenses of crop species.

Worker brigades from schools and factories help provide farm labor during harvest season. In addition to state farms and rural communes, urban gardening provides a much-needed supplement to city diets. Individual gardens are encouraged, but community or institutional gardens—schools, factories, and mass organizations—also produce large amounts of food.

Although food supplies in Cuba still are limited and diets are austere, the crisis wasn't as bad as many feared. In some ways, this draconian transition is fortunate. Cuba is now on a sustainable path and is a world leader in sustainable agriculture. It could serve as a model for others who surely will face a similar transition when their supplies of fossil fuels run out.

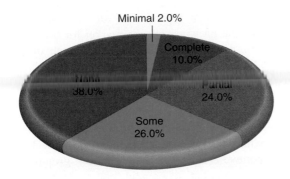

Minimal 2.0%

Complete 10.0%

Mann 38.0%

Partial 24.0%

Some 26.0%

FIGURE 10.22 Level of testing for health effects of 3,350 active and inert pesticide ingredients. Complete testing includes a battery of ten tests for both chronic and acute toxicity and genetic effects.
Source: National Academy of Sciences.

Protection Agency (EPA), the Food and Drug Administration (FDA), and the Department of Agriculture (USDA). The EPA regulates the sale and use of pesticides under the Federal Insecticide, Fungicide, and Rodenticide Act (FIFRA), which mandates the "registration" (licensing) of all pesticide products. Based on scientific studies, the EPA determines which pesticides will not pose significant risks to human health or the environment when used according to label directions. Under the Federal Food, Drug, and Cosmetic Act (FFDCA), the EPA also sets "tolerance levels" or limits for the amount of pesticide residues that lawfully may remain in or on foods marketed in the United States, whether grown domestically or imported from abroad. The FDA and USDA enforce pesticide use and tolerance levels set by the EPA. These agencies can seize and destroy food shipments found to contain pesticide residues in violation of limits set by the EPA.

In Canada, the Pest Control Products Act and Regulations, administered by Agriculture Canada, performs similar functions. Individual Canadian provinces also set local standards. The Ontario Ministry of Agriculture and Food, for instance, introduced a comprehensive program in 1988 that sets a goal of 50 percent reduction in pesticide use by the year 2002. Nonchemical pest control methods, on-farm education, and biotechnology and the development of pest-resistant crop varieties are all planned to play a role in this program. Savings on chemical costs are expected to be about $100 million per year.

In 1996, as a result of growing evidence of pesticide risks to children, Congress passed the Food Quality Protection Act. Key provisions of this act direct the EPA to set pesticide tolerance levels based on potential aggregate pesticide exposures. Benefits of pesticides can be considered when evaluating tolerances for carcinogens in adults, but not in calculating risks on reproduction and prenatal development or on exposure levels for infants and children. This act requires that by 2006 the EPA reassess the risks of over 9,700 pesticides, reevaluate the allowable levels of pesticides that can remain in or on food, and apply a safety factor of 10 where comprehensive and complete information is not available regarding the cumulative impacts on children.

For the first time, under this act, the EPA is required to examine exemptions given to "inert" pesticide ingredients used to dilute or carry the active agents. Of the 2,500 substances added to pesticides but not usually named on product labels, more than 650 have been identified as hazardous by federal, state, or international agencies. Nearly half of these "inert" ingredients have been classified as carcinogens, occupation hazards, or air and water pollutants. Naphthalene, for instance, which is commonly described as an inert ingredient, is designated a hazardous air pollutant under the Clean Air Act and a priority pollutant under the Clean Water Act.

Based on these new rules, the EPA recently banned the use of methyl parathion on all fruits and many vegetables. It also limited the quantity of azinphos methyl that can be used on foods common in children's diets, such as apples, peaches, and pears. Similarly, the EPA has prohibited most uses of chlorpyrifos, commonly known by its trade name Dursban. Found in more than 1,000 commercial products, this organophosphate represented 20 percent of all residential insecticide applications. It was shown, however, to cause brain damage in developing rat pups, and was judged an unacceptable risk to children. Some 36 other organophosphates are still being reviewed. The EPA is also studying hormone-disrupting chemicals and will probably regulate products on this basis in the near future.

Wood preservatives are another example of pesticide use especially dangerous to children. Thirty years ago, the principal wood preservatives were petroleum-based compounds like tar and creosote. These highly toxic biocides prevented growth of molds, invasion by termites, and just about all other living organisms, but also were dangerous to humans. Many Superfund sites are contaminated with these wood preservatives. As understanding of the lingering toxicity of creosote and its chemical cousins emerged, consumers switched to CCA (chromated copper arsenate) pressure-treated lumber. We now know that the arsenic in CCA-treated wood is a serious hazard, especially to workers exposed to sawdust or children who come in contact with the lumber in play equipment, decks, sidewalks, and other outdoor settings (see What Do You Think? p. 160). CCA-treated wood was banned for virtually all residential uses in the United States in 2003, but will the substitutes be safer?

With globalization, our food comes increasingly from other countries where rules governing food safety may be less strict than we would like. The U.S. General Accounting Office reports that the volume of imported food has doubled over the last five years, while food inspections have decreased during that same time. Currently, about 38 percent of the fruits and 12 percent of the vegetables consumed by Americans are imported. The Food and Drug Administration (FDA) inspects only about 2 percent of all food shipments. Less than 0.2 percent of imported fruits and vegetables are examined for microbial contamination, and even a smaller fraction is tested for pesticides. In studies of a wide range of foods collected by the USDA, the State of California, and the Consumers Union between 1994 and 2000, 73 percent of conventionally grown food had residue from at least one pesticide and were six times as likely as organic foods to contain multiple pesticide residues. Only 23 percent of the organic samples of the same groups had any residues. Using these data, the Environmental Working Group has assembled a list of the fruits and vegetables most commonly contaminated with pesticides (table 10.1).

TABLE 10.1	The Twelve Most Contaminated Foods

RANK	FOOD
1.	Strawberries
2.	Bell peppers
3.	Spinach
4.	Cherries (U.S.)
5.	Peaches
6.	Cantaloupe (Mexican)
7.	Celery
8.	Apples
9.	Apricots
10.	Green beans
11.	Grapes (Chilean)
12.	Cucumbers

Source: *Environmental Working Group, 2002.*

In 2001, the Canadian Supreme Court ruled that pesticides can be excluded from residential areas. In response, the province of Quebec announced that it would immediately ban the use of 30 highly toxic pesticides on public lands including parks, schools, day-care centers and hospitals. By 2005, the province will prohibit the use of most nonfarm pesticides. "People's health is more important than a perfect lawn," said Quebec Environment Minister Andre Boisclair. "I enjoin Quebecers to no longer use pesticides."

Is Organic the Answer?

Many farmers and consumers are turning to organic agriculture as a way to reduce pesticide exposures. Numerous studies have shown that organic, sustainable agriculture is more eco-friendly and leaves soils healthier than intensive, chemical-based monoculture cropping. A Swiss study spanning two decades found that average yields on organic plots were 20 percent less than adjacent fields farmed by conventional methods. Costs also were lower and prices paid for organic produce were higher, so that net returns were actually higher with organic crops. Energy use, for example, was 56 percent less per unit of yield in organic farming than conventional approaches. In addition, root fungi were 40 percent higher, earthworms were three times as abundant, and spiders and other pest-eating predators were doubled in the organic plots. Both the organic farmers and their families reported better health and greater satisfaction than their conventional neighbors. A study of food quality in Sweden reported that organic food contained more cancer-fighting polyphenolics and antioxidants than pesticide-treated produce. As mentioned in chapter 9, farms using sustainable techniques can have up to 400 times less erosion after heavy rains than monoculture row crops.

Currently, less than 1 percent of all American farmland is devoted to organic growing, but the market for organic products may stimulate more conversion to this approach in the future. Organic food is much more popular in Europe than in North America. Tiny Liechtenstein is probably the leader among industrialized nations with 18 percent of its land in certified organic agriculture. Sweden is second with 11 percent of its land in organic production.

Considerable anguish and confusion have existed in the United States over what constitutes organic agriculture. How do you know if you can believe claims that food has been grown safely (fig. 10.23)? With the market for organic food generating $11 billion per year, some farmers and marketers are likely to try to pass off foods grown inexpensively with pesticides as more valuable organic produce. In 2002, the U.S. Department of Agriculture passed rules to clarify food content. According to these rules, products labeled "100 percent organic" must be produced without hormones, antibiotics, pesticides, synthetic fertilizers, or genetic modification. "Organic" means that at least 95 percent of the ingredients must be organic. "Made with organic ingredients" must contain at least 70 percent organic contents. Products containing less than 70 percent organic ingredients can list them individually. Organic animals must be raised on organic feed, given access to the outdoors, given no steroidal growth hormones, and treated with antibiotics only to treat diseases.

Many who endorse the concept of organic food are disappointed by the limited scope of these definitions. They point out that there are three goals for organic, sustainable agriculture. One is growing food in harmony with nature—a nonindustrial approach to food production that treats animals humanely and avoids the use of chemical pesticides. Another is a food distribution system based on co-ops, farmer's markets, community supported agriculture, and

FIGURE 10.23 These strawberries were grown organically, but how could you tell? The USDA reports more pesticides in commercial strawberries than any other crop. © William P. Cunningham.

FIGURE 10.24 Your local farmer's market is a good source of locally grown, organic produce. *Source:* William P. Cunningham.

local production rather than crops flown and trucked from far-away places where consumers have no connection with producers (fig. 10.24). The third goal is simple, wholesome, nutritious food. It's true that the USDA rules eliminate many of the pesticides, but a myriad of highly processed ingredients are allowed in organic food. You might find yourself someday buying organic Coca-Cola to go with your organic TV dinner. You can already buy organic, intercontinental grapes in which thousands of calories of jet and diesel fuel were consumed to transport every calorie of food energy from Chile to your supermarket. Some of the best farmers are refusing to apply for organic certification, regarding it as a meaningless term under these rules.

Other people doubt that organic growers can produce enough—even within these limited definitions—to feed everyone. We need the efficiency of chemical pesticides and large-scale, chemical-intensive farming to provide food for the 8 to 9 billion people expected by the middle of this century. According to Vaclav Smil, a Canadian geographer, without synthetic fertilizer we could feed only 2 to 3 billion people. Dennis Avery, director of global food issues for the conservative Hudson Institute calculates that if we were to depend entirely on animal manure to fertilize crops, America would need to increase its cattleherd ninefold and convert almost the entire U.S. landmass to pastureland to support all those animals.

Furthermore, eating organic food doesn't protect you from all risks. Dr. Elsa Murano, undersecretary for food safety in the U.S. Department of Agriculture warns that consumers should be wary of organic food. "We must remember that bacteria and parasites are also all-natural," she said. "As a microbiologist, I know that preservatives are used in foods for a reason . . . to preserve food against the growth of microorganisms." Supporting her point, a Danish research group found more campylobacteria—one of the most common causes of diarrhea—in organic chickens than in standard chickens. Some natural pesticides allowed in organic farming have been linked to health problems in both producers or consumers. Rotenone, for example, has been associated with Parkinson's disease in farmers, while *Bacillus thuringiensis* (Bt) toxin has been found to cause allergies and asthma in crop pickers and handlers.

Biochemist Bruce Ames, who is famous for developing the most widely used cancer screening test, argues that pesticides confer a public-health advantage that far outweighs any risks posed by their toxicity or hormone-disrupting effects. "If the EPA eliminates all pesticides," he says, "all it's going to do is increase the price of fruits and vegetables, and there will be more cancer." It's not that he argues that pesticides are safe, but rather that we're ignoring the real risks we face like smoking, poor nutrition, lack of health care, and accidents. We could save more lives, he claims, by focusing on the factors that kill the vast majority of us than by spending so much time and money worrying about relatively minor hazards.

A Personal Plan

Even though pesticide exposure may not be the greatest threat that most of us face, why not take reasonable steps to reduce your risks? There are many things each of us can do to minimize pesticides in our food and to encourage sustainable agriculture.

- Wash and scrub all fresh fruits and vegetables thoroughly under running water.
- Peel fruits and vegetables when possible. Throw away the outer leaves of leafy vegetables such as lettuce and cabbage.
- Store food carefully so it doesn't get moldy or pick up contaminants from other foods. Use food as soon as possible to ensure freshness.
- Cook or bake foods that you suspect have been treated with pesticides to break down chemical residues.
- Trim the fat from meat, chicken, and fish. Eat lower on the food chain where possible to reduce bioaccumulated chemicals.
- Don't pick and eat berries or other wild foods that grow on the edges of roadsides where pesticides may have been sprayed.
- Grow your own fruits and vegetables without using pesticides or with minimal use of dangerous chemicals.
- Ask for organically grown food at your local grocery store or shop at a farmer's market or co-op where you can get such food.

Summary

- Biological pests are organisms that reduce the availability, quality, or value of resources useful to humans. Pesticides are chemicals intended to kill or drive away pests. Of the millions of species in the world, only about 100 kinds of animals, plants, fungi, and microbes cause most crop damage.

- Humans have probably always known of ways to protect themselves from annoying creatures, but our war against pests entered a new phase with the invention of synthetic organic chemicals such as DDT. These chemicals have brought several important benefits, including increased crop production and control of disease-causing organisms.

- Indiscriminate and profligate pesticide use also has caused many problems, such as killing nontarget species, creating new pests of organisms that were previously not a problem, and causing widespread pesticide resistance among pest species.

- The U.S. EPA estimates that several million metric tons of pesticides are used worldwide each year. U.S. use of conventional pesticides is about 1.24 billion pounds, or about 20 percent of the global total for this category.

- Often highly persistent and mobile in the environment, many pesticides move through air, water, and soil and bioaccumulate or bioconcentrate in food chains causing serious ecological and human health problems.

- A number of good alternatives offer ways to reduce our dependence on dangerous chemical pesticides. Among these are behavioral changes such as crop rotation, cover crops, mechanical cultivation, and planting mixed polycultures rather than vast monoculture fields.

- Consumers may have to learn to accept less than perfect fruits and vegetables. Biological controls such as insect predators, pathogens, or natural poisons specific for a particular pest can help reduce chemical use.

- Genetic breeding and biotechnology can produce pest-resistant crop and livestock strains, as well. Integrated pest management (IPM) combines all of these alternative methods together with judicious use of synthetic pesticides under precisely controlled conditions.

- Regulating pesticide use is a controversial subject. Many people fear that we are exposed to far too many dangerous chemicals. Industry claims that it could not do business without these materials.

- The USDA recently promulgated a definition for organic food grown without the use of synthetic pesticides or fertilizers, under conditions that treat animals humanely and avoid use of antibiotics and growth hormones.

- Many of the procedures and approaches suggested for agriculture and industry also work at home to protect us from pests and toxic chemicals alike. By using a little common sense, we can have a healthier diet, lifestyle, and environment.

Questions for Review

1. What is a pest and what are pesticides? What is the difference between a biocide, a herbicide, an insecticide, and a fungicide?

2. What is DDT and why was it considered a "magic bullet"? Why was it listed among the "dirty dozen" persistent organic pollutants (POPs)?

3. How much pesticide is used worldwide and which of the general categories accounts for the greatest use? Why are "inert" ingredients of concern?

4. Describe fumigants, botanicals, chlorinated hydrocarbons, organophosphates, carbamates, and microbial pesticides.

5. What are endocrine disrupters and why are they dangerous?

6. Explain why pests often resurge or rebound after treatment with pesticides and how they become pesticide resistant. What is a pesticide treadmill?

7. Identify three major categories of alternatives to synthetic pesticides and describe, briefly, how each one works.

8. What is IPM, and how is it used in pest control?

9. Why are children more susceptible to pesticides than adults are? What is being done to protect children?

10. List eight things you could do to reduce your dietary exposure to pesticides.

Questions for Critical Thinking

1. In retrospect, do you think Paul Müller should have received a Nobel prize for discovering the insecticidal properties of DDT?

2. If you were a public health official in a country in which malaria, filariasis, or onchocerciasis were rampant, would you spray DDT to eradicate vector organisms? Would you spray it in your own house?

3. Pesticide treadmill, "dirty dozen" pesticides, and environmental estrogens are all highly emotional terms. Why would some people choose to use or not use these terms? Can you suggest alternative terms for the same phenomena that convey different values?

4. Many farmworkers who suffer from pesticide poisoning are migrants or minorities. Is this evidence of environmental racism? What evidence would you look for to determine whether environmental justice is being served?

5. Suppose that a developing country believes that it needs a pesticide banned in the United States or Canada to feed or protect the health of its people. Are we right to refuse to sell that pesticide?

6. How much extra would you pay for organically grown food? How would you define organic in this context?

7. If alternative pest control methods are so effective and so much safer, why aren't farmers and consumers adopting them more rapidly?

8. Do you believe we could grow enough food for everyone if we switched to organic, low-input, sustainable agriculture? How would you design a research program to test this question?

9. What would you personally consider a "negligible" risk? Would you eat grapes if you knew they had a measurable amount of some pesticide? How small would the amount have to be?

10. A number of sources have been quoted in this chapter. Identify five of them and briefly describe what you think the worldview or political/economic agenda of each might be.

Key Terms

biocide 197
biological controls 207
biological pests 196
carbamates 199
chlorinated hydrocarbons 198
economic thresholds 208
fumigants 198
fungicides 197
herbicides 197
inorganic pesticides 198
insecticides 197
integrated pest management (IPM) 208
microbial agents 199
natural organic pesticides 198
organophosphates 199
persistent organic pollutants (POPs) 204
pesticide 197
pesticide treadmill 201
pest resurgence 201

Further Readings

Ames, B. N., et al. 1997. Environmental pollution, pesticides. In *The Standard Handbook of Environmental Science, Health and Technology*. J. Lehr, ed. McGraw-Hill.

Benbrook, Charles. 2001. Do GM crops mean less pesticide use? *Pesticide Outlook* 5:204–207.

Carson, Rachael. 1962. *Silent Spring*. Riverside Press.

Colborn, Theo, et al. 1996. *Our Stolen Future: How We Are Threatening Our Fertility, Intelligence, and Survival—A Scientific Detective Story*. Dutton Book.

Gurunathan, S., et al. 1998. Accumulation of chlorpyrifos on residential surfaces and toys accessible to children. *Environmental Health Perspectives* 106(1):9–16.

Hayes, T. B., et al. 2002. Hermaphroditic, demasculinized frogs after exposure to the herbicide atrazine at low ecologically relevant doses. *Proceedings of the National Academy of Sciences* 99:5476–80.

Lewis, W. J., et al. 1997. A total system approach to sustainable pest management. *Proceedings of the National Academy of Sciences* 94(23):12243–48.

National Research Council. 1996. *Ecologically Based Pest Management: New Solutions for a New Century*. National Academy Press.

Russell, Edmund. 2001. *War and Nature: Fighting Humans and Insects with Chemicals from World War I to Silent Spring (Studies in Environment and History)*. Cambridge Univ. Press.

Welcome to McGraw-Hill's Online Learning Center

Location: http://www.mhhe.com/environmentalscience

WEB EXERCISES

What Is Integrated Pest Management?

Integrated pest management uses biocontrol measures, biotechnology, cultural practices, and precision application of pesticides to protect crops, ornamentals, livestock, and people from harmful insects and other pests. The U.S. Department of Agriculture maintains a National IPM Network to link a cooperating group of universities, government agencies, and other organizations and to provide up-to-date, accurate information for pest management. You can learn about the NIPMN at www.ipmcenters.org/. Case studies and other information are organized by crop, pest, location, or tactics. Click on the region button and then on your own state on the map to find out about issues where you live. What are the top insect problems in your region and what can be done about them in an environmentally friendly way? You can find more information at www.epa.gov/pesticides/food/ipm.html.

Which Pesticides Are Used in Your State?

The National Center for Food and Agriculture Policy lists pesticide use by state or crop. Go to http://pestdata.ncsu.edu/ncfap/search.cfm, and under the *Pesticide Type* drop-down menu, select an option, then click *Send Year* and *Region*. Now you will have an option of selecting a pesticide (or herbicide or fungicide), a crop, or a state. Select your state and leave the other options blank in order to view the chemicals used, and the crops they are used on, in your state. You should get a table listing the chemicals used for your state. (If the results are returned in an illegible table, try using another type of browser—for example, Netscape instead of Internet Explorer.) How many compounds are used? Which crops use the most pesticides/herbicides/fungicides? Which compounds are used in the greatest volume?

Pesticides from Natural Sources

The EPA monitors pesticide use and development. Much information about pesticides is available at the website www.epa.gov/pesticides/. Go to this site and look in the index for *Biopesticides*, with the subheadings *What are Biopesticides?*, *Regulatory Activity*, and *Active Ingredients*. Look at the explanation of what biopesticides are; then look at the list of active ingredients in the *Biopesticide Fact Sheets*. This is a long list with many unfamiliar terms, but take a few minutes to try to identify what sorts of compounds there are and what their uses are. How many active ingredients are listed? Choose several and identify the source, how it works, and what risks are associated with each.

Now look at the regulatory activity page for biopesticides, www.epa.gov/pesticides/biopesticides/regactall.htm. This page lists recent and pending decisions on applications for approval of new biopesticides or for new uses for existing pesticides, and for exemptions from current controls. Look at the list of recent actions (entries indicating approval or exemptions established, or rejection). What is the approximate ratio of rejections to approvals? Discuss with other students how this ratio can best be explained.

Biodiversity

The first rule of intelligent tinkering is to save all the pieces.

Aldo Leopold

OBJECTIVES

After studying this chapter, you should be able to:

- define *biodiversity* and *species*.
- report on the total number and relative distribution of living species on the earth.
- summarize some of the benefits we derive from biodiversity.
- describe the ways humans cause biodiversity losses.
- evaluate the effectiveness of the Endangered Species Act and CITES in protecting endangered species.
- understand how gap analysis, habitat conservation plans, and captive breeding can contribute to preserving biological resources.
- propose ways we could protect endangered habitats and communities through large-scale, long-range, comprehensive planning.

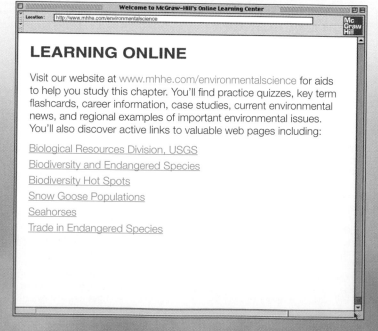

LEARNING ONLINE

Visit our website at www.mhhe.com/environmentalscience for aids to help you study this chapter. You'll find practice quizzes, key term flashcards, career information, case studies, current environmental news, and regional examples of important environmental issues. You'll also discover active links to valuable web pages including:

Biological Resources Division, USGS
Biodiversity and Endangered Species
Biodiversity Hot Spots
Snow Goose Populations
Seahorses
Trade in Endangered Species

Photo: Student interns weed experimental biodiversity plots at Cedar Creek Natural History Area in Minnesota. Courtesy Dr. David Tilman.

Saving Seahorses

Of all the amazing creatures in the world, seahorses surely are among the most unusual and captivating. Although they look like miniature, scaly dragons with prehensile tails, they are members of the pipefish family. Their life cycle is even more remarkable than their appearance. It's thought that seahorses mate for life. Couples greet each other every morning with a complex dance lasting up to four hours that involves synchronized swimming and dramatic body color changes. After an elaborate courtship ritual, the female deposits her eggs inside the male's brood pouch, where they are fertilized and incubated for several weeks. Eventually, the male goes into labor and gives birth to a brood of babies.

Adult seahorses can range from 2 to 25 cm tall (depending on the species) and can weigh as little as a few grams to more than 150 g. Seahorses have small stomachs and high metabolic rates, so they need to eat constantly. A two-week-old seahorse can consume 3,600 shrimp larvae in a single day—up to 25 times its body weight. Raising them in captivity is challenging because they won't eat unless the food is the right color, texture, and moves properly. They are sensitive to changes in water temperature and chemistry as well as the aquatic plants that make up their habitat.

Once found along coastlines and shoals from the tropics to the arctic, seahorses are thought to be declining rapidly nearly everywhere. The best-studied population in the central Philippines, for instance, is reported to have declined 70 percent between 1985 and 1995. Of the 32 known seahorse species, 20 appear on the World Conservation Union (IUCN) Red List of Threatened Species. Traditional Chinese medicine represents the largest threat to seahorses. An estimated 25 million seahorses (more than 70 metric tons) are consumed yearly in Asia to treat a variety of illnesses, including asthma, impotence, and general lethargy and pain. Additional millions of seahorses are caught for the aquarium trade or dried and made into tourist curios. They also are threatened by pollution, habitat destruction, and unwanted by-catch by shrimp trawlers.

China's rapid economic growth is probably the main reason for a recent surge in seahorse demand. Traders report a tenfold increase in the Chinese traditional medicine market in the 1990s, because more people can afford expensive cures, and because other species like tigers, bears, turtles, and snakes that once were used for medicine are becoming rare. Furthermore, small-scale fishers, whose livelihoods are threatened as marine fish stocks decline, increasingly target seahorses. A kilogram of dried seahorses can bring up to $1,900 (U.S.) in Chinese markets, although little of that profit goes to the fishers who caught them (fig. 11.1).

FIGURE 11.1 Dried seahorses for sale in a Chinese apothecary shop. © William P. Cunningham.

In 2002, the 160 countries making up the Convention on International Trade in Endangered Species of Wild Fauna and Flora (CITES) voted to list all 32 species of seahorses on Appendix II of the convention, which means that the international trade in these animals must be regulated to ensure it is not detrimental to the survival of wild populations. This listing will prohibit most trade except for captive-reared animals. Of all wildlife trade issues under international conservation management, seahorses will represent the greatest volume when the listing takes effect in 2004, and will be the first fully marine fish species of commercial importance to receive this type of protection. If this ban is to be effective, much must be done to help subsistence fishing communities to find alternative livelihoods and to develop the skills and legal authority to manage resources sustainably and prevent poaching.

Unfortunately, this story represents a widespread trend. Throughout the world, humans are threatening biodiversity through overexploitation, habitat destruction, pollution, introduction of exotic species, and a variety of other damaging activities. Finding ways to halt or reverse biodiversity losses is one of our most important environmental challenges. In this chapter, we'll look at other examples of threats to wild flora and fauna as well as what's being done to protect and preserve our biological legacy.

BIODIVERSITY AND THE SPECIES CONCEPT

From the driest desert to the dripping rainforests, from the highest mountain peaks to the deepest ocean trenches, life on earth occurs in a marvelous spectrum of sizes, colors, shapes, life cycles, and interrelationships. Think for a moment how remarkable, varied, abundant, and important the other living creatures are with whom we share this planet (fig. 11.2). How will our lives be impoverished if this biological diversity diminishes?

What Is Biodiversity?

Previous chapters of this book have described some of the fascinating varieties of organisms and complex ecological relationships that give the biosphere its unique, productive characteristics. Three kinds of **biodiversity** are essential to preserve these ecological systems: (1) *genetic diversity* is a measure of the variety of different versions of the same genes within individual species; (2) *species diversity* describes the number of different kinds of organisms within individual communities or ecosystems; and (3) *ecological*

FIGURE 11.2   A red-eyed tree frog (*Agalychnis callidryas*) rests on a fern frond in the forest. What will be lost if creatures like this disappear? © Corbis/Volume 6.

diversity assesses the richness and complexity of a biological community, including the number of niches, trophic levels, and ecological processes that capture energy, sustain food webs, and recycle materials within this system.

Within species diversity, we can distinguish between *species richness* (the total number of species in a community) and *species evenness* (the relative abundance of individuals within each species). To illustrate this difference, imagine two ecological communities, each with 10 species and 100 individual plants or animals. Suppose that one community has 82 individuals of one species and two each of nine other species. In the other community, all 10 species are equally abundant, meaning they have 10 individuals each. Although the species richness is the same, if you were to walk through these communities, you'd have the impression that the second is much more diverse because you'd be much more likely to encounter a greater variety.

What Are Species?

As you can see, the concept of species is fundamental in defining biodiversity, but what, exactly, do we mean by the term? In chapter 3, we defined species in terms of *reproductive isolation;* that is, all the organisms potentially able to breed in nature and produce fertile offspring. As we pointed out, this definition has some serious problems, especially among plants and protists, many of which either reproduce asexually or regularly make fertile hybrids.

Another definition favored by some taxonomists is the *phylogenetic species concept* (PSC), which emphasizes the branching (or cladistic) relationships among species or higher taxa regardless of whether organisms can breed successfully. Increasingly, DNA sequencing and other molecular biology techniques are giving us insights into these relationships. Some populations that were formerly thought to be different species on morphological criteria have been shown to be very closely related genetically or biochemically. Others that appeared to be a single species are being divided and even reassigned to different genera or families based on molecular evidence.

A third definition, favored by some conservation biologists, is the *evolutionary species concept* (ESC), which defines species in evolutionary and historic terms rather than reproductive potential. The advantage of this definition is that it recognizes that there can be several "evolutionarily significant" populations within a genetically related group of organisms. Unfortunately, we rarely have enough information about a population to judge what its evolutionary importance or fate may be. Paul Ehrlich and Gretchen Daily calculate that species average 220 evolutionarily significant populations. This calculation could mean that there are up to 10 billion different populations in total. Deciding which ones we should protect becomes an even more daunting prospect.

Perhaps the best compromise between these different positions is a *pluralistic species concept,* which argues that species definitions should vary with the taxa being studied. Thus, there might be paleospecies for fossils, sibling species in cladistics, morphospecies for museum studies, hybrid species among certain groups, and a host of other types depending on the information available and our reasons for studying particular groups.

How Many Species Are There?

At the end of the great exploration era of the nineteenth century, some scientists confidently declared that every important kind of living thing on earth would soon be found and named. Most of those explorations focused on charismatic species such as birds and mammals. Recent studies of less conspicuous organisms such as insects and fungi suggest that millions of new species and varieties remain to be studied scientifically.

The 1.7 million species presently known (table 11.1) probably represent only a small fraction of the total number that exist. Based on the rate of new discoveries by research expeditions—especially in the tropics—taxonomists estimate that there may be somewhere between 3 million and 50 million different species alive today. In fact, some taxonomists estimate that there are 30 million species of tropical insects alone (fig. 11.3). The upper limits for these estimates assume a high degree of ecological specialization among tropical insects. A recent study in New Guinea, however, found that 51 plant species were host to 900 species of herbivorous insects. This evidence would suggest no more than 4 to 6 million insect species worldwide.

About 70 percent of all known species are invertebrates (animals without backbones, such as insects, sponges, clams, worms, etc.). This group probably makes up the vast majority of organisms yet to be discovered and may constitute 95 percent of all species. What constitutes a species in bacteria and viruses is even less certain than for other organisms, but there are large numbers of physiologically or genetically distinct varieties of these organisms.

Biodiversity Hot Spots

Of all the world's currently identified species, only 10 to 15 percent live in North America and Europe. The greatest concentration of different organisms tends to be in the tropics, especially in tropical rainforests and coral reefs. Norman Myers, Russell Mittermeier, and others have identified **biodiversity hot spots** that have exceptionally high numbers of endemics (species that occur nowhere else). Using plants and land-based vertebrates as indicators, they have proposed 25 hot spots that represent a high-priority for conservation because they have both high biodiversity and a high risk of disruption by human activities (fig. 11.4). Although they occupy only 1.4 percent of the world's land area, these hot spots account for 44 percent of all known higher plant species and 35 percent of all terrestrial vertebrate species. The hottest of these hot spots tend to be tropical islands such as Madagascar, Indonesia, and the Philippines, where geographic isolation has resulted in large numbers of unique plants and animals. Special climatic conditions such as those found in South Africa, California, and the Mediterranean Basin also produce highly distinctive flora and fauna.

Some areas with high biodiversity—such as Amazonia, New Guinea, and the Congo basin, for example—aren't included in this hot spot map because most of their land area is relatively undisturbed. Other groups prefer different criteria for identifying important conservation areas. Aquatic biologists, for example, point out that marine reefs, estuaries, and shoals host some of the most diverse wildlife communities in the world, and warn that freshwater species are more highly endangered than terrestrial ones. Other scientists worry that the hot spot approach neglects many rare species and major groups that live in less biologically rich areas (cold spots). Nearly two-thirds of all terrestrial vertebrates, after all, aren't represented in Myers's hot spots. Focusing on a few hot spots also doesn't recognize the importance of certain species and ecosystems to human beings. Wet-lands, for instance, may contain just a few, common plant species but perform valuable ecological services, such as filtering water, regulating floods, and serving as nurseries for fish. Some conservationists argue that we should concentrate on saving important biological communities or landscapes rather than rare species.

Anthropologists point out that regions with high biodiversity are also often home to high cultural diversity as well (see fig. 1.20). It isn't a precise correlation; some countries, like Madagascar, New Zealand, and Cuba, with a high percentage of endemic species, have only a few cultural groups. Often, however, the varied habitat and high biological productivity of places like Indonesia, New Guinea, and the Philippines that allow extensive species specialization also have fostered great cultural variety. By preserving some of the 7,200 recognized language groups in the world—more than half of which are projected to disappear in this century—we might also protect some of the natural setting in which those cultures evolved.

HOW DO WE BENEFIT FROM BIODIVERSITY?

We benefit from other organisms in many ways, some of which we don't appreciate until a particular species or community disappears. Even seemingly obscure and insignificant organisms can play irreplaceable roles in ecological systems or be the source of genes or drugs that someday may be indispensable.

TABLE 11.1	Approximate Numbers of Known Living Species by Taxonomic Group	
Bacteria and cyanobacteria		4,000
Protozoa (single-celled animals)		31,000
Algae (single-celled plants)		40,000
Fungi (molds, mushrooms)		72,000
Multicellular plants		270,000
Sponges		5,000
Jellyfish, corals, anemones		10,000
Flatworms (tapeworms, flukes)		12,000
Roundworms (nematodes, earthworms)		25,000
Clams, snails, slugs, squids, octopuses		70,000
Insects		1,025,000
Mites, ticks, spiders, crabs, shrimp, centipedes, other noninsect arthropods		110,000
Starfish, sea urchins		6,000
Fish and sharks		27,000
Amphibians		4,000
Reptiles		7,150
Birds		9,700
Mammals		4,650
Total		1,733,000

Source: *Norman Myers, 2000.*

FIGURE 11.3 The development of flowering plants—and the insects that pollinate them—vastly increased the diversity of living things on earth. © Corbis/Volume 46.

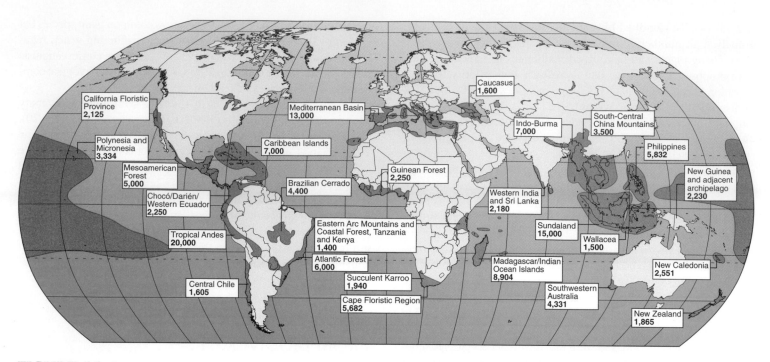

FIGURE 11.4 Biodiversity "hot spots" identified by Conservation International tend to be in tropical or Mediterranean climates and on islands, coastlines, or mountains where many habitats exist and physical barriers encourage speciation. *Source:* Conservation International.

Food

All of our food comes from other organisms. Many wild plant species could make important contributions to human food supplies either as new crops or as a source of genetic material to provide disease resistance or other desirable traits to current domestic crops. Norman Myers estimates that as many as 80,000 edible wild plant species could be utilized by humans. Villagers in Indonesia, for instance, are thought to use some 4,000 native plant and animal species for food, medicine, and other valuable products. Few of these species have been explored for possible domestication or more widespread cultivation. A 1975 study by the National Academy of Science (U.S.) found that Indonesia has 250 edible fruits, only 43 of which have been cultivated widely (fig. 11.5).

Drugs and Medicines

Living organisms provide us with many useful drugs and medicines (table 11.2). More than half of all modern medicines are either derived from or modeled on natural compounds from wild species. The United Nations Development Programme estimates the value of pharmaceutical products derived from developing world plants, animals, and microbes to be more than $30 billion per year. Indigenous communities that have protected and nurtured the biodiversity on which these products are based are rarely acknowledged—much less compensated—for the resources extracted from them. Many consider this expropriation "biopiracy" and call for royalties to be paid for folk knowledge and natural assets.

Consider the success story of vinblastine and vincristine. These anticancer alkaloids are derived from the Madagascar peri-

FIGURE 11.5 Mangosteens from Indonesia have been called the world's best-tasting fruit, but they are practically unknown beyond the tropical countries where they grow naturally. There may be thousands of other traditional crops and wild food resources that could be equally valuable but are threatened by extinction. © William P. Cunningham.

winkle (*Catharanthus roseus*) (fig. 11.6). They inhibit the growth of cancer cells and are very effective in treating certain kinds of cancer. Before these drugs were introduced, childhood leukemias were invariably fatal. Now the remission rate for some childhood leukemias is 99 percent. Hodgkin's disease was 98 percent fatal a few years ago, but is now only 40 percent fatal, thanks to these compounds. The total value of the periwinkle crop is roughly $15 million per year, although Madagascar gets little of those profits.

TABLE 11.2	Some Natural Medicinal Products	
PRODUCT	**SOURCE**	**USE**
Penicillin	Fungus	Antibiotic
Bacitracin	Bacterium	Antibiotic
Tetracycline	Bacterium	Antibiotic
Erythromycin	Bacterium	Antibiotic
Digitalis	Foxglove	Heart stimulant
Quinine	Chincona bark	Malaria treatment
Diosgenin	Mexican yam	Birth-control drug
Cortisone	Mexican yam	Anti-inflammation treatment
Cytarabine	Sponge	Leukemia cure
Vinblastine, vincristine	Periwinkle plant	Anticancer drugs
Reserpine	Rauwolfia	Hypertension drug
Bee venom	Bee	Arthritis relief
Allantoin	Blowfly larva	Wound healer
Morphine	Poppy	Analgesic

FIGURE 11.6 The rosy periwinkle from Madagascar provides anticancer drugs that now make childhood leukemias and Hodgkin's disease highly remissible. You may recognize it as a popular flowering ornamental plant. © William P. Cunningham.

Pharmaceutical companies are actively prospecting for useful products in many tropical countries. Merck, the world's largest biomedical company, paid $1.4 million (U.S.) to the Instituto Nacional de Biodiversidad (INBIO) of Costa Rica for plant, insect, and microbe samples to be screened for medicinal applications. INBIO, a public/private collaboration, trained native people as practical "parataxonimists" to locate and catalog all the native flora and fauna—between 500,000 and 1 million species—in Costa Rica (fig. 11.7). This effort may be a good model both for scientific information gathering and as a way for developing countries to share in the profits from their native resources.

Ecological Benefits

Human life is inextricably linked to ecological services provided by other organisms. Soil formation, waste disposal, air and water purification, nutrient cycling, solar energy absorption, and management of biogeochemical and hydrological cycles all depend on the biodiversity of life (chapter 3). Total value of these ecological services is at least $33 trillion per year, or more than double total world GNP.

There has been a great deal of controversy about the role of biodiversity in ecosystem stability. It seems intuitively obvious that having more kinds of organisms would make a community better able to withstand or recover from disturbance, but few empirical studies show an unequivocal relationship. One such study is being carried out at a long-term ecological site in Minnesota (Exploring Science, p. 222).

Because we don't fully understand the complex interrelationships between organisms, we often are surprised and dismayed at the effects of removing seemingly insignificant members of biological communities. For instance, wild insects provide a valuable

FIGURE 11.7 Costa Rican taxonomists study insect collections as part of an ambitious project to identify and catalog all the species in this small, but highly diverse, tropical country. The knowledge gained may contribute toward valuable commercial products that will provide funds to help preserve biodiversity. Courtesy INBIO, San Jose, Costa Rica.

but often unrecognized service in suppressing pests and disease-carrying organisms. It is estimated that 95 percent of the potential pests and disease-carrying organisms in the world are controlled by other species that prey upon them or compete with them in some way. Many unsuccessful efforts to control pests with synthetic chemicals (chapter 10) have shown that biodiversity provides essential pest control services.

Aesthetic and Cultural Benefits

The diversity of life on this planet brings us many aesthetic and cultural benefits, and cultural diversity is inextricably linked to biodiversity (chapter 1). Millions of people enjoy hunting, fishing,

Diversity and Ecological Stability

Are highly diverse communities more resilient and resistant to environmental stress than those with fewer species? For many years, this has been an ongoing debate among ecologists. A number of pioneering scientists, including Charles Darwin (1859), Charles Elton (1927), Robert McArthur (1955), and Eugene Odum (1959), suggested that communities with more species should be more stable and better able to withstand disturbance than those with less diversity. In spite of the importance of this debate in biological conservation, however, direct evidence from field studies about the link between diversity and stability has been ambiguous because natural environments have so many simultaneously changing variables.

In the 1970s theoretical ecologists, such as Robert May, showed that simple, mathematical models of interspecific competition became less stable—as measured by their ability to return to an equilibrium after perturbation—as the number of interacting species increased. Explaining this result, May suggested that a community of a few generalist species with a wide environmental tolerance might be better able to withstand change than a community with many more specialized organisms. Many ecologists concluded that diversity may not necessarily be associated with ecological stability. Others, however, objected that mathematical models might be misleading in terms of real-world relationships.

An important experimental test of the diversity/stability hypothesis has been car-ried out in a series of long-term studies by David Tilman and his colleagues at the University of Minnesota. Since 1982, the Tilman team has been studying the effects of nitrogen fertilization on grassland plots. Initially Tilman's goal was to study the effect of fertilizer on regenerating grass on abandoned farm fields and on natural oak savannas at the Cedar Creek Natural History Area in central Minnesota. The experimental design was to randomly assign different fertilizer treatments to a set of sample plots on old fields and natural savannas. Study plots varied considerably in species diversity: some contained only one species, some had as many as 26. Using the same set of sample plots for nearly two decades, Tilman and his colleagues—and platoons of student assistants—have carefully gathered, counted, and weighed all the plants growing on the plots (see opening photograph, this chapter). The result has been a long record of highly detailed data.

In a fortuitous turn of events, severe environmental stress occurred several years into the experiment. The summer of 1988 was the hottest, driest summer in 50 years. When researchers tallied up the total plant productivity on the study plots that year, they found that the plots with the most species suffered much less than those with few species. By contrast, the species-poor plots had only about one-eighth their pre-drought productivity. And in subsequent years, the species-poor plots also took longer to recover from the drought. Resilience—the ability to recover from stress—is an important component of ecosystem stability.

Other ecologists have criticized Tilman's conclusions, arguing that plots with higher species numbers came out better after the drought simply because they were more likely to contain larger species that produce more biomass. Tilman counters that he can control for this possibility and still show a strong effect of diversity on stability. But the debate still rages over whether it's the *number* of species or the *type* of species that's crucial to ecosystem health. Nevertheless, experiments like Tilman's, with long-term data gathering on a set of controlled sample plots, provide an important type of evidence to the debate. Experimental studies such as this are better controlled—reducing the number of simultaneously changing variables—than natural systems, but they are more realistic (and often more convincing) than computer models.

This work also shows the value of careful, long-term record keeping in science. Cedar Creek is one of eighteen Long-Term Ecological Research sites, funded by the National Science Foundation precisely to allow long-term monitoring of ecological systems. Tilman's work also shows the importance of attention to unexpected implications of research. As Louis Pasteur said, "chance favors the prepared mind." In this case, the 1988 drought provided an unexpected opportunity to explore critical questions about stability and resilience.

camping, hiking, wildlife watching, and other outdoor activities based on nature. These activities keep us healthy by providing invigorating physical exercise. Contact with nature also can be psychologically and emotionally restorative. In some cultures, nature carries spiritual connotations, and a particular species or landscape may be inextricably linked to a sense of identity and meaning. Many moral philosophies and religious traditions hold that we have an ethical responsibility to care for creation and to save "all the pieces" as far as we are able (chapter 2).

Nature appreciation is economically important. The U.S. Fish and Wildlife Service estimates that Americans spend $104 billion every year on wildlife-related recreation (fig. 11.8). This compares to $81 billion spent each year on new automobiles. Forty percent of all adults enjoy wildlife, including 39 million who hunt or fish and 76 million who watch, feed, or photograph wildlife. Ecotourism can be a good form of sustainable economic development, although we have to be careful that we don't abuse the places and cultures we visit (chapter 13).

FIGURE 11.8 Birdwatching and other wildlife observation contribute more than $29 million each year to the U.S. economy.
© William P. Cunningham.

For many people, the value of wildlife goes beyond the opportunity to shoot or photograph, or even see, a particular species. They argue that **existence value,** based on simply knowing that a species exists, is reason enough to protect and preserve it. We contribute to programs to save bald eagles, redwood trees, whooping cranes, whales, and a host of other rare and endangered organisms because we like to know they still exist somewhere, even if we may never have an opportunity to see them.

WHAT THREATENS BIODIVERSITY?

Extinction, the elimination of a species, is a normal process of the natural world. Species die out and are replaced by others, often their own descendants, as part of evolutionary change. In undisturbed ecosystems, the rate of extinction appears to be about one species lost every decade. In this century, however, human impacts on populations and ecosystems have accelerated that rate, causing hundreds or perhaps even thousands of species, subspecies, and varieties to become extinct every year. If present trends continue, we may destroy *millions* of kinds of plants, animals, and microbes in the next few decades. In this section, we will look at some ways we threaten biodiversity.

Natural Causes of Extinction

Studies of the fossil record suggest that more than 99 percent of all species that ever existed are now extinct. Most of those species were gone long before humans came on the scene. Species arise through processes of mutation and natural selection and disappear the same way (chapter 4). Often, new forms replace their own parents. The tiny *Hypohippus,* for instance, has been replaced by the much larger modern horse, but most of its genes probably still survive in its distant offspring.

Periodically, mass extinctions have wiped out vast numbers of species and even whole families (table 11.3). The best studied of these events occurred at the end of the Cretaceous period when dinosaurs disappeared, along with at least 50 percent of existing genera and 15 percent of marine animal families. An even greater disaster occurred at the end of the Permian period about 250 million years ago when 95 percent of all marine species and nearly half of all plant and animal families died out over a period of about 10,000 years—a short time by geological standards. Current theories suggest that these catastrophes were caused by climate changes, perhaps triggered when large asteroids struck the earth. Many ecologists worry that global climate change caused by our release of "greenhouse" gases in the atmosphere could have similarly catastrophic effects (chapter 15).

Human-Caused Reductions in Biodiversity

The rate at which species are disappearing appears to have increased dramatically over the last 150 years. Between A.D. 1600 and 1850, human activities appear to have been responsible for the extermination of two or three species per decade. By some estimates, we are now losing species at thousands of times natural rates. If present trends continue, the United Nations Environment Program warns, between 22 and 47 percent of all known plant species—and the animals dependent on them—could be extinct in the next 50 years. The eminent biologist E. O. Wilson says the impending biodiversity crash could be more abrupt than any previous mass extinction. Some biologists call this the sixth mass extinction, but note that this time it's not asteroids or volcanoes, but human impacts that are responsible.

TABLE 11.3	Mass Extinctions	
HISTORIC PERIOD	**TIME (BEFORE PRESENT)**	**PERCENT OF SPECIES EXTINCT**
Ordovician	444 million	85
Devonian	370 million	83
Permian	250 million	95
Triassic	210 million	80
Cretaceous	65 million	76
Quaternary	Present	33–66

Source: *W. W. Gibbs, 2001.*

Accurate predictions of biodiversity losses are difficult when many species probably haven't yet been identified. A number of scientists said 30 years ago that 20 to 50 percent of all existing species would be extinct by 2000. So far, these warnings seem to have been premature. Most predictions of anthropogenic mass extinction are based on an assumption that habitat area and species abundance are tightly correlated. E. O. Wilson calculates, for example, that if you cut down 90 percent of a forest, you'll eliminate at least half of the species originally present. In some of the best studied biological communities, however, this seems not to be true. More than 90 percent of Costa Rica's dry seasonal forest, for instance, has been converted to pasture land. Yet entomologist Dan Janzen reports that no more than 10 percent of the original flora and fauna appear to have been permanently lost. Wilson and others respond that remnants of the native species may be hanging on temporarily, but that in the long run they're doomed without sufficient habitat.

Still, it's clear that habitat is being destroyed in many places, and that numerous species are less abundant than they once were. Shouldn't we try to protect and preserve as much as we can? E. O. Wilson summarizes human threats to biodiversity with the acronym **HIPPO,** which stands for Habitat destruction, Invasive species, Pollution, Population (human), and Overharvesting. Let's look in more detail at each of these issues.

Habitat Destruction

The most important extinction threat for most species—especially terrestrial ones—is habitat loss. Perhaps the most obvious example of habitat destruction is clear-cutting of forests and conversion of grasslands to crop fields (fig. 11.9). Over the past 10,000 years, humans have transformed billions of hectares of former forests and grasslands to croplands, cities, roads, and other uses. These human-dominated spaces aren't devoid of wild organisms, but they generally favor weedy species adapted to coexist with us.

Today, forests cover less than half the area they once did, and only around one-fifth of the original forest retains its old-growth characteristics. Species like the northern spotted owl (*Strix occidentalis caurina*) that depend on the varied structure and resources of old-growth forest vanish as their habitat disappears (chapter 12). Grasslands currently occupy about 4 billion ha (roughly equal to the area of closed-canopy forests). Much of the most highly productive and species-rich grasslands—for example, the tallgrass prairie that once covered the United States corn belt—has been converted to cropland. Much more may need to be used as farmland or pasture if human populations continue to expand.

Sometimes we destroy habitat as side effects of resource extraction such as mining, dam-building, and indiscriminate fishing methods. Surface mining, for example, strips off the land covering along with everything growing on it. Waste from mining operations can bury valleys and streams with toxic material (see Mountaintop Removal, chapter 14). Dam-building floods vital stream habitat under deep reservoirs and eliminates food sources and breeding habitat for some aquatic species. Our current fishing methods are highly unsustainable. One of the most destructive fishing techniques is bottom trawling, in which heavy nets are dragged across the ocean floor, scooping up every living thing and crushing the bottom structure to lifeless rubble (chapter 9).

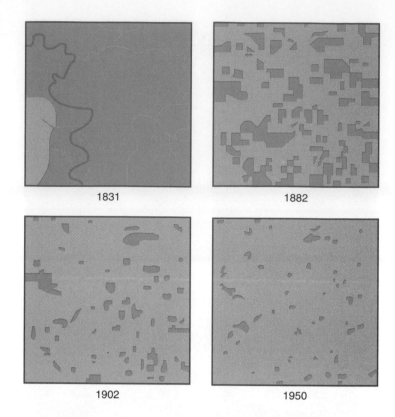

1831 1882
1902 1950

FIGURE 11.9 Decrease in wooded area of Cadiz Township in southern Wisconsin during European settlement. Green areas represent the amount of land in forest each year.

Preserving small, scattered areas of habitat often isn't sufficient to maintain a complete species collection. Large mammals, like tigers or wolves, need large expanses of contiguous range relatively free of human incursion. Even species that occupy less space individually suffer when habitat is fragmented into small, isolated pieces. If the intervening areas create a barrier to migration, isolated populations become susceptible to environmental catastrophes such as bad weather or disease epidemics. They also can become inbred and vulnerable to genetic flaws (chapter 6).

Invasive Species

A major threat to native biodiversity in many places is from accidentally or deliberately introduced species. Called a variety of names—alien, exotic, non-native, non-indigenous, unwanted, disruptive, or pests—**invasive species** are organisms that thrive in new territory where they are free of predators, diseases, or resource limitations that may have controlled their population in their native habitat. Although humans have probably transported organisms into new habitats for thousands of years, the rate of movement has increased sharply in recent years with the huge increase in speed

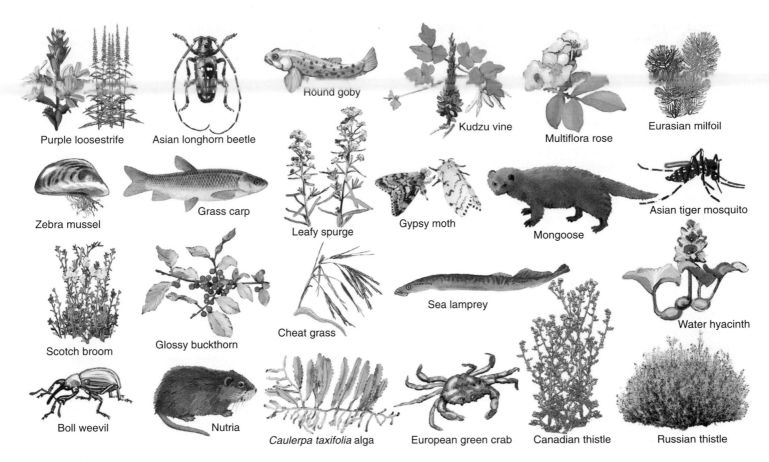

FIGURE 11.10 A few of the approximately 50,000 invasive species in North America. Do you recognize any that occur where you live? What others can you think of?

and volume of travel by air, water, and land. We move species around the world in a variety of ways. Some are deliberately released because people believe they will be aesthetically pleasing or economically beneficial. Others hitch a ride in ship ballast water, in the wood of packing crates, inside suitcases or shipping containers, in the soil of potted plants, even on people's shoes.

Over the past 300 years, approximately 50,000 non-native species have become established in the United States. Many of these introductions such as corn, wheat, rice, soybeans, cattle, poultry, and honeybees have proved to be both socially and economically beneficial. At least 4,500 of these species have established free-living populations, of which 15 percent cause environmental or economic damage (fig. 11.10). Invasive species are estimated to cost the United States some $138 billion annually and are forever changing a variety of many ecosystems.

A few important examples of invasive species include the following:

- Eurasian milfoil (*Myriophyllum spicatum* L.) is an exotic aquatic plant native to Europe, Asia, and Africa. Scientists believe that milfoil arrived in North America during the late nineteenth century in shipping ballast. It grows rapidly and tends

to form a dense canopy on the water surface, which displaces native vegetation, inhibits water flow, and obstructs boating, swimming and fishing. Humans spread the plant between water body systems from boats and boat trailers carrying the plant fragments. Herbicides and mechanical harvesting are effective in milfoil control but can be expensive (up to $5,000 per hectare per year). There is also concern that the methods may harm non-target organisms. A native milfoil weevil, *Euhrychiopsis lecontei,* is being studied as an agent for milfoil biocontrol.

- European green crab (*Carcinils maenav*), a native of the Atlantic coasts of Europe and Africa, was introduced into the waters of Massachusetts in the mid-1800s and has spread both north and south. A voracious predator of mussels, oysters, clams, scallops, and juvenile crabs, this exotic invader is believed to be a major contributor to the demise of the New England soft clam fishery. In the 1990s, this species was found on the Pacific coast, posing a potential threat to the region's clam- and oyster-growing industries and to the Dungeness crab fishery. Despite its name, the crab isn't always green; it can change color to orange and red with yellow patches on its abdomen during its mating cycle.

- Cheatgrass (*Bromus tectorum* L.) is rapidly becoming one of the biggest problem plants in the western United States. Cheatgrass is an annual grass native to Eurasia and the Mediterranean. It was probably introduced to the United States in the 1880s in impure grain seed. By 2002, it was reported to be the dominant plant on more than 50 million ha (123 million acres) in the western United States. The flammability of cheatgrass greatly increases fire size and frequency, eliminating native shrubs, forbs, and perennial grasses. Cheatgrass seeds are borne on sharp awls that burrow into the skin. Both cattle and native wildlife find cheatgrass unpalatable, so it degrades the value of grazing lands as much as 80 percent.

- Water hyacinth (*Eichhornia crassipes*) is a free-floating aquatic plant that grows up to a meter in height. It has thick, waxy, dark green leaves with bulbous, spongy stalks. It grows a tall spike of lovely flowers with blue, lavender, or pink petals, some of which have yellow splotches. This South American native was introduced into the United States in the 1880s. Its growth rate is among the highest of any plant known: hyacinth populations can double in as little as 12 days. Many lakes and ponds are covered from shore to shore with up to 500 tons of hyacinths per hectare. Besides blocking boat traffic and preventing swimming and fishing, water hyacinth infestations also prevent sunlight and oxygen from getting into the water. Thus, water hyacinth infestations reduce fisheries, shade-out submersed plants, crowd-out immersed plants, and diminish biological diversity. Water hyacinth is controlled by using herbicides, machines, and biocontrol insects.

- Kudzu vine (*Pueraria lobata*) has blanketed large areas of the southeastern United States. Long cultivated in Japan for edible roots, medicines, and fibrous leaves and stems used for paper production, kudzu was introduced by the U.S. Soil Conservation Service in the 1930s to control erosion. Unfortunately, it succeeded too well. In the ideal conditions of its new home, kudzu can grow 18 to 30 m in a single season. Smothering everything in its path, it kills trees, pulls down utility lines, and causes millions of dollars in damage every year.

- Asian tiger mosquitoes (*Aedes albopictus*) are unusually aggressive species that now infest many coastal states in the United States. These species have apparently arrived on container ships carrying used tires, a notorious breeding habitat for mosquitoes. Asian tiger mosquitoes spread West Nile virus (another species introduced with the mosquitoes), which is deadly to many wild birds and occasionally to people and livestock.

- Purple loosestrife (*Lythrum salicaria*) grows in wet soil. Originally cultivated by gardeners for its bright purple flower spikes, this tall wetland plant escaped into New England marshes about a century ago. Spreading rapidly across the Great Lakes, it now fills wetlands across much of the northern United States and southern Canada. Because it crowds out indigenous vegetation and has few native predators or symbionts, it tends to reduce biodiversity wherever it takes hold.

- Zebra mussels (*Dreissena polymorpha*) probably made their way from their home in the Caspian Sea to the Great Lakes in ballast water of transatlantic cargo ships, arriving sometime around 1985. Attaching themselves to any solid surface, zebra mussels reach enormous densities—up to 70,000 animals per square meter—covering fish spawning beds, smothering native mollusks, and clogging utility intake pipes. Found in all the Great Lakes, zebra mussels have recently moved into the Mississippi River and its tributaries. Public and private costs for zebra mussel removal now amount to some $400 million per year. On the good side, mussels have improved water clarity in Lake Erie at least fourfold by filtering out algae and particulates.

Disease organisms, or pathogens, may also be considered predators. To be successful over the long term, a pathogen must establish a balance in which it is vigorous enough to reproduce, but not so lethal that it completely destroys its host. When a disease is introduced into a new environment, however, this balance may be lacking and an epidemic may sweep through the area.

The American chestnut was once the heart of many Eastern hardwood forests. In the Appalachian Mountains, at least one of every four trees was a chestnut. Often over 45 m (150 ft) tall, 3 m (10 ft) in diameter, fast growing, and able to sprout quickly from a cut stump, it was a forester's dream. Its nutritious nuts were important for birds (like the passenger pigeon), forest mammals, and humans. The wood was straight grained, light, rot-resistant and used for everything from fence posts to fine furniture, and its bark was used to tan leather. In 1904, a shipment of nursery stock from China brought a fungal blight to the United States, and within 40 years, the American chestnut had all but disappeared from its native range. Efforts are now underway to transfer blight-resistant genes into the few remaining American chestnuts that weren't reached by the fungus or to find biological controls for the fungus that causes the disease.

An infection called whirling disease is decimating trout populations in many western states. It is caused by an exotic microorganism, *Myxobolus cerebralis,* which destroys cartilage in young fish causing them to swim erratically. The parasite is thought to have come into the United States in 1956 in a shipment of frozen fish.

The politics of invasive species are complex and contentious. We spend millions of dollars every year and often use very toxic chemicals or damaging mechanical treatments to remove non-native organisms. Widespread aerial pesticide spraying to control gypsy moths or fire ants, for example, can have deleterious effects on other organisms. Some ecologists question whether it's wise to carry out such massive and expensive campaigns. In some cases, introduced species have been around so long that people regard them fondly. Efforts in California, for example, to protect inconspicuous indigenous species by eradicating Australian eucalyptus have been met with angry opposition from local residents who love the pungent aroma of the eucalyptus and don't know, or don't care, what the native vegetation once was.

The flow of organisms isn't just into America; we also send exotic species to other places. The Leidy's comb jelly, for example, which is native to the western Atlantic coast, has devastated the Black Sea, now making up more than 90 percent of all biomass at certain times of the year. Similarly, the bristle worm from North America has invaded the coast of Poland and now is almost the only thing living on the bottom of some bays and lagoons. A tropical seaweed named *Caulerpa taxifolia,* originally grown for the aquarium trade, has escaped into the northern Mediterranean, where it covers the shallow seafloor with a dense, meter-deep shag carpet from Spain to Croatia. Producing more than 5,000 leafy fronds per square meter, this aggressive weed crowds out everything in its path. Rarely growing in more than scattered clumps less than 25 cm (10 in.) high in its native habitat, this alga was transformed by aquarium breeding into a supercompetitor that grows over everything and can withstand a wide temperature range. Getting rid of these alien species once they dominate an ecosystem is difficult if not impossible.

Pollution

We have known for a long time that toxic pollutants can have disastrous effects on local populations of organisms. Pesticide-linked declines of fish-eating birds and falcons was well documented in the 1970s (fig. 11.11). Marine mammals, alligators, fish, and other declining populations suggest complex interrelations between pollution and health (chapter 8). Mysterious, widespread deaths of thousands of seals on both sides of the Atlantic in recent years are thought to be linked to an accumulation of persistent chlorinated hydrocarbons, such as DDT, PCBs, and dioxins, in fat, causing weakened immune systems that make animals vulnerable to infections. Similarly, mortality of Pacific sea lions, beluga whales in the St. Lawrence estuary, and striped dolphins in the Mediterranean are thought to be caused by accumulation of toxic pollutants.

Lead poisoning is another major cause of mortality for many species of wildlife. Bottom-feeding waterfowl, such as ducks, swans, and cranes, ingest spent shotgun pellets that fall into lakes and marshes. They store the pellets, instead of stones, in their gizzards and the lead slowly accumulates in their blood and other tissues. The U.S. Fish and Wildlife Service (USFWS) estimates that

FIGURE 11.11 Brown pelicans, and other bird species at the top of the food chain, were decimated by DDT in the 1960s. Pelicans and other species have largely recovered since DDT was banned in the United States. © Corbis/Volume 5.

3,000 metric tons of lead shot are deposited annually in wetlands and that between 2 and 3 million waterfowl die each year from lead poisoning.

Population

Human population growth represents a threat to biodiversity in several ways. If our consumption patterns remain constant, with more people, we will need to harvest more timber, catch more fish, plow more land for agriculture, dig up more fossil fuels and minerals, build more houses, and use more water. All of these demands impact wild species. Unless we find ways to dramatically increase the crop yield per unit area, it will take much more land than is currently domesticated to feed everyone if our population grows to 8 to 10 billion as current projections predict. This will be especially true if we abandon intensive (but highly productive) agriculture and introduce more sustainable practices. The human population growth curve is leveling off (chapter 7), but it remains unclear whether we can reduce global inequality and provide a tolerable life for all humans while also preserving healthy natural ecosystems and a high level of biodiversity.

Overharvesting

Overharvesting is responsible for depletion or extinction of many species. A classic example is the extermination of the American passenger pigeon (*Ectopistes migratorius*). Even though it inhabited only eastern North America, 200 years ago this was the world's most abundant bird with a population of between 3 and 5 billion animals (fig. 11.12). It once accounted for about one-quarter of all birds in North America. In 1830, John James Audubon saw a single flock of birds estimated to be ten miles wide, hundreds of miles long, and thought to contain perhaps a billion birds. In spite of this vast abundance, market hunting and habitat destruction caused the entire population to crash in only about 20 years between 1870 and 1890. The last known wild bird was shot in 1900 and the last existing passenger pigeon, a female named Martha, died in 1914 in the Cincinnati Zoo.

At about the same time that passenger pigeons were being extirpated, the American bison or buffalo (*Bison bison*) was being hunted to near extinction on the Great Plains. In 1850, some 60 million bison roamed the western plains. Many were killed only for their hides or tongues, leaving millions of carcasses to rot. Much of the bison's destruction was carried out by the U.S. Army to deprive native peoples who depended on bison for food, clothing, and shelter of these resources, thereby forcing them onto reservations. After 40 years, there were only about 150 wild bison left and another 250 in captivity.

Fish stocks have been seriously depleted by overharvesting in many parts of the world. A huge increase in fishing fleet size and efficiency in recent years has led to a crash of many oceanic populations (fig. 11.13). Worldwide, 13 of 17 principal fishing zones are now reported to be commercially exhausted or in steep decline. At least three-quarters of all commercial oceanic species are overharvested. Canadian fisheries biologists estimate that only 10 percent of the top predators such as swordfish, marlin, tuna, and shark

FIGURE 11.12 A pair of stuffed passenger pigeons (*Ectopistes migratorius*). The last member of this species died in the Cincinnati Zoo in 1914. Courtesy Bell Museum/U. MN.

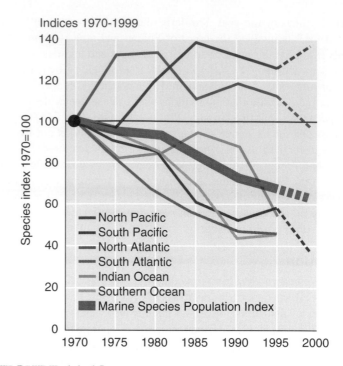

FIGURE 11.13 Changes in abundance of 217 species of marine fish, birds, mammals, and reptiles, from 1970 to 1999, by continent. *Source:* World Wide Fund for Nature (WWF), 2000.

remain in the Atlantic Ocean. Groundfish, such as cod, flounder, halibut, and hake, also are severely depleted. You can avoid adding to this overharvest by eating only abundant, sustainably harvested varieties (What Can You Do? p. 229).

Perhaps the most destructive example of harvesting terrestrial wild animal species today is the African bushmeat trade. Wildlife biologists estimate that 1 million tons of bushmeat, including antelope, elephants, primates, and other animals, are sold in African markets every year (fig. 11.14). For many poor Africans, this is the only source of animal protein in their diet. If we hope to protect the animals targeted by bushmeat hunters, we will need to help them find alternative livelihoods and replacement sources of high-quality protein. The emergence of SARS in 2003 (chapter 8) resulted from the wild food trade in China and Southeast Asia, where millions of civets, monkeys, snakes, turtles, and other animals are consumed each year as luxury foods.

Commercial Products and Live Specimens

In addition to harvesting wild species for food, we also obtain a variety of valuable commercial products from nature. Much of this represents sustainable harvest, but some forms of commercial exploitation are highly destructive, however, and represent a serious threat to certain rare species (fig. 11.15). Despite international bans on trade in products from endangered species, smuggling of furs, hides, horns, live specimens, and folk medicines amounts to millions of dollars each year.

Developing countries in Asia, Africa, and Latin America with the richest biodiversity in the world are the main sources of wild animals and animal products, while Europe, North America, and some of the wealthy Asian countries are the principal importers. Japan, Taiwan, and Hong Kong buy three-quarters of all cat and snake skins, for instance, while European countries buy a similar percentage of live birds. The United States imports 99 percent of all live cacti and 75 percent of all orchids sold each year.

The profits to be made in wildlife smuggling are enormous. Tiger or leopard fur coats can bring $100,000 in Japan or Europe. The population of African black rhinos dropped from approximately 100,000 in the 1960s to about 3,000 in the 1980s because of a demand for their horns. In Asia, where it is prized for its supposed medicinal properties, powdered rhino horn fetches $28,000 per kg. In Yemen, a rhino horn dagger handle can sell for up to $1,000.

Plants also are threatened by overharvesting. Wild ginseng has been nearly eliminated in many areas because of the Asian demand for roots that are used as an aphrodisiac and folk medicine. Cactus "rustlers" steal cacti by the ton from the American southwest and Mexico. With prices ranging as high as $1,000 for rare specimens, it's not surprising that many are now endangered.

The trade in wild species for pets is an enormous business. Worldwide, some 5 million live birds are sold each year for pets, mostly in Europe and North America. Currently, pet traders import (often illegally) into the United States some 2 million reptiles, 1 million amphibians and mammals, 500,000 birds, and 128 million tropical fish each year. About 75 percent of all saltwater tropical

FIGURE 11.14 More than 1 million tons of wild animals are sold every year in the African bushmeat trade. Courtesy Dr. Mitch Eaton.

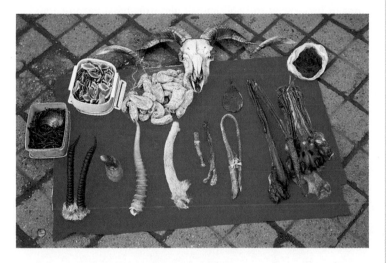

FIGURE 11.15 Parts from rare and endangered species for sale on the street in China. Use of animal products in traditional medicine and prestige diets is a major threat to many species. © William P. Cunningham.

aquarium fish sold come from coral reefs of the Philippines and Indonesia.

Many of these fish are caught by divers using plastic squeeze bottles of cyanide to stun their prey (fig. 11.16). Far more fish die with this technique than are caught. Worst of all, it kills the coral animals that create the reef. A single diver can destroy all of the life on 200 m^2 of reef in a day. Altogether, thousands of divers currently destroy about 50 km^2 of reefs each year. Net fishing would prevent this destruction, and it could be enforced if pet owners would insist on net-caught fish. More than half the world's coral reefs are potentially threatened by human activities, with up to 80 percent at risk in the most populated areas.

What can you do?

Don't Buy Endangered Species Products

You probably are not shopping for a fur coat from an endangered tiger, but there might be other ways you are supporting unsustainable harvest and trade in wildlife species. To be a sustainable consumer, you need to learn about the source of what you buy. Often plant and animal products are farm-raised, not taken from wild populations. But some commercial products are harvested in unsustainable ways. Here are a few products about which you should inquire before you buy:

Seafood includes many top predators that grow slowly and reproduce only when many years old. Despite efforts to manage many fisheries, the following have been severely, sometimes tragically, depleted:

- Top predators: swordfish, marlin, shark, bluefin tuna, albacore ("white") tuna.

- Groundfish and deepwater fish: orange roughy, Atlantic cod, haddock, pollack (source of most fish sticks, artificial crab, generic fish products), yellowtail flounder, monkfish.

- Other species, especially shrimp, yellowfin tuna, and wild sea scallops, are often harvested with methods that destroy other species or habitats.

- Farm-raised species such as shrimp and salmon can be contaminated with PCBs, pesticides, and antibiotics used in their rearing. In addition, aquaculture operations often destroy coastal habitat, pollute surface waters, and deplete wild fish stocks to stock ponds and provide fish meal.

Pets and plants are often collected from wild populations, some sustainably and others not:

- Aquarium fish (often harvested by stunning with dynamite and squirts of cyanide, which destroy tropical reefs and many fish).

- Reptiles: snakes and turtles, especially, are often collected in the wild.

- Plants: orchids and cacti are the best-known, but not the only, group collected in the wild.

Herbal products such as wild ginseng, wild echinacea (purple coneflower), should be investigated before purchasing.

Do buy some of these sustainably harvested products:

- Shade-grown (or organic) coffee, nuts, and other sustainably harvested forest products.

- Pets from the Humane Society, which works to protect stray animals.

- Organic cotton, linen, and other fabrics.

- Fish products that have relatively little environmental impact or fairly stable populations: farm-raised catfish or tilapia, wild-caught salmon, mackerel, Pacific pollack, dolphinfish (mahimahi), squids, crabs, and crayfish.

- Wild freshwater fish like bass, sunfish, pike, catfish, and carp, which are usually better managed than most ocean fish.

FIGURE 11.16 A diver uses cyanide to stun tropical fish being caught for the aquarium trade. Many fish are killed by the method itself, while others die later during shipment. Even worse is the fact that cyanide kills the coral reef itself. © Lynn Funkhouser/Peter Arnold, Inc.

Predator and Pest Control

Some animal populations have been greatly reduced, or even deliberately exterminated, because they are regarded as dangerous to humans or livestock or because they compete with our use of resources. Every year, U.S. government animal control agents trap, poison, or shoot thousands of coyotes, bobcats, prairie dogs, and other species considered threats to people, domestic livestock, or crops.

This animal control effort costs about $20 million in federal and state funds each year and kills some 700,000 birds and mammals, about 100,000 of which are coyotes. Defenders of wildlife regard this program as cruel, callous, and mostly ineffective in reducing livestock losses. Protecting flocks and herds with guard dogs or herders or keeping livestock out of areas that are home range of wild species would be a better solution they believe (fig. 11.17). Ranchers and trappers, on the other hand, argue that without predator control, western livestock operations would be uneconomical.

ENDANGERED SPECIES MANAGEMENT AND BIODIVERSITY PROTECTION

Over the years, we have gradually become aware of the harm we have done—and continue to do—to wildlife and biological resources. Slowly, we are adopting national legislation and international treaties to protect these irreplaceable assets. Parks, wildlife refuges, nature preserves, zoos, and restoration programs have been established to protect nature and rebuild depleted populations. There has been encouraging progress in this area, but much remains to be done. While most people favor pollution control or protection of favored species such as whales or gorillas, surveys show that few understand what biological diversity is or why it is important.

Hunting and Fishing Laws

In 1874, a bill was introduced in the United States Congress to protect the American bison, whose numbers were already falling to dangerous levels. This initiative failed, however, because most legislators believed that all wildlife—and nature in general—was so abundant and prolific that it could never be depleted by human activity. As we discussed earlier in this chapter, however, by the end of the nineteenth century, bison had plunged from some 60 million to only a few hundred animals.

By the 1890s most states had enacted some hunting and fishing restrictions. The general idea behind these laws was to conserve the resource for future human use rather than to preserve wildlife for its own sake. The wildlife regulations and refuges established since that time have been remarkably successful for many species. At the turn of the century, there were an estimated half a million white-tailed deer in the United States; now there are some 14 million—more in some places than the environment can support. Wild turkeys and wood ducks were nearly all gone 50 years ago. By restoring habitat, planting food crops, transplanting breeding stock, building shelters or houses, protecting these birds during breeding season, and other conservation measures, populations of these beautiful and interesting birds have been restored to several million each. Snowy egrets, which were almost wiped out by plume hunters 80 years ago, are now common again.

The Endangered Species Act

Establishment of the U.S. Endangered Species Act (ESA) of 1973 and the Committee on the Status of Endangered Wildlife in Canada (COSEWIC) in 1976 represented powerful new approaches to wildlife protection. Where earlier regulations had been focused almost

FIGURE 11.17 A guard dog watches attentively over a goat herd.
© Mary Ann Cunningham.

Key Concepts

- Biodiversity includes the genetic, species, and ecological variety in populations and communities.

- The main human threats to biodiversity can be summarized by HIPPO: habitat destruction, invasive species, pollution, population growth (human), and overharvesting.

- Endangered species are those considered in imminent danger of extinction, while threatened species are those likely to become endangered in the foreseeable future.

- The Endangered Species Act is one of the most effective environmental laws in United States history, but also is among the most disliked because it can limit property rights and economic benefits where species are endangered.

exclusively on "game" animals, these programs seek to identify all endangered species and populations and to save as much biodiversity as possible, regardless of its usefulness to humans. **Endangered species** are those considered in imminent danger of extinction, while **threatened species** are those that are likely to become endangered—at least locally—within the foreseeable future. Bald eagles, gray wolves, brown (or grizzly) bears, and sea otters, for instance, together with a number of native orchids and other rare plants are considered to be locally threatened even though they remain abundant in other parts of their former range. **Vulnerable species** are naturally rare or have been locally depleted by human activities to a level that puts them at risk. They often are candidates for future listing. For vertebrates, protected categories include species, subspecies, and local races or ecotypes.

The ESA regulates a wide range of activities involving endangered species including "taking" (harassing, harming, pursuing, hunting, shooting, trapping, killing, capturing, or collecting) either accidentally or on purpose; importing into or exporting out of the United States; possessing, selling, transporting, or shipping; and selling or offering for sale any endangered species. Prohibitions apply to live organisms, body parts, and products made from endangered species. Violators of the ESA are subject to fines up to $100,000 and one year imprisonment. Vehicles and equipment used in violations may be subject to forfeiture. In 1995 the Supreme Court ruled that critical habitat—habitat essential for a species' survival—must be protected whether on public or private land.

Currently, the United States has 1,300 species on its endangered and threatened species lists and about 250 candidate species waiting to be considered. The number of listed species in different taxonomic groups reflects much more about the kinds of organisms that humans consider interesting and desirable than the actual number in each group. In the United States, invertebrates make up about three-quarters of all known species but only 9 percent of those deemed worthy of protection. Worldwide, the International Union for Conservation of Nature and Natural Resources (IUCN) lists a total of 16,496 endangered and threatened species, including nearly one-quarter of all known bird varieties (table 11.4).

Listing of new species in the United States has been very slow, generally taking several years from the first petition to final determination. Limited funding, political pressures, listing moratoria, and changing administrative policies have created long delays. In December 2000, flooded by lawsuits over species protection and badly underfunded, the USFWS stopped reviewing new cases altogether. Hundreds of species are classified as "warranted (deserving of protection) but precluded" for lack of funds or local support. At least 18 species have gone extinct since being nominated for protection. Advocates for listing insist that early action is far cheaper than last-ditch rescue efforts. Both environmentalists and property-rights advocates, however, tend to believe that the "single-species" approach is flawed and that efforts should be focused on multiple species and habitat protection.

When Congress passed the original ESA, it probably intended to protect only a few charismatic species like birds and big game animals. Sheltering obscure species such as the Delhi Sands flower-loving fly, the Coachella Valley fringe-toed lizard, Mrs. Furbisher's lousewort, or the orange-footed pimple-back mussel most likely never occurred to those who voted for the bill. This raises some interesting ethical questions about the rights and values of seemingly minor species. Although uncelebrated, these species may be indicators of environmental health. Protecting them usually preserves habitat and a host of unlisted species.

Recovery Plans

Once a species is officially listed as endangered, the Fish and Wildlife Service is required to prepare a recovery plan detailing how populations will be rebuilt to sustainable levels. It usually takes years to reach agreement on specific recovery plans. Among the difficulties are costs, politics, interference on local economic interests, and the fact that once a species is endangered, much of its habitat and ability to survive is likely compromised. The total cost of recovery plans for all currently listed species is estimated to be nearly $5 billion.

The United States currently spends about $150 million per year on endangered species protection and recovery. About half

TABLE 11.4	Endangered and Threatened Species, Worldwide
Mammals	2,133
Birds	2,123
Reptiles	454
Amphibians	231
Fish	1,159
Insects and other invertebrates	3,374
Total fauna	9,474
Plants	7,022

Source: *IUCN Red List, 2000.*

FIGURE 11.18 The California condor is recovering from near extinction. This is one of a few species that have received the bulk of restoration dollars. © Tom McHugh/Photo Researchers, Inc.

that amount is spent on a dozen charismatic species like the California condor, and the Florida panther and grizzly bear, which receive around $13 million per year (fig. 11.18). By contrast, the 137 endangered invertebrates and 532 endangered plants get less than $5 million per year altogether. Our funding priorities often are based more on emotion and politics than biology. A variety of terms are used for rare or endangered species thought to merit special attention:

- *Keystone species* are those with major effects on ecological functions and whose elimination would affect many other members of the biological community; examples are prairie dogs (*Cynomys ludovicianus*) or bison (*Bison bison*).

- *Indicator species* are those tied to specific biotic communities or successional stages or set of environmental conditions. They can be reliably found under certain conditions but not others; an example is brook trout (*Salvelinus fontinalis*).

- *Umbrella species* require large blocks of relatively undisturbed habitat to maintain viable populations. Saving this habitat also benefits other species. Examples of umbrella species are the northern spotted owl (*Strix occidentalis caurina*) and tiger (*Panthera tigris*) (fig. 11.19).

- *Flagship species* are especially interesting or attractive organisms to which people react emotionally. These species can motivate the pubic to preserve biodiversity and contribute to conservation; an example is the giant panda (*Ailuropoda melanoleuca*).

Some recovery plans have been gratifyingly successful. The American alligator was listed as endangered in 1967 because hunting (for meat, skins, and sport) and habitat destruction had reduced populations to precarious levels. Protection has been so effective that the species is now plentiful throughout its entire southern range. Florida alone estimates that it has at least 1 million alligators.

Twenty years ago, due mainly to DDT poisoning, only 800 bald eagles (*Haliaeetus leucocephalis*) remained in the contiguous United States. In 1994, the population had rebounded to more than 8,000 birds and the eagle's status was reduced from endangered to threatened. Bald eagles have been proposed for complete delisting, but disagreement over how they will be managed has delayed this step. Similarly, peregrine falcons, which had been down to only 39 breeding pairs in the 1970s, had rebounded to 1,650 pairs in 1999 and also were taken off the endangered species list. Declaring the ESA a success, former Interior Secretary Bruce Babbit announced that 29 species, including mammals, fish, reptiles, birds, plants, and even one insect (the Tinian monarch) in addition to eagles and falcons have been removed or downgraded from the endangered species list.

Opponents of the ESA have repeatedly tried to require that economic costs and benefits be incorporated into endangered species planning. An important test of the ESA occurred in 1978 in Tennessee where construction of the Tellico Dam threatened a tiny fish called the snail darter. As a result of this case, a federal committee (the so-called "God Squad") was given power to override the ESA for economic reasons. Another important example is the case of the northern spotted owl (chapter 12), whose protection depends on preserving old-growth forest in the Pacific Northwest. Timber-industry economists estimate that saving a population of 1,600 to 2,400 owls would cost $33 billion, with most of the losses borne by local companies and residents of Washington and Oregon. Conservationists dispute these numbers and claim the owl is an umbrella species whose protection would aid many other organisms and resources (fig. 11.20).

An even more costly recovery program may be required for Columbia River salmon and steelhead endangered by hydropower dams and water storage reservoirs that block their migration to the sea. Opening floodgates to allow young fish to run downriver and

FIGURE 11.19 Umbrella or flagship species generally are highly charismatic and can mobilize support to protect large blocks of habitat that benefit other creatures as well. © Corbis/Volume 6.

"DAMN SPOTTED OWL!"

TIMBER
MINING
CO.

©1990 HERBLOCK

FIGURE 11.20 Endangered species often serve as a barometer for the health of an entire ecosystem and as surrogate protector for a myriad of less well-known creatures. Copyright 1990 by Herblock in the *Washington Post.*

adults to return to spawning grounds would have high economic costs to barge traffic, farmers, and electric rate payers who have come to depend on abundant water and cheap electricity. On the other hand, commercial and sport fishing for salmon is worth $1 billion per year and employs about 60,000 people directly or indirectly.

Private Land and Critical Habitat

Private land is essential in endangered species protection. Eighty percent of the habitat for more than half of all listed species is on nonpublic property. The Supreme Court has ruled that destroying habitat is as harmful to endangered species as directly taking (killing) them. Many people, however, resist restrictions on how they use their own property to protect what they perceive to be insignificant or worthless organisms. This is especially true when the land has potential for economic development. If property is worth millions of dollars as the site of a housing development or shopping center, most owners don't want to be told they have to leave it undisturbed to protect some rare organism. Landowners

may be tempted to "shoot, shovel, and shut up," if they discover endangered species on their property. Many feel they should be compensated for lost value caused by ESA regulations.

Recently, to avoid crises like the northern spotted owl, the Fish and Wildlife Service has been negotiating agreements called **habitat conservation plans (HCP)** with private landowners. Under these plans, landowners are allowed to harvest resources or build on part of their land as long as the species benefits overall. By improving habitat in some areas, funding conservation research, removing predators and competitors, or other steps that benefit the endangered species, developers are allowed to destroy habitat or even "take" endangered organisms.

In 2003, the Interior Department released $70 million to purchase critical habitat and to support HCPs. Scientists and environmentalists often are critical of HCPs, claiming these plans often are based more on politics than biology, and that the potential benefits are frequently overstated. Defenders argue that by making the ESA more landowner-friendly, HCPs benefit wildlife in the long run.

Among the more controversial proposals for HCPs are the so-called Safe Harbor and No-Surprises Policies. Under the Safe Harbor clause, any increase in an animal's population resulting from a property owner's voluntary good stewardship would not increase their responsibility or affect future land-use decisions. As long as the property owner complies with the terms of the agreement, he or she can make any use of the property. The No-Surprises provision says that the property owner won't be faced with new requirements or regulations after entering into an HCP. Scientists warn that change, uncertainty, dynamics, and flux are characteristic of all ecosystems. We can't say that natural catastrophes or environmental events won't make it necessary to modify conservation plans in the future.

Reauthorizing the Endangered Species Act

The ESA officially expired in 1992. Since then, Congress has debated many alternative proposals ranging from outright elimination to substantial strengthening of the act. Perhaps no other environmental issue divides Americans more strongly than the ESA. In the western United States, where traditions of individual liberty and freedom are strong and the federal government is viewed with considerable suspicion and hostility, the ESA seems to many to be a diabolical plot to take away private property and trample on individual rights. Many people believe that the law puts the welfare of plants and animals above that of humans. Farmers, loggers, miners, ranchers, developers, and other ESA opponents repeatedly have tried to scuttle the law or greatly reduce its power. Environmentalists, on the other hand, see the ESA as essential to protecting nature and maintaining the viability of the planet. They regard it as the single most effective law in their arsenal and want it enhanced and improved.

Proposals for a new ESA generally fall into one of two general categories. Environmentalists have repeatedly introduced versions that encourage an ecosystem and habitat protection approach rather than focusing on individual species. Reinforcing critical habitat protection and placing deadlines for listing and recovery plan completion, these bills would require that proposed HCPs be independently peer-reviewed by qualified scientists. They also would expand citizen ability to enforce the law and increase civil and criminal penalties.

On the other side, ESA opponents want to reinforce "safe harbor" and "no surprises" policies and to allow exceptions to critical habitat designation. All recovery plan teams in their versions of the ESA would be required to include individuals with vested economic interests. Under these proposals, only the least costly, most cost effective or least burdensome measures would be taken to protect endangered organisms. Federal agencies would be allowed to do "self-consultation"—that is to avoid consulting wildlife specialists from the Fish and Wildlife Service before undertaking projects. ESA opponents also would require notification and hearings in all states with resident candidate species and would give small-parcel landowners expedited permits for HCPs.

Habitat Protection

Over the past decade, growing numbers of scientists, land managers, policymakers, and developers have been making the case that it is time to focus on a rational, continent-wide preservation of ecosystems that support maximum biological diversity rather than a species-by-species battle for the rarest or most popular organisms. By focusing on populations already reduced to only a few individuals, we spend most of our conservation funds on species that may be genetically doomed no matter what we do. Furthermore, by concentrating on individual species we spend millions of dollars to breed plants or animals in captivity that have no natural habitat where they can be released. While flagship species such as mountain gorillas or Indian tigers are reproducing well in zoos and wild animal parks, the ecosystems that they formerly inhabited have largely disappeared.

A leader of this new form of conservation is J. Michael Scott, who was project leader of the California condor recovery program in the mid-1980s and had previously spent ten years working on endangered species in Hawaii. In making maps of endangered species, Scott discovered that even Hawaii, where more than 50 percent of the land is federally owned, has many vegetation types completely outside of natural preserves (fig. 11.21). The gaps between protected areas may contain more endangered species than are preserved within them.

This observation has led to an approach called **gap analysis** in which conservationists and wildlife managers look for unprotected landscapes that are rich in species. Computers and geographical information systems (GIS) make it possible to store, manage, retrieve, and analyze vast amounts of data and create detailed, high-resolution maps relatively easily. This broad-scale, holistic approach seems likely to save more species than a piecemeal approach.

Conservation biologist, R. E. Grumbine suggests four remanagement principles for protecting biodiversity in a large-scale, long-range approach:

1. Protect enough habitat for viable populations of all native species in a given region.
2. Manage at regional scales large enough to accommodate natural disturbances (fire, wind, climate change, etc.).

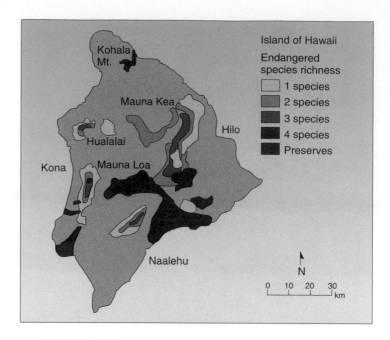

FIGURE 11.21   An example of the biodiversity maps produced by J. Michael Scott and the U.S. Fish and Wildlife Service. Notice that few of the areas of endangered species richness are protected in preserves, which were selected more for scenery or recreation than for biology.

3. Plan over a period of centuries so that species and ecosystems may continue to evolve.
4. Allow for human use and occupancy at levels that do not result in significant ecological degradation.

International Wildlife Treaties

The 1975 Convention on International Trade in Endangered Species (CITES) was a significant step toward worldwide protection of endangered flora and fauna. It regulated trade in living specimens and products derived from listed species, but has not been foolproof. Species are smuggled out of countries where they are threatened or endangered, and documents are falsified to make it appear they have come from areas where the species are still common. Investigations and enforcement are especially difficult in developing countries where wildlife is disappearing most rapidly. Still, eliminating markets for endangered wildlife is an effective way of stopping poaching. Appendix I of CITES lists 700 species threatened with extinction by international trade.

CAPTIVE BREEDING AND SPECIES SURVIVAL PLANS

Breeding programs in zoos and botanical gardens are one way to attempt to save severely threatened species. Institutions like the Missouri Botanical Garden and the Bronx Zoo's Wildlife Conservation Society sponsor conservation and research programs. Botanical gardens, such as the Kew Gardens in England, and research

FIGURE 11.22 Rare plants can be preserved and studied in botanical gardens. © The McGraw-Hill Companies, Inc./Barry W. Barker, photographer.

stations, such as the International Rice Institute in the Philippines, are repositories for rare and endangered plant species that sometimes have ceased to exist in the wild. Valuable genetic traits are preserved in these collections, and in some cases, plants with unique cultural or ecological significance may be reintroduced into native habitats after being cultivated for decades or even centuries in these gardens and seed banks (fig. 11.22).

Zoos can also help preserve wildlife, and act as repositories for genetic diversity. Until fairly recently, zoos depended on primarily wild-caught animals for most of their collections. This was a serious drain on wild populations, because up to 80 percent of the animals caught died from the trauma of capture and shipping. With better understanding of reproductive biology and better breeding facilities, most mammals in North American zoos now are produced by captive breeding programs.

Some zoos now participate in programs that reintroduce endangered species to the wild. The California condor (see fig. 11.18) is one of the best known cases of successful captive breeding. In 1986, only nine of these birds existed in their native habitat. Fearing the loss of these last condors, biologists captured them and brought them to the San Diego and Los Angeles zoos, which had begun breeding programs in the 1970s. By 2003 the population had reached 202 birds, with 73 reintroduced to the wild. In 2001, the first condor egg was laid in the wild since 1986, demonstrating that captive-bred condors can breed in the wild.

Such breeding programs have limitations, however. Bats, whales, and many reptiles rarely reproduce in captivity and still come mainly from the wild. Furthermore, we will never be able to protect the complete spectrum of biological variety in zoos. According to one estimate, if all the space in U.S. zoos were used for captive breeding, only about 100 species of large mammals could be maintained on a long-term basis (fig. 11.23).

These limitations lead to what is sometimes called the "Noah question": how many species can or should we save? How much are we willing to invest to protect the slimy, smelly, crawly things? Would you favor preserving disease organisms, parasites, and vermin or should we use our limited resources to protect only beautiful, interesting, or seemingly useful organisms?

FIGURE 11.23 Primates endangered in the wild may be successfully preserved in zoos. Less charismatic species are rarely included in breeding programs, however. © The McGraw-Hill Companies, Inc./Barry W. Barker, photographer.

Even given adequate area and habitat conditions to perpetuate a given species, continued inbreeding of a small population in captivity can lead to the same kinds of fertility and infant survival problems described earlier for wild populations. To reduce genetic problems, zoos often exchange animals or ship individuals long distances to be bred. It sometimes turns out, however, that zoos far distant from each other unknowingly obtained their animals from the same source. Computer databases operated by the International Species Information System located at the Minnesota Zoo, now keep track of the genealogy of many species. This system can tell the complete reproductive history of every animal in every zoo in the world for some species. Comprehensive species survival plans based on this genealogy help match breeding pairs and project resource needs.

The ultimate problem with captive breeding, however, is that natural habitat may disappear while we are busy conserving the species itself. Large species such as tigers or apes are sometimes called "umbrella species." As long as they persist in their native habitat, many other species survive as well.

Saving Rare Species in the Wild

Renowned zoologist George Schaller says that ultimately "zoos need to get out of their own walls and put more effort into saving the animals in the wild." An interesting application of this principle is a partnership between the Minnesota Zoo and the Ujung Kulon National Park in Indonesia, home to the world's few remaining Javanese rhinos. Rather than try to capture rhinos and move them to Minnesota, the zoo is helping to protect them in their native habitat by providing patrol boats, radios, housing, training, and salaries for Indonesian guards (fig. 11.24). There are no plans to bring any rhinos to Minnesota and chances are very slight that any of us will ever see one, but we can gain satisfaction that, at least for now, a few Javanese rhinos still exist in the wild.

FIGURE 11.24 The *KM Minnesota* anchored in Tamanjaya Bay in west Java. Funds raised by the Minnesota Zoo paid for local construction of this boat, which allows wardens to patrol Ujung Kulon National Park and protect rare Javanese rhinos from poachers.
Courtesy Dr. Ronald Tilson, Minnesota Zoo.

Summary

- Biodiversity includes the genetic, species, and ecological variety in populations and communities.

- Approximately 1.7 million species have been named, so far, but biologists estimate that somewhere between 3 million to 50 million may exist, 95 percent of which are plants and invertebrate animals (those without backbones).

- Biodiversity provides food, medicines, ecological services, and aesthetic and cultural benefits to humans.

- The main human threats to biodiversity can be summarized by HIPPO: habitat destruction, invasive species, pollution, population growth (human), and overharvesting.

- The Endangered Species Act is one of the most effective environmental laws in United States history, but also is among the most disliked because it can limit property rights and economic benefits where species are threatened or endangered.

- Endangered species are those considered in imminent danger of extinction, while threatened species are those likely to become endangered in the foreseeable future.

- A few charismatic, flagship, or politically powerful species get the vast bulk of funding for protection and recovery programs, while ordinary endangered species get little attention.

- Reauthorization of the endangered species act has been highly controversial. Environmentalists generally want to see it strengthened, while property rights advocates and resource developers want to see it weakened or eliminated entirely.

- Zoos can be educational and entertaining while still serving important wildlife conservation and scientific functions. Still, there are limits to the numbers and types of organisms we could maintain under captive conditions. Saving biodiversity in the wild may be the most economical and effective conservation strategy.

Questions for Review

1. What is the range of estimates of the total number of species on the earth? Why is the range so great?

2. What group of organisms has the largest number of species?

3. Define *extinction*. What is the natural rate of extinction in an undisturbed ecosystem?

4. What are rosy periwinkles and what products do we derive from them?

5. Describe some foods we obtain from wild plant species.

6. Define *HIPPO* and describe what it means for biodiversity conservation.

7. What is the current rate of extinction and how does this compare to historic rates?

8. Compare the scope and effects of the Endangered Species Act and CITES.

9. Define *endangered* and *threatened*. Give an example of each.

10. What is gap analysis and how is it related to ecosystem management and design of nature preserves?

Questions for Critical Thinking

1. One reviewer said that this chapter is the most biased in this book. Do you agree? How much moral outrage is appropriate in an issue such as this? Does emotion interfere with rational analysis or effective communication? What is the proper balance between emotion and objectivity in a subject such as this?

2. Many ecologists would like to move away from protecting individual endangered species to concentrate on protecting whole communities or ecosystems. Others fear that the public will only respond to and support glamorous "flagship" species such as gorillas, tigers, or otters. If you were designing conservation strategy, where would you put your emphasis?

3. Put yourself in the place of a fishing industry worker. If you continue to catch many species they will quickly become economically extinct if not completely exterminated. On the other hand, there are few jobs in your village and welfare will barely keep you alive. What would you do?

4. Only a few hundred grizzly bears remain in the contiguous United States, but populations are healthy in Canada and Alaska. Should we spend millions of dollars for grizzly recovery and management programs in Yellowstone National Park and adjacent wilderness areas?

5. How could people have believed a century ago that nature is so vast and fertile that human actions could never have a lasting impact on wildlife populations? Are there similar examples of denial or misjudgment occurring now?

6. Suppose you're having dinner with a friend who orders swordfish. What would you say? What are the ethical and biological arguments for or against eating endangered species?

7. In the past, mass extinction has allowed for new growth, including the evolution of our own species. Should we assume that another mass extinction would be a bad thing? Could it possibly be beneficial to us? To the world?

8. Some captive breeding programs in zoos are so successful that they often produce surplus animals that cannot be released into the wild because no native habitat remains. Plans to euthanize surplus animals raise storms of protests from animal lovers. What would you do if you were in charge of the zoo?

9. Debate with a friend or classmate the ethics of keeping animals captive in a zoo. After exploring the subject from one side, debate the issue from the opposite perspective. What do you learn from this exercise?

Key Terms

biodiversity 217
biodiversity hot spots 219
endangered species 231
existence value 223
extinction 223
gap analysis 234

habitat conservation plans
 (HCP) 233
HIPPO 224
invasive species 224
overharvesting 227
threatened species 231
vulnerable species 231

Further Readings

Baskin, Yvonne. 2003. *A Plague of Rats and Rubbervines: The Growing Threat of Species Invasions.* Island Press.

Ellis, Richard. 2003. *The Empty Ocean: Plundering the World's Marine Life.* Island Press.

Gibbs, W. Wayt. 2001. On the termination of species. *Scientific American* 285(5):40–49.

Lehman, Clarence L., and David Tilman. 2000. Biodiversity, stability, and productivity in competitive communities. *The American Naturalist* 156:534–52.

MacArthur, R. H., and E. O. Wilson. 1963. An equilibrium theory of insular zoogeography. *Evolution* 17:373–87.

May, Robert M. 1972. Will a large complex system be stable? *Nature* 238:413–14.

Myers, N., et al. 2000. Biodiversity hotspots for conservation priorities. *Nature* 403:853.

Poani, K., B. D. Richter, M. G. Anderson, and H. E. Richter. 2002. Biodiversity conservation at multiple scales: Functional sites, landscapes, and networks. *Bioscience* 50(2):133–46.

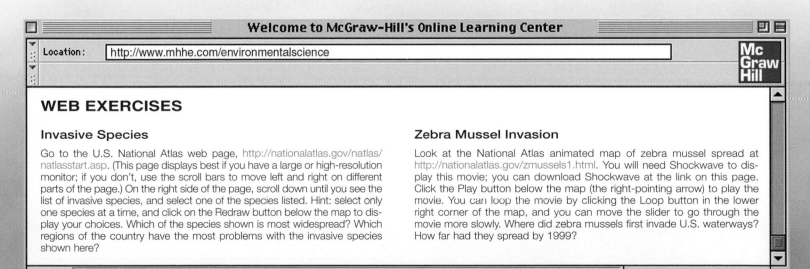

Welcome to McGraw-Hill's Online Learning Center

Location: http://www.mhhe.com/environmentalscience

WEB EXERCISES

Invasive Species

Go to the U.S. National Atlas web page, http://nationalatlas.gov/natlas/natlasstart.asp. (This page displays best if you have a large or high-resolution monitor; if you don't, use the scroll bars to move left and right on different parts of the page.) On the right side of the page, scroll down until you see the list of invasive species, and select one of the species listed. Hint: select only one species at a time, and click on the Redraw button below the map to display your choices. Which of the species shown is most widespread? Which regions of the country have the most problems with the invasive species shown here?

Zebra Mussel Invasion

Look at the National Atlas animated map of zebra mussel spread at http://nationalatlas.gov/zmussels1.html. You will need Shockwave to display this movie; you can download Shockwave at the link on this page. Click the Play button below the map (the right-pointing arrow) to play the movie. You can loop the movie by clicking the Loop button in the lower right corner of the map, and you can move the slider to go through the movie more slowly. Where did zebra mussels first invade U.S. waterways? How far had they spread by 1999?

12

Land Use: Forests and Grasslands

A tree is a tree—how many more do you need to look at?

Ronald Reagan

OBJECTIVES

After studying this chapter, you should be able to:

- discuss the major world land uses and how human activities impact these areas.
- summarize some forest types and the products we derive from them.
- report on how and why tropical forests are being disrupted as well as how they might be better used.
- understand the major issues concerning forests in more highly developed countries such as the United States and Canada.
- outline the extent, location, and state of grazing lands around the world.
- describe how overgrazing causes desertification of rangelands.
- explain why land reform and recognition of indigenous rights are essential for social justice as well as environmental protection.

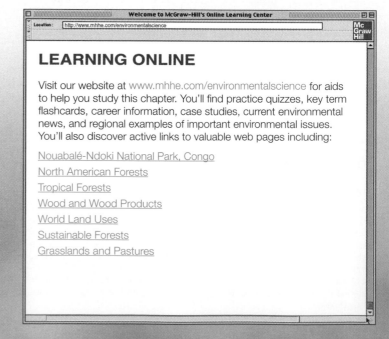

LEARNING ONLINE

Visit our website at www.mhhe.com/environmentalscience for aids to help you study this chapter. You'll find practice quizzes, key term flashcards, career information, case studies, current environmental news, and regional examples of important environmental issues. You'll also discover active links to valuable web pages including:

Nouabalé-Ndoki National Park, Congo

North American Forests

Tropical Forests

Wood and Wood Products

World Land Uses

Sustainable Forests

Grasslands and Pastures

Photo: Elephants gather in a clearing in Nouabalé-Ndoki National Park, Congo.
© Michael Nick Nichols/National Geographic.

Saving an African Eden

When conservation biologist Paul Elkan first entered the Goualougo tract in remote northern Republic of Congo on a wildlife reconnaissance expedition in 1997, he was astonished at how little fear the animals showed. Generally, when wild monkeys or chimpanzees spot humans, they shriek and run away. Conditioned by years of hunting, wild animals fear humans. In the pristine forests of Goualougo, however, Elkan found that the animals followed him for hours, seemingly curious about this strange new species in their forest home. Bordered by two untamed rivers and many kilometers of dense, flooded forests, the Goualougo shows no signs of human intrusion. Even the local Bambendjelle people (called forest Pygmies by Europeans) had no memories of ever hunting there. "This is one of the last great wild places in the world," Elkan said.

The Congo basin holds about one-fourth of the world's tropical forests and is the largest stretch of lowland rainforest in the world, aside from the Amazon. Those forests are home to some of the most important wildlife populations remaining in Central Africa, including forest elephants (*Loxodonta africana cyclotis*), lowland gorillas (*Gorilla gorilla*), chimpanzees (*Pan troglodytes troglodytes*), bongo (*Tragelaphus eurycerus*), buffalo (*Syncerus caffer nanus*), leopard (*Panthera pardus*), six species of small antelope, and eight species of monkeys. Logging and hunting are a growing threat to the forests and the wildlife they shelter. It has long been common practice for loggers to subsist on the local wildlife, or bushmeat. And logging roads provide access deep into the forest for hunters and settlers. Up to 1 million tons of bushmeat may be consumed in Central Africa each year.

In 1982, all the lands in the northern Republic of Congo were divided into large logging concessions. In the late 1980s, wildlife explorations revealed that the Nouabalé-Ndoki region held one of the densest assemblages of wildlife and the richest collection of primates on earth. Recognizing its ecological importance, the government of Congo upgraded the Noubalé-Ndoki forest from a forest management unit to national park in 1993. Adjacent parks in the Central African Republic and Cameroon create a contiguous area of irreplaceable lowland tropical rainforest (fig. 12.1).

Unfortunately, when the parks were being established, the most pristine area was cut in two, and half of it, Goualougo triangle, was slated for logging. Logging rights were sold to a German-owned company, Congolaise Industrielle des Bois (C.I.B.). Biologist/explorer Mike Fay, who publicized the region's plight with his "megatransect" across the Congo basin, calls the Gualougo triangle the Inner Sanctum of this great forest. Elkan and others have confirmed the richness of this African Eden.

In July 2002, C.I.B., the Congolese government, and the Wildlife Conservation Society (WCS) announced that a 26,000 ha (100 mi²) tract of the Goualougo triangle would be added to the Nouabalé-Ndoki National Park. This was the first time any logging company in the entire Congo Basin had voluntarily given up rights to valuable timberland. The C.I.B. estimated that it would forgo logging timber worth an estimated $40 million (U.S.). In addition, C.I.B. promised to practice sustainable forest management on its remaining timber concessions and to participate in a wildlife protection program run by the WCS.

C.I.B. promises to prohibit hunting and transporting of bushmeat by company employees and promises to develop alternative protein sources (imported beef, fish, and poultry farms). The company will provide garden supplies and training to local farmers and a conservation awareness program for villagers and company employees. The forest outside the park will be zoned with community hunting areas near villages and no-hunting zones immediately surrounding the park.

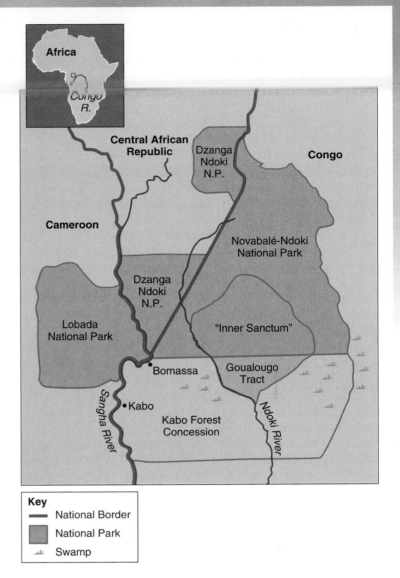

FIGURE 12.1 Adjacent parks in Cameroon, Central African Republic, and Congo protect an irreplaceable expanse of tropical rainforest.

Protected areas will be patrolled by six government officers and 40 local ecoguards funded and managed by the WCS.

Why did the C.I.B. give up such lucrative logging concessions? Biologists played a vital role by revealing the incredible richness of the wildlife and the absence of human encroachment. Consumers who voiced concern about tropical forests were also key. More and more, consumers of paper products, furniture, and lumber are recognizing their links to tropical forests, and public pressure often leads to better policies. European environmental groups provided leadership and helped educate the public. C.I.B. gave up the Goualougo triangle as part of an effort to get its products certified as sustainably harvested, so that it could retain its European market.

Land-use policies are central to environmental quality, and they are influenced by decisions of national governments, corporations, citizen organizations, and individuals. In this chapter, we'll look at land uses, especially on forests and grasslands, and we'll consider how each of us plays a role, as individuals or as members of a community, in land-use decisions affecting our common environment.

WORLD LAND USES

The earth's total land area is about 133 million km2 (56 million mi²), or about 29 percent of the surface of the globe. Forests, grasslands, and agriculture cover about two-thirds of this area (fig. 12.2). Much of the land that falls into the residual "other" category is tundra, marsh, desert, scrub forest, urban areas, bare rock, and ice or snow. About one-third of this land is so barren that it lacks plant cover altogether. While deserts and other unproductive lands are generally unsuitable for intensive human use, they play an important role in biogeochemical cycles and as a refuge for biological diversity. Presently, nearly 10 percent of the world's land surface is formally protected in parks, wildlife refuges, and nature preserves although enforcement of protection varies (chapter 13).

Notice that approximately 11 percent of the earth's landmass is now used for crops. Some agricultural experts claim that as much as half of the present forests and grazing lands—especially in Africa and South America—could be converted to crop production, given the proper inputs of water, fertilizer, erosion control, and mechanical preparation. Although this land could feed a vastly larger human population (perhaps ten times the present number), sustained intensive agriculture could result in serious environmental and social problems (chapter 9).

Rapidly increasing human populations and expanding forestry and agriculture have brought about extensive land-use changes throughout the world. We now dominate most areas with moderate climates and good soils. Globally, the biomes most extensively altered by humans are temperate (midlatitude) forests and grasslands. Over the past 10,000 years, since the spread of agriculture, billions of hectares of forests have been cleared for firewood or timber or converted to cropland. Once-extensive grasslands have been converted to farm fields or greatly altered by grazing livestock. Overharvesting has converted dry forests to scrub and sometimes to desert. In some regions, such as Europe and central Asia, widespread deforestation and grassland conversion occurred centuries ago. Elsewhere, as in the rainforests of Southeast Asia and Amazonia, land conversion has expanded mainly in the past 20 or 30 years. Resulting losses in biodiversity are a matter of great concern (chapter 11), as are effects on climate and air quality (chapters 15 and 16). In this chapter we'll discuss dominant land uses and how those activities affect land cover and ecosystem health around the world.

Clearing forests or plowing grasslands have immediate and obvious destructive impacts on landscapes and wildlife. Given enough time, however, nature can be surprisingly resilient. New England, for example, lost much of its native forest in the nineteenth century, but today the region is largely reforested. Vermont, which had only 35 percent of its original forest in 1850, is now 80 percent forested. Biological communities in these second-growth forests differ from those of the primeval forests, but many ecosystem functions, such as water storage and soil development, are restored. Long-diminished populations of large animals such as moose and bears are resurging. While this recovery doesn't give us a license to mistreat natural landscapes, it does show that damage can be repaired.

WORLD FORESTS

Forests play a vital role in regulating climate, controlling water runoff, providing shelter and food for wildlife, and purifying the air. They produce valuable materials, such as wood and paper pulp. They also have scenic, cultural, and historic values that deserve to be protected. In this section, we will look at forest distribution, use, and management.

How Much Forest Is There?

Estimating original forest and the extent of forest destruction is difficult, because definitions vary (do scattered trees and plantations count?) and forests are so widely distributed. The United Nations calculates that more than half of the world's original forests have been converted to cropland, pasture, settlements, or wasteland. Some of this conversion happened thousands of years ago, but most of it occurred within the past two hundred years. The remaining 3.8 billion ha of forest and woodland covers about 29 percent of the earth's land surface (fig. 12.2). About four-fifths of this land is classified as **closed canopy** (with tree crowns spread over 20 percent or more of the ground). The rest is **open canopy** forest or *woodland,* in which tree crowns cover less than 20 percent of the ground.

The distribution of remaining forest types is shown in fig. 12.3. Tropical rainforests and humid forests, mainly in South America, Central Africa, and Southeast Asia, make up 43 percent of the world's standing forests (fig. 12.4). Another 30 percent are boreal forests, mainly in Siberia and Canada. Among the forests of greatest concern are **old-growth forests,** remnants of ancient forests that still contain much of the world's biodiversity, endangered species, and indigenous human cultures (chapter 11). Generally, old growth is considered to mean forests that are large enough and undisturbed enough that a tree can live out its natural life cycle, and ecosystem processes are not significantly modified by human activity. That doesn't mean that all trees need be enormous or thou-

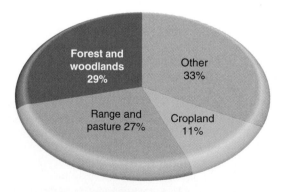

FIGURE 12.2 World land use. The "other" category includes tundra, desert, wetlands, and urban areas. *Source:* Food and Agriculture Organization, (FAO), 1999.

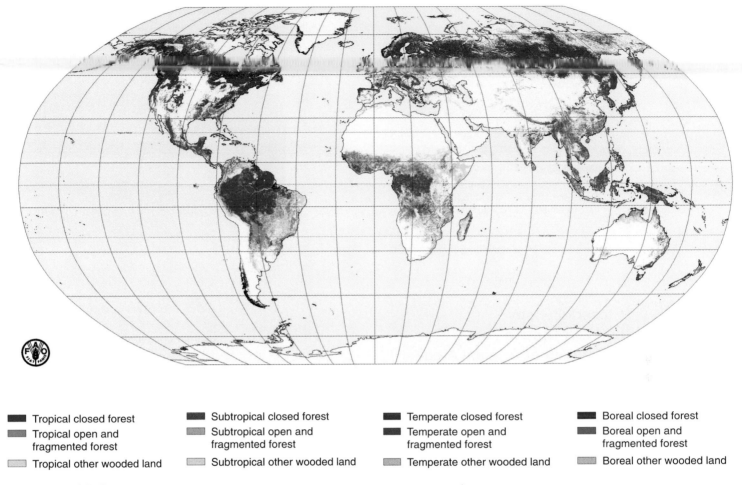

■	Tropical closed forest	■	Subtropical closed forest	■	Temperate closed forest	■	Boreal closed forest
■	Tropical open and fragmented forest	■	Subtropical open and fragmented forest	■	Temperate open and fragmented forest	■	Boreal open and fragmented forest
▢	Tropical other wooded land	▢	Subtropical other wooded land	▢	Temperate other wooded land	▢	Boreal other wooded land

FIGURE 12.3   Major forest types. Note that some of these forests are dense; others may have only 10–20 percent actual tree cover.
Source: United Nations Food and Agriculture Organization, 2002.

sands of years old. In some old-growth forests most trees may live less than a century before being killed by disease or some natural disturbance like a fire. Nor does it mean that humans have never been present. Where human occupation entails relatively little impact, an old-growth forest may have been inhabited by people for a very long time.

Less than half of forests still retain old-growth features. The largest remaining areas of old-growth are in Russia, North America (mainly Canada), South America (mainly Brazil), and Oceania (mainly Papua New Guinea) (table 12.1). Together these four regions account for more than three-quarters of all relatively undisturbed forests in the world. In general, remoteness rather than laws protect those forests. Unfortunately, as global trade expands markets for timber, distance from civilization offers less and less protection. The ancient forests of eastern Siberia, Central Africa, and Indonesia are increasingly vulnerable to both legal and illegal logging. According to the World Wildlife Fund, 50 percent of logging in Siberia is illegal and uncontrolled, 70 percent in Indonesia, and 80 percent in Brazil. These problems are discussed in detail in the next section.

Forest Products

Wood plays a part in more activities of the modern economy than does any other commodity. There is hardly any industry that does not use wood or wood products somewhere in its manufacturing and marketing processes. Think about the amount of junk mail, newspapers, photocopies, and other paper products that each of us in developed countries handles, stores, and disposes of in a single day. Global wood consumption has nearly doubled in the past 50 years, to 3.4 billion m^3 per year (fig. 12.5). Wood harvesting is expected to increase another 50 percent by 2050.

Industrial timber and roundwood (unprocessed logs) are used to make lumber, plywood, veneer, particleboard, and chipboard. Together, they account for about one-half of worldwide wood consumption. This exceeds the use of steel and plastics combined. International trade in wood and wood products amounts to more than $100 billion each year. Developed countries produce less than half of all industrial wood but account for about 80 percent of its consumption. Less-developed countries, mainly in the tropics, produce more than half of industrial wood but use only 20 percent.

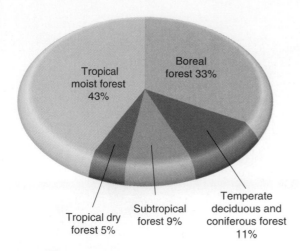

FIGURE 12.4 Relative distribution of remaining forests world-wide.

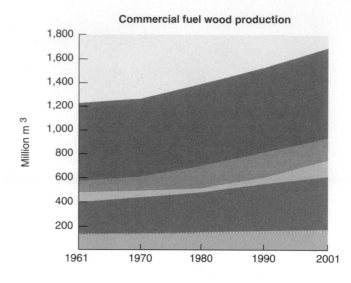

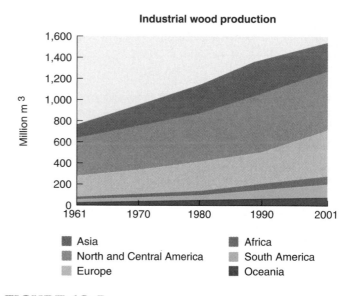

FIGURE 12.5 Commercial wood production by region, according to the UN Food and Agriculture Organization (FAO), 2002. Industrial wood production has grown by 75 percent in 40 years, fuelwood by 50 percent.

TABLE 12.1	Closed-Canopy and Old-Growth Forest as Percent of Original Area	
REGION	REMAINING FOREST (%)	REMAINING OLD-GROWTH (%)
Africa	33.9	7.8
Asia	28.2	5.3
North America	77.3	34.1
Central America	54.5	9.7
South America	69.1	45.6
Europe	30.5	0.1
Russia	68.7	29.3
Oceania*	64.9	22.3
World	53.6	21.7

*Oceania consists of Papua New Guinea, Australia, New Zealand, and many small-island nations.

Source: World Resources Institute 1998–99, *World Resources Institute.*

The United States, Russia, and Canada are the largest producers of both industrial wood and paper pulp. Although old-growth, virgin forest with trees large enough to make plywood or clear furniture lumber is diminishing everywhere, much of the industrial logging in North America and Europe occurs in managed forests where cut trees are replaced by new seedlings. In contrast, tropical hardwoods in Southeast Asia, Africa, and Latin America are being cut at an unsustainable rate, mostly from virgin forests.

Japan is by far the world's largest net importer of wood, purchasing about 46 million m³ per year or 96 percent of net Asian imports. Ironically, the United States is both a major exporter and importer of wood because we buy wood and paper pulp from Canada and finished wood products from Asia at the same time that we sell raw logs, rough lumber, and waste paper to Japan and other countries.

At least one-third of the people in the world depend on firewood or charcoal as their principal source of heating and cooking fuel (fig. 12.6). Consequently, **fuelwood** accounts for about half of all wood harvested worldwide. Unfortunately, burgeoning populations and dwindling forests are causing wood shortages in many less-developed countries. About 1.5 billion people who depend on fuelwood as their primary energy source have less than they need. At present rates of population growth and wood consumption, the annual deficit is expected to increase from 400 million m³ in 1995, to 2,600 m³ in 2025. At that point, the demand will be twice the available fuelwood supply. The average amount of wood used for cooking and heating in 63 less-developed countries is about 1 m³

FIGURE 12.6 Firewood accounts for almost half of all wood harvested worldwide and is the main energy source for one-third of all humans. © William P. Cunningham.

per person per year, roughly equal to the amount of wood that each American consumes as paper products alone.

Many people in poorer countries cook over open fires that deliver only about one-tenth of the available heat to cooking pots. Inexpensive metal stoves can double this efficiency, while locally made ceramic stoves can be four times as efficient as open fires (fig. 12.7). These stoves can save up to 20 percent of household income for urban families.

Non-Timber Forest Benefits

Forests are, of course, far more than simply timber-producing plantations. For many people, particularly in developing countries, forests are a source of mushrooms, nuts, fruits, fodder for livestock, and materials such as rattan and cane used in building and crafts. Forests also provide habitat for wildlife, recreational opportunities, and havens for biodiversity and evolution. They produce oxygen and sequester carbon (thus reducing global warming). In a World Bank study of 105 major world cities, more than one-third, including New York, Tokyo, Barcelona, and Melbourne, get much of their water from protected forests. Preserving these forests reduces landslides, erosion, and sediment runoff. It also improves water purity by filtering pollutants. Many natural forests capture and store water. Protecting them is the most cost-effective way to provide clean drinking water, the World Bank says.

FIGURE 12.7 Simple, inexpensive stoves like these in Guatemala use as little as one-fourth as much wood as an open cooking fire. This simple technology can help save forests and reduce air pollution and household expenses. © William P. Cunningham.

Forest Management

Approximately 25 percent of the world's forests are managed scientifically for wood production. **Forest management** involves planning for sustainable harvests, with particular attention paid to forest regeneration (Case Study, p. 244). Fires, insects, and diseases damage up to one-quarter of the annual growth in temperate forests. Recently, reduced forest growth and sudden die-off of certain tree species in industrialized countries have caused great concern. It is thought that long-range transport of air pollutants (chapter 16) is contributing to this sudden forest death, but not all the causes and solutions are yet understood.

Most countries replant far less forest than is harvested or converted to other uses, but there are some outstanding examples of successful reforestation. China, for instance, cut down most of its forests 1,000 years ago and has suffered centuries of erosion and terrible floods as a consequence. Recently, however, a massive

Forestry for the Seventh Generation

Until the nineteenth century, the Menominee Nation occupied nearly half of what is now Wisconsin and northern Michigan. A woodland people, the Menominee hunted, fished, and gathered wild rice. Their name for themselves, "Mano'min ini'niwuk," means wild rice people. In 1854, besieged by smallpox, alcohol, and pressure from land-hungry European settlers, the tribe was forced onto a reservation representing less than 3 percent of their ancestral lands. Although the reservation—which lies along the Wolf River about 50 miles northwest of Green Bay, WI—is in the poorest county of the state, it represents a unique treasure. The forests covering 98 percent of the land make up the densest, most diverse woodlands in the Great Lakes region and the longest-running operation for sustained-yield forestry in the country.

Timber harvesting started in 1854, when 20 million board feet were cut for lumber, planks, firewood, and fence rails. Greedy lumber barons tried to gain control of valuable white pine holdings but the tribe resisted. While the rest of the state was clear-cut, burned over, and turned into farmland, the Menominee insisted on careful, selective cutting of individual trees. By 1890, the tribe had built its own sawmill and carried out the first sustainable harvest management plan in the country. When first inventoried in 1890, the 89,000 ha (234,000 acres) reservation contained 1.3 billion board feet of lumber. Today, after 107 years in which a total of 2.25 billion board feet were harvested, the forest stock has increased to 1.7 billion board feet.

Ironically, the successful forestry operations almost brought about an end to the tribe. By 1959, the Menominee had accumulated a $10 million surplus. The Bureau of Indian affairs declared them too wealthy for continued protected status. Congress officially terminated the tribe in 1960, distributing trust funds and land allotments to individual tribal members. Forestry and mill operations were turned over to a private corporation, which immediately began liquidating reserves and racking up debts. Tribal leaders fought termination and successfully restored reservation status in 1973. Although the forest remains largely undivided, tribal enterprises still are plagued by debts incurred under privatization.

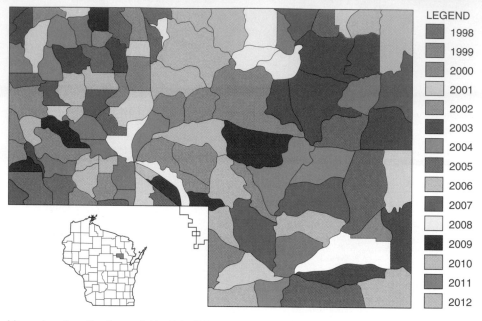

Menominee forestlands are divided into 109 compartments based on species composition, topography, stand history, and management goals. Legend shows planned year for harvest in 15-year cycle. Courtesy of Menominee Tribal Enterprises. Reprinted by permission.

LEGEND
1998
1999
2000
2001
2002
2003
2004
2005
2006
2007
2008
2009
2010
2011
2012

Wise elders set up a simple forest management plan when they began operations a century ago. Rather than manage for short-term yields, as is the case for lands around them, the tribe aims for maximum quantity and quality of native species. They say, "instead of cutting the best, we cut the worst first. We're managing our resources to last forever." They have one of the few lumber operations in the country that preserves old-growth characteristics.

The forest is northern hardwood type with a mixture of sugar maple, beech, hemlock, basswood, yellow birch, white pine, jack pine, and aspen. The heart of the management plan is a continuous forest inventory to determine optimum growth, species balance, ecological health, and cutting cycles. Land is divided into 109 compartments based on 11 species combinations, topography, stand history, and management goals. Two-thirds of the forest is managed for mixed species and ages, with selective cutting on a 15-year cycle. About 20 percent is devoted to aspen and jack pine in even-age (clear-cut) stands of no more than 12 ha (30 acres) each. Nearly 400 ha (1,000 acres) of white pine utilize a two-step shelterwood program that mimics the natural fire-succession sequence by artificially manipulating the balance of sunlight, competition, and soil disturbance. Judicial use of herbicides, prescribed burns, selective cutting, and rock raking maintain optimum growth and regeneration of this valuable species, which has largely disappeared elsewhere in the Great Lakes forest.

Some 300 people are now employed in the tribal forestry and sawmill operations. Logging is carried out by both Indian and non-Indian private contractors. Current harvest levels are about 30 million board feet per year. Lumber from the tribal mill is certified by the Green Cross organization as "good wood" harvested in a socially and environmentally responsible manner.

As Aldo Leopold said in *Sand County Almanac,* the best definition of conservation "is written not with a pen, but with an axe. It is a matter of what a man thinks about while chopping, or while deciding what to chop." He was describing the kind of stewardship practiced on the Menominee reservation. In addition to welcome economic returns, the sustainable harvesting has brought them an aesthetically pleasing forest, spiritual rejuvenation, clean water, and a sense of pride in being Menominee.

reforestation campaign has been started. An average of 4.5 million ha per year were replanted during the last decade. South Korea also has had very successful forest restoration programs. After losing nearly all its trees during the civil war 30 years ago, the country is now about 70 percent forested again.

In spite of being the world's largest net importer of wood, Japan has increased forests to approximately 68 percent of its land area. Strict environmental laws and constraints on the harvesting of local forests encourage imports so that Japan's forests are being preserved while it uses those of its trading partners. It is estimated that two-thirds of all tropical hardwoods cut in Asia are shipped to Japan.

Many reforestation projects involve large plantations of a single species, such as eucalyptus or hybrid poplar, in a single-use, intensive cropping called **monoculture forestry.** Although this produces high profits, a dense, single-species stand encourages pest and disease infestations. This type of management lends itself to mechanized clear-cut harvesting, which saves money and labor but tends to leave soil exposed to erosion. Monocultures eliminate habitat for many woodland species and often disrupt ecological processes that keep forests healthy and productive. When profits from these forest plantations go to absentee landlords or government agencies, local people have little incentive to prevent fires or keep grazing animals out of newly planted areas. In some countries, such as the Philippines, Israel, and El Salvador, government reforestation projects have been targets for destruction by antigovernment forces, with devastating environmental impacts.

Promising alternative agroforestry plans are being promoted by conservation and public service organizations such as The New Forest Fund and Oxfam. These groups encourage planting of mixed species, community woodlots including fruit and nut trees as well as fast-growing, multipurpose trees such as *Leucaena*. Millions of seedlings have been planted in hundreds of self-help projects in Asia, Africa, and Latin America. *Leucaena* is a legume, so it fixes nitrogen and improves the soil. Its nutritious leaves are good livestock fodder. It can grow up to 3 meters per year and quickly provides shade, forage for livestock, firewood, and good lumber for building. *Leucaena* can be an aggressive, weedy exotic, however, if it escapes from cultivation. As in most environmental and social solutions, a combination of species and methods are needed. Community woodlots can be planted on wasteland or along roads or slopes too steep to plow so they do not interfere with agriculture. They protect watersheds, create windbreaks, and, if composed of mixed species, also provide useful food and forest products such as fruits, nuts, mushrooms, or materials for handicrafts on a sustained-yield basis.

TROPICAL FORESTS

The richest and most diverse terrestrial ecosystems on the earth are the tropical forests. Although they now occupy less than 10 percent of the earth's land surface, these forests are thought to contain more than two-thirds of all higher plant biomass and at least one-half of all plant, animal, and microbial species in the world.

Diminishing Forests

Worldwide, we lost between 9 million and 12 million ha of forest per year from 1990 to 2000. The greatest losses were in tropical Africa (5.3 million ha/yr) and South America (4.4 million ha/yr), according to the United Nations Food and Agriculture Organization (FAO). The FAO estimates that about 0.8 percent of the remaining tropical forest is cleared every year (fig. 12.8). The species extinctions due to this loss of habitat are discussed in chapter 13.

There is considerable uncertainty about the rate of deforestation in the tropics because detecting and interpreting change is difficult in vast, remote regions. Satellite data often provide the best available estimates, although there is room for error in these data when second-growth and old-growth forests are hard to distinguish from space. Uncertainty also exists because definitions of deforestation differ. Some scientists and politicians consider an area deforested only when it is completely converted to agriculture, urban areas, or desert. Others call any logged area deforested, even if logging was selective and regrowth will be rapid. Countries also have political and economic reasons to exaggerate or understate the extent of their activities. Consequently, estimates of total tropical forest losses range from about 5 million ha to more than 20 million ha per year. The FAO recently revised its estimate from 14 million ha to 9 million ha per year by including open savannas and cultivated plantations in its definition of forests.

The Congo and Amazon river basins have the highest rates of deforestation in the world. Coastal forests in Central Africa have been logged for decades, but logging in the Congo basin is accelerating. The Congo region now loses some 4 million ha of forests

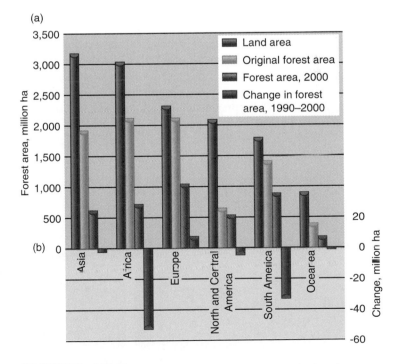

FIGURE 12.8 Regional distribution of original and current forests (a) and regional forest losses, 1990 to 2000 (b). Europe includes Russia's Siberian forests. *Source:* United Nations Food and Agriculture Organization.

FIGURE 12.9 Cattle graze on recently cleared tropical rainforest land. Millions of hectares have been cleared for pasture and cropland in recent decades. Unfortunately, the soil is poorly suited to grazing or farming, and new land must be cleared in a few years.
© William P. Cunningham.

per year. Much of this loss results from logging to produce wood products in Europe, but expanding settlements and farms are also responsible. In Brazil, satellite images from August 2002 indicated that Amazon rainforest area had shrunk by 26,000 km² (2.6 million ha) in the previous 12 months. Alarmingly, this rate was up 40 percent from the previous year. About 16 percent of Brazil's original forest is gone. This clearing results from logging, especially for prize tropical hardwoods, from settlement by land-hungry farmers (fig. 12.9), and more recently from an expansion of soybean farming for livestock and for export (fig. 12.10).

A variety of causes lead to deforestation. Indonesia loses an estimated 2.1 million ha of rainforest per year, and has already lost 80 percent of its lowland forests, mainly to make a way for oil palm and fruit plantations. Forests there are also frequently burned to disguise illegal logging activities. During dry seasons, fires in Borneo and Sumatra can make the air unbreathable in Singapore, Malaysia, and the Philippines. In Central America, nearly two-thirds of the original moist tropical forest has been destroyed, mostly within the past 30 years and primarily due to conversion of

forest to banana plantations and cattle range (fig. 12.11). In West Africa, Colombia, and Ecuador, forests are threatened mainly by oil and gas exploration, which involve extensive road building and land clearing. Settlers and farmers quickly follow as roads are built into the forests.

A number of tropical countries have already lost most of their original forests. Haiti was once 80 percent forested, but today the country is nearly all deforested. India, Burma, Cambodia, Thailand, and Vietnam retain little of their once-expansive forests.

Swidden Agriculture

Indigenous forest people often are blamed for tropical forest destruction because they carry out shifting agriculture that requires forest clearing. Actually, this ancient farming technique can be an ecologically sound way of obtaining a sustained yield from fragile tropical soils if it is done carefully and in moderation. This practice is sometimes called "slash and burn" by people who don't realize how complex and carefully balanced this method of farming actually can be when practiced sustainably. The preferred terms of **milpa** or **swidden agriculture** are taken from local names for *field*.

In this system, farmers clear a new plot of about a hectare (2.5 acres) each year. Small trees are felled and large trees are killed by girdling (cutting away a ring of bark) so that sunlight can penetrate through to the ground. After a few weeks of drying, the branches, leaf litter, and fallen trunks are burned to prepare a rich seedbed of ashes. Fast-growing crops, such as bananas and papayas, are planted immediately to control erosion and to shade root crops, such as cassava and sweet potato, which anchor the soil. Maize, rice, and up to 80 other crops are planted in a riotous profusion. Although they would not recognize the terms, these indigenous people are practicing **mixed perennial polyculture.**

The diversity of the milpa plot mimics that of the jungle itself, even though the species representation is more restricted. When managed well, the soil is covered with vegetation. This variety means that crops mature in a staggered sequence and there is almost always something to eat from the plot. It also helps prevent eruptive insect infestations that would plague a monoculture crop. Annual yields from a single hectare can be as high as 6 tons of

1975

1986

1992

FIGURE 12.10 Roads built for logging encourage further clearance for settlements and farming. These satellite images of Rondonia, Brazil, show the classic "fishbone" pattern of expanding clearance. *Source:* U.S. Geological Survey Earth Shots.

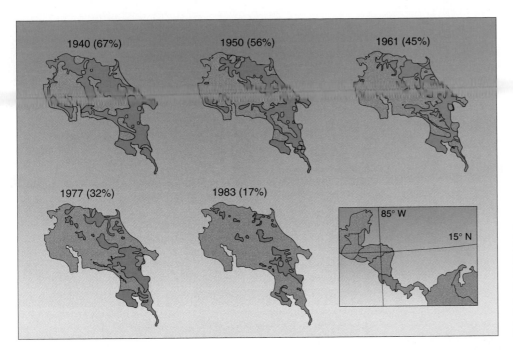

FIGURE 12.11 Loss of primary forest in Costa Rica, 1940 to 1983. Percentages show forestland as a proportion of total land area. Costa Rica now protects most of its remaining primary forests in national parks or other protected areas.

Sources: Sader and Joyce, "Deforestation Rates and Trends in Costa Rica, 1940–1983," *Biotropica* 20:11–19; and T. C. Whitmore and G. T. Prance, *Biogeography and Quaternary History in Tropical America,* 1987, Clarendon Press, Oxford.

grain (maize or rice) and another 5 tons of roots, vegetable crops, nuts, and berries. This yield is comparable to the best results with intensive row cropping, and about 1,000 times as much food is produced from the same land as when it is converted to cattle pasture.

After a year or two, the forest begins to take over the garden plot again. The farmer will continue to harvest perennial crops for a while and will hunt for small animals that are attracted to the lush vegetation. Ideally, the land then will be allowed to remain covered with jungle vegetation for 10 to 15 years while nutrients accumulate before it is cleared and replanted again. In many places, population growth and displacement of farmers from other areas have forced shifting agriculturalists to reuse their traditional plots on shorter and shorter rotation. When plots are farmed every year or two, nutrients are lost faster than they can be replaced, the forest doesn't regrow as vigorously as it once did, and additional species are lost. Eventually, erosion and overuse reduce productivity so much that the land becomes practically useless.

Logging and Land Invasions

The other major source of forest destruction is usually the result of logging and subsequent invasion by land-hungry people from other areas. The loggers often are interested only in "creaming" the most valuable hardwoods, such as teak, mahogany, sandalwood, or ebony. Although only one or two trees per hectare might be taken, widespread devastation usually results. Because the canopy of tropical forests is usually strongly linked by vines and interlocking branches, felling one tree can easily bring down a dozen others. Tractors dragging out logs damage more trees, and construction of roads takes large land areas. Insects and infections invade wounded trees. Tropical trees, which usually have shallow root systems, are easily toppled by wind and erosion when they are no longer supported by their neighbors. Up to three-fourths of the canopy may be destroyed for the sake of a few logs. Obviously, the complex biological community of the layered canopy (chapter 4) is severely disrupted by this practice.

What happens next? Bulldozed roads make it possible for large numbers of immigrants to move into the forest in search of land to farm. People with little experience or understanding of the complex rainforest ecosystem try to turn it into farms and ranches (fig. 12.12). Too often, the result is ecological disaster. Rains wash away the topsoil and the tropical sun bakes the exposed subsoil into an impervious hardpan that is nearly useless for farming.

Degradation of rivers is another disastrous result of forest clearing. Tropical rivers carry two-thirds of all freshwater runoff in the world. In an undisturbed forest, rivers are usually clear, clean, and flow year-round because of the "sponge effect" of the thick root mat created by the trees. When the forest is disrupted and the thin forest soil is exposed, erosion quickly carries away the soil, silting river bottoms, filling reservoirs, ruining hydroelectric and irrigation projects, filling estuaries, and smothering coral reefs offshore. In Malaysia, sediment yield from an undisturbed primary forest was about 100 m^3 per km^2 per year. After forest clearing, the same river carried 2,500 m^3 per km^2 per year.

Forest Protection

What can be done to stop this destruction and encourage careful management? While much of the news is discouraging, there are some hopeful signs for tropical forest conservation. Many tropical countries have realized that forests represent a valuable resource and they are taking steps to protect them. Indonesia has announced plans to preserve 100,000 km^2, one-tenth of its original forest. Zaire and Brazil each plan to protect 350,000 km^2 (about the size of Norway) in parks and forest preserves. In the past several years, Gabon, Congo, South Africa, and other African countries have established new parks and forest preserves (see opening story).

FIGURE 12.12 Logging roads open up access to landless settlers, miners, and hunters who drive away native species and indigenous people. There often is a marked difference between new migrants and indigenous swidden agriculturists. © The McGraw-Hill Companies, Inc./Barry W. Barker, photographer.

People on the grassroots level also are working to protect and restore forests. Reforestation projects build community pride while also protecting the land (fig. 12.13). India, for instance, has a long history of nonviolent, passive resistance to protest unfair government policies. The *satyagrahas* go back to the beginning of Indian culture and often have been associated with forest preservation. Gandhi drew on this tradition in his protests of British colonial rule in the 1930s and 1940s. During the 1970s, commercial loggers began large-scale tree felling in the Garhwal region in the state of Uttar Pradesh in northern India. Landslides and floods resulted from stripping the forest cover from the hills. The firewood on which local people depended was destroyed, and the way of life of the traditional forest culture was threatened. In a remarkable display of courage and determination, the village women wrapped their arms around the trees to protect them, sparking the *Chipko Andolan* movement (literally, movement to hug trees). They prevented logging on 12,000 km² of sensitive watersheds in the Alakanada basin. Today, the *Chipko Andolan* movement has grown to more than 4,000 groups working to save India's forests.

FIGURE 12.13 Schoolchildren plant trees in a community reforestation project in Haiti. Grassroots efforts, such as this, benefit both local people and their environment. Many top-down development projects benefit only the ruling elite and multinational corporations. Courtesy Andy White, University of Minnesota, College of Agriculture.

Debt-for-Nature Swaps

Those of us in developed countries can make a contribution toward saving tropical forests as well. Financing nature protection is often a problem in developing countries where the need is greatest. One promising approach is called **debt-for-nature swaps.** Banks, governments, and lending institutions now hold nearly $1 trillion in loans to developing countries. There is little prospect of ever collecting much of this debt, and banks are often willing to sell bonds at a steep discount—perhaps as little as ten cents on the dollar. Conservation organizations buy debt obligations on the secondary market at a discount and then offer to cancel the debt if the debtor country will agree to protect or restore an area of biological importance.

There have been many such swaps. Conservation International, for instance, bought $650,000 of Bolivia's debt for $100,000—an 85 percent discount. In exchange for canceling this debt, Bolivia agreed to protect nearly 1 million ha (2.47 million acres) around the Beni Biosphere Reserve in the Andean foothills. Ecuador and Costa Rica have had a different kind of debt-for-nature swap. They exchanged debt for local currency bonds that are used to fund activities of local private conservation organizations in the country. This has the dual advantage of building and supporting indigenous environmental groups while protecting the land.

Agreements have been reached with Madagascar and Zambia to swap debts for nature, and negotiations are underway with Peru, Mexico, and Tanzania. Critics charge that these swaps compromise national sovereignty and that they will do little to reduce the debts of developing countries or change the situations that led to environmental destruction in the first place. In 2000, Conservation

International made an effort to remedy this problem. By buying long-term logging concessions for 81,000 ha (200,000 acres) of tropical rainforest in Guyana, the organization is bringing the same economic benefits that the government would get if the land were actually logged—without losing the natural resources of the forest.

TEMPERATE FORESTS

Tropical countries are not unique in harvesting forests at unsustainable rates and in an ecologically damaging fashion. Northern countries have a long history of liquidating forest resources that continues today in many places. Perhaps the largest and most destructive harvest in the world today is taking place in Eastern Russia. Siberia is larger than Amazonia and contains one-fourth of the world's timber reserves. Four million ha (10 million acres) of Siberian taiga and deciduous forests are being felled annually—primarily by Korean loggers—to shore up Russia's faltering economy. China has announced plans for a giant hydroelectric dam on the Amur River, which forms its border with Siberia. This power source will allow China to move 100 million settlers into the region, which is home to the endangered Siberian tiger and several species of cranes. The results are likely to be similar to those we have just discussed in reference to tropical forests.

In the United States and Canada, the two main issues in timber management are (1) cutting of the last remnants of old-growth forest and (2) methods used in timber harvest.

Ancient Forests of the Pacific Northwest

Only a century ago, most of the coastal ranges of Washington, Oregon, northern California, British Columbia, and southeastern Alaska were clothed in a lush forest of huge, ancient trees (fig. 12.14). The moist, mild climate and rich soil of the lowland valleys nurtured magnificent stands of redwood in California and of western red cedar, Douglas fir, western hemlock, and Sitka spruce along the rest of the coast. Everyone knows that redwoods can be huge and very old, but did you know that these other species can reach 3 to 4 m (9 to 12 ft) in diameter, 90 m in height (as high as a 20-story building), and 1,000 or more years in age? These temperate rainforests are probably second only to tropical rainforests in terms of terrestrial biodiversity, and they accumulate more total biomass in standing vegetation per unit area than any other ecosystem on earth.

These old-growth temperate rainforests are extremely complex ecologically. Only in recent years have we begun to realize how many different species live there and how interrelated their life cycles are. Many endemic species such as the northern spotted owl (fig. 12.15), Vaux's swift, and the marbled murrelet are so highly adapted to the unique conditions of these ancient forests that they live nowhere else.

Before loggers and settlers arrived, there were probably 12.5 million ha (31 million acres) of virgin temperate rainforest in the Pacific Northwest. Less than 10 percent of that forest in the United States still remains, and 80 percent of what is left is scheduled to be cut down in the near future. British Columbia has felled at least

FIGURE 12.14 This forest in Oregon's Opal Creek Valley is an excellent example of the old-growth temperate rainforests of the Pacific Northwest. © William P. Cunningham.

FIGURE 12.15 Only about 2,000 pairs of northern spotted owls remain in the old-growth forests of the Pacific Northwest. Cutting old-growth forests threatens the endangered species, but reduced logging threatens the jobs of many timber workers. © Greg Vaughn/Tom Stack & Associates.

60 percent of its richest and most productive ancient forests and is now cutting some 240 million ha (600 million acres) annually, about ten times the rate of old-growth harvest in the United States. At these rates, the only remaining ancient forests in North America in 50 years will be a fringe around the base of the mountains in a few national parks.

Wilderness and Wildlife Protection

Many environmentalists would like to save all remaining virgin forest in the United States as a refuge for endangered wildlife, a laboratory for scientific study, and a place for recreation and spiritual renewal. Economic pressures to harvest the valuable giant trees are considerable, however. The forest products industry employs about 150,000 people in the Pacific Northwest (2–3 percent of jobs) and adds nearly $7 billion annually to the economy. In Oregon, forest products account for one-fifth of the gross state product and many small towns depend almost entirely on logging for their economic life. On the other hand, an economic development study of Washington and Oregon suggests that recreation could provide 16 jobs for every one lost by logging. The biggest problem is how people who are currently employed in the timber industry can make a living during a transition period.

In 1989, environmentalists sued the U.S. Forest Service over plans to clear-cut most of the remaining old-growth forest, arguing that spotted owls are endangered and must be protected under the Endangered Species Act. Federal courts agreed and ordered some 1 million ha (2.5 million acres) of ancient forest set aside to preserve the last 1,000 pairs of owls. This is about half the remaining virgin forest in Washington and Oregon. The timber industry claims that 40,000 jobs have been lost, although environmentalists dispute this number.

Environmentalists also point out that logging jobs are disappearing because of mechanization, a naturally dwindling resource base, and the shipping of raw logs to Japan. They argue that the big trees are disappearing anyway. The question is when to stop cutting—now while a few remain, or in a few years when they are all gone? The workers will have to be retrained anyway; why not sooner rather than later? In spite of complaints that logging restrictions on public land hurt business, companies with private land holdings have benefited from rising timber prices. Weyerhauser, for instance, reported a $90 million (81 percent) increase in the value of its timber holdings in 1992 as a result of the reduced cut in federal forests.

After more than a decade of public meetings and studies, a compromise forest management plan was developed for the Pacific Northwest in 2000. This plan would allow some continued cutting but also would protect a high percentage of prime ancient forests. The plan would preserve nearly a million ha (2.2 million acres) of riparian (streamside) corridors to protect salmon and other aquatic wildlife, as well as 2 million ha (4.9 million acres) of "old-growth" reserves on national forest lands. In 2003, however, the Bush administration largely rewrote this plan, eliminating wildlife surveys and reducing environmental review for logging. The ultimate fate of these forests remains uncertain.

Although current plans protect only a fraction of the original forest, local residents in many areas resent *any* restrictions on their access to resources on public or private lands. "Wise Use" groups have brought lawsuits and introduced federal, state, and local ordinances to prohibit programs that limit land use or to require repayment of losses that result from land-use regulations (What Do You Think? p. 251).

Harvest Methods

Most lumber and pulpwood in the United States and Canada currently is harvested by **clear-cutting,** in which every tree in a given area is cut regardless of size (fig. 12.16). This method enables large

FIGURE 12.16 This huge clear-cut in Washington's Gifford Pinchot National Forest threatens species dependent on old-growth forest and exposes steep slopes to soil erosion. Restoring something like the original forest will take hundreds of years. © Gary Braasch/Stone/Getty.

machines to fell, trim, and skid logs rapidly, but wastes many small trees, exposes soil to erosion, and eliminates habitat for many forest species. Early successional species, such as Douglas fir, jack pine, lodgepole pine, or loblolly pine, on the other hand, flourish after clear-cutting. In a forest managed for these trees, or for raspberries, blueberries, deer, or grouse, this is a good method.

Size and shape of clear-cuts vary depending on topography and management policies. Clear-cuts can be in individual strips, alternating rows, small scattered patches, or areas as large as thousands of acres. Natural regeneration is better in small patches or strips that resemble natural forest openings than in huge clear-cuts. Small patches also are less disruptive for wildlife than are huge denuded spaces. It was once thought that good forest management required immediate removal of all dead trees and logging residue. Research has shown, however, that standing snags and coarse woody debris play important ecological roles, including soil protection, habitat for a variety of organisms, and nutrient recycling.

Other harvest practices offer variations on, or substitutes to, clear-cutting. *Coppicing* is used to encourage stump sprouts from species such as aspen, red oak, beech, or short-leaf pine and is usually accomplished by clear-cutting. In *seed tree harvesting,* some mature trees (generally two to five trees per hectare) are left standing to serve as a seed source in an otherwise clear-cut patch. For many forest types, the least disruptive harvest method is **selective cutting,** in which only a small percentage of the mature trees are taken in each 10- or 20-year rotation. Ponderosa pines, for example, are usually selectively cut to thin stands and to improve growth of the remaining trees. A forest managed by selective cutting can retain many of the characteristics of age distribution and ground cover of a mature, old-growth forest.

What do you think?

Regulations and Property Rights

Should private property owners be entitled to compensation when environmental regulations restrict how their land may be used? That is the subject of an ongoing debate in Congress and across the country. The issue is referred to as "takings." The matter is rooted in the Fifth Amendment to the United States Constitution, which states that "... nor shall private property be taken for public use, without just compensation."

This requirement imposed on the government was designed to make sure that when private land was needed for a public purpose, such as a highway, the government could not simply seize the land, but had to pay for it. A growing number of landowners are claiming that environmental restrictions reduce the potential economic gain from their land. They argue that this lost potential is comparable to "taking" their property, and the Fifth Amendment prohibits this "without just compensation."

Over the years, landowners have brought a variety of lawsuits seeking compensation for lost profits. Pennsylvania passed a mine safety act requiring coal companies to leave part of the coal seams in place underground beneath homes and other surface users to prevent highways,

buildings, and gas pipelines from collapsing. A coal company claimed in court that this constituted a "taking" because of the profits lost on unmined coal, and that they should be compensated for these losses.

A rancher in Nevada sued the government to recover the value of grass and water eaten on public land by wild elk herds, claiming this use deprived his cattle of these resources. In California, agricultural interests are concerned about plans to use water to reestablish salmon runs. They are preparing to sue if this reduces subsidized irrigation water available to them. An Arkansas tavern owner even sued the state for income lost because its drunk driver checkpoints caused people to drink less!

Often, courts have rejected these suits. Property owners have never been viewed as having unrestricted rights of use. Court decisions have held that property owners do not have a right to use property in ways that harm the community interest as a whole, that the Fifth Amendment does not guarantee owners the right to use land in the most profitable way conceivable, and that use restrictions may be imposed for a wide range of environmental and land-use purposes. Newly conservative judges, however, have ruled against state and federal governments. The balance of private and public rights is increasingly open to question.

Mining, logging, and land development interests have also sought legislation to restrict the scope of government regulations. Some of these bills would require government agencies to do a thorough economic analysis of any environmental regulations that would affect either the use or value of privately owned property.

These efforts are opposed by many environmental, consumer, civil rights, and labor groups. They contend that many of the proposals go far beyond constitutional provisions and would paralyze the government's ability to protect the public interest in a clean environment and healthful surroundings. They fear that the bills being proposed would prevent enforcement of many antipollution laws, zoning ordinances, and even obscenity laws (one dial-a-porn company sued the federal government for damages due to rules that restrict the access of children to its services).

In many respects, the controversy over takings represents a classic conflict over two kinds of freedoms: the freedom to act as you like and the right to be protected from the damaging actions of others.

Where should the balance point be between these two types of conflicting rights? To what extent should property rights include the right to negatively impact the rights of others? How would you decide?

The lush Douglas fir and redwood forests of the rainy Pacific Coast Range provide a vivid illustration of why clear-cutting in old-growth forests is controversial. The rich soil and mild, moist climate of the region promote rapid regeneration after clear-cutting, but many of these forests are on steep slopes where erosion is a serious problem. Hillsides are stripped of soil, which fills streams and smothers aquatic life. British Columbia has felled at least 60 percent of its richest and most productive ancient forests. About 5 percent of the province—but only 2 percent of the temperate rainforest—is preserved in parks, ecological reserves, or recreational areas.

Canadian First Nation people have blocked roads and brought lawsuits to protest destruction of traditional lands and subsistence ways of life. People concerned about commercial and sport fishing have joined the battle, both in British Columbia and in the United States. They argue that salmon spawning depends on the clear cold streams of the native forests. The income from a single year's

salmon run can outweigh all the profits from timber harvesting, and the salmon return year after year, while 1,000-year-old trees will never be seen again.

U.S. FOREST MANAGEMENT

Forest management is a subject of perennial, often rancorous debate in U.S. politics. Since its inception, the U.S. Forest Service's primary mission has been to provide cheap timber for home builders. On this principle, the federal government (that is, tax payers) pay to manage forests, build roads to extract trees, and clean up streams and logging debris. All these subsidies help make logging operations profitable—in order to ensure that there is a steady supply of wood in the marketplace. In its effort to make sure that logging operations remain profitable, though, the Forest Service frequently sells logging rights in an area for much less than the

FIGURE 12.17 Many communities depend on timber for survival. Logging can threaten other sources of income, such as salmon fishing, however. © William P. Cunningham.

cost of managing that area or building roads to extract the timber. Below-cost timber sales have cost the federal government billions of dollars since the 1970s. The nonpartisan U.S. Congressional Research Service reported that in 1997 alone, the net loss on timber sales was $1.2 billion, not counting the costs of flood damage, lost wildlife habitat, and lost tourism revenue. Of the 104 national forests, 83 lost money on timber sales. The timber industry insists that this system is necessary and that it provides high-paying jobs in rural communities. Environmentalists point out that we could end all timber sales on federal land and provide $30,000 per year for retraining or forest restoration for every public land timber worker and still be ahead financially.

Part of the reason subsidized logging is so expensive is that it requires a dense network of roads (fig. 12.17). The Forest Service builds and manages more roads than any other agency in the United States. Since the 1960s, it has built more than half a million miles of roads, ten times the length of the entire U.S. interstate highway system. Wilderness advocates argue that this road-building causes soil erosion, clogs streams with debris, destroys wildlife habitat, spreads invasive weeds, and encourages people to build homes in the woods—where they are vulnerable to forest fires. Forestry advocates insist the roads provide public access to otherwise inaccessible country.

Forest policy is increasingly controversial because the agency has evolved new missions that conflict with its traditional job of providing lumber and jobs. Now we also expect forests to provide recreation, including wilderness hiking and camping, and wildlife habitat. In fact, 75 percent of all USFS jobs involve recreation, compared to just 3 percent of jobs involved in logging. Both wilderness designation and back-country recreation require roadless areas, in direct contradiction of pro-road forest policy since the 1960s. In 2000, the Forest Service's first-ever nonindustry chief, Mike Dombeck, initiated several major changes in forest policy. He banned new roads on 58 million acres of "de facto" wilderness, that is, roadless areas that are not officially designated as wilderness areas. And he blocked new sales of old-growth forests. He also directed forest planners to consider ecosystem health and sustainability in their management plans, in addition to timber yield. Although the public was overwhelmingly in favor of those new policies, most have been reversed.

Fire Management

Another focus of controversy in forest management is fires. Since the 1930s, forest managers, with the help of Smokey Bear, have made it a priority to prevent fires in national forests. It has become common wisdom that forest fires are dangerous, wasteful of resources, and destructive of natural beauty. The public demands fire protection, including protection for homes built deep in the forest and far from towns. In 2002, the United States spent about $1.6 billion to fight forest fires—$1 billion more than had been budgeted. Images in the news of sobbing homeowners reinforce public enthusiasm for fire-fighting, as do images of helicopters and National Guard members called in to wage a military-style battle against the flames. For 2004, the budget for fire-fighting is $2.2 billion, four times the previous ten-year average. The cost of fire-fighting has mushroomed partly because there are more remote homes to protect, partly because public expectations have grown to demand more fire suppression, and partly because there have been some very big fires in recent dry summers. In 2002, one of the biggest fire years on record, wildfires burned (to some extent) nearly 2.8 million ha of forests and grasslands in the United States.

Ecologists increasingly recognize that fires are a normal part of a healthy forest. In past centuries, most forests endured periodic fires. Most would burn near the ground, clearing the undergrowth but leaving mature trees with little damage. Fire suppression has eliminated these small fires, but it has increased the chances of huge, catastrophic fires by allowing understory fuel to accumulate. Historically only 20 percent of forest fires were large enough to kill mature trees, according to a study in Arizona; that figure has grown to 40 to 50 percent of fires now (fig. 12.18).

After a fire, many industry advocates encourage **salvage logging,** removal of dead or dying trees from forests damaged by fire or disease. In 1995, Congress passed a budget bill containing a rider (a hidden, unrelated rule slipped into a large, urgent piece of legislation) allowing salvage logging to bypass ordinary environmental reviews or logging limits. Proponents argue that salvage logging is an emergency action necessary to protect forest health in the aftermath of disease or fire, to prevent waste of valuable timber, and to replace jobs lost to restrictions such as the northern spotted owl protection plan. Opponents of salvage logging were outraged by the rider. They insisted that standing dead trees provide essential habitat for wildlife and that forests regenerate rapidly when undisturbed. Bowing to local pressure for economic activity, the Forest Service quickly approved hundreds of sales under this special rider. Many are in the old-growth forests that would otherwise be off-limits, and many healthy trees have been cut in the name of salvage.

Forest Policy and the Bush Administration

The election of George W. Bush in 2000 brought about major changes in forest management. Road building, fire suppression, old-growth extraction, and salvage logging, all reduced during the previous Clinton administration, have been pursued with a new vigor. Protection was quickly removed from the 23 million ha (58 million acres) of "de facto" wilderness declared off-limits to road building

FIGURE 12.18 By suppressing fires and allowing fuel to accumulate, we make major fires such as this more likely. The safest and most ecologically sound management policy for some forests may be to allow natural or prescribed fires, that don't threaten property or human life, to burn periodically. Courtesy John McColgan, Alaska Fires Service/Bureau of Land Management.

in 1998. The court-ordered Northwest Forest Plan, which required that the Forest Service examine the impacts of old-growth logging on wildlife in the Pacific Northwest, was abandoned. At the same time, Interior Secretary Gale Norton directed the Bureau of Land Management (BLM) to abandon its plan to protect 2.4 million ha of roadless wilderness in Utah. The BLM has been directed to undertake no more studies of possible wilderness areas, and these lands are to be made available for logging, mining, and motorized recreation.

The Bush administration has been very supportive of the logging, road building, and mining industries on forest lands. Salvage logging and forest "thinning"—removing trees in advance of possible fires—have been central to the administration's forest policies. Normally, forest thinning, like other logging operations, is subject to standard environmental reviews under the National Environmental Policy Act (NEPA), and there is a public comment period so that citizens can weigh in on contract decisions. A major Bush change of 2003, however, was the "Forest Health Initiative." Among other things, this rule speeds up "thinning" operations by directing federal agencies to identify lands in danger of fire, eliminating environmental review, and abandoning public comment periods. To ensure that these operations are profitable for timber companies, loggers can harvest large, valuable, and fire-resistant trees deep in the back country, rather than just fire-prone deadwood near settlements.

Environmental groups complain that these new rules eliminate public oversight. Even if opponents have the money to sue over a forest-thinning contract, the project will be finished before a complaint gets to court. They argue that thinning contracts are simply logging in disguise, allowing loggers to remove ecologically valuable trees in remote areas otherwise off-limits to harvest. A demonstration project in the Coconino National Forest shows that thinning operations, as currently planned, create open space that will quickly fill in with dense, fire-prone shrubs (fig. 12.19). Ironically, this management may make forests more fire-prone, not less. Researchers for the Sierra Nevada Ecosystem Report wrote, "Timber harvest,

FIGURE 12.19 Demonstration forest thinning (or fuel reduction) project in the Coconino National Forest, Arizona. "Exactly this kind of treatment has to happen across the western U.S.," said Forest Service chief Dale Bosworth. "We have only 20 years to treat 30 million acres." © Martos Hoffman, photographer.

through its effects on forest structure, local microclimate, and fuel accumulation, has increased fire severity more than any other recent activity, including fire suppression." Many forest ecologists maintain that small prescribed burns (fires carefully planned and carried out on humid, windless days) can reduce fuel and produce healthy forests more effectively than commercial logging.

Other research questions the main justification for thinning: protecting houses that have proliferated around the edges of national forests. The Fire Sciences Laboratory in Missoula, Montana, the premier laboratory studying fire behavior, recently found that even in a tinder-dry forest, the most effective fire protection was clearing just 60 m (200 ft) immediately around buildings. Building metal roofs and clearing pine needles and brush around the house are also helpful. Forest thinning beyond 60 m is essentially irrelevant.

Lowering Our Forest Impacts

Americans throw away 30 million trees' worth of newspaper every year. Your habits and purchases affect the health of world forests. Here are some ways you can make a difference.

- Reuse and recycle paper. Make double-sided copies. Save office paper and use the back for scratch paper. Buy recycled paper.

- Use e-mail. Store information in digital form, and avoid printing messages.

- If you build, conserve wood. Use wafer board, particle board, laminated beams or other composites rather than plywood and timbers made from old-growth trees.

- Buy products made from "good wood" or other certified sustainably harvested wood.

- Don't patronize fast-food restaurants that purchase beef from cattle grazing on deforested rainforest land. Don't buy coffee, bananas, pineapples or other cash crops if their production contributes to forest destruction.

- Do buy Brazil nuts, cashews, mushrooms, rattan furniture, and other non-timber forest products harvested sustainably by local people from intact forests. Learn about sustainability certification from the Forest Stewardship Council (www.fscus.org).

- Stay informed about resource and land-use policies, and let your elected representatives know what you think.

Opponents of forest thinning also worry that it is another disguise for below-cost timber sales. A recent Forest Service study found that the cost of thinning the 1.6 million acres of forest in the Klamath Mountains of southeastern Oregon would be $2.7 billion, more than $1,685 per acre, and more than the entire fire-fighting budget for 2004.

Sustainable Forestry and Non-Timber Forest Products

Creative solutions to forest management problems are available. In both temperate and tropical regions, scores of certification programs are being developed to identify sustainably produced wood products. One organization that is currently active in 40 countries is the Forest Stewardship Council (FSC). The FSC works to set standards for certification. SmartWood, a program of the Rainforest Alliance, is the most extensive certification program. This organization works with both tropical and temperate forest products companies. One of the promising movements in North American forestry is the development of cooperatives and networks among private landowners. In the United States alone, there are more than 9 million owners of small (less than 100 acres) forest lands. Groups such as the Community Forestry Resource Center are sharing information and resources to assist in sustainable management of small working forests like these.

Consumer preferences play a role in forest protection (see this chapter's opening story). In 2003, Home Depot adopted a policy of buying wood products only from suppliers committed to environmentally friendly logging and lumber practices. The retailer sells about $5 billion of wood products each year. The number of vendors providing Home Depot with products certified by the FSC grew from 5 in 1999 to 40 in 2000. In addition, Home Depot says that nearly all of the cedar it now buys comes from second- or third-generation forests, rather than old-growth. It also has cut purchases of Indonesian luan wood by 70 percent, because much of it is illegally logged. Staples, an office supply retailer with more than 1,000 stores, announced that it would increase the average amount of recycled content in its paper products from less than 10 percent to more than 30 percent.

Logging is not the only way to make a living in a forest. Increasingly, non-timber forest products are seen as an alternative to timber production. In the United States alone, a $3 billion natural plants industry depends on healthy forests. Non-timber forest products have been around for centuries: latex (rubber), chicle (gum), nuts, and many other products have long been gathered sustainably from tropical forests (fig. 12.20). Medicinal plants, fruits,

FIGURE 12.20 Non-timber forest products, such as natural rubber, can provide an income without destroying forests. © The McGraw-Hill Companies, Inc./Barry W. Barker, photographer.

rattan (vines), and other products can be developed from both temperate and tropical forests. The principal challenge is developing markets for these new products—something that is becoming easier with modern communication technology and niche marketing methods.

GRASSLANDS

After forests, grasslands are the environment most extensively used and altered by human activity. Although grasslands often appear monotonous to us, they can hold great biodiversity, and they support the world's great populations of grazing animals—from American bison to Africa's vast herds of wildebeest and many species of antelope. Grasslands also feed humans, through the milk and meat of livestock, or through the rich soils that make prime farmland.

Worldwide, grasslands naturally occur where rainfall is sufficient for drought-tolerant grasses and flowering plants but too slight, or too irregular, to support forests (fig. 12.21). Grasslands currently cover about 4 billion ha (12 million mi^2), or about 27 percent of the earth's land surface (see fig. 5.3). This represents about two-thirds of the grasslands that existed before the spread of agriculture.

Pasture (enclosed or managed grasslands) and **rangelands** (unfenced, natural prairie and open woodlands) make up about twice the area of the world's croplands. More than 3 billion cattle, sheep, goats, camels, buffalo, and other domestic animals on these lands turn plants into protein-rich human nutrition. Grasslands remain healthy and support rich herds of grazers when those herds are migratory. Grasslands deteriorate when herds become too abundant, or when they cannot move on and let plants recover. Overgrazing, soil erosion, and gradual degradation to scrub or desert eliminate millions of hectares of grasslands each year. We hear a great deal about tropical forest losses, but you might be surprised to learn that each year we lose nearly six times as much grasslands as tropical forest!

FIGURE 12.21 Grasslands are expansive, open environments that can support surprising biodiversity. © William P. Cunningham.

Range Management

By carefully monitoring the numbers of animals and the condition of the range, ranchers and pastoralists (people who live by herding animals) can adjust to variations in rainfall, seasonal plant conditions, and nutritional quality of forage to keep livestock healthy and avoid overusing any particular area. Conscientious management can actually improve the quality of the range.

Some nomadic pastoralists who follow traditional migration routes and animal management practices produce admirable yields from harsh and inhospitable regions. They can be ten times more productive than dryland farmers in the same area and come very close to maintaining the ecological balance, diversity, and productivity of wild ecosystems on their native range. Nomadic herding requires large open areas, however. Wars, political disputes, travel restrictions, incursions by agriculturalists, growing populations, and changing climatic conditions on many traditional ranges have combined to disrupt an ancient and effective way of life. The social and environmental consequences often are tragic.

Overgrazing and Land Degradation

About one-third of the world's range is severely degraded by overgrazing, making this the largest cause of soil degradation. Among the countries with the most damage and the greatest area at risk are Pakistan, Sudan, Zambia, Somalia, Iraq, and Bolivia. Usually, the first symptom of improper range management is elimination of the most palatable herbs and grasses. Grazing animals tend to select species they prefer and leave the tougher, less tasty plants. When native plant species are removed from the range, weedy invaders move in. Gradually, the nutritional value of the available forage declines. As overgrazing progresses, hungry animals strip the ground bare and their hooves pulverize the soil, hastening erosion (fig. 12.22).

The process of denuding and degrading a once-fertile land initiates a desert-producing cycle that feeds on itself and is called **desertification.** With nothing to hold back surface runoff, rain drains off quickly before it can soak into the soil to nourish plants or replenish groundwater. Springs and wells dry up. Trees and bushes not killed by browsing animals or humans scavenging for firewood or fodder for their animals die from drought. When the earth is denuded, the microclimate near the ground becomes inhospitable to seed germination. The dry barren surface reflects more of the sun's heat, changing wind patterns, driving away moisture-laden clouds, and leading to further desiccation. Because of this process, deserts have been called the footprints of civilization.

This process is ancient, but in recent years it has been accelerated by expanding populations and political conditions that force people to overuse fragile lands. Those places that are most severely affected by drought are the desert margins, where rainfall is the single most important determinant in success or failure of both natural and human systems (fig. 12.23). In good years, herds and farms prosper and the human population grows. When drought comes, there is no reserve of food or water and starvation and suffering are widespread. Can we reverse this process? In some places, people are reclaiming deserts and repairing the effects of neglect and misuse.

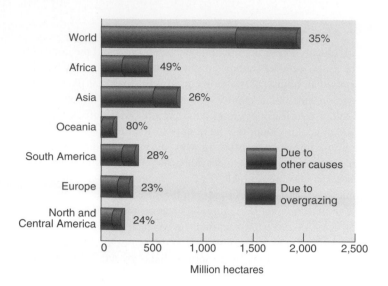

FIGURE 12.22 Rangeland soil degradation due to overgrazing and other causes. Notice that in Europe, Asia, and the Americas, farming, logging, mining, urbanization, etc., are responsible for about three-quarters of all soil degradation. In Africa and Oceania, where more grazing occurs and desert or semiarid scrub make up much of the range, grazing damage is higher. *Source:* World Resource Institute.

As is the case in tropical forests, estimates of the extent of desertification around the world vary widely. Many arid lands are prone to highly variable rainfall with long periods of drought in which natural vegetation disappears and the land lies barren for months or years, making it difficult to distinguish between human-caused changes and normal climatic variations. A recent survey by the United Nations Environment Program found that about half of Africa, for instance, has become significantly drier over the past half century and that arid and hyperarid areas—which are most susceptible to desertification—have increased by about 50 million hectares, while humid and semiarid lands have decreased by the same amount. According to the International Soil Reference and Information Centre in the Netherlands, nearly three-quarters of all rangelands in the world show signs of either degraded vegetation or soil erosion. The highest percentage of moderate, severe, and extreme land degradation is in Mexico and Central America.

Forage Conversion by Domestic Animals

Ruminant animals, such as cows, sheep, goats, buffaloes, camels, and llamas, are especially efficient at turning plant material into protein because bacterial digestion in their multiple stomachs allows them to utilize cellulose and other complex carbohydrates that many mammals (including humans) cannot digest. As a result, they can forage on plant material from which we could otherwise extract little food value. Different grazers have distinct feeding preferences and habits. Often the most effective use of rangelands is to maintain small mixed-species herds so that all vegetation types are utilized equally and none is overgrazed. Cattle and sheep, for instance, prefer grass and herbaceous plants, goats will browse on low woody shrubs, and camels can thrive on tree leaves and larger woody plants.

FIGURE 12.23 Sheep graze on land in Guatemala that was once tropical forest. Soils in the tropics are often thin and nutrient-poor. When forests are felled, heavy rains carry away the soil, and fields quickly degrade to barren scrubland. In some areas, land clearing has resulted in climate changes that have turned once lush forests into desert. © Eastcott/Momatiuk/The Image Works.

Worldwide, 85 percent of the forage for ruminants comes from native rangelands and pasture. In the United States, however, only 15 percent of livestock feed comes from native grasslands. The rest is made up of crops grown specifically for feed, particularly alfalfa, corn, and oats. Grain surpluses and our taste for well-marbled (fatty) meat have shifted livestock growing in the developed countries to feedlot confinement and high-quality diets. In the United States, roughly 90 percent of our total grain crop is used for livestock feed.

Harvesting Wild Animals

A few people in the world still depend on wild animals for a substantial part of their food. About one-half of the meat eaten in Botswana, for instance, is harvested from the wild. There are good reasons to turn even more to native species for a meat source. On the African savanna, researchers are finding that springbok, eland,

FIGURE 12.24 Native species, such as these American bison, are a valuable part of grassland ecosystems; they also may be the best way to harvest biomass for human consumption. Native species often forage more efficiently, resist harsh climates, and are more pest- and disease-resistant than domestic livestock. © Corbis Website.

impala, kudu, gnu, oryx, and other native animals forage more efficiently, resist harsh climates, are more pest- and disease-resistant, and fend off predators better than domestic livestock. Native species also are members of natural biological communities and demonstrate niche diversification, spreading feeding pressure over numerous plant populations in an area. A study by the U.S. National Academy of Sciences concluded that the semiarid lands of the African Sahel can support only about 20 to 28 kg (44 to 62 lbs) of cattle per hectare but can produce nearly three times as much meat from wild ungulates (hooved mammals) in the same area. In the United States, native bison and elk are raised as "novelty" meat sources; might there be a greater future role for ranching of wild animals on our own rangelands (fig. 12.24)?

Rangelands in the United States

The United States has approximately 319 million ha (788 million acres) of rangeland. Most of this rangeland is in the West, and about 60 percent is privately owned. Of the 120 million cattle and 20 million sheep in the United States, only about 2 percent of the cattle and 10 percent of the sheep graze on public rangelands. Federal lands are not very important, overall, in livestock production, but they do have important local economic and environmental ramifications. The Bureau of Land Management (BLM) controls 84 million ha (200 million acres) of grazing lands and the U.S. Forest Service manages about one-fourth as much.

The BLM manages more land than any other agency in the United States, but it is little known outside of the western states where most of its lands are located. The agency was created in 1946 by a merger of the Grazing Service and the General Land Office, whose task had been to distribute public lands to settlers. This merger committed the government to manage lands for resource use. The BLM has such a strong inclination toward resource utilization that critics claim the initials really stand for the "bureau of livestock and mining." While only 25 percent of BLM land is considered suitable for grazing, its policies have an important effect on the economy and environment of western states.

State of the Range

The health of most public grazing lands in the United States is not good. Political and economic pressures encourage managers to increase grazing allotments beyond the carrying capacity of the range. Lack of enforcement of existing regulations and limited funds for range improvement have resulted in overgrazing, loss of native forage species, and erosion. The Natural Resources Defense Council claims that only 30 percent of public rangelands are in fair condition, and 55 percent are poor or very poor (fig. 12.25).

Overgrazing has allowed populations of unpalatable or inedible species, such as sage, mesquite, cheatgrass, and cactus, to build up on both public and private rangelands. Furthermore, competing herbivores, such as grasshoppers, jackrabbits, prairie dogs, and feral (wild) burros and horses, further damage the range and reduce the available forage.

Several wildlife conservation groups regard cattle grazing as the most ubiquitous form of ecosystem degradation and the greatest threat to endangered species in the southwestern United States. They call for a ban on cattle and sheep grazing on all public lands, noting that it provides only 2 percent of the total forage consumed by beef cattle and supports only 2 percent of all livestock producers.

FIGURE 12.25 On the right side of this fence is a lightly grazed pasture consisting of a dense growth of grasses and small, broad-leaved plants. On the left side is overgrazed rangeland, where the vegetation has become sparse and bare soil is exposed. © Francois Gohier/Photo Researchers, Inc.

FIGURE 12.26 Rotational grazing reduces impacts of grazing in the Sheyenne National Grasslands, North Dakota. © Mary Ann Cunningham.

Key Concepts

- Forests occupy about 29 percent of the world's land surface, about half of their preagricultural extent.

- Most remaining forests are in extreme climate areas—boreal forests and tropical rainforests.

- Wood consumption has doubled in the past 60 years; it is expected to increase 10 percent per year for the next 50 years.

- Grasslands cover about one-third of the earth's surface. A primary threat to grasslands is overgrazing, leading to desertification and erosion.

Grazing Fees

Another controversial aspect of federal range management is that fees charged for grazing on public lands are far below market value and represent an enormous hidden subsidy to western ranchers. The 1999 minimum charge for grazing permits on BLM or USFS land was $1.35 per cow per month. Comparable private land rents for an average of $11.10 per cow per month. According to the Congressional Research Service, it costs $3.21 per animal unit month (AUM) just to administer grazing on USFS or BLM land. Add to this $40 million a year in the USDA Livestock Assistance Program, as well as additional range, road, and watershed improvements, and this "cow welfare" system amounts to $100 million a year in taxpayer subsidies, according to a report in the San Jose *Mercury News.*

In 1993, the Clinton administration proposed raising grazing fees gradually to a minimum of $3.96/AUM. This proposal was defeated by a group of western legislators and rates were rolled back to 1986 levels. Just 10 percent of permit holders on BLM land account for almost two-thirds of the animals allowed to graze those lands; 6 percent of the permit holders on Forest Service lands are responsible for about half of the livestock. The Office of Management and Budget concluded that increasing grazing fees to a fair market value would reduce the livestock herd less than 10 percent and would result in removing animals only from the most marginal areas, which probably should not be grazed anyway. Thus far, however, there have not been significant changes in fees or stocking levels.

Rotational Grazing

Where a small number of livestock are free to roam a large area, they generally eat the tenderest, best-tasting grasses and forbs first (wouldn't you?) leaving the tough, unpalatable species to flourish and gradually dominate the vegetation. In some places, farmers and ranchers find that short-term, intensive grazing helps to maintain forage quality. As South African range specialist, Allan Savory, observed, wild ungulates (hoofed animals) like gnus or zebras in Africa or bison (buffalo) in America often tend to form dense herds that graze briefly but intensively in a particular location before moving on to the next area. Short duration, **rotational grazing**—confining animals to a small area for a short time (often only a day or

two) before shifting them to a new location—simulates the effects of wild herds. Forcing livestock to eat everything equally, to trample the ground thoroughly, and to fertilize heavily with manure before moving on helps to keep weeds in check and encourages growth of more desirable forage species.

In some U.S. national grasslands and forestlands, grazing permits now require rotation. This strategy can help keep grasslands healthy while allowing sustainable economic activity (fig. 12.26). This approach doesn't work everywhere, however. Many plant communities in the U.S. desert Southwest, for example, apparently evolved in the absence of large, hoofed animals and can't withstand intensive grazing.

LANDOWNERSHIP AND LAND REFORM

Many of the problems discussed in this chapter have their roots in landownership and public policies concerning resource use. Some land tenure patterns have supported or even created inequality, ignorance, and environmental degradation, while other systems of land use have promoted equality, liberty, and progress. In this final section, we will look at how fair access to the land and its benefits can help bring about better conditions for humans and their environment.

Who Owns How Much?

Whatever political and economic system prevails in a given area, the largest landowners usually wield the most power and reap the most benefits, while the landless and near landless suffer at the bottom of the scale. The World Bank estimates that nearly 800 million people live in "absolute poverty . . . at the very margin of existence." Three-fourths of these people are rural poor who have too little land

FIGURE 12.27 Landless rural peasants still work most of the land in Asia, Latin America, and Africa while they live in poverty. © The McGraw-Hill Companies, Inc./Barry W. Barker, photographer.

to support themselves. Most of the other 200 million absolute poor are urban slum dwellers, many of whom migrated to the city after being forced out of rural areas (chapter 22).

In many countries, inequitable landownership is a legacy of colonial estate systems in which landless peasants still work in virtual serfdom (fig. 12.27). For instance, the United Nations reports that only 7 percent of the landowners in Latin America own or control 93 percent of the productive agricultural land. By far the largest share of landless poor in the world are in South Asia. In Haiti, the poorest and most environmentally degraded country in the Western Hemisphere, 1 percent of the population owns 90 percent of the land.

The 86 million landless households in India comprise at least 500 million people; they are between half and two-thirds of both India's population and about half of all the landless people in the world. Not surprisingly, the countries with the widest disparity in wealth are often the most troubled by social unrest and political instability. As Adam Smith said in *The Wealth of Nations* (1776), "No society can surely be flourishing and happy, of which the far greater part of the members are poor and miserable."

Land tenure is not just a question of social justice and human dignity. Political, economic, and ecological side effects of inequitable land distribution affect those of us far from the immediate location of the problem. For example, people in developing countries forced from their homes by increasing farm mechanization and cash crop production often try to make a living on marginal land that should not be cultivated. The local ecological damage caused by their struggle for land can, collectively, have a global impact. The contribution of concentrated landownership to these environmental problems, however, receives less attention than do the threatening ecological trends themselves.

Land Reform

Throughout history, some **land reform** movements seeking redistribution of landownership have been successful and others have not. Significant land reforms have occurred in many countries only after violent revolutionary movements or civil wars. An important factor in revolutions in Cuba, Mexico, China, Peru, the former Soviet Union, and Nicaragua was the breaking of feudal or colonial landownership patterns. This struggle continues in many countries. Hundreds of large and small peasant movements demand land redistribution, economic security, and political autonomy in Latin America, Africa, and Asia. Entrenched powers respond with economic reprisals, intimidation, "disappearances," and even open warfare. Millions of people have been killed in such struggles in recent centuries alone.

Consider Brazil: 340 rich landowners possess 46.8 million ha (117 million acres) of cropland, while 9.7 million families (70 percent of all rural people) are landless or nearly landless. Studies have shown that 13 percent of the land on the biggest estates is completely idle, while another 76 percent is unimproved pasture. These inequities and the tensions they create are important factors in rainforest invasions described earlier in this chapter.

In some countries, land reform has been relatively peaceful and also successful. In Taiwan, only 33 percent of the farm families owned the land they worked in 1949, whereas nearly 60 percent of rural families farm their own land today. South Korea also has carried out sweeping land redistribution. In the 1950s, more than half of all farmers were landless, but now 90 percent of all South Korean farmers own at least part of the land they till. In general, countries that have had successful land reform also have stabilized population growth and have begun industrialization and development.

What are the ecological implications of landownership? Absentee landlords have little personal contact with the land, may not know or care about what happens to it, and often won't let sharecroppers cultivate the same land from year to year for fear they may lay claim to it. This gives tenant farmers little incentive to protect or improve the land because they can't count on reaping any long-term benefits. In fact, if tenants do improve the land or increase their yields, their rents may be raised. Also, where land is aggregated into collective farms or large estates worked by landless peasants, productivity and health of the soil tend to suffer.

Far from being a costly concession to the idea of equality, land reform often can provide a key to agricultural modernization. In many countries, the economic case for land reform rivals the social case for redistributive policies. Many studies have shown that the productivity of owner-operated farms is significantly higher than that of corporate or absentee landowner farms. Independent farmers, especially those who have only a few hectares to grow crops, tend to lavish a great deal of effort and attention on their small plots, and their yields per hectare often are twice those of larger farms.

Indigenous Lands

Indigenous (native or tribal) people make up about 10 percent of the world's population but occupy about 25 percent of the land. Many of the approximately 5,000 indigenous cultures that remain today possess ecological knowledge of their ancestral lands that is of vital importance to all of us. According to author Alan Durning, "encoded in indigenous languages, customs, and practices may be as much understanding of nature as is stored in the libraries of modern science." And indigenous people continue to play an important role as guardians of wildlife and forests. The Kuna Indians of Panama say, "Where there are forests there are indigenous people, and where there are indigenous people there are forests."

Nearly every country in the world has a sad history of decimation of indigenous cultures and exploitation or annexation of ancestral lands. Time after time, native people have been stripped of their rights through economic pressure, cunning and deceit, or outright theft (fig. 12.28). In the past, colonial governments appropriated tribal lands under the excuse that sparsely populated or uncultivated land was free for the taking. Many nations still fail to recognize indigenous rights. In the Philippines, Cameroon, and Tanzania, for example, the government claims ownership of all forest lands, and indigenous people are considered squatters, even on their ancestral territories.

Our appetite for natural resources and land puts both native people and the ecosystems in which they live at risk. Decimated by plagues, violence, and corruption—and overwhelmed by the onrush of materialistic Western culture—at least half of the world's indigenous cultures are at risk of disappearing within the next century. Consider the case of the 9,000 Yanomami people who live along the border between Brazil and Venezuela (fig. 12.29). In 1991, Brazil recognized Yanomami claims to 17.6 million ha of land, but invasions by *garimpeiros* (miners) into this territory have

FIGURE 12.29   Native tribal people, such as these Yanomami from Brazil, are threatened by tropical forest destruction. These people have lived in harmony with nature for thousands of years. If their culture is lost, valuable knowledge about the forest will be lost as well. © G. Prance/ Visuals Unlimited.

introduced malaria, tuberculosis, flu, and respiratory diseases to which native people have no immunity. Placer mining and mercury effluents kill fish and poison the rivers. Wildlife is killed or driven off by miners, and hundreds of the Yanomami themselves have been hunted down and shot.

From the Philippines to Labrador, indigenous people are fighting for their ancestral territories. Some countries with large native populations, such as Papua New Guinea, Ecuador, Canada, and Australia, have acknowledged indigenous titles or rights to extensive areas. Often, years of history and modern development complicate these indigenous claims. One aboriginal band in Australia, for instance, claims ownership of Circular Quay, some of the most valuable land in the heart of Sidney. What would be a fair compensation after 150 years of occupation and building?

In April 1999, a Canadian territory called Nunavut (meaning "our home" in the native Inuktitut language) was created. Spanning nearly 2 million km^2 or almost 20 percent of Canada's landmass, this vast area of arctic tundra and glaciated mountains will be managed by its 25,000 residents, 90 percent of whom are Inuit. Nunavut has some of the harshest weather anywhere in the world. Ellesmere National Park at the northern extreme of the territory typically has one week of spring, one week of summer, and one week of fall. The rest of the year is winter, and five months are completely dark. With a population spread among 25 widely scattered villages and only one short paved road in the entire territory, Nunavut faces some difficult challenges. Foremost among these is the difficulty of preserving indigenous culture, skills, and values while also becoming part of the modern world. But the right to self-rule and control over their ancestral territory has brought a new sense of energy and purpose to native people. This settlement might be a good model for resolution of native land claims elsewhere in the world.

FIGURE 12.28   George Gillette, chairman of the Fort Berthold Indian Tribe Business Council, weeps as he watches the signing of a 1948 contract to sell the tribe's best land along the Missouri River for the Garrison Dam and Reservoir Project. After signing the contract, Gillette said, "Right now, the future does not look good to us." © AP/Wide World Photos.

Summary

- Land uses of key importance are forests (about 29 percent of global lands), grasslands (about 27 percent), and agriculture (11 percent). Exact statistics vary because definitions and methods of tabulating these areas differ.

- Forests are declining worldwide, but the greatest forest losses are in Africa, South America, and Asia. In general, tropical developing countries are exploiting their forests most rapidly; however, forests in Siberia, Canada, and other northern countries are also experiencing relatively rapid depletion.

- Forests are lost most often to settlement, conversion to agriculture or pasture, exploitation for wood products, or fire. Traditional swidden agriculture can be highly productive and sustainable if forests are allowed to regenerate; increased pressure from migrants and lost land rights often make this type of farming unsustainable, however.

- In parts of North America and Europe, forests are expanding due to replanting or a decline in agriculture. Debt-for-nature swaps and other conservation efforts are slowing forest declines in some developing countries.

- Open canopy forests have tree crowns covering less than 20 percent of ground surface; closed canopy forests have tree crowns over more than 20 percent of the ground. Old-growth forests are considered those in which trees can live out a natural life-cycle without interruption by humans. Worldwide, old-growth forests are biologically important but severely threatened.

- About half of world wood consumption is for fuel, and half is industrial timber or paper pulp. More than half the world's people depend on wood and charcoal for cooking. Wealthy countries produce less than half of industrial wood products, but they consume more than 80 percent.

- In the United States, forest management is often controversial. Issues of greatest disagreement include old growth harvesting, road building, below cost timber sales, fire management, salvage logging, forest thinning (or fuel reduction), and habitat or wilderness preservation. Public forests have traditionally been an important source of timber and jobs, but the number of logging-related jobs has declined because of industry automation, international trade, and conservation efforts.

- Options for sustainable forest use include alternative harvesting methods, reliance on plantation forests for wood products, and non-timber forest products. Thousands of alternative products can be produced from forests, and often these products are highly valuable. Finding markets for alternative products is often a challenge, however.

- Grasslands are extremely important to human livelihoods: they support grazing animals, both domestic and wild, and they have widely been converted to agriculture. Grasslands cover about 30 percent of the world's land surface.

- Grasslands support animals and humans best when herds are allowed to move frequently and pastures are not overstocked. Grasslands are being lost to degradation and desertification, as well as to agriculture. Although grassland losses are noted less than forest losses, they are occurring at a much faster rate.

- In the United States, overgrazing on public rangelands is a significant problem for both environmental health and resource management. Subsidized grazing permits are widely blamed for range degradation. Reducing herds is extremely controversial, however. Rotational grazing is an environmentally sound option that can help restore pasture and range health.

- Land ownership for the poor can help reduce overexploitation of land. Although it is a difficult process, land reform has been attempted in many places to restore land rights to peasants and indigenous people.

Questions for Review

1. Which type of land use occupies the greatest land area?

2. Which kinds of forests are most extensive? Which regions are losing forest at the highest rate?

3. What are some results of deforestation?

4. What are clear-cutting and below-cost timber sales, and why are they controversial?

5. What are salvage logging and forest thinning? What are some arguments for and against these practices?

6. What are the advantages and disadvantages of monoculture forestry?

7. Describe milpa (or swidden) agriculture. Why are these techniques better for some fragile rainforest ecosystems than other types of agriculture?

8. How does overgrazing encourage undesirable forage species to flourish?

9. Why can grazing animals generate food on land that would otherwise be unusable for human occupation?

10. Give some examples of recognition of indigenous land titles.

Questions for Critical Thinking

1. Much forestland and rangeland would be suitable as cropland. Do you think it should be converted? Why or why not?

2. Conservationists argue that we could reduce or redirect the demand for wood products; timber companies claim that continued production is essential for jobs and the economy. What evidence would you need to make a decision in this case and where might you get it?

3. Brazil needs cash to pay increasing foreign debts and to fund needed economic growth. Why shouldn't it harvest its forests and mineral resources to gain the foreign currency it wants and needs? If we want Brazil to save its forests, what can or should we do to encourage conservation?

4. Thousands of landless peasants are mining gold or cutting trees to create farms in officially protected Brazilian rainforest. What might the government do to stop these practices? Should it try to stop them?

5. What lessons do you think milpa (or swidden) agriculture has to offer large-scale commercial farming?

6. The U.S. government has kept timber sale prices and grazing leases low to maintain low prices of lumber and meat for consumers and to help support rural communities and traditional ways of life. How would you weigh those human interests against the ecological values of forests and rangelands?

7. Loggers and ranchers feel deeply threatened by environmental initiatives. What ideas or steps might help preservationists and resource users discuss these issues productively?

8. Native Americans often were cheated out of their land or paid ridiculously low prices for it. Present owners, however, may not have been a part of earlier land deals and may have invested a great deal to improve the land. What would be a fair and reasonable settlement of these competing land claims?

9. There is considerable uncertainty about the extent of desertification of grazing lands or destruction of tropical rainforests.

Put yourself in the place of a decision maker evaluating these data. What evidence would you want to see, or how would you appraise conflicting evidence?

Key Terms

clear-cutting 250
closed canopy 240
debt-for-nature swap 248
desertification 255
forest management 243
fuelwood 242
industrial timber 241
land reform 259
milpa agriculture 246
mixed perennial
    polyculture 246

monoculture forestry 245
old-growth forests 240
open canopy 240
pasture 255
rangelands 255
rotational grazing 258
salvage logging 252
selective cutting 250
swidden agriculture 246

Further Readings

Harrison, Paul, and Fred Pierce. 2000. *AAAS Atlas of Population and Environment.* University of California Press.

Hessl, Amy. 2003. Aspen, elk, and fire: The effects of human institutions on ecosystem processes. *Bioscience* 52(11):1011–23.

Miller, Char. 2001. *Gifford Pinchot and the Making of Modern Environmentalism.* Shearwater Books.

Sharp, Ben R., and Robert J. Whittaker. 2003. The irreversible cattle-driven transformation of a seasonally flooded Australian savanna. *Journal of Biogeography* 30(5):783–802.

UN Food and Agriculture Organization. 2001. Global forest resources assessment 2000. FAO Forestry Paper 140. United Nations (www.fao.org/forestry/fo/fra/main/).

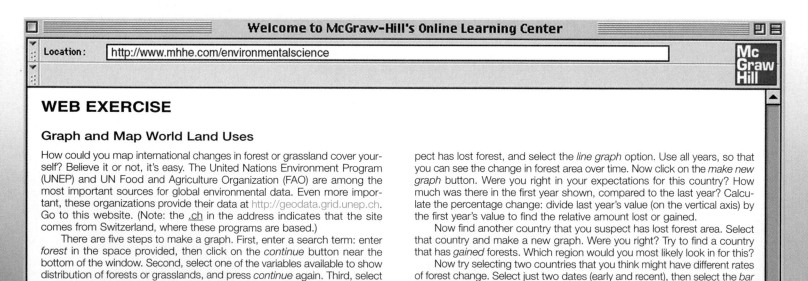

Welcome to McGraw-Hill's Online Learning Center

Location: http://www.mhhe.com/environmentalscience

WEB EXERCISE

Graph and Map World Land Uses

How could you map international changes in forest or grassland cover yourself? Believe it or not, it's easy. The United Nations Environment Program (UNEP) and UN Food and Agriculture Organization (FAO) are among the most important sources for global environmental data. Even more important, these organizations provide their data at http://geodata.grid.unep.ch. Go to this website. (Note: the .ch in the address indicates that the site comes from Switzerland, where these programs are based.)

There are five steps to make a graph. First, enter a search term: enter *forest* in the space provided, then click on the *continue* button near the bottom of the window. Second, select one of the variables available to show distribution of forests or grasslands, and press *continue* again. Third, select all the years available, and press *continue* again. Fourth, select the *draw graph* option. Finally, you'll have to select a country and a graph type. Since you're interested first in change over time, choose a country that you sus-

pect has lost forest, and select the *line graph* option. Use all years, so that you can see the change in forest area over time. Now click on the *make new graph* button. Were you right in your expectations for this country? How much was there in the first year shown, compared to the last year? Calculate the percentage change: divide last year's value (on the vertical axis) by the first year's value to find the relative amount lost or gained.

Now find another country that you suspect has lost forest area. Select that country and make a new graph. Were you right? Try to find a country that has *gained* forests. Which region would you most likely look in for this?

Now try selecting two countries that you think might have different rates of forest change. Select just two dates (early and recent), then select the *bar graph* option, and make a new graph. Were you right in your expectations?

Note that this web page lets you make maps or graphs of many variables. Explore and experiment a bit to see what you can learn.

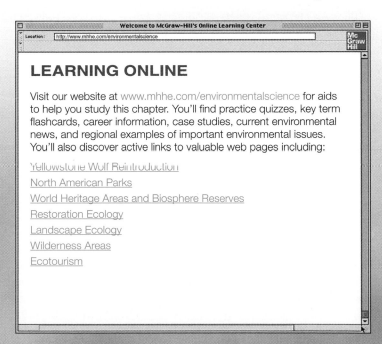

13

Preserving and Restoring Nature

Never doubt that a small group of thoughtful, committed people can change the world;
indeed it is the only thing that can.

Margaret Mead

OBJECTIVES

After studying this chapter, you should be able to:

- understand the origins and current problems of national parks in America and other countries.
- recount some of the current problems in national parks in the United States and elsewhere in the world.
- evaluate the tension between conservation and economic development and how the Man and Biosphere program and ecotourism projects address this tension.
- explain the need for, and problems with, wildlife refuges and wilderness areas in the United States.
- explain the principles and practices of landscape ecology and ecosystem management.
- evaluate the pros and cons of restoring, replacing, or substituting ecosystems and resources for those we have damaged.

Welcome to McGraw-Hill's Online Learning Center

Location: http://www.mhhe.com/environmentalscience

LEARNING ONLINE

Visit our website at www.mhhe.com/environmentalscience for aids to help you study this chapter. You'll find practice quizzes, key term flashcards, career information, case studies, current environmental news, and regional examples of important environmental issues. You'll also discover active links to valuable web pages including:

Yellowstone Wolf Reintroduction
North American Parks
World Heritage Areas and Biosphere Reserves
Restoration Ecology
Landscape Ecology
Wilderness Areas
Ecotourism

Photo: The Florida Everglades, often described as a "river of grass," is threatened by water pollution and diversion projects. © Jeffrey Greenberg/Photo Researchers, Inc.

The World's Biggest Restoration Project

The Florida Everglades have been celebrated as a life-giving river of grass and maligned as an impenetrable, mosquito-infested and alligator-ridden swamp. They have been drained for farming and urban development, yet Floridians depend on their steady flow of water for drinking, fishing, and irrigation. In 2000, the Everglades also became the subject of what may be the world's largest-ever restoration project, involving at least $8 billion worth of engineering, pumps and channels, modeling, and water management.

Everglades National Park (fig. 13.1) receives over a million visitors each year. People come to see the alligators, the fabulous flocks of egrets, ibises, and herons, and sometimes the elusive manatees that graze in the warm, brackish water of the coastal mangroves and lagoons. The ecosystem survives on a steady sheet-flow of water that spills over from Lake Okeechobee to the north. Slowly moving water provides breeding and feeding grounds for fish, invertebrates, birds, alligators, even the rare Florida panther.

Today's Everglades is a shadow of its former glory, though. The wetlands are diminished, and 90 percent of the birds have disappeared, mainly because of habitat loss. The reason for the ecosystem's gradual collapse is channelization and drainage. Since the park was established, the U.S. Army Corps of Engineers has built 1,600 km of canals, 1,000 km of levees, and 200 water control structures to drain farmlands, prevent flooding, and stop the natural sheet-flow of water through the region. Much of this work was done to benefit a growing sugar industry, built largely by Cuban sugar magnates who fled the communist takeover in Cuba. But other agriculturists have benefited, too, as have residents of nearby Miami, one of the country's fastest-growing metropolitan areas.

Today, straightened rivers and a network of canals send much of the region's water out to the ocean. Growing cities demand more and more land—and more water. Meanwhile, agricultural chemicals from the sugar fields upstream contaminate surface water, groundwater, and soils in the area. The national park is vulnerable to all these changes because it occupies—and controls—only the downstream end of the much larger Everglades ecosystem.

To resolve these complex problems, Florida and the Army Corps of Engineers have proposed the Comprehensive Everglades Restoration Plan (CERP). The plan aims to reengineer the water systems, returning some water to the Everglades, yet retaining control and preventing flooding. The Corps will remove nearly 400 km of levees and canals. New reservoirs will store water currently lost to the ocean, and 500 million liters of water per day will be pumped into underground

Historic flow | Current flow | Planned flow

FIGURE 13.1 Planned results of the Comprehensive Everglades Restoration Plan. Red dashed outline shows National Park boundary.

aquifers for later release. New canals and control structures will restore some of the original sheet-flow. Rivers will be de-channelized. Agricultural wastewater will be cleaned by special treatment plants and 14,000 ha of new filtration wetlands.

All these steps are, of course, controversial. Sugar interests have delayed and reduced water cleanup measures. Groundwater storage may contaminate aquifers. Ecologists doubt that the ecosystem will fully recover without restoration of seasonal flooding. And growing urban and suburban demands for water are likely to crowd out the urgent needs of the wetlands.

Despite the controversy and debate, the Comprehensive Everglades Restoration Plan is remarkable because it represents a huge commitment on the part of Congress, and the state of Florida, to preserving this suffering ecosystem. Pragmatic concerns for human livelihoods (fishing, tourism, clean water sources) help justify this commitment. But the restoration project is also driven by a widespread concern for the loss of a remarkable and beautiful ecological area. Park managers are excited about the possibility of seeing the park reconnected to the larger Everglades ecosystem.

In this chapter, we'll investigate efforts to preserve natural environments. Why do we do it? How do we do it? What works, and what doesn't? By now, you are well aware of the damage humans can do to their environment, but the steps we're beginning to take toward preservation and restoration are also promising—and they can make an important difference in preserving biodiversity and environmental health.

PARKS AND NATURE PRESERVES

Most land uses we've discussed thus far are purely utilitarian farmland, timberland, and grazing lands. But we also conserve lands for aesthetic reasons, for recreation, and for wildlife. Many of our highest ideals and values can be seen in parks and preserves. In this chapter, we'll examine the state of parks and wilderness areas, and we'll investigate ways of managing and restoring ecosystems and landscapes.

Park Origins and History

Since ancient times, sacred groves have been set aside for religious purposes and hunting preserves or pleasuring grounds for royalty. As such, they have been reserved primarily for elite members of society. The imperial retreat of the Han emperors of China, for example, built in the second century B.C. near their capital Ch'ang-an, is the earliest landscaped park of which we have detailed description. Large enough to encompass mountains, forests, and marshes as well as palaces and formal gardens, the park reflected Taoist beliefs about the ideal landscape and our place in the cosmos. Great towers and mountaintop pavilions served as retreats from which the emperor could contemplate nature in tranquility. Not all was peace and serenity, however; the emperor and his entourage also enjoyed hunting herds of wildlife maintained for their enjoyment. Although much of the park appeared natural, it required enormous feats of engineering and earth moving to achieve and maintain this wild appearance.

Perhaps the first public parks open to ordinary citizens were the grand esplanades and the tree-sheltered agora that served as a gathering place in the planned Greek city. Central Park in New York City is an important successor to both this democratic ideal and the naturalistic principles of romantic landscaping. Promoted in 1844 by newspaper editor William Cullen Bryant as a "pleasuring ground in the open air for the benefit of all," Central Park was to provide healthful open space and contact with nature for the crowded masses of the city (fig. 13.2). A worldwide competition for design of the park was won by Frederick Law Olmstead, who became the father of landscape architecture in the United States.

Olmstead left New York in 1864 to supervise the first area set aside to protect wild nature in the United States: Yosemite Park in California. Yosemite was authorized by President Abraham Lincoln in the midst of the Civil War to protect its resources from the unbridled exploitation by frontier settlers. Because there was no mechanism for running a park at the national level at that time, Yosemite was deeded to the State of California. It was transferred back to the federal government as a national park in 1890.

In 1872, President Ulysses S. Grant established the first *national* park in the world, Yellowstone National Park. The purpose of the park was to protect the spectacular "curiosities" and natural "wonders" of the geysers, hot springs, and canyons (fig. 13.3). But with 800,000 ha (almost 2 million acres) of land, the park was large enough to encompass and preserve real wilderness.

As it became apparent that wild nature and places of scenic beauty and cultural importance were rapidly disappearing with the closing of the North American frontier, the drive to set aside more national parks accelerated. Canada's Banff National Park was established in 1885. In the United States, Mount Rainier was authorized in 1899, Crater Lake in 1902, Mesa Verde in 1906, Grand Canyon in 1908, Glacier Park in 1910, and Rocky Mountain National Park in 1915.

The U.S. national park system has grown to more than 280,000 km² (108,000 mi²) in 388 parks, monuments, historic sites, and recreation areas. In addition, there are 264 wilderness areas totaling 50,000 km². Each year, about 300 million visitors enjoy these lands. The most heavily visited sites are urban recreation areas, parkways, and historic sites. The jewels of the park system, however, and what most people imagine when they think of a national park, are the great wilderness parks of the West, such as Yellowstone, Yosemite, and Denali National Parks.

Canada has 1,470 parks and protected areas occupying about 150,000 km². Among these are national parks, provincial parks, outdoor recreation parks, and historic parks. They range in size from the small cultural heritage sites to the vast wilderness of Wood Buffalo National Park in northern Alberta (nearly 45,000 km²) or

FIGURE 13.2 Central Park in New York City was one of the first large urban parks designed to provide contact with nature and healthful recreation for common people. © William P. Cunningham.

FIGURE 13.3 Yellowstone National Park, established in 1872, is regarded as the first national park in the world. Although focused initially on the spectacle of natural curiosities and wonders, it has come to be appreciated for its beauty and wilderness values. Courtesy David E. Wieprecht,USGS/CVO.

Nunavut's Quttinirpaaq (Ellesmere Island) National Park Reserve. One of the world's largest park preserves is composed of Kluane National Park in the Yukon, British Columbia's new Tatshenshini-Alsek Wilderness, and Alaska's Glacier Bay and Wrangell-St. Elias National Parks. While Canada enforces controlled access to many reserves, others allow intensive recreation, hunting, logging, and mining.

The United States and Canada have the greatest amount of protected areas, followed by Australia, Greenland, and Saudi Arabia. Note that all these countries have vast, thinly populated, or resource-poor lands that are relatively easy to designate as protected areas (see Web Exercises at the end of this chapter).

Trouble in Our Parks and Monuments

National parks and monuments serve many purposes. They teach us about our natural and cultural heritage, they provide sanctuaries for nature, they offer space for recreation and for solitude. They are intended to provide recreation and entertainment, but for some visitors, recreation involves solitude and contemplation; for others, recreation means snowmobiles and Jetskis. There are also valuable natural resources that can be extracted from many national monument lands. Inevitably, these activities come into conflict on our public lands.

Originally, the great wilderness parks of Canada and the United States were distant from development and protected from most human impacts. This has changed in many national parks. Forests are clear-cut right up to the edges of some parks. Mine drainage contaminates streams and groundwater. The Bush administration opened 13 national monuments to oil and gas drilling, including Texas's Padre Island, the only breeding ground in the United States for endangered Kemps Ridley sea turtles. Pollution and noise from snowmobiles continues to put wildlife, vegetation, and park workers at risk in Yellowstone and other national parks (fig. 13.4). Most two-cycle snowmobile engines produce as much

FIGURE 13.4 Air pollution and noise pollution cause ongoing debates over vehicle use in parks. Are off-road vehicles a public nuisance or a basic right and freedom? Courtesy National Park Service.

FIGURE 13.5 Off-road vehicles are fun to ride, but they cause lasting damage to soils, vegetation, wetlands, and streams. © Scott T. Smith Photography.

air pollution in 7 hours as a passenger car does in 100,000 miles. Personal watercraft (Jetskis and Waverunners) produce similar emissions, as well as leaking oil and fuel into water and making waves that erode shorelines. Off-road vehicles also cause erosion, as well as soil compaction (fig. 13.5).

Although the number of park visitors has risen by one-third over the past decade, U.S. park budgets have fallen by 25 percent. On popular weekends, Zion National Park takes in 7,000 cars, and in Yosemite up to 25,000 visitors crowd into the 18 km^2 (7 mi^2) valley (fig. 13.6). You can buy a pizza, play video games, do laundry, play golf or tennis, and shop for curios in Yosemite, but you are unlikely to experience the solitude extolled by John Muir. Aging buildings and roads are deteriorating, and park staff are overworked. The General Accounting Office estimates that national parks need $6 billion to $8 billion just for overdue repairs to infrastructure.

Air pollution and acid rain, originating from coal-fired power plants, threaten Shenandoah National Park, the Great Smoky Mountains, and Acadia National Park in Maine. Even the desert air of the Grand Canyon, where visitors could see mountains 150 km away, is often too smoggy to allow visibility across the canyon. Power plants supplying energy to Los Angeles, Phoenix, and other cities are the cause. Photochemical smog is damaging the giant redwoods of California's Sequoia National Park. In addition, **inholdings** (land already in private ownership when a park was established) are a threat to the environment in parks, as land owners work to develop hotels, mines, and oil and gas wells deep in parks.

Proposed solutions to traffic congestion include buses, high permit fees for cars, and limits to the number of cars allowed to enter parks. These measures are controversial, of course, because most visitors are reluctant to leave their cars behind—even though traffic jams reduce the convenience of driving. In response to air pollution, national park proponents are lobbying to strengthen the

FIGURE 13.6 Visitors crowd popular trails in Yosemite National Park on summer weekends, making quiet and solitude impossible to find. © John Gerlach/Visuals Unlimited.

FIGURE 13.7 Wild animals have always been one of the main attractions in national parks. Many people lose all common sense when interacting with big, dangerous animals. This is not a petting zoo. © American Heritage Center, University of Wyoming.

Clean Air Act, which would reduce power plant emissions. In recent years, however, this act has been diminished, rather than strengthened. The most controversial efforts have been to restrict snowmobiles and personal watercraft. Industry and user groups have fought the Park Service over this for years.

New Parks and Monuments

One solution to congestion and overuse of parks is creating new ones. One of the most recent additions to the U.S. parks and preserve system is the Grand Staircase-Escalante National Monument, a 660,000 ha tract of desert canyonlands in southern Utah. Established in 1997 by presidential decree, the monument is a remote wilderness of deep, twisting canyons, rock spires, and lofty arches sculpted from brilliantly colored sandstone. Environmentalists were delighted by the protection of this area, which had long been a high priority for wilderness status. Local residents, however, were outraged that their access to the area's resources should be limited. Beneath the brilliantly colored stone may be large deposits of coal, natural gas, and oil. As soon as the national monument was announced, opponents began to build roads, fight for declassification, and remove funding for management. Clearly, although we love our national parks and monuments, it's not easy to establish a new one.

Shortly before the end of his presidency, Bill Clinton established 1.1 million ha of national monuments, and expanded national parks and underwater marine preserves. Most of these designations were controversial, however, and after Clinton left office, the new secretary of the Interior, Gale Norton, began to gradually dismantle many of the new protected areas.

Wildlife in Parks

Many arguments about national parks focus on wildlife. As wildlife habitat has diminished outside parks, many people also want parks to be a haven for wildlife. Often, this means maintaining a healthy, functioning ecosystem—at least this is the goal. Ecologists would like to see naturally balanced populations of predators—mountain lions, wolves, and grizzlies—as well as photogenic deer, elk, and black bears (fig. 13.7). But residents surrounding parks often dislike the idea of grizzlies and wolves in the neighborhood. Efforts to stabilize biological communities by reintroducing predators have produced decades-long debates (What Do You Think? p. 268).

Another reason wildlife is controversial is that public lands occupy only a portion of the **biogeographical area** (regional ecosystem) within which parks occur (fig. 13.8). Wildlife would naturally range far beyond park boundaries, but now the lands beyond parks are owned or used by private citizens, who may not want public wildlife wandering onto their lands. In Yellowstone National Park, unusually cold weather and deep snow in 1996–97 drove bison beyond park boundaries and into National Forest lands in search of food. Ranchers holding grazing permits in the National Forest called in state animal control officials, claiming that the bison carried brucellosis, a disease that could spread to cattle (although this has never been documented). Two-thirds of the park herd, 2,000 animals, either died or were shot. The following year, the United States spent over $500,000 to build fences and control bison. Ironically, the grazing fees protected by these measures earn only $15,000 per year.

WILDERNESS AREAS AND WILDLIFE REFUGES

When Europeans first explored most of North America, the land was thinly populated, and it seemed a vast, empty wilderness. One of the reasons the land seemed empty was that smallpox and other diseases

What do you think?

Reintroducing Wolves to Yellowstone

A little more than a century agao, an estimated 100,000 gray, or timber, wolves (*Canis Lupis*) roamed the western United States. As farmers and ranchers moved west, however, wolves were poisoned, shot, and trapped wherever they could be found. The last wolves were eliminated from the northern Rocky Mountains in the early 1900s. Without a predator to restrain their numbers, elk and deer populations expanded rapidly. In Yellowstone National Park, for instance, the elk herd grew to some 25,000 animals, probably four or five times the vegetation's carrying capacity. Vegetation was overgrazed, and populations of smaller animals, such as ground squirrels, declined. For several decades, ecologists urged that wolves be reintroduced to control prey populations. These proposals brought howls of angry protest from local ranchers, who see wolves as sinister killers that threaten children, pets, livestock, and the ranching way of life. It took more than 20 years to get approval for wolf reintroduction.

In 1995, wolves were trapped in western Canada and relocated to Yellowstone. Once in the park, wolves became established surprisingly quickly. Taking advantage of the abundant food supply, they tripled their population in just three years. The effects on the ecosystem were immediate and striking. Biodiversity increased noticeably. Fewer elk, deer, and moose meant more food for squirrels, gophers, voles, and mice. Abundant small prey, in turn, led to increased numbers of eagles, hawks, fox, pine martens, and weasels. Large animal carcasses left by the wolves provided a feast for scavengers, such as bears, ravens, and magpies. Nearly half the coyotes, which had become common in the wolf's absence, were killed by their larger cousins. This helped small mammals that once were coyote prey. Plants such as grasses, forbs, willows, and aspen flourished in the absence of grazing and browsing pressure. Rangers and naturalists were delighted that the ecosystem was back in balance again, while tourists were thrilled to catch a glimpse of a wolf or to hear them howl.

By the summer of 2000, researchers counted about 125 wolves in nine packs in Yellowstone, just one pack short of the recovery goal of ten packs for three consecutive years. Not everyone was happy, however, with the reintroduction program. Ranchers continue to regard wolves as a threat to their way of life. And some animal rights groups object to a policy that allows any wolf to be shot if it strays from the park and attacks livestock.

Wolf reintroduction raises some important questions about the purposes of parks. Are they recreational areas or refuges for nature? What responsibilities do parks have to their neighbors (and vice versa)? What role do you think science should play in this

Are wolves beautiful, thrilling symbols of wild nature or ruthless killers? Reintroduction of these top predators into Yellowstone National Park has enthusiastic support from environmental groups but passionate opposition from local ranchers and hunters. © Gary Milburn/Tom Stack & Associates.

dispute? Suppose you were a wildlife biologist charged with designing an acceptable wolf management program in Yellowstone. What strategies would you use, and where would you start?

---

had wiped out up to 90 percent (by some estimates) of the resident Native American population. Nonetheless, the American wilderness became celebrated as part of the national character. Wilderness was not only a source of wealth, but a wellspring of independence, self-reliance, democracy, social progress, and national identity. Numerous authors, from Henry David Thoreau to Aldo Leopold, Sigurd Olson, Edward Abbey, and Barry Lopez, have written of the physical, mental, and social benefits of wilderness (fig. 13.9).

Wilderness Areas

The United States has established a system of 264 wilderness areas encompassing 40 million ha. The 1964 Wilderness Act, which is the basis for much of this system, defined **wilderness** as "an area of undeveloped land affected primarily by the forces of nature, where man is a visitor who does not remain; it contains ecological, geological, or other features of scientific or historic value; it possesses outstanding opportunities for solitude or a primitive and unconfined type of recreation; and it is an area large enough so that continued use will not change its unspoiled, natural conditions." Most of the areas meeting these standards are in the western states and Alaska.

Additional wilderness areas continue to be evaluated for protected status. Conservation groups would like to see another 80 million ha added to the wilderness system, half of it in Alaska. For more than 30 years, the Forest Service has been working to identify roadless forest lands that are "de facto" wilderness but not officially designated wilderness. Excluding all lands with any history of roads or development, the Forest Service has found that about one-fourth of its 23 million ha of roadless areas qualified for pro-

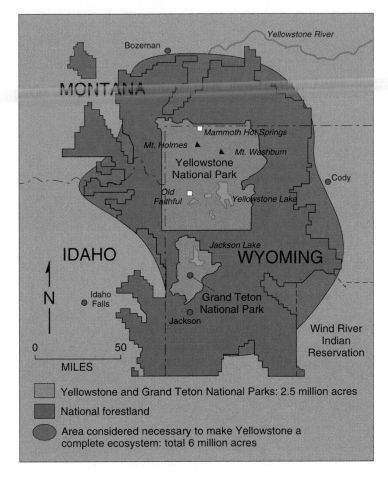

FIGURE 13.8 This map shows the Yellowstone ecosystem complex or biogeographical region, which extends far beyond the park boundaries. Park managers and ecologists believe that it is necessary to manage the entire region if the park itself is to remain biologically viable.

Map labels: Yellowstone River, Bozeman, MONTANA, Mammoth Hot Springs, Mt. Holmes, Mt. Washburn, Yellowstone National Park, Cody, Old Faithful, Yellowstone Lake, IDAHO, Jackson Lake, WYOMING, Idaho Falls, Grand Teton National Park, Jackson, Wind River Indian Reservation

Legend:
- Yellowstone and Grand Teton National Parks: 2.5 million acres
- National forestland
- Area considered necessary to make Yellowstone a complete ecosystem: total 6 million acres

0   50 MILES

FIGURE 13.9 These hikers traveled thousands of miles to experience wilderness in Nunavut's Ellesmere Island. © William P. Cunningham.

tection. The fate of these roadless areas remains an open question. Environmentalists want more wilderness. Loggers, miners, and ranchers want less wilderness. The arguments for saving wilderness are that it provides (1) a refuge for endangered wildlife, (2) an opportunity for solitude and primitive recreation, (3) a baseline for ecological research, and (4) an area where we have chosen simply to leave things in their natural state. The arguments against more wilderness are that timber, energy resources, and critical minerals contained on these lands are essential for economic development.

To people who live in remote areas, jobs, personal freedom, and local control of resources seem more important than abstract values of wilderness. They often see themselves as an embattled minority trying to protect an endangered, traditional way of life against a wealthy elite who want to lock up huge areas for recreation or aesthetic purposes. Wilderness proponents point out that 96 percent of the country already is open for resource exploitation; the remaining 4 percent is mostly land that developers didn't want anyway.

The last large areas still being studied for wilderness preservation in the United States are on Bureau of Land Management (BLM) land. Until 1997, only 1.7 million acres out of a total of 276 million

acres managed by the BLM had been designated as wilderness. Establishment of the Grand Staircase-Escalante National Monument doubled that total. Still, environmental groups argue that ten times as much should be given wilderness protection.

For many people, especially those in developing countries, the idea of pristine wilderness untouched by humans is regarded as neither very important nor very interesting. In most places, all land is occupied fully—if sparsely—by indigenous people. To them the area is home no matter how empty it may look to outsiders. From this perspective, preserving biological diversity, scenic beauty, and other natural resources may be a good idea, but excluding humans and human features from the land does not necessarily make it more valuable. In fact, saving cultural heritage, working landscapes, and historical evidence of early human occupation can often be among the most important reasons to protect an area.

Wildlife Refuges

In 1901 President Teddy Roosevelt established 51 national **wildlife refuges,** the first in an important but troubled system for wildlife preservation in the United States (fig. 13.10). There are now 540 wildlife refuges in this system, encompassing nearly 40 million ha of land and water and representing every major biome in North America.

Major additions to the refuge system were made by President Franklin D. Roosevelt and Harold Ickes, his secretary of the Interior, who took advantage of low prices during the Depression to add additional areas. The largest additions of land protected for wildlife came in 1980 when President Jimmy Carter signed the Alaska National Interest Land Act and added 22 million ha of new refuges to the 12.5 million ha already existing, about two-thirds of which was in Alaska.

FIGURE 13.10 Wildlife refuges have contributed to the recovery of both game and nongame wildlife populations. In some cases, such as these elk on the National Elk Refuge near Jackson Hole, Wyoming, protected species become so numerous that surplus population must be harvested or translocated. © Francois Gohier/Photo Researchers, Inc.

Refuge Management

Although refuges were originally intended to be sanctuaries in which wildlife would be protected from hunting and other disturbances, a 1948 compromise allowed hunting in refuges in exchange for an agreement by hunters to purchase special duck stamps to raise money for wetland protection. Although a refuge that allows hunting seems like an oxymoron, and in spite of the fact that only 10 percent of refuge lands have been acquired with duck stamp funds, hunting has become firmly established in most units (fig. 13.11).

Over the years, a number of improbable and incompatible uses have become accepted in wildlife refuges including oil drilling, cattle grazing, snowmobiling, motorboating, off-road vehicle use, timber harvesting, hay cutting, trapping, and camping. One Nevada refuge has a bombing and gunnery range, while another is the site of a brothel. A General Accounting Office report found that 60 percent of all refuges allow activities that are harmful to wildlife.

Refuges also face threats from external activities. More than three-quarters of all refuges in the United States have water pollution problems, two-thirds of which are serious enough to affect wildlife.

The biggest current battle over wildlife refuges concerns proposals for oil and gas drilling in the Arctic National Wildlife Refuge (ANWR) on the north slope of Alaska's Brooks Range. Environmentalists fear that drilling would destroy the fragile ecosystem, pollute expansive wetlands, and drive away calving caribou. Proponents of drilling argue that we need the oil, or at least its revenue, since much could be sold overseas. Many Alaskans would like to have high-paying industry jobs, and they point out that we really don't know how badly calving caribou would be harmed.

FIGURE 13.11 Hunters on their way to duck blinds in a wildlife refuge. Although it seems a contradiction, half of all refuges allow hunting. © Jeff Foott Productions.

WORLD PARKS AND PRESERVES

Global parks and preserves face innumerable challenges, but the statistics have been improving in recent years (fig. 13.12). The idea of setting aside nature preserves has spread rapidly over the past 50 years, as people around the world have become aware of the growing scarcity of wildlife and wild places. There are several reasons for a global increase in protected areas. International nongovernmental organizations have developed a range of tactics, including debt-for-nature swaps (chapter 12) to establish preserves. Consumer pressure has forced logging and mining companies to collaborate in preserving forests. And governments are increasingly interested in slowing resource depletion and in gaining status by protecting lands.

Within the past several years, Brazil has pledged to protect 12 percent of its Amazon rainforest, some 500,000 km², and both Brazil and Bolivia are working to protect the Pantanal region, the world's largest freshwater wetland complex. Mexico has increased protection for its biosphere reserve at Laguna San Ignacio, the breeding ground for gray whales. Mozambique, despite extreme poverty, has created two new national parks to protect dugongs, sea turtles, elephants, and other wildlife.

Regions with the most dramatic increases have been Asia, North America, and Latin America (fig. 13.13). Among the individual countries with the most admirable plans to protect natural resources are Costa Rica, Tanzania, Rwanda, Botswana, Benin, Senegal, Central Africa Republic, Zimbabwe, Butan, and Switzerland, each of which has designated 10 percent or more of its land as ecological protectorates. Brazil has even more ambitious plans, calling for some 231,600 km², or 18 percent of the country to be protected in nature preserves. So far, however, many of these areas are parks in name only. Lacking guards, visitor centers, administrative personnel, or even boundary fences, many are vulnerable to poaching by hunters and loggers, as well as to encroachment by settlers.

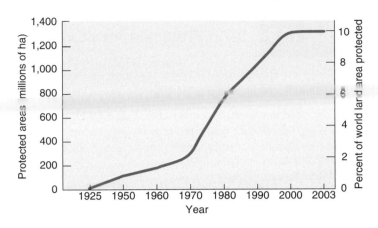

FIGURE 13.12 Global protected lands, 1925 to 2003.
Source: UN Environment Program, 2003.

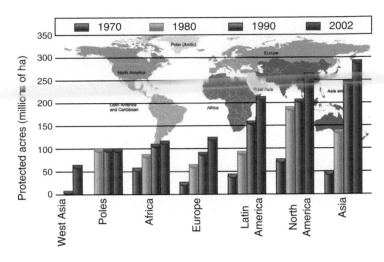

FIGURE 13.13 Change in amount of protected areas by region.
Source: UN Environment Program, 2003.

Paper Parks?

While the expansion of protected areas is promising, in many countries it is easier to declare a new park than to protect and manage it. Often parks are created where people already live, farm, and hunt. The needs of these people must be accounted for if a protected area is to work on the ground, rather than just on paper.

Even parks with systems in place for protection and management are not always safe from exploitation or changes in political priorities. In Greece, the Pindos National Park is threatened by plans to build a hydroelectric dam in the center of the park. Furthermore, excessive stock grazing and forestry exploitation in the peripheral zone are causing erosion and loss of wildlife habitat. In Colombia, the Paramillo National Park also is threatened by dam building. Oil exploration along the border of the Yasuni National Park in Ecuador pollutes water supplies, while miners and loggers in Peru have invaded portions of Huascaran National Park. In Palau, coral reefs identified as a potential biosphere reserve are damaged by dynamiting, while on some "protected" beaches in Indonesia every egg laid by endangered sea turtles is taken by egg hunters. These are just a few of the many problems in parks around the world. Often countries with the most important biomes lack funds, trained personnel, and experience to manage some of the areas under their control.

The IUCN has developed a **world conservation strategy** for natural resources that includes the following three objectives: (1) to maintain essential ecological processes and life-support systems (such as soil regeneration and protection, recycling of nutrients, and cleansing of waters) on which human survival and development depend; (2) to preserve genetic diversity, which is the foundation of breeding programs necessary for protection and improvement of cultivated plants and domesticated animals; (3) to ensure that any utilization of species and ecosystems is sustainable.

These goals are further elaborated in the ecological plan of action adopted by the IUCN and shown in table 13.1. A promising approach for financing these objectives is debt-for-nature swaps (chapter 12).

TABLE 13.1	**IUCN Ecological Plan of Action**

1. Launch a consciousness-raising exercise to bring the issue of biological resources to the attention of policymakers and the public at large.
2. Design national conservation strategies that take explicit account of the values at stake.
3. Expand our network of parks and preserves to establish a comprehensive system of protected areas.
4. Undertake a program of training in the fields relevant to biological diversity to improve the scientific skills and technological grasp of those charged with its management.
5. Work through conventions and treaties to express the interest of the community of nations in the collective heritage of biological diversity.
6. Establish a set of economic incentives to make species conservation a competitive form of land use.

Source: *International Union for Conservation and National Resources.*

Conservation and Economic Development

Many of the most seriously threatened species and ecosystems of the world are in the developing countries, especially in the tropics. This situation concerns us all because these countries are the guardians of biological resources that may be vital to all of us. Unfortunately, where political and economic systems fail to provide people with land, jobs, and food, disenfranchised citizens turn to legally protected lands, plants, and animals for their needs. Immediate human survival always takes precedence over long-term environmental goals. Clearly the struggle to save species and unique ecosystems cannot be divorced from the broader struggle to achieve a new world order in which the basic needs of all are met.

FIGURE 13.14 Ecotourism can be a sustainable resource use. If local communities share in the revenue, it gives them an incentive to value and protect biodiversity and natural beauty. © The McGraw-Hill Companies, Inc./Barry W. Barker, photographer.

Ia. Strict Nature Reserve: for science or environmental monitoring; outstanding or representative ecosystems.

Ib. Wilderness Area: for wilderness protection. Large, mostly unmodified areas without permanent inhabitants.

II. National Park: for ecosystem protection and recreation.

III. Natural Monument: natural features of outstanding natural or cultural value.

IV. Habitat/Species Management Area: for active management of habitats or species.

V. Protected Landscape: for conservation and recreation.

VI. Resource Protected Area: for sustainable use, serving both biological diversity and community needs.

Source: *UNEP-WCMC, 2003.*

The tropics are suffering the greatest destruction and species loss in the world, especially in humid forests and coastal ecosystems. People in some of the affected countries are beginning to realize that the biological richness of their environment may be their most valuable resource and that its preservation is vital for sustainable development. Ecotourism can be more beneficial to many of these countries over the long term than extractive industries such as logging and mining (fig. 13.14 and Case Study, p. 274).

In many cases, sustainable production of food, fiber, medicines, and water in rural areas depends on ecosystem services derived from adjacent conservation reserves. Tourism associated with wildlife watching and outdoor recreation can be a welcome source of income for underdeveloped countries. If local people share in the benefits of saving wildlife, they probably will cooperate and the programs will be successful. To reformulate Thoreau's famous dictum, "In broadly shared economic progress is preservation of the wild."

Because the uses and populations of resources vary so much, the IUCN has established categories of protected areas, from pristine to heavily used (table 13.2). This system acknowledges the many diverse needs for protected areas worldwide.

Indigenous Communities and Biosphere Reserves

Areas chosen for nature preservation are often traditional lands of indigenous people who cannot simply be ordered out. Finding ways to integrate human needs with those of wildlife is essential for local acceptance of conservation goals in many countries. In 1986, UNESCO initiated its **Man and Biosphere (MAB) program** that encourages division of protected areas into zones with different purposes. Critical ecosystem functions and endangered wildlife are protected in a central core region where limited scientific study is the only human access allowed. Ecotourism and research facilities are located in a relatively pristine buffer zone around the core, while sustainable resource harvesting and permanent habitation are allowed in multiple-use peripheral regions (fig. 13.15).

Mexico's 545,000-hectare (2,100 mi²) Sian Ka'an Reserve on the Caribbean coast is a good example of a MAB reserve. The core area includes 528,000 ha (1.3 million acres) of coral reef and adjacent bays, marshes, and lowland tropical forest. More than 335 bird species have been observed within the reserve, along with endangered manatees, five types of jungle cats, spider and howler monkeys, and four species of increasingly rare sea turtles. Approximately 25,000 people live in communities in peripheral regions around the reserve, and the resort developments of Cancun are located just to the north. In addition to tourism, the economic base of the area includes lobster fishing, small-scale farming, and coconut cultivation.

The Amigos de Sian Ka'an, a local community organization, played a central role in establishing the reserve and is working to protect the resource base while it improves living standards for local people. New intensive farming techniques and sustainable harvesting of forest products enable people to make a living without destroying the resource base. Better lobster harvesting techniques developed at the reserve have improved the catch without depleting native stocks. Local people now see the preserve as a benefit rather than an imposition from outside. Unfortunately, the government has very limited funds to develop or patrol the reserve.

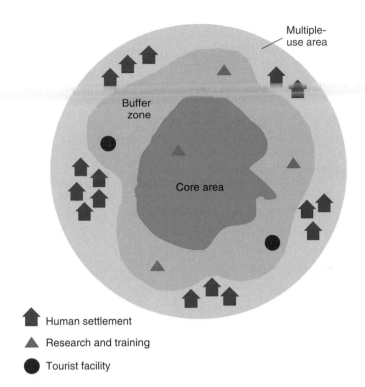

Legend:
- **Human settlement**
- **Research and training**
- **Tourist facility**

Buffer zone

Multiple-use area

Core area

FIGURE 13.15 A model biosphere reserve. Traditional parks and wildlife refuges have well-defined boundaries to keep wildlife in and people out. Biosphere reserves, by contrast, recognize the need for people to have access to resources. Critical ecosystem is preserved in the core. Research and tourism are allowed in the buffer zone, while sustainable resource harvesting and permanent habitations are situated in the multiple-use area around the perimeter.

International Wildlife Preserves

As we have seen, most developing countries rarely have separate systems of parks and wildlife refuges. Many nature preserves are set up primarily to protect wildlife, however. An outstanding example of both the promise and the problems in managing parks in the less-developed countries is seen in the Serengeti ecosystem in Kenya and Tanzania. This area of savanna, thorn woodland, and volcanic highland lying between Lake Victoria and the Great Rift Valley in East Africa is home to the highest density of ungulates (hoofed grazing animals) in the world. Over 1.5 million wildebeests (or gnus) graze on the savanna in the wet season, when grass is available, and then migrate through the woodlands into the northern highlands during the dry season. The ecosystem also supports hundreds of thousands of zebras, gazelles, impalas, giraffes, and other beautiful and intriguing animals. The herbivores, in turn, support lions and a variety of predators and scavengers, such as leopards, hyenas, cheetahs, wild dogs, and vultures. This astounding diversity and abundance is surely one of the greatest wonders of the world.

Tanzania's Serengeti National Park was established in 1940 to protect 15,000 km², an area about the size of Connecticut, or twice as big as Yellowstone Park. It is bordered on the east by the much smaller Ngorongoro Conservation Area and Lake Manyara National Park. Kenya's Masai Mara National Reserve borders the Serengeti

FIGURE 13.16 A guard protects black rhinoceroses from poachers. His ancient rifle is no match, however, for the powerful automatic weapons with which the poachers are now armed. © Steve Raymer/National Geographic Society.

on the north. Rapidly growing human populations push against the boundaries of the park on all sides. Herds of domestic cattle compete with wild animals for grass and water. Agriculturalists clamor for farmland, especially in the temperate highlands along the Kenya-Tanzania border. So many tourists flock to these parks that the vegetation is ground to dust by hundreds of sight-seeing vans, and wildlife find it impossible to carry out normal lives.

Perhaps the worst problem in Africa is **poachers,** illegal hunters who massacre wildlife for valuable meat, horns, and tusks. Where there once were about 1 million rhinos in Africa, the population had dropped by the mid-1980s to less than 10,000 animals. Antipoaching efforts have allowed the population to recover to about 15,000 currently (fig. 13.16).

Elephants are under a similar assault. Thirty years ago there were no elephants in the Serengeti, but perhaps 3 million in all of Africa. Since then, about 80 percent of the African elephants have been killed—mainly for their ivory—at a rate of 100,000 each year. The 2,000 elephants now in Serengeti National Park have been driven there by hunting pressures elsewhere.

The poachers continue to pursue the elephants and rhinos, even in the park. Armed with high-powered rifles and even machine guns and bazookas from the many African wars in the last decade, the poachers take a terrible toll on the wildlife. Park rangers try to stop the carnage, but they often are outgunned by the poachers. The parks themselves are beginning to resemble war zones, with fierce, lethal firefights rather than peace and tranquility.

Transboundary Peace Parks

International boundary regions, especially in the developing world, are often lightly populated, lawless areas. Many border areas have remnants of important ecosystems, but many are also regions of

Case Study

Ecotourism on the Roof of the World

Rising dramatically from the steamy southern jungles of the Ganges River Valley to the icy peaks of the Himalayan mountains on the Tibetan border, Nepal is one of the most scenic countries in the world. Tourists savor the exotic culture of Katmandu or Namche Bazar or hike through lush mountain forests of rhododendron and pine. Offering spectacular scenery, friendly people, and low prices, this charismatic country has become a premiere destination for adventure travelers.

With an annual per capita income of only $170, Nepal is among the poorest countries in the world. The phenomenal increase in visitors over the past 20 years has brought much-needed income but also has caused severe environmental degradation. Forests along popular trekking trails have been decimated to provide firewood for cooking and heating of water for the numerous wealthy outsiders, while tons of garbage and discarded gear litter popular campsites.

One of the most popular Nepalese trekking routes is a three- to four-week circuit of the Annapurna Range in the center of the Himalayan Range. Crossing rushing rivers on swaying suspension bridges, passing between the 8,167 m Dhaulagiri and the 8,091 m Annapurna I (the seventh and tenth highest mountains in the world, respectively), this ancient pilgrim trail follows the Kali Gandaki Valley to holy shrines at Muktinath. First opened to foreigners in 1977, this trail now attracts over 45,000 visitors each year.

Most Nepalese benefit very little from tourists who congest their villages, consume resources, and snap photographs incessantly, but the Annapurna region is different. An innovative project was launched in 1985 to alleviate the destructive impact of masses of trekkers and to maximize the income-generating potential of ecotourism. The Annapurna Conservation Area Project (ACAP) is a

The Annapurna Conservation Area Project directs money from visitors into development and environmental programs directed by, and of benefit to, local residents. This may be a model for preserving nature in other developing countries. © Galen Rowell, 1987 Mountain Light Photography.

2,590 km^2 (1,000 mi^2) biosphere reserve that serves as an encouraging model for conservation in developing nations.

Far different from Western ideals of parks composed of empty, virgin land, the ACAP is home to more than 100,000 people who continue to use resources in traditional ways. The area is divided into five different zones: intensive farming lands around the periphery, protected forest and seasonal grazing areas in the foothills, special management zones along tourist routes, protected regions with high biological or cultural richness, and wilderness areas in the high peaks.

Recognizing that there can be no meaningful conservation without the active involvement of local people, the ACAP returns visitor fees directly to residents to manage the preserve. About $500,000 per year finances a variety of conservation, education, and

development projects. More than 700 local entrepreneurs have been trained in lodge management, hygiene, and marketing. Forest guards have been hired, latrines built, trails repaired, and schools and clinics built for local people. Trekkers now are required to use kerosene rather than wood. Local tree nurseries provide stock for reforestation projects. Solar panels and water turbines provide renewable energy for both tourists and residents. The area is cleaner, healthier, and more enjoyable for everyone.

This successful experiment gives us a different view of the meaning and purpose of parks and nature preserves than the ideal of pristine, unchanging nature conveyed by most American national parks. It also provides a model of how competing human and nonhuman needs might be balanced in other developing countries.

perennial conflict. A movement to change that pattern is developing in southern Africa. Governments, environmental organizations, and international donors are teaming up to develop peace parks, or transfrontier conservation areas (TFCAs). In late 2002, Africa's biggest park was established in the Great Limpopo Transfrontier Park. With a series of undeveloped corridors, this park merges Mozambique's Limpopo National Park, South Africa's Kruger National Park, and Zimbabwe's Gonarezhou National Park. Border fences will be removed, and historic wildlife migrations, long blocked by those fences, can resume. Other parks are planned: one would merge Rwanda's war-torn and severely threatened Kagera National Park with Tanzania's Biharmulo Game Reserve. Another

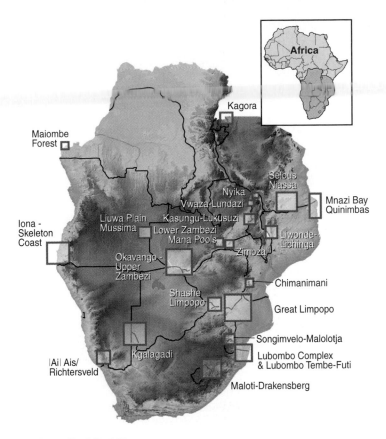

FIGURE 13.17 Existing (*red*) and proposed (*green*) transfrontier conservation areas (TCFAs) in southern Africa.

would link Namibia's and Angola's Skeleton Coast region, protecting breeding colonies of Cape fur seals and other wildlife. A proposed Okovango-Upper Zambezi park would join 36 protected areas in Angola, Botswana, Namibia, Zambia, and Zimbabwe, including the renowned Victoria Falls (fig. 13.17).

These proposed parks are fraught with difficulties. The Great Limpopo park requires cooperation between Africa's wealthiest country and one of its poorest. Political tensions may undermine Zimbabwe's commitment to the park. Each country brings different amounts of money and staff and different priorities to the project. Most TCFAs are justified largely by anticipated income from tourism, but developing this industry requires economic, social, and political stability. Some of the countries involved in the new peace parks, including Rwanda and Angola, have seen little but civil war in a generation. All these parks must meet the needs of subsistence residents as well as political leaders if they are to persist.

Despite these challenges, Peace parks have been widely embraced. They offer a hopeful strategy for cooperation in long-tense regions. They promise to preserve the wild areas and wildlife that are so important to the region's cultural heritage, and they may empower impoverished people in rural areas.

Peace parks are in place or planned elsewhere; for example, Glacier (U.S.) and Waterton (Canada) National Parks are recognized as a joint international peace park. Others are proposed along the Mexico-U.S. border.

PRESERVING FUNCTIONAL ECOSYSTEMS AND LANDSCAPES

Increasingly, we are learning that effective conservation of functioning ecosystems requires understanding the environment at a landscape scale. Communities and ecosystem processes do not exist in isolation; instead, living things interact in heterogeneous, patchy environments. Often these patchy landscapes are referred to as landscape mosaics. **Landscape ecology** is the study of how ecological processes shape these diverse environments, and how mosaic landscapes, in turn, shape ecosystem processes.

For example, in the forests of Yellowstone National Park, fires burn periodically, opening up patches of forest and leaving others standing. The openings allow new trees to grow, and they allow sun-loving plants—and herbivores—to flourish for a time. Blueberries grow, and bears grow fat. Gradually trees grow up, but another opening occurs somewhere else. Wandering bears and birds scatter seeds in the forest to grow as new openings appear. This shifting mosaic maintains biodiversity in the area. The mixture of mature forest, edges, regrowing shrubs, and fresh burns provides habitat for a wide range of birds, insects, amphibians, and rodents. The details of these interactions and landscape patterns are what landscape ecologists study.

Landscape ecology often considers human actions as important elements of landscape processes (Exploring Science, p. 276). Traditionally, ecologists have tried to study ecosystems unaffected by humans. Few places on earth, however, are devoid of human impacts, and few ecosystem processes are free of human interference. Understanding the role of humans in a landscape, therefore, is a useful task. Human disturbances include farming, hunting, timber harvesting, grazing, pollution, introducing invasive species, changes to air and water quality, and even climate change. Human influences can also include ecosystem restoration and management. In developing countries, especially, we can see that humans have been part of the natural landscape for thousands of years. There it is especially important to explore human influences on ecosystem processes.

Patchiness and Heterogeneity

We often imagine the Amazon rainforests as a continuous, unchanging expanse of trees. On closer examination, these forests are made up of shifting patches of different tree types, with different understory species, epiphytes, invertebrates, and birds. A tree falls, creating an opening, and this opening allows another type of tree, and another ecological community, to grow. At a larger scale, the Amazon rainforest is a complex assemblage of upland and wetland trees and soils, trees of differing ages and types that change continually (although very slowly).

Landscape ecologists point out that if we look closely, all landscapes consist of mosaics of different abiotic and biotic conditions. Often a predominant cover type acts as a matrix in which other patch types appear to be embedded. Like individual habitats in the rainforest, these patches are dynamic and change components or functions over time. Among the causes of this patchiness are human

Landscape ecology investigates spatial processes and patterns in ecological systems, such as habitat fragmentation, landscape diversity, and the effects of surrounding land uses on waterways. How can we identify and monitor those processes and patterns? Geographic information systems (GIS) are an important tool for investigating spatial questions in ecology.

What is a GIS? It is software that makes maps from spatial data. Spatial data might be elevation data, locations of animal observations, or boundaries of habitat areas. By mapping these data, scientists can investigate relationships among them. You may have used a GIS: online mapping services such as MapQuest use digital data representing roads, cities, landmarks, and addresses to make maps on demand. You might make a map showing the distance, direction, and driving routes between your location and a destination. An ecologist would use GIS to show the number of animal observations in a habitat type, to measure the mean size of habitat fragments, or to monitor movements of animals among habitat patches.

Landscape ecologists are often explicitly interested in human roles in ecological systems. The Baltimore Ecosystem Study (BES) is a multidisciplinary landscape ecology project aimed at understanding how human-dominated ecosystems differ from, and are similar to, natural ecosystems. Using a watershed in Baltimore, Maryland, researchers have mapped and measured the effects of settlement density on biodiversity, water quality, wetland habitats, and other environmental conditions.

In one study, BES researchers mapped suburban, urban, and undeveloped areas in subwatersheds, then calculated fertilizer inputs to streams from each type of land use. Fertilizer is important because excess nutrients, especially nitrogen and phosphorus, are important pollutants in river systems. By mapping elevations, they could calculate slopes in sub-watersheds; using digital maps of land cover, scientists could measure the amount of impervious (paved or built) surfaces, then calculate the rate and volume of storm runoff from streets to waterways. Contrary to expectations, they are finding that urban neighborhoods introduce significantly less nitrogen into streams than suburbs do. In some cases, urban nutrient inputs are not much greater than those from an undisturbed forest. Sediment runoff (such as sand and silt) is greater in urban areas, though, leading to faster erosion of streambeds and destabilization of stream-side vegetation.

There is growing interest in how landscape features and landscape diversity affect

Land cover, topography, and streams can be overlaid in a GIS to analyze landscape-scale ecology in the Gwynns Falls watershed near Baltimore, Maryland.

ecosystems such as the streams in Baltimore. GIS data recording land uses around stream reaches were central to this study. Many other questions about the interaction of features in landscapes, and the effects of human developments on natural processes, are best answered with GIS data.

and natural disturbances as well as underlying ecological factors such as successional processes.

Landscape heterogeneity can exist across a wide range of scale, from burned patches measuring thousands of hectares in Yellowstone National Park to the effects of soil crumb size and insect burrows in a few square centimeters of soil. A basic question in landscape ecology is whether a given phenomenon appears or applies across many different scales or is restricted to a particular scale. In other words, do patterns emerge, change, or disappear if we look at a landscape up close or from afar?

Landscape Dynamics

Time and space are of special concern in landscape ecology, and for that matter, to the rest of us as well. As humorist Garrison Keillor says, "Time exists so everything doesn't happen at once; space exists so everything doesn't happen to us." How does this apply to ecology? A physiological ecologist might want to study the effects of temperature on the rate of photosynthesis in plants. To make the study manageable, she or he might ask, "What is the photosynthetic rate in this particular *plant*, at this *exact* time, at these *specific* temperatures?"

A landscape ecologist, in contrast, might ask, "What is the effect of temperature on photosynthesis in this *place,* given its unique combination of history, composition, and characteristics?" Developing a deep relationship with, and understanding of, a particular landscape is sometimes referred to as having a "sense of place."

The boundaries between habitat patches are considered especially significant by landscape ecologists. Edges can induce, inhibit, or regulate movement of materials, energy, or organisms across a landscape. Thus the dynamics between patches may be of greater overall importance than what happens within each patch. As chapter 4 shows, wildlife managers are beginning to be very aware of edge effects as well as the size, shape, and distribution of habitat patches in fragmented landscapes.

This focus on interactions between neighboring communities is quite different from classical ecological focus on the structure and functions of discrete communities, populations, or ecosystems. In its interest in complexity and emergent properties of systems, landscape ecology draws on theories of chaos and complexity from mathematics and physics.

There also are many similarities between landscape ecology and the equally new discipline of conservation biology. Among their shared tenets are: (1) evolutionary change is a central feature of natural systems, (2) nonequilibrium dynamics and uncertainty are more characteristics of nature than are stability and determinism, (3) heterogeneity and diversity are good and ought to be maintained, (4) the context of the surrounding landscape is important, and (5) human needs, desires, abilities, and potential must be considered in efforts to understand and protect nature.

Size and Design of Nature Preserves

What is the optimum size and shape of a wildlife preserve? For many years, conservation biologists have disputed whether it is better to have a *single large or several small* reserves (the SLOSS debate). Ideally, a reserve should be large enough to support viable populations of endangered species, keep ecosystems intact, and isolate critical core areas from damaging external forces (see chapter 11). But as gaps are opened in habitat by human disturbance (fig. 13.18), and eventually areas are fragmented into isolated islands, edge effects may eliminate core characteristics everywhere.

For some species with small territories, several small isolated refuges can support viable populations and provide insurance against a disease or other calamity that might wipe out a single population. But small preserves can't support species such as elephants or tigers that need large amounts of space. Given human needs and pressures, however, big preserves aren't always possible. Establishing **corridors** of natural habitat to allow movement of species from one area to another (fig. 13.19) can help maintain genetic exchange and prevent the high extinction rates often characteristic of isolated and fragmented areas.

An interesting experiment funded by the World Wildlife Fund and the Smithsonian Institution is being carried out in the Brazilian rainforest to determine the effects of shape and size on biological reserves. Some 23 test sites, ranging in size from one hectare to 10,000 hectares have been established. Some areas are sur-

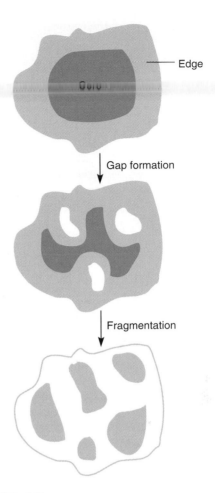

FIGURE 13.18   A patch or "island" of habitat becomes reduced and fragmented as vegetation gaps, or other human disturbances, expand until only small fragments remain. Note that core regions disappear early in this process. Finally, only edges are left in a matrix of disturbed habitat.

rounded by clear-cuts and newly created pastures (fig. 13.20), while others remain connected to the surrounding forest. Selected species are regularly inventoried to monitor their dynamics after a disturbance. As was expected, some species disappear very quickly, especially from small areas. Sun-loving species flourish in the newly created forest edges, but deep-forest, shade-loving species move out, particularly when size or shape reduces the distance from the edge to the center below a certain minimum. This demonstrates the importance of surrounding some reserves with buffer zones that maintain the balance of edge and shade species.

RESTORATION ECOLOGY

Can the lessons of Aldo Leopold's Sand County farm be applied to disrupted and damaged landscapes elsewhere in the world? Advances in ecological understanding coupled with practical knowledge gained through hands-on field work have given birth to a new discipline of **restoration ecology** that seeks to repair or reconstruct ecosystems damaged by humans or natural forces.

- Nature preserves, parks, and other conservation areas serve many important purposes and often face conflicting demands.
- Globally, the extent of protected areas has grown in recent years. Ensuring effective protection is often a challenge.
- Restoration ecology aims to restore, reclaim, or re-create lost habitat.
- Wetland mitigation is an important, and often controversial, aspect of restoration ecology.
- Ecosystem management aims at large-scale, long-term, wholistic management of natural systems.

Although land stewardship and husbandry have long been practiced by people in many land-based cultures, the field of ecological restoration suddenly captured public attention in 1988 when a conference called Restoring the Earth, held in Berkeley, California, drew an overflow crowd of more than 800 scientists, policymakers, and activists to share ideas and experiences from restoration projects. There is now a society for, and journal of, restoration ecology. This fledgling science has been given a boost in recent years as courts and regulatory agencies increasingly have required companies and developers to restore or replace illegally damaged wetlands or habitats of endangered species.

Defining Some Terms

Restoration generally means to bring something back to a former condition. Ecological restoration involves active manipulation of nature to re-create species composition and ecosystem processes as close as possible to the state that existed before human disturbance (fig. 13.21). Although there may be many similarities between restoration and conservation, stewardship, or management, the former often entails more direct intervention to achieve a predetermined end than do these other fields.

Rehabilitation refers to attempts to rebuild elements of structure or function in an ecological system without necessarily achieving complete restoration to its original condition. Often rehabilitation means to bring an area back to a useful state for human purposes rather than to a truly natural state. It aims to reverse deterioration of a resource even if it cannot be restored fully.

Remediation is a process of cleaning chemical contaminants from a polluted area by physical or biological methods as a first step toward protecting human and ecosystem health. Incineration, for instance, often is a cost-effective method of cleaning oil-contaminated soils. Volatile organic solvents in aquifers can be removed by pumping out groundwater and aerating it or passing it through absorbent materials.

Living organisms are highly effective cleaning agents for many contaminants. Water hyacinths, for example, have a great capacity for absorbing heavy metals and other toxins from polluted water. Microorganisms (bacteria and fungi) can be found in nature

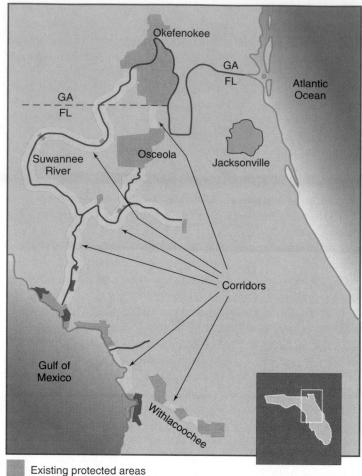

Existing protected areas

Already formally proposed acquisition

Presently proposed to establish functional system

FIGURE 13.19   Corridors serve as routes of migration, linking isolated populations of plants and animals in scattered nature preserves. Although individual preserves may be too small to sustain viable populations, connecting them through river valleys and coastal corridors can facilitate interbreeding and provide an escape route if local conditions become unfavorable. *Source:* R.F. Noss and L.D. Harris, "Nodes, Networks and MUMs: Preserving Diversity at All Levels," in *Environmental Management*, vol. 10:299–309, 1986.

or engineered in the laboratory to destroy many dangerous chlorinated compounds. Chopped-up horseradish roots are very effective in removing phenols from industrial effluents. And common locoweeds can extract selenium from contaminated soils. Sometimes simply adding fertilizer to encourage plant and microbial growth is the best way to clean up surface pollutants.

Reclamation typically is used to describe chemical or physical manipulations carried out in severely degraded sites, such as open-pit mines or large-scale construction. The Surface Mining Control and Reclamation Act (SMCRA), for example, requires mine operators to use reclamation techniques to restore the shape of the land to the original contour and revegetate it to minimize impacts on local surface and ground waters.

FIGURE 13.20 How small can a nature preserve be? In an ambitious research project, scientists in the Brazilian rainforest are carefully tracking wildlife in plots of various sizes, either connected to existing forests or surrounded by clear-cuts. As you might expect, the largest and most highly specialized species are the first to disappear. Courtesy R. O. Bierregaard.

FIGURE 13.21 Volunteers plant native plant species in a landscape restoration project. © William P. Cunningham.

Historically, reclamation meant irrigation projects that brought wetlands and deserts (considered useless wastelands) into agricultural production. Thus, the Bureau of Reclamation and the Army Corps of Engineers, in the early part of this century dredged, diked, drained, and provided irrigation water to convert millions of acres of wild lands into farmlands. Many of those projects were highly destructive to natural landscapes, and we are now using reclamation techniques to restore formerly "reclaimed" land to a more natural state.

Re-creation attempts to construct a new biological community on a site so severely disturbed that there is virtually nothing left to restore. The new system may be modeled on what we think was there before human disturbance or it may be something that never existed on that site but that we think suits current conditions. Often developers and government are required to mitigate damage caused in one area by re-creating a comparable biological community in another place.

In what is probably the most ambitious ecological restoration project in history, the U.S. Army Corps of Engineers is now trying to reverse 50 years of draining, ditching, and damming in south Florida in an effort to repair damage to the Everglades (see opening story; fig. 13.22). Artificial lagoons and underground aquifers will store millions of cubic meters of water for use during dry seasons. This will not only benefit wild landscapes but will also provide a much-needed water source for rapidly growing urban centers. Thousands of hectares of former wetlands that had been drained and farmed are now being turned back into wetlands. Years of accumulated pesticides and fertilizers in the soil, however, are

FIGURE 13.22 The naturally meandering Kissimmee River (*right channel*) was straightened and drained by the Army Corps of Engineers (*left*) for flood control 30 years ago. Now the Corps is attempting to reverse its actions and restore the Kissimmee and its associated wetlands to their original state. Courtesy South Florida Water Management District.

solubilized from flooded fields, and massive bird, fish, and reptile mortalities have been reported, especially around Lake Apopka in central Florida.

Many similar experiences make environmentalists suspicious of some mitigation and restoration projects. "We can't just move ecosystems around like furniture that we put wherever we want," says author Seth Zuckerman. And John Berger, organizer of the first Restoring the Earth Conference, said, "The purpose of restoration ecology is to repair previous damage, not to legitimize further destruction."

Tools of Restoration

Some of the earliest examples of restoration have involved labor-intensive horticultural or animal control methods. The Curtis Prairie at the University of Wisconsin in Madison, for instance, was reclaimed by Civilian Conservation Corps workers (fig. 13.23) and student volunteers, starting in 1934. Native species from the area were collected from remnant prairies along railroad rights-of-way and in pioneer cemeteries, and then hand-planted and cultivated on old, abandoned farm fields. Prairie plants initially had difficulty getting established and competing against exotics until it was recognized that periodic fires were essential for maintaining this biological community. The prairie is now flourishing and serves as a seed source for other restoration projects.

Sometimes the key to rebuilding a community is to remove alien intruders. In the Coachella Valley of California, for example, thousands of exotic salt cedars that dry out riparian habitats and crowd out native vegetation are being removed to protect the endangered Coachella Valley fringe-toed lizard. In Hawaii, feral pigs that root out native plant species and eat native birds are being hunted down and removed. Similarly, on the Galápagos Islands, feral goats and rats are being shot, trapped, and poisoned to protect native species.

A less expensive approach has been adopted in restoring an oak savanna near Chicago. Seeds collected from remnant prairies

are simply broadcast rather than being carefully planted. This is not only faster and cheaper, it is more like the natural process of wind-dispersal of seeds. Fire is used to discourage invasions of exotic species (fig. 13.24). This is called successional restoration because it depends on natural succession to determine the outcome.

In the Flint Hills of southern Kansas, the Nature Conservancy has reintroduced both fire and native bison (buffalo) to its 16,000 ha (40,000 acres) tall-grass prairie preserve. The bison disperse seeds and create disturbances necessary for survival of many other native species. And in Costa Rica, an almost vanished dry tropical forest is being rebuilt in the Guanacaste Conservation Area by preventing fires and allowing domestic livestock to act as seed dispersal agents since the native grazing animals have completely disappeared.

Letting Nature Heal Itself

Sometimes all we have to do to reestablish a healthy ecosystem is simply to walk away and let organisms recolonize an area. The demilitarized zone (DMZ) between North Korea and South Korea was totally devastated in 1953 when the Korean War ended. Shattered tree stumps and ruined villages lay strewn across a barren landscape pockmarked by bomb craters and littered with the debris of war. Because neither side was allowed to move back into this political buffer zone, the area has become a wildlife refuge and a luxuriant oasis for nature in an otherwise densely populated countryside. Scarred slopes have been reclothed with a dense hardwood forest where deer, lynx, and an occasional tiger find shelter. Former rice paddies have reverted to marshland in which waterfowl flourish and the trumpeting cries of the endangered Manchurian cranes are heard once again.

Similarly, parts of the political no-man's-land between the former East and West Berlin became a haven for wildlife during the 40 years that the Berlin Wall separated the two parts of the city. This is a heartening example of the regenerative power of nature. Now that the city is reunified, the question arises about what to do with this area. Should it be allowed to remain an unkempt urban wilderness or should it be turned into a park?

(a)

(b)

FIGURE 13.23 In 1934, workers from the Civilian Conservation Corps dug up old farm fields and planted native prairie seeds for the University of Wisconsin's Curtis Prairie (*a*). The restored prairie (*b*) has taught ecologists much about ecological resilience. *Sources:* (*a*) Courtesy University of Wisconsin-Madison Archives. (*b*) Courtesy University of Wisconsin Arboretum.

FIGURE 13.24 A burn-crew technician sets a back fire to control the prescribed fire that will restore a native prairie. Courtesy State of Minnesota, Department of Natural Resources. Reprinted with permission.

Back to What?

When humans take charge of ecosystems, questions sometimes are raised about what our goals should be. Suppose natural forces, such as hurricanes or fires, disrupt a wilderness area; should we use the principles of restoration ecology to tidy up or improve an area or leave it to natural processes? In some cases where humans have altered an area, there may be more than one historic state to which we could restore it. In the Nature Conservancy's Hassayampa River Preserve near Phoenix, for instance, pollen grains preserved in sediments reveal that 1,000 years ago the area was grassy marsh that was unique in the surrounding desert landscape. Some ecologists would try to rebuild a similar marsh. Corn pollen in the same sediments, however, show that 500 years ago Native Americans began farming the marsh. Which is more important, the natural or early agricultural landscape?

Unfortunately, it may not be possible to return to conditions of either 500 or 1,000 years ago since climate changes and evolution may have made the communities existing at that time incompatible with current conditions. This creates a difficult question for restoration ecologists. If change is natural and inevitable, who is to say that present conditions—whatever they are or however they have come about—are bad? How should we distinguish between desirable and undesirable changes? If it is simply a matter of human preference, some people might prefer a golf course or a shopping mall on the site. What would you say to someone who claims that since humans are a part of nature, whatever changes we make to a landscape also are natural?

Creating Artificial Ecosystems

As we learn more about how ecosystems work and about the valuable ecological services they provide, we are coming to realize that we can use some natural principles in human-designed systems. The Arcata, California, marsh and wildlife sanctuary is a famous example of using nature to solve a human problem. In 1974, Arcata

FIGURE 13.25 Arcata, California, built an artificial marsh as a low-cost, ecologically based treatment system for sewage effluent. © William P. Cunningham.

was faced with the prospect of spending $56 million to upgrade its municipal sewage treatment plant and reduce dumping of inadequately treated wastewater into Humboldt Bay. City residents and faculty from Humboldt State University devised a low-cost alternative that might serve as a model for many other communities.

An abandoned dump near the existing treatment plant was turned into an artificial wetland (fig. 13.25). Wastewater from conventional treatment processes flows through oxidation ponds where sunlight and air kill pathogenic microbes. The effluent is then filtered through the artificial marsh, where aquatic plants filter out nutrients, and zooplankton eat microbes, and fish eat the plankton. Finally, the water trickles into a saltwater slough where pelicans and cormorants gather to feed on the plentiful fish and where oysters perform a final filtering step before the "polished" water enters the bay. The marsh has become a favorite bird-watching and recreational site. It also saves millions of dollars and has turned what was formerly an eyesore into a source of civic pride.

PRESERVING ECOSYSTEM SERVICES: WETLANDS AND FLOODPLAINS

Wetlands and floodplains provide irreplaceable ecological services. They are often highly productive and provide food and habitat for a wide variety of species. Although wetlands currently occupy less than 5 percent of the land in the United States, the Fish and Wildlife

FIGURE 13.26 Louisiana has nearly 40 percent of the remaining coastal wetlands in the United States, but diversion of river sediments that once replenished these marshes and swamps, together with channels and boat wakes, are causing the loss of about 50 mi² annually. A third of these wetlands may be gone in 50 years. © Chris Harris/Gamma Liaison International.

Service estimates that one-third of all endangered species spend at least part of their lives in wetlands. Wetland storage of flood waters is worth an estimated $3 billion to $4 billion per year. Wetlands also improve water quality by acting as natural water purification systems, removing silt and absorbing nutrients and toxins.

Throughout much of our history, however, wetlands have been considered disagreeable, dangerous, and useless swamps and mires. This attitude was reflected in public policies such as the U.S. Swamp Lands Act of 1850, which allowed individuals to buy swamps and marshes for as little as 10 cents per acre. Until recently, federal, state, and local governments encouraged wetland drainage and filling, including dumping waste into wetlands in order to create more land for development. As a result of these policies, about 66 percent of original wetlands in the lower 48 states were destroyed, leaving about 40 million ha (see fig. 5.26). Coastal Louisiana, which has some 40 percent of remaining coastal wetlands in the United States (outside of Alaska), continues to lose about 13,000 ha (50 mi²) per year (fig. 13.26).

Wetland Conservation and Mitigation

The 1972 Clean Water Act began protecting wetlands by requiring discharge permits (called Section 404 permits) for discharging waste into surface waters. In 1977, federal courts interpreted this rule to prohibit both pollution and filling of wetlands (but not drainage). The 1985 Farm Bill went further with a "swampbuster" provision that blocked agricultural subsidies to farmers who drain, fill, or damage wetlands. The Clean Water Act and Farm Bill rules substantially reduced the rate of wetland destruction. In the 1990s, the United States lost an average of 24,000 ha (58,000 acres) per year—80 percent less than in the previous decade.

These laws are not necessarily enforced effectively: Section 404 permit applications, allowing filling and draining, are almost always approved, and exemptions to swampbuster provisions are almost always granted. Even so, these rules are hated by many landowners, farmers, and development, mining, and other groups. In 2001, the Supreme Court sided with developers and drastically reduced wetland protection. The court declared that the Clean Water Act has no jurisdiction over isolated wetlands and ponds or tidal mudflats. This ruling, which settled a suit by Chicago suburbs wanting to use wetlands as waste dumps, eliminated legal protection for about 20 percent of U.S. wetlands.

Because the ecological and economic benefits of wetlands are gaining recognition, **wetland mitigation** (creating wetlands to restore those lost to development) is now one of the most active areas of restoration ecology. Reestablishing both ecological and hydrologic functions in a wetland is a huge and difficult task. Despite decades of experience, we often know too little about how wetlands function. In part, wetlands mitigation is often unsuccessful because the letter of the law may allow a holding pond to replace a natural wetland (fig. 13.27).

Many wetlands ecologists argue that we can never really recreate a functional wetland. We might create something that has some of the right vegetation, but wetland ecosystems are too complex to be fabricated effectively.

Furthermore, mitigation often produces the wrong kind of wetland. According to Joy Zedler of the University of Wisconsin, most wetlands lost to development are wet meadows and wooded swamps. Nearly all mitigation wetlands, though, are ponds. People like these ponds, but their ecological functions are much different than those of the lost wetlands.

Public policy is gradually changing to give incentives for preserving wetlands, rather than just penalties for destroying them. A number of states have established wetland reserve programs that buy conservation easements from farmers (pay farmers not to drain their wetlands). The Department of Agriculture's Wetland Reserve Program and Conservation Reserve Enhancement Program have

FIGURE 13.27 These artificial ponds were created as mitigation for natural wetlands destroyed by this housing project. © William P. Cunningham.

reestablished more than 85,000 ha of wetlands on formerly cultivated fields. A number of fish and wildlife organizations are working to buy and restore rural wetlands. In Minnesota, where the flat Red River valley floods almost yearly, it has been proposed that buying land and rebuilding wetlands would be cheaper and more effective than continuing to clean up flooded cities and ruined crops every few years. Thus far, though, this proposal has made little progress toward implementation.

Floodplains and Flood Control

Not only wetlands, but also floodplains, are responsible for holding floodwaters. **Floodplains,** low-lying land along river banks and lakes subjected to periodic inundation, are often seasonally flooded forests, rich breeding grounds for fish and water birds. They also have rich soils, built up from centuries of silty flood deposits. Because they are flat, they make prime farmland, and they are convenient for building houses and cities.

Floodplains, of course, are subject to flooding. In some cases, flooding is more frequent because farmlands are drained and ditched, sending rainfall and snowmelt quickly into rivers. Expanding impervious surfaces in cities also sends water faster into a river, and straightened channels and levees concentrate flood waters in what floodplains remain. The Mississippi River, for example, has had two "hundred year floods," floods so big they should occur only once in a century, in less than a decade. Increasingly, policy analysts and environmentalists are wondering about the possibility of restoring floodplains in order to reclaim the ecosystem services of floodwater management.

Public flood-control efforts on the Mississippi began in 1879, when construction was started on levees (earthen berms along the river banks) to protect towns and low-lying farm fields. After a particularly disastrous flood in 1927, the Army Corps of Engineers began work on an extensive system of locks, dams, and channel dredging to improve navigation and provide further flood control (fig. 13.28).

The $25 billion river-control system built on the Mississippi and its tributaries has protected many communities over the past century. For others, however, this elaborate system has helped turn a natural phenomenon into a natural disaster. Deprived of the ability to spill out over floodplains, the river is pushed downstream to create faster currents and deeper floods until eventually a levee gives way somewhere. Hydrologists calculate that the Mississippi floods of 1993 were about 3 m (10 ft) higher than they would have been given the same rainfall in 1900, before the control structures were in place.

ECOSYSTEM MANAGEMENT

As you probably already know, there is a growing sense among ecologists, and the public at large, that our environmental situation is rapidly reaching a crisis point. Many of our management policies, while well-meaning, have made matters worse rather than better. **Ecosystem management** is a relatively new discipline in

FIGURE 13.28 Locks, dams, levees, and other flood control and navigational structures such as this complex on the Mississippi River at Dubuque, Iowa, have been a godsend for commercial shippers but have also been disastrous for riparian ecosystems, backwater sloughs, marshes, and the wildlife that depends on them. © The McGraw-Hill Companies, Inc./Bob Coyle, photographer.

environmental science that attempts to integrate ecological, economic, and social goals in a unified, systems approach. It recognizes that we cannot have sustained progress toward social goals in a deteriorating environment or economy, and vice versa. Each of these domains affects, and is affected by, the others. There are many definitions of ecosystem management. Guidelines suggested for U.S. Federal Agencies are shown in table 13.3.

A Brief History of Ecosystem Management

While the term "ecosystem management" is new, a few visionary ecologists such as Aldo Leopold had the foresight to advocate many specific elements of this science 50 years ago. Leopold's attempts to restore his Sand County farm to ecological health and beauty foreshadowed many of the ideas now widely espoused by ecosystem managers. Another pioneer in this field is environmental policy expert Lynton Caldwell, who wrote in 1970 that we should use ecosystems as the basis for public land policy. He understood that to do so "would require that the conventional [political] matrix be unraveled and rewoven in a new pattern." Unfortunately, although land-use planning was a high priority of the burgeoning U.S. environmental movement in the 1970s, the forces of inertia and private interest thwarted attempts to manage ecologically.

Most federal and many state natural resource agencies in the United States are attempting to identify endangered landscapes and implement ecosystem management as their guiding policy. The U.S. Forest Service, the Bureau of Land Management, and the National Park Service, for instance, all have adopted versions of ecosystem management. This is a marked improvement in many cases over past practices. Traditionally, many of these agencies emphasized

TABLE 13.3	Principles of Ecosystem Management

- Managing across whole landscapes, watersheds, or regions over an ecological time scale.
- Considering human needs and promoting sustainable economic development and communities.
- Maintaining biological diversity and essential ecosystem processes.
- Utilizing cooperative institutional arrangements.
- Integrating science and management.
- Generating meaningful stakeholder and public involvement and facilitating collective decision making.
- Adapting management over time, based on conscious experimentation and routine monitoring.

commodity production and the commercial or recreational use of natural resources as their first priority. Management objectives often were designed to expedite the development, extraction, and/or production of resources on public lands above all other uses. Wildlife and fish habitats, endangered species, and cultural, scenic, historic, or aesthetic values tended to be viewed mainly as inconveniences or hindrances to economic uses.

Principles and Goals of Ecosystem Management

Conservation biologist E. R. Grumbine, in a seminal article on ecosystem management, pointed out several important differences between this integrative approach and the traditional policies of the past:

1. *Manage at multiple scales:* A focus on any one level of the biodiversity hierarchy (genes, species, populations, ecosystems, landscapes) is insufficient. Ecosystem managers must see interconnections between all levels.

2. *Use ecological boundaries:* Rather than divide administrative units by political boundaries, the watersheds, ecosystems, or other natural units should be managed in an integrated fashion.

3. *Monitor the ecosystem:* To function correctly, ecosystem management requires ongoing research and data collection so that successes or failures may be recognized and evaluated (fig. 13.29).

4. *Use adaptive management:* Ecosystem management assumes that scientific knowledge is provisional and regards management plans as learning processes or continuous experiments where incorporating the results of previous actions allows managers to remain flexible and adapt to uncertainty.

5. *Allow organizational change:* Implementing ecosystem management requires changes in agency structure and ways of

FIGURE 13.29 Ecosystem management requires a high level of routine monitoring and data collection to assess what works and what doesn't. Adaptive management means that every program is regarded as experimental and subject to change as new information becomes available. Courtesy Jennifer Stelton-Kivioja.

doing business. Accomplishing meaningful changes in entrenched bureaucracies may be the most challenging aspect of this new approach.

6. *Consider humans in nature:* People cannot be separated from nature. Humans inescapably affect ecological patterns and processes and are in turn affected by them.

7. *Identify values:* Regardless of the role of scientific knowledge, human values play a dominant role in ecosystem management goals.

Some commonly reported goals for ecosystem management are presented in table 13.4. In general, the criteria listed by Aldo Leopold, as the basis of environmental ethics—integrity, stability, and beauty of biotic communities—are good benchmarks for ecosystem management.

TABLE 13.4 Ecosystem Management Goals

- Maintain viable populations of native species *in situ*.

- Represent, within protected areas, all native ecosystem types across their natural range of variation.

- Protect essential ecological processes such as nutrient cycles, succession, hydrologic processes, etc.

- Manage over long enough time periods to sustain the evolutionary potential of species and ecosystems.

- Accommodate human use and occupancy within these constraints.

Conflicting Views of Restoration and Ecosystem Management

Is restoration a good strategy for protecting nature? Preservationists often argue that when we assume that we can restore ecosystems, we justify their continued destruction. Just as our reclamation efforts are often misguided, our efforts to restore, rehabilitate, and remediate damaged systems are often unsuccessful and almost always incomplete. Some preservationists say we are arrogant to assume that we have the ability to restore natural systems. Instead, we need to prevent environmental destruction in the first place.

Restorationists counter that we are unlikely to preserve more than small areas of pristine parks and nature reserves. These areas are important, but they aren't sufficient to maintain biological diversity. Isolated preserves tend to deteriorate over time, so preserves are also often unsuccessful and incomplete. Some restorationists point out that current ecological theory holds that ecosystems change naturally. If restored ecosystems differ from original systems, is that necessarily unnatural? Furthermore, many ecosystems are already damaged but could be restored, at least partially. Restorationists hope that increasing knowledge of ecosystem dynamics, wetland hydrology, water chemistry, and other topics will improve the effectiveness of restoration efforts. Finally, some restorationists say that since we have an unavoidable influence on nature, we should work to improve it while we extract its resources.

What do you think? Should we assume that we can rely on restoration to repair damaged systems? Or is this assumption a dangerous justification for further environmental destruction? Can we ensure that restoration serves nature, rather than just human needs?

Similar questions arise regarding ecosystem management, which presumes that we understand ecosystems well enough to manage them. Many ecologists argue that we probably cannot understand the complex, chaotic, and unpredictable processes of natural systems. Thus it is arrogant to suppose we can direct them.

Furthermore, ecosystem management demands ongoing investment in data gathering and adaptive management. In developing countries, political pressures and immediate needs of poor people make even basic environmental protection difficult and expensive. Ecosystem management could be a dangerously ambitious goal in impoverished regions. Advocates of ecosystem management counter that their approach is necessary for long-term environmental health, even in developing areas. Perhaps ecosystem management is also the best approach for protecting impoverished people, as well, since they often depend on healthy ecosystems for subsistence—hunting, fishing, and farming.

The debate over ecosystem management reflects a deep division that runs through much of environmental science. The cleavage between preservation and management, between conservative and progressive policies, and between pessimism and optimism all are based on different worldviews about what is possible and what is likely to occur. Where do you stand in this spectrum of opinions? Do you think we can—or should even try to—manage ecosystems; restore, reclaim, or rehabilitate landscapes; or create new and imaginative forms of nature? Is it possible for us to do what is necessary to preserve the integrity, stability, and beauty of our environment?

Summary

- Parks, wildlife refuges, wilderness areas, and nature preserves occupy a small percentage of our land area, but they protect valuable cultural and biological resources. Parks have evolved from elite pleasure grounds to popular recreation areas. The number of North American monuments and wildlife areas has grown in recent years, but many new conservation areas are controversial and are being challenged by industry interests.

- Wildlife lands and parks serve multiple, often contradictory demands, including many forms of recreation, wildlife habitat, and extraction of timber, minerals, oil, and gas.

- Global protected areas have expanded nearly tenfold in the past 50 years, to nearly 10 percent of world land area. These lands must meet diverse needs, however, including hunting and farming, and actual protection is rarely as secure as it looks on paper.

- In developing countries, wildlife lands have often suffered poaching and encroachment from farmers and loggers. Recent efforts, such as the Man and Biosphere program, have attempted to compromise between human and ecological needs.

- Landscape ecology is the study of the reciprocal effects of spatial patterns and ecological processes. History and human actions are important considerations in understanding most ecological functions at a landscape scale.

- Patchiness and heterogeneity are characteristics of most landscapes. Movement of organisms and materials among patches helps regulate conditions in many ecosystems.

- Ecosystem management involves managing whole landscapes over long time scales, taking human needs into account, and maintaining biodiversity and ecological systems.

- Restoration ecology is the science of restoring, rehabilitating, or re-creating ecological systems. Wetland mitigation, grassland restoration, and species conservation are all important themes of restoration ecology.

- The goals of restoration ecology include maintaining original species in their original locations. Just how authentic restoration can be, or should be, is a major topic of debate.

- Floodplains and wetlands provide essential ecosystem services. Reversing the decline of wetlands and floodplains is an important issue in environmental restoration. In theory, floodplain and wetland restoration should be cost-effective, as well, since they could reduce the costs of flood damage.

Questions for Review

1. Why is Yosemite called the first park to preserve wilderness, while Yellowstone is the first national park?

2. List some problems and threats from inside and outside our national parks.

3. Why is the reintroduction of wolves into Yellowstone a controversial issue?

4. Which countries have the most protected lands? How has the amount of global protected lands changed in the past 50 years?

5. Draw a diagram of an ideal MAB reserve. What activities would be allowed in each zone?

6. What is the legal definition of wilderness in the United States?

7. Describe some ecological values of wetlands and how they are threatened by human activities.

8. Which major biomes have been most heavily disturbed by human activities?

9. Define a landscape. Describe the major ecological and cultural features of the landscape in which you live.

10. Define *restoration, rehabilitation, remediation, reclamation, ecological re-creation,* and *mitigation.* Give an example of each.

Questions for Critical Thinking

1. Is "contrived" naturalness a desirable feature in parks and nature preserves? How much human intervention do you think is acceptable in trying to make nature more beautiful, safe, comfortable, or attractive to human visitors? Think of some specific examples that you would or would not accept.

2. Suppose you were superintendent of Yellowstone National Park. How would you determine the carrying capacity of the park for elk? How would you weigh having more elk or more ground squirrels? If there are too many elk, how would you thin the herd?

3. What are the differences and similarities between indigenous rights to park resources in developing countries and local rights to timber and minerals in conservation areas in the United States or Canada?

4. Why do you suppose that dry tropical forests and tundra are well represented in protected areas, whereas grasslands and wetlands are rarely protected? Consider social, cultural, and economic as well as biogeographical reasons in your answer.

5. Suppose that preserving healthy populations of grizzly bears and wolves requires that we set aside some large fraction of our national parks as a zone into which no humans will ever again be allowed to enter for any reason. Would you support protecting bears even if no one ever sees them or could even be sure that they still existed?

6. Suppose that you had trespassed into the bear sanctuary and were attacked by a bear. Should the rangers shoot the bear or let it eat you?

7. Explain why there might be differences in philosophy or worldview between preservationists and restorationists. Which approach do you prefer?

8. Authenticity is a contentious issue in restoration. Is it necessary, or even desirable, to try to create an exact replica of an original ecosystem? Could the changes that humans make be considered part of natural change?

9. Some environmentalists worry that the science of restoration ecology may give us the arrogant attitude that we can do anything we want now because we can repair the damage later. How would you respond to that concern?

Key Terms

Further Readings

Bojorquez-Tapia, et al. 2003. Mapping expert knowledge: Redesigning the Monarch Butterfly Biosphere Reserve. *Conservation Biology* 17(2):367–79.

Campbell, G. S., P. G. Blackwell, and F. I. Woodward. 2002. Can landscape-scale characteristics be used to predict plant invasions along rivers? *Journal of Biogeography* 29(4):535–44.

Gutzwiller, K. 2002. *Applying Landscape Ecology in Biological Conservation.* Springer.

Holl, Karen D., Elizabeth E. Crone, and Cheryl B. Schultz. 2003. Landscape restoration: Moving from generalities to methodologies. *Bioscience* 53(5):491–502.

Hutchings, M. J., E. A. John, and A. J. A. Stewart. 2000. *The Ecological Consequences of Environmental Heterogeneity.* Blackwell.

Purcell, A. H., C. Friedrich, and V. H. Resh. 2002. An assessment of a small urban stream restoration project in northern California. *Restoration Ecology* 10(4):685–94.

Turner, Monica G. 1989. Landscape ecology: The effect of pattern on process. *Annual Review of Ecology and Systematics* 20:171–97.

Welcome to McGraw-Hill's Online Learning Center

Location: http://www.mhhe.com/environmentalscience

McGraw Hill

WEB EXERCISES

World Protected Areas

The United Nations Environment Program (UNEP) is one of the main sources of information on the global environment. You can map (or graph) protected areas on the UNEP website. Go to http://geodata.grid.unep.ch/. You'll see an option to "search the GEO Database." In the search space, enter *protected areas.* The search should return a list of variables, such as extent of protected mangroves (by number of sites or by total area), number or area of marine protected areas, and so on. Some of these variables are grouped by country, some by region (e.g., Africa), and some by subregion (e.g., southern Africa).

Select *Mangroves, Forest Extent* (at the national level). Then click *continue.* The next window asks you to select a year to map, but you have only one option, so click *continue* again. Next, select the *Draw Map* option. Look at the legend below the map, showing ranges in extent of protected mangroves. Which countries have the most protected mangroves?

Zoom in on Indonesia (using the *Basic Tools, Zoom In* option). As you zoom in, you should see the shape of the ocean floor. Can you see the flat shelves around Indonesia where mangroves might flourish?

Now check which countries have the maximum amount of protected mangroves: click on the *show extremes* box, then click on the map to redraw. Which five countries have the most?

Go back to the Geo Data Portal website, and you can return to the list of data sets by clicking on *2. Dataset,* just above the *Draw Map* option. Select another variable to map. Protected Areas (IUCN Categories I–VI) is a good option. After you've made your map, you can use the *show extremes* option again. Which countries have the most protected areas? Which have the least? Can you think of some explanations for these extremes?

Note that you can make bar graphs to compare selected countries, you can look at (and download) data tables, and if you select a variable with multiple dates, you can draw a line graph showing change over time.

Parks and Preserves in Australia

Many countries have national park and nature preserve systems. Australia provides information on these areas on the Web. Go to www.ea.gov.au/parks/, and choose a national park or nature preserve by clicking on a name in the list in the left frame. Look at the web page for that site. What is the site like? Where is it? What characteristics make it worth preserving? Is there anything like it near where you live, or have you ever been to a place like this one?

Wetland Losses in the United States

Wetland losses represent habitat losses for a huge variety of wildlife, as well as a reduction of water storage capacity on land. Look at the Environmental Protection Agency's map of wetland losses: this map is halfway down the page at www.epa.gov/wateratlas/geo/maplist.html. To see a larger version of this map, click on the map image. How severe is wetland loss in your area? What reasons can you think of to explain wetland losses in your area? Where are wetland losses most severe? What land-use activities might explain losses in those areas? Now look at the map of aquatic/wetland species at risk, on the same page. What is the date on this map? Do you suppose conditions have improved or worsened since that date?

14

Geology and Earth Resources

Nothing hurries geology.

Mark Twain

OBJECTIVES

After studying this chapter, you should be able to:

- understand some basic geologic principles, including how tectonic plate movements affect conditions for life on the earth.
- explain how the three major rock types are formed and how the rock cycle works.
- summarize economic mineralogy and strategic minerals.
- discuss the environmental effects of mining and mineral processing.
- recognize the geologic hazards of earthquakes, volcanoes, and tsunamis.

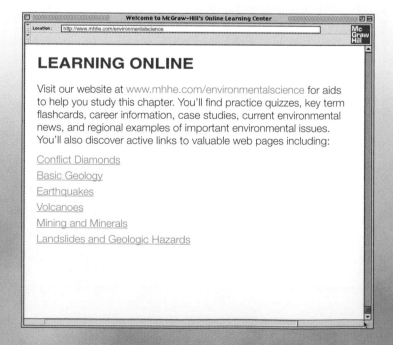

LEARNING ONLINE

Visit our website at www.mhhe.com/environmentalscience for aids to help you study this chapter. You'll find practice quizzes, key term flashcards, career information, case studies, current environmental news, and regional examples of important environmental issues. You'll also discover active links to valuable web pages including:

Conflict Diamonds

Basic Geology

Earthquakes

Volcanoes

Mining and Minerals

Landslides and Geologic Hazards

Photo: Geologic resources lead to some of our worst environmental problems and some of our greatest sources of wealth. Courtesy of Kennecott.

Conflict Diamonds

Someday soon, you may find yourself giving or receiving a diamond ring. Or perhaps you have a diamond earring or necklace. Are there hidden environmental and social costs in the jewelry so many of us wear?

Mineral wealth has always been both a blessing and a curse. Economies and civilizations are built on precious metals, gems, and other mineral resources. But where political and social conditions are unstable, competition to extract diamonds, gold, and other resources can become a free-for-all, a race to plunder the earth's wealth. In Sierra Leone, Angola, the Democratic Republic of Congo, Liberia, and Rwanda, diamond mining has paid for weapons and ammunition on both sides (or sometimes several sides) of brutal civil wars. When Sierra Leone's Revolutionary United Front invaded from Liberia in 1991, its main goal was to capture diamond fields. This army became infamous for its practice of raping and maiming innocent civilians, hacking off hands and arms with machetes, and abducting children for slave labor in the diamond fields. More than half of Sierra Leone's population was displaced, and 75,000 people died in the rebellion. Liberia's infamous Charles Taylor supported the invasion in order to share in the profits, and his armies may have pursued similar tactics of brutality and slave labor. In a similar fashion, Rwanda invaded Congo to gain control over diamonds, gold, and other minerals. In Angola, a quarter-century of civil war was fueled mainly by oil and diamonds. Environmental costs of these wars are incalculable: they include decimated wildlife, poached timber, polluted waters, and devastated farmlands.

Diamonds are portable and valuable, and tracing their point of origin has always been next to impossible. Since 2000, importing countries have banned trade in diamonds from Sierra Leone, Angola, and other countries where the link between the gems and brutality was especially clear. These bans proved only a slight obstacle. Disguising sources of origin is an easy task in lawless regions, and smuggling diamonds across porous borders, or between corrupt governments, is relatively easy. Small diamond pro-

© Corbis Royalty-Free Website.

ducers, such as the Central African Republic and Ivory Coast, soon began to export vastly more diamonds than ever before—and more than they could have produced. In 2003, as the international community struggled to decide how to deal with Liberia's civil war, international calls went out to slow the slaughter by stopping all sales of Liberian diamonds, gold, and other raw resources.

These conflicts are not only local. There is evidence that the Islamic terrorist organization al Qaeda financed much of its work through the diamond trade. Al Qaeda mined and traded diamonds through operatives in Kenya, Tanzania, and Sierra Leone, one of the first internationally banned exporters.

The majority of world diamond trade, between 80 percent and 96 percent, comes from legal sources. Legitimate traders worry, though, that illicit diamonds will undermine the entire market. Stakes are high: the global diamond trade is worth $7 billion per year in rough diamonds alone. Two-thirds of the $50 billion market in diamond jewelry sells in the United States. To protect the market, more than 70 countries have signed an agreement to require certification of origin for all diamonds traded internationally. The plan is to make certificates that are tightly controlled and hard to forge. (Ironically, this effort is being led by South Africa, whose racist apartheid regime was financed by diamonds and other precious minerals up to the 1990s.)

Gold and silver jewelry also have environmental and social costs. Gold (also mined almost exclusively for jewelry), copper, and other precious metals are responsible for some of the world's worst water pollution, habitat destruction, and air pollution. Precious metals also fuel, or incite, the same brutalities as diamonds do in Africa, South America, and Asia.

These minerals are also the basis of the modern global, technological society. In a world of free trade, do all of us share responsibility for financing distant disasters? How can we reduce the damage closer to home, where mining is also an essential industry and a cause of environmental devastation? We'll ask you to consider these and other questions in environmental geology as you read this chapter.

A DYNAMIC PLANET

Although we think of the ground under our feet as solid and stable, the earth is a dynamic and constantly changing structure. Titanic forces stir inside the earth, causing continents to split, drift apart, and then crash into each other in slow but inexorable collisions. In this section, we will look at the structure of our planet and some of the forces that shape it.

A Layered Sphere

The **core,** or interior of the earth (fig. 14.1), is composed of a dense, intensely hot mass of metal—mostly iron—thousands of kilometers in diameter. Solid in the center but more fluid in the

outer core, this immense mass generates the magnetic field that envelops the earth.

Surrounding the molten outer core is a hot, pliable layer of rock called the **mantle.** The mantle is much less dense than the core because it contains a high concentration of lighter elements, such as oxygen, silicon, and magnesium.

The outermost layer of the earth is the cool, lightweight, brittle **crust** of rock that floats on the mantle something like the "skin" on a bowl of warm chocolate pudding. The oceanic crust, which forms the seafloor, has a composition dominated by iron, magnesium, and silicon. Continents are thicker, lighter regions of crust rich in calcium, sodium, potassium, and aluminum. The continents rise above both the seafloor and the ocean surface. Table 14.1 compares the composition of the whole earth (dominated by the dense core) and the crust.

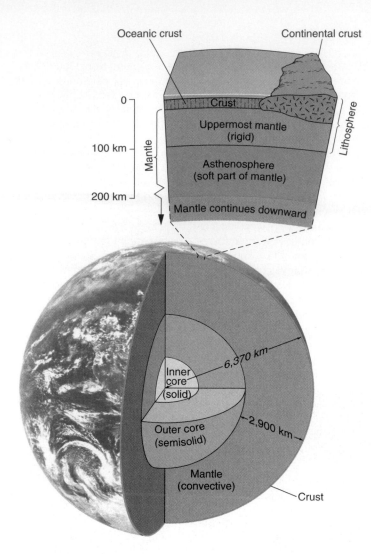

FIGURE 14.1 Earth's cross-section. Slow convection in the mantle causes the thin, brittle crust to move.

TABLE 14.1	Eight Most Common Chemical Elements (percent)		
WHOLE EARTH		**CRUST**	
Iron	33.3	Oxygen	45.2
Oxygen	29.8	Silicon	27.2
Silicon	15.6	Aluminum	8.2
Magnesium	13.9	Iron	5.8
Nickel	2.0	Calcium	5.1
Calcium	1.8	Magnesium	2.8
Aluminum	1.5	Sodium	2.3
Sodium	0.2	Potassium	1.7

Tectonic Processes and Shifting Continents

Hot enough to flow, the upper layer of the mantle has huge convection currents that break the overlying crust into a mosaic of huge blocks called **tectonic plates** (fig. 14.2). Convection in the mantle moves these plates across the earth's surface, breaking apart in some areas and ponderously crushing into each other in other places (fig. 14.3). Continents separate to form ocean basins, or they merge to create new continents. Where oceanic plates separate (at *divergent* plate boundaries), **magma** (molten rock) forces up through cracks to form mid-ocean ridges. These ridges make up the largest mountain range in the world, snaking around the earth for 74,000 km. Although concealed beneath the ocean, these ridges have higher peaks and deeper canyons than any continental mountains. This is because they are sheltered from the erosion that goes on above the water's surface.

When an oceanic plate collides with a continental landmass (at *convergent* boundaries), the continental plate rides up over the seafloor, and the oceanic plate is subducted, or pushed down into

the mantle. As it sinks, the crust material melts. Often it rises back to the surface as magma (fig. 14.3). Deep ocean trenches mark these subduction zones, and volcanoes form where the magma erupts through vents and fissures in the overlying crust. All around the Pacific Ocean rim from Indonesia to Japan to Alaska and down the West Coast of the Americas is the so-called "ring of fire" where oceanic plates are being subducted under the continental plates. This ring is the source of more earthquakes and volcanic activity than any other place on the earth.

Earthquakes are caused by the grinding and jerking as plates slide past each other or as they converge or diverge. Mountain ranges are pushed up at the margins of colliding continental plates. The Atlantic Ocean is growing slowly as Europe and Africa drift away from the Americas. The Himalayas are still rising as the Indian subcontinent smashes into Asia. Southern California is slowly sailing north toward Alaska. In about 30 million years, Los Angeles will pass San Francisco, if either still exists by then.

The speed of this movement is sometimes compared to the rate at which your fingernails grow. Actually, plate movement varies a good deal, from 1 cm per year to as much as 18 cm per year at seafloor spreading zones.

Over millions of years, the drifting continents can move long distances. Antarctica and Australia once were connected to Africa, for instance, somewhere near the equator, and supported luxuriant forests. Over time, continents have formed supercontinents such as Pangaea (fig. 14.4). Ocean circulation changed as continents gathered and disbursed, causing profound changes in global climates. These climate changes may help explain the periodic mass extinctions of organisms marking the divisions between many major geologic periods (fig. 14.5).

ROCKS AND MINERALS

A **mineral** is a naturally occurring, inorganic, solid element or compound with a definite chemical composition and a regular internal crystal structure. Naturally occurring means not created by humans (or synthetic). Organic materials, such as coal, produced by living organisms or biological processes are generally not min-

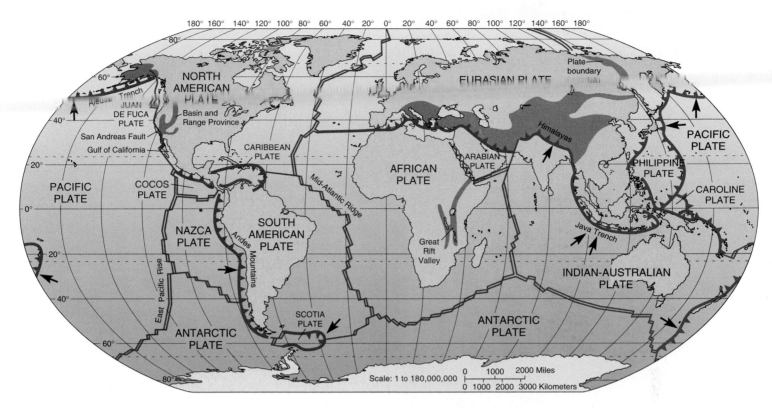

FIGURE 14.2   Map of tectonic plates. Plate boundaries are dynamic zones, characterized by earthquakes and volcanism and the formation of great rifts and mountain ranges. Arrows indicate direction of subduction where one plate is diving beneath another. These zones are sites of deep trenches in the ocean floor and high levels of seismic and volcanic activity. *Sources:* U.S. Department of the Interior, U.S. Geological Survey.

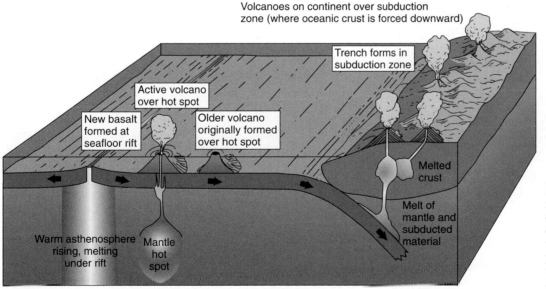

FIGURE 14.3   Tectonic plate movement. Where thin, oceanic plates diverge, upwelling magma forms mid-ocean ridges. A chain of volcanoes, like the Hawaiian Islands, may form as plates pass over a "hot spot." Where plates converge, melting can cause volcanoes, such as the Cascades.

erals. The two fundamental characteristics of a mineral that distinguish it from all other minerals are its chemical composition and its crystal structure. No two minerals are identical in both respects. Once purified, metals such as iron, aluminum, or copper lack a crystal structure, and thus are not minerals. The ores from which they are extracted, however, are minerals and make up an important part of economic mineralogy.

A **rock** is a solid, cohesive, aggregate of one or more minerals. Within the rock, individual mineral crystals (or grains) are mixed together and held firmly in a solid mass. The grains may be large or small, depending on how the rock was formed, but each grain retains its own unique mineral qualities. Each rock type has a characteristic mixture of minerals (and therefore of different chemical elements), grain sizes, and ways in which the grains are

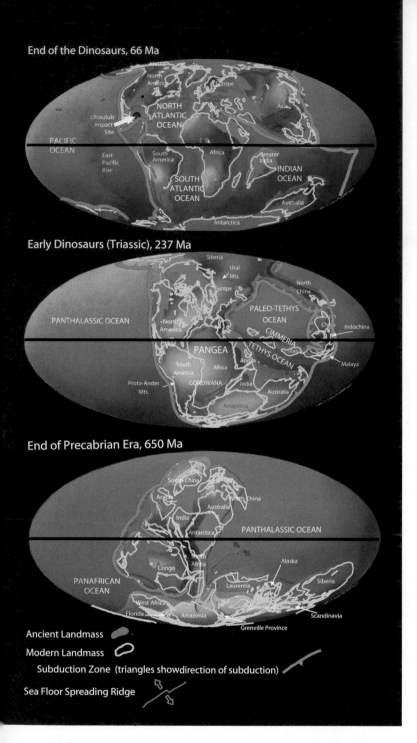

End of the Dinosaurs, 66 Ma

Early Dinosaurs (Triassic), 237 Ma

End of Precabrian Era, 650 Ma

Ancient Landmass

Modern Landmass

Subduction Zone (triangles showdirection of subduction)

Sea Floor Spreading Ridge

FIGURE 14.4 Continents have shifted constantly over geologic time. These three snapshots show their position when multicellular life became abundant, when dinosaurs appeared, and when mammals became abundant. Dates are shown in millions of years (Ma).

Source: Paleogeographic maps by Christopher R. Scotese, © 2003 PALEOMAP Project (www.scotese.com).

mixed and held together. Granite, for example, is a mixture of quartz, feldspar, and mica crystals. Different kinds of granite have distinct percentages of these minerals and particular grain sizes depending on how quickly the rock solidified. These minerals, in turn, are made up of a few elements such as silicon, oxygen, potassium, and aluminum.

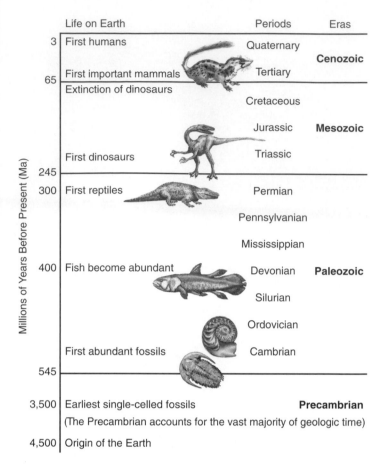

FIGURE 14.5 Periods and eras in geologic time, and major life-forms that mark some periods.

Rock Types and How They Are Formed

What could be harder and more permanent than rocks? Like the continents they create, rocks are also part of a relentless cycle of formation and destruction. They are made and then torn apart, cemented together by chemical and physical forces, crushed, folded, melted, and recrystallized by dynamic processes related to those that shape the large-scale features of the crust. We call this cycle of creation, destruction, and metamorphosis the **rock cycle** (fig. 14.6). Understanding something of how this cycle works helps explain the origin and characteristics of different types of rocks, as well as how they are shaped, worn away, transported, deposited, and altered by geologic forces.

There are three major rock classifications: igneous, sedimentary, and metamorphic. In this section, we will look at how they are made and some of their properties.

Igneous Rocks

The most common rock-type in the earth's crust is solidified from magma, welling up from the earth's interior. These rocks are classed as **igneous rocks** (from *igni,* the Latin word for fire). Magma extruded to the surface from volcanic vents cools quickly to make

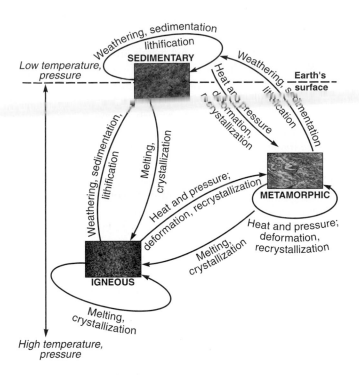

FIGURE 14.6   The rock cycle includes a variety of geologic processes that can transform any rock.

FIGURE 14.7   Weathering slowly reduces an igneous rock to loose sediment. Here, exposure to moisture expands minerals in the rock, and frost may also force the rock apart. © David McGeary.

basalt, rhyolite, andesite, and other fine-grained rocks. Magma that cools slowly in subsurface chambers or is intruded between overlying layers makes granite, gabbro, or other coarse-grained crystalline rocks, depending on its specific chemical composition.

Weathering and Sedimentation

Most of these crystalline rocks are extremely hard and durable, but exposure to air, water, changing temperatures, and reactive chemical agents slowly breaks them down in a process called **weathering** (fig. 14.7). *Mechanical weathering* is the physical breakup of rocks into smaller particles without a change in chemical composition of the constituent minerals. You have probably seen mountain valleys scraped by glaciers or river and shoreline pebbles that are rounded from being rubbed against one another as they are tumbled by waves and currents. *Chemical weathering* is the selective removal or alteration of specific components that leads to weakening and disintegration of rock. Among the more important chemical weathering processes are oxidation (combination of oxygen with an element to form an oxide or hydroxide mineral) and hydrolysis (hydrogen atoms from water molecules combine with other chemicals to form acids). The products of these reactions are more susceptible to both mechanical weathering and to dissolving in water. For instance, when carbonic acid (formed when CO_2 and H_2O combine) percolates through porous limestone layers in the ground, it dissolves the calcium carbonate (limestone) and creates caves.

Particles of rock are transported by wind, water, ice, and gravity until they come to rest again in a new location. The deposition of these materials is called **sedimentation.** Waterborne particles from sediments cover ocean continental shelves and fill valleys and plains. Most of the American Midwest, for instance, is covered with a layer of sedimentary material hundreds of meters thick in the form of glacierborne till (rock debris deposited by glacial ice), windborne loess (fine dust deposits), riverborne sand and gravel, and ocean deposits of sand, silt, and clay. Deposited material that remains in place long enough, or is covered with enough material to compact it, may once again become rock. Some examples of **sedimentary rock** are shale (compacted mud), sandstone (cemented sand), tuff (volcanic ash), and conglomerates (aggregates of gravel, sand, silt, and clay).

Sedimentary rocks are also formed from crystals that precipitate out of, or grow from a solution. An example is rock salt, made of the mineral halite, which is the name for ordinary table salt (sodium chloride). Salt deposits often form when a body of salt water dries up and salt crystals are left behind.

Sedimentary formations often have distinctive layers that show different conditions when they were laid down. Sedimentary rocks such as sandstone can be shaped by erosion into striking features (fig. 14.8). Geomorphology is the study of the processes that shape the earth's surface and the structures they create.

Humans have become a major force in shaping landscapes. Geomorphologist Roger Hooke, of the University of Maine, looking only at housing excavations, road building, and mineral production, estimates that we move somewhere around 30 to 35 gigatons (billion tons or Gt) per year worldwide. When combined with the 10 Gt each year that we add to river sediments through erosion, our earth-moving prowess is comparable to, or greater than, any other single geomorphic agent except plate tectonics.

Metamorphic Rocks

Preexisting rocks can be modified by heat, pressure, and chemical agents to create new forms called **metamorphic rock.** Deeply buried strata of igneous, sedimentary, and metamorphic rocks are

FIGURE 14.8 Soft sedimentary rock layers can be sculpted by erosion into remarkable shapes such as Rainbow Bridge in Utah.
© William P. Cunningham.

subjected to great heat and pressure by deposition of overlying sediments or while they are being squeezed and folded by tectonic processes. Chemical reactions can alter both the composition and structure of the rocks as they are metamorphosed. Some common metamorphic rocks are marble (from limestone), quartzite (from sandstone), and slate (from mudstone and shale). Metamorphic rocks are often the host rock for economically important minerals such as talc, graphite, and gemstones.

ECONOMIC GEOLOGY AND MINERALOGY

Economic mineralogy is the study of minerals that are valuable for manufacturing and are, therefore, an important part of domestic and international commerce. Most economic minerals are metal-bearing ores (table 14.2). Nonmetallic economic minerals are mostly graphite, some feldspars, quartz crystals, diamonds, and many other crystals that are valued for their usefulness, beauty, and/or rarity. Metals have been so important in human affairs that major epochs of human history are commonly known by the dominant materials and the technology to use them (Stone Age, Bronze Age, Iron Age, etc.). The mining, processing, and distribution of these materials have broad and varied implications both for culture and our environment. Most economically valuable crustal resources exist everywhere in small amounts; the important thing is to find them concentrated in economically recoverable levels.

Public policy in the United States has encouraged mining on public lands as a way of boosting the economy and utilizing natural resources. Today many people think these laws seem outmoded and in need of reform (What Do You Think? p. 295).

Metals

How has the quest for minerals and metals affected global development? We will focus first on world use of metals, earth resources that always have received a great deal of human attention. The availability of metals and the methods to extract and use them have determined technological developments, as well as economic and political power for individuals and nations. We still are strongly dependent on the unique lightness, strength, and malleability of metals.

The metals consumed in greatest quantity by world industry include iron (740 million metric tons annually), aluminum (40 million metric tons), manganese (22.4 million metric tons), copper and chromium (8 million metric tons each), and nickel (0.7 million metric tons). Most of these metals are consumed in the United States, Japan, and Europe, in that order. They are produced in many countries, often in mountainous areas, where heat and pressure of mountain building tend to concentrate ores (fig. 14.9). The worldwide mineral trade has become crucially important to the economic and social stability of all nations involved.

Nonmetal Mineral Resources

Nonmetal minerals are a broad class that covers resources from silicate minerals (gemstones, mica, talc, and asbestos) to sand, gravel, salts, limestone, and soils. Sand and gravel production comprise by far the greatest volume and dollar value of all nonmetal mineral resources and a far greater volume than all metal ores. Sand and gravel are used mainly in brick and concrete construction, paving, as loose road filler, and for sandblasting. High-purity silica sand is our source of glass. These materials usually are retrieved from surface pit mines and quarries, where they have been deposited by glaciers, winds, or ancient oceans.

TABLE 14.2 	Primary Uses of Some Major Metals Consumed in the United States
METAL	**USE**
Aluminum	Packaging foods and beverages (38%), transportation, electronics
Chromium	High-strength steel alloys
Copper	Building construction, electric and electronic industries
Iron	Heavy machinery, steel production
Lead	Leaded gasoline, car batteries, paints, ammunition
Manganese	High-strength, heat-resistant steel alloys
Nickel	Chemical industry, steel alloys
Platinum-group	Automobile catalytic converters, electronics, medical uses
Gold	Medical, aerospace, electronic uses; accumulation as monetary standard
Silver	Photography, electronics, jewelry

What do you think?

Should We Revise Mining Laws?

In 1872, the U.S. Congress passed the General Mining Law intended to encourage prospectors to open up the public domain and promote commerce. This law, which has been in effect more than a century, allows miners to stake an exclusive claim anywhere on public lands and to take—for free—any minerals they find. Claim holders can "patent" (buy) the land for $2.50 to $5 per acre (0.4 hectares) depending on the type of claim. Once the patent fee is paid, the owners can do anything they want with the land, just like any other private property. Although $2.50 per acre may have been a fair market value in 1872, many people regard it as ridiculously low today, amounting to a scandalous give-away of public property.

In Nevada, for example, a mining company paid $9,000 for federal land that contains an estimated $20 billion worth of precious metals. Similarly, Colorado investors bought about 7,000 ha (17,000 acres) of rich oil-shale land in 1986 for $42,000 and sold it a month later for $37 million. You don't actually have to find any minerals to patent a claim. A Colorado company paid a total of $400 for 65 ha (160 acres) it claimed would be a gold mine. Almost 20 years later, no mining has been done, but the property—which just happens to border the Keystone Ski Area—is being subdivided for condos and vacation homes.

According to the Bureau of Land Management (BLM), some $4 billion in minerals are mined each year on U.S. public lands. Under the 1872 law, mining companies don't pay a penny for the ores they take. Furthermore, they can deduct a depletion allowance from taxes on mineral profits. Former Senator Dale Bumpers of Arkansas, who calls the 1872 mining law "a license to steal," has estimated that the government could derive $320 million per year by charging an 8 percent royalty on all minerals and probably could save an equal amount by requiring a bond to be posted to clean up after mining is finished.

On the other hand, mining companies argue they would be forced to close down if they had to pay royalties or post bonds. Many people would lose jobs and the economies of western mining towns would collapse if mining becomes uneconomic. We provide subsidies and economic incentives to many industries to stimulate economic growth. Why not mining for metals essential for our industrial economy? Mining is a risky and expensive business. Without subsidies, mines would close down and we would be completely dependent on unstable foreign supplies.

Mining critics respond that other resource-based industries have been forced to pay royalties on materials they extract from public lands. Coal, oil, and gas companies pay 12.5 percent royalties on fossil fuels obtained from public lands. Timber companies—although they don't pay the full costs of the trees they take—have to bid on logging sales and clean up when they are finished. Even gravel companies pay for digging up the public domain. Ironically, we charge for digging up gravel, but give gold away free.

Over the past decade, numerous mining bills have been introduced in Congress. Those supported by environmental groups generally would require companies mining on federal lands to pay a higher royalty on their production. They also would eliminate the patenting process, impose stricter reclamation requirements, and give federal managers authority to deny inappropriate permits. In contrast, bills offered by Western legislators,

Thousands of abandoned mines on public lands poison streams and groundwater with acid, metal-laced drainage. This old mine in Montana drains into the Blackfoot River, the setting of Norman Maclean's book A River Runs Through It.
© Bryan F. Peterson.

and enthusiastically backed by mining supporters, tend to leave most provisions of the 1872 bill in place. They would charge a 2-percent royalty, but only after exploration, production, and other costs were deducted. Permit processes would consider local economic needs before environmental issues in this version. What do you think we should do about this mining law? How could we separate legitimate public-interest land use from private speculation and profiteering? Are current subsidies necessary and justifiable or are they just a form of corporate welfare?

Limestone, like sand and gravel, is mined and quarried for concrete and crushed for road rock. It also is cut for building stone, pulverized for use as an agricultural soil additive that neutralizes acidic soil, and roasted in lime kilns and cement plants to make plaster (hydrated lime) and cement.

Evaporites (materials deposited by evaporation of chemical solutions) are mined for halite, gypsum, and potash. These are often found at or above 97 percent purity. Halite, or rock salt, is used for water softening and melting ice on winter roads in some northern areas. Refined, it is a source of table salt. Gypsum (calcium sulfate) now makes our plaster wallboard, but it has been used for plaster ever since the Egyptians plastered the walls of their frescoed tombs along the Nile River some 5,000 years ago. Potash is an evaporite composed of a variety of potassium chlorides and

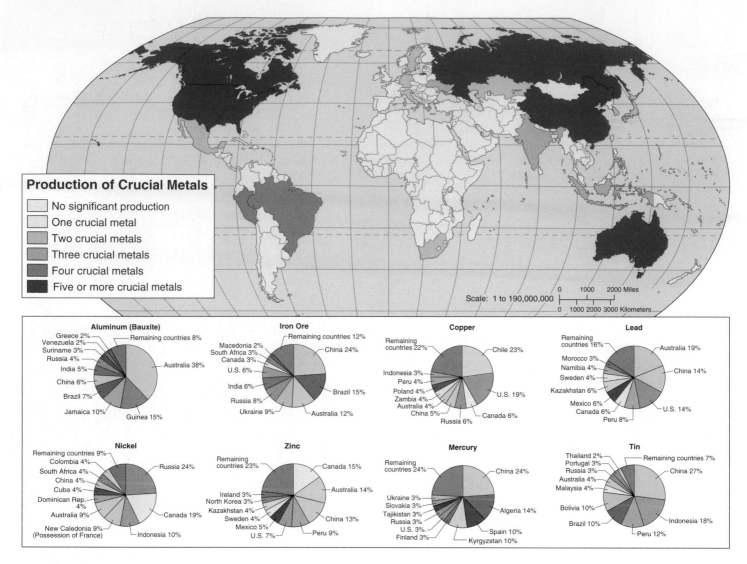

FIGURE 14.9 World production of metals most essential for an industrial economy. Principal consumers are the United States, Western Europe, Japan, and China.

potassium sulfates. These highly soluble potassium salts have long been used as a soil fertilizer.

Sulfur deposits are mined mainly for sulfuric acid production. In the United States, sulfuric acid use amounts to more than 200 lbs per person per year, mostly because of its use in industry, car batteries, and some medicinal products.

ENVIRONMENTAL EFFECTS OF RESOURCE EXTRACTION

Each of us depends daily on geologic resources mined from sites around the world. We use scores of metals and minerals, many of which we've never heard of, in our lights, computers, watches, fertilizers, and cars. Mining and purifying all these resources can have severe environmental and social consequences. The most obvious

effect of mining is often the disturbance or removal of the land surface. Farther-reaching effects, though, include air and water pollution. The EPA lists more than 100 toxic air pollutants, from acetone to xylene, released from U.S. mines every year. Nearly 80,000 metric tons of particulate matter (dust) and 11,000 tons of sulfur dioxide are released from nonmetal mining alone. Chemical- and sediment-runoff pollution is a major problem in many local watersheds.

Resource extraction can affect water quality for several reasons. Gold and other metals are often found in sulfide ores. Sulfide minerals often occur in hydrothermal (hot water) deposits, where sulfur and other elements are mobilized, then deposited as they are heated by deep magma chambers. When sulfur-bearing minerals are exposed to air and water, they produce sulfuric acid, which is highly mobile and strongly acidic. In addition, metal elements often occur in very low concentrations—10 to 20 parts per billion may be economically extractable for gold, platinum, and other metals. Consequently, vast quantities of ore must be crushed and washed

to extract metals. Cyanide, mercury, and other toxic substances are used to chemically separate metals from the minerals that contain them, and these substances can easily contaminate lakes and streams. Further, a great deal of water is used in washing crushed ore with cyanide and other solutions. In arid Nevada, the USGS estimates that mining consumes about 230,000 m³ (60 million gal) of fresh water per day. After use in ore processing, much of this water contains sulfuric acid, arsenic, heavy metals, and other contaminants. Mine runoff leaking into lakes and streams damages or destroys aquatic ecosystems.

Mining

There are many techniques for extracting geologic materials. The most common methods are open-pit mining, strip mining, and underground mining. An ancient method of accumulating gold, diamonds, and coal is placer mining, in which pure nuggets are washed from stream sediments. Since the California gold rush of 1849, placer miners have used water cannons to blast away hillsides. This method, which chokes stream ecosystems with sediment, is still used in Alaska, Canada, and many other regions. Another ancient, and much more dangerous, method is underground mining. Ancient Roman, European, and Chinese miners tunneled deep into tin, lead, copper, coal, and other mineral seams. Mine tunnels occasionally collapse, and natural gas in coal mines can explode. Water seeping into mine shafts also dissolves toxic minerals. Contaminated water seeps into groundwater; it is also pumped to the surface, where it enters streams and lakes.

In underground coal mines, another major environmental risk is fires. Hundreds of coal mines smolder in the United States, China, Russia, India, South Africa, and Europe. The inaccessibility and size of these fires make many impossible to extinguish or control. One mine fire in Centralia, Pennsylvania, has been burning since 1962; control efforts have cost at least $40 million, but the fire continues to expand. China, which depends on coal for much of its heating and electricity, has hundreds of smoldering mine fires; one has been burning for 400 years. According to a recent study from the International Institute for Aerospace Survey in the Netherlands, these fires consume up to 200 million tons of coal every year and emit as much carbon dioxide as all the cars in the United States. Toxic fumes, explosive methane, and other hazardous emissions are also released from these fires.

Open-pit mines are used to extract massive beds of metal ores and other minerals. The size of modern open pits can be hard to comprehend. The Bingham Canyon mine, near Salt Lake City, Utah (see opening photo of this chapter), is 800 m (2,640 ft) deep and nearly 4 km (2.5 mi) wide at the top. More than 5 billion tons of copper ore and waste material have been removed from the hole since 1906. A chief environmental challenge of open-pit mining is that groundwater accumulates in the pit. In metal mines, a toxic soup results. No one yet knows how to detoxify these lakes, which endanger wildlife and nearby watersheds. (See the related essay "Death in a Mine Pit" on the Online Learning Center.)

Half the coal used in the United States comes from strip mines (fig. 14.10). Since coal is often found in expansive, horizontal beds, the entire land surface can be stripped away to cheaply and quickly expose the coal. The overburden, or surface material, is placed back into the mine, but usually in long ridges called spoil banks. Spoil banks are very susceptible to erosion and chemical weathering. Since the spoil banks have no topsoil (the complex organic mixture that supports vegetation—see chapter 9), revegetation occurs very slowly.

The 1977 federal Surface Mining Control and Reclamation Act (SMCRA) requires better restoration of strip-mined lands, especially where mines replaced prime farmland. Since then, the record of strip-mine reclamation has improved substantially. Complete mine restoration is expensive, often more than $10,000 per hectare. Restoration is also difficult because the developing soil is usually acidic and compacted by the heavy machinery used to reshape the land surface.

Bitter controversy has grown recently over mountaintop removal, a coal mining method mainly practiced in Appalachia.

FIGURE 14.10 Some giant mining machines stand as tall as a 20-story building and can scoop up thousands of cubic meters of rock per hour. © James P. Blair/National Geographic Image Collection.

FIGURE 14.11 Mountaintop removal mining is a relatively new, and deeply controversial, method of extracting Appalachian coal. © AP Photo/Bob Bird.

Long, sinuous ridge-tops are removed by giant, 20-story-tall shovels to expose horizontal beds of coal (fig. 14.11). Up to 215 m (700 ft) of ridge-top is pulverized and dumped into adjacent river valleys. The debris can be laden with selenium, arsenic, coal, and other toxic substances. At least 900 km (560 mi) of streams have been buried in West Virginia alone. Environmental lawyers have sued to stop the destruction of streams, arguing that it violates the Clean Water Act (chapter 24). In response, the Bush administration issued a "clarification" of the Clean Water Act, rewriting the law to allow stream filling by any sort of mining or industrial debris, rather than forbidding any stream destruction. Environmentalists charge that the new wording subverts the Clean Water Act, giving a green light to all sorts of stream destruction. West Virginians are deeply divided about mountaintop removal because they live in affected stream valleys but also depend on a coal economy. (See related essay, "Mining a Tropical Paradise" on the Online Learning Center.)

The Mineral Policy Center in Washington, D.C., estimates that 19,000 km (12,000 mi) of rivers and streams in the United States are contaminated by mine drainage. The EPA estimates that cleaning up impaired streams, along with 550,000 abandoned mines in the U.S., may cost $70 billion. Worldwide, mine closing and rehabilitation costs are estimated in the trillions of dollars. Because of the volatile prices of metals and coal, many mining companies have gone bankrupt before restoring mine sites, leaving the public responsible for cleanup. In 2002, more than 500 leading mine executives and their critics convened at the Global Mining Initiative, a meeting in Toronto aimed at improving the sustainability of mining. Executives acknowledged that in the future they will increasingly be held liable for environmental damages, and they said they were seeking ways to improve the industry's social and environmental record. Jay Hair, secretary general of the International Council on Mining and Metals, stated that "environmental protection and social responsibility are important," and that mining companies were interested in participating in sustainable development. Mine executives also recognized that, increasingly, big cleanup bills will cut into company values and stock prices. Finding creative ways to keep mines cleaner from the start will make good economic sense, even if it's not easy to do.

Processing

Metals are extracted from ores by heating or with chemical solvents. Both processes release large quantities of toxic materials that can be even more environmentally hazardous than mining. **Smelting**—roasting ore to release metals—is a major source of air pollution. One of the most notorious examples of ecological devastation from smelting is a wasteland near Ducktown, Tennessee. In the mid-1800s, mining companies began excavating the rich copper deposits in the area. To extract copper from the ore, they built huge, open-air, wood fires, using timber from the surrounding forest. Dense clouds of sulfur dioxide released from sulfide ores poisoned the vegetation and acidified the soil over a 50 mi^2 (13,000 ha) area. Rains washed the soil off the denuded land, creating a barren moonscape.

Sulfur emissions from Ducktown smelters were reduced in 1907 after Georgia sued Tennessee over air pollution. In the 1930s the Tennessee Valley Authority (TVA) began treating the soil and replanting trees to cut down on erosion. Recently, upwards of $250,000 per year has been spent on this effort. While the trees and other plants are still spindly and feeble, more than two-thirds of the area is considered "adequately" covered with vegetation. Similarly, smelting of copper-nickel ore in Sudbury, Ontario, a century ago caused widespread ecological destruction that is slowly being repaired following pollution-control measures (see fig. 16.19).

Chemical extraction is used to dissolve or mobilize pulverized ore, but it uses and pollutes a great deal of water. A widely used method is **heap-leach extraction,** which involves piling crushed ore in huge heaps and spraying it with a dilute alkaline-cyanide solution (fig. 14.12). The solution percolates through the pile and dissolves gold. The gold-containing solution is then pumped to a processing plant that removes the gold by electrolysis. A thick clay pad and plastic liner beneath the ore heap is supposed to keep the poisonous cyanide solution from contaminating surface or groundwater, but leaks are common.

Once all the gold is recovered, mine operators may simply walk away from the operation, leaving vast amounts of toxic effluent in open ponds behind earthen dams. A case in point is the Summitville Mine near Alamosa, Colorado. After extracting $98 million in gold, the absentee owners declared bankruptcy in 1992, abandoning millions of tons of mine waste and huge leaking ponds of cyanide. The Environmental Protection Agency may spend more than $100 million trying to clean up the mess and keep the cyanide pool from spilling into the Alamosa River.

CONSERVING GEOLOGIC RESOURCES

Conservation offers great potential for extending our supplies of economic minerals and reducing the effects of mining and processing. The advantages of conservation are also significant: less waste to dispose of, less land lost to mining, and less consumption of money, energy, and water resources.

Recycling

Some waste products already are being exploited, especially for scarce or valuable metals. Aluminum, for instance, must be extracted from bauxite ore by electrolysis, an expensive, energy-intensive process. Recycling waste aluminum, such as beverage cans, on the other hand, consumes one-twentieth of the energy of extracting new aluminum. Today, nearly two-thirds of all aluminum beverage cans in the United States are recycled, up from only 15 percent 20 years ago. The high value of aluminum scrap ($650 a ton versus $60 for steel, $200 for plastic, $50 for glass, and $30 for paperboard) gives consumers plenty of incentive to deliver their cans for collection. Recycling is so rapid and effective that half of all the aluminum cans now on a grocer's shelf will be made into another can within two months. The energy cost of extracting other materials is shown in table 14.3.

Platinum, the catalyst in automobile catalytic exhaust converters, is valuable enough to be regularly retrieved and recycled from used cars (fig. 14.13). Other metals commonly recycled are gold, silver, copper, lead, iron, and steel. The latter four are readily available in a pure and massive form, including copper pipes, lead batteries, and steel and iron auto parts. Gold and silver are valuable enough to warrant recovery, even through more difficult means. See chapter 21 for further discussion of this topic.

TABLE 14.3 Energy Requirements in Producing Various Materials from Ore and Raw Source Materials

PRODUCT	ENERGY REQUIREMENT (MJ/KG)	
	NEW	FROM SCRAP
Glass	25	25
Steel	50	26
Plastics	162	n.a.
Aluminum	250	8
Titanium	400	n.a.
Copper	60	7
Paper	24	15

Source: *E. T. Hayes*, Implications of Materials Processing, *1997.*

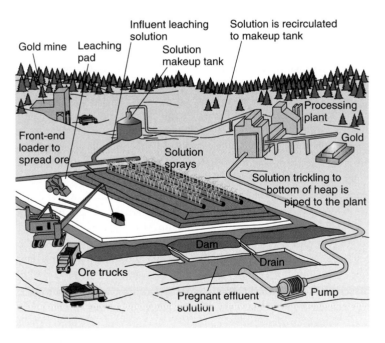

FIGURE 14.12 In a heap-leach operation, huge piles of low-grade ore are heaped on an impervious pad and sprayed continuously with a cyanide solution. As the leaching solution trickles through the crushed ore, it extracts gold and other precious metals. The "pregnant" effluent solution is then pumped to a processing plant where metals are extracted and purified. This technique is highly profitable but carries large environmental risks. *Source:* George Laycock, *Audubon Magazine,* vol. 91(7), July 1989, p. 77.

FIGURE 14.13 The richest metal source we have—our mountains of scrapped cars—offers a rich, inexpensive, and ecologically beneficial resource that can be "mined" for a number of metals. © Joseph Nettis/Photo Researchers, Inc.

Steel and Iron Recycling: Minimills

While total U.S. steel production has fallen in recent decades—largely because of inexpensive supplies from new and efficient Japanese steel mills—a new type of mill subsisting entirely on a readily available supply of scrap/waste steel and iron is a growing industry. Minimills, which remelt and reshape scrap iron and steel, are smaller and cheaper to operate than traditional integrated mills that perform every process from preparing raw ore to finishing iron and steel products. Minimills produce steel at between $225 and $480 per metric ton, while steel from integrated mills costs $1,425 to $2,250 per metric ton on average. The energy cost is likewise lower in minimills: 5.3 million BTU/ton of steel compared to 16.08 million BTU/ton in integrated mill furnaces. Minimills now produce about half of all of U.S. steel production. Recycling is slowly increasing as raw materials become more scarce and wastes become more plentiful.

Substituting New Materials for Old

Mineral and metal consumption can be reduced by new materials or new technologies developed to replace traditional uses. This is a long-standing tradition; for example, bronze replaced stone technology and iron replaced bronze. More recently, the introduction of plastic pipe has decreased our consumption of copper, lead, and steel pipes. In the same way, the development of fiber-optic technology and satellite communication reduces the need for copper telephone wires.

Iron and steel have been the backbone of heavy industry, but we are now moving toward other materials. One of our primary uses for iron and steel has been machinery and vehicle parts. In automobile production, steel is being replaced by polymers (long-chain organic molecules similar to plastics), aluminum, ceramics, and new, high-technology alloys. All of these reduce vehicle weight and cost, while increasing fuel efficiency. Some of the newer alloys that combine steel with titanium, vanadium, or other metals wear much better than traditional steel. Ceramic engine parts provide heat insulation around pistons, bearings, and cylinders, keeping the rest of the engine cool and operating efficiently. Plastics and glass fiber-reinforced polymers are used in body parts and some engine components.

Electronics and communications (telephone) technology, once major consumers of copper and aluminum, now use ultrahigh-purity glass cables to transmit pulses of light, instead of metal wires carrying electron pulses. Once again, this technology has been developed for its greater efficiency and lower cost, but it also affects consumption of our most basic metals.

GEOLOGIC HAZARDS

Earthquakes, volcanoes, floods, and landslides are normal earth processes, events that have made our earth what it is today. However, when they occur in proximity to human populations, they can be devastating, especially where people live in large but poorly built structures. For thousands of years people have been watching and recording these hazards, trying to understand them and learn how to avoid them.

Earthquakes

Earthquakes are sudden movements in the earth's crust that occur along faults (planes of weakness) where one rock mass slides past another one. When movement along faults occurs gradually and relatively smoothly, it is called creep or seismic slip and may be undetectable to the casual observer. When friction prevents rocks from slipping easily, stress builds up until it is finally released with a sudden jerk. The point on a fault at which the first movement occurs during an earthquake is called the epicenter.

Earthquakes have always seemed mysterious, sudden, and violent, coming without warning and leaving in their wake ruined cities and dislocated landscapes (table 14.4). Cities such as Kobe, Japan, or Mexico City, parts of which are built on soft landfill or poorly consolidated soil, usually suffer the greatest damage from earthquakes (fig. 14.14). Water-saturated soil can liquify when shaken. Buildings sometimes sink out of sight or fall down like a row of dominoes under these conditions.

Earthquakes frequently occur along the edges of tectonic plates, especially where one plate is subducting, or diving down, beneath another. Earthquakes also occur in the centers of continents, however. One of the largest earthquakes ever recorded in North America, with an estimated magnitude of 8.8, struck the area around New Madrid, Missouri, in 1811. This event triggered a series of very large earthquakes that continued for several weeks.

The worst death toll from an earthquake was in China in 1976, when an estimated 242,000 people died in collapsing buildings. Today contractors in earthquake zones are attempting to prevent damage and casualties by constructing buildings that can withstand tremors. The primary methods used are heavily reinforced structures, strategically placed weak spots in the building that can absorb vibration from the rest of the building, and pads or floats beneath the building on which it can shift harmlessly with ground motion. The problem of planning and building for earthquakes is at the center of a persistent dispute about nuclear waste storage at Yucca Mountain, Nevada (Case Study, p. 302).

One of the most deadly effects of earthquakes is the **tsunami.** These giant seismic sea swells (sometimes improperly called tidal waves) can move at 1,000 km/hr (600 mph), or faster, away from the center of an earthquake. When these swells approach the shore, they can easily reach 15 m or more and some can be as high as 65 m (nearly 200 ft). A 1960 tsunami coming from a Chilean earthquake still caused 7-meter breakers when it reached Hawaii fifteen hours later. Tsunamis also can be caused by underwater volcanic explosions or massive seafloor slumping. The eruption of the Indonesian volcano Krakatoa in 1883 created a tsunami 40 m (130 ft) high that killed 30,000 people on nearby islands.

While most earthquakes occur in known earthquake-prone areas, sometimes they strike in unexpected places. In the United States, major quakes have occurred in nearly every state.

TABLE 14.4 Worldwide Frequency and Effects of Earthquakes of Various Magnitudes

RICHTER SCALE MAGNITUDE*	DESCRIPTION	AVERAGE NUMBER PER YEAR	OBSERVABLE EFFECTS
2–2.9	Unnoticeable	300,000	Detected by instruments, but not usually felt by people.
3–3.9	Smallest felt	49,000	Hanging objects swing, vibrations like passing of light truck felt.
4–4.9	Minor earthquake	6,200	Dishes rattle, doors swing, pictures move, walls creak.
5–5.9	Damaging earthquake	800	Difficult to stand up. Windows, dishes break. Plaster cracks, loose bricks and tile fall, small slides along sand or gravel banks.
6–6.9	Destructive earthquake	120	Chimneys and towers fall, masonry walls damaged, frame houses move on foundations, small cracks in ground. Broken gas pipes start fires.
7–7.9	Major earthquake	18	General panic. Frame houses split and fall off foundations, some masonry buildings collapse, underground pipes break, large cracks in the ground.
8–8.9	Great earthquake	1 or 2	Catastrophic damage. Most masonry and frame structures destroyed. Roadways, dams, dikes collapse. Large landslides. Rails twist and bend. Underground pipes rupture.

*For every unit increase in the Richter scale, ground displacement increases by a factor of 10, while energy release increases by a factor of 30. There is no upper limit to the scale, but the largest earthquakes recorded have been 8.9.

Source: B. Gutenberg in Earth by F. Press and R. Siever, 1978, W. H. Freeman & Company.

FIGURE 14.14 An elevated freeway collapsed and buckled as a result of the 1995 earthquake in Kobe, Japan. © Reuter/Sankei Shimbun.

Volcanoes

Volcanoes and undersea magma vents produce much of the earth's crust. Over hundreds of millions of years, gaseous emissions from these sources formed the earth's earliest oceans and atmosphere. Many of the world's fertile soils are weathered volcanic materials. Volcanoes have also been an ever-present threat to human populations. One of the most famous historic volcanic eruptions was that of Mount Vesuvius in southern Italy, which buried the cities of Herculaneum and Pompeii in A.D. 79. The mountain had been giving signs of activity before it erupted, but many citizens chose to stay and take a chance on survival. On August 24, the mountain buried the two towns in ash. Thousands were killed by the dense, hot, toxic gases that accompanied the ash flowing down from the volcano.

Nuees ardentes (French for "glowing clouds") are deadly, denser-than-air mixtures of hot gases and ash like those that inundated Pompeii and Herculaneum. Temperatures in these clouds may exceed 1,000°C, and they move at more than 100 km/hour (60 mph). Nuees ardentes destroyed the town of St. Pierre on the Caribbean island of Martinique on May 8, 1902. Mount Pelee released a cloud of nuees ardentes that rolled down through the town, killing somewhere between 25,000 and 40,000 people within a few minutes. All the town's residents died except a single prisoner being held in the town dungeon.

Mudslides are also disasters sometimes associated with volcanoes. The 1985 eruption of Nevado del Ruiz, 130 km (85 mi) northwest of Bogotá, Colombia, caused mudslides that buried most of the town of Armero and devastated the town of Chinchina. An estimated 25,000 people were killed. Heavy mudslides also accompanied the eruption of Mount St. Helens in Washington in 1980. Sediments mixed with melted snow and the waters of Spirit

Case Study

Radioactive Waste Disposal at Yucca Mountain

In July 2002 the U.S. Senate voted to designate Yucca Mountain, Nevada, as the permanent resting place for 77,000 metric tons of high-level nuclear waste, most of it spent fuel from the nation's 107 nuclear power plants. Nuclear power proponents celebrated: after twenty years of research and lobbying, the site was finally approved. With a repository on the horizon, plans for new nuclear plants could proceed, and existing plants could look forward to clearing out 50,000 tons of spent nuclear fuel currently in temporary, sometimes inadequate, storage at reactor sites. The state of Nevada, meanwhile, immediately filed a lawsuit to stop the plan. Nevada Governor Kenny Guinn charged that the Department of Energy (DOE) had lowered its scientific standards for evaluating the geological integrity of the site. Opponents charged that heavy-handed eastern politics, not sound geology, was behind the decision to put the entire nation's nuclear waste in Nevada.

The stakes at Yucca Mountain are high. The DOE and the nuclear power industry have invested more than $4 billion in research, testing, and promotion of the site. The lack of permanent storage has been an obstacle to construction of nuclear power plants since the 1980s. Further, cleanup is beginning at old military installations, such as Hanford, Washington; and Rocky Flats, Colorado. These are some of the nation's worst toxic waste sites, and permanent storage is needed for their radioactive materials.

Geology is the central problem in siting a waste repository of any kind. For high-level radioactive wastes, the DOE needed to find a site where a labyrinth of deep tunnels could keep extremely dangerous materials isolated and secure for 10,000 years, more than twice the length of recorded human history. (The waste will remain highly radioactive for more than 500,000 years, but the DOE considers itself unlikely to be responsible for the site for that long.) This time span requires a geologically stable area—no active faults like those in California, no volcanoes like those in Washington and Oregon. Bedrock needs to be relatively impermeable, not riddled with underground channels and sinkholes as in Florida or parts of Texas. There must be no groundwater, which would soak storage casks and mobilize radioactive materials, possibly allowing tainted water to reach the surface. This rules out the central Great Plains and most of the eastern United States, where plentiful rainfall keeps water tables high and aquifers full.

Yucca Mountain (see photo) fits these requirements better than most places in the United States. But there is conflicting evidence, and conflicting interpretations of evidence, about whether even Yucca Mountain is good enough. DOE geologists insist that the only aquifers in the area are 300 m below the site; other geologists say there is evidence in the rocks that aquifers have risen in the past, suggesting that they might rise again to the level of the repository tunnels. Critics of the site point out that the area has more than 30 known faults. Geologists disagree on how long these faults have been

Yucca Mountain, Nevada (long ridge in center of photo), has been chosen as the United States' first permanent high-level nuclear waste storage site. Source: *Department of Energy.*

dormant, and on how long it will be before they shift again. In addition, there are seven dormant volcanoes in the area. While these are likely to remain dormant for 10,000 years, their activity in several hundred thousand years is hard to predict. Opponents of the site also worry about the potential dangers of shipping waste, potentially 28,000 truckloads and 10,000 rail cars over the site's 30-year lifespan. Defenders point out that high-level waste is currently stored at more than 130 temporary sites, and those sites are clearly unsafe for the long term.

Do you think Nevada should be the repository for the nation's nuclear waste? Does the availability of a Yucca Mountain storage facility justify further nuclear power development? Would our economy be different today if we had waited to build nuclear power plants until a permanent storage site was available?

Lake at the mountain's base and flowed many kilometers from their source. Extensive damage was done to roads, bridges, and property, but because of sufficient advance warning, there were few casualties.

Volcanic eruptions often release large volumes of ash and dust into the air (fig. 14.15). Mount St. Helens expelled 3 km³ of dust and ash, causing ash fall across much of North America. This was only a minor eruption. An eruption in a bigger class of volcanoes was that of Tambora, Indonesia, in 1815, which expelled 175 km² of dust and ash, more than 58 times that of Mount St. Helens. These dust clouds

circled the globe and reduced sunlight and air temperatures enough so that 1815 was known as the year without a summer.

It is not just a volcano's dust that blocks sunlight. Sulfur emissions from volcanic eruptions combine with rain and atmospheric moisture to produce sulfuric acid (H_2SO_4). The resulting droplets of H_2SO_4 interfere with solar radiation and can significantly cool the world climate. In 1991, Mt. Pinatubo in the Philippines emitted 20 million tons of sulfur dioxide that combined with water to form tiny droplets of sulfuric acid. This acid aerosol reached the stratosphere where it circled the globe for two years. This thin haze

FIGURE 14.15 Pyroclastic flows spill down the slopes of Mayon volcano in the Philippines in this September 23, 1984, image. Because more than 73,000 people evacuated danger zones, this eruption caused no casualties. Courtesy Chris G. Newhall/U.S. Geological Survey.

cooled the entire earth by 1°C and postponed global warming for several years. It also caused a 10 to 15 percent reduction in stratospheric ozone, allowing increased ultraviolet light to reach the earth's surface. One of several theories about the extinction of the dinosaurs 65 million years ago is that they were killed by acid rain and climate changes caused by massive volcanic venting in the Deccan Plateau of India.

Mass Wasting

Gravity constantly pulls downward on every material everywhere on earth, causing a variety of phenomena collectively termed **mass wasting** or mass movement, in which geological materials are moved downslope from one place to another. The resulting movement is often slow and subtle, but some slope processes such as rockslides, avalanches, and land slumping can be swift, dangerous, and very obvious. Landslide is a general term for rapid downslope movement of soil or rock. In the United States alone, over $1 billion in property damage is done every year by landslides and related mass wasting.

In some areas, active steps are taken to control landslides or to limit the damage they cause. On the other hand, many human activities such as road construction, forest clearing, agricultural cultivation, and building houses on steep, unstable slopes increase both the frequency and the damage done by landslides. In some

FIGURE 14.16 Parts of an expensive house slide down the hillside in Pacific Palisades, California. Building at the edge of steep slopes made of unconsolidated sediment in an earthquake-prone region is a risky venture. © Al Seib/LA Times.

cases, people are unaware of the risks they face by locating on or under unstable hillsides. In other cases, they simply deny clear and obvious danger. Southern California, where people build expensive houses on steep hills of relatively unconsolidated soil, is often the site of large economic losses from landslides. Chaparral fires expose the soil to heavy winter rains. Resulting mudslides carry away whole neighborhoods and bury downslope areas in debris flows (fig. 14.16). Generally these processes are slow enough that few lives are lost, but the property damage can be high.

Summary

- The earth seems stable to us, but it is a layered, constantly moving system. Tectonic plates, fragments of the earth's crust, move continuously, at a rate of 1–18 cm per year.

- Where tectonic plates diverge, rifts occur. Mid-ocean ridges form where upwelling magma breaks through the crust at rift zones. Subduction occurs where dense, thin oceanic plates collide with less dense, brittle, continental crust. Subducted material often melts and rises, producing volcanoes.

- Rocks are classified by mineral content, grain size, and origin. Three general classes of rock are igneous, metamorphic, and sedimentary. Each can be transformed into the others by processes such as weathering, sedimentation, heat and pressure, or melting.

- Economically important concentrations of minerals occur in many parts of the world. Ores often occur as a result of hydrothermal (hot water) activity belowground.

- Major mining methods include placer mining (of stream sediments), underground mining, open-pit mining, and strip mining. Strip mining is especially important for coal production.

- Water contamination is one of the main environmental costs of mining. Pollutants include acids, cyanide, mercury, heavy metals, and sediment. Air pollution, especially acidic fumes from smelting ores, is also an important consequence of mining and mineral processing.

- Worldwide, only a small percentage of minerals are recycled. Recycling saves energy and reduces environmental damage caused by mining and processing ore. Material substitution, such as replacing copper telephone lines with fiber-optic lines, also reduces resources consumption.

- Earthquakes, tsunamis, landslides, and volcanoes are among the geologic hazards that take many lives. Big earthquakes are the most deadly geologic hazards, especially where building construction is poor.

Questions for Review

1. Describe the layered structure of the earth.

2. Define mineral and rock.

3. What are tectonic plates and why are they important to us?

4. Why are there so many volcanoes and earthquakes along the "ring of fire" that rims the Pacific Ocean?

5. Describe the rock cycle and name the three main rock types that it produces.

6. Figure 14.9 maps sources of some of our most important metals. What are they, and where do they come from?

7. Give some examples of nonmetal mineral resources and describe how they are used.

8. Describe some ways we recycle metals and other mineral resources.

9. What are some environmental hazards associated with mineral extraction?

10. Describe some of the leading geologic hazards and their effects.

Questions for Critical Thinking

1. Look at the walls, floors, appliances, interior, and exterior of the building around you. How many earth materials were used in their construction?

2. What is the geologic history of your town or county?

3. Is your local bedrock igneous, metamorphic, or sedimentary? If you don't know, who might be able to tell you?

4. What would life be like without the global mineral trade network? Can you think of advantages as well as disadvantages?

5. Suppose you live in a small, mineral-rich country, and a large, foreign mining company proposes to mine and market your minerals. How should revenue be divided between the company, your government, and citizens of the country? Who bears the greatest costs? How should displaced people be compensated? Who will make sure that compensation is fairly distributed?

6. Geologic hazards affect a minority of the population who build houses on unstable hillsides, in flood-prone areas, or on faults. What should society do to ensure safety for these people and their property?

7. A persistent question in this chapter is how to reconcile our responsibility as consumers with the damage from mineral extraction and processing. How responsible are you? What are some steps you could take to reduce this damage?

8. If gold jewelry is responsible for environmental and social devastation, should we stop wearing it? Should we worry about the economy in producing areas if we stop buying gold and diamonds? What further information would you need to decide these questions?

9. How could, or should, our government encourage more recycling and more efficient use of earth materials?

10. In Greek mythology, Pluto ("the rich") and Hades ("the unseen") were two names for the god of the underworld. Do these descriptions have relevance for geologic resources today?

Key Terms

core 289
crust 289
earthquakes 300
heap-leach extraction 298
igneous rocks 292
magma 290
mantle 289
mass wasting 303
metamorphic rock 293
mineral 290

rock 291
rock cycle 292
sedimentary rock 293
sedimentation 293
smelting 298
tectonic plates 290
tsunami 300
volcanoes 301
weathering 293

Further Readings

Bawden, G. W., et al. 2001. Tectonic contraction across Los Angeles after removal of groundwater pumping effects. *Nature* 412:812–15.

Gould, Stephen Jay. *Wonderful Life: The Burgess Shale and the Nature of History*. W. W. Norton and Company.

Kennedy, Danny. 1997. U.S. mine gouges for gold. *Earth Island Journal* 12(2):24.

Knoo, Andrew H. 2003. *When Life Nearly Died: The Greatest Mass Extinction of All Time*. Thames & Hudson.

McPhee, John. 2000. *Annals of a Former World*. Farrar Strauss Giroux.

Rudwick, Martin J. S. 1995. *Scenes from Deep Time: Early Pictorial Representations of the Prehistoric World*. University of Chicago Press.

Winchester, Simon. 2002. *The Map That Changed the World: William Smith and the Birth of Modern Geology*. Perennial Press.

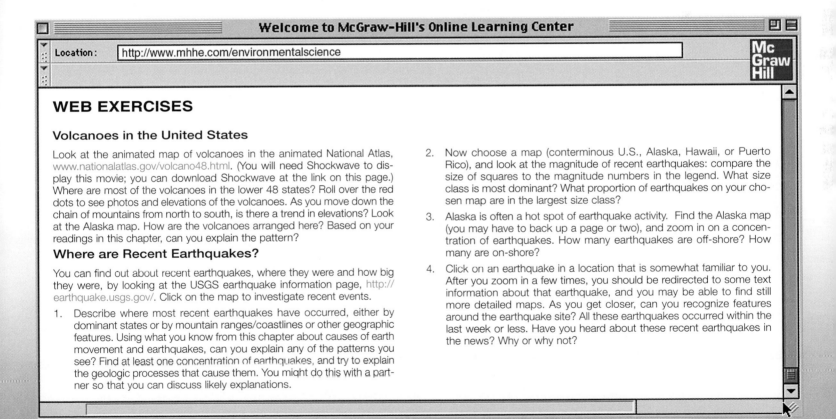

Welcome to McGraw-Hill's Online Learning Center

Location: http://www.mhhe.com/environmentalscience

WEB EXERCISES

Volcanoes in the United States

Look at the animated map of volcanoes in the animated National Atlas, www.nationalatlas.gov/volcano48.html. (You will need Shockwave to display this movie; you can download Shockwave at the link on this page.) Where are most of the volcanoes in the lower 48 states? Roll over the red dots to see photos and elevations of the volcanoes. As you move down the chain of mountains from north to south, is there a trend in elevations? Look at the Alaska map. How are the volcanoes arranged here? Based on your readings in this chapter, can you explain the pattern?

Where are Recent Earthquakes?

You can find out about recent earthquakes, where they were and how big they were, by looking at the USGS earthquake information page, http://earthquake.usgs.gov/. Click on the map to investigate recent events.

1. Describe where most recent earthquakes have occurred, either by dominant states or by mountain ranges/coastlines or other geographic features. Using what you know from this chapter about causes of earth movement and earthquakes, can you explain any of the patterns you see? Find at least one concentration of earthquakes, and try to explain the geologic processes that cause them. You might do this with a partner so that you can discuss likely explanations.

2. Now choose a map (conterminous U.S., Alaska, Hawaii, or Puerto Rico), and look at the magnitude of recent earthquakes: compare the size of squares to the magnitude numbers in the legend. What size class is most dominant? What proportion of earthquakes on your chosen map are in the largest size class?

3. Alaska is often a hot spot of earthquake activity. Find the Alaska map (you may have to back up a page or two), and zoom in on a concentration of earthquakes. How many earthquakes are off-shore? How many are on-shore?

4. Click on an earthquake in a location that is somewhat familiar to you. After you zoom in a few times, you should be redirected to some text information about that earthquake, and you may be able to find still more detailed maps. As you get closer, can you recognize features around the earthquake site? All these earthquakes occurred within the last week or less. Have you heard about these recent earthquakes in the news? Why or why not?

15

Air, Weather, and Climate

Climate is an angry beast and we are poking it with sticks.

Wallace Broecker, Lamont-Doherty Earth Observatory

OBJECTIVES

After studying this chapter, you should be able to:

- analyze human contributions to global climate change and what effects our modifications are having on physical and biological systems.
- summarize the structure and composition of the atmosphere.
- explain how the jet streams, prevailing winds, and frontal systems determine local weather.
- describe how tornadoes and cyclonic storms form and why they are dangerous.
- comprehend how El Niño cycles change ocean surface temperatures and affect continental climate.
- understand the driving forces thought to bring about normal climatic change.
- debate policy options for responding to the threats of global climate change.

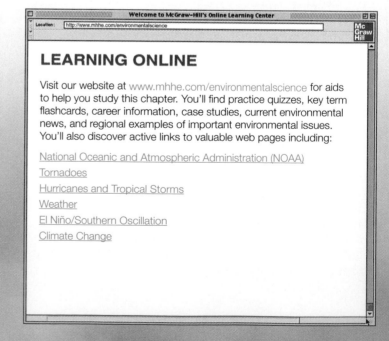

LEARNING ONLINE

Visit our website at www.mhhe.com/environmentalscience for aids to help you study this chapter. You'll find practice quizzes, key term flashcards, career information, case studies, current environmental news, and regional examples of important environmental issues. You'll also discover active links to valuable web pages including:

National Oceanic and Atmospheric Administration (NOAA)

Tornadoes

Hurricanes and Tropical Storms

Weather

El Niño/Southern Oscillation

Climate Change

Photo: Towering ice sheets float in the ocean around Antarctica. Their demise could lead to an accelerated melting of Antarctic glaciers and a catastrophic sea-level rise. © Bryan and Cherry Alexander Photography.

Is Antarctica Melting?

In February 2002, a huge section of the Larsen B Ice Shelf along the Antarctic Peninsula coast suddenly disintegrated into thousands of icebergs (fig. 15.1). Over a 30-day period, about 3,250 km² (an area larger than Rhode Island) shattered in the largest single event in a 30-year series of ice shelf retreats in the peninsula. The total volume of ice released was 720 billion metric tons, or the equivalent of 29 trillion bags of party ice. Climatologists warn that this dramatic collapse could be a signal of global climate change and an omen of catastrophic events to come.

Ice shelves are thick slabs of ice that float on the ocean around much of Antarctica (see opening photograph, this chapter). They are fed by glaciers sliding down off inland mountains. The Larsen B shelf was about 220 m (720 ft) thick, or about as tall as a 50-story building. Based on studies of ice flow and sediment thickness beneath the ice shelf, scientists believe that it existed for hundreds, or perhaps thousands, of years.

Although much of Antarctica has gotten cooler in recent years, the Antarctic Peninsula, which juts out toward the tip of South America, has warmed about 2.5°C over the last half century. This warming has destabilized ice sheets around the peninsula, which have lost about 13,500 km² (5,200 m²) over the past three decades. The Larsen A Ice Shelf, which lay just to the north of Larsen B, collapsed equally suddenly in 1995. Glaciologists speculate that pools of meltwater observed on the surface of the ice worked their way into cracks and crevasses, widening the gaps through freezing and thawing until the whole shelf abruptly collapsed. The next shelf to the south, the Larsen C, is now beginning to show similar meltwater ponds, and the giant Ross and Ronne Ice Shelves (each roughly the size of France) may be near the temperature limit for stability as well.

Ice shelf breakup has little direct consequence for global sea levels because floating ice displaces as much water as it contains. However, the shelves act as buttresses or braking systems for land-based glaciers. The Ross Ice Shelf, in particular, is the main outlet for several major glaciers draining the West Antarctic Ice Sheet, which contains enough water to raise sea level 5 m, if it were all to melt. As ice shelves disintegrate, glaciers slide down off the continent at an ever

February 23, 2002   March 2, 2002

FIGURE 15.1 In early 2002, a Rhode Island-sized section of the Larsen B ice shelf on the Antarctic Peninsula disintegrated into shards. Satellite images show 3,250 km² that collapsed into the ocean.
Source: National Snow and Ice Data Center/University of Colorado/NOAA.

quickening pace, raising worries that rising sea levels could flood coastal areas home to billions of people.

Are increasing temperatures in Antarctica and disintegrating ice shelves proof that human actions are causing global climate change? We can't be sure. It's possible that we're now witnessing merely a normal climatic variation and ice shelf retreat that occurs on a longer time scale than our records cover. As you'll learn in this chapter, however, events in Antarctica are only part of an ever increasing body of evidence that suggests we're having a significant impact on global climate. Many scientists regard this issue to be the single most important environmental problem of our age.

To understand what all these observations may have to say about our future and how we should respond to the threat of changing climate, we have to know something about our atmosphere and how it produces local weather and global climate. We also need to be acquainted with the international politics of this crucial topic. In this chapter, we'll look at how atmospheric systems work, examine the driving forces behind climate modification, and describe, in more detail, the ways climate affects human and natural systems.

THE ATMOSPHERE AND CLIMATE

We live at the bottom of a virtual ocean of air that extends upward about 500 km (300 mi). In the lowest 10 to 12 km, a layer known as the troposphere, the air moves ceaselessly, flowing and swirling, and continually redistributing heat and moisture from one part of the globe to another. The composition and behavior of the troposphere and other layers control our **weather** (daily temperature and moisture conditions in a place) and our **climate** (long-term weather patterns).

The earth's earliest atmosphere probably consisted mainly of hydrogen and helium. Over billions of years, most of that hydrogen and helium diffused into space. Volcanic emissions added carbon, nitrogen, oxygen, sulfur, and other elements to the atmosphere. Virtually all of the molecular oxygen (O_2) we breathe was probably produced by photosynthesis in blue-green bacteria, algae, and green plants.

Clean, dry air is mostly nitrogen and oxygen (table 15.1). Water vapor concentrations vary from near zero to 4 percent, depending on air temperature and available moisture. Minute particles and liquid droplets—collectively called **aerosols**—also are suspended in the air (fig. 15.2). Atmospheric aerosols play important roles in the earth's energy budget and in producing rain.

The atmosphere has four distinct zones of contrasting temperature, due to differences in absorption of solar energy (fig. 15.3). The layer of air immediately adjacent to the earth's surface is called the **troposphere** (*tropein* means to turn or change, in Greek). Within the troposphere, air circulates in great vertical and horizontal **convection currents,** constantly redistributing heat and moisture around the globe. The troposphere ranges in depth from about 18 km (11 mi) over the equator to about 8 km (5 mi) over the poles, where air is cold and dense. Because gravity holds most air molecules close to the earth's surface, the troposphere is much more dense than the other layers: it contains about 75 percent of the

TABLE 15.1	Present Composition of the Lower Atmosphere*	
GAS	SYMBOL OR FORMULA	PERCENT BY VOLUME
Nitrogen	N_2	78.08
Oxygen	O_2	20.94
Argon	Ar	0.934
Carbon dioxide	CO_2	0.035
Neon	Ne	0.00182
Helium	He	0.00052
Methane	CH_4	0.00015
Krypton	Kr	0.00011
Hydrogen	H_2	0.00005
Nitrous oxide	N_2O	0.00005
Xenon	Xe	0.000009

*Average composition of dry, clean air

FIGURE 15.2   The atmospheric processes that purify and redistribute water, moderate temperatures, and balance the chemical composition of the air are essential in making life possible. To a large extent, living organisms have created, and help to maintain, the atmosphere on which we all depend. © William P. Cunningham.

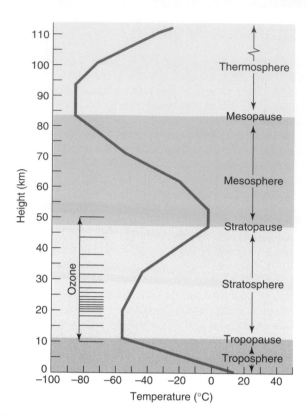

FIGURE 15.3   Temperatures change dramatically in different layers of the atmosphere. Bars in the ozone graph represent relative concentrations of the "ozone layer" with altitude.
Source: Courtesy of Dr. William Culver, St. Petersburg Junior College.

total mass of the atmosphere. Air temperature drops rapidly with increasing altitude in this layer, reaching about –60°C (–76°F) at the top of the troposphere. A sudden reversal of this temperature gradient creates a sharp boundary called the tropopause, which limits mixing between the troposphere and upper zones.

The **stratosphere** extends from the tropopause up to about 50 km (31 mi). It is vastly more dilute than the troposphere, but it has similar composition—except that it has almost no water vapor and nearly 1,000 times more **ozone** (O_3). This ozone absorbs some wavelengths of ultraviolet solar radiation, known as UV-B (290–330 nm,

see fig. 3.9). This absorbed energy makes the atmosphere warmer toward the top of the stratosphere. Since UV radiation damages living tissues, this UV absorption in the stratosphere also protects life on the surface. Recently discovered depletion of stratospheric ozone, especially over Antarctica, is allowing increased amounts of UV radiation to reach the earth's surface. If observed trends continue, this radiation could cause higher rates of skin cancer, genetic mutations, crop failures, and disruption of important biological communities, as you will see later in this chapter.

Unlike the troposphere, the stratosphere is relatively calm. There is so little mixing in the stratosphere that volcanic ash or human-caused contaminants can remain in suspension there for many years.

Above the stratosphere, the temperature diminishes again, creating the mesosphere, or middle layer. The thermosphere (heated layer) begins at about 80 km. This is a region of highly ionized (electrically charged) gases, heated by a steady flow of high-energy solar and cosmic radiation. In the lower part of the thermosphere, intense pulses of high-energy radiation cause electrically charged particles (ions) to glow. This phenomenon is what we know as the *aurora borealis* and *aurora australis,* or northern and southern lights.

No sharp boundary marks the end of the atmosphere. Pressure and density decrease with distance from the earth until they become indistinguishable from the near-vacuum of interstellar space.

Energy and the "Greenhouse Effect"

The sun supplies the earth with an enormous amount of energy, but that energy is not evenly distributed over the globe. Incoming solar radiation (insolation) is much stronger near the equator than at high latitudes. Of the solar energy that reaches the outer atmosphere, about one-quarter is reflected by clouds and atmospheric gases, and another quarter is absorbed by carbon dioxide, water vapor, ozone, methane, and a few other gases (fig. 15.4). This energy absorption warms the atmosphere slightly. About half of incoming solar radiation (insolation) reaches the earth's surface. Some of this energy is reflected by bright surfaces, such as snow, ice, and sand. The rest is absorbed by the earth's surface and by water. Surfaces that *reflect* energy have a high **albedo** (reflectivity). Most of these surfaces appear bright to us because they reflect light as well as other forms of radiative energy. Surfaces that *absorb* energy have a low albedo and generally appear dark. Black soil, asphalt pavement, and dark green vegetation, for example, have low albedos (table 15.2).

Absorbed energy heats the absorbing surface (such as an asphalt parking lot in summer), evaporates water, or provides the energy for photosynthesis in plants. Following the second law of thermodynamics, absorbed energy is gradually reemitted as lower-quality heat energy. A brick building, for example, absorbs energy in the form of light and reemits that energy in the form of heat.

The change in energy quality is very important because the atmosphere selectively absorbs longer wavelengths. Most solar energy comes in the form of intense, high-energy light or near-infrared wavelengths. This short-wavelength energy passes relatively easily through the atmosphere to reach the earth's surface. Energy re-released from the earth's warmed surface ("terrestrial energy") is lower-intensity, longer-wavelength energy in the far-infrared part of the spectrum. Atmospheric gases, especially carbon

TABLE 15.2 — Albedo (Reflectivity) of Surfaces

SURFACE	ALBEDO (%)
Fresh snow	80–85
Dense clouds	70–90
Water (low sun)	50–80
Sand	20–30
Water (sun overhead)	5
Forest	5–10
Black soil	3
Earth/atmosphere average	30

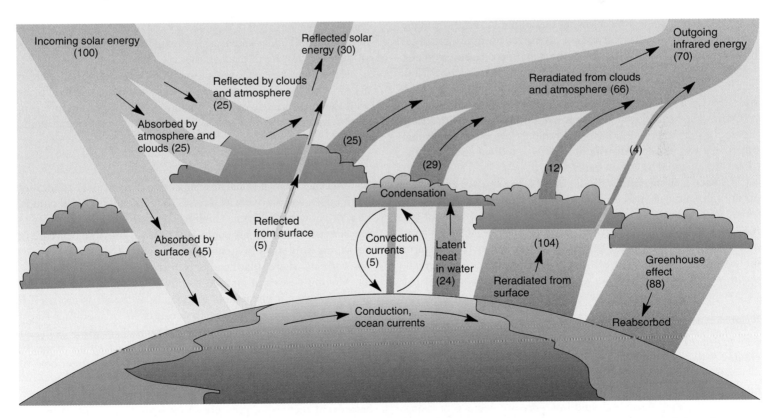

FIGURE 15.4 Energy balance between incoming and outgoing radiation. The atmosphere absorbs or reflects about half of the solar energy reaching the earth. Most of the energy reemitted from the earth's surface is long-wave, infrared energy. Most of this infrared energy is absorbed by aerosols and gases in the atmosphere and is reradiated toward the planet, keeping the surface much warmer than it would otherwise be. This is known as the greenhouse effect. The numbers shown are arbitrary units. Note that for 100 units of incoming solar energy, 100 units are reradiated to space, but more than 100 units are radiated from the earth's surface because of the greenhouse effect.

dioxide and water vapor, absorb much of this long-wavelength energy, re-releasing it in the lower atmosphere and letting it leak out to space only slowly. This terrestrial energy provides most of the heat in the lower atmosphere. If the atmosphere were as transparent to infrared radiation as it is to visible light, the earth's average surface temperature would be about –18°C (0°F) to 33°C (59°F) colder than it is now.

This phenomenon is called the **greenhouse effect** because the atmosphere, loosely comparable to the glass of a greenhouse, transmits sunlight while trapping heat inside. The greenhouse effect is a natural atmospheric process that is necessary for life as we know it. However, too strong a greenhouse effect, caused by burning of fossil fuels and deforestation, may cause harmful environmental change.

Convection and Atmospheric Pressure

Much of the incoming solar energy is used to evaporate water. Every gram of evaporating water absorbs 580 calories of energy as it transforms from liquid to gas. Globally, water vapor contains a huge amount of stored energy, known as **latent heat.** When water vapor condenses, returning from a gas to a liquid form, the 580 calories of heat energy are released. Imagine the sun shining on the Gulf of Mexico in the winter. Warm sunshine and plenty of water allow continuous evaporation that converts an immense amount of solar (light) energy into latent heat stored in evaporated water. Now imagine a wind blowing the humid air north from the Gulf toward Canada. The air cools as it moves north (especially if it encounters cold air moving south). Cooling causes the water vapor to condense. Rain (or snow) falls as a consequence. Note that it is not only water that has moved from the Gulf to the Midwest: 580 calories of heat have also moved with every gram of moisture. The heat and water have moved from a place with strong incoming solar energy to a place with much less solar energy and much less water. The redistribution of heat and water around the globe are essential to life on earth.

Uneven heating, with warm air close to the equator and colder air at high latitudes, also produces pressure differences that cause wind, rain, storms, and everything else we know as weather. As the sun warms the earth's surface, the air nearest the surface warms and expands, becoming less dense than the air above it. The warm air must then rise above the denser air. Vertical convection currents result, which circulate air from warm latitudes to cool latitudes and vice versa. These convection currents can be as small and as localized as a narrow column of hot air rising over a sun-heated rock, or they can cover huge regions of the earth. At the largest scale, the convection cells are described by a simplified model known as Hadley cells, which redistribute heat globally (fig. 15.5).

Where air rises in convection currents, air pressure at the surface is low. Where air is sinking, or subsiding, air pressure is high. On a weather map these high and low pressure centers, or rising and sinking currents of air, move across continents. In most of North America, they generally move from west to east. Rising air tends to cool with altitude, releasing latent heat that causes further rising. Very warm and humid air can rise very vigorously, espe-

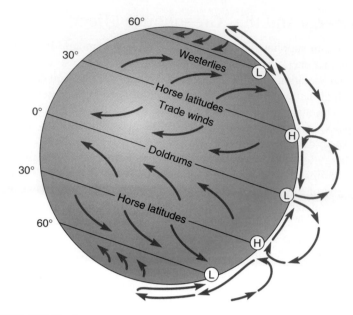

FIGURE 15.5 General circulation patterns. The actual boundaries of the circulating convection cells vary from day to day and season to season, as do the local directions of surface winds. Surface topography also complicates circulation patterns, but within these broad regions, winds usually have a predominant and predictable direction. L and H represent low and high pressure.

cially if it is rising over a mass of very cold air. As water vapor carried aloft cools and condenses, it releases energy that fuels violent storms, which we will discuss later.

Pressure differences are an important cause of wind. There is always someplace with sinking (high pressure) air and someplace with low pressure (rising) air. Air moves from high-pressure centers toward low-pressure areas, and we call this movement wind.

Why Does It Rain?

To understand why it rains, remember two things: water condenses as air cools, and air cools as it rises. Any time air is rising, clouds, rain, or snow might form. Cooling occurs because of changes in pressure with altitude: air cools as it rises (as pressure decreases); air warms as it sinks (as pressure increases). Air rises in convection currents where solar heating is intense, such as over the equator. Moving masses of air also rise over each other and cool. Air also rises when it encounters mountains. If the air is moist (if it has recently come from over an ocean or an evaporating forest region, for example), condensation and rainfall are likely as the air is lifted (fig. 15.6). Regions with intense solar heating, frequent colliding air masses, or mountains tend to receive a great deal of precipitation.

Where air is sinking, on the other hand, it tends to warm because of increasing pressure. As it warms, available moisture evaporates. Rainfall occurs relatively rarely in areas of high pressure. High pressure and clear, dry conditions occur where convection currents are sinking. High pressure also occurs where air sinks after flowing over mountains. Figure 15.5 shows sinking, dry air

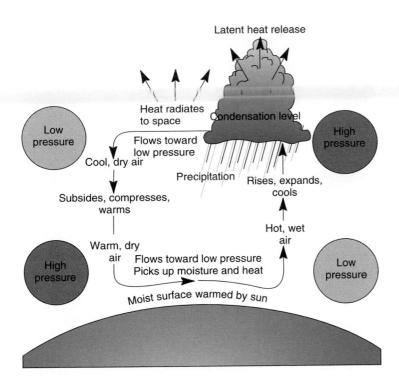

FIGURE 15.6 Convection currents and latent energy cause atmospheric circulation and redistribute heat and water around the globe.

at about 30° north and south latitudes. If you look at a world map, you will see a band of deserts at approximately these latitudes.

Another ingredient is usually necessary to initiate condensation of water vapor: condensation nuclei. Tiny particles of smoke, dust, sea salts, spores, and volcanic ash all act as condensation nuclei. These particles form a surface on which water molecules can begin to coalesce. Without them even supercooled vapor can remain in gaseous form. Even apparently clear air can contain large numbers of these particles, which are generally too small to be seen by the naked eye.

The Coriolis Effect and Jet Streams

Large-scale winds tend not to move in a straight line across the earth's surface. In the Northern Hemisphere, they generally bend clockwise (right), and in the Southern Hemisphere, they bend counterclockwise (left). This curving pattern results from the fact that the earth rotates in an eastward direction as the winds move above the surface. The apparent curvature of the winds is known as the **Coriolis effect.** On a global scale, this effect produces steady, reliable wind patterns, such as the trade winds and the midlatitude Westerlies (see fig. 15.5). Ocean currents similarly curve clockwise in the Northern Hemisphere and counterclockwise in the south. On a regional scale, the Coriolis effect produces cyclonic winds, or wind movements controlled by the earth's spin. Cyclonic winds spiral clockwise out of an area of high pressure in the Northern Hemisphere and counterclockwise into a low-pressure zone. If you look at a weather map in the newspaper, can you find this counterclockwise spiral pattern?

Why does this curving or spiraling motion occur? Imagine you were looking down on the North Pole of the rotating earth. Now imagine that the earth was a merry-go-round in a playground, with the North Pole at its center and the equator around the edge. As it spins counterclockwise (eastward), the spinning edge moves very fast (a full rotation, 39,800 km, every 24 hours for the real earth, or more than 1,600 km/hour). Near the center, though, there is very little eastward velocity. If you threw a ball from the edge toward the center, it would be traveling faster (edge speed) than the middle. It would appear, to someone standing on the merry-go-round, to curve toward the right. If you threw the ball from the center toward the edge, it would start out with no eastward velocity, but the surface below it would spin eastward, making the ball end up, to a person on the merry-go-round, west of its starting point. Winds move above the earth's surface much as the ball does. If you were looking down at the South Pole, you would see the earth spinning clockwise, and winds—or thrown balls—would appear to bend left. Incidentally, this effect does not apply to drains in your house. Their movement is far too small to be affected by the spinning of the earth.

At the top of the troposphere are **jet streams,** hurricane-force winds that circle the earth. These powerful winds follow an undulating path approximately where the vertical convection currents known as the Hadley and Ferrell cells meet. The approximate path of one jet stream over the Northern Hemisphere is shown in fig. 15.7. Although we can't perceive jet streams on the ground, they are important to us because they greatly affect weather patterns. Sometimes jet streams dip down near the top of the world's highest mountains, exposing mountain climbers to violent, brutally cold winds.

Ocean Currents

Warm and cold ocean currents strongly influence climate conditions on land. Surface ocean currents result from wind pushing on the ocean surface, as well as from the Coriolis effect. As surface water moves, deep water wells up to replace it, creating deeper ocean currents. Differences in water density—depending on the temperature and saltiness of the water—also drive ocean circulation. Huge cycling currents called gyres carry water north and south, redistributing heat from low latitudes to high latitudes (see fig. 17.5). The Alaska current, flowing from Alaska southward to California, keeps San Francisco cool and foggy during the summer. The Gulf Stream, one of the best known currents, carries warm Caribbean water north past Canada's maritime provinces to northern Europe. This current is immense, some 800 times the volume of the Amazon, the world's largest river. The heat transported from the Gulf keeps Europe much warmer than it should be for its latitude. As the warm Gulf Stream passes Scandinavia and swirls around Iceland, the water cools and evaporates, becomes dense and salty, and plunges downward, creating a strong, deep, southward current.

Ocean circulation patterns were long thought to be unchanging, but now oceanographers believe that currents can shift abruptly. For example, melting ice at the North Pole could create a flood of cold, fresh water, disrupting the Gulf Stream and causing rapid, drastic

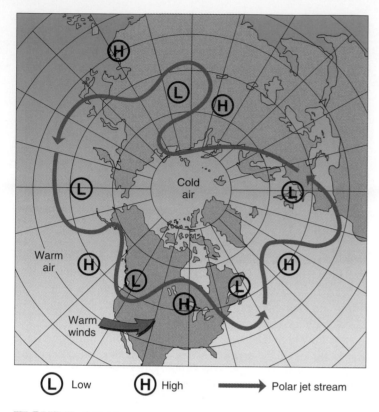

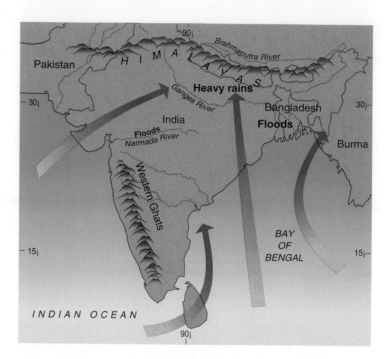

FIGURE 15.8 Summer monsoon air flows over the Indian subcontinent. Warming air rises over the plains of central India in the summer, creating a low-pressure cell that draws in warm, wet oceanic air. As this moist air rises over the Western Ghats or the Himalayas, it cools and heavy rains result. These monsoon rains flood the great rivers, bringing water for agriculture, but also causing much suffering.

FIGURE 15.7 A typical pattern of the arctic circumpolar vortex. This large, circulating mass of cold air sends "fingers," or lobes, across North America and Eurasia, spreading storms in their path. If the vortex becomes stalled, weather patterns stabilize, causing droughts in some areas and excess rain elsewhere.

cooling in Europe. Sudden changes of this sort appear to have happened in geologic history, and many climatologists fear it could occur again if global warming melts arctic ice again.

Seasonal Winds and Monsoons

While figure 15.5 shows regular global wind patterns, large parts of the world, especially the tropics, receive seasonal winds and rainy seasons that are essential for sustaining both ecosystems and human life. Sometimes these seasonal rains are extreme, causing disastrous flooding. In 2003, immense floods in China forced 100 million people from their homes. Many more millions elsewhere in Asia were similarly affected.

Sometimes the rains fail, causing crop failures and famine. The most regular seasonal winds and rains are known as **monsoons.** In India and Bangladesh, monsoon rains come when seasonal winds blow hot, humid air from the Indian Ocean (fig. 15.8). Strong convection currents lift this air, causing heavy rain across the subcontinent. When the rising air reaches the Himalayas, it rises even further, creating some of the heaviest rainfall in the world. During the 5-month rainy season of 1970, a weather station in the foothills of the Himalayas recorded 25 m (82 ft) of rain!

Tropical and subtropical regions around the world have seasonal rainy and dry seasons (see the discussion of tropical biomes,

chapter 5). The main reason for this variable climate is that the region of most intense solar heating and evaporation shifts through the year. Remember that the earth's axis of rotation is at an angle. In December and January the sun is most intense just south of the equator; in June and July the sun is most intense just north of the equator. Wherever the sun shines most directly, evaporation and convection currents—and rainfall and thunderstorms—are very strong. As the earth orbits the sun, the tilt of its axis creates seasons with varying amounts of wind, rain, and heat or cold. Seasonal rains support seasonal tropical forests, and they fill some of the world's greatest rivers, including the Ganges and the Amazon. As the year shifts from summer to winter, solar heating weakens, the rainy season ends, and little rain may fall for months.

During the 1970s and 1980s, rains failed repeatedly in parts of the Sahel in northern Africa. Although rains have often been irregular in this region, increasing populations and civil disorder made these droughts result in widespread starvation and death (fig. 15.9).

Frontal Weather

The boundary between two air masses of different temperature and density is called a front. When cooler air displaces warmer air, we call the moving boundary a **cold front.** Since cold air tends to be more dense than warm air, a cold front will hug the ground and push under warmer air as it advances. As warm air is forced upward, it cools adiabatically (without loss or gain of energy), and its cargo of water vapor condenses and precipitates. Upper layers

FIGURE 15.9 Failure of monsoon rains brings drought, starvation, and death to both livestock and people in the Sahel desert margin of Africa. Although drought is a fact of life in Africa, many governments fail to plan for it, and human suffering is much worse than it needs to be.
© Vic Englebert/Photo Researchers, Inc.

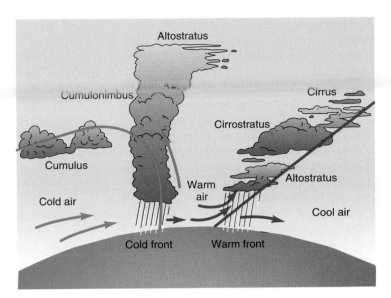

FIGURE 15.10 A cold front assumes a bulbous, "bull-nose" appearance because ground drag retards forward movement of surface air. As warm air is lifted up over the advancing cold front, it cools, producing precipitation. When warm air advances, it slides up over cooler air in front and produces a long, wedge-shaped zone of clouds and precipitation. The high cirrus clouds that mark the advancing edge of the warm air mass may be 1,000 km and 48 hours ahead of the front at ground level.

of a moving cold air mass move faster than those in contact with the ground because of surface friction or drag, so the boundary profile assumes a curving, "bull-nose" appearance (fig. 15.10, left). Notice that the region of cloud formation and precipitation is relatively narrow. Cold fronts generate strong convective currents and often are accompanied by violent surface winds and destructive storms. An approaching cold front generates towering clouds called thunderheads that reach into the stratosphere where the jet stream pushes the cloud tops into a characteristic anvil shape. The weather after the cold front passes is usually clear, dry, and invigorating.

If the advancing air mass is warmer than local air, a **warm front** results. Since warm air is less dense than cool air, an advancing warm front will slide up over cool, neighboring air parcels, creating a long, wedge-shaped profile with a broad band of clouds and precipitation (fig. 15.10, right). Gradual uplifting and cooling of air in the warm front avoids the violent updrafts and strong convection currents that accompany a cold front. A warm front will have many layers of clouds at different levels. The highest layers are often wispy cirrus (mare's tail) clouds that are composed mainly of ice crystals. They may extend 1,000 km (621 mi) ahead of the contact zone with the ground and appear as much as 48 hours before any precipitation. A moist warm front can bring days of drizzle and cloudy skies.

Cyclonic Storms

Few people experience a more powerful and dangerous natural force than cyclonic storms spawned by low-pressure cells over warm tropical oceans. As we discussed earlier in this chapter, low pressure is generated by rising warm air. Winds swirl into this low-pressure area, turning counterclockwise in the Northern Hemisphere due to the Coriolis effect. When rising air is laden with water vapor, the latent energy released by condensation intensifies convection currents and draws up more warm air and water vapor. As long as a temperature differential exists between air and ground and a supply of water vapor is available, the storm cell will continue to pump energy into the atmosphere.

Called **hurricanes** in the Atlantic and eastern Pacific, typhoons in the western Pacific, or cyclones in the Indian Ocean, winds near the center of these swirling air masses can reach hundreds of kilometers per hour and cause tremendous suffering and destruction. Often hundreds of kilometers across, these giant storms can generate winds as high as 320 km/hr (200 mph) and push walls of water called storm surges far inland (fig. 15.11). In 1970, a killer typhoon brought torrential rains, raging winds, and a storm surge up to 5 m high that flooded thousands of square kilometers of the flat coastal area of Bangladesh, drowning more than a half million people.

Hurricane Mitch, which swept across Central America in October 1998, was the most deadly storm to strike the Western Hemisphere in 200 years. Stretching some 1,200 km (745 mi) across the Caribbean Ocean, it had sustained winds above 155 knots (180 mph) and gusts well over 200 mph. More than two meters (about six feet) of rain fell in four days as the storm moved slowly across the mountains of Central America. Runoff from this one storm may have doubled the pesticide load in the Atlantic. At least 11,000 people died and more than 3 million were left homeless by the floods and mudslides triggered by the intense rain. Total damage in Honduras, Nicaragua, Guatemala, Belize, and El Salvador was estimated to be around $5 billion. This about equals the average annual cost for hurricane or flood damage for the entire United States and is about five times the annual U.S. losses from tornadoes.

Tornadoes, swirling funnel clouds that form over land, also are considered to be cyclonic storms although their rotation isn't generated by Coriolis forces. While never as large or powerful as hurricanes, tornadoes can be just as destructive in the limited areas where they touch down (fig. 15.12). Some meteorologists consider smaller, less destructive phenomena such as waterspouts, dust devils, and smoke funnels over forest fires to be miniature tornadoes, but others think they have very different origins.

Tornadoes are generated on the American Great Plains by giant "supercell" frontal systems where strong, dry air cold fronts from Canada collide with warm humid air moving north from the Gulf of Mexico. Greater air temperature differences cause more powerful storms, which is why most tornadoes occur in the spring, when arctic cold fronts penetrate far south over the warming plains. As warm air rises rapidly over dense, cold air, intense vertical convection currents generate towering thunderheads with anvil-shaped leading edges and domed tops up to 20,000 m (65,000 ft) high (fig. 15.13). Water vapor cools and condenses as it rises, releasing latent heat and accelerating updrafts within the supercell. Penetrating into the stratosphere, the tops of these clouds can encounter jet streams, which help create even stronger convection currents.

But what sets the air mass spinning? Scientists debate this point, but one theory is that differential air speeds called shear forces—fast-moving above but slower near the ground—roll the air ahead of the advancing front in invisible horizontal tubes (or vortices) much as you might roll modeling clay between your hands. As these rolling vortex tubes are sucked into the vertical convective currents, they tilt upright and set the air column in motion, creating a spiral structure called a mesocyclone. Initially, the swirling air mass may be many kilometers wide, but cold downdrafts from the rear flank of the storm can spiral around the mesocyclone, increasing its velocity and narrowing its diameter much as a skater spins faster as she pulls in her arms. Winds at the edge of the rapidly spinning air column can reach speeds up to 510 km/hr (318 mph) that rip trees out of the ground and drive straws through telephone poles. In addition, low pressure in the center of the funnel cloud implodes houses, sucks people out of windows, and helps hurl large objects across the countryside.

Sometimes a supercell isn't organized enough to spawn tornadoes, but strong downdrafts, called **downbursts,** can generate straight-line winds well over 160 km/hr (100 mph). On July 4,

FIGURE 15.11   Hurricane Floyd approaches Florida in September 1999. The hole in the center is the "eye" of the storm. Note the counterclockwise circulation. How wide is the main body of this storm? Images produced by Hal Pierce, Laboratory for Atmospheres, NASA Goddard Space Flight Center/NOAA.

FIGURE 15.12   Tornadoes are local cyclonic storms caused by rapid mixing of cold, dry air and warm, wet air. Wind speeds in the swirling funnel can reach 320 km/hr (200 mph). © Corbis Royalty Free Website.

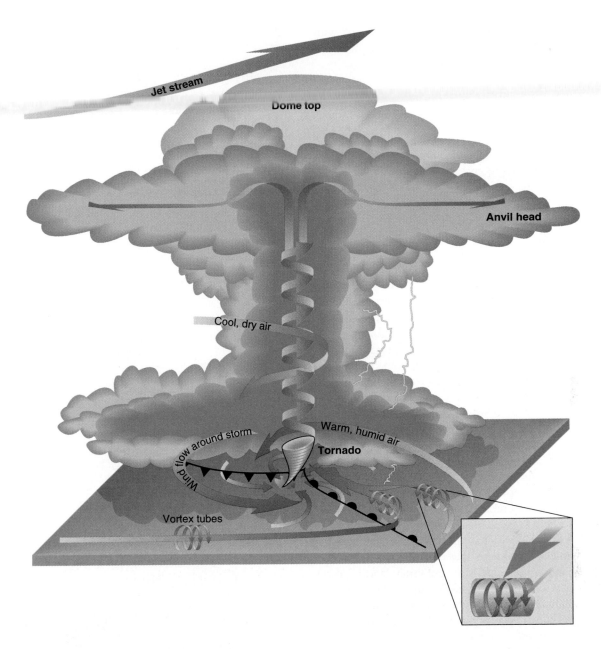

FIGURE 15.13 Tornadoes are generated by rapidly rising thermal updrafts where warm humid air is drawn up into a cold supercell thunderstorm. Rolling vortex tubes generated by differential air currents are drawn into the updraft and start it rotating. Cool, dry air spiraling down from the rear tightens the spiral and intensifies its speed. *Source:* National Center for Atmospheric Research.

1999, a powerful straight-line wind flattened about 25 million trees in a swath 16 km by 48 km across the heart of the Boundary Waters Canoe Area in northern Minnesota. Witnesses said that as the winds hit, the forest began rippling like a field of wheat, and then in seconds, was knocked down as if by a steamroller.

The same strong updrafts that create tornadoes can also generate hail. Falling water droplets rise on powerful updrafts, freeze, fall again, and rise again. Droplets accumulate water, or ice, as they rise and fall, until they become heavy enough to resist being lofted by updrafts. If they remain frozen all the way to the ground, we call them hail. If you cut open large hail stones, you often can see accretion rings where successive layers of ice were added.

There seems to be a connection between strong monsoon winds over West Africa and the frequency of killer hurricanes in the Caribbean and along the Atlantic coast of North and Central America. In years when the Atlantic surface temperatures are high, a stronger than average monsoon trough forms over Africa. This pulls in moist maritime air that brings rain (and life) to the Sahel (the margin of the Sahara). This trough gives rise to tropical depressions (low-pressure storms) that follow one another in regular waves across the Atlantic. The weak trade winds associated with "wet" years allow these storms to organize and gain strength. Evaporation from the warm surface water provides energy.

During the 1970s and 1980s when the Sahel had devastating droughts, the weather was relatively quiet in North America. Only one killer hurricane (winds over 70 km/hr or 112 mph) reached the United States in two decades. By contrast, in 1995 when ocean surface temperatures were high and rain returned to the Sahel, seven hurricanes and six named tropical storms were spawned in the Atlantic. During the summer of 1996, satellite images of the Atlantic showed five active hurricanes roaring across the Caribbean at the same time. Are we witnessing a long-term climatic change or merely a temporary aberration?

Weather Modification

As author Samuel Clemens (Mark Twain) said, "Everybody talks about the weather, but nobody does anything about it." People probably always have tried to influence local weather through religious ceremonies, dancing, or sacrifices. During the drought of the 1930s in the United States, "rainmakers" fleeced desperate farmers of thousands of dollars with claims that they could bring rain.

Some recent developments appear to be effective in local weather modification, at least in some circumstances, but they are not without drawbacks and controversy. Seeding clouds with dry ice or ionized particles, such as silver iodide crystals, can initiate precipitation if water vapor is present and air temperatures are near the condensation point (chapter 17). Dry ice also is effective at dispersing cold fog (where supercooled water droplets are present). Warm fog (air temperatures above freezing) and ice fog (ice crystals in the air) are not usually amenable to weather modification. Hail suppression by cloud seeding also can be effective, but dissipation of the clouds that generate hail diverts rain from areas that need it, as well. There are concerns that materials used in cloud seeding could cause air, ground, and water pollution. North Dakota recently agreed to pay for an environmental impact assessment to convince Montana farmers that North Dakota is not stealing rain by seeding clouds. Similarly, Wyoming and Idaho have been arguing over cloud seeding to increase winter snowpack in the Tetons that would provide summer irrigation water downstream in Idaho.

CLIMATE

If weather is a description of physical conditions in the atmosphere (humidity, temperature, pressure, wind, and precipitation), then climate is the *pattern* of weather in a region over long time periods. The interactions of atmospheric systems are so complex that climatic conditions are never exactly the same at any given location from one time to the next. While it is possible to discern patterns of average conditions over a season, year, decade, or century, complex fluctuations and cycles within cycles make generalizations doubtful and forecasting difficult (fig. 15.14). We always wonder whether anomalies in local weather patterns represent normal variations, a unique abnormality, or the beginnings of a shift to a new regime.

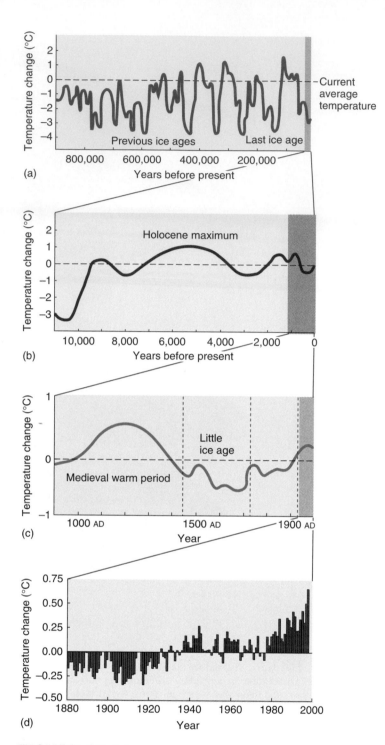

FIGURE 15.14 Global temperature variations over four time periods: (*a*) the past 900,000 years, (*b*) the past 10,000 years, (*c*) the past 1,000 years, and (*d*) the most recent 120 years. The dotted lines in each represent the mean temperature for the twentieth century. Note that both horizontal and vertical scales change in each figure.
Source: (*d*) National Climate Data Center/NESDIS/NOAA, 1998.

Climatic Catastrophes

Major climatic changes, such as those of the Ice Ages, have drastic effects on living organisms. When climatic change is gradual, species may have time to adapt or migrate to more suitable locations. Where climatic change is relatively abrupt, many organisms are unable to respond before conditions exceed their tolerance limits. Whole communities may be destroyed, and if the climatic change is widespread, many species may become extinct.

Perhaps the most well-studied example of this phenomenon is the great die-off that occurred about 65 million years ago at the end of the Cretaceous period. Most dinosaurs—along with 75 percent of all previously existing plant and animal species—became extinct, apparently as a result of sudden cooling of the earth's climate. Geologic evidence suggests that this catastrophe was not an isolated event. There appear to have been several great climatic changes, perhaps as many as a dozen, in which large numbers of species were exterminated (see table 11.3, p. 223).

Driving Forces and Patterns in Climatic Changes

What causes catastrophic climatic changes? There are many explanations. One of these is that changes in solar energy associated with 11-year sunspot cycles or 22-year solar magnetic cycles might play a role. Another theory is that a regular 18.6-year cycle of shifts in the angle at which our moon orbits the earth alters tides and atmospheric circulation in a way that affects climate. A theory that has received attention in recent years suggests that orbital variations as the earth rotates around the sun might be responsible for cyclic weather changes.

Milankovitch cycles, named after Serbian scientist Milutin Milankovitch, who first described them in the 1920s, are periodic shifts in the earth's orbit and tilt (fig. 15.15). The earth's elliptical orbit stretches and shortens in a 100,000-year cycle, while the axis of rotation changes its angle of tilt in a 40,000-year cycle. Furthermore, over a 26,000-year period, the axis wobbles like an out-of-balance spinning top. These variations change the distribution and intensity of sunlight reaching the earth's surface and, consequently, global climate. Bands of sedimentary rock laid in the oceans (fig. 15.16) seem to match both these Milankovitch cycles and the periodic cold spells associated with worldwide expansion of glaciers every 100,000 years or so.

A historical precedent for global climate change that had disastrous effects on humans was the "little ice age" that began in the 1400s. Temperatures dropped so that crops failed repeatedly in parts of northern Europe that once were good farmland. Scandinavian settlements in Greenland founded during the warmer period around A.D. 1000 lost contact with Iceland and Europe as ice blocked shipping lanes. It became too cold to grow crops, and fish that once migrated along the coast stayed farther south. The settlers slowly died out, perhaps in battles with Inuit people who were driven south from the high Arctic by colder weather.

Evidence from ice cores drilled in the Greenland ice cap suggests that world climate may change much more rapidly than pre-

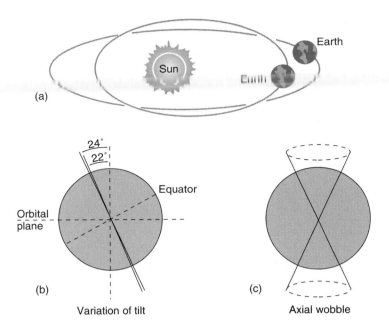

(a)

24°
22°

Equator

Orbital plane

(b) Variation of tilt

(c) Axial wobble

FIGURE 15.15 Milankovitch cycles, which may affect long-term climate conditions: (*a*) changes in the elliptical shape of the earth's orbit, (*b*) shifting tilt of the axis, and (*c*) wobble of the earth.

FIGURE 15.16 Milankovitch cycle periodicity? The light and dark bands in these seafloor sediments were laid down under different climatic regimes, each lasting 20,000 to 40,000 years. Now exposed along the French coast, these layers suggest the regularity of Milankovitch cycles. Courtesy Peter Gilday.

viously thought. During the last major interglacial period 135,000 to 115,000 years ago, it appears that temperatures flipped suddenly from warm to cold or vice versa over a period of years or decades rather than centuries. One possible explanation is that ocean currents transporting heat from the equator to the North Atlantic may have suddenly stopped or even reversed course when surges of fresh water diluted salty ocean waters. Volcanic eruptions may have played a role as well. When Mt. Pinatubo in the Philippines erupted in 1991, it ejected enough dust, aerosols, and gas into the atmosphere to cool the climate by at least 1°C. Larger volcanoes may have had greater effects.

El Niño/Southern Oscillations

What do sea surface temperatures near Indonesia, anchovy fishing off the coast of Peru, and the direction of tropical trade winds have to do with rainfall and temperatures over the United States? Strangely enough, they all may be interrelated. If the terms El Niño, La Niña, and Southern Oscillation are not yet a part of your vocabulary, perhaps they should be. They describe a connection between the ocean and atmosphere that appears to affect these factors as well as weather patterns throughout the world.

The core of this climatic system is a huge pool of warm surface water in the Pacific Ocean that sloshes slowly back and forth between Indonesia and South America like water in a giant bathtub. Most years, this pool is held in the western Pacific by steady equatorial trade winds pushing ocean surface currents westward (fig. 15.17). These surface winds are generated by a huge low-pressure cell formed by upwelling convection currents of moist air warmed by the ocean. Towering thunderheads created by rising air bring torrential summer rains to the tropical rainforests of Northern Australia and Southeast Asia. Winds high in the troposphere carry a return flow back to the eastern Pacific where dry subsiding currents create deserts from Chile to southern California. Surface waters driven westward by the trade winds are replaced by upwelling of cold, nutrient-rich, deep waters off the west coast of South America that support dense schools of anchovies and other finfish.

Every three to five years, for reasons that we don't fully understand, the Indonesian low collapses and the mass of warm surface water surges back east across the Pacific. One theory is that the high cirrus clouds atop the cloud columns block the sun and cool

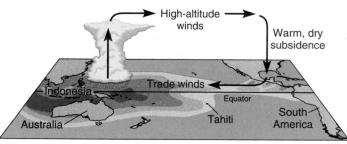

La Niña

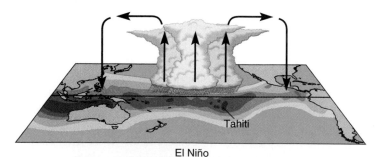

El Niño

FIGURE 15.17

The El Niño/La Niña Southern Oscillation Cycle. During El Niño years, surface trade winds that normally push warm water westward toward Indonesia weaken and allow this pool of water to flow eastward, bringing storms to the Americas.

the ocean surface enough to reverse trade winds and ocean surface currents so they flow eastward rather than westward. Another theory is that eastward-flowing deep currents called baroclinic waves periodically interfere with coastal upwelling, warming the sea surface off South America and eliminating the temperature gradient across the Pacific. At any rate, the shift in position of the tropical depression sets off a chain of events lasting a year or more with repercussions in weather systems across North and South America and perhaps around the world.

Peruvian fishermen were the first to notice irregular cycles of rising ocean temperatures that resulted in disappearance of the anchovy schools on which they depended. They named these events **El Niño** (Spanish for the Christ child) because they often occur around Christmas time. We have come to call the intervening years **La Niña** (or little girl). Together, this cycle is called the El Niño Southern Oscillation (ENSO).

Actually, sea surface temperature measurements in the central Pacific near Tahiti are the best measurement of this giant oscillation. In May 1998, as an unusually strong El Niño was ending, surface temperatures in this region plunged 7°C (12.5°F) in a little over a week, a dramatic event considering the enormous heat-storing capacity of ocean waters.

How does the ENSO cycle affect us? During an El Niño year, the northern jet stream—which normally is over Canada—splits and is drawn south over the United States. This pulls moist air from the Pacific and Gulf of Mexico inland, bringing intense storms and heavy rains from California across the midwestern states. The intervening La Niña years bring hot, dry weather to these same areas. An exceptionally long El Niño event from 1991 to 1995 broke the seven-year drought over the western United States and resulted in floods of the century in the Mississippi Valley. Oregon, Washington, and British Columbia, on the other hand, tend to have warm, sunny weather in El Niño years rather than their usual rain. Droughts in Australia and

Key Concepts

- Most solar energy is intense, high-energy light to which our atmosphere is relatively transparent. Most of the energy re-released from the earth's surface is longer wavelength, infrared energy that can be absorbed by water vapor, carbon dioxide, and other gases. This "greenhouse effect" keeps the earth significantly warmer than it would otherwise be.

- Rising air cools as its pressure decreases. This cooling results in condensation of water vapor, which releases large amounts of latent energy as it changes from gas to liquid form. This energy intensifies the convection currents and draws up more warm air and water vapor. Where there are sharp temperature differentials between air masses and large amounts of water vapor available, powerful storms such as hurricanes and tornadoes are possible.

- A large body of evidence has accumulated suggesting that our global climate is now changing as a result of human-caused releases of CO_2 and other "greenhouse gases" into the atmosphere. This includes more extreme weather events, changes in the habits and ranges of organisms, melting of sea ice and glaciers, and direct measurment of global temperatures.

Indonesia during El Niño episodes cause disastrous crop failures and forest fires, including one in Borneo in 1983 that burned 3.3 million ha (8 million acres). An even stronger El Niño in 1997 spread health-threatening forest fire smoke over much of Southeast Asia.

Are ENSO events becoming stronger or more irregular because of global climate change? Studies of corals up to 130,000 years old suggest this is true. Furthermore, there are signs that warm ocean surface temperatures are spreading. In addition to the pool of warm water in the western Pacific associated with La Niña years, oceanographers recently discovered a similar warm region in the Indian Ocean. High sea surface temperatures spawn larger and more violent storms such as hurricanes and typhoons. On the other hand, increased cloud cover would raise the albedo while upwelling convection currents generated by these storms could pump heat into the stratosphere. This might have an overall cooling effect and act as a safety valve for global warming.

A longer-scale ocean/weather connection called the **Pacific Decadal Oscillation (PDO)** involves a very large pool of warm water that moves back and forth across the North Pacific every 30 years or so. From about 1977 to 1997, surface water temperatures in the middle and western part of the North Pacific Ocean were cooler than average, while waters off the western United States were warmer. During this time, salmon runs in Alaska were bountiful, while those in Washington and Oregon were greatly diminished. In 1997, however, ocean surface temperatures along the coast of western North America turned significantly cooler, perhaps marking a return to conditions that prevailed between 1947 and 1977. Under this cooler regime, Alaskan salmon runs declined while those in Washington and Oregon improved somewhat. This decades-long oscillation can be seen in rainfall patterns shown by tree-ring growth as far east as Yellowstone National Park.

Human-Caused Global Climate Change

Many scientists regard anthropogenic (human-caused) global climate change to be the most important environmental issue of our times. The possibility that humans might alter world climate is not a new idea. In 1895, Svante Arrhenius, who subsequently received a Nobel Prize for his work in chemistry, predicted that CO_2 released by coal burning could cause global warming. Most of Arrhenius' contemporaries dismissed his calculations as impossible, but his work seems remarkably prescient now.

The first evidence that human activities are increasing atmospheric CO_2 came from an observatory on top of the Mauna Loa volcano in Hawaii. The observatory was established in 1957 as part of an International Geophysical Year, and was intended to provide data on air chemistry in a remote, pristine environment. Surprisingly, measurements showed CO_2 levels increasing about 0.5 percent per year, rising from 315 ppm in 1958 to 372 ppm in 2002 (fig. 15.18). This increase isn't a straight line, however. Because a majority of the world's land and vegetation are in the Northern Hemisphere, northern seasons dominate the signal. Every May, CO_2 levels drop slightly as plant growth on northern continents use CO_2 in photosynthesis. During the northern winter, levels rise again as respiration releases CO_2.

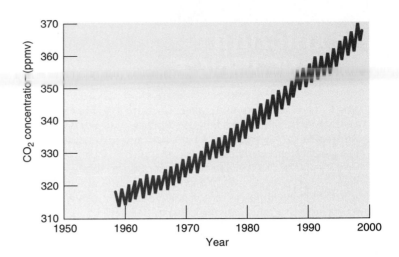

FIGURE 15.18 Carbon dioxide concentrations at Mauna Loa Observatory in Hawaii have increased dramatically since 1958. Annual fluctuations reflect differences in photosynthesis and respiration between winter and summer. *Source:* Dave Keeling and Tim Whorf, Scripps Institute of Oceanography.

In 1988 the United Nations Environment Program and World Meteorological Organization formed the **Intergovernmental Panel on Climate Change (IPCC).** This panel brings together scientists from many nations and a wide variety of fields to assess the current state of knowledge about climate change. The IPCC's First Assessment Report played an important role in adoption of the UN Framework Convention on Climate Change at the Rio Earth Summit in 1992. In February 2001, the IPCC released its third climate report, which said "with a high degree of confidence" that "recent changes in the world's climate have had discernible impacts on physical and biological systems." For this report, more than 700 scientists representing 100 countries reviewed results from some 3,000 scientific studies, which showed changes in 420 different physical and biological systems.

Noting that the earth's surface temperature has risen by about 0.6°C (1.1°F) over the past century, with most of that warming occurring in the past two decades, the IPCC concluded that human activities must be at least partially responsible. Separating the impact of human activity from natural climate variation is extremely difficult (Exploring Science, p. 320). Nonetheless, the IPCC claims, the observed global warming is unlikely to be the result of natural variability alone. "We have altered the chemical composition of the atmosphere through the buildup of greenhouse gases—primarily carbon dioxide, methane, and nitrous oxide," the report declared. "The heat-trapping property of these gases is undisputed although uncertainties exist about exactly how earth's climate responds to them."

Climate Skeptics

Not everyone agrees with the IPCC conclusions or its method of reaching them. Dr. Frederick Seitz, president emeritus of Rockefeller University and former president of the U.S. National Academy

Uncertainty in Climate Change

Why is there so much disagreement about whether human activities are causing global climate change? A big part of the problem is that weather and climate are complex and highly variable. Making predictions about what we might expect over the long term is difficult even in normal conditions. Many scientists agree, for example, that mean global temperatures will probably warm 1.5°–6°C (2.7°–11°F) over the next century. They can't say, however, whether particular regions will be warmer or cooler; and for many regions they are even unable to state whether a wetter or a drier climate is more likely. Virtually all published estimates of how the climate could change are the results of computer models of the atmosphere known as "general circulation models." These complicated mathematical models are able to simulate many features of the climate, but they are still not accurate enough to provide reliable forecasts of specific climatic conditions. In fact, different models often yield contradictory results. Given the unreliability of these models, researchers trying to understand the future impacts of climate change generally analyze different scenarios from several different climate models. The hope is that, by using a wide variety of different climate models, one's analysis can include the entire range of scientific uncertainty. Climatologists warn us that projections of climate change in specific areas are not forecasts but are reasonable examples of what might happen.

The U.S. EPA lists several reasons for caution in interpreting climate models:

1. The cooling effects of sulfate aerosols (from volcanoes and burning fossil fuels) could make a big difference in local conditions. Introducing aerosol effects into some models, for instance, changes precipitation predictions from a decline of 15 percent to an increase of 18 percent.

2. Most models focus on a doubling of CO_2, but we may very well be forced to cope with three or four times the current CO_2 concentrations over the longer term.

3. Nature may produce major surprises. If ocean currents are disrupted, for example, some parts of the world may actually need to deal with a substantially cooler climate rather than warmer.

4. Climate may not change in a "linear" fashion. This assumed gradualness underlies the assumption that society will have time to adjust. Some evidence suggests that climate has changed in a rather abrupt and chaotic fashion in the past.

5. Climate model simulations of today's regional climates are often inaccurate. This could be due simply to a lack of computational power to allow for local variations, or it could indicate a serious flaw in the models.

In the face of scientific uncertainty, interpretations of the costs and benefits of actions will vary. If you think major climate change is highly probable, then inaction is perilous. If you think that changes, if any, are likely to be small and gradual, then changing current policies may be unacceptable. For example, some economists (using uncertain models themselves) project that strict emission controls would cost the U.S. economy 10,000 jobs and $30 billion per year. It would be cheaper, they argue, to adapt to climate change than to try to prevent it. Ecological economists, meanwhile, argue that the costs of action need not be so high. They point out that we could cut emissions by 30 percent simply by improving efficiency and using newer technology. Further, leading the development of efficient technology will create jobs. These calculations also involve uncertainty, of course; economic models differ just as widely in their predictions as climate models do.

It's important to think about who bears the costs and who enjoys the benefits of any particular policy. Mandating energy conservation would be good for those selling renewable energy, but painful for those whose livelihoods depend on fossil fuels. And assuming that it will be cheaper to adapt to climate changes in the future rather than change our habits now, means shifting the costs to future generations. Is that fair?

If you were in Congress and had to vote on whether or not to mandate greenhouse gas reductions, how would you use scenarios from climate models? What additional information would you need to make a decision in this issue? How would you weigh the risks of inaction compared to the immediate pain of making changes in our society and technology?

of Sciences, for example, charged that bureaucrats and politicians altered the report in ways that didn't represent the consensus agreed upon by scientists. Others claim we don't know enough yet to make predictions about what may be happening to our global climate. While most dissenting views come from industry lobbying groups, some prominent scientists also hold that dire predictions about climate change may be overblown. John Cristy of the University of Alabama, for example, says that the upper trophosphere isn't warming as fast as the earth's surface, perhaps showing that our current ideas about global climate are incorrect.

Richard Lindzen of MIT contends that current computer models don't adequately account for cloud effects. Higher evaporation rates, he believes, could produce more clouds that reflect sunlight and balance the greenhouse effect. And Vincent Gray, a climate consultant from New Zealand, contends that CO_2 concentrations measured at Mauna Loa have increased at a remarkably constant rate over the past 42 years, despite the fact that worldwide fossil fuel consumption has increased almost 50 percent during that time. Something is missing from our models, he claims.

Sources of Greenhouse Gases

Since preindustrial times atmospheric concentrations of CO_2, CH_4 and N_2O have climbed by over 31 percent, 151 percent, and 17 percent, respectively. Carbon dioxide is by far the most important cause of anthropogenic climate change (table 15.3). Burning fossil fuels, making cement, burning forests and grasslands, and other human activities release nearly 30 billion tons of CO_2 every year, on average, containing some 8 billion tons of carbon (fig. 15.19). About 3 billion tons of this excess carbon is taken up by terrestrial ecosystems, and around 2 billion tons are absorbed by the oceans, leaving an annual atmospheric increase of some 3 billion tons per year. If current trends continue, CO_2 concentrations could reach about 500 ppm (approaching twice the preindustrial level of 280 ppm) by the end of the twenty-first century.

Although rarer than CO_2, methane absorbs 20 to 30 times as much infrared and is accumulating in the atmosphere about twice as fast as CO_2. Methane is released by ruminant animals, wet-rice paddies, coal mines, landfills, and pipeline leaks. Chloroflurocarbons (CFCs) also are powerful infrared absorbers. CFC releases have declined since many of their uses were banned, but the CFCs already in the atmosphere will persist for many years. Nitrous oxide is produced by burning organic material and by soil denitrification. As table 15.3 shows, CFCs and N_2O together are thought to account for only about 17 percent of human-caused global warming.

Aerosols have a tendency to counteract the effects of greenhouse gases, on local and sometimes global scales. Sulfate aerosols and soot produced by some of the same activities that release CO_2 can shade urban air, for example, and reduce air temperatures by as much or more than greenhouse gases increase them. The 1991 eruption of Mt. Pinatubo in the Philippines ejected enough ash and sulfate particles to cool the global climate about 1°C, but this cooling lasted for only about a year.

Current Evidence of Climate Change

Increasingly, evidence from around the world suggests that global climate is already changing as a result of human actions. The World Meteorological Organisation (WMO) reports that global average surface temperature rose 0.6°C in the twentieth century, which appears to be the largest increase in any century during the past 1,000 years. Furthermore, the rate of warming since 1976 is

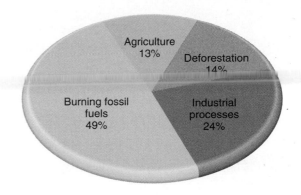

FIGURE 15.19   Contributions to global warming by different types of human activities. *Source:* World Resources Institute.

roughly three times that for the whole period. The 10 hottest years in the 143 years for which we have instrument readings have all been since 1990, with the three hottest being 1998, 2002, and 2001. The WMO also notes that extreme weather events, such as record high and low tempertures, intense rainfall, and severe storms, have increased dramatically in the past decade. As global temperatures continue to warm due to climate change, the WMO warns, we can expect the number and severity of extreme weather events to rise.

Polar regions are changing even faster than the rest of the world. Meteorologists report that average summer temperatures in Alaska are 2.7°C higher than the historic average, while winter temperatures have risen 5.4°C. Permafrost is melting, houses and trees are tipping over, and a beetle infestation (made possible by warmer temperatures) has killed 25 million spruce trees on the Kenai Peninsula. Coastal Inupiat villages such as Shishmaref, Point Hope, and Barrow are having to relocate inland as increasingly severe storms erode the arctic shoreline. As the opening story of this chapter describes, a similar warming trend on the Antarctic Peninsula is causing rapid disintegration of coastal ice shelves. While melting of this floating ice doesn't change sea levels, unblocking of inland antarctic glaciers and ice sheets could result in calamitous sea-level increases.

Satellite images and surface temperature data show that growing seasons are now as much as three weeks longer in a band across northern Eurasia than they were 30 years ago (see fig. 1.9). Arctic sea ice is 40 percent thinner now, and the edge of the Arctic sea ice pack averages about 500 km farther north than it was a century ago. As sea ice forms later in the fall and melts earlier in the spring, polar bears have a shorter seal-hunting season (fig. 15.20). Hudson's Bay polar bears now weigh as much as 100 kg (220 lbs) less than in the 1960s. Surprisingly, warmer winters are hurting some land-based mammals as well. On Canada's remote Ellesmere Island, winters are usually so cold and dry that the little snow that falls quickly blows away exposing moss and lichens on which the musk oxen, Perry caribou, and arctic hares depend. Now, warmer winter temperatures allow freezing rains that coat vegetation with an impervious ice coating. Animals on the verge of starvation are pushed over the edge when they can't scrape through the ice.

Alpine glaciers everywhere are retreating rapidly (fig. 15.21). Mt. Kilimanjaro has lost 85 percent of its famous ice cap since 1915

TABLE 15.3	Proportion of Global Warming Caused by Four Greenhouse Gases

GAS	PERCENTAGE
Carbon dioxide (CO₂)	64
Methane (CH₄)	19
Chlorofluorocarbons (CFCs)	11
Nitrous oxide (N₂O)	6
Sulfur hexafluoride	0.4

FIGURE 15.20 Receding arctic sea ice prevents polar bears from hunting seals, their main food source. © Corbis/Volume 244.

FIGURE 15.21 Alpine glaciers are retreating rapidly everywhere in the world. For the billions of people whose drinking water comes from glacier-fed rivers, this could be a serious threat. © William P. Cunningham.

Current range

Potential range

FIGURE 15.22 Change in suitable range for sugar maple, according to the climate-change model.
Source: Margaret Davis and Catherine Zabriskie in *Global Warming and Biological Diversity,* ed. by Peters and Lovejoy, 1991, Yale University Press.

or both. Given enough time and a route for migration, these species might adapt to new conditions, but we now are forcing them to move at least ten times the rate many achieved at the end of the last ice age (fig. 15.22). The disappearance of amphibians such as the beautiful golden toads from the cloud forests of Costa Rica or western toads from Oregon's Cascade Range are thought to have been caused at least in part by changing weather patterns.

Around the world, coral reefs are bleaching because of higher water temperatures. Rivaling tropical rainforests in their biological diversity and net productivity, coral reefs already are threatened by human activities such as dynamite fishing and limestone excavation. If sea temperatures continue to rise, most coral reefs will be wiped out in the next 50 years. Similarly, the Cape Floral Kingdom of South Africa, one of the most unique terrestrial ecosystems in the world, could disappear due to changing weather patterns.

Winners and Losers

Local climate changes could well have severe effects on human societies, agriculture, and natural ecosystems. In many cases, both organisms and human infrastructure will not be able to move or adapt quickly enough to accommodate the rates at which climate seems to be changing. Tropical areas will probably not get much warmer than they are now, but some middle- and high-latitude areas could experience unaccustomed heat (fig. 15.23).

Does anyone win in these scenarios? Residents of northern Canada, Siberia, and Alaska probably would enjoy warmer temperatures, longer growing seasons, and longer ice-free shipping seasons from their ports. Stronger monsoons bringing more moisture to most of Africa and South Asia could increase crop yields. Already, welcome rains are greening the Sahel at the edge of the Saharan desert. Areas that were disastrously dry in the 70s and 80s are once again suitable for agriculture.

(see opening photo, chapter 1). By 2015 all permanent ice on the mountaintop is expected to be gone. In 1972, Venezuela had six glaciers; now it has only two. All the glaciers in Montana's Glacier National Park will be gone by 2070, if current trends continue. Over the next 50 years, at least half of all alpine glaciers in the world could disappear. Without this resource, agriculture, industry, power generation, and drinking-water supplies will suffer. About 1.7 billion people now live in areas where water supplies are tight. This number could increase to 5.4 billion by 2025. Sea level has risen worldwide approximately 15–20 cm (6–8 in.) in the past century. About one-quarter of this increase is due to melting alpine glaciers; roughly half is caused by thermal expansion of ocean water.

Many wild plant and animal species are being forced out of their current ranges as the climate warms. In Europe and North America, for example, 57 butterfly species have either died out at the southern end of their range, or extended their northern limits,

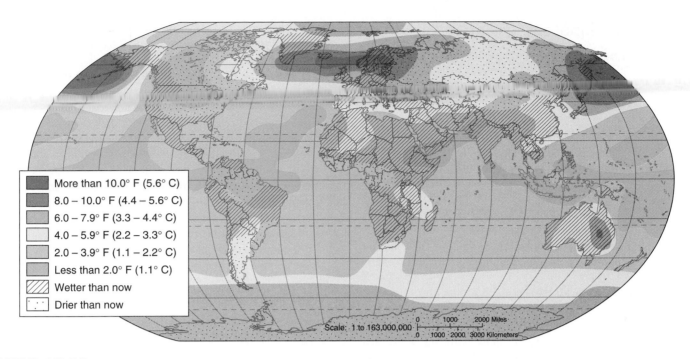

FIGURE 15.23 Potential global temperature and precipitation change by 2050. Note that most of the United States and Canada are expected to be drier (stippled) while most tropical countries will probably be wetter (diagonal lines).

Carbon dioxide is a fertilizer for plants, and higher CO_2 levels bring faster growth for many plant species. Sherwood Idso of the U.S. Department of Agriculture says that increased CO_2 concentrations could trigger lush plant growth that would make crops abundant. But soils in some areas may not be suitable for crops no matter how agreeable the climate.

Enough water is stored in glacial ice caps in Greenland and Antarctica to raise sea levels around 100 meters (300 feet) if they all melt. About one-third of the world's population now live in areas that would be flooded if that happens. Even the 75 cm (30 in.) sea-level rise expected by 2050 will flood much of south Florida, Bangladesh, Pakistan, and many other low-lying coastal areas. Most of the world's largest urban areas are on coastlines. Wealthy cities such as New York or London can probably afford to build dikes to keep out rising seas, but poorer cities such as Jakarta, Calcutta, or Manila might simply be abandoned as residents flee to higher ground. Several small island countries such as the Maldives, the Bahamas, Kiribati, and the Marshall Islands could become uninhabitable if sea levels rise a meter or more. The South Pacific nation of Tuvalu has already announced that it is abandoning its island homeland. All 11,000 residents will move to New Zealand, perhaps the first of many climate change refugees.

Insurance companies worry that the $2 trillion in insured property along U.S. coastlines could be at risk from a combination of high seas and catastrophic storms. Some 87,000 homes in the United States within 150 m (500 ft) of a shoreline are in danger of coastal erosion or flooding in the next 50 years. Accountants warn that loss of land and structures to flooding and coastal erosion together with damage to fishing stocks, agriculture, and water supplies could raise worldwide insurance claims from about $40 billion in 2000 (already ten times more than they had been 50 years earlier) to more than $300 billion in 2050. Some of this increase in insurance claims is that more people are living in dangerous places, but extra-severe storms only exacerbate this problem.

Infectious diseases are likely to increase as the insects and rodents that carry them spread to new areas. Already we have seen diseases such as malaria, dengue fever, and West Nile virus appear in parts of North America where they had never before been reported. Coupled with the movement of hundreds of millions of environmental refugees and greater crowding as people are forced out of areas made uninhabitable by rising seas and changing climates, the spread of epidemic diseases will also increase.

An ominous consequence of arctic permafrost melting might be the release of vast stores of methane hydrate now locked in frozen ground and in sediments in the ocean floor. Together with the increased oxidation of high-latitude peat lands potentially caused by warmer and dryer conditions, release of these carbon stores could add as much CO_2 to the atmosphere as all the fossil fuels ever burned. We could trigger a disastrous positive feedback loop in which the effects of warming cause even more warming.

Another potentially catastrophic outcome of global climate change is that greater terrestrial runoff from increased rainfall might suddenly change ocean circulation patterns that now moderate northern climates (see fig. 17.4). On the other hand, increased ocean evaporation might intensify snowfall at high latitudes so that arctic glaciers and snow pack would increase rather than decrease. Ironically, the increased albedo of colder, snow-covered surfaces might then trigger a new ice age. Clearly, we're disturbing a complex system and facing consequences that we don't fully understand.

International Climate Negotiations

One of the centerpieces of the 1992 United Nations "Earth Summit" meeting in Rio de Janeiro was the Framework Convention on Climate Change, which set an objective of stabilizing greenhouse gas emissions to reduce the threats of global warming. At a follow-up conference in Kyoto, Japan in 1997, 160 nations agreed to roll back CO_2, methane, and nitrous oxide emissions about 5 percent below 1990 levels by 2012. Three other greenhouse gases, hydrofluorocarbons, perfluorocarbons, and sulfur hexafluoride, would also be reduced, although from what level was not decided. Known as the **Kyoto Protocol,** this treaty sets different limits for individual nations depending on their output before 1990. Poorer nations like China and India were exempted from emission limits to allow development to increase their standard of living. Wealthy countries created the problem, the poorer nations argue; and the wealthy should deal with it.

In 2002 all the nations of the European Union plus Japan ratified the Kyoto agreement. When Russia also ratifies, as it has promised to do, the 55-percent threshold will be reached and the protocol will become binding. The United States, however, has refused so far to comply with the agreement it helped broker. In spite of his campaign promises, President Bush has remained adamant that he will not honor U.S. promises at Kyoto. He contends that reducing emissions would be too costly for the U.S. economy. "We're going to put the interests of our own country first and foremost," he said. This go-it-alone attitude provoked a stunned and angry reaction among other countries, some of which began to talk about trade sanctions as a way of forcing U.S. cooperation. Rather than reduce emissions, however, the United States has continued to increase them. If current energy consumption patterns persist, government scientists calculate, the United States will be 43 percent above 1990 emission levels by 2020 rather than 7 percent below them as was promised.

Currently, the United States, with less than 5 percent of the world's population, produces 28 percent of all anthropogenic CO_2 (fig. 15.24). China, with 1.3 billion people, is second in total CO_2 emissions, but fourteenth in per capita production. Japan and Western Europe, which by most measures have standards of living at least equal to the United States, produce only about half as much CO_2 per person.

In 2002 the U.S. EPA released a report essentially agreeing with the IPCC that greenhouse gases produced by human activities are causing global warming. It noted that climate changes and rising sea levels are especially threatening to sensitive landscapes such as mountain meadows and coastal barrier islands. President Bush, however, insisted that many questions remain about the science of climate change. Rather than impose uniform limits on industry, President Bush offered a voluntary program to reduce "greenhouse intensity"—the ratio of of emissions to economic output—by 18 percent over the next decade. Not surprisingly, many businesses love this plan, while environmental groups describe it as "a total charade," and "fiddling while Rome burns." As the economy grows over the next decade, they maintain, the reduction in greenhouse intensity will really result in about a 19-percent increase in actual emissions.

Controlling Greenhouse Emissions

In spite of U.S. intransigence, progress already is being made in many places toward reducing greenhouse emissions. The United Kingdom, for example, had already rolled CO_2 emissions back to 1990 levels by 2000, and vowed to reduce them 60 percent more by 2050. Britain already has started to substitute natural gas for coal, promote energy efficiency in homes and industry, and raise its already high gasoline tax. Plans are to "decarbonize" British society and to decouple GNP growth from CO_2 emissions. A revenue-neutral carbon levy is expected to lower CO_2 releases and trigger a transition to renewable energy over the next five decades.

Germany, also, has reduced its CO_2 emissions at least 10 percent by switching from coal to gas and by encouraging energy efficiency throughout society. Atmospheric scientist Steve Schneider calls this a "no regrets" policy; even if we didn't need to stabilize our climate, many of these steps save money, conserve resources, and have other environmental benefits. Nuclear power also is being promoted as an alternative to fossil fuels. It's true that nuclear reactions don't produce greenhouse gases, but security worries and unresolved problems of how to store wastes safely make this option unacceptable to many people.

Renewable energy sources offer a better solution to climate problems, many people believe. Chapter 20 discusses options for conserving energy and switching to renewable sources such as solar, wind, geothermal, biomass, and fuel cells. Denmark, the world's leader in wind power, now gets 20 percent of its electricity from wind generators. Plans are to generate half of the nation's electricity from offshore wind farms by 2030. Even China reduced its CO_2 emissions 20 percent between 1997 and 2003 through greater efficiency in coal burning and industrial energy use. The U.S. energy plan, meanwhile, is to burn more coal, drill for oil in the Arctic National Wildlife Refuge, and continue driving gas-guzzling trucks and SUVs.

In addition to reducing its output, there are options for capturing and storing CO_2. Planting trees can be effective if they're

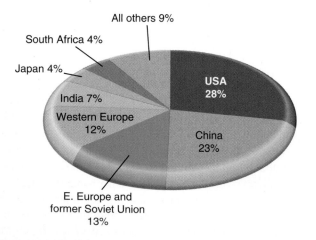

All others 9%

South Africa 4%

Japan 4%

India 7%

Western Europe 12%

USA 28%

China 23%

E. Europe and former Soviet Union 13%

FIGURE 15.24   Sources of anthropogenic CO_2 by country or region in 2000. *Source:* World Resources Institute, 2001.

allowed to mature to old-growth status or are made into products like window frames and doors that will last for many years before they're burned or recycled. Farmland also can serve as a carbon sink if farmers change their crop mixture and practice minimum-till cultivation that keeps carbon in the soil.

One of the provisions of the Kyoto agreement is to establish a greenhouse gas emissions trading system. Under this plan, rather than spend money to build new, efficient facilities, a company could buy pollution credits from a firm that can cut its emissions more cheaply or that will plant trees or sequester carbon in some other manner. Economists estimate the global greenhouse emissions market at about $100 billion by 2010. The European Union has already instituted such a trading system. The primary recipients of payments, so far, are states of the former USSR, which have many dirty, antiquated, inefficient factories and power plants that can (and should) be inexpensively updated to greatly reduce CO_2 and other air pollutants. Some tropical countries also have received payments for agreeing to plant new forests. U.S. businesses bemoan the fact that they are excluded from this lucrative market because their government refuses to ratify the Kyoto agreement.

In 1988 the late John H. Martin suggested that phytoplankton growth in the ocean may be limited by iron deficiency. This theory was tested in 2000, when researchers spread about 3.5 tons of dissolved iron over 75 km^2 (about 30 mi^2) of the South Pacific. Monitored by satellite, a tenfold rise in chlorophyll concentration eventually spread over about 1,700 km^2 (660 mi^2) and was calculated to have pulled several thousand tons of CO_2 out of the air. Still, some ecologists are wary about such large-scale tinkering. We don't know whether the carbon fixed by phytoplankton growth will be stored in sediments or simply eaten by predators and returned immediately to the atmosphere. It's possible that dying algae might create an anoxic zone that would devastate oceanic food webs.

Another way to store CO_2 is to inject it into underground rock layers or deep ocean waters. Since 1996, Norway's Statoil has pumped more than 1 million metric tons of CO_2 into an aquifer 1,000 m (3,000 ft) below the seafloor at a North Sea gas well. It is economical to do so because otherwise the company would have to pay a $50 per ton carbon tax on its emissions. Around the world, deep, briny aquifers could store a century worth of CO_2 output at current fossil fuel consumption rates. It might also be possible to pump liquid CO_2 into deep ocean trenches, where it would form lakes contained by enormous water pressures. There are worries, however, about what this might do to deep ocean fauna and what might happen if earthquakes or landslides caused a sudden release of this CO_2. In Canada, an Alberta power plant is injecting CO_2 into a coal seam too deep to be mined. The CO_2 releases natural gas, which is burned to produce electricity and more CO_2. Proponents of **carbon management,** as these various projects are called, argue that it may be cheaper to clean up fossil fuel effluents than to switch to renewable energy sources.

Most attention is focused on CO_2 because it lasts in the atmosphere, on average, for about 120 years. Methane and other greenhouse gases are much more powerful heat absorbers but remain in the air for a much shorter time. NASA's Dr. James Hansen made a controversial suggestion, however, that the best short-term attack

What can you do?

Reducing Carbon Dioxide Emissions

Individuals can help reduce global warming. Although some actions may cause only a small impact, collectively they add up. Many of these options save money in the long run and have other environmental benefits such as reducing air pollution and resource consumption.

1. Drive less, walk, bike, take public transportation, carpool, or buy a vehicle that gets at least 30 mpg (12.6 km/l). Average annual CO_2 reduction: about 20 lbs for each gallon of gasoline saved. Increasing your mileage by 10 mpg, for example, eliminates about 2,500 lbs (1.1 metric tons) of CO_2 per year. Becoming entirely self-propelled would save, on average, about 12,000 lbs (5.45 metric tons) per year.

2. Plant trees to shade your house during the summer, and paint your house a light color if you live in a warm climate or a dark color in a cold climate. Average annual CO_2 reduction: about 5,000 lbs (2.2 metric tons).

3. Insulate your house and seal all drafts. Average annual CO_2 reduction for the highest efficiency insulation, weatherstripping, and windows: about 5,000 lbs (2.2 metric tons).

4. Replace old appliances with new, energy-efficient models. Average annual CO_2 reduction for the most efficient refrigerator, for instance: about 3,000 lbs (1.4 metric tons).

5. Produce less waste. Buy minimally packaged goods and reusable products. Recycle. Average annual CO_2 reduction: about 1,000 lbs (0.45 metric tons) for 25 percent less garbage.

6. Turn down your thermostat in the winter and turn it up in the summer. Average annual CO_2 reduction: about 500 lbs (0.23 metric tons) for every 1°C (1.8°F) change.

7. Replace standard light bulbs with long-lasting compact fluorescent ones. Average annual CO_2 reduction: about 500 lbs (0.23 metric tons) for every bulb.

8. Wash laundry in warm or cold water, not hot. Average annual CO_2 reduction: about 500 lbs (0.23 metric tons) for two loads per week.

9. Set your water heater thermostat no higher than 48.9°C (120°F). Average annual CO_2 reduction: about 500 lbs (0.23 metric tons) for each 5°C (9°F) temperature change.

10. Buy renewable energy from your local utility if possible. Potential annual CO_2 reduction: about 30,000 lbs (13.5 metric tons).

on warming might come by focusing not on carbon dioxide but on methane. Reducing gas pipeline leaks would conserve this valuable resource as well as help the environment. Methane from landfills, oil wells, and coal mines that once would have been simply vented into the air is now being collected and used to generate electricity. Rice paddies are a major methane source. Changing flooding schedules and fertilization techniques can reduce anoxia (oxygen depletion) that produces marsh gas. Finally, ruminant animals (like cows, camels, and buffalo) create large amounts of digestive system gas. Modified diets can reduce flatulence significantly.

Soot, while not mentioned in the Kyoto Protocol, might also be important in global warming. Dark, airborne particles absorb both ultraviolet and visible light, converting them to heat energy. According to some calculations, reducing soot from diesel engines, coal-fired generators, forest fires, and wood stoves might reduce net global warming by 40 percent within three to five years. Curbing soot emissions would also have beneficial health effects.

Some individual cities and states have announced their own plans to combat global warming. Among the first of these were Toronto, Copenhagen, and Helsinki, which pledged to reduce CO_2 emissions 20 percent from 1990 levels by 2010. Portland, Oregon, was the first U.S. city to implement a CO_2-reduction strategy. States with plans to limit greenhouse gas emissions include California, New York, and New Hampshire. Some corporations are following suit. British Petroleum has set a goal of cutting CO_2 releases from all its facilities by 10 percent before 2010. BP, Alcan, DuPont, and other companies joined with the Environmental Defense Fund to launch a partnership for climate action, pledging to meet or exceed Kyoto requirements. Each of us can make a contribution in this effort (What Can You Do? p. 325). Chapter 20 discusses ways we can cut our individual energy consumption, the principal cause of greenhouse gas emissions.

Summary

- Global climate change may well be *the* most momentous issue in environmental science today. To understand why this is happening and what we can do about it, we need to know something about atmospheric processes.

- Weather is a description of local conditions; climate describes long-term weather patterns.

- The atmosphere and living organisms have evolved together so that the present chemical composition of the air is both suitable for, and largely the result of, biological processes.

- The atmosphere is relatively transparent to visible light that warms the earth's surface and is captured by photosynthetic organisms and stored as potential energy in organic chemicals. Heat is lost from the earth's surface as infrared radiation, but fortunately for us, carbon dioxide and water vapor naturally present in the air capture the radiation and keep the atmosphere warmer than it would otherwise be.

- When air is warmed by conduction or radiation of heat from the earth's surface, it expands and rises, creating convection currents. These vertical updrafts carry water vapor aloft and initiate circulation patterns that redistribute energy and water from areas of surplus to areas of deficit. Pressure gradients created by this circulation drive great air masses around the globe and generate winds that determine both immediate weather and long-term climate.

- The earth's rotation causes wind deflection called the Coriolis effect, which makes air masses circulate in spiraling patterns.

- Strong cyclonic convection currents fueled by temperature and pressure gradients and latent energy in water vapor can create devastating storms, including hurricanes.

- Tornadoes, while classified as cyclonic storms, are not set spinning by Coriolis forces. Instead, shear forces caused by differential wind speeds, together with rapidly rising warm convection currents and cold downdrafts are thought to create intensely focused spinning vortices. Although top wind speeds in tornadoes can be higher than those in hurricanes, total damage in the former is usually smaller because the area covered is smaller.

- Other sources of storms are the seasonal winds, or monsoons, generated by temperature differences between the ocean and a landmass. Monsoons often bring torrential rains and disastrous floods, but they also bring needed moisture to farmlands that feed a majority of the world's population.

- The El Niño/Southern Oscillation is a complex interaction between oceans and atmosphere that has far-reaching climatic, ecological, and social effects. ENSO cycles can affect things as widely different as forest fires in Indonesia, anchovy fishing in Peru, rainfall in the Sahara Desert, and how the corn grows in Iowa. Knowing something about how weather works can be helpful in our everyday life.

- Although some scientists disagree about the causes, timing, and consequences of global climate change, an overwhelming majority now agree that there are unmistakable signals of human impacts on the world's climate.

- Melting of arctic and alpine glaciers could raise sea levels to disastrous heights, while also drying up water sources on which billions of people depend. Many wild animal and plant species are being driven out of their current ranges as the climate changes. Some, like polar bears and musk oxen, have no place to go, and may become extinct. Some low-lying, small island nations such as Tuvalu are becoming uninhabitable.

- In 1997, 160 nations meeting in Kyoto, Japan, signed an agreement to reduce greenhouse gas emissions. Every industrial nation in the world—except the United States—has ratified this agreement, and many are beginning to implement it. Despite evidence from his own scientists about the reality and seriousness of global warming, President Bush continues to call for more study and voluntary efforts to combat this threat.

- Many American states, cities, and companies have begun programs to limit greenhouse gas emissions. There are ways that you, too, can lower your impacts on our global climate.

Questions for Review

1. What are weather and climate? How do they differ?
2. Name and describe the four layers of air in the atmosphere.
3. What is the greenhouse effect?
4. What are the ENSO cycle and the PDO?
5. What are the jet streams? How do they influence weather patterns?
6. How do tornadoes form and why are they so destructive?
7. Describe the Coriolis effect. What causes it?
8. Explain some of the observed or expected effects of climate change.
9. Summarize some actions we could take, collectively and as individuals, to combat global climate change.
10. Why has the United States not ratified the Kyoto Protocol?

Questions for Critical Thinking

1. In most places, weather changes in highly variable, seemingly chaotic ways. How would you distinguish between meaningful patterns and random variations?
2. Many scientists say global warming is real; others dispute this claim. How can we decide what to do when the science is uncertain and the risk is great?
3. Has there been a major change recently in the weather where you live? Can you propose any reasons for such changes?
4. What forces determine the climate in your locality? Are they the same for neighboring states?
5. What was the weather like when your parents were young? Do you believe the stories they tell you? Why or why not?
6. Have you ever experienced a tornado or hurricane? What was it like? What omens or warnings told you it was coming?
7. What would you do to adapt to a permanent drought? What effects would it have on your life?
8. What should we do about coastal cities threatened by rising oceans—rebuild, enclose in dikes, or just move?

9. Would you favor building nuclear power plants to reduce CO_2 emissions? Why or why not?

Key Terms

aerosols 307	Kyoto Protocol 324
albedo 309	La Niña 318
carbon management 325	latent heat 310
climate 307	Milankovitch cycles 317
cold front 312	monsoon 312
convection currents 307	ozone 308
Coriolis effect 311	Pacific Decadal Oscillation
downbursts 314	(PDO) 319
El Niño 318	stratosphere 308
greenhouse effect 310	tornadoes 314
hurricanes 313	troposphere 307
Intergovernmental Panel on	warm front 313
Climate Change (IPCC) 319	weather 307
jet streams 311	

Further Readings

Beardsley, T. 2000. Dissecting a hurricane. *Scientific American* 282(3):80–85.

Binschadler, R. A., and C. R. Bentley. 2002. On thin ice? *Scientific American* 287(6):98–105.

Chavez, F. P., et al. 2003. From anchovies to sardines and back: Multidecadal change in the Pacific Ocean. *Science* 299:217–21.

De Leo, G. A., et al. 2001. Carbon emissions: The economic benefits of the Kyoto Protocol. *Nature* 413:478–79.

Herzog, Howard, et al. 2000. Capturing greenhouse gases. *Scientific American* 282(2):72–89.

Intergovernmental Panel on Climate Change (IPCC). 2001. *Climate Change IPCC Third Assessment Report.* Available as pdf files at: www.grida.no/climate/ipcc_tar/.

Schneider, S., et al., eds. 2002. *Climate Change Policy: A Survey.* Island Press.

Upgren, Arthur, and Jurgen Stock. 2001. *Weather: How It Works and Why It Matters.* Freeman.

Welcome to McGraw-Hill's Online Learning Center

Location: http://www.mhhe.com/environmentalscience

McGraw Hill

WEB EXERCISE

Calculate Your Climate Impacts

Global climate change may well be the single most important environmental problem your generation will face. Each of us plays a role in creating this problem in the ways we live, the things we buy, and, most importantly, the way we use energy. You can calculate how much CO_2 and other greenhouse gases your activities create. Go to www.greentagsusa.org/GreenTags/calculator_intro.cfm, a site maintained by the Bonneville Environmental Fund, that educates individuals about their climate impacts. Follow the steps in their carbon calculator to see how you compare. If you don't know specifics for your consumption patterns, use average figures supplied on each page.

1. How much greenhouse gas did your activities account for last year?
2. How do your activities compare to U.S. averages?
3. Why is there so much emphasis on energy use in this calculator? (Hint: this question has both a scientific and an economic answer.)
4. What does it mean to be climate neutral?
5. What are "green tags" or "green power" options?
6. How much would you be willing to pay to avoid a climate catastrophe?

16

Air Pollution

The only thing we have to fear is fear itself.

Franklin D. Roosevelt

OBJECTIVES

After studying this chapter, you should be able to:

- describe the major categories and sources of air pollution.
- distinguish between conventional or "criteria" pollutants and unconventional types as well as explain why each is important.
- analyze the origins and dangers of some indoor air pollutants.
- relate why atmospheric temperature inversions occur and how they affect air quality.
- understand the difference between ambient and stratospheric ozone, and appreciate the danger of statospheric ozone loss.
- understand how air pollution damages human health, vegetation, and building materials.
- compare different approaches to air pollution control and report on clean air legislation.
- judge how air quality around the world has improved or degraded in recent years and suggest what we might do about problem areas.

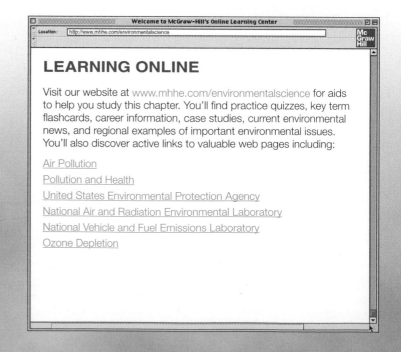

LEARNING ONLINE

Visit our website at www.mhhe.com/environmentalscience for aids to help you study this chapter. You'll find practice quizzes, key term flashcards, career information, case studies, current environmental news, and regional examples of important environmental issues. You'll also discover active links to valuable web pages including:

Air Pollution
Pollution and Health
United States Environmental Protection Agency
National Air and Radiation Environmental Laboratory
National Vehicle and Fuel Emissions Laboratory
Ozone Depletion

Photo: Energy facilities, like this oil refinery, often are major air pollution sources. © Corbis/Volume 12.

Killer Smog

London was cold and foggy on December 5, 1952. Damp, chilly air from the English Channel blanketed the city, trapping a dense stagnant layer just above ground level. As the 8.0 million Londoners stoked coal furnaces that heated most buildings and fueled most industry in the city, smoke mingled with the fog to form a dark, acrid smog. By midday, visibility dropped to a few meters. Traffic slowed to a standstill, and pedestrians, unable to see landmarks, got lost only blocks from home. Hospitals overflowed with people suffering from respiratory distress and cardiovascular problems. With all beds occupied, patients on stretchers filled hallways.

As the smog lingered for three more days, visibility dropped until people couldn't see their own feet as they walked down the street. Abandoned cars littered the roads. People huddled in their homes, stuffing wet rags around windows and doors trying to keep out the choking smog. Prize cows at the Earl's Court Cattle Show suddenly dropped dead, their lungs black with coal smoke. Humans, also, began to die in alarming numbers. Undertakers ran out of coffins. Several temporary morgues were set up to deal with the sudden influx of corpses. Many of those killed were elderly, or already weak or ill, but young, apparently healthy people also collapsed and died after only a few hours exposure to the toxic cloud.

By the time winds finally swept away the smog on December 9, more than 4,700 people had died—three times the number for the same period the previous year. The first government reports correctly attributed the deaths to air pollution. Worried, however, that the public might demand costly pollution controls or cleaner-burning fuel, the government later blamed the deaths on an influenza epidemic, even though medical records show no increase in flu diagnoses. In a recent study of historic documents, epidemiologists Devra Davis and Michelle Bell conclude that death rates in London continued to be abnormally high for at least three months after the 1952 episode. Altogether, they calculate, at least 12,000 early deaths occurred because of this killer smog, and hundreds of thousands of people suffered from asthma, heart attacks, and other conditions aggravated by polluted air. This would make London's killer smog the greatest air pollution disaster in recorded history.

Dirty air wasn't new to London. Until the twelfth century, most Londoners burned wood for fuel. As the city grew and the forests shrank, wood became scarce and expensive. Most people switched to abundant supplies of low-quality, bituminous coal for fuel. In 1272, Edward I forbade burning coal in the city and threatened to execute anyone caught breaking his ban. Lacking affordable firewood, however, most people ignored this royal proclamation and continued to use coal. In 1578, Queen Elizabeth I complained about the foul air of London, and in 1661, John Evelyn published *Fumifugium or the Inconvenience of the Air and Smoke of London Dissipated,* in which he deplored the "clouds of smoke and sulfur so full of stink and darkness."

Still, as the population grew, air pollution worsened. A fog in 1879 lasted from November to March, four long months of sunless gloom. Residents described the air as "thick as pea soup." They complained about the bitter smoke and darkness, but most people assumed that smoky urban air was just an inconvenience or the cost of progress. The high number of deaths in 1952, however, changed attitudes toward air pollution. In 1956, Parliament enacted a Clean Air Act restricting coal use and requiring filters and scrubbers on industrial smokestacks. Subsequently, most other industrial countries have passed similar legislation, and air quality in the developed world has increased dramatically. Still, air pollution is probably responsible for many health problems. In megacities of the developing world, poor air quality remains a major health threat.

In this chapter, we'll look at the main sources and impact of air pollution, as well as ways to control and reduce this important environmental threat. We'll also examine some of the politics of air pollution and the controversy about how best to ensure a healthy environment.

THE AIR AROUND US

How does the air taste, feel, smell, and look in your neighborhood? Chances are that wherever you live, the air is contaminated to some degree. Smoke, haze, dust, odors, corrosive gases, noise, and toxic compounds are present nearly everywhere, even in the most remote, pristine wilderness. Air pollution is generally the most widespread and obvious kind of environmental damage. According to the Environmental Protection Agency (EPA), some 147 million metric tons of air pollution (not counting carbon dioxide or wind-blown soil) are released into the atmosphere each year in the United States by human activities. Total worldwide emissions of these pollutants are around 2 billion metric tons per year.

Over the past 20 years, however, air quality has improved appreciably in most cities in Western Europe, North America, and Japan. Many young people might be surprised to learn that a generation ago most American cities were much dirtier than they are today. This is an encouraging example of improvement in environmental conditions. Our success in controlling some of the most serious air pollutants gives us hope for similar progress in other environmental problems.

While developed countries have been making progress, however, air quality in the developing world has been getting much worse. Especially in the burgeoning megacities of rapidly industrializing countries (chapter 22), air pollution often exceeds World Health Organization standards most of the time. In many Chinese cities, for example, airborne dust, smoke, and soot often are ten times higher than levels considered safe for human health (fig. 16.1).

NATURAL SOURCES OF AIR POLLUTION

It is difficult to give a simple, comprehensive definition of pollution. The word comes from the Latin *pollutus,* which means made foul, unclean, or dirty. Some authors limit the use of the term to damaging materials that are released into the environment by human activities. There are, however, many natural sources of air quality degradation. Volcanoes spew out ash, acid mists, hydrogen sulfide, and other toxic gases (fig. 16.2). Sea spray and decaying vegetation are major sources of reactive sulfur compounds in the air. Trees and bushes emit millions of tons of volatile organic compounds (terpenes and isoprenes), creating, for example, the blue

FIGURE 16.1   On a smoggy day in Shanghai (*left*) visibility is less than 1 km. Twenty-four hours later, after a rainfall (*right*), the air has cleared dramatically. © William P. Cunningham.

haze that gave the Blue Ridge Mountains their name. Pollen, spores, viruses, bacteria, and other small bits of organic material in the air cause widespread suffering from allergies and airborne infections. Storms in arid regions raise dust clouds that transport millions of tons of soil and can be detected half a world away. Bacterial metabolism of decaying vegetation in swamps and of cellulose in the guts of termites and ruminant animals is responsible for as much as two-thirds of the methane (natural gas) in the air.

Does it make a difference whether smoke comes from a natural forest fire or one started by humans? In many cases, the chemical compositions of pollutants from natural and human-related sources are identical, and their effects are inseparable. Sometimes, however, materials in the atmosphere are considered innocuous at naturally occurring levels, but when humans add to these levels, overloading of natural cycles or disruption of essential processes can occur. While the natural sources of suspended particulate material in the air outweigh human sources at least tenfold worldwide, in many cities more than 90 percent of the airborne particulate matter is anthropogenic (human-caused).

HUMAN-CAUSED AIR POLLUTION

What are the major types of anthropogenic air pollutants and where do they come from? In this section, we will define some general categories and sources of air pollution and survey the

FIGURE 16.2   Natural pollution sources, such as volcanoes, can be important health hazards. Courtesy of U.S. Geological Survey.

characteristics and emission levels of the seven conventional pollutants regulated by the Clean Air Act.

Primary and Secondary Pollutants

Primary pollutants are those released directly from the source into the air in a harmful form (fig. 16.3). **Secondary pollutants,** by contrast, are modified to a hazardous form after they enter the air or are formed by chemical reactions as components of the air mix and interact. Solar radiation often provides the energy for these reactions. Photochemical oxidants and atmospheric acids formed by these mechanisms are probably the most important secondary pollutants in terms of human health and ecosystem damage. We will discuss several important examples of such pollutants in this chapter.

Fugitive emissions are those that do not go through a smokestack. By far the most massive example of this category is dust from soil erosion, strip mining, rock crushing, and building construction (and destruction). In the United States, natural and anthropogenic sources of fugitive dust add up to some 100 million metric tons per year. The amount of CO_2 released by burning fossil fuels and biomass is nearly equal in mass to fugitive dust. Fugitive industrial emissions are also an important source of air pollution. Leaks around valves and pipe joints contribute as much as 90 percent of the hydrocarbons and volatile organic chemicals emitted from oil refineries and chemical plants.

Conventional or "Criteria" Pollutants

The U.S. Clean Air Act of 1970 designated seven major pollutants (sulfur dioxide, carbon monoxide, particulates, hydrocarbons, nitrogen oxides, photochemical oxidants, and lead) for which maximum **ambient air** (air around us) levels are mandated. These seven **conventional** or **criteria pollutants** contribute the largest volume of air-quality degradation and also are considered the most serious threat of all air pollutants to human health and welfare. Figure 16.4 shows the major sources of the first six criteria pollutants. Table 16.1 shows an estimate of the total annual worldwide emissions of some important air pollutants. Now let's look more closely at the sources and characteristics of each of these major pollutants.

Sulfur Compounds

Natural sources of sulfur in the atmosphere include evaporation of sea spray, erosion of sulfate-containing dust from arid soils, fumes from volcanoes and fumaroles, and biogenic emissions of hydrogen sulfide (H_2S) and organic sulfur-containing compounds, such as dimethylsulfide, methyl mercaptan, carbon disulfide, and carbonyl sulfide. Total yearly emissions of sulfur from all sources amount to some 114 million metric tons (fig. 16.5). Worldwide, anthropogenic sources represent about two-thirds of the total sulfur flux, but in most urban areas they contribute as much as 90 percent of the sulfur in the air. The predominant form of anthropogenic sulfur is sulfur dioxide (SO_2) from combustion of sulfur-containing fuel (coal and oil), purification of sour (sulfur-containing) natural gas or oil, and industrial processes, such as smelting of sulfide ores.

FIGURE 16.3 Primary pollutants are released directly from a source into the air. A point source is a specific location of highly concentrated discharge, such as this smokestack. © William P. Cunningham.

China and the United States are the largest sources of anthropogenic sulfur, primarily from coal burning.

Sulfur dioxide is a colorless corrosive gas that is directly damaging to both plants and animals. Once in the atmosphere, it can be further oxidized to sulfur trioxide (SO_3), which reacts with water vapor or dissolves in water droplets to form sulfuric acid (H_2SO_4), a major component of acid rain. Very small solid particles or liquid droplets can transport the acidic sulfate ion (SO_4^{-2}) long distances through the air or deep into the lungs where it is very damaging. Sulfur dioxide and sulfate ions are probably second only to smoking as causes of air pollution-related health damage. Sulfate particles and droplets reduce visibility in the United States as much as 80 percent. Some of the smelliest and most obnoxious air pollutants are sulfur compounds, such as hydrogen sulfide from pig manure lagoons or mercaptans (organosulfur thiols) from papermills (fig. 16.6).

Nitrogen Compounds

Nitrogen oxides are highly reactive gases formed when nitrogen in fuel or combustion air is heated to temperatures above 650°C (1,200°F) in the presence of oxygen, or when bacteria in soil or

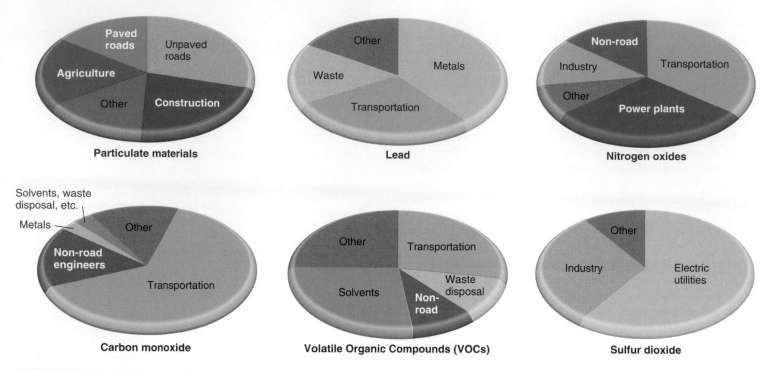

FIGURE 16.4 Anthropogenic sources of six of the primary "criteria" air pollutants in the United States. *Source:* Joyce E. Penner, "Atmospheric Chemistry and Air Quality" in W. B. Meyer and B. L. Turner (eds.), *Changes in Land Use and Land Cover: A Global Perspective,* 1994. Cambridge University Press and UNEP, 1999.

TABLE 16.1 Estimated Fluxes of Pollutants and Trace Gases to the Atmosphere

SPECIES	SOURCES	APPROXIMATE ANNUAL FLUX (MILLIONS OF METRIC TONS/YR)	
		NATURAL	ANTHROPOGENIC
CO_2 (carbon dioxide)	Respiration, fossil fuel burning, land clearing, industrial processes	370,000	29,600*
CH_4 (methane)	Rice paddies and wetlands, gas drilling, landfills, animals, termites	155	350
CO (carbon monoxide)	Incomplete combustion, CH_4 oxidation, biomass burning, plant metabolism	1,580	930
NMHC (nonmethane hydrocarbons)	Fossil fuels, industrial uses, plant isoprenes and other biogenics	860	92
NO_x (nitrogen oxides)	Fossil fuel burning, lightning, biomass burning, soil microbes	90	140
SO_x (sulfur oxides)	Fossil fuel burning, industry, biomass burning, volcanoes, oceans	35	79
SPM (suspended particulate materials)	Biomass burning, dust, sea salt, biogenic aerosols, gas to particle conversion	583	362

Only 27.3 percent of this amount—or 8 billion tons—is carbon.

Source: *Joyce E. Penner, "Atmospheric Chemistry and Air Quality" in W. B. Meyer and B. L. Turner (eds.), Changes in Land Use and Land Cover: A Global Perspective, 1994. Cambridge University Press and UNEP, 1999.*

water oxidize nitrogen-containing compounds. The initial product, nitric oxide (NO), oxidizes further in the atmosphere to nitrogen dioxide (NO_2), a reddish brown gas that gives photochemical smog its distinctive color. Because of their interconvertibility, the general term NO_x is used to describe these gases. Nitrogen oxides combine with water to make nitric acid (HNO_3), which is also a major component of atmospheric acidification.

The total annual emissions of reactive nitrogen compounds into the air are about 230 million metric tons worldwide (see table 16.1). Anthropogenic sources account for 60 percent of these emissions (fig. 16.7). About 95 percent of all human-caused NO_x in the United States is produced by fuel combustion in transportation and electric power generation. Nitrous oxide (N_2O) is an intermediate in soil denitrification that absorbs ultraviolet light and plays an

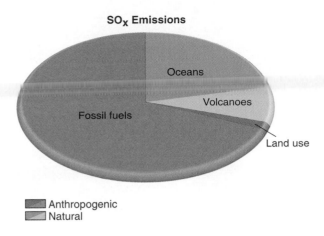

SO$_X$ Emissions

Oceans

Volcanoes

Fossil fuels

Land use

Anthropogenic
Natural

FIGURE 16.5   Sulfur fluxes into the atmosphere.

Source: Joyce E. Penner, "Atmospheric Chemistry and Air Quality" in W. B. Meyer and B. L. Turner (eds.), *Changes in Land Use and Land Cover: A Global Perspective,* 1994. Cambridge University Press and UNEP, 1999.

important role in climate modification (chapter 15). Excess nitrogen is causing fertilization and eutrophication of inland waters and coastal seas. It also may be adversely affecting terrestrial plants both by excess fertilization and by encouraging growth of weedy species that crowd out native varieties.

Carbon Oxides

The predominant form of carbon in the air is carbon dioxide (CO_2). It is usually considered nontoxic and innocuous, but increasing atmospheric levels (about 0.5 percent per year) due to human activities appear to be causing a global climate warming that may have disastrous effects on both human and natural communities. As table 16.1 shows, more than 90 percent of the CO_2 emitted each year is from respiration (oxidation of organic compounds by plant and animal cells). These releases are usually balanced, however, by an equal uptake by photosynthesis in green plants.

Anthropogenic (human-caused) CO_2 releases are difficult to quantify because they spread across global scales. The best current estimate from the Intergovernmental Panel on Climate Change (IPCC) is that between 7 and 8 billion tons of carbon (in the form of CO_2) are released each year by fossil fuel combustion and that another 1 to 2 billion tons are released by forest and grass fires, cement manufacturing, and other human activities. Typically, terrestrial ecosystems take up about 3 billion tons of this excess carbon every year, while oceanic processes take up another 2 billion tons. This leaves an average of at least 3 billion tons to accumulate in the atmosphere. The actual releases and uptakes vary greatly, however, from year to year. Some years almost all anthropogenic CO_2 is reabsorbed; in other years, almost none of it is. The ecological processes that sequester CO_2 depend strongly on temperature, nutrient availability, and other environmental factors.

United States negotiators at the Global Climate meetings claim that forests and soils in North America act as carbon sinks—that is, they take up more carbon than is released by other activities. Over the past decade, CO_2 levels in air coming ashore on the

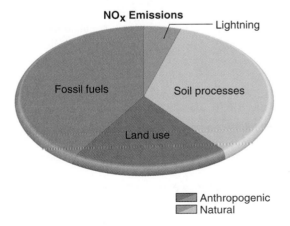

FIGURE 16.6   The most annoying pollutants from this paper mill are pungent organosulfur thiols and sulfides. Chlorine bleeching can also produce extremely dangerous organochlorines, such as dioxins.
© William P. Cunningham.

NO$_X$ Emissions

Lightning

Fossil fuels

Soil processes

Land use

Anthropogenic
Natural

FIGURE 16.7   Worldwide sources of reactive nitrogen gases in the atmosphere.

Source: Joyce E. Penner, "Atmospheric Chemistry and Air Quality" in W. B. Meyer and B. L. Turner (eds.), *Changes in Land Use and Land Cover: A Global Perspective,* 1994. Cambridge University Press and UNEP, 1999.

U.S. West Coast have averaged about 2 ppm higher than air leaving from the East Coast. If we assume that there isn't a major inflow of CO_2-depleted air entering from Canada or Mexico, this would mean that somewhere between 1.6 and 2.2 billion tons of CO_2 are being taken up every year than are being released in the United States. Other countries doubt these measurements, however, and refuse to give the United States credit for this large carbon sequestration.

Carbon monoxide (CO) is a colorless, odorless, nonirritating but highly toxic gas produced by incomplete combustion of fuel (coal, oil, charcoal, or gas), incineration of biomass or solid waste, or partially anaerobic decomposition of organic material. CO inhibits respiration in animals by binding irreversibly to hemoglobin. About 1 billion metric tons of CO are released to the atmosphere each year, half of that from human activities. In the United States, two-thirds of the CO emissions are created by internal combustion engines in transportation. Land-clearing fires and cooking fires also are major sources. About 90 percent of the CO in the air is consumed in photochemical reactions that produce ozone.

Particulate Material

An **aerosol** is any system of solid particles or liquid droplets suspended in a gaseous medium. For convenience, we generally describe all atmospheric aerosols, whether solid or liquid, as **particulate material.** This includes dust, ash, soot, lint, smoke, pollen, spores, algal cells, and many other suspended materials. Anthropogenic particulate emissions amount to about 362 million metric tons per year worldwide. Wind-blown dust, volcanic ash, and other natural materials may contribute considerably more suspended particulate material.

Particulates often are the most apparent form of air pollution since they reduce visibility and leave dirty deposits on windows, painted surfaces, and textiles. Respirable particles smaller than 2.5 micrometers are among the most dangerous of this group because they can be drawn into the lungs, where they damage respiratory tissues. Asbestos fibers and cigarette smoke are among the most dangerous respirable particles in urban and indoor air because they are carcinogenic.

Diesel fumes also are highly toxic because they contain both fine particulates and chemicals such as benzene, dioxins, and mercury. The EPA has proposed new rules to require low-sulfur fuel and antipollution devices, particularly for off-road engines such as bulldozers, tractors, pumps, and generators. Epidemiologists estimate that these new standards will prevent more than 360,000 asthma attacks and 8,300 premature deaths annually. Diesel owners protest that expenses will be exorbitant, but Europe has had standards ten times more stringent than the EPA proposes for several years.

In the 1930s, America experienced terrible soil erosion known as the dust bowl. Poor agricultural practices and policies, coupled with years of drought, left soil on 5 million ha of the southern plains exposed to the wind. Billowing clouds of dust darkened the skies for days and reached as far as Washington, D.C. In 1935, at the peak of the drought, an estimated 850 million tons of topsoil blew away in "black blizzards" (fig. 16.8).

FIGURE 16.8 A cloud of dust more than 1 km thick rolls across the southern plains during the 1930s dust bowl. Courtesy of USDA/National Resource Conservation Service.

Metals and Halogens

Many toxic metals are mined and used in manufacturing processes or occur as trace elements in fuels, especially coal. These metals are released to the air in the form of metal fumes or suspended particulates by fuel combustion, ore smelting, and disposal of wastes. Worldwide atmospheric lead emissions amount to about 2 million metric tons per year, or two-thirds of all metallic air pollution. Most of this lead is from leaded gasoline. Lead is a metabolic poison and a neurotoxin that binds to essential enzymes and cellular components and inactivates them.

Banning leaded gasoline is one of the most successful pollution-control measures in American history. Since 1986, when the ban was enforced, children's average blood lead levels have dropped 90 percent and average IQs have risen three points. Now, 50 nations have renounced leaded gasoline. The global economic benefit of this step is estimated to be more than $200 billion per year.

Mercury is another dangerous neurotoxin that is widespread in the environment. The two largest sources of atmospheric mercury appear to be coal-burning power plants and waste incinerators. Mercuric fungicides in house paint were once a major source of this deadly pollutant but now are restricted. Long-range transport of lead and mercury through the air is causing bioaccumulation in aquatic ecosystems far from the emission sources. It is now dangerous to eat fish from some once-pristine lakes and rivers because of toxic metal contamination.

Other toxic metals of concern are nickel, beryllium, cadmium, thallium, uranium, cesium, and plutonium. Some 780,000 tons of arsenic, a highly toxic metalloid, are released from metal smelters, coal combustion, and pesticide use each year. Halogens (fluorine, chlorine, bromine, and iodine) are highly reactive and generally toxic in their elemental form. Chlorofluorocarbons (CFCs) have been banned for most uses in industrialized countries, but about 600 million tons of these highly persistent chemical compounds are used annually worldwide in spray propellants, refrigeration compressors, and for foam blowing. They diffuse into the stratosphere where they release chlorine and fluorine atoms that destroy the ozone shield that protects the earth from ultraviolet radiation. We'll return to this topic later in this chapter.

Volatile Organic Compounds

Volatile organic compounds (VOCs) are organic chemicals that exist as gases in the air. Plants are the largest source of VOCs, releasing an estimated 350 million tons of isoprene (C_5H_8) and 450 million tons of terpenes ($C_{10}H_{15}$) each year (fig. 16.9). About 400 million tons of methane (CH_4) are produced by natural wetlands and rice paddies and by bacteria in the guts of termites and ruminant animals. These volatile hydrocarbons are generally oxidized to CO and CO_2 in the atmosphere.

In addition to these natural VOCs, a large number of other synthetic organic chemicals, such as benzene, toluene, formaldehyde, vinyl chloride, phenols, chloroform, and trichloroethylene, are released into the air by human activities. About 28 million tons of these compounds are emitted each year in the United States, mainly unburned or partially burned hydrocarbons from transportation, power plants, chemical plants, and petroleum refineries. These chemicals play an important role in the formation of photochemical oxidants.

Photochemical Oxidants

Photochemical oxidants are products of secondary atmospheric reactions driven by solar energy (fig. 16.10). One of the most important of these reactions involves formation of singlet (atomic) oxygen by splitting nitrogen dioxide (NO_2). This atomic oxygen then reacts with another molecule of O_2 to make **ozone** (O_3). Ozone formed in the stratosphere provides a valuable shield for the biosphere by absorbing incoming ultraviolet radiation. In ambient air, however, O_3 is a strong oxidizing reagent and damages vegetation, building materials (such as paint, rubber, and plastics), and sensitive tissues (such as eyes and lungs). Ozone has an acrid, biting odor that is a distinctive characteristic of photochemical smog.

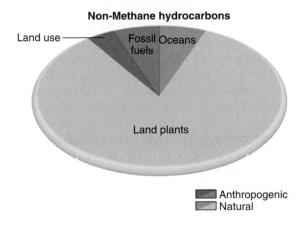

FIGURE 16.9   Sources of non-methane hydrocarbons in the atmosphere.

Source: Joyce E. Penner, "Atmospheric Chemistry and Air Quality" in W. B. Meyer and B. L. Turner (eds.), *Changes in Land Use and Land Cover: A Global Perspective,* 1994. Cambridge University Press and UNEP, 1999.

Hydrocarbons in the air contribute to accumulation of ozone by removing NO in the formation of compounds, such as peroxyacetyl nitrate (PAN), which is another damaging photochemical oxidant.

Air Toxins

Although most air contaminants are regulated because of their potential adverse effects on human health or environmental quality, a special category of toxins is monitored by the U.S. EPA because they are particularly dangerous. Called **hazardous air pollutants (HAPs),** these chemicals include carcinogens, neurotoxins, mutagens, teratogens, endocrine system disrupters, and other highly toxic compounds (chapter 8). Twenty of the most "persistent bioaccumulative toxic chemicals" (see table 8.3) require special reporting and management because they remain in ecosystems for long periods of time, and accumulate in animal and human tissues. Most of these chemicals are either metal compounds, chlorinated hydrocarbons, or volatile organic compounds.

Only about 50 locations in the United States regularly measure concentrations of HAPs in ambient air. Often the best source of information about these chemicals is the **Toxic Release Inventory (TRI)** collected by the EPA as part of the community right to know program. Established by Congress in 1986, the TRI requires 23,000 factories, refineries, hard rock mines, power plants, and chemical manufacturers to report on toxin releases (above certain minimum amounts) and waste management methods for 667 toxic chemicals. Although this total is less than 1 percent of all chemicals registered for use, and represents a limited range of sources, the TRI is widely considered the most comprehensive source of information about toxic pollution in the United States.

In 2002, U.S. industries released 6.6 billion pounds (2.8 million metric tons) of toxic chemicals into the environment and disposed of roughly four times that much through waste management and recycling methods. This represents a 21-percent reduction from releases in 1999. Of the environmental releases, 65 percent was discharged on land, 27 percent (810,000 metric tons) was released into the air, 4 percent was discharged into water, and 4 percent was injected into deep wells. Twenty chemicals accounted for 88 percent of the total releases, with metals and mining waste comprising a vast majority of that amount. Over 200,000 metric tons of

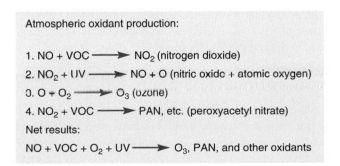

FIGURE 16.10   Secondary production of urban smog oxidants by photochemical reactions in the atmosphere.

Source: B. J. Finlayson-Pitts and J. N. Pitts, *Atmospheric Chemistry,* 1986, John Wiley & Sons, Inc.

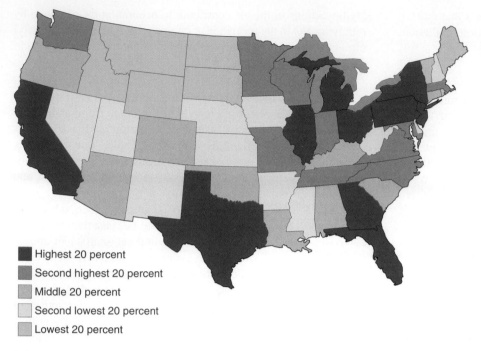

Highest 20 percent

Second highest 20 percent

Middle 20 percent

Second lowest 20 percent

Lowest 20 percent

FIGURE 16.11 Number of people living in areas where the estimated cancer risk from HAPs is greater than 1 in 10,000. *Source:* Environmental Defense Fund based on EPA data, 2003.

persistent bioaccumulative chemicals were emitted in 2002, with lead and lead compounds comprising 97 percent of that total.

While most HAP releases are decreasing, discharges of mercury and dioxins—both of which are bioaccumulative and toxic at extremely low levels—have increased in recent years. Mercury is released primarily by coal-burning power plants, while dioxins are created mainly by burning plastics and medical waste containing chlorine. The EPA reports that 100 million Americans live in areas where the cancer rate from HAPs exceeds 10 in 1 million or ten times the normally accepted standard for action (fig. 16.11). Benzene, formaldehyde, acetaldehyde, and 1,3 butadiene are responsible for most of this HAP cancer risk. Furthermore, twice that many (70 percent of the U.S. population) live in areas where the non-cancer risk of death exceeds 1 in 1 million. To help residents track local air quality levels, the EPA recently estimated the concentration of HAPs in localities across the continental United States (over 60,000 census tracts). You can access this information on the Environmental Defense Fund web page at www.scorecard.org/env-releases/hap/ or the EPA web exercise at the end of this chapter.

Unconventional Pollutants

In addition to toxic air pollutants, some other unconventional forms of air pollution deserve mention. **Aesthetic degradation** includes any undesirable changes in the physical characteristics or chemistry of the atmosphere. Noise, odors, and light pollution are examples of atmospheric degradation that may not be life-threatening but reduce the quality of our lives. This is a very subjective category. Odors and noise (such as loud music) that are offensive to some may be attractive to others. Often the most sensitive device for odor

detection is the human nose. We can smell styrene, for example, at 44 parts per billion (ppb). Trained panels of odor testers often are used to evaluate air samples. Factories that emit noxious chemicals sometimes spray "odor maskants" or perfumes into smokestacks to cover up objectionable odors.

In most urban areas, it is difficult or impossible to see stars in the sky at night because of dust in the air and stray light from buildings, outdoor advertising, and streetlights. This light pollution has become a serious problem for astronomers.

Indoor Air Pollution

We have spent a considerable amount of effort and money to control the major outdoor air pollutants, but we have only recently become aware of the dangers of indoor air pollutants. The EPA has found that indoor concentrations of toxic air pollutants are often higher than outdoors. Furthermore, people generally spend more time inside than out and therefore are exposed to higher doses of these pollutants.

Smoking is without doubt the most important air pollutant in the United States in terms of human health. The Surgeon General estimates that more than 400,000 people die each year in the United States from emphysema, heart attacks, strokes, lung cancer, or other diseases caused by smoking. These diseases are responsible for 20 percent of all mortality in the United States, or four times as much as infectious agents. Lung cancer has now surpassed breast cancer as the leading cause of cancer deaths for U.S. women. Advertising aimed at making smoking appear stylish and liberating has resulted in a 600 percent increase in lung cancer among women since 1950.

Total costs for early deaths and smoking-related illnesses in the United States are estimated to be $100 billion per year. Eliminating smoking probably would save more lives than any other pollution-control measure. Smoking restrictions in many places have resulted in dramatic declines in second-hand smoke exposure to nonsmokers, EPA data show. In just a decade after indoor smoking bans were passed, levels of tobacco by-products in nonsmokers' blood dropped 75 percent.

In some cases, indoor air in homes has concentrations of chemicals that would be illegal outside or in the workplace. The EPA has found that concentrations of such compounds as chloroform, benzene, carbon tetrachloride, formaldehyde, and styrene can be seventy times higher in indoor air than in outdoor air. "Green design" principles can make indoor spaces both healthier and more pleasant (Case Study, p. 338).

In the less-developed countries of Africa, Asia, and Latin America where such organic fuels as firewood, charcoal, dried dung, and agricultural wastes make up the majority of household energy, smoky, poorly ventilated heating and cooking fires represent the greatest source of indoor air pollution (fig. 16.12). The World Health Organization (WHO) estimates that 2.5 billion people—nearly half

FIGURE 16.12 Smoky cooking and heating fires may cause more ill health effects than any other source of indoor air pollution except tobacco smoking. Some 2.5 billion people, mainly women and children, spend hours each day in poorly ventilated kitchens and living spaces where carbon monoxide, particulates, and cancer-causing hydrocarbons often reach dangerous levels. © Guang Hui/Image Bank.

the world's population—are adversely affected by pollution from this source. Women and small children spend long hours each day around open fires or unventilated stoves in enclosed spaces. The levels of carbon monoxide, particulates, aldehydes, and other toxic chemicals can be 100 times higher than would be legal for outdoor ambient concentrations in the United States. Designing and building cheap, efficient, nonpolluting energy sources for the developing countries would not only save shrinking forests but would make a major impact on health as well.

CLIMATE, TOPOGRAPHY, AND ATMOSPHERIC PROCESSES

Topography, climate, and physical processes in the atmosphere play an important role in transport, concentration, dispersal, and removal of many air pollutants. Wind speed, mixing between air layers, precipitation, and atmospheric chemistry all determine whether pollutants will remain in the locality where they are produced or go elsewhere. In this next section, we will survey some environmental factors that affect air pollution levels.

Inversions

Temperature inversions occur when a stable layer of warmer air overlays cooler air, reversing the normal temperature decline with increasing height and preventing convection currents from dispersing pollutants. Several mechanisms create inversions. When a cold front slides under an adjacent warmer air mass or when cool air subsides down a mountain slope to displace warmer air in the valley below, an inverted temperature gradient is established. These inversions are usually not stable, however, because winds accompanying these air exchanges tend to break up the temperature gradient fairly quickly and mix air layers.

The most stable inversion conditions are usually created by rapid nighttime cooling in a valley or basin where air movement is restricted. Los Angeles is a classic example of the conditions that create temperature inversions and photochemical smog (fig. 16.13). The city is surrounded by mountains on three sides and the climate is dry and sunny. Millions of automobiles and trucks create high pollution levels. Skies are generally clear at night, allowing rapid radiant heat loss, and the ground cools quickly. Surface air layers are cooled by conduction, while upper layers remain relatively warm. Density differences retard vertical mixing. During the night, cool, humid, onshore breezes slide in under the contaminated air, squeezing it up against the cap of warmer air above and concentrating the pollutants accumulated during the day.

Morning sunlight is absorbed by the concentrated aerosols and gaseous chemicals of the inversion layer. This complex mixture quickly cooks up a toxic brew of hazardous compounds. As the

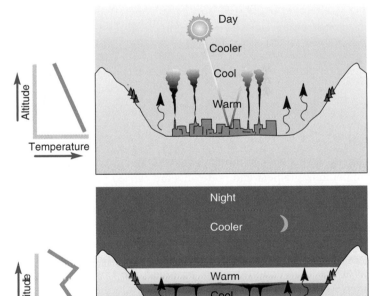

FIGURE 16.13 Atmospheric temperature inversions occur where ground level air cools more quickly than upper levels. This temperature differential prevents mixing and traps pollutants close to the ground.

Indoor Air

How safe is the air in your home, office, or school room? As we decrease air-infiltration into buildings to conserve energy, we often trap indoor air pollutants within spaces where most of us spend the vast majority of our time. In what has come to be known as "sick building syndrome," people complain of headaches, fatigue, nausea, upper-respiratory problems, and a wide variety of allergies from workplace or home exposure to airborne toxins. While these symptoms often are vague and difficult to verify scientifically, the U.S. Environmental Protection Agency estimates that sick building syndrome may cost $60 billion a year in medical expenses, absenteeism, and reduced productivity.

What might be making us sick? Mold spores are probably the greatest single cause of allergic reactions in indoor air. Moisture trapped in air-tight houses often accumulates in walls where molds flourish. Air ducts provide both a good environment for growth of pathogens such as Legionnaire's disease bacteria as well as a path for their dispersal. Legionnaire's pneumonia is much more prevalent than most people realize in places like California and Australia where air-conditioning is common. Uranium-bearing rocks and sediment are widespread across North America.

When uranium decays, it produces carcinogenic radon gas that can seep into buildings. The EPA warns that one home in ten in the United States may exceed the recommended maximum radon concentration of 4 picocuries per liter.

In addition, we are exposed to a variety of synthetic chemicals emitted from carpets, wall coverings, building materials, and combustion gases (see fig. 8.9). You might be surprised to learn how many toxic, synthetic compounds are used to construct buildings and make furniture. Formaldehyde, for instance is a component of more than 3,000 products, including building materials such as particle board, waferboard, and urea-formaldehyde foam insulation. Vinyl chloride is used in plastic plumbing pipe, floor and wall coverings, and countertops. Volatile organic solvents make up as much as half the volume of some paint. New carpets and drapes typically contain up to two dozen chemical compounds designed to kill bacteria and molds, resist stains, bind fibers, and retain colors.

What can you do if you suspect that your living spaces are exposing you to materials that make you sick? Probably few students will be in a position anytime soon to build a new house with nontoxic materials, but there are some principles from the emerging field of "green design" that you might apply if you're house hunting, redecorating your apartment, or interviewing for a job (chapter 8). Low-volatile paint is now available for indoor use. Nontoxic, formaldehyde-free plywood, particle board, and insulation can be used in new construction. Nonallergenic carpets, drapes, and wall coverings are available, but some architects recommend natural wood, stone, and plaster surfaces that are easier to clean and less allergenic than any fabric.

High rates of air exchange can help rid indoor air of moisture, odors, mold spores, radon, and toxins. Does that mean energy inefficiency? Not necessarily. Air-to-air heat exchangers keep heat in during the winter and out during the summer, while still providing a healthy rate of fresh-air flow. Bathrooms and kitchens should have outdoor vents. Gas or oil furnaces should be checked for carbon monoxide production. Although many cooks prefer gas stoves because they heat quickly, they can produce toxic carbon monoxide and nitrogen oxides. Contact your city housing authority or county extension service for further tips on how to make your home, work, or study environment healthier. No matter what your situation, there are things that each of us can do to make our indoor air cleaner and safer.

---

ground warms later in the day, convection currents break up the temperature gradient and pollutants are carried back down to the surface where more contaminants are added. Nitric oxide (NO) from automobile exhaust is oxidized to nitrogen dioxide. As nitrogen oxides are used up in reactions with unburned hydrocarbons, the ozone levels begin to rise. By early afternoon, an acrid brown haze fills the air, making eyes water and throats burn. In the 1970s, before pollution controls were enforced, the Los Angeles basin often would reach 0.34 ppm or more by late afternoon and the pollution index could be 300, the stage considered a health hazard.

Dust Domes and Heat Islands

Even without mountains to block winds and stabilize air layers, many large cities create an atmospheric environment quite different from the surrounding conditions. Sparse vegetation and high levels of concrete and glass in urban areas allow rainfall to run off quickly and create high rates of heat absorption during the day and radiation at night. Tall buildings create convective updrafts that sweep pollutants into the air. Temperatures in the center of large cities are frequently 3° to 5°C (5° to 9°F) higher than the surrounding countryside. Stable air masses created by this "heat island" over the city concentrate pollutants in a "dust dome." Rural areas downwind from major industrial areas often have significantly decreased visibility and increased rainfall (due to increased condensation nuclei in the dust plume) compared to neighboring areas with cleaner air. In the late 1960s, for instance, areas downwind from Chicago and St. Louis reported up to 30 percent more rainfall than upwind regions.

Aerosols and dust in urban air seem to trigger increased cloud-to-ground lightning strikes. Houston and Lake Charles, Louisiana, for instance, which have many petroleum refineries, have among the highest number of lightning strikes in the United States and twice as many as nearby areas with similar climate but cleaner air.

Long-Range Transport

Dust and fine aerosols can be carried great distances by the wind. Pollution from the industrial belt between the Great Lakes and the Ohio River Valley regularly contaminates the Canadian Maritime Provinces, and sometimes can be traced as far as Ireland (fig. 16.14). Similarly, dust storms from China's Gobi and Takla Makan Deserts routinely close schools, factories, and airports in Japan and Korea, and often reach western North America. In one particularly severe dust storm in 1998, chemical analysis showed that 75 percent of the particulate pollution in Seattle, Washington, air came from China.

Studies of air pollutants over southern Asia reveal a 3 km thick toxic cloud of ash, acids, aerosols, dust, and photochemical reactants regularly covers the entire Indian subcontinent for much of the year. Nobel laureate Paul Crutzen estimates that up to 2 million people in India alone die each year from atmospheric pollution. Produced by forest fires, the burning of agricultural wastes, and dramatic increases in the use of fossil fuels, the Asian smog layer cuts the amount of solar energy reaching the earth's surface beneath it by up to 15 percent. Meteorologists suggest that the cloud—80 percent of which is human-made—could disrupt monsoon weather patterns and cut rainfall over northern Pakistan, Afghanistan, western China, and central Asia by up to 40 percent. Shifting monsoon flows may also have contributed to catastrophic floods in Nepal, Bangladesh, and eastern India that killed at least 1,000 people in 2002, and left more than 25 million homeless.

When this "Asian Brown Cloud" drifts out over the Indian Ocean at the end of the monsoon season, it cools sea temperatures and may be changing El Niño/Southern Oscillation patterns in the Pacific Ocean as well (chapter 15). As UN Environment Programme executive director, Klaus Töpfer, said, "There are global implications because a pollution parcel like this, which stretches three km high, can travel half way round the globe in a week."

Increasingly sensitive monitoring equipment has begun to reveal industrial contaminants in places usually considered among the cleanest in the world. Samoa, Greenland, and even Antarctica and the North Pole, all have heavy metals, pesticides, and radioactive elements in their air. Since the 1950s, pilots flying in the high Arctic have reported dense layers of reddish-brown haze clouding the arctic atmosphere. Aerosols of sulfates, soot, dust, and toxic heavy metals such as vanadium, manganese, and lead travel to the pole from the industrialized parts of Europe and Russia.

In a process called "grasshopper" transport, or atmosphere distillation, volatile compounds evaporate from warm areas, travel through the atmosphere, then condense and precipitate in cooler regions (fig. 16.15). Over several years, contaminants accumulate in the coldest places, generally at high latitudes where they bioaccumulate in food chains. Whales, polar bears, sharks, and other top carnivores in polar regions have been shown to have dangerously high levels of pesticides, metals, and other HAPs in their bodies. The Inuit people of Broughton Island, well above the Arctic Circle, have higher levels of polychlorinated biphenyls (PCBs) in their blood than any other known population, except victims of industrial accidents. Far from any source of this industrial by-product, these people accumulate PCBs from the flesh of fish, caribou, and other animals they eat.

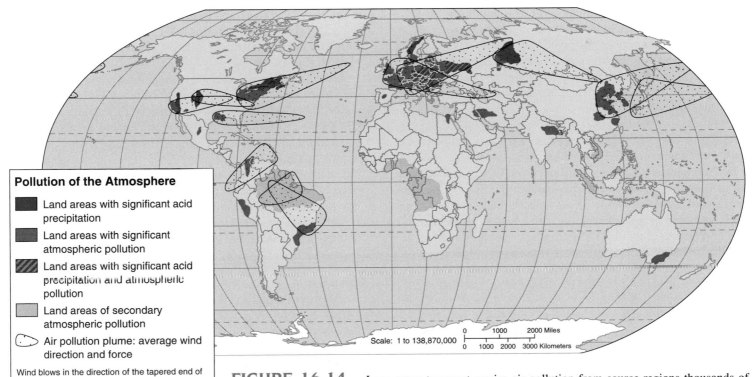

Pollution of the Atmosphere

- ▪ Land areas with significant acid precipitation
- ▪ Land areas with significant atmospheric pollution
- ▨ Land areas with significant acid precipitation and atmospheric pollution
- ▪ Land areas of secondary atmospheric pollution
- ⬭ Air pollution plume: average wind direction and force

Wind blows in the direction of the tapered end of the air pollution plume and the force of the wind is indicated by the size of the plume.

Scale: 1 to 138,870,000

0   1000   2000 Miles

0   1000   2000   3000 Kilometers

FIGURE 16.14 Long-range transport carries air pollution from source regions thousands of kilometers away into formerly pristine areas. Secondary air pollutants can be formed by photochemical reactions far from primary emissions sources.

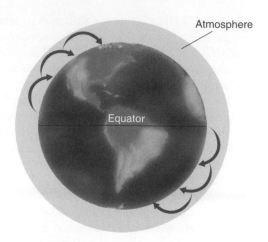

FIGURE 16.15 Air pollutants evaporate from warmer areas then condense and precipitate in cooler regions. Eventually, this "grasshopper" redistribution leads to accumulation in the Arctic and Antarctic.

Stratospheric Ozone

In 1985, the British Antarctic Atmospheric Survey announced a startling and disturbing discovery: **stratospheric ozone** levels over the South Pole were dropping precipitously during September and October every year as the sun reappears at the end of the long polar winter (fig. 16.16). This ozone depletion has been occurring at least since the 1960s but was not recognized because earlier researchers programmed their instruments to ignore changes in ozone levels that were presumed to be erroneous.

The 2000 ozone "hole" over Antarctia was the largest ever recorded, covering 44.3 million km² (17.1 million mi², or larger than all of North America) in which all the ozone between 14 and 20 kilometers altitude was destroyed. Ominously, this phenomenon is now spreading to other parts of the world as well. About 10 percent of all stratospheric ozone worldwide was destroyed during the spring of 1998 and levels over the Arctic averaged 40 percent below normal.

Why are we worried about stratospheric ozone? At ground level, ozone is a harmful pollutant, damaging plants, building materials, and human health; in the the upper atmosphere, however, where it screens out dangerous ultraviolet (UV) rays from the sun, ozone is an irreplaceable resource. Without this shield, organisms on the earth's surface would be subjected to life-threatening radiation burns and genetic damage. A 1 percent loss of ozone results in a 2 percent increase in UV rays reaching the earth's surface and could result in about a million extra human skin cancers per year worldwide if no protective measures are taken. Thus it is urgent that we learn what is attacking the ozone layer and find ways to reverse these trends if possible.

The exceptionally cold temperatures (–85° to –90°C) in Antarctica play a role in ozone losses. During the long, dark winter months, the strong circumpolar vortex (chapter 15) isolates Antarctic air and allows stratospheric temperatures to drop low enough to create ice crystals at high altitudes—something that rarely happens elsewhere over the world. Ozone and chlorine-

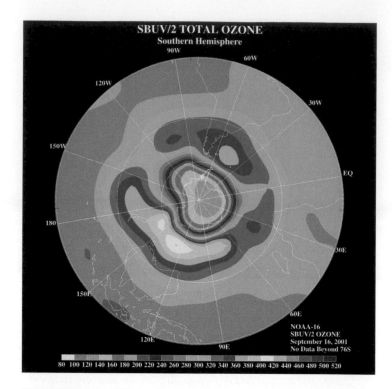

FIGURE 16.16 This satellite image from September 16, 2001, shows ozone depletion over the South Pole. Outside the polar vortex, green and yellow areas show elevated ozone levels. Over Antarctica (outlined in white), however, stratospheric ozone levels are reduced by 80 percent or more from normal levels. Courtesy Ronald Nagatani/National Centers for Environmental Prediction/NOAA.

containing molecules are absorbed on the surfaces of these ice particles. When the sun returns in the spring and provides energy to liberate chlorine ions, destructive chemical reactions proceed quickly (fig. 16.17).

Humans release a variety of chlorine-containing molecules into the atmosphere. The ones suspected of being most important in ozone losses are **chlorofluorocarbons (CFCs)** and halon gases. CFCs were invented in 1928 by Thomas Midgley, Jr., who also discovered the anti-knock properties of tetraethyl lead. Commonly known by the trade name Freon, CFCs were regarded as wonderful compounds. They are nontoxic, nonflammable, chemically inert, cheaply produced, and useful in a wide variety of applications.

Because these molecules are so stable, however, they persist for decades or even centuries once released. When they diffuse out into the stratosphere, the intense UV irradiation releases chlorine atoms that destroy ozone. Since the chlorine atoms are not themselves consumed in these reactions, they continue to destroy ozone for years until they finally precipitate or are washed out of the air.

Until 1978, aerosol spray cans used more CFCs than any other product. Although we didn't know about the special conditions in the Antarctic at that time, it was suspected that CFCs might threaten stratospheric ozone, so laws were passed in the United States, Canada, and some European countries to ban nonessential uses. Still, CFCs continue to be used in the developing world as refrigerants, solvents, spray propellants, and foam-blowing agents. Smuggling of

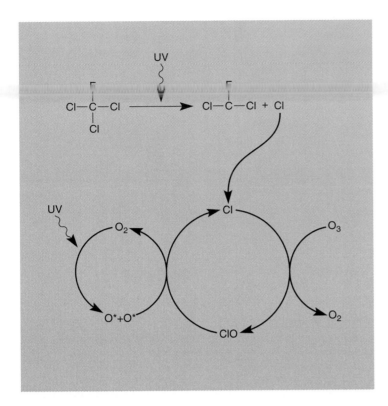

FIGURE 16.17 Destruction of stratospheric ozone by chlorine atoms. When exposed to UV radiation, CFCs release highly reactive chlorine atoms that react with ozone to produce chlorine monoxide. Reactive oxygen atoms, also produced by UV irradiation, react with ClO to start the cycle over again. A single chlorine atom can destroy thousands of ozone molecules before precipitating or drifting off to space.
*O = reactive oxygen atoms

CFCs into the United States from Mexico and other countries where they are still legal now rivals drugs in total value of contraband.

The discovery of stratospheric ozone losses has brought about a remarkably quick international response. At a 1987 conference in Montreal, 81 nations agreed to phase out all CFCs (halons, carbon tetrachloride, and methyl chloroform). Later, in Helsinki, a $500 million fund was established to assist poorer countries to switch to non-CFC technologies. Fortunately, alternatives to CFCs for most uses already exist. The first substitutes are hydrochlorofluorocarbons (HCFCs), which release much less chlorine per molecule. Eventually, we hope to develop halogen-free molecules that work just as well and are no more expensive than CFCs.

There is some evidence that the CFC ban is already having an effect. CFC production in industrialized countries has fallen nearly 80 percent since enactment of the Montreal Protocol in 1989. Chlorine in the stratosphere has shown a steady decline, and satellite measurements have recently detected lower rates of ozone destruction. Models created by Japanese researchers suggest that stratospheric ozone levels could return to normal by 2040.

In 1995, chemists Sherwood Rowland, Mario Molina, and Paul Crutzen shared the Nobel Prize for their work on atmospheric chemistry and stratospheric ozone. This was the first Nobel Prize for an environmental issue.

EFFECTS OF AIR POLLUTION

So far we have looked primarily at the major types and sources of air pollutants. Now we will focus more closely on the effects of those pollutants on human health, physical materials, ecosystems, and global climate.

Human Health

The World Health Organization estimates that some 5 to 6 million people die prematurely every year from illnesses related to air pollution. Heart attacks, respiratory diseases, and lung cancer all are significantly higher in people who breathe dirty air, compared to matching groups in cleaner environments. Residents of the most polluted cities in the United States, for example, are 15 to 17 percent more likely to die of these illnesses than those in cities with the cleanest air. This can mean as much as a 5- to 10-year decrease in life expectancy if you live in the worst parts of Los Angeles or Baltimore, compared to a place with clean air. Of course your likelihood of suffering ill health from air pollutants depends on the intensity and duration of exposure as well as your age and prior health status. You are much more likely to be at risk if you are very young, very old, or already suffering from some respiratory or cardiovascular disease. Some people are super-sensitive because of genetics or prior exposure. And those doing vigorous physical work or exercise are more likely to succumb than more sedentary folks.

Conditions are often much worse in other countries than Canada or the United States. The United Nations estimates that at least 1.3 billion people around the world live in areas where outdoor air is dangerously polluted. In the "black triangle" region of Poland, Hungary, the Czech Republic, and Slovakia, for example, respiratory ailments, cardiovascular diseases, lung cancer, infant mortality, and miscarriages are as much as 50 percent higher than in cleaner parts of those countries. In Madrid, Spain, smog is estimated to shave one-half year off the life of each resident. This adds up to more than 50,000 years lost annually for the whole city. In China, city dwellers are four to six times more likely than country folk to die of lung cancer. As mentioned earlier, the greatest air quality problem is often in poorly ventilated homes in poorer countries where smoky fires are used for cooking and heating. Billions of women and children spend hours each day in these unhealthy conditions. The World Health Organization estimates that 2 million children under age 5 die each year from acute respiratory diseases exacerbated by air pollution.

How does air pollution cause these health effects? The most common route of exposure to air pollutants is by inhalation, but direct absorption through the skin or contamination of food and water also are important pathways. Because they are strong oxidizing agents, sulfates, SO_2, NO_x, and O_3 act as irritants that damage delicate tissues in the eyes and respiratory passages. Fine suspended particulate materials (less than 2.5 µm) penetrate deep into the lungs and are irritants in their own right, as well as carrying metals and other HAPs on their surfaces. Inflammatory responses set in motion by these irritants impair lung function and trigger cardiovascular problems as the heart tries to compensate for lack of oxygen by pumping faster and harder. If the irritation is really severe, so much fluid seeps into lungs through damaged tissues that the victim actually drowns.

Carbon monoxide binds to hemoglobin and decreases the ability of red blood cells to carry oxygen. Asphyxiants such as this cause headaches, dizziness, heart stress, and can even be lethal if concentrations are high enough. Lead also binds to hemoglobin and reduces oxygen-carrying capacity at high levels. At lower levels, lead causes long-term damage to critical neurons in the brain that results in mental and physical impairment and developmental retardation.

Some important chronic health effects of air pollutants include bronchitis and emphysema.

Bronchitis is a persistent inflammation of bronchi and bronchioles (large and small airways in the lung) that causes mucus buildup, a painful cough, and involuntary muscle spasms that constrict airways. Severe bronchitis can lead to emphysema, an irreversible **chronic obstructive lung disease** in which airways become permanently constricted and alveoli are damaged or even destroyed. Stagnant air trapped in blocked airways swells the tiny air sacs in the lung (alveoli), blocking blood circulation. As cells die from lack of oxygen and nutrients, the walls of the alveoli break down, creating large empty spaces incapable of gas exchange (fig. 16.18). Thickened walls of the bronchioles lose elasticity and breathing becomes more difficult. Victims of emphysema make a characteristic whistling sound when they breathe. Often they need supplementary oxygen to make up for reduced respiratory capacity.

Irritants in the air are so widespread that about half of all lungs examined at autopsy in the United States have some degree of alve-

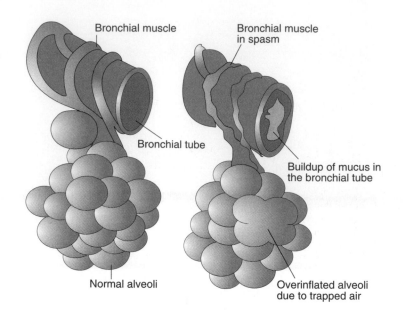

FIGURE 16.18 Bronchitis and emphysema can result in constriction of airways and permanent damage to tiny, sensitive air sacs called alveoli where oxygen diffuses into blood vessels.

olar deterioration. The Office of Technology Assessment (OTA) estimates that 250,000 people suffer from pollution-related bronchitis and emphysema in the United States, and some 50,000 excess deaths each year are attributable to complications of these diseases, which are probably second only to heart attack as a cause of death.

Smoking is undoubtedly the largest cause of obstructive lung disease and preventable death in the world. The World Health Organization says that tobacco kills some 3 million people each year. This makes it rank with AIDS as one of the world's leading killers. Because of cardiovascular stress caused by carbon monoxide in smoke and chronic bronchitis and emphysema, about twice as many people die of heart failure as die from lung cancer associated with smoking.

Plant Pathology

In the early days of industrialization, fumes from furnaces, smelters, refineries, and chemical plants often destroyed vegetation and created desolate, barren landscapes around mining and manufacturing centers. The copper-nickel smelter at Sudbury, Ontario, is a spectacular and notorious example of air pollution effects on vegetation and ecosystems. In 1886, the corporate ancestors of the International Nickel Company (INCO) began open-bed roasting of sulfide ores at Sudbury. Sulfur dioxide and sulfuric acid released by this process caused massive destruction of the plant community within about 30 km of the smelter. Rains washed away the exposed soil, leaving a barren moonscape of blackened bedrock. Super-tall, 400 m smokestacks were installed in the 1950s and sulfur scrubbers were added 20 years later. Emissions were reduced by 90 percent and the surrounding ecosystem is beginning to recover. The area near the factory is still a grim, empty wasteland, however (fig. 16.19). Similar destruction occurred at many

FIGURE 16.19 Sulfur dioxide emissions and acid precipitation from the International Nickel Company copper smelter (*background*) killed all vegetation over a large area near Sudbury, Ontario. Even the pink granite bedrock has burned black. The installation of scrubbers has dramatically reduced sulfur emissions. The ecosystem farther away from the smelter is slowly beginning to recover. © William P. Cunningham.

FIGURE 16.20 Soybean leaves exposed to 0.8 parts per million sulfur dioxide for 24 hours show extensive chlorosis (chlorophyll destruction) in white areas between leaf veins. © William P. Cunningham.

FIGURE 16.21 An open-top chamber tests air pollution effects on plants under normal conditions for rain, sun, field soil, and pest exposure. © Volume 44/PhotoDisc.

other sites during the nineteenth century. Copperhill, Tennessee; Butte, Montana; and the Ruhr Valley in Germany are some well-known examples, but these areas also are showing signs of recovery since corrective measures were taken.

There are two probable ways that air pollutants damage plants. They can be directly toxic, damaging sensitive cell membranes much as irritants do in human lungs. Within a few days of exposure to toxic levels of oxidants, mottling (discoloration) occurs in leaves due to chlorosis (bleaching of chlorophyll), and then necrotic (dead) spots develop (fig. 16.20). If injury is severe, the whole plant may be killed. Sometimes these symptoms are so distinctive that positive identification of the source of damage is possible. Often, however, the symptoms are vague and difficult to separate from diseases or insect damage.

Another mechanism of action is exhibited by chemicals, such as ethylene, that act as metabolic regulators or plant hormones and disrupt normal patterns of growth and development. Ethylene is a component of automobile exhaust and is released from petroleum refineries and chemical plants. The concentration of ethylene around highways and industrial areas is often high enough to cause injury to sensitive plants. Some scientists believe that the devastating forest destruction in Europe and North America may be partly due to volatile organic compounds.

Certain combinations of environmental factors have **synergistic effects** in which the injury caused by exposure to two factors together is more than the sum of exposure to each factor individually. For instance, when white pine seedlings are exposed to subthreshold concentrations of ozone and sulfur dioxide individually, no visible injury occurs. If the same concentrations of pollutants are given together, however, visible damage occurs. In alfalfa, however, SO_2 and O_3 together cause less damage than either one alone. These complex interactions point out the unpredictability of future effects of pollutants. Outcomes might be either more or less severe than previous experience indicates.

Pollutant levels too low to produce visible symptoms of damage may still have important effects. Field studies using open top chambers (fig. 16.21) and charcoal-filtered air show that yields in some sensitive crops, such as soybeans, may be reduced as much as 50 percent by currently existing levels of oxidants in ambient air. Some plant pathologists suggest that ozone and photochemical oxidants are responsible for as much as 90 percent of agricultural, ornamental, and forest losses from air pollution. The total costs of this damage may be as much as $10 billion per year in North America alone.

Acid Deposition

Most people in the United States became aware of problems associated with **acid precipitation** (the deposition of wet acidic solutions or dry acidic particles from the air) within the last decade or so, but English scientist Robert Angus Smith coined the term "acid rain" in his studies of air chemistry in Manchester, England, in the 1850s. By the 1940s, it was known that pollutants, including atmospheric acids, could be transported long distances by wind currents. This was thought to be only an academic curiosity until it was shown that precipitation of these acids can have far-reaching ecological effects.

pH and Atmospheric Acidity

We describe acidity in terms of pH (the negative logarithm of the hydrogen ion concentration in a solution). The pH scale ranges from 0 to 14, with 7, the midpoint, being neutral (chapter 3). Values below 7 indicate progressively greater acidity, while those above 7 are progressively more alkaline. Since the scale is logarithmic, there is a tenfold difference in hydrogen ion concentration for each pH unit. For instance, pH 6 is 10 times more acidic than pH 7; likewise, pH 5 is 100 times more acidic, and pH 4 is 1,000 times more acidic than pH 7.

Normal, unpolluted rain generally has a pH of about 5.6 due to carbonic acid created by CO_2 in air. Volcanic emissions, biological decomposition, and chlorine and sulfates from ocean spray can drop the pH of rain well below 5.6, while alkaline dust can raise it above 7. In industrialized areas, anthropogenic acids in the air usually far outweigh those from natural sources. Acid rain is only one form in which acid deposition occurs. Fog, snow, mist, and dew also trap and deposit atmospheric contaminants. Furthermore, fallout of dry sulfate, nitrate, and chloride particles can account for as much as half of the acidic deposition in some areas.

Aquatic Effects

It has been known for about 30 years that acids—principally H_2SO_4 and HNO_3—generated by industrial and automobile emissions in northwestern Europe are carried by prevailing winds to Scandinavia where they are deposited in rain, snow, and dry precipitation. The thin, acidic soils and oligotrophic lakes and streams in the mountains of southern Norway and Sweden have been severely affected by this acid deposition. Some 18,000 lakes in Sweden are now so acidic that they will no longer support game fish or other sensitive aquatic organisms.

Generally, reproduction is the most sensitive stage in fish life cycles. Eggs and fry of many species are killed when the pH drops to about 5.0. This level of acidification also can disrupt the food chain by killing aquatic plants, insects, and invertebrates on which fish depend for food. At pH levels below 5.0, adult fish die as well. Trout, salmon, and other game fish are usually the most sensitive. Carp, gar, suckers, and other less desirable fish are more resistant. There are several ways acids kill fish. Acidity alters body chemistry, destroys gills and prevents oxygen uptake, causes bone decalcification, and disrupts muscle contraction. Another dangerous effect (for us as well as fish) is that acid water leaches toxic metals, such as mercury and aluminum, out of soil and rocks.

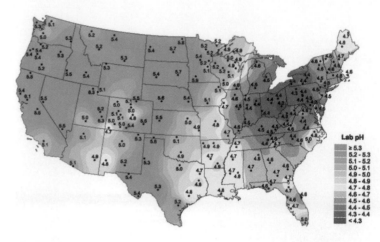

FIGURE 16.22   Acid precipitation over the United States.
Source: National Atmospheric Deposition Program/National Trends Network, 2000. http:// nadp.sws.uiuc.edu.

In the early 1970s, evidence began to accumulate suggesting that air pollutants are acidifying many lakes in North America. Studies in the Adirondack Mountains of New York revealed that about half of the high-altitude lakes (above 1,000 m or 3,300 ft) are acidified and have no fish. Areas showing lake damage correlate closely with average pH levels in precipitation (fig. 16.22). Some 48,000 lakes in Ontario are endangered and nearly all of Quebec's surface waters, including about 1 million lakes, are believed to be highly sensitive to acid deposition.

Much of the western United States has relatively alkaline bedrock and carbonate-rich soil, which counterbalance acids from the atmosphere. Recent surveys of the Rocky Mountains, the Sierra Nevadas in California, and the Cascades in Washington, however, have shown that many high mountain lakes and streams have very low buffering capacity (ability to resist pH change) and are susceptible to acidification.

Sulfates account for about two-thirds of the acid deposition in eastern North America and most of Europe, while nitrates contribute most of the remaining one-third. In urban areas, where transportation is the major source of pollution, nitric acid is equal to or slightly greater than sulfuric acids in the air. A vigorous program of pollution control has been undertaken by both Canada and the United States. Although SO_2 and NO_x emissions have decreased dramatically over the past three decades over much of Europe and eastern North America as a result of pollution control measures, rain falling in these areas remains acidic. Damage to natural ecosystems also continues to be greater than scientists expected. Apparently, alkaline dust that would once have neutralized acids in air has been depleted by years of acid rain and is now no longer effective. This may also lead to a loss of cations such as calcium, magnesium, sodium, and potassium essential to plant growth.

Forest Damage

In the early 1980s, disturbing reports appeared of rapid forest declines in both Europe and North America. One of the earliest was a detailed ecosystem inventory on Camel's Hump Mountain in

FIGURE 16.23 A Fraser fir forest on Mount Mitchell, North Carolina, killed by acid rain, insect pests, and other stressors. © William P. Cunningham.

Vermont. A 1980 survey showed that seedling production, tree density, and viability of spruce-fir forests at high elevations had declined about 50 percent in 15 years. By 1990, almost all the red spruce, once the dominant species on the upper part of the mountain, were dead or dying. A similar situation was found on Mount Mitchell in North Carolina where almost all red spruce and Fraser fir above 2,000 m (6,000 ft) are in a severe decline. Nearly all the trees are losing needles and about half of them are dead (fig. 16.23).

Similar catastrophic declines have been reported in Europe. In Germany, for example, foresters claimed in 1985 that more than 4 million ha (about half the total forest) was diseased or dying. Czechoslovakia, Poland, Austria, and Switzerland found similar problems. Again, high-elevation forests are most severely affected. This is very dangerous for mountain villages in the Alps that depend on forests to prevent avalanches in the winter.

Many plant pathologists now believe that dire predictions of forest death in Europe were greatly exaggerated, but air pollution and deposition of atmospheric acids are still thought to be important causes of forest destruction in many areas. In the longest-running forest-ecosystem monitoring record in North America, researchers at the Hubbard Brook Experimental Forest in New Hampshire have shown that forest soils have become depleted of natural buffering reserves of basic cations such as calcium and magnesium through years of exposure to acid rain. Replacement of these cations by hydrogen and aluminum ions seems to be one of the main causes of plant mortality.

Other mechanisms may also play a role in forest decline. Overfertilization by nitrogen compounds may make trees sensitive to early frost. Toxic metals, such as aluminum, may be solubilized by acidic groundwater. Plant pathogens and insect pests may damage trees or attack trees debilitated by air pollution. Fungi that form essential mutualistic associations (called mycorrhizae) with tree roots may be damaged by acid rain. Other air pollutants, such as sulfur dioxide, ozone, or toxic organic compounds may damage trees. Repeated harvesting cycles in commercial forests may remove nutrients and damage ecological relationships essential for healthy tree growth. Perhaps the most likely scenario is that all these environmental factors act cumulatively but in different combinations in the deteriorating health of individual trees and entire forests.

Buildings and Monuments

In cities throughout the world, some of the oldest and most glorious buildings and works of art are being destroyed by air pollution. Smoke and soot coat buildings, paintings, and textiles. Limestone and marble are destroyed by atmospheric acids at an alarming rate. The Parthenon in Athens, the Taj Mahal in Agra, the Colosseum in Rome, frescoes and statues in Florence, medieval cathedrals in Europe (fig. 16.24), and the Lincoln Memorial and Washington Monument in Washington, D.C., are slowly dissolving and flaking away because of acidic fumes in the air. Medieval stained glass windows in Cologne's gothic cathedral are so porous from etching by atmospheric acids that pigments disappear and the glass literally crumbles away. Restoration costs for this one building alone are estimated at three to four billion German marks ($1.5 to $2 billion, U.S.).

On a more mundane level, air pollution also damages ordinary buildings and structures. Corroding steel in reinforced concrete weakens buildings, roads, and bridges. Paint and rubber deteriorate due to oxidization. Limestone, marble, and some kinds of sandstone flake and crumble. The Council on Environmental Quality estimates that U.S. economic losses from architectural damage caused by air pollution amount to about $4.8 billion in direct costs and $5.2 billion in property value losses each year.

Visibility Reduction

We have realized only recently that pollution affects rural areas as well as cities. Even supposedly pristine places like our national parks are suffering from air pollution. Grand Canyon National Park, where maximum visibility used to be 300 km, is now so smoggy on some winter days that visitors can't see the opposite rim only 20 km across the canyon. Mining operations, smelters, and power plants (some of

FIGURE 16.24 Atmospheric acids, especially sulfuric and nitric acids, have almost completely eaten away the face of this medieval statue. Each year, the total losses from air pollution damage to buildings and materials amounts to billions of dollars. © John D. Cunningham/Visuals Unlimited.

which were moved to the desert to improve air quality in cities like Los Angeles) are the main culprits. Similarly, the vistas from Shenandoah National Park just outside Washington, D.C., are so hazy that summer visibility is often less than 1.6 km because of smog drifting in from nearby urban areas.

Historical records show that over the past four or five decades human-caused air pollution has spread over much of the United States. John Trijonis of the Santa Fe Corporation reports that a gigantic "haze blob" as much as 3,000 km across covers much of the eastern United States in the summer, cutting visibility as much as 80 percent. Smog and haze are so prevalent, Trijonis says, that it's hard for people to believe that the air once was clear. Studies indicate, however, that if all human-made sources of air pollution were shut down, the air would clear up in a few days and there would be about 150 km visibility nearly everywhere rather than the 15 km to which we have become accustomed.

AIR POLLUTION CONTROL

"Dilution is the solution to pollution" was one of the early approaches to air pollution control. Tall smokestacks were built to send emissions far from the source, where they became unidentifiable and largely untraceable. But dispersed and diluted pollutants are now the source of some of our most serious pollution problems. We are finding that there is no "away" to which we can throw our waste products. While most of the discussion in this section focuses on industrial solutions, each of us can make important personal contributions to this effort (What Can You Do? p. 346).

Reducing Production

The most effective strategy for controlling pollution is to minimize polluting activities. Since most air pollution in the developed world is associated with transportation and energy production, the most effective strategy would be conservation: reducing electricity consumption, insulating homes and offices, and developing better public transportation could all greatly reduce air pollution in the United States, Canada, and Europe. Alternative energy sources, such as wind and solar power, produce energy with little or no pollution, and these and other technologies are becoming economically competitive (chapter 20). In addition to conservation, pollution can be controlled by technological innovation.

Particulate removal involves filtering air emissions. Filters trap particulates in a mesh of cotton cloth, spun glass fibers, or asbestos-cellulose. Industrial air filters are generally giant bags 10 to 15 m long and 2 to 3 m wide. Effluent gas is blown through the bag, much like the bag on a vacuum cleaner. Every few days or weeks, the bags are opened to remove the dust cake. Electrostatic precipitators are the most common particulate controls in power plants. Ash particles pick up an electrostatic surface charge as they pass between large electrodes in the effluent stream. Charged particles then collect on an oppositely charged collecting plate. These precipitators consume a large amount of electricity, but maintenance is relatively simple, and collection efficiency can be as high

What can you do?

Saving Energy and Reducing Pollution

- Conserve energy: carpool, bike, walk, use public transport, buy compact fluorescent bulbs, and energy-efficient appliances (see chapter 20 for other suggestions).

- Don't use polluting two-cycle gasoline engines if cleaner four-cycle models are available for lawn mowers, boat motors, etc.

- Buy refrigerators and air conditioners designed for CFC alternatives. If you have old appliances or other CFC sources, dispose of them responsibly.

- Plant a tree and care for it (every year).

- Write to your Congressional representatives and support a transition to an energy-efficient economy.

- If green-pricing options are available in your area, buy renewable energy.

- If your home has a fireplace, install a high-efficiency, clean-burning, two-stage insert that conserves energy and reduces pollution up to 90 percent.

- Have your car tuned every 10,000 miles (16,000 km) and make sure that its antismog equipment is working properly. Turn off your engine when waiting longer than one minute. Start trips a little earlier and drive slower—it not only saves fuel but it's safer, too.

- Use latex-based, low-volatile paint rather than oil-based (alkyd) paint.

- Avoid spray can products. Light charcoal fires with electric starters rather than petroleum products.

- Don't top off your fuel tank when you buy gasoline; stop when the automatic mechanism turns off the pump. Don't dump gasoline or used oil on the ground or down the drain.

- Buy clothes that can be washed rather than dry-cleaned.

as 99 percent. The ash collected by both of these techniques is a solid waste (often hazardous due to the heavy metals and other trace components of coal or other ash source) and must be buried in landfills or other solid-waste disposal sites.

Sulfur removal is important because sulfur oxides are among the most damaging of all air pollutants in terms of human health and ecosystem viability. Switching from soft coal with a high sulfur content to low-sulfur coal is the surest way to reduce sulfur emissions. High-sulfur coal is frequently politically or economically expedient, however. In the United States, Appalachia, a region of chronic economic depression, produces most high-sulfur coal. In China, much domestic coal is rich in sulfur. Switching to cleaner oil or gas would eliminate metal effluents as well as sulfur. Cleaning fuels is an alternative to switching. Coal can be crushed, washed, and gasified to remove sulfur and metals before combustion. This improves heat content and firing properties, but may replace air pollution with solid-waste and water pollution problems; furthermore, these steps are expensive.

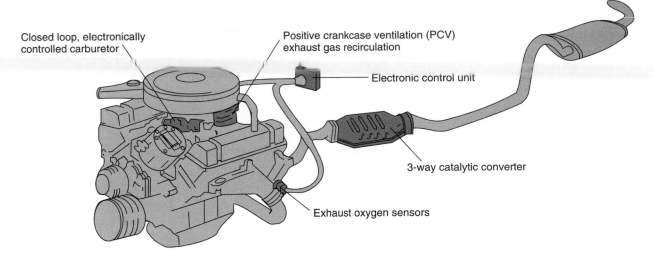

FIGURE 16.25 Elements of a modern automobile emission-control system. A closed-loop, electronically controlled carburetor or fuel-injector carefully meters fuel-air ratios to optimize combustion. Exhaust oxygen sensors measure completeness of fuel burning. Positive crankcase ventilation captures oil "blow-by" and unburned exhaust gases and recycles them to the cylinder.

Labels in figure:
- Closed loop, electronically controlled carburetor
- Positive crankcase ventilation (PCV) exhaust gas recirculation
- Electronic control unit
- 3-way catalytic converter
- Exhaust oxygen sensors

Sulfur can also be removed to yield a usable product instead of simply a waste disposal problem. Elemental sulfur, sulfuric acid, or ammonium sulfate can all be produced using catalytic converters to oxidize or reduce sulfur. Markets have to be reasonably close and fly ash contamination must be reduced as much as possible for this procedure to be economically feasible.

Nitrogen oxides (NO_x) can be reduced in both internal combustion engines and industrial boilers by as much as 50 percent by carefully controlling the flow of air and fuel. Staged burners, for example, control burning temperatures and oxygen flow to prevent formation of NO_x. The catalytic converter on your car uses platinum-palladium and rhodium catalysts to remove up to 90 percent of NO_x, hydrocarbons, and carbon monoxide at the same time (fig. 16.25).

Hydrocarbon controls mainly involve complete combustion or controlling evaporation. Hydrocarbons and volatile organic compounds are produced by incomplete combustion of fuels or by solvent evaporation from chemical factories, paints, dry cleaning, plastic manufacturing, printing, and other industrial processes. Closed systems that prevent escape of fugitive gases can reduce many of these emissions. In automobiles, for instance, positive crankcase ventilation (PCV) systems collect oil that escapes from around the pistons and unburned fuel and channels them back to the engine for combustion. Controls on fugitive losses from industrial valves, pipes, and storage tanks can have a significant impact on air quality. Afterburners are often the best method for destroying volatile organic chemicals in industrial exhaust stacks.

Fuel Switching and Fuel Cleaning

Switching from soft coal with a high sulfur content to low-sulfur coal can greatly reduce sulfur emissions. This may eliminate jobs, however, in such areas as Appalachia that are already economically depressed. Changing to another fuel, such as natural gas or nuclear energy, can eliminate all sulfur emissions as well as those of particulates and heavy metals. Natural gas is more expensive and more difficult to ship and store than coal, however, and many people prefer the sure dangers of coal pollution to the uncertain dangers of nuclear power (chapter 19). Alternative energy sources, such as wind and solar power, are preferable to either fossil fuel or nuclear power, and are becoming economically competitive (chapter 20) in many areas. In the interim, coal can be crushed, washed, and gassified to remove sulfur and metals before combustion. This improves heat content and firing properties but may replace air pollution with solid waste and water pollution problems.

CLEAN AIR LEGISLATION

Throughout history, countless ordinances have prohibited emission of objectionable smoke, odors, and noise. Air pollution traditionally has been treated as a local problem, however. The Clean Air Act of 1963 was the first national legislation in the United States aimed at air pollution control. The act provided federal grants to states to combat pollution but was careful to preserve states' rights to set and enforce air quality regulations. It soon became obvious that some pollution problems cannot be solved on a local basis.

In 1970, an extensive set of amendments essentially rewrote the Clean Air Act. These amendments identified the "criteria" pollutants discussed earlier in this chapter, and established primary and secondary standards for ambient air quality. Primary standards (table 16.2) are intended to protect human health, while secondary standards are set to protect materials, crops, climate, visibility, and personal comfort.

Since 1970 the Clean Air Act has been modified, updated, and amended. The most significant amendments were in the 1990 update. Amendments have involved acrimonious debate, with bills sometimes languishing in Congress from one session to the next because of disputes over burdens of responsibility and cost and definitions of risk. Some of the principal problems that have been addressed in amendments include the following:

- Acid rain: reductions in sulfur dioxide and nitrogen oxides

- Urban smog: lower vehicle emissions, alternative fuels in the smoggiest cities

- Toxic air pollutants: controls on 188 airborne toxins from about 250 types of sources

- Ozone protection: phasing out CFCs, carbon tetrachloride, and other compounds (Smuggling from developing countries such as Mexico, where CFCs are still cheap and legal, has become a serious problem.)

- Marketing pollution rights: allowing corporations to avoid pollution reduction by buying, selling, and "banking" pollution rights from other factories

- Fugitive emissions of volatile organic compounds (VOCs): detailed standards for manufacturing, use, and storage methods

- Ambient ozone, soot, and dust: reduced-sulfur gasoline, cleaner smokestacks and cars

- NO_x emissions: improved standards for sport-utility vehicles, personal watercraft, and other polluting vehicles

A 2002 report concluded that simply by enforcing existing clean air legislation, the United States could save at least another 6,000 lives per year and prevent 140,000 asthma attacks.

Clear Skies

In 2002, President Bush announced his "Clear Skies" plan, calling it the "most aggressive initiative in American history to cut power plant emissions." Supporters applauded his declared intent to eliminate "outdated and obstructive features" of the Clean Air Act, and to establish a market-based approach for controlling three key air pollutants (sulfur dioxide, nitrogen oxide, and mercury). Opponents, however, criticized the plan as "the biggest rollback of the Clean Air Act in history."

The most controversial aspect of Bush's plan is elimination of the "new source review," which was established in 1977. This provision was originally adopted because industry argued that it would be intolerably expensive to install new pollution-control equipment on old power plants and factories that were about to close down anyway. Congress agreed to "grandfather" or exempt existing equipment from new pollution limits with the stipulation that when they were upgraded or replaced, more stringent rules would apply (fig. 16.26). The result was that owners kept old facilities operating precisely because they were exempted from pollution control. In fact, corporations poured millions into aging power plants and factories, expanding their capacity rather than build new ones. A quarter of a century later, most of those grandfathered plants are still going strong, and continue to be among the biggest contributors to smog and acid rain.

In 1999, the Clinton administration sued 51 power plants and dozens of other industrial facilities for illegally making major upgrades without installing the necessary pollution controls. In

POLLUTANT	PRIMARY (HEALTH-BASED) AVERAGING TIME	STANDARD CONCENTRATION
TSP[a]	Annual geometric mean[b]	50 µg/m³
	24 hours	150 µg/m³
SO₂	Annual arithmetic mean[c]	80 µg/m³ (0.03 ppm)
	24 hours	120 µg/m³ (0.14 ppm)
CO	8 hours	10 mg/m³ (9 ppm)
	1 hour	40 mg/m³ (35 ppm)
NO₂	Annual arithmetic mean	80 µg/m³ (0.05 ppm)
O₃	Daily max 8 hour avg.	157 µg/m³ (0.08 ppm)
Lead	Maximum quarterly avg.	1.5 µg/m³

[a]Total suspended particulate material.

[b]The geometric mean is obtained by taking the nth root of the product of n numbers. This tends to reduce the impact of a few very large numbers in a set.

[c]An arithmetic mean is the average determined by dividing the sum of a group of data points by the number of points.

response, utilities in Ohio, Virginia, and Florida agreed to make billion-dollar investments to reduce pollution. Two years later, President Bush said that determining which facilities are new, and which are not, represents a cumbersome and unreasonable imposition on industries (many of which contributed generously to his election campaign). The EPA subsequently announced it would abandon new source reviews, depending instead on voluntary emissions controls and a trading program for air pollution allowances. Several of the utilities that had already agreed to install new pollution controls immediately put their plans on hold.

Environmental groups generally agree that cap-and-trade (which sets maximum amounts for pollutants, and then lets facilities facing costly cleanup bills to pay others with lower costs to reduce emissions on their behalf) has worked fairly well for sulfur dioxide. When trading began in 1990, economists estimated that eliminating 10 million tons of sulfur dioxide would cost $15 billion per year. Left to find the most economical ways to reduce emissions, however, utilities have been able to reach clean air goals for one-tenth that price. A serious shortcoming of this approach is that while trading has resulted in overall pollution reduction, some local "hot spots" remain where owners have found it cheaper to pay someone else to reduce pollution than to do it themselves. Knowing that the average person is enjoying cleaner air isn't much comfort if you're living in one of the persistently dirty areas.

Critics also complained that while the Bush administration moved quickly to abandon the new source review, legislation for replacing it with market-based controls was slow in coming. Nine northeastern states, which find themselves on the receiving end of acid rain and air pollution from grandfathered power plants and factories in the Midwest, sued the EPA for abandoning one of the

FIGURE 16.26 Should old power plants be required to install costly pollution-control equipment? This is the critical issue in the "new source review" under the Clean Air Act. © Corbis Royalty Free Website.

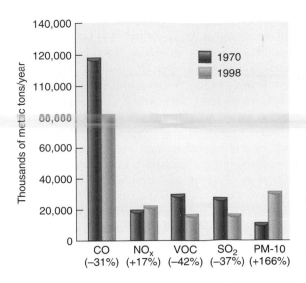

FIGURE 16.27 Air pollution trends in the United States, 1970 to 1998. Although population and economic activity increased during this period, emissions of all "criteria" air pollutants, except for nitrogen oxides and particulate matter, decreased significantly.
Source: Environmental Protection Agency, 2002.

few effective programs for combating dirty air. In 2003, the EPA agreed to reconsider its revisions of the new source review. Many environmentalists would also like any market-based plan to take a "four pollutant" approach, adding carbon dioxide to sulfur dioxide, nitrogen oxide, and mercury regulations. President Bush, who continues to maintain that global warming remains an uncertain theory, resists calls for expensive greenhouse gas restrictions.

CURRENT CONDITIONS AND FUTURE PROSPECTS

Although the United States has not yet achieved the Clean Air Act goals in many parts of the country, air quality has improved dramatically in the last decade in terms of the major large-volume pollutants. For 23 of the largest U.S. cities, the number of days each year in which air quality reached the hazardous level is down 93 percent from a decade ago. Of 97 metropolitan areas that failed to meet clean air standards in the 1980s, 41 were in compliance in 1991 to 1992. For many cities, this was the first time they met air quality goals in 20 years. Still, the EPA reports that nearly half of all Americans live in counties with unhealthy smog levels, and that

at least 200 million people are exposed to hazardous air pollutants in indoor or outside air.

The EPA estimates that between 1970 and 1998, lead fell 98 percent, SO_2 declined 35 percent, and CO shrank 32 percent (fig. 16.27). Filters, scrubbers, and precipitators on power plants and other large stationary sources are responsible for most of the particulate and SO_2 reductions. Catalytic converters on automobiles are responsible for most of the CO and O_3 reductions.

The only conventional "criteria" pollutants that have not dropped significantly are particulates and NO_x. Because automobiles are the main source of NO_x, cities, such as Los Angeles, where pollution comes largely from traffic, still have serious air quality problems. Particulate matter (mostly dust and soot) is produced by agriculture, fuel combustion, metal smelting, concrete manufacturing, and other activities. Industrial cities, such as Baltimore, Maryland, and Baton Rouge, Louisiana, also have continuing problems. In 1999, for the first time in many years, Houston passed Los Angeles as having the worst air quality in the country. Eighty-five other urban areas are still considered nonattainment regions. In spite of these local failures, however, 80 percent of the United States now meets the National Ambient Air Quality Standards. This improvement in air quality is perhaps the greatest environmental success story in our history.

Air Pollution in Developing Countries

The outlook is not so encouraging in other parts of the world. The major metropolitan areas of many developing countries are growing at explosive rates to incredible sizes (chapter 22), and environmental quality is abysmal in many of them. Mexico City remains notorious for bad air. Pollution levels exceed WHO health standards 350 days per year, and more than half of all city children have lead levels in their blood high enough to lower intelligence and retard development. Mexico City's 131,000 industries and 2.5 million vehicles spew out more than 5,500 tons of air pollutants daily. Santiago, Chile, averages 299 days per year on which suspended particulates exceed WHO standards of 90 mg/m³.

While there are few statistics on China's pollution situation, it is known that many of China's 400,000 factories have no air pollution controls. Experts estimate that home coal burners and factories emit 10 million tons of soot and 15 million tons of sulfur dioxide annually and that emissions have increased rapidly over the past 20 years. Sheyang, an industrial city in northern China, is thought to have the world's worst continuing particulate problem, with peak winter concentrations over 700 mg/m³ (nine times U.S. maximum standards). Airborne particulates in Sheyang exceed WHO standards on 347 days per year. Beijing, Xi'an, and Guangzhou are nearly as bad. The high incidence of cancer in Shanghai is thought to be linked to air pollution (see fig. 16.1).

As political walls came down across Eastern Europe and the Soviet Union at the end of the 1980s, horrifying environmental conditions in these centrally planned economies were revealed. Inept industrial managers, a rigid bureaucracy, and lack of democracy have created ecological disasters. Where governments own, operate, and regulate industry, there are few checks and balances or incentives to clean up pollution. Much of the Eastern bloc depends heavily on soft, brown coal for its energy, and pollution controls are absent or highly inadequate.

Southern Poland, the northern Czech Republic, and Slovakia are covered most of the time by a permanent cloud of smog from factories and power plants. Acid rain is eating away historic buildings and damaging already inadequate infrastructures. Home gardening has been banned in Katowice, Poland, because vegetables raised there have unsafe levels of lead and cadmium. For miles around the infamous Romanian "black town" of Copsa Mica, the countryside is so stained by soot that it looks as if someone poured black ink over everything (fig. 16.28). Birth defects afflict 10 percent of infants in northern Bohemia. Workers in factories there get extra hazard pay—burial money, they call it. Life expectancy in these industrial towns is as much as 10 years less than the national average.

Signs of Hope

Not all is pessimistic, however. There have been some spectacular successes in air pollution control. Sweden and West Germany (countries affected by forest losses due to acid precipitation) cut their sulfur emissions by two-thirds between 1970 and 1985. Austria and Switzerland have gone even further, regulating even motorcycle emissions. The Global Environmental Monitoring System (GEMS) reports declines in particulate levels in 26 of 37 cities worldwide. Sulfur dioxide and sulfate particles, which cause acid rain and respiratory disease, have declined in 20 of these cities.

Ten years ago, Cubatao, Brazil, was described as the "Valley of Death," one of the most dangerously polluted places in the

FIGURE 16.28 Soot from factories and refineries has turned buildings, streets, and even this shepherd's sheep black in Copsa Mica, Romania. © James Nachtwey/Magnum Photos Inc.

world. A steel plant, a huge oil refinery, and fertilizer and chemical factories churned out thousands of tons of air pollutants every year. Trees died on the surrounding hills. Birth defects and respiratory diseases were alarmingly high. Since then, however, the citizens of Cubatao have made remarkable progress in cleaning up their environment. The end of military rule and restoration of democracy allowed residents to publicize their complaints. The environment became an important political issue. The state of São Paulo invested about $100 million and the private sector spent twice as much to clean up most pollution sources in the valley. Particulate pollution was reduced 75 percent, ammonia emissions were reduced 97 percent, hydrocarbons that cause ozone and smog were cut 86 percent, and sulfur dioxide production fell 84 percent. Fish are returning to the rivers, and forests are regrowing on the mountains. Progress is possible! We hope that similar success stories will be obtainable elsewhere.

California has gone further than the federal government in making specific plans for air pollution control. The South Coast Air Quality Management District has adopted 160 rules to clean the air in the Los Angeles Basin. If these measures are successful, smog-causing emissions could be reduced by 70 percent. The number of days when the air is considered hazardous to breathe would decrease from 150 per year to none.

Reaching these goals will require substantial lifestyle changes for most Californians. Aerosol hair sprays, deodorants, charcoal lighter fluid, gasoline-powered lawnmowers, and drive-through burger stands could be banned. More than 3,000 consumer products, including automotive polishes, spot removers, herbicides, lubricants, and floor-wax strippers, would need to meet new pollution limits. Paints and cleaning solutions would have to contain fewer volatile solvents.

Summary

- Air pollution is physical or chemical changes brought about by natural processes or human activities that result in air quality degradation.

- Primary pollutants are released directly into the air in a harmful form. Secondary pollutants are created or converted to a hazardous form after they enter the atmosphere, usually by photochemical reactions. Fugitive emissions come from non-point sources such as soil erosion, mining, or building deconstruction.

- The Clean Air Act of 1970 designated seven major "criteria" pollutants (sulfur dioxide, carbon monoxide, particulates, volatile hydrocarbons, nitrogen oxides, photochemical oxidants, and lead) because these were regarded as the greatest threat to human health. Subsequently, some 660 other hazardous air pollutants have been added to the regulatory list.

- More than 200 million Americans live in areas where the risk of death from air pollution effects is greater than the acceptable level of 1 in 1 million. Indoor air is a major part of this health risk, with pollution levels often greater than those outdoors.

- Globally, some 2.5 billion people (mostly women and children) are exposed to hazardous levels of smoke from poorly ventilated heating and cooking fires. Altogether, air pollution probably contributes to more deaths per year than any single infectious disease.

- Aerosols and air toxins can be carried long distances by wind currents. Dust from Chinese deserts, for example, often falls out on the American West. Through a process of sequential evaporation and precipitation, hazardous air pollutants are accumulating in the Arctic and Antarctic, where they concentrate through food chains to reach dangerous levels in both humans and top predators like polar bears and whales.

- Chloroflurocarbons (CFCs) and other long-lasting chlorine-containing compounds migrate into the stratosphere where they destroy the ozone layer that protects us from harmful ultraviolet solar radiation. The Montreal Protocol, which called for a phase-out of these chemicals, is one of the best examples of international cooperation to fight global air pollution.

- Sulfur and nitrogen oxides react in the air to form sulfuric and nitric acids, which fall to earth as acid rain, snow, or dry precipitation. These acids pollute surface waters, kill aquatic organisms, harm vegetation, destroy building materials, and reduce visibility. Pollution controls have significantly reduced these emissions, but more remains to be done.

- Clean Air Acts passed in most developed countries and in many developing nations are among the central tools for environmental protection. They have played a major role in dramatic air pollution reduction in many nations. A highly controversial provision of the U.S. Clean Air Act is the "new source review," which requires that modern pollution-control equipment be installed when old power plants or factories are expanded or upgraded.

- Market-based approaches to pollution control have been proposed as efficient alternatives to government mandates for specific equipment requirements or emission limits. Determining the best way to regulate air quality remains a controversial question.

- Air quality has improved dramatically over the past 30 years in most developed countries. Air pollution remains a grave issue in many poorer countries, especially in the megacities of the developing world and the former Soviet Union.

Questions for Review

1. Define primary and secondary air pollutants.

2. What are the seven "criteria" pollutants in the original Clean Air Act? Why were they chosen? How many more hazardous air toxins have been added?

3. What pollutants in indoor air may be hazardous to your health? What is the greatest indoor air problem globally?

4. What is acid deposition? What causes it?

5. What is an atmospheric inversion and how does it trap air pollutants?

6. What is the difference between ambient and stratospheric ozone? What is destroying stratospheric ozone?

7. What is long-range air pollution transport? Give two examples.

8. What is "new source review," and why is it controversial?

9. Which of the conventional pollutants has decreased most in the recent past and which has decreased least?

10. Give one example of current air quality problems in a developing country.

Questions for Critical Thinking

1. Some authors limit pollution to human-caused materials. If a natural source releases the same toxic chemicals as a factory, is one pollution and the other not?

2. What might be done to improve indoor air quality? Should the government mandate such changes? What values or world views are represented by different sides of this debate?

3. Why do you suppose that air pollution is so much worse in Eastern Europe than in the West?

4. Suppose air pollution causes a billion dollars in crop losses each year but controlling the pollution would also cost a billion dollars. Should we insist on controls?

5. Utility managers once claimed that it would cost $1,000 per fish to control acid precipitation in the Adirondack lakes and that it would be cheaper to buy fish for anglers than to put scrubbers on power plants. Suppose that was true. Does it justify continuing pollution?

6. Developing nations claim that richer countries created global warming and stratospheric ozone depletion, and therefore should bear responsibility for fixing these problems. How would you respond?

7. Is it possible to have zero emissions of pollutants? What does zero mean in this case?

8. If there are thresholds for pollution effects (at least as far as we know now), is it reasonable or wise to depend on environmental processes to disperse, assimilate, or inactivate waste products?

9. How would you choose between government "command and control" regulations versus market-based trading programs for air pollution control? Are there situations where one approach would work better than the other?

10. If you were mayor of Shanghai or Mexico City, what steps would you take to improve air quality?

Key Terms

acid precipitation 344
aerosol 334
aesthetic degradation 336
ambient air 331
bronchitis 342
carbon monoxide 334
chlorofluorocarbons (CFCs) 340
chronic obstructive lung disease 342
conventional or criteria pollutants 331
fugitive emissions 331
hazardous air pollutants (HAPs) 335
nitrogen oxides 331
ozone 335
particulate material 334
photochemical oxidants 335
primary pollutants 331
secondary pollutants 331
stratospheric ozone 340
sulfur dioxide 331
synergistic effects 343
temperature inversions 337
Toxic Release Inventory (TRI) 335
volatile organic compounds 335

Further Readings

Boubel, Richard W., et al. 1994. *Fundamentals of Air Pollution.* 3rd ed. Academic Press.

Davis, Devra. 2002. *When Smoke Ran Like Water: Tales of Environmental Deception and the Battle against Pollution.* Basic Books.

Krupa, Sagar V. 1997. *Air Pollution, People, and Plants: An Introduction.* The American Phytopathological Society.

Lee, Shun Cheng, et al. 2002. Investigation of indoor air quality at residential homes in Hong Kong: A case study. *Atmospheric Environment* 36(2):225–37.

May, Jeffrey C. 2001. *My House Is Killing Me! The Home Guide for Families with Allergies and Asthma.* Johns Hopkins Univ. Press.

Nagashima, T., et al. 2002. Future development of the ozone layer calculated by a general circulation model with fully interactive chemistry. *Geophysical Research Letters* 29:31–34.

Yli-Tuomi, T., et al. 2003. Composition of the Finnish arctic aerosol: Collection and analysis of historic filter samples. *Atmospheric Environment* 37(17):2355–64.

Welcome to McGraw-Hill's Online Learning Center

Location: http://www.mhhe.com/environmentalscience

McGraw Hill

WEB EXERCISE

Finding Air Pollution Sources in Your Neighborhood

The U.S. EPA has a wealth of general information on environmental quality at www.epa.gov/. From there, you can follow links to air information and "where you live" to find data about your local air quality, or you can go directly to www.epa.gov/air/data/index.html and click on the icon for reports and maps. You can either zoom in on your state and neighborhood by clicking on the interactive map, or you can get there a little faster by going to the geographic selection page and choosing your city in the MSA (metropolitan statistical area) list, or entering your zip code on the "list only" page and clicking go.

1. Within the emissions section, click on the facility/monitor locator map to find major sources of pollutants in your area. You can choose a specific pollutant, or, hold down the shift key to choose multiple pollutants. Select a year for your report, specify geographical features, and then click on *generate map*. What is the closest pollution source to your home? Were you aware of it before this exercise? If you can't recognize sources from locations on the map, you can click on *view data* at the bottom of the page to generate a list of pollution sources. What's the largest source of SO_2 emissions in your neighborhood? Are you surprised to learn about it?

2. Go back to *select geographic area* and select your whole state, then choose Air Quality Index Summary under Air Monitoring. Choose a year and either county or MSA and generate a report of air quality levels. How many days in the year you selected had air quality unhealthy for sensitive people? What pollution type was highest in your area? What do you think might be done to improve air quality where you live?

Water Use and Management

*I tell you gentlemen, you are piling up a heritage of conflict and litigation of water rights,
for there is not sufficient water to supply the land.*

John Wesley Powell

OBJECTIVES

After studying this chapter, you should be able to:

- summarize how the hydrologic cycle delivers fresh water to terrestrial ecosystems and how the cycle balances over time.
- contrast the volume and residence time of water in the earth's major compartments.
- describe the important ways we use water and distinguish between withdrawal, consumption, and degradation.
- appreciate the causes and consequences of water shortages around the world and what they mean in people's lives in water-poor countries.
- debate the merits of proposals to increase water supplies and manage demand.
- apply some water conservation methods in your own life.

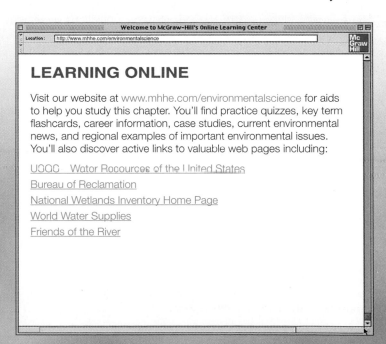

Welcome to McGraw-Hill's Online Learning Center

Location: http://www.mhhe.com/environmentalscience

LEARNING ONLINE

Visit our website at www.mhhe.com/environmentalscience for aids to help you study this chapter. You'll find practice quizzes, key term flashcards, career information, case studies, current environmental news, and regional examples of important environmental issues. You'll also discover active links to valuable web pages including:

USGS — Water Resources of the United States
Bureau of Reclamation
National Wetlands Inventory Home Page
World Water Supplies
Friends of the River

Photo: By 2025, experts predict that two-thirds of the world's population will live in countries suffering from serious water shortages. © William P. Cunningham.

China's South-to-North Water Diversion Project

Beijing, soon to be the seat of the world's largest economy, is desperately thirsty. This city of 13 million is part of the country's fast-growing northeastern industrial region. Already, Beijing has sucked nearby rivers dry. Groundwater is overpumped and running out. Even the great Yellow River is largely consumed before it reaches the sea. Now, Beijing and neighboring Tianjin are eying the far-distant Yangtze River, China's longest and largest river.

The Yangtze, the world's fourth largest river, is already a subject of international debate. The world's largest, most expensive, and perhaps most controversial dam, the Three Gorges Dam, has recently been built on the river. More than 600 km of the river has become the world's largest reservoir. Now that the dam is nearly complete, China is beginning work on the South-to-North Water Diversion Project. With an estimated cost of at least $58 billion (U.S.), this project will be the world's largest water movement project. It will be twice as expensive as the Three Gorges Dam. Pumps will lift Yangtze water over several mountain ranges. Three major canal systems will move 45 billion m³ of water a year 1,300 km north (fig. 17.1). Aqueducts will carry water across or below dozens of rivers. Some 300 million people will benefit, some of them farmers and most urban residents or industries in the north. New reservoirs will displace at least 350,000 people. Portions of the route follow the 1,500-year-old Grand Canal, built by Zhou and Sui emperors to provide transportation from Beijing to southern cities.

Planners have waited a lifetime to see this project move forward. Revolutionary leader Mao Zedong proposed the project 50 years ago. Momentum has grown as Beijing and Tianjin have expanded, and as China's economy has boomed, growing at 8 percent per year. Officials hope to have the eastern route under way by 2005. Water should reach Beijing by 2010. The most difficult and expensive parts of the plan, the far-western sections, will not be complete for at least 50 years.

Water shortages are endemic in China. Four hundred of the country's 670 large cities lack sufficient water. Farmers have rioted over scarce supplies, and more than half the population drinks contaminated water. Desperate rural conditions have increased migration to major cities, where rapidly growing populations increase pressure on scarce water resources.

Chinese environmentalists worry that the project could dry up the country's greatest river, eliminate ecosystems and species, and

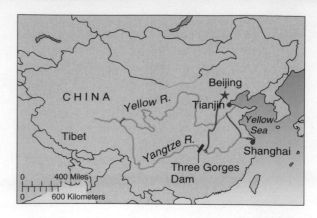

FIGURE 17.1 With the Yellow River nearly depleted by overuse, northern China now plans canals (in red) to deliver Yangtze water to Beijing.

possibly even alter ocean circulation and the climate along China's eastern coast. Perhaps the greatest uncertainty is how to keep the water clean. Canals will pass through heavily industrialized regions between Shanghai and Tianjin. Southern water will mix into rivers flooded with untreated urban waste, sewage, factory effluent, and contaminants from shipping traffic. Opponents of the project argue that less ambitious projects, together with conservation, would be a safer alternative.

China's south-to-north water project is huge, but it is not unique. Southern California continues to covet northern Californian water; Canadian entrepreneurs would like to sell some of British Columbia's water wealth; growing cities of the southern United States thirstily eye the Great Lakes; Russian developers would like to send water from the Ob and Yenisey Rivers to the dry cotton fields of the Aral Sea region. Most of these ideas are decades old, but growing populations, growing economies, and possibly changing climates all add urgency to the proposals.

As you will read in this chapter, water shortages increasingly threaten economies, societies, and our environment. Conflicts have already arisen over shrinking supplies—between the United States and Mexico, between Israel and Jordan, between Turkey and Iraq. Many experts see water as a chief cause of international conflict in the coming century. How to conserve and allocate water resources is a central theme in environmental science. In this chapter, we will survey water supplies, water use, and strategies for water conservation.

WATER RESOURCES

Water is a marvelous substance—flowing, rippling, swirling around obstacles in its path, seeping, dripping, trickling, constantly moving from sea to land and back again. Water can be clear, crystalline, icy green in a mountain stream, or black and opaque in a cypress swamp. Water bugs skitter across the surface of a quiet lake; a stream cascades down a stairstep ledge of rock; waves roll endlessly up a sand beach, crash in a welter of foam, and recede. Rain falls in a gentle mist, refreshing plants and animals. A violent thunderstorm floods a meadow, washing away stream banks. Water is a most beautiful and precious resource.

Water is also a great source of conflict. Some 2 billion people, a third of the world's population, live in countries with insufficient clean water. Some experts estimate this number could double in 25 years. To understand this resource, let's first ask, where does our water come from, and why is it so unevenly distributed?

The Hydrologic Cycle

The hydrologic cycle (water cycle) describes the circulation of water as it evaporates from land, water, and organisms; enters the atmosphere; condenses and is precipitated to the earth's surfaces; and moves underground by infiltration or overland by runoff into rivers, lakes, and seas (see fig. 3.18). This cycle supplies fresh water to the landmasses, maintains a habitable climate, and moderates world temperatures. Movement of water back to the sea in rivers and glaciers is a major geological force that shapes the land and redistributes material. Plants play an important role in the hydrologic cycle, absorbing groundwater and pumping it into the atmosphere by **transpiration** (transport plus evaporation). In tropical forests, as much as 75 percent of annual precipitation is returned to the atmosphere by plants. Much of this moisture falls again, keeping the rainforest humid.

Solar energy drives the hydrologic cycle by evaporating surface water. **Evaporation** is the process in which a liquid is changed to vapor (gas phase) at temperatures well below its boiling point. Water also can move between solid and gaseous states without ever becoming liquid in a process called **sublimation.** On bright, cold, windy winter days, when the air is very dry, snowbanks disappear by sublimation, even though the temperature never gets above freezing. This is the same process that causes "freezer burn" of frozen foods.

In both evaporation and sublimation, molecules of water vapor enter the atmosphere, leaving behind salts and other contaminants and thus creating purified fresh water. This is essentially distillation on a grand scale.

The amount of water vapor in the air is called humidity. Warm air can hold more water than cold air. When a volume of air contains as much water vapor as it can at a given temperature, we say that it has reached its **saturation point. Relative humidity** is the amount of water vapor in the air expressed as a percentage of the maximum amount (saturation point) that could be held at that particular temperature.

When the saturation concentration is exceeded, water molecules begin to aggregate in the process of **condensation.** If the temperature at which this occurs is above 0°C, tiny liquid droplets result. If the temperature is below freezing, ice forms. For a given amount of water vapor, the temperature at which condensation occurs is the **dew point.** Tiny particles, called **condensation nuclei,** float in the air and facilitate this process. Smoke, dust, sea salts, spores, and volcanic ash all provide such particles. Even apparently clear air can contain large numbers of these particles, which are generally too small to be seen by the naked eye. Sea salt is an excellent source of such nuclei, and heavy, low clouds frequently form in the humid air over the ocean.

A cloud, then, is an accumulation of condensed water vapor in droplets or ice crystals. Normally, cloud particles are small enough to remain suspended in the air, but when cloud droplets and ice crystals become large enough, gravity overcomes uplifting air currents, and precipitation occurs.

Balancing the Water Budget

Everything about global hydrological processes is awesome in scale. Each year, the sun evaporates approximately 496,000 km³ of water from the earth's surface (table 17.1). More water evaporates in the tropics than at higher latitudes, and more water evaporates over the oceans than over land. Although the oceans cover about

TABLE 17.1 | **Some Units of Water Measurement**

One cubic kilometer (km³) equals 1 billion cubic meters (m³), 1 trillion liters, or 264 billion gallons.

One acre-foot is the amount of water required to cover an acre of ground 1 foot deep. This is equivalent to 325,851 gallons, or 1.2 million liters, or 1,234 m³, about the amount consumed annually by a family of four in the United States.

One cubic foot per second of river flow equals 28.3 liters per second or 449 gallons per minute.

70 percent of the earth's surface, they account for 86 percent of total evaporation. Ninety percent of the water evaporated from the ocean falls back on the ocean as rain. The remaining 10 percent is carried by prevailing winds over the continents where it combines with water evaporated from soil, plant surfaces, lakes, streams, and wetlands to provide a total continental precipitation of about 111,000 km³.

What happens to the surplus water on land—the difference between what falls as precipitation and what evaporates? Some of it is incorporated by plants and animals into biological tissues. A large share of what falls on land seeps into the ground to be stored for a while (from a few days to many thousands of years) as soil moisture or groundwater. Eventually, all the water makes its way back downhill to the oceans. The 40,000 km³ carried back to the ocean each year by surface runoff or underground flow represents the renewable supply available for human uses and sustaining freshwater-dependent ecosystems.

Evaporation and condensation help regulate the earth's climate. As warm, humid air travels from the tropics to cooler latitudes, it transports heat as well as moisture. The ocean also stores heat, releasing it slowly.

Without oceans to absorb and store heat, and wind currents to redistribute that heat in the latent energy of water vapor, the earth would undergo extreme temperature fluctuations like those of the moon, where it is 100°C (212°F) during the day and –130°C (–200°F) at night. Water performs this vital function because of its unique properties in heat absorption and energy of vaporization (chapter 3).

Regions of Plenty and Regions of Deficit

Rain falls unevenly over the planet (fig. 17.2). Some places get almost no precipitation, and some receive heavy rain almost daily. At Iquique, in Chile's Atacama Desert, no rain has fallen in recorded history. At the other end of the scale, Cherrapunji, in northeastern India, received nearly 23 m (897 in.) of rain in a single year in 1861.

Three principal factors control these global water deficits and surpluses. First, global atmospheric circulation creates regions of persistent high air pressure and low rainfall about 20° to 40° north and south of the equator (chapter 15). Second, where prevailing

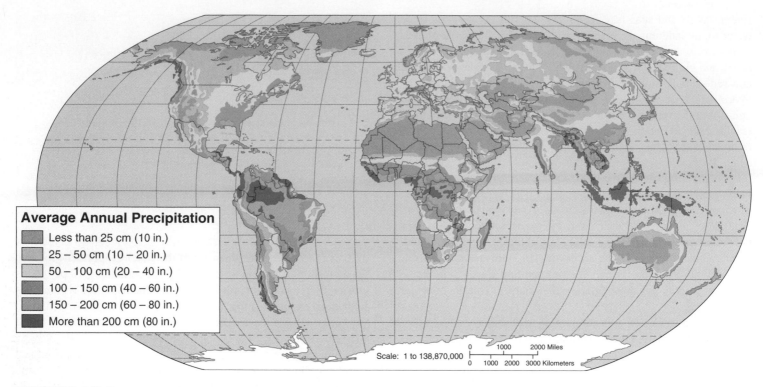

Average Annual Precipitation

■	Less than 25 cm (10 in.)
■	25 – 50 cm (10 – 20 in.)
■	50 – 100 cm (20 – 40 in.)
■	100 – 150 cm (40 – 60 in.)
■	150 – 200 cm (60 – 80 in.)
■	More than 200 cm (80 in.)

Scale: 1 to 138,870,000

FIGURE 17.2 Average annual precipitation. Note wet areas that support tropical rainforests occur along the equator, while the major world deserts occur in zones of dry, descending air between 20° and 40° north and south.

winds come over oceans, they bring moisture to land. Areas far from oceans—in a windward direction—are usually relatively dry.

A third factor in water distribution is topography. Mountains act as both cloud formers and rain catchers. As air sweeps up the windward side of a mountain, air pressure decreases and air cools. As the air cools, it reaches the saturation point, and moisture condenses as either rain or snow. Thus the windward side of a mountain range, as in the Pacific Northwest, is usually wet much of the year. Precipitation leaves the air drier than it was on its way up the mountain. As the air passes the mountaintop and descends the other side, air pressure rises, and the already-dry air warms, increasing its ability to hold moisture. Descending, warming air rarely produces any rain or snow. Places in the **rain shadow,** the dry, leeward side of a mountain range, receive little precipitation. A striking example of the rain shadow effect is that of Mount Waialeale, on the island of Kauai, Hawaii (fig. 17.3). The windward side of the island receives nearly 12 m of rain per year, while the leeward side, just a few kilometers away, receives just 46 cm.

Usually a combination of factors affects precipitation. In Cherrapunji, India, atmospheric circulation sweeps moisture from the warm Indian Ocean toward the high ridges of the Himalayas. Iquique, Chile, lies in the rain shadow of the Andes and in a high-pressure desert zone. Prevailing winds are from the east, so even though Iquique lies near the ocean, it is far from the winds' moisture source—the Atlantic. In the American Southwest, Australia, and the Sahara, high-pressure atmospheric conditions tend to keep the air and land dry. The global map of precipitation represents a complex combination of these forces of atmospheric circulation, prevailing winds, and topography. Can you guess which are most important where you live?

Human activity also explains some regions of water deficit. As noted earlier, plant transpiration recycles moisture and produces rain. When forests are cleared, falling rain quickly enters streams and returns to the ocean. In Greece, Lebanon, parts of Africa, the Caribbean, South Asia, and elsewhere, desert-like conditions have developed since the original forests were destroyed.

MAJOR WATER COMPARTMENTS

The distribution of water often is described in terms of interacting compartments in which water resides for short or long times. Table 17.2 shows the major water compartments in the world.

Oceans

Together, the oceans contain more than 97 percent of all the *liquid* water in the world. (The water of crystallization in rocks is far larger than the amount of liquid water.) Oceans are too salty for most human uses, but they contain 90 percent of the world's living biomass. While the ocean basins really form a continuous reservoir, shallows and narrows between them reduce water exchange, so they have different compositions, climatic effects, and even different surface elevations.

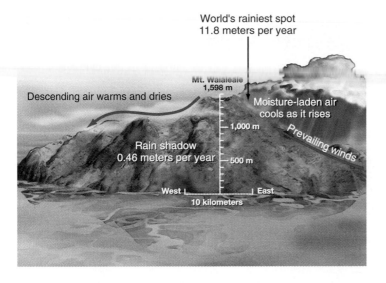

World's rainiest spot
11.8 meters per year

Mt. Waialeale
1,598 m

Descending air warms and dries

Moisture-laden air
cools as it rises

1,000 m

Rain shadow
0.46 meters per year

Prevailing winds

500 m

West   East
10 kilometers

FIGURE 17.3 Rainfall on the east side of Mount Waialeale in Hawaii is more than 20 times as much as on the west side. Prevailing trade winds bring moisture-laden sea air onshore. The air cools as it rises up the flanks of the mountain and the water it carries precipitates as rain—11.8 m (38 ft) per year!

Oceans play a crucial role in moderating the earth's temperature (fig. 17.4). Vast river-like currents transport warm water from the equator to higher latitudes, and cold water flows from the poles to the tropics (fig. 17.5). The Gulf Stream, which flows northeast from the coast of North America toward northern Europe, flows at a steady rate of 10–12 km per hour (6–7.5 mph) and carries more than 100 times more water than all rivers on earth put together.

In tropical seas, surface waters are warmed by the sun, diluted by rainwater and runoff from the land, and aerated by wave action. In higher latitudes, surface waters are cold and much more dense. This dense water subsides or sinks to the bottom of deep ocean basins and flows toward the equator. Warm surface water of the tropics stratifies or floats on top of this cold, dense water as currents carry warm water to high latitudes. Sharp boundaries form between different water densities, different salinities, and different temperatures, retarding mixing between these layers.

While parts of the hydrologic cycle occur on a time scale of hours or days, other parts take centuries. The average **residence time** of water in the ocean (the length of time that an individual molecule spends circulating in the ocean before it evaporates and starts through the hydrologic cycle again) is about 3,000 years. In the deepest ocean trenches, movement is almost nonexistent and water may remain undisturbed for tens of thousands of years.

Glaciers, Ice, and Snow

Of the 2.4 percent of all water that is fresh, nearly 90 percent is tied up in glaciers, ice caps, and snowfields (fig. 17.6). Glaciers are really rivers of ice flowing downhill very slowly (fig. 17.7). They now occur only at high altitudes or high latitudes, but as recently as 18,000 years ago about one-third of the continental landmass was covered by glacial ice sheets. Most of this ice has now melted and the largest remnant is in Antarctica. As much as 2 km (1.25 mi) thick, the Antarctic glaciers cover all but the highest mountain peaks and contain nearly 85 percent of all ice in the world.

A smaller ice sheet on Greenland, together with floating sea ice around the North Pole, makes up another 10 percent of the world's frozen water reservoirs. Mountain snow pack and ice constitute the remaining 5 percent.

 TABLE 17.2	Earth's Water Compartments—Estimated Volume of Water in Storage, Percent of Total, and Average Residence Time		
	VOLUME (THOUSANDS OF KM³)	% TOTAL WATER	AVERAGE RESIDENCE TIME
Total	1,403,377	100	2,800 years
Ocean	1,370,000	97.6	3,000 years to 30,000 years*
Ice and snow	29,000	2.07	1 to 16,000 years*
Groundwater down to 1 km	4,000	0.28	From days to thousands of years*
Lakes and reservoirs	125	0.009	1 to 100 years*
Saline lakes	104	0.007	10 to 1,000 years*
Soil moisture	65	0.005	2 weeks to a year
Biological moisture in plants and animals	65	0.005	1 week
Atmosphere	13	0.001	8 to 10 days
Swamps and marshes	3.6	0.003	From months to years
Rivers and streams	1.7	0.0001	10 to 30 days

Depends on depth and other factors.
Source: *Data from U.S. Geological Survey.*

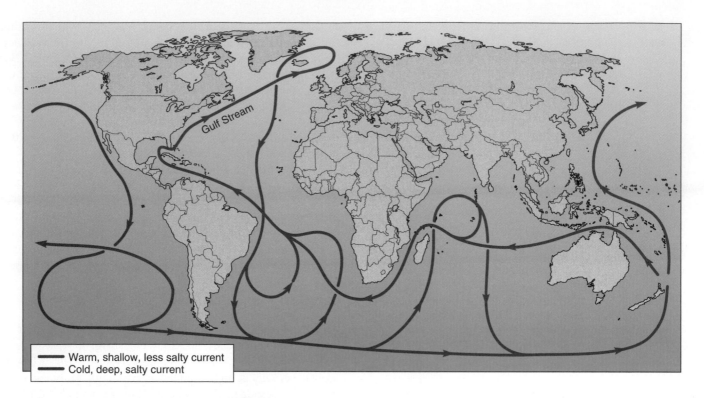

FIGURE 17.4 Ocean currents act as a global conveyor system, redistributing warm and cold water around the globe. These currents moderate our climate. For example, the Gulf Stream keeps northern Europe much warmer than northern Canada.

FIGURE 17.5 Ocean currents, such as the warm Gulf Stream, redistribute heat as they flow around the globe. Here, orange and yellow indicate warm water temperatures (25°–30°C); blue and green are cold (0°–5°C). Courtesy of NASA.

Groundwater

After glaciers, the next largest reservoir of fresh water is held in the ground as **groundwater.** Precipitation that does not evaporate back into the air or run off over the surface percolates through the soil and into fractures and spaces of permeable rocks in a process called **infiltration** (fig. 17.8). Upper soil layers that hold both air and water make up the **zone of aeration.** Moisture for plant growth comes primarily from these layers. Depending on rainfall amount, soil type, and surface topography, the zone of aeration may be very shallow or quite deep. Lower soil layers where all spaces are filled with water make up the **zone of saturation.** The top of this zone is the **water table.** The water table is not flat, but undulates according to the surface topography and subsurface structure. Nor is it stationary through the seasons, rising and falling according to precipitation and infiltration rates.

Porous layers of sand, gravel, or rock lying below the water table are called **aquifers.** Aquifers are always underlain by relatively impermeable layers of rock or clay that keep water from seeping out at the bottom.

Folding and tilting of the earth's crust by geologic processes can create shapes that generate water pressure in confined aquifers (those trapped between two impervious, confining rock layers). When a pressurized aquifer intersects the surface, or if it is penetrated by a pipe or conduit, an **artesian** well or spring results from which water gushes without being pumped.

Areas in which infiltration of water into an aquifer occurs are called **recharge zones** (fig. 17.9). The rate at which most aquifers

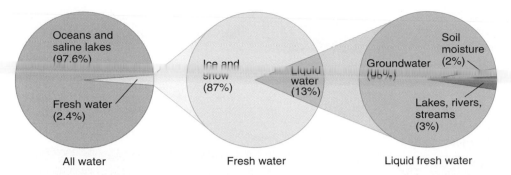

FIGURE 17.6   The easily accessible water in lakes, rivers, and streams represents only 3 percent of all liquid fresh water, which is 13 percent of all fresh water, which is 2.4 percent of all water.

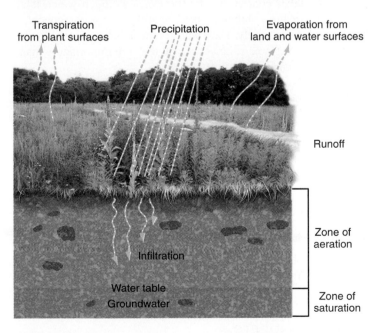

FIGURE 17.7   Glaciers are rivers of ice sliding very slowly downhill. Together, polar ice sheets and alpine glaciers contain more than three times as much fresh water as all the lakes, ponds, streams, and rivers in the world. The dark streaks on the surface of this Alaskan glacier are dirt and rocks marking the edges of tributary glaciers that have combined to make this huge flow.
© William P. Cunningham.

FIGURE 17.8   Precipitation that does not evaporate or run off over the surface percolates through the soil in a process called infiltration. The upper layers of soil hold droplets of moisture between air-filled spaces. Lower layers, where all spaces are filled with water, make up the zone of saturation, or groundwater.

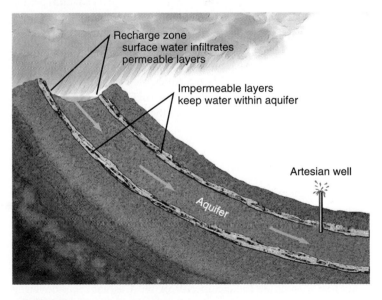

FIGURE 17.9   An aquifer is a porous, water-bearing layer of sand, gravel, or rock. This aquifer is confined between layers of rock or clay and bent by geological forces, creating hydrostatic pressure. A break in the overlying layer creates an artesian well or spring.

are refilled is very slow, however, and groundwater presently is being removed faster than it can be replenished in many areas. Urbanization, road building, and other development often block recharge zones and prevent replenishment of important aquifers. Contamination of surface water in recharge zones and seepage of pollutants into abandoned wells have polluted aquifers in many places, making them unfit for most uses (chapter 18). Many cities protect aquifer recharge zones from pollution or development, both as a way to drain off rainwater and as a way to replenish the aquifer with pure water.

Some aquifers contain very large volumes of water. The groundwater within 1 km of the surface in the United States is more than 30 times the volume of all the freshwater lakes, rivers, and reservoirs on the surface. While water can flow through limestone caverns in underground rivers, most movement in aquifers is a dispersed and almost imperceptible trickle through tiny fractures and spaces. Depending on geology, it can take anywhere from a few hours to several years for contaminants to move a few hundred meters through an aquifer.

Rivers and Streams

Precipitation that does not evaporate or infiltrate into the ground runs off over the surface, drawn by the force of gravity back toward the sea. Rivulets accumulate to form streams, and streams join to form rivers. Although the total amount of water contained at any one time in rivers and streams is small compared to the other water reservoirs of the world (see table 17.2), these surface waters are vitally important to humans and most other organisms. Most rivers, if they were not constantly replenished by precipitation, meltwater from snow and ice, or seepage from groundwater, would begin to diminish in a few weeks.

We measure the size of a river in terms of its **discharge,** the amount of water that passes a fixed point in a given amount of time. This is usually expressed as liters or cubic feet of water per second. The 16 largest rivers in the world carry nearly half of all surface runoff on earth. The Amazon is by far the largest river in the world (table 17.3), carrying roughly ten times the volume of the Mississippi. Several Amazonian tributaries such as the Maderia, Rio Negro, and Ucayali would be among the world's top rivers in their own right.

Lakes, Ponds, and Wetlands

Ponds are generally considered to be small temporary or permanent bodies of water shallow enough for rooted plants to grow over most of the bottom. Lakes are inland depressions that hold standing fresh water year-round. Maximum lake depths range from a few meters to over 1,600 m (1 mi) in Lake Baikal in Siberia. Surface areas vary in size from less than one-half hectare (one acre) to large inland seas, such as Lake Superior or the Caspian Sea, covering hundreds of thousands of square kilometers. Both ponds and lakes are relatively temporary features on the landscape because they eventually fill with silt or are emptied by cutting of an outlet stream through the barrier that creates them.

TABLE 17.3	Major Rivers of the World	
RIVER	COUNTRIES IN RIVER BASIN	AVERAGE ANNUAL DISCHARGE AT MOUTH (CUBIC METERS/SECOND)*
Amazon	Brazil, Peru	175,000
Orinoco	Venezuela, Colombia	45,300
Congo	Congo	39,200
Yangtze	Tibet, China	28,000
Bramaputra	Tibet, India, Bangladesh	19,000
Mississippi	United States	18,400
Mekong	China, Laos, Burma, Thailand, Cambodia, Vietnam	18,300
Parana	Paraguay, Argentina	18,000
Yenisey	Russia	17,200
Lena	Russia	16,000
Irrawaddy	Burma	13,000
Ob	Russia	12,000
Ganges	Nepal, India, Bangladesh	11,600
Amur	China, Russia	11,000
St. Lawrence	Canada, United States	10,700
Mackenzie	Canada	9,600

*One cubic meter equals 264 gallons.
Source: *World Resources Institute.*

While lakes contain nearly 100 times as much water as all rivers and streams combined, they are still a minor component of total world water supply. Their water is much more accessible than groundwater or glaciers, however, and they are important in many ways for humans and other organisms.

Wetlands play a vital and often unappreciated role in the hydrologic cycle. Their lush plant growth stabilizes soil and holds back surface runoff, allowing time for infiltration into aquifers and producing even, year-long stream flow. In the United States, about 20 percent of the 1 billion ha of land area was once wetland. In the past 200 years, more than one-half of those wetlands have been drained, filled, or degraded. Agricultural drainage accounts for the bulk of the losses.

When wetlands are disturbed, their natural water-absorbing capacity is reduced and surface waters run off quickly, resulting in floods and erosion during the rainy season and dry, or nearly dry, stream beds the rest of the year. This has a disastrous effect on biological diversity and productivity, as well as on human affairs (chapter 13).

The Atmosphere

The atmosphere is among the smallest of the major water reservoirs of the earth in terms of water volume, containing less than 0.001 percent of the total water supply. It also has the most rapid turnover rate. An individual water molecule resides in the atmosphere for about ten days, on average. While water vapor makes up only a small amount (4 percent maximum at normal temperatures) of the total volume of the air, movement of water through the atmosphere provides the mechanism for distributing fresh water over the landmasses and replenishing terrestrial reservoirs.

WATER AVAILABILITY AND USE

Clean, fresh water is essential for nearly every human endeavor. Perhaps more than any other environmental factor, the availability of water determines the location and activities of humans on earth (fig. 17.10). **Renewable water supplies** are made up, in general, of surface runoff plus the infiltration into accessible freshwater aquifers. About two-thirds of the water carried in rivers and streams every year occurs in seasonal floods that are too large or violent to be stored or trapped effectively for human uses. Stable runoff is the dependable, renewable, year-round supply of surface water. Much of this occurs, however, in sparsely inhabited regions or where technology, finances, or other factors make it difficult to use it productively. Still, the readily accessible, renewable water supplies are very large, amounting to some 1,500 km³ (about 400,000 gal) per person per year worldwide.

Water-Rich and Water-Poor Countries

As you can see in figure 17.2, South America, West Central Africa, and South and Southeast Asia all have areas of very high rainfall. Brazil and the Democratic Republic of Congo, because they have high precipitation levels and large land areas, are among the most water-rich countries on earth. Canada and Russia, which are both very large, also have some of the highest total annual water supplies of any country. The highest per capita water supplies generally occur in countries with moist climates and low population densities (table 17.4). Iceland, for example, has about 160 million gallons per person per year. In contrast, Kuwait, where temperatures are extremely high and rain almost never falls, has less than 3,000 gallons per person per year from renewable natural sources (table 17.5). Almost all of Kuwait's water comes from imports and desalinized seawater. Egypt, in spite of the fact that the Nile River flows through it, has only about 11,000 gallons of water annually per capita, or about 15,000 times less than Iceland. Singapore is an unusual case. Although it has a humid climate, this small, island nation has very little land area for its 3.5 million people, and must depend on its neighbor, Malaysia, for almost all its water.

Periodic droughts create severe regional water shortages. Droughts are most common and often most severe in semiarid zones where moisture availability is the critical factor in determining plant and animal distribution. Undisturbed ecosystems often survive

FIGURE 17.10 Water has always been the key to survival. Who has access to this precious resource and who doesn't has long been a source of tension and conflict. © Ray Ellis/Photo Researchers, Inc.

TABLE 17.4	Water-Rich Countries
COUNTRY	RENEWABLE ANNUAL SUPPLY[1] (CUBIC METERS/CAPITA)[2]
Iceland	606,500
Surinam	452,000
Guyana	281,500
Papua New Guinea	174,000
Gabon	140,000
Solomon Islands	107,000
Canada	94,000
Norway	88,000
Panama	52,000
Brazil	31,000

[1]Renewable supplies include both surface runoff and groundwater replenishment.
[2]One cubic meter equals 264 gallons.
Source: World Resources, 1998–99.

TABLE 17.5 Water-Poor Countries

COUNTRY	RENEWABLE ANNUAL SUPPLY[1] (CUBIC METERS/CAPITA)[2]
Kuwait	11
Egypt	43
United Arab Emirates	63
Malta	85
Jordan	114
Saudi Arabia	119
Singapore	172
Moldavia	225
Israel	289
Oman	393

[1]Renewable supplies include both surface runoff and groundwater replenishment.
[2]One cubic meter equals 264 gallons.
Source: World Resources, 1998–99.

extended droughts with little damage, but introduction of domestic animals and agriculture disrupts native vegetation and undermines natural adaptations to low moisture levels.

Droughts are often cyclic, and land-use practices exacerbate their effects. In the United States, the cycle of drought seems to be about 30 years. There were severe dry years in the 1870s, 1900s, 1930s, 1950s, and 1970s. The worst of these in economic and social terms were the 1930s. Poor soil conservation practices and a series of dry years in the Great Plains combined to create the "dust bowl." Wind stripped topsoil from millions of hectares of land, and billowing dust clouds turned day into night. Thousands of families were forced to leave farms and migrate to cities. The El Niño, Southern Oscillation (ENSO) system plays an important role in droughts in North America and elsewhere. There now is a great worry that the greenhouse effect (see chapter 15) will bring about major climatic changes and make droughts both more frequent and more severe than in the past in some places.

Types of Water Use

Most water we use eventually returns to rivers and streams. Therefore, it is important to distinguish between withdrawal and consumption. **Withdrawal** is the total amount of water taken from a lake, river, or aquifer for any purpose. Much of this water is employed in nondestructive ways and is returned to circulation in a form that can be used again. **Consumption** is the fraction of withdrawn water that is lost in transmission, evaporation, absorption, chemical transformation, or otherwise made unavailable for other purposes as a result of human use. Note that much water that is withdrawn but not consumed may be **degraded**—polluted or heated so that it is unsuitable for other uses.

Many societies have always treated water as if there is an inexhaustible supply. It has been cheaper and more convenient for most people to dump all used water and get a new supply than to determine what is contaminated and what is not. The natural cleansing and renewing functions of the hydrologic cycle do replace the water we need if natural systems are not overloaded or damaged. Water is a renewable resource, but renewal takes time. The rate at which many of us are using water now may make it necessary to conscientiously protect, conserve, and replenish our water supply.

Quantities of Water Used

Human water use has been increasing about twice as fast as population growth over the past century (fig. 17.11). Water use is stabilizing in industrialized countries, but demand will increase in developing countries where supplies are available. The average amount of water withdrawn worldwide is about 646 m³ (170,544 gal) per person per year. This overall average hides great discrepancies in the proportion of annual runoff withdrawn in different areas. As you might expect, those countries with a plentiful water supply and a small population withdraw a very small percentage of the water available to them. Canada, Brazil, and the Congo, for instance, withdraw less than 1 percent of their annual renewable supply.

By contrast, in countries such as Libya and Israel, where water is one of the most crucial environmental resources, groundwater and surface water withdrawal together amount to more than

FIGURE 17.11 Growth of water withdrawal and consumption, by sector, with projected levels to 2025. *Source:* UNEP, 2002.

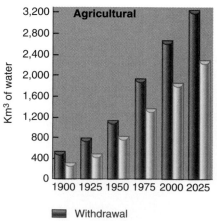
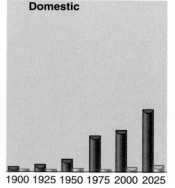
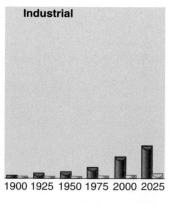

100 percent of their renewable supply. They are essentially "mining" water—extracting groundwater faster than it is being replenished. Obviously, this is not sustainable in the long run.

The total annual renewable water supply in the United States amounts to an average of about 9,000 m³ (nearly 2.4 million gal) per person per year. We now withdraw about one-fifth of that amount, or some 5,000 l (1,300 gal) per person per day, including industrial and agricultural water. By comparison, the average water use in Haiti is less than 30 l (8 gal) per person per day.

Agricultural Water Use

We can divide water use into three major sectors: agricultural, domestic, and industrial. Of these, agriculture accounts for by far the greatest use and consumption. Worldwide, crop irrigation is responsible for two-thirds of water withdrawal and 85 percent of consumption. Evaporation and seepage from unlined irrigation canals are the principal consumptive water losses. Agricultural water use varies greatly, of course. Over 90 percent of water used in India is agricultural; in Kuwait, where water is especially precious, only 4 percent is used for crops. In the United States, which has both a large industrial sector and a highly urbanized population, about half of all water withdrawal, and about 80 percent of consumption, is agricultural. These withdrawals are unevenly distributed (fig. 17.12). Can you identify crops that might be produced in some of the dominant irrigation regions shown in fig. 17.12?

Irrigation can be extremely inefficient. Traditionally, the main method has been flood or furrow irrigation, in which water floods a field (fig. 17.13). As much as half of this water can be lost directly through evaporation. Much of the rest runs off before it is used by plants. In arid lands, flood irrigation is needed to help remove toxic salts from soil, but these salts contaminate streams, lakes, and wetlands downstream. Repeated flood irrigation also waterlogs the soil, reducing crop growth. Sprinkler systems are another notoriously inefficient form of irrigation (fig. 17.14). Water spraying high in the air quickly evaporates, rather than watering crops. In recent years, growing pressure on water resources has led to more efficient sprinkler systems that hang low over crops to reduce evaporation (see fig. 9.21).

Drip irrigation (fig. 17.15) is a promising technology for reducing irrigation water use. These systems release carefully regulated amounts of water just above plant roots, so that nearly all water is used by plants. Only about 1 percent of the world's croplands currently use these systems, however.

Irrigation infrastructure, such as dams, canals, pumps, and reservoirs, is expensive. Irrigation is also the economic foundation of many regions. In the United States, the federal government

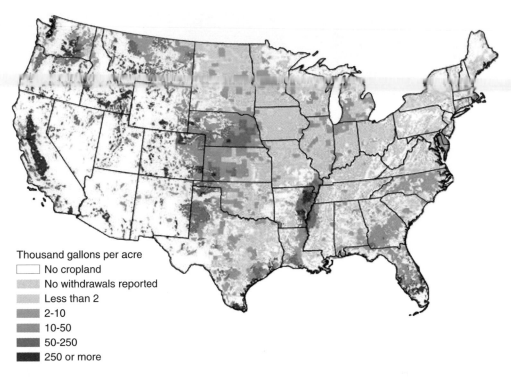

Thousand gallons per acre
- No cropland
- No withdrawals reported
- Less than 2
- 2-10
- 10-50
- 50-250
- 250 or more

FIGURE 17.12 Distribution of groundwater withdrawals for irrigation. *Source:* USDA.

FIGURE 17.13 Furrow irrigation uses extravagant amounts of water, but it also washes toxic minerals and salts from arid-land soils. *Source:* U.S. Department of Agriculture, NRCS.

has taken responsibility for providing irrigation for nearly a century. The argument for doing so is that irrigated agriculture is a public good that cannot be provided by individual farmers. A consequence of this policy has frequently been heavily subsidized crops whose costs, in water and in dollars, far outweigh their value (Case Study, p. 365).

Many resource analysts argue that state-subsidized water encourages and rewards wasteful water use. One recent dispute has

FIGURE 17.14 Rolling sprinklers allow farmers to irrigate crops on uneven terrain. In some areas, irrigation consumes 90 percent of all available water. © Corbis/Volume 102.

FIGURE 17.15 Drip irrigation delivers measured amounts of water exactly where the plants need and can use it. This technique can save up to 90 percent of irrigation water usage and reduces salt buildup. It is more expensive to install and operate, however, than simple flood irrigation. © Sylvan H. Wittwer/Visuals Unlimited.

focused on Arkansas rice growers. After depleting available groundwater in just a few decades, farmers are now asking the federal government to spend $200 million, or about $300,000 per farm, to build pumps and aqueducts to deliver river water to their fields. Ironically, this water would maintain rice crops whose value is less than half the guaranteed price paid by federal price supports. Proponents of the project say that farming is the lifeblood of the region, and that family farms and communities would wither and disappear without public support. Opponents of the project say this amounts to rewarding farmers for overpumping the regional

aquifer, and that they should have taken steps to conserve water years ago. Lowered river levels could reduce fishing and hunting opportunities, as well as tourism revenue. Much of the debate is about precedents. Aquifers are being overpumped nationwide, and other farmers are watching this contest carefully. What do you think? Is this Arkansas project rewarding profligate water wasters, or is it protecting an essential economic activity?

Domestic and Industrial Water Use

Worldwide, domestic water use accounts for about one-fifth of water withdrawals. Because little of this water evaporates or seeps into the ground, consumptive water use is slight, about 10 percent on average. Where sewage treatment is unavailable, water can be badly degraded by urban uses, however. In wealthy countries, each person uses about 500 to 800 l per day (180,000 to 280,000 l per year), far more than in developing countries (30 to 150 l per day). In North America, the largest single user of domestic water is toilet flushing (fig. 17.16). Each person in the United States uses about 50,000 l (13,000 gal) of drinking-quality water annually to flush toilets. Bathing accounts for nearly a third of water use, followed by laundry and washing. In western cities such as Denver and Phoenix, lawn watering is also a major water user.

Urban and domestic water use have grown approximately in proportion with urban populations, about 50 percent between 1960 and 2000. Although individual water use seems slight on the scale of world water withdrawals, the cumulative effect of inefficient appliances, long showers, liberal lawn-watering, and other uses is enormous. California has established increasingly stringent standards for washing machines, toilets, and other appliances, in order to reduce urban water demands (see Case Study, p. 365). Many other cities and states are following this lead to reduce domestic water use.

Industry accounts for 20 percent of global freshwater withdrawals. Industrial use rates range from 70 percent in industrialized parts of Europe to less than 5 percent in countries with little industry. Power production, including hydropower, nuclear, and thermoelectric power, make up 50 to 70 percent of industrial uses, and industrial processes make up the remainder. As with domestic

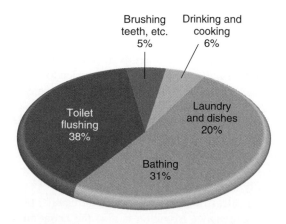

FIGURE 17.16 Typical household water use in the United States.

Water Wars on the Klamath

Until a century ago, the Klamath River was the third most productive salmon fishery south of the Canadian border. The 35 km (22 mi) long Upper Klamath Lake, the largest lake in the Pacific Northwest, once contained great schools of C'wam and Qapdo (Lost River and shortnose suckers). Millions of hectares of marshes, lakes, and dry steppe in the Klamath Basin teemed with waterfowl and wildlife. At least 80 percent of the birds following the Pacific Flyway stopped to feed in the area. Native American tribes, including the Klamath and Paiute in the upper basin, and the Yurok and Hupa downstream, depended on the abundant fish and wildlife for their survival.

In 1905 the newly formed United States Bureau of Reclamation was directed to "reclaim the sunbaked prairies and worthless swamps" in the upper Klamath Basin. Spending $50 million, the bureau built 7 dams, 45 pumping stations, and more than 1,600 km (1,000 mi) of canals and ditches. The project drained three-quarters of the wetlands in the upper basin, and provided irrigation water to 90,000 ha (220,000 acres) of cropland. Promises of cheap land and subsidized water lured some 1,400 farmers to the valley to grow potatoes, alfalfa, sugar beets, mint, onions, and cattle. In the 1990s, irrigators in the Klamath Basin used almost 1 million acre-feet (325 billion gallons or 1.2 trillion liters) of surface water per year.

These water diversions left salmon stranded in dried-up streams, while oxygen-depleted lake water was contaminated with agricultural runoff and clogged with algae. Downstream, Native American tribes and commercial fishermen, who once had brought in about 500,000 kg (roughly a million pounds) of salmon per year, saw their catches decline by as much as 90 percent. More than 7,000 jobs were lost when the fisheries collapsed. In 1997, the C'wam and Qapdo were declared endangered, and Coho salmon were listed as threatened. A coalition of commercial and sports fishermen, environmentalists, and native people sued the government for damaging fish and wildlife resources. A federal judge agreed, and ordered the Bureau

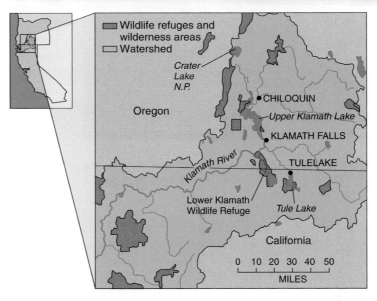

The Klamath River rises in the Cascade Mountains and flows about 400 km (250 mi) across Oregon and Washington to the Pacific Ocean.

of Reclamation to reduce irrigation flow and to maintain minimum water levels in rivers and lakes.

A severe drought in 2001 precipitated a crisis. For the first time in its history, the bureau closed the gates on its irrigation canals and cut off water to the farmers. Outraged at being denied the water they had come to believe was their inalienable right, angry farmers broke open headgate locks and let water flow to their drying fields. News media flocked to the site to film the confrontation. Conservative politicians condemned government actions and pledged to uphold property rights and to eliminate the Endangered Species Act. Progressives supported fishing and tribal rights. And environmentalists called for more water for the Lower Klamath Wildlife Refuge, the first in the United States established to protect migratory waterfowl, and home to the largest population of wintering bald eagles in the lower 48 states.

Sympathizing with the farmers, Interior Secretary Gale Norton went to Upper Klamath Lake and personally opened the gates to give them more water. Favoring a market-

based approach, she suggested a "water bank," in which farmers could voluntarily sell surplus water to Indian tribes or the Fish and Wildlife Service. The tribes and others objected that, while this approach might work when water was abundant, there probably wouldn't be any surplus in dry years.

Conservationists argue that two-thirds of Klamath County's irrigated acreage is watered with inefficient flood irrigation in which about half the water is lost to evaporation. Farmers could save water by installing efficient drip or sprinkler systems, or they could abandon water-demanding crops such as alfalfa and potatoes and convert to dryland agriculture. Farmers demanded compensation for the "taking" of water they considered promised to them in perpetuity.

This case illustrates some of the complexity and importance of water resources. With growing demands on dwindling supplies, it becomes increasingly difficult to satisfy all the needs we have for this precious commodity. If you were Secretary of the Interior, how would you weigh the competing claims for water in the Klamath Basin?

FIGURE 17.17 Hoover Dam powers Las Vegas, Nevada. Lake Mead, behind the dam, loses about 1.3 billion m³ per year to evaporation. Courtesy USDA/NRCS, photo by Lynn Betts.

water, little of this water is made unavailable after use, but it is often degraded by defouling agents, chlorine, or heat when it is released to the environment. The greatest industrial producer of degraded water is mining. Ores must be washed and treated with chemicals such as mercury and cyanide (chapter 14). As much as 80 percent of water used in mining and processing is released with only minimal treatment. In developed countries, industries have greatly improved their performance in recent decades, however. Water withdrawal and consumption have both fallen relative to industrial production.

Most water used for hydroelectricity and for irrigation is stored in reservoirs. Evaporation from large reservoirs in arid regions is now the second greatest consumptive "use" of water. In 2000, evaporation from the world's reservoirs exceeded domestic and industrial water consumption by more than 30 percent. Seepage into porous bedrock also contributes to water losses in reservoirs. In the United States, Lake Powell, on the Colorado River, loses more than 1 billion m³ per year to evaporation and seepage. Lake Mead, just downstream, loses more than 1.3 billion m³ per year to evaporation (fig. 17.17). Egypt's Lake Nassar, on the Nile, loses about 15 percent of its water to evaporation each year. Large dams are generally considered water reclamation and conservation projects, but their net effect can be an extravagant loss of water in thirsty regions.

FRESHWATER SHORTAGES

In 1977, the United Nations Water Conference declared that all people, regardless of their social or economic conditions, have a right to the clean drinking water and basic sanitation needed to prevent communicable diseases and provide for basic human dignity. A quarter century later, however, an estimated 1.5 billion people lack access to adequate quantity and quality of drinking water, while nearly 3 billion, or half the world's population, do not have acceptable sanitation. As populations grow, more people move into

cities, and agriculture and industry compete for increasingly scarce water supplies, water shortages are expected to become even worse in the future. A country in which consumption exceeds more than 20 percent of the available, renewable supply is considered vulnerable to **water stress.**

Globally, water supplies are abundant but unevenly distributed, and the capital needed to collect, store, purify, and distribute water is unavailable in many developing countries. Worldwide, water consumption has increased sixfold over the past century, or about twice as fast as population growth. With easily accessible water already exploited in most places, the World Bank estimates that the financial and environmental costs of tapping new supplies will be two to three times more expensive than current water projects. If present trends continue, the UN cautions, some two-thirds of the world's population will live in countries experiencing water shortages by 2025.

A Precious Resource

The World Health Organization considers 1,000 m³ (264,000 gal) of water per person per year to be the minimum level below which most countries are likely to experience chronic shortages on a scale that will impede development and harm human health. Currently, some 45 countries—most of them in Africa or the Middle East—are considered to have serious water stress, and cannot meet the minimum essential water needs of all their citizens. In some countries the problem is not so much total water supply as access to clean water. In Mali, for example, 88 percent of the population lacks clean water; in Ethiopia it is 94 percent. Rural people often have less access to clean water than do city dwellers. In the 33 worst affected countries, 60 percent of urban people can get clean water as opposed to only 20 percent of those in the country.

More than two-thirds of the world's households have to fetch water from outside the home (fig. 17.18). For many people, this is time-consuming and heavy work; it may take up to 2 hours per day to gather a meager supply. Women and children do most of this work. Improved public systems bring many benefits to poor families. In Mozambique, for example, the World Bank reports that the average time women spent carrying water decreased from 2 hours per day to only 25 minutes when village wells were installed.

Key Concepts

- Water withdrawal is all water used. Consumption is water made unavailable for later use. Degradation is contamination that reduces water's usefulness.

- Water is unevenly divided: people in rainy, thinly populated, or wealthy regions can afford to use up to 5,000 l per day; those in dry, impoverished, or densely populated countries may use less than 100 l per day.

- Water shortages are increasingly urgent as populations, cities, industries, and agriculture expand.

- Water conservation is key to reducing water stress.

The time saved could be used to garden, tend livestock, trade in the market, care for children, or even rest!

The reasons for water shortages are many. In some cases, deficits are caused by natural forces: the rains fail; hot winds dry up reservoirs that normally would carry people through the dry season; rivers change their courses, leaving villages stranded. In other cases, shortages are human in origin: too many people compete for the resource; urbanization, overgrazing, and inappropriate agricultural practices allow water to run off before it can be captured; a lack of adequate sewage systems causes contamination of local supplies. Without money for wells, storage reservoirs, delivery pipes, and other infrastructure, people can't use the resources available to them.

Safe, clean water is available in most countries—for those who can pay the price. Water from vendors often is the only source for many in the crowded slums and shanty towns around the major cities of the developing world. Although the quality is often questionable, this water generally costs about ten times more than a piped city supply. Naturally, sanitation levels decline when water is so expensive. A typical poor family in Lima, Peru, for instance, uses one-sixth as much water as a middle-class American family but pays three times as much for it. Following government recommendations that all water be boiled to prevent cholera would take up to one-third of the total income for such a poor family.

Investments in rural development have brought significant improvements over the past decade. Since 1990, nearly 800 million people—about 13 percent of the world population—have gained access to clean water (see "Rural Water Programs in Malawi" at the Online Learning Center). The percentage of rural families with safe drinking water has risen from less than 10 percent to nearly 75 percent. Still, many people suffer from a lack of this essential resource.

Depleting Groundwater

Groundwater is the source of nearly 40 percent of the fresh water for agricultural and domestic use in the United States. Nearly half of all Americans and about 95 percent of the rural population depend on groundwater for drinking and other domestic purposes. Overuse of these supplies causes several kinds of problems, including drying of wells, natural springs, and disappearance of surface water features such as wetlands, rivers, and lakes.

In many areas of the United States, groundwater is being withdrawn from aquifers faster than natural recharge can replace it. On a local level, this causes a cone of depression in the water table, as is shown in figure 17.19. A heavily pumped well can lower the local water table so that shallower wells go dry. On a broader scale, heavy pumping can deplete a whole aquifer. The Ogallala Aquifer, for example, underlies eight states in the arid high plains between Texas and North Dakota. As deep as 400 m (1,200 ft) in its center, this porous bed of sand, gravel, and sandstone once held more water than all the freshwater lakes, streams, and rivers on earth.

FIGURE 17.18 Drawing water from a village well is a daily chore in many developing countries. © The McGraw-Hill Companies, Inc./Barry W. Barker, photographer.

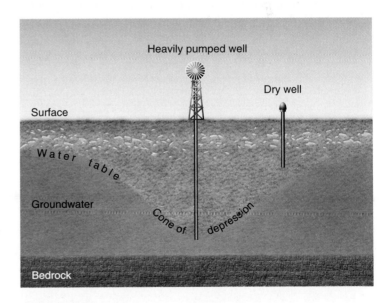

FIGURE 17.19 A cone of depression forms in the water table under a heavily pumped well. This may dry up nearby shallow wells or make pumping so expensive that it becomes impractical.

Excessive pumping for irrigation and other uses has removed so much water that wells have dried up in many places, and farms, ranches, even whole towns are being abandoned.

Many aquifers have slow recharge rates, so it will take thousands of years to refill them once they are emptied. Much of the groundwater we now are using probably was left there by the glaciers thousands of years ago. It is fossil water, in a sense. When we pump water out of a reservoir such as the Ogallala that cannot be refilled in our lifetime, we essentially are mining a nonrenewable resource. Covering aquifer recharge zones with urban development or diverting runoff that once replenished reservoirs ensures that they will not refill.

Withdrawal of large amounts of groundwater causes porous formations to collapse, resulting in **subsidence** or settling of the surface above. The U.S. Geological Survey estimates that the San Joaquin Valley in California has sunk more than 10 m in the last 50 years because of excessive groundwater pumping. Around the world, many cities are experiencing subsidence. Many are coastal cities, built on river deltas or other unconsolidated sediments. Flooding is frequently a problem as these coastal areas sink below sea level. Some inland areas also are affected by severe subsidence. Mexico City is one of the worst examples. Built on an old lake bed, it has probably been sinking since Aztec times. In recent years, rapid population growth and urbanization (chapter 22) have caused groundwater overdrafts. Some areas of the city have sunk as much as 8.5 m (25.5 ft). The Shrine of Guadalupe, the cathedral, and many other historic monuments are sinking at odd and perilous angles.

Sinkholes form when the roof of an underground channel or cavern collapses, creating a large surface crater (fig. 17.20). Drawing water from caverns and aquifers accelerates the process of collapse. Sinkholes can form suddenly, dropping cars, houses, and trees without warning into a gaping crater hundreds of meters across. Subsidence and sinkhole formation generally represent permanent loss of an aquifer. When caverns collapse or the pores between rock particles are crushed as water is removed, it is usually impossible to restore their water-holding capacity.

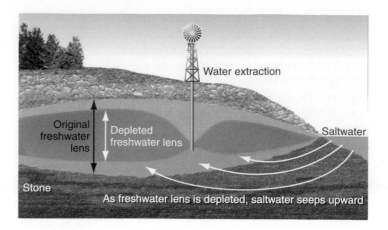

FIGURE 17.21 Saltwater intrusion into a coastal aquifer as the result of groundwater depletion. Many coastal regions of the United States are losing freshwater sources due to saltwater intrusion.

A widespread consequence of aquifer depletion is **saltwater intrusion.** Along coastlines and in areas where saltwater deposits are left from ancient oceans, overuse of freshwater reservoirs often allows saltwater to intrude into aquifers used for domestic and agricultural purposes (fig. 17.21).

INCREASING WATER SUPPLIES

Where do present and impending freshwater shortages leave us now? On a human time scale, the amount of water on the earth is fixed, for all practical purposes, and there is little we can do to make more water. There are, however, several ways to increase local supplies.

Seeding Clouds and Towing Icebergs

In the dry prairie states of the 1800s and early 1900s, desperate farmers paid self-proclaimed "rainmakers" in efforts to save their withering crops. Centuries earlier, Native Americans danced and prayed to rain gods. We still pursue ways to make rain. Seeding clouds with dry ice or potassium iodide particles has been tested for many years with mixed results. Recently, researchers have been having more success using hygroscopic salts that seem to significantly increase rainfall amounts. This technique is being tested in Mexico, South Africa, and the western United States. There is a concern, however, that rain induced to fall in one area decreases the precipitation somewhere else. Furthermore, there are worries about possible contamination from the salts used to seed clouds.

Another scheme that has been proposed for supplying fresh water to arid countries is to tow icebergs from the Arctic or Antarctic. Icebergs contain tremendous quantities of fresh water, but whether they can be towed without excessive melting to places where water is needed is not clear. Nor is it clear whether the energy costs to move an iceberg thousands of miles (if that were, in fact, possible) would be worthwhile.

FIGURE 17.20 Sinkhole in Winter Park, Florida. Notice the cars and truck sliding into the hole. © M. Timothy O'Keefe/Tom Stack & Associates.

Desalination

A technology that might have great potential for increasing freshwater supplies is **desalination** of ocean water or brackish saline lakes and lagoons. The most common methods of desalination are distillation (evaporation and recondensation) or reverse osmosis (forcing water under pressure through a semipermeable membrane whose tiny pores allow water to pass but exclude most salts and minerals). In 2000, the global capacity was about 23 million m³ per day, less than 0.1 percent of all freshwater withdrawals worldwide. Middle Eastern oil-rich states produce about 60 percent of desalinated water. Saudi Arabia is the largest single producer, at about 34 percent of world total. The United States is second, at 20 percent. Although desalination is still three to four times more expensive than most other sources of fresh water, it provides a welcome water supply in such places as Oman and Bahrain where there is no other access to fresh water. If a cheap, inexhaustible source of energy were available, however, the oceans could supply all the water we would ever need.

Dams, Reservoirs, Canals, and Aqueducts

People have been moving water around for thousands of years. Some of the great civilizations (Sumeria, Egypt, China, and the Inca culture of South America) were based on large-scale irrigation systems that brought river water to farm fields. In fact, some historians argue that organizing people to carry out large-scale water projects was the catalyst for the emergence of civilization. Roman aqueducts built 2,000 years ago are still in use. Those early water engineers probably never even dreamed of moving water on a scale that is being proposed and, in some cases, being accomplished now.

Environmental Costs

There has been much controversy about huge projects like the California Water Plan. Some of the water now being delivered to southern California is claimed by other parts of the state, neighboring states, and even Mexico. Environmentalists claim that this water transfer upsets natural balances of streams, lakes, estuaries, and terrestrial ecosystems. Fishing enthusiasts, whitewater boaters, and others who enjoy the scenic beauty of free-running rivers mourn the loss of rivers drowned in reservoirs or dried up by diversion projects. These projects also have been criticized for using public funds to increase the value of privately held farmland and for encouraging agricultural development and urban growth in arid lands where other uses might be more appropriate.

Ecosystem Losses

Mono Lake, a salty desert lake in California just east of Yosemite National Park, is an example of environmental consequences of interbasin water transfers and is an important legal symbol as well (fig. 17.22). Diversion of tributary rivers has shrunk the surface area of this lake by one-third, threatening millions of resident and migratory wading birds, ducks, and gulls that seek shelter and food there. After years of legal wrangling, the California Water Resources Con-

FIGURE 17.22   Mono Lake in eastern California. Diversion of tributary rivers to provide water for Los Angeles has shrunk the lake's surface area by one-third, threatening migratory bird flocks that feed here. These formations were created underwater where calcium-rich springs entered the brine-laden lake. © John Gerlach/Visuals Unlimited.

trol Board ruled in 1994 that Los Angeles must restrict diversions and allow the lake's surface to rise 17 ft to 6,391 ft above sea level. This is less than the 1941 level, but should improve conditions for birds and aquatic life.

About three-fourths of California's Trinity River is diverted to the arid Central Valley. This leaves barely enough water for a few remaining salmon. One of the final acts of the outgoing Clinton administration in 2000 was to propose reapportioning the water, but the current water users are unlikely to give up the precious resource easily. West Africa's Lake Chad, once the world's sixth-largest lake, used to support important fisheries. Water diversion for irrigation, coupled with years of drought, have reduced Lake Chad to less than one-tenth of its size four decades ago. Towns now stand on the old lake bed.

Canada has been the site of decades of protest by First Nations people over flooding of ancestral lands for hydroelectric projects. The James Bay project built by Hydro-Quebec has diverted three major rivers flowing west into Hudson Bay and created huge lakes that flooded more than 10,000 km² (4,000 mi²) of forest and tundra, to generate 26,000 megawatts of electrical power. In 1984, shortly after Phase I of this project was completed, 10,000 caribou drowned when trying to follow migration routes across the newly flooded land. The loss of traditional hunting and fishing sites has been culturally devastating for native Cree people. In addition, mercury leeched out of rocks in newly submerged lands has entered the food chain, and many residents show signs of mercury poisoning. In a similar, but less well-known case, Manitoba Hydro has diverted the Churchill River from the west shore of Hudson Bay so that it now flows into the Nelson River and then to Lake Winnipeg. More than 125,000 km² (50,000 mi²) has been inundated, much of it on permafrost, which is slowly melting and causing reservoir banks to expand continually. Here also, Cree people are being displaced against their will. As is the case in Quebec, most of the electricity generated by these dams is exported to the United States. In some places, however, dams are being removed (What Do You Think? p. 370).

What do you think?

Should We Remove Dams?

For nearly a century, the U.S. Army Corps of Engineers and the Bureau of Reclamation have constructed dams to store water and to generate electricity. Building these dams provided cheap electricity, created jobs for workers, stimulated regional economic development, and allowed farming on lands that would otherwise be too dry. But not everyone agrees that big dam projects are an unmitigated benefit. Their storage reservoirs drown free-flowing rivers and often submerge towns, farms, cemeteries, and historic sites. They block fish migrations and change aquatic habitats essential for endemic species.

The tide may be turning, however, against dam building. In 1998, the Army Corps announced that it would no longer be building large dams and diversion projects. In the few remaining sites where dams might be built, public opposition is so great that getting approval for projects is unlikely. Instead, the new focus may be on removing existing dams and restoring natural habitats. In 1999, Interior Secretary Bruce Babbitt said, "Of the 75,000 large dams in the United States, most were built a long time ago and are now obsolete, expensive, and unsafe. They were built with no consideration of the environmental costs. As operating licenses come up for renewal, removal and restoration to original stream flows will be one of the options to be considered."

The first active hydroelectric dam in the United States to be breached against the wishes of its owners was the 162-year-old Edwards Dam, on the Kennebec River in Augusta, Maine. For many years, the U.S. Fish and Wildlife Service has advocated removal of this dam, which prevented migra-

tion of salmon, shad, sturgeon, and other species up the river. In a precedent-setting decision, the Federal Energy Regulatory Commission ordered the dam removed after concluding that the environmental and economic benefits of a free-flowing river outweigh the electricity generated by the dam. In July, 1999, the dam was breached and restoration work began on wetlands and stream banks long underwater.

The next dams likely to be taken down are the Elwha and Glines Dams on the Elwha River in Olympic National Park in Washington. Built nearly a century ago to provide power to lumber and paper mills in the town of Port Angeles, these dams blocked access to upstream spawning beds for six species of salmon on what once was one of the most productive salmon rivers in the world. Where 50 kg (110 lb) King salmon once fought their way up thundering waterfalls, there now are only concrete walls holding back still water and deep beds of silty sediment. Simply breaching the dams won't restore the salmon, however. Removing the sediment, uncovering gravel spawning beds, and finding suitable breeding stock to rebuild the population is a daunting task. And it won't be cheap. Congress will have to appropriate somewhere around $300 million to $400 million to remove and rehabilitate these two relatively small dams.

Fishing advocates, encouraged by these examples, have begun to talk about much more ambitious projects. Four giant dams on the Snake River in Washington State, for example, might be removed to restore salmon and steelhead runs in the Columbia River. The Hetch Hetchy Dam in Yosemite National Park might be taken down to reveal what John Muir called a valley "just as beautiful and worthy of preservation as the

Protesters invoke the spirit of John Muir as they paint a symbolic crack on the Hetch Hetchy Dam in Yosemite National Park and call for restoration of the Tuolumne River. Courtesy Patagonia/David Cross, photographer.

majestic Yosemite." Some groups have even suggested breaching the Glen Canyon Dam on the Colorado. In each of these cases, powerful interests stand in opposition. These dams generate low-cost electricity and store water that is needed for agriculture and industry. Local economies, domestic water supplies, and certain types of recreation all would be severely impacted by destruction of these dams.

What do you think? How would you go about weighing the many different economic, cultural, and aesthetic considerations for removing or not removing these dams? Do certain interests, such as the rights of native people, or the continued existence of endemic species of fish or wildlife take precedence over economic factors; or should this be a utilitarian calculation of the greatest good for the greatest number? Does that number include only humans or do other species count as well? If you were Secretary of the Interior, how many dams would you remove?

Displacement of People

In China, huge dams are being built in the Three Gorges section of the Yangtze River. More than a million people are being forced to relocate as the huge reservoir (185 m, or 600 ft, deep and 600 km, or 375 mi, long) inundates some 150 cities and villages. Because the dam is being built on an active geological fault, there is a worry that a strong earthquake could rupture the dam and drown millions of people living downstream.

Many other huge dam and water diversion projects have generated controversy and protest around the world. In Turkey, construction of a network of 22 large dams and reservoirs for the Southeast Anatolia Project on the Tigris and Euphrates Rivers is regarded as a serious threat by Turkey's downstream neighbors, Syria and Iraq, which are now at the mercy of Turkey to allow the rivers to continue to flow. In India, the Sardar Sarovar Dam on the sacred Narmada River has been the focus of decades of protest.

Many of the 1 million villagers and tribal people being displaced by this project have engaged in mass resistance and civil disobedience when police try to remove them forcibly. Some have vowed to drown rather than leave their homes. In Nepal, construction of the 240 m (850 ft) high Tehri Dam on the Bhagirathi River has stirred fears that a strong earthquake in this active seismic region might cause the dam to collapse and cause a catastrophic flood downstream. The Tehri Dam is only one of 17 high dams that Nepal and India plan for the Himalayan mountains.

Evaporation, Leakage, and Siltation

The main problem with dams is inefficiency. Some dams built in the western United States lose so much water through evaporation and seepage into porous rock beds that they waste more water than they make available. The evaporative loss from Lake Mead and Lake Powell on the Colorado River is more than 2 billion m³ per year. The salts left behind by evaporation and agricultural runoff nearly double the salinity of the river and make its water unusable when it reaches Mexico. To compensate, the United States has built a $350 million desalination plant at Yuma, Arizona, to improve water quality.

As the turbulent Colorado River slows in the reservoirs created by Glen Canyon and Boulder Dams, it drops its load of suspended material. More than 10 million metric tons of silt per year collect behind these dams. Imagine a line of twenty thousand dump trucks backed up to Lake Mead and Lake Powell every day, dumping dirt into the water. Within as little as 100 years, these reservoirs could be full of silt and useless for either water storage or hydroelectric generation (fig. 17.23).

The accumulating sediments that clog reservoirs and make dams useless also represent a loss of valuable nutrients. The Aswan High Dam in Egypt was built to supply irrigation water to make agriculture more productive. Although thousands of hectares are being irrigated, the water available is only about half that anticipated because of evaporation in Lake Nasser behind the dam, and seepage losses in unlined canals which deliver the water. Controlling the annual floods of the Nile also has stopped the deposition of nutrient-rich silt on which farmers depended for fertilizing their fields. This silt is being replaced with commercial fertilizer costing more than $100 million each year. Furthermore, the nutrients carried by the river once supported a rich fishery in the Mediterranean that was a valuable food source for Egypt. After the dam was installed, sardine fishing declined 97 percent. To make matters worse, growth of snail populations in the shallow permanent canals that distribute water to fields has led to an epidemic of schistosomiasis, a debilitating disease caused by parasitic flatworms in irrigation canals. In some areas, 80 percent of the residents are infected (chapter 8).

Loss of Free-Flowing Rivers

Many water projects drown or drain free-flowing rivers, often turning them into linear, sterile irrigation canals. Conservation history records many battles between those who want to preserve wild rivers and those who would benefit from development (fig. 17.24).

One of the first and most divisive of these battles was over the flooding of the Hetch Hetchy Valley in Yosemite National Park. In the early 1900s, San Francisco wanted to dam the Tuolumne River to produce hydroelectric power and provide water for the city water system. This project was supported by many persons at San Francisco citizens because it represented an opportunity for both clean water and municipal power. Leader of the opposition was John Muir, founder of the Sierra Club and protector of Yosemite Park. Muir said that Hetch Hetchy Valley rivaled Yosemite itself in beauty and grandeur and should be protected. After a prolonged and bitter fight, the developers won and the dam was built. Hard feelings from this controversy persisted for many years. Similar impoundment schemes for the remaining free stretches of the Tuolumne River are being proposed.

FIGURE 17.23 This dam is now useless because its reservoir has filled with silt and sediment. Courtesy Tim McCabe, Soil Conservation Service, USDA.

FIGURE 17.24 The recreational and aesthetic values of free-flowing wild rivers and wilderness lakes may be their greatest assets. Competition between *in situ* values and extractive uses can lead to bitter fights and difficult decisions. © Josh Baker/Colorado Whitewater Photography.

WATER MANAGEMENT AND CONSERVATION

Watershed management and conservation are often more economical and environmentally sound ways to prevent flood damage and store water for future use than building huge dams and reservoirs.

Watershed Management

A **watershed,** or catchment, is all the land drained by a stream or river. It has long been recognized that retaining vegetation and ground cover in a watershed helps hold back rainwater and lessens downstream floods. In 1998, Chinese officials acknowledged that unregulated timber cutting upstream on the Yangtze contributed to massive floods that killed 30,000 people.

Similarly, after disastrous floods in the upper Mississippi Valley in 1993, it was suggested that, rather than allowing residential, commercial, or industrial development on floodplains, these areas should be reserved for water storage, aquifer recharge, wildlife habitat, and agriculture (chapter 13). Sound farming and forestry practices can reduce runoff. Retaining crop residue on fields reduces flooding, and minimizing plowing and forest cutting on steep slopes protects watersheds. Wetlands conservation preserves natural water storage capacity and aquifer recharge zones. A river fed by marshes and wet meadows tends to run consistently clear and steady rather than in violent floods.

A series of small dams on tributary streams can hold back water before it becomes a great flood. Ponds formed by these dams provide useful wildlife habitat and stock-watering facilities. They also catch soil where it could be returned to the fields. Small dams can be built with simple equipment and local labor, eliminating the need for massive construction projects and huge dams.

In 1998, the U.S. Forest Service chief, Mike Dombeck, announced a major shift in his agency's priorities. "Water," he said, "is the most valuable and least appreciated resource the National Forests provide. More than 60 million people in 33 states obtain their drinking water from national forest lands. Protecting watersheds is far more economically important than logging or mining, and will be given the highest priority in forest planning."

Domestic Conservation

We could probably save as much as half of the water we now use for domestic purposes without great sacrifice or serious changes in our lifestyles. Simple steps, such as taking shorter showers, stopping leaks, and washing cars, dishes, and clothes as efficiently as possible, can go a long way toward forestalling the water shortages that many authorities predict. Isn't it better to adapt to more conservative uses now when we have a choice than to be forced to do it by scarcity in the future?

The use of conserving appliances, such as low-volume shower heads and efficient dishwashers and washing machines, can reduce water consumption greatly (What Can You Do? p. 374). If you live in an arid part of the country, you might consider whether you really need a lush green lawn that requires constant watering, feed-

FIGURE 17.25   By using native plants in a natural setting, residents of Phoenix save water and fit into the surrounding landscape. © William P. Cunningham.

ing, and care. Planting native ground cover in a "natural lawn" or developing a rock garden or landscape in harmony with the surrounding ecosystem can be both ecologically sound and aesthetically pleasing (fig. 17.25). There are about 30 million ha (75 million acres) of cultivated lawns, golf courses, and parks in the United States. They receive more water, fertilizer, and pesticides per hectare than any other kind of land.

Our largest domestic water use is toilet flushing (see fig. 17.16). There are now several types of waterless or low-volume toilets. The Swedish-made Clivus Multrum (fig. 17.26) digests both human and kitchen wastes by aerobic bacterial action, producing a rich, nonoffensive compost that can be used as garden fertilizer. There are also low-volume toilets that use recirculating oil or aqueous chemicals to carry wastes to a holding tank, from which they are periodically taken to a treatment plant. Anaerobic digesters use bacterial or chemical processes to produce usable methane gas from domestic wastes. These systems provide valuable energy and save water but are more difficult to operate than conventional toilets. Few cities are ready to mandate waterless toilets, but in 1988 a number of cities (including Los Angeles, California; Orlando, Florida; Austin, Texas; and Phoenix, Arizona) ordered that water-saving toilets, showers, and faucets be installed in all new buildings. The motivation was twofold: to relieve overburdened sewer systems and to conserve water.

Significant amounts of water can be reclaimed and recycled. In California, water recovered from treated sewage constitutes the fastest growing water supply, growing about 30 percent per year. Despite public squeamishness, purified sewage effluent is being used for everything from agricultural irrigation to flushing toilets (fig. 17.27). In a statewide first, San Diego is currently moving toward piping water from the local sewage plant directly into a drinking-water reservoir. Already, California uses more than 555 million m³ (450,000 acre-feet) of recycled water annually. That's equivalent to about two-thirds of the water consumed by Los Angeles every year. Other places with growing populations and limited water supplies may soon find that it pays to follow California's example.

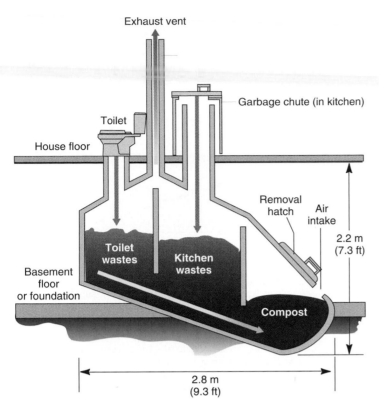

FIGURE 17.26 The Clivus Multrum waterless toilet. Wastes decompose and compost into an odorless, safe, rich fertilizer as they slowly move down to the bottom compartment.

Recycling and Water Conservation

In many developing countries as much as 70 percent of all the agricultural water used is lost to leaks in irrigation canals, application to areas where plants don't grow, runoff, and evaporation. Better farming techniques, such as leaving crop residue on fields and ground cover on drainage ways, intercropping, use of mulches, and low-volume irrigation, could reduce these water losses dramatically.

Nearly half of all industrial water use is for cooling of electric power plants and other industrial facilities. Some of this water use could be avoided by installing dry cooling systems similar to the radiator of your car. In many cases, cooling water could be reused for irrigation or other purposes in which water does not have to be drinking quality. The waste heat carried by this water could be a valuable resource if techniques were developed for using it.

Price Mechanisms and Water Policy

Through most of U.S. history, water policies have generally worked against conservation. In the well-watered eastern United States, water policy was based on riparian usufructuary (use) rights—those who lived along a river bank had the right to use as much water as they liked as long as they didn't interfere with its quality or availability to neighbors downstream. It was assumed that the supply would always be endless and that water had no value until it was used. In the drier western regions where water often is a limiting resource, water law is based primarily on the Spanish system of

FIGURE 17.27 Recycled water is being used in California and Arizona for everything from agriculture, to landscaping, to industry. Some cities even use treated sewage effluent for human drinking-water supplies. © William P. Cunningham.

prior appropriation rights, or "first in time are first in right." Even if the prior appropriators are downstream, they can legally block upstream users from taking or using water flowing over their property. But the appropriated water had to be put to "beneficial" use by being consumed. This creates a policy of "use it or lose it." Water left in a stream, even if essential for recreation, aesthetic enjoyment, or to sustain ecological communities, is not being appropriated or put to "beneficial" (that is, economic) use. Under this system, water rights can be bought and sold, but water owners frequently are reluctant to conserve water for fear of losing their rights.

In most federal "reclamation" projects, customers were charged only for the immediate costs of water delivery. The costs of building dams and distribution systems was subsidized, and the potential value of competing uses was routinely ignored. Farmers in California's Central Valley, for instance, for many years paid only about one-tenth of what it cost the government to supply water to them. This didn't encourage conservation. Subsidies created by underpriced water amounted to as much as $500,000 per farm per year in some areas. Many cities also have long had unreasonably low prices for water. In New York City, for example, water was supplied to homes and businesses for many decades at a flat rate. There were no meters because it was considered more expensive to install and read them than the water was worth. With no incentive to restrict water use or repair leaks, it is estimated that 750,000 m³ (200 million gal) of water was wasted each year from leaky faucets, toilets, and water pipes. The drought of 1988 convinced the city to begin a ten-year, $290 million program to install meters and reduce waste.

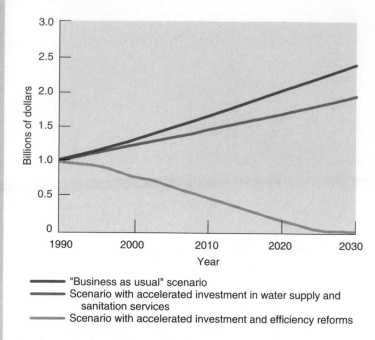

What can you do?

Saving Water and Preventing Pollution

Each of us can conserve much of the water we use and avoid water pollution in many simple ways.

- Don't flush every time you use the toilet. Take shorter showers; don't wash your car so often.

- Don't let the faucet run while washing hands, dishes, food, or brushing your teeth. Draw a basin of water for washing and another for rinsing dishes. Don't run the dishwasher when half full.

- Dispose of used motor oil, household hazardous waste, batteries, etc., responsibly. Don't dump anything down a storm sewer that you wouldn't want to drink.

- Avoid using toxic or hazardous chemicals for simple cleaning or plumbing jobs. A plunger or plumber's snake will often unclog a drain just as well as caustic acids or lye. Hot water and soap will clean brushes more safely than organic solvents.

- If you have a lawn, use water sparingly. Water your grass and garden at night, not in the middle of the day. Consider planting native plants, low-maintenance ground cover, a rock garden, or some other xeriphytic landscaping.

- Use water-conserving appliances: low-flow showers, low-flush toilets, and aerated faucets.

- Use recycled (gray) water for lawns, house plants, car washing.

- Check your toilet for leaks. A leaky toilet can waste 50 gallons per day. Add a few drops of dark food coloring to the tank and wait 15 minutes. If the tank is leaking, the water in the bowl will change color.

FIGURE 17.28 Three scenarios for government investments on clean water and sanitation services, 1990 to 2030.

Source: World Bank estimates based on research paper by Dennis Anderson and William Cavendish, "Efficiency and Substitution in Pollution Abatement: Simulation Studies in Three Sectors."

Growing recognition that water is a precious and finite resource has changed policies and encouraged conservation across the United States. Despite a growing population, the United States is now saving some 144 million l (38 million gal) per day—or enough water to fill Lake Erie in a decade—compared to per capita consumption rates of 20 years ago. With 37 million more people in the United Sates now than there were in 1980, we get by with 10 percent less water. New requirements for water-efficient fixtures and low-flush toilets in many cities help to conserve water on the home front. More efficient irrigation methods on farms also are a major reason for the downward trend. Instead of old systems that sprayed water up into the air, where it could evaporate or be blown away by the wind before reaching crops, new systems have small spray heads just a foot or so above the plant tops, and apply water much more directly. Even better is drip irrigation (see fig. 17.15), which applies water directly to plant roots. California and Florida farmers currently water about 500,000 ha (1.3 million acres) with this technique. Unfortunately, these efficient systems are too expensive for farms in most countries and are used on less than 1 percent of all farmland worldwide.

Charging a higher proportion of real costs to users of public water projects has helped encourage conservation, and so have water marketing policies that allow prospective users to bid on water rights. Both the United States and Australia have had effective water pricing and allocation policies that encourage the most socially beneficial uses and discourage wasteful water uses. Market mechanisms for water allotment can be sensitive, however, in developing countries where farmers and low-income urban residents could be outbid for irreplaceable water supplies.

Even in countries where water is abundant, many countries are reluctant to begin trading water as a commodity. Canada, for instance, which has 20 percent of all the world's known freshwater resources, believes that it has none to spare. For years, Canada has resisted schemes to sell water to the United States. In 1999, the Canadian government passed federal legislation banning any bulk water export. At issue was a plan to ship supertankers of water from Grand Le Pierre on the southern coast of Newfoundland to thirsty Asian cities. Claiming that water is "too precious" to be treated like other commodities, the federal government is also urging provinces and territories to institute similar bans. Whether these policies will hold up under NAFTA and WTO trade policies remains to be seen.

It will be important, as water markets develop, to be sure that environmental, recreational, and wildlife values are not sacrificed to the lure of high-bidding industrial and domestic uses. Given prices based on real costs of using water and reasonable investments in public water supplies, pollution control, and sanitation, the World Bank estimates that everyone in the world could have an adequate supply of clean water by the year 2030 (fig. 17.28). We will discuss the causes, effects, and solutions for water pollution in chapter 18.

Summary

- Water shortages and water stress affect at least a third of the world's population. Water resources are expected to be a major source of regional and international conflict in coming decades.

- Distribution of water around the globe depends mainly on climate factors, including high-pressure zones and prevailing winds, and topography. Human activities, such as deforestation, also affect regional water supplies.

- The hydrologic cycle is the movement of water between the ocean, atmosphere, land, and living things. All water on land originates as evaporation, mostly from oceans, which produces rain and snow. Major water compartments involved in the hydrologic cycle include oceans; glaciers, ice, and snow; groundwater; lakes, rivers, and wetlands; and the atmosphere. Residence time in compartments ranges from thousands of years in the ocean to minutes or days in the atmosphere.

- Aquifers are porous rock formations that hold water. Water enters aquifers through recharge zones. A confined aquifer is one whose saturated layers are capped (overlain) by impermeable rock layers.

- Pumping water from an aquifer produces a cone of depression, often drying out shallower wells. Risks of overpumping aquifers—extracting water faster than it is recharged—include subsiding ground levels, sinkholes, and saltwater intrusion, as well as depletion of water supplies.

- Water withdrawal refers to all water taken for use. Consumption refers to water lost to direct use, usually through evaporation or seepage into the ground. Degradation is water contamination by pollutants, salts, or heat, which reduces its utility for later uses.

- Worldwide, two-thirds of water withdrawn is used for agriculture. Agriculture accounts for 85 percent of water consumption, mainly through evaporative losses and seepage from unlined canals. Improved irrigation methods are beginning to reduce some consumptive losses.

- Industrial and domestic water uses are increasing, although not as fast as agricultural use. Consumption and degradation have fallen somewhat due to conservation and increased efficiency in households and industries.

- Water stress occurs when consumption exceeds 20 percent of available, renewable water supplies. At least 45 countries have serious water stress and cannot supply minimum essential water needs for citizens.

- Many strategies have been attempted to increase available water supplies. Desalination and water diversions are the principal methods that increase supplies in arid regions. Dams and reservoirs are controversial. They provide essential power and irrigation, but they also have great environmental, economic, and social costs, including ecosystem losses, displaced human populations, and water loss through evaporation.

- Water conservation is often the cheapest and most effective way to increase water supplies. Watershed management involves coordinated planning to improve resource allocation and reduce water loss. Efficient household appliances, such as toilets, shower heads, and laundry machines, have greatly reduced per capita consumption in many cities. Dry landscaping (xeriscaping) is required in some southwestern communities of the United States. Drip irrigation and other agricultural practices can reduce irrigation demands. Price mechanisms can provide incentives for more efficient water use. Water policy—laws regarding water rights and use—is a key factor in conservation.

Questions for Review

1. What is the difference between withdrawal, consumption, and degradation of water?

2. Describe the changes in water withdrawal and consumption by sector shown in figure 17.11.

3. Figure 17.12 shows water use for irrigation in the United States. Which regions have the greatest rate of use?

4. Describe some problems associated with dam building and water diversion projects.

5. Describe the path a molecule of water might follow through the hydrologic cycle from the ocean to land and back again.

6. Where are the five largest rivers in the world (table 17.3)?

7. How do mountains affect rainfall distribution? Does this affect your part of the country?

8. Identify and explain three consequences of overpumping aquifers.

9. How much water is fresh (as opposed to saline) and where is it?

10. Define *aquifer*. How does water get into an aquifer?

Questions for Critical Thinking

1. What changes might occur in the hydrologic cycle if our climate were to warm or cool significantly?

2. Why does it take so long for the deep ocean waters to circulate through the hydrologic cycle? What happens to substances that contaminate deep ocean water or deep aquifers in the ground?

3. Where would you most like to spend your vacations? Does availability of water play a role in your choice? Why?

4. Why do we use so much water? Do we need all that we use?

5. Are there ways you could use less water in your own personal life? What obstacles prevent you from taking these steps?

6. Should we use up underground water supplies now or save them for some future time?

7. How much should the United States invest to provide clean water to people in less-developed countries?

8. How should we compare the values of free-flowing rivers and natural ecosystems with the benefits of flood control, water diversion projects, hydroelectric power, and dammed reservoirs?

9. Would it be feasible to change from flush toilets and using water as a medium for waste disposal to some other system? What might be the best way to accomplish this?

10. How does water differ from other natural liquids? How do the properties of water make the hydrologic cycle, and life, possible? (You may need to review chapter 2 to answer this question.)

Key Terms

aquifer 358
artesian 358
condensation 355
condensation nuclei 355
consumption 362
degraded 362

desalination 369
dew point 355
discharge 360
evaporation 355
groundwater 358
infiltration 358

rain shadow 356
recharge zones 358
relative humidity 355
renewable water supplies 361
residence time 357
saltwater intrusion 368
saturation point 355
sinkholes 368
sublimation 355

subsidence 368
transpiration 355
watershed 372
water stress 366
water table 358
withdrawal 362
zone of aeration 358
zone of saturation 358

Further Readings

Gleick, Peter H., et al. 2002. *World's Water 2002–2003: The Biennial Report on Freshwater Resources.* Island Press.

Glennon, Robert J. 2002. *Water Follies: Groundwater Pumping and the Fate of America's Fresh Waters.* Island Press.

Gollehon, Noel, and William Quinby. 2000. Irrigation in the American West: Area, water and economic activity. *Water Resources Development* (special issue) 16:2.

Holland, Marjorie M., Elizabeth R. Blood, and Lawrence R. Shaffer, eds. 2003. *Achieving Sustainable Freshwater Systems.* Island Press.

Reisner, Marc. 1993. *Cadillac Desert: The American West and Its Disappearing Water.* Penguin Books.

United Nations Educational, Scientific, and Cultural Organization (UNESCO). 2003. *World Water Development Report—Water for People, Water for Life.* United Nations.

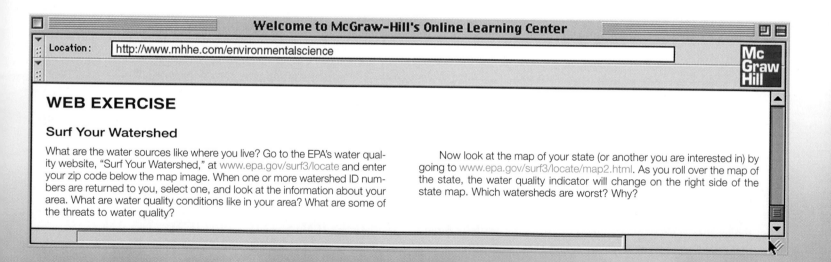

Welcome to McGraw-Hill's Online Learning Center

Location: http://www.mhhe.com/environmentalscience

WEB EXERCISE

Surf Your Watershed

What are the water sources like where you live? Go to the EPA's water quality website, "Surf Your Watershed," at www.epa.gov/surf3/locate and enter your zip code below the map image. When one or more watershed ID numbers are returned to you, select one, and look at the information about your area. What are water quality conditions like in your area? What are some of the threats to water quality?

Now look at the map of your state (or another you are interested in) by going to www.epa.gov/surf3/locate/map2.html. As you roll over the map of the state, the water quality indicator will change on the right side of the state map. Which watersheds are worst? Why?

18

Water Pollution

Water, water, everywhere; nor any drop to drink.

Samuel Taylor Coleridge

OBJECTIVES

After studying this chapter, you should be able to:

- define *water pollution* and describe the sources and effects of some major types.
- appreciate why access to sewage treatment and clean water are important to people in developing countries.
- discuss the status of water quality in developed and developing countries.
- delve into groundwater problems and suggest ways to protect this precious resource.
- fathom the causes and consequences of ocean pollution.
- weigh the advantages and disadvantages of different human waste disposal techniques.
- judge the impact of water pollution legislation and differentiate between best available and best practical technology, and total maximum daily pollution loads.

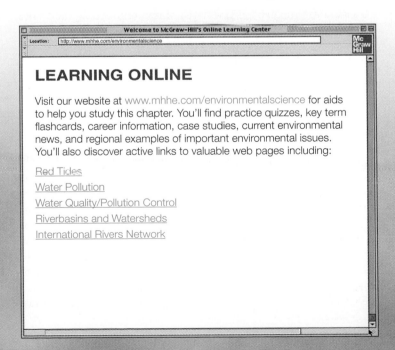

Welcome to McGraw-Hill's Online Learning Center

Location: http://www.mhhe.com/environmentalscience

LEARNING ONLINE

Visit our website at www.mhhe.com/environmentalscience for aids to help you study this chapter. You'll find practice quizzes, key term flashcards, career information, case studies, current environmental news, and regional examples of important environmental issues. You'll also discover active links to valuable web pages including:

Red Tides
Water Pollution
Water Quality/Pollution Control
Riverbasins and Watersheds
International Rivers Network

Photo: Students sample aquatic indicator organisms in a stream. © William P. Cunningham.

A Flood of Pigs

When Hurricane Floyd roared ashore at Cape Fear in the early morning hours of September 16, 1999, it brought much more than just strong winds to North Carolina. More than 50 cm (20 in) of rain poured down in less than 24 hours on ground already saturated from previous storms. The eastern part of North Carolina is a low, coastal plain drained by an intricate web of sluggish creeks and meandering rivers. The deluge filled these myriad channels until they spilled over and merged to create an enormous lake more than 300 km (nearly 200 mi) across, covering towns, farms, factories, and forests. The Great Dismal Swamp in the northeast corner of the state became truly soggy and dismal. Worst of all were the thousands of hog and poultry operations and their open manure lagoons submerged by the rising water.

In the 1990s, North Carolina became the leading turkey-producing state in the United States, and second to Iowa in pork production, with more than 45 million turkeys and 10 million pigs (but only 7 million people), mostly raised in huge, factory operations. Manure is washed out of giant confinement barns and stored in huge ponds, each typically about the size of several football fields and holding around 40,000 m^3 (10 million gal) of liquid waste. The heart of corporate animal-farming is along the sluggish Neuse River as it winds across the flat coastal plain to its estuary on Pamlico Sound (fig. 18.1). Even now, no flood protection measures are required for large-scale animal operations. Environmentalists have warned for years that storage of such massive amounts of manure in a storm-prone floodplain is an accident waiting to happen. A 1996 editorial in the Raleigh *News and Observer* complained: "Are we so foolish as to think the winds won't blow, the rains won't fall, and the rivers won't rise again?"

We'll probably never know the entire damage from Hurricane Floyd. Fifty-one people died in eastern North Carolina, about 8,000 homes were damaged or destroyed, and a million ha (roughly 2.5 million acres) of crops were destroyed. Property losses were several billion dollars. Environmental damage is harder to measure. Millions of dead pigs, chickens, turkeys, and cattle littered the landscape or washed out to sea. Flooded junkyards, factories, and chemical plants oozed a toxic mix of gasoline, oil, and other chemicals. Twenty municipal wastewater treatment plants were swamped. Environmentalists claim that at least 250 overflowing manure ponds disgorged around 10 million m^3 (roughly 2.5 billion gal) of hog and poultry waste into the flood. The Carolina Pork Council, on the other hand, says that only three of 4,000 waste lagoons overflowed. The fecal coliform bacteria now found in wells, industry claims, probably came from humans

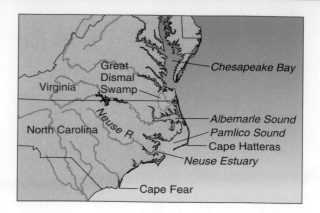

FIGURE 18.1   Heavy rains from Hurricane Floyd flooded a large area of eastern North Carolina in 1999, and washed millions of gallons of hog and turkey manure into Pamlico and Albemarle Sounds.

or wild Canadian geese. At the same time, North Carolina farmers are asking the federal government for $1 billion in flood relief to rebuild damaged barns and waste lagoons.

Whatever the source, decomposing sewage and dead organisms have created an anerobic "dead zone" in the Pamlico Sound, second largest estuary system in the United States. Nutrients also are thought to contribute to blooms of toxic microorganisms such as the dinoflagellate *Pfiesteria*, which kills both finfish and shellfish, and has long-term adverse health effects on humans who eat seafood, get water on their skin, or even breathe toxic fumes from contaminated sources. When the floods finally receded, tens of thousands of dead animals were buried in shallow, watery graves, where they continue to contaminate groundwater.

Few environmental science issues are as widespread or emotionally charged as water pollution. According to the EPA, the vast majority of Americans—some 218 million—live within 16 km (10 mi) of a degraded water body. The dismal example of the flood of dead pigs introduces many of the topics we'll cover in this chapter: point and nonpoint water pollution, oxygen depletion and ocean dead zones, polluted groundwater, toxic and infectious aquatic organisms, waste disposal, and watershed management. We'll look at water quality in the United States and other countries, ocean pollution, stream and river health, and methods of water pollution control, as well as some important legislation and water policy issues that you, as a student of environmental science, should know something about. Let's get started.

WATER POLLUTION

Most students today are too young to appreciate that water in most industrialized countries was once far more polluted and dangerous than it is now. Thirty years ago, Lake Erie was on the brink of ecological collapse. The Cuyahoga River, choked with oil and floating debris, burned regularly. Factories and cities routinely dumped untreated chemicals, metals, oil, solvents, and sewage into rivers and lakes. Toxic solvents and organic chemicals were commonly dumped or buried in the ground, poisoning groundwater that we're now paying billions to clean up. In 1972, President Nixon signed the Clean Water Act, which has been called the United States' most

successful and popular environmental legislation. This act established a goal that all the nation's waters should be "fishable and swimmable." While this goal is far from being achieved, the Clean Water Act remains popular because it protects public health (thus saving taxpayer dollars), as well as reducing environmental damage. In addition, water has an aesthetic appeal: the view of a clean lake, river, or seashore makes people happy, and water provides for recreation, so many people feel their quality of life has improved as water quality has been restored.

We still have a long way to go in improving water quality. While pollution from factory pipes has been vastly reduced in the past 30 years, erosion from farm fields, construction sites, and streets has, in many areas, gotten worse since 1972. Airborne mercury, sulfur, and other substances are increasingly contaminating lakes and wetlands. Concentrated livestock production, as you have just read, threatens underground water as well as surface water systems. Increasing industrialization in developing countries has led to widespread water pollution in impoverished regions with little environmental regulation.

What Is Water Pollution?

Any physical, biological, or chemical change in water quality that adversely affects living organisms or makes water unsuitable for desired uses can be considered pollution. There are natural sources of water contamination, such as poison springs, oil seeps, and sedimentation from erosion, but in this chapter we will focus primarily on human-caused changes that affect water quality or usability.

Pollution-control standards and regulations usually distinguish between point and nonpoint pollution sources. Factories, power plants, sewage treatment plants, underground coal mines, and oil wells are classified as **point sources** because they discharge pollution from specific locations, such as drain pipes, ditches, or sewer outfalls (fig. 18.2). These sources are discrete and identifiable, so they are relatively easy to monitor and regulate. It is generally possible to divert effluent from the waste streams of these sources and treat it before it enters the environment.

In contrast, **nonpoint sources** of water pollution are scattered or diffuse, having no specific location where they discharge into a particular body of water. Nonpoint sources include runoff from farm fields and feedlots (fig. 18.3), golf courses, lawns and gardens, construction sites, logging areas, roads, streets, and parking lots. Whereas point sources may be fairly uniform and predictable throughout the year, nonpoint sources are often highly episodic. The first heavy rainfall after a dry period may flush high concentrations of gasoline, lead, oil, and rubber residues off city streets, for instance, while subsequent runoff may have lower levels of these pollutants. Spring snowmelt carries high levels of atmospheric acid deposition into streams and lakes in some areas. The irregular timing of these events, as well as their multiple sources and scattered location, makes them much more difficult to monitor, regulate, and treat than point sources.

Perhaps the ultimate in diffuse, nonpoint pollution is **atmospheric deposition** of contaminants carried by air currents and precipitated into watersheds or directly onto surface waters as rain, snow, or dry particles. The Great Lakes, for example, have been found to be accumulating industrial chemicals such as PCBs and dioxins, as well as agricultural toxins such as the insecticide toxaphene that cannot be accounted for by local sources alone. The nearest sources for many of these chemicals are sometimes thousands of kilometers away (chapter 16).

Amounts of these pollutants can be quite large. It is estimated that there are 600,000 kg of the herbicide atrazine in the Great Lakes, most of which is thought to have been deposited from the atmosphere. Concentration of persistent chemicals up the food chain can produce high levels in top predators. Several studies have indicated health problems among people who regularly eat fish from the Great Lakes.

FIGURE 18.2 Sewer outfalls, industrial effluent pipes, acid draining out of abandoned mines, and other point sources of pollution are generally easy to recognize. © Simon Fraser/SPL/Photo Researchers, Inc.

FIGURE 18.3 This bucolic scene looks peaceful and idyllic, but allowing cows to trample stream banks is a major cause of bank erosion and water pollution. Nonpoint sources such as this have become the leading unresolved cause of stream and lake pollution in the United States. © Joe McDonald/Animals Animals/Earth Scenes.

Ironically, lakes can be pollution sources as well as recipients. In the past 12 years, about 26,000 metric tons of PCBs have "disappeared" from Lake Superior. Apparently, these compounds evaporate from the lake surface and are carried by air currents to other areas where they are redeposited.

TYPES AND EFFECTS OF WATER POLLUTION

Although the types, sources, and effects of water pollutants are often interrelated, it is convenient to divide them into major categories for discussion (table 18.1). Let's look more closely at some of the important sources and effects of each type of pollutant.

TABLE 18.1 Major Categories of Water Pollutants

CATEGORY	EXAMPLES	SOURCES
A. CAUSES HEALTH PROBLEMS		
1. Infectious agents	Bacteria, viruses, parasites	Human and animal excreta
2. Organic chemicals	Pesticides, plastics, detergents, oil, and gasoline	Industrial, household, and farm use
3. Inorganic chemicals	Acids, caustics, salts, metals	Industrial effluents, household cleansers, surface runoff
4. Radioactive materials production, natural sources	Uranium, thorium, cesium, iodine, radon	Mining and processing of ores, power plants, weapons
B. CAUSES ECOSYSTEM DISRUPTION		
1. Sediment	Soil, silt	Land erosion
2. Plant nutrients	Nitrates, phosphates, ammonium	Agricultural and urban fertilizers, sewage, manure
3. Oxygen-demanding wastes	Animal manure and plant residues	Sewage, agricultural runoff, paper mills, food processing
4. Thermal	Heat	Power plants, industrial cooling

Infectious Agents

The most serious water pollutants in terms of human health worldwide are pathogenic organisms (chapter 8). Among the most important waterborne diseases are typhoid, cholera, bacterial and amoebic dysentery, enteritis, polio, infectious hepatitis, and schistosomiasis. Malaria, yellow fever, and filariasis are transmitted by insects that have aquatic larvae. Altogether, at least 25 million deaths each year are blamed on these water-related diseases. Nearly two-thirds of the mortalities of children under 5 years old are associated with water-borne diseases.

The main source of these pathogens is from untreated or improperly treated human wastes. Animal wastes from feedlots or fields near waterways and food processing factories with inadequate waste treatment facilities also are sources of disease-causing organisms.

In developed countries, sewage treatment plants and other pollution-control techniques have reduced or eliminated most of the worst sources of pathogens in inland surface waters. Furthermore, drinking water is generally disinfected by chlorination so epidemics of waterborne diseases are rare in these countries. The United Nations estimates that 90 percent of the people in developed countries have adequate (safe) sewage disposal, and 95 percent have clean drinking water.

The situation is quite different in less-developed countries (fig. 18.4). The United Nations estimates that at least 2.5 billion people in these countries lack adequate sanitation, and that about half these people also lack access to clean drinking water. Conditions are especially bad in remote, rural areas where sewage treatment is usually primitive or nonexistent, and purified water is either unavailable or too expensive to obtain (fig. 18.5). The World Health Organization estimates that 80 percent of all sickness and disease in less-developed countries can be attributed to waterborne infectious agents and inadequate sanitation.

If everyone had pure water and satisfactory sanitation, the World Bank estimates that 200 million fewer episodes of diarrheal

FIGURE 18.4 Village water supplies in Ghana.
Courtesy National Renewable Energy Laboratory/NREL/PIX.

illness would occur each year, and 2 million childhood deaths would be avoided. Furthermore, 450 million people would be spared debilitating roundworm or fluke infections. Surely these are goals worth pursuing.

Detecting specific pathogens in water is difficult, time-consuming, and costly; thus, water quality control personnel usually analyze water for the presence of **coliform bacteria,** any of the many types that live in the colon or intestines of humans and other animals. The most common of these is *Eschericha coli* (or *E. coli*). Many strains of bacteria are normal symbionts in mammals, but some, such as *Shigella, Salmonella,* or *Lysteria* can cause fatal diseases. It is usually assumed that if any coliform bacteria are present in a water sample, infectious pathogens are present also.

To test for coliform bacteria, a water sample (or a filter through which a measured water sample has passed) is placed in a dish

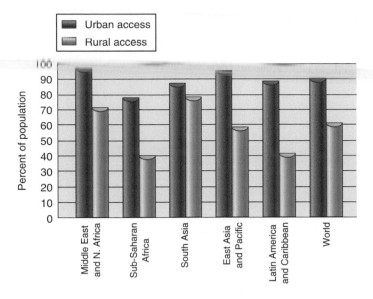

FIGURE 18.5 Proportion of people in developing regions with access to safe drinking water. *Source:* UNESCO, 2002.

FIGURE 18.6 Our national goal of making all surface waters in the United States "fishable and swimmable" has not been fully met, but scenes like this have been reduced by pollution-control efforts.
© Roger A. Clark, Jr./Photo Researchers, Inc.

containing a nutrient medium that supports bacterial growth. After 24 hours in an incubator, living cells will have produced small colonies. If *any* colonies are found in drinking water samples, the U.S. Environmental Protection Agency considers the water unsafe and requiring disinfection. The EPA-recommended maximum coliform count for swimming water is 200 colonies per 100 ml, but some cities and states allow higher levels. If the limit is exceeded, the contaminated pool, river, or lake usually is closed to swimming (fig. 18.6).

Oxygen-Demanding Wastes

The amount of oxygen dissolved in water is a good indicator of water quality and of the kinds of life it will support. Water with an oxygen content above 6 parts per million (ppm) will support game fish and other desirable forms of aquatic life. Water with less than 2 ppm oxygen will support mainly worms, bacteria, fungi, and other detritus feeders and decomposers. Oxygen is added to water by diffusion from the air, especially when turbulence and mixing rates are high, and by photosynthesis of green plants, algae, and cyanobacteria. Oxygen is removed from water by respiration and chemical processes that consume oxygen.

The addition of certain organic materials, such as sewage, paper pulp, or food-processing wastes, to water stimulates oxygen consumption by decomposers. The impact of these materials on water quality can be expressed in terms of **biochemical oxygen demand (BOD):** a standard test of the amount of dissolved oxygen consumed by aquatic microorganisms over a five-day period. An alternative method, called the chemical oxygen demand (COD), uses a strong oxidizing agent (dichromate ion in 50 percent sulfuric acid) to completely break down all organic matter in a water sample. This method is much faster than the BOD test, but normally gives much higher results because it oxidizes compounds not ordinarily metabolized by bacteria. A third method of assaying pollution levels is to measure **dissolved oxygen (DO) content** directly, using an oxygen electrode. The DO content of water depends on factors other than pollution (for example, temperature and aeration), but it is usually more directly related to whether aquatic organisms survive than is BOD.

The effects of oxygen-demanding wastes on rivers depends to a great extent on the volume, flow, and temperature of the river water. Aeration occurs readily in a turbulent, rapidly flowing river, which is, therefore, often able to recover quickly from oxygen-depleting processes. Downstream from a point source, such as a municipal sewage plant discharge, a characteristic decline and restoration of water quality can be detected either by measuring dissolved oxygen content or by observing the flora and fauna that live in successive sections of the river.

The oxygen decline downstream is called the **oxygen sag** (fig. 18.7). Upstream from the pollution source, oxygen levels support normal populations of clean-water organisms. Immediately below the source of pollution, oxygen levels begin to fall as decomposers metabolize waste materials. Rough fish, such as carp, bullheads, and gar, are able to survive in this oxygen-poor environment where they eat both decomposer organisms and the waste itself. Further downstream, the water may become so oxygen-depleted that only the most resistant microorganisms and invertebrates can survive. Eventually, most of the nutrients are used up, decomposer populations are smaller, and the water becomes oxygenated once again. Depending on the volumes and flow rates of the effluent plume and the river receiving it, normal communities may not appear for several miles downstream.

Plant Nutrients and Cultural Eutrophication

Water clarity (transparency) is affected by sediments, chemicals, and the abundance of plankton organisms, and is a useful measure of

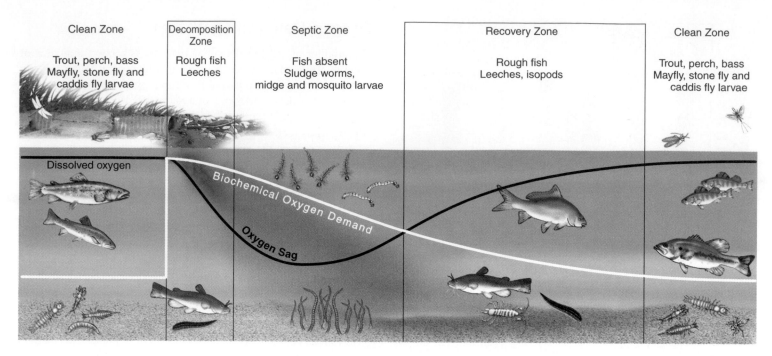

Clean Zone

Trout, perch, bass
Mayfly, stone fly and
caddis fly larvae

Decomposition
Zone

Rough fish
Leeches

Septic Zone

Fish absent
Sludge worms,
midge and mosquito larvae

Recovery Zone

Rough fish
Leeches, isopods

Clean Zone

Trout, perch, bass
Mayfly, stone fly and
caddis fly larvae

Dissolved oxygen

Biochemical Oxygen Demand

Oxygen Sag

FIGURE 18.7   Oxygen sag downstream of an organic source. A great deal of time and distance may be required for the stream and its inhabitants to recover.

water quality and water pollution. Rivers and lakes that have clear water and low biological productivity are said to be **oligotrophic** (oligo = little + trophic = nutrition). By contrast, **eutrophic** (eu + trophic = truly nourished) waters are rich in organisms and organic materials. Eutrophication is an increase in nutrient levels and biological productivity. Some amount of eutrophication is a normal part of successional changes (chapter 5) in most lakes. Tributary streams bring in sediments and nutrients that stimulate plant growth. Over time, most ponds or lakes tend to fill in, eventually becoming marshes. The rate of eutrophication and succession depends on water chemistry and depth, volume of inflow, mineral content of the surrounding watershed, and the biota of the lake itself.

Human activities can greatly accelerate eutrophication. An increase in biological productivity and ecosystem succession caused by human activities is called **cultural eutrophication.** Cultural eutrophication can result from increased nutrient flows, higher temperatures, more sunlight reaching the water surface, or a number of other changes. Increased productivity in an aquatic system sometimes can be beneficial. Fish and other desirable species may grow faster, providing a welcome food source.

Often, however, eutrophication has undesirable results. Elevated phosphorus and nitrogen levels stimulate "blooms" of algae or thick growths of aquatic plants (fig. 18.8). Bacterial populations also increase, fed by larger amounts of organic matter. The water often becomes cloudy or turbid and has unpleasant tastes and odors. In extreme cases, plants and algae die and decomposers deplete oxygen in the water. Collapse of the aquatic ecosystem can result.

Eutrophication also occurs in marine ecosystems, especially in nearshore waters and partially enclosed bays or estuaries. Partially enclosed seas such as the Black Sea, the Baltic, and the

FIGURE 18.8   Eutrophic lake. Nutrients from agriculture and domestic sources have stimulated growth of algae and aquatic plants. This reduces water quality, alters species composition, and lowers recreational and aesthetic values of the lake. © William P. Cunningham.

Mediterranean tend to be in especially critical condition. During the tourist season, the coastal population of the Mediterranean, for example, swells to 200 million people. Eighty-five percent of the effluents from large cities go untreated into the sea. Beach pollution, fish kills, and contaminated shellfish result. Extensive "dead zones" often form where rivers dump nutrients into estuaries and shallow seas. The largest in the world occurs during summer months in the Gulf of Mexico at the mouth of the Mississippi

River. This hypoxic zone (less than 2 mg oxygen per liter) can cover 18,000 km^2 (7,000 mi^2, or about as big as New Jersey). The hypoxic zone in Pamlico Sound following Hurricane Floyd was only about one-tenth this size.

Toxic Tides

According to the Bible, the first plague to afflict the Egyptians when they wouldn't free Moses and the Israelites was that the water in the Nile turned into blood. All the fish died and the people were unable to drink the water, a terrible calamity in a desert country. Some modern scientists believe this may be the first recorded history of a **red tide** or a bloom of deadly aquatic microorganisms called dinoflagellates. Red tides—and other colors, depending on the species involved—have become increasingly common in slow-moving rivers, brackish lagoons, estuaries, and bays, as well as nearshore ocean waters where nutrients and wastes wash down our rivers.

One of the most feared of these organisms is *Pfiesteria piscicida,* an extraordinarily poisonous dinoflagellate that only recently has been recognized as a killer of finfish and shellfish in polluted rivers and estuaries such as North Carolina's Pamlico Sound, where it recently has wiped out hundreds of thousands to millions of fish every year. Dinoflagellates are peculiar organisms with complex life cycles and many different shapes. They are unicellular, photosynthetic microorganisms distinct from both plants and animals. They typically swim with two slender, whiplike flagella, one of which vibrates in a shallow grove in the cell surface and causes the organism to rotate, while the other propels it through the water. *Pfiesteria* can change into at least two dozen distinct forms and sizes depending on water temperature, turbulence, and food supply available (fig. 18.9). In calm, warm, brackish water, rich in sewage or other nutrients, *Pfiesteria* assumes a spherical, nontoxic, but predatory, swimming form that preys on bacteria and plankton.

Under the right conditions, a population explosion can produce a dense bloom of these cells, which can reproduce either by binary fission or sexual fusion of small gametes. If fish blunder into this profuse swarm, *Pfiesteria* quickly turn to a toxic, swimming form that attacks with soluble poisons. These toxins paralyze fish, so they can't escape, and produce skin lesions that expose the fish to infections by pathogenic bacteria and fungi. The predatory zoospores feed on substances that leak from the sores, and also eat skin and flesh directly. When the fish die, zoospores change into star-shaped or lobose (lobed) amoebae that crawl along the bottom and feed on dead carcasses. If the water is cold and calm, these amoebae can attack fish directly; or if it is very turbulent, they spend their time crawling on the bottom or burrowing into the mud looking for something to eat. If the environment becomes adverse, any of these forms can turn into resistant cysts that can withstand extremely harsh conditions.

Perhaps the scariest thing about *Pfiesteria* is the recently recognized fact that it can be extremely poisonous to humans, either through eating contaminated seafood, or by breathing aerosols (airborne mist or dust) containing *Pfiesteria* cells or their secretions. The symptoms of *Pfiesteria* poisoning include headaches, blurred vision, aching joints, burning muscles, difficulty breathing, disori-

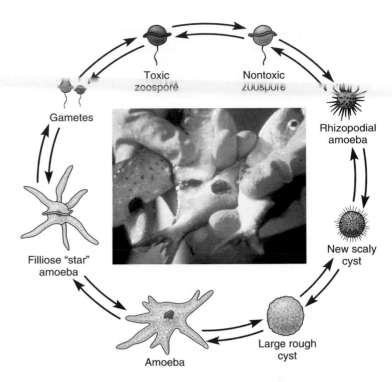

FIGURE 18.9 *Pfiesteria piscicida* life cycle. This toxic, predatory dinoflagellate has at least two dozen different sizes and cell shapes depending on nutrient supply and environmental conditions. They attack fish, causing large skin lesions that often are lethal.
Source: Adapted from J. M. Burkholder (1999), *Scientific American,* 281(2): pp. 42–49. Photo: Courtesy Howard Glasgow and JoAnn M. Burkholder, NCSU Botany.

entation, memory loss, and long-term damage to the brain, liver, pancreas, kidneys, and immune system. Just being on a boat for a few minutes in the midst of *Pfiesteria* bloom can result in debilitating symptoms and long-term injury. This does seem like a plague, indeed, and it appears to be provoked primarily by human actions. Runoff of agricultural wastes, fertilizer, human sewage, and other nutrients into shallow, near-shore water, seems to be the main cause of this terrible affliction.

Inorganic Pollutants

Some toxic inorganic chemicals are released from rocks by weathering, are carried by runoff into lakes or rivers, or percolate into groundwater aquifers. This pattern is part of natural mineral cycles (chapter 3). Humans often accelerate the transfer rates in these cycles thousands of times above natural background levels through the mining, processing, using, and discarding of minerals.

In many areas, toxic, inorganic chemicals introduced into water as a result of human activities have become the most serious form of water pollution. Among the chemicals of greatest concern are heavy metals, such as mercury, lead, tin, and cadmium. Supertoxic elements, such as selenium and arsenic, also have reached hazardous levels in some waters. Other inorganic materials, such as acids, salts, nitrates, and chlorine, that normally are not toxic at low concentrations may become concentrated enough to lower water quality or adversely affect biological communities.

Metals

Many metals such as mercury, lead, cadmium, and nickel are highly toxic. Levels in the parts per million range—so little that you cannot see or taste them—can be fatal. Because metals are highly persistent, they accumulate in food chains and have a cumulative effect in humans.

A mercury-poisoning disaster appears to be in process in South America. Since the mid-1980s, a gold rush has been under way in Brazil, Ecuador, and Bolivia. Forty thousand *garimperios,* or prospectors, have invaded the jungles along the Amazon River and its tributaries to pan for gold. They use mercury to trap the gold and separate it from sediments. Then, the mercury is boiled off with a blow torch. Miners and their families suffer nerve damage from breathing the toxic fumes. Estimates are that 130 tons of mercury per year are deposited in the Amazon, which will be impossible to clean up.

We have come to realize that other heavy metals released as a result of human activities also are concentrated by hydrological and biological processes so that they become hazardous to both natural ecosystems and human health. A condition known as Itai-Itai (literally ouch-ouch) disease that developed in the Japanese living near the Jintsu River was traced to cadmium poisoning from mining and smelting waste-water discharges. Bacteria forming methylated tin have been found in sediments in Chesapeake Bay, leading to worries that this toxic metal may be entering the food web. Tin compounds used as antifouling agents on ship bottoms have been banned because of their toxic effects.

Lead poisoning has been known since Roman times to be dangerous to human health. Lead pipes remain a source of drinking-water pollution, especially in older homes or in areas where water is acidic and therefore leaches more lead from pipes. Even lead solder in pipe joints and metal containers can be hazardous. In 1990, the EPA lowered the maximum limit for lead in public drinking water from 50 parts per billion (ppb) to 20 ppb. Some public health officials argue that lead is neurotoxic at any level, and the limits should be less than 10 ppb.

Mine drainage and leaching of mining wastes are serious sources of metal pollution in water (chapter 14). A survey of water quality in eastern Tennessee found that 43 percent of all surface streams and lakes and more than half of all groundwater used for drinking supplies was contaminated by acids and metals from mine drainage. In some cases, metal levels were 200 times higher than what is considered safe for drinking water.

Nonmetallic Salts

Desert soils often contain high concentrations of soluble salts, including toxic selenium and arsenic (Case Study, p. 385). You have probably heard of poison springs and seeps in the desert where these compounds are brought to the surface by percolating groundwater. Irrigation and drainage of desert soils mobilize these materials on a larger scale and can result in serious pollution problems, as in Kesterson Marsh in California where selenium poisoning killed thousands of migratory birds in the 1980s.

Such salts as sodium chloride (table salt) that are nontoxic at low concentrations also can be mobilized by irrigation and concentrated by evaporation, reaching levels that are toxic for plants and animals. Salt levels in the San Joaquin River in central California rose from 0.28 gm/l in 1930 to 0.45 gm/l in 1970 as a result of agricultural runoff. Salinity levels in the Colorado River and surrounding farm fields have become so high in recent years that millions of hectares of valuable croplands have had to be abandoned. The United States has built a huge desalinization plant at Yuma, Arizona, to reduce salinity in the river. In northern states, millions of tons of sodium chloride and calcium chloride are used to melt road ice in the winter. The corrosive damage to highways and automobiles and the toxic effects on vegetation are enormous. Leaching of road salts into surface waters may have a similarly devastating effect on aquatic ecosystems.

Acids and Bases

Acids are released as by-products of industrial processes, such as leather tanning, metal smelting and plating, petroleum distillation, and organic chemical synthesis. Coal mining is an especially important source of acid water pollution. Sulfur compounds in coal react with oxygen and water to make sulfuric acid. Thousands of kilometers of streams in the United States have been acidified by acid mine drainage, some so severely that they are essentially lifeless.

Coal and oil combustion also leads to formation of atmospheric sulfuric and nitric acids (chapter 16), which are disseminated by long-range transport processes and deposited via precipitation (acidic rain, snow, fog, or dry deposition) in surface waters. Where soils are rich in such alkaline material as limestone, these atmospheric acids have little effect because they are neutralized. In high mountain areas or recently glaciated regions where crystalline bedrock is close to the surface and lakes are oligotrophic, however, there is little buffering capacity (ability to neutralize acids) and aquatic ecosystems can be severely disrupted. These effects were first recognized in the mountains of northern England and Scandinavia about 30 years ago.

In recent years, aquatic damage due to acid precipitation has been reported in about 200 lakes in the Adirondack Mountains of New York State and in several thousand lakes in eastern Quebec, Canada. Game fish, amphibians, and sensitive aquatic insects are generally the first to be killed by increased acid levels in the water. If acidification is severe enough, aquatic life is limited to a few resistant species of mosses and fungi. Increased acidity may result in leaching of toxic metals, especially aluminum, from soil and rocks, making water unfit for drinking or irrigation, as well.

Organic Chemicals

Thousands of different natural and synthetic organic chemicals are used in the chemical industry to make pesticides, plastics, pharmaceuticals, pigments, and other products that we use in everyday life. Many of these chemicals are highly toxic (chapter 8). Exposure to very low concentrations (perhaps even parts per quadrillion in the case of dioxins) can cause birth defects, genetic disorders, and cancer. Some can persist in the environment because they are resistant to degradation and toxic to organisms that ingest them. Contamination of surface waters and groundwater by these chemicals is a serious threat to human health.

Case Study

Arsenic in Drinking Water

When we think of water pollution, we usually visualize sewage or industrial effluents pouring out of a discharge pipe, but there are natural toxins that threaten us as well. One of these is arsenic, a common contaminate in drinking water that may be poisoning millions of people around the world. Arsenic has been known since the fourth century B.C. to be a potent poison. It has been used for centuries as a rodenticide, insecticide, and weed killer, as well as a way of assassinating enemies. Because it isn't metabolized or excreted from the body, arsenic accumulates in hair and fingernails, where it can be detected long after death. Napoleon Bonaparte was recently found to have high enough levels of arsenic in his body to suggest he was poisoned.

Perhaps the largest population to be threatened by naturally occurring groundwater contamination by arsenic is in West Bengal, India, and adjacent areas of Bangladesh. Arsenic occurs naturally in the sediments that make up the Ganges River delta (see map). Rapid population growth, industrialization, and intensification of agricultural irrigation, however, have put increasing stresses on the limited surface-water supplies. Most surface water is too contaminated to drink, so groundwater has all but replaced other water sources for most people in this region.

In the 1960s, thousands of deep tube wells were sunk throughout the region to improve water supplies. Much of this humanitarian effort was financed by loans from the World Bank. At first, villagers were suspicious of well water, regarding it as unnatural and possibly evil. But as surface-water supplies

West Bengal and adjoining areas of Bangladesh have hundreds of millions of people who may be exposed to dangerous arsenic levels in well water.

diminished and populations grew, Bangladesh became more and more dependent on this new source of supposedly fresh, clean water. By the late 1980s, health workers became aware of widespread signs of chronic arsenic poisoning among villagers. Symptoms include watery and inflamed eyes, gastrointestinal cramps, gradual loss of strength, scaly skin and skin tumors, anemia, confusion, and, eventually, death.

Why is arsenic poisoning appearing now? Part of the reason is increased dependence on well water, but some villages have had wells for centuries with no problem. One theory is that excessive withdrawals now lower the water table during the dry season, exposing arsenic-bearing minerals to air, which converts normally insoluble salts to soluble oxides. When aquifers are refilled during the next rainy season, dissolved arsenic can be pumped out. Health workers estimate that the total number of potential victims in India and Bangladesh may exceed

200 million people. But with no other source of easily accessible or affordable water, few of the poorest people have much choice.

There are worries that millions of Americans also are exposed to dangerously high levels of arsenic. In 1942, the U.S. Government set the acceptable level of arsenic in drinking water at 50 parts per billion or ppb. A 1999 study by the National Academy of Sciences found a one in 100 risk of cancer from drinking water with that level of arsenic for a lifetime. This is 10,000 times the normally accepted risk level. Following years of heated debate, the U.S. limit was revised in 2002 to meet the World Health Organization standard of 10 ppb. Local officials and private water supply owners argued that it would cost too much to upgrade their systems.

In the end, public outrage over tainted water, combined with the enormous public health costs of chronic arsenic poisoning, convinced the federal government to enforce stricter standards.

The two most important sources of toxic organic chemicals in water are improper disposal of industrial and household wastes and runoff of pesticides from farm fields, forests, roadsides, golf courses, and other places where they are used in large quantities (fig. 18.10). The U.S. EPA estimates that about 500,000 metric tons of pesticides are used in the United States each year. Much of this material washes into the nearest waterway, where it passes through ecosystems and may accumulate in high levels in certain nontarget organisms. The bioaccumulation of DDT in aquatic ecosystems was one of the first of these pathways to be understood. Dioxins, and other chlorinated hydrocarbons (hydrocarbon molecules that contain chlorine atoms) have been shown to accumulate to dangerous levels in the fat of salmon, fish-eating birds, and humans and to cause health problems similar to those resulting from toxic metal compounds (fig. 18.11).

Hundreds of millions of tons of hazardous organic wastes are thought to be stored in dumps, landfills, lagoons, and underground tanks in the United States (chapter 21). Many, perhaps most, of these

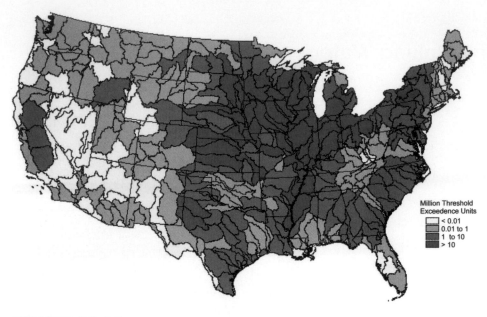

FIGURE 18.10 Watersheds at risk from pesticide runoff. Threshold Exceedence Units represent the number of acres, and amount of chemicals contributing to high rates of pesticide runoff. *Source:* Natural Resources Conservation Service.

Million Threshold
Exceedence Units
☐ < 0.01
☐ 0.01 to 1
☐ 1 to 10
■ > 10

FIGURE 18.11 The deformed beak of this young robin is thought to be due to dioxins, DDT, and other toxins in its mother's diet. © Rob & Melissa Simpson/Valan Photos.

sites have leaked toxic chemicals into surface waters or groundwater or both. The EPA estimates that about 26,000 hazardous waste sites will require cleanup because they pose an imminent threat to public health, mostly through water pollution.

Countless additional organic compounds enter our water unmonitored. In 2002, the USGS released the first-ever study of pharmaceuticals and hormones in streams. Scientists sampled 130 streams, looking for 95 contaminants, including antibiotics, natural and synthetic hormones, detergents, plasticizers, insecticides, and fire retardants. All these substances were found, usually in low concentrations. One stream had 38 of the compounds tested. Drinking water standards exist for only 14 of the 95 substances. What are the effects of these widely used chemicals, on our environment or on people consuming the water? Nobody knows. This study is a first step toward filling huge gaps in our knowledge about their distribution, though.

Sediment

Rivers have always carried sediment to the oceans, but erosion rates in many areas have been greatly accelerated by human activities. As chapters 11 and 19 describe, some rivers carry astounding loads of sediment. Erosion and runoff from croplands contribute about 25 billion metric tons of soil, sediment, and suspended solids to world surface waters each year. Forests, grazing lands, urban construction sites, and other sources of erosion and runoff add at least 50 billion additional tons. This sediment fills lakes and reservoirs, obstructs shipping channels, clogs hydroelectric turbines, and makes purification of drinking water more costly. Sediments smother gravel beds in which insects take refuge and fish lay their eggs. Sunlight is blocked so that plants cannot carry out photosynthesis, and oxygen levels decline. Murky, cloudy water also is less attractive for swimming, boating, fishing, and other recreational uses (fig. 18.12).

Sediment also can be beneficial. Mud carried by rivers nourishes floodplain farm fields. Sediment deposited in the ocean at river mouths creates valuable deltas and islands. The Ganges River, for instance, builds up islands in the Bay of Bengal that are eagerly colonized by land-hungry people of Bangladesh. Louisiana's coastal wetlands require constant additions of sediment from the muddy Mississippi to counteract coastal erosion. These wetlands are now disappearing at a disastrous rate: levees now channel the river and its load straight out to the Gulf of Mexico, where sediments are dumped beyond the continental shelf.

Thermal Pollution and Thermal Shocks

Raising or lowering water temperatures from normal levels can adversely affect water quality and aquatic life. Water temperatures are usually much more stable than air temperatures, so aquatic organisms tend to be poorly adapted to rapid temperature changes. Lowering the temperature of tropical oceans by even one degree can be lethal to some corals and other reef species. Raising water temperatures can have similar devastating effects on sensitive organisms. Oxygen solubility in water decreases as temperatures increase, so species requiring high oxygen levels are adversely affected by warming water.

Humans cause thermal pollution by altering vegetation cover and runoff patterns, as well as by discharging heated water directly into rivers and lakes.

FIGURE 18.12 Sediment and industrial waste flow from this drainage canal into Lake Erie. © Lawrence Lowry/Photo Researchers, Inc.

The cheapest way to remove heat from an industrial facility is to draw cool water from an ocean, river, lake, or aquifer, run it through a heat-exchanger to extract excess heat, and then dump the heated water back into the original source. A **thermal plume** of heated water is often discharged into rivers and lakes, where raised temperatures can disrupt many processes in natural ecosystems and drive out sensitive organisms. Nearly half the water we withdraw is used for industrial cooling. Electric power plants, metal smelters, petroleum refineries, paper mills, food-processing factories, and chemical manufacturing plants all use and release large amounts of cooling water.

To minimize thermal pollution, power plants frequently are required to construct artificial cooling ponds or cooling towers in which heat is released into the atmosphere and water is cooled before being released into natural water bodies.

Some species find thermal pollution attractive. Warm water plumes from power plants often attract fish, birds, and marine mammals that find food and refuge there, especially in cold weather. This artificial environment can be a fatal trap, however. Organisms dependent on the warmth may die if they leave the plume or if the flow of warm water is interrupted by a plant shutdown. Endangered manatees in Florida, for example, are attracted to the abundant food and warm water in power plant thermal plumes and are enticed into spending the winter much farther north than they normally would. On several occasions, a midwinter power plant breakdown has exposed a dozen or more of these rare animals to a sudden thermal shock that they could not survive.

WATER QUALITY TODAY

Surface-water pollution is often both highly visible and one of the most common threats to environmental quality. In more developed countries, reducing water pollution has been a high priority over the past few decades. Billions of dollars have been spent on control programs and considerable progress has been made. Still much

remains to be done. In developed countries, poor water quality often remains a serious problem. In this section, we will look at progress as well as continuing obstacles in this important area.

Surface Waters in the United States and Canada

Like most developed countries, the United States and Canada have made encouraging progress in protecting and restoring water quality in rivers and lakes over the past 40 years. In 1948, only about one-third of Americans were served by municipal sewage systems, and most of those systems discharged sewage without any treatment or with only primary treatment (the bigger lumps of waste are removed). Most people depended on cesspools and septic systems to dispose of domestic wastes.

Areas of Progress

The 1972 Clean Water Act established a National Pollution Discharge Elimination System (NPDES), which requires an easily revoked permit for any industry, municipality or other entity dumping wastes in surface waters. The permit requires disclosure of what is being dumped and gives regulators valuable data and evidence for litigation. As a consequence, only about 10 percent of our water pollution now comes from industrial or municipal point sources. One of the biggest improvements has been in sewage treatment.

Since the Clean Water Act was passed in 1972, the United States has spent more than $180 billion in public funds and perhaps ten times as much in private investments on water pollution control. Most of that effort has been aimed at point sources, especially to build or upgrade thousands of municipal sewage treatment plants. As a result, nearly everyone in urban areas is now served by municipal sewage systems and no major city discharges raw sewage into a river or lake except as overflow during heavy rainstorms.

This campaign has led to significant improvements in surface-water quality in many places. Fish and aquatic insects have returned to waters that formerly were depleted of life-giving oxygen. Swimming and other water-contact sports are again permitted in rivers, lakes, and at ocean beaches that once were closed by health officials.

Key Concepts

- Point source water pollution is discharged from specific locations; non-point sources are scattered or diffuse, such as field or road runoff.

- Major types of water pollutants include infectious agents; oxygen-demanding wastes; nutrients; inorganic salts, metals, acids and bases; organic chemicals; sediment; and thermal pollution.

- Key problems in water quality now are sediment, nutrients, and pathogens, especially from nonpoint sources.

- Water pollution control strategies include waste reduction, source reduction, and reclamation; improved management of farms and construction sites; land-use planning; and sewage treatment.

The Clean Water Act goal of making all U.S. surface waters "fishable and swimmable" has not been fully met, but in 1999 the EPA reported that 91.4 percent of all monitored river miles and 87.5 percent of all assessed lake acres are suitable for their designated uses. This sounds good, but you have to remember that not all water bodies are monitored. Furthermore, the designated goal for some rivers and lakes is merely to be "boatable." Water quality doesn't have to be very high to be able to put a boat on it. Even in "fishable" rivers and lakes, there isn't a guarantee that you can catch anything other than rough fish like carp or bullheads, nor can you be sure that what you catch is safe to eat. Even with billions of dollars of investment in sewage treatment plants, elimination of much of the industrial dumping and other gross sources of pollutants, and a general improvement in water quality, the EPA reports that 21,000 water bodies still do not meet their designated uses. According to the EPA, an overwhelming majority of the American people—almost 218 million—live within 16 km (10 mi) of an impaired water body.

In 1998, a new regulatory approach to water quality assurance was instituted by the EPA. Rather than issue standards on a river by river approach or factory by factory permit discharge, the focus is being changed to watershed-level monitoring and protection. Some 4,000 watersheds are monitored for water quality (fig. 18.13). You can find information about your watershed at www.epa.gov/owow/tmdl/. The intention of this program is to give the public more and better information about the health of their watersheds. In addition, states will have greater flexibility as they identify impaired water bodies and set priorities, and new tools will be used to achieve goals. States are required to identify waters not meeting water quality goals and to develop **total maximum daily loads (TMDL)** for each pollutant and each listed water body. A TMDL is the amount of a particular pollutant that a water body can receive from both point and nonpoint sources. It considers seasonal variation and includes a margin of safety.

By 1999, all 56 states and territories had submitted TMDL lists, and the EPA had approved most of them. Of the 3.5 million mi (5.6 million km) of rivers monitored, only 300,000 mi (480,000 km) fail to meet their clean water goals. Similarly, of 40 million lake acres (99 million ha), only 12.5 percent (in about 20,000 lakes) failed to meet their goal. To give states more flexibility in planning, the EPA has proposed new rules that include allowances for reasonably foreseeable increases in pollutant loadings to encourage "Smart Growth." In the future, TMDLs also will include load allocations from all nonpoint sources, including air deposition and natural background levels.

An encouraging example of improved water quality is seen in Lake Erie. Although widely regarded as "dead" in the 1960s, the lake today is promoted as the "walleye capital of the world." Bacteria counts and algae blooms have decreased more than 90 percent

Percent of Impaired Waters

- No data
- No waters listed
- < 5%
- 5 — 10%
- 10 — 25%
- > 25%

FIGURE 18.13 Percent of impaired rivers by watershed. *Source:* Environmental Protection Agency, August 1999.

since 1962. Water that once was murky brown is now clear. Interestingly, part of the improved water quality is due to immense numbers of exotic zebra mussels, which filter the lake water very efficiently. Swimming is now officially safe along 96 percent of the lake's shoreline. Nearly 40,000 nesting pairs of double-crested cormorants nest in the Great Lakes region, up from only about 100 in the 1970s. Anglers now complain that the cormorants eat too many fish. In 1998 wildlife agents found 800 cormorants shot to death in a rookery on Galloo Island at the east end of Lake Ontario.

Canada's 1970 Water Act has produced comparable results. Seventy percent of all Canadians in towns over 1,000 population are now served by some form of municipal sewage treatment. In Ontario, the vast majority of those systems include tertiary treatment. After ten years of controls, phosphorus levels in the Bay of Quinte in the northeast corner of Lake Ontario have dropped nearly by half, and algal blooms that once turned waters green are less frequent and less intense than they once were. Elimination of mercury discharges from a pulp and paper mill on the Wabigoon-English River system in western Ontario has resulted in a dramatic decrease in mercury contamination. Twenty years ago this mercury contamination was causing developmental retardation in local residents. Extensive flooding associated with hydropower projects has raised mercury levels in fish to dangerous levels elsewhere, however.

Remaining Problems

The greatest impediments to achieving national goals in water quality in both the United States and Canada are sediment, nutrients, and pathogens, especially from nonpoint discharges of pollutants. These sources are harder to identify and to reduce or treat than are specific point sources. About three-fourths of the water pollution in the United States comes from soil erosion, fallout of air pollutants, and surface runoff from urban areas, farm fields, and feedlots. In the United States, as much as 25 percent of the 46,800,000 metric tons (52 million tons) of fertilizer spread on farmland each year is carried away by runoff (fig. 18.14).

Cattle in feedlots produce some 129,600,000 metric tons (144 million tons) of manure each year, and the runoff from these sites is rich in viruses, bacteria, nitrates, phosphates, and other contaminants. A single cow produces about 30 kg (66 lb) of manure per day, or about as much as that produced by ten people. Some feedlots have 100,000 animals with no provision for capturing or treating runoff water. Imagine drawing your drinking water downstream from such a facility. Pets also can be a problem. It is estimated that the wastes from about a half million dogs in New York City are disposed of primarily through storm sewers, and therefore do not go through sewage treatment.

Loading of both nitrates and phosphates in surface water have decreased from point sources but have increased about fourfold since 1972 from nonpoint sources. Fossil fuel combustion has become a major source of nitrates, sulfates, arsenic, cadmium, mercury, and other toxic pollutants that find their way into water. Carried to remote areas by atmospheric transport, these combustion products now are found nearly everywhere in the world. Toxic organic compounds, such as DDT, PCBs, and dioxins, also are transported long distances by wind currents.

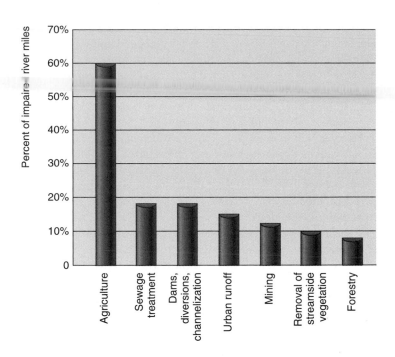

FIGURE 18.14 Percentage of impaired river miles in the United States by source of damage. Totals add up to more than 100 percent because one river can be affected by many sources.

Source: USDA and Natural Resources Conservation Service, *America's Private Land: A Geography of Hope,* USDA, 1996.

Surface Waters in Other Countries

Japan, Australia, and most of western Europe also have improved surface-water quality in recent years. Sewage treatment in the wealthier countries of Europe generally equals or surpasses that in the United States. Sweden, for instance, serves 98 percent of its population with at least secondary sewage treatment (compared with 70 percent in the United States), and the other 2 percent have primary treatment. Poorer countries have much less to spend on sanitation. Spain serves only 18 percent of its population with even primary sewage treatment. In Ireland, it is only 11 percent, and in Greece, less than 1 percent of the people have even primary treatment. Most of the sewage, both domestic and industrial, is dumped directly into the ocean.

The fall of the "iron curtain" in 1989 revealed appalling environmental conditions in much of the former Soviet Union and its satellite states in eastern and central Europe. The countries closest geographically and socially to western Europe, the Czech Republic, Hungary, East Germany, and Poland, have made massive investments and encouraging progress toward cleaning up environmental problems. Parts of Russia itself, however, along with former socialist states in the Balkans and Central Asia, remain some of the most polluted places on earth. In Russia, for example, only about half the tap water is fit to drink. In cities like St. Petersburg, even boiling and filtering isn't enough to make municipal water safe. About one-third of all Russians live in regions where air pollution levels are ten times higher than World Health Organization safety standards. Life expectancies for Russian men have plummeted from about 72 years in 1980 to 59 years in 2003. Deaths now exceed births in Russia by about 1 million per year. Only about one-quarter of all Russian children are considered healthy. In heavily industrialized cities like Magnitogorsk, a steel-manufacturing center in the Ural Mountains, nine out of ten children suffer from pollution-related illnesses and birth defects.

High levels of radioactivity from the Chernobyl accident in 1986 remain in many former Eastern Bloc countries. The Danube River, which originates in Bavaria and Austria, and flows through Slovakia, Poland, Hungary, Serbia, Romania, and Bulgaria before emptying into the Black Sea, illustrates some of the problems besetting the area (fig. 18.15). Draining a landscape with a long history of unregulated mining and heavy industry, the Danube carries chrome, copper, mercury, lead, zinc, and oil to the Black Sea at 20 times the levels these materials flow into the North Sea. Just one city, Bratislava, the capital of Slovakia, dumps 73 million m³ (about 2 billion gal) of industrial and municipal wastes into the river each year. With a population of about 100 million people in its catchment area, the beautiful blue Danube isn't blue anymore. Bombing in Serbia during the Kosovo War in 1999 further exacerbated the situation by releasing oil, industrial chemicals, agricultural fertilizers, pesticides, and other toxic materials into the Danube.

There are also some encouraging pollution-control stories. In 1997, Minamata Bay in Japan, long synonymous with mercury poisoning, was declared officially clean again. Another important success is found in Europe, where one of its most important rivers has been cleaned up significantly through international cooperation. The Rhine, which starts in the rugged Swiss Alps and winds 1,320 km through five countries before emptying through a Dutch delta into the North Sea, has long been a major commercial artery into the heart of Europe. More than 50 million people live in its catchment basin and nearly 20 million get their drinking water from the river or its tributaries. By the 1970s, the Rhine had become so polluted that dozens of fish species disappeared and swimming was discouraged along most of its length.

Efforts to clean up this historic and economically important waterway began in the 1950s, but a disastrous fire at a chemical warehouse near Basel, Switzerland, in 1986 provided the impetus for major changes. Through a long and sometimes painful series of international conventions and compromises, land-use practices, waste disposal, urban runoff, and industrial dumping have been changed and water quality has significantly improved. Oxygen concentrations have gone up fivefold since 1970 (from less than 2 mg/l to nearly 10 mg/l or about 90 percent of saturation) in long stretches of the river. Chemical oxygen demand has fallen fivefold during this same period, and organochlorine levels have decreased as much as tenfold. Many species of fish and aquatic invertebrates have returned to the river. In 1992, for the first time in decades, mature salmon were caught in the Rhine.

The less-developed countries of South America, Africa, and Asia have even worse water quality than do the poorer countries of Europe. Sewage treatment is usually either totally lacking or woefully inadequate. In urban areas, 95 percent of all sewage is discharged untreated into rivers, lakes, or the ocean. Low technological capabilities and little money for pollution control are made even

FIGURE 18.15 Two Gypsy children from Copsa Mica, Romania, play in a polluted river across from the carbon factory that turns everything in town black. Eventually this filthy water will be flushed down the Danube and into the Black Sea, which has lost nearly all its commercial fisheries in only 25 years. © James Nachtwey/Magnum Photos.

worse by burgeoning populations, rapid urbanization, and the shift of much heavy industry (especially the dirtier ones) from developed countries where pollution laws are strict to less-developed countries where regulations are more lenient.

Appalling environmental conditions often result from these combined factors (fig. 18.16). Two-thirds of India's surface waters are contaminated sufficiently to be considered dangerous to human health. The Yamuna River in New Delhi has 7,500 coliform bacteria per 100 ml (37 times the level considered safe for swimming in the United States) *before* entering the city. The coliform count increases to an incredible 24 *million* cells per 100 ml as the river leaves the city! At the same time, the river picks up some 20 million liters of industrial effluents every day from New Delhi. It's no wonder that disease rates are high and life expectancy is low in this area. Only 1 percent of India's towns and cities have any sewage treatment, and only eight cities have anything beyond primary treatment.

In Malaysia, 42 of 50 major rivers are reported to be "ecological disasters." Residues from palm oil and rubber manufacturing, along with heavy erosion from logging of tropical rainforests, have destroyed all higher forms of life in most of these rivers. In the Philippines, domestic sewage makes up 60 to 70 percent of the total volume of Manila's Pasig River. Thousands of people use the river not only for bathing and washing clothes but also as their source of drinking and cooking water. China treats only 2 percent of its sewage. Of 78 monitored rivers in China, 54 are reported to be seriously polluted. Of 44 major cities in China, 41 use "contaminated" water supplies, and few do more than rudimentary treatment before it is delivered to the public.

Groundwater and Drinking Water Supplies

About half the people in the United States, including 95 percent of those in rural areas, depend on underground aquifers for their drinking water. This vital resource is threatened in many areas by overuse and pollution and by a wide variety of industrial, agricultural, and domestic contaminants. For decades it was widely assumed that groundwater was impervious to pollution because soil would bind chemicals and cleanse water as it percolated through. Springwater or artesian well water was considered to be the definitive standard of water purity, but that is no longer true in many areas.

One of the serious sources of groundwater pollution throughout the United States is MTBE (methyl tertiary butyl ether), a suspected carcinogen. MTBE is a gasoline additive that has been used since the 1970s to reduce the amount of carbon monoxide and ozone in vehicle exhaust. By the time the health dangers of MTBE were confirmed in the late 1990s, aquifers across the country had been contaminated—mainly from leaking underground storage tanks at gas stations. About 250,000 of these tanks are leaking MTBE into groundwater nationwide. In one U.S. Geological Survey (USGS) study, 27 percent of shallow urban wells tested contained MTBE. The additive is being phased out, but plumes of tainted water will continue to move through aquifers for decades to come. (Surface waters have also been contaminated, especially by two-stroke engines, such as those on personal watercraft.)

Treating MTBE-laced aquifers is expensive but not impossible. Douglas MacKay of the University of Waterloo in Ontario suggests that if oxygen could be pumped into aquifers, then naturally occurring bacteria could metabolize (digest) the compound. It could take decades or even centuries for natural bacteria to eliminate MTBE from a water supply, however. Water can also be pumped out of aquifers, reducing the flow and spread of contamination. Thus far, little funding has been invested in finding cost-effective remedies, however.

The U.S. EPA estimates that every day some 4.5 trillion l (1.2 trillion gal) of contaminated water seep into the ground in the United States from septic tanks, cesspools, municipal and industrial landfills and waste disposal sites, surface impoundments, agricul-

FIGURE 18.16 Ditches in this Haitian slum serve as open sewers into which all manner of refuse and waste are dumped. The health risks of living under these conditions are severe. © Les Stone/Sygma/Corbis.

tural fields, forests, and wells (fig. 18.17). The most toxic of these are probably waste disposal sites. Agricultural chemicals and wastes are responsible for the largest total volume of pollutants and area affected. Because deep underground aquifers often have residence times of thousands of years, many contaminants are extremely stable once underground. It is possible, but expensive, to pump water out of aquifers, clean it, and then pump it back.

In farm country, especially in the Midwest's corn belt, fertilizers and pesticides commonly contaminate aquifers and wells. Herbicides such as atrazine and alachlor are widely used on corn and soybeans and show up in about half of all wells in Iowa, for example. Nitrates from fertilizers often exceed safety standards in rural drinking water. These high nitrate levels are dangerous to infants (nitrate combines with hemoglobin in the blood and results in "blue-baby" syndrome). They also are transformed into cancer-causing nitrosamines in the human gut. In Florida, 1,000 drinking water wells were shut down by state authorities because of excessive levels of toxic chemicals, mostly ethylene dibromide (EDB), a pesticide used to kill nematodes (roundworms) that damage plant roots.

Although most of the leaky, single-walled underground storage tanks once common at filling stations and factories have now been removed and replaced by more modern ones, a great deal of soil in American cities remains contaminated by previous careless storage and disposal of petroleum products. Considering that a single gallon (3.8 l) of gasoline can make a million gallons of water undrinkable, soil contamination remains a serious problem.

In addition to groundwater pollution problems, contaminated surface waters and inadequate treatment make drinking water unsafe in many areas. A 1996 survey concluded that nearly 20,000 public drinking water systems in the United States expose consumers to contaminants such as lead, pesticides, and pathogens at levels that violate EPA rules (fig. 18.18). A vast majority of these

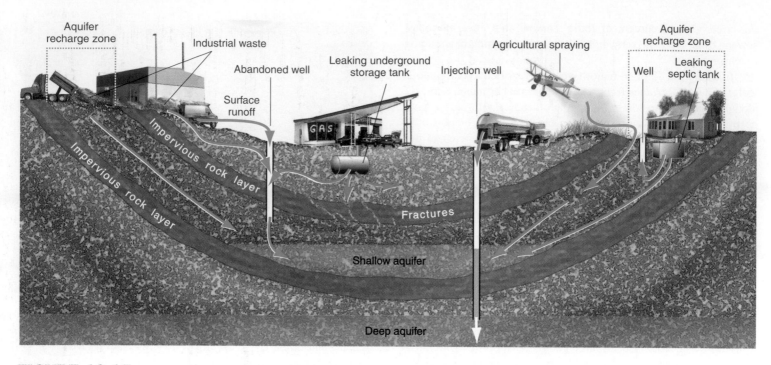

FIGURE 18.17   Sources of groundwater pollution. Septic systems, landfills, and industrial activities on aquifer recharge zones leach contaminants into aquifers. Wells provide a direct route for injection of pollutants into aquifers.

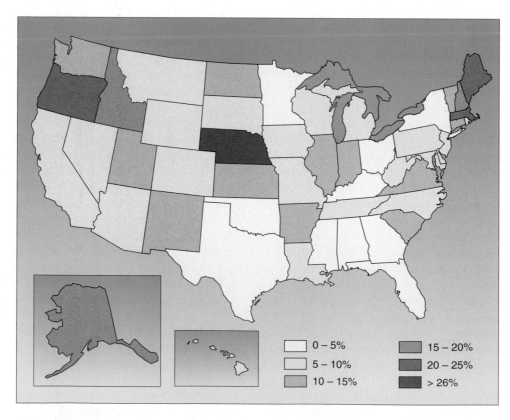

FIGURE 18.18   Percentage of drinking-water systems within states with violations of EPA health standards in 2000. *Source:* EPA Safe Drinking Water Information System 2001.

systems are small, serving fewer than 3,000 people, but altogether some 50 million people are sometimes at risk. Problems often occur because small systems can't afford modern purification and distribution equipment, regular testing, and trained operators to bring water quality up to acceptable standards.

Every year epidemiologists estimate that around 1.5 million Americans fall ill from infections caused by fecal contamination. In 1993, for instance, a pathogen called cryptosporidium got into the Milwaukee public water system, making 400,000 people sick and killing at least 100 people. The total costs of these diseases amount to billions of dollars per year. Preventive measures such as protecting water sources and aquifer recharge zones, providing basic treatment for all systems, installing modern technology and distribution networks, consolidating small systems, and strengthening the Clean Water Act and the Safe Drinking Water Act would cost far less. Unfortunately, in the present climate of budget-cutting and anti-regulation, these steps seem unlikely.

Ocean Pollution

Coastal zones, especially bays, estuaries, shoals, and reefs near large cities or the mouths of major rivers, often are overwhelmed by human-caused contamination. Suffocating and sometimes poisonous blooms of algae regularly deplete ocean waters of oxygen and kill enormous numbers of fish and other marine life. High levels of toxic chemicals, heavy metals, disease-causing organisms, oil, sediment, and plastic refuse are adversely affecting some of the most attractive and productive ocean regions. The potential losses caused by this pollution amount to billions of dollars each year.

Discarded plastic flotsam and jetsam are lightweight and non-biodegradable. They are carried thousands of miles on ocean currents and last for years (fig. 18.19). Even the most remote beaches of distant islands are likely to have bits of polystyrene foam containers or polyethylene packing material that were discarded half a world away. It has been estimated that some 6 million metric tons of plastic bottles, packaging material, and other litter are tossed from ships every year into the ocean where they ensnare and choke seabirds, mammals (fig. 18.20), and even fish. Sixteen states now require that six-pack yokes be made of biodegradable or photodegradable plastic, limiting their longevity as potential killers. In one day, volunteers in Texas gathered more than 300 tons of plastic refuse from Gulf Coast beaches.

Few coastlines in the world remain uncontaminated by oil or oil products. Figure 18.21 shows locations where visible oil slicks have been reported. Oceanographers estimate that somewhere between 3 million and 6 million metric tons of oil are discharged into the world's oceans each year from both land- and sea-based operations. About half of this amount is due to maritime transport. Most oil spills result not from catastrophic, headliner accidents, but from routine open-sea bilge pumping and tank cleaning. These procedures are illegal but are easily carried out once ships are beyond sight of land. Much of the rest comes from land-based municipal and industrial runoff or from atmospheric deposition of residues from refining and combustion of fuels.

FIGURE 18.19 Beach pollution, including garbage, sewage, and contaminated runoff, is a growing problem associated with ocean pollution. © Steve Austin/Paplio/Corbis.

FIGURE 18.20 A deadly necklace. Marine biologists estimate that castoff nets, plastic beverage yokes, and other packing residue kill hundreds of thousands of birds, mammals, and fish each year. © Frans Lanting/Photo Researchers, Inc.

The transport of huge quantities of oil creates opportunities for major oil spills through a combination of human and natural hazards. Military conflict in the Middle East and oil drilling in risky locations, such as the notoriously rough North Sea and the Arctic Ocean, make it likely that more oil spills will occur. Plans to drill for oil along the seismically active California and Alaska coasts have been controversial because of the damage that oil spills could cause to these biologically rich coastal ecosystems.

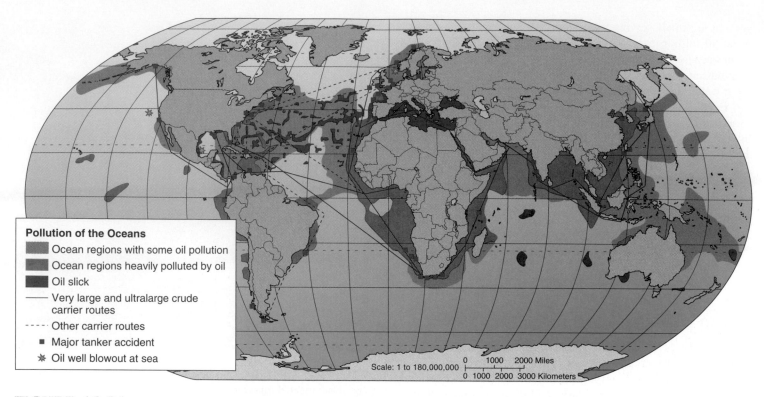

FIGURE 18.21 Location of oil pollution in the ocean. Major shipping routes of large oil tankers are very likely to be contaminated with oil.

WATER POLLUTION CONTROL

Appropriate land-use practices and careful disposal of industrial, domestic, and agricultural wastes are essential for control of water pollution.

Source Reduction

The cheapest and most effective way to reduce pollution is usually to avoid producing it or releasing it to the environment in the first place. Elimination of lead from gasoline has resulted in a widespread and significant decrease in the amount of lead in surface waters in the United States. Studies have shown that as much as 90 percent less road deicing salt can be used in many areas without significantly affecting the safety of winter roads. Careful handling of oil and petroleum products can greatly reduce the amount of water pollution caused by these materials. Although we still have problems with persistent chlorinated hydrocarbons spread widely in the environment, the banning of DDT and PCBs in the 1970s has resulted in significant reductions in levels in wildlife (fig. 18.22).

Modifying agricultural practices in headwater streams in the Chesapeake Bay watershed and the Catskill Mountains of New York have had positive and cost-effective impacts on downstream water quality (Case Study, p. 395).

Industry can reduce pollution by recycling or reclaiming materials that otherwise might be discarded in the waste stream. Both of these approaches usually have economic as well as environmental benefits. It turns out that a variety of valuable metals can be recovered from industrial wastes and reused or sold for other purposes. The company benefits by having a product to sell, and the municipal sewage treatment plant benefits by not having to deal with highly toxic materials mixed in with millions of gallons of other types of wastes.

Nonpoint Sources and Land Management

Among the greatest remaining challenges in water pollution control are diffuse, nonpoint pollution sources. Unlike point sources, such as sewer outfalls or industrial discharge pipes, which represent both specific locations and relatively continuous emissions, nonpoint sources have many origins and numerous routes by which contaminants enter ground and surface waters. It is difficult to identify—let alone monitor and control—all these sources and routes. Some main causes of nonpoint pollution are:

- *Agriculture:* The EPA estimates that 60 percent of all impaired or threatened surface waters are affected by sediment from eroded fields and overgrazed pastures; fertilizers, pesticides, and nutrients from croplands; and animal wastes from feedlots.

- *Urban runoff:* Pollutants carried by runoff from streets, parking lots, and industrial sites contain salts, oily residues, rubber, metals, and many industrial toxins. Yards, golf courses, parklands, and urban gardens often are treated with far more fertilizers and pesticides per unit area than farmlands. Excess chemicals are carried by storm runoff into waterways.

Case Study

Watershed Protection in the Catskills

New York City has long been proud of its excellent municipal drinking water. Drawn from the rugged Catskill Mountains 100 km (60 mi) north of the city, stored in hard-rock reservoirs, and transported through underground tunnels, the city water is outstanding for so large an urban area. Yielding 450,000 m³ (1.2 billion gal) per day, and serving more than 9 million people, this is the largest surface-water storage and supply complex in the world. As the metropolitan agglomeration has expanded, however, people have moved into the area around the Catskill Forest Preserve, and water quality is not as high as it was a century ago.

When the 1986 U.S. Safe Drinking Water Act mandated filtration of all public surface-water systems, the city was faced with building an $8 billion water treatment plant that would cost up to $500 million per year to operate. In 1989, however, the EPA ruled that the city could avoid filtration if it could meet certain minimum standards for microbial contaminants such as bacteria, viruses, and protozoan parasites. In an attempt to avoid the enormous cost of filtration, the city proposed land-use regulations for the five counties (Green, Ulster, Sullivan, Schoharie, and Delaware) in the Catskill/Delaware watershed from which it draws most of its water.

With a population of 50,000 people, the private land within the 520 km² (200 mi²) watershed is mostly devoted to forestry and small dairy farms, neither of which are highly profitable. Among the changes the city called for was elimination of storm water runoff from barnyards, feedlots, or grazing

areas into watersheds. In addition, farmers would be required to reduce erosion and surface runoff from crop fields and logging operations. Property owners objected strenuously to what they regarded as onerous burdens that would cost enough to put many of them out of business. They also bristled at having the huge megalopolis impose rules on them. It looked like a long and bitter battle would be fought through the courts and the state legislature.

To avoid confrontation, a joint urban/rural task force was set up to see if a compromise could be reached, and to propose alternative solutions to protect both the water supply and the long-term viability of agriculture in the region. The task force agreed that agriculture is the "preferred land use" on private land, and that agriculture has "significant present and future environmental benefits." In addition, the task force proposed a voluntary, locally developed and administered program of "whole farm planning and best management approaches" very similar to ecosystem-based, adaptive management (chapter 9).

This grass-roots program, financed mainly by the city, but administered by local farmers themselves, attempts to educate landowners, and provides alternative marketing opportunities that help protect the watershed. Economic incentives are offered to encourage farmers and foresters to protect the water supply. Collecting feedlot and barnyard runoff in infiltration ponds together with solid conservation practices such as terracing, contour plowing, strip farming, leaving crop residue on fields, ground cover on waterways, and culti-

Investing in soil conservation and water-quality protection on small dairy farms and agro-forestry programs within the Catskill/Delaware watershed has saved New York City billions of dollars in filtration costs and has also improved community relations. Source: *William P. Cunningham.*

vation of perennial crops such as orchards and sugarbush have significantly improved watershed water quality. As of 1999, about 400 farmers—close to the 85 percent participation goal—have signed up for the program. The cost, so far, to the city has been about $50 million—or less than 1 percent of constructing a treatment plant.

In addition to saving billions of dollars, this innovative program has helped create good will between the city and its neighbors. It has shown that upstream cleanup, prevention, and protection are cheaper and more effective than treating water after it's dirty. Farmers have learned they can be part of the solution, not just part of the problem. And we have learned that watershed planning through cooperation is effective when local people are given a voice and encouraged to participate.

- *Construction sites:* New buildings and land development projects such as highway construction affect relatively small areas but produce vast amounts of sediment, typically ten to twenty times as much per unit area as farming (fig. 18.23).

- *Land disposal:* When done carefully, land disposal of certain kinds of industrial waste, sewage sludge, and biodegradable garbage can be a good way to dispose of unwanted materials. Some poorly run land disposal sites, abandoned dumps, and leaking septic systems, however, contaminate local waters.

Generally, soil conservation methods (see chapter 9) also help protect water quality. Applying precisely determined amounts of

fertilizer, irrigation water, and pesticides saves money and reduces contaminants entering the water. Preserving wetlands that act as natural processing facilities for removing sediment and contaminants helps protect surface and groundwaters.

In urban areas, reducing materials carried away by storm runoff is helpful. Citizens can be encouraged to recycle waste oil and to minimize use of fertilizers and pesticides. Regular street sweeping greatly reduces contaminants. Runoff can be diverted away from streams and lakes. Many cities are separating storm sewers and municipal sewage lines to avoid overflow during storms.

A good example of watershed management is seen in Chesapeake Bay, the United States' largest estuary. Once fabled for its

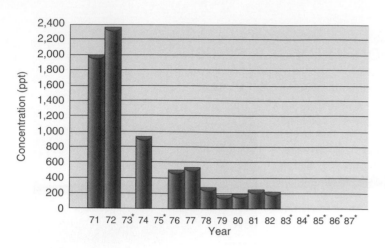

FIGURE 18.22 Dioxin concentrations in herring gull eggs on Scotch Bonnet Island in Lake Ontario have fallen sharply since 1971. Discontinued production and use of the herbicide 2,4,5-T are major reasons for this decline. *Source: State of Canada's Environment,* Minister of the Environment. *Data not available.

FIGURE 18.23 Erosion on construction sites produces a great deal of sediment and is a major cause of nonpoint water pollution. Builders are generally required to install barriers to contain sediments, but these measures are often ineffectual. © The McGraw-Hill Companies, Inc./Bob Coyle, photographer.

abundant oysters, crabs, shad, striped bass, and other valuable fisheries, the Bay had deteriorated seriously by the early 1970s. Citizens' groups, local communities, state legislatures, and the federal government together established an innovative pollution-control program that made the Bay the first estuary in America targeted for protection and restoration.

Among the principal objectives of this plan is reducing nutrient loading through land-use regulations in the six watershed states to control agricultural and urban runoff. Pollution prevention measures such as banning phosphate detergents also are important, as are upgrading wastewater treatment plants and improving compliance with discharge and filling permits. Efforts are underway to replant thousands of hectares of seagrasses and to restore wetlands that filter out pollutants. Since the 1980s, annual phosphorous discharges into the bay dropped 40 percent. Nitrogen levels, however, have remained constant or have even risen in some tributaries. Although progress has been made, the goals of reducing both nitrogen and phosphate levels by 40 percent and restoring viable fish and shellfish populations are still decades away. Still, as former EPA Administrator Carol Browner says, it demonstrates the "power of cooperation" in environmental protection.

Human Waste Disposal

As we have already seen, human and animal wastes usually create the most serious health-related water pollution problems. More than 500 types of disease-causing (pathogenic) bacteria, viruses, and parasites can travel from human or animal excrement through water. In this section, we will look at how to prevent the spread of these diseases.

Natural Processes

In the poorer countries of the world, most rural people simply go out into the fields and forests to relieve themselves as they have

always done. Where population densities are low, natural processes eliminate wastes quickly, making this an effective method of sanitation. The high population densities of cities make this practice unworkable, however. Even major cities of many less-developed countries are often littered with human waste which has been left for rains to wash away or for pigs, dogs, flies, beetles, or other scavengers to consume. This is a major cause of disease, as well as being extremely unpleasant. Studies have shown that a significant portion of the airborne dust in Mexico City is actually dried, pulverized human feces.

Where intensive agriculture is practiced—especially in wet rice paddy farming in Asia—it has long been customary to collect "night soil" (human and animal waste) to be spread on the fields as fertilizer. This waste is a valuable source of plant nutrients, but it is also a source of disease-causing pathogens in the food supply. It is the main reason that travelers in less-developed countries must be careful to surface sterilize or cook any fruits and vegetables they eat. Collecting night soil for use on farm fields was common in Europe and America until about 100 years ago when the association between pathogens and disease was recognized.

Until about 50 years ago, most rural American families and quite a few residents of towns and small cities depended on a pit toilet or "outhouse" for waste disposal. Untreated wastes tended to seep into the ground, however, and pathogens sometimes contaminated drinking water supplies. The development of septic tanks and properly constructed drain fields represented a considerable improvement in public health (fig. 18.24). In a typical septic system, wastewater is first drained into a septic tank. Grease and oils rise to the top and solids settle to the bottom, where they are subject to bacterial decomposition. The clarified effluent from the septic tank is channeled out through a drainfield of small perforated

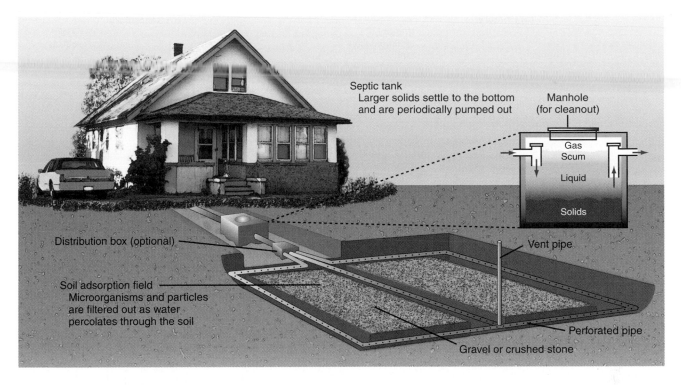

Septic tank
Larger solids settle to the bottom and are periodically pumped out

Manhole
(for cleanout)

Gas
Scum
Liquid
Solids

Distribution box (optional)

Soil adsorption field
Microorganisms and particles are filtered out as water percolates through the soil

Vent pipe

Perforated pipe

Gravel or crushed stone

FIGURE 18.24 A domestic septic tank and drain field system for sewage and wastewater disposal. To work properly, a septic tank must have healthy microorganisms, which digest toilet paper and feces. For this reason, antimicrobial cleaners and chlorine bleach should never be allowed down the drain.

pipes embedded in gravel just below the surface of the soil. The rate of aeration is high in this drainfield so that pathogens (most of which are anaerobic) will be killed, and soil microorganisms can metabolize any nutrients carried by the water. Excess water percolates up through the gravel and evaporates. Periodically, the solids in the septic tank are pumped out into a tank truck and taken to a treatment plant for disposal.

Where land is available and population densities are not too high, this can be an effective method of waste disposal. It is widely used in rural areas, but with urban sprawl, groundwater pollution often becomes a problem, indicating the need to shift to a municipal sewer system.

Municipal Sewage Treatment

Over the past 100 years, sanitary engineers have developed ingenious and effective municipal wastewater treatment systems to protect human health, ecosystem stability, and water quality. This topic is an important part of pollution control, and is a central focus of every municipal government; therefore, let's look more closely at how a typical municipal sewage treatment facility works.

Primary treatment is the first step in municipal waste treatment. It physically separates large solids from the waste stream. As raw sewage enters the treatment plant, it passes through a metal grating that removes large debris (fig. 18.25a). A moving screen then filters out smaller items. Brief residence in a grit tank allows sand and gravel to settle. The waste stream then moves to the primary sedimentation tank where about half the suspended, organic solids settle to the bottom as sludge. Many pathogens remain in

the effluent and it is not yet safe to discharge into waterways or onto the ground.

Secondary treatment consists of biological degradation of the dissolved organic compounds. The effluent from primary treatment flows into a trickling filter bed, an aeration tank, or a sewage lagoon. The trickling filter is simply a bed of stones or corrugated plastic sheets through which water drips from a system of perforated pipes or a sweeping overhead sprayer. Bacteria and other microorganisms in the bed catch organic material as it trickles past and aerobically decompose it.

Aeration tank digestion is also called the activated sludge process. Effluent from primary treatment is pumped into the tank and mixed with a bacteria-rich slurry (fig. 18.25b). Air pumped through the mixture encourages bacterial growth and decomposition of the organic material. Water flows from the top of the tank and sludge is removed from the bottom. Some of the sludge is used as an inoculum for incoming primary effluent. The remainder would be valuable fertilizer if it were not contaminated by metals, toxic chemicals, and pathogenic organisms. The toxic content of most sewer sludge necessitates disposal by burial in a landfill or incineration. Sludge disposal is a major cost in most municipal sewer budgets (fig. 18.26). In some communities this is accomplished by land farming, composting, or anaerobic digestion, but these methods don't inactivate metals and some other toxic materials.

Where space is available for sewage lagoons, the exposure to sunlight, algae, aquatic organisms, and air does the same job more slowly but with less energy costs. Effluent from secondary treatment processes is usually disinfected with chlorine, UV light,

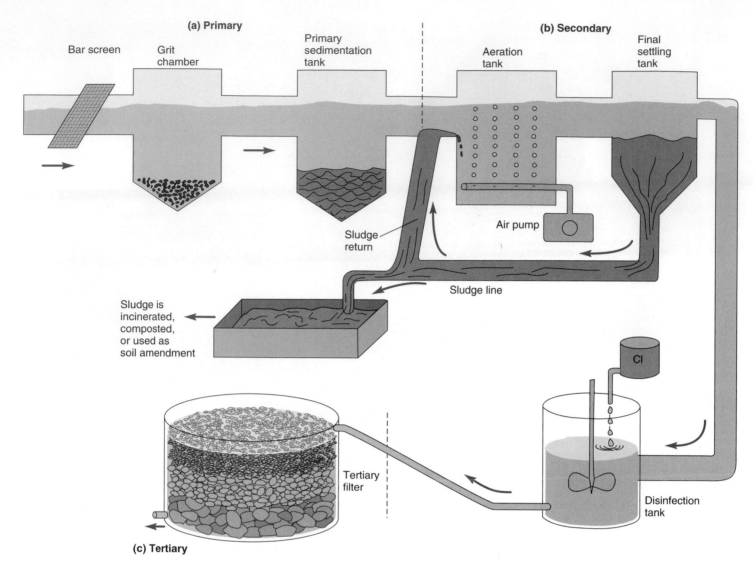

(a) Primary

Bar screen Grit chamber Primary sedimentation tank

(b) Secondary

Aeration tank Final settling tank

Sludge return

Air pump

Sludge line

Sludge is incinerated, composted, or used as soil amendment

Cl

Tertiary filter

Disinfection tank

(c) Tertiary

FIGURE 18.25 (*a*) Primary sewage treatment removes only solids and suspended sediment. (*b*) Secondary treatment, through aeration of activated sludge (or biosolids), followed by sludge removal and chlorination of effluent, kills pathogens and removes most organic material. (*c*) During tertiary treatment, passage through a trickling bed evaporator and/or a lagoon or marsh further removes inorganic nutrients, oxidizes any remaining organics, and reduces effluent volume.

or ozone to kill harmful bacteria before it is released to a nearby waterway.

Tertiary treatment removes plant nutrients, especially nitrates and phosphates, from the secondary effluent. Although wastewater is usually free of pathogens and organic material after secondary treatment, it still contains high levels of inorganic nutrients, such as nitrates and phosphates. When discharged into surface waters, these nutrients stimulate algal blooms and eutrophication. To preserve water quality, these nutrients also must be removed. Passage through a wetland or lagoon can accomplish this. Alternatively, chemicals often are used to bind and precipitate nutrients (see fig. 18.25*c*).

In many American cities, sanitary sewers are connected to storm sewers, which carry runoff from streets and parking lots. Storm sewers are routed to the treatment plant rather than discharged into surface waters because runoff from streets, yards, and industrial sites generally contains a variety of refuse, fertilizers,

pesticides, oils, rubber, tars, lead (from gasoline), and other undesirable chemicals. During dry weather, this plan works well. Heavy storms often overload the system, however, causing bypass dumping of large volumes of raw sewage and toxic surface runoff directly into receiving waters. To prevent this overflow, cities are spending hundreds of millions of dollars to separate storm and sanitary sewers. These are huge, disruptive projects. When they are finished, surface runoff will be diverted into a river or lake and cause another pollution problem.

Low-Cost Waste Treatment

The municipal sewage systems used in developed countries are often too expensive to build and operate in the developing world where low-cost, low-tech alternatives for treating wastes are needed. One option is **effluent sewerage,** a hybrid between a traditional septic tank and a full sewer system. A tank near each dwelling collects

FIGURE 18.26 "Well, if *you* can't use it, do you know anyone who *can* use 3,000 tons of sludge every day?" © 2001 by Sidney Harris.

and digests solid waste just like a septic system. Rather than using a drainfield, however, to dispose of liquids—an impossibility in crowded urban areas—effluents are pumped to a central treatment plant. The tank must be emptied once a year or so, but because only liquids are treated by the central facility, pipes, pumps, and treatment beds can be downsized and the whole system is much cheaper to build and run than a conventional operation.

Another alternative is to use natural or artificial wetlands to dispose of wastes. Arcata, California, for instance, needed an expensive sewer plant upgrade. By transforming a 65-hectare (160-acre) garbage dump into a series of ponds and marshes that serve as a simple, low-cost waste treatment facility, the city saved millions of dollars and improved the environment simultaneously. Sewage is piped to holding ponds where solids settle out and are digested by bacteria and fungi. Effluent flows through marshes where it is filtered and cleansed by aquatic plants and microorganisms. The marsh is a haven for wildlife and has become a prized recreation area for the city (chapter 13). Eventually, the purified water flows into the bay where marine life flourishes.

Similar wetland waste treatment systems are now operating in many developing countries. Effluent from these operations can be used to irrigate crops or raise fish for human consumption if care is taken to first destroy pathogens (fig. 18.27). Usually 20 to 30 days of exposure to sun, air, and aquatic plants is enough to make the water safe. These systems make an important contribution to human food supplies. A 2,500-hectare (6,000-acre) waste-fed aquaculture facility in Calcutta, for example, supplies about 7,000 metric tons of fish annually to local markets. The World Bank estimates that some 3 billion people will be without sanitation services by the middle of the next century under a business-as-usual scenario (fig. 18.28). With investments in innovative programs, however, sanitation could be provided to about half those people and a great deal of misery and suffering could be avoided.

FIGURE 18.27 In India, a poplar plantation thrives on raw sewage water piped directly from nearby homes. Innovative solutions like this can make use of nutrients that would pollute water systems.
Courtesy FAO, photo by I. De Borjegyi.

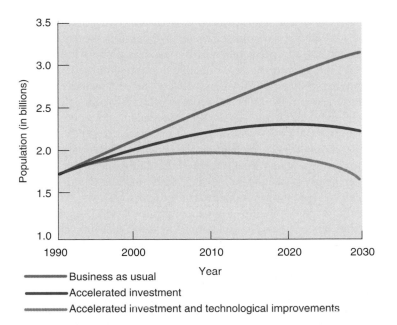

FIGURE 18.28 World population without adequate sanitation—three scenarios in the year 2030. If business as usual continues, more than 3 billion people will lack safe sanitation. Accelerated investment in sanitation services could lower this number. Higher investment, coupled with technological development, could keep the number of people without adequate sanitation from growing even though the total population increases.
Source: World Bank estimates based on research paper by Dennis Anderson and William Cavendish, "Efficiency and Substitution in Pollution Abatement: Simulation Studies in Three Sectors."

Water Remediation

Remediation means finding remedies for problems. Just as there are many sources for water contamination, there are many ways to clean it up. New developments in environmental engineering are providing promising solutions to many water pollution problems.

Containment methods confine or restrain dirty water or liquid wastes *in situ* (in place) or cap the surface with an impermeable layer to divert surface water or groundwater away from the site and to prevent further pollution. Where pollutants are buried too deeply to be contained mechanically, materials sometimes can be injected to precipitate, immobilize, chelate, or solidify them. Bentonite slurries, for instance, can effectively stabilize liquids in porous substrates. Similarly, straw or other absorbent material is spread on surface spills to soak up contaminants.

Extraction techniques pump out polluted water so it can be treated. Many pollutants can be destroyed or detoxified by chemical reactions that oxidize, reduce, neutralize, hydrolyze, precipitate, or otherwise change their chemical composition. Where chemical techniques are ineffective, physical methods may work. Solvents and other volatile organic compounds, for instance, can be stripped from solution by aeration and then burned in an incinerator. Some contaminants can be removed by semipermeable membranes or resin filter beds that bind selectively to specific materials. Some of the same techniques used to stabilize liquids *in situ* can also be used *in vitro* (in a reaction vessel). Metals, for instance, can be chelated or precipitated in insoluble, inactive forms.

Often, living organisms can be used effectively and inexpensively to clean contaminated water. We call this bioremediation (chapter 21). Restored wetlands, for instance, along stream banks or lake margins can be very effective in filtering out sediment and removing pollutants. They generally cost far less than mechanical water treatment facilities and provide wildlife habitat as well.

Lowly duckweed (*Lemna* sp.), the green scum you often see covering the surface of eutrophic ponds, grows fast and can remove large amounts of organic nutrients from water. Under optimal conditions, a few square centimeters of these tiny plants can grow to cover nearly a hectare (about 2.5 acres) in four months. Large duckweed lagoons are being used as inexpensive, low-tech sewage treatment plants in developing countries. Where conventional wastewater purification typically costs $300 to $600 per person served, a duckweed system can cost one-tenth as much. The duckweed can be harvested and used as feed, fuel, or fertilizer. Up to 35 percent of its dry mass is protein—about twice as much as alfalfa, a popular animal feed.

Where space for open lagoons is unavailable, bioremediation can be carried out in reaction vessels. This has the advantage of controlling conditions more precisely and doesn't release organisms into the environment. Some of the most complex, holistic systems for water purification are designed by Ocean Arks International (OAI) in Falmouth, Massachusetts. Their "living machines" combine living organisms—chosen to perform specific functions—in contained environments. In a typical living machine, water flows through a series of containers, each with a distinct ecological community designed for a particular function. Wastes generated by the inhabitants of one vessel become the food for inhabitants of another. Sunlight provides the primary source of energy.

OAI has created or is in the process of building water treatment plants in a dozen states and foreign countries. Designs range from remediating toxic wastes from Superfund sites to simply treating domestic wastes. Starting with microorganisms in aerobic and anaerobic environments where different kinds of wastes are metabolized or broken down, water moves through a series of containers containing hundreds of different kinds of plants and animals,

FIGURE 18.29 In-house wastewater treatment in Oberlin College's Environmental Studies building. Constructed wetlands outside, and tanks inside, allow water plants to filter water and use nutrients.
Courtesy of National Renewable Energy Lab.

including algae, rooted aquatic plants, clams, snails, and fish, each chosen to provide a particular service. Technically, the finished water is drinkable, although few people feel comfortable doing so. More often, the final effluent is used to flush toilets or for irrigation. Called ecological engineering, this novel approach can save resources and money as well as clean up our environment and serve as a valuable educational tool (fig. 18.29).

WATER LEGISLATION

Water pollution control has been among the most broadly popular and effective of all environmental legislation in the United States. It has not been without controversy, however. In this section, we will look at some of the major issues concerning water quality laws and their provisions (table 18.2).

The Clean Water Act

Passage of the Clean Water Act of 1972 was a bold, bipartisan step determined to "restore and maintain the chemical, physical, and biological integrity of the Nation's waters" that made clean water

TABLE 18.2 Some Important U.S. and International Water Quality Legislation

1. *Federal Water Pollution Control Act* (1972). Established uniform nationwide controls for each category of major polluting industries.
2. *Marine Protection Research and Sanctuaries Act* (1972). Regulates ocean dumping and established sanctuaries for protection of endangered marine species.
3. *Ports and Waterways Safety Act* (1972). Regulates oil transport and the operation of oil handling facilities.
4. *Safe Drinking Water Act* (1974). Requires minimum safety standards for every community water supply. Among the contaminants regulated are bacteria, nitrates, arsenic, barium, cadmium, chromium, fluoride, lead, mercury, silver, pesticides; radioactivity and turbidity also regulated. This act also contains provisions to protect groundwater aquifers.
5. *Resource Conservation and Recovery Act* (RCRA) (1976). Regulates the storage, shipping, processing, and disposal of hazardous wastes and sets limits on the sewering of toxic chemicals.
6. *Toxic Substances Control Act* (TOSCA) (1976). Categorizes toxic and hazardous substances, establishes a research program, and regulates the use and disposal of poisonous chemicals.
7. *Comprehensive Environmental Response, Compensation, and Liability Act* (CERCLA) (1980) and *Superfund Amendments and Reauthorization Act* (SARA) (1984). Provide for sealing, excavation, or remediation of toxic and hazardous waste dumps.
8. *Clean Water Act* (1985) (amending the 1972 Water Pollution Control Act). Sets as a national goal the attainment of "fishable and swimmable" quality for all surface waters in the United States.
9. *London Dumping Convention* (1990). Calls for an end to all ocean dumping of industrial wastes, tank washing effluents, and plastic trash. The United States is a signatory to this international convention.

a national priority. Along with the Endangered Species Act and the Clean Air Act, this is one of the most significant and effective pieces of environmental legislation ever passed by the U.S. Congress. It also is an immense and complex law, with more than 500 sections regulating everything from urban runoff, industrial discharges, and municipal sewage treatment to land-use practices and wetland drainage.

The ambitious goal of the Clean Water Act was to return all U.S. surface waters to "fishable and swimmable" conditions. For specific "point" sources of pollution such as industrial discharge pipes or sewage outfalls, the act requires discharge permits and **best practicable control technology (BPT).** It sets national goals of **best available, economically achievable technology (BAT),** for toxic substances and zero discharge for 126 priority toxic pollutants. As we discussed earlier in this chapter, these regulations have had a positive effect on water quality. While not yet swimmable or fishable everywhere, surface-water quality in the United States has significantly improved on average over the past quarter century. Perhaps the most important result of the act has been investment of $54 billion in federal funds and more than $128 billion in state and local funds for municipal sewage treatment facilities.

Not everyone, however, is completely happy with the Clean Water Act. Industries, state and local governments, farmers, land developers, and others who have been forced to change their operations or spend money on water protection often feel imposed upon. One of the most controversial provisions of the act has been Section 404, which regulates draining or filling of wetlands. Although the original bill only mentions wetlands briefly, this section has evolved through judicial interpretation and regulatory policy to become one of the principal federal tools for wetland protection. Many people applaud the protection granted to these ecologically important areas that were being filled in or drained at a rate of about half a million hectares per year before the passage of the Clean Water Act. Farmers, land developers, and others who

are prevented from converting wetlands to other uses often are outraged by what they consider "taking" of private lands.

Another sore point for opponents of the Clean Water Act are what are called "unfunded mandates," or requirements for state or local governments to spend money that is not repaid by Congress. You will notice that the $128 billion already spent by cities to install sewage treatment and stormwater diversion to meet federal standards far exceeds the $54 billion in congressional assistance for these projects. Estimates are that local units of government could be required to spend another $130 billion to finish the job without any further federal funding. Small cities that couldn't afford or chose not to participate in earlier water quality programs, in which the federal government paid up to 90 percent of the costs, are especially hard hit by requirements that they upgrade municipal sewer and water systems. They now are faced with carrying out those same projects entirely on their own funds.

Clean Water Act Reauthorization

Opponents of federal regulation have tried repeatedly to weaken or eliminate the Clean Water Act. They regard restriction of their "right" to dump toxic chemicals and waste into wetlands and waterways to be an undue loss of freedom. They resent being forced to clean up municipal water supplies, and call for cost/benefit analysis that places greater weight on economic interests in all environmental planning. Most of all, they view any limitation on use of private property to be a "taking" for which they should be fully compensated.

Even those who support the Clean Water Act in principle would like to see it changed and strengthened. Among these proposals are a shift from "end-of-the-pipe" focus on removing specific pollutants from effluents to more attention to changing industrial processes so toxic substances won't be produced in the first place. Another important issue is nonpoint pollution from agricultural runoff and urban areas, which has become the largest source of surface-water degradation in the United States. Regulating these sources remains a difficult problem.

Environmentalists also would like to see stricter enforcement of existing regulations, mandatory minimum penalties for violations, more effective community right-to-know provisions, and increased powers for citizen lawsuits against polluters. Studies have found that, in practice, polluters are given infrequent and light fines for polluting. Under the current law, using data that polluters themselves are required to submit, groups such as the Natural Resources Defense Council and the Citizens for a Better Environment have won million-dollar settlements in civil lawsuits (the proceeds generally are applied to clean-up projects) and some transgressors have even been sent to jail. Not surprisingly, environmentalists want these powers expanded, while polluters find them very disagreeable.

Other Important Water Legislation

In addition to the Clean Water Act, several other laws help to regulate water quality in the United States and abroad. Among these is the Safe Drinking Water Act, which regulates water quality in commercial and municipal systems. Critics complain that standards and enforcement policies are too lax, especially for rural water districts and small towns. Some researchers report pesticides, herbicides, and lead in drinking water at levels they say should be of concern (fig. 18.30). Atrazine, for instance, a widely used herbicide, was detected in 96 percent of all surface-water samples in one study of 374 communities across 12 states. Remember, however, that simply detecting a toxic compound is not the same as showing dangerous levels.

The Superfund program for remediation of toxic waste sites was created in 1980 by the Comprehensive Environmental Response,

What can you do?

Steps You Can Take to Improve Water Quality

Individual actions have important effects on water quality. Here are some steps you can take to make a difference.

- Compost your yard waste and pet waste. Nutrients from decayed leaves, grass, and waste are a major urban water pollutant. Many communities have public compost sites available.

- Don't fertilize your lawn or apply lawn chemicals. Untreated grass can be just as healthy, and it won't poison your pets or children.

- Make sure your car doesn't leak fluids, oil, or solvents on streets and parking lots, from which contaminants wash straight into rivers and lakes.Recycle motor oil at a gas station or oil change shop.

- Don't buy lawn mowers, personal watercraft, or other vehicles with two-cycle engines, which release abundant fuel and oil into air and water. Instead, buy more efficient, four-stroke engines.

- Visit your local sewage treatment plant. Often public tours are available or group tours can be arranged, and these sites can be fascinating.

- Keep informed about water policy debates at local and federal levels. Policies change often, and public input is important.

Compensation, and Liability Act (CERCLA) and was amended by the Superfund Amendments and Reauthorization Act (SARA) of 1984. This program is designed to provide immediate response to emergency situations and to provide permanent remedies for abandoned or inactive sites. These programs provide many jobs for environmental science majors in monitoring and removal of toxic wastes and landscape restoration. A variety of methods have been developed for remediation of problem sites.

Among the most important international agreements on water quality is the 1972 Great Lakes Water Quality Agreement between Canada and the United States. The agreement has produced encouraging progress in cleaning up the world's largest freshwater system. Another major international agreement was the 1990 London Dumping Convention, which called for phasing out all ocean dumping of industrial waste, tank-washing effluent, and plastic trash by 1995. The 64 nations that have signed the Law of the Sea Treaty are bound by this agreement. The United States has passed legislation to support its provisions. Whether this can be enforced remains to be seen, however.

Is it safe to assume that we are well on our way to solving our water pollution problems? We are better off in terms of legislation, policy, and practice than we were in 1960. Laws, however, are only as good as (1) the degree to which they are not weakened by subsequent amendments and exceptions and (2) the degree to which they are funded for research and enforcement. Economic interests cause continued pressure on both of these points, so that the importance of an overriding national attitude to maintain the intent of protective legislation must continually be stressed and retaught.

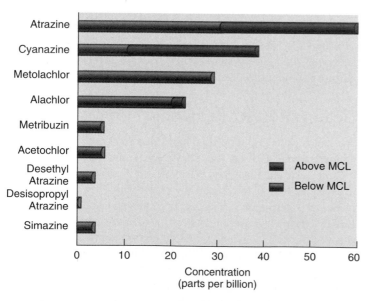

FIGURE 18.30 Nine herbicide active ingredients and metabolites found in drinking water samples in Fort Wayne, Indiana, Spring 1995, compared to federal Maximum Concentration Levels (MCL). The use of these chemicals, and their presence in water sources, has increased in subsequent years. *Source:* B. Cohen and E. Bondoc, "Weed Killers by the Glass," Environmental Working Group 1995.

Summary

- Water pollution control laws have greatly reduced the worst water pollution in most industrialized countries. In many developing countries, water pollution is getting worse, but access to safe drinking water is improving slowly, especially in urban areas.

- Water pollution is any physical, biological, or chemical change in water quality that adversely affects organisms or makes it unsuitable for other uses. Pollution is classed as point source, originating at a specific location such as a drain pipe, ditch, or sewer outfall, or nonpoint source, which is widespread runoff from fields, streets, or cities. Nonpoint sources are usually more irregular and harder to control than point sources. Sediment, untreated human waste, and nutrient enrichment are the most widespread pollutants.

- Major types of water pollution include infectious agents; oxygen-demanding wastes; nutrients; inorganic salts, metals, acids and bases; organic chemicals; sediment; and thermal pollution. In developing countries, infectious agents in drinking water are a primary cause of disease and debilitation.

- Nutrient enrichment and warming cause eutrophication, which involves rapid plant and algae growth, and often results in rapid decomposition and oxygen depletion in aquatic systems. Oligotrophic lakes and streams, by contrast, are cold, and oxygen-rich, but have low biological productivity. In marine environments, excess nutrients can produce dead zones and toxic tides—concentrations of poisonous microorganisms.

- The U.S. Clean Water Act requires that the EPA monitor water quality and regulate discharge. Total maximum daily loads (TMDLs) are set for major pollutants. Only 10 percent of water pollution in the United States now comes from point sources. Nonpoint sources, including feedlots, manure lagoons, farm fields, construction sites, and cities, remain important problems.

- Groundwater can be contaminated by waste in recharge zones through abandoned wells, and by buried waste. Leaking underground storage tanks release contaminants, such as MTBE, into groundwater.

- Ocean pollution comes from many sources, including garbage, sewage, nutrients, and oil spills. About half of oil releases come from routine releases and tank cleaning.

- The cheapest and most effective way to reduce water pollution is source reduction. Often, industrial wastes can be recycled or reclaimed rather than released. Agricultural practices can reduce field runoff, and sediment barriers at construction sites can reduce sediment releases. Land-use planning can greatly reduce pollution production.

- Sewage treatment is one of the most important steps in maintaining clean drinking water. Primary treatment removes, strains, and settles out solids. Secondary treatment, including aeration, digestion, and chlorination, removes pathogens and organic material. Tertiary treatment removes inorganic nutrients and oxidizes remaining organics. A variety of low-cost methods, such as constructed wetlands, can be used to purify water.

- Water legislation is credited with radically improving water quality. Legislation, including the Clean Water Act, remains controversial: costs can be high and are largely borne by producers; it can be difficult to identify the best, most affordable, or best practicable technology; and proponents of greater controls object that stricter rules are needed and that enforcement is too often lax.

Questions for Review

1. Define *water pollution*.
2. List eight major categories of water pollutants and give an example for each category.
3. Describe eight major sources of water pollution in the United States. What pollution problems are associated with each source?
4. What is *Pfiesteria* and why is it dangerous?
5. What is eutrophication? What causes it?
6. What are the origins and effects of siltation?
7. Describe primary, secondary, and tertiary processes for sewage treatment. What is the quality of the effluent from each of these processes?
8. Why do combined storm and sanitary sewers cause water quality problems? Why does separating them also cause problems?
9. What pollutants are regulated by the Clean Water Act? What goals does this act set for abatement technology?
10. Describe remediation techniques and how they work.

Questions for Critical Thinking

1. Cost is the greatest obstacle to improving water quality. How would you decide how much of the cost of pollution control should go to private companies, government, or individuals?
2. How would you define *adequate* sanitation? Think of some situations in which people might have different definitions for this term.
3. Humans can survive relatively unclean water. How much effort should we make to reduce pollution to benefit other species?
4. What sorts of information would you need to make a judgment about whether water quality in your area is getting better

or worse? How would you weigh different sources, types, and effects of water pollution?

5. Imagine yourself in a developing country with a severe shortage of clean water. What would you miss most if your water supply were suddenly cut by 90 percent?

6. Proponents of deep well injection of hazardous wastes argue that it will probably never be economically feasible to pump water out of aquifers more than 1 kilometer below the surface. Therefore, they say, we might as well use those aquifers for hazardous waste storage. Do you agree? Why or why not?

7. Suppose that part of the silt in a river is natural and part is human-caused. Is one pollution but the other not?

8. Arsenic contamination in Bangladesh results from geological conditions, World Bank and U.S. aid, poverty, government failures, and other causes. Who do you think is responsible for finding a solution? Why? Would you answer differently if you were a poor villager in Bangladesh?

Key Terms

atmospheric deposition 379
best available, economically achievable technology (BAT) 401
best practicable control technology (BPT) 401
biochemical oxygen demand (BOD) 381
coliform bacteria 380
cultural eutrophication 382
dissolved oxygen (DO) content 381
effluent sewerage 398
eutrophic 382
nonpoint sources 379
oligotrophic 382
oxygen sag 381
point sources 379
primary treatment 397
red tide 383
secondary treatment 397
tertiary treatment 398
thermal plume 387
total maximum daily loads (TMDL) 388

Further Readings

Adler, Robert W., et al. 1993. *Clean Water Act Twenty Years Later.* Island Press.

Forster, D. L. 2000. Public policies and private decisions: Their impacts on Lake Erie water quality and farm economy. *Journal of Soil and Water Conservation* 309:322–26.

Harvell, C. D., et al. 1999. Emerging marine diseases—Climate links and anthropogenic factors. *Science* 285:1505–10.

Paul , Michael J., and Judy L. Meyer. 2001. Streams in the urban landscape. *Annual Review of Ecology and Systematics* 32:333–65.

Pickett, S. T. A., et al. 2001. Urban ecological systems: Linking terrestrial, ecological, physical, and socioeconomic components of metropolitan areas. *Annual Review of Ecology and Systematics.* 32:127–57.

Shiva, Vandana. 2002. *Water Wars: Privatization, Pollution, and Profit.* South End Press.

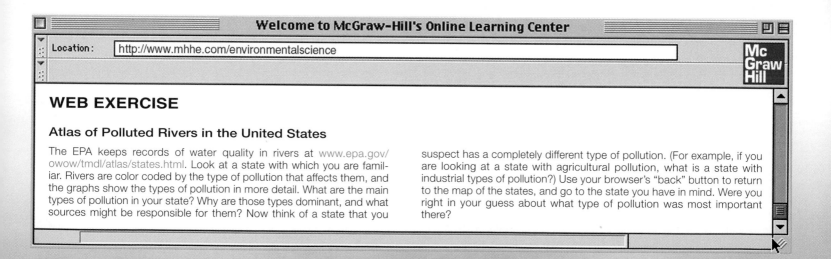

Welcome to McGraw-Hill's Online Learning Center

Location: http://www.mhhe.com/environmentalscience

WEB EXERCISE

Atlas of Polluted Rivers in the United States

The EPA keeps records of water quality in rivers at www.epa.gov/owow/tmdl/atlas/states.html. Look at a state with which you are familiar. Rivers are color coded by the type of pollution that affects them, and the graphs show the types of pollution in more detail. What are the main types of pollution in your state? Why are those types dominant, and what sources might be responsible for them? Now think of a state that you suspect has a completely different type of pollution. (For example, if you are looking at a state with agricultural pollution, what is a state with industrial types of pollution?) Use your browser's "back" button to return to the map of the states, and go to the state you have in mind. Were you right in your guess about what type of pollution was most important there?

19

Conventional Energy

The complex problems we face today will not be solved by the same level of thinking that created them.

Albert Einstein

OBJECTIVES

After studying this chapter, you should be able to:

- summarize our current energy sources and explain briefly how energy use has changed through history.
- compare our energy consumption with that of other people in the world.
- itemize the ways we use energy.
- analyze the resources and reserves of fossil fuels in the world.
- evaluate the costs and benefits of using coal, oil, and natural gas.
- understand how nuclear reactors work, why they are dangerous, and how they might be made safer.
- discuss our options for radioactive waste storage and disposal.
- summarize public opinion about nuclear power and how you feel about this controversial energy source.

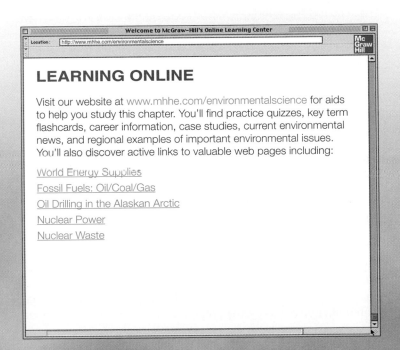

LEARNING ONLINE

Visit our website at www.mhhe.com/environmentalscience for aids to help you study this chapter. You'll find practice quizzes, key term flashcards, career information, case studies, current environmental news, and regional examples of important environmental issues. You'll also discover active links to valuable web pages including:

World Energy Supplies
Fossil Fuels: Oil/Coal/Gas
Oil Drilling in the Alaskan Arctic
Nuclear Power
Nuclear Waste

Photo: Fossil fuels provide about 86 percent of our commercial energy, but also cause air pollution, global warming, and vulnerability to supply disruptions. © Corbis Royalty Free Website.

What a Tangled Web We Weave

At 4:09 P.M. Thursday, August 14, 2003, erratic electrical currents surged through the complex electrical grid circling Lake Erie (fig. 19.1). Like falling dominoes, high-voltage transmission lines, power plants, and distribution systems failed in a furious avalanche that eventually shut off power across much of the eastern United States and Canada. Within a few minutes, at least 100 large generating stations—20 of them nuclear power plants—disconnected from the grid, and 50 million people were without electricity. New York City, Cleveland, Detroit, Ottawa, and Toronto went dark, as did a wide swath reaching from New Jersey to western Michigan and eastern Ontario.

Office workers watched their computer monitors blink off without warning. Worried residents poured out onto city streets, asking anxiously if this was a terrorist attack. Many were dumbfounded to find that their cell phones—the modern lifeline—wouldn't work. With traffic lights out, city streets soon turned into gridlock. Volunteers tried to direct traffic with mixed results. Soon hospitals and government buildings were switching on backup generators to keep essential equipment operating, and the police began evacuating thousands of people trapped in elevators. The 350,000 people on the New York City subways when the lights went out, began walking through black, muddy tunnels to find escape ladders. Some had to climb the equivalent of four stories to reach street level. New York streets became a sea of pedestrians as millions of commuters started walking home.

The lost power closed factories and retail stores across eight states and much of eastern Canada. As the power failure wore on through Friday and Saturday, at least a dozen major airports canceled or diverted flights, triggering a chain reaction of interruptions across the continent. Grocery stores reported that millions of dollars worth of frozen goods were thawing and spoiling. The total economic impact may be as high as $5 billion. People in high-rise buildings had no water without pumps to lift it. Metropolitan sewer systems stopped functioning. On one of the hottest weekends of the summer, air conditioners were dead and even fans weren't working. People cooked on outdoor grills, and ate by candlelight on their front steps or porches. Some city residents saw the stars for the first time in their lives.

The huge blackout provides a springboard for various factions to advance their ideas about energy policy. Utility companies immediately called for billions of dollars in rate increases and subsidies to build new power plants and to upgrade the transmission grid. They blamed the failure on environmental regulations and public opposition that prevents them from siting new power plants and transmission lines.

Conservationists, on the other hand, say that rather than commit ourselves to huge nuclear and fossil fuel-burning plants and high-voltage transmission lines, we should focus instead on conservation and renewable energy. The best way to prevent future energy bottle-

FIGURE 19.1   Power lines in Ontario form one leg of the "Lake Erie Loop" that failed catastrophically in August 2003.
© William P. Cunningham.

necks and grid overload, they argue, is to increase the efficiency of our buildings, homes, factories, and appliances, in addition to our transmission lines. They also call for speedier development of dispersed, renewable energy, such as wind and solar power, as well as small, decentralized on-site power sources such as fuel cells, wind turbines, rooftop solar electric cells, and microgenerators that furnish both heat and electricity.

Many industry observers believe that a decade of government deregulation made the blackout inevitable. No single authority is in charge of the grid, and utilities now have little incentive to invest the money needed to improve its reliability. In addition, the sheer complexity of the system may have led to its downfall. Being able to buy power from an idle plant half a continent away saves consumers money in the short run, but it exposes us to cascading failures precisely because of its interconnectivity.

Within 48 hours, power had been restored to most of the stricken eastern interconnection, and life returned slowly to normal. This massive power outage reminds us, however, of how much we have come to depend on uncertain energy supplies. In this chapter, we'll look at conventional energy sources such as fossil fuels and nuclear power. Chapter 20 continues with a consideration of conservation and renewable energy sources such as wind, solar, biomass, and geothermal energy.

WHAT IS ENERGY AND WHERE DO WE GET IT?

Work is the application of force through a distance. **Energy** is the capacity to do work. **Power** is the rate of flow of energy, or the rate at which work is done. The energy that we use to move our muscles, to think, and to carry out metabolic functions comes from stored chemical energy (or potential energy) in our food. Food energy is generally measured in calories (cal). One calorie is the amount of energy to heat 1 gram of water 1°C. A kilocalorie (or food Calorie) is 1,000 calories. In physics, the basic metric unit of force is a newton, which is the force necessary to accelerate 1 kilogram 1 meter per second. A **joule (J)** is the amount of work done when a force of 1 newton is exerted over 1 meter or 1 amp per second flows through 1 ohm. One J equals 0.238 cal. Some other common energy units are presented in table 19.1.

TABLE 19.1	Some Energy Units

1 joule (J) = the force exerted by a current of 1 amp per second flowing through a resistance of 1 ohm

1 watt (W) = 1 joule (J) per second

1 kilowatt-hour (kWh) = 1 thousand (10^3) watts exerted for 1 hour

1 megawatt (MW) = 1 million (10^6) watts

1 gigawatt (GW) = 1 billion (10^9) watts

1 petajoule (PJ) = 1 quadrillion (10^{15}) joules

1 PJ = 947 billion BTU, or 0.278 billion kWh

1 British thermal unit (BTU) = energy to heat 1 lb of water 1°F

1 standard barrel (bbl) of oil = 42 gal (160 l) or 5.8 million BTU

1 metric ton of standard coal = 27.8 million BTU or 4.8 bbl oil

FIGURE 19.2 Sudden price shocks in the 1970s caused by anticipated oil shortages showed Americans that our energy-intensive lifestyles may not continue forever. © Owen Franken/Stock Boston.

A Brief Energy History

Fire was probably the first human energy technology. Charcoal from fires has been found at sites occupied by our early ancestors 1 million years ago. Muscle power provided by domestic animals has been important at least since the dawn of agriculture some 10,000 years ago. Wind and water power have been used nearly as long. The invention of the steam engine, together with diminishing supplies of wood in industrializing countries, caused a switch to coal as our major energy source in the nineteenth century. Coal, in turn, has been replaced by oil in this century due to the ease of shipping, storing, and burning liquid fuels.

World oil use peaked in 1979, when daily production passed 66 million barrels per day. An Arab oil embargo in 1973 and the Iranian revolution in 1979 caused crude oil prices to rise nearly tenfold during the 1970s. These sudden price shocks were a major source of the crushing debt burdens that still hold back many developing countries. The results were less disruptive in richer countries, but Americans waiting in long lines at gas stations became aware for the first time that much of our luxurious lifestyle is dependent on a limited, unstable oil supply (fig. 19.2).

The early 1980s saw an increased concern about conservation and development of renewable energy resources. This concern didn't last long, unfortunately, as reduced demand and increased production in the mid-1980s caused an oil glut that made prices fall almost as rapidly as they had risen a decade earlier. Through most of the 1990s, the price of a benchmark barrel of Texas light crude hovered around $15 when adjusted for inflation. Since then, production curbs instituted by the Organization of Petroleum Exporting Countries (OPEC) pushed prices above $30 per barrel, still less, in real dollars, than prices 30 years earlier. The United States now imports more than half of its oil. These imports cost some $50 billion each year, more than one-half our total foreign trade deficit.

Current Energy Sources

Fossil fuels (petroleum, natural gas, and coal) now provide about 86 percent of all commercial energy in the world (fig. 19.3). Hydroelectric dams supply about 7 percent of our commercial power,

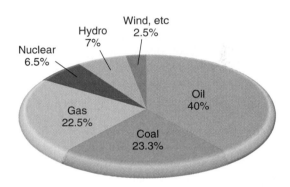

FIGURE 19.3 Worldwide commercial energy consumption. This does not include energy collected for personal use or traded in informal markets. Courtesy of British Petroleum, 2003.

while wind, solar, geothermal, and biomass together make up less than 3 percent. Although not included in these data, private systems (individual houses, businesses, industries) may capture as much renewable energy as that reported in commercial transactions. Similarly, this survey probably underrepresents the importance of biomass because these fuels often are gathered by the people using them, or they are sold in the informal economy and not reported internationally. In many poorer countries, such as Haiti, Bhutan, and Malawi, biomass supplies more than 90 percent of the energy used for heating and cooking. This is an important energy source for poor people, but can be a serious cause of forest destruction and soil loss (chapter 12).

Nuclear power is roughly equal to hydroelectricity, worldwide (i.e., 6.5 percent of all commercial energy), but it makes up about 20 percent of all electric power in more developed countries such as Canada and the United States. We have enough nuclear fuel to produce power for a long time, and it has the benefit of not emitting greenhouse gases (chapter 15), but as we discuss later in this chapter, safety concerns make this option unacceptable to most people.

Per Capita Consumption

Perhaps the most important facts about fossil fuel consumption are that those of us in the 20 richest countries consume nearly 80 percent of the natural gas, 65 percent of the oil, and 50 percent of the coal produced each year. Although we make up less than *one-fifth* of the world's population, we use more than *one-half* of the commercial energy supply. The United States and Canada, for instance, constitute only 5 percent of the world's population but consume about one-quarter of the available energy. To be fair, however, some of that energy goes to grow crops or manufacture goods that are later shipped to developing countries.

How much energy do you use every year? Most of us don't think about it much, but maintaining the lifestyle we enjoy requires an enormous energy input. On average, each person in the United States and Canada uses more than 300 GJ (equivalent to about 60 barrels of oil) per year. By contrast, in the poorest countries of the world, such as Ethiopia, Kampuchea, Nepal, and Bhutan, each person generally consumes less than one GJ per year. This means that each of us consumes, on average, almost as much energy in a single day as a person in one of these countries consumes in a year. On the other hand, Saudi Arabia, where water often is more expensive than oil, uses ten times as much energy per person as North America to maintain a luxurious, modern lifestyle.

Clearly, having an abundant external energy supply contributes to the comfort and convenience of our lives (fig. 19.4). Those of us in the richer countries enjoy many amenities not available to most people in the world. The linkage is not absolute, however. Several European countries, including Sweden, Denmark, and Switzerland, have higher standards of living than does the United States by almost any measure but use about half as much energy as we do. These countries have had effective energy conservation programs for many years. Japan, also, has a far lower energy consumption rate than might be expected for its industrial base and income level. Because Japan has few energy resources of its own, it has developed very efficient energy conservation measures.

One measure of energy efficiency is the amount used to produce a dollar of economic output. As figure 19.4 shows, while most parts of the world have made progress in energy conservation over the past decade or so, large regional differences persist. The former USSR, for instance, still uses more than twice as much energy per unit of Gross Domestic Product (GDP) as does Western Europe. This presents a great opportunity for energy conservation, and explains why most carbon dioxide trading (see chapter 15) is taking place in Eastern Europe.

HOW ENERGY IS USED

The largest share (33 percent) of the energy used in the United States is consumed by industry (fig. 19.5). Mining, milling, smelting, and forging of primary metals consume about one-quarter of the industrial energy share. The chemical industry is the second largest industrial user of fossil fuels, but only half of its use is for energy generation. The remainder is raw material for plastics, fertilizers, solvents, lubricants, and hundreds of thousands of organic chemicals in commercial use. The manufacture of cement, glass, bricks, tile, paper, and processed foods also consumes large amounts of energy. Residential and commercial buildings use some 20 percent of the primary energy consumed in the United States, mostly for space heating, air conditioning, lighting, and water heating. Small motors and electronic equipment take an increasing share of residential and commercial energy. In many office buildings, computers, copy machines, lights, and human bodies liberate enough waste heat so that a supplementary source is not required, even in the winter.

Transportation consumes about 27 percent of all energy used in the United States each year. About 98 percent of that energy comes from petroleum products refined into liquid fuels, and the remaining 2 percent is provided by natural gas and electricity. Almost three-quarters of all transport energy is used by motor vehicles. Nearly 3 trillion passenger miles and 600 billion ton miles of freight are carried annually by motor vehicles in the United States. Much of our transportation is extremely inefficient. Driving a 2,200 kg vehicle to take one 60 kg person a few kilometers for shopping or work might be considered a questionable use of resources. About 75 percent of all freight traffic in the United States is carried by trains, barges, ships, and pipelines, but because they are very efficient, they use only 12 percent of all transportation fuel.

Finally, analysis of how energy is used has to take into account waste and loss of potential energy. About *half* of all the energy in primary fuels is lost during conversion to more useful forms, while it is being shipped to the site of end use, or during its use. Electricity, for instance, is generally promoted as a clean, efficient source of energy because when it is used to run a resistance heater or an electrical appliance almost 100 percent of its energy is converted to useful work and no pollution is given off.

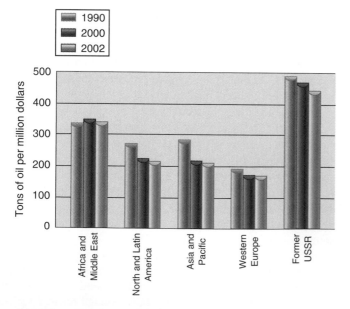

FIGURE 19.4 Energy consumption (tons of oil equivalent) per million dollars of GDP by region, 1990, 2000, and 2002. *Source:* Statistical Yearbook, 2003.

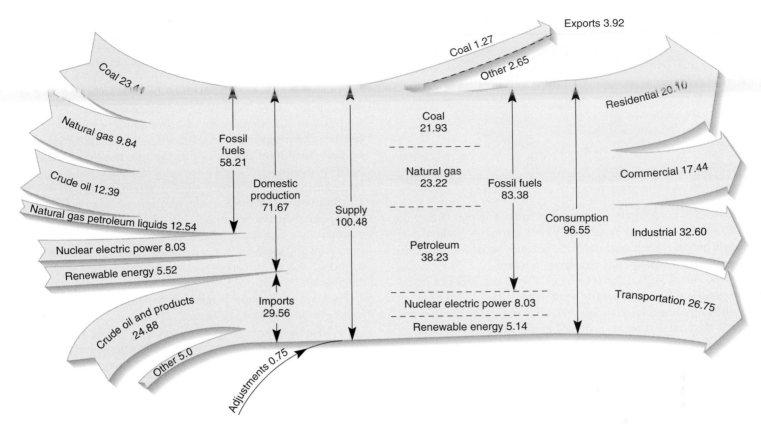

FIGURE 19.5 U.S. Energy Flow, 2002. Quantities are in quadrillion BTUs (quads). *Source:* U.S. Energy Flow, U.S. Department of Energy, 2003.

What happens before then, however? We often forget that huge amounts of pollution are released during mining and burning of the coal that fires power plants. Furthermore, nearly two-thirds of the energy in the coal that generated that electricity was lost in thermal conversion in the power plant. About 10 percent more is lost during transmission and stepping down to household voltages. Similarly, about 75 percent of the original energy in crude oil is lost during distillation into liquid fuels, transportation of that fuel to market, storage, marketing, and combustion in vehicles.

Natural gas is our most efficient fuel. Only 10 percent of its energy content is lost in shipping and processing since it moves by pipelines and usually needs very little refining. Ordinary gas-burning furnaces are about 75 percent efficient, and high-economy furnaces can be as much as 95 percent efficient. Because natural gas has more hydrogen per carbon atom than oil or coal, it produces about half as much carbon dioxide—and therefore half as much contribution to global warming—per unit of energy.

COAL

Coal is fossilized plant material preserved by burial in sediments and altered by geological forces that compact and condense it into a carbon-rich fuel. Coal is found in every geologic system since the Silurian Age 400 million years ago, but graphite deposits in very old rocks suggest that coal formation may date back to Pre-cambrian times. Most coal was laid down during the Carbonifer-

ous period (286 million to 360 million years ago) when the earth's climate was warmer and wetter than it is now. Because coal takes so long to form, it is essentially a nonrenewable resource.

Coal Resources and Reserves

World coal deposits are vast, ten times greater than conventional oil and gas resources combined. Coal seams can be 100 m thick and can extend across tens of thousands of square kilometers that were vast swampy forests in prehistoric times. The total resource is estimated to be 10 trillion metric tons. If all this coal could be extracted, and if coal consumption continued at present levels, this would amount to several thousand years' supply. At present rates of consumption, these proven-in-place reserves—those explored and mapped but not necessarily economically recoverable—will last about 200 years. Note that "known reserves" have been identified but not thoroughly mapped. **"Proven reserves"** have been mapped, measured, and shown to be economically recoverable. Ultimate reserves include unknown as well as known resources.

Where are these coal deposits located? They are not evenly distributed throughout the world (fig. 19.6). North America, Europe, and Asia contain more than 90 percent of the world's coal, and five nations (United States, Russia, China, India, and Australia) account for three-quarters of that amount. In part, countries with large land areas are more likely to have coal deposits, but this resource is very rare in Africa, the Middle East, or Central and South America (fig. 19.7). Both China and India plan to greatly

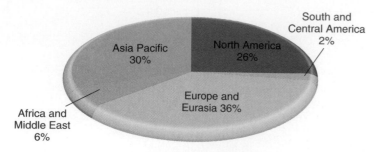

FIGURE 19.6 Proven-in-place coal reserves by region, 2002.

increase coal consumption to raise standards of living. If they do so, CO_2 released could exacerbate global warming. Antarctica is thought to have large coal deposits, but they would be difficult, expensive, and ecologically damaging to mine.

It would seem that the abundance of coal deposits is a favorable situation. But do we really want to use all of the coal? What are the environmental and personal costs of coal extraction and use? In the next section, we will look at some of the disadvantages and dangers of mining and burning coal.

Mining

Coal mining is a hot, dirty, and dangerous business (fig. 19.8). Underground mines are subject to cave-ins, fires, accidents, and accumulation of poisonous or explosive gases (carbon monoxide, carbon dioxide, methane, hydrogen sulfide). Between 1870 and 1950, more than 30,000 coal miners died of accidents and injuries in Pennsylvania alone, equivalent to one man per day for 80 years.

Untold thousands have died of respiratory diseases. In some mines, nearly every miner who did not die early from some other cause was eventually disabled by **black lung disease,** inflammation and fibrosis caused by accumulation of coal dust in the lungs or airways (chapter 8). Few of these miners or their families were compensated for their illnesses by the companies for which they worked. The U.S. Department of Labor now compensates miners and their dependents under the Black Lung Benefits Program, but workers often have difficulty proving that health problems are occupationally related. Even when compensation is provided, black lung remains a wretched occupational hazard of coal mining.

Strip mining or surface mining is cheaper than underground mining but often makes the land unfit for any other use. Mine reclamation is now mandated in the United States, but efforts often are superficial and ineffective. Coal mining also contributes to water pollution. Sulfur and other water soluble minerals make mine drainage and runoff from coal piles and mine tailings acidic and highly toxic. Thousands of miles of streams in the United States have been poisoned by coal-mining operations.

Perhaps the most egregious type of strip mining is "mountaintop removal," practiced mainly in Appalachia, where the tops of mountain ridges are scraped off and dumped into valleys below to get at coal seams (see fig. 14.11). Streams, farms, even whole towns are buried by this practice under hundreds of meters of toxic rubble.

Air Pollution

Many people aren't aware that coal burning releases radioactivity and many toxic metals. Uranium, arsenic, lead, cadmium, mercury, rubidium, thallium, and zinc—along with a number of other

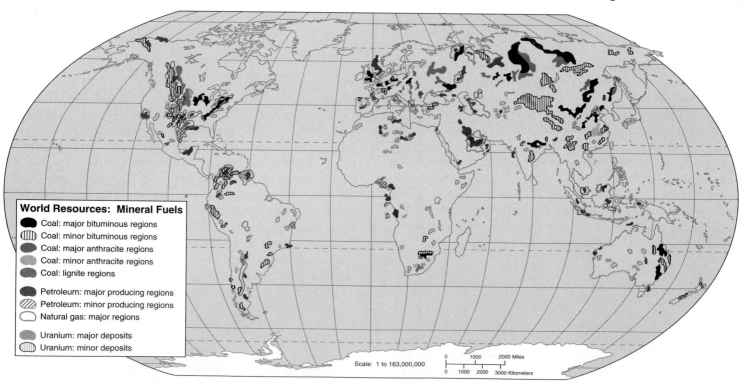

FIGURE 19.7 Where are fossil fuels and uranium located? North America, Europe, the Middle East, and parts of Asia are richly endowed. Africa, most of South America, and island states, like Japan, generally lack these fuels, greatly limiting their economic development.

FIGURE 19.8 A loader fills an enormous truck with coal in Australia. Coal mining is a dangerous and dirty job, and often leaves unrepairable scars on the land. © Volume 31/PhotoDisc.

FIGURE 19.9 Coal delivery in China. In parts of Guizhou Province, coal can have 35,000 ppm arsenic, and peppers dried over smokey coal fires can have 50,000 times as much arsenic as is considered safe. © William P. Cunningham.

elements—are absorbed by plants and concentrated in the process of coal formation. These elements are not destroyed when the coal is burned, instead they are released as gases or concentrated in fly ash and bottom slag. You are likely to get a higher dose of radiation living next door to a coal-burning power plant than a nuclear plant under normal (nonaccident) conditions. Coal combustion is responsible for about 25 percent of all atmospheric mercury pollution in the United States.

In China alone, several hundred million people commonly burn coal in unvented stoves that permeate homes with high levels of toxic metals including arsenic, mercury, selenium, and fluorine (fig. 19.9). Coal samples from Guizhou Province in south central China, for example, have been found to have as much as 35,000 ppm arsenic. Chili peppers dried over coal fires (as they typically are in this area) can have 500 ppm arsenic, or 50,000 times the acceptable level for drinking water in the United States. At least 3,000 people from Guizhou were found in one survey to be suffering from severe arsenic poisoning, as much as 80 percent of which is believed to come from contaminated food.

Coal also contains up to 10 percent sulfur (by weight). Unless this sulfur is removed by washing or flue-gas scrubbing, it is released during burning and oxidizes to sulfur dioxide (SO_2) or sulfate (SO_4) in the atmosphere. The high temperatures and rich air mixtures ordinarily used in coal-fired burners also oxidize nitrogen compounds (mostly from the air) to nitrogen monoxide, dioxide, and trioxide. Every year the 900 million tons of coal burned in the United States (83 percent for electric power generation) releases 18 million metric tons of SO_2, 5 million metric tons of nitrogen oxides (NO_x), 4 million metric tons of airborne particulates, 600,000 metric tons of hydrocarbons and carbon monoxide, and close to a trillion metric tons of CO_2. This is about three-quarters of the SO_2, one-third of the NO_x, and about half of the industrial CO_2 released in the United States each year. Coal burning is the largest single source of acid rain in many areas.

These air pollutants have many deleterious effects, including human health costs, injury to domestic and wild plants and animals, and damage to buildings and property (chapter 16). Total losses from air pollution are estimated to be between $5 billion and $10 billion per year in the United States alone. By some accounts, at least 5,000 excess human deaths per year can be attributed to coal production and burning.

Sulfur can be removed from coal before it is burned, or sulfur compounds can be removed from the flue gas after combustion. Formation of nitrogen oxides during combustion also can be minimized. Perhaps the ultimate limit to our use of coal as a fuel will be the release of carbon dioxide into the atmosphere. As we discussed in chapter 15, carbon traps heat in the atmosphere and is a major contributor to global warming. Trapping and storing CO_2 produced by stationary facilities such as power plants is beginning to be explored, but this would be difficult to do with mobile sources such as automobiles. Coal could be used, however, to make pure hydrogen or methane for fuel cells. So-called "clean coal" technologies might make this valuable resource available to us without disastrous environmental damage.

OIL

Like coal, petroleum is derived from organic molecules created by living organisms millions of years ago and buried in sediments where high pressures and temperatures concentrated and transformed them into energy-rich compounds. Depending on its age and history, a petroleum deposit will have varying mixtures of oil, gas, and solid tarlike materials. Some very large deposits of heavy oils and tars are trapped in porous shales, sandstone, and sand deposits in the western areas of Canada and the United States.

Liquid and gaseous hydrocarbons can migrate out of the sediments in which they formed through cracks and pores in surrounding rock layers. Oil and gas deposits often accumulate under layers of shale or other impermeable sediments, especially where folding and deformation of systems create pockets that will trap upward-moving hydrocarbons (fig. 19.10). Contrary to the image implied by its name, an oil pool is not usually a reservoir of liquid in an open cavern but rather individual droplets or a thin film of liquid permeating spaces in a porous sandstone or limestone, much like water saturating a sponge.

Pumping oil out of a reservoir is much like sucking liquid out of a sponge. The first fraction comes out easily, but removing subsequent fractions requires increasing effort. We never recover all the oil in a formation; in fact, a 30 to 40 percent yield is about average. There are ways of forcing water or steam into the oil-bearing formations to "strip" out more of the oil, but at least half the total

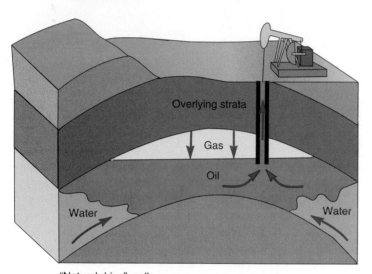

"Natural drive" well

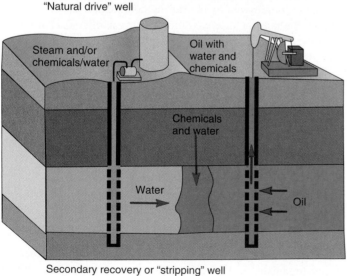

Secondary recovery or "stripping" well

FIGURE 19.10   Recovery process for petroleum. In a "natural drive" well, water or gas pressure forces liquid petroleum into the well. In an enhanced recovery or "stripping" well, steam and/or water mixed with chemicals are pumped down a well behind the oil pool, forcing oil up the extraction well. *Source:* World Resources Institute, 1998–99.

deposit usually remains in the ground at the point at which it is uneconomical to continue pumping. Methods for squeezing more oil from a reservoir are called **secondary recovery techniques** (fig. 19.10).

Oil Resources and Reserves

The total amount of oil in the world is estimated to be about 4 trillion barrels (600 billion metric tons), half of which is thought to be ultimately recoverable. Some 465 billion barrels of oil already have been consumed. In 2003, the proven reserves were roughly 1 trillion bbls, enough to last only 40 years at the current consumption rate of 25 billion barrels per year. It is estimated that another 800 billion barrels either remain to be discovered or are not recoverable at current prices with present technology. As oil resources become depleted and prices rise, it probably will become economical to find and bring this oil to the market unless alternative energy sources are developed. This estimate of the resource does not take into account the very large potential from unconventional liquid hydrocarbon resources, such as shale oil and tar sands, which might double the total reserve if they can be mined with acceptable social, economic, and environmental impacts.

By far the largest supply of proven-in-place oil is in Saudi Arabia, which has 250 billion barrels, about one-fourth of the total proven world reserve (fig. 19.11). Kuwait had more than 10 percent of the proven world oil reserves before Iraq invaded in 1990. Some 600 wells were blown up and set on fire by the retreating Iraqis and at least 5 billion barrels of Kuwait's oil was burned, spilled, or otherwise lost. Together, the Persian Gulf countries in the Middle East contain nearly two-thirds of the world's proven petroleum supplies. With our insatiable appetite for oil (some would say addiction), it is not difficult to see why this volatile region plays such an important role in world affairs.

Note that this discussion has been of *proven* reserves. Oil companies estimate that reservoirs around and under the Caspian Sea may hold 200 billion barrels. If true, this would make Azerbaijan and Kazakhstan second only to Saudi Arabia in oil riches.

Although oil consumption in Europe and North America has been relatively constant in recent years, Asian demand has been growing rapidly as economies there expand and people become more affluent. China, for instance, is expanding oil use 10 percent per year. If current trends continue, about half of all oil produced by the Organization of Petroleum Exporting Countries (OPEC) will be going to Asia.

Oil Imports and Domestic Supplies

Originally, the United States is estimated to have had about 200 billion barrels of recoverable oil, or about as much as is thought to still be in Central Asia. This large supply of relatively accessible oil played an important part in development of America's economic and industrial power during the twentieth century. Until 1947, the United States was the leading oil export country in the world. By 1998, the United States was importing nearly 10 million barrels per day, more than half of total consumption. The major suppliers of this oil were Mexico and Venezuela in the 1970s but shifted to

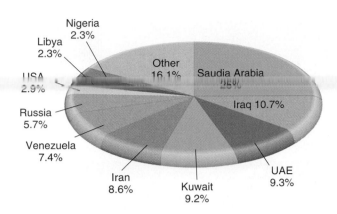

FIGURE 19.11   Proven oil reserves. Ten countries account for nearly 84 percent of all known recoverable oil. *Source:* British Petroleum.

FIGURE 19.12   Alaska's Arctic National Wildlife Refuge is home to one of the world's largest caribou herds as well as 200 other wildlife and plant species. It may also contain North America's last big unexploited oil and gas deposit. Can we extract fossil fuels without irreparably damaging the fragile tundra ecosystem? © Mary Ann Cunningham.

Saudi Arabia and other Middle Eastern countries in the 1990s, making the United States vulnerable to the unstable political situations in that volatile region.

Altogether, the United States has already used about 40 percent of its original recoverable petroleum resource. Of the 120 billion barrels thought to remain, about 58 billion barrels are proven-in-place. If we stopped importing oil and depended exclusively on indigenous supplies, our proven reserves would last only ten years at current rates of consumption.

By some estimates, the cost of supporting regimes that ensure us access to oil supplies, and maintaining military forces to keep that oil flowing smoothly, amounts to a subsidy of at least $50 billion per year. Adding the environmental and health costs of refining and using that oil shows that the real cost of gasoline is probably four or five times what we pay at the pump.

The largest remaining untapped oil field in the United States is thought to be in the Arctic National Wildlife Refuge (ANWR) in northwestern Alaska. Specifically, the narrow coastal plain where the 130,000 caribou of the Porcupine herd migrate in the summer to bear their young (fig. 19.12) is targeted by oil companies for exploration. Geologists claim there may be 16 billion barrels of oil under the refuge, and that it could be extracted without undue harm to the land or wildlife. Conservationists disagree, claiming that the economically recoverable petroleum resource may be no more than 1 million barrels per day or less than 4 percent of U.S. daily oil consumption. They also believe that wells, roads, pipelines, air strips, and other facilities needed to get this oil to market would cause irreparable harm. Raising the average fuel efficiency of all cars and light trucks in America by just one mile per gallon would save more oil than is ever likely to be recovered from ANWR. This topic has been one of the most controversial environmental issues in the United States for the past decade. Check with your instructor or legislators to learn about its current status.

Oil Shales and Tar Sands

Estimates of our total oil supply usually do not reflect the very large potential from unconventional oil resources, such as shale oil and tar sands, which might double the total reserve if they

FIGURE 19.13   Canada has huge reserves of tar sands such as the one being mined here in Alberta. By 2010, Canada could become the largest supplier of U.S. oil. Courtesy American Petroleum Institute.

can be extracted with reasonable social, economic, and environmental costs. Canada, for example, has an estimated 270 billion cubic meters of bituminous **tar sands,** mostly in northern Alberta (fig. 19.13). Liquid petroleum can be extracted from these sands with hot water, chemicals, or other stripping processes.

The oil content of Canadian tar sand is thought to be 40 times as much as that under ANWR. By 2010, Alberta expects to be shipping 2 million barrels of oil per day to the United States, or about twice as much as the Trans-Alaska pipeline can carry at peak capacity. This would make Canada surpass Saudi Arabia as the United States' biggest oil supplier. There are severe environmental costs, however, to producing this oil. A typical plant producing 125,000 barrels of oil per day creates about 15 million m³ of toxic sludge and consumes or contaminates billions of liters of water each year. Some Canadians worry about becoming an energy colony for the United States.

Similarly, vast deposits of oil shale occur in the western United States. Actually, **oil shale** is neither oil nor shale but a fine-grained sedimentary rock rich in solid organic material called kerogen. When heated to about 480°C (900°F), the kerogen liquefies and can be extracted from the stone. Oil shale beds up to 600 m (1,800 ft) thick occur in the Green River Formation in Colorado, Utah, and Wyoming, and lower-grade deposits are found over large areas of the eastern United States. If these deposits could be extracted at a reasonable price and with acceptable environmental impacts, they might yield the equivalent of several trillion barrels of oil.

Mining and extracting shale oil also creates many problems. It is expensive; it uses vast quantities of water, a scarce resource in the arid west; it has a high potential for air and water pollution; and it produces enormous quantities of waste. In the early 1980s, when the search for domestic oil supplies was at fever pitch, serious discussions occurred about filling whole canyons, rim to rim, with oil shale waste. One experimental mine used a nuclear explosion to break up the oil shale. All the oil shale projects dried up when oil prices fell in the mid-1980s, however. What do you think? Should we declare some western states sacrifice areas so we can have cheap gasoline?

NATURAL GAS

Natural gas is the world's third largest commercial fuel (after oil, and coal), making up 23 percent of global energy consumption. It is the most rapidly growing energy source because it is convenient, cheap, and clean burning. Because natural gas produces only half as much CO_2 as an equivalent amount of coal, substitution could help reduce global warming (chapter 15). Natural gas is difficult to ship across oceans or to store in large quantities. North America is fortunate to have an abundant, easily available supply of gas and a pipeline network to deliver it to market. Many developing countries cannot afford such a pipeline network and don't have access to this versatile fuel. In remote areas the natural gas produced in conjunction with oil pumping often is simply burned, a terrible waste of a valuable resource (fig. 19.14). The World Bank estimates that 100 billion m³ of gas are flared every year, or about 1.5 times the amount used annually in Africa.

Natural Gas Resources and Reserves

The republics of the former Soviet Union have 31 percent of known natural gas reserves (mostly in Siberia and the Central Asian republics) and account for about 40 percent of all production. Both

FIGURE 19.14   In remote areas, like this offshore oil platform, natural gas is flared off as an unwanted by-product. © Corbis/Volume 160.

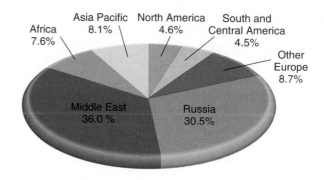

FIGURE 19.15   Proven natural gas reserves by region, 2002. *Source:* British Petroleum, 2003.

eastern and western Europe buy substantial quantities of gas from these wells. Figure 19.15 shows the distribution of proven natural gas reserves in the world.

The total ultimately recoverable natural gas resources are estimated to be 10,000 trillion ft³, corresponding to about 80 percent as much energy as the recoverable reserves of crude oil. The proven world reserves of natural gas are 5,500 trillion ft³ (125,000 million metric tons). Because gas consumption rates are only about half of those for oil, current gas reserves represent roughly a 60-year supply at present usage rates. Proven reserves in North America are about 250 trillion ft³, or 4.6 percent of the world total. This is a ten-year supply at current rates of consumption. Known reserves are more than twice as large.

Across the western United States, a fierce battle is underway over plans to drill thousands of natural gas wells into underground coal seams (What Do You Think? p. 415). The Arctic National Wildlife Refuge was the biggest energy controversy of the 1990s, and coal-bed methane may be our next big environmental struggle.

What do you think?

Coal-Bed Methane

Vast deposits of coal, oil, and gas lie under the sage scrub and arid steppe of North America's intermountain West. Geologists estimate that at least 346 trillion ft^3 of "technically recoverable" natural gas and 62 billion barrels of petroleum liquids occur in five intermountain basins stretching from Montana to New Mexico. These deposits would provide a 15-year supply of gas at present usage rates, and at least four times as much oil as the most optimistic estimates for the Arctic National Wildlife Refuge. About half of that gas and oil is in or around relatively shallow coal seams, which makes it vastly cheaper to extract than most other gas supplies. Drilling a typical offshore gas well costs tens of millions of dollars, and a deep conventional gas well costs several million dollars, but a coal-bed methane well is generally less than $100,000. The total value of the methane and petroleum liquids from the Rocky Mountains could be as much as $200 billion over the next decade.

Most coal-bed methane is held in place by pressure from overlying aquifers. Pumping the water out these aquifers releases the gas, but creates phenomenal quantities of effluent that often is contaminated with salt and other minerals. A typical coal-bed well produces 75,000 liters of water per day. Dumping it on the surface can poison fields and pastures, erode stream banks, contaminate rivers, and harm fish and wildlife. Drawing down aquifers depletes the wells on which many ranches depend, and also dries up natural springs and wetlands essential for wildlife. Ranchers complain that livestock and wildlife are killed by traffic and poisoned by discarded toxic waste around well sites. "It may be a clean fuel," says one rancher, "but it's a dirty business."

Another objection to coal-bed methane extraction is simply the enormous scope of the enterprise. In Wyoming's Power River Basin, energy companies have already installed 12,000 wells and have proposed 39,000 more. Eventually, this area could contain as many as 140,000 wells, together with the sprawling network of roads, pipelines, compressor stations, and waste water pits necessary for such a gargantuan undertaking. The Green River Basin and the San Juan Basin, with three to five times as much potential gas and oil as Powder River, have even greater probability for environmental damage.

In 2002, the U.S. Environmental Protection Agency (EPA) gave its worst possible rating to the Environmental Impact Statement (EIS) for the proposed Powder River wells because of concerns over waste water disposal. Nevertheless, the Bush administration approved the plan and ordered federal land managers across the Rockies to look for ways to remove or reduce environmental restrictions on gas drilling. This order came in spite of a federal study finding that 63 percent of the natural gas in the five basins was completely open to drilling, 25 percent had some restrictions, and only 12 percent of the gas was totally protected, which was about what conservationists had been saying.

An unlikely coalition of ranchers, hunters, anglers, conservationists, water users, and renewable energy activists have banded together to fight against coal-bed gas extraction, calling on Congress to protect private property rights, preserve water quality, and conserve sensitive public lands. Lifelong Republicans, who once looked with suspicion and disgust at environmentalists, suddenly find themselves banding together with tree-huggers and off-the-grid hippie communes to protect their way of life.

On the other side, producers argue that they are helping to preserve the American way of life. If we want to be independent of foreign energy sources, they point out, we need to develop our own energy sources. A decade ago, environmentalists were promoting natural gas as a clean energy alternative because it produces far less carbon dioxide, and air pollutants such as particulates and sulfur oxides, than does burning

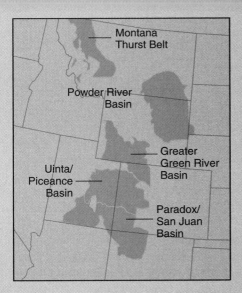

Coal-bed methane deposits occur in five intermountain basins in the western United States.

coal. Disagreeable as gas well effluents may be, they aren't nearly as toxic and long lasting as nuclear waste.

What do you think? Does having access to cleaner fuels justify the social and environmental costs of their extraction? If you had a vote in this issue, what restrictions would you impose on the companies carrying out these projects? Could renewable energy sources, such as wind or solar, substitute for coal-bed methane (chapter 20).

Ethical Issues

What responsibility do those of us who consume fuel have for the way it's produced? How would you weigh the rights of a minority (a few thousand ranchers) against the needs of tens of millions of urban residents who would benefit from cheaper energy prices and cleaner air? Much of the land involved is arid and appears barren to those of us used to more verdant landscapes, yet it is loved and treasured by those who live there. How would you set a value on the solitude, history, and harsh beauty of these places?

Unconventional Gas Sources

Natural gas resources have been less extensively investigated than petroleum reserves. There may be extensive "unconventional" sources of gas in unexpected places. Prime examples are recently discovered methane hydrate deposits in arctic permafrost and beneath deep ocean sediments. **Methane hydrate** is composed of small bubbles or individual molecules of natural gas trapped in a crystalline matrix of frozen water. At least 50 oceanic deposits and a dozen land deposits are known. Altogether, they are thought to hold some 10,000 gigatons (10^{13} tons) of carbon or twice as much as the combined amount of all coal, oil, and conventional natural gas. This could be a valuable energy source but would be difficult to extract, store, and ship. If climate change causes melting of these deposits, it could trigger a catastrophic spiral of global warming because methane is ten times as powerful a greenhouse gas as CO_2. Japan plans exploratory extraction of methane hydrate in the next few years, first on land near Prudhoe Bay, Alaska, and then in Japanese waters.

Methane also can be produced by digesting garbage or manure. Some U.S. cities collect methane from landfills and sewage sludge digestion. Because methane is ten times more potent than CO_2 as a greenhouse gas, stopping leaks from pipelines and other sources is important in preventing global warming. In developing countries, small-scale manure digesters provide a valuable, renewable source of gas for heating, lighting, and cooking (chapter 20).

NUCLEAR POWER

In 1953, President Dwight Eisenhower presented his "Atoms for Peace" speech to the United Nations. He announced that the United States would build nuclear-powered electrical generators to provide clean, abundant energy. He predicted that nuclear energy would fill the deficit caused by predicted shortages of oil and natural gas. It would provide power "too cheap to meter" for continued industrial expansion of both the developed and the developing world. It would be a supreme example of "beating swords into plowshares." Technology and engineering would tame the evil genie of atomic energy and use its enormous power to do useful work.

Glowing predictions about the future of nuclear energy continued into the early 1970s. Between 1970 and 1974, American utilities ordered 140 new reactors for power plants (fig. 19.16). Some advocates predicted that by the end of the century there would be 1,500 reactors in the United States alone. In 1970, the International Atomic Energy Agency (IAEA) projected worldwide nuclear power generation of at least 4.5 million megawatts (MW) by the year 2000, 18 times more than our current nuclear capacity and twice as much as present world electrical capacity from all sources.

Rapidly increasing construction costs, declining demand for electric power, and safety fears have made nuclear energy much less attractive than promoters expected. Electricity from nuclear plants was about half the price of coal in 1970 but twice as much by 1990. Wind energy is already cheaper than nuclear power in

FIGURE 19.16   Two nuclear reactors (domes) at the San Onofre Nuclear Generating Station sit between the beach and Interstate 5, the major route between Los Angeles and San Diego. © Corbis Royalty Free Website.

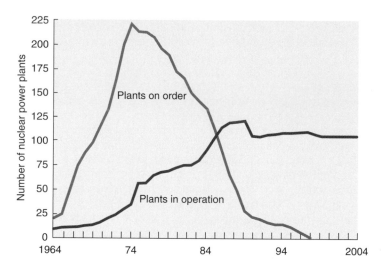

FIGURE 19.17   The changing fortunes of nuclear power in the United States are evident in this graph showing the number of nuclear plants on order and plants in operation. *Source:* World Resources Institute.

many areas and solar power or hydropower are becoming cheaper as well (chapter 20).

The United States led the world into the nuclear age, and it appears that we are leading the world out of it. After 1975, only 13 orders were placed for new nuclear reactors in the United States, and all of those orders subsequently were canceled (fig. 19.17). In fact, 100 of the 140 reactors on order in 1975 were canceled. It began to look as if the much-acclaimed nuclear power industry might have been a very expensive wild goose chase that would never produce enough energy to compensate for the amount invested in research, development, mining, fuel preparation, and waste storage.

How Do Nuclear Reactors Work?

The most commonly used fuel in nuclear power plants is U^{235}, a naturally occurring radioactive isotope of uranium. Ordinarily, U^{235} makes up only about 0.7 percent of uranium ore, too little to sustain a chain reaction in most reactors. It must be purified and concentrated by mechanical or chemical procedures (fig. 19.18). Mining and processing uranium to create nuclear fuel is even more dirty and dangerous than coal mining. In some uranium mines 70 percent of the workers—most of whom were Native Americans—have died from lung cancer caused by high radon and dust levels. In addition, mountains of radioactive tailings and debris have been left around fuel preparation plants.

When the U^{235} concentration reaches about 3 percent, the uranium is formed into cylindrical pellets slightly thicker than a pencil and about 1.5 cm long. Although small, these pellets pack an amazing amount of energy. Each 8.5-gram pellet is equivalent to a ton of coal or four barrels of crude oil.

The pellets are stacked in hollow metal rods approximately 4 m long. About 100 of these rods are bundled together to make a **fuel assembly.** Thousands of fuel assemblies containing 100 tons of uranium are bundled together in a heavy steel vessel called the reactor core. Radioactive uranium atoms are unstable—that is, when struck by a high-energy subatomic particle called a neutron, they undergo **nuclear fission** (splitting), releasing energy and more neutrons. When uranium is packed tightly in the reactor core, the neutrons released by one atom will trigger the fission of another uranium atom and the release of still more neutrons (fig. 19.19). Thus a self-sustaining **chain reaction** is set in motion and vast amounts of energy are released.

The chain reaction is moderated (slowed) in a power plant by a neutron-absorbing cooling solution that circulates between the fuel rods. In addition, **control rods** of neutron-absorbing material, such as cadmium or boron, are inserted into spaces between fuel assemblies to shut down the fission reaction or are withdrawn to allow it to proceed. Water or some other coolant is circulated between the fuel rods to remove excess heat.

The greatest danger in one of these complex machines is a cooling system failure. If the pumps fail or pipes break during operation, the nuclear fuel quickly overheats and a "meltdown" can

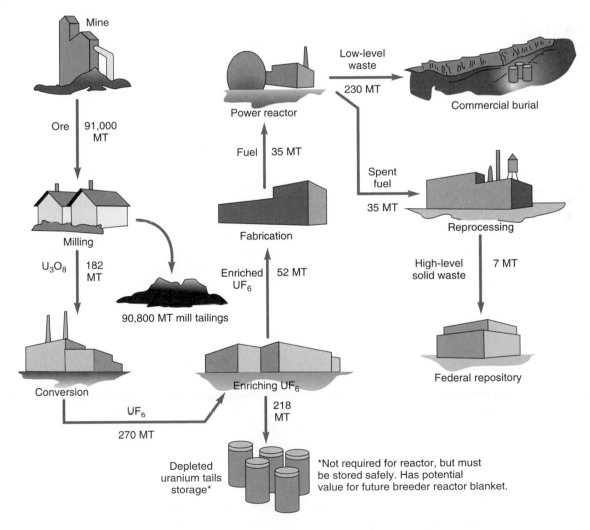

FIGURE 19.18   The nuclear fuel cycle. Quantities represent the average annual fuel requirements for a typical 1,000 MW light water reactor (MT = metric tons). About 35 MT or one-third of the reactor fuel is replaced every year. Reprocessing is not currently done in the United States.

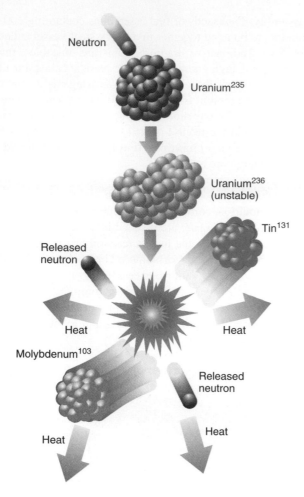

FIGURE 19.19 The process of nuclear fission is carried out in the core of a nuclear reactor. In the sequence shown here, the unstable isotope, uranium-235, absorbs a neutron and splits to form tin-131 and molybdenum-103. Two or three neutrons are released per fission event and continue the chain reaction. The total mass of the reaction product is slightly less than the starting material. The residual mass is converted to energy (mostly heat).

result that releases deadly radioactive material. Although nuclear power plants cannot explode like a nuclear bomb, the radioactive releases from a worst-case disaster like the 1986 fire at Chernobyl in the Ukraine are just as devastating as a bomb.

Kinds of Reactors in Use

Seventy percent of the nuclear plants in the United States and in the world are pressurized water reactors (PWR) (fig. 19.20). Water is circulated through the core, absorbing heat as it cools the fuel rods. This primary cooling water is heated to 317°C (600°F) and reaches a pressure of 2,235 psi. It then is pumped to a steam generator where it heats a secondary water-cooling loop. Steam from the secondary loop drives a high-speed turbine generator that produces electricity. Both the reactor vessel and the steam generator are contained in a thick-walled concrete and steel containment building that prevents radiation from escaping and is designed to

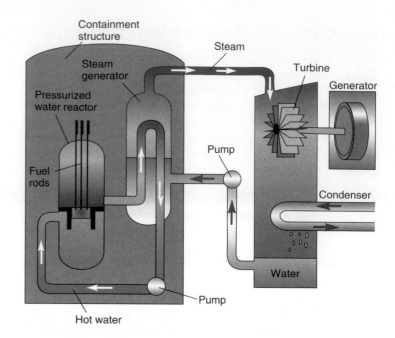

FIGURE 19.20 Pressurized water nuclear reactor. Water is superheated and pressurized as it flows through the reactor core. Heat is transferred to nonpressurized water in the steam generator. The steam drives the turbogenerator to produce electricity.

withstand high pressures and temperatures in case of accidents. Engineers operate the plant from a complex, sophisticated control room containing many gauges and meters to tell them how the plant is running.

Overlapping layers of safety mechanisms are designed to prevent accidents, but these fail-safe controls make reactors very expensive and very complex. A typical nuclear power plant has 40,000 valves compared to only 4,000 in a fossil fuel-fired plant of similar size. In some cases, the controls are so complex that they confuse operators and cause accidents rather than prevent them. Under normal operating conditions, a PWR releases very little radioactivity and is probably less dangerous for nearby residents than a coal-fired power plant.

A simpler, but dirtier and more dangerous reactor design is the boiling water reactor (BWR). In this model, water from the reactor core boils to make steam, which directly drives the turbine-generators. This means that highly radioactive water and steam leave the containment structure. Controlling leaks is difficult, and the chances of releasing radiation in an accident are very high.

Canadian nuclear reactors use heavy water containing deuterium (H^2 or ^2H, the heavy, stable isotope of hydrogen) as both a cooling agent and a moderator. These *Can*adian *deu*terium (CANDU) reactors operate with natural, unconcentrated uranium (0.7 percent U^{235}) for fuel, eliminating expensive enrichment processes. "Heavy water," deuterium-containing water (^2H_2O), is expensive, however, and these reactors are susceptible to overheating and meltdown if cooling pumps fail.

In Britain, France, and the former Soviet Union, a common reactor design uses graphite, both as a moderator and as the structural material for the reactor core. In the British MAGNOX design (named after the magnesium alloy used for its fuel rods), gaseous carbon dioxide is blown through the core to cool the fuel assemblies and carry heat to the steam generators. In the Soviet design, called RBMK (the Russian initials for a graphite-moderated, water-cooled reactor), low-pressure cooling water circulates through the core in thousands of small metal tubes.

These designs were originally thought to be very safe because graphite has high capacity for both capturing neutrons and dissipating heat. Designers claimed that these reactors could not possibly run out of control; unfortunately, they were proven wrong. The small cooling tubes are quickly blocked by steam if the cooling system fails and the graphite core burns when exposed to air. The two most disastrous reactor accidents in the world, so far, involved fires in graphite cores that allowed the nuclear fuel to melt and escape into the environment. In 1956, a fire at the Windscale Plutonium Reactor in England released roughly 100 million curies of radionuclides and contaminated hundreds of square kilometers of countryside. Similarly, burning graphite in the Chernobyl nuclear plant in Ukraine made the fire much more difficult to control than it might have been in another reactor design.

Concerns about nuclear plant safety were heightened in 2002 when inspectors found that leaking boric acid had eaten nearly all the way through the reactor vessel lid of First Energy's Davis-Besse plant near Toledo, Ohio. A hole about 50 cm wide penetrated almost all the way through the 15 cm thick steel cap. Only a 1 cm thick stainless steel liner prevented a rupture of the high-pressure reactor vessel and a catastrophic loss of cooling fluid. If the vessel had ruptured, it may well have been as disastrous as the calamity at Chernobyl. First Energy, which owns power plants in Ohio, Pennsylvania, and New Jersey, has been cited numerous times for maintenance errors, health and safety problems, and pollution violations. Powerlines owned by First Energy have been identified as the source of power surges that triggered a massive blackout in 2003 that left 50 million people in Canada and the United States without electricity (see p. 406). The terrorist attacks of September 11, 2001, also reminds those living near nuclear plants of their risks. Utility engineers claim that reactor containment structures (the large cement domes over the reactor) can withstand a direct hit by a loaded jumbo jet. Critics doubt this claim, which has not been tested empirically.

Alternative Reactor Designs

Several other reactor designs are inherently safer than the ones we now use. Among these are the modular High-Temperature, Gas-Cooled Reactor (HTGCR) and the Process-Inherent Ultimate-Safety (PIUS) reactor.

In HTGCR, which is sometimes called a "pebble-bed reactor," uranium is encased in tiny ceramic-coated pellets; gaseous helium blown around these pellets is the coolant. If the reactor core is kept small, it cannot generate enough heat to melt the ceramic coating, even if all coolant is lost; thus, a meltdown is impossible and operators could walk away during an accident without risk of a fire or radioactive release. Fuel pellets are loaded into the reactor from

the top, shuffle through the core as the uranium is consumed, and emerge from the bottom as spent fuel. This type of reactor can be reloaded during operation. Since the reactors are small, they can be added to a system a few at a time, avoiding the costs, construction time, and long-range commitment of large reactors. Only two of these reactors have been tried in the United States: the Brown's Ferry reactor in Alabama and the Fort St. Vrain reactor near Loveland, Colorado. Both were continually plagued with problems (including fires in control buildings and turbine-generators), and both were closed without producing much power.

A much more successful design has been built in Europe by General Atomic. In West German tests, a HTGCR was subjected to total coolant loss while running at full power. Temperatures remained well below the melting point of fuel pellets and no damage or radiation releases occurred. These reactors might be built without expensive containment buildings, emergency cooling systems, or complex controls. They would be both cheaper and safer than current designs.

The PIUS design features a massive, 60-meters-high pressure vessel of concrete and steel, within which the reactor core is submerged in a very large pool of boron-containing water (fig. 19.21). As long as the primary cooling water is flowing, it keeps the borated water away from the core. If the primary coolant pressure is lost, however, the surrounding water floods the core, and the boron poisons the fission reaction. There is enough secondary water in the pool to keep the core cool for at least a week without any external power or cooling. This should be enough time to resolve the problem. If not, operators can add more water and evaluate conditions further.

The Canadian "slow-poke" is a small-scale version of the PIUS design that doesn't produce electricity but might be a useful substitute for coal, oil, or gas burners in district heating plants. The core of this "mini-nuke" is only about half the size of a kitchen stove and generates only 1/10 to 1/100 as much power as a conventional reactor. The fuel sits in a large pool of ordinary water, which it heats to just below boiling and sends to a heat exchanger. A secondary flow of hot water from the exchanger is pumped to

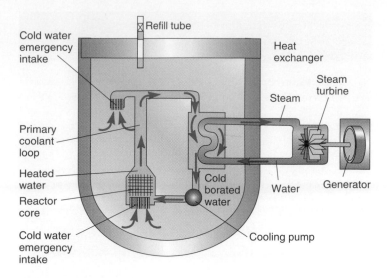

FIGURE 19.21 A PIUS reactor consists of a core and primary cooling system immersed in a very large pool of cold, borated water. As long as the reactor is operating, the cool water is excluded from the cooling circuit by the temperature and pressure differential. Any failure of the primary cooling system would allow cold water to flood the core and shut down the nuclear reaction. The large volume of the pool would cool the reactor for several days without any replenishment.

nearby buildings for space heating. Promoters claim that a runaway reaction is impossible in this design and that it makes an attractive and cost-efficient alternative to fossil fuels. Despite a widespread aversion to anything nuclear, Switzerland and Germany are developing similar small nuclear heating plants.

If these reactors had been developed initially, the history of nuclear power might have been very different. Neither reactor type is suitable, however, for mobile power plants, such as nuclear submarines, which tells you something about the history of nuclear power and the motivation for its development. Aside from being inherently safer in case of coolant pressure loss, neither of these alternative reactor designs eliminates other problems, such as waste disposal, that we will discuss later in this chapter. Still, one of these alternate forms of nuclear power might make it an attractive energy source some day.

Breeder Reactors

For more than 30 years, nuclear engineers have been proposing high-density, high-pressure, **breeder reactors** that produce fuel rather than consume it. These reactors create fissionable plutonium and thorium isotopes from the abundant, but stable, forms of uranium (fig. 19.22). The starting material for this reaction is plutonium reclaimed from spent fuel from conventional fission reactors. After about ten years of operation, a breeder reactor would produce enough plutonium to start another reactor. Sufficient uranium currently is stockpiled in the United States to produce electricity for 100 years at present rates of consumption, if breeder reactors can be made to work safely and dependably.

Several problems have held back the breeder reactor program in the United States. One problem is the concern about safety. The

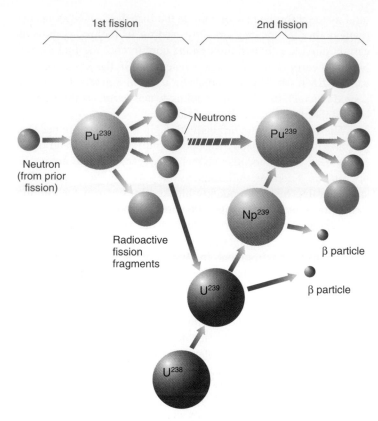

FIGURE 19.22 Reactions in a "breeder" fission process. Neutrons from a plutonium fission change U^{238} to U^{239} and then to Pu^{239} so that the reactor creates more fuel than it uses.

reactor core of the breeder must be at a very high density for the breeding reaction to occur. Water does not have enough heat capacity to carry away the high heat flux in the core, so liquid sodium generally is used as a coolant. Liquid sodium is very corrosive and difficult to handle. It burns with an intense flame if exposed to oxygen, and it explodes if it comes into contact with water. Because of its intense heat, a breeder reactor will melt down and self-destruct within a few seconds if the primary coolant is lost, as opposed to a few minutes for a normal fission reactor.

Another very serious concern about breeder reactors is that they produce excess plutonium that can be used for bombs. It is essential to have a spent-fuel reprocessing industry if breeders are used, but the existence of large amounts of weapons-grade plutonium in the world would surely be a dangerous and destabilizing development. The chances of some of that material falling into the hands of terrorists or other troublemakers are very high. Japan planned to purchase 30 tons of this dangerous material from France and ship it half way around the world through some of the most dangerous and congested shipping lanes on the planet to fuel a breeder program. In 1995, a serious accident at Japan's Moju breeder reactor caused reevaluation of the whole program.

A proposed $1.7 billion breeder-demonstration project in Clinch River, Tennessee, has been on and off for 15 years. At last estimate, it would cost up to five times the original price if it is ever completed. In 1986, France put into operation a full-sized commercial breeder reactor, the SuperPhenix, near Lyons. It cost three times the original estimate to build and produces electricity at twice the cost per kilowatt of conventional nuclear power. After only a year of operation, a large crack was discovered in the inner containment vessel of the SuperPhenix, and in 1997 it was shut down permanently.

RADIOACTIVE WASTE MANAGEMENT

One of the most difficult problems associated with nuclear power is the disposal of wastes produced during mining, fuel production, and reactor operation. How these wastes are managed may ultimately be the overriding obstacle to nuclear power.

Ocean Dumping of Radioactive Wastes

Until 1970, the United States, Britain, France, and Japan disposed of radioactive wastes in the ocean. Dwarfing all these dumps, however, are those of the former Soviet Union, which has seriously—and some fear permanently—contaminated the Arctic Ocean. Rumors of Soviet nuclear waste dumping had circulated for years, but it was not until after the collapse of the Soviet Union that the world learned the true extent of what happened. Starting in 1965, the Soviets disposed of eighteen nuclear reactors—seven loaded with nuclear fuel—in the Kara Sea off the eastern coast of Novaya Zemlya island, and millions of liters of liquid waste in the nearby Barents Sea. Two other reactors were sunk in the Sea of Japan.

Altogether, the former Soviet Union dumped 2.5 million curies of radioactive waste into the oceans, more than twice as much as the combined amounts that 12 other nuclear nations have reported dumping over the past 45 years. In 1993, despite protests from Japan, Russia dumped 900 tons of additional radioactive waste into the Sea of Japan.

Land Disposal of Nuclear Waste

Enormous piles of mine wastes and abandoned mill tailings in all uranium-producing countries represent serious waste disposal problems. Production of 1,000 tons of uranium fuel typically generates 100,000 tons of tailings and 3.5 million liters of liquid waste. There now are approximately 200 million tons of radioactive waste in piles around mines and processing plants in the United States. This material is carried by the wind and washes into streams, contaminating areas far from its original source. Canada has even more radioactive mine waste on the surface than does the United States.

In addition to the leftovers from fuel production, there are about 100,000 tons of low-level waste (contaminated tools, clothing, building materials, etc.) and about 15,000 tons of high-level (very radioactive) wastes in the United States. The high-level wastes consist mainly of spent fuel rods from commercial nuclear power plants and assorted wastes from nuclear weapons production. For the past 20 years, spent fuel assemblies from commercial reactors have been stored in deep water-filled pools at the power plants. These pools were originally intended only as temporary storage until the wastes were shipped to reprocessing centers or permanent disposal sites.

With internal waste storage pools now full but neither reprocessing nor permanent storage available, a number of utility companies are beginning to store nuclear waste in large metal dry casks placed outside power plants (fig. 19.23). These projects are meeting with fierce opposition from local residents who fear the casks will leak. Most nuclear power plants are built near rivers, lakes, or seacoasts. Extremely toxic radioactive materials could spread quickly over large areas if leaks occur. A hydrogen gas explosion and fire in a dry storage cask at Wisconsin's Point Beach nuclear plant intensified opponents' suspicions about this form of waste storage.

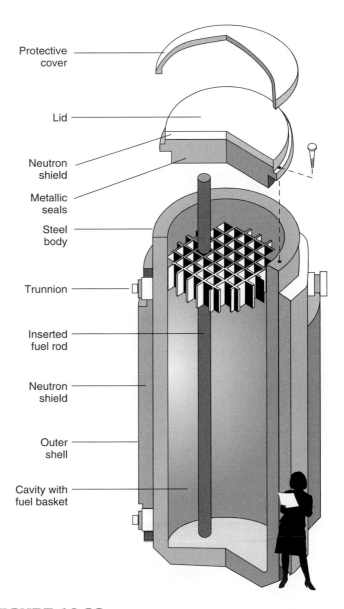

Protective cover

Lid

Neutron shield

Metallic seals

Steel body

Trunnion

Inserted fuel rod

Neutron shield

Outer shell

Cavity with fuel basket

FIGURE 19.23 Dry cask storage for nuclear waste. Each cask is 17 ft tall, 8.5 ft in diameter, has steel walls 10 in. thick, and weighs 122 tons when fully loaded with 40 fuel assemblies. The casks are expected to last 40 years and cost $700,000 each. Critics are worried that the casks will leak, but proponents point out that they can be monitored easily and replaced if necessary.

In 1987, the U.S. Department of Energy announced plans to build the first **high-level waste repository** on a barren desert ridge near Yucca Mountain, Nevada. Intensely radioactive wastes are to be buried deep in the ground where it is hoped that they will remain unexposed to groundwater and earthquakes for the thousands of years required for the radioactive materials to decay to a safe level (fig. 19.24). Although the area is very dry now, we can't be sure that it will always remain that way (see Case Study, p. 304). A billion dollars already has been spent at Yucca Mountain and total costs now are expected to be $10 billion to $35 billion. Although the facility was supposed to open in 1998, the earliest possible date is now 2010.

Several states have nuclear power plants that will be forced to close if waste storage isn't approved soon. They have been pushing for temporary storage in Nevada until Yucca Mountain is ready. Nevada has threatened to close its borders if such a plan is approved. Meanwhile several Native American tribes have offered to store nuclear waste if the price is right. What do you think, is it ethical—or safe—to let them do so?

Russia has offered to store nuclear waste from other countries. Plans are to transport wastes to the Mayak in the Ural mountains. The storage site is near Chelyabinsk, where an explosion at a waste facility in 1957 contaminated about 24,000 km^2 (9,200 mi^2). The region is now considered the most radioactive on earth, so the Russians feel it can't get much worse. They expect that storing 20,000 tons of nuclear waste should pay about $20 billion.

Some nuclear experts believe that **monitored, retrievable storage** would be a much better way to handle wastes. This method involves holding wastes in underground mines or secure surface facilities where they can be watched. If canisters begin to leak, they could be removed for repacking. Safeguarding the wastes would be expensive and the sites might be susceptible to wars or terrorist attacks. We might need a perpetual priesthood of nuclear guardians to ensure that the wastes are never released into the environment.

Decommissioning Old Nuclear Plants

Old power plants themselves eventually become waste when they have outlived their useful lives. Most plants are designed for a life of only 30 years. After that, pipes become brittle and untrustworthy because of the corrosive materials and high radioactivity to which they are subjected. Plants built in the 1950s and early 1960s already are reaching the ends of their lives. You don't just lock the door and walk away from a nuclear power plant; it is much too dangerous. It must be taken apart, and the most radioactive pieces have to be stored just like other wastes. This includes not only the reactor and pipes but also the meter-thick, steel-reinforced concrete containment building. The pieces have to be cut apart by remote-control robots because they are too dangerous to be worked on directly.

Only a few plants have been decommissioned so far, but it has generally cost two to ten times as much to tear them down as

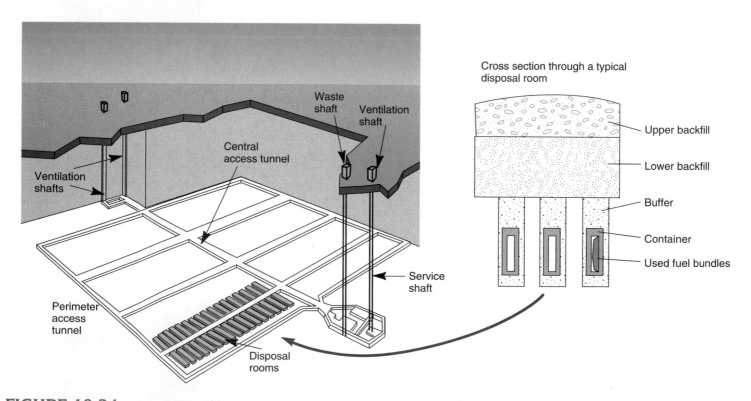

FIGURE 19.24   One proposal for the permanent disposal of used fuel is to bury the used fuel containers 500 to 1,000 m deep in the stable rock formations. Spent fuel, in the form of ceramic fuel pellets, is sealed inside the corrosion-resistant containers. Glass beads are compacted around the used fuel bundles and into the spaces between the fuel and the container shell. The containers are then buried in specially built vaults that are packed and sealed with special materials to further retard the migration of radioactive material. Redrawn with permission of Atomic Energy of Canada Limited.

to build them in the first place. One example is Unit 1 of California's San Onofre Nuclear Plant (see fig. 19.16). After having operated for 24 years, the reactor was shut down in 1992. It took a decade to dismantle the reactor and shell and to pack them in a 930-ton shipping assembly, which is scheduled to be shipped 18,000 km around the tip of South America to Barnwell, South Carolina, for permanent storage. The Panama Canal refused passage to this cargo, and Chile may prohibit it as well. Another option is to go west from California, around the world to whichever East Coast port chooses to accept it. At the time of this writing, none has.

Altogether, the 103 reactors now in operation might cost somewhere between $200 billion and $1 trillion to decommission. No one knows how much it will cost to store the debris for thousands of years or how it will be done. However, we would face this problem, to some degree, even without nuclear electric power plants. Plutonium production plants and nuclear submarines also have to be decommissioned. Originally, the Navy proposed to just tow old submarines out to sea and sink them. The risk that the nuclear reactors would corrode away and release their radioactivity into the ocean makes this method of disposal unacceptable, however.

CHANGING FORTUNES OF NUCLEAR POWER

Although promoted originally as a new wonder of technology that could open the door to wealth and abundance, nuclear power has long been highly controversial. Around the world, many countries have decided to forego this power source. Germany, for example, has decided to close all its nuclear power plants and to depend, instead, on wind energy. In 2003, the first of 19 German nuclear power plants was shut down. About 500 million euros ($590 million U.S.) will be spent to dismantle this facility, four times as much as to build it. What to do with the contaminated waste, however, remains contentious (fig. 19.25).

For many U.S. environmental organizations, opposition to nuclear power is a perennial priority. Antinuclear groups have organized mass protest rallies featuring popular actors, musicians, and celebrities who help raise funds and attract attention to their cause. Protests and civil disobedience at some sites have gone on for decades, and plants that were planned—or in some cases, already built—have been abandoned or modified to burn fossil fuels because of public opposition.

On the other side, workers, consumers, utility officials, and others who stand to benefit from this new technology rally in support of nuclear power. They argue that abandoning this energy source is foolish since a great deal of money already has been invested and the risks may be lower than is commonly believed.

Public opinion about nuclear power has fluctuated over the years. Before the Three-Mile Island accident in 1978, two-thirds of Americans supported nuclear power. By the time Chernobyl exploded in 1985, however, less than one-third of Americans favored this power source. More recently, however, memories of these ear-

FIGURE 19.25 German protesters block a shipment of nuclear waste being sent to a permanent storage site. © Associated Press.

lier incidents have faded. Now about half of all Americans support atomic energy, and about one-quarter say they wouldn't mind having a nuclear plant within 16 km (10 mi) from their home.

The 103 nuclear reactors now operating in 31 states produce about 20 percent of all electricity consumed in the Untied States. Vermont, which gets 85 percent of its electricity from nuclear power, leads the nation in this category. With natural gas prices soaring and electrical shortages (mostly contrived by suppliers, critics charge) in California, the Bush administration is once again promoting nuclear reactors as clean and environmentally friendly because they don't emit greenhouse gases.

Still, critics regard the problems of reactor safety and waste disposal reasons to abandon this technology as quickly as possible. We've entered into a Faustian bargain that our descendents will regret for thousands of years to come, they claim. Over the past 50 years the U.S. government has supported nuclear power with more than $150 billion in subsidies. During that same time, less than $5 billion was spent on renewable energy research and development. Where might we be now if that ratio had been reversed? What do you think? Is the energy from nuclear power worth the costs? Should we build new reactors and allow existing ones to continue to operate in order to reduce our dependence on fossil fuels? How would you evaluate the risks and benefits of this technology?

NUCLEAR FUSION

Fusion energy is an alternative to nuclear fission that could have virtually limitless potential. **Nuclear fusion** energy is released when two smaller atomic nuclei fuse into one larger nucleus. Nuclear fusion reactions, the energy source for the sun and for hydrogen bombs, have not yet been harnessed by humans to produce useful net energy. The fuels for these reactions are deuterium and tritium, two heavy isotopes of hydrogen.

It has been known for 40 years that if temperatures in an appropriate fuel mixture are raised to 100 million degrees Celsius and pressures of several billion atmospheres are obtained, fusion of deuterium and tritium will occur. Under these conditions, the electrons are stripped away from atoms and the forces that normally keep nuclei apart are overcome. As nuclei fuse, some of their mass is converted into energy, some of which is in the form of heat. There are two main schemes for creating these conditions: magnetic confinement and inertial confinement.

Magnetic confinement involves the containment and condensation of plasma, a hot, electrically neutral gas of ions and free electrons in a powerful magnetic field inside a vacuum chamber. Compression of the plasma by the magnetic field should raise temperatures and pressures enough for fusion to occur. The most promising example of this approach, so far, has been a Russian design called *tokomak* (after the Russian initials for "torodial magnetic chamber"), in which the vacuum chamber is shaped like a large donut (fig. 19.26).

Inertial confinement involves a small pellet (or a series of small pellets) bombarded from all sides at once with extremely high-intensity laser light. The sudden absorption of energy causes an implosion (an inward collapse of the material) that will increase densities by 1,000 to 2,000 times and raise temperatures above the critical minimum. So far, no lasers powerful enough to create fusion conditions have been built.

In both of these cases, high-energy neutrons escape from the reaction and are absorbed by molten lithium circulating in the walls of the reactor vessel. The lithium absorbs the neutrons and transfers heat to water via a heat exchanger, making steam that drives a turbine generator, as in any steam power plant. The advantages of fusion reactions, if they are ever feasible, include production of fewer radioactive wastes, the elimination of fissionable products that could be made into bombs, and a fuel supply that is much larger and less hazardous than uranium.

Despite 50 years of research and a $25 billion investment, fusion reactors never have reached the breakeven point at which they produce more energy than they consume. A major setback occurred in 1997, when Princeton University's Tokomak Fusion Test Reactor was shut down. Three years earlier, this reactor had set a world's record by generating 10.7 million watts for one second, but researchers conceded that the project was still decades away from self-sustaining power generation. Proponents of fusion power urge that research be continued, maintaining that success could be just around the corner. Opponents view this technology as just another expensive wild-goose chase and predict that it will never generate enough energy to pay back the fortune spent on its development.

U.S. ENERGY POLICY

In light of the dangers of air pollution, global climate change, and other environmental problems associated with burning of fossil fuels and nuclear power, what energy policy is the United States pursuing? Speaking on behalf of the energy task force he headed, Vice Presi-

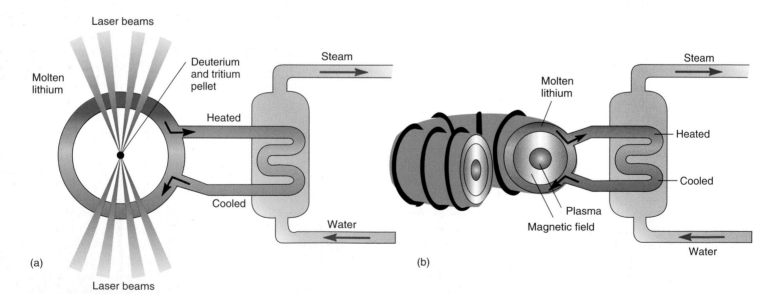

FIGURE 19.26 Nuclear fusion devices. (*a*) Inertial confinement is created by laser beams that bombard and ignite fuel pellets. Molten lithium absorbs energy from the fusion and transfers it as heat to a steam generator. (*b*) A powerful magnetic field confines the plasma and compresses it so that critical temperatures and pressures are reached. Again, molten lithium absorbs energy and transfers it to a steam generator.

dent Dick Cheney said that oil, coal, and natural gas would remain the United States' primary energy resources for "years down the road" and the Bush administration's energy strategy would aim mainly to increase supply of fossil fuels, rather than limit demand. Mr. Cheney dismissed as "1970s-era thinking" the notion that "we could simply conserve or ration our way out" of an energy crisis. "Conservation," he said, "may be a sign of personal virtue, but it is not a sufficient basis for a sound, comprehensive energy policy."

Mr. Cheney, who ran Halliburton, Inc., an oil-services company, before becoming vice president, offered a supply-oriented energy philosophy that emphasized opening protected lands to oil and gas exploration, while rolling back environmental rules that inhibit the burning of coal and the construction of pipelines and refineries. He claimed that the United States needs to build one large electric-generating plant a week for 20 years. He also strongly supported the use of nuclear power, saying that the most environ-mentally friendly way to increase energy supplies was to extend the life of existing nuclear plants and grant permits to build new ones, because they had no emissions of greenhouse gases. Speaking of nuclear power, he said, "If we are serious about environmental protection, then we must seriously question the wisdom of backing away from what is, as a matter of record, a safe, clean and plentiful energy source."

Although federal courts have ordered Mr. Cheney to reveal how his task force reached these conclusions, he has steadfastly refused to comply. Still, it is widely known that the committee consisted entirely of executives and lobbyists from the energy industry. It had no representatives from consumer or environmental groups, many of whom prefer the conservation options and sustainable energy sources described in chapter 20. What do you think? What questions would you ask the Vice President if you had an opportunity to debate this issue?

Summary

- Energy is the capacity to do work. Power is the rate of doing work. The huge blackout of 2003 reminds us of how dependent we are on energy.

- Worldwide, about 86 percent of all commercial energy is generated by fossil fuels, about 40 percent coming from petroleum. Next are coal, with 23.3 percent, and natural gas (methane), with 22.5 percent. Coal supplies will last several more centuries at present rates of usage, but it appears that the fossil fuel age will have been a rather short episode in the total history of humans. Nuclear power provides about 6.5 percent of all commercial energy, but about 20 percent of the electricity in the United States.

- Energy is essential for most activities of modern society. Its use generally correlates with standard of living, but there are striking differences between energy use per capita in countries with relatively equal standards of living. The United States, for instance, consumes nearly twice as much energy per person as does Switzerland, which is higher in many categories that measure quality of life. This difference is based partly on level of industrialization and partly on policies, attitudes, and traditions in the United States that encourage extravagant or wasteful energy use.

- The largest share of energy used in the United States is for industry. Transportation consumes about 27 percent of the U.S. energy supply, almost entirely from petroleum. One measure of energy efficiency is the amount used per unit of economic output. While efficiency has increased in most areas in recent years, Europe still uses less than half as much energy per unit of GDP as the former Soviet Union.

- The environmental damage caused by mining, shipping, processing, and using fossil fuels may necessitate cutting back on our use of these energy sources. Coal is a dirty and dangerous fuel, at least as we currently obtain and use it. Some new coal treatment methods remove contaminants, reduce emissions, and make its use more efficient (so less will be used). Coal combustion is a major source of acid precipitation that is suspected of being a significant cause of environmental damage in many areas. We now recognize that CO_2 buildup in the atmosphere has the potential to trap heat and raise the earth's temperature to catastrophic levels.

- Natural gas is a convenient, clean-burning fuel that produces less CO_2 per BTU than coal. Russia and the Middle East have nearly two-thirds of the world's proven supply of gas. Large, easily accessible gas deposits are found in coal seams in the western United States, but the environmental and social costs of extracting this fuel are high.

- Nuclear energy offers an alternative to many of the environmental and social costs of fossil fuels, but it introduces serious problems of its own. In the 1950s, there was great hope that these problems would be overcome and that nuclear power plants would provide energy "too cheap to meter." Recently, however, much of that optimism has been waning. No new reactors have been started in the United States since 1975. Many countries are closing down existing nuclear power plants, and a growing number have pledged to remain or become "nuclear-free."

- The greatest worry about nuclear power is the danger of accidents that release hazardous radioactive materials into the environment. Several accidents, most notably the "meltdown" at the Chernobyl plant in the Soviet Ukraine in 1986, have convinced many people that this technology is too risky to pursue.

- Other major worries about nuclear power include where to put the waste products of the nuclear fuel cycle and how to ensure that it will remain safely contained for the thousands of years required for "decay" of the radioisotopes to nonhazardous lev-

els. Yucca Mountain, Nevada, was chosen for a high-level waste repository, but many experts believe that burying these toxic residues in nonretrievable storage is a mistake.

- None of our current major energy sources appear to offer security in terms of stable supply or environmental considerations. Neither coal nor nuclear power is a good long-term energy source with our present level of technology. We urgently need to develop alternative sources of sustainable energy.

Questions for Review

1. What is energy? What is power?

2. What are the major sources of commercial energy worldwide and in the United States? Why are data usually presented in terms of commercial energy?

3. How does energy use in the Untied States compare with that in other countries?

4. How much coal, oil, and natural gas are in proven reserves worldwide? Where are those reserves located?

5. What is coal-bed methane, and why is it controversial?

6. What are the most important health and environmental consequences of our use of fossil fuels?

7. Describe how a nuclear reactor works and why reactors can be dangerous.

8. What are the four most common reactor designs? How do they differ from each other?

9. What are the advantages and disadvantages of the breeder reactor?

10. Describe methods proposed for storing and disposing of nuclear wastes.

Questions for Critical Thinking

1. We have discussed a number of different energy sources and energy technologies in this chapter. Each has advantages and disadvantages. If you were an energy policy analyst, how would you compare such different problems as the risk of a nuclear accident versus air pollution effects from burning coal?

2. If your local utility company were going to build a new power plant in your community, what kind would you prefer? Why?

3. The nuclear industry is placing ads in popular magazines and newspapers claiming that nuclear power is environmentally friendly since it doesn't contribute to the greenhouse effect. How do you respond to that claim?

4. Our energy policy effectively treats some strip-mine and well-drilling areas as national sacrifice areas knowing they will

never be restored to their original state when extraction is finished. How do we decide who wins and who loses in this transaction?

5. Storing nuclear wastes in dry casks outside nuclear power plants is highly controversial. Opponents claim the casks will inevitably leak. Proponents claim they can be designed to be safe. What evidence would you consider adequate or necessary to choose between these two positions?

6. Since the environmental impact of a nuclear power plant accident cannot be limited to national boundaries, should there be international regulations for power plants? Who would define these regulations? How could they be enforced?

7. Although we have wasted vast amounts of energy resources in the process of industrialization and development, some would say that it was a necessary investment to get to a point at which we can use energy more efficiently and sustainably. Do you agree? Might we have followed a different path?

8. The policy of the United States has always been to make energy as cheap and freely available as possible. Most European countries charge three to four times as much for gasoline as we do. Think about the ramifications of our energy policy. Why is it so different from that in Europe? Who benefits and who or what loses in these different approaches? How have our policies shaped our lives? What does existing policy tell you about how our government works?

Key Terms

black lung disease 410
breeder reactor 420
chain reaction 417
control rods 417
energy 406
fossil fuels 407
fuel assembly 417
high-level waste
  repository 422
joule (J) 406
methane hydrate 416

monitored, retrievable
  storage 422
nuclear fission 417
nuclear fusion 424
oil shale 414
power 406
proven reserves 409
secondary recovery
  techniques 412
tar sands 413
work 406

Further Readings

Banerjee, Subhankar. 2003. *Arctic National Wildlife Refuge: Seasons of Life and Land.* Mountaineers Books.

Freese, Barbara. 2003. *Coal. A Human History.* Perseus Books.

Heinberg, Richard. 2003. *The Party's Over: Oil, War and the Fate of Industrial Societies.* New Society Publishers.

Lake, James A., et al. 2002. Next-generation nuclear power. *Scientific American* 286(1):72–81.

Liu, Jie, et al. 2002. Chronic arsenic poisoning from burning high-arsenic-containing coal in Guizhou, China. *Environmental Health Perspectives* 110(2):119–23.

Orr, Robert Bent Lloyd. 2002. *Energy: Science Policy and the Pursuit of Sustainability.* Island Press.

Shcherbak, Yuri M. 1996. Ten years of the Chernobyl era. *Scientific American* 274(4):44–49.

Suess, Erwin, et al. 1999. Flammable ice. *Scientific American* 281(5):76–83.

Williams, Ted. 2001. Mountain madness. *Audubon* 103(3):36–43.

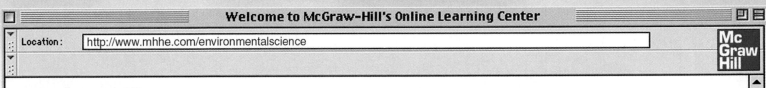

Welcome to McGraw-Hill's Online Learning Center

Location: http://www.mhhe.com/environmentalscience

WEB EXERCISES

Energy Resources in Canada

A Canadian atlas of resources is available online at http://ccrs-gad1.cgdi.gc.ca/resources/EngNRAtlas.html. Look at the oil and gas reserves in Canada. Read the directions for drawing a map in the lower left frame; then, in the upper left "Topics" frame, select "Resource Development" from the drop-down menu. Click on the boxes for all energy-related layers. Note: to see where these layers are, be sure to click on the Canadian Boundaries at the bottom of the list, too. Then click on the DRAW MAP button below the list of map layers. Note that the map legend will appear below the map. Where are most of the energy sources in Canada? Are they clustered or widely distributed? Did you turn on the hydro sources as well? If not, do so now. Are they distributed in the same areas as oil and gas, or are they in different areas? Add population density to the map. Are people located where the energy is? Does this matter?

Where in the World Are Our Fossil Fuels?

The U.S. Department of Energy maintains an excellent data set on energy issues. Go to www.eia.doe.gov/fueloverview.html to view a comprehensive, alphabetical list of energy information and then answer the following questions:

1. How much did total U.S. energy consumption increase between 1949 and 2000? (Hint: look at the GNP data.)

2. How much did total U.S. GNP increase during that time?

3. What is the ratio between answers 2 and 1? What's the significance of these numbers?

4. Where are the four Liquid Natural Gas ports in the U.S.? (Hint: look for LNG imports.)

5. From information in your text, why does this matter?

6. In 1995, how did U.S. electricity prices compare to those in Japan, Germany, China, and India? What effect do you think prices may have had on conservation and renewable energy sources?

7. What ten countries have the largest coal reserves? (Hint: scroll down to World Energy Data.)

8. Where does the U.S. rank in this list?

9. What percentage of total world coal reserves are thought to be in the U.S.?

10. What percentage of total world coal consumption occurs in the U.S.?

11. According to the *Oil and Gas Journal,* where are the ten largest natural gas reserves in the world?

12. What percentage of the total world gas supplies does the U.S. have?

13. What percentage of the total world gas supplies does the U.S. consume each year?

14. What ten countries have the world's largest oil reserves (as far as we know)?

15. What percentage of known oil is in the U.S.?

16. What percentage of all oil does the U.S. consume?

17. What do the answers in questions 15 and 16 suggest to you?

20

Sustainable Energy

Two roads diverged in a wood, and I—I took the one less traveled by, And that has made all the difference.

Robert Frost

OBJECTIVES

After studying this chapter, you should be able to:

- appreciate the opportunities for energy conservation available to us.
- evaluate the transportation options open to us and what energy trade-offs they offer.
- understand how active and passive systems capture solar energy and how photovoltaic collectors generate electricity.
- comprehend why diminishing fuelwood supplies are a crisis in less-developed countries.
- evaluate the use of dung, crop residues, energy crops, and peat as potential energy sources.
- explain how hydropower, wind, and geothermal energy contribute to our power supply.
- describe how tidal and wave energy and ocean thermal gradients can be used to generate electrical energy.

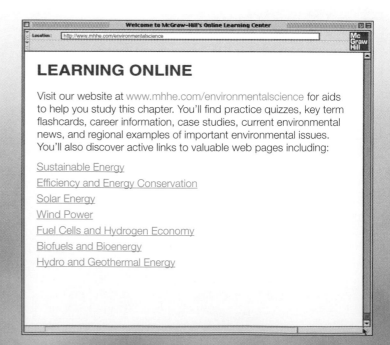

Photo: Windmills, each taller than the Statue of Liberty, stand on a shoal off the Swedish coast.
© Yttre Stengrund/Windpowerphotos.com.

Sea Power

In 2003, British Prime Minister Tony Blair and Swedish Prime Minister Goran Persson surprised the world (and many of their constituents) by pledging to reduce carbon dioxide emissions by 60 percent by 2050. Calling climate change "unquestionably the most urgent environmental challenge, we face," Blair said that issues of global poverty and environmental degradation—especially global warming—are just as devastating in their potential impact as terrorism or weapons of mass destruction. Noting that his country is "well on the way" to meeting its greenhouse gas reduction target of 12.5 percent under the United Nations Kyoto Protocol, Blair said that this goal is not enough to avert global warming. He said that "to stop further damage to the climate, we need a reduction of 60 percent worldwide." Furthermore, he said, "I believe we can achieve that target at reasonable cost."

To reach this ambitious goal, both the United Kingdom and Sweden are promoting a portfolio of conservation and renewable energy projects including windmills, fuel cells, wave power, solar energy, and cogeneration (combined production of heat and electricity) that, it is hoped, will produce at least 20 percent of each country's electricity by 2020.

Of the new energy sources, offshore wind power is likely to be the largest for these countries. Coinciding with Blair's promise for greenhouse gas reductions, the British government began a new round of licensing seabed locations for thousands of giant windmills (see photo, p. 428). Noting that the U.K. has among the best wind resources in Europe and great expertise in offshore development from North Sea oil, the Minister of Trade predicted that wind could provide up to half of Britain's renewable energy, while also providing up to 20,000 jobs. The U.K. could become a net energy exporter in 2010 for the first time in 30 years, rather than importing 75 percent of its supply as is predicted under a business-as-usual scenario.

Offshore wind projects already underway in the U.K. are expected to provide as much electricity as six large nuclear power plants. Some people object to the visual impact of windmills up to 127 meters tall (417 ft) from the water line to the tip of the blades. Others, however, see these wind farms as welcome alternatives to our present energy sources. As American author Bill McKibben says, "The choice is not between windmills and untouched nature. It's between windmills and the destruction of the planet's biology." A spokesperson for Greenpeace U.K. said, "For 30 years Greenpeace has opposed the pollution of our oceans but can today fully support this massive commitment to harness wind power at sea."

Wind energy is the fastest-growing generation source in the world (although admittedly it's growing from a small base). Worldwide, wind power now exceeds 35,000 MW of installed capacity, generating enough electricity to power 3.5 million average American homes or twice that many in Europe. Over the next five years, this capacity is expected to double. Europe has the fastest growth in wind power, with installations up 33 percent in 2002. Germany and Denmark currently lead in wind power, already obtaining about 20 percent of their electricity from wind. And Spain—the home of Don Quixote—increased its wind power capacity 44 percent in 2002.

This example illustrates both the need for, and promise of, renewable energy sources if we are to live in a sustainable world. In this chapter we'll look in more detail at the options for conservation and environmentally friendly energy supplies.

CONSERVATION

As the previous chapter and the opening story of this chapter suggest, we urgently need to move toward sustainable, environmentally friendly, affordable, politically progressive energy sources. One of the easiest ways to avoid energy shortages and to relieve environmental and health effects of our current energy technologies is simply to use less. Conservation offers many benefits both to society and to the environment.

Utilization Efficiencies

Much of the energy we consume is wasted. This statement is not a simple admonishment to turn off lights and turn down furnace thermostats in winter; it is a technological challenge. Our ways of using energy are so inefficient that most potential energy in fuel is lost as waste heat, becoming a form of environmental pollution. Of the energy we do extract from primary resources, however, much is used for frankly trivial or extravagant purposes. As chapter 19 shows, several European countries have higher standards of living than the United States, and yet use 30 to 50 percent less energy.

Many conservation techniques are relatively simple and highly cost effective. Compact fluorescent bulbs, for example, produce four times as much light as an incandescent bulb of the same wattage, and last ten times as long (fig. 20.1). Although they cost more initially, total lifetime savings can be $30 to $50 per fluorescent bulb. Light-emitting diodes (LEDs) also are extremely efficient, consuming 90 percent less energy and lasting hundreds of times as long as ordinary light bulbs. They can produce millions of colors and be adjusted in brightness to suit ambient conditions. They are being used now in everything from flashlights and Christmas lights, to advertising signs, brake lights, exit signs, and street lights.

Many improvements in domestic energy efficiency have occurred in the past decade. Today's average new home uses one-half the fuel required in a house built in 1974, but much more can be done. Household energy losses can be reduced by one-half to three-fourths through better insulation, double or triple glazing of windows, thermally efficient curtains or window coverings, and by sealing cracks and loose joints. Reducing air infiltration is usually the cheapest, quickest, and most effective way of saving energy because it is the largest source of losses in a typical house. It doesn't take much skill or investment to caulk around doors, windows, foundation joints, electrical outlets, and other sources of air leakage.

According to national standards passed in 2001, all new washing machines will have to use 35 percent less water in 2007 than current models. This will add an estimated $240 to the cost of a washer, but will pay back in seven years. It also will cut water use in the United States by 40 trillion liters (10.5 trillion gallons) per

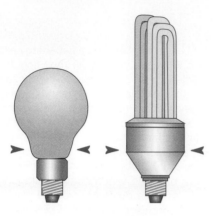

FIGURE 20.1   A 15-watt compact fluorescent bulb (*right*) produces as much light as a 60-watt incandescent bulb (*left*) but lasts ten times as long.

FIGURE 20.2   Earth-sheltered homes take advantage of the stable temperatures and insulating qualities of the earth. This house has south-facing windows for maximum solar gain, and high celerestory windows that give light to the back of the house as well as summer ventilation. Courtesy National Renewable Energy Laboratory/NREL/PIX.

year and save four times as much electricity every year as is used to light all the homes in the United States. Air conditioners will also be required to be about 20 percent more efficient.

For even greater savings, new houses can be built with extra thick superinsulated walls, air-to-air heat exchangers to warm incoming air, and even double-walled sections that create a "house within a house." The R-2000 program in Canada details how energy conservation can be built into homes. Special double-glazed windows that have internal reflective coatings and that are filled with an inert gas (argon or xenon) have an insulation factor of R11, the same as a standard 4-inch thick insulated wall or ten times as efficient as a single-pane window. Superinsulated houses now being built in Sweden require 90 percent less energy for heating and cooling than the average American home.

Orienting homes so that living spaces have passive solar gain in the winter and are shaded by trees or roof overhang in the summer also helps conserve energy. Earth-sheltered homes built into the south-facing side of a slope or protected on three sides by an earth berm are exceptionally efficient energy savers because they maintain relatively constant subsurface temperatures (fig. 20.2). Sod roofs provide good insulation, prevent rain runoff, and last longer than asphalt shingles. Because they are heavier, however, they need stronger supports.

Straw-bale construction offers both high insulating qualities and a renewable, inexpensive building material that can be assembled by amateurs (fig. 20.3). This isn't a new technique. Settlers on the Great Plains built straw-bale houses a century ago because they didn't have wood. Some of those houses are still standing. The bales are strong and will support the roof without any additional timber framing. They must be thoroughly waterproofed, however, with stucco, adobe, or plaster both inside and out so the straw doesn't decay. It's also important to seal them so mice and other vermin can't take up residence. The thick walls are terrific sound insulators as well as highly energy efficient. The cost can be far less than a conventionally built home.

One of the most direct and immediate ways that individuals can save energy is to turn off appliances. Few of us realize how much electricity is used by appliances in a standby mode. You may think you've turned off your TV, DVD player, cable box, or printer,

FIGURE 20.3   Carolyn Roberts and her sons build a straw bale house near Tucson, AZ. Courtesy Carolyn Roberts/A House of Straw.

but they're really continuing to draw power in an "instant-on" mode. For the average home, standby appliances can represent up to 25 percent of the monthly electric bill. Home office equipment including computers, printers, cable modems, copiers, etc., usually are the biggest energy consumers (fig. 20.4). Putting your computer to sleep saves about 90 percent of the energy it uses when fully on, but turning it completely off is even better.

Industrial energy savings are another important part of our national energy budget. More efficient electric motors and pumps, new sensors and control devices, advanced heat-recovery systems, and material recycling have reduced industrial energy requirements significantly. In the early 1980s, U.S. businesses saved $160 billion per year through conservation. When oil prices collapsed, however, many businesses returned to wasteful ways.

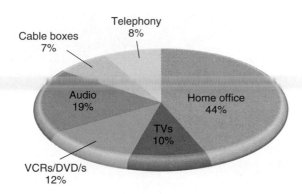

FIGURE 20.4   Typical standby energy consumption by household electrical appliances. *Source:* U.S. Department of Energy.

Energy Conversion Efficiencies

Energy efficiency is a measure of energy produced compared to energy consumed. Table 20.1 shows the typical energy efficiencies of a variety of energy-conversion devices. Thermal-conversion machines, such as steam turbines in coal-fired or nuclear power plants, can turn no more than 40 percent of the energy in their primary fuel into electricity or mechanical power because of the need to reject waste heat. Does this mean that we can never increase the efficiency of fossil fuel use? No. Some waste heat can be recaptured and used for space heating, raising the net yield to 80 or 90 percent. In another kind of process, fuel cells convert the chemical energy of a fuel directly into electricity without an intermediate combustion cycle. Since this process is not limited by waste heat elimination, its efficiencies can approach 80 percent with such fuel as hydrogen gas or methane.

Transportation

One of the areas in which most of us can accomplish the greatest energy conservation is in our transportation choices. You may not be able to build an energy-efficient house or persuade your utility company to switch from coal or nuclear to solar energy, but you can decide every day how you travel to school, to work, or for shopping or entertainment. Automobiles and light trucks account for 40 percent of the U.S. oil consumption and produce one-fifth of its carbon dioxide emissions. According the U.S. EPA, raising the average fuel efficiency of the passenger fleet by 3 miles per gallon (approx. 1.4 l/100 km), would save American consumers about $25 billion a year in fuel costs, reduce carbon dioxide emissions by 140 million metric tons per year, and save more oil than the maximum expected production from Alaska's Arctic National Wildlife Refuge.

The Bureau of Transportation Statistics reports that there are now more vehicles in the United States (214 million) than licensed drivers (190 million). More importantly, those vehicles are used for an average of 1 billion trips per day. Many of us drive now for errands or short shopping trips that might have previously been made on foot. Some of that is due to the design of our cities (chapter 22). Suburban subdivisions have replaced compact downtown

TABLE 20.1   Typical Net Efficiencies of Energy-Conversion Devices

	YIELD (PERCENT)
ELECTRIC POWER PLANTS	
Hydroelectric (best case)	90
Combined-cycle steam	90
Fuel cell (hydrogen)	80
Coal-fired generator	38
Oil-burning generator	38
Nuclear generator	30
Photovoltaic generation	10
TRANSPORTATION	
Pipeline (gas)	90
Pipeline (liquid)	70
Waterway (no current)	65
Diesel-electric train	40
Diesel-engine automobile	35
Gas-engine automobile	30
Jet-engine airplane	10
SPACE HEATING	
Electric resistance	99*
High-efficiency gas furnace	90
Typical gas furnace	70
Efficient wood stove	65
Typical wood stove	40
Open fireplace	−10
LIGHTING	
Sodium vapor light	60*
Fluorescent bulb	25*
Incandescent bulb	5*
Gas flame	1

*Note that 60–70 percent of the energy in the original fuel is lost in electric power generation.
Source: *U.S. Department of Energy.*

centers in most cities. Shopping areas are surrounded by busy streets and vast parking lots that are highly pedestrian unfriendly. But sometimes we use fuel inefficiently simply because we haven't thought about alternatives. The Census Bureau reports that three-quarters of all workers commute alone in private vehicles. Less than 5 percent use public transportation or carpool, and a mere 0.38 percent walk or travel by bicycle.

In response to the 1970s oil price shocks, automobile gas-mileage averages in the United States more than doubled from 13.3 mpg in 1973 to 25.9 mpg in 1988. Unfortunately, falling fuel prices of the 1990s discouraged further conservation. By 2002, the average fuel economy of America's passenger fleet had dropped to 20.4 miles a gallon, the lowest since 1980. Most of this decrease is due to the popularity of SUVs and light trucks, which now account for slightly more than half of all passenger vehicle sales in the United States. According to the Environmental Protection Agency (EPA), in 2002, SUVs averaged 17.3 miles per gallon (mpg), and pickups averaged 16.5 mpg, while cars averaged 24.4 mpg.

Often people choose SUVs or trucks in the mistaken belief that they're safer than small cars. In fact, studies have shown that SUVs and pickup trucks are actually more dangerous—both to those in them and to those in vehicles they may hit—than most midsize or large automobiles, and no safer than many compact or even some subcompact cars. Because they're heavier and taller than most other vehicles, SUVs and pickups are harder to control in emergencies, and more likely to roll over.

What can you do if you want to be environmentally responsible? The cheapest, least environmentally damaging, and healthiest alternative for short trips is walking. You need to get some exercise every day, why not make walking part of it? Next, in terms of minimal expense and environmental impact, is an ordinary bicycle. For trips less than 2 km, it's often quicker to go by bicycle than to find a parking space for your car. While many cities have downgraded their mass transit systems, you might be surprised at the places you can go with this option. A number of electric-powered (fig. 20.5) or electric-assisted (fig. 20.6) vehicles are available, but many are expensive specialty items.

Hybrid gasoline-electric vehicles have the highest efficiency rating and lowest emissions available in the United States (What Do You Think? p. 433). Diesel-powered vehicles are popular in Europe because they can be 35 to 50 percent more fuel efficient than similar sized gas-powered vehicles. Some European models

FIGURE 20.6   The Twike (or twin bike) is a fully enclosed, two-seat, three-wheeled, pedal-assisted, electric bicycle made in Switzerland. Capable of traveling 65 mph and 45 miles between battery charging (more if you pedal harder), it costs about $20,000 (U.S.). © William P. Cunningham.

get up to 78 mpg. Many Americans think of diesels as noisy, smoke-belching, smelly beasts, but clean-diesel technologies have made them much quieter and cleaner than they were 20 years ago. The diesel-powered Volkswagen Beetle, for example, gets a combined city/highway average of 45.5 mpg. The diesel fuel available in the United States, however, is ten times as high in acid-rain-forming sulfur as that available in Europe. There also are worries about the health hazards of diesel soot. You might be surprised to learn that some ordinary gasoline-powered automobiles like the Toyota Echo get a combined city/highway average as high as 39 mpg.

Both the United States and the European Union have announced plans to spend billions of dollars on research and development of hydrogen fuel cell-powered vehicles. Using hydrogen gas for fuel, these vehicles would produce water as their only waste product. We'll discuss how fuel cells work in more detail later in this chapter. Although prototype fuel cell vehicles are already being tested in several places, even the most optimistic predictions are that it will take at least 20 years for this technology to be mass produced at a reasonable cost. Although hydrogen fuel could be produced with electricity from remote wind or solar facilities, providing a convenient and inexpensive way to get surplus energy to market, most hydrogen currently is created from natural gas, making it no cleaner or more efficient than simply burning the gas directly. While not calling for an end to fuel cell research, conservation groups are urging the government not to abandon other useful technologies, such as hybrid engines and conventional pollution control, while waiting for fuel cells.

Another way to look at energy efficiency is to consider the total **net energy yield** from energy-conversion devices (table 20.2). Net energy yield is based on the total useful energy produced during the lifetime of an entire energy system minus the energy

FIGURE 20.5   The Segway Human Transporter is a dynamically stabilized, self-balancing, electric vehicle capable of navigating most walkable areas. © William P. Cunningham.

What do you think?

Hybrid Automobile Engines

In 1990, the California Air Resources Board shocked the automobile industry by ordering it to start producing emission-free vehicles or face stiff penalties. It's not surprising that California was the first state to get tough on automobile emissions. Auto exhaust counts for 90 percent of the state's carbon monoxide, 77 percent of its nitrogen oxides, and 55 percent of its smog-producing hydrocarbons. At the time this order was issued, however, only battery-powered electric vehicles were available. Although several manufacturers built all-electric autos, the batteries were heavy, expensive, and required more frequent recharging than customers would accept. Even though 90 percent of all daily commutes are less than 80 km (50 mi), most people want the capability to take a long road trip of several hundred kilometers without needing to stop for fuel or recharging.

An alternative that appears to have much more customer appeal is the hybrid gas-electric vehicle. The first of these hybrid cars to be marketed in the United States is the two-seat Honda Insight. Its 3-cylinder, 1.0-liter gas engine is the main power source. A 7 hp electric motor helps during acceleration and hill climbing. When the small battery pack begins to run down, it is recharged by the gas engine, so that the vehicle never needs to be plugged in. More electricity is captured during "regenerative" braking, further increasing efficiency. With a streamlined, lightweight, plastic and aluminum body, the Insight gets 72 mpg (30.3 km/liter) in highway driving and has low-enough emissions to satisfy California requirements. Quick acceleration and nimble handling make the Insight fun to drive. The cost is about $20,000.

Both Toyota and Honda have also introduced 5 passenger hybrids. The Toyota is called a Prius, and the Honda is a Civic hybrid. During most city driving, they depend only on a quiet, emission-free, battery-powered, electric motor. The 1.5-liter gas engine kicks in to help accelerate or when the batteries need recharging. With a com-

The hybrid gas-electric Toyota Prius seats five adults, gets 52 mpg (22 km/l) in city driving, and produces 90 percent less smog-forming exhaust gas than current U.S. standards. Courtesy Toyota.

bined city/highway average of about 50 mpg (22 km/liter) in city driving, both models are among the most efficient cars on the road and can travel more than 625 mi (1,000 km) without refueling. Some drivers are unnerved by the noiseless electric motor of hybrids. Sitting at a stoplight, it makes no sound at all. You might think it was dead, but when the light changes, you glide off silently and smoothly. Both the hybrid Civic and Prius sell for about $20,000 and qualify for a $2,000 Federal Clean-Fuel Vehicle tax deduction. The Sierra Club estimates that in 100,000 mi (160,000 km), a Prius will generate 27 tons of CO_2, a Ford Taurus will generate 64 tons, while the Ford Excursion SUV will produce 134 tons. In 1999 the Sierra Club awarded both the Insight and the Prius an "excellence in engineering" award, the first time this organization has ever endorsed commercial products.

In 2003, Toyota introduced a new version of the Prius with more interior space, better performance, and more distinctive styling. A more aerodynamic shape, better batteries, and higher voltage are expected

to give the Prius II slightly better fuel efficiency. Although Toyota also has an experimental fuel cell car, it expects hybrids to be the best automobile choice for the next 20 years at least. Toyota is also working on a diesel hybrid for the European market.

Most American automobile makers are skeptical of hybrid designs and are putting most of their research into fuel cells, which may be years from commercial use but have the promise of being truly zero-emission. Ford and General Motors have prototype hybrid vehicles with 42 V generators (versus 500 V in the Prius). These "mild" hybrids use the electrical system primarily to run CD players, dashboard computers, extra lights, and other accessories. Mileage is improved very little.

What do you think? Would you buy a vehicle with a hybrid engine system? Would you take a chance on this new technology to get a quiet, clean, efficient, environmentally friendly means of transportation? Do you think that automobile makers are wise to wait for fuel cells, or should they be producing hybrid vehicles as well?

TABLE 20.2 Typical Net Useful Energy Yields

ENERGY SOURCE	YIELD/COST RATIO
NONRENEWABLE RESOURCES	
Coal (space or process heat)	20/1
Natural gas (as heat source)	10/1
Gasoline and fuel oil	7/1
Coal gasification (combined cycle)	5/1
Oil shale (as liquid fuel)	1/1
Nuclear (excluding waste disposal)	2/1*
RENEWABLE RESOURCES	
Hydroelectric (best case)	20/1**
Wind (electric generation)	2/1
Biomass methane	2/1
Solar electric (10 percent efficient)	1/1
Solar electric (20 percent efficient)	2/1

*Decommissioning of old nuclear plants and perpetual storage of wastes may consume more energy than nuclear power has produced.

**Hydropower yield depends on availability of water and life expectancy of dam and reservoir. Most dams produce only about 40 percent of rated capacity due to lack of water. Some dams have failed or silted up without producing any net energy yield.

What can you do?

Some Things You Can Do to Save Energy

1. Drive less: make fewer trips, use telecommunications and mail instead of going places in person.
2. Use public transportation, walk, or ride a bicycle.
3. Use stairs instead of elevators.
4. Join a car pool or drive a smaller, more efficient car; reduce speeds.
5. Insulate your house or add more insulation to the existing amount.
6. Turn thermostats down in the winter and up in the summer.
7. Weatherstrip and caulk around windows and doors.
8. Add storm windows or plastic sheets over windows.
9. Create a windbreak on the north side of your house; plant deciduous trees or vines on the south side.
10. During the winter, close windows and drapes at night; during summer days, close windows and drapes if using air conditioning.
11. Turn off lights, television sets, and computers when not in use.
12. Stop faucet leaks, especially hot water.
13. Take shorter, cooler showers; install water-saving faucets and showerheads.
14. Recycle glass, metals, and paper; compost organic wastes.
15. Eat locally grown food in season.
16. Buy locally made, long-lasting materials.

required to make useful energy available. To make comparisons between different energy systems easier, the net energy yield is often expressed as a *ratio* between the output of useful energy and the energy costs for construction, fuel extraction, energy conversion, transmission, waste disposal, etc.

Nuclear power is a good case for net energy yield studies. We get a large amount of electricity from a small amount of fuel in a nuclear plant, but it also takes a great deal of energy to extract and process nuclear fuel and to build, operate, and eventually dismantle power plants. As chapter 19 points out, we may never reach a break-even point where we get back more energy from nuclear plants than we put into them, especially considering the energy that may be required to decommission nuclear plants and guard their waste products in secure storage for thousands of years.

Net yields and overall conversion efficiencies are not the only considerations when we compare different energy sources. The yield/cost ratio and conversion-cycle efficiency is much higher for coal burning, for instance, than for photovoltaic electrical production, making coal appear to be a better source of energy than solar radiation. Solar energy, however, is free, renewable, and nonpolluting. If we can use solar energy to get electrical energy, it doesn't matter how *efficient* the process is, as long as we get more out of it than we put in.

Negawatt Programs

Utility companies are finding it much less expensive to finance conservation projects than to build new power plants. Rather than buy megawatts of new generating capacity, power companies are investing in "negawatts" of demand avoidance. Each of us can help save energy in everyday life (What Can You Do? p. 434). Pacific Gas and Electric in California and Potomac Power and Light in Washington, D.C., both have instituted large conservation programs. They have found that conservation costs about $350 per kilowatt (kW) saved. By contrast, a new nuclear power plant costs between $3,000 and $8,000 per kW of installed capacity. New coal-burning plants with the latest air pollution-control equipment cost at least $1,000 per kW. By investing $200 million to $350 million in public education, home improvement loans, and other efficiency measures, a utility can avoid building a new power plant that would cost a billion dollars or more. Furthermore, conservation measures don't consume expensive fuel or produce pollutants.

Can application of this approach help alleviate energy shortages in other countries as well? Yes. For example, Brazil could cut its electricity consumption an estimated 30 percent with an investment of $10 billion (U.S.). It would cost $44 billion to build new power plants to produce that much electricity. South Korea has instituted a comprehensive conservation program with energy-saving building standards, efficiency labels on new household appliances, depreciation allowances, reduced tariffs on energy-conserving equipment, and loans and tax breaks for upgrading homes and businesses. It also forbids some unnecessary uses, such as air conditioning or the use of elevators between the first and third floors.

Cogeneration

One of the fastest growing sources of new energy is **cogeneration,** the simultaneous production of both electricity and steam or hot water in the same plant. By producing two kinds of useful energy in the same facility, the net energy yield from the primary fuel is increased from 30–35 percent to 80–90 percent. In 1900, half the electricity generated in the United States came from plants that also provided industrial steam or district heating. As power plants became larger, dirtier, and less acceptable as neighbors, they were forced to move away from their customers. Waste heat from the turbine generators became an unwanted pollutant to be disposed of in the environment. Furthermore, long transmission lines, which are unsightly and lose up to 20 percent of the electricity they carry, became necessary.

By the 1970s, cogeneration had fallen to less than 5 percent of our power supplies, but interest in this technology is being renewed. The capacity for cogeneration more than doubled in the 1980s to about 30,000 megawatts (MW). District heating systems are being rejuvenated, and plants that burn municipal wastes are being studied. New combined-cycle coal-gasification plants or "mini-nukes" (chapter 19) offer high efficiency and clean operation that may be compatible with urban locations. Small neighborhood- or apartment building-sized power-generating units are being built that burn methane (from biomass digestion), natural gas, diesel fuel, or coal (fig. 20.7). The Fiat Motor Company makes a small

FIGURE 20.7 A technician adjusts a gas microturbine that produces on-site heat and electricity for businesses, industry, or multiple housing units. Courtesy Capstone Micro Turbines.

generator for about $10,000 that produces enough electricity and heat for four or five energy-efficient houses. These units are especially valuable for facilities like hospitals or computer centers that can't afford power outages.

TAPPING SOLAR ENERGY

The sun serves as a giant nuclear furnace in space, constantly bathing our planet with a free energy supply. Solar heat drives winds and the hydrologic cycle. All biomass, as well as fossil fuels and our food (both of which are derived from biomass), results from conversion of light energy (photons) into chemical bond energy by photosynthetic bacteria, algae, and plants.

A Vast Resource

The average amount of solar energy arriving at the top of the atmosphere is 1,330 watts per square meter. About half of this energy is absorbed or reflected by the atmosphere (more at high latitudes than at the equator), but the amount reaching the earth's surface is some 10,000 times all the commercial energy used each year. However, this tremendous infusion of energy comes in a form that, until this century, has been too diffuse and low in intensity to be used except for environmental heating and photosynthesis. Figure 20.8 shows solar energy levels over the United States for a typical summer and winter day.

Passive Solar Heat

Our simplest and oldest use of solar energy is **passive heat absorption,** using natural materials or absorptive structures with no moving parts to simply gather and hold heat. For thousands of years, people have built thick-walled stone and adobe dwellings that slowly collect heat during the day and gradually release that heat at night (fig. 20.9). After cooling at night, these massive building materials maintain a comfortable daytime temperature within the house, even as they absorb external warmth.

A modern adaptation of this principle is a glass-walled "sunspace" or greenhouse on the south side of a building (fig. 20.10). Incorporating massive energy-storing materials, such as brick walls, stone floors, or barrels of heat-absorbing water into buildings also collects heat to be released slowly at night. An interior, heat-absorbing wall called a Trombe wall is an effective passive heat collector. Some Trombe walls are built of glass blocks enclosing a water filled space or water-filled circulation tubes so heat from solar rays can be absorbed and stored, while light passes through to inside rooms.

Active Solar Heat

Active solar systems generally pump a heat-absorbing, fluid medium (air, water, or an antifreeze solution) through a relatively small collector, rather than passively collecting heat in a stationary medium like masonry. Active collectors can be located adjacent to

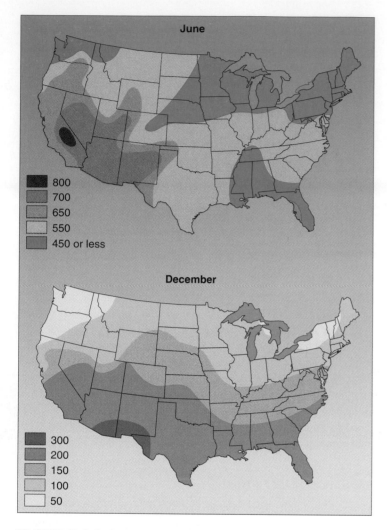

FIGURE 20.8 Average daily solar radiation in the United States in June and December. One langley, the unit for solar radiation, equals 1 calorie per square centimeter of earth surface (3.69 Btu/square foot).
Source: National Weather Bureau, U.S. Department of Commerce.

FIGURE 20.9 Taos Pueblo in northern New Mexico uses adobe construction to keep warm at night and cool during the day.
© William P. Cunningham.

FIGURE 20.10 The Adam Joseph Lewis Center for Environmental Studies at Oberlin College is designed to be self-sustaining even in northern Ohio's cool, cloudy climate. Large, south-facing windows let in sunlight, while 370 m² of solar panels on the roof generate electricity. A constructed wetland outside and a living machine inside (see fig. 18.29) purify wastewater. National Renewable Energy Lab.

or on top of buildings rather than being built into the structure. Because they are relatively small and structurally independent, active systems can be retrofitted to existing buildings (fig. 20.11).

A flat black surface sealed with a double layer of glass makes a good solar collector. A fan circulates air over the hot surface and into the house through ductwork of the type used in standard forced-air heating. Alternatively, water can be pumped through the collector to pick up heat for space heating or to provide hot water. Water heating consumes 15 percent of the United States' domestic energy budget, so savings in this area alone can be significant. A simple flat panel with about 5 m² of surface can reach 95°C (200°F) and can provide enough hot water for an average family of four almost anywhere in the United States. In California, 650,000 homes now heat water with solar collectors. In Greece, Italy, Israel, and other countries where fuels are more expensive, up to 70 percent of domestic hot water comes from solar collectors.

Sunshine doesn't reach us all the time, of course. How can solar energy be stored for times when it is needed? There are a number of options. In a climate where sunless days are rare and seasonal variations are small, a small, insulated water tank is a good solar energy storage system. For areas where clouds block the sun for days at a time or where energy must be stored for winter use, a large, insulated bin containing a heat-storing mass, such as stone, water, or clay, provides solar energy storage. During the summer months, a fan blows the heated air from the collector into the storage medium. In the winter, a similar fan at the opposite end of the bin

FIGURE 20.11 A roof-top solar water heater (top dark panel) sits above eight photovoltaic electric-generating panels on an eco-tourist villa at Maho Bay in the U.S. Virgin Islands. © William P. Cunningham.

blows the warm air into the house. During the summer, the storage mass is cooler than the outside air, and it helps cool the house by absorbing heat. During the winter, it is warmer and acts as a heat source by radiating stored heat. In many areas, six or seven months' worth of thermal energy can be stored in 10,000 gallons of water or 40 tons of gravel, about the amount of water in a very small swimming pool or the gravel in two average-sized dump trucks.

HIGH-TEMPERATURE SOLAR ENERGY

Parabolic mirrors are curved reflecting surfaces that collect light and focus it into a concentrated point. There are two ways to use mirrors to collect solar energy to generate high temperatures. One technique uses long curved mirrors focused on a central tube, containing a heat-absorbing fluid (fig. 20.12). Fluid flowing through the tubes reaches much higher temperatures than possible in a basic flat panel collector. Another high-temperature system uses thousands of smaller mirrors arranged in concentric rings around a tall central tower. The mirrors, driven by electric motors, track the sun and focus its light on a heat absorber at the top of the "power tower" where molten salt is heated to temperatures as high as 500°C (1,000°F), which then drives a steam-turbine electric generator.

Under optimum conditions, a 50 ha (130 acres) mirror array should be able to generate 100 MW of clean, renewable power. The only power tower in the United States is Southern California Edison's Solar II plant in the Mojave Desert east of Los Angeles. Its 2,000 mirrors focused on a 100 m (300 ft) tall tower generates 10 MW or enough electricity for 5,000 homes at an operating cost far below that of nuclear power or oil. We haven't had enough experience with these facilities to know how reliable the mirrors, motors, heat absorbers, and other equipment will be over the long run.

If the entire U.S. electrical output came from such central tower solar steam generators, 60,000 km² of collectors would be

FIGURE 20.12 Parabolic mirrors focus sunlight on steam-generating tubes at this power plant in the California desert. © The McGraw-Hill Companies, Inc./Doug Sherman, photographer.

needed. This is an area about half the size of South Dakota. It is less land, however, than would be strip mined in a 30-year period if all our energy came from coal or uranium. In contrast with windmill farms, which can be used for grazing or farming while also producing energy, mirror arrays need to be carefully protected and are not compatible with other land uses.

Solar Cookers

Parabolic mirrors have been tested for home cooking in tropical countries where sunshine is plentiful and other fuels are scarce. They produce such high temperatures and intense light that they are dangerous, however. A much cheaper, simpler, and safer alternative is the solar box cooker (fig. 20.13). An insulated box costing only a few dollars, with a black interior and a glass or clear plastic lid, serves as a passive solar collector. Several pots can be placed inside at the same time. Temperatures only reach about 120°C (250°F) so cooking takes longer than an ordinary oven. Fuel is free, however, and the family saves hours each day usually spent hunting for firewood or dung. These solar ovens help reduce tropical forest destruction and reduce the adverse health effects of smoky cooking fires.

FIGURE 20.13 A simple box of wood or cardboard, plastic, and foil can help reduce tropical deforestation, improve women's lives, and avoid health risks from smoky fires in developing countries. These inexpensive solar cookers could revolutionize energy use in developing tropical countries. © William P. Cunningham.

Promoting Renewable Energy

Energy policies in some states include measures to encourage conservation and alternative energy sources. Among these are: (1) "distributional surcharges" in which a small per kWh charge is levied on all utility customers to help renewable energy finance research and development, (2) "renewables portfolio" standards to require power suppliers to obtain a minimum percentage of their energy from sustainable sources, and (3) **green pricing** that allows utilities to profit from conservation programs and charge premium prices for energy from renewable sources. Perhaps your state has some or all of these in place.

Iowa, for example, has a Revolving Loan Fund supported by a surcharge on investor-owned gas and electric utilities. This fund provides low-interest loans for renewable energy and conservation. One of the first to provide green pricing is Colorado, where 1,000 customers have agreed to pay $2.50 per month above their regular electric rates to help finance a 10 MW wind farm being built on the Colorado-Wyoming border. Buying a 100 kW "block" of wind power provides the same environmental benefits as planting a half acre of trees or not driving an automobile 4,000 km (2,500 mi) per year. Not all green pricing plans are as straightforward as this, however. Some utilities collect the premium rates for facilities that already exist or for renewable energy bought from other utilities at much lower prices.

Interestingly, some nonutility companies are investing in sustainable energy. In 1997, British Petroleum PLC broke ranks with other fossil fuel corporations and put $20 million into a solar cell manufacturing facility located in Fairfield, CA. Although fossil fuel interests have generally tried to obstruct renewables, British Petroleum believes that the threat of global climate change requires us to search for new types of energy. Similarly, two European insurance companies, concerned about potential losses from storms and rising sea levels caused by global warming, are investing $5 million in Sunlight Power, a U.S. company that makes and services solar power systems for remote regions of developing countries where electric service is unavailable.

Photovoltaic Solar Energy

The photovoltaic cell offers an exciting potential for capturing solar energy in a way that will provide clean, versatile, renewable energy. This simple device has no moving parts, negligible maintenance costs, produces no pollution, and has a lifetime equal to that of a conventional fossil fuel or nuclear power plant.

Photovoltaic cells capture solar energy and convert it directly to electrical current by separating electrons from their parent atoms and accelerating them across a one-way electrostatic barrier formed by the junction between two different types of semiconductor material (fig. 20.14). The photovoltaic effect, which is the basis of these devices, was first observed in 1839 by French physicist Alexandre-Edmond Becquerel, who also discovered radioactivity. His discovery didn't lead to any useful applications until 1954, when researchers at Bell Laboratories in New Jersey learned how to carefully introduce impurities into single crystals of silicon.

These handcrafted single-crystal cells were much too expensive for any practical use until the advent of the U.S. space program. In 1958, when *Vanguard I* went into orbit, its radio was powered by six palm-sized photovoltaic cells that cost $2,000 per peak watt of output, more than 2,000 times as much as conventional energy at the time. Since then, prices have fallen dramatically. In

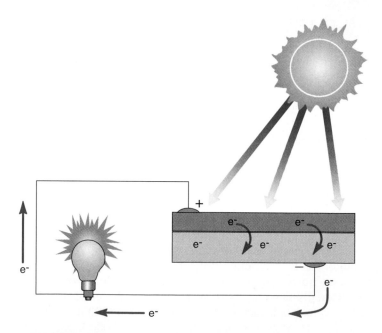

FIGURE 20.14 The operation of a photovoltaic cell. Boron impurities incorporated into the upper silicon crystal layers cause electrons (e-) to be released when solar radiation hits the cell. The released electrons move into the lower layer of the cell, thus creating a shortage of electrons, or a positive charge, in the upper layer and an oversupply of electrons, or negative charge, in the lower layer. The difference in charge creates an electric current in a wire connecting the two layers.

1970, they cost $100 per watt; in 2003 they were approaching $5 per watt. This makes solar energy cost-competitive with other sources in remote areas (more than 1 km from a power line).

During the last 25 years, the efficiency of energy capture by photovoltaic cells has increased from less than 1 percent of incident light to more than 10 percent under field conditions and over 75 percent in the laboratory. Promising experiments are under way using exotic metal alloys, such as gallium arsenide, and semiconducting polymers of polyvinyl alcohol, which are more efficient in energy conversion than silicon crystals. Every year college students from all over North America race solar-powered cars across the country to test solar technology and publicize its potential (fig. 20.15). Perhaps you and your classmates would like to build a solar car and enter the race.

One of the most promising developments in photovoltaic cell technology in recent years is the invention of amorphous silicon collectors. First described in 1968 by Stanford Ovshinky, a self-taught inventor from Detroit, these noncrystalline silicon semiconductors can be made into lightweight, paper-thin sheets that require much less material than conventional photovoltaic cells. They also are vastly cheaper to manufacture and can be made in a variety of shapes and sizes, permitting ingenious applications. Roof tiles with photovoltaic collectors layered on their surface already are available (fig. 20.16). Even flexible films can be coated with amorphous silicon collectors. Silicon collectors already are providing power to places where conventional power is unavailable, such as lighthouses, mountaintop microwave repeater stations, villages on remote islands, and ranches in the Australian outback.

You probably already use amorphous silicon photovoltaic cells. They are being built into light-powered calculators, watches, toys, photosensitive switches, and a variety of other consumer products. Japanese electronic companies presently lead in this field, having foreseen the opportunity for developing a market for photovoltaic cells. This market is already more than $700 million per year. Japanese companies now have home-roof arrays capable of providing all the electricity needed for a typical home at prices in some areas competitive with power purchased from a utility. By the end of this century, Japan plans to meet a considerable part of its energy requirements with solar power, an important goal for a country lacking fossil and nuclear fuel resources.

The world market for solar energy is expected to grow rapidly in the near future, especially in remote places where conventional power isn't available. Altogether, some 2 billion people around the world now have no access to electricity. Most would like to have a modern power source if it were affordable.

Think about how solar power could affect your future energy independence. Imagine the benefits of being able to build a house anywhere and having a cheap, reliable, clean, quiet source of energy with no moving parts to wear out, no fuel to purchase, and little equipment to maintain. You could have all the energy you need without commercial utility wires or monthly energy bills. Coupled with modern telecommunications and information technology, an independent energy source would make it possible to live in the countryside and yet have many of the employment and entertainment opportunities and modern conveniences available in a metropolitan area.

Storing Electrical Energy

Electrical energy is difficult and expensive to store. This is a problem for photovoltaic generation as well as other sources of electric power. Traditional lead-acid batteries are heavy and have low energy densities; that is, they can store only moderate amounts of energy per unit mass or volume. Acid from batteries is corrosive and lead from smelters or battery manufacturing is a serious health hazard for workers who handle these materials. A typical lead-acid battery array sufficient to store several days of electricity for an average home would cost about $5,000 and weigh 3 or 4 tons. All the components for an electric car are readily available, but, as mentioned earlier, battery technology limits how far they can go between charges. Still, some communities are encouraging use of electric vehicles because they produce zero emissions.

FIGURE 20.15 The Aurora IV was built by students at the University of Minnesota for the 1999 Sunrayce. It was traveling about 130 km/hr (80 mph) when this photo was taken. © Charles Habermann.

FIGURE 20.16 Solar roof tiles (shiny area) can generate enough electricity for a house full of efficient appliances. On sunny days, this array can produce a surplus to sell back to the utility company, making it even more cost efficient. Courtesy National Renewable Energy Laboratory/NREL/PIX.

Other types of batteries also have drawbacks. Metal-gas batteries, such as the zinc-chloride cell, use inexpensive materials and have relatively high-energy densities, but have shorter lives than other types. Sodium-sulfur batteries have considerable potential for large-scale storage. They store twice as much energy in half as much weight as lead-acid batteries. They require an operating temperature of about 300°C (572°F) and are expensive to manufacture. Alkali-metal batteries have a high storage capacity but are even more expensive. Lithium batteries have very long lives and store more energy than other types, but are the most expensive of all. Recent advances in thin-film technology may hold promise for future energy storage but often require rare, toxic elements that are both costly and dangerous.

Another strategy is to store energy in a form that can be turned back into electricity when needed. Pumped-hydro storage involves pumping water to an elevated reservoir at times when excess electricity is available. The water is released to flow back down through turbine generators when extra energy is needed. Using a similar principle, pressurized air can be pumped into such reservoirs as natural caves, depleted oil and gas fields, abandoned mines, or special tanks. An Alabama power company currently uses off-peak electricity to pump air at night into a deep salt mine. By day, the air flows back to the surface through turbines, driving a generator that produces electricity. Cool night air is heated to 1,600°F by compression plus geothermal energy, increasing pressure and energy yield.

An even better way to use surplus electricity is in electrolytic decomposition of water to H_2 and O_2. These gases can be liquefied (like natural gas) at –252°C (–423°F), making them easier to store and ship than most forms of energy. They are highly explosive, however, and must be handled with great care. They can be burned in internal combustion engines, producing mechanical energy, or they can be used to power fuel cells, which produce more electrical energy. There is a concern that if hydrogen escapes, it could destroy statospheric ozone.

Flywheels are the subject of current experimentation for energy storage. Massive, high-speed flywheels, spinning in a nearly friction-free environment, store large amounts of mechanical energy in a small area. This energy is convertible to electrical energy. It is difficult, however, to find materials strong enough to hold together reliably when spinning at high speed. Flywheels have a disconcerting tendency to fail explosively and unexpectedly, sending shrapnel flying in all directions. Still, this might represent a useful technology if these problems can be overcome.

FUEL CELLS

Rather than store and transport energy, another alternative would be to generate it locally, on demand. **Fuel cells** are devices that use ongoing electrochemical reactions to produce an electric current. They are very similar to batteries except that rather than recharging them with an electrical current, you add more fuel for the chemical reaction.

Fuel cells are not new; the basic concept was recognized in 1839 by William Grove, who was studying the electrolysis of water. He suggested that rather than use electricity to break apart water and produce hydrogen and oxygen gases, it should be possible to reverse the process by joining oxygen and hydrogen to produce water and electricity. The term "fuel cell" was coined in 1889 by Ludwig Mond and Charles Langer, who built the first practical device using a platinum catalyst to produce electricity from air and coal gas. The concept languished in obscurity until the 1950s when the U.S. National Aeronautics and Space Administration (NASA) was searching for a power source for spacecraft. Research funded by NASA eventually led to development of fuel cells that now provide both electricity and drinkable water on every space shuttle flight. The characteristics that make fuel cells ideal for space exploration—small size, high efficiency, low emissions, net water production, no moving parts, and high reliability—also make them attractive for a number of other applications.

All fuel cells consist of a positive electrode (the cathode) and a negative electrode (the anode) separated by an electrolyte, a material that allows the passage of charged atoms, called ions, but is impermeable to electrons (fig. 20.17). In the most common systems, hydrogen or a hydrogen-containing fuel is passed over the anode while oxygen is passed over the cathode. At the anode, a reactive catalyst, such as platinum, strips an electron from each hydrogen atom, creating a positively charged hydrogen ion (a proton). The hydrogen ion can migrate through the electrolyte to the cathode, but the electron is excluded. Electrons pass through an external circuit, and the electrical current generated by their passage can be used to do useful work. At the cathode, the electrons and protons are reunited and combined with oxygen to make water.

The fuel cell provides direct-current electricity as long as it is supplied with hydrogen and oxygen. For most uses, oxygen is provided by ambient air. Hydrogen can be supplied as a pure gas, but

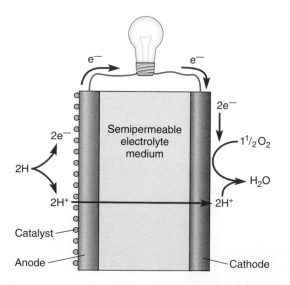

FIGURE 20.17 Fuel cell operation. Electrons are removed from hydrogen atoms at the anode to produce hydrogen ions (protons) that migrate through a semipermeable electrolyte medium to the cathode, where they reunite with electrons from an external circuit and oxygen atoms to make water. Electrons flowing through the circuit connecting the electrodes create useful electrical current.

storing hydrogen gas is difficult and dangerous because of its volume and explosive nature. Liquid hydrogen takes far less space than the gas, but must be kept below –250°C (–400°F), not a trivial task for most mobile applications. The alternative is a device called a **reformer** or converter that strips hydrogen from fuels such as natural gas, methanol, ammonia, gasoline, ethanol, or even vegetable oil. Many of these fuels can be derived from sustainable biomass crops. Even methane effluents from landfills and wastewater treatment plants can be used as a fuel source. Where a fuel cell can be hooked permanently to a gas line, hydrogen can be provided by solar, wind, or geothermal facilities that use electricity to hydrolyze water.

A fuel cell run on pure oxygen and hydrogen produces no waste products except drinkable water and radiant heat. When a reformer is coupled to the fuel cell, some pollutants are released (most commonly carbon dioxide), but the levels are typically far less than conventional fossil fuel combustion in a power plant or automobile engine. Although the theoretical efficiency of electrical generation of a fuel cell can be as high as 70 percent, the actual yield is closer to 40 or 45 percent. This is not much better than a very good fossil fuel power plant or a gas turbine electrical generator. On the other hand, the quiet, clean operation and variable size of fuel cells make them useful in buildings where waste heat can be captured for water heating or space heating. A new 45-story office building at 4 Times Square, for example, has two 200-kilowatt fuel cells on its fourth floor that provide both electricity and heat. This same building has photovoltaic panels on its façade, natural lighting, fresh air intakes to reduce air conditioning, and a number of other energy conservation features.

The current from a fuel cell is proportional to the size (area) of the electrodes, while the voltage is limited to about 1.23 volts per cell. A number of cells can be stacked together until the desired power level is reached. A fuel cell stack that provides almost all of the electricity needed by a typical home (along with hot water and space heating) would be about the size of a refrigerator. A 200 kilowatt unit fills a medium-size room and provides enough energy for 20 houses or a small factory (fig. 20.18). Tiny fuel cells running on

FIGURE 20.18 The Long Island Power Authority has installed 75 stationary fuel cells to provide reliable backup power.
Courtesy Long Island Power Authority.

methanol may soon be used in cell phones, pagers, toys, computers, videocameras, and other appliances now run by batteries. Rather than buy new batteries or spend hours recharging spent ones, you might just add an eyedropper of methanol every few weeks to keep your gadgets operating.

Fuel Cell Types

Several different electrolytes can be used in fuel cells, each with advantages and disadvantages (table 20.3). The design being developed for use in cars, buses, and trucks is called a proton exchange membrane (PEM). The membrane is a thin semipermeable layer of an organic polymer containing sulfonic acid groups that facilitate passage of hydrogen ions but block electrons and oxygen. The surface of the membrane is dusted with tiny particles of platinum

TABLE 20.3	Fuel Cell Types			
TYPE	**PROTON ELECTROLYTE MEMBRANE**	**PHOSPHORIC ACID**	**MOLTEN CARBONATE**	**SOLID OXIDE**
Electrolyte	Semipermeable organic polymer	Phosphoric acid	Liquid carbonate	Solid-oxide ceramic
Charge carrier	H^+	H^+	$CO_3^=$	$O^=$
Catalyst	Platinum	Platinum	Nickel	Perovskites (calcium titanate)
Operating temperature, °C	80	200	650	1,000
Cell material	Carbon or metal-based	Graphite-based	Stainless steel	Ceramic
Efficiency (percent)	Less than 40	40 to 50	50 to 60	More than 60
Heat cogeneration	None	Low	High	High
Status	Demonstration systems	Commercially available	Demonstration systems	Under development

Source: Alan C. Lloyd, Scientific American, July 1999, vol. 281 (1) p. 83.

catalyst. These cells have the advantage of being lightweight and operating at a relatively low temperature (80°C or 176°F). The fuel efficiency of PEM systems is typically less than 40 percent. Buses equipped with PEM stacks have been demonstrated in Chicago, Miami, and Vancouver, BC. In 1999, DaimlerChrysler unveiled their 5-passenger NECAR-4. A 500 kg (1,100 lb) PEM stack that costs $30,000 (compared to about $3,000 for a conventional gasoline engine) produces performance comparable to most passenger cars, but takes up considerable interior room.

For stationary electrical generation, the most common fuel cell design uses phosphoric acid immobilized in a porous ceramic matrix as the electrolyte. Because this system operates at higher temperatures than PEM cells, less platinum is needed for the catalyst. It has a higher efficiency, 40 to 50 percent, but is heavier and larger than PEM cells. It also is less sensitive to carbon dioxide contamination than other designs. Hundreds of 200 kilovolt phosphoric acid fuel cells have been installed around the world. Some have run for decades. They supply dependable electricity in remote locations without the spikes and sags and risk of interruption common in utility grids. The largest fuel cell ever built was an 11 MW unit in Japan, that provides enough electricity for a small town.

In 1999, the Central Park police station in New York City was equipped with a 200 kilowatt phosphoric acid fuel cell. The station is located in the middle of the park, so bringing in new electric lines would have cost $1.2 million and would have disrupted traffic and park use for months. A diesel generator was ruled out as too noisy and polluting. Solar photovoltaic panels were thought to be too obtrusive. A small, silent fuel cell provided just the right solution.

Carbonate fuel cells use an inexpensive nickel catalyst and operate at 650°C (1,200°F). The electrode is a very hot (thus the name molten carbonate) solution trapped in a porous ceramic. The charge carrier is carbonate ion, which is formed at the cathode where oxygen and carbon dioxide react in the presence of a nickel oxide catalyst. Migrating through the electrolyte, the carbonate ion

reacts at the anode with hydrogen and carbon monoxide to release electrons. The high operating temperature of this design means that it can reform fuels internally and ionize hydrogen without expensive catalysts. Heat cogeneration is very good, but the high temperature makes these units more difficult to operate. It takes hours for carbonate fuel cells to get up to operating temperature, so they aren't suitable for short-term, quick-response uses.

The least developed of the fuel cell design is called solid oxide, because it uses a coated zirconium ceramic as an electrolyte. Oxygen ions formed by the titanium catalyst carry the charge across the electrolyte. Operating temperatures are 1,000°C (1,800°F). They have the highest fuel efficiency of any current design, but mass production of components has not yet been mastered, and these cells are still in the experimental stage.

ENERGY FROM BIOMASS

Photosynthetic organisms have been collecting and storing the sun's energy for more than 2 billion years. Plants capture about 0.1 percent of all solar energy that reaches the earth's surface. That kinetic energy is transformed, via photosynthesis, into chemical bonds in organic molecules (chapter 3). A little more than half of the energy that plants collect is spent in such metabolic activities as pumping water and ions, mechanical movement, maintenance of cells and tissues, and reproduction; the rest is stored in biomass.

The magnitude of this resource is difficult to measure. Most experts estimate useful biomass production at 15 to 20 times the amount we currently get from all commercial energy sources. It would be ridiculous to consider consuming all green plants as fuel, but biomass has the potential to become a prime source of energy. It has many advantages over nuclear and fossil fuels because of its renewability and easy accessibility. Renewable energy resources account for about 18 percent of total world energy use, and biomass makes up three-quarters of that renewable energy supply. Biomass resources used as fuel include wood, wood chips, bark, branches, leaves, starchy roots, and other plant and animal materials.

Burning Biomass

Wood fires have been a primary source of heating and cooking for thousands of years. As recently as 1850, wood supplied 90 percent of the fuel used in the United States. Wood now provides less than 1 percent of the energy in the United States, but in many of the poorer countries of the world, wood and other biomass fuels provide up to 95 percent of all energy used (fig. 20.19). The 1,500 million cubic meters of fuelwood collected in the world each year is about half of all wood harvested.

In northern industrialized countries, wood burning has increased since 1975 in an effort to avoid rising oil, coal, and gas prices. Most of these northern areas have adequate wood supplies to meet demands at current levels, but problems associated with wood burning may limit further expansion of this use. Inefficient and incomplete burning of wood in fireplaces and stoves produces smoke laden with fine ash and soot and hazardous amounts of car-

FIGURE 20.19 Firewood is an important resource in this Siberian village near Lake Baikal. © William P. Cunningham.

bon monoxide (CO) and hydrocarbons. In valleys where inversion layers trap air pollutants, the effluent from wood fires can present a major source of air quality degradation and health risk. Polycyclic aromatic compounds produced by burning are especially worrisome because they are carcinogenic (cancer-causing).

In Oregon's Willamette Valley or in the Colorado Rockies, where woodstoves are popular and topography concentrates contaminants, as much as 80 percent of air pollution on winter days is attributable to wood fires. Several resort towns, such as Vail, Aspen, and Telluride, Colorado, have banned installation of new woodstoves and fireplaces because of high pollution levels. Oregon, Colorado, and Vermont now have emission standards for new woodstoves. The Environmental Protection Agency ranks wood burners high on a list of health risks to the general population, and standards are being considered to regulate the use of woodstoves nationwide and to encourage homeowners to switch to low-emission models.

Highly efficient and clean-burning woodstoves are available but expensive. Running exhaust gases through heat exchangers recaptures heat that would escape through the chimney, but if flue temperatures drop too low, flammable, tarlike creosote deposits can build up, increasing fire risk. A better approach is to design combustion chambers that use fuel efficiently, capturing heat without producing waste products. Brick-lined fireboxes with afterburner chambers to burn gaseous hydrocarbons do an excellent job. Catalytic combusters, similar to the catalytic converters on cars, also are placed inside stovepipes. They burn carbon monoxide and hydrocarbons to clean emissions and to recapture heat that otherwise would have escaped out the chimney in the chemical bonds of these molecules.

Wood chips, sawdust, wood residue, and other plant materials (fig. 20.20) are being used in some places in the United States and Europe as a substitute for coal and oil in industrial boilers. In Vermont, for instance, where fossil fuels are expensive and 76 percent of the land is covered by forest, 250,000 cords of unmarketable cull wood are burned annually to fuel a 50 MW power plant in Burlington. Michigan's Public Service Commission estimated that the state's surplus forest growth could provide 1,300 MW of electricity each year. Pollution-control equipment is easier to install and maintain in a central power plant than in individual home units. Wood burning also contributes less to acid precipitation than does coal. Because wood has little sulfur, it produces few sulfur gases and burns at lower temperatures than coal; thus, it produces fewer nitrogen oxides. Burning wood as a renewable crop doesn't produce any net increase in atmospheric carbon dioxide (and, therefore, doesn't add to global warming) because all the carbon released by burning biomass was taken up from the atmosphere when the biomass was grown.

Fuelwood Crisis in Less-Developed Countries

Two billion people—about 40 percent of the total world population—depend on firewood and charcoal as their primary energy source. Of these people, three-quarters (1.5 billion) do not have an adequate, affordable supply. Most of these people are in the less-developed countries where they face a daily struggle to find enough fuel to warm their homes and cook their food. The problem is intensifying because rapidly growing populations in many developing countries create increasing demands for firewood and charcoal from a diminishing supply.

As firewood becomes increasingly scarce, women and children, who do most of the domestic labor in many cultures, spend more and more hours searching for fuel (fig. 20.21). In some places, it now takes eight hours, or more, just to walk to the nearest fuelwood supply and even longer to walk back with a load of sticks and branches that will only last a few days.

FIGURE 20.20 Harvesting marsh reeds (*Phragmites* sp.) in Sweden as a source of biomass fuel. In some places, biomass from wood chips, animal manure, food-processing wastes, peat, marsh plants, shrubs, and other kinds of organic material make a valuable contribution to energy supplies. Care must be taken, however, to avoid environmental damage in sensitive areas. Courtesy Dr. Douglas Pratt.

For people who live in cities, the opportunity to scavenge firewood is generally nonexistent and fuel must be bought from merchants. This can be ruinously expensive. In Addis Ababa, Ethiopia, 25 percent of household income is spent on wood for cooking fires. A circle of deforestation has spread more than 160 km (100 mi) around some major cities in India, and firewood costs up to ten times the price paid in smaller towns.

The poorest countries such as Ethiopia, Bhutan, Burundi, and Bangladesh depend on biomass for 90 percent of their energy. Often, the harvest is sustainable, consisting of deadwood, branches, trimmings, and shrubs. In Pakistan, for example, some 4.4 million tons of twigs and branches and 7.7 million tons of shrubs and crop residue are consumed as fuel each year with destruction of very few living trees.

In countries where fuel is scarce, however, desperate people often chop down anything that will burn. In Haiti, for instance, more than 90 percent of the once-forested land has been almost completely denuded and people cut down even valuable fruit trees to make charcoal they can sell in the marketplace. It is estimated that the 1,700 million tons of fuelwood now harvested each year globally is at least 500 million tons less than is needed. By 2025 the worldwide demand for fuelwood is expected to be about twice current harvest rates while supplies will not have expanded much beyond current levels. Some places will be much worse than this average. In some African countries such as Mauritania, Rwanda, and the Sudan, firewood demand already is ten times the sustainable yield. Reforestation projects, agroforestry, community woodlots, and inexpensive, efficient, locally produced woodstoves could help alleviate expected fuelwood shortages in many places.

Dung and Methane as Fuels

Where wood and other fuels are in short supply, people often dry and burn animal manure. This may seem like a logical use of waste biomass, but it can intensify food shortages in poorer countries. Not putting this manure back on the land as fertilizer reduces crop production and food supplies. In India, for example, where fuelwood supplies have been chronically short for many years, a limited manure supply must fertilize crops and provide household fuel. Cows in India produce more than 800 million tons of dung per year, more than half of which is dried and burned in cooking fires (fig. 20.22). If that dung were applied to fields as fertilizer, it could boost crop production of edible grains by 20 million tons per year, enough to feed about 40 million people.

When cow dung is burned in open fires, more than 90 percent of the potential heat and most of the nutrients are lost. Compare that to the efficiency of using dung to produce methane gas, an excellent fuel. In the 1950s, simple, cheap methane digesters were designed for villages and homes, but they were not widely used. In China, 6 million households use biogas for cooking and lighting. Two large municipal facilities in Nanyang will soon provide fuel for more than 20,000 families. Perhaps other countries will follow China's lead.

Methane gas is the main component of natural gas. It is produced by anaerobic decomposition (digestion by anaerobic bacte-

FIGURE 20.21 A charcoal market in Ghana. Firewood and charcoal provide the main fuel for billions of people. Forest destruction results in wildlife extinction, erosion, and water loss. Courtesy National Renewable Energy Laboratory/NREL/PIX.

FIGURE 20.22 In many countries, animal dung is gathered by hand, dried, and used for fuel. This girl in Mali is carrying home fuel for cooking. Using dung as fuel deprives fields of nutrients and reduces crop production. © Wolfgang Kaehler.

ria) of any moist organic material. Many people are familiar with the fact that swamp gas is explosive. Swamps are simply large methane digesters, basins of wet plant and animal wastes sealed from the air by a layer of water. Under these conditions, organic materials are decomposed by anaerobic (oxygen-free) rather than aerobic (oxygen-using) bacteria, producing flammable gases instead of carbon dioxide. This same process may be reproduced artificially by placing organic wastes in a container and providing warmth and water (fig. 20.23). Bacteria are ubiquitous enough to start the culture spontaneously.

Burning methane produced from manure provides more heat than burning the dung itself, and the sludge left over from bacterial digestion is a rich fertilizer, containing healthy bacteria as well as most of the nutrients originally in the dung. Whether the manure is of livestock or human origin, airtight digestion also eliminates some health hazards associated with direct use of dung, such as exposure to fecal pathogens and parasites.

How feasible is methane—from manure or from municipal sewage—as a fuel resource in developed countries? Methane is a clean fuel that burns efficiently. Any kind of organic waste material: livestock manure, kitchen and garden scraps, and even municipal garbage and sewage can be used to generate gas. In fact, municipal landfills are active sites of methane production, contributing as much as 20 percent of the annual output of methane to the atmosphere. This is a waste of a valuable resource and a threat to the environment because methane absorbs infrared radiation and contributes to the greenhouse effect (chapter 15). About 300 landfills in the United States currently burn methane and generate enough electricity together for a million homes. Another 600 landfills have been identified as potential sources for methane development. Hydrologists worry, however, that water will be pumped into landfills to stimulate fermentation, thus increasing the potential for groundwater contamination.

Cattle feedlots and chicken farms in the United States are a tremendous potential fuel source. Collectible crop residues and feedlot wastes each year contain 4.8 billion gigajoules (4.6 quadrillion BTUs) of energy, more than all the nation's farmers use. The Haubenschild farm in central Minnesota, for instance, uses manure from 850 Holsteins to generate all the power needed for their dairy operation and still have enough excess electricity for an additional 80 homes. In January 2001, the farm saved 35 tons of coal, 1,200 gallons of propane, and made $4,380 from electric sales.

Municipal sewage treatment plants also routinely use anaerobic digestion as a part of their treatment process, and many facilities collect the methane they produce and use it to generate heat or electricity for their operations. Although this technology is well-developed, its utilization could be much more widespread.

Alcohol from Biomass

Ethanol (grain alcohol) and methanol (wood alcohol) are produced by anaerobic digestion of plant materials with high sugar content, mainly grain and sugarcane. Ethanol can be burned directly in automobile engines adapted to use this fuel, or it can be mixed with gasoline (up to about 10 percent) to be used in any normal automobile engine. A mixture of gasoline and ethanol is often called gasohol or **biofuel.** Ethanol in gasohol raises octane ratings and is a good substitute for lead antiknock agents, the major cause of lead pollution. It also helps reduce carbon monoxide emissions in automobile exhaust. Gas stations in U.S. cities that exceed EPA air quality standards have been ordered to sell so-called oxygenated fuels containing ethanol, methanol, or methyltertiarybutyl ether (MTBE). The latter, however, is a suspected carcinogen that has contaminated many lakes and aquifers and is being banned in most states.

Ethanol production could be a solution to grain surpluses and bring a higher price for grain crops than the food market offers. It also offers promise for reduced dependence on gasoline, which is refined from petroleum. However, it's not clear, given the way crops are currently grown, harvested, dried and fermented, that ethanol represents a net energy gain. Local residents also complain about odors from ethanol plants. Brazil has instituted an ambitious national program to substitute crop-based ethanol for imported petroleum. In 1995, the Brazilian sugar harvest produced 2.5 billion gal of ethanol, but falling oil prices have undercut plans to expand ethanol production to 4 billion gal per year. Both methanol and ethanol make good fuels for fuel cells.

Crop Residues, Energy Crops, and Peat

Crop residues, such as cornstalks, corncobs, or wheat straw, can be used as a fuel source, but they are expensive to gather and often are better left on the ground as soil protection (chapter 9). The residue accumulated in food processing, however, can be a useful fuel. Hawaiian sugar growers burn bagasse, a fibrous sugarcane residue, to produce the state's second-largest energy source, surpassed only by petroleum.

Some crops are being raised specifically as an energy source. Fast-growing trees, such as eucalyptus or poplar and shrubs, such

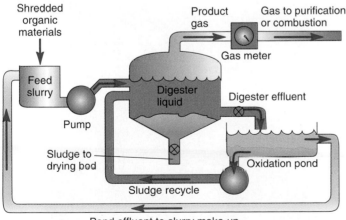

FIGURE 20.23 Continuous unit for converting organic material to methane by anaerobic fermentation. One kilogram of dry organic matter will produce 1–1.5 m³ of methane, or 2,500–3,600 million calories per metric ton.

Source: Solar Energy as a National Energy Resource, NSF/NASA Solar Energy Panel, National Science Foundation, December 1972.

FIGURE 20.24 These fast-growing hybrid poplars are less than two years old. They make a good energy crop in some circumstances. Courtesy National Renewable Energy Laboratory/NREL/PIX.

as alder and willow, are grown to provide wood for energy (fig. 20.24). Milkweeds, sedges, marsh grasses, cattails, and other bio-mass crops also might become useful energy sources, grown on land that is otherwise unsuitable for crops. Unfortunately, there may be unacceptable environmental costs from disruption of wet-lands and forests that are converted to energy crop plantations. Some plant species produce hydrocarbons that can be used directly as fuel in internal combustion engines without having to be fer-mented or transformed to hydrogen or methane gas. Sunflower oil, for instance, can be burned directly in diesel engines. In most cases, these oils and other high-molecular-weight hydrocarbons are more valuable for other purposes, such as food, plastics, and industrial chemicals; however, they might be an attractive energy source in some circumstances.

Several other forms of biomass are actual or proposed sources of energy. People in northern climates have been digging up peat (partially decomposed plant residues) from bogs and marshes for centuries. There has been a resurgence of interest in exploiting the vast Minnesota and Canadian peat bogs, which some people regard as wastelands. After mining some or all of the peat, the land could be used for energy crops, such as cattails. Critics of these schemes point out the damaging effects of bog draining on ecosystems,

watersheds, and wildlife. Burning of municipal garbage ("waste to watts") is a popular proposal; however, many people are worried about air pollution from these facilities (chapter 21). Garbage may contain hazardous, volatile, or explosive materials that are difficult or dangerous to sort out. Effects on air quality from all forms of direct combustion of biomass are a serious concern.

ENERGY FROM THE EARTH'S FORCES

The winds, waves, tides, ocean thermal gradients, and geothermal areas are renewable energy sources. Although available only in selected locations, these sources could make valuable contributions to our total energy supply.

Hydropower

Falling water has been used as an energy source since ancient times. The invention of water turbines in the nineteenth century greatly increased the efficiency of hydropower dams. By 1925, falling water generated 40 percent of the world's electric power. Since then, hydroelectric production capacity has grown 15-fold, but fossil fuel use has risen so rapidly that water power is now only 20 percent of total electrical generation. Still, many countries produce most of their electricity from falling water (fig. 20.25). Norway, for instance, depends on hydropower for 99 percent of its electricity. Currently, total world hydroelectric production is about 3,000 terrawatt hours (10^{12} Whr). Six countries, Canada, Brazil, the United States, China,

FIGURE 20.25 Hydropower dams produce clean renewable energy but can be socially and ecologically damaging. The McGraw-Hill Companies, Inc./Barry Barker, photographer.

Russia, and Norway, account for more than half that total. Approximately two-thirds of the economically feasible potential remains to be developed. Untapped hydro resources are still abundant in Latin America, Central Africa, India, and China.

Much of the hydropower development in recent years has been in enormous dams. There is a certain efficiency of scale in giant dams, and they bring pride and prestige to the countries that build them, but they can have unwanted social and environmental effects. The largest hydroelectric dam in the world at present is the Itaipu Dam on the Parana River between Brazil and Paraguay. Designed to generate 12,600 MW of power, this dam should produce as much energy as 13 large nuclear power plants when completed. The lake that it is creating already has flooded about 1,300 km² (500 mi²) of tropical rainforest and displaced many thousands of native people and millions of other creatures. An even larger dam that will produce twice as much electrical power is now under construction at the Three Gorges site on China's Yangtze River.

There are other problems with big dams, besides human displacement, ecosystem destruction, and wildlife losses. Dam failure can cause catastrophic floods and thousands of deaths. Sedimentation often fills reservoirs rapidly and reduces the usefulness of the dam for either irrigation or hydropower. In China, the Sanmenxia Reservoir silted up in only four years, and the Laoying Reservoir filled with sediment before the dam was even finished. Rotting vegetation in artificial impoundments can have disastrous effects on water quality. When Lake Brokopondo in Suriname flooded a large region of uncut rainforest, underwater decomposition of the submerged vegetation produced hydrogen sulfide that killed fish and drove out villagers over a wide area. Acidified water from this reservoir ruined the turbine blades, making the dam useless for power generation.

Floating water hyacinths (rare on free-flowing rivers) have already spread over reservoir surfaces behind the Tucurui Dam on the Amazon River in Brazil, impeding navigation and fouling machinery. Herbicides sprayed to remove aquatic vegetation have contaminated water supplies. Herbicides used to remove forests before dam gates closed caused similar pollution problems. Schistosomiasis, caused by parasitic flatworms called blood flukes (chapter 8), is transmitted to humans by snails that thrive in slow-moving, weedy tropical waters behind these dams. It is thought that 14 million Brazilians suffer from this debilitating disease.

As mentioned before, dams displace indigenous people. The Narmada Valley project in India will drown 150,000 ha of tropical forest and displace 1.5 million people, mostly tribal minorities and low-caste hill people. China's Three Gorges project has displaced more than a million people. The Akosombo Dam built on the Volta River in Ghana nearly 20 years ago displaced 78,000 people from 700 towns. Few of these people ever found another place to settle, and those still living remain in refugee camps and temporary shelters. The Cree First Nations people of Canada also have been displaced by massive hydroelectric projects, most notably the James Bay project in Quebec and the Churchill Nelson River diversion in Manitoba.

In tropical climates, large reservoirs often suffer enormous water losses. Lake Nasser, formed by the Aswan High Dam in Egypt, loses 15 billion m³ each year to evaporation and seepage. Unlined canals lose another 1.5 billion m³. Together, these losses represent one-half of the Nile River flow, or enough water to irrigate 2 million ha of land. The silt trapped by the Aswan Dam formerly fertilized farmland during seasonal flooding and provided nutrients that supported a rich fishery in the Delta region. Farmers now must buy expensive chemical fertilizers, and the fish catch has dropped almost to zero. As in South America, schistosomiasis is an increasingly serious problem.

If big dams—our traditional approach to hydropower—have so many problems, how can we continue to exploit the great potential of hydropower? Fortunately, there is an alternative to gigantic dams and destructive impoundment reservoirs. Small-scale, **low-head hydropower** technology can extract energy from small headwater dams that cause much less damage than larger projects. Some modern, high-efficiency turbines can even operate on **run-of-the-river flow.** Submerged directly in the stream and small enough not to impede navigation in most cases, these turbines don't require a dam or diversion structure and can generate useful power with a current of only a few kilometers per hour. They also cause minimal environmental damage and don't interfere with fish movements, including spawning migration. **Micro-hydro generators** operate on similar principles but are small enough to provide economical power for a single home. If you live close to a small stream or river that runs year-round and you have sufficient water pressure and flow, hydropower is probably a cheaper source of electricity for you than solar or wind power (fig. 20.26).

FIGURE 20.26 Solar collectors capture power only when the sun shines, but hydropower is available 24 hours a day. Small turbines such as this one can generate enough power for a single-family house with only 15 m (50 ft) of head and 200 l (50 g) per minute flow. The turbine can have up to four nozzles to handle greater water flow and generate more power. Courtesy Burkhardt Turbines.

Small-scale hydropower systems also can cause abuses of water resources. The Public Utility Regulatory Policies Act of 1978 included economic incentives to encourage small-scale energy projects. As a result, thousands of applications were made to dam or divert small streams in the United States. Many of these projects have little merit. All too often, fish populations, aquatic habitat, recreational opportunities, and the scenic beauty of free-flowing streams and rivers are destroyed primarily to provide tax benefits for wealthy investors.

Wind Energy

As this chapter's opening story shows, wind power offers an enormous potential for renewable energy. The World Meteorological Organization estimates that 20 million MW of wind power could be developed economically worldwide. This would be ten times the total current global electrical generating capacity. Wind has a number of advantages over most other power sources. Wind farms have much shorter planning and construction times than fossil fuel or nuclear power plants. Wind generators are modular (more turbines can be added if loads grow) and they have no fuel costs or air emissions (fig. 20.27). Since 1995, total wind generating capacity has increased nearly tenfold making it the fastest growing energy source in the world. With 35,000 MW of installed capacity, wind power is making a valuable contribution to reducing global warming. Wind does have limitations, however. Like solar energy, it is an intermittent source. Furthermore, not every place has strong enough or steady enough wind to make this an economical resource. Although modern windmills are more efficient than those of a few years ago, it takes a wind velocity between 7 m per second (16 mph) and 27 m per second (60 mph) to generate useful amounts of electricity.

In places like the Netherlands, which has winds above 16 mph an average of 245 days per year, windmills have long been recognized as a valuable energy source for pumping water and grinding grain. Wind power also played a crucial role in the settling of the American West, much of which has abundant underground aquifers, but little surface water. The strong, steady winds blowing across the prairies provided the energy to pump water that allowed ranchers and farmers to settle the land. By the end of the nineteenth century, nearly every farm or ranch west of the Mississippi River had at least one windmill, and the manufacture, installation, and repair of windmills was a major industry. The Rural Electrification Act of 1935 brought many benefits to rural America, but it effectively killed wind power development, and shifted electrical generation to large dams and fossil fuel-burning power plants. It's interesting to speculate what the course of history might have been if we had not spent trillions of dollars on fossil fuels and nuclear power, but instead had invested that money on small-scale, renewable energy systems.

The oil price shocks of the 1970s spurred a renewed interest in wind power. In the 1980s, the United States was a world leader in wind technology, and California hosted 90 percent of all wind power generators in the world. Poor management, technical flaws, and overdependence on subsidies, however, led to bankruptcy of

FIGURE 20.27 Fifty-meter-tall towers (about the height of a 12-story building) stand on the Minnesota prairie. Each one, with a 750 kW generator about the size of a minivan, can generate enough electricity for about 75 homes. Courtesy National Renewable Energy Laboratory/NREL/PIX.

many of the most important companies of that era, including Kenetech, once the world's largest manufacturer of wind generators. Now European companies dominate this $1 billion (U.S.) per year market.

Modern wind machines are far different from those employed a generation ago. Out of commission for maintenance only about three days per year, many can produce power 90 percent of the time. Theoretically up to 60 percent efficient, modern windmills typically produce about 35 percent of peak capacity under field conditions. Currently, wind farms are the cheapest source of *new*

TABLE 20.4 — Jobs and Land Required for Alternative Energy Sources

TECHNOLOGY	LAND USE (m² PER GIGAWATT-HOUR FOR 30 YEARS)	JOBS (PER TERAWATT-HOUR PER YEAR)
Coal	3,642	116
Photovoltaic	3,237	175
Solar thermal	3,561	248
Wind	1,335	542

Source: *Lester R. Brown, et al. Saving the Planet, 1991, W. W. Norton & Co., Inc.*

FIGURE 20.28   A 23-meter-long, 1.6-meter-wide fiberglass blade waits to be attached to a new wind generator. © William P. Cunningham.

power generation, costing as little as 3 cents/kWh compared to 4 to 5 cents/kWh for coal and five times that much for nuclear fuel. As table 20.4 shows, when the land consumed by mining is taken into account, wind power takes about one-third as much area and creates about five times as many jobs to create the same amount of electrical energy as coal.

While Europe, with its high population density, is focusing most attention to offshore wind farms, the bulk of North America's wind potential is situated across the Great Plains. Compared to offshore installations, which are costly because of the need to operate in deep water and withstand storms and waves, wind tower construction on land is relatively simple and cheap. There also is growing demand for wind projects from farmers, ranchers, and rural communities because of the economic benefits that wind energy brings. One thousand megawatts of wind power (equivalent to one large nuclear or fossil fuel plant) can create more than 3,000 permanent jobs, while paying about $4 million in rent to landowners and $3.6 million in tax payments to local governments. Seven Midwestern states—North Dakota, Texas, Kansas, South Dakota, Montana, Iowa, and Minnesota—all now have at least 1,000 MW of wind power, but wind-energy experts estimate this is less than 1 percent of their ultimate potential (fig. 20.28).

With each tower taking only about a 0.1 ha (0.25 acre) of cropland, farmers find that they can continue to cultivate 90 percent of their land while getting $2,000 or more in annual rent for each wind machine. An even better return results if the landowner builds and operates the wind generator, selling the electricity to the local utility. Annual profits can be as much as $100,000 per turbine, a wonderful bonus for use of 10 percent of your land. Cooperatives are springing up to help landowners finance, build, and operate their own wind generators. About 20 Native American tribes, for example, have formed a coalition to study wind power. Together, their reservations (which were sited in the windiest, least productive parts of the Great Plains) could generate at least 350,000 MW of electrical power, equivalent to about half of the total U.S. installed capacity.

There are problems with wind energy. In some places, high bird mortality has been reported around wind farms. This seems to be particularly true in California, where rows of generators were placed

at the summit of mountain passes where wind velocities are high but where migrating birds were likely to fly into rotating blades. New generator designs and more careful tower placement seems to have reduced this problem in most areas. As mentioned in this chapter's opening story, some people object to the sight of large machines looming on the horizon. To others, however, windmills offer a welcome alternative to nuclear or fossil fuel-burning plants.

Many of the places most appropriate for wind development are far from the urban centers where power is needed (a major reason why people didn't settle in extremely windy places, is the wind). This means that some method is needed to transfer wind-generated power to the market. Currently, the only way to do that is high-voltage power lines (fig. 20. 29). There may be more resistance to building thousands of kilometers of new power lines than to the wind farms themselves. An attractive alternative is to use the electricity from wind power to split water into hydrogen and oxygen gas. The gas could then be pumped through underground gas pipes to the city, where it could be used in fuel cells.

Small windmills can generate enough electricity for a single home or farm, and are generally the least expensive form of renewable energy, if the winds are strong and steady enough in your area. The problem is how to store energy for windless days. Buying enough batteries to provide power for several days can be exorbitantly expensive, not to mention bulky and difficult to maintain. If you're able to hook up to the utility grid, the best solution to this problem is what's called reverse metering, in which you sell electricity back to your utility when you have a surplus and then draw from them when you have a shortage. If your system is sized right, it should pay for itself in about 5 years if wind speeds in your area are favorable.

Geothermal Energy

The earth's internal temperature can provide a useful source of energy in some places. High-pressure, high-temperature steam fields exist below the earth's surface. Around the edges of continental

FIGURE 20.29   As we come to depend more on wind, solar, and hydropower energy, we will probably need more transmission lines to get that energy to market. © Corbis/Volume 160.

plates or where the earth's crust overlays magma (molten rock) pools close to the surface, this energy is expressed in the form of hot springs, geysers, and fumaroles (chapter 14). Yellowstone National Park is the largest geothermal region in the United States. Iceland, Japan, and New Zealand also have high concentrations of geothermal springs and vents. Depending on the shape, heat content, and access to groundwater, these sources produce wet steam, dry steam, or hot water.

Until recently, the main use of this energy source was in baths built at hot springs. More recently, **geothermal energy** has been used in electric power production, industrial processing, space heating, agriculture, and aquaculture. In Iceland, most buildings are heated by geothermal steam. A few communities in the United States, including Boise, Idaho, use geothermal home heating. Recently, geothermal steam also has been developed for electrical generation.

California's Geysers project is the world's largest commercial geothermal electric-generating complex, with 200 steam wells that provide some 1,300 MW of power. This system functions much like other heat-driven generators. A shaft sunk into the subsurface steam reservoir brings pressurized steam up to a turbine at the surface. The steam spins the turbine to produce electricity and then is piped off to a condensing unit and then to a cooling pond. The remaining brine (water heavily laden with dissolved minerals) is pumped back down to the reservoir. When the steam source has high mineral concentrations that would corrode delicate turbine blades, a heat exchanger is used to generate clean steam as is done in a nuclear power plant.

Among the advantages of geothermal generators are a reasonably long life span (at least several decades), no mining or transporting of fuels, and little waste disposal. The main disadvantages are the potential danger of noxious gases in the steam and noise problems from steam-pressure relief valves.

While few places in the world have geothermal steam, our planet is a huge heat sink that can serve as a valuable energy resource. Nearly everywhere, the earth just below ground level maintains a nearly constant temperature between 10° and 20°C (50° and 70°F). Using a set of closed-loop, underground tubes, homeowners can utilize this stable temperature to help with space heating and cooling (fig. 20.30). In the winter, water or another heat exchange fluid circulating inside the loop absorbs heat from the earth and carries it to a geothermal heat pump, which heats air or water for circulation through the heating system. In the summer the system reverses, drawing heat from the house and cooling the transfer liquid by circulating it through the earth loops.

Several different loop configurations are available. If space permits, the easiest way to install the earth loops is generally in horizontal trenches dug about 2 meters below the ground surface. Coiling the tubes in a "slinky" spiral gives a greater surface area for heat exchange in a smaller area. In small lots, vertical shafts can be drilled to accommodate the heat exchange tubes. If a pond or other water body is nearby, it, too, may serve as a heat sink. Water wells may also serve this purpose. A typical geothermal system will often reduce home heating and cooling costs by about half and can pay for itself in about 5 years, while also reducing your environmental impacts.

Tidal and Wave Energy

Ocean tides and waves contain enormous amounts of energy that can be harnessed to do useful work. Tidal power exploitation is not new. The earliest recorded tide-powered mills were built nearly 1,000 years ago in England. One built in Woodbridge, England, in 1170 functioned productively for 800 years. Through the seventeenth century, similar mills were built by the Dutch in the Netherlands.

The Rance River Power Station in France, in operation since 1966, was the first large tidal electric generation plant, producing 160 MW. A **tidal station** works like a hydropower dam, with its turbines spinning as the tide flows through them (fig. 20.31). It requires a high-tide low-tide differential of several meters to spin the turbines. Unfortunately, the tidal period of $13\frac{1}{2}$ hours causes problems in integrating the plant into the electric utility grid, as it seldom coincides with peak-use hours. Nevertheless, demand has kept the plant running for more than two decades.

The first North American tidal generator, producing 20 MW, was completed in 1984 at Annapolis Royal, Nova Scotia. A much larger project has been proposed to dam the Bay of Fundy and produce 5,000 MW of power on the Bay's 17-meter tides. The total flow at each tide through the Bay of Fundy theoretically could gen-

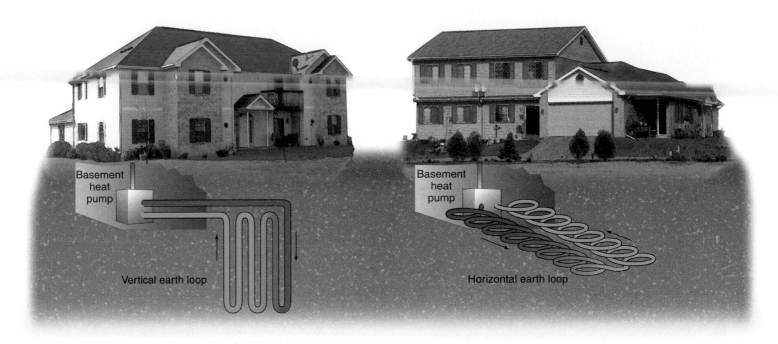

FIGURE 20.30   Geothermal energy can cut heating and cooling costs by half in many areas. In summer (shown here), warm water is pumped through buried tubing (earth loops) where it is cooled by constant underground temperatures. In winter, the system reverses and the relatively warm soil helps heat the house. Where space is limited (left), earth loops can be vertical. If more space is available (right) the tubing can be laid in shallow horizontal trenches.

erate energy equivalent to the output of 250 large nuclear power plants. The environmental consequences of such a gargantuan project, however, may prevent its ever being built. The main worries are saltwater flooding of freshwater aquifers when seawater levels rise behind the dam and the flooding and destruction of rich shoals and salt flats, breeding grounds for aquatic species and a vital food source for millions of migrating shorebirds. There also would be heavy siltation, as well as scouring of the seafloor as water shoots through the dam. In both California and Norway, plans are being considered for turbines that would sit on the ocean floor and spin as tides flow in and out of bays. This would avoid many of the problems associated with building dams to harness tidal flows.

Ocean wave energy can easily be seen and felt on any seashore. The energy that waves expend as millions of tons of water are picked up and hurled against the land, over and over, day after day, can far exceed the combined energy budget for both solar energy and wind power in localized areas. Captured and turned into useful forms, that energy could make a very substantial contribution to meeting local energy needs.

Dutch researchers estimate that 20,000 km of ocean coastline are suitable for harnessing wave power. Among the best places in the world for doing this are the west coasts of Scotland, Canada, the United States (including Hawaii), South Africa, and Australia. Wave energy specialists rate these areas at 40 to 70 kW per meter of shoreline. Altogether, it's calculated, if the technologies being studied today become widely used, wave power could amount to as much as 16 percent of the world's current electrical output. Some of the designs being explored include oscillating water columns that push or pull air through a turbine, and a variety of floating buoys, barges, and cylinders that bob up and down as waves pass, using a generator to convert mechanical motion into electricity. In 2001, Scotland installed the first commercial wave-power station in the world on the island of Islay. This facility generates about 500 kW, or enough electricity for some 400 island homes.

Ocean Thermal Electric Conversion

Temperature differentials between upper and lower layers of the ocean's water also are a potential source of renewable energy. In a closed-cycle **ocean thermal electric conversion (OTEC)** system, heat from sun-warmed upper ocean layers is used to evaporate a working fluid, such as ammonia or Freon, which has a low boiling point. The pressure of the gas produced is high enough to spin turbines to generate electricity. Cold water then is pumped from the the ocean depths to condense the gas.

As long as a temperature difference of about 20°C (36°F) exists between the warm upper layers and cooling water, useful amounts of net power can, in principle, be generated with one of these systems. This differential corresponds, generally, to a depth of about 1,000 m in tropical seas. The places where this much temperature difference is likely to be found close to shore are islands that are the tops of volcanic seamounts, such as Hawaii, or the edges of continental plates along subduction zones (chapter 14) where deep trenches lie just offshore. The west coast of Africa, the south coast of Java, and a number of South Pacific islands, such as Tahiti, have usable temperature differentials for OTEC power.

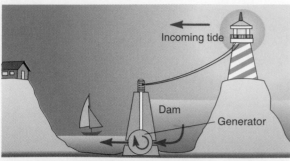

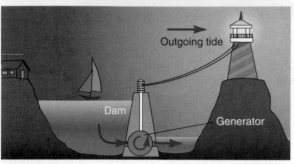

FIGURE 20.31   Tidal power station. Both incoming and outgoing tides are held back by a dam. The difference in water levels generates electricity in both directions as water runs through reversible turbogenerators.

Although their temperature differentials aren't as great as the ocean, deep lakes can have very cold bottom water. Ithaca, New York, has recently built a system to pump cold water out of Lake Cayuga to provide natural air conditioning during the summer. Cold water discharge from a Hawaiian OTEC system has been used to cool the soil used to grow cool-weather crops such as strawberries.

WHAT'S OUR ENERGY FUTURE?

None of the renewable energy sources discussed in this chapter are likely to completely replace fossil fuels and nuclear power in the near future. They could, however, make a substantial collective contribution toward providing us with the conveniences we crave in a sustainable, environmentally friendly manner. They could also

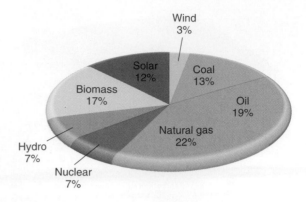

FIGURE 20.32   Idealized "ecological" scenario for cumulative world energy consumption, 2000 to 2100. *Source:* World Energy Council, 2002.

make us energy independent and balance our international payment deficit.

The World Energy Council projects that renewables could provide about 40 percent of world cumulative energy consumption under an idealized "ecological" scenario assuming that political leaders take global warming seriously and pass taxes to encourage conservation and protect the environment (fig. 20.32). This scenario also envisions measures to shift wealth from the north to south, and to enhance economic equity. By the end of the twenty-first century, renewable sources could provide all our energy needs if we take the necessary steps to make this happen.

Rising fuel prices and increasing dependence on imported oil have prompted demands for a U.S. energy policy. Environmentalists point to the dangers of air pollution, global climate change, and other environmental problems associated with burning of fossil fuels. Businesses stress the importance of a reliable energy supply for economic growth. While both call for a new policy, they disagree on what it should contain. Conservatives tend to favor increasing production and easing regulations on power plant operation and transmission-line siting, rather than limiting demand. Progressives, on the other hand, prefer conservation measures such as forcing automakers to increase average fuel efficiency of cars and light trucks and providing heating bill assistance for low-income households.

What path do you think we should take to achieve an ideal energy future?

Summary

- More efficient energy use, together with a greater reliance on renewable sources, could reduce or even eliminate our dependence on fossil fuels and nuclear energy.

- One of the best ways to relieve environmental problems resulting from our energy use is to simply use less. There are many opportunities to conserve energy through greater efficiencies in lighting, space and water heating, insulation, better industrial motors and controls, and simply turning off appliances when not in use. Cogeneration of heat and electricity can double the efficiency of home or business energy use. Affordable systems provide stable, efficient backup energy for many applications.

- Transportation is one of our biggest energy uses as well as a great opportunity for conservation. Raising the average mileage of U.S. automobiles and light trucks by 3 mpg would save consumers about $25 billion per year, reduce CO_2 emissions by 140 tons annually, and save more oil every year than the maximum expected flow from the Arctic National Wildlife Refuge. Hybrid gasoline-electric vehicles have the highest efficiency rating and lowest emissions of any commercially available option.

- Solar energy is a vast resource. People have used passive space heating for thousands of years. Active solar systems collect and store solar energy, and move it mechanically to where it's needed. One of the most promising solar technologies is photovoltaic cells that convert sunlight into electricity. Prices of these cells have come down dramatically, and they are now competitive with other electrical sources in many remote locations. A rooftop array could provide all the power you need without any moving parts. Storage of this energy can be a problem. For homes connected to a utility grid, the best solution may be to sell surplus energy to the utility and then buy back what's needed when the sun doesn't shine. For remote, large-scale solar arrays, the best way to transport energy may be to use it to generate hydrogen gas that can be shipped to markets via underground pipelines.

- Fuel cells are devices that use electrochemical reactions and semipermeable membranes to generate an electrical current. Ideally, their only waste product is clean water. A number of different fuel cell designs are available and several are now available at competitive prices.

- Biomass has long been used as a fuel for heat and light. In many places, this has led to deforestation and air pollution. Supplies are becoming scarce for the billions of people who depend on biomass. Firewood, dung, and charcoal are the main forms of biomass used in most places. More efficient stoves or digesters that convert dung to methane could help conserve energy.

- Hydropower once produced most electricity in the world, but has largely been replaced by fossil fuels. Large hydropower resources remain untapped, but there are severe social and environmental problems associated with many large dam projects. Small-scale, run-of-the-river turbines may offer a better solution in some cases.

- Wind power is the fastest growing energy source in the world and offers a huge potential for clean, renewable power. Wind farms already supply about 35,000 MW, and this capacity is expected to double in 5 years. Wind could easily supply all the electricity we need and free us from the economic, social, and environmental problems associated with fossil fuels and nuclear power.

- Geothermal energy is a valuable resource in places like Iceland and New Zealand that have large geothermal features. Nearly everywhere, the constant temperature of the earth a few meters below the surface can provide a useful heat sink for heating or cooling. Tidal and wave energy and steep ocean temperature gradients also can be a valuable resource in some locations.

- Although none of the renewable sources discussed in this chapter are likely to completely replace fossil fuels in the near future, a combination of alternative approaches could make a significant difference in our energy uses.

Questions for Review

1. Describe five ways that we could conserve energy individually or collectively.
2. Explain the principle of net energy yield. Give some examples.
3. What is the difference between active and passive solar energy?
4. How do photovoltaic cells generate electricity?
5. What is a fuel cell and how does it work?
6. Describe some problems with wood burning in both industrialized nations and developing nations.
7. How is methane made? Give an example of a useful methane source.
8. What are some advantages and disadvantages of large hydroelectric dams?
9. What are some examples of biomass fuel other than wood?
10. Describe how tidal power or ocean wave power generate electricity.

Questions for Critical Thinking

1. What alternative energy sources are most useful in your region and climate? Why?

2. What can you do to conserve energy where you live? In personal habits? In your home, dormitory, or workplace?

3. What massive heat storage materials can you think of that could be attractively incorporated into a home?

4. Do you think building wind farms in remote places, parks, or scenic wilderness areas would be damaging or unsightly?

5. If you were the energy czar of your state, where would you invest your budget?

6. What could (or should) we do to help developing countries move toward energy conservation and renewable energy sources? How can we ask them to conserve when we live so wastefully?

7. Can you think of environmental consequences associated with tidal or geothermal energy? If so, how can they be mitigated?

8. You are offered a home solar energy system that costs $10,000 but saves you $1,000 a year. Will you take it at this rate? If the cost were higher and the payoff time longer, what is the threshold at which you would not buy the system?

Key Terms

active solar systems 435
biofuel 445
cogeneration 435
energy efficiency 431
fuel cells 440
geothermal energy 450
green pricing 438
hybrid gasoline-electric
 vehicles 432
low-head hydropower 447

micro-hydro generators 447
net energy yield 432
ocean thermal electric
 conversion (OTEC) 451
passive heat absorption 435
photovoltaic cell 438
reformer 441
run-of-the-river flow 447
tidal station 450

Further Readings

Asmus, Peter. 2000. *Reaping the Wind: How Mechanical Wizards, Visionaries, and Profiteers Helped Shape Our Energy Future.* Island Press.

Ayres, Robert U. 2001. The energy we overlook. *World Watch* 14(6):30–39.

DeCicco, John. 2001. *Aceee's Green Book: The Environmental Guide to Cars and Trucks.* American Council for an Energy-Efficient Economy.

Geller, Howard. 2002. *Energy Revolution: Policies for a Sustainable Future.* Island Press.

Hoffmann, Peter. 2001. *Tomorrow's Energy: Hydrogen, Fuel Cells, and the Prospects for a Cleaner Planet.* MIT Press.

Jurgen, Ronald K., ed. 2002. *Electric and Hybrid-Electric Vehicles.* Society of Automotive Engineers.

Roberts, Carolyn. 2002. *A House of Straw.* Chelsea Green Publishers.

Schaeffer, J., and D. Pratt. 1999. *Solar Living Sourcebook,* 10th ed. Chelsea Green Publishers.

Smith, Douglas. 2001. Big plans for ocean power hinge on funding and additional research. *Power Engineering* 105(11):91–95.

Tromp, Tracey K., et al. 2003. Potential environmental impact of a hydrogen economy on the stratosphere. *Science* 300(5626): 1740–42.

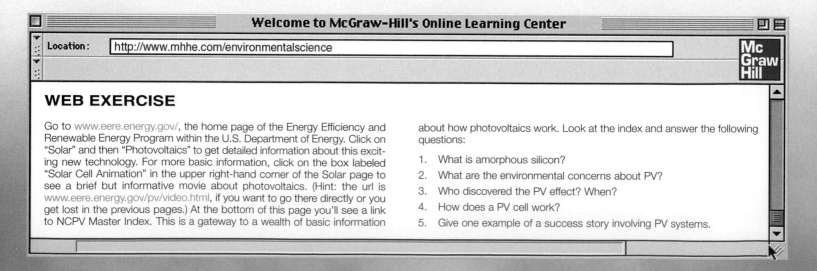

Welcome to McGraw-Hill's Online Learning Center

Location: http://www.mhhe.com/environmentalscience

WEB EXERCISE

Go to www.eere.energy.gov/, the home page of the Energy Efficiency and Renewable Energy Program within the U.S. Department of Energy. Click on "Solar" and then "Photovoltaics" to get detailed information about this exciting new technology. For more basic information, click on the box labeled "Solar Cell Animation" in the upper right-hand corner of the Solar page to see a brief but informative movie about photovoltaics. (Hint: the url is www.eere.energy.gov/pv/video.html, if you want to go there directly or you get lost in the previous pages.) At the bottom of this page you'll see a link to NCPV Master Index. This is a gateway to a wealth of basic information

about how photovoltaics work. Look at the index and answer the following questions:

1. What is amorphous silicon?

2. What are the environmental concerns about PV?

3. Who discovered the PV effect? When?

4. How does a PV cell work?

5. Give one example of a success story involving PV systems.

21

Solid, Toxic, and Hazardous Waste

We have no knowledge, so we have stuff; but stuff without knowledge is never enough.

Greg Brown

OBJECTIVES

After studying this chapter, you should be able to:

- identify the major components of the waste stream and describe how wastes have been—and are being—disposed of in North America and around the world.
- explain the differences between dumps, sanitary landfills, and modern secure landfills.
- summarize the benefits, problems, and potential of recycling and reusing wastes.
- analyze some alternatives for reducing the waste we generate.
- understand what hazardous and toxic wastes are and how we dispose of them.
- evaluate the options for hazardous waste management.
- outline some ways we can destroy or permanently store hazardous wastes.

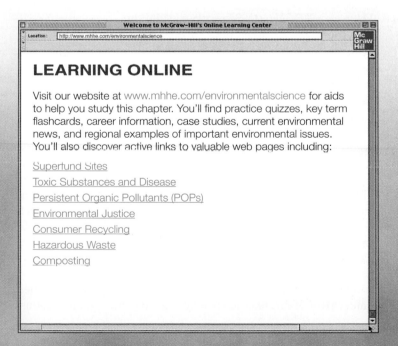

Welcome to McGraw-Hill's Online Learning Center

Location: http://www.mhhe.com/environmentalscience

LEARNING ONLINE

Visit our website at www.mhhe.com/environmentalscience for aids to help you study this chapter. You'll find practice quizzes, key term flashcards, career information, case studies, current environmental news, and regional examples of important environmental issues. You'll also discover active links to valuable web pages including:

Superfund Sites

Toxic Substances and Disease

Persistent Organic Pollutants (POPs)

Environmental Justice

Consumer Recycling

Hazardous Waste

Composting

Photo: Fresh Kills, the largest landfill in the United States, closed in 2001 for lack of space.
© Ray Pfortner/Peter Arnold, Inc.

South Africa's "National Flower"?

You've seen them in ditches, caught in fences, fluttering in trees on a windy day. Plastic shopping bags, an item of convenience world-over, are increasingly an eyesore and a nuisance. Some South Africans have begun referring to these ubiquitous bags as the country's national flower because they seem to bloom everywhere. Now the government is trying to make them disappear.

South African shops hand out about 8 billion light-weight, single-use plastic bags per year. Most may be disposed of properly (buried in landfills or burned), but many end up blowing along public streets and across the countryside. This mobile litter is not just an aesthetic nuisance: bags clog sewers and streams and threaten wildlife, as well. The principal problem is extremely fine bags, averaging 17 microns in thickness. (A micron is one-thousandth of a millimeter.) These thin bags are cheap enough to give away with groceries and other goods, but they are too fragile for reuse.

In an effort to make bags less disposable and more reusable, the South African government proposed a mandatory minimum thickness of 80 microns. Stores would be more likely to charge for these sturdier bags—and consumers would reconsider before throwing away bags they had paid for. Trade unions and plastics manufacturers insisted that existing equipment couldn't produce these bags, and they threatened job losses and factory closures. Eventually, unions and the government reached a compromise: there is now a 30 micron minimum, thick enough for reuse, and the government promises new jobs in recycling industries.

The effort to control disposable bags is just part of a wider South African effort to reduce litter and encourage recycling. Deposits have been proposed for tires, bottles, cans, and other products. Local governments are enthusiastic about these steps because litter is a chronic problem in many cities. Central Johannesburg alone is flooded with more than 200 tons of litter per day and spends nearly 50 million rand (nearly $7 million U.S.) a year cleaning up this debris.

South Africa is not the only country worried about this problem. Taiwan initiated a rule in 2003 that restaurants and supermarkets must charge customers for plastic bags and utensils. Australia is considering a tax on disposable bags. British supermarkets pass out some 10 billion bags each year, and the government there is considering a 9-penny-per-bag tax, which should force stores to charge for bags. Reportedly, stores support such a move. Currently, they spend £1 billion per year on bags that they give away for free. Ireland imposed such a tax, to be used for environmental cleanup, in 2002. A survey in Country Durham, England, found that 70 percent of residents favored a system of paying for bags in order to encourage reuse. County Durham dumps more than 600 tons of plastic shopping bags in landfills every year, at a cost of about £20,000 ($32,000 U.S.).

By most estimates, charging customers for bags would reduce consumption by 40 to 50 percent. Germany, Norway, and others have charged shoppers for years, and as a result, reusable cloth or plastic bags are widely used.

Are disposable paper bags a better choice? They decompose or burn more readily than plastic, but they also require logging, bleaching, and waste disposal. Proponents of thicker plastic bags hope that consumers will start carrying their own shopping bags. Perhaps the best answer to the question, "paper or plastic?" is "I've brought my own, thanks."

Waste management is a growing and global problem. Often, waste production depends on our individual choices, to buy highly packaged goods, to buy unnecessary items, to reuse, recycle, or dispose of goods. Our personal choices are also complicated by cultural expectations and economic policies. In this chapter, we'll review the state of waste production, including solid waste and hazardous waste, and study efforts to reduce waste production and its environmental effects.

SOLID WASTE

Waste is everyone's business. We all produce wastes in nearly everything we do. According to the Environmental Protection Agency, the United States produces 11 billion tons of solid waste each year. About half of that amount consists of agricultural waste, such as crop residues and animal manure, which are generally recycled into the soil on the farms where they are produced. They represent a valuable resource as ground cover to reduce erosion and fertilizer to nourish new crops, but they also constitute the single largest source of nonpoint air and water pollution in the country. More than one-third of all solid wastes are mine tailings, overburden from strip mines, smelter slag, and other residues produced by mining and primary metal processing. Road and building construction debris is another major component of solid waste. Much of this material is stored in or near its source of production and isn't mixed with other kinds of wastes. Improper disposal practices, however, can result in serious and widespread pollution.

Industrial waste—other than mining and mineral production—amounts to some 400 million metric tons per year in the United States. Most of this material is recycled, converted to other forms, destroyed, or disposed of in private landfills or deep injection wells. About 60 million metric tons of industrial waste falls in a special category of hazardous and toxic waste, which we will discuss later in this chapter.

Municipal waste—a combination of household and commercial refuse—amounts to more than 200 million metric tons per year in the United States (fig. 21.1). That's approximately two-thirds of a ton for each man, woman, and child every year—twice as much per capita as Europe or Japan, and five to ten times as much as most developing countries.

The Waste Stream

Does it surprise you to learn that you generate that much garbage? Think for a moment about how much we discard every year. There are organic materials, such as yard and garden wastes, food wastes, and sewage sludge from treatment plants; junked cars; worn out furniture; and consumer products of all types. Newspapers, magazines, advertisements, and office refuse make paper one of our major wastes (fig. 21.2). In spite of recent progress in recycling, many of the 200 *billion* metal, glass, and plastic food and bever-

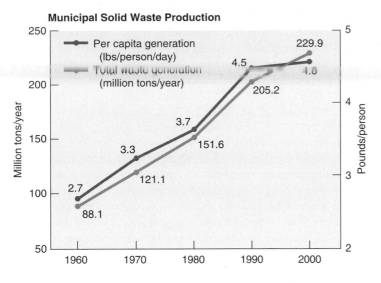

Municipal Solid Waste Production

Municipal Solid Waste Recycling

FIGURE 21.1 Bad news and good news in solid waste production. Per capita waste has risen steadily to more than 2 kg per person per day. Recycling rates are also rising, however.
Source: Environmental Protection Agency, 2001.

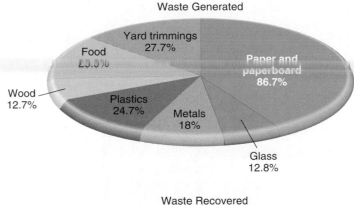

Waste Generated

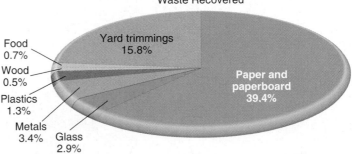

Waste Recovered

FIGURE 21.2 Composition of domestic waste in the United States by weight (*top*) and of waste recovered for recycling (*bottom*).
Source: Environmental Protection Agency, 2003.

of miscellaneous garbage. This mixing makes the disposal or burning of what might have been rather innocuous stuff a difficult, expensive, and risky business. Spray paint cans, pesticides, batteries (zinc, lead, or mercury), cleaning solvents, smoke detectors containing radioactive material, and plastics that produce dioxins and PCBs when burned are mixed willy-nilly with paper, table scraps, and other nontoxic materials. The best thing to do with household toxic and hazardous materials is to separate them for safe disposal or recycling, as we will see later in this chapter.

WASTE DISPOSAL METHODS

Where are our wastes going now? In this section, we will examine some historic methods of waste disposal as well as some future options. Notice that our presentation begins with the least desirable—but most commonly used—measures and proceeds to discuss some preferable options. Keep in mind as you read this that modern waste management reverses this order and stresses the "three R's" of reduction, reuse, and recycling before destruction or, finally, secure storage of wastes.

Open Dumps

For many people, the way to dispose of waste is to simply drop it someplace. Open, unregulated dumps are still the predominant method of waste disposal in most developing countries. The giant developing world megacities have enormous garbage problems.

age containers used every year in the United States end up in the trash. Wood, concrete, bricks, and glass come from construction and demolition sites, dust and rubble from landscaping and road building. All of this varied and voluminous waste has to arrive at a final resting place somewhere.

The **waste stream** is a term that describes the steady flow of varied wastes that we all produce, from domestic garbage and yard wastes to industrial, commercial, and construction refuse. Many of the materials in our waste stream would be valuable resources if they were not mixed with other garbage. Unfortunately, our collecting and dumping processes mix and crush everything together, making separation an expensive and sometimes impossible task. In a dump or incinerator, much of the value of recyclable materials is lost.

Another problem with refuse mixing is that hazardous materials in the waste stream get dispersed through thousands of tons

FIGURE 21.3 Scavengers sort through the trash at "Smoky Mountain," one of the huge metropolitan dumps in Manila, Philippines. Some 20,000 people live and work on these enormous garbage dumps. The health effects are tragic. In many cities of the developing world, communities subsist by picking and recycling waste. © Fred McConnaughey/Photo Researchers, Inc.

Mexico City, one of the largest cities in the world, generates some 10,000 tons of trash *each day.* Until recently, most of this torrent of waste was left in giant piles, exposed to the wind and rain, as well as rats, flies, and other vermin. Manila, in the Philippines, generates a similar amount of waste, half of which goes to a giant, constantly smoldering dump called "Smoky Mountain." Over 20,000 people live and work on this mountain of refuse, scavenging for recyclable items or edible food scraps (fig. 21.3). In July 2000, torrential rains spawned by Typhoon "Kai Tak" caused part of the mountain to collapse, burying at least 215 people. The government would like to close these dumps, but how will the residents be housed and fed? Where else will the city put its garbage?

Most developed countries forbid open dumping, at least in metropolitan areas, but illegal dumping is still a problem. You have undoubtedly seen trash accumulating along roadsides and in vacant, weedy lots in the poorer sections of cities. Is this just a question of aesthetics? Consider the problem of waste oil and solvents. An estimated 200 million liters of waste motor oil are poured into the sewers or allowed to soak into the ground every year in the United States. This is about five times as much as was spilled by the *Exxon Valdez* in Alaska in 1989! No one knows the volume of solvents and other chemicals disposed of by similar methods.

Increasingly, these toxic chemicals are showing up in the groundwater supplies on which nearly half the people in America depend for drinking (chapter 18). An alarmingly small amount of oil or other solvents can pollute large quantities of drinking or irrigation water. One liter of gasoline, for instance, can make a million liters of water undrinkable. The problem of illegal dumping is likely to become worse as acceptable sites for waste disposal become more scarce and costs for legal dumping escalate. We clearly need better enforcement of antilittering laws as well as a change in our attitudes and behavior.

Ocean Dumping

The oceans are vast, but not so large that we can continue to treat them as carelessly as has been our habit. Every year some 25,000 metric tons (55 million lbs) of packaging, including half a million bottles, cans, and plastic containers, are dumped at sea. Beaches, even in remote regions, are littered with the nondegradable flotsam and jetsam of industrial society (fig. 21.4). About 150,000 tons (330 million lbs) of fishing gear—including more than 1,000 km (660 mi) of nets—are lost or discarded at sea each year. Environmental groups estimate that 50,000 northern fur seals are entangled in this refuse and drown or starve to death every year in the North Pacific alone.

Until recently, many cities in the United States dumped municipal refuse, industrial waste, sewage, and sewage sludge in the ocean. Federal legislation now prohibits this dumping. New York City, the last to stop offshore sewage sludge disposal, finally ended

FIGURE 21.4 Garbage dumping at sea is a persistent problem. Here an elephant seal rests amid washed-up plastic flotsam. © Corbis/Royalty Free Website.

this practice in 1992. Still, 60 million to 80 million m³ of dredge spoil—much of it contaminated with heavy metals and toxic organic compounds—are disposed of at sea. Some people claim that the deep abyssal plain is the most remote, stable, and innocuous place to dump our wastes. Others argue that we know too little about the values of these remote places or the rare species that live there to smother them with sludge and debris.

Landfills

Over the past 50 years most American and European cities have recognized the health and environmental hazards of open dumps. Increasingly, cities have turned to **sanitary landfills,** where solid waste disposal is regulated and controlled. To decrease smells and litter and to discourage insect and rodent populations, landfill operators are required to compact the refuse and cover it every day with a layer of dirt (fig. 21.5). This method helps control pollution, but the dirt fill also takes up as much as 20 percent of landfill space. Since 1994, all operating landfills in the United States have been required to control such hazardous substances as oil, chemical compounds, toxic metals, and contaminated rainwater that seeps through piles of waste. An impermeable clay and/or plastic lining underlies and encloses the storage area. Drainage systems are installed in and around the liner to catch drainage and to help monitor chemicals that may be leaking. Modern municipal solid-waste landfills now have many of the safeguards of hazardous waste repositories described later in this chapter.

More careful attention is now paid to the siting of new landfills. Sites located on highly permeable or faulted rock formations are passed over in favor of sites with less leaky geologic foundations. Landfills are being built away from rivers, lakes, floodplains, and aquifer recharge zones rather than near them, as was often done in the past. More care is being given to a landfill's long-term effects so that costly cleanups and rehabilitation can be avoided.

Historically, landfills have been a convenient and relatively inexpensive waste-disposal option in most places, but this situation is changing rapidly. Rising land prices and shipping costs, as well as increasingly demanding landfill construction and maintenance requirements, are making this a more expensive disposal method. The cost of disposing a ton of solid waste in Philadelphia went from $20 in 1980 to more than $100 in 1990. Union County, New York, experienced an even steeper price rise. In 1987, it paid $70 to get rid of a ton of waste; a year later, that same ton cost $420, or about $10 for a typical garbage bag. In the past decades, costs have continued to rise steadily, though not as sharply. The United States now spends about $10 billion per year to dispose of trash. A decade from now, it may cost Americans $100 billion per year to dispose of their garbage.

Suitable places for waste disposal are becoming scarce in many areas. Other uses compete for open space. Citizens have become more concerned and vocal about health hazards, as well as aesthetics. It is difficult to find a neighborhood or community willing to accept a new landfill. Since 1984, when stricter financial and environmental protection requirements for landfills took effect, more than 1,200 of the 1,500 existing landfills in the United States

FIGURE 21.5 In a sanitary landfill, trash and garbage are crushed and covered each day to prevent accumulation of vermin and spread of disease. A waterproof lining is now required to prevent leaching of chemicals into underground aquifers.

have closed. Many major cities are running out of local landfill space. They export their trash, at enormous expense, to neighboring communities and even other states. More than half the solid waste from New Jersey goes out of state, some of it up to 800 km (500 mi) away.

A positive trend in landfill management is methane recovery. Methane, or natural gas, is a natural product of decomposing garbage deep in a landfill. It is also an important "greenhouse gas." Normally methane seeps up to the landfill surface and escapes. At 300 U.S. landfills, the methane is being collected and burned. Cumulatively, these landfills could provide enough electricity for a city of a million people. Three times as many landfills could be recovering methane. Tax incentives could be developed to encourage this kind of resource recovery.

Exporting Waste

Although most industrialized nations in the world have agreed to stop shipping hazardous and toxic waste to less-developed countries, the practice still continues. In 1999, for example, 3,000 tons of incinerator waste from a plastics factory in Taiwan was unloaded from a ship in the middle of the night and dumped in a field near the small coastal Cambodian village of Bet Trang. The village residents thought they had been blessed with a windfall. They emptied out chunks of crumbling residue so they could use the white plastic shipping bags as bedding and roofing material. They rinsed out bags to use for rice storage, and they ripped them open with their teeth to get string to use as clotheslines and lashing for their oxcarts. Children played happily on the big pile of dusty, white material.

In the following weeks, the villagers discovered that rather than a treasure, they had a calamity. The first sign of trouble was when one of the dock workers who unloaded the waste died and five others were hospitalized with symptoms of nerve damage and respiratory distress. Villagers also began to complain of a variety

Labels on figure: Compacted waste filling trench; Original terraine; Daily 6-inch earth cover

of illnesses. The village was evacuated, and about 1,000 residents of the nearby city of Sihanoukville fled in panic. Subsequent analysis found high levels of mercury and other toxic metals in the residue. The Formosa Plastics Corp., which shipped the waste, admitted paying a $3 million bribe to Cambodian officials to permit its dumping. They said they couldn't dispose of it in Taiwan because of a threat of public protest. Following an international uproar, the plastics company agreed to go back and pick up the waste. But the villagers who handled the toxic wastes face an uncertain future. Is it safe to reinhabit their homes? Is it wise to have children? Will they suffer long-term health effects from exposure to this material?

As we will discuss later in this chapter, "garbage imperialism" also operates within richer countries as well. Poor neighborhoods and minority populations are much more likely than richer ones to be the recipients of dumps, waste incinerators, and other locally unwanted land uses (LULUs). In recent years, attention has turned, in the United States, to Indian reservations, which are exempt from some state and federal regulations concerning waste disposal. Virtually every tribe in America has been approached with schemes to store wastes on their reservation (What Do You Think? p. 462).

Another method of disposing of toxic wastes is to "recycle" them as asphalt or concrete filler for building highways. This is considered a beneficial use, but what happens to the toxins as the roadway is slowly worn away by traffic? Similar waste products are "land farmed" or sold as fertilizer and soil amendments. There are no safety standards for fertilizer composition because it's not intended for human consumption, but much of it is used on crops that humans will eat, or will be fed to livestock that are part of our food chain. For example, Florida has about a billion m^3 of phosphogypsum, a waste product of phosphate mining that producers want to market as a soil amendment. While it's true that phosphate is an essential plant nutrient, this particular product is also radioactive. In another case in Oregon, metal-rich dust and ash from steel mills is classified as hazardous waste when it leaves the mill. After being mixed with other minerals, however, it becomes fertilizer that will be spread on farm fields. Manufacturers are required to report the "active" ingredient—things like nitrogen, phosphorus, and phosphate—content of their product, but much can go unreported as "inert" matter.

Incineration and Resource Recovery

Landfilling is still the disposal method for the majority of municipal waste in the United States (fig. 21.6). Faced with growing piles of garbage and a lack of available landfills at any price, however, public officials are investigating other disposal methods. The method to which they frequently turn is burning. Another term commonly used for this technology is **energy recovery,** or waste-to-energy, because the heat derived from incinerated refuse is a useful resource. Burning garbage can produce steam used directly for heating buildings or generating electricity. Internationally, well over 1,000 waste-to-energy plants in Brazil, Japan, and western Europe generate much-needed energy while also reducing the amount that needs to be landfilled. In the United States, more than 110 waste

incinerators burn 45,000 tons of garbage daily. Some of these are simple incinerators; others produce steam and/or electricity.

Types of Incinerators

Municipal incinerators are specially designed burning plants capable of burning thousands of tons of waste per day. In some plants, refuse is sorted as it comes in to remove unburnable or recyclable materials before combustion. This is called **refuse-derived fuel** because the enriched burnable fraction has a higher energy content than the raw trash. Another approach, called **mass burn,** is to dump everything smaller than sofas and refrigerators into a giant furnace and burn as much as possible (fig. 21.7). This technique avoids the expensive and unpleasant job of sorting through the garbage for nonburnable materials, but it often causes greater problems with air pollution and corrosion of burner grates and chimneys.

In either case, residual ash and unburnable residues representing 10 to 20 percent of the original volume are usually taken to a landfill for disposal. Because the volume of burned garbage is reduced by 80 to 90 percent, disposal is a smaller task. However, the residual ash usually contains a variety of toxic components that make it an environmental hazard if not disposed of properly. Ironically, one worry about incinerators is whether enough garbage will be available to feed them. Some communities in which recycling has been really successful have had to buy garbage from neighbors to meet contractual obligations to waste-to-energy facilities. In other places, fears that this might happen have discouraged recycling efforts.

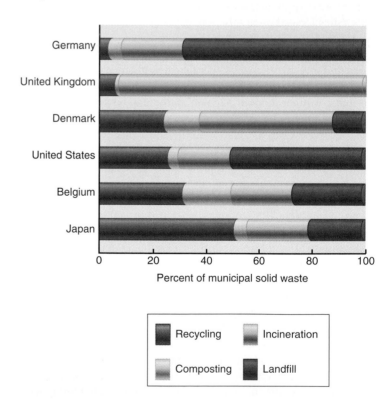

FIGURE 21.6 Percentage of municipal solid waste recycled, composted, incinerated, and landfilled in selected developed countries.
Source: Eurostat, UNEP, 2003.

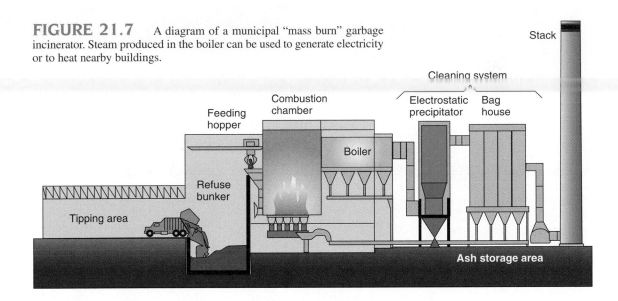

FIGURE 21.7 A diagram of a municipal "mass burn" garbage incinerator. Steam produced in the boiler can be used to generate electricity or to heat nearby buildings.

Incinerator Cost and Safety

The cost-effectiveness of garbage incinerators is the subject of heated debates. Initial construction costs are high—usually between $100 million and $300 million for a typical municipal facility. Tipping fees at an incinerator, the fee charged to haulers for each ton of garbage dumped, are often much higher than those at a landfill. As landfill space near metropolitan areas becomes more scarce and more expensive, however, landfill rates are certain to rise. It may pay in the long run to incinerate refuse so that the lifetime of existing landfills will be extended.

Environmental safety of incinerators is another point of concern. The EPA has found alarmingly high levels of dioxins, furans, lead, and cadmium in incinerator ash. These toxic materials were more concentrated in the fly ash (lighter, airborne particles capable of penetrating deep into the lungs) than in heavy bottom ash. Dioxin levels can be as high as 780 parts per billion. One part per billion of TCDD, the most toxic dioxin, is considered a health concern. All of the incinerators studied exceeded cadmium standards, and 80 percent exceeded lead standards. Proponents of incineration argue that if they are run properly and equipped with appropriate pollution-control devices, incinerators are safe to the general public. Opponents counter that neither public officials nor pollution-control equipment can be trusted to keep the air clean. They argue that recycling and source reduction efforts are better ways to deal with waste problems.

The EPA, which generally supports incineration, acknowledges the health threat of incinerator emissions but holds that the danger is very slight. The EPA estimates that dioxin emissions from a typical municipal incinerator may cause one death per million people in 70 years of operation. Critics of incineration claim that a more accurate estimate is 250 deaths per million in 70 years.

One way to reduce these dangerous emissions is to remove batteries containing heavy metals and plastics containing chlorine before wastes are burned. Bremen, West Germany, is one of several European cities now trying to control dioxin emissions by keeping all plastics out of incinerator waste. Bremen is requiring households to separate plastics from other garbage. This is expected to eliminate nearly all dioxins and other combustion by-products and prevent the expense of installing costly pollution-control equipment that otherwise would be necessary to keep the burners operating. Several cities have initiated a recycling program for the small "button" batteries used in hearing aids, watches, and calculators in an attempt to lower mercury emissions from its incinerator.

SHRINKING THE WASTE STREAM

Having less waste to discard is obviously better than struggling with disposal methods, all of which have disadvantages and drawbacks. In this section we will explore some of our options for recycling, reuse, and reduction of the wastes we produce.

Recycling

The term *recycling* has two meanings in common usage. Sometimes we say we are *recycling* when we really are *reusing* something, such as refillable beverage containers. In terms of solid waste management, however, **recycling** is the reprocessing of discarded materials into new, useful products (fig. 21.8). Some recycling processes reuse materials for the same purposes; for instance, old aluminum cans and glass bottles are usually melted and recast into new cans and bottles. Other recycling processes turn old materials into entirely new products. Old tires, for instance, are shredded and turned into rubberized road surfacing. Newspapers become cellulose insulation, kitchen wastes become a valuable soil amendment, and steel cans become new automobiles and construction materials.

The high value of aluminum scrap ($700 per ton in 1999) has spurred a large percentage of aluminum recycling nearly

What do you think?

Environmental Justice

Who do you suppose lives closest to toxic waste dumps, Superfund sites, or other polluted areas in your city or county? If you answered poor people and minorities, you are probably right. Everyday experiences tell us that minority neighborhoods are much more likely to have high pollution levels and unpopular industrial facilities such as toxic waste dumps, landfills, smelters, refineries, and incinerators than are middle- or upper-class, white neighborhoods.

One of the first systematic studies showing this inequitable distribution of environmental hazards based on race in the United States was conducted by Robert D. Bullard in 1978. Asked for help by a predominantly black community in Houston that was slated for a waste incinerator, Bullard discovered that all five of the city's existing landfills and six of eight incinerators were located in African-American neighborhoods. In a book entitled *Dumping on Dixie,* Bullard showed that this pattern of risk exposure in minority communities is common throughout the United States.

In 1987, the Commission for Racial Justice of the United Church of Christ published an extensive study of environmental racism. Its conclusion was that race is the most significant variable in determining the location of toxic waste sites in the United States. Among the findings of this study are:

- three of the five largest commercial hazardous waste landfills accounting for about 40 percent of all hazardous waste

Native Americans march in protest of toxic waste dumping on tribal lands.
© Barbara Gauntt/The Clarion-Ledger, Jackson, MS.

disposal in the United States are located in predominantly black or Hispanic communities.

- 60 percent of African Americans and Latinos and nearly half of all Asians, Pacific Islanders, and Native Americans live in communities with uncontrolled toxic waste sites.

- The average percentage of the population made up by minorities in communities without a hazardous waste facility is 12 percent. By contrast, communities with one hazardous waste facility have, on average, twice as high (24 percent) a minority population, while those with two or more such facilities average three times as high a minority population (38 percent) as those without one.

- The "dirtiest" or most polluted zip codes in California are in riot-torn South Central Los Angeles where the population is predominantly African American or Latino. Three-quarters of all blacks and half of all Hispanics in Los Angeles live in these polluted areas, while only one-third of all whites live there.

Race is claimed to be the strongest determinant of who is exposed to environmental hazards. Where whites can often "vote with their feet" and move out of polluted and dangerous neighborhoods, minorities are restricted by color barriers and prejudice to less desirable locations. In some areas, though, class or income also are associated with environmental hazards. The difference between *environmental racism* and other kinds of *environmental injustice* can be hard to define. Economic opportunity is often closely tied to race and cultural background in the United States.

Racial inequities also are revealed in the way the government cleans up toxic waste sites and punishes polluters. White communities see faster responses and get better results once toxic wastes are discovered than do minority communities. Penalties

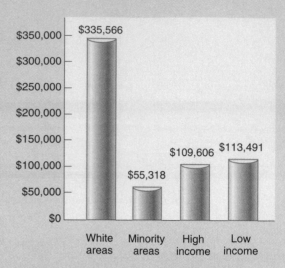

Hazardous waste law enforcement. The average fines or penalties per site for violation of the Resource Conservation and Recovery Act vary dramatically with racial composition of the communities where waste was dumped.
Source: M. Lavelle and M. Coyle, The National Law Journal, Vol. 15: 52–56, No. 3, September 21, 1992.

assessed against polluters of white communities average six times higher than those against polluters of minority communities. Cleanup is more thorough in white communities as well. Most toxic wastes in white communities are removed or destroyed. By contrast, waste sites in minority neighborhoods are generally only "contained" by putting a cap over them, leaving contaminants in place to potentially resurface or leak into groundwater at a later date. The growing environmental justice movement works to combine civil rights and social justice with environmental concerns to call for a decent, livable environment and equal environmental protection for everyone.

Ethical Considerations

What are the ethical considerations in waste disposal? Does everyone have a right to live in a clean environment or only a right to buy one if they can afford it? What would be a fair way to distribute the risks of toxic wastes? If you had to choose between an incinerator, a secure landfill, or a composting facility for your neighborhood, which would you take?

FIGURE 21.8 Trucks with multiple compartments pick up residential recyclables at curbside, greatly reducing the amount of waste that needs to be buried or burned. For many materials, however, collection costs are too high and markets are lacking for recycling to be profitable. © William P. Cunningham.

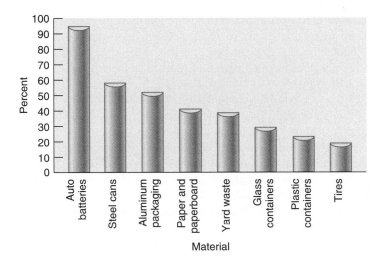

FIGURE 21.9 Recycling rates of selected materials in the United States. *Source:* Environmental Protection Agency, 2003.

everywhere (fig. 21.9). About two-thirds of all aluminum beverage cans are now recycled; up from only 15 percent in 1970. Aluminum recycling is so rapid that half of the cans now on grocery shelves will be made into another can within two months. A big problem for recyclers is the wild fluctuation in market prices for commodities. Newsprint, for example, which peaked at $160 a ton in 1995, dropped to $42 per ton in 1999. One day, it's so valuable that people are stealing it off the curb; the next day it's literally down in the dumps. It's hard to build a recycling program when you can't count on a stable price for your product.

Another problem in recycling is contamination. Most of the 24 billion plastic soft drink bottles sold every year in the United States are made of PET (polyethylene terephthalate), which can be melted and remanufactured into carpet, fleece clothing, plastic-strapping, and nonfood packaging. However, even a smidgen of vinyl—a single PVC (polyvinyl chloride) bottle in a truckload, for example—can make PET useless. Although most bottles are now marked with a recycling number, it's hard for consumers to remember which is which. A looming worry is the prospect of single-use, plastic beer bottles. Already being test marketed, these bottles are made of PET but are amber colored to block sunlight and have a special chemical coating to keep out oxygen, which would ruin the beer. The special color, interior coating, and vinyl cap lining will make these bottles incompatible with regular PET, and it will probably cost more to remove them from the waste stream than the reclaimed plastic is worth. Plastic recycling already is down 50 percent from a decade ago because so many soft drink bottles are sold and consumed on the go, and never make it into recycling bins. Throw-away beer bottles are a looming threat to this industry.

Benefits of Recycling

Recycling is usually a better alternative to either dumping or burning wastes. It saves money, energy, raw materials, and land space, while also reducing pollution. Recycling also encourages individual awareness and responsibility for the refuse produced. Curbside pickup of recyclables costs around $35 per ton, as opposed to the $80 paid to dispose of them at an average metropolitan landfill. Many recycling programs cover their own expenses with materials sales and may even bring revenue to the community.

Another benefit of recycling is that it could cut our waste volumes drastically and reduce the pressure on disposal systems. Philadelphia is investing in neighborhood collection centers that will recycle 600 tons a day, enough to eliminate the need for a previously planned, high-priced incinerator. New York City closed its last remaining landfill, Fresh Kills, in 2001. The city now exports its 11,000 tons per day of waste by truck, train, and barge, to New Jersey, Pennsylvania, Virginia, and Ohio. New York has set ambitious recycling goals of 50 percent waste reduction, but still the city recycles less than 20 percent of its household and office waste. In contrast, Minneapolis and Seattle recycle nearly 60 percent of domestic waste, Los Angeles and Chicago over 40 percent. In 2002, New York Mayor Michael Bloomberg raised a national outcry by canceling most of the city's recycling program. He argued that the program didn't pay for itself and the money should be spent to balance the city's budget. A year later, Bloomberg relented, and recycling was reinstated for nearly all recyclable materials. The speed of this reversal indicates widespread support for recycling, as well as the rising costs of exporting garbage.

Japan probably has the most successful recycling program in the world (fig. 21.6). Half of all household and commercial wastes in Japan are recycled while the rest is about equally incinerated or landfilled. By comparison, the United States landfills more than 60 percent of all solid waste. Japanese families diligently separate wastes into as many as seven categories, each picked up on a different day. Would we do the same (fig. 21.10)? Some authors say that Americans are too lazy to recycle. North Stonington, Connecticut, however, faced with escalating disposal costs, reduced its

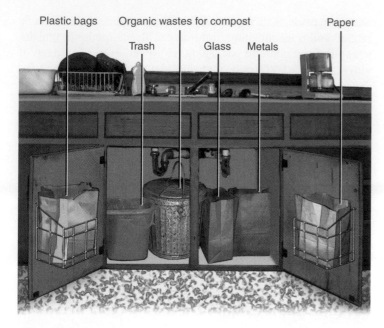

Plastic bags Organic wastes for compost Paper
Trash Glass Metals

FIGURE 21.10 Source separation in the kitchen—the first step in a strong recycling program.

waste volume by two-thirds in just two years. Between 1980 to 2000, landfilling rates in the United States remained nearly constant, while recycling and composting almost doubled (fig. 21.11).

Recycling lowers our demands for raw resources. In the United States, we cut down 2 million trees every day to produce newsprint and paper products, a heavy drain on our forests. Recycling the print run of a single Sunday issue of the *New York Times* would spare 75,000 trees. Every piece of plastic we make reduces the reserves supply of petroleum and makes us more dependent on foreign oil. Recycling 1 ton of aluminum saves 4 tons of bauxite (aluminum ore) and 700 kg (1,540 lb) of petroleum coke and pitch, as well as keeping 35 kg (77 lb) of aluminum fluoride out of the air.

Recycling also reduces energy consumption and air pollution. Plastic bottle recycling could save 50 to 60 percent of the energy needed to make new ones. Making new steel from old scrap offers up to 75 percent energy savings. Producing aluminum from scrap instead of bauxite ore cuts energy use by 95 percent, yet we still throw away more than a million tons of aluminum every year. If aluminum recovery were doubled worldwide, more than a million tons of air pollutants would be eliminated every year.

Reducing litter is an important benefit of recycling. Ever since disposable paper, glass, metal, foam, and plastic packaging began to accompany nearly everything we buy, these discarded wrappings have collected on our roadsides and in our lakes, rivers, and oceans. Without incentives to properly dispose of beverage cans, bottles, and papers, it often seems easier to just toss them aside when we have finished using them. Litter is a costly as well as unsightly problem. We pay an estimated 32 cents for each piece of litter picked up by crews along state highways, which adds up to $500 million every year. "Bottle-bills" requiring deposits on bottles and cans have reduced littering in many states.

Creating a Market for Recycling

In many communities, citizens have done such a good job of collecting recyclables that a glut has developed. Mountains of some waste materials accumulate in warehouses (fig. 21.12) because there are no markets for them. Too often, wastes that we carefully separate for recycling end up being mixed together and dumped in a landfill or incinerator, because a market for these resources is not well developed.

Our present public policies often tend to favor extraction of new raw materials. Energy, water, and raw materials are often sold to industries below their real cost to create jobs and stimulate the economy. For instance, in 1999, a pound of recycled clear PET, the material in most soft drink bottles, sold for about 40¢. By contrast, a pound of off-grade, virgin PET cost 25¢. Setting the prices of natural resources at their real cost would tend to encourage efficiency and recycling. State, local, and national statutes requiring

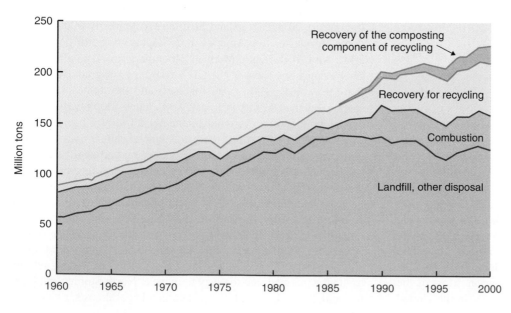

FIGURE 21.11 Disposal of municipal solid waste from 1960 to 2000. Landfills remain the dominant destination, but recycling and composting are increasing.
Source: Environmental Protection Agency, 2003.

FIGURE 21.12 In some places, mountains of paper, plastic, and glass have accumulated in warehouses because collection programs have been more successful than material marketing efforts. Federal and state governments need to encourage use of recycled materials. Each of us can help by buying recycled products. © Mike Brisson.

government agencies to purchase a minimum amount of recycled material have helped create a market for used materials. Each of us can play a role in creating markets, as well. If we buy things made from recycled materials—or ask for them if they aren't available—we will help make it possible for recycling programs to succeed.

Composting

Pressed for landfill space, many cities have banned yard waste from municipal garbage. Rather than bury this valuable organic material, they are turning it into a useful product through **composting:** biological degradation or breakdown of organic matter under aerobic (oxygen-rich) conditions. The organic compost resulting from this process makes a nutrient-rich soil amendment that aids water retention, slows soil erosion, and improves crop yields. A home compost pile is an easy and inexpensive way to dispose of organic waste in an interesting and environmentally friendly way.

Increasingly, cities are composting yard waste and other organic material. This quickly reduces the waste stream, and rich compost can be sold to gardeners for an extra profit.

Energy from Waste

Every year we throw away the energy equivalent of 80 million barrels of oil in organic waste in the United States. In developing countries, up to 85 percent of the waste stream is food, textiles, vegetable matter, and other biodegradable materials. Worldwide, at least one-fifth of municipal waste is organic kitchen and garden refuse. In a landfill, much of this matter is decomposed by microorganisms generating billions of cubic meters of methane ("natural gas"), which contributes to global warming if allowed to escape into the atmosphere (chapter 15). Many cities are drilling methane wells in their landfills to capture this valuable resource. Fuel cells (chapter 20) are a good way to use this fuel.

This valuable organic material can be burned in an incinerator rather than being buried in landfills, but there are worries about air pollution from incineration. Organic wastes also can be decomposed in large, oxygen-free digesters to produce methane under more controlled conditions than in a landfill and with less air pollution than mass garbage burning.

Anaerobic digestion also can be done on a small scale. Millions of household methane generators provide fuel for cooking and lighting for homes in China and India (chapter 20). In the United States, some farmers produce all the fuel they need to run their farms—both for heating and to run trucks and tractors—by generating methane from animal manure.

Demanufacturing

Demanufacturing is the disassembly and recycling of obsolete products, such as TV sets, computers, refrigerators, and air conditioners. Electronics and appliances are one of the fastest-growing components of the global waste stream. Most are large appliances, such as refrigerators and stoves, and **e-waste:** computers, cell phones, TVs, and printers. Americans throw away about 54 million household appliances, 12 million computers, and uncounted cell phones each year. Most office computers are used only 3 years; televisions last 5 years or so; refrigerators last longer, an average of 12 years. In the United States, an estimated 300 to 500 million computers will be discarded by 2007, and about three-fourths of all computers ever sold in this country await disposal in storage rooms and garages.

Demanufacturing is key to reducing the environmental costs of e-waste and appliances. A single personal computer can contain 700 different chemical compounds, including toxic metals (mercury, lead, gallium, germanium, nickel, palladium, beryllium, selenium, arsenic, and others), and valuable metals (gold, silver, copper), as well as brominated fire retardants and plastics. A typical personal computer has about $6 worth of gold, $5 worth of copper, and $1 of silver. Approximately 40 percent of lead entering U.S. landfills, and 70 percent of heavy metals, comes from e-waste. Batteries and electronic switches in toys and electronics make up another 10 to 20 percent of heavy metals in our waste stream. These contaminants can enter groundwater if computers are landfilled, or worse, enter the air if they are incinerated. When collected, these materials can become a valuable resource—and an alternative to newly mined materials.

To reduce these environmental hazards, the European Union now requires cradle-to-grave responsibility for electronic products. By 2005, manufacturers will have to accept used products or fund independent collectors. An extra $20 (less than one percent of the price of most computers) will be added to the purchase price to pay for collection and demanufacturing. Manufacturers selling computers, televisions, refrigerators, and other appliances in Europe must also phase out many of the toxic compounds used in production. Japan is rapidly adopting European environmental standards, and some U.S. companies are following suit, in order to maintain their international markets. In the United States, at least 29 states have passed, or are considering, legislation to control disposal of appliances and computers, in order to protect groundwater and air quality.

Unfortunately, about 80 percent of American e-waste is currently sent to Asia for disposal or recycling. Cheap labor, including child labor, and weak environmental regulations make this expedient. This waste export is legal because the United States has not ratified the 1989 Basel Convention banning trade in hazardous waste to developing countries. As a result, demanufacturing areas have highly contaminated soil, groundwater, and surface water. Many companies and states are working to implement alternative, safer strategies for demanufacturing.

Reuse

Even better than recycling or composting is cleaning and reusing materials in their present form, thus saving the cost and energy of remaking them into something else. We do this already with some specialized items. Auto parts are regularly sold from junkyards, especially for older car models. In some areas, stained glass windows, brass fittings, fine woodwork, and bricks salvaged from old houses bring high prices. Some communities sort and reuse a variety of materials received in their dumps (fig. 21.13).

FIGURE 21.13 Reusing discarded products is a creative and efficient way to reduce wastes. This recycling center in Berkeley, California, is a valuable source of used building supplies and a money saver for the whole community. Courtesy Urban Ore, Inc., Berkeley, CA.

In many cities, glass and plastic bottles are routinely returned to beverage producers for washing and refilling. The reusable, refillable bottle is the most efficient beverage container we have. This is better for the environment than remelting and more profitable for local communities. A reusable glass container makes an average of 15 round-trips between factory and customer before it becomes so scratched and chipped that it has to be recycled. Reusable containers also favor local bottling companies and help preserve regional differences.

Since the advent of cheap, lightweight, disposable food and beverage containers, many small, local breweries, canneries, and bottling companies have been forced out of business by huge national conglomerates. These big companies can afford to ship food and beverages great distances as long as it is a one-way trip. If they had to collect their containers and reuse them, canning and bottling factories serving large regions would be uneconomical. Consequently, the national companies favor recycling rather than refilling because they prefer fewer, larger plants and don't want to be responsible for collecting and reusing containers. In some circumstances, life-cycle assessment shows that washing and decontaminating containers takes as much energy and produces as much air and water pollution as manufacturing new ones.

In less affluent nations, reuse of all sorts of manufactured goods is an established tradition. Where most manufactured products are expensive and labor is cheap, it pays to salvage, clean, and repair products. Cairo, Manila, Mexico City, and many other cities have large populations of poor people who make a living by scavenging. Entire ethnic populations may survive on scavenging, sorting, and reprocessing scraps from city dumps.

Producing Less Waste

What is even better than reusing materials? Generating less waste in the first place. The "What Can You Do?" box on page 467 describes some contributions you can make to reducing the volume of our waste stream. Industry also can play an important role in source reduction. The 3M Company saved over $500 million since 1975 by changing manufacturing processes, finding uses for waste products, and listening to employees' suggestions. What is waste to one division is a treasure to another.

Excess packaging of food and consumer products is one of our greatest sources of unnecessary waste. Paper, plastic, glass, and metal packaging material make up 50 percent of our domestic trash by volume. Much of that packaging is primarily for marketing and has little to do with product protection (fig. 21.14). Manufacturers and retailers might be persuaded to reduce these wasteful practices if consumers ask for products without excess packaging. Canada's National Packaging Protocol (NPP) recommends that packaging minimize depletion of virgin resources and production of toxins in manufacturing. The preferred hierarchy is (1) no packaging, (2) minimal packaging, (3) reusable packaging, and (4) recyclable packaging.

Where disposable packaging is necessary, we still can reduce the volume of waste in our landfills by using materials that are compostable or degradable. **Photodegradable plastics** break down

FIGURE 21.14 How much more do we need? Where will we put what we already have? Reprinted with special permission of King Features Syndicate.

when exposed to ultraviolet radiation. **Biodegradable plastics** incorporate such materials as cornstarch that can be decomposed by microorganisms. Several states have introduced legislation requiring biodegradable or photodegradable six-pack beverage yokes, fast-food packaging, and disposable diapers. These degradable plastics often don't decompose completely; they only break down to small particles that remain in the environment. In doing so, they can release toxic chemicals into the environment. And in modern, lined landfills they don't decompose at all. Furthermore, they make recycling less feasible and may lead people to believe that littering is okay.

Some environmental groups are beginning to think that we have put too much emphasis on recycling. Many people think that if they recycle aluminum cans and newspapers they are doing everything they can for the environment. In surveys conducted by the New Jersey-based Environmental Research Associates, only two in ten people surveyed could define waste reduction. Companies that make "throw away" or heavily packaged products generally favor recycling because it allows them to continue business-as-usual. While recycling is an important part of waste management, we have to remember that it is actually the third "R" in the waste hierarchy. The two preferred methods—reduction and reuse—get lost in our enthusiasm for recycling.

HAZARDOUS AND TOXIC WASTES

The most dangerous aspect of the waste stream we have described is that it often contains highly toxic and hazardous materials that are injurious to both human health and environmental quality. We now produce and use a vast array of flammable, explosive, caustic, acidic, and highly toxic chemical substances for industrial, agricultural, and domestic purposes (fig. 21.15). According to the EPA, industries in the United States generate about 265 million metric tons of *officially* classified hazardous wastes each year, slightly more than 1 ton for each person in the country. In addition, considerably more toxic and hazardous waste material is generated by

What can you do?

industries or processes not regulated by the EPA. Shockingly, at least 40 million metric tons (22 billion lbs) of toxic and hazardous wastes are released into the air, water, and land in the United States each year. The biggest source of these toxins are the chemical and petroleum industries (fig. 21.16).

What Is Hazardous Waste?

Legally, a **hazardous waste** is any discarded material, liquid or solid, that contains substances known to be (1) fatal to humans or laboratory animals in low doses, (2) toxic, carcinogenic, mutagenic, or teratogenic to humans or other life-forms, (3) ignitable with a flash point less than 60°C, (4) corrosive, or (5) explosive or highly reactive (undergoes violent chemical reactions either by itself or when mixed with other materials). Notice that this definition includes both toxic and hazardous materials as defined in chapter 8. Certain compounds are exempt from regulation as hazardous waste if they are accumulated in less than 1 kg (2.2 lb) of commercial chemicals or 100 kg of contaminated soil, water, or debris. Even larger amounts (up to 1,000 kg) are exempt when stored at an approved waste treatment facility for the purpose of being beneficially used, recycled, reclaimed, detoxified, or destroyed.

Hazardous Waste Disposal

Most hazardous waste is recycled, converted to nonhazardous forms, stored, or otherwise disposed of on-site by the generators—chemical companies, petroleum refiners, and other large industrial facilities—so that it doesn't become a public problem. Still, the hazardous waste that does enter the waste stream or the environment

FIGURE 21.15 According to the U.S. Environmental Protection Agency, industries produce about one ton of hazardous waste per year for every person in the United States. Responsible handling and disposal is essential. © Michael Greenlar/The Images Works.

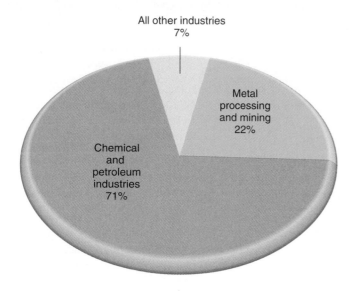

FIGURE 21.16 Producers of hazardous wastes in the United States.

represents a serious environmental problem. And orphan wastes left behind by abandoned industries remain a serious threat to both environmental quality and human health. For years, little attention was paid to this material. Wastes stored on private property, buried, or allowed to soak into the ground were considered of little concern to the public. An estimated 5 billion metric tons of highly poisonous chemicals were improperly disposed of in the United States between 1950 and 1975 before regulatory controls became more stringent.

Federal Legislation

Two important federal laws regulate hazardous waste management and disposal in the United States. The Resource Conservation and Recovery Act (RCRA, pronounced "rickra") of 1976 is a com-

prehensive program that requires rigorous testing and management of toxic and hazardous substances. A complex set of rules require generators, shippers, users, and disposers of these materials to keep meticulous account of everything they handle and what happens to it from generation (cradle) to ultimate disposal (grave) (fig. 21.17).

The Comprehensive Environmental Response, Compensation and Liability Act (CERCLA or Superfund Act), passed in 1980 and modified in 1984 by the Superfund Amendments and Reauthorization Act (SARA), is aimed at rapid containment, cleanup, or remediation of abandoned toxic waste sites. This statute authorizes the Environmental Protection Agency to undertake emergency actions when a threat exists that toxic material will leak into the environment. The agency is empowered to bring suit for the recovery of its costs from potentially responsible parties such as site owners, operators, waste generators, or transporters.

SARA also established (under title III) community right to know and state emergency response plans that give citizens access to information about what is present in their communities. One of the most useful tools in this respect is the **Toxic Release Inventory,** which requires 20,000 manufacturing facilities to report annually on releases of more than 300 toxic materials. You can find specific information there about what is in your neighborhood.

The government does not have to prove that anyone violated a law or what role they played in a Superfund site. Rather, liability under CERCLA is "strict, joint, and several," meaning that anyone associated with a site can be held responsible for the entire cost of cleaning it up no matter how much of the mess they made. In some cases, property owners have been assessed millions of dollars for removal of wastes left there years earlier by previous owners. This strict liability has been a headache for the real estate and insurance businesses.

CERCLA was amended in 1995 to make some of its provisions less onerous. In cases where treatment is unavailable or too costly and it is likely that a less-costly remedy will become available within a reasonable time, interim containment is now allowed. The EPA also now has the discretion to set site-specific cleanup levels rather than adhere to rigid national standards.

Superfund Sites

The EPA estimates that there are at least 36,000 seriously contaminated sites in the United States. The General Accounting Office (GAO) places the number much higher, perhaps more than 400,000 when all are identified. By 1997, some 1,400 sites had been placed on the National Priority List (NPL) for cleanup with financing from the federal Superfund program. The **Superfund** is a revolving pool designed to (1) provide an immediate response to emergency situations that pose imminent hazards, and (2) to clean up or remediate abandoned or inactive sites. Without this fund, sites would languish for years or decades while the courts decided who was responsible to pay for the cleanup. Originally a $1.6 billion pool, the fund peaked at $3.6 billion. From its inception, the fund was financed by taxes on producers of toxic and hazardous wastes. Industries opposed this "polluter pays" tax, because current man-

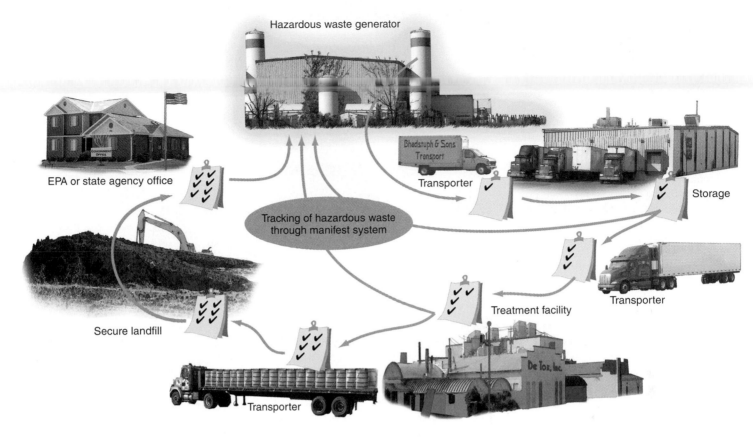

FIGURE 21.17 Toxic and hazardous wastes must be tracked from "cradle to grave" by detailed shipping manifests.

ufacturers are often not the ones responsible for the original contamination. In 1995, Congress agreed to let the tax expire. Since then the Superfund has dwindled, and the public has picked up an increasing share of the bill. In the 1980s the public covered less than 20 percent of the Superfund. In 2003 the figure was nearly 80 percent, and by 2005 the industry share will be zero.

Total costs for hazardous waste cleanup in the United States are estimated between $370 billion and $1.7 trillion, depending on how clean sites must be and what methods are used. For years, Superfund money was spent mostly on lawyers and consultants, and cleanup efforts were often bogged down in disputes over liability and best cleanup methods. During the 1990s, however, progress improved substantially, with a combination of rule adjustments and administrative commitment to cleanup. From 1993 to 2000, the number of completed NPL cleanups jumped from 155 to 757, just over half the list's 1,500 sites (fig. 21.18). Since 2000, progress has slowed again, due to underfunding and a lower priority among administrators.

What qualifies a site for the NPL? These sites are considered to be especially hazardous to human health and environmental quality because they are known to be leaking or have a potential for leaking supertoxic, carcinogenic, teratogenic, or mutagenic materials (chapter 8). The ten substances of greatest concern or most commonly detected at Superfund sites are lead, trichloro-

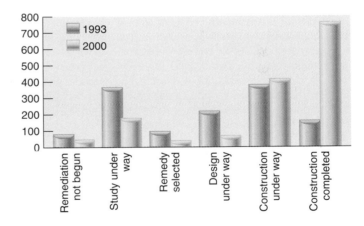

FIGURE 21.18 Progress on Superfund National Priority List (NPL) sites. After years of little progress, the number of completed sites jumped from 155 in 1993 to 757 in 2000. Over 90 percent of the 1,500 NPL sites are under construction or completed.
Source: Environmental Protection Agency, 2001.

ethylene, toluene, benzene, PCBs, chloroform, phenol, arsenic, cadmium, and chromium. These and other hazardous or toxic materials are known to have contaminated groundwater at 75 percent of the sites now on the NPL. In addition, 56 percent of these sites

Getting contaminants out of soil and groundwater is one of the most widespread and persistent problems in waste cleanup. Once leaked into the ground, solvents, metals, radioactive elements, and other contaminants are dispersed and difficult to collect and treat. The main method of cleaning up contaminated soil is to dig it up, then decontaminate it or haul it away and store it in a landfill in perpetuity. At a single site, thousands of tons of tainted dirt and rock may require incineration or other treatment. Cleaning up contaminated groundwater usually entails pumping vast amounts of water out of the ground—hopefully extracting the contaminated water faster than it can spread through the water table or aquifer. In the United States alone, there are tens of thousands of contaminated sites on factories, farms, gas stations, military facilities, sewage treatment plants, landfills, chemical warehouses, and other types of facilities. Cleaning up these sites is expected to cost at least $700 billion.

Recently, a number of promising alternatives have been developed using plants, fungi, and bacteria to clean up our messes. *Phytoremediation* (remediation, or cleanup, using plants) can include a variety of strategies for absorbing, extracting, or neutralizing toxic compounds. Certain types of mustards and sunflowers can extract lead, arsenic, zinc, and other metals (*phytoextraction*). Poplar trees can absorb and break down toxic organic chemicals (*phytodegradation*). Reeds and other water-loving plants can filter water tainted with sewage, metals, or other contaminants. Natural bacteria in groundwater, when provided with plenty of oxygen, can neutralize contaminants in aquifers, minimizing or even eliminating the need to extract

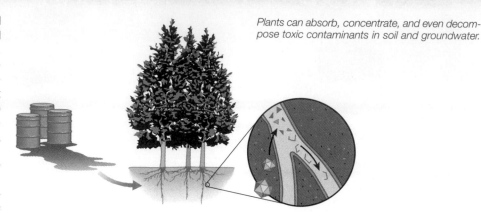

Plants can absorb, concentrate, and even decompose toxic contaminants in soil and groundwater.

and treat water deep in the ground. Radioactive strontium and cesium have been extracted from soil near the Chernobyl nuclear power plant using common sunflowers.

How do the plants, bacteria, and fungi do all this? Many of the biophysical details are poorly understood, but in general, plant roots are designed to efficiently extract nutrients, water, and minerals from soil and groundwater. The mechanisms involved may aid extraction of metallic and organic contaminants. Some plants also use toxic elements as a defense against herbivores—locoweed, for example, selectively absorbs elements such as selenium, concentrating toxic levels in its leaves. Absorption can be extremely effective. Braken fern growing in Florida was found to contain arsenic at concentrations more than 200 times higher than the soil in which it was growing.

Genetically modified plants are also being developed to process toxins. Poplars have been grown with a gene borrowed from bacteria that transform a toxic compound of mercury into a safer form. In another experiment, a gene for producing mammalian liver

enzymes, which specialize in breaking down toxic organic compounds, was inserted into tobacco plants. The plants succeeded in producing the liver enzymes and breaking down toxins absorbed through their roots.

These remediation methods are not without risks. As plants take up toxins, insects could consume leaves, allowing contaminants to enter the food web. Some absorbed contaminants are volatilized, or emitted in gaseous form, through pores in plant leaves. Once toxic contaminants are absorbed into plants, the plants themselves are usually toxic and must be landfilled. But the cost of phytoremediation can be less than half the cost of landfilling or treating toxic soil, and the volume of plant material requiring secure storage ends up being a fraction of a percent of the volume of the contaminated dirt.

Cleaning up hazardous and toxic waste sites will be a big business for the foreseeable future, both in the United States and around the world. Innovations such as phytoremediation offer promising prospects for business growth as well as for environmental health and saving taxpayers' money.

have contaminated surface waters, and airborne materials are found at 20 percent of the sites.

Where are these thousands of hazardous waste sites, and how did they get contaminated? Old industrial facilities such as smelters, mills, petroleum refineries, and chemical manufacturing plants are highly likely to have been sources of toxic wastes. Regions of the country with high concentrations of aging factories

such as the "rust belt" around the Great Lakes or the Gulf Coast petrochemical centers have large numbers of Superfund sites (fig. 21.19). Mining districts also are prime sources of toxic and hazardous waste. Within cities, factories and places such as railroad yards, bus repair barns, and filling stations where solvents, gasoline, oil, and other petrochemicals were spilled or dumped on the ground often are highly contaminated.

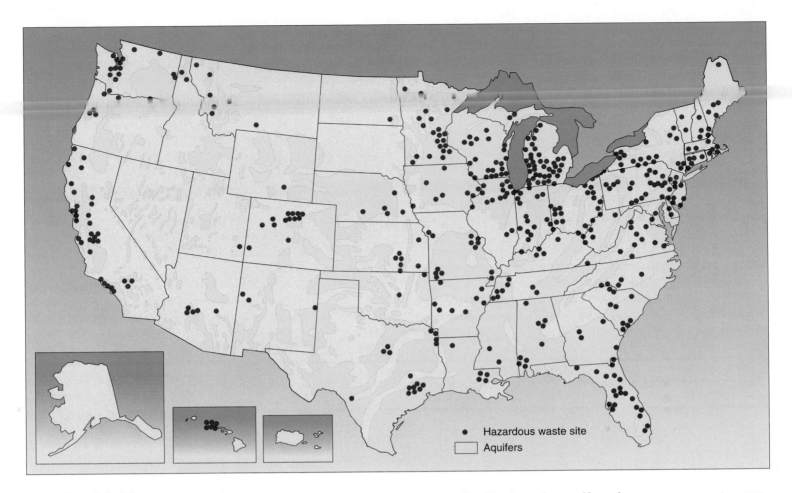

FIGURE 21.19   Some of the hazardous waste sites on the EPA priority cleanup list. Sites located on aquifer recharge zones represent an especially serious threat. Once groundwater is contaminated, cleanup is difficult and expensive. In some cases, it may not be possible.

Source: Environmental Protection Agency.

Some of the most infamous toxic waste sites were old dumps where many different materials were mixed together indiscriminately. For instance, Love Canal in Niagara Falls, New York, was an open dump used by both the city and nearby chemical factories as a disposal site. More than 20,000 tons of toxic chemical waste was buried under what later became a housing development. Another infamous example occurred in Hardeman County, Tennessee, where about a quarter of a million barrels of chemical waste were buried in shallow pits that subsequently leaked toxins into the groundwater. In other sites, liquid wastes were pumped into open lagoons or abandoned in warehouses.

Studies of who lives closest to Superfund and toxic release inventory sites reveal that minorities often are overrepresented in these neighborhoods. Charges of environmental racism have been made, but this is difficult to show conclusively (What Do You Think? p. 462).

How Clean Is Clean?

Among the biggest problems in cleaning up hazardous waste sites are questions of liability and the degree of purity required. In many cities, these problems have created large areas of contaminated properties known as **brownfields** that have been abandoned or are not being used up to their potential because of real or suspected pollution. Up to one-third of all commercial and industrial sites in the urban core of many big cities fall in this category. In heavy industrial corridors the percentage typically is higher.

For years, no one was interested in redeveloping brownfields because of liability risks. Who would buy a property knowing that they might be forced to spend years in litigation and negotiations and be forced to pay millions of dollars for pollution they didn't create? Even if a site has been cleaned to current standards, there is a worry that additional pollution might be found in the future or that more stringent standards might be applied.

In many cases, property owners complain that unreasonably high levels of purity are demanded in remediation programs. Consider the case of Columbia, Mississippi. For many years a 35 ha (81 acre) site in Columbia was used for turpentine and pine tar manufacturing. Soil tests showed concentrations of phenols and other toxic organic compounds exceeding federal safety standards. The site was added to the Superfund NPL and remediation was ordered. Some experts recommended that the best solution was to simply cover the surface with clean soil and enclose the property with a fence to keep people out. The total costs would have been about $1 million.

Instead, the EPA ordered Reichhold Chemical, the last known property owner, to excavate more than 12,500 tons of soil and haul it to a commercial hazardous waste dump in Louisiana at a cost of some $4 million. The intention is to make the site safe enough to be used for any purpose, including housing—even though no one has proposed building anything there. According to the EPA, the dirt must be clean enough for children to play in it—even eat it every day for 70 years—without risk.

Similarly, in places where contaminants have seeped into groundwater, the EPA generally demands that cleanup be carried to drinking water standards. Many critics believe that these pristine standards are unreasonable. Former Congressman Jim Florio, a principal author of the original Superfund Act, says, "It doesn't make any sense to clean up a rail yard in downtown Newark so it can be used as a drinking water reservoir." Depending on where the site is, what else is around it, and what its intended uses are, much less stringent standards may be perfectly acceptable.

Recognizing that reusing contaminated properties can play a significant role in rebuilding old cities, creating jobs, increasing the tax base, and preventing needless destruction of open space at urban margins, programs have been established at both federal and state levels to encourage brownfield recycling. Adjusting purity standards according to planned uses and providing liability protection for nonresponsible parties gives developers and future purchasers confidence that they won't be unpleasantly surprised in the future with further cleanup costs. In some communities, former brownfields are being turned into "eco-industrial parks" that feature environmentally friendly businesses and bring in much needed jobs to inner-city neighborhoods.

Options for Hazardous Waste Management

What shall we do with toxic and hazardous wastes? In our homes, we can reduce waste generation and choose less toxic materials. Buy only what you need for the job at hand. Use up the last little bit or share leftovers with a friend or neighbor. Many common materials that you probably already have make excellent alternatives to commercial products (What Can You Do? p. 473). Dispose of unneeded materials responsibly (fig. 21.20).

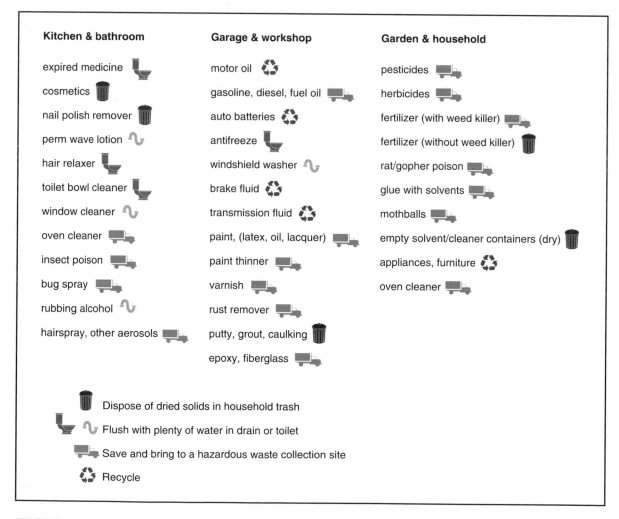

Kitchen & bathroom

- expired medicine
- cosmetics
- nail polish remover
- perm wave lotion
- hair relaxer
- toilet bowl cleaner
- window cleaner
- oven cleaner
- insect poison
- bug spray
- rubbing alcohol
- hairspray, other aerosols

Garage & workshop

- motor oil
- gasoline, diesel, fuel oil
- auto batteries
- antifreeze
- windshield washer
- brake fluid
- transmission fluid
- paint, (latex, oil, lacquer)
- paint thinner
- varnish
- rust remover
- putty, grout, caulking
- epoxy, fiberglass

Garden & household

- pesticides
- herbicides
- fertilizer (with weed killer)
- fertilizer (without weed killer)
- rat/gopher poison
- glue with solvents
- mothballs
- empty solvent/cleaner containers (dry)
- appliances, furniture
- oven cleaner

- Dispose of dried solids in household trash
- Flush with plenty of water in drain or toilet
- Save and bring to a hazardous waste collection site
- Recycle

FIGURE 21.20 Household waste disposal guide. Solvent-containing products have the words "flammable," "combustible," or "petroleum distillates" on the labels. These materials should *not* be disposed of in a drain or toilet. Never mix products containing bleach with those containing ammonia. A toxic gas can form!

What can you do?

Alternatives to Hazardous Household Chemicals

Chrome cleaner: Use vinegar and nonmetallic scouring pad.

Copper cleaner: Rub with lemon juice and salt mixture.

Floor cleaner: Mop linoleum floors with 1 cup vinegar mixed with 2 gallons of water. Polish with club soda.

Brass polish: Use Worcestershire sauce.

Silver polish: Rub with toothpaste on a soft cloth.

Furniture polish: Rub in olive, almond, or lemon oil.

Ceramic tile cleaner: Mix 1/4 cup baking soda, 1/2 cup white vinegar, and 1 cup ammonia in 1 gallon warm water (good general purpose cleaner).

Drain opener: Use plunger or plumber's snake, pour boiling water down drain.

Upholstery cleaner: Clean stains with club soda.

Carpet shampoo: Mix 1/2 cup liquid detergent in 1 pint hot water. Whip into stiff foam with mixer. Apply to carpet with damp sponge. Rinse with 1 cup vinegar in 1 gal water. Don't soak carpet—it may mildew.

Window cleaner: Mix 1/3 cup ammonia, 1/4 cup white vinegar in 1 quart warm water. Spray on window. Wipe with soft cloth.

Spot remover: For butter, coffee, gravy, or chocolate stains: Sponge up or scrape off as much as possible immediately. Dab with cloth dampened with a solution of 1 teaspoon white vinegar in 1 quart cold water.

Toilet cleaner: Pour 1/2 cup liquid chlorine bleach into toilet bowl. Let stand for 30 minutes, scrub with brush, flush.

Pest control: Spray plants with soap-and-water solution (3 tablespoons soap per gallon water) for aphids, mealybugs, mites, and whiteflies. Interplant with pest repellent plants such as marigolds, coriander, thyme, yarrow, rue, and tansy. Introduce natural predators such as ladybugs or lacewings.

Indoor pests: Grind or blend 1 garlic clove and 1 onion. Add 1 tablespoon cayenne pepper and 1 quart water. Add 1 tablespoon liquid soap.

Moths: Use cedar chips or bay leaves.

Ants: Find where they are entering house, spread cream of tartar, cinnamon, red chili pepper, or perfume to block trail.

Fleas: Vacuum area, mix brewer's yeast with pet food.

Mosquitoes: Brewer's yeast tablets taken daily repel mosquitoes.

Note: test cleaners in small, inconspicuous area before using.

Produce Less Waste

As with other wastes, the safest and least expensive way to avoid hazardous waste problems is to avoid creating the wastes in the first place. Manufacturing processes can be modified to reduce or eliminate waste production. In Minnesota, the 3M Company reformulated products and redesigned manufacturing processes to eliminate more than 140,000 metric tons of solid and hazardous wastes,

4 billion l (1 billion gal) of wastewater, and 80,000 metric tons of air pollution each year. They frequently found that these new processes not only spared the environment but also saved money by using less energy and fewer raw materials.

Recycling and reusing materials also eliminates hazardous wastes and pollution. Many waste products of one process or industry are valuable commodities in another. Already, about 10 percent of the wastes that would otherwise enter the waste stream in the United States are sent to surplus material exchanges where they are sold as raw materials for use by other industries. This figure could probably be raised substantially with better waste management. In Europe, at least one-third of all industrial wastes are exchanged through clearinghouses where beneficial uses are found. This represents a double savings: the generator doesn't have to pay for disposal, and the recipient pays little, if anything, for raw materials.

Convert to Less Hazardous Substances

Several processes are available to make hazardous materials less toxic. *Physical treatments* tie up or isolate substances. Charcoal or resin filters absorb toxins. Distillation separates hazardous components from aqueous solutions. Precipitation and immobilization in ceramics, glass, or cement isolate toxins from the environment so that they become essentially nonhazardous. One of the few ways to dispose of metals and radioactive substances is to fuse them in silica at high temperatures to make a stable, impermeable glass that is suitable for long-term storage.

Incineration is applicable to mixtures of wastes. A permanent solution to many problems, it is quick and relatively easy but not necessarily cheap—nor always clean—unless it is done correctly. Wastes must be heated to over 1,000°C (2,000°F) for a sufficient period of time to complete destruction. The ash resulting from thorough incineration is reduced in volume up to 90 percent and often is safer to store in a landfill or other disposal site than the original wastes. Nevertheless, incineration remains a highly controversial topic (fig. 21.21).

Several sophisticated features of modern incinerators improve their effectiveness. Liquid injection nozzles atomize liquids and mix air into the wastes so they burn thoroughly. Fluidized bed burners pump air from the bottom up through burning solid waste as it travels on a metal chain grate through the furnace. The air velocity is sufficient to keep the burning waste partially suspended. Plenty of oxygen is available, and burning is quick and complete. Afterburners add to the completeness of burning by igniting gaseous hydrocarbons not consumed in the incinerator. Scrubbers and precipitators remove minerals, particulates, and other pollutants from the stack gases.

Chemical processing can transform materials so they become nontoxic. Included in this category are neutralization, removal of metals or halogens (chlorine, bromine, etc.), and oxidation. The Sunohio Corporation of Canton, Ohio, for instance, has developed a process called PCBx in which chlorine in such molecules as PCBs is replaced with other ions that render the compounds less toxic. A portable unit can be moved to the location of the hazardous waste, eliminating the need for shipping them.

Biological waste treatment or **bioremediation** taps the great capacity of microorganisms to absorb, accumulate, and detoxify a

FIGURE 21.21 Actor Martin Sheen joins local activists in a protest in East Liverpool, Ohio, site of the largest hazardous waste incinerator in the United States. About 1,000 people marched to the plant to pray, sing, and express their opposition. Involving celebrities draws attention to your cause. A peaceful, well-planned rally builds support and acceptance in the broader community. © Piet Van Lier.

variety of toxic compounds. Bacteria in activated sludge basins, aquatic plants (such as water hyacinths or cattails), soil microorganisms, and other species remove toxic materials and purify effluents. Biotechnology offers exciting possibilities for finding or creating organisms to eliminate specific kinds of hazardous or toxic wastes. By using a combination of classic genetic selection techniques and high-technology gene-transfer techniques, for instance, scientists have recently been able to generate bacterial strains that are highly successful at metabolizing PCBs. There are concerns about releasing such exotic organisms into the environment, however (chapter 11). It may be better to keep these organisms contained in enclosed reaction vessels and feed contaminated material to them under controlled conditions.

Store Permanently

Inevitably, there will be some materials that we can't destroy, make into something else, or otherwise cause to vanish. We will have to store them out of harm's way. There are differing opinions about how best to do this.

Retrievable Storage. Dumping wastes in the ocean or burying them in the ground generally means that we have lost control of them. If we learn later that our disposal technique was a mistake, it is difficult, if not impossible, to go back and recover the wastes. For many supertoxic materials, the best way to store them may be in **permanent retrievable storage.** This means placing waste storage containers in a secure building, salt mine, or bedrock cavern where they can be inspected periodically and retrieved, if necessary, for repacking or for transfer if a better means of disposal is developed. This technique is more expensive than burial in a landfill because the storage area must be guarded and monitored continuously to prevent leakage, vandalism, or other dispersal of toxic materials. Remedial measures are much cheaper with this technique, however, and it may be the best system in the long run.

Secure Landfills. One of the most popular solutions for hazardous waste disposal has been landfilling. Although, as we saw earlier in this chapter, many such landfills have been environmental disasters, newer techniques make it possible to create safe, modern **secure landfills** that are acceptable for disposing of many hazardous wastes.

The first line of defense in a secure landfill is a thick bottom cushion of compacted clay that surrounds the pit like a bathtub (fig. 21.22). Moist clay is flexible and resists cracking if the ground shifts. It is impermeable to groundwater and will safely contain wastes. A layer of gravel is spread over the clay liner and perforated drain pipes are laid in a grid to collect any seepage that escapes from the stored material. A thick polyethylene liner, protected from punctures by soft padding materials, covers the gravel bed. A layer of soil or absorbent sand cushions the inner liner and the wastes are packed in drums, which then are placed into the pit, separated into small units by thick berms of soil or packing material.

When the landfill has reached its maximum capacity, a cover much like the bottom sandwich of clay, plastic, and soil—in that order—caps the site. Vegetation stabilizes the surface and improves its appearance. Sump pumps collect any liquids that filter through the landfill, either from rainwater or leaking drums. This leachate is treated and purified before being released. Monitoring wells check groundwater around the site to ensure that no toxins have escaped.

Most landfills are buried below ground level to be less conspicuous; however, in areas where the groundwater table is close to the surface, it is safer to build above-ground storage. The same protective construction techniques are used as in a buried pit. An advantage to such a facility is that leakage is easier to monitor because the bottom is at ground level.

Transportation of hazardous wastes to disposal sites is of concern because of the risk of accidents. Emergency preparedness officials conclude that the greatest risk in most urban areas is not nuclear war or natural disaster but crashes involving trucks or trains carrying hazardous chemicals through densely packed urban corridors. Another worry is who will bear financial responsibility for abandoned waste sites. The material remains toxic long after the businesses that created it are gone. As is the case with nuclear wastes (chapter 19), we may need new institutions for perpetual care of these wastes.

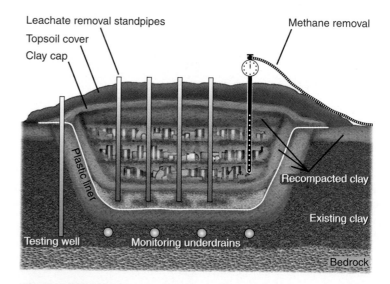

FIGURE 21.22 A secure landfill for toxic waste. A thick plastic liner and two or more layers of impervious compacted clay enclose the landfill. A gravel bed between the clay layers collects any leachate, which can then be pumped out and treated. Well samples are tested for escaping contaminants and methane is collected for combustion.

Summary

- Global waste production is a critical issue in environmental quality. Global waste production is rapidly growing, as nonbiodegradable materials grow in a waste stream.

- Solid waste includes domestic, commercial, industrial, agricultural, and mining wastes that are primarily nontoxic. About 60 percent of North American domestic and industrial wastes are deposited in landfills; most of the rest is incinerated or recycled. Old landfills were often leaky and messy, and they have been an important source of groundwater contamination. Modern landfills must have impermeable liners that prevent seepage to groundwater, and they must be covered with soil and monitored for gas emissions and water contamination.

- Incineration is our second most important method of waste disposal. Incineration can destroy organic compounds, and it can be used to produce energy. Airborne contaminants can result from burning, however, especially if incinerators are not operated at optimal temperatures or if toxic substances are burned.

- Recycling and composting (of yard waste and other organic materials) are growing in North America and globally. The growing cost of waste disposal makes collection increasingly cost-effective, and the rapid depletion of landfill space makes recycling attractive. A principal obstacle to recycling is weak or unstable markets for recycled plastic, paper, and other materials. The growing use of disposable plastic beverage containers poses another important problem to widespread recycling. Japan is a world leader in recycling, with about 60 percent of domestic waste being recycled.

- Reusing, demanufacturing, and reducing material consumption are important additional strategies for reducing the waste stream.

- Hazardous and toxic wastes are waste materials that cause health problems, including birth defects, neurological disorders, reduced resistance to infection, and cancer. Environmental costs of hazardous and toxic waste include contamination of water supplies, poisoning of soil, and destruction of habitat.

- The major categories of hazardous wastes are ignitable, corrosive, reactive, explosive, and toxic materials. Some materials of greatest concern are heavy metals, solvents, and synthetic organic chemicals such as halogenated hydrocarbons, organophosphates, and phenoxy herbicides.

- Disposal practices for solid and hazardous wastes have often been unsatisfactory. Thousands of abandoned, often unknown waste disposal sites still leak toxic materials into the environment. Techniques for controlling hazardous wastes include not making the material in the first place; incineration; secure landfills; and physical, chemical, or biological treatment to detoxify or immobilize wastes.

- The Superfund is a revolving fund to finance cleanup of some of our worst hazardous waste sites. The Superfund was established by the Comprehensive Environmental Response, Compensation, and Liability Act (CERCLA) of 1980. Many sites on the Superfund's National Priority List are abandoned factories or dumps.

- Dangerous wastes are often removed from wealthy countries or neighborhoods to poorer ones, and cleanup efforts may be faster and more complete in wealthier areas. Proponents of environmental justice try to identify these patterns and rectify them.

- People often resist having transfer facilities, storage sites, disposal operations, or transportation of hazardous materials (often referred to as Locally Unwanted Land Uses, or LULUs) near where they live. Safe handling and liability remain unanswered in solid and hazardous waste disposal.

Questions for Review

1. What are solid wastes and hazardous wastes? What is the difference between them?

2. Describe the difference between an open dump, a sanitary landfill, and a modern, secure, hazardous waste disposal site.

3. Why are landfill sites becoming limited around most major urban centers in the United States? What steps are being taken to solve this problem?

4. Describe some concerns about waste incineration.

5. List some benefits and drawbacks of recycling wastes. What are the major types of materials recycled from municipal waste and how are they used?

6. What is composting, and how does it fit into solid waste disposal?

7. Describe some ways that we can reduce the waste stream to avoid or reduce disposal problems.

8. List ten toxic substances in your home and how you would dispose of them.

9. What are brownfields and why do cities want to redevelop them?

10. What societal problems are associated with waste disposal? Why do people object to waste handling in their neighborhoods?

Questions for Critical Thinking

1. A toxic waste disposal site has been proposed for the Pine Ridge Indian Reservation in South Dakota. Many tribal members oppose this plan, but some favor it because of the jobs and income it will bring to an area with 70 percent unemployment. If local people choose immediate survival over long-term health, should we object or intervene?

2. There is often a tension between getting your personal life in order and working for larger structural changes in society. Evaluate the trade-offs between spending time and energy sorting recyclables at home compared to working in the public arena on a bill to ban excess packaging.

3. Should industry officials be held responsible for dumping chemicals that were legal when they did it but are now known to be extremely dangerous? At what point can we argue that they *should* have known about the hazards involved?

4. Look at the discussion of recycling or incineration presented in this chapter. List the premises (implicit or explicit) that underlie the presentation as well as the conclusions (stated or not) that seem to be drawn from them. Do the conclusions necessarily follow from these premises?

5. Suppose that your brother or sister has decided to buy a house next to a toxic waste dump because it costs $20,000 less than a comparable house elsewhere. What do you say to him or her?

6. Is there an overall conceptual framework or point of view in this chapter? If you were presenting a discussion of solid or hazardous waste to your class, what would be your conceptual framework?

7. Is there a fundamental difference between incinerating municipal, medical, or toxic industrial waste? Would you oppose an incinerator for one type of waste in your neighborhood but not others? Why, or why not?

8. The Netherlands incinerates much of its toxic waste at sea by a shipborne incinerator. Would you support this as a way to dispose of our wastes as well? What are the critical considerations for or against this approach?

Key Terms

biodegradable plastics 467
bioremediation 473
brownfields 471
composting 465
demanufacturing 465
e-waste 465
energy recovery 460
hazardous waste 467
mass burn 460
permanent retrievable
   storage 474

photodegradable plastics 466
recycling 461
refuse-derived fuel 460
sanitary landfills 459
secure landfills 474
Superfund 468
Toxic Release Inventory 468
waste stream 457

Further Readings

Bergman, B. J. 1999. The hidden life of computers. *Sierra* 84(4): 32–33.

Blumberg, Louis, and Robert Gottlieb. 1989. *War on Waste: Can America Win Its Battle with Garbage?* Island Press.

Kazuhiro, Ueta, and Harumi Koizumi. 2001. Reducing household waste: Japan learns from Germany. *Environment* 43(9):20–32.

McDonough, William, and Michael Braungart. 2002. *Cradle to Cradle: Remaking the Way We Make Things.* North Point Press.

Rathje, William, and Cullen Murphy. 1992. *Rubbish! The Archaeology of Garbage.* HarperCollins.

Taylor, David. 1999. Talking trash: The economic and environmental issues of landfills. *Environmental Health Perspectives* 107(8):A404–A409.

Warren-Rhodes, Kimberley, and Albert Koenig. 2001. Escalating trends in the urban metabolism of Hong Kong: 1971–1997. *AMBIO: A Journal of the Human Environment* 30(7):429–38.

Welcome to McGraw-Hill's Online Learning Center

Location: http://www.mhhe.com/environmentalscience

McGraw Hill

WEB EXERCISE

Recycling Challenge

The Internet Consumer Recycling Guide, www.idiom.com/~bryce/recycle/, provides a detailed list of recyclable materials and where to send them. Before you look at this website, take five minutes and write down at least 25 different types of items in the room around you. These items can be small (paper clips, hair clips, books, food packaging, etc.) or large (carpets, furniture, computers, light fixtures, plumbing, etc.). Once you have made the list, mark all those that are recyclable and note how you would recycle them.

Now look at the Recycling Guide web page. Look at all three "guide" pages—The World's Shortest Comprehensive Recycling Guide, the Guide to Recycling Common Materials, and Guide to Hard-to-Recycle Materials. How many additional items on your list can be recycled, according to these guides? How many could be recycled in principle but would be hard to recycle in your community? Why? What recycled products could be made by recycling the materials on your list?

How many of the items on your list might be available with recycled content? How many of them could you have avoided acquiring in the first place?

Mapping Superfund Sites

The National Priority List (NPL) is a list of contaminated sites scheduled for cleanup using money from the so-called Superfund. To be placed on this list, sites must be nominated and studied. A final decision can take years. Remediation can take many years more. How many "Superfund" sites are in your state (or any state you choose)? Go to www.epa.gov/superfund/sites/npl/npl.htm, where you'll find a map of the United States. Click on a state to see a state map. How many sites are there in your state, or in the one you chose? As you look at the state map, can you identify whether sites are clustered on a river, on a coastline, or on some other natural features? Can you think of any reasons why sites are placed as they are?

Click on the site nearest you, and you'll see a list of sites in that state. Click on the name of the site, and you should go to a description of the facility, its dates, and the types of contamination. Have you heard of this site? What type of site is it? Has it been cleaned up?

Click on several of the sites in your state. What is the approximate range of ages of these sites? How many were polluted before you were born? Do you think you should be responsible for cleaning them up? If not you, then who?

Urbanization and Sustainable Cities

The problems that overwhelm us today are precisely those we failed to solve decades ago.

Mostafa K. Tolba

OBJECTIVES

After studying this chapter, you should be able to:

- distinguish between a rural village, a city, and a megacity.
- recognize the push and pull factors that lead to urban growth.
- report on the growth rate of giant metropolitan urban areas such as Mexico City, as well as the problems this growth engenders.
- understand the causes and consequences of urban sprawl.
- explain the principles of smart growth, garden cities, and conservation designs.
- evaluate the ways that cities can be ecologically, socially, and economically sustainable.

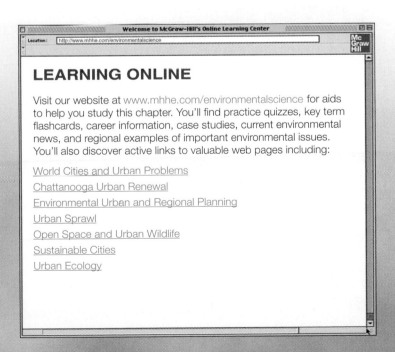

Welcome to McGraw-Hill's Online Learning Center

Location: http://www.mhhe.com/environmentalscience

LEARNING ONLINE

Visit our website at www.mhhe.com/environmentalscience for aids to help you study this chapter. You'll find practice quizzes, key term flashcards, career information, case studies, current environmental news, and regional examples of important environmental issues. You'll also discover active links to valuable web pages including:

World Cities and Urban Problems
Chattanooga Urban Renewal
Environmental Urban and Regional Planning
Urban Sprawl
Open Space and Urban Wildlife
Sustainable Cities
Urban Ecology

Photo: Cities like Hong Kong are powerful engines for cultural, economic, and political change. They also consume resources and create environmental problems. © William P. Cunningham.

Chattanooga, A Model Sustainable City

Nestled in the Tennessee River valley, at the foot of the Cumberland Plateau, Chattanooga, Tennessee, is a medium-size industrial city that provides an inspiring example of participatory city planning that has produced a clean urban environment, a vibrant economy, and strong social capital. Twenty-five years ago, the U.S. Department of Health, Education, and Welfare named Chattanooga America's dirtiest city. Factory smoke trapped between the mountains gave city air the highest particulate count in the nation—dirtier, even, than Los Angeles. Per capita tuberculosis rates were three times the national average. Chattanooga Creek was so polluted by toxic waste from coke ovens and steel mills that the EPA designated 4 km (2.5 mi) of it a Superfund site.

Social conditions declined along with environmental quality. As industries moved out, and the upper class fled to the suburbs, the city was left with abandoned factories, boarded-up buildings, a declining tax base, unemployment, crime, and despair. Most of the worst pollution and highest unemployment were in predominantly African-American neighborhoods, where residents had the least resources to improve their situation. How could the city reinvent itself and turn these trends around?

In 1984, the city council, led by David Crockett—the great nephew of the famous frontiersman—brought together a variety of city residents to talk about their vision for the future. A series of town meetings drew more than 1,700 people who brainstormed what their city might be like. Among the values mentioned most often were clean air and water, jobs, affordable housing, personal safety, a place to fish, and somewhere to walk along the river. Forty goals were identified, ranging from cleaning up the river to building low-cost housing. By giving everyone a voice, the process built community spirit and a broad base of support. Suddenly people with energy and good ideas from every part of the city got involved.

Public-private partnerships led to nearly $800 million in investments (two-thirds from the private sector) and created 8,000 new permanent and temporary jobs. The Chattanooga Neighborhood Enterprise Corporation invested $60 million in affordable housing. More than 1,000 first-time home buyers obtained low-interest loans, while other residents got home-improvement help. By 1990, Chattanooga was in compliance with national air quality standards. As part of urban redevelopment, the city encouraged use of public transport by building satellite parking lots

FIGURE 22.1   Chattanooga's waterfront has been cleaned up and has become an asset rather than a liability. © Bob Boyer.

at the city outskirts and providing free, convenient, rapid bus service to the city center. The electric buses designed for this system are now being sold to other cities and provide jobs and revenue for the city.

The riverfront has been cleaned up. Old warehouses that blocked river access have been removed and public fishing piers built. A 35 km (22 mi) long riverfront park with walking and bike trails has been built along both sides of the river through downtown (fig. 22.1). Fountains, flowers, shade trees, street musicians, and outdoor festivals now fill the park. A world-class freshwater aquarium attracts tourists. What had been an embarrassment of derelict buildings and weedy, vacant lots, has become the most desirable location in the city. Property values have shot up, and yuppies are moving back to the city center to take advantage of the recreational and cultural opportunities.

Chattanooga is now considered the "granddaddy" of the sustainable cities movement. Using principles of urban ecology, the city has created a livable environment, a strong sense of community, and a flourishing economy. As David Crockett says, it is "a city that is cleaner, greener, and safer; that values human and natural resources; and that provides an economic system that will keep our children here."

URBANIZATION

Since their earliest origins, cities have been centers of education, religion, commerce, record keeping, communication, and political power. As cradles of civilization, cities have influenced culture and society far beyond their proportion of the total population (fig. 22.2). Until recently, however, only a small percentage of the world's people lived permanently in urban areas and even the greatest cities of antiquity were small by modern standards. The vast majority of humanity has always lived in rural areas where farming, fishing, hunting, timber harvesting, animal herding, mining, or other natural resource-based occupations provided support.

Since the beginning of the Industrial Revolution some 300 years ago, however, cities have grown rapidly in both size and power. In every developing country, the transition from an agrarian society to an industrial one has been accompanied by **urbanization,** an increasing concentration of the population in cities and a transformation of land use and society to a metropolitan pattern of organization. Industrialization and urbanization bring many benefits—especially to the top members of society—but they also cause many problems, as we will detail in this chapter.

Nearly half the people in the world now live in urban areas. Demographers predict that by the end of the twenty-first century, 80 or 90 percent of all humans will live in cities and that some

FIGURE 22.2 Since their earliest origins, cities have been centers of education, religion, commerce, politics, and culture. Unfortunately, they have also been sources of pollution, crowding, disease, and misery. © Corbis/Royalty Free Website.

FIGURE 22.3 This village in Chiapas, Mexico, is closely tied to the land through culture, economics, and family relationships. While the timeless pattern of life here gives a great sense of identity, it can also be stifling and repressive. © William P. Cunningham.

giant interconnecting metropolitan areas could have hundreds of millions of residents. In this chapter, we will look at how cities came into existence, why people live there, and what the environmental conditions of cities have been, are now, and might be in the future. Some of the most severe urban problems in the world are found in the giant megacities of the developing countries. Far more lives may be threatened by the desperate environmental conditions in these cities than by any other issue we have studied. We also will look at a few of those problems and possible solutions that could improve the urban environment.

What Is a City?

Just what makes up an urban area or a city? Definitions differ. The U.S. Census Bureau considers any incorporated community to be a city, regardless of size, and defines any city with more than 2,500 residents as urban. More meaningful definitions are based on *functions*. In a **rural area,** most residents depend on agriculture or other ways of harvesting natural resources for their livelihood. In an **urban area,** by contrast, a majority of the people are not directly dependent on natural resource-based occupations.

A **village** is a collection of rural households linked by culture, custom, family ties, and association with the land (fig. 22.3). A **city,** by contrast, is a differentiated community with a population and resource base large enough to allow residents to specialize in arts, crafts, services, or professions rather than natural resource-based occupations. While the rural village often has a sense of security and connection, it also can be stifling. A city offers more freedom to experiment, to be upwardly mobile, and to break from restrictive traditions, but it can be harsh and impersonal (fig. 22.4).

Beyond about 10 million inhabitants, an urban area is considered a supercity or **megacity.** Megacities in many parts of the world have grown to enormous size. Chongqing, China, having annexed a large part of Sichuan province and about 30 million people, claims to be the biggest city in the world. In the United States,

FIGURE 22.4 A city is a differentiated community with a large enough population and resource base to allow specialization in arts, crafts, services, and professions. Although there are many disadvantages in living so closely together, there are also many advantages. Cities are growing rapidly, and most of the world's population will live in urban areas if present trends continue. © The McGraw-Hill Companies, Inc./Barry W. Barker, photographer.

urban areas between Boston and Washington, D.C., have merged into a nearly continuous megacity (sometimes called Bos-Wash) containing about 35 million people. The Tokyo-Yokohama-Osaka-Kobe corridor contains nearly 50 million people. Because these agglomerations have expanded beyond what we normally think of as a city, some geographers prefer to think of urbanized **core regions** that dominate the social, political, and economic life of most countries (fig. 22.5). Architect and city planner C. A. Doxiadis predicts that as core regions of adjacent countries merge, the sea coasts and major river valleys of most continents will be covered with continuous strip cities, which he calls ecumenopolises, each containing billions of people.

World Urbanization

The United States underwent a dramatic rural-to-urban shift in the nineteenth and early twentieth centuries (fig. 22.6). Now many developing countries are experiencing a similar demographic movement. In 1850, only about 2 percent of the world's population lived in cities. By 2000, 47 percent of the people in the world were urban. Only Africa and South Asia remain predominantly rural, but people there are swarming into cities in ever-increasing numbers. About three-fourths of the people in Europe, North America, and Latin America already live in cities (table 22.1). Some urbanolo-

gists predict that by 2100, the whole world will be urbanized to the levels now seen in developed countries.

As figure 22.7 shows, 90 percent of the population growth over the next 25 years is expected to occur in the less-developed countries of the world. Most of that growth will be in the already overcrowded cities of the least affluent countries, such as India, China, Mexico, and Brazil. The combined population of these cities is projected by the Population Reference Bureau to jump from its present 6 billion to more than 8 billion by the year 2025. Meanwhile, rural populations in these countries are expected to remain constant or even decline somewhat as rural people migrate into the cities.

Recent urban growth has been particularly dramatic in the largest cities, especially those of the developing world. In 1900, 13 cities had populations over 1 million; all except Tokyo and Peking were in Europe or North America (table 22.2). By 2000, there were 241 metropolitan areas of more than 1 million people but only three of the largest cities were in developed countries.

A century ago, London was the only city with more than 5 million people; now 30 cities have populations larger than that. Some

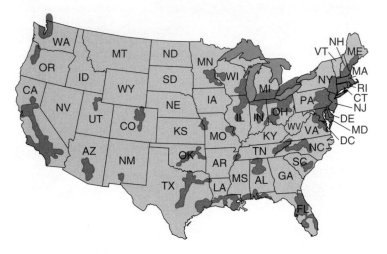

FIGURE 22.5 Urban core agglomerations (lavender areas) are forming megalopolises in many areas. While open space remains in these areas, the flow of information, capital, labor, goods, and services links each into an interacting system. *Source:* U.S. Census Bureau, 1998.

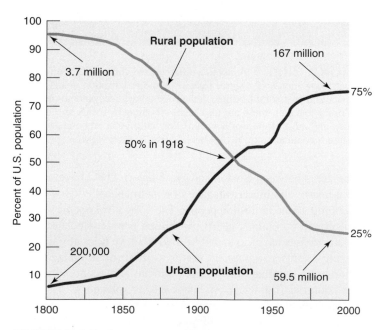

FIGURE 22.6 United States rural-to-urban shift. Percentages of rural and urban populations, 1800 to 2000. *Source:* U.S. Bureau of the Census.

TABLE 22.1	Urban Share of Total Population (Percentage)		
	1950	2000	2030*
Africa	18.4	40.6	57.0
Asia	19.3	43.8	59.3
Europe	56.0	75.0	81.5
Latin America	40.0	70.3	79.7
North America	63.9	77.4	84.5
Oceania	32.0	49.5	60.7
World	38.3	59.4	70.5

*Projected
Source: *United Nations Population Division, 2003.*

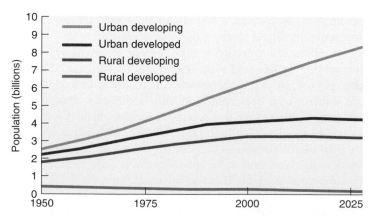

FIGURE 22.7 Urban and rural growth in developed and developing countries. *Source:* United Nations (U.N.) Population Division, *World Urbanization Prospects (The 1996 Revision),* on diskette (U.N. New York, 1996).

TABLE 22.2 The World's Largest Urban Areas (Populations in Millions)

1900		2015**	
London, England	6.6	Tokyo, Japan	31.0
New York, USA	4.2	New York, USA	29.9
Paris, France	3.3	Mexico City	21.0
Berlin, Germany	2.4	Seoul, Korea	19.8
Chicago, USA	1.7	São Paolo, Brazil	18.5
Vienna, Austria	1.6	Osaka, Japan	17.6
Tokyo, Japan	1.5	Jakarta, Indonesia	17.4
St. Petersburg, Russia	1.4	Delhi, India	16.7
Philadelphia, USA	1.4	Los Angeles, USA	16.6
Manchester, England	1.3	Beijing, China	16.0
Birmingham, England	1.2	Cairo, Egypt	15.5
Moscow, Russia	1.1	Manila, Philippines	13.5
Peking, China*	1.1	Buenos Aires, Brazil	12.9

*Now spelled Beijing.

**Projected

Source: *T. Chandler,* Three Thousand Years of Urban Growth, *1974, Academic Press; and World Gazetter, 2003.*

futurists predict that by 2025 at least 400 cities will have populations of 1 million or more, and 93 supercities each will have 5 million or more residents. Three-fourths of those cities will be in the developing world (fig. 22.8). In just the next 25 years, Mumbai, India; Delhi, India; Karachi, Pakistan; Manila, Philippines; and Jakarta, Indonesia all are expected to grow by at least 50 percent.

The actual population of many of these cities is uncertain. Where to draw city boundaries and how to count everyone is controversial. If you live near a large metropolitan area, you probably have experienced how the city gradually merges into the countryside. Many people live beyond the official city boundaries but depend on the urban economy for their livelihoods. In megacities of the developing world, as much as half of urban populations are transient workers or residents of unplanned slums and shantytowns. These people are hard to count. Beijing, for example, has a population according to the official census of about 10 million residents, but has an additional 3 to 4 million undocumented, transitory workers.

Can cities function with 20 or 25 million people? Can they supply the public services necessary to sustain a civilized life? Adding 750,000 new people annually to Mexico City amounts to building a new city the size of Baltimore or San Francisco every year. This growth is occurring, as is most urban growth in the world, in a country with a sagging economy, an unstable government, and a high foreign debt load.

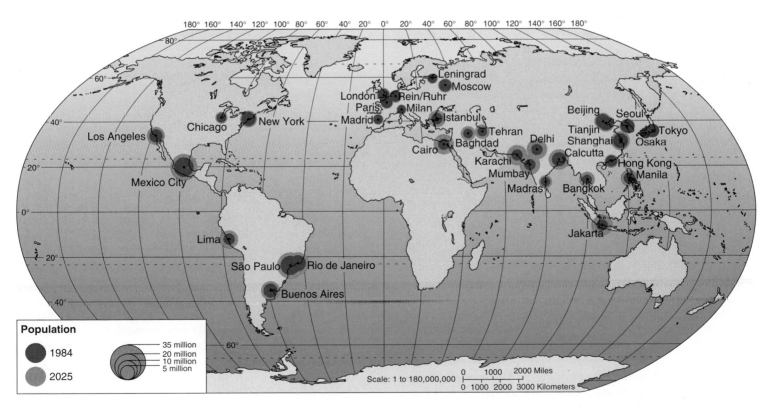

FIGURE 22.8   By 2025, at least 400 cities will have populations of 1 million or more, and 93 supercities will have populations above 5 million. Three-fourths of the world's largest cities will be in developing countries that already have trouble housing, feeding, and employing their people.

CAUSES OF URBAN GROWTH

Urban populations grow in two ways: by natural increase (more births than deaths) and by immigration. Natural increase is fueled by improved food supplies, better sanitation, and advances in medical care that reduce death rates and cause populations to grow both within cities and in the rural areas around them (chapter 7). In Latin America and East Asia, natural increase is responsible for two-thirds of urban population growth. In Africa and West Asia, immigration is the largest source of urban growth. Immigration to cities can be caused both by **push factors** that force people out of the country and by **pull factors** that draw them into the city.

Immigration Push Factors

People migrate to cities for many reasons. In some areas, the countryside is overpopulated and simply can't support more people. The "surplus" population is forced to migrate to cities in search of jobs, food, and housing. Not all rural-to-urban shifts are caused by overcrowding in the country, however. In some places, economic forces or political, racial, or religious conflicts drive people out of their homes. The countryside may actually be depopulated by such demographic shifts. The United Nations estimated that in 2002 at least 19.8 million people fled their native country and that about another 20 million were internal refugees within their own country, displaced by political, economic, or social instability. Many of these refugees end up in the already overcrowded megacities of the developing world.

Land tenure patterns and changes in agriculture also play a role in pushing people into cities. The same pattern of agricultural mechanization that made farm labor largely obsolete in the United States early in this century is spreading now to developing countries. Furthermore, where land ownership is concentrated in the hands of a wealthy elite, subsistence farmers are often forced off the land so it can be converted to grazing lands or monoculture cash crops. Speculators and absentee landlords let good farmland sit idle that otherwise might house and feed rural families.

Immigration Pull Factors

Even in the largest and most hectic cities, many people are there by choice, attracted by the excitement, vitality, and opportunity to meet others like themselves. Cities offer jobs, housing, entertainment, and freedom from the constraints of village traditions. Possibilities exist in the city for upward social mobility, prestige, and power not ordinarily available in the country. Cities support specialization in arts, crafts, and professions for which markets don't exist elsewhere.

Modern communications also draw people to cities by broadcasting images of luxury and opportunity. An estimated 90 percent of the people in Egypt, for instance, have access to a television set. The immediacy of television makes city life seem more familiar and attainable than ever before. We generally assume that beggars and homeless people on the streets of developing nations' teeming cities have no other choice of where to live, but many of these people want to be in the city. In spite of what appears to be dismal conditions, living in the city may be preferable to what the country had to offer.

Government Policies

Government policies often favor urban over rural areas in ways that both push and pull people into the cities. Developing countries commonly spend most of their budgets on improving urban areas (especially around the capital city where leaders live), even though only a small percentage of the population lives there or benefits directly from the investment. This gives the major cities a virtual monopoly on new jobs, housing, education, and opportunities, all of which bring in rural people searching for a better life. In Peru, for example, Lima accounts for 20 percent of the country's population, but has 50 percent of the national wealth, 60 percent of the manufacturing, 65 percent of the retail trade, 73 percent of the industrial wages, and 90 percent of all banking in the country. Similar statistics pertain to São Paulo, Mexico City, Manila, Cairo, Lagos, Bogotá, and a host of other cities.

Governments often manipulate exchange rates and food prices for the benefit of more politically powerful urban populations but at the expense of rural people. Importing lower-priced food pleases city residents, but local farmers then find it uneconomical to grow crops. As a result, an increased number of people leave rural areas to become part of a large urban workforce, keeping wages down and industrial production high. Zambia, for instance, sets maize prices below the cost of local production to discourage farming and to maintain a large pool of workers for the mines. Keeping the currency exchange rate high stimulates export trade but makes it difficult for small farmers to buy the fuels, machinery, fertilizers, and seeds that they need. This depresses rural employment and rural income while stimulating the urban economy. The effect is to transfer wealth from the country to the city.

CURRENT URBAN PROBLEMS

Large cities in both developed and developing countries face similar challenges in accommodating the needs and by-products of dense populations. The problems are most intense, however, in rapidly growing cities of developing nations.

The Developing World

As figure 22.8 shows, 90 percent of the human population growth in the next century is expected to occur in the developing world, mainly in Africa, Asia, and South America. Almost all of that growth will occur in cities—especially the largest cities—which already have trouble supplying food, water, housing, jobs, and basic services for their residents. The unplanned and uncontrollable growth of those cities causes tragic urban environmental problems. Consider as you study urban conditions what responsibilities we in the richer countries have to help others who are less fortunate and how we might do so.

Traffic and Congestion

A first-time visitor to a supercity—particularly in a less-developed country—is often overwhelmed by the immense crush of pedestrians and vehicles of all sorts that clog the streets. The noise,

FIGURE 22.9 Motorized rickshaws, motor scooters, bicycles, street vendors, and pedestrians all vie for space on the crowded streets of Jakarta. But in spite of the difficulties of living here, people work hard and have hope for the future. © William P. Cunningham.

FIGURE 22.10 This tidal canal in Jakarta serves as an open sewer. By some estimates, about half of the 10 million residents of this city have no access to modern sanitation systems. © William P. Cunningham.

congestion, and confusion of traffic make it seem suicidal to venture onto the street. Jakarta, for instance, is one of the most densely populated cities in the world (fig. 22.9). Traffic is chaotic almost all the time. People commonly spend three or four hours each way commuting to work from outlying areas. Bangkok also has monumental traffic problems. The average resident spends the equivalent of 44 days a year sitting in traffic jams. About 20 percent of all fuel is consumed by vehicles standing still. Hours of work lost each year are worth at least $3 billion.

Air Pollution

The dense traffic (commonly old, poorly maintained vehicles), smoky factories, and use of wood or coal fires for cooking and heating often create a thick pall of air pollution in the world's supercities. Lenient pollution laws, corrupt officials, inadequate testing equipment, ignorance about the sources and effects of pollution, and lack of funds to correct dangerous situations usually exacerbate the problem. What is its human toll? An estimated 60 percent of Calcutta's residents are thought to suffer from respiratory diseases linked to air pollution. Lung cancer mortality in Shanghai is reported to be four to seven times higher than rates in the countryside. Mexico City, which sits in a high mountain bowl with abundant sunshine, little rain, high traffic levels, and frequent air stagnation, has one of the highest levels of photochemical smog (chapter 16) in the world.

Sewer Systems and Water Pollution

Few cities in developing countries can afford to build modern waste treatment systems for their rapidly growing populations. The World Bank estimates that only 35 percent of urban residents in developing countries have satisfactory sanitation services (fig. 22.10). The situation is especially desperate in Latin America, where only 2 percent of urban sewage receives any treatment. In Egypt, Cairo's sewer system was built about 50 years ago to serve a population of 2 million people. It is now being overwhelmed by more than 10 million people. Less than one-tenth of India's 3,000 towns and cities have even partial sewage systems and water treatment facilities. Some 150 million of India's urban residents lack access to sanitary sewer systems. In Colombia, the Bogotá River, 200 km (125 mi) downstream from Bogotá's 5 million residents, still has an average fecal bacteria count of 7.3 million cells per liter, more than 700,000 times the safe drinking level and 3,500 times higher than the limit for swimming.

Some 400 million people, or about one-third of the population, in developing world cities do not have safe drinking water, according to the World Bank. Although city dwellers are somewhat more likely than rural people to have clean water, this still represents a large problem. Where people have to buy water from merchants, it often costs 100 times as much as piped city water and may not be any safer to drink. Many rivers and streams in developing countries are little more than open sewers, and yet they are all that poor people have for washing clothes, bathing, cooking, and—in the worst cases—for drinking. Diarrhea, dysentery, typhoid, and cholera are widespread diseases in these countries, and infant mortality is tragically high (chapter 8).

Housing

The United Nations estimates that at least 1 billion people—20 percent of the world's population—live in crowded, unsanitary slums of the central cities and in the vast shantytowns and squatter settlements that ring the outskirts of most developing world cities. Around 100 million people have no home at all. In Mumbay, India, for example, it is thought that half a million people sleep on the streets, sidewalks, and traffic circles because they can find no other place to live (fig. 22.11). In São Paulo, perhaps 1 million "street kids" who have run away from home or been abandoned by their parents live however and wherever they can. This is surely a symptom of a tragic failure of social systems.

Slums are generally legal but inadequate multifamily tenements or rooming houses, either custom built to rent to poor people or converted from some other use. The chals of Mumbay, India, for example, are high-rise tenements built in the 1950s to house immigrant workers. Never very safe or sturdy, these dingy, airless buildings are already crumbling and often collapse without warning. Eighty-four percent of the families in these tenements live in a single room; half of those families consist of six or more people. Typically, they have less than 2 square meters of floor space per person and only one or two beds for the whole family. They may share kitchen and bathroom facilities down the hall with 50 to 75 other people. Even more crowded are the rooming houses for mill workers where up to 25 men sleep in a single room only 7 meters square. Because of this crowding, household accidents are a common cause of injuries and deaths in developing world cities, especially to children. Charcoal braziers or kerosene stoves used in crowded homes are a routine source of fires and injuries. With no place to store dangerous objects beyond the reach of children, accidental poisonings and other mishaps are a constant hazard.

Shantytowns are settlements created when people move onto undeveloped lands and build their own houses. Shacks are built of corrugated metal, discarded packing crates, brush, plastic sheets, or whatever building materials people can scavenge. Some shantytowns are simply illegal subdivisions where the landowner rents land without city approval. Others are spontaneous or popular settlements or **squatter towns** where people occupy land without the owner's permission. Sometimes this occupation involves thousands of people who move onto unused land in a highly organized, overnight land invasion, building huts and laying out streets, markets, and schools before authorities can root them out. In other cases, shantytowns just gradually "happen."

Called *barriads, barrios, favelas,* or *turgios* in Latin America, *bidonvillas* in Africa, or *bustees* in India, shantytowns surround every megacity in the developing world (fig. 22.12). They are not an exclusive feature of poor countries, however. Some 250,000 immigrants and impoverished citizens live in the *colonias* along the southern Rio Grande in Texas. Only 2 percent have access to adequate sanitation. Many live in conditions as awful as you would see in any developing world city.

About three-quarters of the residents of Addis Ababa, Ethiopia, or Luanda, Angola, live in squalid refugee camps. Two-thirds of the population of Calcutta live in unplanned squatter settlements and nearly half of the 20 million people in Mexico City live in uncon-

FIGURE 22.11 In Mumbay, as many as half a million people sleep on the streets because they have no other place to live. Ten times as many live in crowded, dangerous slums and shantytowns throughout the city. © Louis Psihoyos/Matrix International, Inc.

FIGURE 22.12 Shantytowns called *favelas* perch on hillsides above the Amazon in Manaus, Brazil. Courtesy Kathleen Smith.

trolled, unauthorized shantytowns. Many governments try to clean out illegal settlements by bulldozing the huts and sending riot police to drive out the settlers, but the people either move back in or relocate to another shantytown.

These popular but unauthorized settlements usually lack sewers, clean water supplies, electricity, and roads. Often the land on which they are built was not previously used because it is unsafe or unsuitable for habitation. In Bhopal, India, and Mexico City, for example, squatter settlements were built next to deadly industrial sites. In Rio de Janeiro, La Paz (Bolivia), Guatemala City, and Caracas (Venezuela), they are perched on landslide-prone hills. In Lima (Peru), Khartoum (Sudan), and Nouakchott (Mauritania), shantytowns have spread onto sandy deserts. In Manila, thousands of people live in huts built on towering mounds of garbage and burning industrial waste in city dumps (see fig. 21.3).

As desperate and inhumane as conditions are in these slums and shantytowns, many people do more than merely survive there. They keep themselves clean, raise families, educate their children, find jobs, and save a little money to send home to their parents. They learn to live in a dangerous, confusing, and rapidly changing world and have hope for the future. The people have parties; they sing and laugh and cry. They are amazingly adaptable and resilient. In many ways, their lives are no worse than those in the early industrial cities of Europe and America a century ago. Perhaps continuing development will bring better conditions to cities of the developing world as it has for many in the developed world.

The Developed World

For the most part, the rapid growth of central cities that accompanied industrialization in nineteenth- and early twentieth-century Europe and North America has now slowed or even reversed. London, for instance, once the most populous city in the world, has lost nearly 2 million people, dropping from its high of 8.6 million in 1939 to about 6.7 million now. While the greater metropolitan area surrounding London has been expanding to about 10 million inhabitants, the city itself is now only the twelfth largest city in the world.

Many of the worst urban environmental problems of the more developed countries have been substantially reduced in recent years. Improved sanitation and medical care have reduced or totally eliminated many of the communicable diseases that once afflicted urban residents. Air and water quality have improved dramatically as heavy industry such as steel smelting and chemical manufacturing have moved to developing countries. In consumer and information economies, workers no longer need to be concentrated in central cities. They can live and work in dispersed sites. Automobiles now make it possible for much of the working class to enjoy amenities such as single-family homes, yards, and access to recreation that once were available only to the elite.

In the United States, old, dense manufacturing cities such as Philadelphia and Detroit have lost population as industry has moved to developing countries. In a major demographic shift, both businesses and workers have moved west and south. Some of the most rapidly growing metropolitan areas like Phoenix, Arizona; Boulder, Colorado; Austin, Texas; and San Jose, California, are centers for high-technology companies located in landscaped suburban office parks. These cities often lack a recognizable downtown, being organized instead around low-density housing developments, national-chain shopping malls, and extensive freeway networks. For many high-tech companies, being located near industrial centers and shipping is less important than a good climate, ready access to air travel, and amenities such as natural beauty and open space.

Urban Sprawl

While the move to suburbs and rural areas has brought many benefits to the average citizen, it also has caused numerous urban problems. Cities that once were compact now spread over the landscape, consuming open space and wasting resources. This pattern of urban growth is known as **sprawl.** While there is no universally accepted definition of the term, sprawl generally includes the characteristics outlined in table 22.3. As former Maryland Governor Parris N. Glendening said, "In its path, sprawl consumes thousands of acres of forests and farmland, woodlands and wetlands. It requires government to spend millions extra to build new schools, streets, and water and sewer lines." And Christine Todd Whitman, former New Jersey governor and head of the Environmental Protection Agency, said, "Sprawl eats up our open space. It creates traffic jams that boggle the mind and pollute the air. Sprawl can make one feel downright claustrophobic about our future."

In most American metropolitan areas, the bulk of new housing is in large, tract developments that leapfrog out beyond the edge of the city in a search for inexpensive rural land with few restrictions on land use or building practices (fig. 22.13). The U.S. Department of Housing and Urban Development estimates that urban sprawl consumes some 200,000 ha (roughly 500,000 acres) of farmland each year. Because cities often are located in fertile river valleys or shorelines, much of that land would be especially valuable for producing crops for local consumption. But with planning authority divided among many small, local jurisdictions, metropolitan areas have no way to regulate growth or provide for rational, efficient resource use. Small towns and township or county officials generally welcome this growth because it profits local landowners and business people. Although the initial price of tract homes often is less than comparable urban property, there are external costs in the form of new roads, sewers, water mains, power lines, schools, and shopping centers and other extra infrastructure required by this low-density development.

TABLE 22.3	Characteristics of Urban Sprawl

1. Unlimited outward extension.
2. Low-density residential and commercial development.
3. Leapfrog development that consumes farmland and natural areas.
4. Fragmentation of power among many small units of government.
5. Dominance of freeways and private automobiles.
6. No centralized planning or control of land uses.
7. Widespread strip malls and "big-box" shopping centers.
8. Great fiscal disparities among localities.
9. Reliance on deteriorating older neighborhoods for low-income housing.
10. Decaying city centers as new development occurs in previously rural areas.

Source: *Excerpt from speech by Anthony Downs at the CTS Transportation Research Conference, as appeared on Website by Planners Web, Burlington, VT, 2001.*

FIGURE 22.13 Sprawl consumes natural areas, farmland, and open space. It wastes resources, results in traffic congestion, and requires large investments to replace schools, roads, parks, and shopping that is abandoned in the inner city. © William P. Cunningham.

Landowners, builders, real estate agents, and others who profit from this crazy-quilt development pattern generally claim that growth benefits the suburbs in which it occurs. They promise that adding additional residents will lower the average taxes for everyone, but in fact, the opposite often is true. In a study titled *Better Not Bigger,* author Eben Fodor analyzed the costs of medium-density and low-density housing. In suburban Washington, D.C., for instance, each new house on a quarter acre (0.1 ha) lot cost $700 more than it paid in taxes. A typical new house on a 5 acre (2 ha) lot, however, cost $2,200 more than it paid in taxes because of higher expenses for infrastructure and services. Ironically, people who move out to rural areas to escape from urban problems such as congestion, crime, and pollution often find that they have simply brought those problems with them. A neighborhood that seemed tranquil and remote when they first moved in, soon becomes just as crowded, noisy, and difficult as the city they left behind as more people join them in their rural retreat.

In a study of 213 American urban areas, author and former mayor of Albuquerque, David Rusk, found that between 1960 and 1990 total population grew 47 percent while land use increased by 107 percent. This means that urban density decreased by 28 percent. Atlanta, Georgia, exemplifies urban sprawl. Between 1990 and 2000, the Atlanta population grew 32 percent while the total metropolitan area increased by 300 percent. In 1990, the metro area was about 100 km (62 miles) across and encompassed about 7,700 km^2 (3,000 mi^2). By 2000, the metro area had expanded to about 175 km (110 mi) across and covered about 25,000 km^2 (9,500 mi^2). By far the fastest growing metropolitan region in the United States is Las Vegas, Nevada, which doubled its population but quadrupled its size in the 1990s (fig. 22.14 *a* and *b*).

Because many Americans live far from where they work, shop, and recreate, they consider it essential to own a private automobile. The average U.S. driver spends about 443 hours per year behind a steering wheel. This means that for most people, the equivalent of one full 8-hour day per week is spent sitting in an automobile. Of the 5.8 billion barrels of oil consumed each year in the United States (60 percent of which is imported), about two-thirds is burned in cars and trucks. As chapter 16 shows, about two-thirds of all carbon monoxide, one-third of all nitrogen oxides, and one-quarter of all volatile organic compounds emitted each year from human-caused sources in the United States are released by automobiles, trucks, and buses.

Building the roads, parking lots, filling stations, and other facilities needed for an automobile-centered society takes a vast amount of space and resources. In some metropolitan areas it is estimated that one-third of all land is devoted to the automobile. To

(a)

(b)

FIGURE 22.14 Satellite images of Las Vegas, Nevada, in 1972 (*a*) and 1992 (*b*) show how the metropolitan area has grown over two decades. *Source:* U.S. Geological Survey.

FIGURE 22.15 Building new freeways to reduce congestion is like trying to solve a weight problem by loosening your belt.
© William P. Cunningham.

make it easier for suburban residents to get from their homes to jobs and shopping, we provide an amazing network of freeways and highways. At a cost of several trillion dollars to build, the interstate highway system was designed to allow us to drive at high speeds from source to destination without ever having to stop. As more and more drivers clog the highways, however, the reality is far different. In Los Angeles, for example, which has the worst congestion in the United States, the average speed in 1982 was 58 mph (93 km/hr), and the average driver spent less than 4 hours per year in traffic jams. In 2000, the average speed in Los Angeles was only 35.6 mph (57.3 km/hr), and the average driver spent 82 hours per year waiting for traffic. Although new automobiles are much more efficient and cleaner operating than those of a few decades ago, the fact that we drive so much farther today and spend so much more time idling in stalled traffic means that we burn more fuel and produce more pollution than ever before.

Altogether, it is estimated that traffic congestion costs the United States $78 billion per year in wasted time and fuel. Some people argue that the existence of traffic jams in cities shows that more freeways are needed. Often, however, building more traffic lanes simply encourages more people to drive farther than before. Rather than ease congestion and save fuel, more freeways can exacerbate the problem (fig. 22.15).

Sprawl also creates problems in central cities from which residents and businesses have fled. With a reduced tax base and fewer civic leaders living or working in downtown areas, the city is unable to maintain its infrastructure. Streets, parks, schools, and civic buildings fall into disrepair at the same time that these facilities are being built at great expense in new suburbs. The poor who are left behind when the upper and middle classes abandon the city center often can't find jobs where they live and have no way to commute to the suburbs where jobs are now located. The low-density development of suburbs is racially and economically exclusionary because it provides no affordable housing and makes it impractical to design a viable public transit system.

Sprawl also is bad for your health. By encouraging driving and discouraging walking, sprawl promotes a sedentary lifestyle that contributes to heart attacks and diabetes, among other problems. In Atlanta, for example, the lowest-density neighborhoods (suburbs) tend to have significantly higher rates of overweight residents than the highest-density neighborhoods.

Finally, sprawl fosters uniformity and alienation from local history and natural environment. Housing developments often are based on only a few standard housing styles, while shopping centers and strip malls everywhere feature the same national chains. You could drive off the freeway in the outskirts of almost any big city in America and see exactly the same brands of fast-food restaurants, motels, stores, filling stations, and big-box shopping centers.

Smart Growth

Are there alternatives to unplanned sprawl and wasteful resource use? One option proposed by many urban planners is **smart growth** that makes efficient and effective use of land resources and existing infrastructure by encouraging in-fill development that avoids costly duplication of services and inefficient land use. Smart growth aims to provide a mix of land uses to create a variety of affordable housing choices and opportunities. It also attempts to provide a variety of transportation choices including pedestrian friendly neighborhoods. This approach to planning also seeks to maintain a unique sense of place by respecting local cultural and natural features.

By making land-use planning open and democratic, smart growth makes urban expansion fair, predictable, and cost-effective. All stakeholders are encouraged to participate in creating a vision for the city and to collaborate rather than confront each other. Goals are established for staged and managed growth in urban transition areas with compact development patterns. This approach is not opposed to growth. It recognizes that the goal is not to block growth but to channel it to areas where it can be sustained over the long term. It strives to enhance access to equitable public and private resources for everyone and to promote the safety, livability, and revitalization of existing urban and rural communities.

Smart growth protects environmental quality. It attempts to reduce traffic and to conserve farmlands, wetlands, and open space. This may mean restricting land use, but it also means finding economically sound ways to reuse polluted industrial "brownfields" within the city (What Do You Think? p. 489). As cities grow and transportation and communications enable communities to interact more, the need for regional planning becomes both more possible and more pressing. Community and business leaders need to make decisions based on a clear understanding of regional growth needs and how infrastructure can be built most efficiently and for the greatest good.

One of the best examples of successful urban land-use planning in the United States is Portland, Oregon, which has rigorously enforced a boundary on its outward expansion, requiring, instead, that development be focused on in-filling unused space within the city limits. Because of its many urban amenities, Portland is considered one of the best cities in America. Between 1970 and 1990, the Portland population grew by 50 percent but its total land area grew only 2 percent. During this time, Portland property taxes decreased 29 percent and vehicle miles traveled increased only 2 percent. By contrast, Atlanta, which had similar population growth, experienced an explosion of urban sprawl that increased its land area three-fold, drove up property taxes 22 percent, and increased traffic miles by 17 percent. A result of this expanding traffic and increasing congestion was that Atlanta's air pollution increased by 5 percent, while Portland's, which has one of the best public transit systems in the nation, decreased by 86 percent. Portland shares many of the same goals as Chattanooga, described at the beginning of this chapter and in table 22.4. In some cities, like Boulder, Colorado, rigid limits on city boundaries have driven up land prices and limited the amount of affordable housing. Additional steps need to be taken to ensure a diverse community.

Garden Cities and New Towns

The twentieth century has seen numerous experiments in building **new towns** for society at large that try to combine the best features of the rural village and the modern city. One of the most influential of all urban planners was Ebenezer Howard (1850–1929), who not only wrote about ideal urban environments but also built real

TABLE 22.4	Goals for Smart Growth

1. Create a positive self-image for the community.
2. Make the downtown vital and livable.
3. Alleviate substandard housing.
4. Solve problems with air, water, toxic waste, and noise pollution.
5. Improve communication between groups.
6. Improve community member access to the arts.

Source: *Vision 2000, Chattanooga, TN.*

cities to test his theories. In his *Garden Cities of Tomorrow,* written in 1898, Howard proposed that the congestion of London could be relieved by moving whole neighborhoods to **garden cities** separated from the central city by a greenbelt of forests and fields.

In the early 1900s, Howard worked with architect Raymond Unwin to build Letchworth and Welwyn Garden just outside of London. Interurban rail transportation provided access to these cities. Houses were clustered in "superblocks" surrounded by parks, gardens, and sports grounds. Streets were curved. Safe and convenient walking paths and overpasses protected pedestrians from traffic. Businesses and industries were screened from housing areas by vegetation. Each city was limited to about 30,000 people to facilitate social interaction. Housing and jobs were designed to create a mix of different kinds of people and to integrate work, social activities, and civic life. Trees and natural amenities were carefully preserved and the towns were laid out to maximize social interactions and healthful living. Care was taken to meet residents' psychological needs for security, identity, and stimulation.

Letchworth and Welwyn Garden each have 70 to 100 people per acre. This is a true urban density, about the same as New York City in the early 1800s and five times as many people as most suburbs today. By planning the ultimate size in advance and choosing the optimum locations for housing, shopping centers, industry, transportation, and recreation, Howard believed he could create a hospitable and satisfying urban setting while protecting open space and the natural environment. He intended to create parklike surroundings that would preserve small-town values and encourage community spirit in neighborhoods.

Letchworth and Welwyn Garden were the first of 32 new towns established in Great Britain, which now house about 1 million people. The Scandinavian countries have been especially successful in building garden cities. The former Soviet Union also built about 2,000 new towns. Some are satellites of existing cities, and others are entirely new communities far removed from existing urban areas.

Planned communities also have been built in the United States following the theories of Ebenezer Howard, but most plans have been based on personal automobiles rather than public transit. In the 1920s, Lewis Mumford, Clarence Stein, and Henry Wright drew up plans that led to the establishment of Radburn, New Jersey, and Chatham Village (near Pittsburgh). Reston, Virginia, and

What do you think?

People for Community Recovery

The Lake Calumet Industrial District on Chicago's far South Side is an environmental disaster area. A heavily industrialized center of steel mills, oil refineries, railroad yards, coke ovens, factories, and waste disposal facilities, much of the site is now a marshy wasteland of landfills, toxic waste lagoons, and slag dumps, around a system of artificial ship channels.

At the southwest corner of this degraded district sits Altgeld Gardens, a low-income public housing project built in the late 1940s by the Chicago Housing Authority. The 2,000 units of "The Gardens" or "The Projects," as they are called by the largely minority residents, are low-rise rowhouses, many of which are vacant or in poor repair. But residents of Altgeld Gardens are doing something about their neighborhood. People for Community Recovery (PCR) is a grassroots citizen's group organized to work for a clean environment, better schools, decent housing, and job opportunities for the Lake Calumet neighborhood.

PCR was founded in 1982 by Mrs. Hazel Johnson, an Altgeld Gardens resident whose husband died from cancer that may have been pollution-related. PCR has worked to clean up more than two dozen waste sites and contaminated properties in their immediate vicinity. Often this means challenging authorities to follow established rules and enforce existing statutes. Public protests, leafleting, and community meetings have been effective in public education about the dangers of toxic wastes and have helped gain public support for cleanup projects.

PCR's efforts successfully blocked construction of new garbage and hazardous waste landfills, transfer stations, and incinerators in the Lake Calumet district. Pollution prevention programs have been established at plants still in operation. And PCR helped set up a community monitoring program to stop illegal dumping and to review toxic inventory data from local companies.

Education is an important priority for PCR. An environmental education center

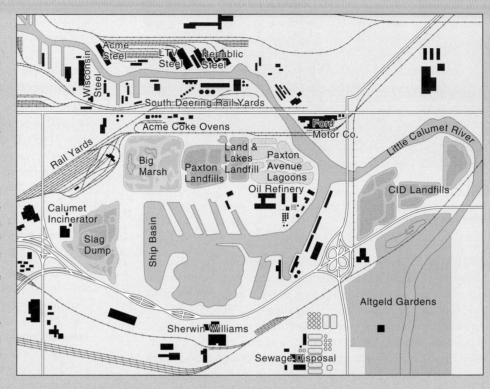

The Calumet industrial district in South Chicago.

administered by community members organizes workshops, seminars, fact sheets, and outreach for citizens and local businesses. A public health education and screening program has been set up to improve community health. Partnerships have been established with nearby Chicago State University to provide technical assistance and training in environmental issues.

PCR also works on economic development. Environmentally responsible products and services are now available to residents. Jobs that are being created as "green" businesses are brought into the community. Wherever possible, local people and minority contractors from the area are hired to clean up waste sites and restore abandoned buildings. Job training for youth and adults as well as retraining for displaced workers is a high priority. Funding for these projects

has come from fines levied on companies for illegal dumping.

PCR and Mrs. Johnson have received many awards for their fight against environmental racism and despair. In 1992, PCR was the recipient of the President's Environmental and Conservation Challenge Award. PCR is the only African-American grassroots organization in the country to receive this prestigious award.

Although Altgeld Gardens is far from clean, much progress has been made. Perhaps the most important accomplishment is community education and empowerment. Residents have learned how and why they need to work together to improve their living conditions. Could these same lessons be useful in your city or community? What could you do to help improve urban environments where you live?

FIGURE 22.16 By clustering buildings together, Kentlands, Maryland, has maintained a high percentage of open space as public commons. © William P. Cunningham.

Columbia, Maryland, both were founded in the early 1960s and are widely regarded as the most successful attempts to build new towns of their era. Another movement to build new towns according to Howard's principles has sprung up in the 1990s. Towns such as Seaside in northern Florida and Kentlands outside Washington, D.C., cluster houses to save open space and create a sense of community (fig. 22.16). Commercial centers are located within a few minutes walk of most houses, and streets are designed to encourage pedestrians and to provide places to gather and visit.

New Urbanist Movement

Rather than abandon the cultural history and infrastructure investment in existing cities, a group of architects and urban planners is attempting to redesign metropolitan areas to make them more appealing, efficient, and livable. European cities such as Stockholm, Sweden; Helsinki, Finland; Leichester, England; and Neerlands, the Netherlands, have a long history of innovative urban planning. In the United States, Andres Duany, Elizabeth Plater-Zyberk, Peter Calthorpe, and Sym Van Der Ryn have been leaders in this movement. Sometimes called a neo-traditionalist approach, these designers attempt to recapture some of the best features of small towns and the best cites of the past. They are designing urban neighborhoods that integrate houses, offices, shops, and civic buildings. Ideally, no house should be more than a five-minute walk from a neighborhood center with a convenience store, a coffee shop, a bus stop, and other amenities. A mix of apartments, townhouses, and detached houses in a variety of price ranges insures that neighborhoods will include a diversity of ages and income levels. Some design principles of this movement include:

- Limit city size or organize them in modules of 30,000 to 50,000 people, large enough to be a complete city but small enough to be a community. A greenbelt of agricultural and recreational land around the city limits growth while promoting efficient land use. By careful planning and cooperation with neighboring regions, a city of 50,000 people can have real urban amenities such as museums, performing arts centers, schools, hospitals, etc.

- Determine in advance where development will take place. This protects property values and prevents chaotic development in which the lowest uses drive out the better ones. It also recognizes historical and cultural values, agricultural resources, and such ecological factors as impact on wetlands, soil types, groundwater replenishment and protection, and preservation of aesthetically and ecologically valuable sites.

- Turn streets and shopping malls into areas that invite people to stroll, meet friends, or listen to a debate or a street musician (fig. 22.17). If there aren't 100 places for an impromptu celebration, a place isn't a real city. Another test of a city is a vital nightlife. Design city spaces with sidewalk cafes, pocket parks, courtyards, balconies, and porticoes that shelter pedestrians, bring people together, and add life and security to the street. Restaurants, theaters, shopping areas, and public entertainment that draw people to the streets generate a sense of spontaneity, excitement, energy, and fun.

- Locate everyday shopping and services so people can meet daily needs with greater convenience, less stress, less automobile dependency, and less use of time and energy. This might be accomplished by encouraging small-scale commercial development in or close to residential areas. Perhaps we should once again have "mom and pop" stores on street corners or in homes.

- Increase jobs in the community by locating offices, light industry, and commercial centers in or near suburbs, or by enabling work at home via computer terminals. These alternatives save commuting time and energy and provide local jobs. There are

FIGURE 22.17 Many cities have redesigned core shopping areas to be more "user friendly." Walking streets, such as this one in Amsterdam, create space for entertainment, dining, and chance encounters and provide opportunities for building a sense of community. © Peter Halden.

FIGURE 22.18 Bicycles provide low-cost, nonpolluting, energy-efficient urban transportation. Guangzhou (Canton), China, a city of 4 million people, is said to have 3 million bicycles. © William P. Cunningham.

also concerns, however, about work-at-home employees being exploited in low-paying "sweatshop" conditions by unscrupulous employers. Some safeguards may be needed.

- Encourage walking or the use of small, low-speed, energy-efficient vehicles (microcars, motorized tricycles, bicycles, etc.) for many local trips now performed by full-size automobiles. Creating special traffic lanes, reducing the number or size of parking spaces, or closing shopping streets to big cars might encourage such alternatives (fig. 22.18).

- Promote more diverse, flexible housing as alternatives to conventional, detached single-family houses. "In-fill" building between existing houses saves energy, reduces land costs, and might help provide a variety of living arrangements. Allowing owners to turn unused rooms into rental units provides space for those who can't afford a house and brings income to retired people who don't need a whole house themselves. Allowing single-parent families or groups of unrelated adults to share housing and to use facilities cooperatively also provides alternatives to those not living in a traditional nuclear family. One of the great "discoveries" of urban planning is that mixing various types of housing—individual homes, townhouses, and high-rise apartments—can be attractive if buildings are aesthetically arranged in relation to one another.

- Create housing "superblocks" that use space more efficiently and foster a sense of security and community. Widen peripheral arterial streets and provide pedestrian overpasses so traffic flows smoothly around residential areas; then reduce interior streets within blocks to narrow access lanes with speed bumps and barriers to through traffic so children can play more safely. The land released from streets can be used for gardens, linear parks, playgrounds, and other public areas that will foster community spirit and encourage people to get out and walk. Cars can be parked in remote lots or parking ramps, especially where people have access to public transit and can walk to work or shopping.

FIGURE 22.19 Communal gardens in Seattle, Washington, use once-vacant land to grow food and give local residents a chance to experience nature in a highly urban setting. © Kevin R. Morris/Stone/Getty.

- Make cities more self-sustainable by growing food locally, recycling wastes and water, using renewable energy sources, reducing noise and pollution, and creating a cleaner, safer environment (fig. 22.19). Reclaimed inner-city space or a greenbelt of agricultural and forestland around the city provides food and open space as well as such valuable ecological services as purifying air, supplying clean water, and protecting wildlife habitat and recreation land.

- Rooftop gardens and natural plantings can absorb up to 70 percent of rain water. They provide habitat for birds and insects, and ameliorate climate. They save energy and provide contact with nature for building residents. Many German cities now require that at least half of all new development must be covered with vegetation. The least expensive way to do this is with green roofs on buildings and parking structures.

- Invite public participation in decision making as was done in Chattanooga. Emphasize local history, culture, and environment to create a sense of community and identity. Create local networks in which residents take responsibility for crime prevention, fire protection, and home care of children, the elderly, sick, and disabled. Coordinate regional planning through metropolitan boards that cooperate with but do not supplant local governments.

Ecologists and conservation biologists have, in the past, tended to study pristine nature, untouched—as much as possible—by human influences. In recent years, however, we have come to recognize that cities are ecological systems too. If ecology is the relationship among organisms and their environment, what could be more ecological than studying how humans impact other species and our environment? And for a vast majority of us, that means studying urban ecosystems where we are the keystone species. Rather than merely seeing humans as disturbing factors in these ecosystems, some scientists are welcoming people as participants in research and project-based service learning.

Two of the most important examples of urban ecology research are Long-Term Ecological Research (LTER) sites in Phoenix and Baltimore. Funded by the National Science Foundation for an initial period of six years at nearly a million dollars per year, but with an understanding that the research will go on much beyond that, these projects use modern technology to study every aspect of urban ecology. How does the city shape its own weather? What plants and animals live in the city, and where, and how? What ecological processes cycle materials and energy? How do human activities influence those processes? One of the first things that scientists are discovering is that cities are not homogenous; like other ecosystems, they have patches at varying scales that change over time. Using computer models and sophisticated mapping techniques, ecologists can explore how patches shift around in space and time.

Environmental justice concerns are at the forefront of many urban ecology research projects. Where are toxic and hazardous materials generated, stored, and released in the city? How do they move around and where do they accumulate? In Detroit, for example, a group of students worked with experts to map data on more than 5,000 children with elevated blood lead levels. Not surprisingly, they found a correlation between low income, old housing, incidence of poisoning, and concentration of special-education students. Sometimes public awareness of a problem is one of the best outcomes of this research. Another large group of students in the Detroit area used geographic information systems together with chemical and biological analysis to prepare a detailed study of water quality and ecosystem health in the Rouge River. Participatory planning coupled with citizen science enable some cities to identify indicators of urban sustainability (table 22.5).

Designing for Open Space

Traditional suburban development typically divides land into a checkerboard layout of nearly identical 1 to 5 ha parcels with no designated open space (fig. 22.20, *top*). The result is a sterile landscape consisting entirely of house lots and streets. This style of development, which is permitted—or even required—by local zoning and ordinances, consumes agricultural land and fragments wildlife habitat. Many of the characteristics that people move to the

TABLE 22.5	Urban Sustainability Indicators

Children in poverty
Violent crime
Access to health care
Air and water quality, litter
Vacant or deteriorating housing
Participation in neighborhood organizations
Money earned and spent in neighborhood
Access to public transportation
Shopping and services within walking distance
Quality of schools
Cultural and recreational opportunities

Source: *MN Citizen's Environment Action Committee, 1999.*

country to find—space, opportunities for outdoor recreation, access to wild nature, a rural ambience—are destroyed by dividing every acre into lots that are "too large to mow but too small to plow."

An interesting alternative known as **conservation development,** cluster housing, or open space zoning preserves at least half of a subdivision as natural areas, farmland, or other forms of open space. Among the leaders in this design movement are landscape architects Ian McHarg, Frederick Steiner, and Randall Arendt. They have shown that people who move to the country don't necessarily want to own a vast acreage or to live miles from the nearest neighbor; what they most desire is long views across an interesting landscape, an opportunity to see wildlife, and access to walking paths through woods or across wildflower meadows.

By carefully clustering houses on smaller lots, a conservation subdivision can provide the same number of buildable lots as a conventional subdivision and still preserve 50 to 70 percent of the land as open space (fig. 22.20, *bottom*). This not only reduces development costs (less distance to build roads, lay telephone lines, sewers, power cables, etc.) but also helps foster a greater sense of community among new residents. Walking paths and recreation areas get people out of their houses to meet their neighbors. Home owners have smaller lots to care for and yet everyone has an attractive vista and a feeling of spaciousness.

Some good examples of this approach are Farmview near Yardley, Pennsylvania, and Hawksnest in Delafield Township, Wisconsin. In Farmview, 332 homes are clustered in six small villages set in a 160 ha (414 acre) rural landscape, more than half of which is dedicated as permanent farmland. House lots and villages were strategically placed to maximize views, helping the development to lead its county in sales for upscale developments. Hawksnest is situated in dairy-farming country outside of Waukesha, Wisconsin. Seventy homes are situated amid 70 ha (180 acres) of meadows, ponds, and woodlands. Restored prairies, neighborhood recreational facilities, and connections to a national scenic trail have proved to be valuable marketing assets for this subdivision.

Urban habitat can make a significant contribution toward saving biodiversity. In a ground-breaking series of habitat conservation plans triggered by the need to protect the endangered

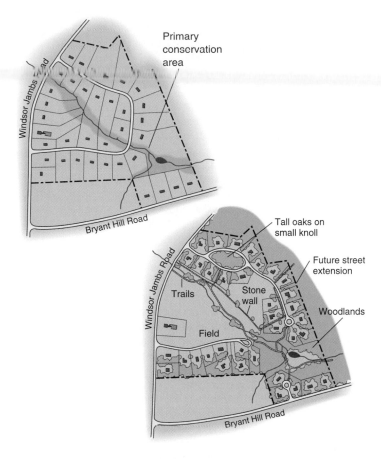

FIGURE 22.20 Conventional subdivision (*top*) and an open space plan (*bottom*). Although both plans provide 36 home sites, the conventional development allows for no open space. Cluster housing on smaller lots in the open space design preserves at least half the area as woods, prairie, wetlands, farms, or other conservation lands, while providing residents with more attractive vistas and recreational opportunities than a checkerboard development.

California gnatcatcher, some 85,000 ha (210,000 acres) of coastal scrub near San Diego was protected as open space within the rapidly expanding urban area. This is an area larger than Yosemite Valley, and will benefit many other species as well as humans.

SUSTAINABLE DEVELOPMENT IN POORER COUNTRIES

What can be done to improve conditions in developing world cities? Curitiba, Brazil, is an outstanding example of what can be done, even in relatively undeveloped countries, to improve transportation, protect central cities, and create a sense of civic pride (Case Study, p. 494). Other cities have far to go, however, before they reach this standard. Among the immediate needs are housing, clean water, sanitation, food, education, health care, and basic transportation for their residents. The World Bank estimates that

FIGURE 22.21 In this *colonia* on the outskirts of Mexico City, residents work with the government to bring in electricity, water, and sewers to shantytowns and squatter settlements. Like many developing world countries, Mexico recognizes that helping people help themselves is the best way to improve urban living. © William P. Cunningham.

interventions to improve living conditions in urban households in the developing world could average the annual loss of almost 80 million "disability-free" years of life. This is about twice the feasible benefit estimated from all other environmental programs studied by the World Bank.

Some countries, recognizing the need to use vacant urban land, are redistributing unproductive land or closing their eyes to illegal land invasions. Indonesia, Peru, Tanzania, Zambia, and Pakistan have learned that squatter settlements make a valuable contribution to meeting national housing needs. Squatters' rights are being upheld in some cases, and such services as water, sewers, schools, and electricity are being provided to the settlements (fig. 22.21). Some countries intervene directly in land distribution and land prices. Tunisia, for instance, has a "rolling land bank" to buy and sell land. This strong and effective program controls urban land prices and reduces speculation and unproductive land ownership.

Many planners argue that social justice and sustainable economic development are answers to the urban problems we have discussed in this chapter. If people have the opportunity and money to buy better housing, adequate food, clean water, sanitation, and other things they need for a decent life, they will do so. Democracy, security, and improved economic conditions help in slowing population growth and reducing rural-to-city movement. An even more important measure of progress may be institution of a social welfare safety net guaranteeing that old or sick people will not be abandoned and alone.

Some countries have accomplished these goals even without industrialization and high incomes. Sri Lanka, for instance, has lessened the disparity between the core and periphery of the country. Giving all people equal access to food, shelter, education, and health care eliminates many incentives for interregional migration.

Curitiba: An Environmental Showcase

Curitiba, a Brazilian city of about 2 million people located on the Atlantic coast about 650 km (400 mi) southwest of Rio de Janeiro, has acquired a worldwide reputation for its innovative urban planning and environmental protection policies. Tree-lined streets, clean air, smoothly flowing traffic, and absence of litter and garbage make this one of Latin America's most liveable cities.

The architect of this remarkable program is mayor Jaime Lerner, who began in 1962 as a student leader protesting the proposed destruction of Curitiba's historic downtown center. Elected mayor nine years later, he has worked for more than two decades to preserve the city and make it a more beautiful and habitable place. Now Governor of the State of Parana, Lerner's success has thrust the city into the international spotlight as a shining example of what conservation and good citizenship can do to improve urban environments, even in developing world cities.

The heart of Curitiba's environmental plan is education for both children and adults. Signs posted along roadways proclaim "50 kg of paper equals one tree" and "recycle; it pays." School children study ecology along with Portuguese and math. With the assistance of children who encourage their parents, the city has successfully instituted a complex recycling plan that requires careful separation of different kinds of materials. The city calculates that 1,200 trees per day are saved by paper recycled in this program. "Imagine if the whole of Brazil did this with an urban population 60 times greater than Curitiba," Lerner exclaims. "We could save 26 million trees per year!" More than 70 percent of the residents now recycle. Food and bus coupons are given out in exchange for recyclables, providing an alternative to welfare while also protecting the environment.

Another area in which Curitiba is setting an example is transportation. Faced with a population that tripled in two decades, bringing increasing levels of traffic congestion and air pollution, Curitiba had a transportation dilemma. The choices were either to bulldoze freeways through the historic heart of the city or to institute mass transit. The city chose mass transit.

Special articulated buses each carry up to 300 passengers on dedicated transit lanes. You pay your fare before entering the platform waiting area. When the bus pulls in, multiple doors open flush with the platform so that many people can enter or exit at once (even in wheelchairs) and each stop takes only seconds. A feeder network of smaller buses and vans collects riders from neighborhoods, so that everyone has convenient, frequent, and inexpensive transit service. Now more than three-quarters of the city's population leave their automobiles at home every day and take special express buses to work. The system is so successful that ridership has increased from 25,000 passengers a day 20 years ago to more than 1.5 million per day now, or 70 percent of all trips in the city. The result is not only less congestion and pollution but major energy savings.

Other measures to clear the air and reduce congestion include a limit on building height and construction of an industrial park outside city boundaries. Maximum use is made of all materials and buildings. Worn-out buses become city training centers, an old military fort is a cultural center, and a gunpowder depot is now a theater. Water and energy conservation are practiced widely. Even litter—an ubiquitous component of most Brazilian cities—is absent in Curitiba. More than one million trees have been planted and the green area ratio of 52 square meters of park per capita is higher than most American cities.

People are so imbued with city pride that they keep their surroundings spotless. To further beautify their city, civic volunteers have planted 1.5 million trees—more than any other place in Brazil.

Although many residents initially were skeptical of this environmental plan, now a remarkable 99 percent of the city's inhabitants would not want to live anywhere else. The World Bank uses Curitiba as an example of what can be done through civic leadership and public participation to clean up the urban environment. Some people claim that Curitiba, with its cool climate and high percentage of European immigrants, may be a special case among Brazilian cities. Lerner claims that Curitiba has no special features except concern, creativity, and communal efforts to care for its environment. Could you start a similar program in your hometown?

Both population growth and city growth have been stabilized, even though the per capita income is only $800 per year. Could we help other countries do something similar?

Whether sustained, environmentally sound economic development is possible for a majority of the world's population remains one of the most important and most difficult questions in environmental science. The unequal relationship between the richer "Northern" countries and their impoverished "Southern" neighbors is a major part of this dilemma. Some people argue that the best hope for developing countries may be to "delink" themselves from the established international economic systems and develop direct south-south trade based on local self-sufficiency, regional cooperation, barter, and other forms of nontraditional exchange that are not biased in favor of the richer countries.

Summary

- A rural area is one in which a majority of residents are supported by methods of harvesting natural resources. An urban area is one in which a majority of residents are supported by manufacturing, commerce, or services. A village is a rural community. A city is an urban community with sufficient size and complexity to support economic specialization and to require a higher level of organization and opportunity than is found in a village.

- Urbanization in the United States over the past 200 years has caused a dramatic demographic change. A similar shift is now occurring in most parts of the world. Only Africa and South Asia remain predominantly rural, but cities are growing rapidly there as well. In 2000, for the first time in history, more than half the world's people lived in urban areas. Most future urban growth in the next century will be in the supercities of the developing world. A century ago only 13 cities had populations above 1 million; now there are 241 such cities. By 2050 that number will probably double again, and three-fourths of those cities will be in the developing world.

- Cities grow by natural increase (births) and migration. People move into the city because they are "pushed" out of rural areas or because they are "pulled" in by the advantages and opportunities of the city.

- Huge, rapidly growing cities in the developing world often have appalling environmental conditions. Among the worst problems faced in these cities are traffic congestion, air pollution, inadequate or nonexistent sewers and waste disposal systems, water pollution, and housing shortages. Millions of people live in slums and shantytowns where conditions are frightful, yet these people raise families, educate their children, learn new jobs and new ways of living, and have hope for the future.

- The problems of developed world cities tend to be associated with urban sprawl around the outskirts and decay and blight in the core. Unlimited expansion into rural areas, leapfrog development, and lack of coordinated land-use planning lead to loss of farmlands and open space, traffic congestion, air and water pollution, and numbingly uniform housing tracts and shopping centers. Sprawl also requires local government to spend millions of dollars to replace roads, sewers, water lines, schools, parks, power grids, and other infrastructure being abandoned in the inner city.

- Still, there are ways that we can improve cities in both the developed and the developing world to make them healthier, safer, and more environmentally sound, socially just, and culturally fulfilling than they are now. Smart growth, garden cities, new traditionalist urban movements, and conservation development are among the ideas advanced for improving our cities. Curitiba, Brazil, is an encouraging example of how these principles can be applied in the developing world.

Questions for Review

1. What is the difference between a city and a village and between rural and urban?

2. How many people now live in cities, and how many live in rural areas worldwide?

3. What changes in urbanization are predicted to occur in the next 30 years, and where will that change occur?

4. Identify the 13 largest cities in the world. Has the list changed in the past century? Why?

5. When did the United States pass the point at which more people live in the city than the country? When will the rest of the world reach this point?

6. Describe the current conditions in a typical megacity of the developing world. What forces contribute to its growth?

7. Describe the difference between slums and shantytowns.

8. Why are urban areas in U.S. cities decaying?

9. How has transportation affected the development of cities? What have been the benefits and disadvantages of freeways?

10. Describe some ways that American cities and suburbs could be redesigned to be more ecologically sound, socially just, and culturally amenable.

Questions for Critical Thinking

1. Picture yourself living in a rural village or a developing world city. What aspects of life there would you enjoy? What would be the most difficult for you to accept?

2. Why would people move to one of the megacities of the developing world if conditions are so difficult there?

3. A city could be considered an ecosystem. Using what you learned in chapters 3 and 4, describe the structure and function of a city in ecological terms.

4. Look at the major urban area(s) in your state. Why were they built where they are? Are those features now a benefit or drawback?

5. Who benefits from urban sprawl and who suffers? Why is this process so powerful and persistent?

6. Weigh the costs and benefits of automobiles in modern American life. Is there a way to have the freedom and convenience of a private automobile without its negative aspects?

7. Boulder, Colorado, has been a leader in controlling urban growth. One consequence is that housing costs have skyrocketed and poor people have been driven out. If you lived in Boulder, would you vote for additional population limits? What do you think is an optimum city size?

8. A number of proposals are presented in this chapter for urban redesign. Which of them would be appropriate or useful for your community? Try drawing up the ideal plan for your neighborhood.

Key Terms

city 479
conservation development 492
core regions 479
garden cities 488
megacity 479
new towns 488
pull factors 482
push factors 482
rural area 479

shantytowns 484
slums 484
smart growth 487
sprawl 485
squatter towns 484
urban area 479
urbanization 478
village 479

Further Readings

Arendt, R. 1999. *Growing Greener: Putting Conservation into Local Plans and Ordinances.* Island Press.

Beatly, T. 2000. *Green Urbanism: Learning from European Cities.* Island Press.

Bullard, Robert D., et al. 2000. *Sprawl City: Race, Politics, and Planning in Atlanta.* Island Press.

Calthorpe, Peter, and William Fulton. 2000. *The Regional City: Planning for the End of Sprawl.* Island Press.

Chen, Donald T. 2000. The science of smart growth. *Scientific American* 283(6):84–91.

Freeman, C., and O. Buck. 2003. Development of an ecological mapping methodology for urban areas in New Zealand. *Landscape and Urban Planning* 63(3):161–73.

Hall, Kenneth B., and Gerald A. Porterfield. 2001. *Community by Design: New Urbanism for Suburbs and Small Communities.* McGraw-Hill Co.

Kuhn, M. 2003. Greenbelt and green heart: Separating and integrating landscapes in European city regions. *Landscape and Urban Planning* 64(1–2):19–27.

Motavalli, Jim. 1998. Chattanooga on a roll: From America's dirtiest city to one of its greenest. *E Magazine,* 9(2):14–15.

Pickett, S. T. A., et al. 2001. Urban ecological systems: Linking terrestrial ecological, physical, and socioeconomic components of metropolitan areas. *Annual Review of Ecology and Systematics* 32:127–57.

Rabinovitch, Jonas, and Josef Leitman. 1996. Urban planning in Curitiba. *Scientific American* 274(3):46–54.

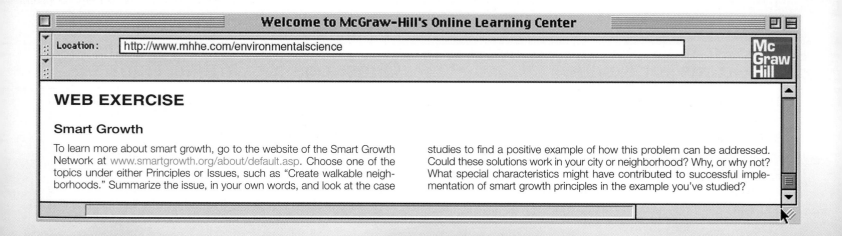

Welcome to McGraw-Hill's Online Learning Center

Location: http://www.mhhe.com/environmentalscience

WEB EXERCISE

Smart Growth

To learn more about smart growth, go to the website of the Smart Growth Network at www.smartgrowth.org/about/default.asp. Choose one of the topics under either Principles or Issues, such as "Create walkable neighborhoods." Summarize the issue, in your own words, and look at the case studies to find a positive example of how this problem can be addressed. Could these solutions work in your city or neighborhood? Why, or why not? What special characteristics might have contributed to successful implementation of smart growth principles in the example you've studied?

23

Ecological Economics

Is it progress if a cannibal uses a fork?

Stanislaw Lee

OBJECTIVES

After studying this chapter, you should be able to:

- explain the difference between neoclassical and ecological economics and how each discipline views ecological processes and natural resources.
- distinguish between different types and categories of resources.
- understand how resource supply and demand affect price and technological progress.
- develop a position on limits to growth and economic carrying capacity of our environment.
- discuss internal and external costs, market approaches to pollution control, and cost-benefit analysis.
- define GNP and explain some alternative ways to measure values of natural resources and real social progress.
- analyze the role of business and some possible strategies for achieving future sustainability.

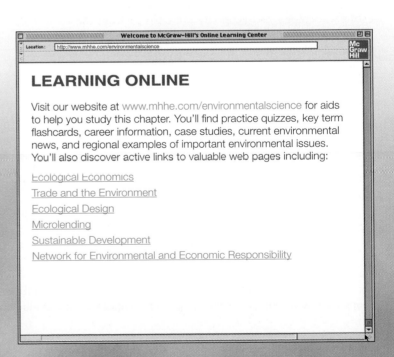

Welcome to McGraw-Hill's Online Learning Center

Location: http://www.mhhe.com/environmentalscience

LEARNING ONLINE

Visit our website at www.mhhe.com/environmentalscience for aids to help you study this chapter. You'll find practice quizzes, key term flashcards, career information, case studies, current environmental news, and regional examples of important environmental issues. You'll also discover active links to valuable web pages including:

Ecological Economics

Trade and the Environment

Ecological Design

Microlending

Sustainable Development

Network for Environmental and Economic Responsibility

Photo: Creating and maintaining an artificial environment in Biosphere 2 turned out to be very difficult and expensive. © John Miller/AP Wide World.

Creating Another Earth

In September 1993, eight tired, hungry people emerged from two years of isolation in a high-tech greenhouse in the desert near Oracle, Arizona. Designed as an ecotechnological model for space exploration and colonization, this bioengineered facility was intended to grow food, cleanse the air, and recirculate and purify water for its eight inhabitants without exchange of materials (including atmospheric gases) with the outside world. Called Biosphere 2 (Biosphere 1 being the earth), the 13,000 square meter airtight complex consisted of living quarters and greenhouses containing food crops and small, but photogenic, areas representing deserts, rainforests, savannas, and an ocean.

Almost from its beginning, Biosphere 2 experienced controversy and problems. By the end of the first year of their mission, the Biospherians reported deteriorating air and water quality. Oxygen concentrations in the air had fallen from 21 percent to 14 percent. This is equivalent to oxygen levels at an elevation of 5,300 m (17,500 ft) and was barely sufficient to keep the occupants alive and functioning. At the same time, carbon dioxide concentrations were undergoing large daily and seasonal variations and nitrous oxide (N_2O) in the air had reached mind-numbing levels. In January 1993, fresh air was pumped in to replenish the dome's atmosphere and rescue the inhabitants.

Subsequent investigations showed that the missing oxygen was being consumed by microbes in the excessively rich soil in which food crops were being grown. At the same time, fresh concrete used in construction was absorbing carbon dioxide released by microbial metabolism. If this carbon dioxide sink hadn't been available, the air would have become unbreathable long before it did. Water systems also became polluted with excess nutrients, degrading aquatic habitats and contaminating drinking water supplies. Species losses were much higher than originally anticipated. Of the 25 introduced vertebrate species, for example, 18 became extinct. All insect pollinators also died, so that most plants were unable to produce seeds and food supplies dwindled to alarming levels. Weedy vines, particularly morning glories (*Ipomoea hederacea*), flourished in the carbon-rich atmosphere and threatened to choke out more desirable plants. Although the majority of the insects disappeared, ants, cockroaches, and katydids thrived and overran everything. Biosphere 2 cost nearly $200 million to build, with an additional cost of about $1 million per year for fossil fuel energy to keep all the systems running. That averages out to about $25 million per human occupant for the two-year experiment. How much more would it cost to produce a truly balanced ecological system that could maintain life indefinitely? Although the experiment failed to meet its original objectives, it illustrates the value of ecological services provided by the natural world that most of us take for granted. What will we have lost and what might it cost to replace resources and services on which we now depend if we deplete or destroy the natural world? In this chapter, we will examine how ecological economists assign prices and values to natural resources and ecological services.

ECONOMIC WORLDVIEWS

Economy is the management of resources to meet our needs in the most efficient manner possible. It typically deals with choices and alternatives. Assuming we don't have unlimited ability to produce, distribute, or consume goods and services, economists ask, "Should we make bread or bullets? What are the trade-offs between them, and which would result in the greatest benefits?" (fig. 23.1). Interestingly, ecology and economy are derived from the same root words and concerns. *Oikos* (ecos) is the Greek word for household. Economics is the *nomos,* or counting, of the household resources. Ecology is the *logos,* or logic of how the household works. In both disciplines, the household is expanded to include the whole world.

Sustainability has become a central theme of environmental science and natural resource use. Although the idea of sustainability has many facets, the central idea is that we should use our assets in ways that will make them last. Resources and natural amenities, including wildlife, natural beauty, and open space, should be conserved so that future generations can have lifestyles at least as healthy and happy as ours—or perhaps better.

Can sustainability be achieved, and if so how? And how can we measure genuine progress toward achieving this goal? These are among the core questions in environmental science, because human impacts will continue to damage our environment unless we find intelligent alternatives and solutions. Key problems of unsustainability have woven through much of this book. In this chapter we will look at ways that economists measure and evalu-

FIGURE 23.1   Bread or bullets? What are the costs and benefits of each? And what are the trade-offs between them? © William P. Cunningham.

 TABLE 23.1	Goals for Sustainable Natural Resource Use

- Harvest rates for renewable resources (those like organisms that regrow or those like fresh water that are replenished by natural processes) should not exceed regeneration rates.

- Waste emissions should not exceed the ability of nature to assimilate or recycle those wastes.

- Nonrenewable resources (such as minerals) may be exploited by humans, but only at rates equal to the creation of renewable substitutes.

FIGURE 23.2 Informal markets such as this one in Iraq may be the purest example of willing sellers and buyers setting prices based on supply and demand. © The McGraw-Hill Companies, Inc./Barry W. Barker, photographer.

ate resources and the benefits we obtain from them. Table 23.1 suggests some goals for a sustainable economy. But as you will see in this chapter, different people have widely varying views of what these goals mean and how best we might attain them.

Classical Economics

Classical economics originally was a branch of moral philosophy concerned with how individual interest and values intersect with larger social goals. The founder of modern Western economics, Adam Smith (1723–1790), was an ethicist concerned with individual freedom of choice. Smith's landmark book *Inquiry into the Nature and Causes of the Wealth of Nations,* published in 1776, argued that

> Every individual endeavors to employ his capital so that its produce may be of the greatest value. He generally neither intends to promote the public interest, nor knows how much he is promoting it. He intends only his own security, only his own gain. And he is in this led by an *invisible hand* to promote an end which was no part of his intention. By pursuing his own interests he frequently promotes that of society more effectually than when he really intends to.

This statement often is taken as justification for the capitalist system, in which market competition between willing sellers and buyers is believed to bring about the greatest efficiency of resource use, the optimum balance between price and quality, and to be critical for preserving individual liberty (fig. 23.2). As we will discuss later in this chapter, however, *laissez faire* market systems rarely incorporate factors such as social or environmental costs.

David Ricardo (1772–1823), an important contemporary of Smith, introduced a better understanding of the relation between supply and demand in economics. **Demand** is the amount of a product or service that consumers are willing and able to buy at various possible prices, assuming they are free to express their preferences. **Supply** is the quantity of that product being offered for sale at various prices, other things being equal. The inverse relationship between supply and demand is shown in figure 23.3. As prices rise, the supply increases and demand falls. The reverse holds true as price decreases. The difference between the cost of

production and the price buyers are willing to pay, Ricardo called rent. Today we call it profit.

In a free market of independent and intelligent buyers and sellers, supply and demand should come into a **market equilibrium,** represented in figure 23.3 by the intersection of the two curves. In real life, prices are not determined strictly by total supply and demand as much as what economists called **marginal costs and benefits.** Sellers ask themselves, "What would it cost to produce one more unit of this product or service? Suppose I add one more worker or buy an extra supply of raw materials, how much profit could I make?" Buyers ask themselves similar questions, "How much would I benefit and what would it cost if I bought one more widget?" If both buyer and seller find the marginal costs and benefits attractive, a sale is made.

There are exceptions, however, to this theory of supply and demand. Consumers will buy some things regardless of cost. Raising the price of cigarettes, for instance, doesn't necessarily reduce

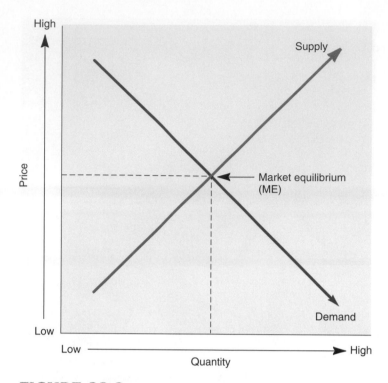

FIGURE 23.3   Classic supply/demand curves. When price is low, supply is low and demand is high. As prices rise, supply increases but demand falls. Market equilibrium is the price at which supply and demand are equal.

demand. We call this price inelasticity. Other items have **price elasticity;** that is, they follow supply/demand curves exactly. When price goes up, demand falls and vice versa.

John Stuart Mill (1806–1873), another important classical philosopher/economist, believed that perpetual growth in material well-being is neither possible nor desirable. Economies naturally mature to a steady state, he believed, leaving people free to pursue nonmaterialistic goals. He didn't regard this equilibrium state to be necessarily one of misery and want, however. In *Principles of Political Economy,* he wrote, "It is scarcely necessary to remark that a stationary condition of capital and population implies no stationary state of human improvement. There would be as much scope as ever for all kinds of mental culture, and moral and social progress; as much room for improving the art of living, and much more likelihood of its being improved when minds cease to be engrossed by the art of getting on." This view has much in common with the concepts of a steady-state economy and sustainable development that we will examine later in this chapter.

Neoclassical Economics

Toward the end of the nineteenth century, the field of economics divided into two broad camps. **Political economy** continued the tradition of moral philosophy and concerned itself with social structures, value systems, and relationships among the classes. This group included reformers such as Karl Marx and E. F. Schumacher, along with many socialists, anarchists, and utopians. The other camp,

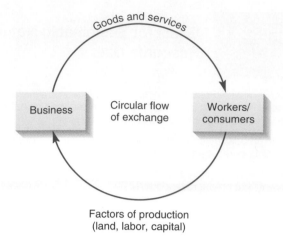

FIGURE 23.4   The neoclassical model of the economy focuses on the flow of goods, services, and factors of production (land, labor, capital) between business and individual workers/consumers. The social and environmental consequences of these relationships are irrelevant in this view.

called **neoclassical economics,** adapted principles of modern science to economic analysis. This approach strives to be mathematically rigorous, noncontextual, abstract, and predictive. Neoclassical economists claim to be objective and value free, leaving social concerns to other disciplines. Like their classical predecessors, they retain an emphasis on scarcity and the interaction of supply and demand in determining prices and resource allocation (fig. 23.4).

Continued economic growth is considered to be both necessary and desirable in the neoclassical worldview. Growth is seen as the only way to maintain full employment and avoid class conflict arising from inequitable distribution of wealth. Natural resources are viewed as merely a factor of production rather than a critical supply of materials, services, and waste sinks by neoclassical economics. Because factors of production are thought to be interchangeable and substitutable, materials and services provided by the environment are not considered indispensable. As one resource becomes scarce, neoclassical economists believe, substitutes will be found.

Natural resource economics shares much in common with neoclassical attitudes but expands the value of nature as an important source of raw materials and a sink for wastes (fig. 23.5). Geological and ecological processes also recycle materials over long or short time scales, but these processes are considered to be external to the economic system. Because the environment and natural resources are seen as relatively abundant compared to "built" or human-made capital, the Riccardian rent for labor, technology, and money are expected to be higher than that for nature.

Ecological Economics

A relatively new discipline called **ecological economics** brings the insights of ecology to economic analysis. In contrast to previous approaches, this is a transdisciplinary field that intends to be holistic, contextual, value-sensitive, and ecocentric. It incorporates the principles of thermodynamics and coevolution into its worldview.

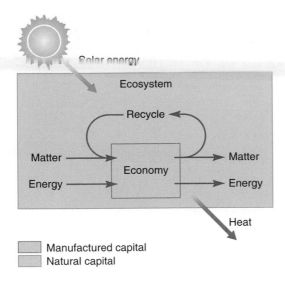

Manufactured capital
Natural capital

FIGURE 23.5   In natural resource economics, the economy is seen as an open subset of the ecosystem. Matter and energy are consumed in the form of raw resources and rejected as waste. Material is recycled by ecological processes, while heat is ejected back to space. Manufactured or "built" capital is regarded as scarce—and therefore valuable—while natural capital is regarded as plentiful and, therefore, cheap.

Source: T. Prugh, *National Capital and Human Economic Survival* 1995, ISEE.

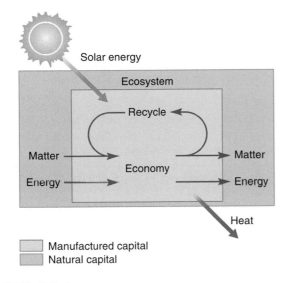

Manufactured capital
Natural capital

FIGURE 23.6   An ecological economic view reverses ideas of scarcity. Because material recycling is considered to be integral to the economy, manufactured or built capital is seen to be large compared to limited supplies of natural capital.

Source: Herman Daly in A. M. Jansson, et al., *Investing in Natural Capital*, ISEE.

A principal concern of ecological economics is the equitable distribution of resources and rights within current generations and between this generation and future ones. It also is mindful of rights of other species. An important difference between this new field and its predecessors is a recognition of the world as an open, dynamic system within which the human economy is inextricably embedded (fig. 23.6).

Ecological economics acknowledges our dependence of the essential life-support services provided by nature. Rather than regard natural and manufactured resources as completely interchangeable, ecological economists regard some aspects of nature as irreplaceable and essential. This places limits on human population growth and resource consumption and raises questions about the carrying capacity of our environment. Seeing nature as limited and relatively fragile compared to the human economy reverses our appraisal of which resources are scarce and, therefore, valuable. Ecological economists regard natural resources as important to our well-being and as worthy of protection and development as are human or technological resources.

The **steady-state economy** advocated by ecological economists is characterized by low human birth and death rates, use of renewable energy sources, material recycling, and emphasis on durability, efficiency, and stability rather than high throughput of materials. Building on the ideas of John Stuart Mill, some pioneers of this concept include the Rumanian economist Nicholas Georgescu-Rogen and American economists Kenneth Boulding and Herman E. Daly, all of whom have explored concepts of sustainability and limits to growth. Much of the rest of this chapter will be devoted to exploring the differences between neoclassical and ecological economics and a discussion of limits to growth.

RESOURCES, CAPITAL, AND RESERVES

A fundamental question in all economic analysis is whether one option or state of the world is better than another. Economics deals with questions of resource allocation or trade-offs either at the "micro" scale of individual decisions or at the "macro" scale of national policies and world trade. How can we determine the available amount and value of a specific resource, given a particular economic system and technology? In this section, we will look at some ways to categorize capital, resources, and reserves.

Resource Types

Capital is any form of wealth available for use in the production of more wealth. Economists distinguish between *natural capital* (goods and services provided by nature), *human or cultural capital* (knowledge, experience, and human enterprise), and *manufactured or built capital* (tools, infrastructure, and technology). Social scientists would add *social capital* (shared values, trust, cooperative spirit, and community organization) to this list. A **resource** is anything with potential use in creating wealth or giving satisfaction. Natural resources can be either renewable or nonrenewable, as well as tangible or intangible. Although we generally define these terms from a human perspective, remember that many resources we desire are important to other species as well.

In general, **nonrenewable resources** are the earth's geological endowment: the minerals, fossil fuels, and other materials present in fixed amounts in the environment (fig. 23.7). Although many of these resources are renewed or recycled by geological or ecological processes, the time scales to do so are so long by human

FIGURE 23.7 The geological resources of the earth's crust, such as oil from this well, are nonrenewable. Often the limit to our use of these resources is not so much the absolute amount available but the energy cost required to extract the resources and the environmental consequences of doing so. © Corbis/Volume 60.

FIGURE 23.8 Biological resources, such as these cashew fruits, are renewable, but if overused and driven to extinction can not be re-created. © The McGraw-Hill Companies, Inc./Barry W. Barker, photographer.

TABLE 23.2	Important Ecological Services

1. *Regulate* global energy balance; chemical composition of the atmosphere and oceans; local and global climate; water catchment and groundwater recharge; production, storage, and recycling of organic and inorganic materials; maintenance of biological diversity.

2. *Provide* space and suitable substrates for human habitation, crop cultivation, energy conversion, recreation, and nature protection.

3. *Produce* oxygen, fresh water, food, medicine, fuel, fodder, fertilizer, building materials, and industrial inputs.

4. *Supply* aesthetic, spiritual, historic, cultural, artistic, scientific, and educational opportunities and information.

Source: *R. S. de Groot*, Investing in Natural Capital, *1994.*

standards that the resource will be gone once present supplies are exhausted. Predictions abound that we are in imminent danger of running out of one or another of these exhaustible resources. The actual available supplies of many commodities—such as metals—can be effectively extended, however, by more efficient use, recycling, substitution of one material for another, or better extraction from dilute or remote sources.

Renewable resources are things that can be replenished or replaced. They include sunlight—our ultimate energy source—and the biological organisms and biogeochemical cycles that provide essential ecological services (fig. 23.8 and table 23.2). This is discussed later in this chapter under nonmarket values. Because biological organisms and ecological processes are self-renewing, we often can harvest surplus organisms or take advantage of ecological services without diminishing future availability, if we do so carefully. Unfortunately, our stewardship of these resources often is less than ideal. Even once vast biological populations such as

passenger pigeons, American bison (buffalo), and Atlantic cod, for instance, were exhausted by overharvesting in only a few years (see chapters 6 and 11). Similarly, we appear to be upsetting climatic systems with potentially disastrous results (see chapter 15). To our dismay, we have found that such mismanagement of many renewable resources may make them more ephemeral and limited than fixed geological resources.

Abstract or **intangible resources** include open space, beauty, serenity, wisdom, diversity, and satisfaction (fig. 23.9). Paradoxically, these resources can be both infinite *and* exhaustible. There is no upper limit to the amount of beauty, knowledge, or compassion that can exist in the world, yet they can be easily destroyed. A single piece of trash can ruin a beautiful vista, or a single cruel remark can spoil an otherwise perfect day. On the other hand, unlike tangible resources that usually are reduced by use or sharing, intangible resources often are increased by use and multiplied by being shared. Nonmaterial assets can be important economi-

FIGURE 23.9 Hikers admire Josephine Lake in Glacier National Park. There is no limit on the amount of beauty, knowledge, or compassion in the world, yet they can be easily destroyed by a single cruel act, rude noise, or careless vandalism. © William P. Cunningham.

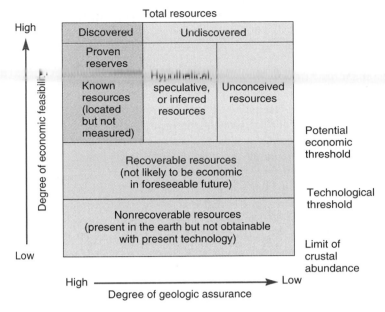

FIGURE 23.10 Categories of natural resources. These categories, based on degree of geological assurance and economic feasibility of recovery, were intended to describe mineral resources. With some modification, these categories can be applied to many other nonrenewable resources.

cally. Information management and tourism—both based on intangible resources—have become two of the largest and most powerful industries in the world.

Economic Resource Categories

Although we have defined a resource as anything useful, we should distinguish between economic usefulness and total abundance or a material or service. Vast supplies of potentially important materials are present in the earth's crust, for instance, but they are useful to us only if they can be recovered in reasonable amounts with available technology and with acceptable environmental and social costs. Within the aggregate total of any natural resource, we can distinguish categories on the basis of economic and technological feasibility, as well as the resource location and quantity (fig. 23.10).

You've probably read about fossil fuel proven reserves or recoverable resources. What do these terms mean? **Proven reserves** of a resource are those that have been thoroughly mapped and are economical to recover at current prices with available technology. **Known resources** are those that have been located but are not completely mapped. Their recovery or development may not be economical now, but they are likely to become economical in the

foreseeable future. **Undiscovered resources** are only speculative or inferred from similarities with known deposits. There also may be unconceived resources—sometimes called "unknown-unknowns"—that we don't even know where to look for.

Recoverable resources are accessible with current technology but are not necessarily economical. Unrecoverable resources are so diffuse or remote that they are not likely ever to be technologically accessible. There can be vast differences between these categories. For instance, the U.S. Geological survey estimates that only 0.01 percent of the minerals in the upper one kilometer of the earth's crust will ever be economically recoverable. That means that 10,000 times as much is not economically available and will probably never be technically feasible to recover. Chapters 14 and 19 give some practical examples of proven, known, and ultimately recoverable resources.

Communal Property Resources

In 1968, biologist Garret Hardin wrote a widely quoted article entitled **"The Tragedy of the Commons"** in which he argued that any commonly held resource inevitably is degraded or destroyed because the narrow self-interests of individuals tend to outweigh public interests. Hardin offered as a metaphor the common woodlands and pastures held by most colonial New England villages. In deciding how many cattle to put on the commons, Hardin explained, each villager would attempt to maximize his or her own personal gain. Adding one more cow to the commons could mean a substantially increased income for an individual farmer. The damage done by overgrazing, however, would be shared among all the farmers. According to Hardin, the only solution would be either to give coercive power to the government or to privatize the resource.

Hardin intended this dilemma, known in economics as the "free-rider" problem, to warn about human overpopulation and resource availability. Other authors have used his metaphor to explain such diverse problems as African famines, air pollution, fisheries declines, and urban crime. What Hardin was really describing, however, was an **open access system** in which there are no rules to manage resource use. In fact, many communal resources have been successfully managed for centuries by cooperative arrangements among users. Some examples include Native American management of wild rice beds and hunting grounds; Swiss village-owned mountain forests and pastures; Maine lobster fisheries; communal irrigation systems in Spain, Bali, and Laos; and nearshore fisheries almost everywhere in the world. A large body of literature in economics and social sciences describes how these cooperative systems work. Among the features shared by **communal resource management systems** are: (1) community members have lived on the land or used the resource for a long time and anticipate that their children and grandchildren will as well, thus giving them a strong interest in sustaining the resource and maintaining bonds with their neighbors; (2) the resource has clearly defined boundaries; (3) the community group size is known and enforced; (4) the resource is relatively scarce and highly variable so that the community is forced to be interdependent; (5) management strategies appropriate for local conditions have evolved over time and are collectively enforced; that is, those affected by the rules have a say in them; (6) the resource and its use are actively monitored, discouraging anyone from cheating or taking too much; (7) conflict resolution mechanisms reduce discord; and (8) incentives encourage compliance with rules, while sanctions for noncompliance keep community members in line.

Rather than being the only workable solution to problems in common pool resources, privatization and increasing external controls often prove to be disastrous. In places where small villages have owned and operated local jointly held forests and fishing grounds for generations, nationalization and commodification of resources generally have led to rapid destruction of both society and ecosystems. Where communal systems once enforced restraint over harvesting, privatization encouraged narrow self-interest and allowed outsiders to take advantage of the weakest members of the community.

A tragic example is the forced privatization of Indian reservations in the United States. Failing to recognize or value local knowledge and forcing local people to participate in a market economy allowed outsiders to disenfranchise native people and resulted in disastrous resource exploitation. Learning to distinguish between open access systems and communal property regimes is important in understanding how best to manage natural resources.

POPULATION, TECHNOLOGY, AND SCARCITY

Are we about to run out of essential natural resources? It stands to reason that if we consume a fixed supply of nonrenewable resources at a constant rate, eventually we'll use up all the economically recoverable reserves. There are many warnings in the environmental literature that our extravagant depletion of non-

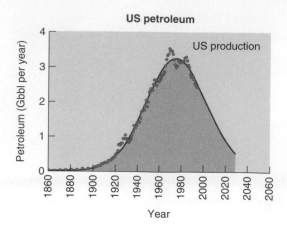

FIGURE 23.11 United States petroleum production, 1860 to 2000. Dots indicate actual production. The bell-shaped curve is a theoretical Hubbert curve for a nonrenewable resource. The shaded area under the curve, representing 220 Gbbl (Gbbl = Gigabarrels or billions of standard 42-gallon barrels), is an estimate of the total economically recoverable resource.

renewable resources sooner or later will result in catastrophe. The dismal prospect of Malthusian diminishing returns and a life of misery, starvation, and social decay inspire many environmentalists to call for an immediate change to voluntary simplicity and lower consumption rates. Models for exploitation rates of nonrenewable resources—called Hubbert curves, after Stanley Hubbert who developed them—often closely match historic experience for natural resource depletion (fig. 23.11).

Many economists, however, contend that neither supply/demand relationships nor economically recoverable reserves are rigidly fixed. Human ingenuity and enterprise often allow us to respond to scarcity in ways that postpone or alleviate the dire effects predicted by modern Jeremiahs. In the next section we will look at some of the arguments for and against limits to growth of the global economy.

Market Efficiencies and Technological Development

In a pioneer or frontier economy, methods for harvesting resources and turning them into useful goods and services tend to be inefficient and wasteful. This may not matter, however, if the supply of resources exceeds the demand for them. The loggers, for example, who swarmed across the Great Lakes Forest at the beginning of the twentieth century, wasted a vast amount of wood. Their inefficiency didn't seem important, however, because the supply of trees appeared infinite, while labor, capital, and means for getting lumber to markets was scarce. As markets and societies develop, however, better technology and more efficient systems allow people to create the same amount of goods and services using far fewer resources. We now produce hundreds of times the crop yield with less labor from the same land that pioneers farmed.

This increasing technological efficiency can dramatically shift supply and demand relationships (fig. 23.12). As technology makes

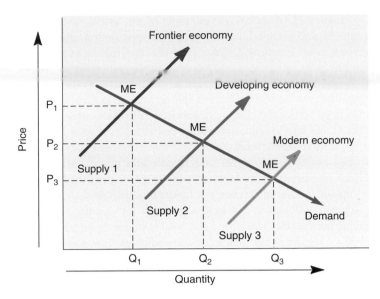

FIGURE 23.12 Supply and demand curves at three different stages of economic development. At each stage there is a market equilibrium point at which supply and demand are in balance. As the economy becomes more efficient, the equilibrium shifts so there is a larger quantity available at a lower price than before. (P = price, Q = quantity, ME = market equilibrium)

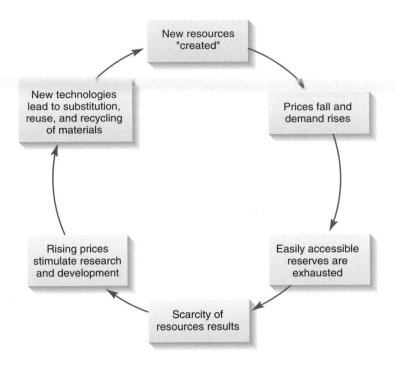

FIGURE 23.13 Scarcity/development cycle. Paradoxically, resource use and depletion of reserves can stimulate research and development, the substitution of new materials, and the effective creation of new resources.

goods and services cheaper to produce, the quantity available at a given price can increase greatly. The market equilibrium or the point at which supply and demand equilibrate will shift to lower prices and higher quantities as a market matures.

Scarcity can actually serve as a catalyst for innovation and change (fig. 23.13). As materials become more expensive and difficult to obtain, it becomes cost-effective to discover new supplies or to use available ones more efficiently. The net effect is as if a new supply of resources has been created or discovered. Several important factors play a role in this cycle of technological development:

- Technical inventions can increase efficiency of extraction, processing, use, and recovery of materials.

- Substitution of new materials or commodities for scarce ones can extend existing supplies or create new ones. For instance, substitution of aluminum for copper, concrete for structural steel, grain for meat, and synthetic fibers for natural ones all remove certain limits to growth.

- Trade makes remote supplies of resources available and may also bring unintended benefits in information exchange and cultural awakening.

- Discovery of new reserves through better exploration techniques, more investment, and looking in new areas becomes rewarding as supplies become limited and prices rise.

- Recycling becomes feasible and accepted as resources become more valuable. Recycling now provides about 37 percent of the iron and lead, 20 percent of the copper, 10 percent of the aluminum, and 60 percent of the antimony that is consumed each year in the United States.

Increasing Environmental Carrying Capacity

Despite repeated warnings that rapidly growing populations and increasing affluence are bound to exhaust natural resources and result in rapid price increases, technological developments of the sort described earlier have resulted in price decreases for most raw materials over the last hundred years. Consider copper for example. Twenty years ago worries about impending shortages led the United States to buy copper and store it in strategic stockpiles. Estimates of the amount of this important metal needed for electric motors, telephone lines, transoceanic cables, and other uses essential for industrialized society far exceeded known reserves. It looked as if severe shortages and astronomical price increases were inevitable. But then aluminum power lines, satellites, fiber optics, integrated circuits, microwave transmission, and other inventions greatly diminished the need for copper. Although prices are highly variable because of world politics and trade policies, the general trend for most materials has been downward in this century. It is as if the carrying capacity of our natural resources—at least in terms of copper—has been increased.

Economists generally believe that this pattern of substitutability and technological development is likely to continue into the future. Ecologists generally disagree. There are bound to be limits, they argue, to how many people our environment can support. An interesting example of this debate occurred in 1980. Ecologist Paul Ehrlich bet economist Julian Simon that increasing human populations and growing levels of material consumption would inevitably lead to price increases for natural resources. They chose a package of five metals—chrome, copper, nickel, tin, and tungsten—priced at the time at $1,000. If, in ten years, the combined price (corrected for inflation) was higher than $1,000, Simon would pay the difference. If the combined price had fallen, Ehrlich would pay. In 1990 Ehrlich sent Simon a check for $576.07; the price for these five metals had fallen 47.6 percent.

Does this prove that resource abundance will continue indefinitely? Hardly. Ehrlich claims that the timing and set of commodities chosen simply were the wrong ones. The fact that we haven't yet run out of raw materials doesn't mean that it will never happen. Many ecological economists now believe that some nonmarket resources such as ecological processes may be more irreplaceable than tangible commodities like metals. What do you think? Are we approaching the limits of nature to support more humans and more consumption? Which resources, if any, do you think are most likely to be limiting in the future?

Economic Models

In the early 1970s, an influential study of resource limitations was funded by the Club of Rome, an organization of wealthy business owners and influential politicians. The study was carried out by a team of scientists from the Massachusetts Institute of Technology headed by the late Donnela Meadows. The results of this study were published in the 1972 book *Limits to Growth*. A complex computer model of the world economy was used to examine various scenarios of different resource depletion rates, growing population, pollution, and industrial output. Given the Malthusian assumptions built into this model, catastrophic social and environmental collapse seemed inescapable.

Figure 23.14 shows one example of the world model. Food supplies and industrial output rise as population grows and resources are consumed. Once past the carrying capacity of the environment, however, a crash occurs as population, food production, and industrial output all decline precipitously. Pollution continues to grow as society decays and people die, but, eventually, it also falls. Notice the similarity between this set of curves and the "boom and bust" population cycles described in chapter 6.

Many economists criticized these results because they discount technological development and factors that might mitigate the effects of scarcity. In 1992, the Meadows group published updated computer models in *Beyond the Limits* that include technological progress, pollution abatement, population stabilization, and new public policies that work for a sustainable future. If we adopt these changes sooner rather than later, the computer shows an outcome like that in figure 23.15, in which all factors stabilize sometime in this century at an improved standard of living for everyone. Of course neither of these computer models shows what will happen, only what some possible outcomes *might* be, depending on the choices we make.

Why Not Conserve Resources?

Even if large supplies of resources are available or the technological advances to mitigate scarcity exist, wouldn't it be better to reduce our use of natural resources so they will last as long as possible? Will anything be lost if we're frugal now and leave more to be used by future generations? Many economists would argue that resources are merely a means to an end rather than an end in themselves. They have to be used to have value.

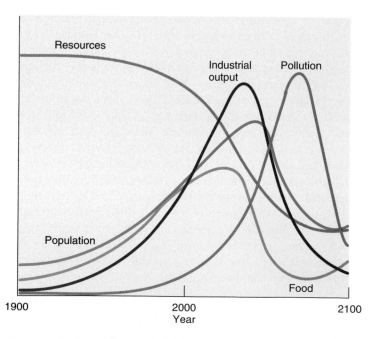

FIGURE 23.14 A run of one of the world models in *Limits to Growth*. This model assumes business-as-usual for as long as possible until Malthusian limits cause industrial society to crash. Notice that pollution continues to increase well after industrial output, food supplies, and population have all plummeted.

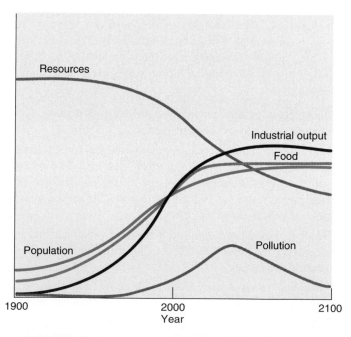

FIGURE 23.15 A run of the world model from *Beyond the Limits*. This model assumes that population and consumption are curbed, new technologies are introduced, and sustainable environmental policies are introduced immediately rather than after resources are exhausted.

TABLE 23.3	Estimated Annual Value of Ecological Services

ECOSYSTEM SERVICES	VALUE (TRILLION $U.S.)
Soil formation	17.1
Recreation	3.0
Nutrient cycling	2.3
Water regulation and supply	2.3
Climate regulation (temperature and precipitation)	1.8
Habitat	1.4
Flood and storm protection	1.1
Food and raw materials production	0.8
Genetic resources	0.8
Atmospheric gas balance	0.7
Pollination	0.4
All other services	1.6
Total value of ecosystem services	**33.3**

Source: *Adapted from R. Costanza et al., "The Value of the World's Ecosystem Services and Natural Capital," Nature, Vol. 387 (1997).*

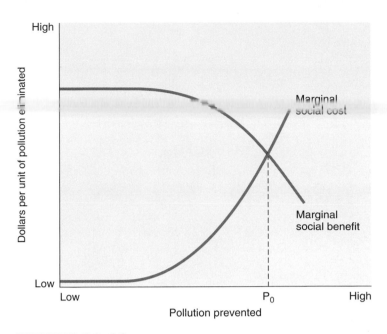

FIGURE 23.18   To achieve maximum economic efficiency, regulations should require pollution prevention up to the optimum point (P_o) at which the costs of eliminating pollution just equal the social benefits of doing so.

nature detoxify them. How much would it cost if we had to do this ourselves? Ecological economists look at everything from recreational beaches to forest lumber to hidden services such as the ocean's regulation of atmospheric carbon dioxide to pollination of crops by insects.

The estimated annual value of all ecological goods and services provided by nature range from $16 trillion to $54 trillion, with a median worth of $33 trillion, or half again the annual GNPs of all national economies in the world (table 23.3). These estimates are probably understated because they omit ecosystem services from several biomes, such as deserts and tundra, that are poorly understood in terms of their economic contributions. The most valuable ecosystems in terms of biological processes are wetlands and coastal estuaries because of their high level of biodiversity and their central role in many biogeochemical cycles.

In 2003, economists from Cambridge University (U.K.) estimated that protecting a series of nature reserves representing samples of all major biomes would cost $45 billion per year, but would preserve ecological services worth between $4.4 trillion to $5.2 trillion annually.

Cost-Benefit Analysis

One way to evaluate public projects is to analyze the costs and benefits they generate in a **cost-benefit analysis (CBA).** This process attempts to assign values to resources as well as to social and environmental effects of carrying out or not carrying out a given undertaking. It tries to find the optimal efficiency point at which the marginal cost of pollution control equals the marginal benefits (fig. 23.18).

CBA is one of the main conceptual frameworks of resource economics and is used by decision makers around the world as a way of justifying the building of dams, roads, and airports, as well as in considering what to do about biodiversity loss, air pollution, and global climate change. Deeply entrenched in bureaucratic practice and administrative culture, this technique has become much more widespread in American public affairs since the Reagan administration's executive orders in the 1980s calling for the application of CBA to all regulatory decisions and legislative proposals. Many conservatives see CBA as a way of eliminating what they consider to be unnecessary and burdensome requirements to protect clean air, clear water, human health, or biodiversity. They would like to add a requirement that all regulations be shown to be cost-effective.

The first step in CBA is to identify who or what might be affected by a particular plan. What are the potential outcomes and results? What alternative actions might be considered? After identifying and quantifying all the contingent factors, an attempt is made to assign monetary costs and benefits to each component. Usually the direct expenses of a project are easy to ascertain. How much will you have to pay for land, materials, and labor? The monetary worth of lost opportunities—to swim or fish in a river, or to see birds in a forest—on the other hand, are much harder to appraise. How would you put a price on good health or a long life? It's also important to ask who will bear the costs and who will reap the benefits of any proposal. Are there consequences that cannot be given a monetary price? What is a bug or a bird worth, for instance, or the opportunity for solitude or inspiration (fig. 23.19)? Eventually, the decision maker compares all the costs and benefits to see whether the project is justified or whether some alternative action might bring more benefits at less cost.

FIGURE 23.19 What is the value of solitude or beauty? How would you assign costs and benefits to a scene such as this?
© The McGraw-Hill Companies, Inc./Barry W. Barker, photographer.

At the same time that some environmentalists are using CBA as a way of obtaining recognition for environmental issues within the terms of mainstream discourse, others are challenging this method of resource accounting as amoral and deeply flawed. Grassroots opponents of roads and hydroelectric dams around the world have repeatedly contested the ways that CBA values land, forests, streams, fisheries and livelihoods, as well as its reliance on unaccountable experts and neglect of equity issues. Ordinary people often refuse to answer questions about how much money they would pay to save a wilderness or how much they would accept to allow it to be destroyed. Developing world delegates to the Intergovernmental Panel on Climate Change angrily rejected a cost-benefit analysis of policy options regarding global warming that assigned a higher value to lives of people in industrialized countries compared to those in less-developed nations.

A study by the Economic Policy Institute of Washington, D.C., found that costs for complying with environmental regulations are almost always less than industry and even governments estimate they will be. For example, electric utilities in the United States claimed that it would cost $4 to $5 billion to meet the 1990 Clean Air Act. But by 1996, utilities were actually saving $150 million per year. Similarly, when CFCs were banned, automobile manufacturers protested it would add $1,200 to the cost of each new car. The actual cost was about $40.

Some other criticisms of CBA include its absence of standards, inadequate attention to alternatives, and the placing of monetary values on intangible and diffuse or distant costs and benefits. Who judges how costs and benefits will be estimated? How can we compare things as different as the economic gain from cheap power with loss of biodiversity or the beauty of a free-flowing river? Critics claim that placing monetary values on everything could lead to a belief that only money and profits count and that any behavior is acceptable as long as you can pay for it. Sometimes speculative or even hypothetical results are given specific numerical values in CBA and then treated as if they are hard facts. Risk-assessment techniques (see chapter 8) may be more appropriate for comparing uncertainties.

Market-Based Mechanisms for Environmental Protection

What is the most efficient and economical way to reduce pollution? Some people argue that we should simply say "Stop it!" to polluters. "Don't dump garbage in our air and water or you'll be punished." Although this approach has a certain moral appeal, it may tend to force all businesses to adopt uniform standards and methods of pollution control regardless of cost or effectiveness. It can also lead to an adversarial climate in which resources are used in litigation rather than pollution control. Furthermore, the "command and control" method can freeze technology by eliminating incentives for continued research and development. Industry is often discouraged—even prohibited—from trying new technologies or alternative production methods. These problems can be overcome, many economists believe, by using market mechanisms to reduce pollution rather than rigid rules and regulations.

Pollution charges are fees assessed per unit of effluent. This approach encourages businesses to do as much pollution control as possible rather than to merely meet a required standard. The more pollution you eliminate, the more you save. Several Scandinavian countries, for example, have passed a carbon tax to discourage CO_2 production and to reduce global climate change. This tax gives industries an incentive to find the most efficient and cost-effective methods of reducing fossil fuel use under their particular set of operating conditions.

Emissions Trading

In **emissions trading,** companies that have cut pollution by more than they're required to can sell "credits" to other companies that still exceed allowed levels. The 1990 Clean Air Act, for example, created a market-based system to reduce emissions of acid-rain-causing sulfur dioxide (SO_2) from power plants and other industrial facilities. A SO_2 target reduction was set for 10 million tons per year, leaving it to industry to find the most efficient way to do this. The government expected that meeting this requirement would cost companies up to $15 billion per year, but the actual cost has been less than one-tenth of that. This program is regarded as a shining example of the benefits of market-based approaches to environmental regulation. There are complaints, however, that while nationwide emissions have come down, "hot spots" remain where local utilities have paid for credits rather than install pollution abatement equipment. If you're living in one of those hot spots and continuing to breathe polluted air, it's not much comfort to know that nationwide average air quality has improved.

Under the Kyoto Protocol, European nations have begun trading carbon emission credits. Prices in 2003 dropped as low as $5.50 (U.S.) per ton, about one-quarter of what had been expected. The main source of the credits are parts of the former Soviet Union, where old, extremely inefficient factories can be upgraded rather cheaply to use less energy and reduce emissions. Ironically, the United States, which insisted on having emissions trading in the protocol, cannot participate in this $10 billion annual market because it has not ratified the treaty.

Intergenerational Justice and Discount Rates

"A bird in the hand is worth two in the bush." All of us are familiar with sayings that suggest it is better to have something now than in the distant future. **Discount rates** are the economists' way to introduce a time factor in accounting. It is a recognition that having something today is worth more to most of us than it will be in the future. In general, discount rates are equivalent to interest rates. When you buy something on credit, for instance, you are deciding that having the use of that item now is worth 10 or 20 percent more to you than the money you paid would be worth in a year.

The choice of discount rates to apply to future benefits becomes increasingly problematic with intangible resources or long time frames. How much will a barrel of oil or a 4,000-year-old redwood tree be worth a couple of centuries from now? Maybe there will be substitutes for oil by then, so the barrel of oil won't be worth much. On the other hand, if we don't find substitutes, that oil or tree may be priceless someday. This forecasting is complicated by the fact that we are making decisions not only for ourselves, but also for future generations.

Although having access to clean groundwater or biological diversity 100 years from now isn't worth much to us—assuming that we will be long gone by that time—those resources might be quite valuable to our descendants. Future citizens will be affected by the choices we make today, but they don't have a vote. Our decisions about how to use resources raise difficult questions about justice between generations. How should we weigh their interests in the future against ours right now?

Which discount rate should be used for future benefits and costs is often a crucial question in large public projects such as dams and airports. Proponents may choose to use low discount rates that make a venture seem attractive while opponents prefer higher rates that show the investment to be questionable. These questions are especially difficult in comparing future environmental costs to immediate financial returns. Environmental activists often need to understand economic nuances when they are fighting environmentally destructive schemes.

Internal and External Costs

Internal costs are the expenses (monetary or otherwise) that are borne by those who use a resource. Often, internal costs are limited to the direct out-of-pocket expenses involved with gaining access to the resource and turning it into a useful product or service.

External costs are the expenses (monetary or otherwise) that are borne by someone other than the individuals or groups who use a resource. External costs often are related to public goods and services derived from nature. Some examples of external costs are the environmental or human health effects of using air or water to dispose of wastes. Since these effects usually are diffuse and difficult to quantify, they do not show on the ledgers of the responsible parties. They are likely to be ignored in private decisions about the costs and benefits of a purchase or a project. One way to use the market system to optimize resource use is to make sure that those who reap the benefits of resource use also bear all the external costs. This is referred to as **internalizing costs.**

TRADE, DEVELOPMENT, AND JOBS

Trade can be a powerful tool in making resources available and raising standards of living. Think of the things you now enjoy that might not be available if you had to live exclusively on the resources available in your immediate neighborhood. Too often, the poorest, least powerful people suffer in this global marketplace. To balance out these inequities, nations can deliberately invest in economic development projects. In this section, we'll look at some aspects of trade, development, business, and jobs that have impacts on our environment and welfare.

International Trade

International issues further complicate questions of resource management. Much of the vast discrepancy between richer and poorer nations is related to past economic and political history, as well as to current international trade relations. According to the economic theory of comparative advantage, each place has goods or services it can supply in better quality or at better prices than its neighbors. International trade allows us to take advantage of all the best products from around the world. The banking and trading systems that regulate credit, currency exchange, shipping rates, and commodity prices were set up by the richer and more powerful nations in their own self-interest. The General Agreement on Tariffs and Trade (GATT) and World Trade Organization (WTO) agreements, for example, negotiated primarily between the largest industrial nations, regulate 90 percent of all international trade.

These systems tend to keep the less-developed countries in a perpetual role of resource suppliers to the more-developed countries. The producers of raw materials, such as mineral ores or agricultural products, get very little of the income generated from international trade (fig. 23.20). Furthermore, they suffer both from low commodity prices relative to manufactured goods and from wild "yo-yo" swings in prices that destabilize their economies and make it impossible to either plan for the future or to accumulate the capital for further development.

Policies of the WTO and the IMF have provoked harsh criticism and mass demonstrations in cities around the world. As a prerequisite for international development loans, the IMF frequently

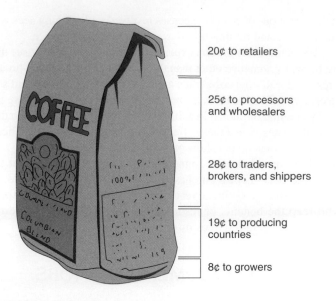

20¢ to retailers

25¢ to processors and wholesalers

28¢ to traders, brokers, and shippers

19¢ to producing countries

8¢ to growers

FIGURE 23.20 What do we really pay for when we purchase a dollar's worth of coffee?

requires debtor nations to adopt harsh "structural adjustment" plans that slash welfare programs and impose cruel hardships on poor people. The WTO has issued numerous rulings that favor international trade over pollution prevention or protection of endangered species. Trade conventions such as the North American Free Trade Agreement (NAFTA) have been accused of encouraging a "race to the bottom" in which companies can play one country against another and move across borders to find the most lax labor and environmental protection standards. See chapter 25 for more on these topics.

International Development

No single institution has more influence on the financing and policies of developing countries than the World Bank. Of some $25 billion loaned each year for developing world projects by multinational development banks, about two-thirds comes from the World Bank. For every dollar invested by the World Bank, two dollars are attracted from other sources. If you want to have an impact on what is happening in the developing countries, it is imperative to understand how this huge enterprise works.

The World Bank was founded in 1945 to provide aid to war-torn Europe and Japan. In the 1950s, its emphasis shifted to development aid for Third World countries. This aid was justified on humanitarian grounds, but providing markets and political support for Western capitalism was an important by-product.

Many World Bank projects have been environmentally destructive and highly controversial. In Botswana, for example, $18 million was provided to increase beef production for export by 20 percent, despite already severe overgrazing on fragile grasslands. The project failed, as did two previous beef production projects in the same area. In India, the sacred Narmada River is being transformed by 30 large dams and 135 medium-sized ones all financed by the World Bank. About 1.5 million hill people and farmers will be displaced as a result.

In response to criticisms of World Bank policies, the U.S. Congress now insists that all loans for international development be reviewed for environmental and social effects before being approved. It asks for assurance that each project (a) uses renewable resources and does not exceed the regenerative capacity of the environment, (b) does not cause severe or irreversible environmental deterioration, and (c) does not displace indigenous people.

In 1999, leaders from the Group of Seven industrial nations (Germany, Japan, France, Canada, Italy, Britain, and the United States) promised to cut the debt burden of the most heavily indebted poor countries in what they described as a decisive step toward alleviating poverty. Altogether, the amount of debt relief could approach $90 billion, and 41 countries (32 of which are African) could benefit from this decision. The share of debt for every man, woman, and child in many of these countries far exceeds the annual per capita income. In Mozambique, for example, where the average income is only $80 per year, the accumulated foreign debt is $10,424 per person. The amount owed to the United States by all the heavily indebted poor countries totals $6.8 billion, or slightly more than the cost of just three of the fleet of 21 B-2 Stealth Bombers.

A number of religious and social justice groups urge the rich countries to go further by declaring a "jubilee for the millennium," in which all debt for foreign assistance or agricultural development programs would be forgiven. Priority should be given to countries that have demonstrated a sustained commitment to poverty alleviation or that have recently suffered a natural disaster, according to these groups. They also would exclude countries that support international terrorism, or which have excessive military spending, gross violations of human rights, or drug trafficking. Under this more generous plan, up to 90 percent of the $127 billion owed by the most heavily indebted countries might be canceled, and the internal spending for education, health, and nutrition by those countries could be quadrupled.

Microlending

An encouraging alternative to the huge development projects financed by the World Bank is provided by small-scale, community-based banking pioneered by economist Muhammad Yunus of Bangladesh. A network of Grammeen (village) Banks makes small loans (averaging $67) directly to the poorest people in the community. Ninety percent of the customers are women who have no collateral and no steady source of income. Loans are enough to buy a used sewing machine, a bicycle, a loom, a cow, some garden tools, or other means of providing needed family income (fig. 23.21).

Compared to a loan repayment rate of only 30 percent in conventional banks, the recovery rate on Grammeen accounts is a remarkable 98 percent. Peer lending is the key to this success. Borrowers are organized into five-member groups that serve both as mutual aid societies and collection agencies. Payments must be made in regular weekly installments. If one member of the group defaults, the others repay the loan. Creating a sense of dignity, respect, and mutual support within the group encourages individual responsibility and enterprise.

FIGURE 23.21   A small loan to buy a used sewing machine can provide a sustainable livelihood for a family in the developing world.
© The McGraw-Hill Companies, Inc./Barry W. Barker, photographer.

Microlending programs similar to the Grammeen Banks are now in place around the world. More than one hundred organizations in the United States assist microenterprises by providing loans, grants, or training. The Women's Self-Employment Project in Chicago, for instance, teaches job skills to single mothers in housing projects. Similarly, "tribal circle" banks on Native American reservations successfully finance microscale economic development projects. Perhaps the biggest banks could learn from the smallest ones how to really help people.

GREEN BUSINESS

During the first Industrial Revolution 200 years ago, raw materials such as lumber, minerals, and clean water seemed inexhaustible, while nature was regarded as something to be tamed and civilized. The "solution to pollution is dilution," was a common approach for disposal of unwanted wastes. Recently, however, some leading industrialists and business executives have realized that business as usual may not be sustainable over the long term. Global pollution, resource depletion, loss of biological diversity, and the threat of global climate change all suggest that change is essential. As the leading sources of much pollution and resource depletion, these leaders realized that they had a role to play in redesigning systems and processes to ensure that their enterprises have a long-term future.

Known by a variety of names, including eco-efficiency, clean production, pollution prevention, industrial ecology, natural capitalism, restorative technology, the natural step, environmentally preferable products, design for the environment, and the next industrial revolution, this movement has had some remarkable successes and presents an encouraging pathway for how we might achieve both environmental protection and social welfare. Some of the leaders in this new approach to business include Paul Hawken, William McDonough, Ray Anderson, Amory Lovins, David Crockett, and John and Nancy Todd.

Operating in a socially responsible manner consistent with the principles of sustainable development and environmental protection, they have shown, can be good for employee morale, public relations, and the bottom line simultaneously. Environmentally conscious or "green" companies such as the Body Shop, Patagonia, Aveda, Malden Mills, Johnson and Johnson, and Interface, Inc. (Case Study, p. 515) consistently earn high marks from community and environmental groups. Conserving resources, reducing pollution, and treating employees and customers fairly may cost a little more initially, but can save money and build a loyal following in the long run.

Paul Hawken's 1993 book, *The Ecology of Commerce,* was a seminal influence in convincing many people to reexamine the role of business and economics in environmental and social welfare. Basing his model for a new industrial revolution on the principles of ecology, Hawken points out that almost nothing is discarded or unused in nature. The wastes from one organism become the food of another. Industrial processes, he argues, should be designed on a similar principle (table 23.4). Rather than a linear pattern in which we try to maximize the throughput of material and minimize labor, products and processes should be designed to

- be energy efficient;
- use renewable materials;
- be durable and reusable or easily dismantled for repair and remanufacture, nonpolluting throughout their entire life cycle;
- provide meaningful and sustainable livelihoods for as many people as possible;
- protect biological and social diversity;
- use minimum and appropriate packaging made of reusable or recyclable materials.

We can do all this and at the same time increase profits, reduce taxes, shrink government, increase social spending, and restore our environment, Hawken claims. Recently, Hawken has served as chairperson for The Natural Step in America, a movement started in Sweden by Dr. K. H. Robert, a physician concerned about the increase in environmentally related cancers. Through a consensus

TABLE 23.4 Goals for an Eco-Efficient Economy

- Introduce no hazardous materials into the air, water, or soil.
- Measure prosperity by how much natural capital we can accrue in productive ways.
- Measure productivity by how many people are gainfully and meaningfully employed.
- Measure progress by how many buildings have no smokestacks or dangerous effluents.
- Make the thousands of complex governmental rules unnecessary that now regulate toxic or hazardous materials.
- Produce nothing that will require constant vigilance from future generations.
- Celebrate the abundance of biological and cultural diversity.
- Live on renewable solar income rather than fossil fuels.

TABLE 23.5 The Natural Step: System Conditions for Sustainability

1. Minerals and metals from the earth's crust must not systematically increase in nature.
2. Materials produced by human society must not systematically increase in nature.
3. The physical basis for biological productivity must not be systematically diminished.
4. The use of resources must be efficient and just with respect to meeting human needs.

process, a group of 50 leading scientists endorsed a description of the living systems on which our economy and lives depend. More than 60 major European corporations and 55 municipalities have incorporated sustainability principles (table 23.5) into their operations.

Another approach to corporate responsibility is called the **triple bottom line.** Rather than reporting only net profits as a measure of success, ethically sensitive corporations include environmental effects and social justice programs as indications of genuine progress.

Corporations committed to eco-efficiency and clean production include such big names as Monsanto, 3M, DuPont, Duracell, and Johnson and Johnson. Following the famous three Rs—reduce, reuse, recycle—these firms have saved money and gotten welcome publicity. Savings can be substantial. Pollution prevention programs at 3M, for example, have saved more than $750 million. Small operations can benefit as well. Stanley Selegut, owner of three eco-tourist resorts in the U.S. Virgin Islands, attributes $5 million worth of business to free press coverage about the resorts' green building features and sustainable operating practices.

Design for the Environment

Our current manufacturing system is incredibly wasteful. On average, for every truckload of products delivered in the United States, 32 truckloads of waste are produced along the way. The automobile is a typical example. Industrial ecologist, Amory Lovins, calculates that for every 100 gallons (380 l) of gasoline burned in your car engine, only one percent (1 gal or 3.8 l) actually moves passengers. All the rest is used to move the vehicle itself. The wastes produced—carbon dioxide, nitrogen oxides, unburned hydrocarbons, rubber dust, heat—are spread through the environment where they pollute air, water, and soil.

Architect William McDonough urges us to rethink design approaches (table 23.6). In the first place, he says, we should question whether the product is really needed. Could we provide the same service in a more eco-efficient manner? According to McDonough, products should be divided into three categories:

1. *Consumables* are products like food, natural fabrics, or paper that can harmlessly go back to the soil as compost.

2. *Service products* are durables such as cars, TVs, and refrigerators. These products should be leased to the customer to provide their intended service, but would always belong to the manufacturer. Eventually they would be returned to the maker, who would be responsible for recycling or remanufacturing the product. Knowing that they will have to dismantle the product at the end of its life will encourage manufacturers to design for easy disassembly and repair.

3. *Unmarketables* are compounds like radioactive isotopes, persistent toxins, and bioaccumulative chemicals. Ideally, no one would make or use these products. But because eliminating their use will take time, McDonough suggests that in the mean time these materials should belong to the manufacturer and be molecularly tagged with the maker's mark. If they are discovered to be discarded illegally, the manufacturer would be held liable.

TABLE 23.6 McDonough Design Principles

Inspired by the way living systems actually work, Bill McDonough offers three simple principles for redesigning processes and products:

1. *Waste equals food.* This principle encourages elimination of the concept of waste in industrial design. Every process should be designed so that the products themselves, as well as leftover chemicals, materials, and effluents, can become "food" for other processes.

2. *Rely on current solar income.* This principle has two benefits: First, it diminishes, and may eventually eliminate, our reliance on hydrocarbon fuels. Second, it means designing systems that sip energy rather than gulping it down.

3. *Respect diversity.* Evaluate every design for its impact on plant, animal, and human life. What effects do products and processes have on identity, independence, and integrity of humans and natural systems? Every project should respect the regional, cultural, and material uniqueness of its particular place.

Eco-Efficient Carpeting from Interface, Inc.

In 1994, in response to customers' concerns about health problems caused by chemical fumes from new carpeting, wall coverings, and other building materials, Ray Anderson, founder and CEO of Interface, Inc., a billion-dollar-a-year interior furnishing company, decided to review company environmental policy. What he found was that the company really didn't have an environmental vision other than to obey all relevant laws and comply with regulations. He also learned that carpeting—of which Interface was the world's third-largest manufacturer—is one of the highest volume and longest lasting components in landfills. A typical carpet is made of nylon embedded in fiberglass and polyvinyl chloride. After a useful life of about 12 years, most carpeting is ripped up and discarded. Every year, more than 770 million m^2 (920 million yd^2) of carpet weighing 1.6 billion kg (3.5 billion lbs) ends up in U.S. landfills. The only recycling that most manufacturers do is to shave off some of the nylon for remanufacture. Everything else is buried in the ground where it will last at least 20,000 years.

At about the same time that Interface was undergoing its environmental audit, Anderson was given a copy of Paul Hawken's book *The Ecology of Commerce*. Reading it, he said, was like "a spear through the chest." He vowed to turn his company around, to make its goal sustainability instead of simply maximizing profits. Rather than sell materials, Interface would focus on selling service. The key is what Anderson calls an "evergreen lease." First of all, the carpet is designed to be completely recyclable. Where most flooring companies merely sell carpet, Interface offers to lease carpets to customers. As carpet tiles wear out, old ones are removed and replaced as part of the lease. The customer pays no installation or removal charges, only a monthly fee for constantly fresh-looking and functional carpeting. Everything in old carpet is used to make new product. Only after many reincarnations as carpet, are materials finally sent to the landfill.

Dramatic changes have been made at Interface's 26 factories. Toxic air emissions have been reduced significantly by changing manufacturing processes and substituting nontoxic materials for more dangerous ones. Solar power is replacing fossil fuel use. Less waste is produced as more material is recycled and products are designed for eco-efficiency. The total savings from pollution prevention and recycling in 1999 was $87 million.

Not only has Interface continued to be an industry leader, it was named one of the "100 Best Companies to Work For in America" by *Fortune* magazine. Ray Anderson has become a popular speaker on the topic of eco-efficiency and clean production. He cochairs the President's Council on Sustainable Development, was named Entrepreneur of the Year by Ernest & Young, and was the Georgia Conservancy's Conservationist of the Year in 1998. Anderson's book, *Mid-Course Correction: Toward a Sustainable Enterprise,* published in 1999 by Chelsea Green, has won critical acclaim.

Ray Anderson. Courtesy Ray Anderson, Interface Inc.

Transforming an industry as large as interior furnishing has not been an overnight success. "Like aircraft carriers," Anderson says, "big businesses don't turn on a dime." Still, he has shown that the principles of sustainability and financial success can coexist and can lead to a new prosperity that includes both environmental and human dividends. His motto, that we should "put back more than we take and do good to the Earth, not just no harm," has become a vision for a new industrial revolution that now is reaching many companies beyond his own.

Ethical Considerations

What responsibilities do businesses have to protect the environment or save resources beyond the legal liabilities spelled out in the law? None whatever, according to conservative economist Milton Friedman. In fact, Friedman argues, it would be unethical for corporate leaders to consider anything other than maximizing profits. To spend time or resources doing anything other than making profits and increasing the value of the company is a betrayal of their duty. What do you think? Should social justice, sustainability, or environmental protection be issues of concern to corporations?

Following these principles, McDonough Bungart Design Chemistry has created nontoxic, easily recyclable materials to use in buildings and for consumer goods. Among some important and innovative "green office" projects designed by the McDonough and Partners architectural firm are the Environmental Defense Fund headquarters in New York City, the Environmental Studies Center at Oberlin College in Ohio (see fig. 20.10), the European Headquarters for Nike in Hilversum, the Netherlands, and the Gap Coporate Offices in San Bruno, California (fig. 23.22). Intended to promote employee well-being and productivity as well as eco-efficiency, the Gap building has high ceilings, abundant skylights, windows that open, a full-service fitness center (including pool),

FIGURE 23.22 The award-winning Gap, Inc. corporate offices in San Bruno, California, demonstrate some of the best features of environmental design. A roof covered with native grasses provides insulation and reduces runoff. Natural lighting, an open design, and careful relation to its surroundings all make this a good place to work. © Mark Luthringer.

and a landscaped atrium for each office bay that brings the outside in. The roof is covered with native grasses. Warm interior tones and natural wood surfaces (all wood used in the building was harvested by certified sustainable methods) give a friendly feeling. Paints, adhesives, and floor coverings are low toxicity and the building is one-third more energy efficient than strict California laws require. A pleasant place to work, the offices help recruit top employees and improve both effectiveness and retention. As for the bottom line, Gap, Inc. estimates that the increased energy and operational efficiency will have a four- to eight-year payback.

Green Consumerism

Consumer choice can play an important role in persuading businesses to produce eco-friendly goods and services (What Can You Do? p. 516). Increasing interest in environmental and social sustainability has caused an explosive growth of green products. The National Green Pages published by Co-Op America currently lists more than 2,000 green companies. You can find ecotravel agencies, telephone companies that donate profits to environmental groups, entrepreneurs selling organic foods, shade-grown coffee, straw-bale houses, geodesic-dome kits, paint thinner made from orange peels, sandals made from recycled auto tires, earthworms for composting, and a plethora of hemp products including burgers, ale, clothing, shoes, rugs, balm, shampoo, and insect repellent. Although these eco-entrepreneurs represent a tiny sliver of the $7 trillion per year U.S. economy, they often serve as pioneers in developing new technologies and offering innovative services.

In some industries eco-entrepreneurs have found profitable niches (organic, naturally colored clothing, for example) within a larger market. In other cases, once a consumer demand has built up, major companies add green products or services to their inventory. Natural foods, for instance, have grown from the domain of a few funky, local co-ops to a $7 billion market segment. Most supermarket chains now carry some organic food choices. Similarly,

What can you do?

Personally Responsible Consumerism

There are many things that each of us can do to lower our ecological impacts and support green businesses through responsible consumerism and ecological economics.

- Practice living simply. Ask yourself if you really need more material goods to make your life happy and fulfilled.

- Rent, borrow, or barter when you can. Do you really need to have your own personal supply of tools, machines, and other equipment that you use, at most, once per year?

- Recycle or reuse building materials: doors, windows, cabinets, appliances. Shop at salvage yards, thrift stores, yard sales, or other sources of used clothes, dishes, appliances, etc.

- Consult the National Green Pages from Co-Op America for a list of eco-friendly businesses. Write one letter each month to a company from which you buy goods or services and ask them what they are doing about environmental protection and human rights.

- Buy green products. Look for efficient, high quality materials that will last and that are produced in the most environmentally friendly manner possible. Subscribe to clean energy programs if they are available in your area. Contact your local utility and ask that they provide this option if they don't now.

- Buy products in bulk if you can or look for the least amount of packaging. Choose locally grown or locally made products made under humane conditions by workers who receive a fair wage.

- Think about the total life-cycle costs of the things you buy, including environmental impacts, service charges, energy use, and disposal costs as well as initial purchase price.

- Stop junk mail. Demand that your name be removed from mass-mailing lists.

- Invest in socially and environmentally responsible mutual funds or green businesses when you have money for investment.

natural-care health and beauty products reached $3 billion in sales in 2003 out of a $33 billion industry. By supporting these products, you can ensure that they will continue to be available and, perhaps, even help expand their penetration into the market.

Jobs and the Environment

For years business leaders and politicians have portrayed environmental protection and jobs as mutually exclusive. Pollution control, protection of natural areas and endangered species, limits on use of nonrenewable resources, they claim, will strangle the economy and throw people out of work. Ecological economists dispute this claim, however. Their studies show that only 0.1 percent of all large-scale layoffs in the United States in recent years were due to government regulations (fig. 23.23). Environmental protection, they argue, is not only necessary for a healthy economic system, it actually creates jobs and stimulates business.

Recycling, for instance, takes much more labor than extracting virgin raw materials. This doesn't necessarily mean that recycled goods are more expensive than those from virgin resources. We're simply substituting labor in the recycling center for energy and huge machines used to extract new materials in remote places.

Japan, already a leader in efficiency and environmental technology, has recognized the multibillion dollar economic potential of "green" business. The Japanese government is investing $4 billion (U.S.) per year on research and development that targets seven areas, ranging from utilitarian projects such as biodegradable plastics and heat-pump refrigerants to exotic schemes such as carbon-dioxide-fixing algae and hydrogen-producing microbes.

A World Wildlife Fund study predicts that over the next decade, energy conservation, renewable energy sources, and other environmental protection programs could result in 750,000 new jobs in the United States if new technologies are embraced and promoted. The net economic impact could be as much as $4.4 billion per year.

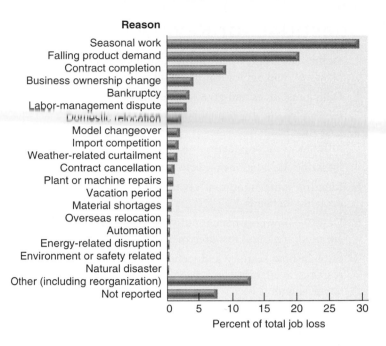

FIGURE 23.23   Although opponents of environmental regulation often claim that protecting the environment costs jobs, studies by economist E. S. Goodstein show that only 0.1 percent of all large-scale layoffs in the United States were the result of environmental laws.
Source: E. S. Goodstein, Economic Policy Institute, Washington, D.C.

Summary

- Neoclassical economics and market-based mechanisms now dominate the world economy but they often do a poor job of incorporating nonmarket resources or protecting the environment.

- Ecological economics brings the insights of ecology into economic analysis. It calls for consideration of the role of natural capital and ecological services in planning and accounting.

- A resource is anything with potential use for creating wealth or giving satisfaction. Minerals and fossil fuels are examples of nonrenewable resources. Biological organisms and ecological processes are renewable and self-reproducing but can be exhausted by overuse. Resources can be defined by their economic and technological feasibility, as well as their location and physical size. We distinguished between known, proven, inferred, and unconceived resources, and between economically important and technically accessible reserves.

- Questions about the scarcity of resources and their effects on economic development are important in determining what kind of society we have. Open-access systems encourage narrow self-interests and resource exploitation. In communal property resource systems, self-governing, locally based community management has successfully sustained natural resources for centuries in many cases.

- Cost-benefit analysis can be a useful tool in environmental management. Ultimately, our aim should be to internalize costs that are now treated as externalities. GNP is used as a measure of economic progress but new measures for human well-being and environmental health are needed. The genuine progress index of sustainable welfare and the human development index are proposed as alternatives.

- Some business leaders believe their only responsibility is to maximize profits. Others regard operating in a socially responsible manner consistent with the principles of sustainable development and environmental protection to be good for employee morale, public relations, and the bottom line.

- The principles of eco-efficiency and design for the environment could help us attain sustainability.

Questions for Review

1. Define economics and distinguish between classical, neo-classical, and ecological economics.

2. Define resources and give some examples of renewable, non-renewable, and intangible resources.

3. List four economic categories of resources and describe the differences among them.

4. Describe the relationship between supply and demand.

5. Identify some important ecological services on which our economy depends.

6. Describe how cost-benefit ratios are determined and how they are used in natural resource management.

7. Explain how scarcity and technological progress can extend resource availability and extend the carrying capacity of the environment.

8. Describe how GNP is calculated and explain why this may fail to adequately measure human welfare and environmental quality. Discuss some alternative measures of national progress.

9. Why does the marketplace sometimes fail to optimally allocate natural resource values?

10. List some of the characteristics of an eco-efficient economic system.

Questions for Critical Thinking

1. If you could retroactively stabilize economic growth or population growth at some point in the past, when would you choose? What assumptions or values shape your choice?

2. When the ecologist warns that we are using up irreplaceable natural resources and the economist rejoins that ingenuity and enterprise will find substitutes for most resources, what underlying premises and definitions shape their arguments?

3. How can intangible resources be infinite and exhaustible at the same time? Isn't this a contradiction in terms? Can you find other similar paradoxes in this chapter?

4. What is the difference between hypothetical and unconceived (or unknown-unknown) resources? How can we plan for resources that we haven't even thought of yet? Are there costs in assuming that there are no unknown-unknowns?

5. What would be the effect on the developing countries of the world if we were to change to a steady-state economic system? How could we achieve a just distribution of resource benefits while still protecting environmental quality and future resource use?

6. Resource use policies bring up questions of intergenerational justice. Suppose you were asked: "What has posterity ever done for me?" How would you answer?

7. If you were doing a cost-benefit study, how would you assign a value to the opportunity for good health or the existence of rare and endangered species in faraway places? Is there a danger or cost in simply saying some things are immeasurable and priceless and therefore off limits to discussion?

8. Why not conserve resources to the maximum extent possible? Discuss the costs and benefits of resource investment versus conservation.

9. Is it right for business to consider ethics and the welfare of future generations? Why or why not?

10. If natural capitalism or eco-efficiency has been so good for some entrepreneurs, why haven't all businesses moved in this direction?

Key Terms

capital 501
classical economics 499
communal resource
  management systems 504
cost-benefit analysis
  (CBA) 509
demand 499
discount rates 511
ecological economics 500
emissions trading 510
external costs 511
genuine progress index
  (GPI) 507
gross domestic product
  (GDP) 507
gross national product
  (GNP) 507
intangible resources 502
internal costs 511

internalizing costs 511
known resources 503
marginal costs and benefits 499
market equilibrium 499
neoclassical economics 500
nonrenewable resources 501
open access system 504
political economy 500
pollution charges 510
price elasticity 500
proven reserves 503
recoverable resources 503
renewable resources 502
resource 501
steady-state economy 501
supply 499
Tragedy of the Commons 503
triple bottom line 514
undiscovered resources 503

Further Readings

Brown, Lester. 2001. *Ecoeconomy: Building an Economy for the Earth.* W. W. Norton & Co.

Collins, J., and J. Porras. 2002. *Built to Last: Successful Habits of Visionary Companies.* Harper Business.

Costanza, Robert, et al. 1997. The value of the world's ecosystem services and natural capital. *Nature* 387:253–60.

Daily, Gretchen C., and Katherine Ellison. 2002. *The New Economy of Nature.* Island Press.

Dunn, Robert H. 1997. Corporate responsibility: The next five years. *The Greenmoney Journal* 6(1–2):7, 21.

Ferraro, P. J., and A. Kiss. 2002. Direct payments to conserve biodiversity. *Science* 298:1718–19.

Hawken, Paul, Amory Lovins, and L. Hunter Lovins. 2000. *Natural Capitalism: Creating the Next Industrial Revolution.* Back Bay Books.

Holliday, Charles O., Stephan Schmidheiny, and Philip Watts. 2002. *Walking the Talk: The Business Case for Sustainable Development.* Greenleaf Publishing

McDonough, William, and Michael Braungart. 2002. *Cradle to Cradle.* North Point Press.

Prugh, T., R. Costanza, and H. Daly. 2000. *The Local Politics of Global Sustainability.* Island Press.

Sachs, Jeffrey D., et al. 2001. The geography of poverty and wealth. *Scientific American* 284(3):71–75.

Wackernagel, Mathis, et al. 2002. Tracking the ecological overshoot of the human economy. *Proc. Natl. Acad. Sci. USA* 99(14):9266–71.

Yunus, Muhammad. 1999. The Grameen bank. *Scientific American* 281(5):114–19.

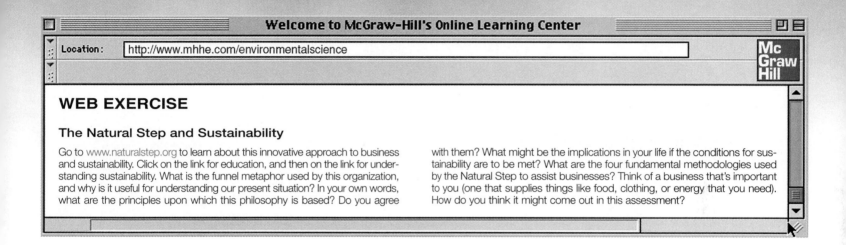

Welcome to McGraw-Hill's Online Learning Center

Location: http://www.mhhe.com/environmentalscience

Mc Graw Hill

WEB EXERCISE

The Natural Step and Sustainability

Go to www.naturalstep.org to learn about this innovative approach to business and sustainability. Click on the link for education, and then on the link for understanding sustainability. What is the funnel metaphor used by this organization, and why is it useful for understanding our present situation? In your own words, what are the principles upon which this philosophy is based? Do you agree with them? What might be the implications in your life if the conditions for sustainability are to be met? What are the four fundamental methodologies used by the Natural Step to assist businesses? Think of a business that's important to you (one that supplies things like food, clothing, or energy that you need). How do you think it might come out in this assessment?

24

Environmental Policy, Law, and Planning

The best lack all conviction, while the worst are full of passionate intensity.

W. B. Yeats

OBJECTIVES

After studying this chapter, you should be able to:

- understand the cycle by which policy is established.
- follow the path of a bill through the legislature.
- explore the differences between civil, criminal, and administrative law.
- judge the effectiveness of litigation in environmental issues.
- consider the reasons that international treaties and global institutions have or have not been successful.
- appreciate the importance of wicked problems, resilience, and adaptive management in environmental planning.
- scrutinize collaborative, community-based planning methods.

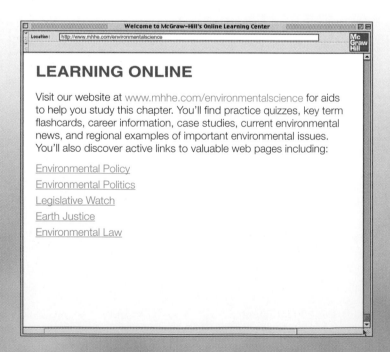

Welcome to McGraw-Hill's Online Learning Center

Location: http://www.mhhe.com/environmentalscience

LEARNING ONLINE

Visit our website at www.mhhe.com/environmentalscience for aids to help you study this chapter. You'll find practice quizzes, key term flashcards, career information, case studies, current environmental news, and regional examples of important environmental issues. You'll also discover active links to valuable web pages including:

Environmental Policy

Environmental Politics

Legislative Watch

Earth Justice

Environmental Law

Photo: The ability of ordinary citizens to petition their government and to participate in public policy formation is a hallmark of democracy. © William P. Cunningham.

Is NEPA an Impediment?

Signed into law by President Nixon in 1970, the **National Environmental Policy Act (NEPA)** is the cornerstone of U.S. environmental policy. Conservationists see this act as a powerful tool for environmental protection, but commercial interests blame it for gridlock and consider it an impediment to business. In 2002 President Bush created a task force to study streamlining NEPA and make it less burdensome for industry. Conservationists worry this is really an effort to weaken the law and return to laissez-faire resource management.

NEPA does three important things: (1) it authorizes the Council on Environmental Quality (CEQ), the oversight board for general environmental conditions; (2) it directs federal agencies to take environmental consequences into account in decision making; and (3) it requires an **Environmental Impact Statement (EIS)** be published for every major federal project likely to have an important impact on environmental quality (fig. 24.1). NEPA doesn't forbid environmentally destructive activities if they comply otherwise with relevant laws, but it demands that agencies admit publically what they plan to do. Once embarrassing information is revealed, however, few agencies will bulldoze ahead, ignoring public opinion. And an EIS can provide valuable information about government actions to public interest groups that wouldn't otherwise have access to these resources.

What kinds of projects require an EIS? The activity must be federal and it must be major, with a significant environmental impact. Evaluations are always subjective as to whether specific activities meet these characteristics. Each case is unique and depends on context, geography, the balance of beneficial versus harmful effects, and whether any areas of special cultural, scientific, or historical importance might be affected. A complete EIS for a project is usually time-consuming and costly. The final document is often hundreds of pages long and generally takes six to nine months to prepare. Sometimes just requesting an EIS is enough to sideline a questionable project. In other cases, the EIS process gives adversaries time to rally public opposition and information with which to criticize what's being proposed. If agencies don't agree to prepare an EIS voluntarily, environmentalists can petition the courts to force them to do so.

Every EIS must contain the following elements: (1) purpose and need for the project, (2) alternatives to the proposed action (including taking no action), (3) a statement of positive and negative environmental impacts of the proposed activities. In addition, an EIS should make clear the relationship between short-term resources and long-term productivity, as well as any irreversible commitment of resources resulting from project implementation.

FIGURE 24.1   Every major federal project in the United States must be preceded by an Environmental Impact Statement.
© Corbis/Volume 160.

Among the areas in which the Bush administration has tried recently to ignore or limit NEPA include forest policy, energy exploration, and marine wildlife protection. The "Healthy Forest Initiative," for example, called for bypassing EIS reviews for logging or thinning projects, and prohibited citizen appeals of forest management plans (chapter 12). Similarly, when the Bureau of Land Management proposed 77,000 coal-bed methane wells in Wyoming and Montana, the administration claimed that water pollution and aquifer depletion associated with this technology didn't require review (chapter 19). And in a court case involving a Navy plan to test sonar devices in the Pacific that might harm marine mammals, the Justice Department argued that environmental laws don't apply in waters more than 3 miles off U.S. shores.

What do these cases mean to you? Perhaps you don't live near a coastline or an area slated for forest thinning or gas drilling, but it's almost certain that federal agencies or corporations that do business with them have projects with potential adverse effects on your local environment. To be informed environmental citizens, we all need to know something about how policies and laws like NEPA are created and applied. In this chapter, we'll look more deeply into environmental policies and how the legislative, legal, and administrative systems work to fulfill or frustrate them.

ENVIRONMENTAL POLICY

The term "policy" is used in many different ways to indicate both formal and informal decisions or intentions at a personal, community, national, or international level. You might have an informal policy never to accept telemarketing calls; your church may have an open-door policy for visitors; and many countries have a policy not to negotiate with terrorists. At the same time, the U.S. Clean Air Act is a formal statement of national policy on acceptable air quality, while the U.N. Convention on Global Climate Change represents the official intentions of many nations to curb greenhouse gases. Interestingly, policy can describe both an actual document as well as a contractual agreement, such as when you buy insurance.

At its core, then, **policy** is a plan or statement of intentions—either written or stated—about a course of action or inaction intended to accomplish some end. Some political scientists limit the term public policy to the principles, laws, executive orders, codes, or goals established by some government body or institution. For the purposes of this chapter, **environmental policy** will be taken as those official rules and regulations concerning the environment that are adopted, implemented, and enforced by some governmental agency as well as general public opinion about environmental issues (fig. 24.2).

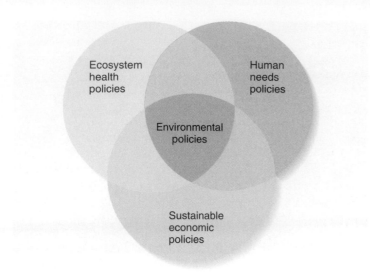

FIGURE 24.2 The best environmental policies incorporate economic, ecological, and social/cultural considerations.
Source: Modified from Ray Grizzle, *Bioscience,* vol. 44(4), April 1994.

Political Decision Making

The policies we establish depend to a great extent on the system within which they operate. For many of us, the ideal political system is one that is open, honest, transparent, reaches the best possible decisions to maximize benefits to everyone. In a pluralistic, democratic society, we aim to give everyone an equal voice in policy making. Ideally, many separate interests put forward their solutions to public problems that are discussed, debated, and evaluated fairly and equally. Facts and access are open to everyone. Policy choices are made democratically but compassionately; implementation is reasonable, fair, and productive. Unfortunately, this isn't always the way our system works. Can you think of some alternate explanations of how we create policy?

Politics as Power

According to some observers, politics is really the struggle among often unequal, competing interest groups as they strive to shape public policy to suit their own agendas. The political system, in this view, manages group conflict by (1) establishing rules to ensure civil competition, (2) encouraging compromises and balancing interests to the extent possible, (3) codifying compromises as public policy, and (4) enforcing laws and rules based on that policy. Where this form of power politics is operational, it often results in a tyranny of a powerful elite over the impotent masses. Those elites manipulate public opinion and give up only as much power and wealth as necessary to maintain overall control.

However, while self-interest and power politics clearly are important forces in American public life, they can't account for all of the civil rights and environmental movements, the war on poverty, and public interest that characterized much of the 1960s and 1970s. The force of ideals, values, and altruism sometimes carry the day even in our winner-take-all system. We seem to go through periods of public spirit, optimism, and openness to change every few decades that give our public life a sense of generosity and good will.

Rational Choice

Another model for public decision making is **rational choice** and science-based management. In this utilitarian approach, no policy should have greater total costs than benefits. In choosing between policy alternatives, we should always prefer those with the greatest cumulative welfare and the least negative impacts. Professional administrators would weigh various options and make an objective, methodical decision that would bring maximum social gain. As chapters 2 and 23 illustrate, there are many arguments against simply applying utilitarian, cost-benefit approaches in public decision making. For example:

- Many conflicting values and needs cannot be compared because they aren't comparable or we don't have perfect information.

- There are few generally agreed-upon broad societal goals but rather benefits to specific groups and individuals, many of which are in conflict.

- Policymakers generally are not motivated to make decisions on the basis of societal goals, but rather to maximize their own rewards: power, status, money, or reelection.

- Large investments in existing programs and policies create "path dependence" and "sunken costs" that prevent policymakers from considering good alternatives foreclosed by previous decisions.

- Uncertainty about consequences of various policy options compels decision makers to stick as closely as possible to previous policies to reduce the likelihood of adverse, calamitous, unanticipated consequences.

- Policymakers, even if well-meaning, don't have sufficient intelligence or adequate data or models to calculate accurate costs and benefits when large numbers of diverse political, social, economic, and cultural values are at stake.

- The segmented nature of policy making in large bureaucracies makes it difficult to coordinate decision making.

The Policy Cycle

How do policy issues and options make their way onto the stage of public debate? In this section, we will look at the **policy cycle** by which problems are identified and acted upon in the public arena (fig. 24.3). The first stage in this process is problem identification. Sometimes the government identifies issues for groups that have no voice or don't recognize problems themselves. In other cases, the public identifies a problem such as loss of biodiversity or health effects exposure to toxic waste and demands redress by the government. In either case, proponents describe the issue—either privately or publicly—and characterize the risks and benefits of their preferred course of action. Seizing the initiative in issue identification often allows leaders to define terms, set the agenda, organize stakeholders, choose tactics, aggregate related issues, and legitimate (or de-legitimate) issues and actors. It can be a great advantage to

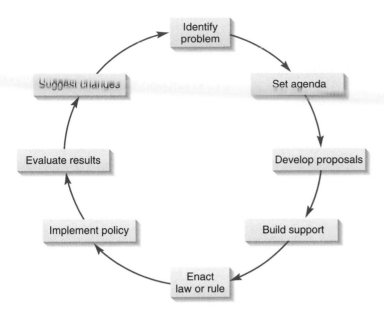

FIGURE 24.3 The policy cycle.

define the terms or choose the location of a debate. Next, stakeholders develop proposals for preferred policy options, often in the form of legislative proposals or administrative rules. Proponents build support for their position through media campaigns, public education, and personal lobbying of decision makers. By following the legislative or administrative process through its many steps, interest groups ensure that their proposals finally get enacted into law or established as a rule or regulation.

The next step is implementation. Ideally, government agencies will faithfully carry out policy directives as they organize bureaucracies, provide services, and enforce rules and regulations, but often it takes continued monitoring to make sure the system works as it should. Evaluating the results of policy decisions is as important as establishing them in the first place. Measuring impacts on target and nontarget populations shows us whether the intended goals, principles, and course of action are being attained. Finally, suggested changes or adjustments are considered that will make the policy fairer or more effective.

There generally are two different routes by which this cycle is carried out. Special economic interest groups such as industry associations, labor unions, or wealthy and powerful individuals don't need (or often want) much public attention or support for their policy initiatives. They generally carry out the steps of issue identification, agenda setting, and proposal development privately because they can influence legislative or administrative processes directly through their contacts with decision makers. Public interest groups, on the other hand, often lack direct access to corridors of power and need to rally broad general support to legitimate their proposals. An important method for getting their interests on the table is to attract media attention. Organizing a dramatic protest or media event can generate a lot of free publicity. Announcing some dire threat or sensational claim is a good way to gain attention. The problem is that it takes ever increasing levels of hysteria and hyperbole to get

yourself heard in the flood of shocking news with which we are bombarded every day. Ironically, many groups that bemoan the overload of rhetorical overstatement that engulfs us, contribute to it in order to be heard above the din.

Environmental Rights

Is a clean, healthy environment a basic human entitlement in the same sense as freedom from discrimination or the right to own property? A long history in international law argues that we all have an inalienable right to a safe, sustainable environment. The 1982 World Charter for Nature, for example, asserts that "man's needs can be met only by ensuring the proper functioning of natural systems," and that it is "an essential human right to means of redress when the human environment has suffered damage or degradation." The 1987 World Commission on Environment and Development (famous for defining sustainable development as meeting the needs of the present without compromising the ability of future generations to meet their own needs) went further in stating, "All human beings have the fundamental right to an environment adequate for their health and well-being."

Of the 191 nations in the world, 109 now have constitutional provisions for protection of the environment and natural resources. One hundred of them specifically recognize the right to a clean and healthy environment and/or the state's obligation to prevent environmental harm. The 1988 American Convention on Human Rights expressly declares, "everyone has a right to live in a healthy environment and to have access to basic public services" and that signatory parties "shall promote the protection, preservation, and improvement of the environment." In spite of having agreed to this Convention, however, the United States has not accepted environmental protection as a fundamental human right. When NEPA was introduced in Congress in 1969, supporters argued that it should be a constitutional amendment. Disagreements about what clean and healthy mean, along with worries about how much it might impede commerce and industry to make such a guarantee, limited NEPA to a statute with more restricted, but still important powers than a constitutional amendment. A statute, however, is much easier to ignore or overturn than a constitutional provision (see the opening story in this chapter). What do you think? Does everyone have an inherent right to a clean, healthy environment? How would you define these terms?

ENVIRONMENTAL LAW

Laws are rules set by authority, society or custom. Church laws, social mores, administrative regulations, and a variety of other codes of behavior can be considered laws if they are backed by some enforcement power. Government laws are established by federal, state, or local legislative bodies or administrative agencies. **Environmental law** constitutes a special body of official rules, decisions, and actions concerning environmental quality, natural resources, and ecological sustainability. Each branch of government plays a role in establishing the rules of law. **Statute**

FIGURE 24.4 What level of force is justified in environmental protection? This park in Singapore takes a draconian stance. © William P. Cunningham.

law consists of formal documents or decrees enacted by the legislative branch of government declaring, commanding, or prohibiting something. It represents the formal will of the legislature. **Case law** is derived from court decisions in both civil and criminal cases. **Administrative law** rises from executive orders, administrative rules and regulations, and enforcement decisions in which statutes passed by the legislature are interpreted in specific applications and individual cases (fig. 24.4). Because every country has different legislative and legal processes, this chapter will focus primarily on the U.S. system in the interest of simplicity and space.

A Brief Environmental History

Before looking at how each of these branches of government functions it might be useful to review how the underlying attitudes toward our environment and the policies that underlie our environmental laws have changed over the past century. For most of its history, U.S. environmental policy has had a laissez-faire or hands-off attitude toward business and private property. Pollution and environmental degradation were regarded as the unfortunate but necessary cost of doing business. If you didn't like the smell of a tannery or the sight of a waste dump, you were free to go somewhere else. While there were some early laws forbidding gross interference with another person's property or rights—the Rivers and Harbors Act of 1899, for example, made it illegal to dump so much refuse in waterways that navigation was blocked—in general, everyone was free to do whatever they wanted on their own property. People were either unaware of, or didn't pay much attention to, the fact that pollutants can move through the air, soil, and water to endanger people distant from the source.

The emergence of the modern environmental movement in the 1960s and 1970s marked a dramatic turning point in our understanding of the dangerous consequences of pollution and our demands to be protected from it. Rachel Carson's *Silent Spring* (1962) and Barry Commoner's *Closing Circle* (1971) alerted the public to the ecological and health risks of pesticides, hazardous

wastes, and toxic industrial effluents. Public activism in the civil rights and antiwar movements was carried over to environmental protests and demands for environmental protection. Emergence of new media—especially television—provided access to environmental news and made events in faraway places seem immediate and important.

The 1969 blowout of an oil well in the Santa Barbara Channel just off the coast of southern California is a good example of how the convergence of actors, events, timing, and media attention can shape public opinion and influence the policy cycle. For many weeks, black, gooey crude oil washed up onto beautiful southern California beaches. The oil spill made a perfect story for TV. The continuing saga was ideal for nightly updates. The setting guaranteed good photos and was readily recognized by the viewing public as an important place (fig. 24.5). National news networks had just developed the capacity for live satellite feeds and were hungry to use their new technology. Los Angeles was one of few locations with reliable uplinks and Santa Barbara was close enough for a film crew to go out every day to get some good footage and be back in the studio in time for the five o'clock news.

Because the story was ongoing, it fit well in the 30-second spots characteristic of TV news. The audience was familiar with both the issue and the images that described it. Reporters didn't have to spend precious seconds explaining what was happening, but could just give a sporting-event-like update on which side was winning today. A policy debate in Congress might be more important, but is too complex to explain in a few seconds and doesn't

FIGURE 24.5 Beach cleanup efforts after the Santa Barbara oil spill in 1969 made excellent media material and had an important role in U.S. environmental policy. © Esther Henderson/Photo Researchers, Inc.

provide exciting visuals. The wealthy residents of Santa Barbara were media-savvy, and had the influence and contacts to publicize the oil spill. While some of the cleanup efforts were not very effective, they made great visual footage. Attractive young people, smudged with oil, trying vainly to sweep gooey crud off a beautiful beach made ideal TV footage. Although the Santa Barbara oil spill wasn't nearly as big as others around the world, it played an important role in mobilizing public opinion and was an major factor in passage of the 1972 U.S. Clean Water Act.

As a result of awakened public concern about environmental issues, more than 27 major federal laws for environmental protection and hundreds of administrative regulations were established in the United States in the environmental decade of the 1970s. The statutes, case law, rules, precedents, and agencies resulting from that period created the foundation on which much of our current environmental protection rests. In the initial phase of this environmental revolution, the main focus was on direct regulation and litigation to force malefactors to obey the law. In recent years, attention has shifted from end-of-the-pipe command and control to pollution prevention and collaborative methods that can provide win/win solutions for all stakeholders. We'll look at some of those alternative dispute resolution approaches later in this chapter. But next, let's look more closely at how environmental law is established and administered.

Statutory Law: The Legislative Branch

Establishing laws at either the state or federal level is one of the most important ways of protecting our environment. Many environmental groups spend a good deal of their time and resources trying to influence the legislative process. In this section we'll look at how that system works.

Federal laws (statutes) are enacted by Congress and must be signed by the president. They originate as legislative proposals called bills, which are usually drafted by the congressional staff, often in consultation with representatives of various interest groups. Thousands of bills are introduced every year in Congress. Some are very narrow, providing funds to build a specific section of road or to help a particular person, for instance. Others are extremely broad, perhaps overhauling the social security system or changing the entire tax code. Similarly, environmental legislation might deal with a very specific local problem or a national or international issue. Often a number of competing bills on a single issue may be introduced as proponents from different sides attempt to incorporate their views into law. A bill may have a single sponsor if it is the pet project of a particular legislator, or it may have 100 or more coauthors if it is an issue of national importance.

A Convoluted Path

After introduction, each bill is referred to a committee or subcommittee with jurisdiction over the issue for hearings and debate. Most hearings take place in Washington, but if the bill is controversial or legislators want to attract publicity for themselves or the issue, they may conduct field hearings closer to the site of the controversy. The public often has an opportunity to give testimony at field hearings (fig. 24.6). Although it's not likely that you will change the opinions of many legislators no matter how fervent or cogent your testimony, these events can be a good place to gain attention and educate the public about a topic. Hearings and debates also build a record of legislative intent that can be valuable in later interpretation and implementation of laws by courts and administrative agencies.

If a bill has sufficient support within the subcommittee, its language will be "marked up" or revised and modified to be more widely acceptable and to improve its chances of passage. At this stage several competing bills might be combined into a single compromise version. If the compromise bill garners sufficient support, it is forwarded to the full committee for more hearings, debate, and a vote. If it fails in the full committee, the bill is sent back to the subcommittee for more work and further compromise. A bill that succeeds in the full committee is reported to the full House or Senate for a floor debate. Often opponents of a bill will attempt to amend it during each of these stages to lessen its impact or to make it so unpalatable that even the original authors can no longer support it. Some amendments add completely unrelated material or even reverse the intent of a bill. As bills move through this convoluted pathway, interested parties can follow their progress in the *Congressional Quarterly Weekly,* a publication both in print and online that keeps track of proposed legislation. Many environmental groups also maintain websites with up-to-date information on events in Congress.

By the time an issue has passed through both the House and Senate, the versions approved by the two bodies are likely to be different. They go then to conference committee to iron out any differences between them. After going back to the House and Senate for confirmation, the final bill goes to the president, who may either sign it into law or veto it. If the president vetoes the bill, it may still become law if two-thirds of the House and Senate vote to override the veto. If the president takes no action within ten days of receiving a bill from Congress, the bill becomes law without his signature.

FIGURE 24.6   Citizens line up to testify at a legislative hearing. By getting involved in the legislative process, you can be informed and have an impact on governmental policy. © Bob Daemmrich/The Image Works.

One exception to this procedure is that if Congress adjourns before the ten-day period elapses, the bill does not become law. The president, by doing nothing, is said to have exercised a "pocket veto."

The Thomas website, maintained by Congress, also has current information about the progress of legislation. You can find out how your senator and representative voted on critical environmental issues by consulting The League of Conservation Voters, which ranks each member of Congress on their voting record. The Defenders of Wildlife maintains a daily e-mail environmental news service called Greenwire that has up-to-date information about what's happening in Washington. Table 24.1 lists some of the most important recent federal environmental legislation.

Legislative Riders

There are two types of legislation: authorizing bills become law, while appropriation bills provide funds for federal agencies and programs. Appropriation bills can have language attached expressing the intent of Congress, but, in theory, at least, are not supposed to make policy, merely fund existing plans and projects. Legislators who can't muster enough votes to pass pet projects through regular channels often will try to add authorizing amendments

called **riders** into completely unrelated funding bills. Even if they oppose the riders, other members of Congress have a difficult time voting against an appropriation package for disaster relief or to fund programs that benefit their districts. Often this happens in conference committee because when the conference report goes back to the House and Senate, the vote is either to accept or reject with no opportunity to debate or amend further.

Starting with the 104th Congress, antienvironmental forces began using this tactic to roll back environmental protections and gain access to natural resources. Environmental groups were outraged, for instance, when riders were attached to 1996 supplemental spending bills that put a moratorium on listing additional species under the Endangered Species Act and exempted "salvage" logging on public lands from environmental laws. In subsequent years, numerous antienvironmental riders have been attached to appropriation bills. The 2004 Omnibus spending bill, for example, included numerous special-interest amendments to prevent administrative appeals and judicial reviews of environmentally destructive government policies, allow increased logging and road building in Alaska's Tongass National Forest, cut funding for land conservation, weaken national organic labeling standards, and

TABLE 24.1	Major U.S. Environmental Laws

LEGISLATION	PROVISIONS
Wilderness Act of 1964	Established the national wilderness preservation system.
National Environmental Policy Act of 1969	Declared national environmental policy, required environmental impact statements, created Council on Environmental Quality.
Clean Air Act of 1970	Established national primary and secondary air quality standards. Required states to develop implementation plans. Major amendments in 1977 and 1990.
Clean Water Act of 1972	Set national water quality goals and created pollutant discharge permits. Major amendments in 1977 and 1996.
Federal Pesticides Control Act of 1972	Required registration of all pesticides in U.S. commerce. Major modifications in 1996.
Marine Protection Act of 1972	Regulated dumping of waste into oceans and coastal waters.
Coastal Zone Management Act of 1972	Provided funds for state planning and management of coastal areas.
Endangered Species Act of 1973	Protected threatened and endangered species, directed FWS to prepare recovery plans.
Safe Drinking Water Act of 1974	Set standards for safety of public drinking-water supplies and to safeguard groundwater. Major changes made in 1986 and 1996.
Toxic Substances Control Act of 1976	Authorized EPA to ban or regulate chemicals deemed a risk to health or the environment.
Federal Land Policy and Management Act of 1976	Charged the BLM with long-term management of public lands. Ended homesteading and most sales of public lands.
Resource Conservation and Recovery Act of 1976	Regulated hazardous waste storage, treatment, transportation, and disposal. Major amendments in 1984.
National Forest Management Act of 1976	Gave statutory permanence to national forests. Directed USFS to manage forests for "multiple use."
Surface Mining Control and Reclamation Act of 1977	Limited strip mining on farmland and steep slopes. Required restoration of land to original contours.
Alaska National Interest Lands Act of 1980	Protected 40 million ha (100 million acres) of parks, wilderness, and wildlife refuges.
Comprehensive Environmental Response, Compensation and Liability Act of 1980	Created $1.6 billion "Superfund" for emergency response, spill prevention, and site remediation for toxic wastes. Established liability for cleanup costs.
Superfund Amendments and Reauthorization Act of 1994	Increased Superfund to $8.5 billion. Shares responsibility for cleanup among potentially responsible parties. Emphasizes remediation and public "right to know."

Source: *N. Vig and M. Kraft,* Environmental Policy in the 1990s, *3rd Congressional Quarterly Press.*

expand forest-thinning projects. Generally, riders are tacked onto completely unrelated bills that legislators will have difficulty voting against. A rider to eliminate critical habitat for endangered species, for example, was hung on a veteran's health care bill. Congressional leaders pledged to end this practice, but little has been done so far to stop it.

Lobbying

Groups or individuals with an interest in pending legislation can often cause a great deal of influence by **lobbying,** or using personal contacts, public pressure, and political action to persuade legislators to vote or act in their favor. The term derives from the habit of partisans and professional lobbyists to lurk in the hallways and lobbies of Congress hoping to snare a passing legislator to urge them to vote in a specific way. We also use the same term to describe efforts to influence administrative agencies.

Most major environmental organizations maintain offices in Washington from which they monitor legislative and administrative programs and policies. Hundreds of professional environmental lobbyists and volunteer activists attend hearings, meet with legislators and agency personnel, draft proposed legislation and administrative rules, and attempt in a variety of ways to shape the national environmental agenda. They join thousands of other amateur and professional lobbyists representing industry and business organizations, workers, property owners, religious groups, ethnic associations, and just about every other kind of interest group that you can imagine. Walking the halls of Congress or those of the House or Senate office buildings, you see an amazing mixture of people attempting to be heard. It's fascinating to be part of this process.

In a survey of professional lobbyists, a majority agreed that personal contacts were the most effective way to influence decision makers. Undoubtedly the best way to make contact is through personal friends of a legislator or someone to whom they owe political allegiance. Having a famous movie star, a person of great power or wealth, or some other celebrity represent your group also can help open doors for your ideas. Your own senator or representative is more likely to be interested in your views than someone with whom you have no connection. They place a high priority on responding to their own constituents who can vote them in (or out) of office. But even if you aren't rich or illustrious or politically connected, you can often get a fair hearing from legislative staff—if not their bosses—if you have a persuasive case concerning an important topic.

What can those of us do to have an impact on legislation who can't afford to go to Washington to be directly involved? Getting involved in local election campaigns can greatly increase your access to legislators. Writing letters or making telephone calls also are highly effective ways to get your message across. You'd be surprised at how few letters or calls legislators receive even on important national issues. Your voice can have an important impact. How to write an effective letter is described further in chapter 25. All legislators now have e-mail addresses, although it isn't clear how much weight this form of communication carries.

Getting media attention can sway the opinions of decision makers. Organizing protests, marches, demonstrations, street

FIGURE 24.7 Making a ruckus on behalf of environmental protection can attract attention to your cause. © William P. Cunningham.

theater, or other kinds of public events can call attention to your issue (fig. 24.7). Public education campaigns, press conferences, TV ads, and a host of other activities can be helpful. Tax-exempt (503c) organizations can't lobby directly or engage in politics, but there is a murky line between educational ads and outright campaigning and lobbying. Joining together with other like-minded groups can greatly increase your clout and ability to get things done. It's hard for a single individual or even a small group to have much impact, but if you can organize a mass movement, you may be very effective.

Case Law: The Judicial Branch

Over the past 30 years, appeals to the judicial system have often been the most effective ways for seeking redress for environmental damage and forcing changes in how things are done. Activist judges and sympathetic juries in both federal and state court systems have been willing to take a stand where legislatures have been too timid or conservative to do so. Many groups spend a great deal of their time and energy bringing lawsuits that will shape environmental policy. The Environmental Defense Fund, for example, operates primarily in this arena. In the early days of the organization, their motto was, "Sue the bastards, that will get their attention." Even if you're not interested in environmental issues, it's worthwhile knowing something about how this system works. You may find yourself in court someday.

The judicial branch of government establishes environmental law by ruling on the constitutionality of statutes and interpreting their meaning. We describe the body of legal opinions built up by many court cases as case law. Often legislation is written in vague and general terms so as to make it widely enough accepted to gain passage. Congress, especially in the environmental area, often leaves it to the courts to "fill in the gaps." As one senator said when Congress was about to pass the Superfund legislation, "All we know is that the American people want these hazardous waste sites cleaned up . . . Let the courts worry about the details." When trying to interpret a law, the courts depend on the legislative record from hearings and debates to determine congressional intent. What was a particular statute meant to do by those who wrote and passed it?

The Court System

The United States is divided into 96 federal court districts, each of which has at least one trial court. Over these district courts are the circuit court of appeals, which hears disputes arising from questions about procedural issues and interpretations of the law in district courts. There are 12 geographic regions for the appeals courts. The federal courts have jurisdiction over federal criminal prosecutions, claims against the federal government, claims arising under federal statutes or treaties, and cases in which defendants or plaintiffs come from two or more states. The residence of the defendant or location of the property in dispute usually determines the venue, or the court in which each case is heard. Each state has its own courts that generally parallel the federal system. These courts have jurisdiction over cases arising from state laws. The U.S. Supreme Court is the court of last resort for appeals for both federal and state court systems.

A trial court judge presides over trials, rules on motions made by attorneys, and decides questions of law, such as what evidence is admissible, and what law applies to the case (fig. 24.8). The judge controls the pace of the proceedings and maintains decorum in the court room. Although Thomas Jefferson said, "Ours is a government of laws, not men," in reality, the judge has tremendous power over the outcome of a trial. Certain courts earn a reputation of being pro- or antienvironmental. Litigants always try to know something about the judge's ideology before bringing a case to trial.

The first judge to hear a case arising from a particular statute or situation has the greatest latitude to interpret the law and set a **precedent** to be used as an example in subsequent trials. These decisions are binding, however, only on those courts on a lower level and in the same system. Precedents from the California courts, for instance, are not binding in Arizona courts. Furthermore, if a judge *distinguishes* a case—determines it is different from

FIGURE 24.8   In a trial court, the judge presides over trials, rules on motions, and interprets the law. © Photodisc/Volume 25.

other cases—she or he is not obliged to adhere to prior precedents. Or a judge can always simply overturn a clearly applicable precedent as a matter of "correcting the law" where technology or changing community values has made prior decisions outdated.

Legal Thresholds

Before a trial can start, the litigants must satisfy certain threshold requirements. The first of these is **standing,** or whether the participants have a right to stand before the bar and be heard. The main criteria for standing is a valid interest in the case. Plaintiffs must show that they are materially affected by the situation they petition the court to redress. This is an important point in environmental cases. Groups or individuals often want to sue a person or corporation for degrading the environment. But unless they can show that they personally suffer from the degradation, courts are likely to deny standing.

In a landmark 1969 case, for example, the Sierra Club challenged a decision of the Forest Service and the Department of the Interior to lease public land in California to Walt Disney Enterprises for a ski resort. The land in question was a beautiful valley that cut into the southern boundary of Sequoia National Park (see fig. 2.4). Building a road into the valley would have necessitated cutting down a grove of giant redwood trees within the park. The Sierra Club argued that it should be granted standing in the case to represent the trees, animals, rocks, and mountains that couldn't defend their own interests in court. After all, the club pointed out corporations—such as Disney Enterprises—are treated as persons and represented by attorneys in the courts. Why not grant trees the same rights? The case went all the way to the Supreme Court, which ruled that the Sierra Club failed to show that it or any of its members would be materially affected by the development.

In addition to standing, plaintiffs must show that their case represents a "live" legal dispute that is likely to result in a final and meaningful judgment, and that there is a present controversy for which a decision is needed. In other words, you can't bring a suit over a hypothetical situation or one that is no longer cogent. For example, suppose you want to stop a development project that would destroy endangered species habitat. If the defendant can show that the species couldn't inhabit the habitat (if it were already extinct, for instance), the case is moot (of no practical importance) and the courts will refuse to hear it.

Criminal Law

Criminal law derives from those federal and state statutes that prohibit wrongs against the state or society, such as arson, rape, murder, and robbery. Serious crimes, like murder or rape, that are punishable by long-term incarceration or heavy fines are called felonies. Lesser crimes, such as shoplifting or vandalism, that result in smaller fines or shorter sentences in a county or city jail are labeled misdemeanors. Definitions vary from state to state. What may be a felony crime in one state may be only a misdemeanor in another.

A criminal case is always initiated by a government prosecutor. Guilt or innocence of the defendant is determined by a jury of peers, but the sentence is imposed, often in consultation with the

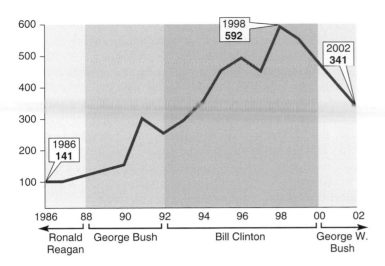

FIGURE 24.9 The number of criminal cases referred each year for prosecution by the EPA has varied with the political climate in Washington.

jury, by the judge. The judge is responsible for keeping order in the hearings and for determining points of law. The jury acts as a fact-finding body that weighs the truth and reliability of the witnesses and evidence.

Violation of many environmental statutes constitutes criminal offenses. In 1975, the U.S. Supreme Court ruled that corporate officers can be held criminally liable for violations of environmental laws if they were grossly negligent, or the illegal actions can be considered willful and knowing violations. In 1982, the EPA created an Office for Criminal Investigation. Under Bill Clinton, prosecutions for environmental crimes rose to nearly 600 per year. They have fallen to roughly half that number under George W. Bush, however (fig. 24.9). In one of the toughest criminal sentences imposed so far, the president of a Colorado company was sentenced, in 1999, to 14 years in prison for knowingly dumping chlorinated solvents that contaminated the water table. The company itself was fined $950,000 and put on probation for ten years.

In 2003, the Colonial Pipeline Company was fined $34 million for oil spills, the largest civil penalty ever under the EPA. Also in 2003, Monsanto and its subsidiaries agreed to an out-of-court settlement of $700 million for dumping PCBs in Anniston, Alabama. Altogether, the EPA forced polluters to pay $3.9 billion in 2003 for pollution controls and cleanup. Too often international environmental crimes go unpunished because of jurisdictional complications. The European Union has called for a Global Environmental Crime Intelligence Unit, much like the renowned Interpol, to investigate illegal logging, waste dumping, and other transborder crimes.

Civil Law

Civil law is defined as a body of laws regulating relations between individuals or between individuals and corporations. Issues such as property rights, and personal dignity and freedom are protected by civil law. In some cases, legislative statutes, such as the Civil Rights Act, establish specific aspects of civil law. In other cases, where no particular statute exists, custom and a body of previous court decisions, collectively called **common law,** establish precedents that constitute a working definition of individual rights and responsibilities. The Woburn case is an example of **tort law** (tort derives from the Middle English word for injury) that seeks compensation for damages. This kind of civil action is usually initiated by the attorney representing the injured party (the plaintiff). The defendant in a civil case has a right to be tried by a jury, but in highly technical issues, this right often is waived and the case is heard only by the judge. Being found guilty of a civil offense can result in financial penalties but not jail time.

In contrast to a criminal case where the burden of proof lies with the prosecution, and defendants are considered innocent until proven guilty, civil cases can be decided on a "preponderance of evidence." This makes civil cases considerably easier to win than criminal cases where the evidence is ambiguous. A number of mitigating factors also are taken into account in determining guilt and assigning penalties in civil cases. Culpability is based on whether the defendant could reasonably have anticipated and avoided the offense. A "good faith effort" to comply or solve the problem can be a factor. The compliance history is important. Is this a first offense or a habitual repeater? Finally, is there evidence of economic benefit to the perpetrator? That is, did the violator gain personally from the action? If so, it is more likely that willful intent was involved.

Most people consider being convicted of a criminal offense much more serious than losing a civil case, because the former can lead to incarceration while the latter only costs money. Civil judgments can be costly, however. A group of Alaskan fishermen won $5 billion from the Exxon oil company for damages caused by the 1989 *Exxon Valdez* oil spill. Civil cases can be brought in both state and federal court. In 2000, the Koch oil company, one of the largest pipeline and refinery operators in the United States, agreed to pay $35 million in fines and penalties to state and federal authorities for negligence in more than 300 oil spills in Texas, Oklahoma, Kansas, Alabama, Louisiana, and Missouri between 1990 and 1997. Koch also agreed to spend more than $1 billion on cleanup and improved operations.

Sometimes the purpose of a civil suit is to seek an injunction or some other form of equitable relief from the actions of an individual, a corporation, or a governmental agency. You might ask the courts, for example, to order the government to cease and desist from activities that are in violation of either the spirit or the letter of the law. This sort of civil action is heard only by a judge; no jury is present. Environmental groups have been very successful in asking courts to stop logging and mining operations, to enforce implementation of the endangered species act, to require agencies to enforce air and water pollution laws, and a host of other efforts to protect the environment and conserve natural resources. Often, rather than sue a corporation directly for environmental damage, it is more effective to sue the government for not enforcing laws that would have prevented the damage. A big corporation with deep pockets can afford squadrons of lawyers and may have the resources and incentive to tie up litigation for years with motions and counter suits. Federal or state agencies may be more inclined to agree that you are right and to be willing to settle the matter quickly.

Adversarial Approaches and SLAPP suits

The American legal system is adversarial, pitting one side against the other in an effort to distinguish right from wrong, or innocence from guilt. In a trial, each side tries to make the strongest possible arguments for its position, and to point out the faults in the opponent's case. The jury, as neutral fact-finder, hears arguments from both sides and makes an objective decision. We believe that this approach gives the best chance for truth to be discovered. It is time-consuming and costly, however, and doesn't always result in justice. Everyone has heard of cases where it seems perfectly obvious that the defendant is guilty, and yet they get off because certain evidence is inadmissible, or they simply have lawyers who can dazzle or confuse the jury with deceptive arguments. This system promotes strife and confrontation. It doesn't encourage compromise and often leaves lasting hatreds that make future cooperation impossible.

Because defending a lawsuit is so expensive, the mere threat of litigation can be a chilling deterrent. Increasingly, environmental activists are being harassed with **Strategic Lawsuits Against Public Participation (SLAPP).** Citizens who criticize businesses that pollute or government agencies that are derelict in their duty to protect the environment are often sued in retaliation. While most of these preemptive strikes are groundless and ultimately dismissed, defending yourself against them can be exorbitantly expensive and take up time that might have been spent working on the original issue. Public interest groups and individual activists—many of whom have little money to defend themselves—often are intimidated from taking on polluters. For example, a West Virginia farmer wrote an article about a coal company's pollution of the Buckhannon River. The company sued him for $200,000 for defamation. Similarly, citizen groups fighting a proposed incinerator in upstate New York were sued for $1.5 million by their own county governments. A Texas woman called a nearby landfill a dump—and her husband was named in a $5 million suit for failing to "control his wife." Of course these suits also are expensive for the company or agency that initiates them, but they may be far cheaper than paying a fine or scrapping a big project.

Administrative Law: The Executive Branch

More than 100 federal agencies and thousands of state and local boards and commissions have environmental oversight. They usually have power to set rules, adjudicate disputes, and investigate misconduct. Federal agencies often delegate power to a matching state agency in order to decentralize authority. The enabling legislation to create each agency is called an "organic" act because it establishes a basic unit of governmental organization. In the federal government, most executive agencies come under the jurisdiction of cabinet-level departments such as Agriculture, Interior, or Justice (fig. 24.10).

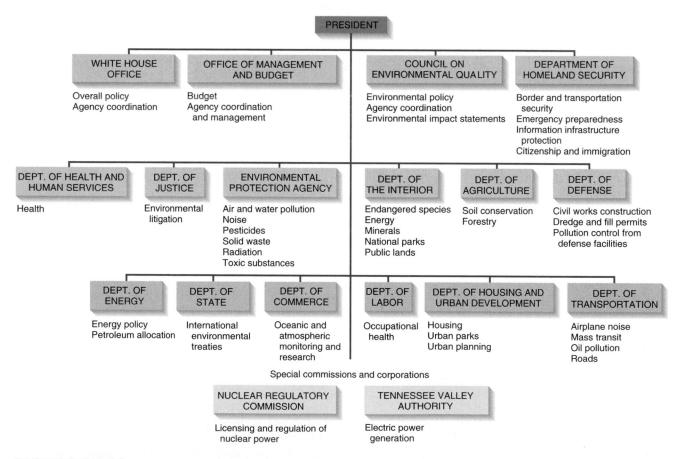

FIGURE 24.10 Major agencies of the Executive Branch of the U.S. federal government with responsibility for resource management and environmental protection. *Source:* U.S. General Accounting Office.

Agency rule-making and standard-setting can be either formal or informal. In an informal case, notice and background for proposed rules are published in the Federal Register. Opportunities for all interested parties to submit comments are provided. This is often an important avenue for environmentally concerned citizens and public interest groups to have an impact on environmental policy. In formal rule making, a public hearing is held with witnesses and testimony much like a civil trial. Witnesses can be cross-examined. A complete transcript is made and final findings are published in the public record. It is generally more difficult for individuals to intervene in a formal hearing, although sometimes there is an opportunity to submit written comments. Rule-making is often a complex, highly technical process that is difficult for citizen groups to understand and monitor. The proceedings are usually less dramatic and colorful than criminal trials, and yet can be very important for environmental protection.

Executive orders also can be powerful agents for change. In 1994, for instance, President Bill Clinton issued Executive Order 12898 requiring all federal agencies to collect data on effects of pollution on minorities, and to develop strategies to promote environmental justice. During his two terms in office, Clinton used the Antiquities Act to establish 22 new national monuments (fig. 24.11). In addition, he expanded dozens of existing national parks and wildlife refuges. Altogether, Clinton ordered protection for about 36 million ha (90 million acres) of nature preserves, the largest of which was the Pacific Ocean reserve composed of 34 million ha of ocean and coral reefs northwest of Hawaii. In addition, the U.S. Forest Service, under Clinton appointee Mike Dombeck, ordered a moratorium on road building and logging on nearly 24 million ha of *de facto* forest wilderness.

Rules and policies made by executive decree in one administration can be quickly undone in the next one. In his first day in office, President George W. Bush ordered all federal agencies to suspend or ignore more than 60 rules and regulations from the Clinton administration. In addition, Mr. Bush called for a sweeping overhaul of environmental laws to ease restrictions on businesses and to speed decisions on development projects. His supporters regard these policies as merely restoring reason and balance to government; critics see this as a radical ideological campaign to roll back all the environmental protections and social progress of the past century (What Do You Think? p. 532). Because most of this agenda has been pursued through agency regulations and executive orders, most Americans, distracted by terrorism, a weak economy, and lingering wars in Afghanistan and Iraq, seem unaware of the magnitude and implications of this abrupt policy shift.

Regulatory Agencies

The EPA is the primary agency with responsibility for protecting environmental quality. Created in 1970, at the same time as NEPA, the EPA is a cabinet-level department, with more than 18,000 employees and ten regional offices. Often in conflict with Congress, other agencies of the Executive Branch, and environmental groups, the EPA has to balance many competing interests and conflicting opinions. Greatly influenced by politics, the agency changes dramatically depending on which party is in power and what attitudes

FIGURE 24.11   In his two terms in office, President Bill Clinton created 22 new national monuments, including the U.S. Virgin Islands Coral Reef Monument shown here. Whether these conservation orders will last remains to be seen. © William P. Cunningham.

toward the environment prevail at any given time. Under the Nixon and Carter administrations, the EPA grew rapidly and enforced air and water quality standards vigorously. It declined sharply during the Reagan administration, then recovered under Bill Clinton.

The Departments of the Interior and Agriculture are to natural resources what the EPA is to pollution. Interior is home to the National Park Service, which is responsible for more than 376 national parks, monuments, historic sites, and recreational areas. It also houses the Bureau of Land Management (BLM), which administers some 140 million ha of land, mostly in the western United States. In addition, Interior is home to the U.S. Fish and Wildlife Service, which operates more than 500 national wildlife refuges and administers endangered species protection.

The Department of Agriculture is home to the U.S. Forest Service, which manages about 175 national forests and grasslands, totaling some 78 million ha. With 39,000 employees, the Forest Service is nearly twice as large as the EPA (fig. 24.12). The Department of Labor houses the Occupational Safety and Health Agency (OSHA), which oversees workplace safety. Research that forms the basis for OSHA standards is carried out by the National Institute for Occupational Safety and Health (NIOSH). In addition, several independent agencies that are not tied to any specific department also play a role in environmental protection and public health. The Consumer Products Safety Commission passes and

What do you think?

Restoring Balance or Promoting Radical Ideology?

In his first term in office, President George W. Bush embarked on a sweeping revision of American environmental policy. His supporters are thrilled by what they see as an effort to restore good sense and fairness to government. Critics, on the other hand, see this as an attempt to roll back most conservation efforts and social progress of the past century.

When campaigning for office, Mr. Bush promised a "New Environmentalism" that would provide incentives for voluntary conservation rather than the rigid "command and control" policies that tell polluters exactly how much to clean up effluents and how to do it. Instead, Mr. Bush promotes "kinder, gentler, more flexible rules" that remove burdensome limitations on developers and resource extractors. Some basic tenets of this agenda are that government is too big and too powerful, and has gone too far in restricting industry and commerce. "At the heart of the New Environmentalism," says Gale Norton, Bush's Secretary of the Interior, "is a conviction that federal lands are underutilized as a supply of natural resources, and that we have reached the limits on what we can do through government regulation and mandates." Previous efforts to safeguard our environment, Mr. Bush argues, have merely produced bureaucratic gridlock and "analysis paralysis." Promoting a kind of libertarian environmentalism, Mr. Bush wants to depend more on market mechanisms and local control to achieve environmental goals.

But what Mr. Bush and his supporters see as common sense, balance, and efficiency, others see as a return to crony capitalism and resource plundering of a century ago. White House initiatives such as "Healthy Forests" and "Clear Skies" are derided as Orwellian double-speak that promote exactly the opposite of what they proclaim. The League of Conservation Voters contends that Bush is well on the way to compiling the worst environmental record of any president in American history. "Across the board," says Senator James M. Jeffords, the Vermont independent who, until recently, was the chairman of the Senate Committee on Environment and Public Works, "we would be better off doing nothing than doing what the Bush administration wants to do, which will make things worse than they already are." Journalist Bill Moyers calls the Bush approach "faith-based resource management" because it is based on an unshakeable belief that humans are meant to dominate the earth, and that free markets always find the optimum solutions for any resource question.

The following are among dozens of policies of the Bush administration that outrage and alarm conservationists:

- Permit logging, mining and road building on 24 million ha of *de facto* wilderness on federal lands.

- Expedite oil drilling in the Arctic National Wildlife Refuge in Alaska.

- Abandon Clean Air rules that require the oldest and dirtiest power plants and refineries to install modern pollution-control equipment when they undergo major repairs or upgrades.

- Reject Clean Water standards that require lower arsenic levels in drinking water.

- Modify or ignore NEPA to speed decisions on resource extraction.

- Refuse to ratify the Kyoto Protocol on climate change.

- Exempt forest-thinning projects from environmental review and citizen appeals.

- Remove protection for up to one-half of the nation's wetlands, small ponds, and headwater streams.

- Allow continued use of snowmobiles, personal watercraft, and off-road vehicles in national parks, national seashores, and other protected areas.

- Cut funds for 33 hazardous waste clean-up projects in 18 states.

- Open millions of hectares of the Tongass National Forest in Alaska to logging and mining.

- Reverse Clean Water Act ban on dumping of mining waste into waterways and wetlands.

- Order the Department of Energy to begin burying high-level nuclear waste at Yucca Mountain, Nevada, despite uncertainties about the safety or stability of the site.

- Promote an energy policy favoring expanded use of fossil fuels and nuclear power and dismissing conservation and renewable energy as impractical.

- Exempt the military from almost all environmental regulations.

- Scrap a $1.5 billion program to build high-efficiency, low-emissions automobiles that would get 60 mpg. Instead, require an increase in average fleet efficiency by 1.5 mpg (from 20.4 to 21.9 mpg) over three years.

What do you think? Would you support these policy changes? If you need more information to decide, you'll find more discussion about many of these issues in prior chapters of this book. Is this a kinder, gentler environmentalism, or return to the ways of robber barons? How much environmental protection or environmental quality is too much, in your opinion?

---

enforces regulations to protect consumers, and the Food and Drug Administration is responsible for the purity and wholesomeness of food and drugs.

All of these agencies have a tendency to be "captured" by the industries they are supposed to be regulating. Many of the people with expertise to regulate specific areas came from the industry or sector of society that their agency oversees. Furthermore, the people they work most closely with and often develop friendships with are those they are supposed to watch. And when they leave the agency to return to private life—as many do when the administration changes—they are likely to go back to the same industry or sector where their experience and expertise lies. The effect is often what's called a "revolving door," where workers move back and forth between industry and government. As a result regulators often become overly sympathetic with and protective of the industry they should be overseeing.

FIGURE 24.12 A forest ranger explains wilderness rules to a hiker. © William P. Cunningham.

Administrative Courts

Administrative regulations and rule-making procedures often are objected to by affected parties. Over the past decade, 80 percent of the rules made by the EPA, for example, were challenged in court. It is very unusual for the courts to overturn agency rules, but the challenge buys time for regulated corporations to continue business as usual for a while longer. **Administrative courts** hear challenges to agency rules and regulations. An administrative judge can consider both the validity of the rule and its application to a specific case. If the parties dispute the judge's findings, they can appeal to a district court. The courts rarely overturn agency rules unless (1) the enabling act is too vague or unconstitutional, (2) the agency has gone beyond the scope of power granted by the legislature, or (3) the agency didn't follow proper procedures.

Administrative courts also hear enforcement cases where an individual or corporation has violated an agency rule or standard. Suppose, for instance, that a factory is found to be exceeding allowable air pollution emissions. After an investigation, a complaint is filed with an administrative judge. A hearing is scheduled and the judge listens to both sides of the case and issues an opinion, which is usually a recommendation to the head of the responsible agency

for a penalty or remediation action. This decision also can be appealed to the circuit court of appeals.

The rules for evidence are usually less strict in an administrative case than in a criminal action. The administrative judge acts as both fact finder and decision maker. There is no jury. The judge can question witnesses and ask for additional evidence. The emphasis is on finding the truth rather than sticking to strict rules of procedure. Administrative courts often recommend relatively small penalties as a way of encouraging early settlement. In one case, a company charged with a violation of EPA standards was assessed a $500 fine. The company appealed the case to the federal district court, which imposed a $10,000 penalty.

INTERNATIONAL TREATIES AND CONVENTIONS

As recognition of the interconnections in our global environment has advanced, the willingness of nations to enter into protective covenants and treaties has grown concomitantly. Table 24.2 lists some major international treaties and conventions, while figure 24.13 shows the number of participating parties in them. Note that the earliest of these conventions has no nations as participants; they were negotiated entirely by panels of experts. Not only the number of parties taking part in these negotiations has grown, but the rate at which parties are signing on and the speed at which agreements take force also have increased rapidly. The Convention on International Trade in Endangered Species (CITES), for example, was not enforced until 14 years after ratification, but the Convention on Biological Diversity was enforceable after just one year, and had 160 signatories only four years after introduction. Over the past 25 years, more than 170 treaties and conventions have been negotiated to protect our global environment. Designed to regulate activities ranging from intercontinental shipping of hazardous waste, to deforestation, overfishing, trade in endangered species, global warming, and wetland protection, these agreements theoretically cover almost every aspect of human impacts on the environment.

Unfortunately, many of these environmental treaties constitute little more than vague, good intentions. In spite of the fact that we

TABLE 24.2	Some Important International Treaties

CBD: Convention on Biological Diversity 1992 (1993)
CITES: Convention on International Trade on Endangered Species of Wild Fauna and Flora 1973 (1987)
CMS: Convention on the Conservation of Migratory Species of Wild Animals 1979 (1983)
Basel: Basel Convention on the Transboundary Movements of Hazardous Wastes and their Disposal 1989 (1992)
Ozone: Vienna Convention for the Protection of the Ozone Layer and Montreal Protocol on Substances that Deplete the Ozone Layer 1985 (1988)
UNFCCC: United Nations Framework Convention on Climate Change 1992 (1994)
CCD: United Nations Convention to Combat Desertification in those Countries Experiencing Serious Drought and/or Desertification, Particularly in Africa 1994 (1996)
Ramsar: Convention on Wetlands of International Importance especially as Waterfowl Habitat 1971 (1975)
Heritage: Convention Concerning the Protection of the World Cultural and Natural Heritage 1972 (1975)
UNCLOS: United Nations Convention on the Law of the Sea 1982 (1994)

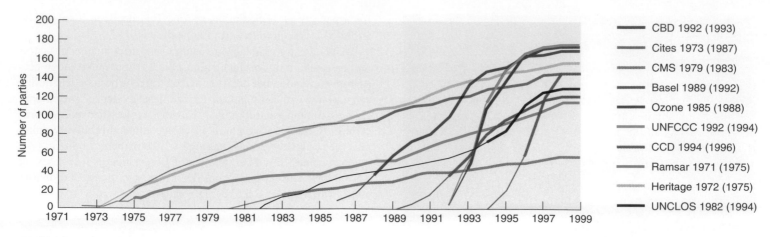

FIGURE 24.13 Number of participating parties in some major international environmental treaties. The thick portion of each line shows when the agreement went into effect (date in parentheses). See table 24.2 for complete treaty names. *Source:* United Nations Environment Programme from Global Environment Outlook-2000.

often call them laws, there is no body that can legislate or enforce international environmental protection. The United Nations and a variety of regional organizations bring stakeholders together to negotiate solutions to a variety of problems but the agreed-upon solutions generally rely on moral persuasion and public embarrassment for compliance. Most nations are unwilling to give up sovereignty. There is an international court, but it has no enforcement power. Nevertheless, there are creative ways to strengthen international environmental protection.

One of the principal problems with most international agreements is the tradition that they must be by unanimous consent. A single recalcitrant nation effectively has veto power over the wishes of the vast majority. For instance, more than 100 countries at the UN Conference on Environment and Development (UNCED), held in Rio de Janeiro in 1992, agreed to restrictions on the release of greenhouse gases. At the insistence of U.S. negotiators, however, the climate convention was reworded so that it only *urged*—but did not require—nations to stabilize their emissions.

As a way of avoiding this problem, some treaties incorporate innovative voting mechanisms. When a consensus cannot be reached, they allow a qualified majority to add stronger measures in the form of amendments that do not need ratification. All members are legally bound to the whole document unless they expressly object. This approach was used in the Montreal Protocol, passed in 1987 to halt the destruction of stratospheric ozone by chlorofluorocarbons (CFCs). The agreement allowed a vote of two-thirds of the 140 participating nations to amend the protocol. Although initially the protocol called for only a 50 percent reduction in CFC production, subsequent research showed that ozone was being depleted faster than previously thought. The protocol was strengthened by amendment to an outright ban on CFC production in spite of the objection of a few countries.

Where strong accords with meaningful sanctions cannot be passed, sometimes the pressure of world opinion generated by revealing the sources of pollution can be effective. NGOs and others can use this information to expose violators. For example, the environmental group Greenpeace discovered monitoring data in

1990 showing that Britain was disposing of coal ash in the North Sea. Although not explicitly forbidden by the Oslo Convention on ocean dumping, this evidence proved to be an embarrassment, and the practice was halted.

Trade sanctions can be an effective tool to compel compliance with international treaties. The Montreal Protocol, for example, bound signatory nations not to purchase CFCs or products made using them from countries that refused to ratify the treaty. Because many products employed CFCs in their manufacture, this stipulation proved to be very effective. On the other hand, trade agreements also can work against environmental protection. The World Trade Organization was established to make international trade more fair and to encourage development. It has been used, however, to subvert national environmental laws. In a ruling in 1998, the WTO forbid the United States from restricting imports of shrimp from Thailand, Malaysia, India, and Pakistan that were caught with nets that trap endangered sea turtles. Similarly, under provisions of the North American Free Trade Agreement, the Methanex Corporation of Canada filed a $1 billion claim against California for banning MTBE (a suspected carcinogen) in gasoline. Methanex claims that their sales of methanol (an ingredient of MTBE) will be harmed by this ban.

Increasingly, private citizens and nongovernmental organizations collect to protest policies of the far-reaching institutions like the World Bank, the International Monetary Fund, and the World Trade Organization (WTO). In 1999, for example, more than 50,000 people gathered in Seattle, Washington, to demonstrate their opposition to the WTO. Including farmers, workers, animal rights activists, environmentalists, and indigenous peoples, the protestors represented a wide range of complaints about how international politics impact their lives (fig. 24.14). Although most demonstrators were peaceful, a small group of anarchists and vandals set fires, broke windows, and looted stores. The police, unprepared for such a large gathering, responded with what many observers regarded as unnecessary force. An even more violent demonstration took place in 2001, when 100,000 protestors converged on a meeting of the Group of Eight Industrialized Nations in Genoa, Italy. Again, a small group of radicals started a riot in which one person was killed and hundreds were injured. Ironically, while many demonstrators called for a return to insular, nationalist policies, others were protesting because global institutions aren't powerful enough to protect them from transnational corporations and inequitable financial arrangements. The 2003 meeting of the WTO in Cancun, Mexico, for example, collapsed when delegates from developing nations insisted that the $300 billion in subsidies paid every year to the world's wealthiest farmers undermined the livelihoods of millions of poor farmers around the world.

Globalization and Environmental Governance

The rapid pace of **globalization**—the revolution in communications, transportation, finances and commerce that has brought about increasing interdependence of national economies—offers both opportunities and challenges to environmental management. Increasingly, we recognize that international cooperation is essential for conserving resources and maintaining a healthy environment. International commissions and conventions are paying ever more attention to how good environmental decisions are made. Do democratic rights and civil liberties contribute to better environmental management? Should local citizens or advocacy groups have the right to appeal a decision they believe harms an ecosystem or is unfair? What is the best way to fight corruption among those who manage our forests, water, parks, and mineral resources? These are all questions about how we make environmental decisions and who makes them—a process called **environmental governance.**

One of the strongest arguments for encouraging better governance is that it requires us to focus on the social dimensions of natural resource use and ecosystem management, in addition to the technical details of how to manage. This includes how we value ecosystems, how we form environmental policies, how we nego-

Key Concepts

- Environmental policies are the goals and rules—whether written or implied—that determine how we manage our natural resources and protect (or fail to protect) environmental quality and ecosystem health.

- Governance is the process by which we implement policy. How we decide and who gets to decide often determines what we decide. Democracy, equity, transparency, and accountability all can facilitate good environmental governance.

- Wicked problems have no clear right or wrong, often because they're nested within other sets of interlocking issues. Many environmental policy questions fall in this category. There are no value-free, objective answers to these dilemmas, only choices that are better or worse, depending on your viewpoint.

tiate trade-offs between conflicting uses or values, and, finally, how we make sure the costs and benefits of our decisions—including impacts on the poor—are equitably shared. A basic principle of governance is that how we decide and who gets to decide often determine what we decide.

The Aarhus Convention of 1988 specifically addresses these issues. First negotiated as a regional agreement among European countries, this document has now been ratified by 40 nations in Europe and Asia, and is open to signature by all nations of the world. In addition to recognizing the basic right of every person of present and future generations to a healthy environment, the convention also specifies how authorities at all levels will provide fair and transparent decision-making processes, access to information, and right to redress of grievances. Individuals don't need to prove legal standing to request information or comment on official decisions that affect their environment. The Aarhus Convention gives citizens, organizations, and governments the right to investigate and seek to curtail pollution caused by public and private entities in other countries that are parties to the treaty. For example, a Dutch public interest group could demand information about air or water emissions from a German factory.

Adopting and implementing the Aarhus Convention could greatly enhance global environmental governance. But while there is growing interest in endorsing the Aarhus principles worldwide, many countries see the treaty's concepts of democratic decision making about the environment as too progressive or threatening to business social relationships. Others worry about giving up national sovereignty and independence. What do you think? Would you support these principles? How would you advise your legislators to vote if the convention is introduced in your country?

FIGURE 24.14 Protestors demonstrate against monetary policies and international institutions that threaten livelihoods and environmental quality. *Source:* Tom Finkle.

DISPUTE RESOLUTION AND PLANNING

The adversarial approach of our current legal system often fails to find good solutions for many complex environmental problems. Identifying an enemy and punishing him for transgressions seems more important to us than finding win/win compromises. Gridlocks occur in which conflicts between adversaries breed mutual suspicion and decision paralysis. The result is continuing ecosystem deterioration, economic stagnation, and growing incivility and confrontation. The complexity of many environmental problems arises from the fact that they are not purely ecological, economic, or social, but a combination of all three. They require an understanding of the interrelations between nature and people. Are there ways to break these logjams and find creative solutions? In this section we will look at some new developments in mediation, dispute resolution, and alternative procedures for environmental decision making.

Wicked Problems and Adaptive Management

Rational choice theories of planning and decision making assume that if we just collect more data, buy faster computers to crunch numbers, build more complex models, and spend more money, any problem should be resolvable. More information, it's assumed, will lead automatically to better management. This "bigger hammer" approach may be effective in problems that are difficult but relatively straightforward. Increasingly, however, we have come to recognize that many of the most important problems we face don't fit this pattern. Questions like what ecosystem health means, or how clean is clean, don't have simple right or wrong answers. They depend on your worldview and how you define these terms. Different people come to different conclusions even if they share the same information.

Environmental scientists describe problems with no simple right or wrong answers as being **wicked problems,** not in the sense of having malicious intent, but rather as obstinate or intractable. These problems often are nested within other sets of interlocking issues. The definition of both the problem and its solutions differ for various stakeholders. There are no value-free, objective answers for these dilemmas, only choices that are better or worse depending on your viewpoint. Wicked problems are important and have serious consequences, but also are complex and have a poor match between who bears the costs and who bears the benefits on any proposed solution (fig. 24.15). They usually can't be solved by simple rules and regulations, more scientific research, or appeals to ethics. Often the best solution comes from community-based planning and consensus building. Inherent uncertainty gives these questions no clear end point. You cannot know when all possible solutions have been explored.

Recent advancement in understanding how ecological systems work gives us some insight into many wicked problems. Like biological organisms, social problems often change and evolve over time. Their history unfolds in complex ways, depending on chance interactions and unpredictable events. Like ecological systems, there may never be a stable equilibrium in many environmental issues. Each involves an assemblage of issues and actors that are unique in time and place. They can't be standardized. There are no good precedents from previous experience. Their solutions are unique, and what may work today, may not be applicable tomorrow. How can we learn to cope with such uncertainty?

One promising approach to solving wicked environmental problems comes from the work of ecologists C. S. Holling and

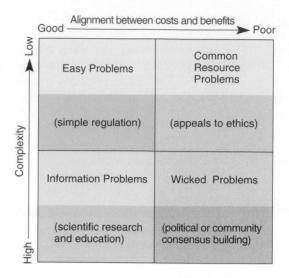

FIGURE 24.15 The difficulty of environmental decision making increases as problems become more complex and the congruence between those who bear costs and enjoy benefits decreases.

K. Sexton, et. al., *Better Environment Decisions.* 1999. Copyright © 1999 Island Press. Reprinted by permission of Alexander Hoyt & Associates.

Lance Gunderson, and planners Steven Light and Kai Lee, among others. Starting with the observation that human understanding of nature is imperfect, this group believes that all human interactions with nature should be experimental. They suggest that environmental policies should incorporate **adaptive management,** or "learning by doing," designed from the outset to test clearly formulated hypotheses about the ecological, social, and economic impacts of the actions being undertaken. Rather than assume that what seemed the best initial policy option will always remain so, we need to carefully monitor how conditions are changing and what effects we are having on both target and nontarget elements of the system. If our policy succeeds, the hypothesis is affirmed. But if the policy fails, an adaptive design still permits learning, so that future decisions can proceed from a better base of understanding. The goal of adaptive management and experimental design is to enable us to live with the unexpected. They aim to yield understanding as much as to produce answers or solutions (table 24.3). This approach to natural resources is similar to—but more explicitly experimental than—ecosystem management (see chapter 13).

Resilience in Ecosystem and Institutions

The great economist Joseph Schumpeter described "waves of creative destruction" that transform economic systems. Another insight from Holling and his collaborators is that similar cycles of destructive creation operate in both ecological systems and in policy institutions (fig. 24.16). This is a familiar process that occurs in secondary succession (chapter 4). The release phase of the cycle occurs when factors such as fires, storms, or pests disturb a biological community, mobilizing nutrients and making space available for new growth. During the reorganization phase, pioneer and opportunist species colonize the new habitat. These species grow

TABLE 24.3 Institutional Conditions for Adaptive Management

1. There is a mandate to take action in the face of uncertainty.
2. Decision makers are aware they are experimenting anyway.
3. Decision makers care about improving outcomes over biological time scales.
4. Preservation of pristine environments is no longer an option, and human intervention cannot produce desired outcomes predictably.
5. Resources are sufficient to measure ecosystem-scale behavior.
6. Theory, models, and field methods are available to estimate and infer ecosystem-scale behavior.
7. Hypotheses can be formulated.
8. Organizational culture encourages learning from experience.
9. There is sufficient stability to measure long-term outcomes; institutional patience is essential.

Kai N. Lee, Compass and Gyroscope, 1993. Copyright (c) 1993 Island Press. Reprinted by permission of Alexander Hoyt & Associates.

rapidly on the accessible carbon, nutrient, and energy sources during the exploitation stage. As the community matures, both the stored capital and connectedness increase until the ecosystem reaches a stage at which the system is poised for some new disturbance that starts the cycle again.

The most important characteristic of natural systems is their **resilience,** or ability to recover from disturbance. This doesn't imply that the ecosystem always returns to the exact condition it was in before the disturbance. It may have a new assemblage of species, or different set of physical conditions, but if it is resilient, the system has the ability to reorganize itself in creative and constructive ways. "Environmental quality is not achieved by attempting to eliminate change or surprises," Holling observed. The goal, instead, is resilience in the face of surprise. Surprise can be counted on. Resilience comes from adaptation to stress, from survival of the fittest in a turbulent environment.

In studying a variety of natural resource management regimes, Holling and others observed that human institutions also follow a similar pattern. In studying a variety of natural resource management issues ranging from restoration of the Florida Everglades, to control of spruce budworm in New England forests, to cattle grazing in South Africa, to protection of salmon in the Pacific Northwest, they observed that every attempt to manage ecological variables one factor at a time inexorably leads to less resilient ecosystems, more rigid management systems, and more dependent societies. Initial success sets the conditions for eventual collapse. Take the example of forest fire suppression. For 70 years, the U.S. Forest Service has had a very effective policy of putting out all forest fires. The result has been that flammable debris has built up in the forests so that major conflagrations are now inevitable. During this time, however, people have felt safe moving to the borders of the forests and now there is a large population with a huge invest-

ment in property that needs to be protected from fire. Furthermore, a big bureaucracy has built up whose raison d'etre is to fight fires. It takes more and more money to forestall a calamity that becomes increasingly likely because of our efforts to prevent it.

What happens in each of these cases is that our goal to control variability in ecological systems leads us to a narrow purpose and to focus exclusively on solving a single problem. But elements of the system change gradually as a consequence of our management success in ways that we did not anticipate. As more homogenous ecosystems develop over a landscape scale, resilience decreases, and it becomes more likely that the system will flip suddenly into a new regime. What can we do to avoid this trap? Table 24.4 suggests some important lessons for ecological managers.

The Precautionary Principle

One response to the uncertainty of wicked problems and chaotic, nonlinear, discontinuous systems is to plan a margin of safety for error or surprises. Drawing on studies of ecological systems, many conservation biologists advocate a **precautionary principle** that says that when an activity raises threats of harm to human health or the environment, precautionary measures should be taken even if some cause and effect relationships are not fully established scientifically. At a meeting at the Wingspread Center in 1998, an international group of scientists, government officials, lawyers, and grassroots environmental activists agreed on four basic tenets of precautionary action:

- People have a duty to take anticipatory steps to prevent harm. If you have a reasonable suspicion that something bad might be going to happen, you have an obligation to try to stop it.

- The burden of proof of carelessness of a new technology, process, activity, or chemical lies with the proponents, not with the general public.

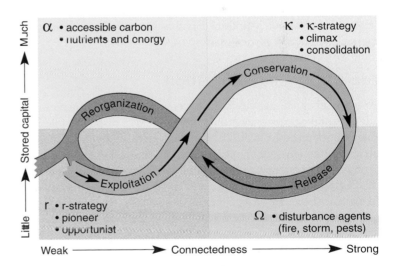

FIGURE 24.16 The creative-destruction cycle. Resilience, or the ability to reorganize and recover from disturbance, is the most important characteristic of both natural and human systems.

Buzz Holling, *Barriers & Bridges,* edited by Gunderson, Holling, and Light, 1995. Columbia University Press. Reprinted by permission of the authors.

| TABLE 24.4 | Planning for Resilience |

1. Interdisciplinary, integrated modes of inquiry are needed for adaptive management of wicked problems.

2. We must recognize that these problems are fundamentally nonlinear and that we need nonlinear approaches to them.

3. Interactions between slow ecological processes such as global climate change or soil erosion in the American cornbelt are difficult to study together with the fast processes that bring creative destruction such as potential collapse of Antarctic ice sheets or appearance of a dead zone in the Gulf of Mexico, but we need to look for connections.

4. The spatial and temporal scales of our concerns are widening. We now need to consider global connections and problems in our planning.

5. Both ecological and social systems are evolutionary and are not amenable to simple solutions based on knowledge of small parts of the whole or on assumptions of constancy or stability of fundamental relationships.

6. We need adaptive management policies that focus on building resilience and the capacity of renewal in both ecosystems and human institutions.

Source: *L. Gunderson, C. Holling, and S. Light,* Barriers and Bridges to the Renewal of Ecosystems and Institutions, *1995, Columbia University Press.*

- Before using a new technology, process, or chemical, or starting a new activity, people have an obligation to examine a full range of alternatives, including the alternative of not using it.

- Decisions applying the precautionary principle must be open, informed, and democratic, and must include the affected parties.

The European Union has adopted this principle as the basis of its environmental policy, but American opponents of this approach claim that it could prevent us from doing anything productive or innovative. What do you think? Is this just common sense or an invitation to decision paralysis?

Arbitration and Mediation

Another set of alternatives to the adversarial nature of litigation and administrative challenges is the growing field of dispute resolution. Increasingly used to avoid the time, expense, and winner-take-all confrontation inherent in tort law, these techniques encourage compromise and workable solutions with which everyone can live.

Arbitration is a formal process of dispute resolution somewhat like a trial. There are stringent rules of evidence, cross-examination of witnesses, and the process results in a legally binding decision. The arbitrator takes a more active role than a judge, however, and is not as constrained by precedent. The arbitrator is more interested in resolving the dispute rather than strict application or interpretation of the law. Arbitrators must have formal training and be certified by the Federal Mediation and Conciliation Service or the American Arbitration Association. Arbitration is usually an attractive prospect if you don't think you could win a formal lawsuit, but why would anyone agree to arbitration if they think they can win the whole enchilada in court? They might take this route just to avoid disagreeable surprises. Juries can be fickle. Furthermore, they might want to get an unpleasant process over with sooner rather than later. In addition, arbitration often is written into contracts so that the disputants have no choice over the matter.

There are disadvantages to arbitration. It doesn't create a legally binding precedent, something that often is the main motivation for a lawsuit. There is less opportunity to appeal if you don't like the decision you get. There also is less protection from self-incrimination, false witnesses, or evidence you didn't expect. You don't generate nearly as much publicity because the proceedings and record are not public. For some litigants, the publicity generated by a trial is more valuable than the settlement itself. Finally, you are less likely to win the whole thing. Some sort of compromise is the most likely outcome.

Mediation is a process in which disputants are encouraged to sit down and talk to see if they can come up with a solution by themselves (fig. 24.17). The mediator makes no final decision but is simply a facilitator of communication. This process is especially useful in complex issues where there are multiple stakeholders with different interests, as is often the case in environmental controversies. For example, a mediation was attempted to work out policy and management disputes about the Boundary Waters Canoe Area Wilderness in northern Minnesota. Local property owners, resort operators, anglers who wanted to use motorboats on wilderness lakes, wildlife protection groups, wilderness advocates, and the U.S. Forest Service all were represented around the table. Each group had its own agenda, although they all quickly coalesced into pro-motor and anti-motor factions. Interestingly, although we usually think of environmentalists as being relatively disadvantaged in these issues, in this case, the local folks felt they were unfairly outgunned by the high-powered lawyers who represented some of the pro-wilderness and wildlife groups. Although the participants agreed on many points and it seemed as if the mediation was about to come up with a workable compromise, at the last minute every-

FIGURE 24.17 Mediation encourages stakeholders to discuss issues and try to find a workable compromise. © Jon Riley/Stone/Getty.

thing fell apart and the groups refused to agree about anything. Each side accused the other of intransigence and bad will, and the issues they disagreed upon will likely end up in court.

This example illustrates both the promises and perils of mediation. It can be quicker and cheaper than court battles. It can lead to compromise and understanding that will lead to further cooperation, and that will solve problems faster than endless appeals. And it may find creative solutions that satisfy multiple parties and interests. On the other hand no one can be forced to mediate or to do so in good faith. Rancorous participants can tie up the process in long, pointless arguments that only make others more angry. Ultimately, there can be a tyranny of the minority. A single person can veto an agreement that everyone else wants. Furthermore, mediation represses or denies certain irreconcilable structural conflicts, giving the impression of equality between disputants when none really exists. Unequal negotiating skills of the participants can lead to unfair outcomes and even more rancor and paranoia than before the mediation was attempted. As is the case in arbitration, mediation doesn't generate the publicity and complete victory that some groups may desire.

Collaborative Approaches to Community-Based Planning

Over the past several decades, natural resource managers have come to recognize the value of holistic, adaptive, multiuse, multivalue approach to planning. Involving all stakeholders and interest groups early in the planning process can help avoid the "train wrecks" in which adversaries become entrenched in non-negotiable positions. Working with local communities can tap into traditional knowledge and gain acceptance for management plans that finally emerge from policy planning. This approach is especially important in nonlinear, nonequilibrium systems and wicked problems. Among the more important reasons to use collaborative approaches are:

- The way wicked problems are formulated depends on your worldview. Incorporating a variety of perspectives early in the process is more likely to lead to the development of acceptable solutions in the end.

- People have more commitment to plans they have helped develop. The first stage is therefore to identify those involved and to engage them in the process.

- There is truth in the old adage that "two heads are better than one." Involving multiple stakeholders and multiple sources of information enriches the process.

- Community-based planning provides access to situation-specific information and experience that can often only be obtained by active involvement of local residents.

- Participation is an important management tool. Project-threatening resistance on the part of certain stakeholders can be minimized by inviting active cooperation of all stakeholders throughout the planning process.

- The knowledge and understanding needed by those who will carry out subsequent phases of a project can only be gained through active participation.

A good example of community-based planning can be seen in the Atlantic Coastal Action Programme (ACAP) in eastern Canada. The purpose of this project is to develop blueprints for the restoration and maintenance of environmentally degraded harbors and estuaries in ways that are both biologically and socially sustainable. Officially established under Canada's Green Plan and supported by Environment Canada, this program created 13 community groups, some rural and some urban, with membership in each dominated by local residents. Federal and provincial government agencies are represented primarily as nonvoting observers and resource people. Each community group is provided with core funding for full-time staff who operate an office in the community and facilitate meetings.

Four of the 13 ACAP sites are in the Bay of Fundy, an important and unique estuary lying between New Brunswick and Nova Scotia. Approximately 270 km long, and with an area of more than 12,000 km^2, the bay, together with the nearby Georges Bank and the Gulf of Maine once formed one of the richest fisheries in the world. With the world's highest recorded tidal range (up to 16 m at maximum spring tide), the bay still sustains a great variety of fishery and wildlife resources, and provides habitat for a number of rare or endangered species. Now home to more than 1 million people, the coastal region is an important agricultural, lumbering, and paper-producing region (fig. 24.18).

Since European settlement began in 1604, the Bay of Fundy region has experienced great changes in population growth, resource use, and human-induced ecosystem change. More than 80 percent of the saltmarshes present in 1604 have been eliminated or degraded. Pollution and sediment damage harbors and biological communities. Overfishing and introduction of exotic species have resulted in endemic species declines. The collapse of cod, halibut, and haddock fishing has had devastating economic effects

FIGURE 24.18 The Bay of Fundy has the greatest tidal range in the world. It is the site of innovative community-based environmental planning process. © Phil Degginger/Animals Animals/Earth Scenes.

on the regional economy and the livelihoods of local residents. Aquaculture is now a more valuable activity than all wild fisheries.

To cope with these complex, intertwined social and biological problems, ACAP is bringing together different stakeholders from around the bay to create comprehensive plans for ecological, economic, and social sustainability. Through citizen monitoring and adaptive management, the community builds social capital (knowledge, cooperative spirit, trust, optimism, working relations), develops a sense of ownership in the planning process, and eliminates some of the fears and sectorial rivalry that often divides local groups, outsiders, and government agents.

On the other hand, giving a greater voice and increased power to local communities could simply result in the foxes guarding the hen house. How would you balance the general public interest with those of local stakeholders?

Green Plans

Several national governments have undertaken integrated environmental planning that incorporates community round-tables for vision development. Canada, New Zealand, Sweden, and Denmark all have so-called **green plans** or comprehensive, long-range national environmental strategies. The best of these plans weave together complex systems, such as water, air, soil, and energy, and mesh them with human factors such as economics, health, and carrying capacity. Perhaps the most thorough and well-thought-out green plan in the world is that of the Netherlands.

Developed in the 1980s through a complex process involving the public, industry, and government, the 400-page Dutch plan contains 223 policy changes aimed at reducing pollution and establishing economic stability. Three important mechanisms have been adopted for achieving these goals: integrated life-cycle management, energy conservation, and improved product quality. These measures should make consumer goods last longer and be more easily recycled or safely disposed of when no longer needed. For example, auto manufacturers are now required to design cars so they can be repaired or recycled rather than being discarded.

Among the guiding principles of the Dutch Green Plan are: (1) the "stand-still" principle that says environmental quality will not deteriorate, (2) abatement at the source rather than cleaning up afterward, (3) the "polluter pays" principle that says users of a resource pay for negative effects of that use, (4) prevention of unnecessary pollution, (5) application of the best practicable means for pollution control, (6) carefully controlled waste disposal, and (7) motivating people to behave responsibly.

The Netherlands have invested billions of guilders in implementing this comprehensive plan. Some striking successes already have been accomplished. Between 1980 and 1990, emissions of sulfur dioxide, nitrogen oxides, ammonia, and volatile organic compounds were reduced 30 percent. By 1995, pesticide use had been reduced 25 percent from 1988 levels, and chlorofluorocarbon use had been virtually eliminated. By 1998, industrial wastewater discharge into the Rhine River was 70 percent less than a decade earlier. Some 250,000 ha (more than 600,000 acres) of former wetlands that had been drained for agriculture are being restored as

FIGURE 24.19 Under the Dutch Green Plan, 250,000 ha (600,000 acres) of drained agricultural land are being restored to wetland and 40,000 ha (99,000 acres) are being replanted as woodland.
© David L. Brown/Tom Stack & Associates.

nature preserves and 40,000 ha (99,000 acres) of forest are being replanted. This is remarkably generous and foresighted in such a small, densely populated country, but the Dutch have come to realize they cannot live without nature (fig. 24.19).

Not all goals have been met so far. Planned reductions in CO_2 emissions failed to materialize when cheap fuel prices encouraged fuel-inefficient cars. Currently a carbon tax is being considered. A sudden population increase caused by immigration from developing countries and Eastern Europe also complicates plan implementation, but the basic framework of the Dutch plan has much to recommend it, nevertheless. Other countries would be more sustainable and less environmentally destructive if they were to adopt a similar plan.

Summary

- Although "policy" can have multiple meanings, environmental policy, in this chapter, is taken to mean public opinion as well as official rules and regulations concerning our environment. The policy cycle describes the steps by which problems are identified and defined, and solutions are proposed, debated, enacted into law, and monitored.

- The National Environmental Policy Act (NEPA) forms the cornerstone of both environmental policy and law in the United States. One of its most important provisions is the requirement of Environmental Impact Statements (EIS) for all major federal projects and programs.

- Laws are rules established through legislation (statutes), judicial decisions (case law), custom (common law), or administrative decisions (administrative law). Over the past century, U.S. policy and law has shifted from a hands-off attitude toward industry and private property, to end-of-the-pipe command and control, to more collaborative, pragmatic approaches.

- Passage of bills in Congress is a convoluted process with many opportunities for amendments and riders to be added. Competing interests "lobby" throughout this process to try to change the outcome. Over the past 30 years, litigation has often been the most effective route for environmental protection. Establishing standing or the right to be heard in court may be the crucial stumbling block in this effort. SLAPP suits, while often baseless, are intended to intimidate or inactivate environmental groups.

- Although many international treaties and conventions have been passed to protect our global environment, many are vague or toothless. Some innovative measures have been devised to compel compliance.

- Globalization presents both the challenge and opportunity to examine the role of democracy, fairness, transparency, and accountability in environmental governance.

- Some alternatives to adversarial litigation include arbitration, mediation, and community-based planning. These techniques are useful in complex, unpredictable, multistakeholder, multivalue issues. We call these wicked problems.

- Learning from ecological systems, we see that some of the main goals for environmental policy and planning should be ecological and institutional resilience and adaptive management. The Netherlands has done one of the best jobs of any nation in using these approaches to formulate a comprehensive "green plan."

Questions for Review

1. What is the policy cycle, and how does it work?

2. Describe the path of a bill through Congress. When are riders and amendments attached?

3. What are the differences and similarities between civil, criminal, and administrative law?

4. List some of the major U.S. environmental laws of the past 30 years.

5. Why have some international environmental treaties and conventions been effective while most have not? Describe two such treaties.

6. Define *globalization* and describe how it impacts environmental quality.

7. What are wicked problems? Why are they difficult?

8. What is resilience? Why is it important?

9. What is collaborative, community-based planning?

10. What is unique about the Dutch Green Plan?

Questions for Critical Thinking

1. In your opinion, how much environmental protection is too much? Think of a practical example in which some stakeholders may feel oppressed by government regulations. How would you justify or criticize these regulations?

2. What role do you believe fairness, power, and rational choice play in our political process? Would our environmental policies be different if we had a different political system?

3. Which is the most important step in the policy cycle? If you were leader of a major environmental group, where would you put your efforts in establishing policy?

4. Do you believe that trees, wild animals, rocks, or mountains should have legal rights and standing in the courts? Why or why not? Are there partial rights or some other form of protection you would favor for nature?

5. It's sometimes difficult to determine whether a lawsuit is retaliatory or based on valid reason. How would you define a SLAPP suit, and differentiate it from a legitimate case?

6. Try creating a list of arguments for and against an international body with power to enforce global environmental laws. Can you see a way to create a body that could satisfy both reasons for and against this power?

7. Think of a familiar example of a wicked environmental problem. What are the most important elements that make it wicked? What institutional changes could we implement to make this issue less wicked?

8. The Holling diagram for the creative-destruction cycle (fig. 24.16) is described in terms of ecological change. Try applying this model to cycles of change in human institutions. Describe the actors, conditions, and forcing factors in each of the four quadrants of the model.

9. Take a current wicked environmental problem. If you were an environmental leader trying to resolve this problem, would you choose litigation, arbitration, or mediation? What are your reasons for favoring or rejecting each one?

10. Would you favor adoption of the Aarhus Convention by your government? Why or why not? Put yourself in the place of someone with an opposite position from your own. What arguments might they make for or against international environmental governance?

Key Terms

adaptive management 536
administrative courts 533
administrative law 524
arbitration 538
case law 524
civil law 529
common law 529
criminal law 528
environmental governance 535
Environmental Impact
    Statement (EIS) 521
environmental law 523
environmental policy 521
globalization 535
green plans 540
lobbying 527
mediation 538

National Environmental Policy
    Act (NEPA) 521
policy 521
policy cycle 522
precautionary principle 537
precedent 528
rational choice 522
resilience 537
riders 526
standing 528
statute law 523
Strategic Lawsuits Against
    Public Participation
    (SLAPP) 530
tort law 529
wicked problems 536

Further Readings

Babiker, M. H., et al. 2002. The evolution of a climate regime: Kyoto to Marrakech and beyond. *Environmental Science and Policy* 5(3):195–206.

Caldwell, Lynton K. 1996. *International Environmental Policy.* Duke University Press.

Chertow, Marian R., and Daniel C. Esty, eds. 1997. *Thinking Ecologically: The Next Generation of Environmental Policy.* Yale University Press.

Clark, Tim W. 2002. *The Policy Process: A Practical Guide for Natural Resources Professionals.* Yale University Press.

Constanza, R., and S. E. Jorgensen, eds. 2002. *Understanding and Solving Environmental Problems in the 21st Century: Toward a New, Integrated Hard Problem Science.* Elsevier Publishers.

Dowie, Mark. 2003. In law we trust: Can environmental legislation protect the commons now? *Orion* 22(4)18–25.

Gunderson, Lance H., and C. S. Holling, eds. 2001. *Panarchy: Understanding Transformations in Systems of Humans and Nature.* Island Press.

Lewicki, Roy J., Barbara Gray, and Michael Elliott, eds. 2002. *Making Sense of Intractable Environmental Conflicts: Concepts and Cases.* Island Press.

Meadowcroft, J. 2002. Politics and scale: Some implications for environmental governance. *Landscape and Urban Planning* 61(2–4):169–79.

Orr, David. 2003. Walking north on a southbound train. *Conservation Biology* 17(2):348–51.

Roberts, Jane. 2002. *Environmental Policy: Theory and Practice.* Routledge Press.

Sexton, Ken, et al. 1999. *Better Environmental Decisions.* Island Press.

Tickner, Joel A. 2002. *Precaution, Environmental Science, and Preventive Public Policy.* Island Press.

Wilkinson, David. 2002. *Environment and Law.* Routledge Press.

Welcome to McGraw-Hill's Online Learning Center

Location: http://www.mhhe.com/environmentalscience

McGraw Hill

WEB EXERCISES

Environmental Scorecard

Find out how the U.S. Congress voted on key environmental issues by going to the web page of the League of Conservation Voters (LCV) at http://scorecard.lcv.org/. You can download the entire scorecard in PDF format or view the full Senate or full House vote online. To find out how your own senators and congressional representative voted, search by state, enter the name of a specific member, or enter your zip code and click on *go*. Look at the descriptions of the specific issues on which the rating was based to learn more about them and to see if you agree with the assessment offered by this scorecard. The LCV may regard voting in a particular way to be pro- or anti-environmental, but do you agree? Note that an

absence during a specific vote is recorded as negative, even if the absence was for illness or some other valid reason. Under the Overview section in the left frame on the scorecard page, you can learn more about the issues discussed by Congress.

To keep up to date on current issues, you can visit the websites of specific groups such as the Natural Resources Defense Council (www.nrdc.org/legislation/legwatch.asp) or the Endangered Species Coalition (www.stopextinction.org/). The Thomas website of the U.S. Library of Congress (http://thomas.loc.gov/) allows you to find the latest action in Congress.

What Then Shall We Do?

All history consists of successive excursions from a single starting point to which man returns again and again to organize yet another search for a durable set of values.

Aldo Leopold

OBJECTIVES

After studying this chapter, you should be able to:

*   be aware of the goals and opportunities in environmental education and environmental careers.
*   recognize opportunities for making a difference through the goods and services we choose as well as the limits of green consumerism.
*   compare the differences between radical and mainline environmental groups and the tactics they employ to bring about social change.
*   summarize some practical suggestions for how to reduce consumption, organize an environmental campaign, communicate effectively with the media, and write to elected officials.
*   appreciate the need for sustainable development and explain how nongovernmental groups work toward this goal.
*   evaluate how green politics and environmental citizenship can help protect the earth.
*   formulate your own philosophy and action plan for what you can and should do to create a better world and a sustainable environment.

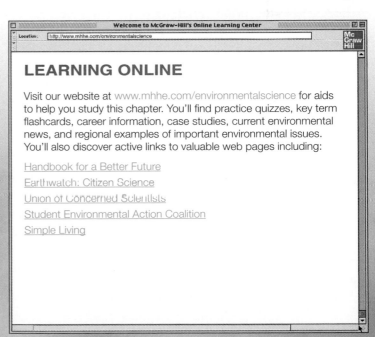

LEARNING ONLINE

Visit our website at www.mhhe.com/environmentalscience for aids to help you study this chapter. You'll find practice quizzes, key term flashcards, career information, case studies, current environmental news, and regional examples of important environmental issues. You'll also discover active links to valuable web pages including:

Handbook for a Better Future
Earthwatch: Citizen Science
Union of Concerned Scientists
Student Environmental Action Coalition
Simple Living

Photo: As developing countries around the world adopt more affluent lifestyles, will there be enough resources for everyone? © William P. Cunningham.

Citizen Science and the Christmas Bird Count

Every Christmas since 1900, dedicated volunteers have counted and recorded all the birds they can find within their team's designated study site (fig. 25.1). This effort has become the largest, longest-running, citizen-science project in the world. For the 100th count, nearly 50,000 participants in about 1,800 teams observed 58 million birds belonging to 2,309 species. Although about 70 percent of the counts in 2000 were made in the United States or Canada, 650 teams in the Caribbean, Pacific Islands, and Central and South America also participated. Participants enter their bird counts on standardized data sheets, or submit their observations over the Internet. Compiled data can be viewed and investigated online, almost as soon as they are submitted.

Frank Chapman, the editor of *Bird-Lore* magazine and an officer in the newly formed Audubon Society, started the Christmas Bird Count in 1900. For years, hunters had gathered on Christmas Day for a competitive hunt, often killing hundreds of birds and mammals as teams tried to outshoot each other. Chapman suggested an alternative contest: to see which team could observe and identify the most birds, and the most species, in a day. The competition has grown and spread. In the 100th annual count, the winning teams was in Monte Verde, Costa Rica, with an amazing 343 species tallied in a single day.

The tens of thousands of birdwatchers participating in the count gather vastly more information about the abundance and distribution of birds than biologists could gather alone. These data provide important information for scientific research on bird migrations, populations, and habitat change. Now that the entire record for a century of bird data is available on the BirdSource website (www.birdsource.org), both professional ornithologists and amateur birdwatchers can study the geographical distribution of a single species over time, or they can examine how all species vary at a single site through the years. Those concerned about changing climate can look for variation in long-term distribution of species. Climatologists can analyze the effects of weather patterns such as El Niño or La Niña on where birds occur. One of the most intriguing phenomena revealed by this continent-wide data collection is irruptive behavior: that is, appearance of massive

FIGURE 25.1 Participating in the Christmas Bird Count is a good way to learn about science and your local environment.
© William P. Cunningham.

numbers of a particular species in a given area in one year, and then their move to other places in subsequent years following weather patterns, food availability, and other factors.

Following the success of the Christmas Bird Count, other citizen-science projects have been initiated. Project Feeder Watch enlists the efforts of schoolchildren and backyard birdwatchers to monitor birds. Farmers have monitored their own pasture and stream health and collaborated with students to explore how management can improve the land. Volunteers are being enlisted to monitor water quality, climate and air quality, forest health, and many other environmental factors. Data collected from such widespread, coordinated projects provide an invaluable source of information for scientific research. How can you learn more about your local environment, and contribute to research, by participating in a citizen-science project? Contact your local Audubon chapter or your state's Department of Natural Resources to find out what you can do.

How does counting birds contribute to sustainability? Citizen-science projects are one way individuals can learn more about the scientific process, become familiar with their local environment, and become more interested in community issues. In this chapter, we'll look at other ways individuals and groups can help protect nature and move toward a sustainable society.

LIVING IN AN UNCERTAIN WORLD

Throughout this book, you've learned about many environmental problems. Evidence of global climate change caused by human actions is increasingly compelling. Forests, wetlands, prairies, and the biodiversity they support are disappearing at an alarming rate. Major ocean fisheries have collapsed abruptly and marine ecosystems have been severely damaged. One-third of all nations don't have enough clean fresh water to meet vital needs. Within 50 years, it's expected that two-thirds of all countries, representing 85 percent of the world's population, will experience water shortages. More than 800 million people are chronically undernourished; 1.5 billion live in abject poverty; and 2.5 billion lack clean drinking water and adequate sanitation. Every year, infectious diseases kill 14 million people, most of whom are among the world's poorest.

At the same time, there also are encouraging signs of movement toward a sustainable society. Population growth is slowing nearly everywhere. Farmers are producing enough food for an adequate diet for everyone if supplies were equitably distributed. Wind farms on the American Great Plains and along Europe's coastlines are providing increasing amounts of clean, renewable energy. Together with wind, other renewable energy sources, such as solar, geothermal, and biomass energy, could provide all the energy we need and eliminate the need for fossil fuels completely. We're finding ways to recycle materials and produce goods and services with fewer toxic chemicals and less air and water pollution. Ecosystems such as the Florida Everglades and the tallgrass prairies of the American West are being cleaned and restored. A growing number of people are using their purchasing power to encourage organic farming, fair trade, fewer toxic chemicals, and more sustainably produced consumer goods.

FIGURE 25.3 Environmental education helps develop awareness and appreciation of ecological systems and how they work. © William P. Cunningham.

FIGURE 25.2 What lives in a tide pool? Learning to appreciate the beauty, richness, and diversity of the natural world is important if we are to protect it. © William P. Cunningham.

How can you contribute to finding solutions to these pressing environmental dilemmas? Continuing your environmental education and helping others understand the challenges we face, in addition to some of our options for solving them, can be a first step. In this chapter, we'll look at other ways that we can act—both individually and collectively—to live sustainably and build a better world. Perhaps most importantly, we all need to know and enjoy some part of nature. How can we protect what we don't know or love? Experiencing first-hand some of the richness, beauty, and diversity of life is essential if we are to be effective in saving it (fig. 25.2). As author Edward Abbey says,

> It is not enough to fight for the land; it is even more important to enjoy it. While you can. While it is still there. So get out there and mess around with your friends, ramble out yonder and explore the forests, encounter the grizz, climb the mountains. Run the rivers, breathe deep of that yet sweet and lucid air, sit quietly for a while and contemplate the precious stillness, that lovely mysterious and awesome space.

ENVIRONMENTAL EDUCATION

In 1990 Congress passed the National Environmental Education Act establishing two broad goals: (1) to improve understanding among the general public of the natural and built environment and the relationships between humans and their environment, including global aspects of environmental problems, and (2) to encourage postsecondary students to pursue careers related to the environment. Specific objectives proposed to meet these goals include developing an awareness and appreciation of our natural and social/cultural environment, knowledge of basic ecological concepts, acquaintance with a broad range of current environmen-

TABLE 25.1 **Outcomes from Environmental Education**

The natural context: An environmentally educated person understands the scientific concepts and facts that underlie environmental issues and the interrelationships that shape nature.

The social context: An environmentally educated person understands how human society is influencing the environment, as well as the economic, legal, and political mechanisms that provide avenues for addressing issues and situations.

The valuing context: An environmentally educated person explores his or her values in relation to environmental issues; from an understanding of the natural and social contexts, the person decides whether to keep or change those values.

The action context: An environmentally educated person becomes involved in activities to improve, maintain, or restore natural resources and environmental quality for all.

Source: A Greenprint for Minnesota, *Minnesota Office of Environmental Education, 1993.*

tal issues, and experience in using investigative, critical-thinking, and problem-solving skills in solving environmental problems (fig. 25.3). Several states, including Arizona, Florida, Maryland, Minnesota, Pennsylvania, and Wisconsin, have successfully incorporated these goals and objectives into their curricula. Table 25.1 presents some guidelines for environmental education.

Environmental Literacy

Speaking to the first of these broad goals, former Environmental Protection Agency administrator William K. Reilly called for broad **environmental literacy** in which every citizen is fluent in the principles of ecology and has a "working knowledge of the basic grammar and underlying syntax of environmental wisdom." Environmental literacy, according to Reilly can help create a stewardship ethic—a sense of duty to care for and manage wisely our natural endowment and our productive resources for the long haul.

TABLE 25.2   The Environmentalist's Bookshelf

What are some of the most influential and popular environmental books? A survey of environmental experts and leaders around the world listed the following among the 500 best books on nature and the environment[1]:

A Sand County Almanac by Aldo Leopold (100)[2]
Silent Spring by Rachel Carson (81)
State of the World by Lester Brown and the Worldwatch Institute (31)
The Population Bomb by Paul Ehrlich (28)
Walden by Henry David Thoreau (28)
Wilderness and the American Mind by Roderick Nash (21)
Small Is Beautiful: Economics as if People Mattered by E. F. Schumacher (21)
Desert Solitaire: A Season in the Wilderness by Edward Abbey (20)
The Closing Circle: Nature, Man, and Technology by Barry Commoner (18)
The Limits to Growth: A Report for the Club of Rome's Project on the Predicament of Mankind by Donella H. Meadows, et al. (17)
The Unsettling of America: Culture and Agriculture by Wendell Berry (16)
Man and Nature by George Perkins Marsh (16)
Encounters with the Archdruid by John McPhee (16)
Man's Role in Changing the Face of the Earth by William Thomas (ed.), with Carl Sauer, Marston Bates, and Lewis Mumford (16)
The Monkey Wrench Gang by Edward Abbey (14)
World Resources by the World Resources Institute (14)
Gaia: An Atlas of Planet Management by Norman Meyers (13)
Soft Energy Paths: Toward a Durable Peace by Amory Lovins (12)
Our Common Future by the World Commission on Environment and Development (12)
The Immense Journey by Loren Eisley (11)
Only One Earth: The Care and Maintenance of a Small Planet by Barbara Ward and Rene Dubos (11)
For the Common Good: Redirecting the Economy Toward Community, the Environment, and a Sustainable Future by Herman Daly and John Cobb (10)
Deep Ecology: Living as if Nature Mattered by Bill Devall and George Sessions (10)
Arctic Dreams: Imagination and Desire in a Northern Landscape by Barry Lopez (9)
Pilgrim at Tinker Creek by Annie Dillard (8)

[1]Robert Merideth, 1992, G. K. Hall/Macmillan Inc.

[2]Indicates number of votes for each book. Because the preponderance of respondents were from the United States (82 percent), American books are probably overrepresented.

Robert Merideth, *The Environmentalist's Bookshelf: A Guide to the Best Books*, 1993, by G. K. Hall, an imprint of Macmillan, Inc. Reprinted by permission.

"Environmental education," he says, "boils down to one profoundly important imperative: preparing ourselves for life in the next century. When the twenty-first century rolls around, it will not be enough for a few specialists to know what is going on while the rest of us wander about in ignorance."

While you've made a great start toward learning about your environment by reading this book and taking a class in environmental science, I hope you will continue to read and expand your horizons after your class is finished. A large body of environmental literature can help you in this quest. Some of the most influential environmental books of all time are listed in table 25.2. To this list we'd add some personal favorites: *The Singing Wilderness* by Sigurd F. Olson, *My First Summer in the Sierra* by John Muir, and *Remaking Society: Pathways to a Green Future* by Murray Bookchin.

Citizen Science

While university classes often tend to be theoretical and abstract, many students are discovering they can make authentic contributions to scientific knowledge through active learning and undergraduate research programs. Internships in agencies or environmental organizations are one way of doing this. Another is to get involved in organized **citizen science** projects in which ordinary people join with established scientists to answer real scientific questions. Community-based research was pioneered in the Netherlands, where several dozen research centers now study environmental issues ranging from water quality in the Rhine River, cancer rates by geographic area, and substitutes for harmful organic solvents. In each project, students and neighborhood groups team with scientists and university personnel to collect data. Their results have been incorporated into official government policies.

Similar research opportunities exist in the United States and Canada. The Audubon Christmas Bird Count, mentioned at the beginning of this chapter, is a good example. Earthwatch offers a much smaller but more intense opportunity to take part in research. Every year hundreds of Earthwatch projects each field a team of a dozen or so volunteers who spend a week or two working on issues ranging from loon nesting behavior to archaeological digs. The American River Watch organizes teams of students to measure water quality. You might be able to get academic credit as well as helpful practical experience in one of these research experiences.

Environmental Careers

The need for both environmental educators and environmental professionals opens up many job opportunities in environmental fields.

WWF estimates, for example, that 750,000 new jobs will be created over the next decade in the renewable energy field alone. Scientists are needed to understand the natural world and the effects of human activity on the environment. Lawyers and other specialists are needed to develop government and industry policy, laws, and regulations to protect the environment. Engineers are needed to develop technologies and products to clean up pollution and to prevent its production in the first place. Economists, geographers, and social scientists are needed to evaluate the costs of pollution and resource depletion and to develop solutions that are socially, culturally, politically, and economically appropriate for different parts of the world. In addition, business will be looking for a new class of environmentally literate and responsible leaders who appreciate how products sold and services rendered affect our environment.

Trained people are essential in these professions at every level, from technical and clerical support staff to top managers. Perhaps the biggest national demand over the next few years will be for environmental educators to help train an environmentally literate populace. We urgently need many more teachers at every level who are trained in environmental education. Outdoor activities and natural sciences are important components of this mission, but environmental topics such as responsible consumerism, waste disposal, and respect for nature can and should be incorporated into reading, writing, arithmetic, and every other part of education.

Can environmental protection and resource conservation—a so-called green perspective—be a strategic advantage in business? Many companies think so. An increasing number are jumping on the environmental bandwagon, and most large corporations now have an environmental department. A few are beginning to explore integrated programs to design products and manufacturing processes to minimize environmental impacts. Often called "design for the environment," this approach is intended to avoid problems at the beginning rather than deal with them later on a case-by-case basis. In the long run, executives believe this will save money and make their businesses more competitive in future markets. The alternative is to face increasing pollution control and waste disposal costs   now estimated to be more than $100 billion per year for all American businesses—as well as to be tied up in expensive litigation and administrative proceedings.

The market for pollution-control technology and know-how is also expected to be huge. Cleaning up the former East Germany is expected to cost some $200 billion, and other parts of the former Soviet Union are even more polluted. Many companies are positioning themselves to cash in on this enormous market. Germany and Japan appear to be ahead of America in the pollution-control field because they have had more stringent laws for many years, giving them more experience in reducing effluents.

The rush to "green up" business is good news for those looking for jobs in environmentally related fields, which are predicted to be among the fastest growing areas of employment during the next few years. The federal government alone projects a need to hire some 10,000 people per year in a variety of environmental disciplines (fig. 25.4). How can you prepare yourself to enter this market? The best bet is to get some technical training: environmental engineering, analytical chemistry, microbiology, ecology, limnol-

FIGURE 25.4   Many interesting, well-paid jobs are opening up in environmental fields. Here an environmental technician takes a sample from a monitoring well for chemical analysis. © William P. Cunningham.

ogy, groundwater hydrology, or computer science all have great potential. Currently, a chemical engineer with a graduate degree and some experience in an environmental field can practically name his or her salary. Some other very good possibilities are environmental law and business administration, both rapidly expanding fields.

For those who aren't inclined toward technical fields there are still opportunities for environmental careers. A good liberal arts education will help you develop skills such as communication, critical thinking, balance, vision, flexibility, and caring that should serve you well. Large companies need a wide variety of people; small companies need a few people who can do many things well.

INDIVIDUAL ACCOUNTABILITY

Some prime reasons for our destructive impacts on the earth are our consumption of resources and disposal of wastes. Technology has made consumer goods and services cheap and readily available in the richer countries of the world. As you already know, we in the industrialized world use resources at a rate out of proportion to our percentage of the population. If everyone in the world were to attempt to live at our level of consumption, given current methods of production, the results would surely be disastrous. In this section we will look at some options for consuming less and reducing our environmental impacts. Perhaps no other issue in this book represents so clear an ethical question as the topic of responsible consumerism.

How Much Is Enough?

A century ago, economist and social critic, Thorstein Veblen, in his book, *The Theory of the Leisure Class,* coined the term **conspicuous consumption** to describe buying things we don't want or need just to impress others. How much more shocked he would be to see

MODERNE MAN

FIGURE 25.5   Is this our highest purpose? © 1990 Bruce von Alten.

current trends. The average American now consumes twice as many goods and services as in 1950. The average house is now more than twice as big as it was 50 years ago, even though the typical family contains half as many people. We need more space to hold all the stuff we buy. Shopping has become the way many people define themselves (fig. 25.5). As Marx predicted, everything has become commodified; getting and spending have eclipsed family, ethnicity, even religion as the defining matrix of our lives. But the futility and irrelevance of much American consumerism leaves a psychological void. Once we possess things, we find they don't make us young, beautiful, smart, and interesting as they promised. So we buy even more stuff to try to fill the empty feeling that we don't have time to have real friends, to cook real food, to have creative hobbies, or to do work that makes us feel we have accomplished something with our lives. Some social critics call this drive to possess stuff "affluenza."

A growing number of people find themselves stuck in a vicious circle: they work frantically at a job they hate, to buy things they don't need, so they can save time to work even longer hours. Seeking a measure of balance in their lives, some opt out of the rat race and voluntarily abandon stressful jobs for simpler, less complex ones, usually at lower pay. They give up big houses and choose smaller, less expensive ones, perhaps moving from busy metro areas to small towns where the pace is slower and the cost of living lower. They strip away the clutter from their lives and deliberately live with less stuff. As Thoreau wrote in *Walden,* "Our life is frittered away by detail . . . simplify, simplify."

Lohas and Cultural Creatives

Marketers and trend spotters have noticed that an increasing number of people in affluent countries are becoming concerned about the effects of pollution and social inequity. There's a name for consumers who worry about the environment, want products to be produced in a fair, sustainable way, and use purchasing power to express their values. They're called Lohas, an acronym for "lifestyles of health and sustainability." Encompassing things like organic food, energy-efficient appliances and automobiles, natural home care and health products, active vacations and eco-tourism,

the total market for this group represents $230 billion per year, according to Natural Business Communications, a company in Colorado that publishes *The Lohas Journal* and is credited with coining the term. Altogether, 68 million Americans—about one-third of the adult population—qualify as Lohas, consumers who take environmental and social issues into account when they make purchases. Ninety percent of this group say they prefer to make purchases from companies that share their values, and many say they are willing to pay a premium for products and services they consider healthier for themselves, their families, society, and the environment. Merchants flock to annual Lohas business conferences to learn how to tap into this important market.

Another name for people who are deeply concerned about nature and want to be involved in creating a new and better way of life is **cultural creatives,** a term introduced by Paul Ray and Sherry Ruth Anderson in a book by the same title. Ray and Anderson describe this group as socially conscious, involved in improving communities, and willing to translate values into action. They are strongly aware of environmental problems and want to do something to remedy them. Most cultural creatives place great importance on helping other people, care intensely about psychological or spiritual development, and volunteer for one or more good causes. They dislike the modern emphasis on wealth, consumerism and power, and enjoy learning about new places and people and alternative ways of life.

Neither cultural creatives nor Lohas are defined by particular demographic characteristics. They work at all sorts of jobs and occupy every economic level. The majority are mainstream in their religious beliefs and are no more liberal or conservative than the U.S. average. One important trait is that about two-thirds of them are women, and many of the values most important to them—relationships, family life, children, education, and responsibility—are traditionally thought of as women's issues. Because women now do a majority of family shopping as well as hold more than half of all personal wealth in America, both businesses and non-profits are beginning to pay attention to these concerns.

In 2003, the United Nations Environment Programme (UNEP) held workshops on sustainable consumption in Paris and Tokyo. Recognizing that making people feel guilty about their lifestyles and purchasing habits isn't working, UNEP is attempting to find ways to make sustainable living something consumers will adopt willingly. The goal is economically, socially, and environmentally viable solutions that allow people to enjoy a good quality of life while consuming fewer natural resources and polluting less. A good example of this approach is a British automaker that provides a mountain bike with every car it sells, urging buyers to use the bike for short journeys. Another example cited by UNEP is European detergent makers who encourage customers to switch to low-temperature washing liquids and powders, not just to save energy but because it's good for their clothes.

Although each of our individual choices may make a small impact, collectively they can be important. The What Can You Do? box on p. 549 offers some suggestions for reducing waste and pollution.

Green Washing and Confusing Choices

Although many people report they prefer to buy products and packaging that are socially and ecologically sustainable, there is a wide gap between what consumers say in surveys about purchasing habits and the actual sales data. Part of the problem is accessibility and affordability. In many areas, green products either aren't available or are so expensive that those on limited incomes (as many living in voluntary simplicity are) can't afford them. Although businesses are beginning to recognize the size and importance of the market for "green" merchandise, the variety of choices and the economies of scale haven't yet made them as accessible as we would like.

Another problem is that businesses, eager to cash in on this premium market, offer a welter of confusing and often misleading claims about the sustainability of their offerings. Consumers must be wary to avoid "green scams" that sound great but are actually only overpriced standard items. Many terms used in advertising are vague and have little meaning. For example:

- "Nontoxic" suggests that a product has no harmful effects on humans. Since there is no legal definition of the term, however, it can have many meanings. How nontoxic is the product? And to whom? Substances not poisonous to humans can be harmful to other organisms.

- "Biodegradable," "recyclable," "reusable," or "compostable" may be technically correct but not signify much. Almost everything will biodegrade *eventually,* but it may take thousands of years. Similarly, almost anything is potentially recyclable or reusable; the real question is whether there are programs to do so in your community. If the only recycling or composting program for a particular material is half a continent away, this claim has little value.

- "Natural" is another vague and often misused term. Many natural ingredients—lead or arsenic, for instance—are highly toxic. Synthetic materials are not necessarily more dangerous or environmentally damaging than those created by nature.

- "Organic" can connote different things in different places. Some states have standards for organic food, but others do not. On items such as shampoos and skin-care products, "organic"

What can you do?

Reducing Your Impact

Purchase Less

Ask yourself whether you really need more stuff.

Avoid buying things you don't need or won't use.

Use items as long as possible (and don't replace them just because a new product becomes available).

Use the library instead of purchasing books you read.

Make gifts from materials already on hand, or give nonmaterial gifts.

Reduce Excess Packaging

Carry reusable bags when shopping and refuse bags for small purchases.

Buy items in bulk or with minimal packaging; avoid single-serving foods.

Choose packaging that can be recycled or reused.

Avoid Disposable Items

Use cloth napkins, handkerchiefs, and towels.

Bring a washable cup to meetings; use washable plates and utensils rather than single-use items.

Buy pens, razors, flashlights, and cameras with replaceable parts.

Choose items built to last and have them repaired; you will save materials and energy while providing jobs in your community.

Conserve Energy

Walk, bicycle, or use public transportation.

Turn off (or avoid turning on) lights, water, heat, and air conditioning when possible.

Put up clotheslines or racks in the backyard, carport, or basement to avoid using a clothes dryer.

Carpool and combine trips to reduce car mileage.

Save Water

Water lawns and gardens only when necessary.

Use water-saving devices and fewer flushes with toilets.

Don't leave water running when washing hands, food, dishes, and teeth.

Based on material by Karen Oberhauser, Bell Museum Imprint, University of Minnesota, 1992. Used by permission.

may have no significance at all. Most detergents and oils are organic chemicals, whether they are synthesized in a laboratory or found in nature. Few of these products are likely to have pesticide residues anyway.

- "Environmentally friendly," "environmentally safe," and "won't harm the ozone layer" are often empty claims. Since there are no standards to define these terms, anyone can use them. How much energy and nonrenewable material are used in manufacture, shipping, or use of the product? How much waste is generated, and how will the item be disposed of when it is no longer functional? One product may well be more environmentally benign than another, but be careful who makes this claim.

Blue Angels and Green Seals

Products that claim to be environmentally friendly are being introduced at 20 times the normal rate for consumer goods. To help consumers make informed choices, several national programs have been set up to independently and scientifically analyze environmental impacts of major products. Germany's Blue Angel, begun in 1978, is the oldest of these programs. Endorsement is highly sought after by producers since environmentally conscious shoppers have shown that they are willing to pay more for products they know have minimum environmental impacts. To date, more than 2,000 products display the Blue Angel symbol. They range from recycled paper products, energy-efficient appliances, and phosphate-free detergents to refillable dispensers.

Similar programs are being proposed in every Western European country as well as in Japan and North America. Some are autonomous, nongovernmental efforts like the United States' new Green Seal program (managed by the Alliance for Social Responsibility in New York). Others are quasigovernmental institutions such as the Canadian Environmental Choice programs (fig. 25.6).

The best of these organizations attempt "cradle-to-grave" **life-cycle analysis** (fig. 25.7) that evaluates material and energy inputs and outputs at each stage of manufacture, use, and disposal of the product. While you need to consider your own situation in making choices, the information supplied by these independent agencies is generally more reliable than self-made claims from merchandisers.

Limits of Green Consumerism

To quote Kermit the Frog, "It's not easy being green." Even with the help of endorsement programs, doing the right thing from an environmental perspective may not be obvious. Often we are faced with complicated choices. Do the social benefits of buying rainforest nuts justify the energy expended in transporting them here, or would it be better to eat only locally grown products? In switching from Freon propellants to hydrocarbons, we spare the stratospheric ozone but increase hydrocarbon-caused smog. By choosing reusable diapers over disposable ones, we decrease the amount of material going to the landfill, but we also increase water pollution, energy consumption, and pesticide use (cotton is one of the most pesticide-intensive crops grown in the United States).

When the grocery store clerk asks you, "Paper or plastic?" you probably choose paper and feel environmentally virtuous, right? Everyone knows that plastic is made by synthetic chemical processes from nonrenewable petroleum or natural gas. Paper from naturally growing trees is a better environmental choice, isn't it? Well, not necessarily. In the first place, paper making consumes water and causes much more water pollution than does plastic man-

(a)   (b)

FIGURE 25.6   American (*a*) and Canadian (*b*) symbols that will be used to indicate products that are "environmentally superior."
(*a*) Reproduced with permission, Green Seal, Inc., Washington, D.C. (*b*) Official mark of Environment Canada.

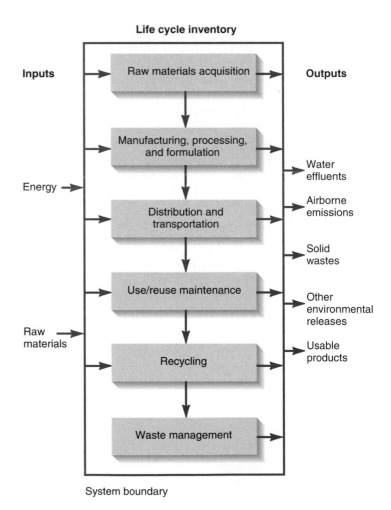

FIGURE 25.7   At each stage in its life cycle, a product receives inputs of materials and energy and produces outputs of materials or energy that move to subsequent phases and wastes that are released into the environment. *Source:* After Society of Environmental Toxicology and Chemistry, 1991.

ufacturing. Paper mills also release air pollutants, including foul-smelling sulfides and captans as well as highly toxic dioxins.

Furthermore, the brown paper bags used in most supermarkets are made primarily from virgin paper. Recycled fibers aren't strong enough for the weight they must carry. Growing, harvesting, and transporting logs from agroforestry plantations can be as environmentally disruptive as oil production. It takes a great deal of energy to pulp wood and dry newly made paper. Paper is also heavier and bulkier to ship than plastic. Although the polyethylene used to make a plastic bag contains many calories, in the end, paper bags are generally more energy-intensive to produce and market than plastic ones.

If both paper and plastic go to a landfill in your community, the plastic bag takes up less space. It doesn't decompose in the landfill, but neither does the paper in an air-tight, water-tight landfill. If paper is recycled but plastic is not, then the paper bag may be the better choice. If you are lucky enough to have both paper and plastic recycling, the plastic bag is probably a better choice since it recycles more easily and produces less pollution in the process. The best choice of all is to bring your own reusable cloth bag.

Complicated, isn't it? We often must make decisions without complete information, but it's important to make the best choices we can. Don't assume that your neighbors are wrong if they reach conclusions different from yours. They may have valid considerations of which you are unaware. The truth is that simple black and white answers often don't exist.

Taking personal responsibility for your environmental impact can have many benefits. Recycling, buying "green" products, and other environmental actions not only set good examples for your friends and neighbors, they also strengthen your sense of involvement and commitment in valuable ways. There are limits, however, to how much we can do individually through our buying habits and personal actions to bring about the fundamental changes needed to save the earth. Green consumerism generally can do little about larger issues of global equity, chronic poverty, oppression, and the suffering of millions of people in the developing world. There is a danger that exclusive focus on such problems as whether to choose paper or plastic bags, or to sort recyclables for which there are no markets, will divert our attention from the greater need to change basic institutions.

COLLECTIVE ACTIONS

While a few exceptional individuals can be effective working alone to bring about change, most of us find it more productive and more satisfying to work with others.

Collective action multiplies individual power (fig. 25.8). You get encouragement and useful information from meeting regularly with others who share your interests. It's easy to get discouraged by the slow pace of change; having a support group helps maintain your enthusiasm. You should realize, however, that there is a broad spectrum of environmental and social action groups. Some will suit your particular interests, preferences, or beliefs more than others. In this section, we will look at some environmental organizations as well as options for getting involved.

Student Environmental Groups

A number of organizations have been established to teach ecology and environmental ethics to elementary and secondary school students, as well as to get them involved in active projects to clean up their local community. Groups such as Kids Saving the Earth or Eco-Kids Corps are an important way to reach this vital audience. Family education results from these efforts as well. In a World Wildlife Fund survey, 63 percent of young people said they "lobby" their parents about recycling and buying environmentally responsible products.

Key Concepts

- Sustainable consumption aims to find ways to enjoy a good quality of life while consuming fewer natural resources and polluting less. Adopting a simpler, less demanding lifestyle can also reduce stress, improve health, and promote social justice.

- Although each of our individual choices may make a small impact, collectively they can be important. Recycling, buying green products, accumulating less stuff, and other personal actions set a good example for friends and neighbors, encourage businesses to provide sustainable goods and services, and strengthen your sense of involvement and commitment.

- Participating in group activities can produce greater impact than working alone. It also provides motivation and information exchange while helping you and others maintain enthusiasm. There are many different types of environmental groups to suit every taste and interest.

- The United Nations Earth Charter provides a good set of principles and goals for economic, social, and ecological sustainability.

FIGURE 25.8 Working together with others can give you energy, inspiration, and a sense of accomplishment. © William P. Cunningham.

FIGURE 25.9 Protests, marches, and public demonstrations can be an effective way to get your message out and to influence legislators. © William P. Cunningham.

TABLE 25.3 Organizing an Environmental Campaign

1. What do you want to change? Are your goals realistic, given the time and resources you have available?
2. What and who will be needed to get the job done? What resources do you have now and how can you get more?
3. Who are the stakeholders in this issue? Who are your allies and constituents? How can you make contact with them?
4. How will your group make decisions and set priorities? Will you operate by consensus, majority vote, or informal agreement?
5. Have others already worked on this issue? What successes or failures did they have? Can you learn from their experience?
6. Who has the power to give you what you want or to solve the problem? Which individuals, organizations, corporations, or elected officials should be targeted by your campaign?
7. What tactics will be effective? Using the wrong tactics can alienate people and be worse than taking no action at all.
8. Are there social, cultural, or economic factors that should be recognized in this situation? Will the way you dress, talk, or behave offend or alienate your intended audience? Is it important to change your appearance or tactics to gain support?
9. How will you know when you have succeeded? How will you evaluate the possible outcomes?
10. What will you do when the battle is over? Is yours a single-issue organization, or will you want to maintain the interest, momentum, and network you have established?

Source: *Based on material from "Grassroots Organizing for Everyone" by Claire Greensfelder and Mike Roselle from* Call to Action, *1990 Sierra Book Club Books.*

Organizations for secondary and college students often are among our most active and effective groups for environmental change. The largest student environmental group in North America is the Student Environmental Action Coalition (SEAC). Formed in 1988 by students at the University of North Carolina at Chapel Hill, SEAC has grown rapidly to more than 30,000 members in some 500 campus environmental groups. SEAC is both an umbrella organization and grassroots network that functions as an information clearinghouse and a training center for student leaders. Member groups undertake a diverse spectrum of activities ranging from politically neutral recycling promotion to confrontational protests of government or industrial projects (fig. 25.9). National conferences bring together thousands of activists who share tactics and inspiration while also having fun. If there isn't a group on your campus, why not look into organizing one?

Another important student organizing group is the network of Public Interest Research Groups active on most campuses in the United States. While not focused exclusively on the environment, the PIRGs usually include environmental issues in their priorities for research. By becoming active, you could probably introduce environmental concerns to your local group if they are not already working on problems of importance to you.

One of the most important skills that you are likely to learn in SEAC or other groups committed to social change is how to organize. Organizing is a dynamic process in which you must constantly adapt to changing conditions. Some basic principles apply in most situations, however (table 25.3). Remember that you are not alone. Others share your concerns and want to work with you to bring about change; you just have to find them. There is power in numbers. As Margaret Mead once said, "Never doubt that a small, highly committed group of individuals can change the world; indeed, it is the only thing that ever has."

Using the communication media to get your message out is an important part of the modern environmental movement. Table 25.4 suggests some important considerations in planning a media campaign.

Schools can be sources of information and inspiration in sustainable living. They have knowledge and expertise to figure out how to do new things, and they have students who have the energy and enthusiasm to do much of the research, and for whom that discovery will be a valuable learning experience. At many colleges and universities, students have undertaken campus audits to examine water and energy use, waste production and disposal, paper consumption, recycling, buying locally produced food, and many other examples of sustainable resource consumption. At more than 100 universities and colleges across America, graduating students have taken a pledge that reads:

> "I pledge to explore and take into account the social and environmental consequences of any job I consider and will try to improve these aspects of any organization for which I work."

Could you introduce something similar at your school?

Campuses often have building projects that can be models for sustainability research and development. Some recent examples of prize-winning sustainable design can be found at Stanford University, Oberlin College in Ohio, and the University of California at Santa Barbara. Stanford's Jasper Ridge building will provide classroom, laboratory, and office space for its biological research station. Stanford students worked with the administration to develop *Guidelines for Sustainable Buildings,* a booklet that cov-

TABLE 25.4	Using the Media to Influence Public Opinion

Shaping opinion, reaching consensus, electing public officials, and mobilizing action are accomplished primarily through the use of the communications media. To have an impact in public affairs, it is essential to know how to use these resources. Here are some suggestions:

1. *Assemble a press list. Learn to write a good press release* by studying books from your public library on press relations techniques. Get to know reporters from your local newspaper and TV stations.

2. *Appear on local radio and TV talk shows.* Get experts from local universities and organizations to appear.

3. *Write letters to the editor, feature stories, and news releases.* You may include black and white photographs. Submit them to local newspapers and magazines. Don't overlook weekly community shoppers and other "freebie" newspapers, which usually are looking for newsworthy material.

4. *Try to get editorial support from local newspapers, radio, and TV stations.* Ask them to take a stand supporting your viewpoint. If you are successful, send a copy to your legislator and to other media managers.

5. *Put together a public service announcement and ask local radio and TV stations to run it* (preferably not at 2 A.M.). Your library or community college may well have audiovisual equipment that you can use. Cable TV stations usually have a public access channel and will help with production.

6. *If there are public figures in your area who have useful expertise, ask them to give a speech or make a statement.* A press conference, especially in a dramatic setting, often is a very effective way of attracting attention.

7. *Find celebrities or media personalities to support your position.* Ask them to give a concert or performance, both to raise money for your organization and to attract attention to the issue. They might like to be associated with your cause.

8. *Hold a media event that is photogenic and newsworthy.* Clean up your local river and invite photographers to accompany you. Picket the corporate offices of a polluter, wearing eye-catching costumes and carrying humorous signs. Don't be violent, abusive, or obnoxious; it will backfire on you. Good humor usually will go farther than threats.

9. *If you hear negative remarks about your issue on TV or radio, ask for free time under the Fairness Doctrine to respond.* Stations have to do a certain amount of public service to justify relicensing and may be happy to accommodate you.

10. *Ask your local TV or newspaper to do a documentary or feature story about your issue or about your organization and what it is trying to do.* You will not only get valuable free publicity, but you may inspire others to follow your example.

ers everything from energy-efficient lighting to native landscaping. With 275 photovoltaic panels to catch sunlight, there should be no need to buy electricity for the building. In fact, it's expected that surplus energy will be sold back to local utility companies to help pay for building operation.

Oberlin's Environmental Studies Center, designed by famed architect Bill McDonough, features 370 m² of photovoltaic panels on its roof, a geothermal well to help heat and cool the building, large south-facing windows for passive solar gain, and a "living machine" for water treatment, including plant-filled tanks in an indoor solarium and a constructed wetland outside (see figs. 18.29 and 20.10).

UCSB's Bren School of Environmental Science and Management looks deceptively institutional but claims to be the most environmentally state-of-the-art structure of its kind in the United States (fig. 25.10). It wasn't originally intended to be a particularly green building, but planners found that some simple features like having large windows that harvest natural light and open to let ocean breezes cool the interior make the building both more functional and more appealing. Motion detectors control light levels and sensors monitor and refresh the air when there is too much CO_2 putting students to sleep. More than 30 percent of interior materials are recycled. Solar panels supply 10 percent of the electricity, and the building exceeds federal efficiency standards by 30 percent. "The overriding and very powerful message is it really doesn't cost any more to do these things," says Dennis Aigner, dean of Bren School.

Mainline Environmental Organizations

Among the oldest, largest, and most influential environmental groups in the United States are the National Wildlife Federation,

FIGURE 25.10 The University of California at Santa Barbara claims its new Bren School of Environmental Science and Management is the most environmentally friendly building of its kind in the United States. © William P. Cunningham.

the World Wildlife Fund, the Audubon Society, the Sierra Club, the Izaak Walton League, Friends of the Earth, Greenpeace, Ducks Unlimited, the Natural Resources Defense Council, and The Wilderness Society. Sometimes known as the "group of 10," these organizations are criticized by radical environmentalists for their tendency to compromise and cooperate with the establishment. Although many of these groups were militant—even extremist— in their formative stages, they now tend to be more staid and conservative. Members are mostly passive and know little about the inner workings of the organization, joining as much for publications or social aspects as for their stands on environmental issues.

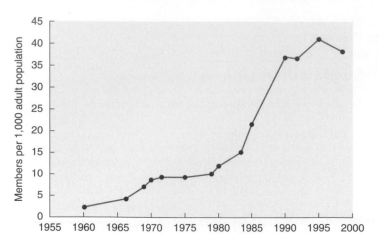

FIGURE 25.11 Growth of national environmental organizations in the United States. *Source:* Robert Putnam, 2000.

Many of these groups had astonishing growth rates in the 1980s and 90s, tripling their collective membership in just a decade (fig. 25.11). Much of this increase, however, was what sociologist Robert Putnam calls an "honorific rhetorical device for fundraising." People make donations in response to direct mail advertising and door-to-door or telemarketing solicitations and are counted as members even though they have little interest in further contact with the organization. Typically, up to half these "members" don't renew in subsequent years, so that organizations dependent on this revenue source send out a never-ending flood of solicitations seeking new donations.

Still, these groups are powerful and important forces in environmental protection. Their mass membership, large professional staffs, and long history give them a degree of respectability and influence not found in newer, smaller groups. The Sierra Club, for instance, with about half a million members and chapters in almost every state, has a national staff of about 400, an annual budget over $20 million, and 20 full-time professional lobbyists in Washington. These national groups have become a potent force in Congress, especially when they band together to pass specific legislation such as the Alaska National Interest Lands Act or the Clean Air Act.

In a survey that asked congressional staff and officials of government agencies to rate the effectiveness of groups that attempt to influence federal policy on pollution control, the top five were national environmental organizations. In spite of their large budgets and important connections, industry groups like the American Petroleum Institute, the Chemical Manufacturers Association, and the Edison Electric Institute ranked far behind these environmental groups in terms of influence.

Although much of the focus of the big environmental organizations is in Washington, Audubon, Sierra Club, and Izaak Walton have local chapters, outings, and conservation projects (fig. 25.12). This can be a good way to get involved. Go to some meetings, volunteer, offer to help. You may have to start out stuffing envelopes or some other unglamourous job, but if you persevere, you may have a chance to do something important and fun. It's a good way to learn and meet people.

FIGURE 25.12 Outings are an important way for environmental groups to build friendships, have fun, and become familiar with threatened natural areas. © William P. Cunningham.

Some environmental groups such as the Environmental Defense Fund (EDF), The Nature Conservancy (TNC), National Resources Defense Council (NRDC), and the Wilderness Society (WS) have limited contact with ordinary members except through their publications. They depend on a professional staff to carry out the goals of the organization through litigation (EDF and NRDC), land acquisition (TNC), or lobbying (WS). Although not often in the public eye, these groups can be very effective because of their unique focus.

Still, there are questions about means and ends of different groups. The Nature Conservancy, for example, is one of the largest and certainly the world's richest environmental group, with 3,200 employees in 30 countries, annual revenues approaching $1 billion, and assets of around $3 billion. Today, it manages 7 million acres in 1,400 nature preserves, which it describes as the world's largest private sanctuary system. Despite its successes, the Conservancy is controversial for management decisions on some of this far-flung system of reserves. For instance, it has logged forests, engineered a $64 million development of opulent houses on fragile coastal grasslands, and drilled for natural gas within some of its preserves.

Today, the million-member Conservancy itself is something of an environmental Leviathan. It is also the leading proponent of a brand of environmentalism that promotes compromise between conservation and corporate America. Its governing board and advisory council now include executives and directors from oil companies, chemical producers, auto manufacturers, mining corporations, logging companies, and electric utilities. In an exposé in the *Washington Post,* TNC was accused of avoiding issues, such as drilling in the Arctic National Wildlife Refuge, that are sensitive to big corporate donors. In response to these criticisms, the Conservancy has promised a thorough review and housecleaning to restore public trust and to return to its original objectives of preserving landscapes and biodiversity.

Broadening the Environmental Agenda

The environmental movement in the United States tends to be overwhelmingly white, middle-class, and suburban. Few blue-collar workers and even fewer minorities are involved, especially in leadership or professional positions. This is unfortunate in several ways. The movement will probably never be truly successful until it has a broad-based coalition of support. Furthermore, as we saw in chapter 2, poor people are often most seriously affected by toxic waste dumps, noise, urban blight, and air and water pollution (fig. 25.13). They should be included in discussions of how to remedy these problems. Unfortunately, most environmental groups have acquired a reputation for caring about plants, animals, and wilderness areas more than about people. Recently, a number of grassroots citizen's groups have begun to demand cleaner local environments. Often organized around tenant councils or other social justice issues, these groups could be a powerful force for general environmental protection.

Deep or Shallow Environmentalism?

Norwegian philosopher Arne Naess criticizes the superficial commitment of environmentalists who claim to be green but are quick to compromise and who do little to bring about fundamental change. He characterizes this as **shallow ecology** and contrasts it with what he calls **deep ecology,** a profoundly biocentric worldview that calls for a fundamental shift in our attitudes and behavior. An equally radical but more humanist philosophy called **social ecology** is proposed by Murray Bookchin, who draws on the communitarian anarchism of the Russian geographer Peter Kropotkin.

Among the tenets of both these "dark green" philosophies are voluntary simplicity; rejection of anthropocentric attitudes; intimate contact with nature; decentralization of power; support for cultural and biological diversity; a belief in the sacredness of nature; and direct personal action to protect nature, improve the environment, and bring about essential societal change. They differ mainly on the necessity for personal freedom and communal interaction.

Both Naess and Bookchin criticize mainstream environmental groups that favor pragmatic work within the broadly agreed upon social agenda and the established political system to bring about incremental, progressive reform rather than radical revolution. Most progressives naturally object to being called shallow, light, or superficial; they prefer instead to characterize their position as moderate rather than radical, or reformist rather than revolutionary. They tend to view the deep ecologists and anarchists as unrealistic zealots who would rather be righteous than effective.

FIGURE 25.13 Mainline environmental organizations tend to concentrate on wildlife and wilderness, paying scant attention to environmental justice and inner-city issues. © Sam Kittner/Greenpeace Photo.

Aside from name-calling, there are some important philosophical questions in this debate. One of the questions is whether we face an environmental crisis that requires radical action or only "problems" that can best be overcome by working within established channels. Another important question is whether it is better to overturn society to explore new ways of living or to work for progressive change within existing political, economic, and social systems. Finally, is it more important to work for personal perfection or collective improvement? There may be no single answer to any of these questions; it's good to have people working in many different ways to find solutions.

Cooperation and Compromise

Rather than use confrontation and denunciation as methods for solving disputes, Dan Dagget, author of *Beyond Rangeland Conflict,* calls for grassroots cooperation and compromise. "Name-calling," Dagget says, "won't solve our environmental problems. Nor does it help to paint the world in shades of guilt or look for someone to blame. I've decided to join the growing number of people who have made a commitment to work together in spite of diverse politics, religious beliefs, and culture to try to improve our common environment . . . This is something each of us uses every day when we work to achieve our goals—to plant a garden, to paint a house, to do our job. Frequently, in these situations, we work with people we don't agree with politically, but we still get the job done. And we get better results than the politicians do when they try to solve our problems for us."

Dagget describes how a diverse group of ranchers, government land managers, wise-users, radical environmentalists, and just plain citizens in central Nevada formed a group called the Toiyabe Wetlands and Watersheds Management Team to work on reversing the effects of more than a century of overgrazing. Although the federal government had plowed loads of money into trying to protect and restore rangelands in this area, they made little progress until local residents began a program of collaborative stewardship. Similarly, on a northern Arizona ranch belonging to the family of former Secretary of the Interior Bruce Babbitt, a team of local environmentalists, ranchers, students, and National Park Service staffers is studying better ways to sustainably manage grasslands of the arid Southwest. And in northeastern California, a group of townsfolk in Quincy got together to try to save their regional economy and to hammer out a management plan for the surrounding fire-prone forests.

In working together, people come to understand perspectives different from their own. They begin to see their opponents as decent individuals rather than faceless members of some vast, evil conspiracy. And when a group has some small, tangible successes, they are more likely to tackle bigger issues. Critics of these efforts, on the other hand, say that working on mundane problems distracts your attention from structural relations that cause those problems in the first place. In the jargon of the 1960s, you become co-opted— too busy with meaningless details to pay attention to what's really important. While personal friendships may facilitate dialogue, they also mute disagreements.

What do you think? Is it better to remain independent and critical, or to join together to try to find workable solutions? How would you decide when to compromise and when to stand firm?

Radical Environmental Groups

A striking contrast to the mainline conservation organizations are the **direct action** groups, such as Earth First!, Sea Shepherd, and a few other groups that form either the "cutting edge" or the "radical fringe" of the environmental movement, depending on your outlook. Often associated with the deep ecology philosophy and bioregional ecological perspective, the strongest concerns of these militant environmentalists tend to be animal rights and protection of wild nature. Their main tactics are civil disobedience and attention-grabbing actions, such as guerrilla street theater, picketing, protest marches, road blockades, and other demonstrations. Many of these techniques are borrowed from the civil rights movement and Mahatma Gandhi's nonviolent civil disobedience. While often more innovative than the mainstream organizations, pioneering new issues and new approaches, the tactics of these groups can be controversial.

Members of Earth First! chained themselves to a giant tree-smashing bulldozer to prevent forest clearing in Texas, and blockaded roads being built into wilderness areas. In one remarkable case, a 25-year-old activist named Julia "Butterfly" Hill spent more than two years sitting on a small, tarp-covered platform 55 m (180 ft) up in a redwood tree in northern California. She was protesting the clear-cutting of the Headwaters forest by the Maxxam/Pacific Lumber Company. She came down from her perch in 1999 when her supporters raised $50,000 to buy the tree from Maxxam. Some protests are humorous and lighthearted, such as skits or street theater that get a point across in a nonthreatening way (fig. 25.14).

A more problematic tactic is **monkey wrenching,** or environmental sabotage, a concept made popular by author Edward Abbey's book *The Monkey Wrench Gang*. Among the actions advocated by some Earth First! members are driving large spikes in trees to protect them from loggers, vandalizing construction equipment, pulling up survey stakes for unwanted developments, and destroying billboards.

A more radical fringe group called Earth Liberation Front claims responsibility for burning or vandalizing millions of dollars' worth of houses, government facilities, and commercial buildings to protest urban sprawl and genetic engineering. Mainstream environmental groups condemn these illegal acts as both unethical and counterproductive. They worry that environmental terrorism will alienate those who might be their allies and could validate matching violence from their opponents. In one case in Arizona, several expensive houses were burned by an arsonist who claimed to be associated with the Earth Liberation Front, but was really just using this as a cover for his own criminal motives.

Sea Shepherd, a radical offshoot from Greenpeace, is determined to stop the killing of marine mammals. Convinced that peaceful protests were not working, the members took direct action by sinking whaling vessels and destroying machinery at whale processing stations in Iceland. Greenpeace criticized this destructive approach, perhaps because of memories of the sinking of its ship *Rainbow Warrior* by the French secret service in New Zealand in 1985. They and other critics of ecotage (ecological sabotage) see

FIGURE 25.14 Street theater can be a humorous, yet effective, way to convey a point in a nonthreatening way. Confrontational tactics get attention, but they may alienate those who might be your allies and harden your opposition. © William P. Cunningham.

What do you think?

Evaluating Extremist Claims

Attaining a sustainable future requires us to change some of the ways we impact the earth. Some problems will likely turn out less serious and others more so, than we currently believe. Undoubtedly new concerns have yet to be uncovered. The issues will continue to stimulate strong feelings on both sides for several reasons. The problems are potentially very serious. Uncertainty is likely to continue to surround many of the issues. And, most importantly, needed changes often will challenge our worldviews.

Sometimes advocates on both sides get carried away and attempt to persuade through the use of biased language instead of relying on objective evidence to influence the thought of others. Critical thinking requires that conclusions be based on the objective evaluation of objective evidence. Bias, a person's inclination to favor or oppose a particular conclusion, is a formidable barrier to critical thinking.

Being able to detect bias enables a person to determine whether a claim or position is being presented objectively. There are a number of strategies to help identify bias. One clue as to whether or not a claim is being presented fairly is the person's choice of language. Many words have positive or negative connotations that can reflect emotion rather than objectivity.

The viewpoints presented below are positions that have appeared in print. Notice that bias can range from obvious to subtle.

Writer A: *Despite the hysterics of a few pseudo-scientists, there is no reason to believe in global warming.*

This writer seeks to persuade by using language to ridicule the view that the global warming threat is real. Would you believe a "*pseudo-scientist,*" especially one using "*hysterics*"? Such emotionally charged language attacks the credibility of the opposing view. The implication is that only kooks think global warming likely. Would the sentence have lost its emotional punch if it had said: "Despite the claims of researchers, there is no reason to believe in global warming"?

Writer B: *With the collapse of Marxism, environmentalism has become the new refuge of socialist thinking. The environment is a great way to advance a political agenda that favors central planning and an intrusive government. What better way to control someone's property than to subordinate one's private property rights to environmental concerns?*

This writer attempts to discredit environmental concerns not with evidence, but by linking them to an unpopular political ideology. *Marxism* and *socialist* are highly negatively charged terms in the minds of most Americans. The writer seeks to influence the reader's opinion by claiming environmental concern is akin to accepting a distasteful political belief.

Writer C: *The battle to feed humanity is over. In the 1970s and 1980s hundreds of millions of people will starve to death in spite of any crash programs embarked upon now. At this late date nothing can prevent a substantial increase in the world death rate, although many lives could be saved through dramatic programs to 'stretch' the carrying capacity of the earth. . . . But these programs will only provide a stay of execution unless they are accompanied by determined and successful efforts at population control.*

Words and phrases conveying bias include: "*The battle . . . is over.*" "Hundreds of millions *will* starve to death. . . . *Crash programs.* . . . will only provide a *stay of execution.* . . ." The writer hopes to persuade the reader through an emotional appeal rather than convince the reader through reason.

All people have biases. They are a part of our conceptual frameworks, which in turn, are rooted in our level of knowledge, past experiences, purposes, and vested interests. Identifying bias in ourselves and others is a necessary skill in the critical-thinking process. What are some other strategies one could use to detect bias in addition to looking at language?

(Writer of A and B: Rush Limbaugh; C: Paul Ehrlich)

---

these acts as environmental terrorism that give all environmentalists a bad name. Sea Shepherd's Paul Watson says, "Our aggressive nonviolence is just doing what must be done, according to the dictates of our own conscience." Earth First!'s Dave Foreman (once described as a kamikaze environmentalist) said, "Extremism in defense of Mother Earth is no vice." Is this merely hyperbole, or a valid rhetorical device? Being able to recognize and evaluate claims like this is an important critical-thinking skill (What Do You Think? p. 557).

Wise Use Movement

At the same time that the American environmental movement has become more militant and more powerful, it is opposed by many people who feel their way of life threatened by increasing restrictions on their access to natural resources. Much of the focus of the mainstream environmental groups has been on wilderness and wildlife issues that appeal to middle- and upper-class urban residents but don't interest a broad section of society. Many people whose livelihood is based on mining, logging, ranching, fishing, farming, and other resource-dependent jobs regard their careers, communities, and values as under attack by outside forces (fig. 25.15). They believe that years of living in a particular place give them special privileges and knowledge neither understood nor appreciated by outsiders. Regulations for resource access, recreational opportunities, and land uses often are seen as insensitive to local needs and a threat to customary uses and property rights.

In their study of environmentalism and cultural creatives, authors Paul Ray and Sherry Anderson found that about one-quarter of all Americans are "traditionalists" who believe that preserving conventional social roles and cultural values are more important than protecting nature. The other half of the population, according to Ray and

FIGURE 25.15   To families dependent on jobs in logging, mining, ranching, or other resource-based occupations, environmental regulations may seem threatening. How do we balance human needs against those of nature? © Bryan F. Peterson.

ductive landscapes and advocate stewardship and constructive use rather than simply leaving nature alone. Wise use proponents often call for fewer governmental regulations, less pollution control, and the precedence of property rights over protection for wild nature or endangered species. Some wise use groups compare in size and influence with the largest national environmental groups. The national Cattlemen's Association and the National Farm Bureau, for example, claim hundreds of thousands of members and each have annual budgets approaching $10 million.

While some wise use groups are genuine grassroots organizations, others are primarily lobbying arms for industry. People for the West, for instance, claims to represent ordinary rural people, but 12 of its 13 board members and almost all of its $1 million annual budget come from big timber or mining corporations. The Mountain States Legal Foundation, which employed both Interior Secretaries James Watt and Gail Norton, is funded by Coors Beer and other corporations and represents primarily business interests in resource and land rights disputes.

In some cases, anti-environmental extremists have been even more violent and destructive than those in the radical fringe of the pro-environmental movement. Survivalists and separatists, for instance, have made death threats against both government employees and environmental activists. In Nevada, militants blew up offices of both the U.S. Forest Service and the Bureau of Land Management. In California, two forest activists were nearly killed when a bomb exploded in their car. Obviously these extreme actions aren't typical of everyone who opposes more environmental regulation, but they show how deep and divisive these differences in worldview and economic interests can be.

Sustainability

As the World Bank says in its *World Development Report,* "Without adequate environmental protection, human development is undermined; without human development, environmental protection will fail." Often, because they have no other options, the poorest people are both the victims and agents of environmental degradation (fig. 25.16). How can we accomplish the twin goals of improving human well-being and protecting our common environment? Sustainability has been a central theme throughout this book, but perhaps we should review its principles and goals one more time.

As the developing countries of the world become more affluent, they are adopting many of the wasteful and destructive lifestyle patterns of the richer countries. Automobile production in China, for example, is increasing at about 19 percent per year, or doubling every 3.7 years. By 2030 there could be more automobiles in China than the United States. What will be the effect on air quality, world fossil fuel supplies, and global climate if that growth rate continues? Already, two-thirds of the children in Shenzhen, China's wealthiest province, suffer from lead poisoning, probably caused by use of leaded gasoline. And, as chapter 8 points out, diseases associated with affluent lifestyles—such as obesity, diabetes, heart attacks, depression, and traffic accidents—are becoming the leading causes of morbidity and mortality worldwide.

Anderson, are "modernists" who are more interested in success, fashion, and a comfortable life than either traditional values or nature appreciation. Traditionalists tend to be religious and social conservatives who aren't necessarily anti-environment, but who see many of the priorities of environmental groups as elitist, unnecessary, and detrimental to their lifestyles. Often residents of small towns or rural areas, traditionalists put a strong emphasis on independence, self-reliance, and personal freedom. They prefer local resource control and are very suspicious of foreigners and outsiders. It might be possible to find areas of agreement and cooperation between environmental activists and traditionalists if there were greater cultural sensitivity and understanding between these groups.

Drawing on the anger and resentment felt by many traditionalists, a number of **wise use groups** have sprung up in recent years. These organizations generally advocate conservation rather than preservation of resources (see the discussion of Gifford Pinchot and John Muir in chapter 2). Many conservationists regard empty wilderness as wasteful and against God's plan. They prefer pro-

FIGURE 25.16 More than 1 billion people—like this Indonesian family—live in acute poverty, lacking access to secure food supplies, safe drinking water, housing, education, health services, and decent jobs. © William P. Cunningham.

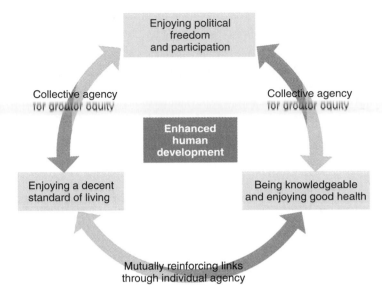

FIGURE 25.17 Human development, democracy, and education are mutually reinforcing. *Source:* UN, 2002.

We would all benefit by helping developing countries access more efficient, less polluting technologies. Education, democracy, and access to information are essential for sustainability (fig. 25.17). It is in our best interest to help finance protection of our common future in some equitable way. Maurice Strong, chair of the Earth Charter Council, estimates that development aid from the richer countries should be some $150 billion per year, while internal investments in environmental protection by developing countries will need to be about twice that amount. Many scholars and social activists believe that poverty is at the core of many of the world's most serious human problems: hunger, child deaths, migrations, insurrections, and environmental degradation. One way to alleviate poverty is to foster economic growth so there can be a bigger share for everyone.

Strong economic growth already is occurring in many places. The World Bank projects that if current trends continue, economic output in developing countries will rise by 4 to 5 percent per year in the next 40 years. Economies of industrialized countries are expected to grow more slowly but could still triple over that period. Altogether, the total world output could be quadruple what it is today.

That growth could provide funds to clean up environmental damage caused by earlier, wasteful technologies and misguided environmental policies. It is estimated to cost $350 billion per year to control population growth, develop renewable energy sources, stop soil erosion, protect ecosystems, and provide a decent standard of living for the world's poor. This is a great deal of money, but it is small compared to the $1 trillion per year spent on wars and military equipment.

While growth simply implies an increase in size, number, or rate of something, development, in economic terms, means a real increase in average welfare or well-being. **Sustainable development** based on the use of renewable resources in harmony with ecological systems is an attractive compromise to the extremes of no growth versus unlimited growth. Perhaps the best definition of this goal is that of the World Commission on Environment and Development, which defined sustainable development in *Our Common Future* as "meeting the needs of the present without compromising the ability of future generations to meet their own needs." Some goals of sustainable development include:

- A demographic transition to a stable world population of low birth and death rates.

- An energy transition to high efficiency in production and use, coupled with increasing reliance on renewable resources.

- A resource transition to reliance on nature's "income" without depleting its "capital."

- An economic transition to sustainable development and a broader sharing of its benefits.

- A political transition to global negotiation grounded in complementary interests between North and South, East and West.

• An ethical or spiritual transition to attitudes that do not separate us from nature or each other.

In the United States, the President's Council on Sustainable Development has recommended 145 specific actions to accomplish three dozen policy goals. At the heart of the Council's findings is a conviction that economic, environmental, and social equity issues are inextricably linked and must be considered together (fig. 25.18). Another core belief is that to achieve sustainability, some things—jobs, productivity, wages, capital, savings, profits, information, knowledge, education, social welfare, and justice—must grow, while others—pollution, waste, and poverty—must not. Ten core goals and indicators of progress suggested by the Council are listed in table 25.5.

International Nongovernmental Organizations

The rise in international environmental organizations in recent years has been phenomenal. At the Stockholm Conference in 1972, only a handful of environmental groups attended, almost all from developed countries. Twenty years later, at the Rio Earth Summit, more than 30,000 individuals representing several thousand environmental groups, many from developing countries, held a global Ecoforum to debate issues and form alliances for a better world. We call these groups working for social change **nongovernmental organizations (NGOs).** They have become a powerful aspect of environmental protection.

Some NGOs are located primarily in the more highly developed countries of the North and work mainly on local issues. Others are headquartered in the North but focus their attention on the problems of developing countries in the South. Still others are truly global, with active groups in many different countries. A few are highly professional, combining private individuals with representatives of government agencies on quasi-governmental boards or standing committees with considerable power. Others are at the fringes of society, sometimes literally voices crying in the wilderness. Many work for political change, more specialize in gathering and disseminating information, and some undertake direct action to protect a specific resource.

Public education and consciousness-raising using protest marches, demonstrations, civil disobedience, and other participatory public actions and media events are generally important tactics for these groups. Greenpeace, for instance, carries out well-publicized confrontations with whalers, seal hunters, toxic-waste dumpers, and others who threaten very specific and visible resources. Greenpeace may well be the largest environmental organization in the world, with some 2.5 million contributing members.

In contrast to these highly visible groups, others choose to work behind the scenes, but their impact may be equally important. Conservation International has been a leader in debt-for-nature swaps to protect areas particularly rich in biodiversity. It also has some interesting initiatives in economic development, seeking products made by local people that will provide income along with environmental protection.

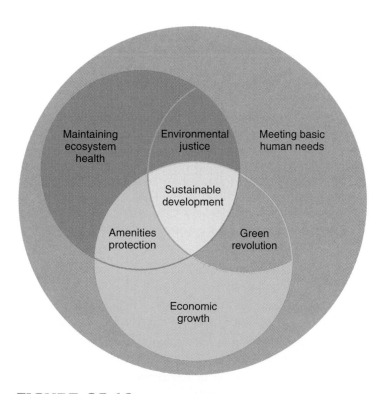

FIGURE 25.18 A model for integrating ecosystem health, human needs, and sustainable economic growth.
(Modified from Raymond Grizzle and Christopher Barrett, personal communication.)

TABLE 25.5 **Goals for Sustainable Development**

1. Ensure every person the benefits of a healthy environment.
2. Sustain a healthy economy that affords the opportunity for a high quality of life.
3. Ensure equity and opportunity for economic, social, and environmental well-being.
4. Protect and restore natural resources for current and future generations.
5. Encourage stewardship.
6. Encourage people to work together to create healthy communities.
7. Create full opportunity for citizens, businesses, and communities to participate in and influence the natural resource, environmental, and economic decisions that affect them.
8. Move toward stabilization of the U.S. population.
9. Lead in developing and carrying out sustainable development policies globally.
10. Ensure access to formal education and lifelong learning that will prepare citizens for meaningful work and a high quality of life, and give them an understanding of concepts involved in sustainable development.

Source: *President's Council on Sustainable Development, 1997.*

GREEN POLITICS

In many countries with a parliamentary form of government, **"green" political parties** based on environmental interests have become a political force in recent years. The largest and most powerful green party is Germany's *Die Grünen*. It is a grassroots, egalitarian, council-style movement, committed to participatory democracy, environmental protection, and a fundamental transformation of society (fig. 25.19). An essential premise is that ceaseless industrial growth destroys both people and the environment and must, therefore, be stopped. The Greens have formed coalitions with antinuclear weapons protesters, feminists, human rights advocates, and other public interest groups, but have struggled over whether to work with or oppose the ruling centrist party. Political realists argue that they could be more effective within a larger coalition; idealists vow never to compromise their principles.

In 2002 federal elections, the Greens won 8.6 percent of the total vote, making them the third largest political party in Germany. Forming a historic "red-green" coalition with the Social Democratic Party, the Greens, for the first time in world history, became a part of the ruling government in an industrialized country. There are strains within the party, however. Some of the far left in the Greens continue to take a hard line on things that most Germans hold as cherished rights, such as the ability to drive at unlimited speeds on the Autobahn, and to take fuel-guzzling vacations via jet airplane to far-away places. Others take a more pragmatic view of what's possible, and are willing to compromise to accomplish important goals.

The majority-rule political system in the United States makes it extremely difficult to start a new political party at the national level. Green candidates have been successful at the local and state levels, however. In 2002, Greens elected 171 local and state officials, including the first state representative (in Maine). One of the more interesting victories was that of Jason West, a 26-year-old house painter and puppeteer, who was elected mayor of New Paltz, New York. With the support of two other Green officials, he now controls the city council.

National groups such as the League of Conservation Voters, and "green committees of correspondents," work to introduce environmental issues into party politics and support candidates who share their environmental concerns. Campus Greens are strong at many colleges and universities. The four key values they espouse are (1) ecological wisdom, (2) peace and social justice, (3) grassroots democracy, and (4) freedom from violence.

In 2001, representatives from 60 countries around the world met in Australia to draft a Global Greens Charter based on the principles of social justice, democracy, nonviolence, and sustainability. "Experience over the past 30 years has shown," said European Greens Secretary General Arnold Cassola, "that being a protest party is not enough; if you want change, you need power." By working together internationally, Greens hope to attain that power.

What Can Individuals Do?

Although there is little likelihood of a Green party gaining power at the United States federal level, there are many opportunities for environmentally concerned citizens to have an influence on government policies. Get to know the positions of your state legislators, congressional representatives, and senators. Be active in party politics. Going to your district, county, and state political conventions gives you a voice in choosing candidates and establishing party platforms. Individuals can also contact legislators directly. Table 25.6 gives some suggestions for how to write effectively. You would probably be surprised at how little mail most legislators get from their constituents. Even on controversial issues, a representative might receive less than 100 personal letters or telephone calls. Since each congressional district includes about a half a million people, the legislator tends to assume that each letter represents the views of 5,000 to 10,000 people. Your voice can have great impact. A single, well-written letter can make a difference in whether important legislation gets passed.

All legislators now have e-mail access, but it isn't clear how many of them read Internet messages or how much attention they pay to them. They may fall in the category of form letters and petitions. Individually they don't count for much, but if a decision maker gets a million of them, they probably will notice. If you don't have time to write, a telephone call to the local or Washington office of your senator or representative can be effective. Follow the same steps in formulating your argument that you would use in a letter. You probably won't talk directly to the senator or representative, but your opinion will be registered by an aide. Often congressional staff are the ones who are really knowledgeable about a specific topic. Their bosses depend on them for advice about how they should vote. A well-organized, factual argument can be very persuasive and may change the position of your legislator. Chapter 24 has further discussion of environmental legislation, planning, policy, and law.

Don't forget that many important decisions also are made at the local level. Your city planning and zoning board, your county commissioners, state departments of natural resources or environmental

FIGURE 25.19 Delegates to the Green Party convention in Germany debate their political, social, economic, and environmental platform. © Photoreporters, Inc./International Press On Line.

TABLE 25.6 How to Write to Your Elected Officials

1. Address your letter properly.
   a. *Your representative:*
      The Honorable _____,
      House Office Building
      Washington, D.C. 20515
      Dear Representative _____,
   b. *Your senators:*
      The Honorable _____,
      Senate Office Building
      Washington, D.C. 20510
      Dear Senator _____,
   c. *The president:*
      The President
      The White House
      1600 Pennsylvania Avenue N.W.
      Washington, D.C. 20500
      Dear Mr. President,
2. Tell who you are and why you are interested in this subject. Be sure to give your return address.
3. Always be courteous and reasonable. You can disagree with a particular position, but be respectful in doing so. You will gain little by being shrill, hostile, or abusive.
4. Be brief. Keep letters to one page or less. Cover only one subject, and come to the point quickly. Trying to cover several issues confuses the subject and dilutes your impact.
5. Write in your own words. It is more important to be authentic than polished. Don't use form letters or stock phrases provided by others. Speak or write from your own personal experiences and interests. Try to show how the issue affects the legislator's own district and constituents.
6. If you are writing about a specific bill, identify it by number (for instance, H.R. 321 or S. 123). You can get a free copy of any bill or committee report by writing to the House Document Room, U.S. House of Representatives, Washington, D.C. 20515 or the Senate Document Room, U.S. Senate, Washington, D.C. 20510.
7. Ask your legislator to vote a specific way, support a specific amendment, or take a specific action. Otherwise you will get a form response that says: "Thank you for your concern. Of course I support clean air, pure water, apple pie, and motherhood."
8. If you have expert knowledge or specifically relevant experience, share it. But don't try to intimidate, threaten, or dazzle your representative. Don't pretend to have vast political influence or power. Legislators quickly see through artifice and posturing; they are professionals in this field.
9. If possible, include some reference to the legislator's past action on this or related issues. Show that you are aware of his or her past record and are following the issue closely.
10. Follow up with a short note of thanks after a vote on an issue that you support. Show your appreciation by making campaign contributions or working for candidates who support issues important to you.
11. Try to meet your senators and representatives when they come home to campaign, or visit their office in Washington if you are able. If they know who you are personally, you will have more influence when you call or write.
12. Join with others to exert your combined influence. An organization is usually more effective than isolated individuals.

---

protection, and a host of other agencies have power to establish many significant policies. These bodies are much more accessible for ordinary citizens. You might even run for a seat on a local board. What better way to ensure that local actions are sustainable than to become a policymaker yourself! Participating in practical environmental projects like litter cleanup or restoration projects can help build a sense of community, be educational, and also do good (fig. 25.20).

Finally, not everyone agrees with all the protest tactics used at the Seattle WTO meeting, but public demonstrations have played an important role in getting issues on the public agenda and influencing public opinion and official policies. You might think about what kinds of protest you believe to be effective and acceptable, and whether you would be willing to participate the next time there is an important issue at stake.

The Earth Charter

The UN Conference on Environment and Development in Rio de Janiero in 1992 was a landmark in global conservation history. More than 30,000 people—including 178 heads of state and representatives from more than 1,000 NGOs—gathered in Rio de Janeiro for this historic occasion. They considered a wide range of international environmental issues, including protecting forests and biodiversity, prevention of global climate change, rights of indigenous and tribal peoples, the role of poverty in environmental problems, sustainable development, and aid for environmental protection programs in poor countries. The final document of the Earth Summit, called Agenda 21, was long on promise but lacking in specific detail about how to accomplish its lofty goals. Still, many important issues were discussed for the first time and the contacts made in this process, both government-to-government and person-to-person networking, may turn out to have been an important contribution.

Following the Rio Earth Summit, a UN Council was established in Costa Rica to draft an **Earth Charter** of principles for environmental protection and sustainable development. Intended to be a statement of fundamental principles of enduring significance widely shared by people of all races, cultures, and religions, this statement draws on traditional wisdom, scientific insights, and a growing world literature on environmental ethics. It also draws on the experience of indigenous and tribal people and others whose cultural practices and belief systems most effectively promote environmental protection and sustainable livelihoods. The Benchmark II draft prepared for the "Millennium" meeting in January 2000 contained 49 principles drawn from hundreds of consultations over a three-year period in many countries and with people from all sectors of society. A summary of some key principles is presented in table 25.7.

FIGURE 25.20 Volunteers cleaned up 25 tons of trash from the banks of the Mississippi River in 2003. Each of us can help improve the quality of our local environment. © William P. Cunningham.

TABLE 25.7	Earth Charter Principles

1. Respect Earth and life in all its diversity.
2. Care for the community of life with understanding, compassion, and love.
3. Build democratic societies that are just, participatory, sustainable, and peaceful.
4. Secure Earth's bounty and beauty for present and future generations.
5. Protect and restore the integrity of Earth's ecological systems, with special concern for biological diversity and the natural processes that sustain life.
6. Prevent harm as the best method of environmental protection and, when knowledge is limited, apply a precautionary approach.
7. Adopt patterns of production, consumption, and reproduction that safeguard Earth's regenerative capacities, human rights, and community well-being.
8. Advance the study of ecological sustainability and promote the open exchange and wide application of the knowledge acquired.
9. Eradicate poverty as an ethical, social, and environmental imperative.
10. Ensure that economic activities and institutions at all levels promote human development in an equitable and sustainable manner.
11. Affirm gender equality and equity as prerequisites to sustainable development and ensure universal access to education, health care, and economic opportunity.
12. Uphold the right of all, without discrimination, to a natural and social environment supportive of human dignity, bodily health, and spiritual well-being, with special attention to the rights of indigenous peoples and minorities.
13. Strengthen democratic institutions at all levels, and provide transparency and accountability in governance, inclusive participation in decision making, and access to justice.
14. Integrate into formal education and life-long learning the knowledge, values, and skills needed for a sustainable way of life.
15. Treat all living beings with respect and consideration.
16. Promote a culture of tolerance, nonviolence, and peace.

Source: *The Earth Council, San Jose, Costa Rica, 2001.*

In the preamble to the Earth Charter, the Earth Council delegates made the following poignant statement that sums up what we have discussed in this book:

"In an increasingly interdependent world, it is imperative that we, the citizens of the Earth, declare our responsibility to one another, the greater community of life, and future generations. The Earth community stands at a defining moment. Environmental degradation, biodiversity loss, and depletion of natural resources threaten to destroy the ecological systems that sustain life. Injustice, poverty, and armed conflict deepen the world's suffering. Fundamental changes in our attitudes, values, and ways of living are necessary. A shared vision of basic values is urgently needed to provide an ethical foundation for the emerging world community. Therefore, in a spirit of human solidarity and kinship with all life, we affirm the following principles for sustainable development, which are interrelated and indivisible. We commit ourselves to implement these principles and to create a global partnership in support of their fulfillment.

As never before in human history, common destiny beckons us to redefine our priorities and to seek a new beginning. Such renewal is the promise of these Earth Charter principles. Fulfillment of this promise requires an inner change—a change of mind and heart. It requires that we take decisive action to adopt, apply, and develop the vision of the Earth Charter. Every individual, family, organization, business enterprise, and government has a critical role to play. Youth are fundamental actors for change. We can, if we will, take advantage of the creative possibilities before us and inaugurate an era of fresh hope."

In the end, although a global perspective is essential for both environmental protection and social justice, we need to, as E. F. Schumacher once said, "think globally but act locally." We hope this book will have given you the inspiration and the tools to do both.

Summary

- Although many serious environmental and social problems continue to confront us, there are encouraging signs of progress in sustainability. And there are ways—both large and small—that we can act individually and collectively, to live with less impact and build a better world.

- The main goals of environmental education are to expand environmental literacy among the general public and to prepare students for environmental careers. Environmental literacy is a basic understanding of the natural and built environments and the relationships between humans and their environment.

- There are many opportunities for careers at a variety of different levels in the environmental field.

- Businesses are rushing to invent and market environmental technology and "green" consumer goods. We have to look carefully at the claims made for "environmentally friendly" products. Some are true but others are not.

- We also must ask ourselves how much we need or have a right to consume. There is much we can do through our individual choices of goods and services.

- Some people, called Lohas or cultural creatives, are opting for voluntary simplicity to reduce stress, create time, and lessen their consumption levels and environmental impacts.

- There are many opportunities for collective action to bring about social change. Student environmental groups offer chances to network with others, learn useful organizing techniques, do good work, and have fun. The large national "mainline" environmental organizations have a degree of respectability, power, and influence unmatched by smaller, independent groups.

- Monkey wrenching, tree-sitting, and other forms of direct action are highly controversial tactics espoused by some radical environmentalists. In some cases, antagonists have been able to set aside their differences to work together on constructive solutions.

- Sustainable development promises to alleviate acute poverty while also protecting the environment. Many nongovernmental organizations (NGOs) as well as intergovernmental treaties and conventions work toward these twin goals. In many countries, green politics espouses four key values of ecological wisdom, peace and social justice, grassroots democracy, and freedom from violence. While it is difficult to introduce new national parties in the United States, there are opportunities to work within the legislative, judicial, and administrative agencies to bring about change.

- The Earth Charter affirms many key values of environmental science and sustainable development, including respect for all life, ecosystem protection and restoration, human rights, equitable sharing of resources, eradication of poverty, democratic decision making, and a sense of shared responsibility for the well-being of the earth community.

Questions for Review

1. List four major environmental problems and four signs of hope for a sustainable future.

2. Summarize the goals of environmental education.

3. Define *affluenza, Lohas,* and *cultural creatives.* Show how these terms are related.

4. Describe some things you can do to reduce resource consumption.

5. List five green marketing claims and describe how they might be misleading.

6. List ten key issues in organizing an environmental campaign or using media to influence public opinion.

7. What are the ten largest, oldest mainline environmental organizations in the United States?

8. Describe and evaluate the tactics used by radical environmental groups. Why is direct action sometimes called monkey wrenching?

9. Identify some core values of traditionalists and explain why these people feel alienated from mainline environmental groups.

10. List the four key values of the Green Party.

11. Define *sustainable development* and identify its six goals.

12. In a few words, summarize the main concept in each of the first four goals of the Earth Charter.

Questions for Critical Thinking

1. How would it change your life if all the principles of the Earth Charter were really taken seriously?

2. Do you agree with the principles presented in table 25.7? Why or why not?

3. Put yourself in the place of someone who might object to one or more of these principles. What would they find challenging or objectionable?

4. The wise use movement appeals to people who fear that their livelihood and way of life are threatened by those who want to "lock up" resources in parks and nature preserves. How might conservationists and preservationists find a common ground?

5. People in many cultures traditionally regard wild nature as disagreeable and dangerous. How would you approach environmental education if you were assigned to such a country?

6. Do you support the conservative, mainstream approach to conservation followed by the mainline environmental groups or do you prefer the more challenging, innovative approaches of more radical groups? What are the advantages and disadvantages of each?

7. Extremists on both sides of the environmental debate sometimes use violence against property—and sometimes against people—to advance their agendas. Would you ever condone taking the law into your own hands for an environmental cause? Under what circumstances and with what limits might you do so?

8. Do you agree that sustainable development has the potential to simultaneously reduce poverty *and* protect the environment? What responsibility do those of us in rich nations bear toward the rest of the world?

9. Choose one of the ambiguous dilemmas presented in this chapter, such as whether it is better to use paper or plastic bags. What additional information would you need to choose between alternatives? How should competing claims or considerations be weighed in an issue such as this?

10. The Talmud says, "If not now, when?" How might this apply to environmental science?

Key Terms

citizen science 546
conspicuous consumption 547
cultural creatives 548
deep ecology 555
direct action 556
Earth Charter 562
environmental literacy 545
green political parties 561

life-cycle analysis 550
monkey wrenching 556
nongovernmental
  organizations (NGOs) 560
shallow ecology 555
social ecology 555
sustainable development 559
wise use groups 558

Further Readings

Bell, S., and S. Morse. 1999. *Sustainability Indicators: Measuring the Immeasurable?* Island Press.

Brophy, Paul C., and Alice Shabecoff. 2000. *A Guide to Careers in Community Development.* Island Press.

Brower, M., and W. Leon. 1999. *The Consumer's Guide to Effective Environmental Choices.* Three Rivers Press.

Corraliza, J. A., and J. Berenguer. 2000. Environmental values, beliefs, and actions: A situational approach. *Environment and Behavior* 32(6):832–49.

Dobson, Andrew. 2000. *Green Political Thought,* 3rd ed. Routledge Press.

Environmental Careers Organization. 1999. *The Complete Guide to Environmental Careers in the 21st Century.* Island Press.

Folke, Carl, et. al. 2002. *Resilience and Sustainable Development: Building Adaptive Capacity in a World of Transformations.* International Council for Science (available online at www.icgu.org/library/WSSD-REP/vol3.pdf).

Hays, Samuel P. 2000. *A History of Environmental Politics Since 1945.* Univ. of Pittsburg Press.

Kellert, Stephen R., and Timothy J. Farnham. 2002. *The Good in Nature and Humanity: Connecting Science, Religion, and Spirituality with the Natural World.* Island Press.

Moyers, Bill, et al. 2001. *Doing Democracy: The MAP Model for Organizing Social Movements.* New Society Publishers.

Orr, David. 2003. Walking north on a southbound train. *Conservation Biology* 17(2):348–51.

Prugh, Thomas, and Erik Assadourian. 2003. What is sustainability, anyway? *Worldwatch* 16(5):10–21.

Putnam, Robert D. 2000. *Bowling Alone: The Collapse and Revival of American Community.* Simon & Schuster.

Ray, Paul, and Sherry Ruth Anderson. 2001. *The Cultural Creatives: How 50 Million People Are Changing the World.* Three Rivers Press.

Wilson, E. O. 2002. The bottleneck. *Scientific American* 286(2): 82–91.

Location: http://www.mhhe.com/environmentalscience

McGraw Hill

WEB EXERCISE

Finding an Environmental Organization

To learn more about environmental organizations, go to www.webdirectory.com/, which describes itself as the "world's biggest search engine." You'll find thousands of listings there. Many of the websites are for institutions and agencies, but you'll also find public groups in almost any environmental issue. Choose a general category that appeals to you and investigate a group that seems interesting. What are the goals and tactics of this group? How long has it been in existence? Do you agree (or disagree) with its principles and practices, as far as you can discern these from what's published on the website?

If you can't find an environmental organization that gives useful information about itself in the categories listed in the Environmental Organization Web directory, try searching for one of the "big ten" organizations described in this chapter. What do they tell you about themselves? Would you consider joining them? Why or why not?

Although the Environmental Working Group no longer updates its Clearinghouse on Environmental Advocacy and Research (CLEAR) website, you still may be able to find useful information about anti-environmental groups at www.ewg.org/pub/home/clear/clear.html. Click on your home state (if you're from the United States) or choose a western state to find out about some of these groups and their activities.

GLOSSARY

A

abiotic Non-living.

abundance The number or amount of something.

acid precipitation Acidic rain, snow, or dry particles deposited from the air due to increased acids released by anthropogenic or natural resources.

acids Substances that release hydrogen ions (protons) in water.

active learner Someone who understands and remembers best by doing things physically.

active solar system A mechanical system that actively collects, concentrates, and stores solar energy.

acute effects Sudden, severe effects.

acute poverty Insufficient income or access to resources needed to provide the basic necessities for life such as food, shelter, sanitation, clean water, medical care, and education.

adaptive management A management plan designed from the outset to "learn by doing," and to actively test hypotheses and adjust treatments as new information becomes available.

administrative courts Courts that hear enforcement cases for agencies or consider appeals to agency rules.

administrative law Executive orders, administrative rules and regulations, and enforcement decisions by administrative agencies and special administrative courts.

aerobic Living or occurring only in the presence of oxygen.

aerosols Minute particles or liquid droplets suspended in the air.

aesthetic degradation Changes in environmental quality that offend our aesthetic senses.

albedo A description of a surface's reflective properties.

allergens Substances that activate the immune system.

ambient air The air immediately around us.

amino acid An organic compound containing an amino group and a carboxyl group; amino acids are the units or building blocks that make peptide and protein molecules.

anaerobic respiration The incomplete intracellular breakdown of sugar or other organic compounds in the absence of oxygen that releases some energy and produces organic acids and/or alcohol.

analytical thinking Asks, how can I break this problem down into its constituent parts?

anemia Low levels of hemoglobin due to iron deficiency or lack of red blood cells.

annual A plant that lives for a single growing season.

anthropocentric The belief that humans hold a special place in nature; being centered primarily on humans and human affairs.

antigens Chemical compounds to which antibodies bind.

appropriate technology Technology that can be made at an affordable price by ordinary people using local materials to do useful work in ways that do the least possible harm to both human society and the environment.

aquifers Porous, water-bearing layers of sand, gravel, and rock below the earth's surface; reservoirs for groundwater.

arbitration A formal process of dispute resolution in which there are stringent rules of evidence, cross-examination of witnesses, and a legally binding decision made by the arbitrator that all parties must obey.

arithmetic growth A pattern of growth that increases at a constant amount per unit time, such as 1, 2, 3, 4 or 1, 3, 5, 7.

artesian well The result of a pressurized aquifer intersecting the surface or being penetrated by a pipe or conduit, from which water gushes without being pumped; also called a spring.

asthma A distressing disease characterized by shortness of breath, wheezing, and bronchial muscle spasms.

atmospheric deposition Sedimentation of solids, liquids, or gaseous materials from the air.

atom The smallest unit of matter that has the characteristics of an element; consists of three main types of subatomic particles: protons, neutrons, and electrons.

atomic number The characteristic number of protons per atom of an element. Used as an identifying attribute.

autotroph An organism that synthesizes food molecules from inorganic molecules by using an external energy source, such as light energy.

B

barrier islands Low, narrow, sandy islands that form offshore from a coastline.

bases Substances that bond readily with hydrogen ions.

BAT *See* best available, economically achievable technology.

Batesian mimicry Evolution by one species to resemble the coloration, body shape, or behavior of another species that is protected from predators by a venomous stinger, bad taste, or some other defensive adaptation.

benthic The bottom of a sea or lake.

best available, economically achievable technology (BAT) The best pollution control available.

best practical control technology (BPT) The best technology for pollution control available at reasonable cost and operable under normal conditions.

beta particles High-energy electrons released by radioactive decay.

bill A piece of legislation introduced in Congress and intended to become law.

bioaccumulation The selective absorption and concentration of molecules by cells.

biocentric preservation A philosophy that emphasizes the fundamental right of living organisms to exist and to pursue their own goods.

biocentrism The belief that all creatures have rights and values; being centered on nature rather than humans.

biochemical oxygen demand A standard test of water pollution measured by the amount of dissolved oxygen consumed by aquatic organisms over a given period.

biocide A broad-spectrum poison that kills a wide range of organisms.

biodegradable plastics Plastics that can be decomposed by microorganisms.

biodiversity The genetic, species, and ecological diversity of the organisms in a given area.

biodiversity hot spots Areas with exceptionally high numbers of endemic species.

Biofuel Fuels such as ethanol, methanol, or vegetable oils from crops.

biogeochemical cycles Movement of matter within or between ecosystems; caused by living organisms, geological forces, or chemical reactions. The cycling of nitrogen, carbon, sulfur, oxygen, phosphorus, and water are examples.

biogeographical area An entire self-contained natural ecosystem and its associated land, water, air, and wildlife resources.

biological community The populations of plants, animals, and microorganisms living and interacting in a certain area at a given time.

biological controls Use of natural predators, pathogens, or competitors to regulate pest populations.

biological or biotic factors Organisms and products of organisms that are part of the environment and potentially affect the life of other organisms.

biological oxygen demand (BOD) A standard test for measuring the amount of dissolved oxygen utilized by aquatic microorganisms.

biological pests Organisms that reduce the availability, quality, or value of resources useful to humans.

biological resources The earth's organisms.

biomagnification Increase in concentration of certain stable chemicals (for example, heavy metals or fat-soluble pesticides) in successively higher trophic levels of a food chain or web.

biomass The total mass or weight of all the living organisms in a given population or area.

biomass fuel Organic material produced by plants, animals, or microorganisms that can be burned directly as a heat source or converted into a gaseous or liquid fuel.

biomass pyramid A metaphor or diagram that explains the relationship between the amounts of biomass at different trophic levels.

biome A broad, regional type of ecosystem characterized by distinctive climate and soil conditions and a distinctive kind of biological community adapted to those conditions.

bioremediation Use of biological organisms to remove or detoxify pollutants from a contaminated area.

biosphere The zone of air, land, and water at the surface of the earth that is occupied by organisms.

biosphere reserves World heritage sites identified by the IUCN as worthy for national park or wildlife refuge status because of high biological diversity or unique ecological features.

biota All organisms in a given area.

biotic Pertaining to life; environmental factors created by living organisms.

biotic potential The maximum reproductive rate of an organism, given unlimited resources and ideal environmental conditions. Compare with environmental resistance.

birth control Any method used to reduce births, including celibacy, delayed marriage, contraception; devices or medication that prevent implantation of fertilized zygotes, and induced abortions.

black lung disease Inflammation and fibrosis caused by accumulation of coal dust in the lungs or airways. *See* respiratory fibrotic agents.

blind experiments Those in which those carrying out the experiment don't know until after data has been gathered and analyzed which was the experimental treatment and which was the control.

blue revolution New techniques of fish farming that may contribute as much to human nutrition as miracle cereal grains but also may create social and environmental problems.

bog An area of waterlogged soil that tends to be peaty; fed mainly by precipitation; low productivity; some bogs are acidic.

boreal forest A broad band of mixed coniferous and deciduous trees that stretches across northern North America (and also Europe and Asia); its northernmost edge, the taiga, intergrades with the arctic tundra.

BPT *See* best practical control technology.

breeder reactor A nuclear reactor that produces fuel by bombarding isotopes of uranium and thorium with high-energy neutrons that convert inert atoms to fissionable ones.

bronchitis A persistent inflammation of bronchi and bronchioles (large and small airways in the lungs).

brownfields Abandoned or underused urban areas in which redevelopment is blocked by liability or financing issues related to toxic contamination.

C

cancer Invasive, out-of-control cell growth that results in malignant tumors.

capital Any form of wealth, resources, or knowledge available for use in the production of more wealth.

captive breeding Raising plants or animals in zoos or other controlled conditions to produce stock for subsequent release into the wild.

carbamates Urethanes such as carbaryl, aldicarb, etc. that are used as pesticides.

carbohydrate An organic compound consisting of a ring or chain of carbon atoms with hydrogen and oxygen attached; examples are sugars, starches, cellulose, and glycogen.

carbon cycle The circulation and reutilization of carbon atoms, especially via the processes of photosynthesis and respiration.

carbon management Storing CO_2 or using it in ways that prevent its release into the air.

carbon monoxide (CO) Colorless, odorless, nonirritating but highly toxic gas produced by incomplete combustion of fuel, incineration of biomass or solid waste, or partially anaerobic decomposition of organic material.

carbon sink Places of carbon accumulation, such as in large forests (organic compounds) or ocean sediments (calcium carbonate); carbon is thus removed from the carbon cycle for moderately long to very long periods of time.

carbon source Originating point of carbon that reenters the carbon cycle; cellular respiration and combustion.

carcinogens Substances that cause cancer.

carnivores Organisms that mainly prey upon animals.

carrying capacity The maximum number of individuals of any species that can be supported by a particular ecosystem on a long-term basis.

case law Precedents from both civil and criminal court cases.

cash crops Crops that are sold rather than consumed or bartered.

catastrophic systems Dynamic systems that jump abruptly from one seemingly steady state to another without any intermediate stages.

cell Minute biological compartments within which the processes of life are carried out.

cellular respiration The process in which a cell breaks down sugar or other organic compounds to release energy used for cellular work; may be anaerobic or aerobic, depending on the availability of oxygen.

chain reaction A self-sustaining reaction in which the fission of nuclei produces subatomic particles that cause the fission of other nuclei.

chaotic systems Systems that exhibit variability, which may not be necessarily random, yet whose complex patterns are not discernible over a normal human time scale.

chaparral Thick, dense, thorny evergreen scrub found in Mediterranean climates.

chemical bond The force that holds atoms together in molecules and compounds.

chemical energy Potential energy stored in chemical bonds of molecules.

chlorinated hydrocarbons Hydrocarbon molecules to which chlorine atoms are attached.

chlorofluorocarbons Chemical compounds with a carbon skeleton and one or more attached chlorine and fluorine atoms. Commonly used as refrigerants, solvents, fire retardants, and blowing agents.

chloroplasts Chlorophyll-containing organelles in eukaryotic organisms; sites of photosynthesis.

chronic effects Long-lasting results of exposure to a toxin; can be a permanent change caused by a single, acute exposure or a continuous, low-level exposure.

chronic food shortages Long-term undernutrition and malnutrition; usually caused by people's lack of money to buy food or lack of opportunity to grow it themselves.

chronic obstructive lung disease Irreversible damage to the linings of the lungs caused by irritants.

chronically undernourished Those people whose diet doesn't provide the 2,200 kcal per day, on average, considered necessary for a healthy productive life.

citizen science Projects in which trained volunteers work with scientific researchers to answer real-world questions.

city A differentiated community with a sufficient population and resource base to allow residents to specialize in arts, crafts, services, and professional occupations.

civil law A body of laws regulating relations between individuals or between individuals and corporations concerning property rights, personal dignity and freedom, and personal injury.

classical economics Modern, western economic theories of the effects of resource scarcity, monetary policy, and competition on supply and demand of goods and services in the marketplace. This is the basis for the capitalist market system.

clear-cut Cutting every tree in a given area, regardless of species or size; an appropriate harvest method for some species; can be destructive if not carefully controlled.

climate A description of the long-term pattern of weather in a particular area.

climax community A relatively stable, long-lasting community reached in a successional series; usually determined by climate and soil type.

closed canopy A forest where tree crowns spread over 20 percent of the ground; has the potential for commercial timber harvests.

cloud forests High mountain forests where temperatures are uniformly cool and fog or mist keeps vegetation wet all the time.

coal gasification The heating and partial combustion of coal to release volatile gases, such as methane and carbon monoxide; after pollutants are washed out, these gases become efficient, clean-burning fuel.

coal washing Coal technology that involves crushing coal and washing out soluble sulfur compounds with water or other solvents.

Coastal Zone Management Act Legislation of 1972 that gave federal money to 30 seacoast and Great Lakes states for development and restoration projects.

co-composting Microbial decomposition of organic materials in solid waste into useful soil additives and fertilizer; often, extra organic material in the form of sewer sludge, animal manure, leaves, and grass clippings are added to solid waste to speed the process and make the product more useful.

coevolution The process in which species exert selective pressure on each other and gradually evolve new features or behaviors as a result of those pressures.

cogeneration The simultaneous production of electricity and steam or hot water in the same plant.

cold front A moving boundary of cooler air displacing warmer air.

coliform bacteria Bacteria that live in the intestines (including the colon) of humans and other animals; used as a measure of the presence of feces in water or soil.

commensalism A symbiotic relationship in which one member is benefited and the second is neither harmed nor benefited.

common law The body of court decisions that constitute a working definition of individual rights and responsibilities where no formal statutes define these issues.

communal resource management systems Resources managed by a community for long-term sustainability.

competitive exclusion A theory that no two populations of different species will occupy the same niche and compete for exactly the same resources in the same habitat for very long.

complexity (ecological) The number of species at each trophic level and the number of trophic levels in a community.

composting The biological degradation of organic material under aerobic (oxygen-rich) conditions to produce compost, a nutrient-rich soil amendment and conditioner.

compound A molecule made up of two or more kinds of atoms held together by chemical bonds.

concept map A two-dimensional representation of the relationship between key ideas. A flow chart or graph of ideas.

conclusion A statement that follows logically from a set of premises.

condensation The aggregation of water molecules from vapor to liquid or solid when the saturation concentration is exceeded.

condensation nuclei Tiny particles that float in the air and facilitate the condensation process.

conifers Needle-bearing trees that produce seeds in cones.

conservation development Consideration of landscape history, human culture, topography, and ecological values in subdivision design. Using cluster housing, zoning, covenants, and other design features, at least half of a subdivision can be preserved as open space, farmland, or natural areas.

conservation of matter In any chemical reaction, matter changes form; it is neither created nor destroyed.

conspicuous consumption A term coined by economist and social critic Thorstein Veblen to describe buying things we don't want or need to impress others.

consumer An organism that obtains energy and nutrients by feeding on other organisms or their remains. *See* also heterotroph.

consumption The fraction of withdrawn water that is lost in transmission or that is evaporated, absorbed, chemically transformed, or otherwise made unavailable for other purposes as a result of human use.

contour plowing Plowing along hill contours; reduces erosion.

control rods Neutron-absorbing material inserted into spaces between fuel assemblies in nuclear reactors to regulate fission reaction.

controlled studies Those in which comparisons are made between experimental and control populations that are identical (as far as possible) in every factor except the one variable being studied.

convection currents Rising or sinking air currents that stir the atmosphere and transport heat from one area to another. Convection currents also occur in water; see spring overturn.

conventional pollutants The seven major pollutants (sulfur dioxide, carbon monoxide, particulates, hydrocarbons, nitrogen oxides, photochemical oxidants, and lead) identified and regulated by the U.S. Clean Air Act.

cool deserts Deserts such as the American Great Basin characterized by cold winters and sagebrush.

coral reefs Prominent oceanic features composed of hard, limy skeletons produced by coral animals; usually formed along edges of shallow, submerged ocean banks or along shelves in warm, shallow, tropical seas.

core The dense, intensely hot mass of molten metal, mostly iron and nickel, thousands of kilometers in diameter at the earth's center.

core region The primary industrial region of a country; usually located around the capital or largest port; has both the greatest population density and the greatest economic activity of the country.

Coriolis effect The influence of friction and drag on air layers near the earth; deflects air currents to the direction of the earth's rotation.

cornucopian fallacy The belief that nature is limitless in its abundance and that perpetual growth is not only possible but essential.

corridor A strip of natural habitat that connects two adjacent nature preserves to allow migration of organisms from one place to another.

cost-benefit analysis An evaluation of large-scale public projects by comparing the costs and benefits that accrue from them.

cover crops Plants, such as rye, alfalfa, or clover, that can be planted immediately after harvest to hold and protect the soil.

creative thinking Asks, how could I do this differently?

criminal law A body of court decisions based on federal and state statutes concerning wrongs against persons or society.

criteria pollutants *See* conventional pollutants.

critical factor The single environmental factor closest to a tolerance limit for a given species at a given time. *See* limiting factors.

critical thinking An ability to evaluate information and opinions in a systematic, purposeful, efficient manner.

croplands Lands used to grow crops.

crude birth rate The number of births in a year divided by the midyear population.

crude death rate The number of deaths per thousand persons in a given year; also called crude mortality rate.

crust The cool, lightweight, outermost layer of the earth's surface that floats on the soft, pliable underlying layers; similar to the "skin" on a bowl of warm pudding.

cultural creatives People who are socially conscious, involve in improving communities and willing to translate values into action.

cultural eutrophication An increase in biological productivity and ecosystem succession caused by human activities.

D

debt-for-nature swap Forgiveness of international debt in exchange for nature protection in developing countries.

deciduous Trees and shrubs that shed their leaves at the end of the growing season.

decline spiral A catastrophic deterioration of a species, community, or whole ecosystem; accelerates as functions are disrupted or lost in a downward cascade.

decomposers Fungi and bacteria that break complex organic material into smaller molecules.

deductive reasoning Deriving testable predictions about specific cases from general principles.

deep ecology A philosophy that calls for a profound shift in our attitudes and behavior based on voluntary simplicity; rejection of anthropocentric attitudes; intimate contact with nature; decentralization of power; support for cultural and biological diversity; a belief in the sacredness of nature; and direct personal action to protect nature, improve the environment, and bring about fundamental societal change.

degradation (of water resource) Deterioration in water quality due to contamination or pollution; makes water unsuitable for other desirable purposes.

Delaney Clause A controversial amendment to the Federal Food, Drug, and Cosmetic Act, added in 1958, prohibiting the addition of any known cancer-causing agent to processed foods, drugs, or cosmetics.

delta Fan-shaped sediment deposit found at the mouth of a river.

demand The amount of a product that consumers are willing and able to buy at various possible prices, assuming they are free to express their preferences.

demanufacturing Disassembly of products so components can be reused or recycled.

demographic bottleneck A population founded when just a few members of a species survive a catastrophic event or colonize new habitat geographically isolated from other members of the same species.

demographic transition A pattern of falling death rates and birthrates in response to improved living conditions; could be reversed in deteriorating conditions.

demography Vital statistics about people: births, marriages, deaths, etc.; the statistical study of human populations relating to growth rate, age structure, geographic distribution, etc., and their effects on social, economic, and environmental conditions.

denitrifying bacteria Free-living soil bacteria that converts nitrates to gaseous nitrogen and nitrous oxide.

dependency ratio The number of nonworking members compared to working members for a given population.

desalinization (or desalination) Removal of salt from water by distillation, freezing, or ultrafiltration.

desert A type of biome characterized by low moisture levels and infrequent and unpredictable precipitation. Daily and seasonal temperatures fluctuate widely.

desertification Denuding and degrading a once-fertile land, initiating a desert-producing cycle that feeds on itself and causes long-term changes in soil, climate, and biota of an area.

detritivore Organisms that consume organic litter, debris, and dung.

dew point The temperature at which condensation occurs for a given concentration of water vapor in the air.

dieback A sudden population decline; also called a population crash.

diminishing returns A condition in which unrestrained population growth causes the standard of living to decrease to a subsistence level where poverty, misery, vice, and starvation makes life permanently drab and miserable. This dreary prophecy has led economics to be called "the dismal science."

direct action Civil disobedience, guerrilla street theater, picketing, protest marches, road blockades, demonstrations, and other techniques borrowed from the civil rights movement and applied to environmental protection.

disability-adjusted life years (DALY) A measure of premature deaths and losses due to illnesses and disabilities in a population.

discharge The amount of water that passes a fixed point in a given amount of time; usually expressed as liters or cubic feet of water per second.

disclimax communities *See* equilibrium community.

discount rates The difference between present value and future value of a resource. Generally equivalent to an interest rate.

disease A deleterious change in the body's condition in response to destabilizing factors, such as nutrition, chemicals, or biological agents.

dissolved oxygen (DO) content Amount of oxygen dissolved in a given volume of water at a given temperature and atmospheric pressure; usually expressed in parts per million (ppm).

diversity (species diversity, biological diversity) The number of species present in a community (species richness), as well as the relative abundance of each species.

DNA Deoxyribonucleic acid; the long, double-helix molecule in the nucleus of cells that contains the genetic code and directs the development and functioning of all cells.

dominant plants Those plant species in a community that provide a food base for most of the community; they usually take up the most space and have the largest biomass.

double-blind design One in which neither the experimenter or the subjects know until after data has been gathered and analyzed which was the experimental treatment and which was the control.

downbursts Sudden, very strong, downdrafts of cold air associated with an advancing storm front.

drip irrigation Uses pipe or tubing perforated with very small holes to deliver water one drop at a time directly to the soil around each plant.

dry alkali injection Spraying dry sodium bicarbonate into flue gas to absorb and neutralize acidic sulfur compounds.

E

earth charter A set of principles for sustainable development, environmental protection, and social justice developed by a council appointed by the United Nations.

earthquakes Sudden, violent movement of the earth's crust.

ecocentric (ecologically centered) A philosophy that claims moral values and rights for both organisms and ecological systems and processes.

ecofeminism A pluralistic, nonhierarchical, relationship-oriented philosophy that suggests how humans could reconceive themselves and their relationships to nature in nondominating ways as an alternative to patriarchal systems of domination.

ecojustice Justice in the social order and integrity in the natural order.

ecological development A gradual process of environmental modification by organisms.

ecological economics Application of ecological insights to economic analysis in a holistic, contextual, value-sensitive, ecocentric manner.

ecological equivalents Different species that occupy similar ecological niches in similar ecosystems in different parts of the world.

ecological niche The functional role and position of a species (population) within a community or ecosystem, including what resources it uses, how and when it uses the resources, and how it interacts with other populations.

ecological succession The process by which organisms occupy a site and gradually change environmental conditions so that other species can replace the original inhabitants.

ecology The scientific study of relationships between organisms and their environment. It is concerned with the life histories, distribution, and behavior of individual species as well as the structure and function of natural systems at the level of populations, communities, and ecosystems.

economic development A rise in real income *per person*; usually associated with new technology that increases productivity or resources.

economic growth An increase in the total wealth of a nation; if population grows faster than the economy, there may be real economic growth, but the share per person may decline.

economic thresholds In pest management, the point at which the cost of pest damage exceeds the costs of pest control.

ecosystem A specific biological community and its physical environment interacting in an exchange of matter and energy.

ecosystem management An integration of ecological, economic, and social goals in a unified systems approach to resource management.

ecosystem restoration To reinstate an entire community of organisms to as near its natural condition as possible.

ecotage Direct action (guerrilla warfare) or sabotage in defense of nature. *See* monkey wrenching.

ecotone A boundary between two types of ecological communities.

ecotourism A combination of adventure travel, cultural exploration, and nature appreciation in wild settings.

edge effects A change in species composition, physical conditions, or other ecological factors at the boundary between two ecosystems.

effluent sewerage A low-cost alternative sewage treatment for cities in poor countries that combines some features of septic systems and centralized municipal treatment systems.

electron A negatively charged subatomic particle that orbits around the nucleus of an atom.

electrostatic precipitators The most common particulate controls in power plants; fly ash particles pick up an electrostatic surface charge as they pass between large electrodes in the effluent stream, causing particles to migrate to the oppositely charged plate.

element A molecule composed of one kind of atom; cannot be broken into simpler units by chemical reactions.

El Niño A climatic change marked by shifting of a large warm water pool from the western Pacific Ocean towards the east. Wind direction and precipitation patterns are changed over much of the Pacific and perhaps around the world.

emergent diseases A new disease or one that has been absent for at least 20 years.

emigration The movement of members from a population.

emission standards Regulations for restricting the amounts of air pollutants that can be released from specific point sources.

emissions trading Programs in which companies that have cut pollution by more than they're required to can sell "credits" to other companies that still exceed allowed levels.

endangered species A species considered to be in imminent danger of extinction.

endemism A state in which species are restricted to a single region.

Endocrine disrupters Chemicals that disrupt normal hormone functions.

energy The capacity to do work (that is, to change the physical state or motion of an object).

energy efficiency A measure of energy produced compared to energy consumed.

energy pyramid A representation of the loss of useful energy at each step in a food chain.

energy recovery Incineration of solid waste to produce useful energy.

environment The circumstances or conditions that surround an organism or group of organisms as well as the complex of social or cultural conditions that affect an individual or community.

environmental ethics A search for moral values and ethical principles in human relations with the natural world.

environmental governance Rules and regulations that govern our impacts on the environment and natural resources.

environmental health The science of external factors that cause disease, including elements of the natural, social, cultural and technological worlds in which we live.

environmental hormones Chemical pollutants that substitute for, or interfere with, naturally occurring hormones in our bodies; these chemicals may trigger reproductive failure, developmental abnormalities, or tumor promotion.

environmental impact statement (EIS) An analysis, required by provisions in the National Environmental Policy Act of 1970, of the effects of any major program a federal agency plans to undertake.

environmental indicators Organisms or physical factors that serve as a gauge for environmental changes. More specifically, organisms with these characteristics are called bioindicators.

environmentalism Active participation in attempts to solve environmental pollution and resource problems.

environmental justice A recognition that access to a clean, healthy environment is a fundamental right of all human beings.

environmental law The special body of official rules, decisions, and actions concerning environmental quality, natural resources, and ecological sustainability.

environmental literacy Fluency in the principles of ecology that gives us a working knowledge of the basic grammar and underlying syntax of environmental wisdom.

environmental policy The official rules or regulations concerning the environment adopted, implemented, and enforced by some governmental agency.

environmental racism Decisions that restrict certain people or groups of people to polluted or degraded environments on the basis of race.

environmental resistance All the limiting factors that tend to reduce population growth rates and set the maximum allowable population size or carrying capacity of an ecosystem.

environmental resources Anything an organism needs that can be taken from the environment.

environmental science The systematic, scientific study of our environment as well as our role in it.

enzymes Molecules, usually proteins or nucleic acids, that act as catalysts in biochemical reactions.

epidemiology The study of the distribution and causes of disease and injuries in human populations.

epiphyte A plant that grows on a substrate other than the soil, such as the surface of another organism.

equilibrium community Also called a **disclimax community;** a community subject to periodic disruptions, usually by fire, that prevent it from reaching a climax stage.

estuary A bay or drowned valley where a river empties into the sea.

eukaryotic cell A cell containing a membrane-bounded nucleus and membrane-bounded organelles.

eutrophic Rivers and lakes rich in organisms and organic material (*eu* = truly; *trophic* = nutritious).

evaporation The process in which a liquid is changed to vapor (gas phase).

evolution A theory that explains how random changes in genetic material and competition for scarce resources cause species to change gradually.

e-waste Discarded electronic equipment such as computers, cell phones, television sets, etc.

exhaustible resources Generally considered the earth's geologic endowment: minerals, nonmineral resources, fossil fuels, and other materials present in fixed amounts in the environment.

existence value The importance we place on just knowing that a particular species or a specific organism exists.

exotic organisms Alien species introduced by human agency into biological communities where they would not naturally occur.

exponential growth Growth at a constant rate of increase per unit of time; can be expressed as a constant fraction or exponent. *See* geometric growth.

external costs Expenses, monetary or otherwise, borne by someone other than the individuals or groups who use a resource.

extinction The irrevocable elimination of species; can be a normal process of the natural world as species out-compete or kill off others or as environmental conditions change.

extirpate To destroy totally; extinction caused by direct human action, such as hunting, trapping, etc.

extreme poverty Living on less than $1 (US) per day.

F

family planning Controlling reproduction; planning the timing of birth and having as many babies as are wanted and can be supported.

famines Acute food shortages characterized by large-scale loss of life, social disruption, and economic chaos.

fauna All of the animals present in a given region.

fecundity The physical ability to reproduce.

fen An area of waterlogged soil that tends to be peaty; fed mainly by upwelling water; low productivity.

feral A domestic animal that has taken up a wild existence.

fermentation (alcoholic) A type of anaerobic respiration that yields carbon dioxide and alcohol.

fertility Measurement of actual number of offspring produced through sexual reproduction; usually described in terms of number of offspring of females, since paternity can be difficult to determine.

fetal alcohol syndrome A tragic set of permanent physical and mental and behavioral birth defects that result when mothers drink alcohol during pregnancy.

fibrosis The general name for accumulation of scar tissue in the lung.

fidelity A principle that forbids misleading or deceiving any creature capable of being mislead or deceived. We are to be truthful in our dealings with others.

filters A porous mesh of cotton cloth, spun glass fibers, or asbestos-cellulose that allows air or liquid to pass through but holds back solid particles.

fire-climax community An equilibrium community maintained by periodic fires; examples include grasslands, chaparral shrubland, and some pine forests.

first law of thermodynamics States that energy is *conserved;* that is, it is neither created nor destroyed under normal conditions.

floodplains Low lands along riverbanks, lakes, and coastlines subjected to periodic inundation.

flora All of the plants present in a given region.

flue-gas scrubbing Treating combustion exhaust gases with chemical agents to remove pollutants. Spraying crushed limestone and water into the exhaust gas stream to remove sulfur is a common scrubbing technique.

fluidized bed combustion High pressure air is forced through a mixture of crushed coal and limestone particles, lifting the burning fuel and causing it to move like a boiling fluid.

food aid Financial assistance intended to boost less-developed countries' standards of living.

food chain A linked feeding series; in an ecosystem, the sequence of organisms through which energy and materials are transferred, in the form of food, from one trophic level to another.

food security The ability of individuals to obtain sufficient food on a day-to-day basis.

food surpluses Excess food supplies.

food web A complex, interlocking series of individual food chains in an ecosystem.

forest management Scientific planning and administration of forest resources for sustainable harvest, multiple use, regeneration, and maintenance of a healthy biological community.

fossil fuels Petroleum, natural gas, and coal created by geological forces from organic wastes and dead bodies of formerly living biological organisms.

founder effect The effect on a population founded when just a few members of a species survive a catastrophic event or colonize new habitat geographically isolated from other members of the same species.

Fourth World A political/economic category describing very poor nations that have neither market economies nor central planning and are either not developing or are developing very slowly. Also used to describe indigenous communities within wealthier nations.

freezing condensation A process that occurs in the clouds when ice crystals trap water vapor. As the ice crystals become larger and heavier, they begin to fall as rain or snow.

fresh water   Water other than seawater; covers only about 2 percent of earth's surface, including streams, rivers, lakes, ponds, and water associated with several kinds of wetlands.

freshwater ecosystems   Ecosystems in which the fresh (nonsalty) water of streams, rivers, ponds, or lakes plays a defining role.

front   The boundary between two air masses of different temperature and density.

fuel assembly   A bundle of hollow metal rods containing uranium oxide pellets; used to fuel a nuclear reactor.

fuel cells   Mechanical devices that use hydrogen or hydrogen-containing fuel such as methane to produce an electric current. Fuel cells are clean, quiet, and highly efficient sources of electricity.

fuel-switching   A change from one fuel to another.

fuelwood   Branches, twigs, logs, wood chips, and other wood products harvested for use as fuel.

fugitive emissions   Substances that enter the air without going through a smokestack, such as dust from soil erosion, strip mining, rock crushing, construction, and building demolition.

fumigants   Toxic gases such as methyl bromine that are used to kill pests.

fungi   One of the five kingdom classifications; consists of nonphotosynthetic, eukaryotic organisms with cell walls, filamentous bodies, and absorptive nutrition.

fungicide   A chemical that kills fungi.

G

Gaia hypothesis   A theory that the living organisms of the biosphere form a single, complex interacting system that creates and maintains a habitable Earth; named after Gaia, the Greek Earth mother goddess.

gamma rays   Very short wavelength forms of the electromagnetic spectrum.

gap analysis   A biogeographical technique of mapping biological diversity and endemic species to find gaps between protected areas that leave endangered habitats vulnerable to disruption.

garden city   A new town with special emphasis on landscaping and rural ambience.

gasohol   A mixture of gasoline and ethanol.

gene   A unit of heredity; a segment of DNA nucleus of the cell that contains information for the synthesis of a specific protein, such as an enzyme.

gene banks   Storage for seed varieties for future breeding experiments.

general fertility rate   Crude birthrate multiplied by the percentage of reproductive age women.

genetic assimilation   The disappearance of a species as its genes are diluted through crossbreeding with a closely related species.

genetic drift   The gradual changes in gene frequencies in a population due to random events.

genetic engineering   Laboratory manipulation of genetic material using molecular biology techniques to create desired characteristics in organisms.

genetically modified organisms (GMOs)   Organisms whose genetic code has been altered by artificial means such as interspecies gene transfer.

genuine progress index (GPI)   An alternative to GNP or GDP for economic accounting that measures real progress in quality of life and sustainability.

geometric growth   Growth that follows a geometric pattern of increase, such as 2, 4, 8, 16, etc. *See* exponential growth.

geothermal energy   Energy drawn from the internal heat of the earth, either through geysers, fumaroles, hot springs, or other natural geothermal features, or through deep wells that pump heated groundwater.

germ plasm   Genetic material that may be preserved for future agricultural, commercial, and ecological values (plant seeds or parts or animal eggs, sperm, and embryos).

global environmentalism   A concern for, and action to help solve, global environmental problems.

globalization   The revolution in communications, transportation, finances and commerce that has brought about increasing inter-dependence of national economies.

grasslands   A biome dominated by grasses and associated herbaceous plants.

greenhouse effect   Gases in the atmosphere are transparent to visible light but absorb infrared (heat) waves that are reradiated from the earth's surface.

green plans   Integrated national environmental plans for reducing pollution and resource consumption while achieving sustainable development and environmental restoration.

green political parties   Political organizations based on environmental protection, participatory democracy, grassroots organization, and sustainable development.

green pricing   Setting prices to encourage conservation or renewable energy. Plans that invite customers to pay a premium for energy from renewable sources.

green revolution   Dramatically increased agricultural production brought about by "miracle" strains of grain; usually requires high inputs of water, plant nutrients, and pesticides.

gross domestic product (GDP)   The total economic activity within national boundaries.

gross national product (GNP)   The sum total of all goods and services produced in a national economy. Gross domestic product (GDP) is used to distinguish economic activity within a country from that of off-shore corporations.

groundwater   Water held in gravel deposits or porous rock below the earth's surface; does not include water or crystallization held by chemical bonds in rocks or moisture in upper soil layers.

gully erosion   Removal of layers of soil, creating channels or ravines too large to be removed by normal tillage operations.

H

habitat   The place or set of environmental conditions in which a particular organism lives.

habitat conservation plans   Agreements under which property owners are allowed to harvest resources or develop land as long as habitat is conserved or replaced in ways that benefit resident endangered or threatened species in the long run. Some incidental "taking" or loss of endangered species is generally allowed in such plans.

Hadley cells   Circulation patterns of atmospheric convection currents as they sink and rise in several intermediate bands.

hazardous   Describes chemicals that are dangerous, including flammables, explosives, irritants, sensitizers, acids, and caustics; may be relatively harmless in diluted concentrations.

hazardous air pollutants (HAPs)   Especially dangerous air pollutants including carcinogens, neurotoxins, mutagens, teratogens, endocrine system disrupters and other highly toxic compounds.

hazardous waste   Any discarded material containing substances known to be toxic, mutagenic, carcinogenic, or teratogenic to humans or other life-forms; ignitable, corrosive, explosive, or highly reactive alone or with other materials.

health   A state of physical and emotional well-being; the absence of disease or ailment.

heap-leach extraction   A technique for separating gold from extremely low-grade ores. Crushed ore is piled in huge heaps and sprayed with a dilute alkaline-cyanide solution, which percolates through the pile to extract the gold, which is separated from the effluent in a processing plant. This process has a high potential for water pollution.

heat   A form of energy transferred from one body to another because of a difference in temperatures.

heat capacity   The amount of heat energy that must be added or subtracted to change the temperature of a body; water has a high heat capacity.

heat of vaporization   The amount of heat energy required to convert water from a liquid to a gas.

herbicide   A chemical that kills plants.

herbivore   An organism that eats only plants.

heterotroph   An organism that is incapable of synthesizing its own food and, therefore, must feed upon organic compounds produced by other organisms.

high-level waste repository   A place where intensely radioactive wastes can be buried and remain unexposed to groundwater and earthquakes for tens of thousands of years.

high-quality energy   Intense, concentrated, and high-temperature energy that is considered high-quality because of its usefulness in carrying out work.

HIPPO   Habitat destruction, Invasive species, Pollution, Population (human), and Overharvesting, the leading causes of extinction.

holistic science   The study of entire, integrated systems rather than isolated parts. Often takes a descriptive or interpretive approach.

homeostasis   Maintaining a dynamic, steady state in a living system through opposing, compensating adjustments.

Homestead Act   Legislation passed in 1862 allowing any citizen or applicant for citizenship over 21 years old and head of a family to acquire 160 acres of public land by living on it and cultivating it for five years.

host organism   An organism that provides lodging for a parasite.

hot desert   Deserts of the American Southwest and Mexico; characterized by extreme summer heat and cacti.

human development index (HDI)   A measure of quality of life using life expectancy, child survival, adult literacy, childhood education, gender equity and access to clean water and sanitation as well as income.

human ecology The study of the interactions of humans with the environment.

human resources Human wisdom, experience, skill, labor, and enterprise.

humus Sticky, brown, insoluble residue from the bodies of dead plants and animals; gives soil its structure, coating mineral particles and holding them together; serves as a major source of plant nutrients.

hurricanes Large cyclonic oceanic storms with heavy rain and winds exceeding 119 km/hr (74 mph).

hybrid gasoline-electric vehicles Automobiles that run on electric power and a small gasoline or diesel engine.

hydrologic cycle The natural process by which water is purified and made fresh through evaporation and precipitation. This cycle provides all the freshwater available for biological life.

hypothesis A provisional explanation that can be tested scientifically.

I

igneous rocks Crystalline minerals solidified from molten magma from deep in the earth's interior; basalt, rhyolite, andesite, lava, and granite are examples.

inbreeding depression In a small population, an accumulation of harmful genetic traits (through random mutations and natural selection) that lowers viability and reproductive success of enough individuals to affect the whole population.

inductive reasoning Inferring general principles from specific examples.

industrial revolution Advances in science and technology that have given us power to understand and change our world.

industrial timber Trees used for lumber, plywood, veneer, particleboard, chipboard, and paper; also called roundwood.

inertial confinement A nuclear fusion process in which a small pellet of nuclear fuel is bombarded with extremely high-intensity laser light.

infiltration The process of water percolation into the soil and pores and hollows of permeable rocks.

informal economy Small-scale family businesses in temporary locations outside the control of normal regulatory agencies.

inherent value Ethical values or rights that exist as an intrinsic or essential characteristic of a particular thing or class of things simply by the fact of their existence.

inholdings Private lands within public parks, forests, or wildlife refuges.

inorganic pesticides Inorganic chemicals such as metals, acids, or bases used as pesticides.

insecticide A chemical that kills insects.

insolation Incoming solar radiation.

instrumental value Value or worth of objects that satisfy the needs and wants of moral agents. Objects that can be used as a means to some desirable end.

intangible resources Factors such as open space, beauty, serenity, wisdom, diversity, and satisfaction that cannot be grasped or contained. Ironically, these resources can be both infinite and exhaustible.

integrated pest management (IPM) An ecologically based pest-control strategy that relies on nat-ural mortality factors, such as natural enemies, weather, cultural control methods, and carefully applied doses of pesticides.

internal costs The expenses, monetary or otherwise, borne by those who use a resource.

internalizing costs Planning so that those who reap the benefits of resource use also bear all the external costs.

interplanting The system of planting two or more crops, either mixed together or in alternating rows, in the same field; protects the soil and makes more efficient use of the land.

interpretive science Explanation based on observation and description of entire objects or systems rather than isolated parts.

interspecific competition In a community, competition for resources between members of *different* species.

intraspecific competition In a community, competition for resources among members of the *same* species.

invasive species Organisms that thrive in new territory where they are free of predators, diseases or resource limitations that may have controlled their population in their native habitat.

ionizing radiation High-energy electromagnetic radiation or energetic subatomic particles released by nuclear decay.

ionosphere The lower part of the thermosphere.

ions Electrically charged atoms that have gained or lost electrons.

irruptive growth *See* Malthusian growth.

island biogeography The study of rates of colonization and extinction of species on islands or other isolated areas based on size, shape, and distance from other inhabited regions.

isotopes Forms of a single element that differ in atomic mass due to a different number of neutrons in the nucleus.

J

J curve A growth curve that depicts exponential growth; called a J curve because of its shape.

jet streams Powerful winds or currents of air that circulate in shifting flows; similar to oceanic currents in extent and effect on climate.

joule A unit of energy. One joule is the energy expended in 1 second by a current of 1 amp flowing through a resistance of 1 ohm.

K

keystone species A species whose impacts on its community or ecosystem are much larger and more influential than would be expected from mere abundance.

kinetic energy Energy contained in moving objects such as a rock rolling down a hill, the wind blowing through the trees, or water flowing over a dam.

known resources Those that have been located but are not completely mapped but, nevertheless, are likely to become economical in the foreseeable future.

kwashiorkor A widespread human protein deficiency disease resulting from a starchy diet low in protein and essential amino acids.

Kyoto Protocol An international agreement to reduce greenhouse gas emissions.

L

La Niña The part of a large-scale oscillation in the Pacific (and, perhaps, other oceans) in which trade winds hold warm surface waters in the western part of the basin and cause upwelling of cold, nutrient-rich, deep water in the eastern part of the ocean.

landfills Land disposal sites for solid waste; operators compact refuse and cover it with a layer of dirt to minimize rodent and insect infestation, wind-blown debris, and leaching by rain.

land reform Democratic redistribution of landownership to recognize the rights of those who actually work the land to a fair share of the products of their labor.

landscape ecology The study of the reciprocal effects of spatial pattern on ecological processes. A study of the ways in which landscape history shapes the features of the land and the organisms that inhabit it as well as our reaction to, and interpretation of, the land.

landslide The sudden fall of rock and earth from a hill or cliff. Often triggered by an earthquake or heavy rain.

latent heat Stored energy in a form that is not sensible (detectable by ordinary senses).

LD50 A chemical dose lethal to 50 percent of a test population.

less-developed countries (LDC) Nonindustrialized nations characterized by low per capita income, high birthrates and death rates, high population growth rates, and low levels of technological development.

life-cycle analysis Evaluation of material and energy inputs and outputs at each stage of manufacture, use, and disposal of a product.

life expectancy The average age that a newborn infant can expect to attain in a particular time and place.

life span The longest period of life reached by a type of organism.

limiting factors Chemical or physical factors that limit the existence, growth, abundance, or distribution of an organism.

lipid A nonpolar organic compound that is insoluble in water but soluble in solvents, such as alcohol and ether; includes fats, oils, steroids, phospholipids, and carotenoids.

liquid metal fast breeder A nuclear power plant that converts uranium 238 to plutonium 239; thus, it creates more nuclear fuel than it consumes. Because of the extreme heat and density of its core, the breeder uses liquid sodium as a coolant.

lobbying Using personal contacts, public pressure, or political action to persuade legislators to vote in a particular manner.

logical learner Someone who understands and remembers best by thinking through a topic and finding logical reasons for statements.

logical thinking Asks, can the rules of logic help understand this?

logistic growth Growth rates regulated by internal and external factors that establish an equilibrium with environmental resources. *See* S curve.

longevity The length or duration of life; compare to survivorship.

low-head hydropower Small-scale hydro technology that can extract energy from small headwater dams; causes much less ecological damage.

low-quality energy Diffuse, dispersed energy at a low temperature that is difficult to gather and use for productive purposes.

LULUs Locally Unwanted Land Uses such as toxic waste dumps, incinerators, smelters, airports, freeways, and other sources of environmental, economic, or social degradation.

M

magma Molten rock from deep in the earth's interior; called lava when it spews from volcanic vents.

magnetic confinement A technique for enclosing a nuclear fusion reaction in a powerful magnetic field inside a vacuum chamber.

malignant tumor A mass of cancerous cells that have left their site of origin, migrated through the body, invaded normal tissues, and are growing out of control.

malnourishment A nutritional imbalance caused by lack of specific dietary components or inability to absorb or utilize essential nutrients.

Malthusian growth A population explosion followed by a population crash; also called irruptive growth.

Man and Biosphere (MAB) program A design for nature preserves that divides protected areas into zones with different purposes. A highly protected core is surrounded by a buffer zone and peripheral regions in which multiple-use resource harvesting is permitted.

mangroves Trees from a number of genera that live in salt water.

mantle A hot, pliable layer of rock that surrounds the earth's core and underlies the cool, outer crust.

marasmus A widespread human protein deficiency disease caused by a diet low in calories and protein or imbalanced in essential amino acids.

marginal costs and benefits The costs and benefits of producing one additional unit of a good or service.

marine Living in or pertaining to the sea.

market equilibrium The dynamic balance between supply and demand under a given set of conditions in a "free" market (one with no monopolies or government interventions).

marsh Wetland without trees; in North America, this type of land is characterized by cattails and rushes.

mass burn Incineration of unsorted solid waste.

mass wasting Mass movement of geologic materials downhill caused by rockslides, avalanches, or simple slumping.

matter Anything that takes up space and has mass.

mediation An informal dispute resolution process in which parties are encouraged to discuss issues openly but in which all decisions are reached by consensus and any participant can withdraw at any time.

Mediterranean climate areas Specialized landscapes with warm, dry summers; cool, wet winters; many unique plant and animal adaptations; and many levels of endemism.

megacity See megalopolis.

megalopolis Also known as a megacity or supercity; megalopolis indicates an urban area with more than 10 million inhabitants.

megawatt (MW) Unit of electrical power equal to 1,000 kilowatts or 1 million watts.

mesosphere The atmospheric layer above the stratosphere and below the thermosphere; the middle layer; temperatures are usually very low.

metabolism All the energy and matter exchanges that occur within a living cell or organism; collectively, the life processes.

metamorphic rock Igneous and sedimentary rocks modified by heat, pressure, and chemical reactions.

metapopulation A collection of populations that have regular or intermittent gene flow between geographically separate units.

methane hydrate Small bubbles or individual molecules of methane (natural gas) trapped in a crystalline matrix of frozen water.

micorrhizal symbiosis An association between the roots of most plant species and certain fungi. The plant provides organic compounds to the fungus, while the fungus provides water and nutrients to the plant.

microbial agents Or biological controls, are beneficial microbes (bacteria, fungi) that can be used to suppress or control pests.

micro-hydro generators Small power generators that can be used in low-level rivers to provide economical power for four to six homes, freeing them from dependence on large utilities and foreign energy supplies.

Milankovitch cycles Periodic variations in tilt, eccentricity, and wobble in the earth's orbit; Milutin Milankovitch suggested that it is responsible for cyclic weather changes.

milpa agriculture An ancient farming system in which small patches of tropical forests are cleared and perennial polyculture agriculture practiced and is then followed by many years of fallow to restore the soil; also called **swidden agriculture.**

mineral A naturally occurring, inorganic, crystalline solid with definite chemical composition and characteristic physical properties.

minimum viable population size The number of individuals needed for long-term survival of rare and endangered species.

mitigation Repairing or rehabilitating a damaged ecosystem or compensating for damage by providing a substitute or replacement area.

mixed perennial polyculture Growing a mixture of different perennial crop species (where the same plant persists for more than one year) together in the same plot.

molecule A combination of two or more atoms.

monitored, retrievable storage Holding wastes in underground mines or secure surface facilities such as dry casks where they can be watched and repackaged, if necessary.

monkey wrenching Environmental sabotage such as driving large spikes in trees to protect them from loggers, vandalizing construction equipment, pulling up survey stakes for unwanted developments, and destroying billboards.

monoculture agroforestry Intensive planting of a single species; an efficient wood production approach, but one that encourages pests and disease infestations and conflicts with wildlife habitat or recreation uses.

monsoon A seasonal reversal of wind patterns caused by the different heating and cooling rates of the oceans and continents.

montane coniferous forests Coniferous forests of the mountains consisting of belts of different forest communities along an altitudinal gradient.

moral agents Beings capable of making distinctions between right and wrong and acting accordingly. Those whom we hold responsible for their actions.

moral extensionism Expansion of our understanding of inherent value or rights to persons, organisms, or things that might not be considered worthy of value or rights under some ethical philosophies.

moral subjects Beings that are not capable of distinguishing between right or wrong or that are not able to act on moral principles and yet are capable of being wronged by others.

morals A set of ethical principles that guide our actions and relationships.

morbidity Illness or disease.

more-developed countries (MDC) Industrialized nations characterized by high per capita incomes, low birth and death rates, low population growth rates, and high levels of industrialization and urbanization.

mortality Death rate in a population; the probability of dying.

Müellerian mimicry Evolution of two species, both of which are unpalatable and, have poisonous stingers or some other defense mechanism, to resemble each other.

mulch Protective ground cover, including both natural products and synthetic materials that protect the soil, save water, and prevent weed growth.

multiple use Many uses that occur simultaneously; used in forest management; limited to mutually compatible uses.

mutagens Agents, such as chemicals or radiation, that damage or alter genetic material (DNA) in cells.

mutation A change, either spontaneous or by external factors, in the genetic material of a cell; mutations in the gametes (sex cells) can be inherited by future generations of organisms.

mutualism A symbiotic relationship between individuals of two different species in which both species benefit from the association.

N

NAAQS National Ambient Air Quality Standard; federal standards specifying the maximum allowable levels (averaged over specific time periods) for regulated pollutants in ambient (outdoor) air.

natality The production of new individuals by birth, hatching, germination, or cloning.

natural history The study of where and how organisms carry out their life cycles.

natural increase Crude death rate subtracted from crude birthrate.

natural organic pesticides "Botanicals" or organic compounds naturally occurring in plants, animals or microbes that serve as pesticides.

natural resources Goods and services supplied by the environment.

natural selection The mechanism for evolutionary change in which environmental pressures cause certain genetic combinations in a population to become more abundant; genetic combinations best adapted for present environmental conditions tend to become predominant.

neo-classical economics A branch of economics that attempts to apply the principles of modern science to economic analysis in a mathematically rigorous, noncontextual, abstract, predictive manner.

neo-Luddites People who reject technology as the cause of environmental degradation and social disruption. Named after the followers of Ned Ludd who tried to turn back the Industrial Revolution in England.

neo-Malthusian A belief that the world is characterized by scarcity and competition in which too many people fight for too few resources. Named for Thomas Malthus, who predicted a dismal cycle of misery, vice, and starvation as a result of human overpopulation.

net energy yield Total useful energy produced during the lifetime of an entire energy system minus the energy used, lost, or wasted in making useful energy available.

neurotoxins Toxic substances, such as lead or mercury, that specifically poison nerve cells.

neutron A subatomic particle, found in the nucleus of the atom, that has no electromagnetic charge.

new towns Experimental urban environments that seek to combine the best features of the rural village and the modern city.

nihilists Those who believe the world has no meaning or purpose other than a dark, cruel, unceasing struggle for power and existence.

NIMBY *Not In My BackYard*: the rallying cry of those opposed to LULUs.

nitrate-forming bacteria Bacteria that convert nitrites into compounds that can be used by green plants to build proteins.

nitrite-forming bacteria Bacteria that combine ammonia with oxygen to form nitrites.

nitrogen cycle The circulation and reutilization of nitrogen in both inorganic and organic phases.

nitrogen-fixing bacteria Bacteria that convert nitrogen from the atmosphere or soil solution into ammonia that can then be converted to plant nutrients by nitrite- and nitrate-forming bacteria.

nitrogen oxides Highly reactive gases formed when nitrogen in fuel or combustion air is heated to over 650°C (1,200°F) in the presence of oxygen or when bacteria in soil or water oxidize nitrogen-containing compounds.

noncriteria pollutants *See* unconventional air pollutants.

nongovernmental organizations (NGOs) A term referring collectively to pressure and research groups, advisory agencies, political parties, professional societies, and other groups concerned about environmental quality, resource use, and many other issues.

nonpoint sources Scattered, diffuse sources of pollutants, such as runoff from farm fields, golf courses, construction sites, etc.

nonrenewable resources Minerals, fossil fuels, and other materials present in essentially fixed amounts (within human time scales) in our environment.

North/South division A description of the fact that most of the world's wealthier countries tend to be in North America, Europe, and Japan while the poorer countries tend to be located closer to the equator.

nuclear fission The radioactive decay process in which isotopes split apart to create two smaller atoms.

nuclear fusion A process in which two smaller atomic nuclei fuse into one larger nucleus and release energy; the source of power in a hydrogen bomb.

nucleic acids Large organic molecules made of nucleotides that function in the transmission of hereditary traits, in protein synthesis, and in control of cellular activities.

nucleus The center of the atom; occupied by protons and neutrons. In cells, the organelle that contains the chromosomes (DNA).

nuées ardentes Deadly, denser-than-air mixtures of hot gases and ash ejected from volcanoes.

numbers pyramid A diagram showing the relative population sizes at each trophic level in an ecosystem; usually corresponds to the biomass pyramid.

O

obese Generally considered to be a body mass greater than 30 kg/m², or roughly 30 pounds above normal for an average person.

ocean shorelines Rocky coasts and sandy beaches along the oceans; support rich, stratified communities.

ocean thermal electric conversion (OTEC) Energy derived from temperature differentials between warm ocean surface waters and cold deep waters. This differential can be used to drive turbines attached to electric generators.

oceanic islands Islands in the ocean; formed by breaking away from a continental landmass, volcanic action, coral formation, or a combination of sources; support distinctive communities.

offset allowances A controversial component of air quality regulations that allows a polluter to avoid installation of control equipment on one source with an "offsetting" pollution reduction at another source.

oil shale A fine-grained sedimentary rock rich in solid organic material called kerogen. When heated, the kerogen liquefies to produce a fluid petroleum fuel.

old-growth forests Forests free from disturbance for long enough (generally 150 to 200 years) to have mature trees, physical conditions, species diversity, and other characteristics of equilibrium ecosystems.

oligotrophic Condition of rivers and lakes that have clear water and low biological productivity (*oligo* = little; *trophic* = nutrition); are usually clear, cold, infertile headwater lakes and streams.

omnivore An organism that eats both plants and animals.

open access system A commonly held resource for which there are no management rules.

open canopy A forest where tree crowns cover less than 20 percent of the ground; also called woodland.

open range Unfenced, natural grazing lands; includes woodland as well as grassland.

open system A system that exchanges energy and matter with its environment.

optimum The most favorable condition in regard to an environmental factor.

orbital The space or path in which an electron orbits the nucleus of an atom.

organic compounds Complex molecules organized around skeletons of carbon atoms arranged in rings or chains; includes biomolecules, molecules synthesized by living organisms.

organophosphates Organic molecules to which phosphate group(s) are attached.

overburden Overlying layers of noncommercial sediments that must be removed to reach a mineral or coal deposit.

overharvesting Harvesting so much of a resource that it threatens its existence.

overnutrition Receiving too many calories.

overshoot The extent to which a population exceeds the carrying capacity of its environment.

oxygen cycle The circulation and reutilization of oxygen in the biosphere.

oxygen sag Oxygen decline downstream from a pollution source that introduces materials with high biological oxygen demands.

ozone A highly reactive molecule containing three oxygen atoms; a dangerous pollutant in ambient air. In the stratosphere, however, ozone forms an ultraviolet absorbing shield that protects us from mutagenic radiation.

P

Pacific Decadal Oscillation (PDO) A large pool of warm water that moves north and south in the Pacific Ocean every 30 years or so and has large effects on North America's climate.

parabolic mirrors Curved mirrors that focus light from a large area onto a single, central point, thereby concentrating solar energy and producing high temperatures.

paradigm A model that provides a framework for interpreting observations.

parasite An organism that lives in or on another organism, deriving nourishment at the expense of its host, usually without killing it.

parsimony If two explanations appear equally plausible, choose the simpler one.

particulate material Atmospheric aerosols, such as dust, ash, soot, lint, smoke, pollen, spores, algal cells, and other suspended materials; originally applied only to solid particles but now extended to droplets of liquid.

parts per billion (ppb) Number of parts of a chemical found in 1 billion parts of a particular gas, liquid, or solid mixture.

parts per million (ppm) Number of parts of a chemical found in 1 million parts of a particular gas, liquid, or solid mixture.

parts per trillion (ppt) Number of parts of a chemical found in 1 trillion (10¹²) parts of a particular gas, liquid, or solid mixture.

passive heat absorption The use of natural materials or absorptive structures without moving parts to gather and hold heat; the simplest and oldest use of solar energy.

pasture   Grazing lands suitable for domestic livestock.

patchiness   Within a larger ecosystem, the presence of smaller areas that differ in some physical conditions and thus support somewhat different communities; a diversity-promoting phenomenon.

pathogen   An organism that produces disease in a host organism, disease being an alteration of one or more metabolic functions in response to the presence of the organism.

peat   Deposits of moist, acidic, semidecayed organic matter.

pelagic   Zones in the vertical water column of a water body.

pellagra   Lassitude, torpor, dermatitis, diarrhea, dementia, and death brought about by a diet deficient in tryptophan and niacin.

peptides   Two or more amino acids linked by a peptide bond.

perennial species   Plants that grow for more than two years.

permafrost   A permanently frozen layer of soil that underlies the arctic tundra.

permanent retrievable storage   Placing waste storage containers in a secure building, salt mine, or bedrock cavern where they can be inspected periodically and retrieved, if necessary.

persistent organic polutants (POPs)   Chemical compounds that persist in the environment and retain biological activity for long times.

pest   Any organism that reduces the availability, quality, or value of a useful resource.

pesticide   Any chemical that kills, controls, drives away, or modifies the behavior of a pest.

pesticide treadmill   A need for constantly increasing doses or new pesticides to prevent pest resurgence.

pest resurgence   Rebound of pest populations due to acquired resistance to chemicals and nonspecific destruction of natural predators and competitors by broadscale pesticides.

pH   A value that indicates the acidity or alkalinity of a solution on a scale of 0 to 14, based on the proportion of H^+ ions present.

phosphorus cycle   The movement of phosphorus atoms from rocks through the biosphere and hydrosphere and back to rocks.

photochemical oxidants   Products of secondary atmospheric reactions. *See* smog.

photodegradable plastics   Plastics that break down when exposed to sunlight or to a specific wavelength of light.

photosynthesis   The biochemical process by which green plants and some bacteria capture light energy and use it to produce chemical bonds. Carbon dioxide and water are consumed while oxygen and simple sugars are produced.

photosynthetic efficiency   The percentage of available light captured by plants and used to make useful products.

photovoltaic cell   An energy-conversion device that captures solar energy and directly converts it to electrical current.

physical or abiotic factors   Nonliving factors, such as temperature, light, water, minerals, and climate, that influence an organism.

phytoplankton   Microscopic, free-floating, autotrophic organisms that function as producers in aquatic ecosystems.

pioneer species   In primary succession on a terrestrial site, the plants, lichens, and microbes that first colonize the site.

plankton   Primarily microscopic organisms that occupy the upper water layers in both freshwater and marine ecosystems.

plasma   A hot, electrically neutral gas of ions and free electrons.

poachers   Those who hunt wildlife illegally.

point sources   Specific locations of highly concentrated pollution discharge, such as factories, power plants, sewage treatment plants, underground coal mines, and oil wells.

policy   A societal plan or statement of intentions intended to accomplish some social good.

policy cycle   The process by which problems are identified and acted upon in the public arena.

political economy   The branch of economics concerned with modes of production, distribution of benefits, social institutions, and class relationships.

pollution   To make foul, unclean, dirty; any physical, chemical, or biological change that adversely affects the health, survival, or activities of living organisms or that alters the environment in undesirable ways.

pollution charges   Fees assessed per unit of pollution based on the "polluter pays" principle.

polycentric complex   Cities with several urban cores surrounding a once dominant central core.

population   A group of individuals of the same species occupying a given area.

population crash   A sudden population decline caused by predation, waste accumulation, or resource depletion; also called a dieback.

population explosion   Growth of a population at exponential rates to a size that exceeds environmental carrying capacity; usually followed by a population crash.

population momentum   A potential for increased population growth as young members reach reproductive age.

postmaterialist values   A philosophy that emphasizes quality of life over acquisition of material goods.

post-modernism   A philosophy that rejects the optimism and universal claims of modern positivism.

potential energy   Stored energy that is latent but available for use. A rock poised at the top of a hill or water stored behind a dam are examples of potential energy.

power   The rate of energy delivery; measured in horsepower or watts.

precautionary principle   The decision to leave a margin of safety for unexpected developments.

precedent   An act or decision that can be used as an example in dealing with subsequent similar situations.

precycling   Making environmentally sound decisions at the store and reducing waste before we buy.

predation   The act of feeding by a predator.

predator   An organism that feeds directly on other organisms in order to survive; live-feeders, such as herbivores and carnivores.

premises   Introductory statements that set up or define a problem. Those things taken as given.

prevention of significant deterioration   A clause of the Clean Air Act that prevents degradation of existing clean air; opposed by industry as an unnecessary barrier to development.

price elasticity   A situation in which supply and demand of a commodity respond to price.

primary pollutants   Chemicals released directly into the air in a harmful form.

primary productivity   Synthesis of organic materials (biomass) by green plants using the energy captured in photosynthesis.

primary standards   Regulations of the 1970 Clean Air Act; intended to protect human health.

primary succession   An ecological succession that begins in an area where no biotic community previously existed.

primary treatment   A process that removes solids from sewage before it is discharged or treated further.

principle of competitive exclusion   A result of natural selection whereby two similar species in a community occupy different ecological niches, thereby reducing competition for food.

producer   An organism that synthesizes food molecules from inorganic compounds by using an external energy source; most producers are photosynthetic.

production frontier   The maximum output of two competing commodities at different levels of production.

productivity   The synthesis of new organic material. That done by green plants using solar energy is called primary productivity.

prokaryotic   Cells that do not have a membrane-bounded nucleus or membrane-bounded organelles.

promoters   Agents that are not carcinogenic but that assist in the progression and spread of tumors; sometimes called cocarcinogens.

pronatalist pressures   Influences that encourage people to have children.

proteins   Chains of amino acids linked by peptide bonds.

proton   A positively charged subatomic particle found in the nucleus of an atom.

proven reserves   *See* proven resources.

proven resources   Those that have been thoroughly mapped and are economical to recover at current prices with available technology.

public trust   A doctrine obligating the government to maintain public lands in a natural state as guardians of the public interest.

pull factors (in urbanization)   Conditions that draw people from the country into the city.

push factors (in urbanization)   Conditions that force people out of the country and into the city.

R

radioactive   An unstable isotope that decays spontaneously and releases subatomic particles or units of energy.

radioactive decay   A change in the nuclei of radioactive isotopes that spontaneously emit high-energy electromagnetic radiation and/or subatomic particles while gradually changing into another isotope or different element.

radionucleides Isotopes that exhibit radioactive decay.

rainforest A forest with high humidity, constant temperature, and abundant rainfall (generally over 380 cm [150 in] per year); can be tropical or temperate.

rain shadow Dry area on the downwind side of a mountain.

rangeland Grasslands and open woodlands suitable for livestock grazing.

rational choice Public decision making based on reason, logic, and science-based management.

recharge zone Area where water infiltrates into an aquifer.

reclamation Chemical, biological, or physical clean-up and reconstruction of severely contaminated or degraded sites to return them to something like their original topography and vegetation.

recoverable resources Those accessible with current technology but not economical under current conditions.

re-creation Construction of an entirely new biological community to replace one that has been destroyed on that or another site.

recycling Reprocessing of discarded materials into new, useful products; not the same as reuse of materials for their original purpose, but the terms are often used interchangeably.

red tide A population explosion or bloom of minute, single-celled marine organisms called dinoflagellates. Billions of these cells can accumulate in protected bays where the toxins they contain can poison other marine life.

reduced tillage systems Systems, such as minimum till, conserve-till, and no-till, that preserve soil, save energy and water, and increase crop yields.

reflective thinking Asks, what does this all mean?

reformer A device that strips hydrogen from fuels such as natural gas, methanol, ammonia, gasoline, or vegetable oil so they can be used in a fuel cell.

refuse-derived fuel Processing of solid waste to remove metal, glass, and other unburnable materials; organic residue is shredded, formed into pellets, and dried to make fuel for power plants.

regenerative farming Farming techniques and land stewardship that restore the health and productivity of the soil by rotating crops, planting ground cover, protecting the surface with crop residue, and reducing synthetic chemical inputs and mechanical compaction.

regulations Rules established by administrative agencies; regulations can be more important than statutory law in the day-to-day management of resources.

rehabilitate land A utilitarian program to make an area useful to humans.

rehabilitation To rebuild elements of structure or function in an ecological system without necessarily achieving complete restoration to its original condition.

relative humidity At any given temperature, a comparison of the actual water content of the air with the amount of water that could be held at saturation.

relativists Those who believe moral principles are always dependent on the particular situation.

remediation Cleaning up chemical contaminants from a polluted area.

renewable resources Resources normally replaced or replenished by natural processes; resources not depleted by moderate use; examples include solar energy, biological resources such as forests and fisheries, biological organisms, and some biogeochemical cycles.

renewable water supplies Annual freshwater surface runoff plus annual infiltration into underground freshwater aquifers that are accessible for human use.

reproducibility Making an observation or obtaining a particular result more than once.

residence time The length of time a component, such as an individual water molecule, spends in a particular compartment or location before it moves on through a particular process or cycle.

resilience The ability of a community or ecosystem to recover from disturbances.

resistance (inertia) The ability of a community to resist being changed by potentially disruptive events.

resource In economic terms, anything with potential use in creating wealth or giving satisfaction.

resource partitioning In a biological community, various populations sharing environmental resources through specialization, thereby reducing direct competition. *See also* ecological niche.

resource scarcity A shortage or deficit in some resource.

restoration To bring something back to a former condition. Ecological restoration involves active manipulation of nature to re-create conditions that existed before human disturbance.

restoration ecology Seeks to repair or reconstruct ecosystems damaged by human actions.

riders Amendments attached to bills in conference committee, often completely unrelated to the bill to which they are added.

rill erosion The removing of thin layers of soil as little rivulets of running water gather and cut small channels in the soil.

risk Probability that something undesirable will happen as a consequence of exposure to a hazard.

risk assessment Evaluation of the short-term and long-term risks associated with a particular activity or hazard; usually compared to benefits in a cost-benefit analysis.

RNA Ribonucleic acid; nucleic acid used for transcription and translation of the genetic code found on DNA molecules.

rock A solid, cohesive, aggregate of one or more crystalline minerals.

rock cycle The process whereby rocks are broken down by chemical and physical forces; sediments are moved by wind, water, and gravity, sedimented and reformed into rock, and then crushed, folded, melted, and recrystallized into new forms.

rotational grazing Confining animals to a small area for a short time (often only a day or two) before shifting them to a new location.

ruminant animals Cud-chewing animals, such as cattle, sheep, goats, and buffalo, with multichambered stomachs in which cellulose is digested with the aid of bacteria.

runoff The excess of precipitation over evaporation; the main source of surface water and, in broad terms, the water available for human use.

run-of-the-river flow Ordinary river flow not accelerated by dams, flumes, etc. Some small, modern, high-efficiency turbines can generate useful power with run-of-the-river flow or with a current of only a few kilometers per hour.

rural area An area in which most residents depend on agriculture or the harvesting of natural resources for their livelihood.

S

S curve A curve that depicts logistic growth; called an S curve because of its shape.

salinity Amount of dissolved salts (especially sodium chloride) in a given volume of water.

salinization A process in which mineral salts accumulate in the soil, killing plants; occurs when soils in dry climates are irrigated profusely.

saltwater intrusion Movement of saltwater into freshwater aquifers in coastal areas where groundwater is withdrawn faster than it is replenished.

salvage logging Harvesting timber killed by fire, disease, or windthrow.

sanitary landfills A landfill in which garbage and municipal waste is buried every day under enough soil or fill to eliminate odors, vermin, and litter.

saturation point The maximum concentration of water vapor the air can hold at a given temperature.

scavenger An organism that feeds on the dead bodies of other organisms.

scientific method A systematic, precise, objective study of a problem. Generally this requires observation, hypothesis development and testing, data gathering, and interpretation.

scientific theory An explanation supported by many tests and accepted by a general consensus of scientists.

secondary pollutants Chemicals modified to a hazardous form after entering the air or that are formed by chemical reactions as components of the air mix and interact.

secondary recovery technique Pumping pressurized gas, steam, or chemical-containing water into a well to squeeze more oil from a reservoir.

secondary standards Regulations of the 1972 Clean Air Act intended to protect materials, crops, visibility, climate, and personal comfort.

secondary succession Succession on a site where an existing community has been disrupted.

secondary treatment Bacterial decomposition of suspended particulates and dissolved organic compounds that remain after primary sewage treatment.

second law of thermodynamics States that, with each successive energy transfer or transformation in a system, less energy is available to do work.

secure landfill A solid waste disposal site lined and capped with an impermeable barrier to prevent leakage or leaching. Drain tiles, sampling wells, and vent systems provide monitoring and pollution control.

sedimentary rock Deposited material that remains in place long enough or is covered with enough material to compact into stone; examples include shale, sandstone, breccia, and conglomerate.

sedimentation The deposition of organic materials or minerals by chemical, physical, or biological processes.

selective cutting Harvesting only mature trees of certain species and size; usually more expensive than clear-cutting, but it is less disruptive for wildlife and often better for forest regeneration.

seriously undernourished Those who receive less than 80 percent of their minimum daily caloric requirements.

shallow ecology A critical term applied to superficial environmentalists who claim to be green but are quick to compromise and who do little to bring about fundamental change.

shantytowns Settlements created when people move onto undeveloped lands and build their own shelter with cheap or discarded materials; some are simply illegal subdivisions where a landowner rents land without city approval; others are land invasions.

sheet erosion Peeling off thin layers of soil from the land surface; accomplished primarily by wind and water.

sick building syndrome Headaches, allergies, chronic fatigue and other symptoms caused by poorly vented indoor air contaminated by pathogens or toxins.

significant numbers Meaningful numbers whose accuracy can be verified.

sinkholes A large surface crater caused by the collapse of an underground channel or cavern; often triggered by groundwater withdrawal.

sludge Semisolid mixture of organic and inorganic materials that settles out of wastewater at a sewage treatment plant.

slums Legal but inadequate multifamily tenements or rooming houses; some are custom built for rent to poor people, others are converted from some other use.

smart growth Efficient use of land resources and existing urban infrastructure.

smelting Heating ores to extract metals.

smog The term used to describe the combination of smoke and fog in the stagnant air of London; now often applied to photochemical pollution products or urban air pollution of any kind.

social ecology A socialist/humanist philosophy based on the communitarian anarchism of the Russian geographer Peter Kropotkin. It shares much with deep ecology except that it is more humanist in its outlook.

social justice Equitable access to resources and the benefits derived from them; a system that recognizes inalienable rights and adheres to what is fair, honest, and moral.

soil A complex mixture of weathered mineral materials from rocks, partially decomposed organic molecules, and a host of living organisms.

soil horizons Horizontal layers that reveal a soil's history, characteristics, and usefulness.

soil profile All the vertical layers or horizons that make up a soil in a particular place.

southern pine forest United States coniferous forest ecosystem characterized by a warm, moist climate.

species A population of morphologically similar organisms that can reproduce sexually among themselves but that cannot produce fertile offspring when mated with other organisms.

species diversity The number and relative abundance of species present in a community.

species recovery plan A plan for restoration of an endangered species through protection, habitat management, captive breeding, disease control, or other techniques that increase populations and encourage survival.

sprawl Unlimited outward extension of city boundaries that lowers population density, consumes open space, generates freeway congestion, and causes decay in central cities.

spring overturn Springtime lake phenomenon that occurs when the surface ice melts and the surface water temperature warms to its greatest density at 4°C and then sinks, creating a convection current that displaces nutrient-rich bottom waters.

squatter towns Shantytowns that occupy land without owner's permission; some are highly organized movements in defiance of authorities; others grow gradually.

stability In ecological terms, a dynamic equilibrium among the physical and biological factors in an ecosystem or a community; relative homeostasis.

stable runoff The fraction of water available year-round; usually more important than total runoff when determining human uses.

Standard Metropolitan Statistical Area (SMSA) An urbanized region with at least 100,000 inhabitants with strong economic and social ties to a central city of at least 50,000 people.

standing The right to take part in legal proceedings.

statute law Formal documents or decrees enacted by the legislative branch of government.

statutory law Rules passed by a state or national legislature.

steady-state economy Characterized by low birth and death rates, use of renewable energy sources, recycling of materials, and emphasis on durability, efficiency, and stability.

stewardship A philosophy that holds that humans have a unique responsibility to manage, care for, and improve nature.

strategic lawsuits against public participation (SLAPP) Lawsuits that have no merit but are brought merely to intimidate and harass private citizens who act in the public interest.

strategic metals and minerals Materials a country cannot produce itself but that it uses for essential materials or processes.

stratosphere The zone in the atmosphere extending from the tropopause to about 50 km (30 mi) above the earth's surface; temperatures are stable or rise slightly with altitude; has very little water vapor but is rich in ozone.

stratospheric ozone The ozone (O_3) occurring in the stratosphere 10 to 50 km above the earth's surface.

stress-related diseases Diseases caused or accentuated by social stresses such as crowding.

strip cutting Harvesting trees in strips narrow enough to minimize edge effects and to allow natural regeneration of the forest.

strip farming Planting different kinds of crops in alternating strips along land contours; when one crop is harvested, the other crop remains to protect the soil and prevent water from running straight down a hill.

strip mining Removing surface layers over coal seams using giant, earth-moving equipment; creates a huge open-pit from which coal is scooped by enormous surface-operated machines and transported by trucks; an alternative to deep mines.

structure (in ecological terms) Patterns of organization, both spatial and functional, in a community.

sublimation The process by which water can move between solid and gaseous states without ever becoming liquid.

subsidence A settling of the ground surface caused by the collapse of porous formations that result from withdrawal of large amounts of groundwater, oil, or other underground materials.

subsoil A layer of soil beneath the topsoil that has lower organic content and higher concentrations of fine mineral particles; often contains soluble compounds and clay particles carried down by percolating water.

sulfur cycle The chemical and physical reactions by which sulfur moves into or out of storage and through the environment.

sulfur dioxide A colorless, corrosive gas directly damaging to both plants and animals.

Superfund A fund established by Congress to pay for containment, cleanup, or remediation of abandoned toxic waste sites. The fund is financed by fees paid by toxic waste generators and by cost-recovery from cleanup projects.

supply The quantity of that product being offered for sale at various prices, other things being equal.

surface mining Some minerals are also mined from surface pits. *See* strip mining.

surface tension A condition in which the water surface meets the air and acts like an elastic skin.

survivorship The percentage of a population reaching a given age or the proportion of the maximum life span of the species reached by any individual.

sustainable agriculture An ecologically sound, economically viable, socially just, and humane agricultural system. Stewardship, soil conservation, and integrated pest management are essential for sustainability.

sustainable development A real increase in well-being and standard of life for the average person that can be maintained over the long-term without degrading the environment or compromising the ability of future generations to meet their own needs.

sustained yield Utilization of a renewable resource at a rate that does not impair or damage its ability to be fully renewed on a long-term basis.

swamp Wetland with trees, such as the extensive swamp forests of the southern United States.

swidden agriculture *See* milpa agriculture.

symbiosis The intimate living together of members of two different species; includes **mutualism, commensalism,** and, in some classifications, **parasitism.**

synergism An interaction in which one substance exacerbates the effects of another. The sum of the interaction is greater than the parts.

synergistic effects When an injury caused by exposure to two environmental factors together is greater than the sum of exposure to each factor individually.

systemic A condition or process that affects the whole body; many metabolic poisons are systemic.

T

taiga The northernmost edge of the boreal forest, including species-poor woodland and peat deposits; intergrading with the arctic tundra.

tailings Mining waste left after mechanical or chemical separation of minerals from crushed ore.

taking Unconstitutional confiscation of private property.

tar sands Sand deposits containing petroleum or tar.

technological optimists Those who believe that technology and human enterprise will find cures for all our problems. Also called **Promethean environmentalism.**

tectonic plates Huge blocks of the earth's crust that slide around slowly, pulling apart to open new ocean basins or crashing ponderously into each other to create new, larger landmasses.

temperate rainforest The cool, dense, rainy forest of the northern Pacific coast; enshrouded in fog much of the time; dominated by large conifers.

temperature A measure of the speed of motion of a typical atom or molecule in a substance.

temperature inversions A stable layer of warm air overlays cooler air, trapping pollutants near ground level.

teratogens Chemicals or other factors that specifically cause abnormalities during embryonic growth and development.

terracing Shaping the land to create level shelves of earth to hold water and soil; requires extensive hand labor or expensive machinery, but it enables farmers to farm very steep hillsides.

territoriality An intense form of intraspecific competition in which organisms define an area surrounding their home site or nesting site and defend it, primarily against other members of their own species.

tertiary treatment The removal of inorganic minerals and plant nutrients after primary and secondary treatment of sewage.

thermal plume A plume of hot water discharged into a stream or lake by a heat source, such as a power plant.

thermocline In water, a distinctive temperature transition zone that separates an upper layer that is mixed by the wind (the epilimnion) and a colder, deep layer that is not mixed (the hypolimnion).

thermodynamics A branch of physics that deals with transfers and conversions of energy.

thermodynamics, first law Energy can be transformed and transferred, but cannot be destroyed or created.

thermodynamics, second law With each successive energy transfer or transformation, less energy is available to do work.

thermosphere The highest atmospheric zone; a region of hot, dilute gases above the mesosphere extending out to about 1,600 km (1,000 mi) from the earth's surface.

Third World Less-developed countries that are not capitalistic and industrialized (First World) or centrally-planned socialist economies (Second World); not intended to be derogatory.

thorn scrub A dry, semi-desert dominated by acacias and other spiny shrubs.

threatened species While still abundant in parts of its territorial range, this species has declined significantly in total numbers and may be on the verge of extinction in certain regions or localities.

tidal station A dam built across a narrow bay or estuary traps tide water flowing both in and out of the bay. Water flowing through the dam spins turbines attached to electric generators.

timberline In mountains, the highest-altitude edge of forest that marks the beginning of the treeless alpine tundra.

tolerance limits *See* limiting factors.

topsoil The first true layer of soil; layer in which organic material is mixed with mineral particles; thickness ranges from a meter or more under virgin prairie to zero in some deserts.

tornado A violent storm characterized by strong swirling winds and updrafts; tornadoes form when a strong cold front pushes under a warm, moist air mass over the land.

tort law Court cases that seek compensation for damages.

total fertility rate The number of children born to an average woman in a population during her entire reproductive life.

total growth rate The net rate of population growth resulting from births, deaths, immigration, and emigration.

total maximum daily loads (TMDL) The amount of particular pollutant that a water body can receive from both point and non-point sources and still meet water quality standards.

toxic colonialism Shipping toxic wastes to a weaker or poorer nation.

toxic release inventory A program created by the Superfund Amendments and Reauthorization Act of 1984 that requires manufacturing facilities and waste handling and disposal sites to report annually on releases of more than 300 toxic materials.

toxins Poisonous chemicals that react with specific cellular components to kill cells or to alter growth or development in undesirable ways; often harmful, even in dilute concentrations.

tradable permits Pollution quotas or variances that can be bought or sold.

tragedy of the commons An inexorable process of degradation of communal resources due to selfish self-interest of "free riders" who use or destroy more than their fair share of common property. *See* open access system.

transitional zone A zone in which populations from two or more adjacent communities meet and overlap.

transpiration The evaporation of water from plant surfaces, especially through stomates.

triple bottom line Corporate accounting that reports social and environmental costs and benefits as well as merely economic ones.

trophic level Step in the movement of energy through an ecosystem; an organism's feeding status in an ecosystem.

tropical rainforests Forests in which rainfall is abundant-more than 200 cm (80 in) per year-and temperatures are warm to hot year-round.

tropical seasonal forest Semievergreen or partly deciduous forests tending toward open woodlands and grassy savannas dotted with scattered, drought-resistant tree species; distinct wet and dry seasons, hot year-round.

tropopause The boundary between the troposphere and the stratosphere.

troposphere The layer of air nearest to the earth's surface; both temperature and pressure usually decrease with increasing altitude.

tsunami Giant seismic sea swells that move rapidly from the center of an earthquake; they can be 10 to 20 meters high when they reach shorelines hundreds or even thousands of kilometers from their source.

tundra Treeless arctic or alpine biome characterized by cold, harsh winters, a short growing season, and potential for frost any month of the year; vegetation includes low-growing perennial plants, mosses, and lichens.

U

unconventional air pollutants Toxic or hazardous substances, such as asbestos, benzene, beryllium, mercury, polychlorinated biphenyls, and vinyl chloride, not listed in the original Clean Air Act because they were not released in large quantities; also called noncriteria pollutants.

unconventional oil Resources such as shale oil and tar sands that can be liquefied and used like oil.

undernourished Those who receive less than 90 percent of the minimum dietary intake over a long-term time period; they lack energy for an active, productive life and are more susceptible to infectious diseases.

undiscovered resources Speculative or inferred resources or those that we haven't even thought about.

universalists Those who believe that some fundamental ethical principles are universal and unchanging. In this vision, these principles are valid regardless of the context or situation.

upwelling Convection currents within a body of water that carry nutrients from bottom sediments toward the surface.

urban area An area in which a majority of the people are not directly dependent on natural resource-based occupations.

urbanization An increasing concentration of the population in cities and a transformation of land use to an urban pattern of organization.

utilitarian conservation A philosophy that resources should be used for the greatest good for the greatest number for the longest time.

utilitarianism *See* utilitarian conservation.

utilitarians Those who hold that an action is right that produces the greatest good for the greatest number of people.

V

values An estimation of the worth of things; a set of ethical beliefs and preferences that determine our sense of right and wrong.

verbal learner Someone who understands and remembers best by listening to the spoken word.

vertical stratification The vertical distribution of specific subcommunities within a community.

village   A collection of rural households linked by culture, custom, and association with the land.

visible light   A portion of the electromagnetic spectrum that includes the wavelengths used for photosynthesis.

visual learner   Someone who understands and remembers best by reading, or looking at pictures and diagrams.

vitamins   Organic molecules essential for life that we cannot make for ourselves; we must get them from our diet; they act as enzyme cofactors.

volatile organic compounds   Organic chemicals that evaporate readily and exist as gases in the air.

volcanoes   Vents in the earth's surface through which gases, ash, or molten lava are ejected. Also a mountain formed by this ejecta.

voluntary simplicity   Deliberately choosing to live at a lower level of consumption as a matter of personal and environmental health.

vulnerable species   Naturally rare organisms or species whose numbers have been so reduced by human activities that they are susceptible to actions that could push them into threatened or endangered status.

W

warm front   A long, wedge-shaped boundary caused when a warmer advancing air mass slides over neighboring cooler air parcels.

waste stream   The steady flow of varied wastes, from domestic garbage and yard wastes to industrial, commercial, and construction refuse.

water cycle   The recycling and reutilization of water on earth, including atmospheric, surface, and underground phases and biological and nonbiological components.

water droplet coalescence   A mechanism of condensation that occurs in clouds too warm for ice crystal formation.

water stress   A situation when residents of a country don't have enough accessible, high-quality water to meet their everyday needs.

water table   The top layer of the zone of saturation; undulates according to the surface topography and subsurface structure.

waterlogging   Water saturation of soil that fills all air spaces and causes plant roots to die from lack of oxygen; a result of overirrigation.

watershed   The land surface and groundwater aquifers drained by a particular river system.

weather   Description of the physical conditions of the atmosphere (moisture, temperature, pressure, and wind).

weathering   Changes in rocks brought about by exposure to air, water, changing temperatures, and reactive chemical agents.

wetland mitigation   Replacing a wetland damaged by development (roads, buildings, etc.) with a new or refurbished wetland.

wetlands   Ecosystems of several types in which rooted vegetation is surrounded by standing water during part of the year. *See also* swamp, marsh, bog, fen.

wicked problems   Problems with no simple right or wrong answer where there is no single, generally agreed-on definition of or solution for the particular issue.

wilderness   An area of undeveloped land affected primarily by the forces of nature; an area where humans are visitors who do not remain.

Wilderness Act   Legislation of 1964 recognizing that leaving land in its natural state may be the highest and best use of some areas.

wildlife   Plants, animals, and microbes that live independently of humans; plants, animals, and microbes that are not domesticated.

wildlife refuges   Areas set aside to shelter, feed, and protect wildlife; due to political and economic pressures, refuges often allow hunting, trapping, mineral exploitation, and other activities that threaten wildlife.

windbreak   Rows of trees or shrubs planted to block wind flow, reduce soil erosion, and protect sensitive crops from high winds.

wind farms   Large numbers of windmills concentrated in a single area; usually owned by a utility or large-scale energy producer.

Wise Use Groups   A coalition of ranchers, loggers, miners, industrialists, hunters, off-road vehicle users, land developers, and others who call for unrestricted access to natural resources and public lands.

withdrawal   A description of the total amount of water taken from a lake, river, or aquifer.

woodland   A forest where tree crowns cover less than 20 percent of the ground; also called open canopy.

work   The application of force through a distance; requires energy input.

world conservation strategy   A proposal for maintaining essential ecological processes, preserving genetic diversity, and ensuring that utilization of species and ecosystems is sustainable.

World Trade Organization (WTO)   An association of 135 nations that meet to regulate international trade.

X

X ray   Very short wavelength in the electromagnetic spectrum; can penetrate soft tissue; although it is useful in medical diagnosis, it also damages tissue and causes mutations.

Y

yellowcake   The concentrate of 70 to 90 percent uranium oxide extracted from crushed ore.

Z

zero population growth (ZPG)   The number of births at which people are just replacing themselves; also called the replacement level of fertility.

zone of aeration   Upper soil layers that hold both air and water.

zone of leaching   The layer of soil just beneath the topsoil where water percolates, removing soluble nutrients that accumulate in the subsoil.

zone of saturation   Lower soil layers where all spaces are filled with water.

INDEX

DDT (dichlor-diphenyl-trichloroethane)
  banning of, 196, 394, 396
  bioaccumulation and biomagnification
    of, 158
  as chlorinated hydrocarbon, 198–99
  discovery, 196
  effectiveness against malaria, 153, 196
  historical overview, 196
  as hormone-disrupting chemical, 202
  persistence, 203–4, 204
  pest resistance and, 153, 203
  as water pollutant, 385, 386
dead zones, 184, 382–83
death rates, 116-117,132–33
debt-for-nature swaps, 248–49
deciduous forests, 96, 98
decisiveness, in critical thinking, 8
decommissioning of old nuclear plants, 423
decomposer organisms, 59
deductive reasoning, 42–43
deep ecology, 555
deer
  chronic wasting disease and, 151–52
  in national parks, 267
  white-tailed, population growth of,
    113, 230
deforestation, 23, 245–46
degradation
  land, 181–82
  water, 362, 366
Delany Clause, to Food and Drug Act
  (1958), 163
Delhi Sands flower-loving fly, 231
demand, in economics, 499, 500
demanufacturing, 464–65
Democratic Republic of Congo
  diamonds, 289
  forest preservation program, 247
demographic bottleneck, 121
demographics
  birth rate, 131–32
  death rates, 132–33
  emigration and immigration, 135–36
  fertility rate, 23, 131–32
  growth rates, 133
  life expectancy, 133–34
  life span, 133–34
  living longer, implications of, 134–35
  mortality, 132–33
demographic transitions, 138–40
  defined, 138, 140
  development and population, 138
  ecojustice view, 140
  Iran, family planning in, 133
  optimistic view, 138
  pessimistic view, 138–39
  social justice view, 139–40
  women's rights and fertility, 140
Denali National Park (Alaska), 265
Denmark
  birth incentives program, 137
  carbon dioxide reduction plan in
    Copenhagen, 326
  per capita energy consumption, 408
  Summit for Social Development
    (1995), 29
  wind power, use of, 324, 429
density-dependent factors regulating
  population growth, 118–19
density-independent factors regulating
  population growth, 118
dentrifying bacteria, 65–66
dependency ratio, 135
Depo-Provera, 141
depression, unipolar, 148
depth, aquatic ecosystems and, 100, 101,
  103, 105
desalination, 361, 369, 384
  plant on Colorado River, 371
Descartes, René, 36, 41
desertification, 255–56
deserts, 95, 96
  human disturbance of, 105–7
  mineral or salt layer in soil, 180
design for the environment, 513, 547
detection limits, toxins and, 163–64

detritivores, 59
deuterium, 50, 418, 424
developed countries
  air pollution, improvement of, 329
  birth dearth, 137
  birth reduction pressures, 137
  demographic transitions and, 138
  versus developing countries, 24–27
  healthcare and, 151
  housing, urban, 484–85
  immigration to, 135–36
  as importers of wildlife and wildlife
    products, 228
  industrial wood and, 241
  population, 130–31
  poverty and, 24–27
  protein-rich foods, production and
    consumption of, 174–75
  sewage treatment in, 380
  smart growth, 487–88
  urban population growth, 480–81
  urban problems, 482–85
  urban sprawl, 485–87
  water pollution, 387–91
  wealth in, 25–27
developing countries
  agricultural land in production, 181
  air pollution in, 329, 349–50
  demographic transitions and, 138
  versus developed countries, 24–27
  economic progress, 27
  food security and, 171–72
  fuelwood consumption, 242–43
  fuelwood crisis in, 443–44
  healthcare and, 151
  indoor air pollution, 336–37
  industrial wood and, 241
  international trade, lack of benefit
    from, 511
  life expectancy, progress in, 134
  pesticide poisoning in, 204–5
  population growth, 130–31, 480
  poverty and, 24–27
  protein-rich foods, production and
    consumption of, 174–75
  sanitation in, 380
  as source of wildlife and wildlife
    products, 228
  species and ecosystems, seriously
    threatened, 271–72
  sustainable development in, 493–94
  toxic colonialism and, 40
  traffic and congestion, 482–83
  urban population growth, 480–81
  urban problems, 482–85
  waste exportation to, 459–60
  water, increasing demand for, 362
  water pollution, 390–91
  wealth in, 25–27
development
  barrier islands, storm damage and
    development of, 102–3
  conservation and, 271–72, 492–93
  human development index, 25–26
  international, 512–13
  land-use planning, 487–88
  smart growth, 487–88
  urban sprawl and, 485–87
dew point, 355
Diamond, Jared, 120
diamonds, 289
diaphragms (birth control), 141
diarrhea, 148, 149, 483
dibromochloropropane, 197, 198
dichlorvos, 199
dieback, population, 113
dieldrin, 198–99, 204, 205
diesel-powered automobiles, 432
diet. See nutrition
dimethyldichlorovinylphosphate (DDVP), 199
dimethylsulfide, 331
dinosaurs, 223
Dioum, Baba, 3
dioxins
  bioaccumulation of, 385
  as hazardous air pollutant, 334, 336

in incinerator ash, 461
  as persistent organic pollutants, 204
direct action environmentalism, 556
disability-adjusted life years (DALYs),
  147–48, 158
discharge, stream, 160, 360
disclimax communities, 88, 89
discount rates, 511
diseases. See also specific diseases
  defined, 147
  ecological, 151–52
  global disease burden, 147–49
  pesticides as controls for, 200
  stress-related, 119
Disney Corporation versus Sierra Club
  (1969), 36, 37, 528
dispute resolution, 535–39
dissolved oxygen (DO) content, 381
distemper virus, 151
distributional surcharges, 438
divergent plate boundaries, 290, 291
diversity, 217–18. See also biodiversity
  in biological communities, 83–84, 89
  respecting, in McDonough design
    principles, 514
DNA (deoxyribonucleic acid), in genetic
  engineering, 187
Doctors Without Borders, 151
dolphins
  domoic acid poisoning and, 152
  immune system depressants, death
    from, 154
  striped, 227
  toxic pollutants in, 227
Dombeck, Mike, 252, 372, 531
Dominican Republic, falling fertility
  rate, 138
domoic acid poisoning, 152
double-blind experiment design, 42
doubling times, population, 112, 126
downbursts (winds), 314–15
Doxiadis, C.A., 479
drip irrigation, 184, 363, 364, 374
droughts
  in African Sahel, 312, 313, 315–16
  cycles, 361–62
drugs, biodiversity and, 220–21
dry cask storage for nuclear waste, 421
Duany, Andres, 490
Dubos, René, 20
ducks
  as biological pest controls, 207
  duck stamps, 270
  lead shot, deaths from, 227
  wood, population restoration of, 230
Ducks Unlimited, 553
duckweed, sewage-treatment by, 400
Dumping on Dixie, 466
dumps, open, 457–58
dung, as fuel, 444–45
Durning, Alan, 30, 260
dust
  long-range transport, 339–40
  as trigger for lightning, 338
dust bowl (1930s), 334
dust domes, 338
dysentery, 380, 483

eagles
  bald, as threatened species, 231
  bald, successful recovery plan for, 232
  DDT and, 196
Earth
  carrying capacity, 112, 505–6
  composition, 290, 291
  layers, 289–90
  Milankovitch cycles, 317
  picture of, 20
  tectonic processes, 290, 291
Earth Charter, 562–63
Earth Charter Council, 559
Earth First!, 556
Earth Liberation Front, 556
earthquakes, 300–301
  Chile (1960), 300
  China (1976), 300

frequency and effects, 300, 301
  Kobe (Japan, 1995), 300, 301
  Krakatoa (Indonesia, 1883), 300
  New Madrid, Missouri (1811), 300
  tectonic processes and, 290, 291, 300
  tsunamis and, 300
  Vesuvius, Mount, (Italy, 79 A.D.), 301
Earth Summit (Brazil, 1992), 20, 319, 324,
  534, 560, 562
Earthwatch, 546
Ebola, 149
eco-efficiency, 513
ecofeminism, 38–39
Eco-Kids Corps, 551
ecological development, 86
ecological diseases, 151–52
ecological diversity, 217–18
ecological economics, 500–501, 507
ecological footprint, 16
ecological niche, 75–77
ecological pyramids, 59–60, 61, 62
ecological services, estimated value of,
  508–9
ecological succession, 86, 88–89
ecology
  biodiversity, ecological benefits of,
    221, 222
  deep ecology, 555
  defined, 498
  landscape (see landscape ecology)
  restoration (see restoration ecology)
  shallow ecology, 555
  social ecology, 555
  systems study of, 49
Ecology, 49
Ecology of Commerce, The, 513, 515
economic minerology, 294–96, 298–300
  conservation of geologic resources,
    298–300
  metals, 294
  mining (see mining)
  nonmetal resources, 294–96
Economic Policy Institute, 510
economics, 498–517
  classical economics, 499–500, 507
  demand, 499, 500
  development, international, 512–13
  ecological economics, 500–501, 507
  economics, defined, 498
  market efficiencies, 504–5
  microlending, 512–13
  models, 506
  natural resource economics, 500, 501
  neoclassical economics, 500
  resources (see natural resources;
    resources)
  scarcity, 504–5
  supply, 499, 500
  technology, effect of, 504–5
  trade, international, 511–12
  worldviews, 498–501
economy, defined, 498
ecosystem management, 283–85
  defined, 283
  historical overview, 283–84
  principles and goals, 284–85
  restoration and, conflicting views
    of, 285
ecosystems
  artificial, creating, 281
  defined, 57–58
  energy exchange in, 55–57
  landscape dynamics, 276–77
  management (see ecosystem
    management)
  marine ecosystems, 100–103
  patchiness and heterogeneity, 275–76
  productivity, 58–60
  pyramids, ecological, 59–60, 61, 62
  resilience in, 536–37, 538
  soil, 179
  water projects, ecosystem losses from,
    369–70
ecotones, 84–85, 86
ecotourism, 272, 274
ecotoxicology, 154